P9-DGG-048

ARCTIC OCEAN

Barents Sea

FINLAND
Isinki
☆
● St. Petersburg
☆ Moscow
aw
☆ Minsk
apest ☆ Kiev
☆ Bucharest
☆ Istanbul
thens ☆ Ankara
TURKEY
ARMENIA
CYPRUS
LEBANON-
ISRAEL
Alexandria
Cairo ☆ JORDAN
☆ Amman
ean Sea

RUSSIA

● Aldan

Sea of Okhotsk

Bering Sea

● Novosibirsk

☆ Astana

KAZAKHSTAN

Ulan Bator ☆
MONGOLIA

● Harbin

Bishkek ☆
☆ Urumqi
Tashkent ☆
UZBEKISTAN
GEORGIA
AZERBAIJAN
Ashgabat ☆
KYRGYZSTAN
TURKMENISTAN TAJIKISTAN
SYRIA
Damascus ☆ Baghdad
IRAQ
Tehran ☆
☆ Mashhad
IRAN
AFGHANISTAN
☆ Kabul
Islamabad ☆
Lahore ●
PAKISTAN
NEPAL
New Delhi ☆
Karachi ●
☆ Kathmandu
BHUTAN
Lhasa ●

CHINA

Shenyang ●
Beijing ☆
Tianjin ●
N. KOREA
Qingdao ●
Tai'an ●
Zaozhuang ●
Xian ●
Chengdu ●
Wuhan ●
Chongqing ●
Shanghai ●
Ningbo ●
Guangzhou
(Canton) ●
BANGLADESH
Dhaka ☆
Kolkata
(Calcutta) ●
INDIA

Sea of Japan

Seoul ☆
S. KOREA
Osaka ●
Yokohama

JAPAN
Tokyo ☆

East China Sea

Taipei ☆
TAIWAN

Hong Kong
(Xianggang) ●

NORTH PACIFIC OCEAN

KUWAIT
BAHRAIN-
Abu
Dhabi ☆
Riyadh ☆ QATAR
SAUDI ARABIA UAE
☆ Muscat
OMAN
Ahmadabad ●
Mumbai
(Bombay) ●
Hyderabad ●

Nay Pyi Taw ☆
BURMA
(MYANMAR)
LAOS
Hanoi ☆

AD
Khartoum ☆
SUDAN
ERITREA
Sana'a ☆
YEMEN
DJIBOUTI
Red Sea
Arabian Sea
Bangalore ●
Chennai
(Madras) ●
Bay of Bengal
Vientiane ☆
THAILAND
Bangkok ☆
VIETNAM
South China Sea

PHILIPPINES
Manila ☆

Philippine Sea

CENTRAL
AFRICAN
REPUBLIC
Bangui ●
SOUTH
SUDAN
Juba ☆
Addis
Ababa ☆
ETHIOPIA
SOMALIA
Colombo ☆
SRI LANKA
CAMBODIA
Phnom Penh ☆
Ho Chi
Minh City ●
BRUNEI

PALAU
FEDERATED STATES
OF MICRONESIA
MARSHALL
ISLANDS

MALDIVES
Kuala Lumpur ☆
MALAYSIA
SINGAPORE ☆

NAURU
KIRIBATI

UGANDA
KENYA
Kampala ☆
● Nairobi
RWANDA
DEM. REP.
OF CONGO
BURUNDI
nshasa
TANZANIA
Dodoma ☆
Dar es
Salaam ●
● Mogadishu
SEYCHELLES

Jakarta ☆
Surabaya ●
INDONESIA
Bandung ●
PAPUA
NEW GUINEA
Port Moresby ●
SOLOMON
ISLANDS

TUVALU

SAMOA

LA
ZAMBIA
MALAWI
Harare ☆
ZIMBABWE
MOZAMBIQUE
COMOROS
Mozambique Channel
MADAGASCAR
Antananarivo ☆
MAURITIUS

INDIAN OCEAN

Timor Sea
TIMOR-LESTE
Darwin ●

Coral Sea
VANUATU
FIJI
TONGA

BOTSWANA
aborone
Pretoria ☆
fontein ☆
Maputo ☆
Mbabane ☆
SWAZILAND
SOUTH
AFRICA
LESOTHO
Maseru ☆

AUSTRALIA

Perth ●

Great Australian Bight

Canberra ☆
Melbourne ●
● Sydney

Tasman Sea

● Auckland

NEW
ZEALAND
● Wellington

RCTICA

NORWAY
Oslo ☆
FINLAND
Stockholm ☆
Helsinki ☆
SWEDEN
Tallinn ☆
ESTONIA
St. Petersburg ●
Riga ☆ LATVIA
LITHUANIA
Vilnius ☆
☆ Minsk
BELARUS
DENMARK
Copenhagen ☆
NETH.
Berlin ☆
Warsaw ☆
POLAND
Dublin ☆
UNITED
KINGDOM
IRELAND
London ☆
Amsterdam ☆
GERMANY
UKRAINE
Kiev ☆
Brussels ☆
BEL.
LUX.
Prague ☆
CZECH.
SLOVAKIA
Paris ☆
Vienna ☆
AUS.
Budapest ☆
MOL.
Chisinau ☆
FRANCE
Bern ☆ SWITZ.
SL.
HUN.
ROMANIA
Bucharest ☆
CROATIA
B&H SERB.
SPAIN
ITALY
MONT.
KO.
BULGARIA
Sofia ☆
Istanbul ☆
PORTUGAL
Madrid ☆
Rome ☆
MAC.
ALB.
GREECE
Ankara ☆
TURKEY
Lisbon ☆
Tunis ☆
ALGERIA
Algiers ☆
TUNISIA
Athens ☆

WORLD REGIONAL GEOGRAPHY

GLOBAL PATTERNS, LOCAL LIVES Seventh Edition

LYDIA MIHELIČ PULSIPHER
Geography Professor Emeritus,
University of Tennessee

ALEX PULSIPHER
Geographer and Independent Scholar

OLA JOHANSSON
Geography Professor,
University of Pittsburgh at Johnstown

with the assistance of

CONRAD "MAC" GOODWIN
Anthropologist/Archaeologist and
Independent Scholar

w. h. freeman
Macmillan Learning
New York

DEDICATED TO SHELAGH LEUTWILER

Vice President, STEM: **Ben Roberts**
Executive Editor: **Andrew Dunaway**
Program Manager: **Jennifer Edwards**
Developmental Editor: **Kharissia Pettus**
Editorial Assistant: **Alexandra Hudson**
Assistant and Supplements Editor: **Kevin Davidson**
Marketing Manager: **Shannon Howard**
Marketing Assistant: **Cate McCaffery**
Director of Digital Production: **Keri deManigold**
Director of Content, Life and Earth Sciences: **Amanda Nietzel**
Director, Content Management Enhancement: **Tracey Kuehn**
Managing Editor, Sciences and Social Sciences: **Lisa Kinne**
Senior Project Editor: **Kerry O'Shaughnessy**
Senior Production Supervisor: **Paul Rohloff**
Senior Media Producer: **Chris Efstratiou**
Media Producer: **Jenny Chiu**
Photo Editor: **Sheena Goldstein**
Photo Researcher: **Lisa Passmore**
Director of Design, Content Management: **Diana Blume**
Cover and Text Designer: **Vicki Tomaselli**
Art Coordinator: **Janice Donnola**
Illustrations: **Dragonfly Media Group**
Maps: **University of Tennessee, Cartographic Services Laboratory, Will Fontanez, Director; Maps.com**
Composition: **codeMantra U.S. LLC**
Printing and Binding: **LSC Communications**
Front cover & title page image: **Jakkree Thampitakkull/Moment/Getty Images**
Back cover images: *top row*: **Niko Guido/E+/Getty Images; Fotosearch/Getty Images; Jumper/Photodisc/Getty Images;**
   bottom row: **SAEED KHAN/AFP/Getty Images; Glowimages/Getty Images**

Library of Congress Control Number: 2016953824

ISBN-13: 978-1-319-04804-4 (with subregions)
ISBN-10: 1-319-04804-8 (with subregions)
ISBN-13: 978-1-319-05976-7 (without subregions)
ISBN-10: 1-319-05976-7 (without subregions)

© 2017, 2014, 2011, 2008 by W. H. Freeman and Company
All rights reserved

Printed in the United States of America
Second printing

W. H. Freeman and Company
One New York Plaza
Suite 4500
New York, NY 10004-1562
www.macmillanlearning.com

Shelley Johansson

Lydia Mihelič Pulsipher is a cultural-historical geographer who studies the landscapes and lifeways of ordinary people through the lenses of archaeology, geography, and ethnography. She has contributed to several geography-related exhibits at the Smithsonian Museum of Natural History in Washington, D.C., including "Seeds of Change," which featured the research she and Conrad Goodwin did in the eastern Caribbean. Lydia Pulsipher has ongoing research projects in the eastern Caribbean (historical archaeology) and in central Europe, where she is interested in various aspects of the post-Communist transition and social inclusion policies in the European Union. Her graduate students have studied human ecology issues in the Caribbean and border issues and issues of national identity and exclusion in several central European countries. She has taught cultural, gender, European, North American, and Mesoamerican geography at the University of Tennessee at Knoxville since 1980; through her research, she has given many students their first experience in fieldwork abroad. Previously she taught at Hunter College and Dartmouth College. She received her B.A. from Macalester College, her M.A. from Tulane University, and her Ph.D. from Southern Illinois University. For personal enjoyment, she works in her gardens, cans the produce, bakes rhubarb pies, and encourages the musical talents of her grandchildren.

Alex Pulsipher is an independent scholar in Knoxville, Tennessee, who has conducted research on vulnerability to climate change, sustainable communities, and the diffusion of green technologies in the United States. In the early 1990s, Alex spent time in South Asia working for a sustainable development research center. He then completed a B.A. at Wesleyan University in Connecticut, writing his undergraduate thesis on the history of Hindu nationalism. Beginning in 1995, Alex contributed to the research and writing of the first edition of *World Regional Geography* with Lydia Pulsipher. Since then he has continued to write and restructure content for subsequent editions, and has created innumerable maps and over 120 photo essays. He has also traveled to South America, Southeast Asia, South Asia, and Europe to collect information in support of the text and for the website. In addition to his work on *World Regional Geography*, Alex developed the first, second, and third editions of *World Regional Geography Concepts*. He has a master's degree in geography from Clark University in Worcester, Massachusetts.

Ola Johansson is a cultural and urban geographer with a particular interest in music. As a cultural geographer, he has studied local music scenes and venues, touring patterns, national identity and cultural hybridity in music, and Swedish popular music. Ola is also co-author of the book *Sound, Society, and the Geography of Popular Music*. As an urban geographer, he is interested in urban policy and governance, the formation of entertainment districts, and place branding. He received both a bachelor's degree and a master's degree in social and economic geography from Lund University in Sweden. He then moved to the United States and graduated with a Ph.D. in geography from the University of Tennessee in 2004. In the past, Ola has been a researcher with the Tennessee Valley Authority in Knoxville and a guest scholar at Linnaeus University in Sweden. He is currently professor of geography at the University of Pittsburgh at Johnstown, Pennsylvania. Ola has been a behind-the-scenes contributor to *World Regional Geography* for several years and is now excited to be a co-author. In his spare time, Ola enjoys playing soccer, traveling, cooking, and going to concerts.

Conrad "Mac" Goodwin assists in the writing and production of *World Regional Geography* and the *World Regional Geography Concepts* version in a variety of ways. Mac, Lydia's husband, is an anthropologist and historical archaeologist with a B.A. in anthropology from the University of California, Santa Barbara; an M.A. in historical archaeology from the College of William and Mary; and a Ph.D. in archaeology from Boston University. He specializes in sites created during the European colonial era in North America, the Caribbean, and the Pacific. He has particular expertise in the archaeology of agricultural systems, gardens, domestic landscapes, and urban spaces, and has written an archaeological management plan for Wilmington, Delaware. In addition to work in archaeology and on the textbook, for the past ten years he has been conducting research on wines and winemaking in Slovenia and delivering papers on these topics at professional geography meetings. In 2013, Knoxville mayor Madeline Rogero appointed him to a four-year term on the Metropolitan Planning Commission as a result of his work with local neighborhood and environmental organizations. For relaxation, Mac works in his organic garden, builds stone walls (including a pizza oven), and is a slow-food chef.

BRIEF CONTENTS

CONTENTS

Ales Beno/Anadolu Agency/Getty Images

CHAPTER 5

Russia and the Post-Soviet States

261

Bruce Yuanyue Bi/Getty Images

PHOTO ESSAYS

CHAPTER 6

North Africa and Southwest Asia 313

PHOTO ESSAYS

CHAPTER 11

Oceania: Australia, New Zealand, and the Pacific   619

Sandra Mu/Getty Images

In this seventh edition, our goal is to show how the activities of ordinary people at the local level are connected to various geographic themes: environment, globalization and development, power and politics, urbanization, and population and gender. We continue to portray the rich diversity of human life across the world and humanize geographic issues by representing the daily lives of women, men, and children in the various regions of the globe. We make a special effort to be current with the data we use and the analysis we offer. In striving to reach these goals, we have made this seventh edition of *World Regional Geography* as up to date, instructive, and visually appealing as possible.

NEW TO THE SEVENTH EDITION

THEMES

Teaching world regional geography is never easy. Many instructors have found that focusing their courses on a few key ideas makes their teaching more effective and helps students retain information. With input from instructors like you, we've woven five common themes throughout each chapter that provide a framework for students to use as they expand their knowledge of the world and each of its regions. These themes are listed here in the order in which they are covered in every chapter:

- **Environment**: What are the basic environmental characteristics of a region? What are the impacts of human use? How do issues of water scarcity, water pollution, and water management affect people and environments in a particular region? How do food production systems affect environments and societies in a region? How might global climate change and changes in food production systems affect soil and water resources? How are places, people, and ecosystems in a particular region vulnerable to shifts in global relationships? Which human activities contribute significant amounts of greenhouse gases?

ENVIRONMENT

GEOGRAPHIC THEME 1

Environment: Deforestation in this region contributes significantly to global climate change. In addition, some areas are experiencing a water crisis related to climate change, inadequate water infrastructure, and the intensified use of water, despite the region's overall abundant water resources.

- **Globalization and Development**: How are lives changing as flows of people, ideas, products, and resources become more global? How has a particular region been affected by globalization, historically and currently? How do shifts in economic, social, and other dimensions of development affect human well-being? What paths have been charted by the so-called developed world, and how are they relevant, or irrelevant, to the rest of the world? What new "homegrown" solutions are emerging from the so-called less developed countries?

- **Power and Politics**: What are the main differences in the ways that power is wielded in societies? Where are authoritarian modes of governance dominant? Where have political freedoms expanded the most? What kind of changes is the expansion of political freedoms bringing to different world regions? How are changes in the geopolitical order affecting current world events?

- **Urbanization**: What forces are driving urbanization in a particular region? How have cities responded to growth? How are regions affected by the changes that accompany urbanization—for example, the growth of slums, changes in access to jobs, education, and health care, and depopulation and decline in rural areas?

- **Population and Gender**: What are the major forces driving population growth or decline in a region? How have changes in gender roles influenced population growth or decline? How are changes in life expectancy, family size, gender balance, and the age of the population influencing population change?

These themes are stated at the beginning of each chapter and tailored to the particular region. They are repeated at relevant points in the text as well as reviewed in "Things to Remember" sections found throughout the chapter and in thought-provoking questions posed in the "Geographic Insights Review and Self-Test" section at the end of each chapter.

RESTRUCTURED CHAPTERS

Each chapter has been restructured so that the discussions to which the Geographic Themes refer now occur in the same order in each chapter. We implemented this change to make it easier for teachers and students to navigate the complex topics of world regional geography. The number of themes has also been condensed from a rotating cast of nine themes in the previous edition to the five themes noted above in this edition. Discussions of climate change, food, and water now come under the heading Environment. Some discussions of population issues have been linked with discussions of changing gender roles.

EPILOGUES: POLAR REGIONS AND SPACE

The epilogue discussing Antarctica has been expanded to include the Arctic because both polar regions are experiencing modifications linked to climate change at an accelerating rate. A new epilogue discusses space, which is an increasingly important geographical topic not only because nations continue to explore space and try to capture its resources but because the parts of space closest to Earth have already become central to our understanding of life on Earth, as well as to the management of global communications and even the conduct of hostilities.

Paul Nicklen/National Geographic Getty Images

CONTINUING IN THE SEVENTH EDITION

ON THE BRIGHT SIDE FEATURES

In light of the often overwhelming and at times worrisome nature of the information presented in any world regional geography course, each chapter has a series of *On the Bright Side* commentaries that explore some of the more hopeful patterns and opportunities emerging within each region, usually at the instigation of local people. New On the Bright Side features have been added or updated in relevant chapters.

ON THE BRIGHT SIDE

In the last decade, Mongolia invested heavily in IT. Today, the national telecom company is upgrading to a 4G network. Most people, including nomads, have at least one cell phone, far out in the steppe. In just a 3-year period (from 2012 to 2015) the number of Internet users in Mongolia more than doubled. *[Source: Laura Pappano, "The Boy Genius of Ulan Bator," New York Times Magazine, September 15, 2013, p. 52.]* ■

VIGNETTES, CASE STUDIES, AND GLOBAL PATTERNS, LOCAL LIVES

These short essays about individual people, families, and groups make geography interesting, immediate, and real to students. In addition, they serve to personalize and make more understandable the effects of abstract geographical processes and trends. Vignettes, Case Studies, and Global Patterns, Local Lives features have been added or updated where appropriate.

LOCAL LIVES PHOTO FEATURES

Three *Local Lives* photo features in each region chapter add further human interest by showing regional customs related to foodways, people and animals, and festivals. Each photo has an extensive caption designed to pique students' curiosity.

NEW PHOTOS

Every photo has been carefully considered for currency in this edition; those not considered current have been replaced. An ongoing

aim of this text has been to awaken students to the circumstances of people around the world, and photos are a powerful way to accomplish this objective. This edition continues our tradition of promoting careful attention to photos by including in Chapter 1 a short lesson on photo interpretation. Students are encouraged to use the skills described as they look at every photo in the text, and instructors are encouraged to use the photos as lecture themes and thus to help generate analytical class discussions.

Each photo was chosen to complement a concept or situation described in the text. All photos are numbered and referenced in the text, making it easier for students to integrate the text with the visuals as they read. Moreover, the photos—like all of the book's graphics, including the maps—have been given significant space and prominence in the page layout. The result is a visually engaging, dynamic, instructive, and up-to-date text.

At the beginning of each regional chapter, a series of photos surrounding the regional map introduces the reader to landscapes within the region. Subsequent figures that are *Photo Essays* illustrate particular themes. For example, each region has a photo essay about urbanization, including a map of national urbanization rates and large cities as well as photos that illustrate various aspects of urban life in the region.

To help instructors make use of all these photo features in their teaching, the many photo essays and photo figures are accompanied by *Thinking Geographically* questions.

UP-TO-DATE CONTENT

Because the world is constantly changing, it is essential that a world regional geography text be as current as possible. Nonetheless, consequential events can come so late in the publishing process that it is impossible to adequately assess likely outcomes. Such is the case with the June 2016 decision in the United Kingdom to leave the European Union (a move popularly known as Brexit). The potential ramifications are enormous, but the exit is likely to take years, so for the foreseeable future the number of countries

FIGURE 3.29 LOCAL LIVES: Festivals in Middle and South America

A A member of a *bateria*, or percussion band, is parading during Rio de Janeiro's Carnival, the largest in the world. Carnival, which is celebrated in much of the United States as Mardi Gras, takes place in the days and weeks before Lent, the 40-day period in Christian traditions of fasting and abstinence that precedes Easter. Elaborate celebrations take place in Barranquilla in Colombia, Port-au-Prince in Haiti, Port of Spain in Trinidad, and Montevideo in Uruguay.

B A procession commemorating Jesus walks on an *alfombra*, or carpet, made of colored sawdust, painstakingly drawn by hand on the streets of Antigua, Guatemala, during *Semana Santa*, or Holy Week, the last week of Lent. The processions will destroy the alfombras. This tradition is an adaptation of the Mayan practice of making elaborate ground designs with flowers and feathers for Mayan royalty to walk on as they made their way to important ceremonies.

C A richly decorated grave in Tijuana, Mexico, during Día de los Muertos—Day of the Dead—a holiday celebrated throughout Mexico on November 1. Families gather in cemeteries, where they build private altars and other decorations dedicated to deceased loved ones. The holiday is a "Christianized" version of an Aztec festival dedicated to the goddess Mictecacihuatl, Queen of the Underworld, whose job is to watch over the bones of the dead.

in the European Union will remain at 28 and internal EU relationships will change only slowly. Likewise, the unexpected upset in the United States presidential and congressional elections in November 2016 is likely to affect many aspects of life in the United States and elsewhere. When we can, we will offer insights on these issues, and others that are bound to arise, on the book's homework site.

Some major content areas of the book that have been updated include:

- Climate change and its environmental, political, and economic implications
- Changes since the global economic recession that began in 2008 and its effect on migrants, labor outsourcing, and job security in importing and exporting countries
- The role of terrorism in the realignment of power globally and locally; the emergence of the Islamic State group
- The growing disparity in wealth in North America and throughout the world
- Domestic and global implications of the U.S. political, economic, and military stances
- Conflict over oil pollution in the Amazon, Caribbean, and West Africa
- Migration (voluntary and forced) and the ways it is changing both sending and receiving countries and regions economically and culturally
- Changing gender roles in both developed and developing countries
- Recent economic and social crises in the European Union, including the unexpected flood of refugees and the consequences for the original EU members, new and potential member states, and the global community
- Growing tensions between Russia, its neighbors, the European Union, Turkey, and the United States
- The conflict in Ukraine
- Political revolutions and conflict (the Arab Spring) in North Africa and Southwest Asia (often referred to as the Middle East)
- Civil war in Syria and new refugee movements
- The new influence of Arabic media outlets, such as Al Jazeera
- Growing tensions over water in a number of regions
- Disputes in East Asia over the islands and reefs in the East and South China Seas
- The end of the one-child policy in China
- Civil change and unrest in Southeast Asia
- The consequences of global climate change on Pacific Island nations

CONSISTENT BASE MAPS

In this edition, we continue to improve what has often been cited as a principal strength of this text: high-quality, relevant, and consistent maps. To help students make conceptual connections and to compare regions, every chapter contains the following:

- Regional map with landscape photos at the beginning of each chapter

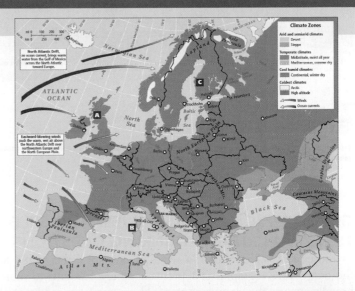

FIGURE 4.4 PHOTO ESSAY: Climates of Europe

- Political map of the region
- Climate map with photos of different climate zones
- Map of the human impacts on the biosphere, with photo essay
- Map of the region's vulnerability to climate change, with photo essay
- Map of regional trends in power and politics, with photo essay
- Urbanization map, with photo essay
- Maps of geographic patterns of human well-being
- Map of population density
- Maps of subregions

VISUAL HISTORIES

These visual timelines for each region use images to illustrate key points in the region's history (see below).

THINGS TO REMEMBER

At the close of every main section, a few concise statements review the important points in the section. The statements emphasize some key themes while encouraging students to think through the ways in which the material illustrates these points. They also review the Geographic Themes that begin each chapter.

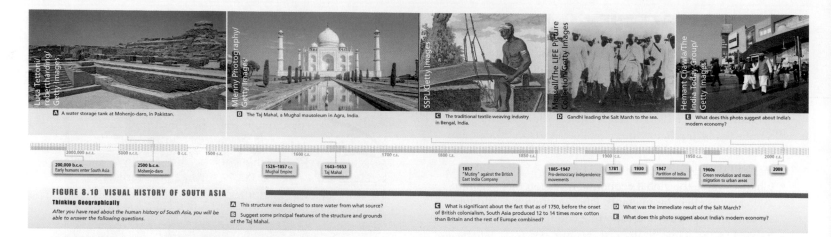

A A water storage tank at Mohenjo-daro, in Pakistan.

B The Taj Mahal, a Mughal mausoleum in Agra, India.

C The traditional textile-weaving industry in Bengal, India.

D Gandhi leading the Salt March to the sea.

E What does this photo suggest about India's modern economy?

200,000 b.c.e. Early humans enter South Asia

2500 b.c.e. Mohenjo-daro

1526–1857 c.e. Mughal Empire

1643–1653 Taj Mahal

1857 "Mutiny" against the British East India Company

1885–1947 Pro-democracy independence movements

1781

1930

1947 Partition of India

1960s Green revolution and mass migration to urban areas

2008

FIGURE 8.10 VISUAL HISTORY OF SOUTH ASIA

Thinking Geographically

After you have read about the human history of South Asia, you will be able to answer the following questions.

A This structure was designed to store water from what source?

B Suggest some principal features of the structure and grounds of the Taj Mahal.

C What is significant about the fact that as of 1750, before the onset of British colonialism, South Asia produced 12 to 14 times more cotton than Britain and the rest of Europe combined?

D What was the immediate result of the Salt March?

E What does this photo suggest about India's modern economy?

GEOGRAPHIC INSIGHTS REVIEW AND SELF-TEST AND CRITICAL THINKING QUESTIONS

At the end of each chapter, a series of questions, many tied to the chapter's Geographic Themes, encourage students to more broadly analyze the chapter content. These questions can be used for assignments, group projects, or class discussion.

MARGINAL GLOSSARY OF KEY TERMS

Terms important to the chapter content are boldfaced on first use and defined on the page on which they first appear. Each term is listed at the end of the chapter, along with the number of the page where the term is defined. The key terms are also listed alphabetically and defined in the glossary at the end of the book.

THE ENDURING VISION: GLOBAL AND LOCAL PERSPECTIVES

THE GLOBAL VIEW

In addition to the new features and enhancements to the text, we retain the hallmark features that have made the first six editions of this text successful for instructors and students. For the seventh edition, we continue to emphasize global trends and the connections between regions that are changing lives throughout the world. The following linkages are explored in every chapter, as appropriate:

- The *multifaceted economic linkages* among world regions. These include (1) the effects of colonialism; (2) trade; (3) the role in the world economy of transnational corporations such as Walmart, Norilsk Nickel, Nike, and Apple; (4) the influence of regional trade organizations such as ASEAN and NAFTA; and (5) the changing roles of the World Bank and the International Monetary Fund as the negative consequences of structural adjustment programs become better understood. These issues are explored primarily in the Globalization and Development section of each chapter.

- *Migration:* Migrants are changing economic and social relationships in virtually every part of the globe. The societies they leave are changed radically by the migrants' absence, just as the host societies are altered by their arrival. The most recent and graphic example is the exodus of millions of refugee asylum seekers from Syria and elsewhere in the Middle East hoping to find refuge in the European Union. The text explores the local and global effects of foreign workers in places such as Japan, Europe, Africa, the Americas, and Southwest Asia, as well as the increasing number of refugees resulting from conflicts around the world. Also discussed are long-standing migrant groups, including the Overseas Chinese and the Indian diaspora. These topics are explored primarily in the Population and Gender section as well as the Sociocultural Issues section of each chapter.

- *Mass communications and marketing techniques* are promoting *world popular culture* across regions. The text integrates coverage of popular culture and its effects in discussions of topics such as tourism in the Caribbean and Southeast Asia; the wide-ranging impact of innovations originating in modernizing economies, such as Ushahidi in Kenya; and the blending of Western and traditional culture in places such as South Africa, Malawi, Japan, and China.

- *Gender issues* are covered in every chapter with the aim of addressing more completely the lives of ordinary people. Gender is intimately connected to other patterns, including internal and global migration, and these connections and other region-wide gender patterns are illustrated in a variety of maps and photos and in vignettes that show gender roles as played out in the lives of individuals. The lives of children, especially with regard to their roles in families, are also covered, often in concert with the treatment of gender issues. These issues are explored primarily in Population and Gender section as well as the Sociocultural Issues section of each chapter.

THE LOCAL LEVEL

We pay special attention in this book to the local scale—a town, a village, a household, an individual. Our hope is, first, that stories of individual people and families will make geography interesting and real to students; and second, that seeing the effects of abstract processes and trends on ordinary lives will dramatize the effects of these developments for students. Reviewers have mentioned that students particularly appreciate the personal vignettes, which are often stories of real people (with names disguised). For each region, we examine the following local responses:

- *Local lives:* We use photo essays to focus on particular regional customs and traditions as they relate to foodways, festivals, and the relationship of animals to the people of a region.

- *Cultural change:* We look closely at changes in the family, gender roles, and social organization in response to urbanization, modernization, and the global economy.

- *Impacts on well-being:* Ideas of what constitutes "well-being" differ from culture to culture, yet broadly speaking, people everywhere try to provide a healthful life for themselves in a community of their choosing. Their success in doing so is affected by local conditions, global forces, and their own ingenuity.

- *Issues of identity:* Paradoxically, as the world becomes more tightly knit through global communications and media, ethnic and regional identities often become stronger. The text examines how modern developments such as the Internet and related technologies are used to reinforce particular cultural identities, often bringing educated emigrants back to their home countries to help with reforms or to facilitate rapid responses in crises.

- *Local attitudes toward globalization:* People often have ambivalent reactions to global forces: they are repelled by the power of these forces, fearing effects on their own lives and livelihoods and on local traditional cultural values, but they are also attracted by the economic opportunities that may emerge from greater global integration. The text looks at how the people of a region react to cultural and economic globalization.

ALSO AVAILABLE

To better serve the different needs of diverse faculty and curricula, two other versions of this textbook are available.

WORLD REGIONAL GEOGRAPHY WITHOUT SUBREGIONS, SEVENTH EDITION
ISBN-13: 978-1-319-05976-7
The briefer version provides essentially the same main text coverage as the version described above, omitting only the subregional sections. This version contains all the pedagogy found in the main version.

WORLD REGIONAL GEOGRAPHY CONCEPTS, THIRD EDITION
ISBN-13: 978-1-4641-1071-9
This streamlined version of *World Regional Geography* is designed to allow instructors to cover all the world regions in a single semester.

FOR THE INSTRUCTOR: A WEALTH OF RESOURCES ONLINE AT SAPLING LEARNING FOR GEOGRAPHY

Engaging students. Empowering instructors. Created and supported by educators, Sapling Learning's instructional online homework drives student success and saves educators time.

With the seventh edition, the authors have created all new content for Sapling Learning, closing the gap between what you do in the classroom and the homework students do on their own. Included in Sapling are the following resources:

- *Google Earth Activities:* Eleven Google Earth tutorials are available with assignable assessment questions.

- *Sapling Tutorials:* Eleven new interactive tutorial assignments (one for each region). Each question includes wrong-answer feedback targeted to students' misconceptions.

- *Chapter Quizzes:* These multiple-choice quizzes help students assess their mastery of each chapter.

- *LearningCurve* is an integral part of Sapling. It is an intuitive, fun, and highly effective formative assessment tool that is based on extensive educational research. Students can use LearningCurve to test their knowledge in a low-stakes environment that helps them improve their mastery of key concepts and prepare for lectures and exams. This adaptive quizzing engine moves students from basic knowledge through critical thinking and synthesis skills as they master content at each level. For a demo, visit *learningcurveworks.com.*

- *Video Library:* Three to four new videos from the BBC have been curated for each chapter. Each video includes assessments, making these an ideal resource for the flipped classroom, as an extension resource, or as a homework assignment.

INSTRUCTOR RESOURCES

- *All text images* are in PowerPoint and JPEG formats, with enlarged labels for better projection quality.

- *PowerPoint lecture outlines* by Nicole C. James. The main themes of each chapter are outlined and enhanced with images from the book, providing a pedagogically sound foundation on which to build personalized lecture presentations.

- *Instructor's Resource Manual* by Jennifer Rogalsky, State University of New York, Geneseo, and Helen Ruth Aspaas, Virginia Commonwealth University, contains suggested lecture outlines, points to ponder for class discussion, and ideas for exercises and class projects. It is offered as chapter-by-chapter Word files to facilitate editing and printing.

- *Test Bank* by Jennifer Rogalsky, State University of New York, Geneseo. The Test Bank is designed to match the pedagogical intent of the text and offers more than 2500 test questions (multiple-choice, short answer, matching, true/false, and essay) in a Word format that makes it easy to edit, add, and resequence questions. A *computerized Test Bank* (powered by Diploma) with the same content is also available.

- *Clicker questions* by Toby Applegate, University of Massachusetts, Amherst. Prepared within the PowerPoint slides, clicker questions allow instructors to jump-start discussions, illuminate important points, and promote better conceptual understanding during lectures.

COURSE MANAGEMENT

All instructor and student resources are also available via *Black-Board, Canvas, Angel, Moodle, Sakai, and Desire2Learn.*

W. H. Freeman offers a course cartridge that populates your LMS (learning management system) site with content tied directly to the book. For access to a specific course cartridge, please contact your W. H. Freeman/Macmillan Learning sales consultant.

FOR STUDENTS: SAPLING FOR GEOGRAPHY

STUDENTS AGREE SAPLING LEARNING GETS RESULTS

Ninety percent of students say they would pay for Sapling Learning the next semester.

- *Google Earth Activities:* Eleven Google Earth tutorials are available.

- *Sapling Tutorials:* Eleven new interactive tutorial assignments (one for each region). Each question includes wrong-answer feedback targeted to students' misconceptions.

- *Chapter Quizzes:* These multiple-choice quizzes help students assess their mastery of each chapter.

- *LearningCurve* is an integral part of Sapling. It is an intuitive, fun, and highly effective formative assessment tool that is based on extensive educational research. Students can use LearningCurve to test their knowledge in a low-stakes environment that helps them improve their mastery of key concepts and prepare for lectures and exams. This adaptive quizzing engine moves students from basic knowledge through critical thinking and synthesis skills as they master content at each level. For a demo, visit *learningcurveworks.com*.

- *Video Library:* Three to four new videos from the BBC have been curated for each chapter. Each video includes assessments, making these an ideal resource for the flipped classroom, as an extension resource, or as a homework assignment.

ALSO AVAILABLE: RAND MCNALLY'S *ATLAS OF WORLD GEOGRAPHY*, 176 PAGES

This atlas, available at a greatly reduced price when bundled with the textbook, contains:

- 52 physical, political, and thematic maps of the world and continents; 49 regional, physical, political, and thematic maps; and dozens of metro-area inset maps

- Geographic facts and comparisons, covering topics such as population, climate, and weather

- A section on common geographic questions, a glossary of terms, and a comprehensive 25-page index

ACKNOWLEDGMENTS

The authors wish to acknowledge the many geographers whose insights and suggestions have informed this book.

WORLD REGIONAL GEOGRAPHY, SEVENTH EDITION

Toby Applegate
University of Massachusetts, Amherst

Michele Barnaby
Pittsburg State University

Marissa Bell
SUNY Buffalo

Kathleen Braden
Seattle Pacific University

Stephen Buchanan
Wilmington University

Dean Butzow
Lincoln Land Community College

Seth Cavello
SUNY Geneseo

John Conley
Orange Coast College

Stentor Danielson
Slippery Rock University

Charles Davis
Mississippi Gulf Coast Community College, Jefferson Davis Campus

Simon Davis
Bronx Community College, CUNY

Bryant Evans
Houston Community College

Caitlin Finlayson
University of Mary Washington

David Fox
Park University

Sunita George
Western Illinois University

Dusty Girard
Brookhaven College

Raymond Greene
Western Illinois University

Nicholas Hill
Greenville Technical College

Gary Howe
Northwestern Michigan College

Pauline Johnson
Mars Hill University

Jeannine Koshear
Fresno City College

Elizabeth Larson
Arizona State University, School of Geographical Sciences and Urban Planning

Julie Lawless
Western Illinois University

James Leonard
Marshall University

John Lindberg
Scott Community College

Enid Lotstein
Bronx Community College, CUNY

Meredith Marsh
Lindenwood University

Blake Mayberry
Red Rocks Community College

Darrel McDonald
Stephen F. Austin State University

Ellen McHale
Utica College

Douglas Munski
University of North Dakota

Tom Mueller
California University of Pennsylvania

Velvet Nelson
Sam Houston State University

Erinn Nicley
Western Governors University

Susan Oldfather
Broward College

Rosann Poltrone
Arapahoe Community College

Dawn Richards
College of Southern Maryland

Jeffrey Richetto
University of Alabama

Kevin Romig
Northwest Missouri State

Scott Roper
Castleton University

Tamar Rothenberg
Bronx Community College, CUNY

Anne Saxe
MiraCosta College

Dudley Shurlds
East Mississippi Community College

Trevor Smith
Mississippi Gulf Coast Community College

John Teeple
Fresno City College

Sam Sweitz
Michigan Technological University

Arthur Viterito
College of Southern Maryland

Susan Walcott
University of North Carolina, Greensboro

Donald Williams
Western New England University

WORLD REGIONAL GEOGRAPHY CONCEPTS, THIRD EDITION

Ola Ahlqvist
Ohio State University

Wayne Brew
Montgomery County Community College

Elizabeth Dudley-Murphy
University of Utah

Chad Garick

Jones County
Junior College

Kari Jensen
Hofstra University

Timothy Kelleher
Florida State University

Mathias Le Bosse
Kutztown University of

Pennsylvania

Fuyuan Liang
Western Illinois University

Chris Post
Kent State University, Stark

Amy Rock
Ohio University

Sarah Smiley
Kent State University, Salem

WORLD REGIONAL GEOGRAPHY, SIXTH EDITION

Victoria Alapo
Metropolitan Community College

Jeff Arnold
Southwestern Illinois College

Shaunna Barnhart
Pennsylvania State University

Dean Butzow
Lincoln Land Community College

Philip Chaney
Auburn University

Christine Hansell
Skyline College

Heidi LaMoreaux
Santa Rosa Junior College

Kent Mathewson
Louisiana State University

Julie Mura
Florida State University

Michael Noll
Valdosta State University

Tim Oakes
University of Colorado, Boulder

Kefa M. Otiso
Bowling Green State University

Sam Sweitz
Michigan Technological University

Jeff Ueland
Bemidji State University

Ben Wolfe
Metropolitan Community College, Blue River

WORLD REGIONAL GEOGRAPHY CONCEPTS, SECOND EDITION

Heike C. Alberts
University of Wisconsin, Oshkosh

John All
Western Kentucky University

Robert G. Atkinson
Tarleton State University

Christopher A. Badurek
Appalachian State University

Bradley H. Baltensperger
Michigan Technological University

Denis A. Bekaert
Middle Tennessee State University

Mikhail Blinnikov
St. Cloud State University

Mark Bonta
Delta State University

Patricia Boudinot
George Mason University

Lara M. P. Bryant
Keene State College

Craig S. Campbell
Youngstown State University

Bruce E. Davis
Eastern Kentucky University

L. Scott Deaner
Owens Community College

James V. Ebrecht
Georgia Perimeter College

Kenneth W. Engelbrecht
Metropolitan State College of Denver

Natalia Fath
Towson University

Alison E. Feeney
Shippensburg University

John H. Fohn II
Missouri State University, West Plains

Stephen Franklin
Coconino Community College

Joy Fritschle
West Chester University

Matthew Gerike
University of Missouri

Carol L. Hanchette
University of Louisville

Ellen R. Hansen
Emporia State University

Katie Haselwood-Weichelt
University of Kansas

Kari Jensen
Hofstra University

Cub Kahn
Oregon State University

Curtis A. Keim
Moravian College

Jeannine Koshear
Fresno City College

Hsiang-te Kung
University of Memphis

Chris Laingen
Eastern Illinois University

Leonard E. Lancette
Mercer University

Heidi Lannon
Santa Fe College

James Leonard
Marshall University

John Lindberg
Scott Community College

Max Lu
Kansas State University

Peter G. Odour
North Dakota State University

Kefa M. Otiso
Bowling Green State University

Lynn M. Patterson
Kennesaw State University

James Penn
Grand Valley State University

Paul E. Phillips
Fort Hays State University

Gabriel Popescu
Indiana University, South Bend

Jennifer Rahn
Samford University

Amanda Rees
Columbus State University

Robert F. Ritchie IV
Liberty University

Ginger L. Schmid
Minnesota State University, Mankato

Cynthia L. Sorrensen
Texas Tech University

Jennifer Speights-Binet
Samford University

Sam Sweitz
Michigan Technological University

Michael W. Tripp
Vancouver Island University

Julie L. Urbanik
University of Missouri, Kansas City

Irina Vakulenko
University of Texas, Dallas

Jean Vincent
Santa Fe College

Linda Wang
University of South Carolina, Aiken

Kelly Watson
Florida State University

Laura A. Zeeman
Red Rocks Community College

Sandra Zupan
University of Kentucky

WORLD REGIONAL GEOGRAPHY CONCEPTS, FIRST EDITION

Gillian Acheson
Southern Illinois University, Edwardsville

Tanya Allison
Montgomery College

Keshav Bhattarai
Indiana University, Bloomington

Leonhard Blesius
San Francisco State University

Jeffrey Brauer
Keystone College

Donald Buckwalter
Indiana University of Pennsylvania

Craig Campbell
Youngstown State University

John Comer
Oklahoma State University

Kevin Curtin
George Mason University

Ron Davidson
California State University, Northridge

Tina Delahunty
Texas Tech University

Dean Fairbanks
California State University, Chico

Allison Feeney
Shippensburg University of Pennsylvania

Eric Fournier
Samford University

Qian Guo
San Francisco State University

Carole Huber
University of Colorado, Colorado Springs

Paul Hudak
University of North Texas

Christine Jocoy
California State University, Long Beach

Ron Kalafsky
University of Tennessee, Knoxville

David Keefe
University of the Pacific

Mary Klein
Saddleback College

Max Lu
Kansas State University

Donald Lyons
University of North Texas

Barbara McDade
University of Florida

Victor Mote
University of Houston

Darrell Norris
State University of New York, Geneseo

Gabriel Popescu
Indiana University, South Bend

Claudia Radel
Utah State University

Donald Rallis
University of Mary Washington

Pamela Riddick
University of Memphis

Jennifer Rogalsky
State University of New York, Geneseo

Tobie Saad
University of Toledo

Charles Schmitz
Towson University

Sindi Sheers
George Mason University

Ira Sheskin
University of Miami

Dmitri Siderov
California State University, Long Beach

Steven Silvern
Salem State College

Ray Sumner
Long Beach City College

Stan Toops
Miami University

Karen Trifonoff
Bloomsburg University of Pennsylvania

Jim Tyner
Kent State University

Michael Walegur
University of Delaware

Scott Walker
Northwest Vista College

Mark Welford
Georgia Southern University

WORLD REGIONAL GEOGRAPHY, FIFTH EDITION

Gillian Acheson
Southern Illinois University, Edwardsville

Greg Atkinson
Tarleton State University

Robert Begg
Indiana University of Pennsylvania

Richard Benfield
Central Connecticut State University

Fred Brumbaugh
University of Houston, Downtown

Deborah Corcoran
Missouri State University

Kevin Curtin
George Mason University

Lincoln DeBunce
Blue Mountain Community College

Scott Dobler
Western Kentucky University

Catherine Doenges
University of Connecticut, Stamford

Jean Eichhorst
University of Nebraska, Kearney

Brian Farmer
Amarillo College

Eveily Freeman
Ohio State University

Hari Garbharran
Middle Tennessee State University

Abe Goldman
University of Florida

Angela Gray
University of Wisconsin, Oshkosh

Ellen Hansen
Emporia State University

Nick Hill
Greenville Technical College

Johanna Hume
Alvin Community College

Edward Jackiewicz
California State University, Northridge

Rebecca Johns
University of South Florida, St. Petersburg

Suzanna Klaf
Ohio State University

Jeannine Koshear
Fresno City College

Brennan Kraxberger
Christopher Newport University

Heidi Lannon
Santa Fe College

Angelia Mance
Florida Community College, Jacksonville

Meredith Marsh
Lindenwood University

Linda Murphy
Blinn Community College

Monica Nyamwange
William Paterson University

Adam Pine
University of Minnesota, Duluth

Amanda Rees
Columbus State University

Benjamin Richason
St. Cloud State University

Amy Rock
Kent State University

Betty Shimshak
Towson University

Michael Siola
Chicago State University

Steve Smith
Missouri Southern State University

Jennifer Speights-Binet
Samford University

Emily Sturgess Cleek
Drury University

Gregory Taff
University of Memphis

Catherine Veninga
College of Charleston

Mark Welford
Georgia Southern University

Donald Williams
Western New England College

Peggy Robinson Wright
Arkansas State University, Jonesboro

WORLD REGIONAL GEOGRAPHY, FOURTH EDITION

Robert Acker
University of California, Berkeley

Joy Adams
Humboldt State University

John All
Western Kentucky University

Jeff Allender
University of Central Arkansas

David L. Anderson
Louisiana State University, Shreveport

Donna Arkowski
Pikes Peak Community College

Jeff Arnold
Southwestern Illinois College

Richard W. Benfield
Central Connecticut University

Sarah A. Blue
Northern Illinois University

Patricia Boudinot
George Mason University

Michael R. Busby
Murray State College

Norman Carter
California State University, Long Beach

Gabe Cherem
Eastern Michigan University

Brian L. Crawford
West Liberty State College

Phil Crossley
Western State College of Colorado

Gary Cummisk
Dickinson State University

Kevin M. Curtin
University of Texas, Dallas

Kenneth Dagel
Missouri Western State University

Jason Dittmer
University College London

Rupert Dobbin
University of West Georgia

James Doerner
University of Northern Colorado

Ralph Feese
Elmhurst College

Richard Grant
University of Miami

Ellen R. Hansen
Emporia State University

Holly Hapke
Eastern Carolina University

Mark L. Healy
Harper College

David Harms Holt
Miami University

Douglas A. Hurt
University of Central Oklahoma

Edward L. Jackiewicz
California State University, Northridge

Marti L. Klein
Saddleback College

Debra D. Kreitzer
Western Kentucky University

Jeff Lash
University of Houston, Clear Lake

Unna Lassiter
California State University, Long Beach

Max Lu
Kansas State University

Donald Lyons
University of North Texas

Shari L. MacLachlan
Palm Beach Community College

Chris Mayda
Eastern Michigan University

Armando V. Mendoza
Cypress College

Katherine Nashleanas
University of Nebraska, Lincoln

Joseph A. Naumann
University of Missouri, St. Louis

Jerry Nelson
Casper College

Michael G. Noll
Valdosta State University

Virginia Ochoa-Winemiller
Auburn University

Karl Offen
University of Oklahoma

Eileen O'Halloran
Foothill College

Ken Orvis
University of Tennessee

Manju Parikh
College of Saint Benedict and St. John's University

Mark W. Patterson
Kennesaw State University

Paul E. Phillips
Fort Hays State University

Rosann T. Poltrone
Arapahoe Community College, Littleton, Colorado

Waverly Ray
MiraCosta College

Jennifer Rogalsky
State University of New York, Geneseo

Gil Schmidt
University of Northern Colorado

Yda Schreuder
University of Delaware

Tim Schultz
Green River Community College, Auburn, Washington

Sinclair A. Sheers
George Mason University

D. James Siebert
North Harris Montgomery Community College, Kingwood

Dean Sinclair
Northwestern State University

Bonnie R. Sines
University of Northern Iowa

Vanessa Slinger-Friedman
Kennesaw State University

Andrew Sluyter
Louisiana State University

Kris Runberg Smith
Lindenwood University

Herschel Stern
MiraCosta College

William R. Strong
University of North Alabama

Ray Sumner
Long Beach City College

Rozemarijn Tarhule-Lips
University of Oklahoma

Alice L. Tym
University of Tennessee, Chattanooga

James A. Tyner
Kent State University

Robert Ulack
University of Kentucky

Jialing Wang
Slippery Rock University of Pennsylvania

Linda Q. Wang
University of South Carolina, Aiken

Keith Yearman
College of DuPage

Laura A. Zeeman
Red Rocks Community College

WORLD REGIONAL GEOGRAPHY, THIRD EDITION

Kathryn Alftine
California State University, Monterey Bay

Donna Arkowski
Pikes Peak Community College

Tim Bailey
Pittsburg State University

Brad Baltensperger
Michigan Technological University

Michele Barnaby
Pittsburg State University

Daniel Bedford
Weber State University

Richard Benfield
Central Connecticut State University

Sarah Brooks
University of Illinois, Chicago

Jeffrey Bury
University of Colorado, Boulder

Michael Busby
Murray State University

Norman Carter
California State University, Long Beach

Gary Cummisk
Dickinson State University

Cyrus Dawsey
Auburn University

Elizabeth Dunn
University of Colorado, Boulder

Margaret Foraker
Salisbury University

Robert Goodrich
University of Idaho

Steve Graves
California State University, Northridge

Ellen Hansen
Emporia State University

Sophia Harmes
Towson University

Mary Hayden
Pikes Peak Community College

R. D. K. Herman
Towson University

Samantha Kadar
California State University, Northridge

James Keese
California Polytechnic State University

Phil Klein
University of Northern Colorado

Debra D. Kreitzer
Western Kentucky University

Soren Larsen
Georgia Southern University

Unna Lassiter
California State University, Long Beach

David Lee
Florida Atlantic University

Anthony Paul Mannion
Kansas State University

Leah Manos
Northwest Missouri State University

Susan Martin
Michigan Technological University

Luke Marzen
Auburn University

Chris Mayda
Eastern Michigan University

Michael Modica
San Jacinto College

Heather Nicol
State University of West Georgia

Ken Orvis
University of Tennessee

Thomas Paradis
Northern Arizona University

Amanda Rees
University of Wyoming

Arlene Rengert
West Chester University of Pennsylvania

B. F. Richason
St. Cloud State University

Deborah Salazar
Texas Tech University

Steven Schnell
Kutztown University

Kathleen Schroeder
Appalachian State University

Roger Selya
University of Cincinnati

Dean Sinclair
Northwestern State University

Garrett Smith
Kennesaw State University

Jeffrey Smith
Kansas State University

Dean Stone
Scott Community College

Selima Sultana
Auburn University

Ray Sumner
Long Beach City College

Christopher Sutton
Western Illinois University

Harry Trendell
Kennesaw State University

Karen Trifonoff
Bloomsburg University

David Truly
Central Connecticut State University

Kelly Victor
Eastern Michigan University

Mark Welford
Georgia Southern University

Wendy Wolford
University of North Carolina, Chapel Hill

Laura A. Zeeman
Red Rocks Community College

WORLD REGIONAL GEOGRAPHY, SECOND EDITION

Helen Ruth Aspaas
Virginia Commonwealth University

Cynthia F. Atkins
Hopkinsville Community College

Timothy Bailey
Pittsburg State University

Robert Maxwell Beavers
University of Northern Colorado

James E. Bell
University of Colorado, Boulder

Richard W. Benfield
Central Connecticut State University

John T. Bowen Jr.
University of Wisconsin, Oshkosh

Stanley Brunn
University of Kentucky

Donald W. Buckwalter
Indiana University of Pennsylvania

Gary Cummisk
Dickinson State University

Roman Cybriwsky
Temple University

Cary W. de Wit
University of Alaska, Fairbanks

Ramesh Dhussa
Drake University

David M. Diggs
University of Northern Colorado

Jane H. Ehemann
Shippensburg University

Kim Elmore
University of North Carolina, Chapel Hill

Thomas Fogarty
University of Northern Iowa

James F. Fryman
University of Northern Iowa

Heidi Glaesel
Elon College

Ellen R. Hansen
Emporia State University

John E. Harmon
Central Connecticut State University

Michael Harrison
University of Southern Mississippi

Douglas Heffington
Middle Tennessee State University

Robert Hoffpauir
California State University, Northridge

Catherine Hooey
Pittsburg State University

Doc Horsley
Southern Illinois University, Carbondale

David J. Keeling
Western Kentucky University

James Keese
California Polytechnic State University

Debra D. Kreitzer
Western Kentucky University

Jim LeBeau
Southern Illinois University, Carbondale

Howell C. Lloyd
Miami University of Ohio

Judith L. Meyer
Southwest Missouri State University

Judith C. Mimbs
University of Tennessee, Chattanooga

Monica Nyamwange
William Paterson University

Thomas Paradis
Northern Arizona University

Firooza Pavri
Emporia State University

Timothy C. Pitts
Edinboro University of Pennsylvania

William Preston
California Polytechnic State University

Gordon M. Riedesel
Syracuse University

Joella Robinson
Houston Community College

Steven M. Schnell
Northwest Missouri State University

Kathleen Schroeder
Appalachian State University

Dean Sinclair
Northwestern State University

Robert A. Sirk
Austin Peay State University

William D. Solecki
Montclair State University

Wei Song
University of Wisconsin, Parkside

William Reese Strong
University of North Alabama

Selima Sultana
Auburn University

Suzanne Traub-Metlay
Front Range Community College

David J. Truly
Central Connecticut State University

Alice L. Tym
University of Tennessee, Chattanooga

WORLD REGIONAL GEOGRAPHY, FIRST EDITION

Helen Ruth Aspaas
Virginia Commonwealth University

Brad Bays
Oklahoma State University

Stanley Brunn
University of Kentucky

Altha Cravey
University of North Carolina, Chapel Hill

David Daniels
Central Missouri State University

Dydia DeLyser
Louisiana State University

James Doerner
University of Northern Colorado

Bryan Dorsey
Weber State University

Lorraine Dowler
Pennsylvania State University

Hari Garbharran
Middle Tennessee State University

Baher Ghosheh
Edinboro University of Pennsylvania

Janet Halpin
Chicago State University

Peter Halvorson
University of Connecticut

Michael Handley
Emporia State University

Robert Hoffpauir
California State University, Northridge

Glenn G. Hyman
International Center for Tropical Agriculture

David Keeling
Western Kentucky University

Thomas Klak
Miami University of Ohio

Darrell Kruger
Northeast Louisiana University

David Lanegran
Macalester College

David Lee
Florida Atlantic University

Calvin Masilela
West Virginia University

Janice Monk
University of Arizona

Heidi Nast
DePaul University

Katherine Nashleanas
University of Nebraska, Lincoln

Tim Oakes
University of Colorado, Boulder

Darren Purcell
Florida State University

Susan Roberts
University of Kentucky

Dennis Satterlee
Northeast Louisiana University

Kathleen Schroeder
Appalachian State University

Dona Stewart
Georgia State University

Ingolf Vogeler
University of Wisconsin, Eau Claire

Susan Walcott
Georgia State University

These world regional geography textbooks have been a family project many years in the making. Lydia Pulsipher came to the discipline of geography at the age of 5, when her immigrant father, Joe Mihelič, hung a world map over the breakfast table in their home in Coal City, Illinois, where he was pastor of the New Hope Presbyterian Church, and challenged her to locate such places as Istanbul, Chicago, and Vienna. The family soon moved to the Mississippi Valley of eastern Iowa, where Lydia's father, then a professor at the Presbyterian theological seminary in Dubuque, continued his geography lessons on the passing landscapes whenever Lydia accompanied him on Sunday trips to small country churches.

Lydia's sons, Anthony and Alex, got their first doses of geography in the bedtime stories she told them. For plots and settings, she drew on Caribbean colonial documents she was then reading for her dissertation. They first traveled abroad and learned about the hard labor of field geography when, at ages 12 and 8 respectively, they were expected to help with the archaeological and ethnographic research conducted by Lydia and her colleagues on the eastern Caribbean island of Montserrat. It was Lydia's brother John Mihelič who first suggested that Lydia, Alex, and Mac write

a book like this one, after he too came to appreciate geography. He has been a loyal cheerleader during the process, as have family and friends in Knoxville, Montserrat, California, Slovenia, and beyond.

For the seventh edition, Ola Johansson became an additional member of the author team. Although it is the first time his name appears on the cover, Ola is not a newcomer to the textbook: he has been a technical reviewer and chapter author during previous editions of both *World Regional Geography* and *World Regional Geography Concepts*. As a graduate of the University of Tennessee geography program, Ola has also known the Pulsiphers for years.

Graduate students and faculty colleagues in the geography department at the University of Tennessee have been generous in their support, serving as helpful impromptu sounding boards for ideas. Ken Orvis, especially, has advised the authors on the physical geography sections of all editions. Yingkui (Philippe) Li provided information on glaciers and climate change; Russell Kirby wrote one of the vignettes based on his research in Vietnam; Toby Applegate, Melanie Barron, Alex Pulsipher (in his capacity as an instructor), Michelle Brym, and Sara Beth Keough helped the authors understand how to better assist instructors; and Derek Alderman, Ron Kalafsky, Tom Bell, Margaret Gripshover, and Micheline Van Riemsdijk chatted with the authors many times on specific and broad issues related to this textbook.

Maps for this edition were conceived by Mac Goodwin and Alex Pulsipher and produced by Will Fontanez and the University of Tennessee cartography shop staff and by Maps.com under the direction of Mike Powers. Alex Pulsipher created and produced the photo essays and chose many of the photos used in the book.

Liz Widdicombe and Sara Tenney at W. H. Freeman were the first to suggest the idea that together we could develop a new direction for *World Regional Geography*, one that included the latest thinking in geography written in an accessible style and well-illustrated with attractive, relevant maps and photos. In accomplishing this goal, we are especially indebted to our first developmental editor, Susan Moran, and to the W. H. Freeman staff for all they have done in the first years and since to ensure that this book is well written, beautifully designed, and well presented to the public.

We would also like to gratefully acknowledge the efforts of the following people at W. H. Freeman for this seventh edition: Andrew Dunaway, executive editor; Jennifer Edwards, program manager; Kharissia Pettus, developmental editor; Kerry O'Shaughnessy, senior project editor; Shannon Howard, marketing manager; Anna Paganelli and Patti Brecht, copyeditors; Vicki Tomaselli, designer; Janice Donnola, art coordinator; Paul Rohloff, senior production supervisor; Sheena Goldstein, photo editor; and Lisa Passmore, photo researcher. We are also grateful to the supplements authors, who have created what we think are unusually useful, up-to-date, and labor-saving materials for instructors who use our book.

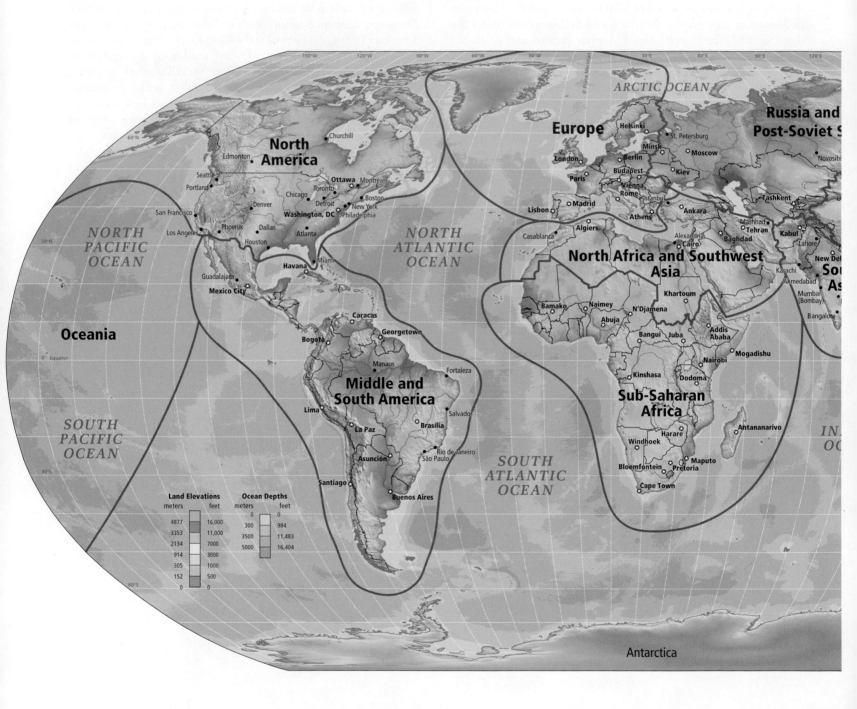

Land Elevations

meters	feet
4877	16,000
3353	11,000
2134	7000
914	3000
305	1000
152	500
0	0

Ocean Depths

meters	feet
0	0
300	984
3500	11,483
5000	16,404

ARCTIC OCEAN

Europe

Russia and Post-Soviet S

North America

NORTH PACIFIC OCEAN

NORTH ATLANTIC OCEAN

North Africa and Southwest Asia

Oceania

Middle and South America

SOUTH PACIFIC OCEAN

Sub-Saharan Africa

SOUTH ATLANTIC OCEAN

IN OC

Sou As

Antarctica

Churchill
Edmonton
Seattle
Portland
San Francisco
Los Angeles
Phoenix
Denver
Dallas
Houston
Atlanta
Chicago
Detroit
Toronto
Ottawa
Montréal
Boston
New York
Philadelphia
Washington, DC
Miami
Havana
Guadalajara
Mexico City
Caracas
Bogotá
Georgetown
Manaus
Fortaleza
Lima
Brasília
Salvador
La Paz
Asunción
Rio de Janeiro
São Paulo
Santiago
Buenos Aires

Helsinki
St. Petersburg
London
Minsk
Moscow
Berlin
Kiev
Budapest
Paris
Vienna
Rome
Istanbul
Lisbon
Madrid
Athens
Ankara
Novosibi
Tashkent
Casablanca
Algiers
Alexandria
Cairo
Mashhad
Tehran
Baghdad
Kabul
Lahore
New Del
Karachi
Ahmedabad
Mumbai (Bombay)
Bangalore
Bamako
Naimey
N'Djamena
Khartoum
Abuja
Bangui
Juba
Addis Ababa
Kinshasa
Nairobi
Mogadishu
Dodoma
Antananarivo
Windhoek
Harare
Bloemfontein
Pretoria
Maputo
Cape Town

0° Prime Meridian
Equator

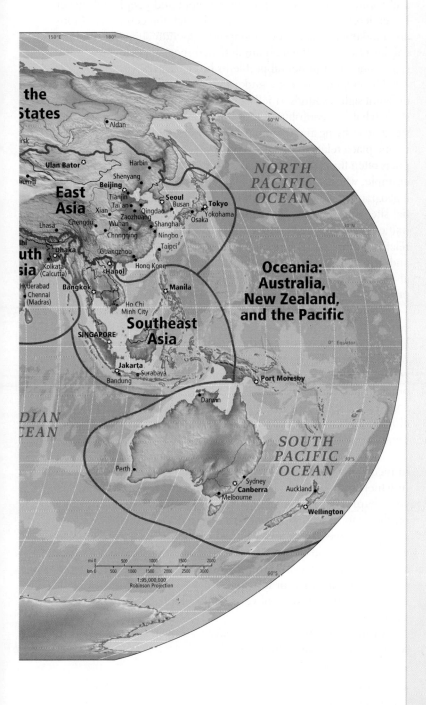

1

Geography: An Exploration of Connections

FIGURE 1.1 Regions of the world.

GEOGRAPHIC THEMES

After you read this chapter, you will be able to discuss the following geographic themes as they relate to the five thematic concepts:

1. Environment: Humans are altering the planet at an unprecedented rate, causing sometimes drastic effects on ecosystems and the climate. Multiple environmental factors often interact to influence the vulnerability of a location to the impacts of climate change. These vulnerabilities have a spatial pattern.

2. Globalization and Development: Global flows of information, goods, and people are transforming patterns of economic development. Local self-sufficiency is giving way to global interdependence as people and places are increasingly becoming connected, sometimes across vast distances.

3. Power and Politics: There are major differences across the world in the ways that power is wielded in societies. Modes of governing that are more authoritarian are based on the power of the state or community leaders. Modes that are based on notions of political freedom and democracy give the general public greater power over themselves and more of a role in deciding how policies are developed and governments are run. There are also many other ways of managing political power.

4. Urbanization: The development of manufacturing and service economies has taken place largely in cities, drawing people from the countryside into urban areas. Many of these new urbanites were displaced by the mechanization of agriculture in rural areas, which drastically reduced the need for agricultural labor and left people with no means of support and a reduced access to food.

5. Population and Gender: Population growth is slowing across the world for a number of reasons. Chief among these are urbanization and the lifestyle changes that it brings; the contraction of agricultural labor as a result of the mechanization of farming; and perhaps most important, the increasing numbers of women who are delaying childbearing as they pursue educational and work opportunities outside the home. The availability of birth control is an important contributing factor in determining success in family planning.

WHERE IS IT? WHY IS IT THERE? WHY DOES IT MATTER?

Where are you? You may be in a house or a library or sitting under a tree on a fine fall afternoon. You are probably in a community (perhaps a college or university), and you are in a country (perhaps the United States) and a region of the world (perhaps North America, Southeast Asia, or the Pacific). Why are you where you are? Some answers are immediate, such as "I have an assignment to read." Other explanations are more complex, such as your belief in the value of an education, your career plans, and your or someone's willingness to sacrifice to pay your tuition. Even past social movements that opened up higher education to more than a fortunate few may help explain why you are where you are.

The questions *where* and *why* are central to geography. Think about a time when you used your cell phone to pinpoint the site of a party on a Saturday night, or the location of the best Greek restaurant, or used a GPS to find the fastest and safest route home. In this age of smart phones, these tasks have become much easier, but the fundamental geographic issues of location, spatial relationships, and connections between the environment and people will never be rendered irrelevant by technology. In fact, the central questions and fundamental issues of geography profoundly shape technological devices, many of which are designed specifically to address the spatial and environmental problems humans face every day.

Geographers seek to understand why different places have different sights, sounds, smells, and arrangements of features. They study what has contributed to the look and feel of a place, to the standard of living and customs of the people, and to the way people in one place relate to people in other places. Furthermore, geographers often think on several scales, from the local to the global. For example, when choosing the best location for a new grocery store, a geographer might consider the physical characteristics of potential sites, the socioeconomic circumstances of the neighborhood, traffic patterns locally and in the city at large, the store's location relative to the main population concentrations for the whole city, and crucially, the location of competitors. She would probably also consider national or even international transportation routes, possibly to determine cost-efficient connections to suppliers.

To make it easier to understand a geographer's many interests, try this exercise. Draw a map of your most familiar childhood landscape. Relax, and recall the objects and experiences that were most important to you there. If the place was your neighborhood, you might start by drawing and labeling your home. Then fill in other places you encountered regularly, such as your backyard, your best friend's home, or your school. **FIGURE 1.2** shows the childhood landscape remembered by Julia Stump in Franklin, Tennessee.

Maps almost always have a depth of meaning that is not immediately apparent. Consider how your map reveals the ways in which your life was structured by space. What is the scale of your map? That is, how much space did you decide to illustrate on the map? The amount of space your map covers may represent the degree of freedom you had as a child, or how aware you were of the world around you. Were there places you were not supposed to go? Does your map reveal, perhaps subtly, such emotions as fear, pleasure, or longing? Does it indicate your sex, your ethnicity, or the makeup of your family? Did you use symbols to show certain features?

In making your map and analyzing it, you have engaged in several aspects of geography:

- Landscape observation and analysis
- Descriptions of Earth's surface and consideration of the natural environment
- Spatial analysis (the study of how people, objects, and ideas are related to one another across space)
- The use of different scales of analysis (your map probably shows the spatial features of your childhood at a detailed *local scale*)
- Cartography (the making of maps)

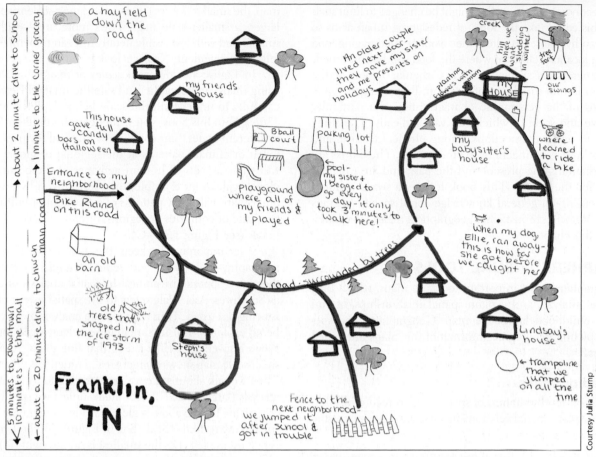

FIGURE 1.2 A childhood landscape map. Julia Stump drew this map of her childhood landscape in Franklin, Tennessee, as an exercise in Dr. Pulsipher's world geography class.

As you progress through this book and this course, you will acquire geographic information and skills that will help you achieve your goals, whatever they are. If you want to travel or work outside your hometown or simply understand local events within the context of world events, knowing how to practice geography will make your task easier and more engaging.

WHAT IS GEOGRAPHY?

Physical Geography and Human Geography

The primary concern of both physical geography and human geography is the study of Earth's surface and the interactive physical and human processes that shape the surface.

Geography is the study of our planet's surface and the processes that shape it. Yet this definition does not begin to convey the fascinating interactions of human and environmental forces that have given Earth its diverse landscapes and ways of life.

Geography as an academic discipline is unique in that it links the physical sciences—such as geology, physics, chemistry, biology, and botany—with the social sciences—such as anthropology, sociology, history, economics, and political science. **Physical geography** generally focuses on how Earth's physical processes work independently of humans, but physical geographers have become increasingly interested in how physical processes may affect humans and how humans affect these processes in return. **Human geography** is the study of the various aspects of

human life that create the distinctive landscapes and regions of the world. Physical geography and human geography are often tightly linked. For example, geographers might try to understand:

- How and why people came to occupy a particular place
- How people use the physical aspects of that place (climate, landforms, and resources) and then modify them to suit their particular needs
- How people may create environmental problems
- How people interact with other places, far and near

Geographers usually specialize in one or more fields of study, or subdisciplines. Some of these particular types of geography are mentioned over the course of the book. Despite their individual specialties, geographers often cooperate in studying **spatial interactions** between people and places and the **spatial distribution** of relevant phenomena. For example, in the face of increasing global warming, climatologists, cultural geographers, and economic geographers work together to understand the spatial distribution of carbon dioxide emissions, as well as the

physical geography the study of Earth's physical processes: how they work and interact, how they affect humans, and how they are affected by humans

human geography the study of patterns and processes that have shaped human understanding, use, and alteration of Earth's surface

spatial interactions the flow of goods, people, services, or information across space and among places

spatial distribution the arrangement of a phenomenon across Earth's surface

cultural and economic practices that might be changed to limit such emissions. This could take the form of redesigning urban areas so that people can live closer to where they work or encouraging food production in locations closer to where the food will be consumed.

Many geographers specialize in a particular region of the world, or even in one small part of a region. Regional geography is the analysis of the geographic characteristics of a particular place, the size and scale of which can vary radically. The study of a region can reveal connections among physical features and ways of life, as well as connections to other places. These links are key to understanding the present and the past, and are essential in planning for the future. This book follows a *world regional* approach, focusing on general knowledge about specific regions of the world. We will see just what geographers mean by *region* a little later in this chapter.

GEOGRAPHERS' VISUAL TOOLS

Among geographers' most important tools are maps, which they use to record, analyze, and explain spatial relationships, as you did on your childhood landscape map. Geographers who specialize in depicting geographic information on maps are called **cartographers.**

Understanding Maps

A map is a visual representation of space used to record, display, analyze, and explain spatial relationships. **FIGURE 1.3** explains the various features of maps.

Legend and Scale The first thing to check on a map is the **legend,** which is usually a small box somewhere on the map that provides basic information about how to read the map, such as the meaning of the symbols and colors used (see parts A–C and the Legend box in Figure 1.3). Sometimes the scale of the map is also given in the legend.

In cartography, *scale* has a slightly different meaning than it does in general geographic analysis. **Scale** on a map refers to the relationship between the size of objects on the map and the actual size they have on the surface of Earth. It is usually represented by a scale bar (see Figure 1.3D–G) but is also sometimes represented by a ratio (for example 1:8000) or a fraction (1/8000), which indicates that one unit of measure on the map equals a particular number of units on the ground. For example, 1:8000 inches means that 1 inch on the map represents 8000 inches (or about an eighth of a mile) on the surface of Earth (1 mile = 63,360 inches divided by 8 = 7,920 inches).

A scale of 1/800 is considered larger than a scale of 1/8000 because the features on a 1/800 scale map are larger and can be shown in more detail. The larger the scale of the

map, the smaller the area it covers. A larger-scale map shows objects larger; a smaller-scale map shows more area—with each object smaller and with less visible detail. You can remember this with the following statement: "Objects look larger on a larger-scale map."

In Figure 1.3, different *scales of imagery* are demonstrated using maps, photographs, and satellite images. Read the captions carefully to understand the scale being depicted in each image. Throughout this book, you will encounter different kinds of maps at different scales. Some will show physical features, such as landforms or climate patterns at the regional or global scale. Others will show aspects of human activities at these same regional or global scales—for example, the routes taken by drug traders. Yet other maps will show patterns of settlement or cultural features at the scale of countries or regions, or cities or even local neighborhoods (see Figure 1.3D–G).

It is important to keep the two types of scales used in geography—*map scale* and *scale of analysis*—distinct, because they have opposite meanings! In spatial analysis of a region, such as Southwest Asia, scale refers to the spatial extent of the area that is being discussed. Thus a large-scale analysis means a large area is being explored. But in cartography, a large-scale map is one that shows a given area enlarged so that fine detail is visible, while a small-scale map shows a larger area in much less detail. In this book, when we talk about scale we are referring to its meaning in spatial analysis (larger scale = larger area), unless we specifically indicate that we are talking about scale as used in cartography (larger scale = smaller area). In the Scale box in Figure 1.3, the largest-scale map is that on the left (D); the smallest is on the right (G).

Longitude and Latitude Most maps contain lines of latitude and longitude, which enable a person to establish a position on the map relative to other points on the globe. Lines of **longitude** (also called *meridians*) run from pole to pole; lines of **latitude** (also called *parallels*) run around Earth parallel to the equator (see Figure 1.3H).

Both latitude and longitude lines describe circles, so there are 360° (the symbol ° refers to degrees) in each circle of latitude and 180° in each pole-to-pole semicircle of longitude. Each degree spans 60 minutes (minutes are designated with the symbol ′), and each minute has 60 seconds (which are designated with the symbol ″). Keep in mind that these are measures of relative linear space on a circle, not measures of time. They do not even represent real distance because the circles of latitude get successively smaller to the north and south of the equator until they become virtual dots at the poles.

The globe is also divided into hemispheres. The Northern and Southern hemispheres are on either side of the equator. The Western and Eastern hemispheres are defined as follows. The prime meridian, 0° longitude, runs from the North Pole through Greenwich, England, to the South Pole. Greenwich was chosen in the seventeenth century as the primary reference point for the global system of longitude because it was the location of the British Royal observatory. As the dominant global colonial power, Britain needed accurate methods of determining the longitude of any particular place. The half of the globe's surface west of the prime meridian is called the Western Hemisphere; the half to the east is called the Eastern Hemisphere. The longitude lines both east and west of the prime meridian are labeled from 1° to 180° by

cartographers geographers who specialize in depicting geographic information on maps

legend a small box somewhere on a map that provides basic information about how to read the map, such as the meaning of the symbols and colors used

scale (of a map) the proportion that relates the dimensions of the map to the dimensions of the area it represents; also, variable-sized units of geographical analysis from the local scale to the regional scale to the global scale

longitude the distance in degrees east and west of Greenwich, England; lines of longitude, also called meridians, run from pole to pole (the line of longitude at Greenwich is 0° and is known as the prime meridian)

latitude the distance in degrees north or south of the equator; lines of latitude run parallel to the equator, and are also called parallels

FIGURE 1.3 Understanding maps.

The Legend

Being able to read a map legend is crucial to understanding the maps in this book. The colors in the legend convey information about different areas on the map. In the population density map below, the lowest density (0–3 persons per square mile), is colored light tan. A part of North America with this density is shown in the map inset to the right of the legend. On the far right is a picture of this area. Two other densities (27–260 per square mile and more than 2600 people per square mile) are also shown in this manner.

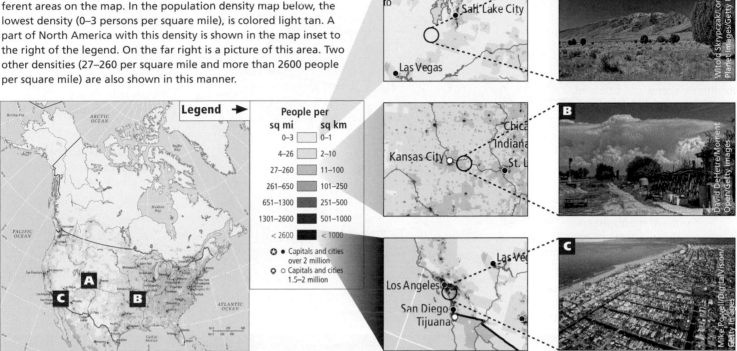

Scale

Maps often display information at different spatial scales, which means that lengths, areas, distances, and sizes can appear dramatically different on otherwise similar maps. This book often combines maps at several different scales with photographs taken by people at Earth's surface and photographs taken by satellites or astronauts in space. All of these visual tools convey information at a spatial scale. Here are some of the map scales you might encounter in this book. The scale is visible below each image.

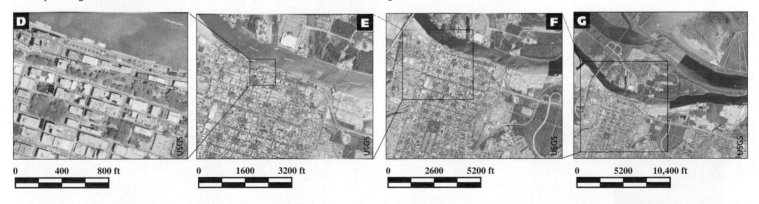

Representation of Scale

Here are some representations of map scale that you may encounter on maps in this book and elsewhere.

(A) This scale bar means that the length of the entire box represents 8000 feet on the ground.

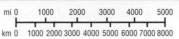

(B) This scale bar works like the one on the left, but also gives lengths in miles and kilometers.

1:8000

(C) This means that 1 unit of measure (an inch, for example) equals 8000 similar units of measure on the ground.

FIGURE 1.3 Understanding maps. *(continued)*

Latitude and Longitude

(**H**) Lines of longitude and latitude form a global scale grid that can be used to designate the location of any place on the planet.

The prime meridian is at zero degrees longitude and passes through Greenwich, England. The counterpart to the prime meridian on the opposite side of the globe is the international date line.

The equator divides the globe into Northern and Southern hemispheres.

All lines of longitude or meridians are of equal length.

The distance between lines of longitude decreases toward the poles.

Lines of latitude decrease in length as they approach the poles.

Lines of latitude and longitude intersect at right angles.

Lines of latitude are parallel to each other.

The equator is at zero degrees latitude.

The half of the globe's surface west of the prime meridian is called the Western Hemisphere; the half to the east is called the Eastern Hemisphere.

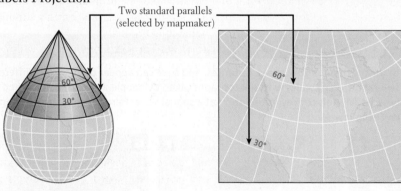

Projections

(**I**) Albers Projection

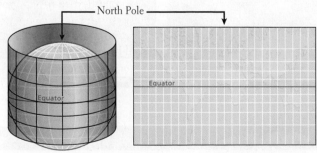

Two standard parallels (selected by mapmaker)

Pros: Minimal distortion near two parallels (lines of latitude)

Cons: Areas farther away from these lines are distorted.

(**J**) Mercator Projection

North Pole

Pros: A straight line between two points on this map gives an accurate compass direction between them. Minimal distortion within 15 degrees of the equator.

Cons: Extreme distortion near the poles, especially above 60° latitude.

(**K**) Robinson Projection

North Pole

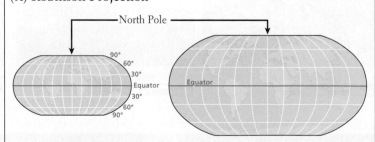

Pros: Uninterrupted view of land and ocean. Less distortion in high latitudes than in the Mercator projection.

Cons: The shapes of landmasses are slightly distorted due to the curvature of the longitude lines.

their direction and distance in degrees from the prime meridian. For example, 20 degrees east longitude would be written as 20° E. The longitude line at 180° runs through the Pacific Ocean and is used roughly as the International Date Line; the calendar day officially begins when midnight falls at this line.

The equator divides the globe into the Northern and Southern hemispheres. Latitude is measured from 0° at the equator to 90° at the North or South poles.

Lines of longitude and latitude form a grid that can be used to designate the location of a place. In Figure 1.3H, notice the dot that marks the location of Khartoum below the 20th parallel in eastern Africa. The position of Khartoum is 15° 35′ 1″ N latitude by 32° 32′ 3″ E longitude.

Map Projections Printed maps must solve the problem of showing the spherical Earth on a flat piece of paper. Imagine drawing a map of Earth on an orange, peeling the orange, and then trying to flatten out the orange-peel map and transfer it exactly to a flat piece of paper. The various ways of showing the spherical surface of Earth on flat paper are called **map projections.** All projections create some distortion. For maps of small parts of Earth's surface, the distortion is minimal. Developing a projection for the whole surface of Earth that minimizes distortion is much more challenging.

For large midlatitude regions of Earth that are mainly east/west in extent (North America, Europe, China, Russia), an *Albers projection* is often used. As you can see in Figure 1.3I, this is a conic, or cone-shaped, projection. The cartographer chooses two standard parallels (lines of latitude) on which to orient the map, and these parallels have no distortion. Areas along and between these parallels display minimal distortion. Areas farther to the north or south of the chosen parallels have more distortion. Although all areas on the map are proportional to areas on the ground, distortion of actual shape is inherent in the projection because, as previously discussed, parts of the globe are being projected onto flat paper.

The *Mercator projection* (see Figure 1.3J) has long been used by the general public and the media, but geographers rarely use this projection because of its gross distortion near the poles. Remember the poles are actually just points (see Figure 1.3H); but to make his flat map, the Flemish cartographer Gerhardus Mercator (1512–1594) stretched out the poles, depicting them as lines equal in length to the equator—a very large distortion! As a result, for example, Greenland appears about as large as Africa, even though it is only about one-fourteenth Africa's size. Nevertheless, the Mercator projection is still useful for navigation because it portrays the shapes of landmasses more or less accurately, and because a straight line between two points on this map gives the compass direction between them. However, actual distance measurements are distorted north and south of the equator.

The *Robinson projection* (see Figure 1.3K) shows the longitude lines bending toward the poles to give an impression of Earth's curvature, and it has the advantage of showing an uninterrupted view of land and ocean; however, as a result, the shapes and sizes of landmasses can be quite distorted (but much less than in Mercator maps) and the poles are still stretched out, not the points they are in reality. In this book we often use the Robinson projection for world maps because they are the lesser of several evils. The reader is advised to keep in mind the spatial distortions of all map projections.

Maps are not politically unbiased. Most currently popular world map projections reflect the European origins of modern cartography. For example, Europe or North America is often placed near the center of the map, where distortion is minimal; other population centers, such as East Asia, are placed at the highly distorted periphery. Therefore, for a less-biased study of the modern world, we need world maps that center on different parts of the globe. Another source of bias in world maps is the convention that north is commonly at the top of the map. Some cartographers think that this can lead to a subconscious assumption that the Northern Hemisphere is somehow superior to the Southern Hemisphere.

Geographic Information Science (GISc)

The acronym **GISc** is now widespread and usually refers to **geographic information science,** the body of science that supports spatial analysis technologies. GISc is multidisciplinary, using techniques from cartography (mapmaking), geodesy (measuring Earth's surface), and photogrammetry (the science of making reliable measurements, especially by using aerial photography). Other sciences, such as cognitive psychology and spatial statistics (geomatics or geoinformatics), are increasingly being used to give more depth and breadth to three-dimensional spatial analysis. GISc, then, can be used in medicine to analyze the human body, in engineering to analyze mechanical devices, in architecture to analyze buildings, in archaeology to analyze sites above and below ground, and in geography to analyze Earth's surface and the space above and below Earth's surface.

GISc is a burgeoning field in geography, with a wide variety of practical applications in government and business and in assessing and improving human and environmental conditions. (*GIS* [without the c] is an older term that refers to geographic information *systems* and describes the computerized analytical systems that are the tools of this newest of spatial sciences.)

The now widespread use of GISc, particularly by governments and corporations, has dramatically increased the amount of information that is collected and stored, and has changed the way those data are analyzed and distributed. These changes create many new possibilities for solving problems; for example, they make it easier for local governments to plan future urban growth. However, these technologies also raise serious ethical questions. What rights do people have over the storage, analysis, and distribution of information about their location and movements, which can now be gathered from their cell phones? Should this information reside in the public domain? Should individuals have the right to have their location-based information suppressed from public view? Should a government or corporation have the right to sell information to anyone, without special permission, about where people spend their time and how frequently they go to particular places? Progress on these societal questions has not kept pace with the technological advances in GISc.

map projections the various ways of showing the spherical Earth on a flat surface

geographic information science (GISc) the body of science that supports multiple spatial analysis technologies and keeps them at the cutting edge

THE DETECTIVE WORK OF PHOTO INTERPRETATION

Most geographers use photographs to help them understand or explain a geographic issue or depict the character of a place. Interpreting a photo to extract its geographical information can sometimes be like detective work. Below are some points to keep in mind as you look at the pictures throughout this book. Try them out first with the photo in **FIGURE 1.4**.

(A) Landforms: Notice the lay of the land and the landform features. Is there any indication of how the landforms and humans have influenced each other? Is environmental stress visible?

(B) Vegetation: Notice whether the vegetation indicates a wet or dry, or warm or cold environment. Can you recognize specific species? Does the vegetation appear to be natural or influenced by human use?

(C) Material culture: Are there buildings, tools, clothing, agricultural products, plantings, or vehicles that give clues about the cultural background, wealth, values, or aesthetics of the people who live where the picture was taken?

(D) People: What does the person in the photo suggest about the situation pictured?

(E) Economy: Can you see evidence of the global economy, such as goods that probably were not produced locally?

(F) Location: From your observations, can you tell where the picture was taken or narrow down the possible locations?

You can use this system to analyze any of the photos in this book and anywhere else. Practice by analyzing the photos in this book before you read their captions. Here is an example of how you could do this with Figure 1.4:

(A) Landforms:
1. The flat horizon suggests a plain or a river delta.
   Environmental stress is visible in several places.
2. This oily liquid doesn't look natural. Could it be crude oil?

What would have caused the landscape transformation? Maybe an oil spill?

(B) Vegetation:
3. This looks like fairly thick and varied vegetation. These could be palm trees and other types of plant life found in tropical climates. There seems to be a lot of dead vegetation in the foreground.
   Must be fairly wet and warm, possibly tropical. Since some vegetation is still green, perhaps the blackened plants were poisoned.

(C) Material culture:
There is not much that is obviously material culture here, except a single person and possibly some spilled oil. The whole area might be abandoned.

(D) People:
4. The clothing on this person doesn't look like he made it. It looks mass-produced.
   This suggests that he has access to goods produced some distance away, maybe in a nearby city. Or he could buy things in a market where imported goods are sold.

(E) Global economy:
See (A2) and (C): Could this be an oil spill? If that is oil, how did it get here? See (D): The person is wearing manufactured clothes, is carrying what look like Crocs, and has some trendy bracelets, all of which indicate that he is participating in the global marketplace.

(F) Location:
This could be somewhere tropical where there could have been an oil spill. *Hint:* Use this book! Look at Figure 6.19 to see the member countries of OPEC (the Organization of the Petroleum Exporting Countries). The combination of the possible oil spill and the vegetation suggests that the photo could be of Venezuela, Ecuador, Nigeria, Angola, the American South, or Indonesia. Suggestion: To further narrow your guesses, quickly look through chapters 3, 7, and 10!

FIGURE 1.4 Oil and the environment. An international development company began extracting products from this area 50 years ago. A recent UN report stated that the area now needs a massive cleanup, which could take up to 30 years and cost more than a billion dollars, making it one of the biggest such efforts in the world. The area has had approximately 300 incidences of oil-related pollution each year since the 1970s, causing an unknown number of deaths. Can you guess where this is?

Pius Utomi Ekpei/AFP/Getty Images

THE REGION AS A CONCEPT

Regions

The concept of *region* is useful to geographers because it allows them to break up the world into manageable units in order to analyze and compare spatial relationships (Figure 1.1). Nonetheless, regions do not have rigid definitions and their boundaries are fluid.

A **region** is a unit of Earth's surface that contains distinct patterns of physical features and/or distinct patterns of human development. It could be a desert region, a region that produces rice, or a region in which there is ethnic violence. Geographers rarely use the same set of attributes to describe any two regions. For example, the region of the southern United States might be defined by its distinctive vegetation, architecture, music, foods, and historical experience, while Siberia, in eastern Russia, could be defined primarily by its climate, vegetation, remoteness, and sparse settlement.

Another issue in defining regions is that they may shift over time. The people and the land they occupy may change so drastically in character that they can no longer be thought of as belonging to a certain region, and become more closely aligned with another, perhaps adjacent, region. Examples of this are countries in Central Europe, such as Poland and Hungary, which, for more than 40 years, were closely aligned with Russia and the Soviet Union, a vast region that stretched across northern Eurasia to the Pacific **(FIGURE 1.5A)**. Poland and Hungary's borders with western Europe were highly militarized and shut to travelers. The Soviet Union collapsed in the early 1990s, bringing drastic political and economic changes to the region, and in 2004, Poland and Hungary became members of the European Union (EU; see Figure 1.5B). Their western borders are now open, while their eastern borders are now more heavily guarded in order to keep unwelcome immigrants and other influences out of the European Union. But through all this change—on the ground, in the border regions—people share cultural features (language, religion, historical connections) even

> **region** a unit of Earth's surface that contains distinct patterns of physical features and/or distinct patterns of human development

FIGURE 1.5 Changing country alliances and relationships in Europe (A) before 1989 and (B) in 2016.

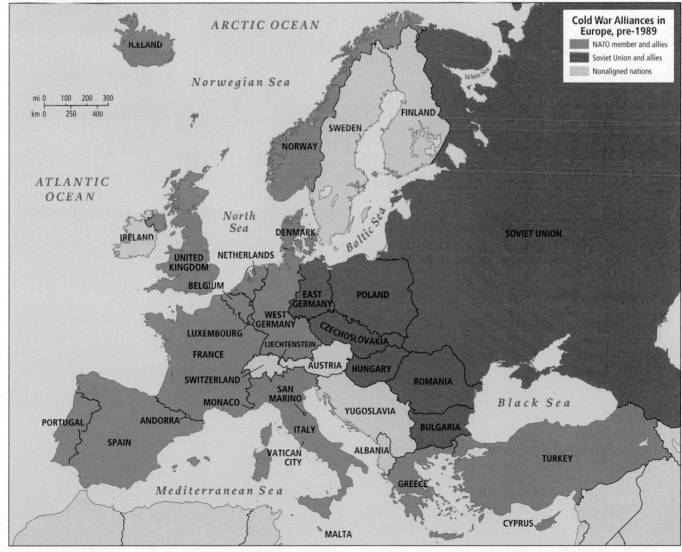

(A) Pre-1989 alignment of countries in Europe and the Soviet Union.

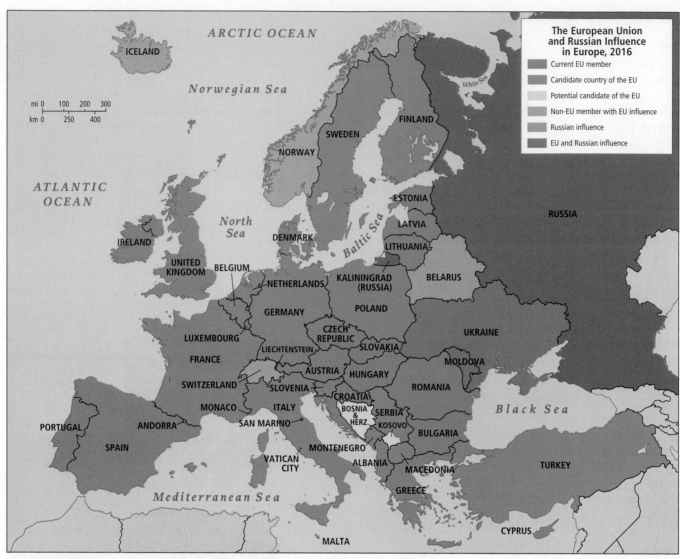

(B) Post-2016 alignments of the European Union and of Russia and the post-Soviet states.

as they may define each other as having very different regional allegiances. For this reason, we say regional borders can be *fuzzy*, meaning that they are hard to determine precisely.

A recurring problem in world regional textbooks is the changing nature of regional boundaries and the fact that on the ground they are not clear lines but linear zones of fuzziness. For example, when we first designed this textbook in the mid-1990s, the changes to Europe were just beginning; its eastern limits were under revision as the Soviet Union disintegrated. There were also hints that as the Soviet Union disappeared, countries of Central Asia, long within the Soviet sphere—places like Kazakhstan, Turkmenistan, Kyrgyzstan, Tajikistan, and Uzbekistan—should perhaps have been defined as constituting a new region of their own. Some suggested that these four, plus Turkey, Iraq, Iran, Saudi Arabia, the United Arab Emirates, Afghanistan, Pakistan, and possibly even western China should become a new post-Soviet world region of Central Asia **(FIGURE 1.6)**. The suggestion was that this region would be defined by what was thought to be a common religious heritage (Islam), plus long-standing cultural and historical ties and a difficult environment marked by water scarcity but also by rich oil and gas resources.

In fact, grouping these countries together in a single region would misrepresent the current situation as well as the past. A world region known as Central Asia may emerge eventually (and there are groups of interested parties discussing that possibility right now), but such a region will be slow to take shape and will be based on criteria very different from common religion, historical experiences, and environmental features. First of all, these supposed uniting features are, on closer inspection, quite diverse. There are many different versions of Islam practiced from the Mediterranean to western China and from southern Russia to the Arabian Sea. Also, to the extent that there are common historical experiences, they are actually linked more to European and Russian colonial exploitation than to an ancient and deeply uniting Central Asian cultural heritage. Finally, while environments generally defined by water scarcity are common to all the countries listed, oil and gas resources are not uniformly distributed at all. It could be that a Central Asian identity will eventually develop, perhaps centered on the leadership of Turkey or Dubai—the latter is trying hard to define itself as the affluent capital of such a region—but thus far, the region has not coalesced. Therefore, in this book, the countries listed

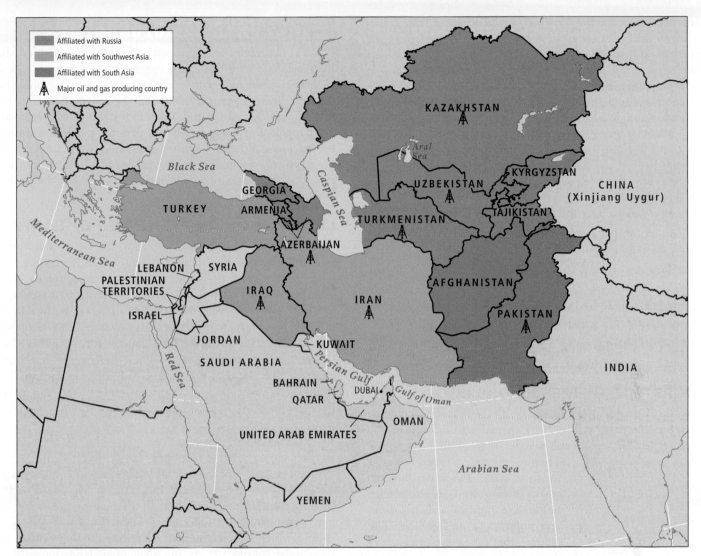

FIGURE 1.6 Hypothetical map of a possible Central Asian world region. A hypothetical region called Central Asia could consist of Kazakhstan, Turkmenistan, Kyrgyzstan, Tajikistan, and Uzbekistan, plus Turkey, Iraq, Iran, the Caucasus (Georgia, Armenia, and Azerbaijan), and also Afghanistan and Pakistan. For cultural reasons, it might even include parts of western China (Xinjiang Uygur), but we will not consider that here. The Caucasus countries have had a long association with Central Asia; economically, they might find this a better association than one with Europe, which is halfheartedly courting them for their oil. Just which region Saudi Arabia, Jordan, Syria, Yemen, and Israel would fall into is debatable—perhaps a region called the Eastern Mediterranean and North Africa. Despite their fringe location on the Arabian Peninsula, the Emirates (especially Dubai) appear to be angling for a leading economic role in a new Central Asian region.

above are still to be found in the three different regions shown in Figure 1.6.

If regions are so difficult to define and describe, why do geographers use them? To discuss the whole world at once would be impossible, so geographers try to find a reasonable way to divide the world into manageable parts. There is nothing sacred about the criteria or the boundaries for the world regions we use. They are just practical aids to learning. In defining each of the world regions for this book, we have considered such factors as physical features, political boundaries, cultural characteristics, history, how the places now define themselves, and what the future may hold. We are constantly reevaluating regional boundaries and, in this edition, have made some changes. For

example, the troubled new country of South Sudan, once part of North Africa and Southwest Asia, is now considered part of sub-Saharan Africa, to which it is more culturally aligned (see Chapter 7).

This book organizes the material into three *regional scales of analysis*: the **global scale,** the *world regional scale*, and the **local scale.** The term *scale of analysis* refers to the relative size of the area under discussion. At the global scale, explored in this chapter, the entire world is treated as a single area—a unity that is more and more relevant as our planet operates as a

global scale the level of geography that encompasses the entire world as a single unified area

local scale the level of geography that describes the space where an individual lives or works; a city, town, or rural area

global system. We use the term *world region* for the largest divisions of the globe, such as East Asia and North America (see Figure 1.1). We have defined ten world regions, each of which is covered in a separate chapter. In each regional chapter, we consider the interactions of human geography and physical geography in relation to the five thematic concepts: environment, globalization and development, power and politics, urbanization, and population and gender. These concepts then allow the reader to make comparisons across regions. Because people live their daily lives at the local scale—in villages, towns, and city neighborhoods—this book shows through vignettes how global or regional patterns affect individuals where they live.

In summary, regions have the following traits:

- A region is a unit of Earth's surface that contains distinct environmental or cultural patterns.

- No two regions are necessarily defined by the same set of attributes.

- Regional definitions and the territory included can change.

- On the ground, the boundaries of regions are usually indistinct and hard to agree upon.

- Regions can vary greatly in size (scale).

GEOGRAPHIC THEMES IN THIS BOOK

Within the world regional framework, this book is also organized around five thematic concepts of special significance in the modern world. These concepts are the focus of the Geographic Themes offered in every chapter in this book: the *environment, globalization and development, power and politics, urbanization,* and *population and gender*. The sections that follow explain each of these five thematic concepts, how they tie in with the main concerns of geographers, and how they are related to each other.

landforms physical features of Earth's surface, such as mountain ranges, river valleys, basins, and cliffs

plate tectonics the scientific theory that Earth's surface is composed of large plates that float on top of an underlying layer of molten rock; the movement and interaction of the plates create many of the large features of Earth's surface, particularly mountains

THINGS TO REMEMBER

- Geographers are interested in questions about where and why.

- Physical geographers study Earth's surface and the processes that shape it. There is a continual interaction between physical and human processes.

- Among geographers' most important tools are maps, which they use to record, analyze, and explain spatial relationships.

- The careful analysis of landscapes (including photographs) can lead to important understandings about places and geographic issues.

- A region is a unit of Earth's surface that has a combination of distinct physical and/or human features; the complex of features can vary from region to region, and regional boundaries are rarely clear or precise.

PHYSICAL GEOGRAPHY

Physical geography is concerned with the processes that shape Earth's landforms, climate, and vegetation. In this sense, physical geographers are similar to scientists from other disciplines who focus on these phenomena, though as geographers they often look at problems spatially and depict the results of their analysis on maps. In this book, physical geography provides a backdrop for the many aspects of human geography we discuss. Physical geography is a fascinating, large, and growing field of study worth exploring in greater detail. What follows are just the basics of landforms and climate.

LANDFORMS: THE SCULPTING OF EARTH

The processes that create the world's varied **landforms**—mountain ranges, continents, and the deep ocean floor—are some of the most powerful and slow-moving forces on the planet. Originating deep beneath Earth's surface, these *internal processes* can move entire continents, often taking hundreds of millions of years to do their work. However, it is *external processes* that form many of Earth's landscape features, such as a beautiful waterfall or a rolling plain. These more rapid and delicate changes take place on the surface of Earth. Geomorphologists study these two processes that constantly shape and reshape Earth's surface.

PLATE TECTONICS

Two key ideas related to internal processes in physical geography are the *Pangaea hypothesis* and *plate tectonics*. The geophysicist Alfred Wegener first suggested the Pangaea hypothesis in 1912. This hypothesis proposes that all the continents were once joined in a single vast continent called Pangaea (meaning "all lands"), which then fragmented over time into the continents we know today **(FIGURE 1.7)**. As one piece of evidence for his theory, Wegener pointed to the neat fit between the west coast of Africa and the east coast of South America.

For decades, most scientists rejected Wegener's hypothesis. We now know, however, that Earth's continents have been assembled into supercontinents a number of times, only to break apart again. All of this activity is made possible by plate tectonics, a process of continental motion discovered in the 1960s, long after Wegener's time.

The study of **plate tectonics** has shown that Earth's surface is composed of large plates that float on top of an underlying layer of molten rock. The plates are composed of two types of "crust," or cooled and hardened rock. Oceanic crust is located under the oceans, where intense pressures created by the large volumes of water makes it dense and relatively thin. Continental crust forms the continents where much lower pressures exerted by the atmosphere make it thicker and less dense. These massive plates drift slowly, driven by the circulation of the underlying molten rock flowing from hot regions deep inside Earth to cooler surface regions and back. The creeping movement of tectonic plates fragmented and separated Pangaea into pieces that are the continents we know today (see Figure 1.7E).

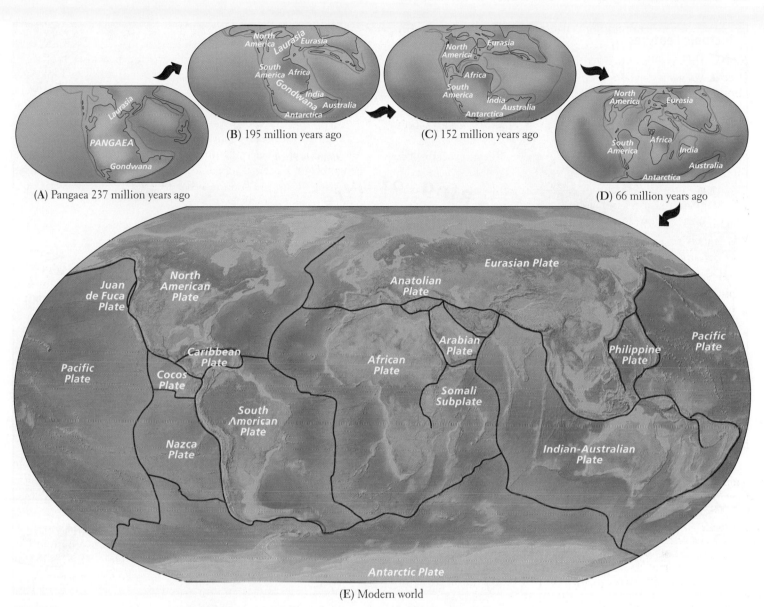

FIGURE 1.7 The breakup of Pangaea. (A) through **(D)** show the changing arrangement of Earth's land surfaces, Pangaea, over 237 million years. The modern world map **(E)** depicts the current boundaries of the major tectonic plates. Pangaea is only the latest of several global configurations that have coalesced and then fragmented over the last billion years. [Source consulted: Frank Press, Raymond Siever, John Grotzinger, and Thomas H. Jordan, *Understanding Earth*, 4th ed. (New York: W. H. Freeman, 2004), pp. 42–43]

Plate movements influence the shapes of major landforms, such as continental shorelines and mountain ranges. Huge mountains have piled up on the leading edges of the continents as the plates carrying them collide with other plates, folding and warping in the process. Plate tectonics accounts for the long, linear mountain ranges that extend from Alaska to Chile in the Western Hemisphere and from Southeast Asia to the European Alps in the Eastern Hemisphere. The highest mountain range in the world, the Himalayas of South Asia, was created when what is now India, situated at the northern end of the Indian-Australian Plate, ground into Eurasia. The only continent that lacks these long, linear mountain ranges is Africa. Often called the "plateau continent," Africa is believed to have been at the center of Pangaea and to have moved relatively little since the breakup. However, as Figure 1.7E shows,

parts of eastern Africa—the Somali Subplate and the Arabian Plate—continue to separate from the continent (the African Plate).

Humans encounter tectonic forces most directly as earthquakes and volcanoes. Plates slipping past each other create the catastrophic shaking of the landscape we know as an earthquake. Plates collide and one may slip under the other in a process called *subduction*. Volcanoes arise at zones of subduction or sometimes in the middle of a plate, where gases and molten rock (called *magma*) can rise to Earth's surface through fissures and holes in the plate. Volcanoes and earthquakes are particularly common around the edges of the Pacific Ocean, an area known as the **Ring of Fire (FIGURE 1.8)**.

Ring of Fire the tectonic plate junctures around the edges of the Pacific Ocean; characterized by volcanoes and earthquakes

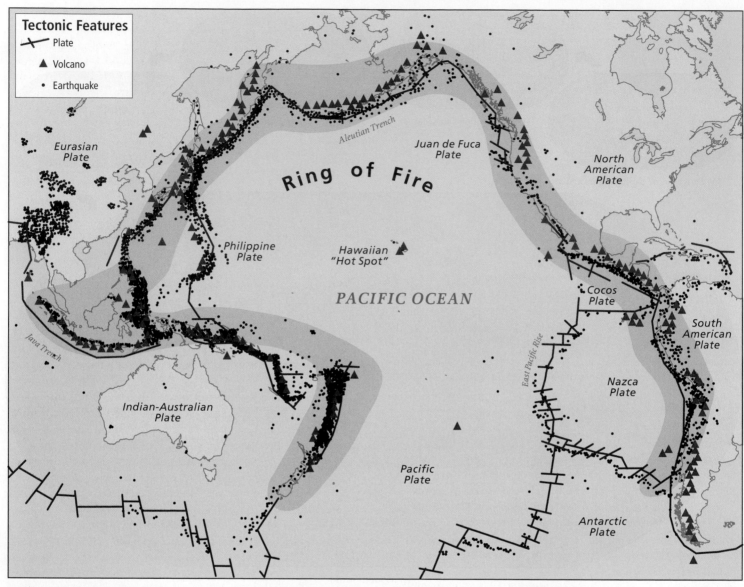

FIGURE 1.8 Ring of Fire. Volcanic formations encircling the Pacific Basin form the Ring of Fire, a zone of frequent earthquakes and volcanic eruptions. [Sources consulted: United States Geological Survey, Active Volcanoes and Plate Tectonics, "Hot Spots" and the "Ring of Fire," at http:/vulcan.wr.usgs.gov/Glossary/PlateTectonics/Maps/map_plate_tectonics_world.html; and Frank Press, Raymond Siever, John Grotzinger, and Thomas H. Jordan, *Understanding Earth*, 4th ed. (New York: W. H. Freeman, 2004), p. 27]

LANDSCAPE PROCESSES

weathering the physical or chemical decomposition of rocks by sunlight, wind, rain, snow, ice, freezing and thawing, and the effects of life-forms

erosion the process by which fragmented rock and soil are moved over a distance, primarily by wind and water

floodplain the flat land along a river where sediment is deposited during flooding

delta the triangular-shaped plain of sediment that forms where a river meets the sea

The landforms created by plate tectonics have been further shaped by external processes, which are more familiar to us because we can observe them daily. One such process is **weathering.** Rock, exposed to the onslaught of sunlight, wind, rain, snow, ice, freezing and thawing, and the effects of life-forms (such as plant roots), fractures and decomposes into tiny pieces. These particles then become subject to another external process, **erosion.**

During erosion, wind and water wear down and carry away rock particles and any associated decayed organic matter and deposit them in new locations. The deposition of eroded material can raise and flatten the land around a river, where periodic flooding spreads huge quantities of silt. As valleys between hills are filled in by silt, a **floodplain** is created. Where rivers meet the sea, floodplains often fan out roughly in the shape of a triangle, creating a **delta.** External processes tend to smooth out the dramatic mountains and valleys created by internal processes.

Human activity often contributes to external landscape processes. By altering the vegetation cover, agriculture and forestry expose Earth's surface to sunlight, wind, and rain. These agents in turn increase weathering and erosion. Flooding becomes

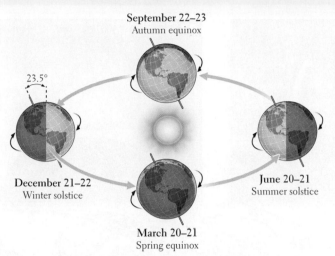

FIGURE 1.9 Seasonal changes in Earth's orientation to the Sun. Diagram of the angle of Earth's orientation to the Sun, showing seasonal changes over the course of one year.

more common because the removal of vegetation limits the ability of Earth's surface to absorb rainwater. As erosion increases, rivers may fill with silt, and deltas may extend into the oceans. When humans dam rivers or build levies to control flooding, they interrupt the natural processes of deposition and removal of silt.

CLIMATE

The processes associated with climate are generally more rapid than those that shape landforms. **Weather,** the short-term and spatially limited expression of climate, can change in a matter of minutes. **Climate** is the long-term balance of temperature and moisture that keeps weather patterns fairly consistent from year to year. By this definition, the last major global climate change took place about 15,000 years ago, when the glaciers of the last ice age began to melt.

Solar energy is the engine of climate. Earth's atmosphere, oceans, and land surfaces absorb large amounts of solar energy, and the differences in the amounts they absorb account for part of the variations in climate we observe. The most intense direct solar energy strikes Earth more or less head-on in a broad band stretching about 30° north and south of the equator. The fact that Earth's axis sits at a 23° angle as it orbits the Sun—an angle that does not change—means that where the band of greatest solar intensity strikes Earth varies in regular sequence over the course of a year, creating seasons. Just how this yearly seasonal pattern works is illustrated by **FIGURE 1.9**. The highest average temperatures on Earth's surface are within this band of the strongest solar intensity. Moving away from the equator, solar energy strikes Earth's surface less directly— at more of an obtuse (wide) angle. This wide angle causes a deflection of some of the Sun's rays, resulting in lower average annual temperatures.

CLIMATE REGION

Geographers have several systems for classifying the world's climates. The systems are based on the patterns of temperature and precipitation just described. This book uses a modification of the widely known *Köppen classification system*, which divides the world into several types of climate regions, labeled (A) through (E) on the climate map in **FIGURE 1.10**. As you look at the regions on this map, examine the photos, and read the accompanying climate descriptions, you will notice the importance of climate to vegetation. Each chapter includes a climate map about the region being discussed; when reading these maps, refer to the written descriptions in Figure 1.10 as necessary. Keep in mind that the sharp boundaries shown on climate maps are in reality much more gradual transitions.

TEMPERATURE AND AIR PRESSURE

The daily wind and weather patterns are largely a result of variations in solar energy absorption that create complex patterns of air temperature and *air pressure*. To understand air pressure, think of air as existing in a particular unit of space—for example, a column of air above a square foot of Earth's surface. Air pressure is the amount of force (due to the pull of gravity) exerted by that column on that square foot of surface. Air pressure and temperature are related: The gas molecules in warm air are relatively far apart and are associated with low air pressure. In cool air, the gas molecules are relatively close together (dense) and are associated with high air pressure.

As the Sun warms a unit of cool air, the molecules move farther apart. The air becomes less dense and exerts less pressure. Air tends to move from areas of higher pressure to areas of lower pressure, creating wind. If you have been to the beach on a hot day, you may have noticed a cool breeze blowing in off the water. This happens because land heats up (and cools down) faster than water, so on a hot day, the air over the land warms, rises, and becomes less dense than the air over the water. This causes the cooler, denser air over the water to flow inland. At night the breeze often reverses direction, blowing from the now cooling land onto the now relatively warmer water.

These air movements have a continuous and important influence on global weather patterns and are closely associated with land and water masses. Because continents heat up and cool off much more rapidly than the oceans that surround them, the wind tends to blow from the ocean to the land during summer and from the land to the ocean during winter. It is almost as if the continents were breathing once a year, inhaling in summer and exhaling in winter.

PRECIPITATION

Perhaps the most tangible way we experience changes in air temperature and pressure is through rain or snow. **Precipitation** (dew, rain, sleet, hail, and snow) occurs primarily because warm air holds more moisture than cool air. When this warmer moist air rises to a higher altitude, its temperature drops, reducing its ability to hold moisture. The moisture condenses into drops that form clouds and may eventually fall as rain or some other form of precipitation.

weather the short-term and spatially limited expression of climate that can change in a matter of minutes

climate the long-term balance of temperature and precipitation that characteristically prevails in a particular region

precipitation dew, rain, sleet, hail, and snow

FIGURE 1.10 PHOTO ESSAY: Climate Regions of the World

A **Tropical humid climates.** In *tropical wet climates*, rain falls predictably every afternoon and usually just before dawn. The *tropical wet/dry climate*, also called a *tropical savanna*, has a wider range of temperatures and a wider range of rainfall fluctuation than the tropical wet climate.

B **Arid and semiarid climates.** *Deserts* generally receive very little rainfall (2 inches or less per year). Most of that rainfall comes in downpours that are extremely rare and unpredictable. *Steppes* have climates similar to those of deserts, but that are more moderate. They usually receive about 10 inches more rain per year than deserts and are covered with grass or scrub.

C **Temperate climates.** *Midlatitude temperate climates* are moist all year and have short, mild winters and long, hot summers. *Subtropical temperate climates* differ from midlatitude climates in that subtropical winters are dry. *Mediterranean climates* have moderate temperatures but are dry in summer and wet in winter.

D **Cool humid climates.** Stretching across the broad interiors of Eurasia and North America are *continental climates*, which either have dry winters (northeastern Eurasia) or are moist all year (North America and north-central Eurasia). Summers in cool humid climates are short but can have very warm days.

E **Coldest climates.** *Arctic and high-altitude climates* are by far the coldest and are also among the driest. Although moisture is present, there is little evaporation because of the low temperatures. The Arctic climate is often called *tundra*, after the low-lying vegetation that covers the ground. The high-altitude version of this climate, which can be far from the Arctic, is more widespread and subject to larger daily fluctuations in temperature. High-altitude microclimates, such as those in the Andes and the Himalayas, can vary tremendously, depending on factors such as available moisture, orientation to the Sun, and vegetation cover. As one ascends in altitude, the changes in climate loosely mimic those found as one moves from lower to higher latitudes. These changes are known as temperature-altitude zones (see Figure 3.6).

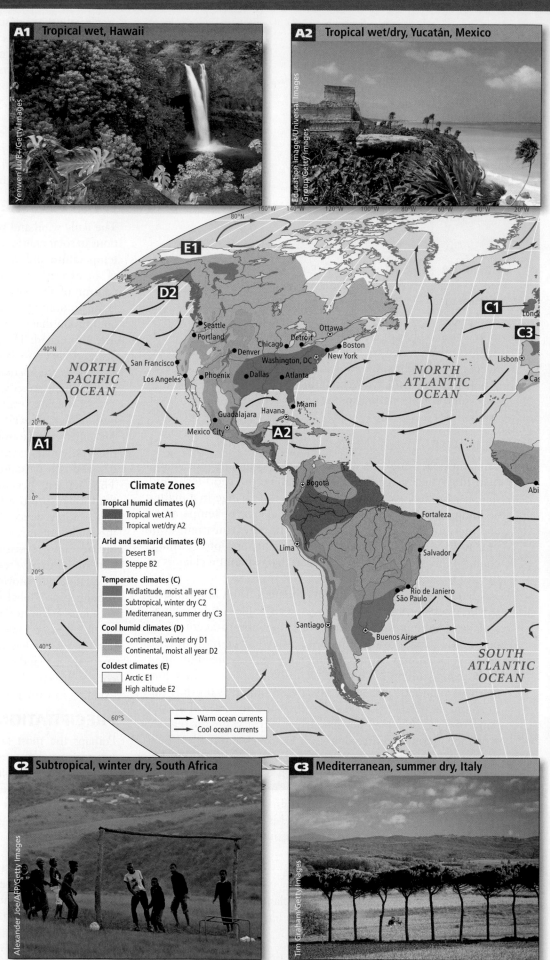

A1 Tropical wet, Hawaii

A2 Tropical wet/dry, Yucatán, Mexico

Climate Zones

Tropical humid climates (A)
- Tropical wet A1
- Tropical wet/dry A2

Arid and semiarid climates (B)
- Desert B1
- Steppe B2

Temperate climates (C)
- Midlatitude, moist all year C1
- Subtropical, winter dry C2
- Mediterranean, summer dry C3

Cool humid climates (D)
- Continental, winter dry D1
- Continental, moist all year D2

Coldest climates (E)
- Arctic E1
- High altitude E2

→ Warm ocean currents
→ Cool ocean currents

C2 Subtropical, winter dry, South Africa

C3 Mediterranean, summer dry, Italy

B1 Desert, Namibia

Hoberman Collection/Universal Images Group/Getty Images

B2 Steppe, Mongolia

Palani Mohan/AsiaPac/Getty Images

C1 Midlatitude, moist all year, United Kingdom

RDImages/Epics/Hulton Archive/Getty Images

E2 High altitude, Tibet

P. Morris/Hulton Archive/Getty Images

D1 Continental, winter dry, Russia

Russian Look/LIG/AGE Fotostock

D2 Continental, moist all year, Alaska

Nikki Kahn/The Washington Post via Getty Images

E1 Arctic tundra, Canada

Paul J. Richards/AFP/Getty Images

ARCTIC OCEAN

NORTH PACIFIC OCEAN

INDIAN OCEAN

SOUTH PACIFIC OCEAN

Helsinki
Berlin
Moscow
Kiev
Paris
Budapest
Madrid
Rome
Istanbul
Athens
Ankara
Tehran
Kabul
Casablanca
Baghdad
Cairo
Tashkent
Lahore
Delhi
Karachi
Kolkata (Calcutta)
Dhaka
Mumbai (Bombay)
Bangalore
Chennai (Madras)
Abuja
Addis Ababa
djan
Kinshasa
Cape Town

Harbin
Shenyang
Beijing
Xian
Chengdu
Chongqing
Shanghai
Guangzhou
Hong Kong
Seoul
Tokyo
Yokohama
Taipei
Manila
Bangkok
Ho Chi Minh City
Singapore
Jakarta
Surabaya
Sydney
Melbourne

B1
C2
E2
B2
D1

80°N
60°N
40°N
20°
0°
20°S
40°S
60°S

0° 20°E 40°E 60°E 80°E 100°E 120°E 140°E 160°E 180°E

Several conditions that encourage moisture-laden air to rise influence the patterns of precipitation around the globe. When moisture-bearing air is forced to rise as it passes over mountain ranges, the air cools, and the moisture condenses to produce rainfall **(FIGURE 1.11)**. This process, known as **orographic rainfall,** is most common in coastal areas where wind blows moist air from above the ocean onto the land and up the side of a coastal mountain range. Most of the moisture falls as rain as the cooling air rises along the coastal side of the range. On the inland side, the descending air warms and ceases to drop its moisture. The drier side of a mountain range is said to be in the *rain shadow.* Rain shadows may extend for hundreds of miles across the interiors of continents, as they do on the Mexican Plateau, or east of California's Pacific coastal ranges, or north of the Himalayas of Eurasia.

A central aspect of Earth's climate is the *rain belt* that exists in equatorial areas (also known as the intertropical convergence zone, or ITCZ; see Chapter 7 and Chapter 8). Near the equator, moisture-laden tropical air is heated by the strong direct sunlight and rises to the point where it releases its moisture as rain. Neighboring nonequatorial areas also receive some of this moisture when seasonally shifting winds (related to the seasonally varying angle of solar radiation striking Earth) move the rain belt north and south of the equator. The huge downpours of the Asian summer monsoon are an example of this equatorial rain belt.

In the summer **monsoon** season, the Eurasian continental landmass heats up, causing the overlying air to expand, become less dense, and rise. The somewhat cooler, yet moist, air of the Indian Ocean is drawn inland. The effect is so powerful that the equatorial rain belt is sucked onto the land (see Figure 8.4). This results in tremendous, sometimes catastrophic, summer rains throughout virtually all of South and Southeast Asia and much of coastal and interior East Asia. The reverse happens in the winter as similar forces pull the equatorial rain belt south during the Southern Hemisphere's summer.

Much of the moisture that falls on North America and Eurasia is *frontal precipitation* caused by the interaction of large air masses of different temperatures and densities. These masses develop when air stays over a particular area long enough to take on the temperature of the land or sea beneath it. Often when we listen to a weather forecast, we hear about

orographic rainfall rainfall produced when a moving moist air mass encounters a mountain range, rises, cools, and releases condensed moisture that falls as rain

monsoon a wind pattern in which in summer months, warm, wet air coming from the ocean brings copious rainfall, and in winter, cool, dry air moves from the continental interior toward the ocean

FIGURE 1.11 Orographic rainfall (and rain shadow diagram).

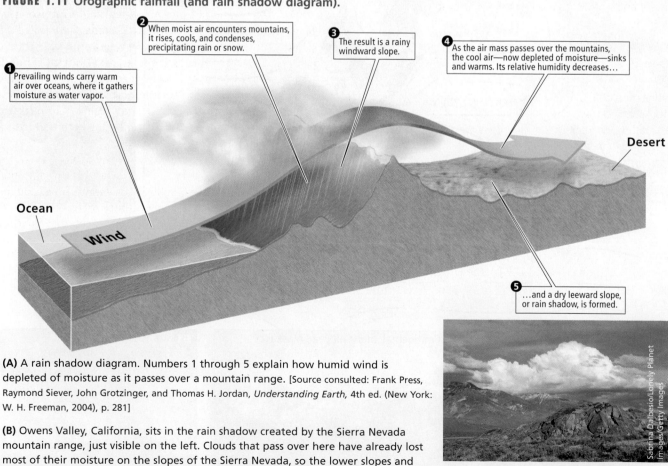

❶ Prevailing winds carry warm air over oceans, where it gathers moisture as water vapor.

❷ When moist air encounters mountains, it rises, cools, and condenses, precipitating rain or snow.

❸ The result is a rainy windward slope.

❹ As the air mass passes over the mountains, the cool air—now depleted of moisture—sinks and warms. Its relative humidity decreases…

❺ …and a dry leeward slope, or rain shadow, is formed.

Ocean

Wind

Desert

(A) A rain shadow diagram. Numbers 1 through 5 explain how humid wind is depleted of moisture as it passes over a mountain range. [Source consulted: Frank Press, Raymond Siever, John Grotzinger, and Thomas H. Jordan, *Understanding Earth,* 4th ed. (New York: W. H. Freeman, 2004), p. 281]

(B) Owens Valley, California, sits in the rain shadow created by the Sierra Nevada mountain range, just visible on the left. Clouds that pass over here have already lost most of their moisture on the slopes of the Sierra Nevada, so the lower slopes and valleys are dry.

Sabrina Dalbesio/Lonely Planet Images/Getty Images

warm fronts or cold fronts. A *front* is the zone where warm and cold air masses come into contact, and it is always named after the air mass whose leading edge is moving into an area. At a front, the warm air tends to rise over the cold air, carrying warm clouds to a higher, cooler altitude. Rain or snow may follow. Much of the rain that falls along the outer edges of a hurricane is the result of frontal precipitation.

THINGS TO REMEMBER

GEOGRAPHIC THEME 1
- **Environment:** Humans are altering the planet at an unprecedented rate, causing sometimes drastic effects on ecosystems and the climate. Multiple environmental factors often interact to influence the vulnerability of a location to the impacts of climate change. These vulnerabilities have a spatial pattern.

- Landforms are shaped by internal processes, such as plate tectonics, and external processes, such as weathering.

- Climate is the long-term balance of temperature and precipitation that keeps weather patterns fairly consistent from year to year.

- Weather is the short-term and spatially limited expression of climate that can change in minutes.

- Seasons occur because Earth's axis is positioned at a consistent angle of 23° as it orbits the Sun over the course of a year.

- Variations in air pressure are caused by heating and cooling and account for phenomena such as wind and precipitation patterns.

- Precipitation occurs primarily because warm air holds more moisture than cool air. When this warmer moist air rises to a higher altitude, its temperature drops, reducing its ability to hold moisture, resulting in rain, snow, sleet, or hail.

ENVIRONMENT

GEOGRAPHIC THEME 1

Environment: Humans are altering the planet at an unprecedented rate, causing sometimes drastic effects on ecosystems and the climate. Multiple environmental factors often interact to influence the vulnerability of a location to the impacts of climate change. These vulnerabilities have a spatial pattern.

HUMAN IMPACT ON THE BIOSPHERE

From the beginning of human life, we have overused resources in trying to improve our own living conditions, sometimes with disastrous consequences. What is new is the scale of human impacts on the planet, which can now be found virtually everywhere. In fact, geoscientists have recently identified a new geologic epoch, called the *Anthropocene*—the time during which humans have had an overwhelming impact on Earth's biosphere. Just when the Anthropocene began is under debate.

As people have grown more aware of their environmental impacts, they have put forth numerous proposals to limit damage to the **biosphere**, defined here as the entirety of Earth's integrated physical and biological systems, with humans and their impacts included as part of nature, not separate from it. Societies have become so transformed by the intensive use of Earth's resources that reversing this level of use is enormously difficult. For example, how possible would it be for you and your entire family to live for even just one day without using any fossil fuels for transportation or home heating or cooling? Would you be able to get to school or work or be comfortable in your home? Intensive per capita resource consumption is now deeply ingrained, especially in wealthy countries that, with just 20 percent of the world's population, consume more than 80 percent of the available world resources.

Human consumption of natural resources is now being examined using the concept of the **ecological footprint.** This is a method of estimating the amount of biologically productive land and sea area needed to sustain a human at the average current standard of living for a given population (country). It is particularly useful for drawing comparisons. For example, the worldwide average of the biologically productive area needed to support one person—this would be one individual's ecological footprint—is about 4.5 acres. Because of the lifestyle in the United States, one person's ecological footprint averages about 24 acres; it is about 18 acres in Canada, and just 4 acres in China. You can calculate your own footprint using the Global Footprint Network's calculator, at http://tinyurl.com/6d2wyl4. A similar concept more closely related to global warming is the *carbon footprint*, which measures the greenhouse gas emissions a person's activities produce. To calculate your family's carbon footprint, use Carbon Footprint's calculator, at http://www.carbonfootprint.com/calculator.aspx.

Because the biosphere is a global ecological system that integrates all living things and their relationships, it is important to raise awareness that actions in widely separated parts of Earth have a cumulative effect on the whole. **FIGURE 1.12** shows a global map of the relative intensity of human biosphere impacts. The map includes photo insets that show particular trouble spots in South America (see Figure 1.12E, F), Europe (see Figure 1.12A), South Asia (see Figure 1.12B), and Southeast Asia (see Figure 1.12C, D). However, to fully appreciate human impacts on the biosphere, some understanding of the underlying physical processes that shape the biosphere is needed.

GLOBAL CLIMATE CHANGE

Planet Earth is continually undergoing **climate change,** a slow shifting of climate patterns caused by

biosphere the entirety of Earth's integrated physical spheres, with humans and other impacts included as part of nature

ecological footprint the amount of biologically productive land and sea area needed to sustain a person at the current average standard of living for a given population

climate change a slow shifting of climate patterns caused by the general cooling or warming of the atmosphere

FIGURE 1.12 PHOTO ESSAY: Human Impacts on the Biosphere

Thinking Geographically

After you have read about human impacts on the biosphere, you will be able to answer the following questions.

A In which sector of the economy is mining?

B What form of pollution does this photo show most directly?

C, **D** What is the evidence that shifting cultivation may be contributing to deforestation in the region depicted?

E, **F** How is logging in Brazil linked to rising CO₂ levels and the global economy? Does your answer to **E** connect to you?

Humans have had enormous impacts on the biosphere. The map and insets show varying levels of these impacts on the biosphere. The map is derived from a synthesis of hundreds of studies. High-impact areas often have roads, railways, agriculture, or other intensive land uses. Low- to medium-impact areas have biodiversity loss and other disturbances related to human activity. [Sources consulted: United Nations Environment Programme, "Human Impact, Year 1700 (approximately)," and "Human Impact, Year 2002," (New York: United Nations Development Program), 2002, 2003, 2004, 2005, 2006, at http://www.grida.no/graphicslib/detail/human-impact-year-1700-approximately_6963 and http://www.grida.no/graphicslib/detail/human-impact-year-2002_157a]

the general cooling or warming of the atmosphere. The current trend of **global warming**—which refers to the observed warming of Earth's climate as atmospheric levels of greenhouse gases increase—is extraordinary because it is happening more quickly than climate changes in the past. Scientific evidence indicates that it is extremely likely that the changes are linked to human agency. **Greenhouse gases (GHG),** which include carbon dioxide (CO₂), methane, water vapor, and other gases, are essential to keeping Earth's incoming and outgoing radiation balanced in such a way that Earth's surface is maintained in a temperature range hospitable to life. Similar to the glass panes of a greenhouse, these gases allow solar radiation to pass through the atmosphere and strike Earth's surface; the gases also allow much of this radiation to bounce back into space as surface radiation. But some radiation is re-reflected back to Earth, keeping Earth's surface (like the interior of a greenhouse) warmer than it would be if incoming and outgoing radiation were in balance. **FIGURE 1.13** shows this system of incoming and outgoing radiation and the role of GHGs in trapping some of the outgoing radiation and sending it back to warm Earth. The current scientific consensus is that the warming trend in Earth's climate is caused by GHGs being released by humans at accelerating rates through the burning of fossil fuels (for example, in car engines and in coal-fired power plants) and by the effects of deforestation, which reduces the amount of carbon dioxide that is absorbed by trees (see Figure 1.12A, C, E). The most recent evidence indicates that Earth is warming very rapidly.

Most scientists now agree that there is an urgent need to reduce GHG emissions to avoid catastrophic climate change in coming years. Climatologists, biogeographers, and other scientists are documenting long-term global warming and cooling trends by examining evidence in tree rings, fossilized pollen and marine creatures, and glacial ice. These data indicate that the twentieth century was the warmest century in 600 years, and that the current decade is the hottest on record. Evidence is mounting that these are not normal fluctuations. Very long-term climate-change patterns indicate that we should be heading into a cooling pattern, but instead it is estimated that, at present rates of warming, by 2100 average global temperatures could rise between 2.5°F and 10°F (about 2°C

global warming the warming of Earth's climate as atmospheric levels of greenhouse gases increase

greenhouse gases (GHG) gases, such as carbon dioxide and methane, released into the atmosphere by human activities; these gases are harmful when released in excessive amounts

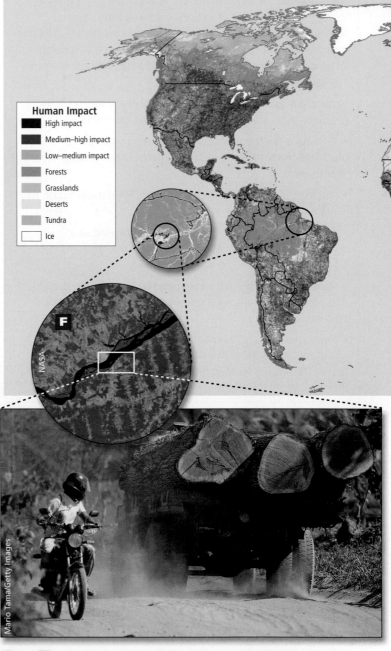

E and **F** **Development and deforestation.** In the Brazilian Amazon, deforestation often is done in regularized spatial patterns, such as the "fishbone" pattern (see satellite image inset **F**). This pattern results from regulations that determine the location of roads used for settlement and logging. Whole logs are brought by road to rivers, where they are put on barges and taken to a port for export.

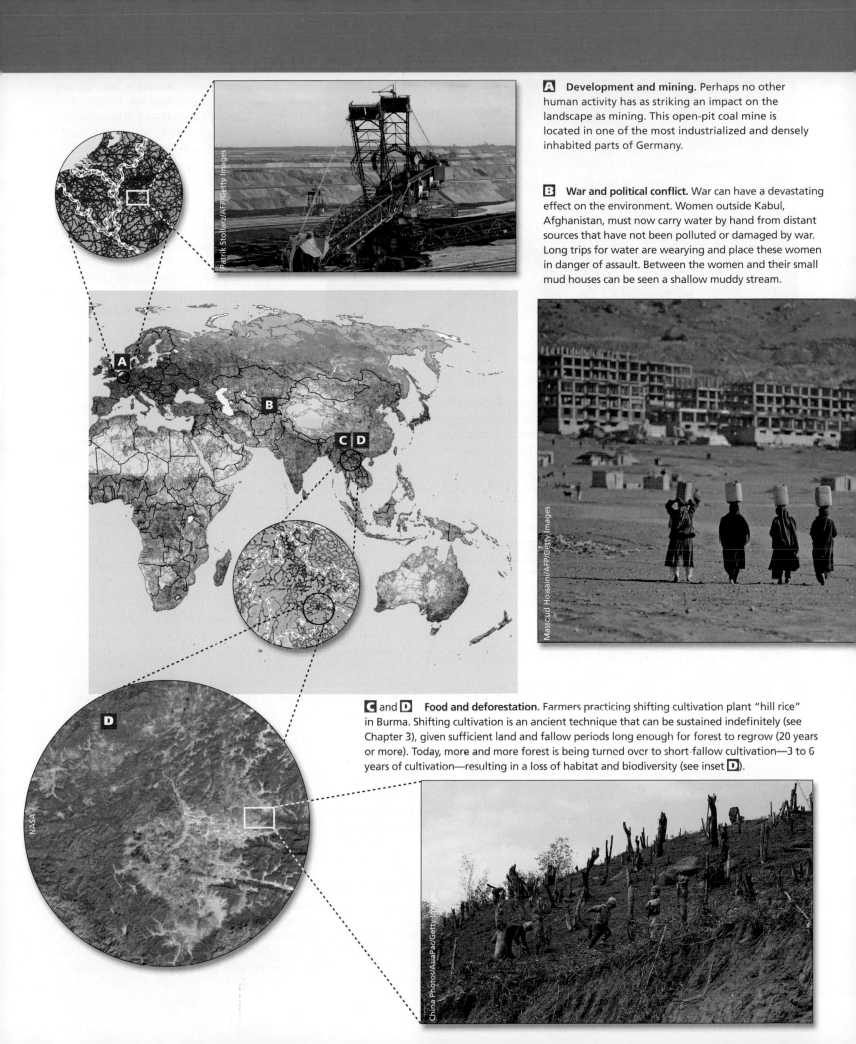

A **Development and mining.** Perhaps no other human activity has as striking an impact on the landscape as mining. This open-pit coal mine is located in one of the most industrialized and densely inhabited parts of Germany.

Patrik Stollarz/AFP/Getty Images

B **War and political conflict.** War can have a devastating effect on the environment. Women outside Kabul, Afghanistan, must now carry water by hand from distant sources that have not been polluted or damaged by war. Long trips for water are wearying and place these women in danger of assault. Between the women and their small mud houses can be seen a shallow muddy stream.

Masscud Hossaini/AFP/Getty Images

NASA

C and **D** **Food and deforestation.** Farmers practicing shifting cultivation plant "hill rice" in Burma. Shifting cultivation is an ancient technique that can be sustained indefinitely (see Chapter 3), given sufficient land and fallow periods long enough for forest to regrow (20 years or more). Today, more and more forest is being turned over to short-fallow cultivation—3 to 6 years of cultivation—resulting in a loss of habitat and biodiversity (see inset **D**).

China Photos/AsiaPac/Getty Images

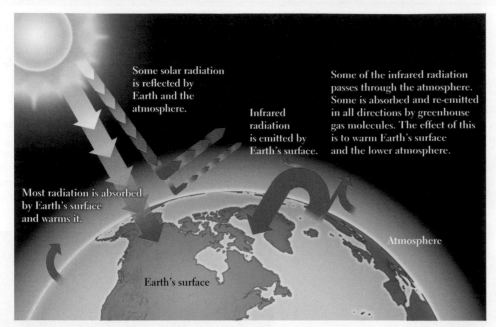

FIGURE 1.13 The balance of incoming and outgoing radiation and the greenhouse effect. To maintain an even temperature, Earth has to balance energy coming in with energy going out. Energy coming in is mostly sunshine, and energy going out is mostly radiant heat.

Summary of diagram:

1. Energy from the Sun arrives at Earth as solar radiation.
2. Some solar radiation is reflected by Earth and the atmosphere. Most radiation is absorbed by Earth's surface and warms it.

Infrared (heat) radiation is emitted by Earth's surface. Some of the infrared radiation passes through the atmosphere. Some is absorbed and re-emitted in all directions by greenhouse gas molecules. The effect of this is to warm Earth's surface and the lower atmosphere. These days, scientists find that extra greenhouse gases released by human activities are causing Earth to retain extra energy—outgoing infrared radiation seems to average nearly 1 watt per square meter (W/m^2) less than incoming solar radiation, so average temperatures on Earth are rising.

to 5°C)—a rise that will have many impacts on the biosphere, such as flooding, desertification, rising sea levels, loss of agricultural land, and loss of species.

A key problem in reducing GHGs is that those most responsible for global warming (the world's wealthiest and most industrialized countries) have the least incentive to reduce emissions because they are the least vulnerable to the changes global warming causes in the physical environment. Those most vulnerable to these changes (poor countries with low levels of human development and with large populations living in low-lying coastal wetlands) are the least responsible for the growth in GHG emissions and have little power in the global geopolitical sphere; hence, they do not have the power (politically and economically) to affect the level of emissions **(FIGURE 1.14)**.

Drivers of Global Climate Change

GHGs exist naturally in the atmosphere. It is their heat-trapping ability that makes Earth warm enough for life to exist. When we increase their levels, as humans are doing now, Earth becomes warmer still.

Electricity generation, vehicles, industrial and construction processes, and the heating/cooling of homes and businesses all burn large amounts of CO_2-producing fossil fuels such as coal, natural gas, and oil. Even the exhalations of ever-larger human populations contribute CO_2, and the commercial raising of grazing animals contributes methane through the animals' flatulence. Unusual quantities of GHGs from these sources are accumulating in Earth's atmosphere, and their presence has already led to significant warming of the planet's climate.

Widespread deforestation worsens the situation. Living forests take in CO_2 from the atmosphere via photosynthesis, releasing oxygen and storing the carbon in their biomass. As more trees are cut down and their wood is used for fuel, more carbon enters the atmosphere, less is taken out, and less is stored. The loss of trees and other forest organisms produces as much as 30 percent of the buildup of CO_2 in the atmosphere. The use of fossil fuels accounts for the other 70 percent.

In 2010, the industrialized countries (the United States, Canada, the European Union, Russia, and Australia) and the large, rapidly developing countries (notably China and India) had the highest percentages of total GHG emissions and that largely remains the case. But by 2014, patterns of CO_2 emissions (the only GHG for which complete global data are available) were beginning to change. Many industrialized countries in Europe, as well as Canada, the United States, and Australia, had reduced emissions in total tons and per capita. Figure 1.14 shows the global patterns in 2014 of total CO_2 emissions in total tons (A) and per capita (B). Note that the United States, despite having made small improvements since 2010, remains among the leaders in both categories. From 1859 to 1995, developed countries produced roughly 80 percent of the GHGs from all types of industrial, home, and transportation sources, and developing countries produced 20 percent. But by 2007, the developing countries were catching up, accounting for nearly 30 percent of total emissions. As developing countries industrialize over the next decades and continue to cut down their forests, they will release more and more GHGs every year. If current patterns hold, by 2040 the developing countries will emit more GHGs than the developed countries.

Climate-Change Impacts

Scientific understanding of the likely impacts of rising global temperatures is growing rapidly, and it is clear that these impacts will not be uniform across the globe. Also, the delay in reducing GHGs has dramatically increased the expected effects of climate change. The glaciers and polar ice caps, for example,

FIGURE 1.14 Carbon dioxide (CO_2) emissions around the world in 2014. [Source consulted: http://edgar.jrc
.ec.europa.eu/overview.php?v=CO2ts1990-2014; http://edgar.jrc.ec.europa.eu/overview.php?v=CO2ts_pc1990-2014]

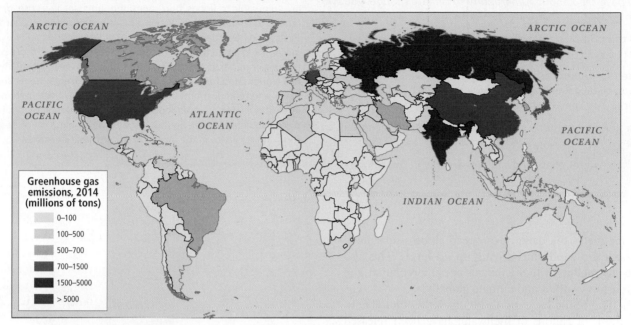

(A) Total emissions per country by millions of tons, 2014. China emits the most tons of CO_2, and its rate has been rising sharply since 2005. The United States emits the second most, having had a small emissions reduction since 2005. India's CO_2 emissions are increasing, and Russia's have slightly decreased, as have Japan's. These top five countries contribute more than 50 percent of the world's CO_2 emissions.

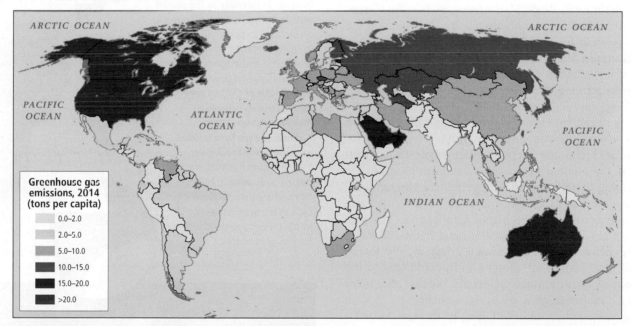

(B) Tons of emissions per capita, 2014. The United States, Canada, Australia, and the Gulf States (Bahrain, Kuwait, Qatar, Saudi Arabia, and the United Arab Emirates) have the highest rates of CO_2 emissions per capita. The United States is eighth, at 16.5 tons per capita. China and India are far down the list—China at 4.40 tons per capita, India at 1.8 tons per capita.

are melting even faster than anticipated, causing a corresponding rise in sea level. In fact, this phenomenon has been observable for several years. Satellite imagery analyzed by scientists at the National Aeronautics and Space Administration (NASA) shows that between 1979 and 2005—just 26 years—the polar ice caps shrank by about 23 percent. The amount of polar ice cap shrinkage wavers from year to year, with the ice caps regaining ice to some extent during winter months, but the overall trend in recent years is 12 percent shrinkage per decade. The polar ice caps normally reflect solar heat back into the atmosphere,

FIGURE 1.15 PHOTO ESSAY: Vulnerability to Climate Change

Thinking Geographically

After you have read about vulnerability to climate change, you will be able to answer the following questions.

A Does this photo relate most to short-term or long-term resilience?

B Of the four thumbnail maps in this graphic, which depicts the information that best explains why Spain and Portugal are so much less vulnerable to climate change than is Morocco?

C What sign of an orderly response to disaster is visible in this picture?

D How can you tell that food supplies are low in this refugee camp?

The map shows overall global patterns of vulnerability to climate change, based on a combination of human and environmental factors. Areas in darkest brown are vulnerable to floods, hurricanes, droughts, sea level rise, or other hazards related to climate change. When a place is exposed to a hazard that it is sensitive to and has little resilience to, it becomes vulnerable. For example, many places are exposed to drought, but generally speaking, those places with the poorest populations are the most sensitive. However, sensitivity to drought can be compensated for if adequate relief and recovery systems, such as emergency water and food distribution systems, are in place. These systems lend an area a level of resilience that can reduce its overall vulnerability to climate change. A place's vulnerability can be thought of as a combination of its sensitivity, exposure, and resilience in the face of multiple climate hazards.

but as the ice melts, the dark, open ocean absorbs solar heat. Warming oceans expand, thus driving up sea levels. This also hastens ice cap melting, which changes ocean circulation, the engine that drives weather and climate globally.

The melting of the ice caps also has several other effects. Already, trillions of gallons of meltwater have been released into the oceans. If this trend continues, at least 60 million people in coastal areas and on low-lying islands could be displaced by rising sea levels. Another issue is the melting of high mountain glaciers in the Himalayas, which are a major source of water for many of the world's large rivers, such as the Ganga, the Indus, the Brahmaputra, the Huang He (Yellow), and the Chang Jiang (Yangtze). Similar melting effects on major rivers are expected in South America. Scientists have monitored mountain glaciers across the globe for more than 30 years; while some are growing, the majority are melting rapidly. Over the short term, melting mountain glaciers will increase the flow of many rivers, but eventually river flows will decrease as mountain glaciers shrink or disappear entirely.

Over time, higher average annual temperatures will shift northward in the Northern Hemisphere and southward in the Southern Hemisphere, bringing warmer climate zones to these regions. In some areas, drought and water scarcity may also become more common since higher temperatures increase evaporation rates from soils, vegetation, and bodies of water. Animal and plant species that cannot adapt rapidly to the changes will disappear. Such climate shifts could lead to the displacement of large numbers of people because the zones where specific crops can grow are likely to change.

Higher temperatures also will lead to stronger tornados and hurricanes because these storms are powered by warm ocean water and warm, rising air (**FIGURE 1.15**). One example is Hurricane Sandy, the unusually large and powerful hurricane that struck the Atlantic Coast of North America in the autumn of 2012; another is Typhoon Haiyan in the Philippines in 2013. In 2015 a record 22 hurricanes (typhoons) reached category 4 or 5 strength in the Northern Hemisphere.

Another effect of global warming is likely to be a shift in ocean currents, caused by warming of the ocean itself. Worries

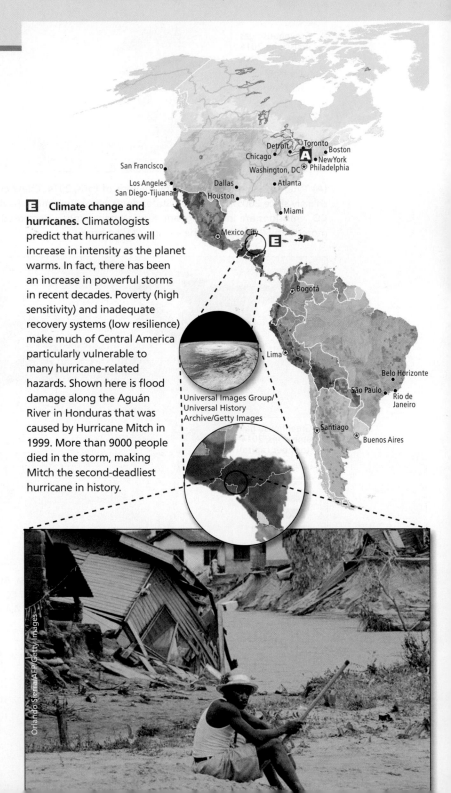

E **Climate change and hurricanes.** Climatologists predict that hurricanes will increase in intensity as the planet warms. In fact, there has been an increase in powerful storms in recent decades. Poverty (high sensitivity) and inadequate recovery systems (low resilience) make much of Central America particularly vulnerable to many hurricane-related hazards. Shown here is flood damage along the Aguán River in Honduras that was caused by Hurricane Mitch in 1999. More than 9000 people died in the storm, making Mitch the second-deadliest hurricane in history.

Universal Images Group/ Universal History Archive/Getty Images

Orlando Sierra/AFP/Getty Images

A **United States: High resilience, low vulnerability.** Effective and well-funded recovery and relief systems give the United States high resilience to climate hazards. This contributes to its generally low vulnerability. Shown here are ambulances in New York City responding to Hurricane Sandy in 2012.

Climate

Population

Vulnerability

Human development

B **Spain and Morocco: The multiple dimensions of vulnerability.** A wide variety of information is used to make the global map of vulnerability shown below. For example, the contrast in vulnerability between Spain and Morocco relates to (among other things) differences in climate, population density and distribution, and human development (which is itself based on many factors).

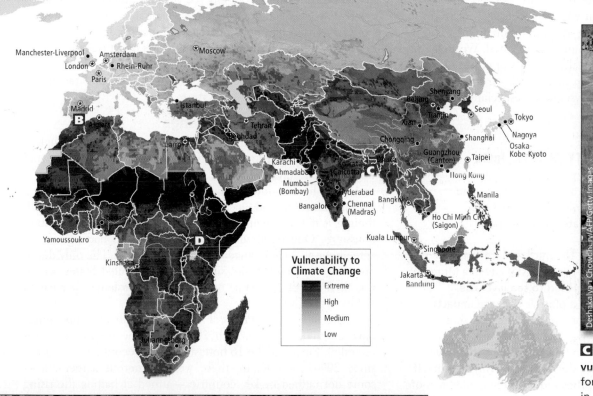

Vulnerability to Climate Change

Extreme
High
Medium
Low

C **India: Moderate resilience, high vulnerability** Rural Indians line up for food and water after Cyclone Aila in 2009. Advances in government-led disaster recovery have increased India's resilience, and many areas no longer have the extreme vulnerability levels that neighboring Pakistan and Afghanistan have. Nevertheless, much of India remains very vulnerable, with high exposure and sensitivity to sea level rise, flooding, hurricanes (cyclones), drought, and other disturbances that climate change can create or intensify.

D **Northern Uganda and South Sudan: High to extreme vulnerability.** The overall situation in these areas is somewhat similar to that of India (C) but for different reasons. Because there is better overall access to water, sensitivity to drought is somewhat lower. However, armed conflict in Uganda has reduced the country's resilience, as many people have been forced to live in refugee camps and are dependent on food and water aid donated by foreigners. Shown here are refugees in northern Uganda, picking up bits of donated grain that has been dropped.

are increasing that the Gulf Stream that warms northwestern Europe and brings rain there is shifting course. Warmer temperatures are also making the oceans expand in size, contributing to sea level rise. The result of all these changes will be more chaotic and severe or even dry weather, especially for places where climates are strongly influenced by ocean currents, such as the entire west coast of the Americas (see Chapter 2) and northwestern Europe (see Chapter 4).

Vulnerability to Climate Change

The vulnerability a place has to climate change can be thought of as the amount of risk its human or natural systems have of being damaged by such climate-related hazards as sea level rise, drought, flooding, or increased storm intensity. Many of these vulnerabilities are water related; others are not. Scientists who study climate change agree that while we can take measures to minimize temperature increases, we can't stop them entirely, much less reverse those that have already occurred. We are going to have to live with and adapt to the impacts of climate change for quite some time. The first step in doing this is to understand how and where humans and ecosystems are especially vulnerable to climate change.

A CASE STUDY OF VULNERABILITY: Mumbai, India

Three concepts are important in understanding a place's vulnerability to climate change: *exposure*, *sensitivity*, and *resilience*. Here we explore them in the context of the vulnerability to water-related climate-change impacts in Mumbai, India.

Exposure refers to the extent to which a place is exposed to climate-change impacts. A low-lying coastal city like Mumbai, India (see Figure 8.1), is highly exposed to sea level rise. *Sensitivity* refers to how sensitive a place is to those impacts. Many of Mumbai's inhabitants, for example, are very sensitive to the impacts of sea level rise because they are extremely poor and can only afford to live in low-lying slums that have no sanitation and therefore have polluted waterways nearby. If these waterways were to flood, they would spread epidemics of waterborne illnesses that could kill millions of people in the city and neighboring areas. *Resilience* refers to a place's ability to bounce back from the disturbances that climate-change impacts create. Mumbai's resilience to sea level rise is bolstered because despite its widespread poverty, it is the wealthiest city in South Asia. This wealth enables it to afford relief and recovery systems that could help it deal with sea level rise over the short and long term.

Over the short term, Mumbai benefits from the fact that it has more and better hospitals and emergency response teams than any other city in South Asia. Over the long term, Mumbai's well-trained municipal planning staff can create and execute plans to help sensitive populations, like people living in slums, adapt to sea level rise. This could be done, for example, through planned

Kyoto Protocol an amendment to a UN treaty on global warming, the Protocol is an international agreement, adopted in 1997 and in force in 2005, that sets binding targets for industrialized countries for reducing emissions of greenhouse gases

relocation to higher ground or by building sea walls and dikes that could keep sea waters out of low-lying slum areas. Of course, Mumbai's overall vulnerability is more complicated than these examples suggest because the city faces many more climate-change impacts than just sea level rise. However, these examples help us understand vulnerability to climate change as a combination of exposure, sensitivity, and resilience.

Figure 1.15 includes a map of vulnerability to climate change and photos of the types of problems that are already being seen around the world (see Figure 1.15A, C–E). One pattern is that places with low levels of human development tend to be more vulnerable to climate change. For example, many of the qualities that make Mumbai more sensitive to sea level rise are less present in urban areas in highly developed countries. New York City has virtually none of the large, unplanned lowland slums found in Mumbai. In addition, numerous world-class hospitals, emergency response teams, and large and well-trained municipal planning staffs boost New York's resilience. However, despite all of New York City's wealth and resources, Hurricane Sandy showed that the metropolitan area is still very vulnerable to a large storm. ■

Responding to Climate Change

In 1997, an agreement known as the **Kyoto Protocol** was adopted; it is an amendment to a United Nations (UN) treaty on global warming. The protocol called for scheduled reductions in CO_2 emissions by the industrialized countries in North America, Europe, East Asia, and Oceania. The agreement also encouraged, though it did not require, developing countries to curtail their emissions. One hundred eighty-three countries had signed the agreement by 2009 (Canada withdrew in 2011). The only developed country that had not signed was the United States, which was then and still remains one of the world's largest per capita producers of CO_2.

Since the Kyoto Protocol was signed, global temperatures have continued to rise, with 2015 being the hottest year ever recorded. Fifteen of the 16 hottest years on record have occurred since 2000. After Kyoto, there were numerous agreements—some not ratified by key countries—aimed at halting the rising concentration of CO_2 in the atmosphere, usually by some point in the future, such as the year 2020. At the 2015 UN Climate Conference in Paris, the industrialized countries that produce the most pollution (the United States, the countries of the EU, and China; see Figure 1.14) agreed to make substantial emission reductions, but the biggest burden of reducing emissions still lies on all other (mostly developing) countries, which will have to adopt emission reductions more than 10 times as aggressive as those of the United States, the EU, and China in order to avoid a 2°C increase by 2030. There has been some progress in helping developing countries create clean-energy economies and otherwise adapt to climate change, but the pace of this has been slow and adequate funds to make these changes have not been allocated.

According to the latest analysis, to avoid a temperature rise of more than 2°C, not only would carbon emissions need to be in decline well before 2020, the global community would have to have implemented strong negative emissions strategies before then—meaning we should be *extracting* more CO_2 from the atmosphere than we add to it. The technology for emission extraction is only just emerging.

ON THE BRIGHT SIDE: Renewable Energy

Renewable energy sources are those that can be replenished in a relatively short amount of time—for example, sunlight, wind, waves, or heat from deep inside Earth. While these sources are relatively underutilized today, many analysts predict a rapid increase in their use in coming decades. This is because the costs of solar and wind power are declining, while the costs of fossil fuels are generally rising. For example, the cost of electricity generated by solar energy has fallen by 99% over the past 25 years and is now cheaper than electricity generated by fossil fuels in Germany, Japan, Spain, Italy, parts of India and China, as well as Southern California and Hawaii. Fossil fuels are relatively expensive in all of these places due to a variety of factors, but by the end of the decade solar power is expected to be cheaper than fossil fuels in most of the developed world.

While the use of renewable energy is expanding quickly, it will take several decades for this growth to translate into significant reductions in the use of fossil fuels. About 80 percent of the energy used throughout the world today comes from fossil fuels (38 percent from petroleum, 26 percent from natural gas, and 16 percent from coal); 8 percent is from nuclear power; and 12 percent is from renewable sources. Hydroelectric energy, ethanol fuels from crops, and wood burning make up the vast majority of current global renewable energy use, with wind and solar power generation each accounting for less than 1 percent of total global energy use. But because the cost of solar and wind technologies continues to decrease, these two technologies together could account for as much as 34 percent of the energy used around the world by 2030. This would significantly reduce greenhouse gas emissions. For further discussion of climate change and renewable energy, watch this video from the Smithsonian, narrated by Bill Nye the Science Guy: http://www.smithsonianmag.com/videos/category/3play_1/climate-change-101-with-bill-nye-the-science/?no-ist. ∎

THINGS TO REMEMBER

• Human activities, such as burning of fossil fuels and deforestation, create large amounts of carbon dioxide, methane, and other greenhouse gases that trap heat in the atmosphere, causing global warming.

• The vulnerability a place has to climate change can be thought of as the amount of risk its human or natural systems have of being damaged by such climate-related hazards as sea level rise, drought, flooding, and increased storm intensity.

• Progress in making global agreements to control CO_2 emissions has been slow and inadequate.

• While renewable resources are relatively underutilized today, many analysts predict a rapid increase in their use in coming decades.

WATER

Water is emerging as the major resource issue of the twenty-first century, but water, energy, and food (and other agricultural products) as resources are intimately linked. Energy is used in the distribution of water, and water and energy are essential to agriculture. Demand for clean water skyrockets as populations grow and people move out of poverty. Irrigated agriculture now accounts for 70 percent of all global freshwater use; this plus increasing meat consumption, industrial demands for water, and greater household use of water are resulting not only in more demand for clean water, but in much higher rates of *water pollution* as well. With fresh, clean water becoming scarce in so many parts of the world, water disputes are proliferating. This is especially true where rivers cross international borders and upstream users take more than their perceived fair share or pollute water for downstream users. Controversy also surrounds the sale of clean water. When water becomes a commodity rather than a free good, as it was before modern times, water prices can increase so much that poor people's access becomes severely reduced, which then negatively affects sanitation and health.

Calculating Water Use per Capita

Humans require an average of 5 to 13 gallons (20 to 50 liters) of clean water per day for basic domestic needs: drinking, cooking, and bathing/cleaning. Per capita domestic water consumption tends to increase as incomes rise; the average person in a wealthy country consumes as much as 20 times the amount of water, per capita, as the average person in a very poor country. Much of this extra water use in developed countries goes for nonessential activities such as landscape watering. However, domestic water consumption is only a fraction of a person's actual water consumption. **Virtual water** is the volume of water required to produce, process, and deliver a good or service that a person consumes. To grow an apple and ship it from the orchard to the consumer, for instance, requires many liters of water. When we add an individual's domestic water consumption to her virtual water consumption, we have that person's total **water footprint**. The more one consumes, the larger one's virtual water footprint. **TABLE 1.1** shows the amounts of water used to produce some commonly consumed products. (As you look at Table 1.1 and read further, note that there are 1000 liters, or 263 gallons, in a cubic meter (m^3).)

Just as domestic water consumption does, personal water footprints vary widely according to physical geography, standards of living, and rates of consumption. Moreover, the amount of water used to produce 1 ton of a specific product varies widely from country to country because of climate conditions as well as agricultural and industrial technology and efficiency. For example, to produce 1 ton of corn in the United States requires 489 m^3 of virtual water, on average, whereas in India the same amount of corn requires 1935 m^3 of virtual water; in Mexico, 1744 m^3; and in the Netherlands, just 408 m^3. In the case of corn, water can be lost to *evapotranspiration* in the field, to the evaporation of standing irrigation water, and to evaporation as

virtual water the water used to produce a product, such as an apple or a pair of shoes

water footprint all the water a person consumes, including both virtual water and the water they consume directly

Table 1.1 The global average virtual water content of everyday products*

Product†	Virtual water content (in liters)
1 potato	25
1 cup tea	35
1 kilogram of bread	1608
1 apple	125
1 glass of beer	75
1 glass of wine	120
1 egg	135
1 cup of coffee	140
1 glass of orange juice	170
1 pound of chicken meat	2000
1 hamburger	2400
1 pound of cheese	2500
1 pair of bovine leather shoes	8000

*Virtual water is the volume of water used to produce a product.

†To see the virtual water content of additional products, go to www.waterfoot
print.org/?page=files/productgallery. [Source consulted: Arjen Y. Hoekstra and
Ashok K. Chapagain, *Globalization of Water—Sharing the Planet's Freshwater Resources*
(Malden, MA: Blackwell, 2008), p. 15, Table 2.2]

water flows to and from the field. Additionally, virtual water that becomes polluted in the production process is lost to further use.

There are several Web sites designed to help people understand water issues. To calculate your individual water footprint, go to Water Footprint's http://www.waterfootprint.org/?page=cal/waterfootprintcalculator_indv or to the *National Geographic* calculator at http://environment.nationalgeographic.com/environment/freshwater/change-the-course/water-footprint-calculator/. For more information on virtual water, look at this UNESCO site: http://www.unesco.org/new/fileadmin/MULTIMEDIA/FIELD/Venice/pdf/special_events/bozza_scheda_DOW04_1.0.pdf.

Water: Who Gets Access to It? Who Owns It?

The availability of water varies greatly across the world, as the map in **FIGURE 1.16** indicates. Countries shaded in blue have levels of access to water that range from adequate to relatively good; those shaded in orange or red are vulnerable or have some amount of stress related to scarcity. This variability has generally been due to natural conditions, but more recently, the availability of water is being affected by human use of water and by population density.

Though many people consider water a human right that should not cost anything to access, water has become the third most valuable commodity, after oil and electricity. Water in wells, streams, and rivers is increasingly being *privatized*. This means that its ownership is being transferred from governments—which can be held accountable for protecting the rights of all citizens to water—to individuals, corporations, and other private entities that manage the water primarily for profit. Governments often

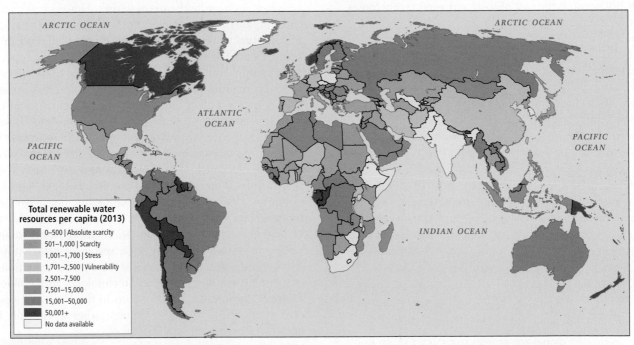

FIGURE 1.16 Map of national renewable water resources, 2013. Access to water varies greatly from place to place. This map shows the total renewable water resources per capita in cubic meters. Blue areas are relatively water secure. Countries that are shown in orange and red are in varying states of vulnerability, with those in red and orange being the most vulnerable, subject to extreme scarcity. The data are for 2013, as reported in the UN Water Report for 2015. [Source: WWAP, with data from the FAO AQUASTAT database: http://www.fao.org/nr/water/aquastat/main/index.stm, using UN-Water category thresholds, aggregate data for all countries except Andorra and Serbia]

privatize water under the rationale that private enterprise will make needed investments that will boost the efficiency of water distribution systems and the overall quality of the water supply. Regardless of whether or not these potential gains are realized, privatization usually raises the cost of water for consumers. This can become quite controversial. For example, in 1999 the city of Cochabamba, Bolivia, sold control of its water to a group of multinational corporations led by the California-based Bechtel, who then raised the price of water beyond what the urban poor could afford. The result was a nationwide series of riots that led finally to the abandonment of privatization. This event inspired the 2008 James Bond movie, *Quantum of Solace*.

Water Quality

About one-sixth of the world's population does not have access to clean drinking water, and dirty water kills more than 6 million people each year. In an average year, more people die this way than in all of the world's armed conflicts. In the poorer parts of cities in the developing world, many people draw water with a pail from a communal spigot, sometimes from shallow wells, or simply from holes dug in the ground (FIGURE 1.17). Usually this water should be boiled before use, even for bathing; but boiling requires fuel, a scarce resource for the poor. These water-quality and access problems help explain why so many people are chronically ill and why 24,000 children under the age of 5 die every day from waterborne diseases.

In Europe and the United States, demand for better water quality has resulted in a $100 billion bottled water industry. However, there are few standards of quality for bottled water; and in addition to generating mountains of plastic bottle waste, the bottled water industry often acquires its water from sources (springs, ponds, deep wells) that are publicly owned. Consumers thus pay for the same high-quality water they could consume as a public commodity for a small fee from the tap in their homes.

Water and Urbanization

Urban development patterns dramatically affect the management of water. In most urban slum areas in poor countries, and even in some places in the United States and Canada, crucial water-management technologies such as sewage treatment systems are sometimes entirely absent. Germ-laden human waste from toilets and kitchens may be deposited in urban gutters that drain into creeks, rivers, and bays, but retrofitting wastewater collection and treatment systems in cities already housing several million inhabitants is often considered prohibitively costly, especially in poor countries.

Independent of sewage, water inevitably becomes polluted in cities as parking lots and rooftops replace areas that were once covered with natural vegetation. Rainwater quickly runs off these hard surfaces, collects in low places, and becomes stagnant instead of being absorbed into the ground. In urban slums, flooding can spread polluted water over wide areas, carrying it into homes and into local freshwater sources, as well as to places where children play (see Figure 1.17). Diseases such as malaria and cholera, carried in this water, can spread rapidly as a result.

FIGURE 1.17 Access to water in water-scarce environments: Mumbai, India. A young girl gathers water from a shallow open well in a slum in Mumbai, where there are vast, low-lying slums with no sanitation and sewage that pollutes nearby waterways. If these waterways were to flood, they would spread deadly epidemics of waterborne illnesses via open wells such as the one shown here.

Sajjad Hussain/AFP/Getty Images

Thinking Geographically
What factors may have led this child to endanger her health by drinking this water?

Fortunately, new technologies and urban planning approaches are being developed that can help cities avoid these problems, but they are not yet in widespread use.

ON THE BRIGHT SIDE: Innovative Water Management in a Dry Land

Namibia is the most arid country in sub-Saharan Africa. The capital, Windhoek, with 322,000 people, lies in the central plateau. The city uses semi-purified sewage effluent to water parks, gardens, and sports fields, lowering the need for potable water. The city has been increasing its use of reclaimed water since 1969 and today recycled sewage supplies more than one quarter of the water needs. This has made it possible for Windhoek to recharge its aquifer, which helped it to withstand 2 years with no rainfall. The wastewater reclamation and the reduction in the use of potable water have also helped save energy that was previously required to transport water and to desalinate seawater. ∎

FOOD

Over time, food production has undergone many changes. It started with hunting and gathering and over the millennia evolved into more labor-intensive, small-scale, subsistence agriculture. For the vast majority of people, subsistence was the mode of food production for many thousands of years, until a series of innovations and ideas changed the way food was produced and goods were manufactured. The **Industrial Revolution,** which took place from about 1750 to 1850, opened the door to the development of what has become modern and mechanized commercial agriculture, involving the intensive use of machinery, fuel, and chemicals. Modern processes of food production, distribution, and consumption have greatly increased the supply and, to some extent, the security of food systems. However, this has come with a cost: it has created environmental pollution that may impede future food production and food security.

Agriculture: Early Human Impacts on the Physical Environment

Agriculture includes animal husbandry, or the raising of animals, as well as the cultivation of plants. The ability to produce food, as opposed to being dependent on hunting and gathering, led to a host of long-term changes. The human population began to grow more quickly, rates of natural resource use increased, permanent settlements eventually developed into towns and cities, and ultimately, human relationships became more formalized. Some would say these are the steps that led to civilization.

Very early humans hunted animals and gathered plants and plant products (seeds, fruits, roots, and fibers) for their food, shelter, and clothing. To successfully use these wild resources, humans developed an extensive folk knowledge of the needs of the plants and animals they favored. The transition from hunting in the wild to tending animals in pens and pastures and from gathering wild plant products to sowing seeds and tending plants in gardens, orchards, and fields probably took place gradually over thousands of years.

Where and when did plant cultivation and animal husbandry first develop? Genetic studies suggest that between 8000 and 20,000 years ago, people in many different places around the world independently learned to develop plants and animals for food and other uses through selective breeding, a process known as **domestication.**

Why did agriculture and animal husbandry develop in the first place? Certainly the desire for more secure food resources played a role, but the opportunity to trade may have been just as important. It is probably not a coincidence that many of the known locations of agricultural innovation lie along early trade routes—for example, along the Tigris and Euphrates rivers and along the Silk Road that runs through Central Asia from the eastern Mediterranean to China. In such locales, people would have had access to new information and new plants and animals brought by traders.

Industrial Revolution a series of innovations and ideas that occurred broadly between 1750 and 1850, which changed the way goods were manufactured

agriculture the practice of producing food and other products through animal husbandry, or the raising of animals, and the cultivation of plants

domestication the process whereby humans used selective breeding to develop plants and animals for use as food, construction materials, and sources of energy, or to improve life in other ways

food security the ability of a society to consistently supply a sufficient amount of basic food to the entire population

Agriculture and Its Consequences

Agriculture made possible the amassing of surplus stores of food (and other agricultural products) for lean times, and allowed some people to specialize in activities other than the procurement of food and other agricultural products. It also may have led to several developments now regarded as problems: rapid population growth, concentrated settlements where diseases could easily spread, environmental degradation, and paradoxically, malnutrition or even famine.

Through the study of human remains, archaeologists have learned that it was not uncommon for the nutritional quality of human diets to decline as people stopped eating diverse wild food species and began to eat primarily one or two species of domesticated plants and animals. Evidence of nutritional stress (shorter stature, malnourished bones and teeth) has been found repeatedly in human skeletons excavated in sites around the world where agriculture was practiced.

Although agriculture could support more people on a given piece of land than could hunting and gathering, as populations expanded and as more land was turned over to agriculture, natural habitats were destroyed, reducing opportunities for hunting and gathering. Furthermore, the storage of agricultural surpluses not only enabled people to trade food and other products, it also made it possible for people to live together in larger communities, which marked the beginning of urban societies, the development of social classes, and the monopolization of wealth. Dense settlements, however, facilitated the spread of disease. Moreover, land clearing increased vulnerability to drought and other natural disasters that could wipe out entire harvests. Thus, as ever-larger populations depended solely on agriculture, episodic famine may have become more common and affected more people.

Modern Food Production and Food Security

For most of human history, people lived in subsistence economies. Over the past five centuries of global interaction and trade, though, people have become ever more removed from their sources of food. Today, occupational specialization means that food is increasingly mass-produced. Far fewer people work in agriculture than did in the past, and now most people know little about cultivation and animal tending. They work for cash to buy food and other necessities.

A side effect of this dependence on money is that the **food security**—the ability of a society to consistently supply a sufficient amount of basic food to the entire population—can be threatened by economic disruptions, even in distant places. As countries become more involved with the global economy, they may import more food, or their own food production may become vulnerable to price swings. For example, a crisis in food security began to develop in 2007 when the world price of corn spiked. Speculators in alternative energy, thinking that corn would be an ideal raw material from which to make ethanol (a substitute for gasoline), invested heavily in it, creating a shortage. As a result, global corn prices rose beyond the reach of those who depended on corn as a food, leading to widespread protests among the food-insecure. Then, between the sharp price rise in oil in 2008 and the recession of 2007–2009, the global cost of many other basic foods rose 17 percent. When oil prices rise, all foods produced and transported with machines get more expensive.

There are many other factors that contribute to food insecurity, some of which will be discussed in later chapters. But in many places around the world, when the global recession—partially caused by rising oil prices—eliminated jobs, migrant workers could no longer send remittances to their families, who then no longer had money with which to buy food. Financially strapped rural farmers were often forced to sell their land for cash. A lot of this land was bought up by corporations to grow crops for export to consumers in East Asia, South Asia, and Europe. This series of events called into question the sustainability of modernized global food production and acquisition systems. In developing countries, some household economies were so ruined that parents sold important assets to feed their children or went without food themselves, to the detriment of their long-term health, and some stopped sending their children to school. By 2015, however, the reversals of 2007–2008 had been partially mitigated, and there were improvements in human well-being. **FIGURE 1.18** identifies countries in which undernourishment remains an ongoing problem and periodic food insecurity is especially intense. Some food security improvement was due to falling oil prices.

Another way that modernized agriculture influences food security is through its reliance on machines, chemical fertilizers, and pesticides. Paradoxically, when the shift to this kind of agriculture was introduced into developing countries like India and Brazil in the 1970s, it was called the **green revolution.** In fact, the green revolution is not "green" in the modern sense of being environmentally savvy. When first implemented, the results of green revolution

agriculture were spectacular: soaring production levels along with high profits for those farmers who could afford the additional investment. In the early years of this movement, it seemed to scientists and developers that the world was literally getting greener. But often, poorer farmers couldn't afford the machinery and chemicals, let alone the expensive new seeds. Also, since greatly increased production leads to lower crop prices on the market, these poorer farmers lost market share. To survive, they often were forced to sell their land and move to crowded cities. Here they joined masses of urban poor people whose access to food was precarious.

Green revolution agriculture can influence food security by damaging the environment. As rains wash fertilizers and pesticides into streams, rivers, and lakes, these bodies of water become polluted. Over time, the pollution destroys fish and other aquatic animals vital to food security. Hormones fed to farm animals to hasten growth may enter the human food chain. Soil degradation can also increase as green revolution techniques (such as mechanical plowing, tilling, and harvesting) leave soils exposed to rains that wash away natural nutrients and the soil itself. Indeed, many of the most agriculturally productive parts of North America, Europe, and Asia have already suffered moderate to serious loss of soil through erosion.

> **green revolution** increases in food production brought about through the use of new seeds, fertilizers, mechanized equipment, irrigation, pesticides, and herbicides

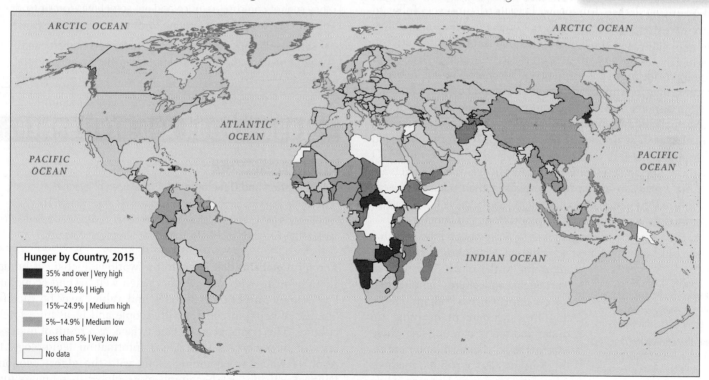

FIGURE 1.18 Global map of undernourishment, 2015. This Food and Agriculture Organization of the United Nations (UNFAO) map is based on 2015 data showing that considerable improvement has been made toward reducing undernourishment. The global recession that began in 2008 put many people back into a state of hunger, and the UN made global food security its primary Millenium Development Goal. The proportion of people suffering from undernourishment—the lack of adequate nutrition to meet their daily needs—has declined in the developing world, most notably in Central and South America, India, and parts of Africa (though data are missing for Libya, Sudan, South Sudan, Uganda, and Somalia). Hunger remains a problem for many in much of sub-Saharan Africa; parts of South Asia; Mongolia and North Korea in East Asia; and in Bolivia in South America and Haiti in the Caribbean. [Source consulted: Food and Agriculture Organization of the United Nations, FAO Statistics Division (Rome, Italy, 2015); available at http://www.fao.org/hunger/en/]

Globally, soil erosion and other problems related to food production affect about 7 million square miles (2000 million hectares), putting at risk the livelihoods of a billion people.

At least in the short term, green revolution agriculture raised the maximum number of people that could be supported on a given piece of land, or its **carrying capacity.** However, it is unclear how sustainable these green revolution gains in food production are. In the 25 years between 1965, when the green revolution was implemented, and 1990, total global food production did rise between 70 and 135 percent (varying from region to region). But, in response, populations also rose quickly during this period. It is now estimated that to feed the population projected for 2050, global food output must increase by another 70 percent. Yet scientists from many disciplines estimate that within the next 50 years, environmental problems such as water scarcity and global climate change will limit, halt, or even reverse increases in food production; indeed, food production by green revolution techniques is already diminishing. How can these discrepancies between expectations and realities be resolved?

The technological advances that could make current agricultural systems more productive are increasingly controversial. In North America, **genetic modification (GMO),** the practice of splicing together the genes from widely divergent species to create particular characteristics, is being used to boost productivity. However, outside of North America, many worry about the side effects of such agricultural manipulation. Europeans have tried (unsuccessfully) to keep GMO food products entirely out of Europe, fearing that they could lead to negative ecological consequences or catastrophic crop failures. The European Commission is carefully crafting legislation that will provide for risk assessment and clear labeling so producers and consumers can make informed choices. They point out that the main advance in GMO agriculture has been the production of seeds that can tolerate high levels of environmentally damaging herbicides, such as Roundup. The use of GMO crops thus could lead to more, not less, environmental degradation. Even if GMO crops prove safe, in developing countries where farmers' budgets are tiny, GMO seeds are much more expensive than traditional seeds and they must be purchased anew each year because GMO plants do not produce viable seeds, as do plants from traditional seeds.

As a result of the uncertainties of GMO crops and the potential negative side effects of new agricultural technologies, many are returning to the much older idea of **sustainable agriculture**—farming that meets human needs without poisoning the environment or using up water and soil resources. Often these systems avoid chemical inputs entirely, as in the case of popular methods of *organic agriculture*. However, while these systems can be productive, they are less so than conventional green revolution systems and often require significantly more human labor, resulting in higher food prices. More dependence on sustainable and organic systems could therefore lead to food insecurity for the less affluent, especially in the world's megacities.

carrying capacity the maximum number of people that a given territory can support sustainably with food, water, and other essential resources

genetic modification (GMO) in agriculture, the practice of splicing together the genes from widely divergent species to create particular desirable characteristics

sustainable agriculture farming that meets human needs without poisoning the environment or using up water and soil resources

globalization the growth of worldwide linkages and the changes these linkages are bringing about in daily life, especially in economies, but also in cultural and biological spheres

According to the UN Food and Agriculture Organization (FAO), one-fifth of humanity subsists on a diet too low in total calories and vital nutrients to sustain adequate health and normal physical and mental development (see Figure 1.18). As we will see in later chapters, this alarming hunger problem is partly due to political instability, corruption, and inadequate distribution systems rather than simple food production. However, when food is scarce for whatever reason, it tends to go to those who have the money to pay for it.

THINGS TO REMEMBER

• Genetic studies suggest that between 8000 and 20,000 years ago, people in many different places around the globe independently learned to use selective breeding to develop plants and animals for food and other agricultural products.

• While modern processes of agriculture and distribution have increased the supply of food and other agricultural products, environmental damage, market disruptions, and population growth could affect food supplies and compromise food security.

• Many farmers are unable to afford the chemicals and machinery required for commercial agriculture or new, genetically modified seeds. Because their production is low, they cannot compete on price and are sometimes forced to give up farming, often migrating to cities.

• Sustainable agriculture is farming that meets human needs without harming the environment or depleting water and soil resources.

GLOBALIZATION AND DEVELOPMENT

GEOGRAPHIC THEME 2

Globalization and Development: Global flows of information, goods, and people are transforming patterns of economic development. Local self-sufficiency is giving way to global interdependence as people and places are increasingly becoming connected, sometimes across vast distances.

The term **globalization** refers to the worldwide changes brought about by many types of flows that reach well beyond economics. For example, a person's attitudes and values may be modified because of being in contact with people and ideas from foreign cultures. This contact can be face to face, as a result of increased rates of migration, or virtual, via information technology, the reach of which is expanding along with the global economy. Political activities are becoming more global in scope as international agreements are necessary to manage an interconnected world. And the causes and impacts of climate change illustrate how local behavior can have implications on a global scale. Globalization is the most complex and far-reaching of the thematic concepts described in this book.

Across the world, local self-sufficiency is giving way to global interdependence and international trade. While for a time there was anticipation that globalization would lead to more prosperity for all, the economic recession that began in 2007 cast a spotlight on the unpredictable and widespread effects of global interdependence. In that year, economic disruptions in the United States and Europe resulted in powerful ripple effects that reached around the world. Foreclosures in the U.S. housing market meant that European banks that had invested in U.S. mortgages faltered and some failed. The governments of some indebted countries, such as Greece and Ireland, could not make loan payments, deepening the economic disruptions in Europe. Consumers in the United States and Europe cut back on shopping and vacations, and as a result, many workers across the world lost their jobs or saw their incomes decline. Recovery is still spotty across Africa, Asia, and Middle and South America, and in 2015, an economic slump in China and a refugee crisis in Europe threatened to bring further global economic instability.

WHAT IS THE ECONOMY?

The *economy* is the forum in which people make their living, and resources are what they use to do so. *Extractive resources* are resources that must be mined from Earth's surface (mineral ores) or grown from its soil (timber and plants). There are also *human resources*, such as skills and brainpower, which are used to transform extractive resources into useful products (such as refrigerators or bread) or bodies of knowledge (such as books or computer software). Economic activities are often divided into three *sectors of the economy*: the **primary sector** is based on **extraction** (mining, fishing, forestry, and agriculture); the **secondary sector** is **industrial production** (processing, manufacturing, and construction); and the **tertiary sector** is **services** (sales, entertainment, and financial). Of late, a fourth, or **quaternary sector,** has been broken off from the tertiary sector and includes advanced intellectual pursuits such as education, research, and IT (information technology) development. Generally speaking, as people in a society shift from extractive activities, such as farming and mining, to industrial and service activities, their material standards of living rise—a process typically known as **development.**

Development is best thought of as taking place on a sliding scale. The development process has several characteristics, one of which is a shift from economies based on extractive resources to those based on human resources. Many countries around the world, especially those that are poorer (often these are referred to as *underdeveloped* or *developing* countries, although the preferred term these days is **less developed countries,** or **LDCs**), have begun to change from labor-intensive and low-wage, often agricultural, economies toward higher-wage but still labor-intensive manufacturing and service economies (including Internet-based economies). Over time, this process often boosts a country's economy, helping it become what is known as moderately developed—incomes are higher and social services, such as health care and education, are much improved. The richest countries, often referred to as *developed* or **more developed countries (MDCs),** are now lessening their dependence on labor-intensive manufacturing and shifting toward more highly skilled mechanized production or knowledge-based service and technology (quaternary) industries. Generally, MDCs provide adequate education, health care, and other social services to help their people contribute to economic development.

WHAT IS THE GLOBAL ECONOMY?

The **global economy** is the worldwide system in which goods, services, and labor are exchanged. Most of us participate in the global economy every day. For example, books like this can be made from trees cut down in Southeast Asia or Siberia and shipped to a paper mill in Oregon. Many books are now printed in Asia because labor costs are lower there. Globalization is not new. At least 2500 years ago, silk and other goods were traded along the Central Asian Silk Road that connected Greece and then Rome in the Mediterranean with distant China, an expanse of more than 4000 miles (6437 km).

European colonization was an early undertaking related to globalization. Starting in about 1500 C.E., European countries began extracting resources from distant parts of the world that they had conquered. The colonizers organized systems to process those resources into higher-value goods to be traded wherever there was a market. Sugarcane, for example, was grown on Caribbean and Brazilian plantations with slave labor from Africa (**FIGURE 1.19**) and made locally into crude sugar, molasses, and rum. It was then shipped to Europe and North America, where it was further refined and sold at considerable profit. The global economy

primary sector economic activity based on extracting resources (mining, forestry, and agriculture)

extraction mining, forestry, and agriculture

secondary sector economic activity based on processing, manufacturing, and construction

industrial production processing, manufacturing, and construction

tertiary sector economic activity based on sales, entertainment, and financial services

services sales, entertainment, and financial services

quaternary sector economic activities based on intellectual pursuits in education, research, and information technology

development an inexact concept typically referring to the shift from primary activities to secondary, tertiary, and quaternary activities

less developed countries (LDCs) countries that have labor-intensive and low-wage, often agricultural, economies

more developed countries (MDCs) countries that have economies often generated by tertiary and quaternary sectors and that generally provide adequate education, health care, and other social services to help their people contribute to economic development

global economy the worldwide system in which goods, services, and labor are exchanged

FIGURE 1.19 A landscape of European colonialism. This sugar plantation on the island of St. Croix was one of thousands in the British Caribbean (and many other parts of the world) where the labor of enslaved people and low-paid workers subsidized the creation of wealth that benefited European colonizers and helped to fund the Industrial Revolution in Europe.

grew as each region produced goods for export rather than just for local consumption. At the same time, regions also became more dependent on imported food, clothing, machinery, energy, and knowledge.

The new wealth derived from the colonies and the ready access to global resources led to Europe's Industrial Revolution. No longer was one woman producing the cotton or wool for clothes or fabric by spinning thread, weaving the thread into cloth, and sewing a garment. Instead, these separate tasks were spread out among many workers, often in distant places, with some people specializing in producing the fiber and others in spinning, weaving, or sewing. These innovations in efficiency were followed by labor-saving improvements such as mechanized reaping, spinning, weaving, and sewing.

This larger-scale mechanized production accelerated globalization, as it created a demand for raw materials and labor and a need for markets in which to sell finished goods. European colonizers managed to integrate the production and consumption of their colonial possessions in the Americas, Africa, and Asia. For example, in the British Caribbean colonies, hundreds of thousands of enslaved Africans wore rough garments made of cheap cloth woven in England from cotton grown in British India. The sugar that slaves produced on British-owned plantations with iron equipment from British foundries was transported to European markets in ships made in the British Isles of trees and resources from the North American colonies and other parts of the world.

Until the early twentieth century, much of the activity of the global economy took place within the colonial empires of a few European nations (Britain, France, the Netherlands, Germany, Belgium, Spain, and Portugal). By the 1960s, global economic and political changes brought an end to these empires, and now almost all colonial territories are independent countries. Nevertheless, the global economy persists in the form of banks and **multinational corporations**—such as Shell, Chevron, IBM, Walmart, Coca-Cola, Bechtel, Apple, British Petroleum, Toyota, Google, and Cisco—that operate across international borders. These corporations extract resources (including intellectual properties, such as knowledge of medicinal plants) from many places, make products in factories located where they can take advantage of cheap labor and transportation facilities, and market their products wherever they can make the most profit. Their global influence, wealth, and importance to local economies enable the multinational companies to influence the economic and political affairs of the countries in which they operate.

multinational corporation a business organization that operates extraction, production, and/or distribution facilities in multiple countries

VIGNETTE Sixty-year-old Olivia lives near Soufrière on St. Lucia, an island in the eastern Caribbean (**FIGURE 1.20A**). She, her daughter Anna, and her three grandchildren live in a wooden house surrounded by a leafy green garden dotted with fruit trees. Anna has a tiny shop at the side of the house, from which she sells various small everyday items and preserves that she and her mother make from the garden fruits.

On days when the cruise ships dock, Olivia heads to the market shed on the beach with a basket of homegrown goods.

FIGURE 1.20 Workers in the global economy.

A Soufrière, St. Lucia, where Olivia and Anna live.

B Setiya, along with other migrant workers, boards a bus that will take him to the construction site in Malacca, Malaysia. Later, during economic recessions, hundreds of thousands of workers, like Setiya, were expelled from the country.

C The trailer at the back of Tanya's lot, where her daughter lives.

Thinking Geographically

A What about this photo might suggest that Olivia and her neighbors share a similar standard of living?

B What about this photo suggests that Setiya is a low-paid worker?

C How does this photo suggest that Tanya's low income will also be experienced by her daughter, Reyna?

She calls out to the passengers as they near the shore, offering her spices and snacks for sale. In a good week she makes U.S.$70. Her daughter makes about U.S.$100 per week in the shop and is constantly looking for other ways to earn a few dollars.

Olivia and Anna support their family of five on about U.S.$170 a week (U.S.$8840 per year). From this income they take care of their bills and other purchases, including school fees for the granddaughter who will go to high school in the capital next year and perhaps college if she succeeds. Their livelihood puts them at or above the standard of living of most of their neighbors.

In Malacca, Malaysia, 30-year-old Setiya, an undocumented immigrant from Tegal, Indonesia, is boarding a bus that will take him to a job site where he is helping build a new tourist hotel (see Figure 1.20B). Like a million other Indonesians attracted by the booming economy, he snuck into Malaysia, risking arrest, because in Malaysia average wages are four times higher than at home.

This is Setiya's second trip to Malaysia. His first trip was to do work legally on a Malaysian oil palm plantation. Upon arrival, however, Setiya found that he would have to work for 3 months just to pay off his boat fare from Indonesia. Not one to give in easily, he quietly went to another city and found a construction job earning U.S.$10 a day (about U.S.$2600 a year), which allowed him to send money home to his family in Indonesia.

Periodically, Malaysia announces it is expelling foreign workers; and in 2008, Setiya was one of them. The country was suffering from rising unemployment due to the global recession, and its leaders wanted to save more jobs for local people by deporting foreign workers. When the recession was alleviated, demand for labor increased, so young fathers from Indonesia (like Setiya) once again risked trips on leaky boats to illegally enter Malaysia and also Singapore in order to support their families.

Fifty-year-old Tanya works at a fast-food restaurant outside of Charleston, South Carolina, making less than U.S.$7.50 an hour. She had been earning U.S.$8.00 an hour sewing shirts at a textile plant until it closed and moved to Indonesia. Her husband is a delivery truck driver for a snack-food company.

Between them, Tanya and her husband make U.S.$30,000 a year, but from this income they must cover all their expenses, including their mortgage, gasoline, and car payments. In addition, they help their daughter, Rayna, who quit school after eleventh grade and married a man who is now out of work. They and their baby live at the back of the lot in an old mobile home (see Figure 1.20C).

With Tanya's now-lower wage (almost U.S.$1000 less a year), there will not be enough money to pay the college tuition for her son, who is in high school. He had hoped to become an engineer, and would have been the first in the family to go to college. For now, he is working at the local gas station.

These people, living worlds apart, are all part of the global economy. Workers around the world are paid startlingly different rates for jobs that require about the same skill level. Varying costs of living and varying local standards of wealth make a difference in how people live and how they perceive their own situation. Though Tanya's family has the highest income by far, they live in poverty compared to their neighbors, and their hopes for the future are dim. Olivia's family, on the other hand, are not well off, but they do not think of themselves as poor because they have what they need, others around them live in similar circumstances, and their children seem to have a future. They can subsist on local resources, and the tourist trade promises continued cash income. But their subsistence depends on circumstances beyond their control; in an instant, the cruise-line companies can choose another port of call. Setiya, by far the poorest, seems trapped by his status as an undocumented worker, which robs him of many of his rights. Still, the higher pay that he can earn in Malaysia offers him a possible way out of poverty. [*Source: From Lydia Pulsipher's and Alex Pulsipher's field notes. For detailed source information, see Text Sources and Credits.*] ∎

THE DEBATE OVER GLOBALIZATION AND FREE TRADE

The term **free trade** refers to the unrestricted international exchange of goods, services, and capital. Free trade is an ideal that has not been achieved and probably never will be. Currently, all governments impose some restrictions on trade to protect their own national economies from foreign competition, although there are far fewer such restrictions than there were in the 1980s. Restrictions take several forms, but among the most important to understand are *tariffs*, *export subsidies*, *import quotas*, *precise product regulations*, and *investment restrictions*. Tariffs are taxes imposed on imported goods that increase the cost of those goods to the consumer, thus giving price advantages to competing, locally made goods. Export subsidies reduce costs for home-country producers so they can compete better with foreign producers in the global market. Import quotas set limits on the amount of a given good that may be imported over a set period of time, curtailing supply and keeping prices high, again to protect local producers. Precise regulations and accompanying paperwork also restrict access to a country's markets. Investment restrictions keep outsiders from building competitive businesses in a particular country.

These and other forms of trade protection are subjects of contention. Proponents of free trade argue that the removal of all such protective measures encourages efficiency, lowers prices, and gives consumers more choices. Companies can sell to larger markets and take advantage of mass-production systems that lower costs further. As a result, businesses can grow faster, thereby providing people with jobs and opportunities to raise their standard of living. These pro–free trade arguments have been quite successful, and in recent decades, restrictions on trade imposed by individual countries have been greatly reduced. Several *regional trade blocs* have been formed; these are associations of neighboring countries that agree to lower trade barriers for one another. The main ones

free trade the unrestricted international exchange of goods, services, and capital

are the North American Free Trade Agreement (NAFTA), the European Union (EU), the Southern Common Market in South America (Mercosur), and the Association of Southeast Asian Nations (ASEAN). Presently under negotiation are the Trans-Pacific Partnership (TPP), which would unite 12 Pacific Rim countries, including the United States but not China, in a free trade agreement; and the Trans-Atlantic Free Trade Agreement (TAFTA) between the European Union and the United States. Both are promoted as enlarging free trade areas, protecting investment, and removing unnecessary regulatory barriers. Critics of both say that the secretive negotiations that produced the draft agreements sidelined public interests and favored corporate interests. Neither has been ratified.

One of the main global institutions that supports the ideal of free trade is the **World Trade Organization (WTO),** whose stated mission is to lower trade barriers and to establish ground rules for international trade. Related institutions, the *World Bank* and the *International Monetary Fund* (IMF), both make loans to countries that need money to pay for economic development or to avoid financial crises. Before approving a loan, the World Bank or the IMF may require a borrowing country to reduce and eventually remove tariffs, import quotas, and other trade-restricting regulations. These requirements are part of larger belt-tightening measures, often called **structural adjustments,** that the IMF imposes on countries seeking loans, such as the requirement to close or privatize government enterprises and to reduce government services, including education and health-care programs that benefit the poor.

Those opposed to free trade argue that it leads to a less regulated global economy that can be chaotic, resulting in rapid cycles of growth and decline that increase the disparity between rich and poor worldwide. They note that the rules of trade have been made by the more powerful countries. That is, free trade is assured on goods and services that benefit more developed countries (MDCs) but is less assured when lesser developed countries (LDCs) have a competitive edge. Labor unions point out that as corporations relocate factories and services to poorer countries where wages are lower, jobs are lost in richer countries, creating poverty there. In the poorer countries, multinational corporations often work with governments to prevent workers from organizing labor unions that could bargain for **living wages,** wages high enough to support a healthy life. Environmentalists have found that in newly industrializing countries, which often lack effective environmental protection laws, multinational corporations tend to use highly polluting and unsafe production methods to lower costs. Many fear that a "race to the bottom" in wages, working conditions, government services,

and environmental quality is underway as countries compete for profits and potential investors. Multinational corporations that have recently agreed to address worker abuses include Apple, Nike, Walmart, and Infosys.

ON THE BRIGHT SIDE: Reforms at the IMF and World Bank

In response to the now widely recognized failures of policies that rely on the power of markets to guide development, the IMF and the World Bank have made some changes. The standard set of loan repayment requirements, known as *structural adjustment programs* (SAPs), have been replaced with *Poverty Reduction Strategy Papers*, or PRSPs. Instead of having a uniform loan repayment policy (structural adjustment) imposed on them, each country in need works with World Bank and IMF personnel to design a broad-based plan for both economic growth *and* poverty reduction. PRSPs still involve market-based solutions and are aimed at reducing the role of government in the economy. These programs remain highly bureaucratic but they do focus more on poverty reduction rather than just on development via structural adjustment. They promote the maintenance of education, health care, social services, and broader participation in civil society. PRSPs also include the possibility that all or some of a country's debt be "forgiven" (written off by the IMF and the World Bank). This alleviates one of the worst problems of SAPs, which was that the belt-tightening measures, instead of making it easier to pay off debt, sometimes sent a country into an economic decline that made loan repayment impossible. ■

Fair trade, proposed as an alternative to free trade, is intended to provide a fair price to producers and to uphold environmental and safety standards in the workplace. Economic relationships surrounding trade are rearranged in order to provide better prices for producers from developing countries. For example, "fair trade" coffee and chocolate are now sold widely in North America and Europe. Prices are somewhat higher for consumers, but the extreme profits of middlemen are eliminated. As a result, growers of coffee and cocoa beans who produce for fair trade companies can receive living wages and work under better conditions.

In evaluating free trade, globalization, and fair trade, consider how many of the things you own or consume were produced in the global economy—your computer, clothes, furniture, appliances, car, and foods. These products are cheaper for you to buy, and your standard of living is higher as a result of lower production costs as well as competition among many global producers. However, you or someone you know may have lost a job because a company moved to another location where labor and resources were cheaper. Underpaid workers (even children) working under harsh conditions that possibly generate high levels of pollution may have made those cheap products. If workers can't earn living wages, often some family members end up migrating, perhaps without the proper papers, to earn a better wage. For example, NAFTA gave the advantage to U.S. corporate farmers who exported crops grown in the United States to Mexico or acquired land in Mexico for corporate farming. In both cases these large producers were able to out-compete small Mexican farmers both

World Trade Organization (WTO) a global institution made up of member countries whose stated mission is to lower trade barriers and to establish ground rules for international trade

structural adjustment belt-tightening measures imposed to force a government to make its economy more open to market forces; often involves closing or privatizing government enterprises and reducing education, health-care programs, and other social services that benefit the poor

living wages minimum wages that are high enough to support a healthy life

fair trade trade that values equity throughout the international trade system, not just market forces; now proposed as an alternative to free trade

on production levels and on price, so the Mexican farmers were forced out of business. These impoverished ex-farmers tried to migrate to the United States to support their families, where they have often encountered contempt and hardship.

THINGS TO REMEMBER

GEOGRAPHIC THEME 2
- **Globalization and Development:** Global flows of information, goods, and people are transforming patterns of economic development. Local self-sufficiency is giving way to global interdependence as people and places are increasingly becoming connected, sometimes across vast distances.

- Workers around the world are paid startlingly different rates for jobs that require about the same skill level.

- Under true free trade, economic transactions would be conducted without interference or regulation in an open market but workers might face hazards in the workplace and consumers might buy unsafe products. Under fair trade, an alternative to free trade, consumers are asked to pay a fair price to producers, and producers are expected to pay living wages and uphold environmental and safety standards in the workplace.

- Trade agreements can result in deep changes in local economies, forcing drastic solutions onto families, such as family members migrating to distant locations to earn a living.

MEASURING ECONOMIC DEVELOPMENT

The most long-standing measure of development has been **gross domestic product (GDP) per capita.** GDP is simply an economic measure that refers to the total market value of all goods and services produced in a country in a given year. A closely related index that is now used more often by international agencies is **gross national income (GNI) per capita,** the sum of a nation's gross domestic product plus net income received from overseas. When GDP or GNI is divided by the number of people in the country, the result is per capita GDP or GNI. This book now uses primarily the GNI per capita statistics.

GNI per capita is often used as a measure of how well people are living in particular places, but it has several disadvantages. First is the matter of wealth distribution. Because GNI per capita is an average, it can hide the fact that a country has a few fabulously rich people and a great mass of abjectly poor people. For example, a GNI per capita of U.S.$50,000 would be meaningless if a few lived on millions per year and most lived on less than U.S.$10,000 per year.

A second problem with GNI and GDP is that they are not indicators of health and well-being in a particular place, although GNI is correlated with other indicators of quality of life, such as infant mortality, life expectancy, and education rates. The third problem with GNI data is that the purchasing power of currency varies widely around the globe. A GNI of U.S.$18,000 per capita in Barbados might represent a middle-class standard of living, whereas that same amount in New York City could not buy even basic food and shelter. Because of these purchasing power variations, in this book GNI (and occasionally GDP) per capita figures have been adjusted for **purchasing power parity (PPP).** PPP, usually indicated in parentheses after GDP or GNI, is the amount that the local currency equivalent of U.S. dollars purchases in a given country. The *Economist* magazine does an annual survey of the purchasing power parity of different currencies, using McDonald's Big Mac because it is a standardized product for sale all over the world. In July of 2015, a Big Mac at McDonald's in the United States cost U.S.$4.79. In the **Euro zone** (those countries in the European Union that use the Euro currency), a Big Mac costs the equivalent of U.S.$4.05, while in India it costs U.S.$1.83. Of course, for the consumer in India, where annual per capita GNI (PPP) is U.S.$5760, this would be a rather expensive meal. On the other hand, at $4.79, the Big Mac would be an economy meal in the United States, where the GNI per capita (PPP) is U.S.$55,860, or in the Euro area, where U.S.$36,280 is the average GNI (PPP) per capita (see "Big Mac Index," at http://www.economist.com/content/big-mac-index).

A third disadvantage of using GDP (PPP) or GNI (PPP) per capita is that both measure only what goes on in the **formal economy**—all the activities that are officially recorded as part of a country's production. Many goods and services are produced outside formal markets, in the **informal economy.** Here, work is often bartered for food, housing, or services, or for cash payments made "off the books"—payments that are not reported to the government as taxable income. It is estimated that one-third or more of the world's work takes place in the informal economy. Examples of workers in this category include anyone who contributes to her/his own or someone else's well-being through bartered or off-the-books services such as housework, gardening, herding, animal care, or elder and child care. *Remittances*, or pay sent home by migrants, become part of the formal economy of the receiving society if they are sent through banks or similar financial institutions, because records are kept and taxes levied. If they are transmitted off the books (perhaps illegally), such as via mail or in cash, they become part of the informal economy.

There is a gender aspect to informal economies. Researchers studying all types of societies and cultures have shown that, on average, women perform about 60 percent of all the work done, and that much of this work is unpaid and in the informal economy. Yet only the work women are paid for in the formal economy appears in the statistics, so economic figures per capita ignore much of the work women do. Statistics also neglect the contributions of millions of men and children who work in the informal economy as street vendors, craftspeople, traders, service workers, or seasonal laborers.

---

gross domestic product (GDP) per capita the total market value of all goods and services produced within a particular country's borders and within a given year, divided by the number of people in the country

gross national income (GNI) per capita the sum of a country's gross domestic product plus all net income received from overseas, divided by the mid-year population

purchasing power parity (PPP) the amount that the local currency equivalent of U.S.$1 purchases in a given country

Euro zone those countries in the European Union that use the Euro currency

formal economy all aspects of the economy that take place in official channels

informal economy all aspects of the economy that take place outside official channels

A fourth disadvantage of GDP (PPP) and GNI (PPP) per capita is that neither takes into consideration whether these levels of income are achieved at the expense of environmental sustainability, human well-being, or human rights.

GEOGRAPHIC PATTERNS OF HUMAN WELL-BEING

Some development experts, such as the Nobel Prize–winning economist Amartya Sen, advocate a broader definition of development that includes measures of **human well-being.** This term generally means a healthy and socially rewarding standard of living in an environment that is safe and sustainable. The following section explores the three measures of human well-being that are used in this book. These measures are discussed and mapped in each regional chapter so that you can make some crude region-to-region and country-to-country comparisons.

Global GNI per capita (PPP) is mapped in **FIGURE 1.21A**. Comparisons between regions and countries are possible, but as discussed above, GNI per capita figures ignore all aspects of development other than economic ones. For example, there is no way to tell from GNI per capita figures how quickly a country is consuming its natural resources, or how well it is educating its young, maintaining its environment, or achieving gender and racial equality. Therefore, along with the traditional GNI per capita figure, geographers often use several other measures of development.

The second measure used in this book is the **United Nations Human Development Index (HDI),** which calculates a country's level of well-being with a formula of factors that considers income adjusted to PPP, data on life expectancy at birth (an indicator of overall health care), and data on educational attainment (see Figure 1.21B).

The third measure of well-being used here, the **United Nations Gender Development Index (GDI),** is a composite measure reflecting the degree to which women and men are equal within a particular country in three dimensions: longevity, education, and income (see Figure 1.21C). Ranks are from most equal (1) to least equal (5). Not all countries are ranked because many did not submit data.

Together, these three measures reveal some of the subtleties and nuances of well-being and make comparisons between countries somewhat more valid. Because the more sensitive indices (HDI and GDI) are also more complex than the purely economic GNI per capita (PPP), they are still being refined by the UN.

human well-being various measures of the extent to which people are able to achieve a healthy and socially rewarding standard of living in an environment that is safe and sustainable

United Nations Human Development Index (HDI) an index that calculates a country's level of well-being, based on a formula of factors that considers income adjusted to PPP, data on life expectancy at birth, and data on educational attainment

United Nations Gender Development Index (GDI) a composite measure of disparity in human development (longevity, education, and income) by gender. The closer the rank is to 1, the smaller the gap between women and men; that is, the more that the genders are tending toward equality

sustainable development the improvement of current standards of living in ways that will not jeopardize those of future generations

political ecology the study of the interactions among development, politics, human well-being, and the environment

A geographer looking at these maps in Figure 1.21 might make the following observations:

- The map of GNI per capita (see Figure 1.21A) shows a wide range of difference across the globe, with very obvious concentrations of high and low GNI per capita (PPP). The most populous parts of the world—China and India—have medium low and low GNI, respectively, and sub-Saharan Africa has the lowest GNI.

- The HDI map (see Figure 1.21B) shows a similar pattern, but Middle and South America, Southeast Asia, China, Italy, and Spain rank a bit higher on HDI than they do on GNI. Several countries in southern Africa, as well as Iran, rank lower on HDI than on GNI, thus illustrating the disconnect between GNI per capita, a measure of average income, and human well-being.

- The GDI map (see Figure 1.21C) shows some strange anomalies, with certain high-income countries, such as the United States (medium high GDI) and Saudi Arabia (very low GDI), ranking far lower on this index than on the HDI index. China (high GDI), which has medium-low GNI and a reputation for gender discrimination, ranks higher than the United States on the GDI scale, in the same category as New Zealand and Ireland. China can rank high in terms of GDI despite having a record of gender discrimination because males and females are close to income parity; although they have relatively low development on the three indices (longevity, education, and income), the genders do tend to be more or less equal. In the early years of the revolution in China, the Communist Party did aim at equality for the whole population and the residual effects of this policy are still evident. As you learn more about actual conditions in these various countries, you will be able to speculate about the causes of the various anomalies in gender development.

SUSTAINABLE DEVELOPMENT AND POLITICAL ECOLOGY

The UN defines **sustainable development** as improving current living standards in ways that will not jeopardize those of future generations. Sustainability has only recently gained widespread recognition as an important goal—well after the developed parts of the world had already achieved high standards of living based on the mass consumption of resources that was accompanied by serious levels of pollution and wasteful resource use. However, an emphasis on sustainability is particularly important for the vast majority of Earth's people who do not yet have an acceptable level of well-being. Without sustainable development strategies, efforts to improve living standards for those who need it most will increasingly be foiled by degraded or scarce resources.

Political ecology is the study of the interactions among development, politics, human well-being, and the environment. Geographers who study these relationships are known for asking the "Development for whom?" question, meaning, "Who is actually benefiting from so-called development projects?" Political ecologists examine how the power relationships

FIGURE 1.21 Global maps of human well-being, showing per capita income (PPP); human development rankings; and gender development rankings. [Sources consulted: The World Bank, Table 1.1, at http://wdi.worldbank.org /table/1.1#; 2015 United Nations Human Development Reports, Table 1, at http://hdr.undp.org/en/composite/HDI; 2015 United Nations Human Development Reports, Table 4, at http://hdr.undp.org/en/composite/GDI]

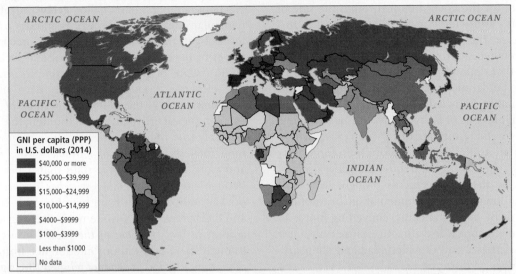

(A) Gross national income (GNI) per capita, adjusted for purchasing power parity (PPP).

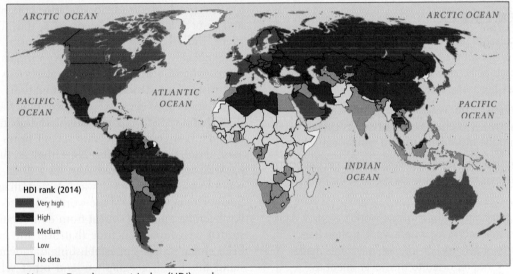

(B) Human Development Index (HDI) rank.

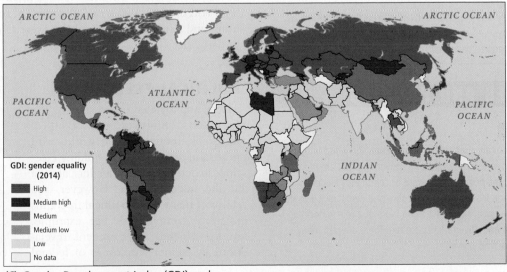

(C) Gender Development Index (GDI) rank.

in a society affect the ways in which development proceeds. For instance, in a Southeast Asian country, the clearing of forests to grow oil palm trees might at first seem to benefit many people. It would create some jobs, earn profits for the growers, and raise tax revenues for the government through the sale of palm oil, an important and widely used edible oil and industrial lubricant. However, these gains must be balanced against the loss of highly biodiverse tropical forest ecosystems and the human cultures that depend on them. Not only are forest dwellers losing their lands and means of livelihood to palm oil agribusiness, valuable knowledge that could be used to develop more sustainable uses of forest ecosystems is being lost as forest dwellers are forced to migrate to crowded cities, where their woodland skills are useless and therefore soon forgotten.

Political ecologists are raising awareness that development should be measured by the improvements brought to overall human well-being and long-term environmental quality, not just by the income created. By these standards, converting forests to oil palm plantations might appear less attractive, since only a few benefit from it and it carries a cost of widespread and often irreversible ecological and social disruption.

THINGS TO REMEMBER

• The term *development* has until recently referred to the rise in material standards of living that usually accompanies the shift from extractive economic activities, such as farming and mining, to industrial and service economic activities.

• Measures of development are being redefined to mean improvements in average per capita income as well as overall average well-being and the movement toward gender equality.

• For development to happen, social services, such as education and health care, are necessary to enable people to contribute to economic growth.

• It is estimated that one-third or more of the world's work takes place in the informal economy, where work is often bartered for food, housing, or services, or for cash payments made "off the books" that are not reported to the government as taxable income.

• Sustainability has only recently gained widespread recognition as an important goal; it involves improving current living standards in ways that will not jeopardize those of future generations.

POWER AND POLITICS

GEOGRAPHIC THEME 3

Power and Politics: There are major differences across the world in the ways that power is wielded in societies. Modes of governing that are more authoritarian are based on the power of the state or community leaders. Modes that are based on notions of political freedom and democracy give the general public greater power over themselves and more of a role in deciding how policies are developed and governments are run. There are also many other ways of managing political power.

For geographers studying globalization, the social and spatial distribution of political power is an area of considerable interest (**FIGURE 1.22**). In recent years, there has been what appears to be a trend toward political systems guided by competitive elections, a process often called **democratization.** This trend is of particular interest because it runs counter to **authoritarianism,** a form of government that subordinates individual freedom to the power of **the state** or elite regional and local leaders. In democratic systems of government, beyond the right to participate in free elections, average individuals have many other **political freedoms,** such as the following: *freedom of speech* (the right to express oneself in public and through the media), *freedom of assembly* (the right to gather together in groups to pursue common interests), *freedom of movement* (the right to travel, live, and work in any part of a state that a person is a citizen of), *freedom from unreasonable searches and seizures* (the right to privacy and protection from searches and seizures of individuals and their property by anyone not possessing a warrant granted by a court of law), *freedom of the press* (the right to communicate through any media without interference from the government or other entities), and *freedom of religion* (the right to practice or not practice any faith or spiritual path).

Geographers do not necessarily conclude that democracy and respect for political freedoms is the best political arrangement. Indeed, many geographers are critical of the imposition of democracy, often by foreign governments or organizations, in places where local people have not chosen democratic systems and where long-standing cultural traditions support other political arrangements. Few would deny that the shift toward more democratic systems of government and greater political freedom over the past century and into recent times is extremely significant, if not always peaceful. Nevertheless, there has been recent backsliding: After the Cold War, for a time, it seemed that there was a logical progression toward democratization, not just in the Soviet Union but around the world. More recently, though, there has been a significant retreat from democratic practice, so much so that we cannot conclude that over the last decade there has been global progress in establishing democracy. To better understand this trend, geographers and other scholars have become particularly interested in the role of political freedoms at the local level and the roles that social movements, international organizations, and a free media play in the exercise of power and politics in particular places.

POLITICAL FREEDOMS

There was a steady expansion in the twentieth century of some political freedoms throughout the world, with more and more countries holding elections to determine their leaders, at least at the national level. However, the status of many other political freedoms is more complicated. For example, in the United States, recent revelations by former employees of federal

democratization the transition toward political systems that are guided by competitive elections

authoritarianism a political system that subordinates individual freedom to the power of the state or of elite regional and local leaders

the state an organized political community in a defined territory and under one government

political freedoms the rights and capacities that support individual and collective liberty and public participation in political decision making

government intelligence agencies have led many to question how well protected some political freedoms are in a country generally considered to be one of the more democratic and "politically free" places on Earth. The revelations disclosed secret mass surveillance programs in which the phone and Internet communications of more than a billion people, including all U.S. citizens, are collected, stored, and analyzed by the federal government and various corporations that it hires. These activities have been criticized as undermining many political freedoms, including the freedom from unreasonable search and seizure and, through intimidation, freedom of the press.

The Arab Spring movements, which commenced in early 2011 in a number of countries in North Africa and Southwest Asia, further highlight the complexity of the expansion of political freedoms. These movements showed that massive public demonstrations can achieve amazing turnovers of power, as was the case in Tunisia, Egypt, and Libya, but the outcomes are more complex and ambiguous than a straightforward expansion of political freedoms. While the demonstrations are an expression of a desire for certain political freedoms (freedom of assembly, freedom of speech) and have resulted in political reforms and elections in some cases, they have also provoked extremely repressive responses from authoritarian states (Egypt and Syria) and similarly forceful tactics from groups that oppose the state, as in the case of the total breakdown of the Libyan state. Although the passage of time may tell a different story, at present the result of the Arab Spring seems to have been widespread violence and constraints on political freedoms.

In Egypt, following demonstrations that precipitated the end of a notoriously corrupt authoritarian government that had ruled for more than 30 years, elections were held in 2012. An Islamist political party with authoritarian leanings, the Muslim Brotherhood, won an easy majority, causing concern that voters had quashed broader democratic reforms and endangered political freedoms. After a year of increasingly authoritarian rule, the Muslim Brotherhood was removed from power by a combination of public protests and a military *coup d'etat* (an overthrow of a government by the military). Supporters of the Muslim Brotherhood held violent protests in reaction, and the military responded with a crackdown in which thousands of demonstrators were killed and an unknown number imprisoned.

The case of Syria (see Figure 1.22B) proved to have more drastic consequences. Syrians held peaceful but urgent Arab Spring demonstrations in 2011 in an attempt to oust a regime that had ruled for more than 40 years. The military and police violently repressed the protests, sparking a civil war and the rise of various factions of rebels against the regime, some of which were antidemocratic in the extreme. More than 470,000 people have died in the ongoing civil war—more than 10 percent of the population—and another 12 million have been displaced from their homes, with several million traveling mostly on foot through Turkey and Greece to seek refuge in West and North Europe. In 2015, a number of terrorist attacks in Europe and the United States appear to have been spawned by the Syrian situation.

What Factors Encourage the Expansion of Political Freedoms?

Here are some of the most widely agreed-upon factors that support the expansion of political freedoms:

- **Peace:** Peace is essential to creating an environment in which people can, among other things, vote in free and fair elections, speak and gather freely, and use print and electronic media to voice their concerns.

- **Broad prosperity:** As a broader segment of the population gains access to more than the bare essentials of life, there is often a shift toward greater political freedom. Whether general prosperity must be in place before this occurs is still widely debated, as is the question of whether prosperity necessarily leads to any expansion of political freedoms.

- **Education:** Better-educated people tend to want a stronger voice in how they are governed. Although democracy has spread to countries with relatively undereducated populations, leaders in such places sometimes become more authoritarian once elected.

- **Civil society: Civil society** is made up of the social groups and traditions that function independently of the state and its institutions to foster a sense of unity and an informed common purpose among the general population. Civil society institutions can include the media, nongovernmental organizations (NGOs; discussed below and in Figure 1.23), political parties, universities, unions, parent/teacher associations, and in some cases, religious organizations (see Figure 1.22C).

GEOPOLITICS

As the map in Figure 1.22 shows, there is a lack of political freedom in many parts of the world. A possible explanation for this is that the expansion of political freedoms is sometimes at odds with **geopolitics,** the strategies that powerful countries use to ensure that their own interests are served in relations with other countries. In the modern era, geopolitics was perhaps most obvious during the *Cold War,* the period from 1946 to the early 1990s when the United States and its allies in western Europe faced off against the Union of Soviet Socialist Republics (USSR) and its allies in eastern Europe and Central Asia. Ideologically, the United States promoted a version of free market **capitalism**—an economic system based on the private ownership of the means of production and distribution of goods, driven by the profit motive and characterized by a competitive marketplace. By contrast, the USSR and its allies favored what was called **communism**

civil society the social groups and traditions that function independently of the state and its institutions to foster a sense of unity and an informed common purpose among the general population

geopolitics the strategies that countries use to ensure that their own interests are served in relations with other countries

capitalism an economic system based on the private ownership of the means of production and distribution of goods, driven by the profit motive and characterized by a competitive marketplace

communism an ideology, based largely on the writings of the German revolutionary Karl Marx, that calls on workers to unite to overthrow capitalism and establish an egalitarian society in which workers share what they produce; as practiced, communism was actually a socialized system of public services and a centralized government and economy in which citizens participated only indirectly through Communist Party representatives

FIGURE 1.22 PHOTO ESSAY: Power and Politics

Thinking Geographically

After you have read about power and politics, you will be able to answer the following questions.

A Why has Juan Manuel Santos's relationship with the Colombian military come under criticism?

B How might the use of cell phone cameras among the demonstrators in the refugee camp contribute to or detract from political freedoms?

C, **D** Of the factors mentioned in the text as necessary for democracy to flourish, which are obviously present in **C** and missing in **D**?

The map and accompanying photo essay show two related trends in political power. Generally speaking, countries with fewer political freedoms and lower levels of democratization also have the most violent conflict. Countries are colored on the map according to their score on a "democracy index" created by the *Economist* magazine, which uses a combination of statistical indicators to capture elements crucial to the process of democratization. These elements include the ability of a country to hold peaceful elections that are accepted as fair and legitimate, people's ability to participate in elections and other democratic processes, the strength of civil liberties (such as a free media and the right to hold political gatherings and peaceful protests), and the ability of governments to enact the will of their citizens and be free of corruption. Also displayed on the map are major conflicts initiated or in progress since 1990 that have resulted in at least 10,000 casualties. Aspects of the connections between democratization and armed conflict are explored in the photo captions.

but what was actually a state-controlled economy and a socialized system of public services and a centralized government in which citizens participated only indirectly through the Communist Party.

The Cold War became a race to attract the loyalties of unallied countries and to arm them. Sometimes the result was that authoritarian rulers were embraced as allies by one side or the other. Eventually, the Cold War influenced the internal and external policies of virtually every country in the world, often causing complex local issues to be oversimplified into a contest of democracy and capitalism versus communism.

In the post–Cold War period of the 1990s, geopolitics shifted. The Soviet Union dissolved, creating many independent states, nearly all of which began to implement some democratic and free market reforms. Globally, countries jockeyed for position in what looked like it might become a new era of trade and amicable prosperity rather than war. But throughout the 1990s, while the developed countries were in a period of unprecedented prosperity, many unresolved political conflicts emerged in southeastern Europe (often referred to as the Balkans), Central and South America, Africa, Southwest Asia, and South Asia. Too often these disputes erupted into bloodshed and the systematic attempt to remove (through **ethnic cleansing**) or exterminate (through **genocide**) all members of a particular ethnic or religious group.

The new geopolitical era ushered in by the terrorist attacks on the United States on September 11, 2001, is still evolving. Because of the size and the global power of the United States, the attacks and the U.S. reactions to them affected virtually every international relationship, public and private. The ensuing adjustments, which will continue for years, are directly or indirectly affecting the daily lives of billions of people around the world. The peace and security scholar, Michael Klare, talks about the "new geography of conflict"—that contemporary geopolitics is shaped around access to vital resources rather than Cold War ideology. In his latest book, *The Race for What's Left* (2012), Klare shows that inaccessible areas—oceans and the polar regions—are the most recent frontier of geopolitical struggle over resources. The resolution of these conflicts will inevitably affect political freedoms, positively or negatively.

Among the geopolitical adjustments taking shape is the still-evolving association between five developing economies: Brazil, Russia, India, China, and South Africa, known collectively as BRICS.

ethnic cleansing the deliberate removal of an ethnic group from a particular area by forced migration

genocide the deliberate destruction of an ethnic, racial, or political group

Democratization and Conflict

Democratization index
- Full democracy
- Flawed democracy
- Hybrid regime
- Authoritarian regime
- No data

Armed conflicts and genocides with high death tolls since 1990
- Ongoing conflict
- ✳ 10,000–100,000 deaths
- ✴ 100,000–1,000,000 deaths
- ✴ More than 1,000,000 deaths

A Government repression and poorly protected political freedoms have frustrated attempts to end the ongoing violence in Colombia between rebel groups, private militias (called *paramilitaries*), and the government. The administration of Colombian president Juan Manuel Santos (shown below) has been criticized for allowing Colombia's military to financially reward its personnel for executing suspected militants, a policy that resulted in thousands of people being killed illegally. Some of the victims may have simply been political activists. Repressive tactics like these help maintain support for violent confrontation (see Chapter 3).

Raul Arboleda/AFP/Getty Images

B Syrian refugees take part in a demonstration at the Zaatari refugee camp in Jordan, near the border with Syria. The Syrians asked in February 2013 that the international community arm the Free Syrian Army. At that time, 83,000 Syrians were in the Zaatari camp and another 300,000 were in camps elsewhere in Jordan. Syria's civil war developed when the government responded to citizen protests with a harsh military crackdown on the civilian population. By 2016, the civil war had led to nearly 500,000 deaths, more than a million war refugees crossing into Europe, and to 12 million displaced people within Syria (see Chapter 6).

Khalil Mazraawi/AFP/Getty Images

Michael Bradley/Getty Images

C A member of the Service and Food Workers Union in New Zealand pickets and solicits support from passing motorists for better wages and working conditions for the union's largely immigrant and female members (see Chapter 11). New Zealand has well-protected political freedoms and is among the world's most democratized nations.

Junior Kannah/AFP/Getty Images

D Political campaigns often become compromised during times of war or conflict. Supporters of the political opposition to Joseph Kabila in September 2015 in Kinshasa, Congo, beat this man, who was suspected of being a pro-government supporter. Kabila was trying to change the constitution so that he could run for a third term in 2016. The country's ongoing civil war has resulted in more than 5.4 million deaths, mostly caused by malnutrition among people displaced by the violence (see Chapter 7).

All five are leaders in their regions, have advanced industrialized economies and show signs of being able to spread general well-being to their populations. None are full democracies: Brazil, China, and India are what are known as flawed democracies, and Russia and South Africa are considered authoritarian regimes (see Figure 1.22). As a group, they would like to be able to counter the persistent international superpower status of the United States and the European Union. Each, however, is dealing with serious internal adjustments to the global economy and with political upheavals, which are covered in the respective regional chapters.

International Cooperation

The trend toward more international cooperation on such things as climate change, peace keeping, and emergency relief in disasters has the potential to expand political freedoms. The prime example of this today is the **United Nations (UN),** an assembly of 193 member states. The member states sponsor programs and agencies that focus on, among other things, economic development, general health and well-being, democratization, peacekeeping, and humanitarian aid. However, the UN rarely challenges a country's *sovereignty*, its right to conduct its internal affairs as it sees fit without interference from outside. Consequently, the UN often can enforce its rulings only through economic sanctions. While the UN does play a role in peacekeeping in troubled areas, there are no true UN military forces. Rather, there are peacekeeping-only troops from member states that wear UN designations on their uniforms and take orders from temporary UN commanders. In some instances, the UN can endorse military action by non-UN troops, via the **Security Council**—a division of the UN with five permanent members (the United States, United Kingdom, China, France, and Russia) and ten rotating members. The Security Council is intended to respond quickly to international crises.

In addition to the UN, **nongovernmental organizations (NGOs)** are an important embodiment of international cooperation. In such associations, individuals, often from widely differing backgrounds and locations, agree on political, economic, social, or environmental goals. For example, some NGOs, such as the World Wildlife Fund, work to protect the environment. Others, such as Doctors Without Borders, provide medical care to those who need it most, especially in conflict zones. The Red Cross and Red Crescent provide emergency relief after disasters, as do Oxfam, Catholic Charities, and Gift of the Givers **(FIGURE 1.23)**.

NGOs can be an important component of civil society, yet there is some concern that the power of huge international NGOs might undermine democratic processes, especially in small countries. Some critics feel that NGO officials are a powerful, do-gooder elite that does not interact sufficiently well with local people. A frequent target of such criticism is Oxfam International (a networked group of 17 NGOs), the world leader in emergency famine relief. Oxfam was a major provider of relief after the 2004 Indian Ocean tsunami, the 2010 Haiti earthquake, and the 2013 cyclone in the Philippines. It has now expanded to work on long-term efforts to reduce poverty and injustice, which Oxfam sees as the root causes of famine. This more politically active role has brought Oxfam into conflict with governments and officials capable of hampering the NGO's ability to achieve many of its goals at the local level.

United Nations (UN) an assembly of 193 member states that sponsors programs and agencies that focus on economic development, general health and well-being, democratization, peacekeeping assistance in hot spots around the world, humanitarian aid, and scientific research

Security Council a division of the UN with five permanent members (the United States, Britain, China, France, and Russia) and ten rotating members

nongovernmental organizations (NGOs) associations outside the formal institutions of government in which individuals, often from widely differing backgrounds and locations, share views and activism on political, social, economic, or environmental issues

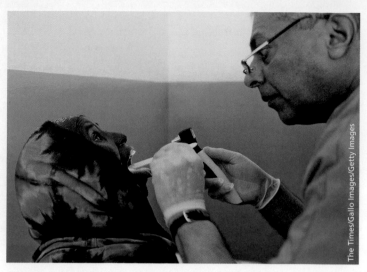

FIGURE 1.23 Nongovernmental organizations in action. A doctor, a volunteer with the South Africa–based NGO Gift of the Givers, examines a patient in Somalia. Gift of the Givers, which is the largest NGO started and staffed completely by Africans, provides medical help and food to famine-stricken parts of Somalia.

The Times/Gallo Images/Getty Images

THINGS TO REMEMBER

GEOGRAPHIC THEME 3 • **Power and Politics:** There are major differences across the world in the ways that power is wielded in societies. Modes of governing that are more authoritarian are based on the power of the state or community leaders. Modes that are based on notions of political freedom and democracy give the general public greater power over themselves and more of a role in deciding how policies are developed and governments are run. There are also many other ways of managing political power.

• There have been expansions over recent decades of some political freedoms throughout the world, with more and more countries holding elections of their leaders, at least at the national level. However, the status of many other political freedoms is more complicated.

• Some of the factors that are most widely agreed upon as supporting the expansion of political freedoms are peace, broad prosperity, education, and civil society.

• The expansion of political freedoms is sometimes at odds with geopolitics, the strategies that countries use to ensure that their own interests are served in relations with other countries.

• NGOs can be an important component of civil society, yet there is some concern that the power of huge international NGOs might undermine democratic processes, especially in small countries.

URBANIZATION

GEOGRAPHIC THEME 4

Urbanization: The development of manufacturing and service economies has taken place largely in cities, drawing people from the countryside into urban areas. Many of these new urbanites were displaced by the mechanization of agriculture in rural areas, which drastically reduced the need for agricultural labor and left people with no means of support and a reduced access to food.

In 1700, fewer than 7 million people, or just 10 percent of the world's total population, lived in cities, and only 5 cities had populations of several hundred thousand people or more. The world we live in today has been transformed by **urbanization,** the process whereby cities, towns, and suburbs grow as populations shift from rural to urban livelihoods. By some measures, a little over half of the world's population now live in urban places, and there are more than 400 cities of more than 1 million people and 28 cities of more than 10 million people **(FIGURE 1.24).**

WHY ARE CITIES GROWING?

Cities are centers of innovation and culture and these features attract new migrants, who are looking not just for jobs but also for excitement (see Figure 1.24A); thus cities are said to *pull* in migrants. Meanwhile, the mechanization of agriculture has drastically reduced the need for rural labor, resulting in people being *pushed* off the land. Increasing the food supply to the point that large nonfarming populations can be fed has made it possible for people to move out of rural areas and into cities, where the development of manufacturing and service economies has created many jobs (see Figure 1.24B). This **push/pull phenomenon of urbanization** from farm to city often is lopsided because the growth of urban jobs does not keep up with in-migration: push forces are stronger than pull forces. One of the push forces is climate change, which can bring extended droughts or floods that force people off the land or otherwise limit rural sustenance. Numerous cities, especially in poorer parts of the world, have been unprepared for the massive inflow of rural migrants, many of whom, now underemployed, live in polluted **slum** areas plagued by natural and human-made hazards, poor housing, out-of-date infrastructures, and inadequate access to food, clean water, education, and social services (see Figure 1.24C). Often a substantial portion of the migrants' meager cash income goes to support their still-rural families.

Patterns of Urban Growth

The most rapidly growing cities are in developing countries (LDCs) in Asia, Africa, and Middle and South America (see the map in Figure 1.24). Countries in Africa and Asia are now between 40 and 48 percent urban, respectively, but are expected to become 56 and 65 percent urban by 2050. The settlement patterns of these cities bear witness to their rapid and often unplanned growth, fueled in part by the steady arrival of masses of poor rural people looking for work. Cities like Mumbai (in India), Cairo (in Egypt), Nairobi (in Kenya), and Rio de Janeiro (in Brazil) sprawl out from a small, affluent core, often the oldest part, where there are upscale businesses, fine old buildings, banks, shopping centers, and residences for wealthy people. Surrounding these elite landscapes are sprawling mixed commercial, industrial, and middle-class residential areas, interspersed with pockets of extremely dense slums. Also known in various locales as *barrios, favelas, hutments, shantytowns, ghettos,* and *tent villages,* these settlements provide housing for the poorest of the poor, who provide low-wage labor for the city. Housing is often self-built out of any materials the residents can find: cardboard, corrugated metal, masonry, scraps of wood and plastic. There are usually no building codes, no toilets with sewer connections, and there is little access to clean water. Electricity is often obtained from illegal and dangerous connections to nearby power lines. Schools are few and overcrowded, and transportation is provided only by informal, nonscheduled van-based services.

The UN estimates that currently more than a billion people live in urban slums, with that number to increase to 2 billion by 2030. Life in these areas can be insecure and chaotic, as criminal gangs often assert control through violence and looting—actions that are especially likely during periods of economic recession and political instability.

ON THE BRIGHT SIDE: Stories of Urban Migrant Success

Slums are only part of the story of urbanization today. Those who are financially able to come to urban areas for education and complete their studies tend to find employment in modern industries and business services. They constitute the new middle class and leave their imprint on urban landscapes via the high-rise apartments they occupy and the shops and entertainment facilities they frequent (see Figure 1.24A and the accompanying map). Cities such as Mumbai in India, São Paulo in Brazil, Cape Town in South Africa, and Shanghai in China are now home to this more educated group of new urban residents, many of whom may have started life on farms and in villages.

In the past, most migrants in cities were young men, but these days, the number of young women migrating is growing. Cities offer women more than better-paying jobs: They provide access to education, better health care, and more personal freedom. Once in cities, women often choose to have fewer children as they delay childbearing in favor of advancing their education or career. Within urban households, people are choosing to have fewer children for a variety of reasons, such as the fact that without farms to run, the labor provided by a large family is no longer needed. Hence, urbanization is linked to slower population growth. ∎

Urban life in more developed countries (MDCs) is quite different from that in LDCs. Because the process of urbanization tends to go hand in hand with economic development, most people in wealthy countries live in cities. In fact, the

urbanization the process whereby cities, towns, and suburbs grow as populations shift from rural to urban livelihoods

push/pull phenomenon of urbanization conditions, such as political instability or economic changes, that encourage (push) people to leave rural areas, and urban factors, such as job opportunities, that encourage (pull) people to move to the urban area

slum densely populated area characterized by crowding, run-down housing, and inadequate access to food, clean water, education, and social services

FIGURE 1.24 PHOTO ESSAY: Urbanization

Thinking Geographically

After you have read about urbanization, you will be able to answer the following questions.

A To what group of urban migrants does this skydiving young man probably belong?

B In what kind of neighborhood of Dhaka would you guess this man lives?

C This photo exemplifies what problem commonly faced by rapidly growing cities?

In the map, the color of the country indicates the percentage of the population living in urban areas. The blue circles represent the populations of the world's largest urban areas in 2015. [Sources consulted: 2015 World Population Data Sheet, Population Reference Bureau, at http://www.prb.org/Publications/Datasheets/2015/2015-world-population-data-sheet.aspx, and https://demographia.wordpress.com/2015/02/02/largest-1000-cities-on-earth-world-urban-areas-2015-edition/]

A Cities have always been centers of innovation, entertainment, and culture, in large part because they draw both money and talented people. This resident of Shanghai is engaging in a current fad: parachuting off one of the new skyscrapers that now dominate the city's skyline.

average urbanization rate among MDCs is close to 80 percent. In MDCs, urban growth is typically slow and comparatively organized in that basic services and housing are adequately provided to newcomers. There has even been some thought that MDC cities are *greener* than LDC cities or rural areas because their very compactness saves energy. People who live in dense urban settlements are less dependent on cars and don't have to commute very far. Also, compact multi-unit housing uses less energy for heating and cooling than do detached single-family units. Dense urban settlement has resulted in reduced carbon emissions in Europe and in some North American cities. Los Angeles, San Francisco, Boston, Washington, DC, and New York City all reduced their per capita emissions between 1980 and 2010 primarily by improving their public transportation systems. But in most of North America, urban sprawl is increasing the spatial extent of cities. The result of lower-density residential and commercial developments in suburbs is that cities are decentralizing and the traditional city center is, at least relatively speaking, losing its dominance as the main hub of economic activity. When public transportation does not keep pace with urban population growth, people drive even farther for work and play, adding pollution to the areas where they live. Since 2000, per capita emissions in Salt Lake City, Atlanta, Houston, Phoenix, Denver, and Detroit all have risen as more suburbanites commute to work.

The spread-out character of cities also means that different ethnic and socioeconomic groups tend to live apart in a segregated urban landscape. Because of economic changes, old industrial districts have declined, resulting in job losses that affect nearby working-class neighborhoods. There are, however, distinct regional differences to these urban patterns. North American cities, for example, are significantly more sprawling and decentralized than European cities. But what they all have in common is the need to knit together the fabric of the city with a transportation system that serves the urban population. Merely expanding highways is not enough to solve existing traffic problems. In fact, new roads can quickly become clogged with traffic; instead, urban planners now seek to reduce the need for cars by developing walkable urban landscapes integrated with public transit (rail and buses) to serve new, denser urban and suburban development.

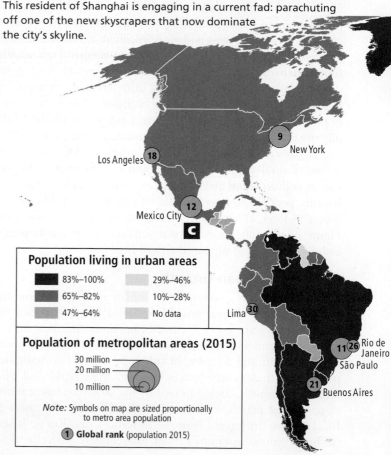

Population living in urban areas

■ 83%–100%	29%–46%
65%–82%	10%–28%
47%–64%	No data

Population of metropolitan areas (2015)

30 million
20 million
10 million

Note: Symbols on map are sized proportionally to metro area population

① **Global rank** (population 2015)

B A recent migrant to Dhaka pulls a cart loaded with goods. Many of the world's fastest-growing cities are attracting more people than they can support with decent jobs, housing, and infrastructure.

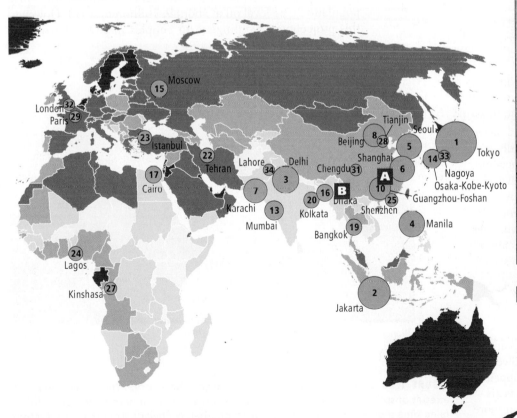

Moscow ⑮
London ㉜
Paris ㉙
㉓ Istanbul
㉒
Tehran
Lahore
Delhi
Chengdu ㉛
⑰
Cairo
Karachi
⑬
Mumbai
Kolkata
Dhaka
Bangkok ⑲
Jakarta ②
Lagos ㉔
Kinshasa ㉗

Tianjin
⑧ ㉘
Beijing
Shanghai
Seoul
① Tokyo
⑤
⑭ ㉝
⑥
Nagoya
Osaka-Kobe-Kyoto
Guangzhou-Foshan
Shenzhen
④ Manila

C Some cities struggle with major environmental problems. Mexico City, currently the world's twelfth-largest city, occasionally suffers from severe flooding because of its location on an old, now-sinking lake bed and its antiquated drainage and sewage infrastructure.

THINGS TO REMEMBER

GEOGRAPHIC THEME 4

• **Urbanization:** The development of manufacturing and service economies has taken place largely in cities, drawing people from the countryside into urban areas. Many of these new urbanites were displaced by the mechanization of agriculture in rural areas, which drastically reduced the need for agricultural labor and left people with no means of support and a reduced access to food.

• Today about half of the world's population live in cities; there are over 400 cities with more than 1 million people and 28 cities of more than 10 million people.

• For some, urbanization means improved living standards, while for others it means being forced into slums with inadequate food, water, and social services.

• There is some hope that with investment in green infrastructure, cities, as they become more dense, will foster lower per capita carbon emissions, especially if they extend public transportation into the suburbs so families can use one car or none.

POPULATION AND GENDER

GEOGRAPHIC THEME 5

Population and Gender: Population growth is slowing across the world for a number of reasons. Chief among these are urbanization and the lifestyle changes that it brings; the contraction of agricultural labor as a result of the mechanization of farming; and perhaps most important, the increasing numbers of women who are delaying childbearing as they pursue educational and work opportunities outside the home. The availability of birth control is an important contributing factor in determining success in family planning.

The explanation lies in the changing relationships between humans and the environment. For most of human history, fluctuating food availability, natural hazards, and disease kept human death rates high, especially for infants. Out of many pregnancies, a couple might raise only one or two children. Also, pandemics such as the Black Death in Europe and Asia in the 1300s killed millions of people in a short time.

An astonishing upsurge in human population began about 1500, at a time when the technological, industrial, and scientific revolutions were beginning in some parts of the world. Human life expectancy increased dramatically, and more and more people lived long enough to reproduce successfully, often many times over. The dispersals of food crops to new areas during the age of European colonization after 1500 and the Industrial Revolution after 1750 accelerated the population growth trend. The result was an exponential pattern of growth (see Figure 1.25). This pattern is often called a *J curve* because the ever-shorter periods between doubling and redoubling of the population cause an abrupt upward swing in the growth line when depicted on a graph.

Today, the human population is growing in most regions of the world, more rapidly in some places than in others. The main reason for this growth is that a very large group of young people has reached the age of reproduction and more will join them shortly. Nevertheless, the *rate* of global population growth is slowing. Per capita, people are having fewer children. In 1968, the global growth rate was 2.1 percent, by 1993 it had dropped to 1.7 percent per year and by 2015 to roughly 1.13 percent per year. If slower growth trends continue, the world population may level off at about 9.5 billion by 2050. However, this projection is contingent on couples in LDCs having the

Over the last several hundred years, the human population has boomed, but growth rates are now slowing in most societies and in a few have even begun to decline (a process called *negative growth*). Much of this pattern has to do with changes in economic development, urbanization, and gender roles that have reduced incentives for large families. It is for these reasons that population and gender are linked here and throughout the book. In those cases where population growth rates are slowing only a little, the explanation is almost always related to women and their status in society and the opportunities afforded them.

GLOBAL PATTERNS OF POPULATION GROWTH

It took between 1 million and 2 million years (at least 40,000 generations) for humans to evolve and to reach a population of 2 billion, which happened around 1945. Then, remarkably, in just 70 years—by October of 2015—the world's population more than tripled to 7.3 billion **(FIGURE 1.25)**. What happened to make the population grow so quickly in such a short time?

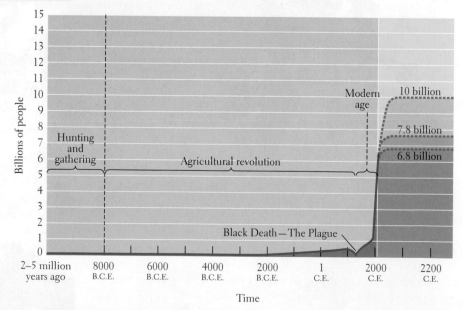

FIGURE 1.25 Exponential population growth: The J curve. The curve's J shape is a result of successive doublings of the population. It starts out nearly flat, but as doubling time shortens, the curve bends ever more sharply upward. Note that B.C.E. (before the common era) is equivalent to B.C. (before Christ); C.E. (common era) is equivalent to A.D. (anno Domini). [Source consulted: G. Tyler Miller, Jr., *Living in the Environment*, 8th ed. (Belmont, CA: Wadsworth, 1994), p. 4]

education and economic security to choose to have smaller families and the ability to access and practice birth control technology.

In a few countries (especially Syria in Southwest Asia; Japan in East Asia; and Ukraine plus Moldova, Bulgaria, and Hungary in Central Europe), the population is actually declining and rapidly aging, due primarily to low birth and death rates but in Syria due to war and migration. This situation could prove problematic as those who are elderly and dependent become more numerous, posing a financial burden on a declining number of working-age people. In addition, HIV/AIDS is affecting population patterns to varying extents in all world regions. In Africa, the epidemic is severe; as a result, several African countries have sharply lowered life expectancies among young and middle-aged adults.

Global Variations in Population Density and Growth

If the more than 7 billion people on Earth today were evenly distributed across the land surface, they would produce an *average population density* of about 121 people per square mile (47 per square kilometer). But people are not evenly distributed (**FIGURE 1.26**). Nearly 90 percent of all people live north of the equator, and most of them live between 20°N and 60°N latitude. Even within that limited territory, people are concentrated on about 20 percent of the available land. They live mainly in zones that have climates warm and wet enough to support agriculture: along rivers, in lowland regions, or fairly close to the sea. In general, people are located where resources are available, but access to resources may be precarious, limited by age, gender, and income.

Usually, the variable that is most important for understanding population growth in a region is the rate of natural increase (often called the *growth rate*). The **rate of natural increase (RNI)** is the relationship in a given population between the number of people being born (the **birth rate**) and the number dying (the **death rate**), without regard to the effects of **migration** (the movement of people from one place or country to another).

The rate of natural increase is expressed as a percentage per year. For example, in 2015, the annual birth rate in Austria (in Europe, population 8.6 million) was 10 per 1000 people, and the death rate was 9 per 1000 people. Therefore, the annual rate of natural increase was 1 per 1000 (10 − 9 = 1), or 0.1 percent.

For comparison, consider Jordan (in Southwest Asia, population 8.1 million). In 2015, Jordan's birth rate was 28 per 1000, and the death rate was 6 per 1000. Thus the annual rate of natural increase was 22 per 1000 (28 − 6 = 22), or 2.2 percent per year. But note that Jordan's death rate was much lower than Austria's. This is due to the fact that, as discussed below, Austria's population at present is much older than Jordan's. In Austria, 18 percent of the people are over the age of 65, while in Jordan, just 3 percent are. Thus, in Austria every year a larger percentage of the population is reaching the natural limit of human life than it is in Jordan.

Total fertility rate (TFR), another term used to indicate trends in population, is the average number of children a woman in a particular population is likely to have during her reproductive years (15–49). The TFR in 2015 for Austrian women was 1.5; for Jordanian women, it was 3.5. As education rates for women increase, and as they postpone childbearing into their late 20s, total fertility rates tend to decline.

Another powerful contributor to population growth is **immigration** (in-migration). In Europe, for example, the rate of natural increase is quite low, but the region's economic power attracts immigrants from throughout the world and in 2015, related to the civil war in Syria and a series of crises in the Middle East, North Africa, Afghanistan, and Pakistan, more than a million migrants and refugees entered Europe, most of them by sea in small boats, but many on foot. They came to Europe because of the economic power of the region and its high standard of living. Even before the massive influx of migrants in 2015, Austria was attracting immigrants at the annual rate of 0.05 percent of its population. Jordan is much more impacted by migration; its population is skewed because in recent decades it has taken in so many Palestinian refugees that today Palestinians outnumber native Jordanians two to one. However, Jordan has granted citizenship to Palestinians, so they are now counted as Jordanians. Jordan's official immigration rate of 0.03 percent of the population per year has been rendered inaccurate by the many refugees who have fled into Jordan from Iraq and Syria over the last few years.

By 2010, international migrants accounted for about 85 percent of the EU's population growth; they were important additions to the labor force in an era of declining births. But when so many refugees from war and unrest came in 2015, even though Europe needed population growth, Europeans worried that they would not be able to care for the large numbers and that the new people, because they were from predominantly Muslim countries, would abruptly change European culture. This situation is further discussed in Chapter 4.

All across the world, people are on the move, trying to improve their circumstances; rarely do people freely choose to leave home. Usually they are fleeing war, natural disasters, economic recessions, and job loss (push forces). Understandably, developed countries are favored destinations (pull forces).

AGE AND SEX STRUCTURES

The age and sex structures of a country's population reflect past and present social conditions and can help predict future population trends. The age distribution, or age structure, of a population is the proportion of the total population in each age group. The sex structure is the proportion of males and females in each age group.

The **population pyramid** (in wealthy developed countries, it looks more like a pillar) is a graph

rate of natural increase (RNI) the rate of population growth measured as the excess of births over deaths per 1000 people per year, without regard to the effects of migration

birth rate the number of births per 1000 people in a given population, per unit of time (usually per year)

death rate the ratio of total deaths to total population in a specified community, usually expressed in numbers per 1000 or in percentages

migration the movement of people from one place or country to another, often for safety or economic reasons

total fertility rate (TFR) the average number of children that women in a particular population are likely to have at the present rate of natural increase

immigration in-migration to a place or country (see also *migration*)

population pyramid a graph that depicts the age and sex structures of a political unit, usually a country

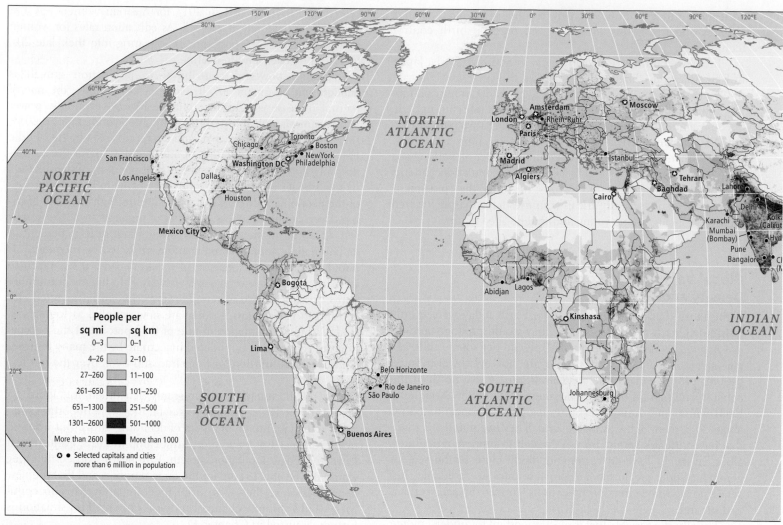

FIGURE 1.26 World population density.

that depicts age and sex structures. Consider the population pyramid/pillars for Austria and Jordan **(FIGURE 1.27A, B)**. Notice that Jordan's (B) is a true pyramid, with a wide bottom, and that the largest groups are in the age categories 0 through 19.

By contrast, Austria's (A) is not a pyramid at all but rather a sort of pillar with an irregular vertical shape that tapers in toward the bottom. The base of the Austrian pillar, which began to narrow about 40 years ago, indicates that there are now fewer people in the youngest age categories than in the teen, young adult, or middle-age groups; those over 70 greatly outnumber the youngest (ages 0 to 4). This age distribution is caused by two emerging trends in developed economies: Many Austrians now live to an advanced age and, in the last several decades, because of the investment of time and money needed to raise children, Austrian couples have chosen to have only one child, or none. If these two trends continue, Austrians will need to support and care for large numbers of elderly people, and those responsible will be an ever-declining group of working-age people. The refugees now entering Austria might turn into valuable assets if they agree to stay and contribute their talents to Austria's economy and tax base, and raise some children.

Population pyramid/pillars also reveal *sex imbalance* within populations. Look closely at the right (female) and left (male) halves of the pyramid/pillars in Figure 1.27. In several age categories, the sexes are not evenly balanced on both sides of the line. In the Austria pyramid, there are more women than men near the top (especially in the age categories of 70 and older). In Jordan, there are more males than females near the bottom (especially for ages 0 to 24).

Demographic research on the reasons behind statistical sex imbalance is relatively new, and many different explanations have been proposed. In Austria, the predominance of elderly women reflects the fact that women live about 5 years longer than men in countries with long life expectancies (a trend that is still poorly understood). But the sex imbalance in younger populations has different explanations. The normal ratio worldwide is for about 95 females to be born for every 100 males. Because baby boys are somewhat more fragile than girls on average, the ratio normally evens out naturally within the first 5 years. However, in many places the ratio is as low as 80 females to 100 males and continues to be so throughout the life cycle. This is due to the widespread cultural preference for boys over girls (discussed in later chapters).

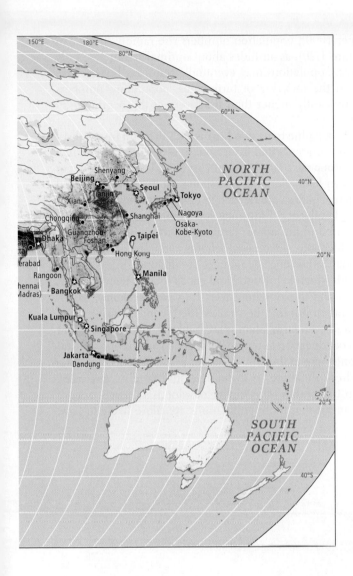

This bias toward boys becomes more observable as couples have fewer children. Some fetuses, if identified as female, are purposely aborted; also, girls and women are sometimes fed less well and receive less health care than males, especially in poverty-stricken areas. Because of this, females are more likely to die, especially in early childhood. The artificially achieved gender imbalance of more males than females can be especially troubling for young adults because the absence of females means that young men will find it hard to find a marriage mate. In some societies this can mean that men seek younger and younger brides; in others, the scarcity of females can result in sex trafficking.

POPULATION GROWTH RATES AND WEALTH

Although there is a wide range of variation, regions with slow population growth rates usually tend to be affluent, and regions with fast growth rates tend to have widespread poverty. The reasons for this difference are complicated; again, Austria and Jordan are useful examples.

In 2014, Austria had an annual gross national income (GNI) per capita of U.S.$45,040 (PPP). This figure represents the total production of goods and services in a country in a given year, divided by the mid-year population; Austria's is among the highest in the world (the comparable figure for the United States is U.S.$55,860). Austria has a very low infant mortality rate of 3 per 1000 live births. Its highly educated population is 100 percent literate, employed largely in technologically sophisticated industries and services. Large amounts of time, effort, and money are required to educate a child to compete in this economy. Austrian women seek higher education and careers, and couples choose to marry late, if at all, and have only one or two children. The cost of raising a child is high and because it is very likely that children will survive to adulthood and thus will be available as companions for aging parents, a small family is considered suitable, even though in single-child families this one offspring may eventually have to care for four grandparents and two parents. But the social welfare system in Austria ensures that citizens will not be impoverished in old age.

In contrast to Austria, Jordan has a GNI (PPP) per capita of U.S.$11,910 and an infant mortality rate of 17 per 1000. Much

FIGURE 1.27 Population pyramids for Austria and Jordan.
[Source consulted: U.S. Census Bureau, International Programs, http://www.census.gov/population/international/data/idb/informationGateway.php]

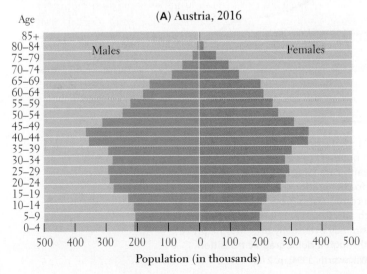

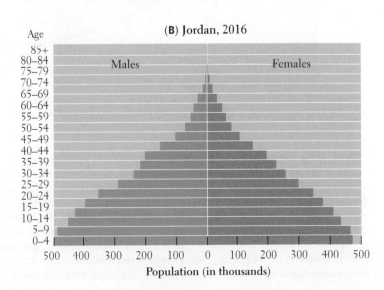

everyday work is still done by hand, so each new child is a potential contributor to the family income from a young age. There is little social welfare for elderly citizens and a much higher risk of children not surviving into adulthood. Having more children helps ensure that someone will be there to provide care for aging parents and grandparents.

However, circumstances in Jordan have changed rapidly over the last 30 years. In 1985, Jordan's per capita GDP was U.S.$993, infant mortality was 77 per 1000, and women had an average of 8 children. Now Jordanian women have an average of only 3.5 children, with the need to educate those children being increasingly recognized. Geographers would say that Jordan is going through a **demographic transition,** meaning that a period of high birth and death rates is giving way to a period of much lower birth and death rates (the red and purple lines, respectively, in **FIGURE 1.28**). In the middle phases, however, when death rates decline more rapidly than birth rates (Jordan's death rate is just 6 per 1000 but its birth

demographic transition the change from high birth and death rates to low birth and death rates that usually accompanies a cluster of other changes, such as change from a subsistence to a cash economy, increased education rates, and urbanization

subsistence economy an economy in which families produce most of their own food, clothing, and shelter

cash economy an economic system that tends to be urban but may be rural, in which skilled workers, well-trained specialists, and even farm laborers are paid in money

rate is 28 per 1000), population numbers rise rapidly (the green line in Figure 1.28), as attitudes about optimal family size only slowly adjust. Populations may eventually stabilize, but because of the lag in the lowering of birth rates, population numbers may be significantly higher than they were at the beginning of the transition.

The demographic transition begins with the shift from subsistence to cash economies and from rural to urban ways of life. In a **subsistence economy,** a family, usually in a rural setting, produces most of its own food, clothing, and shelter, so there is little need for cash. Many children are needed to help perform the work that supports the family, and birth rates are high. Most needed skills are learned around the home, farm, and surrounding places, so expensive educations at technical schools and universities are not needed. Today, subsistence economies are disappearing as people seek cash with which to buy food and goods such as television sets and bicycles, and to pay school fees. In a **cash economy,** which tends to be urban but may be rural, skilled workers, well-trained specialists, and even farm laborers are paid in money. Each child needs years of education to qualify for a good cash-paying job and does not contribute to the family budget while in school. Having many children, therefore, is a drain on the family's resources. Perhaps most important, the higher development levels of cash economies mean that those populations are more likely to have

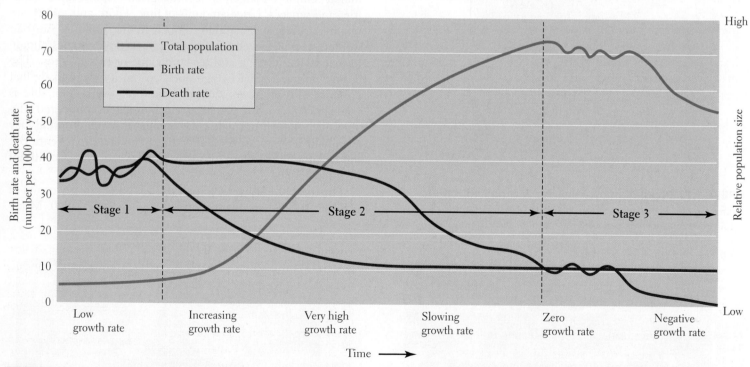

FIGURE 1.28 Demographic transition diagram. In traditional societies (Stage 1), both birth rates and death rates are usually high (red and purple lines, left vertical axis), and population numbers (green line, right vertical axis) remain low and stable. With advances in food production, education, and health care (Stage 2), death rates usually drop rapidly, but cultural values favoring reproduction remain strong, often for generations, so birth rates drop much more slowly, with the result that for decades or longer, the population continues to grow significantly. When changed social and economic circumstances enable most children to survive to adulthood and it is no longer necessary to produce a cadre of family labor (Stage 3), population growth rates slow and may eventually drift into negative growth (decline). At this point, demographers say that the society has gone through a demographic transition.
[Source consulted: G. Tyler Miller, Jr., *Living in the Environment*, 8th ed. (Belmont, CA.: Wadsworth, 1994), p. 218]

better health care, increasing the likelihood that each child will survive to adulthood.

GENDER

Note the difference between the terms *gender* and *sex*. **Gender** indicates how a particular social group defines the differences between the sexes. **Sex** refers to the biological category of male, female, or **intersex,** but does not indicate how people may behave or identify themselves. Gender definitions and accepted behavior for the sexes can vary greatly from one social group to another. Here we consider gender.

For women, the historical and modern global gender picture is puzzlingly negative. In nearly every culture, in every region of the world, and for a great deal of recorded history, women have had (and still have) an inferior status. Exceptions are rare, although the intensity of this second-class designation varies considerably. On average, women have less access to education, medical care, and even food. They start work at a younger age and work longer hours than men. Around the world, people of both sexes still routinely accept the idea that men are more productive and intelligent than women. The puzzling question of how and why women became subordinate to men has not yet been well explored because, oddly enough, few thought the question significant until the last 50 years.

In nearly all cultures, families prefer boys over girls because, as adults, due to social norms, boys will have greater earning capacity **(TABLE 1.2)** and more power in society. This preference for boys has some unexpected side effects. As mentioned above, the preference may lead to fewer girls being born, which then leads to an eventual shortage of marriageable women, leaving many men without the hope of forming a family. Currently, there is concern in several Asian societies that the scarcity of young women, which in some places is extreme (the shortage in China equals the entire population of Canada), is already leading to antisocial behavior on the part of discouraged young men.

Gender Roles

Geographers have begun to pay more attention to **gender roles**—the socially assigned roles for males and females. In virtually all parts of the world, and for at least tens of thousands of years, the biological fact of maleness and femaleness has been translated into specific roles for each sex. The activities assigned to men and to women can vary greatly from culture to culture and from era to era, but they remain central to the ways societies function. The existence of transgender people and social roles for people of a "third gender" are also ancient, and acceptance varies across the globe. Indeed, increasing attention to gender roles by geographers has been driven largely by interest in the shifts that occur when traditional ways are transformed by modernization.

There are some striking consistencies regarding traditional gender roles over time and space. Men are expected to fulfill public roles, while women fulfill private roles and have limited freedom to access public spaces. Certainly there are exceptions in every culture, and customs are changing, especially in wealthier countries. But generally, men work outside the home in positions such as executive jobs, animal herders, hunters, farmers, warriors, or government leaders. Women keep house, bear and rear

Table 1.2 Comparison of male and female incomes for selected countries

Country	HDI rank	Female GNI per capita (PPP U.S.$)	Male GNI per capita (PPP U.S.$)	Female income as percent of male income
Austria	23	29,598	58,826	50.31
Barbados	57	10,245	14,739	69.51
Botswana	106	15,179	18,096	83.88
Canada	9	33,587	50,843	66.06
Japan	20	24,975	49,541	50.41
Jordan	80	3587	18,831	19.04
Kuwait	48	42,292	111,968	37.77
Poland	36	18,423	28,271	65.16
Russia	50	17,269	28,287	61.05
Saudi Arabia	39	20,094	77,044	26.08
Sweden	14	40,222	51,084	78.74
United Kingdom	14	27,259	51,628	52.80
United States	8	43,054	63,156	68.17

Data from the UN Human Development Report 2015, Table 4; see http://report.hdr.undp.org

children, care for the elderly, grow and preserve food, and prepare the meals, among many other tasks. In nearly all cultures, women are defined as dependent on men—their fathers, husbands, brothers, or adult sons—even when the women may produce most of the family sustenance.

Gender Issues

Because their activities are focused on the home, women tend to marry early. One quarter of the girls in developing countries are mothers before they are 18. This is crucial in that pregnancy is the leading cause of death among girls aged 15 to 19 worldwide, primarily because immature female bodies are not ready for the stress of pregnancy and birth. Globally, babies born to women under 18 have a 60 percent higher chance of dying in infancy than do those born to women over 18.

Typically, women also have less access to education than men (around the world, 70 percent of

gender the ways a particular social group defines the differences between the sexes

sex the biological category of male, female, intersex; does not indicate how people may behave or identify themselves

intersex the state of being in variation from typical male or female biology, but this state does not indicate how the individual may behave or identify socially

gender roles the socially assigned roles for males and females although exceptions and variations exist

youth who leave school early are girls). They are less likely to have access to information and paid employment, and so have less access to wealth and political power. When they do work outside the home (as is increasingly the case in every world region), women tend to fill lower-paid positions, often as laborers, service workers, or lower-level professionals. And even when they work outside the home, most women retain their household duties, so they work a *double day*.

The extent to which physical differences between males and females affects their social roles is still being debated. On average, men are stronger and taller, but women have more endurance and pain tolerance relative to men. Women's physical capabilities are somewhat limited during pregnancy and nursing, but from the age of about 45, women are no longer subject to the limitations related to pregnancy, and most contribute in some significant way to the well-being of their adult children and grandchildren, an important social and economic role. A growing number of biologists suggest that the evolutionary advantage of menopause in midlife is that it gives women the time, energy, and freedom to help succeeding generations thrive. This notion—sometimes labeled the *grandmother hypothesis*—seems to have worldwide validity **(FIGURE 1.29)**. Certainly, grandfathers can play nurturing roles, but women tend to live 5 years longer than men.

Perhaps more than for any other culturally defined human characteristic, there is significant agreement that gender is important, but just how gender roles are defined varies greatly across places and over time. Traditional notions of gender roles are now being challenged everywhere. In many countries, including some conservative Muslim countries, women are acquiring education at higher rates than men. Although it will take women a while to catch up in the job market, eventually education should make women competitive with men for jobs and roles in public life as policy makers and government officials, not just as voters. Unless discrimination persists, women should also begin to earn pay equal to that of men (see Table 1.2).

Research data suggest that there is a ripple effect that benefits the whole group when developing countries pay attention to the needs of girls:

- Girls in developing countries who get 7 or more years of education marry 4 years later than average and have 2.2 fewer children.

- An extra year of secondary schooling over the average for her locale boosts a girl's lifetime income by 15 to 25 percent.

- The children of educated mothers are healthier and more likely to finish secondary school.

- When women and girls earn income, 90 percent of their earnings are invested in the family, compared to just 40 percent of males' earnings.

cisgender someone who has a gender identity that aligns with the gender they were assigned at birth

transgender someone who has a gender identity that differs from the gender they were assigned at birth

Considering only women's perspectives on gender, however, misses half the story. Men are also affected by strict gender expectations, often negatively. For most of human history, young men have borne a disproportionate share of

FIGURE 1.29 The grandmother hypothesis. In every culture and community worldwide, grandmothers contribute to the care and education of their grandchildren. This characteristic has played an essential role in human evolution. A grandmother in Xian, China, helps her toddler granddaughter learn to walk.

Tim Graham/Getty Images

Thinking Geographically
What are five things grandmothers in every world region are likely to worry about regarding their grandchildren?

burdensome physical tasks and dangerous undertakings. Until recently, mostly young men left home to migrate to distant, low-paying jobs. Overwhelmingly, it has been young men (the majority of soldiers) who die in wars or suffer physical and psychological injuries from combat.

An emerging body of research suggests that sexual violence is widespread and inextricably linked to gender roles. A recent survey of 10,000 men in six Asian countries indicated that one in four participants had raped a woman, usually a romantic partner. Seventy percent of the men who had committed rape said they felt entitled to do so. This and similar studies done in other countries suggest that rape is a widespread global phenomenon deeply rooted in power relations between genders, and that preventative measures are most effective during childhood and adolescence.

Sexual orientation (straight, gay, etc.) and gender identity (**cisgender, transgender,** etc.) are topics that will be explored in some chapters. Attitudes toward variations in sexual orientation and identity are changing in many modernizing cultures, with homosexuality gaining recognition and acceptance as being biologically determined. Also, there is increasing recognition that gender identity may not necessarily align with the biology of an individual and that a variety of accommodations for transgender people is helpful to their social functioning. This book will return repeatedly to the questions of sexual orientation, gender identity, and disparities because they play such a central role across the world. Careful attention to the roots of gender-based biases and phobias may help to create a more just future for all people.

THINGS TO REMEMBER

GEOGRAPHIC THEME 5 • **Population and Gender:** Population growth is slowing across the world for a number of reasons. Chief among these are urbanization and the lifestyle changes that it brings; the contraction of agricultural labor as a result of the mechanization of farming; and perhaps most important, the increasing numbers of women who are delaying childbearing as they pursue educational and work opportunities outside the home. The availability of birth control is an important contributing factor in determining success in family planning.

• Over the last several hundred years, global population growth has been rapid. Although growth will continue for many years, rates are now slowing in most places and in a few places growth has reversed and the population is even shrinking.

• The circumstances that lead to lower population growth rates also lead to the overall aging of populations; these two phenomena are found nearly everywhere in the world today, but to varying degrees.

• Gender indicates how a particular group defines the social differences between the sexes. Sex is the biological category such as male or female.

• There are some global consistencies in gender disparities: typically, men have public roles and women have private roles; in the present and in every country, the average woman earns less than the average man.

• Variations in sexual identity are increasingly recognized as normal and societies are slowly ending prohibitions to the full participation in civic life of LGBT people.

SOCIOCULTURAL ISSUES

Many geographers are interested in the economic, social, and cultural practices of a people and in the spatial patterns these factors create. Cultural geography focuses on culture as a complex of important distinguishing characteristics of human societies. **Culture** comprises everything people use to live that is not directly part of biological inheritance. Culture is represented by the ideas, materials, methods, and social arrangements that people have invented and passed on to subsequent generations, such as methods of producing food and shelter. Culture includes language, music, tools and technology, clothing, gender roles, belief systems, and moral codes, such as those prescribed in Confucianism, Hinduism, Islam, Judaism, Christianity, and a variety of indigenous belief systems.

ETHNICITY AND CULTURE: SLIPPERY CONCEPTS

A group of people who share a location, a set of beliefs, a way of life, a technology, and usually a common ancestry and sense of common history form an **ethnic group.** The term *culture group* is often used interchangeably with ethnic group. Both of the concepts of culture and ethnicity are imprecise, especially as they are popularly used. For instance, as part of the modern globalization process, migrating people often move well beyond their customary cultural or ethnic boundaries to cities or even distant countries. In these new places they take on many new ways of life and beliefs—their culture actually changes, yet they still may identify with their cultural or ethnic origins.

The Kurds in Southwest Asia are an example of the tenacity of this ethnic or cultural group identity. Long before the U.S. war in Iraq, the Kurds were asserting their right to create their own country in the territory where they have lived as nomadic herders since before the founding of Islam (600 c.e.). Syria, Iraq, Iran, and Turkey all claim parts of the traditional Kurdish area, while Kurds have de facto control over part of this homeland. Many Kurds are now educated urban dwellers, living and working in modern settings in Turkey, Iraq, Iran, or even London and New York, yet they actively support the cause of establishing a permanent Kurdish homeland. Although urban Kurds think of themselves as ethnic Kurds and are so regarded in the larger society, they do not follow the traditional Kurdish way of life. We could argue that these urban Kurds have a new identity within the Kurdish culture or ethnic group. Or they may be in a *transcultural* position, moving from one culture to another. Some scholars, such as Benedict Anderson, argue that a sense of belonging to any community that is not based on regular interaction between individuals is largely grounded in the individual's imagination. This may help explain how a person's beliefs and ways of living can change so profoundly, even while they may still identify with a particular culture or ethnicity.

Another problem with the imprecision of the concept of culture is that it is often applied to a very large group that shares only the most general of characteristics. For example, one often hears the terms American culture, African American culture, or Asian culture. In each case, the group referred to is far too large to share more than a few broad characteristics. It might fairly be said, for example, that U.S. culture is characterized by beliefs that promote individual rights, autonomy, and individual responsibility. But when we look beyond these broad abstractions and get to specifics, just what constitutes the rights and responsibilities of the individual are quite debatable. In fact, U.S. culture encompasses many subcultures that share some of the core set of beliefs but disagree over parts of the core and over a host of other matters. The same is true, in varying degrees, for all other regions of the world. When many culture groups live in close association more or less amicably, the society may be called **multicultural.**

VALUES

Occasionally you will hear someone say, "After all is said and done, people are all alike," or "People ultimately all want the same thing." It is a heartwarming sentiment, but an oversimplification. True, we all want food, shelter, health, love,

---

culture all the ideas, materials, and institutions that people have invented to use to live on Earth that are not directly part of our biological inheritance

ethnic group a group of people who share a common ancestry and sense of common history, a set of beliefs, a way of life, a technology, and usually a common geographic location of origin

multicultural society a society in which many culture groups live in close association and acceptance; but culturally, people are not all alike, and that is one of the qualities that makes the study of geography interesting. We would be wise not to expect or even to want other people to be like us. Cultural diversity has helped humans to be successful and adaptable animals. The various cultures serve as a bank of possible strategies for responding to the social and physical challenges faced by the human species. The reasons for differences in behavior from one culture to the next are usually complex, but they are often related to differences in values. Consider the following vignette that contrasts the values and *norms* (accepted patterns of behavior based on values) held by modern and urban individualistic cultures with those held by rural, community-oriented cultures.

VIGNETTE One recent rainy afternoon, a beautiful 40-something Asian woman walked alone down a fashionable street in Honolulu, Hawaii. She wore high-heeled sandals, a flared skirt that showed off her long legs, and a cropped blouse that allowed a glimpse of her slim waistline. She carried a laptop case and a large fashionable handbag. Her long, shiny black hair was tied back. Everyone noticed and admired her because she exemplified an ideal Honolulu businesswoman: beautiful, self-assured, and rich enough to keep herself well-dressed.

In the village of this woman's grandmother—whether it be in Japan, Korea, Taiwan, or rural Hawaii—the dress that exposed her body to open assessment and admiration by strangers of both sexes would signal that she lacked modesty. The fact that she walked alone down a public street—unaccompanied by her father, husband, or female relatives—might even indicate that she was not a respectable woman. That she at the advanced age of 40 was investing in her own good looks might be assessed as pathetic self-absorption. Thus a particular behavior may be admired when judged by one set of values and norms but may be considered questionable or even disreputable when judged by another. [Source: From Lydia Pulsipher's field notes. For detailed source information, see Text Credit pages.] ∎

If culture groups have different sets of values and standards, does that mean that there are no overarching human values or standards? This question increasingly worries geographers, who try to be sensitive both to the particularities of place and to larger issues of human rights. Those who lean too far toward appreciating difference could end up tacitly accepting inhumane behavior, such as the oppression of minorities or violence against women. Acceptance of difference does not preclude judgments of extreme customs or points of view. Nonetheless, although it is important to take a stand against cruelty of all sorts, deciding when and where to take that stand is rarely easy.

RELIGION AND BELIEF SYSTEMS

The religions of the world are formal and informal institutions that embody value systems. Most have roots deep in history, and many include a spiritual belief in a higher power (such as God, Yahweh, or Allah) as the underpinning for their value systems. Today, religions often focus on reinterpreting age-old values for the modern world. Some formal religious institutions—such as Islam, Buddhism, and Christianity—proselytize; that is, they try to extend their influence by converting others. Others, such as Judaism and Hinduism, accept converts only reluctantly.

Informal religions, often called *belief systems*, have no formal central doctrine and no firm policy on who may or may not be a practitioner. The fact that many people across the world combine informal religious beliefs with their more formal religious practices, of whatever persuasion, accounts for the very rich array of personal beliefs found in the world today.

Religious beliefs are often reflected in the landscape. For example, settlement patterns can demonstrate the central role of religion in community life: village buildings may be grouped around a mosque, a temple, a synagogue, or a church, and the same can be said for urban neighborhoods. In some places, religious rivalry is a major feature of the landscape. Certain spaces may be clearly delineated for the use of one group or another, as in Northern Ireland's Protestant and Catholic neighborhoods.

Religion has often been used to wield power. For example, during the era of European colonization, religion (Christianity) was used as a way to impose a change of attitude on conquered people.

FIGURE 1.30 shows the distribution of the major religious traditions on Earth today; it demonstrates some of the religious consequences of colonization. Note, for instance, the distribution of Roman Catholicism in the parts of the Americas, Africa, and Southeast Asia, all places colonized by European Catholic countries.

Recent research by the Pew Research Center on Religion and Public Life shows some interesting changing patterns for world religions. By 2050, the numbers of Christians and Muslims are predicted to be about equal globally. In Europe, Muslims will make up about 10 percent of the population; and in the United States, Muslims will outnumber Jews. Those who are unaffiliated with religion will rise in developed places (to about 26 percent of the U.S. population) but decline as a share of world population. Many of today's unaffiliated live in countries with low fertility and aging populations (the United States, Europe, China, and Japan). Both Islam and Christianity will grow most rapidly in sub-Saharan Africa. All of these are just predictions based on current trends and could be negated by any number of future developments. [Source: The Future of World Religions: Population Growth Projections, 2010–2050, Pew Research Center, April 2, 2015.]

Religion can also spread through trade contacts. In the seventh and eighth centuries, Islamic people used a combination of trade and political power (and less often, actual conquest) to extend their influence across North Africa, throughout Central Asia, and eventually into South and Southeast Asia (see Figure 6.14).

ON THE BRIGHT SIDE: Altruism

Recognizing all the ills that have emerged from racism and similar prejudices, we need not infer that human history has been marked primarily by conflict and exploitation or that these conditions are inevitable. The science of psychology has shown, through numerous experiments, that humans have strong inclinations toward *altruism*, the willingness to sacrifice one's own well-being for the sake of others. Studies across many scientific disciplines suggest that altruism, especially within groups, has been a strong force supporting the development of human societies since prehistoric times. It could be argued that altruism is the norm, and that one reason inhumane behavior is so distressing is that it is an anomaly. ∎

FIGURE 1.30 PHOTO ESSAY: Major Religions of the World

The small symbols on the map indicate a localized concentration of a particular religion within an area where another religion is predominant.

A Indigenous religion, Mexico

Luis Acosta/AFP/Getty Images

B Islam, Egypt

Yehuda Raizner/AFP/Getty Images

C Christianity, Ethiopia

Peter Delarue/AFP/Getty Images

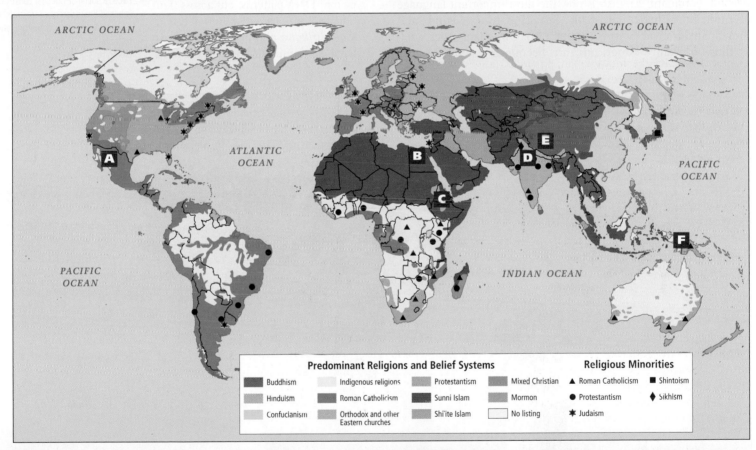

Predominant Religions and Belief Systems

- Buddhism
- Hinduism
- Confucianism
- Indigenous religions
- Roman Catholicism
- Orthodox and other Eastern churches
- Protestantism
- Sunni Islam
- Shi'ite Islam
- Mixed Christian
- Mormon
- No listing

Religious Minorities

- ▲ Roman Catholicism
- ● Protestantism
- ✳ Judaism
- ■ Shintoism
- ◆ Sikhism

D Hinduism, India

Sonu Mehta/Hindustan Times via Getty Images

E Buddhism, Tibet

TPG/Getty Images

F Indigenous religion, West Papua, Indonesia

Romeo Gacad/AFP/Getty Images

LANGUAGE

Language is one of the most important criteria used in delineating cultural regions. The modern global pattern of languages reflects the complexities of human interaction and isolation over several hundred thousand years. Between 2500 and 3500 languages are spoken on Earth today, some by only a few dozen people in isolated places. Many languages have several *dialects*—regional variations in grammar, pronunciation, and vocabulary.

The geographic pattern of languages has continually shifted over time as people have interacted through trade and migration. The pattern changed most dramatically around 1500, when the languages of European colonists began to replace the languages of the people they conquered. For this reason, English, Spanish, Portuguese, or French are spoken in large patches of the Americas, Africa, Asia, and Oceania. Today, because of trade and instantaneous global communication, a few languages have become dominant. English is now the most important language of international trade, but Arabic, Spanish, Chinese, Hindi, and French are also widely used **(FIGURE 1.31)**. Other languages, such as those of Native Americans, are becoming extinct because children no longer learn them within families, nor are these languages used by the media or in school systems.

RACE

Like ideas about gender roles, ideas about race affect human relationships everywhere on Earth. However, while race is of enormous social significance across the world, biologists tell us that from a scientific standpoint, race is a meaningless concept! The characteristics we popularly identify as **race** markers—skin color, hair texture, and face and body shape—have no significance as biological categories. All people now alive in the world are members of one species, *Homo sapiens sapiens*. For any supposed *racial trait*, such as skin color or hair texture or facial features, there are wide variations within human groups. Meanwhile, many invisible biological characteristics, such as blood type and DNA patterns, cut across skin color distributions and other so-called *racial attributes* and are shared across what are commonly viewed as different

race a social or political construct that is based on apparent characteristics such as skin color, hair texture, and face and body shape, but that is of no biological significance

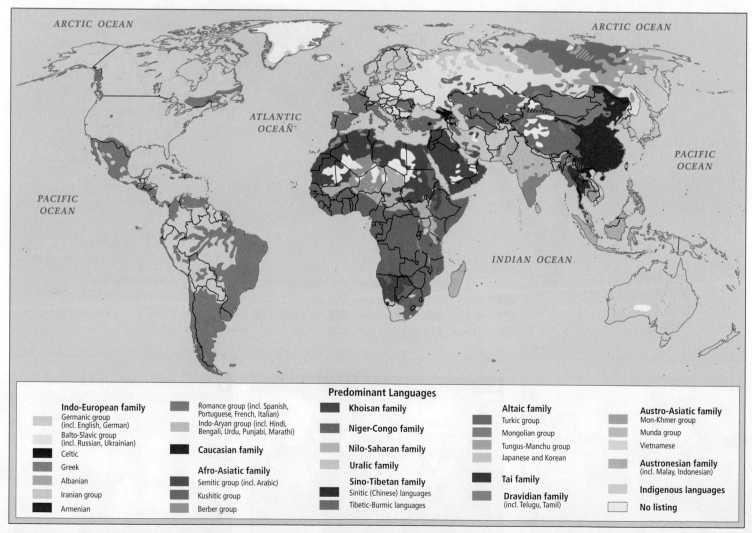

FIGURE 1.31 World's major language families. Distinct languages (Spanish and Portuguese, for example) are part of a larger group (Romance), which in turn is part of a language family (Indo-European).

races. In fact, over the last several thousand years there has been such massive gene flow among moving human populations that no modern group presents a discrete set of biological characteristics. Although any two of us may look quite different, from the biological point of view we are all simply *Homo sapiens sapiens* and are closely related.

Some of the easily visible features of particular human groups evolved to help them adapt to environmental conditions. For example, biologists have shown that people with darker skin (containing a high proportion of protective melanin pigment) evolved in regions close to the equator, where sunlight is most intense (FIGURE 1.32). All humans need the nutrient vitamin D, and sunlight striking the skin helps the body absorb vitamin D. Too much of the vitamin, however, can result in improper kidney functioning. Dark skin absorbs less vitamin D than light skin and thus would be a protective adaptation in equatorial zones. In higher latitudes, where the Sun's rays are more dispersed, light skin facilitates the sufficient absorption of vitamin D; darker-skinned people at these higher latitudes may need to supplement vitamin D to be sure they get enough, since D deficiencies can result in several health risks. Light-skinned people with little protective melanin in their skin—if they live in equatorial or high-intensity sunlit zones (parts of Australia, for example)—need to protect against too much vitamin D. All skin types need to protect against serious sunburn, and skin cancer. Similar correlations have been observed between skin color, sunlight, and another essential vitamin, folate, which if deficient can result in birth defects.

Despite its biological insignificance, over time, race has acquired enormous social and political import as humans from different parts of the world have encountered each other in situations of unequal power. *Racism*—the idea that skin color is a primary determinant of human abilities and even cultural traits—has often been invoked to justify the enslavement of particular groups, or confiscation of their land and resources. Race and its implications in North America will be covered in Chapter 2, and the topic will be discussed in several other world regions as well (see also a TED—Technology, Entertainment, and Design—talk by anthropologist Nina Jablonski, at http://tinyurl.com/l7wg7m).

THINGS TO REMEMBER

• Cultural geographers seek to understand human variability on Earth through a variety of lenses: culture and ethnicity, religion, language, and race.

• A group of people who share a location, a set of beliefs, a way of life, a technology, and usually a common ancestry and sense of common history form an ethnic group.

• Religion has often been used to wield power. For example, during the era of European colonization, religion (Christianity) was used to impose a change of attitude on conquered people.

• Today, with increasing trade and instantaneous global communication, a few languages have become dominant. English is now the most important language of international trade, but Arabic, Spanish, Chinese, Hindi, and French are also widely used. Other languages are becoming extinct because children no longer learn them within families or from the media or in school.

• Race is biologically meaningless, yet it has acquired enormous social and political significance.

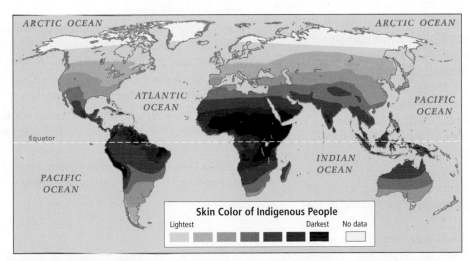

FIGURE 1.32 Skin color map for indigenous people as predicted from multiple environmental factors. Skin color has a biological role completely separate from any other human characteristics. That is, skin plays a twin role with respect to the Sun: protection from excessive UV radiation and absorption of enough sunlight to trigger the production of vitamin D. [Sources consulted: UNEP/GRID-Arendal at http://www.grida.no/graphicslib/detail/skin-colour-map-indigenous-people_8b88, Emmanuelle Bournay, cartographer; and G. Chaplin (2004), "Geographic Distribution of Environmental Factors Influencing Human Skin Coloration," *American Journal of Physical Anthropology, 125,* 292–302; map updated in 2007]

GEOGRAPHIC THEMES: Review and Self-Test

1. Environment: Humans are altering the planet at an unprecedented rate, causing sometimes drastic effects on ecosystems and the climate. Multiple environmental factors often interact to influence the vulnerability of a location to the impacts of climate change. These vulnerabilities have a spatial pattern.

• Describe at least five environmental effects now thought to be related to climate change.

• How is the place where you live vulnerable to climate change?

• Assess your exposure, sensitivity, and resilience to water-related climate-change impacts.

• Describe the environmental and economic impacts on tourism of changes in climate patterns.

• Select one climate zone (temperate, midlatitude, tropical, etc.) and describe the likely effect on food production of climate change.

• Explain the physical and social vulnerabilities of the city of Mumbai to climate change.

2. Globalization and Development: Global flows of information, goods, and people are transforming patterns of economic development. Local self-sufficiency is giving way to global interdependence as people and places are increasingly becoming connected, sometimes across vast distances.

• How is globalization evident in your life—from the clothes you wear, to your favorite foods, to your career plans?

• To what extent does your circle of friends show the effects of globalization?

• Explain how globalization can empower little-known ethnic groups or, alternatively, leave them with curtailed options.

• How has globalization affected such things as human trafficking, for good and for ill?

3. Power and Politics: There are major differences across the world in the ways that power is wielded in societies. Modes of governing that are more authoritarian are based on the power of the state or community leaders. Modes that are based on notions of political freedom and democracy give the general public greater power over themselves and more of a role in deciding how policies are developed and governments are run. There are also many other ways of managing political power.

• Give several examples from recent world events of shifts from authoritarian modes of government to more democratic systems in which individuals have a greater say in the running of their governments.

• Are peace, broad prosperity, education, and civil society features of life in your home country? How do the political freedoms you enjoy shape your daily life?

• Using the concept of geopolitics, explain why and how the 9/11 attacks had a global impact.

4. Urbanization: The development of manufacturing and service economies has taken place largely in cities, drawing people from the countryside into urban areas. Many of these new urbanites were displaced by the mechanization of agriculture in rural areas, which drastically reduced the need for agricultural labor and left people with no means of support and a reduced access to food.

• How far back in your family history would you need to go to find ancestors who lived in a rural area and grew almost all their own food?

• Assuming you do not live in a rural area, how has urban life affected the diet and level of physical activity of you and/or your family?

• When cities expand, agricultural land often contracts. Explain why this happens and describe some economic and social ramifications of this relationship.

• In what ways might it be possible for cities to foster more "green" ways of life? Assess the likelihood of this happening.

5. Population and Gender: Population growth is slowing across the world for a number of reasons. Chief among these are urbanization and the lifestyle changes that it brings; the contraction of agricultural labor as a result of the mechanization of farming; and perhaps most important, the increasing numbers of women who are delaying childbearing as they pursue educational and work opportunities outside the home. The availability of birth control is an important contributing factor in determining success in family planning.

• Explain the supposed interaction between urbanization, gender roles, and fertility rates.

• How is the shift toward more equality between the genders influencing population growth rates, patterns of economic development, and politics?

• How do your plans for a career and a family compare to those of your grandparents and great-grandparents?

• Ask your parents who the first woman in your family was who had a career. How, if at all, did this influence the number of children she had?

• Explain and discuss the most widespread and consistent measurement of gender disparities globally.

Critical Thinking Questions

1. Some people argue that it is acceptable for people in the United States to consume at high levels because their consumerism keeps the world economy going. What are the weaknesses in this idea?

2. What are the causes of the huge global increases in migration, legal and illegal, that have taken place over the last 25 years?

3. What would happen in the global marketplace if all people earned a living wage?

4. What might be some of the expected impacts on society of longer life spans and lower fertility rates?

5. Given the threats posed by global warming, what are the most important steps you are willing to take now?

6. As you read about differing ways of life, values, and perspectives on the world, reflect on the appropriateness of force as a way of resolving conflicts.

7. How might a widespread acceptance of the normalcy of variations in sexual orientation and gender identity change parenting philosophies?

8. Reflect on the reasons why, and impacts of, some people having much while others have little.

9. What would be some of the disadvantages and advantages of abolishing gender roles in any given culture?

Chapter Key Terms

agriculture 30
authoritarianism 40
biosphere 19
birth rate 49
capitalism 41
carrying capacity 32
cartographers 4
cash economy 52
cisgender 54
civil society 41
climate change 19
climate 15
communism 41
culture 55
death rate 49
delta 14
democratization 40
demographic transition 52
development 33
domestication 30
ecological footprint 19
erosion 14
ethnic cleansing 42
ethnic group 55
Euro zone 37
extraction 33
fair trade 36
floodplain 14
food security 30
formal economy 37
free trade 35
gender 53
gender roles 53
genetic modification (GM) 32
genocide 42
geographic information science (GISc) 7
geopolitics 41

global economy 33
global scale 11
global warming 20
globalization 32
green revolution 31
greenhouse gases (GHG) 20
gross domestic product (GDP) per capita 37
gross national income (GNI) per capita 37
human geography 3
human well-being 38
immigration 49
industrial production 33
Industrial Revolution 30
informal economy 37
intersex 53
Kyoto Protocol 26
landforms 12
latitude 4
legend 4
less developed countries (LDC) 33
living wages 36
local scale 11
longitude 4
map projections 7
migration 49
monsoon 18
more developed countries (MDCs) 33
multicultural society 55
multinational corporation 34
nongovernmental organizations (NGOs) 44
orographic rainfall 18
physical geography 3
plate tectonics 12
political ecology 38
political freedoms 40
population pyramid 49

precipitation 15
primary sector 33
purchasing power parity (PPP) 37
push/pull phenomenon of urbanization 45
quaternary sector 33
race 58
rate of natural increase (RNI) 49
region 9
Ring of Fire 13
scale (of a map) 4
secondary sector 33
Security Council 44
services 33
sex 53
slum 45
spatial distribution 3
spatial interaction 3
state, the 40
structural adjustment 36
subsistence economy 52
sustainable agriculture 32
sustainable development 38
tertiary sector 33
total fertility rate (TFR) 49
transgender 54
United Nations (UN) 44
United Nations Gender Development Index (GDI) rank 38
United Nations Human Development Index (HDI) 38
urbanization 45
virtual water 27
water footprint 27
weather 15
weathering 14
World Trade Organization (WTO) 36

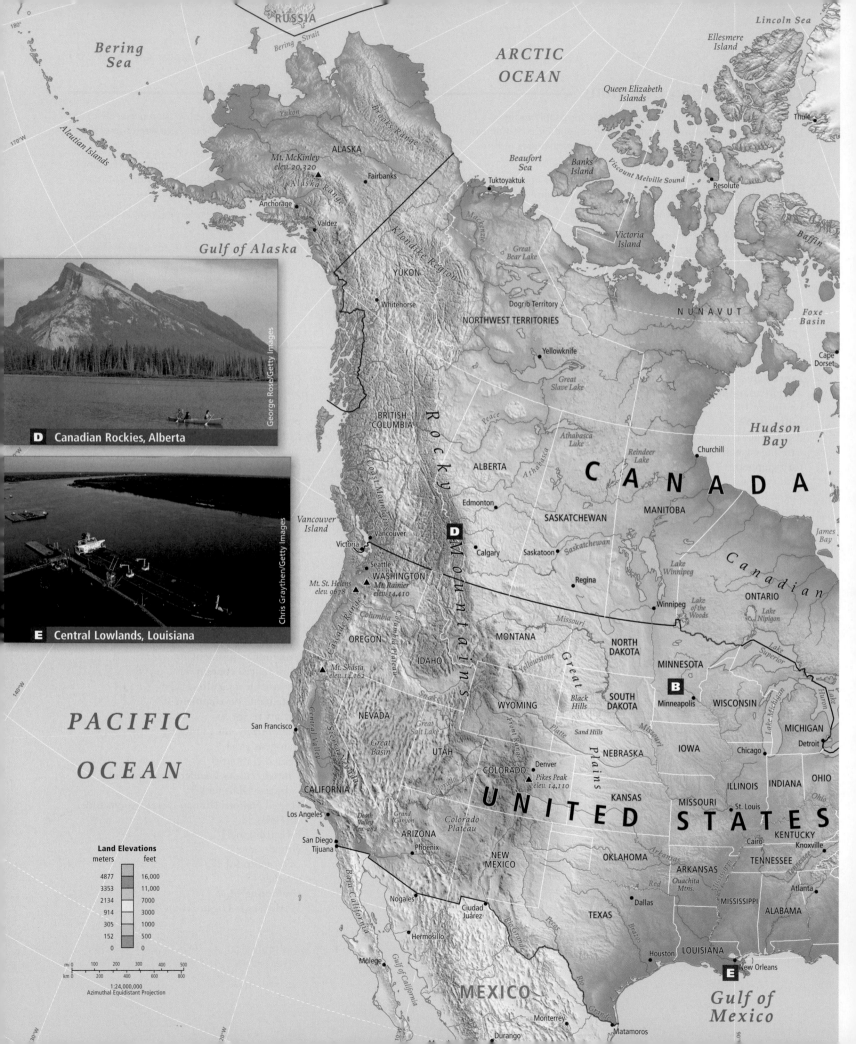

RUSSIA

Bering Strait

Bering Sea

Aleutian Islands

ARCTIC OCEAN

Lincoln Sea

Ellesmere Island

Thule

Queen Elizabeth Islands

ALASKA

Mt. McKinley
elev. 20,320

Fairbanks

Anchorage

Valdez

Gulf of Alaska

Yukon

Brooks Range

Alaska Range

Klondike Region

YUKON

Whitehorse

Beaufort Sea

Tuktoyaktuk

Banks Island

Viscount Melville Sound

Resolute

Victoria Island

Great Bear Lake

Mackenzie

NORTHWEST TERRITORIES

Dogrib Territory

N U N A V U T

Baffin

Foxe Basin

Cape Dorset

Yellowknife

Great Slave Lake

BRITISH COLUMBIA

Rocky

Coast Mountains

Peace

ALBERTA

Athabasca

Athabasca Lake

Reindeer Lake

Churchill

Hudson Bay

George Rose/Getty Images

D Canadian Rockies, Alberta

Vancouver Island

Vancouver

Victoria

Seattle

WASHINGTON

Mt. St. Helens
elev. 9678

Mt. Rainier
elev. 14,410

Edmonton

SASKATCHEWAN

Calgary

Saskatoon

Saskatchewan

MANITOBA

Regina

Winnipeg

Lake Winnipeg

Lake of the Woods

James Bay

C a n a d i a n

ONTARIO

Lake Nipigon

Lake Superior

Chris Graythen/Getty Images

E Central Lowlands, Louisiana

Cascade Range

Columbia

Columbia Plateau

OREGON

Mt. Shasta
elev. 14,162

IDAHO

D

M o u n t a i n s

Snake

MONTANA

Yellowstone

Missouri

Great

WYOMING

Black Hills

NORTH DAKOTA

SOUTH DAKOTA

MINNESOTA

B

Minneapolis

WISCONSIN

Lake Michigan

Lake Huron

MICHIGAN

Detroit

PACIFIC OCEAN

San Francisco

Sierra Nevada

Central Valley

NEVADA

Great Basin

Great Salt Lake

UTAH

Front Range

Platte

Denver

COLORADO

Pikes Peak
elev. 14,110

Plains

Sand Hills

Missouri

NEBRASKA

IOWA

Chicago

ILLINOIS

INDIANA

OHIO

Los Angeles

San Diego

Tijuana

*Death Valley
elev. -282*

Grand Canyon

ARIZONA

Phoenix

Colorado

Colorado Plateau

NEW MEXICO

U N I T E D   S T A T E S

KANSAS

MISSOURI

St. Louis

KENTUCKY

Cairo

Knoxville

Nogales

Baja California

Ciudad Juárez

OKLAHOMA

Arkansas

Red

ARKANSAS

Ouachita Mtns.

TENNESSEE

Tennessee

Atlanta

MISSISSIPPI

ALABAMA

Hermosillo

Rio Grande

Pecos

TEXAS

Dallas

Brazos

Mississippi

LOUISIANA

E

New Orleans

Moleje

Houston

Monterrey

Durango

Gulf of California

MEXICO

Rio Grande

Matamoros

Gulf of Mexico

Land Elevations

meters	feet
4877	16,000
3353	11,000
2134	7000
914	3000
305	1000
152	500
0	0

mi 0   100   200   300   400   500
km 0   200   400   600   800

1:24,000,000
Azimuthal Equidistant Projection

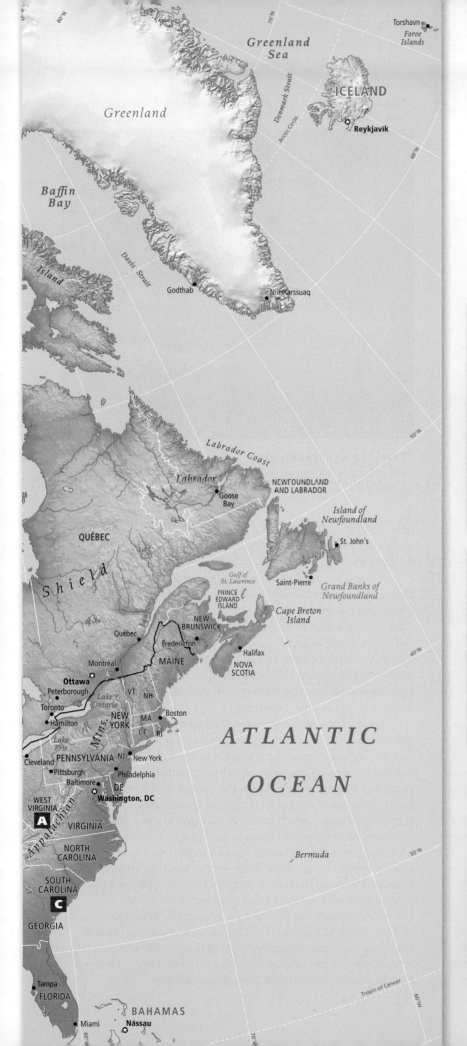

2

North America

A Appalachian Mountains, Virginia

Karen Bleier/AFP/Getty Images

B Central Lowlands, Minnesota

Jeffrey Phelps/Getty Images

C Coastal Lowlands, Charleston, South Carolina

Streeter Lecka/Getty Images

FIGURE 2.1 Regional map of North America.

GEOGRAPHIC THEMES

After you read this chapter, you will be able to discuss the following issues as they relate to the five thematic concepts:

1. Environment: North America's use of resources has enormous environmental impacts. Home to only 5 percent of the world's population, the region produces 17 percent of humanity's greenhouse gas emissions. Other impacts include the depletion and pollution of water resources and fisheries and the destruction of natural habitats.

2. Globalization and Development: Globalization has transformed North America, reorienting it toward knowledge-intensive jobs that require education and training. Income inequality has risen as most manufacturing jobs have been moved to countries with cheaper labor or have been replaced by technology. North America's size and wealth have made it the center of the global economy, and its demand for imported goods and its export of manufacturing jobs make it a major engine of globalization.

3. Power and Politics: Compared to many world regions, North America has relatively high levels of political freedom. However, there is also widespread dissatisfaction with the political process here. Internationally, Canada plays a modest role while the United States has enormous global political influence, although its status as the world's predominant "superpower" is declining.

4. Urbanization: Since World War II, North America's urban areas have become less densely populated, expanding spatially much faster than their populations have increased numerically. This is primarily because of suburbanization and urban sprawl, which have transformed this highly urbanized region.

5. Population and Gender: For more than two centuries, North American birth rates have been in decline. A major factor in this has been women delaying childbearing to pursue education and careers. Declining childbirth rates play a role in the aging of North American populations, which may slow economic growth.

The North American Region

North America (**FIGURE 2.1**) is one of the largest, wealthiest, and most politically powerful regions in the world. It encompasses many environments and a complex array of local cultures and economic activities that interact with each other across wide distances. Many of the trends discussed throughout this book are well advanced in North America.

The five thematic concepts in this book are explored as they arise in the discussion of regional issues, with interactions between two or more themes featured. Vignettes, like the one that follows about Javier Aguilar, illustrate one or more of the themes as they are experienced in individual lives.

VIGNETTE Javier Aguilar, a 39-year-old father of three, has worked as an agricultural laborer in California's Central Valley for 20 years. He and hundreds of thousands like him tend the fields of crops that feed the nation, especially during the winter months (FIGURE 2.2). Aguilar used to make $8.00 an hour, but now he is

FIGURE 2.2 Lettuce harvest in California. Agricultural laborers pick lettuce in California's Central Valley.

Paul Harris/Getty Images

Thinking Geographically
What factors might cause lettuce production to decline during a recession?

unemployed and standing in a church-sponsored food line. "If I don't work, [we] don't live. And here all the work is gone," he says, grimly.

In Mendota, also in California's Central Valley, young Latino men wait on street corners to catch a van to the fields. None come. Unemployment rates in Mendota began to skyrocket in 2009, hitting a high of 45 percent in 2011, and by 2014 the jobless rate was still at 34 percent. Mayor Robert Silva said his community was dying on the vine. He saw the trouble spreading as many small businesses closed. He worried about drug use, alcoholism, family violence, and malnutrition that can accompany severe unemployment in any community.

This level of unemployment in the heart of the nation's biggest producer of fruit, nuts, and vegetables was partly due to drought and partly to a global economic recession. The drought is related to natural dry cycles as well as to global climate change. Water has become more scarce in the Central Valley and access to irrigation water has been cut to force conservation. The global economic recession, which began in 2007 and in some places still continues, reduced overall demand for many of California's fresh fruits and vegetables as families turned to cheaper foods and those grown closer to home. All of these stresses have forced farmers to remove from production as much as 1 million of the 4.7 million acres of land once cultivated and irrigated in the Central Valley. This may eventually result in the loss of as many as 80,000 jobs and as much as U.S.$2.2 billion in California agriculture and related industries. And even though by early 2016 agricultural production in Mendota had partially been revived, unemployment remained over 30 percent, nearly twice the national average.

The confluence of troubles in California's Central Valley has been particularly devastating for low-wage agricultural workers like Javier Aguilar because they have so little to fall back on in terms of savings, education, or skills. Community food banks are essential to keeping Javier's family from going hungry. After years of having their living standards decline, many farmworkers are engaging in political action, such as the "water rallies" that in recent years have pushed the California state government to relax environmental policies that limit the amount of water that can be diverted into the southern portion of the Central Valley via canals and pipelines fed by rivers in much wetter Northern California. However, environmentalists, fishers, and urban water users oppose such moves and so far have kept the government from changing its policies. But for now the situation remains bleak in Mendota, where Mayor Silva observed, "We're supposed to be the cantaloupe capital of the world, but we're the food line capital of the world." [Source: New York Times; PBS Newshour. For detailed source information, see Text Sources and Credits.] ∎

Stories like that of Javier Aguilar are not what people usually expect when they think about this vast region, renowned for its wealth, innovation, and economic opportunities. But Javier's story is one that exemplifies the experiences of the many North Americans whose incomes have stagnated or declined in recent decades as wealth has become more concentrated in the hands of the upper middle class and the rich. With less money going to lower-income people, and with many environments in this region pushed to their limits, more people are having to make do with less. To some extent these shifts reflect broad changes in the global economy. The United States has been the world's largest economy for 140 years and its strongest military power for more than half a century. But after decades of high-paying jobs being moved overseas and multiple financial crises, the U.S. economy has lost much of its momentum. By some measures, the United States is now no longer the world's largest economy; in 2014, China's economy outgrew that of the United States in terms of its *purchasing power*. China will surpass the United States in all measures of economic size within the next decade.

The circumstances that people like Javier Aguilar find themselves in also reflect economic and political changes within the United States, where many economic trends favor the wealthy, and many government policies designed to help poor people have been rolled back in recent decades. And yet there has been renewed interest in helping poor people navigate difficult economic times. For example, health care for low-income people received major government investment with the passage of the Affordable Care Act in 2010. These trends exist to a lesser extent in Canada, due largely to much more stable support for government programs that help poor people.

What Makes North America a Region?

In this book we define North America as Canada and the United States, two countries that are linked because of their geographic proximity, similar history, and many common cultural, economic, and political features. This is a cultural definition of the region, not a physical one. Mexico and much of Central America are physically part of the continent of North America, but culturally they are more connected to Middle and South America.

Terms in This Chapter

The term *North America* is used to refer to both countries. Even though it is common on both sides of the border to call the people of Canada "Canadians" and people in the United States "Americans," this text uses the term *United States*, or *U.S.*, rather than *America*, for the United States. Other terms relate to the cultural diversity of the region. The text uses the term **Latino** to refer to all Spanish-speaking people from Middle and South America, although their ancestors may have been European, African, Asian, or Native American. Native American people have different terms they prefer and those preferences change over time. As of this writing, some in the United States still favor the use of the term *American Indian*. In Canada, *Aboriginal peoples* is the umbrella term, which covers three main groups: *First Nations* (preferred by some), *Nunavut* (85 percent of whom are Inuit, and all of whom are indigenous people of northern Canada and parts of Greenland and Alaska), and *Métis* (people of mixed Native American and European ancestry). In this text we will most often use Native American for those in the United States and Aboriginal peoples for those in Canada.

THINGS TO REMEMBER

• The global economic recession, which began in 2007 had many lingering effects that are still being felt today. It has reduced overall demand for California's fresh fruits and vegetables, as families turned to cheaper foods or those grown closer to home.

• The recession, combined with a drought, forced farmers to remove from production as much as 1 million of the 4.7 million acres of land once cultivated and irrigated in the Central Valley. This may eventually result in a loss of as many as 80,000 jobs and as much as U.S.$2.2 billion in California agriculture and related industries.

• North America consists of the United States and Canada.

PHYSICAL GEOGRAPHY

The continent of North America has almost every type of climate and a wide variety of landforms. Huge expanses of mountain peaks, ridges, and valleys meet expansive plains, long, winding rivers, myriad lakes, and extraordinarily lengthy coastlines. Here the focus is on a few of the most significant landforms.

LANDFORMS

A wide mass of mountains and basins, known as the Rocky Mountain zone, dominates western North America (see Figure 2.1D). It stretches down from the Bering Strait in the far north, through Alaska, and into Mexico. This zone formed about 200 million years ago when, as part of the breakup of the supercontinent Pangaea (see Figure 1.8), the Pacific Plate pushed against the North American Plate, thrusting up mountains. These plates still rub against each other, causing earthquakes along the Pacific coast of North America.

Latino a term used to refer to all Spanish-speaking people from Middle and South America, although their ancestors may have been European, African, Asian, or Native American

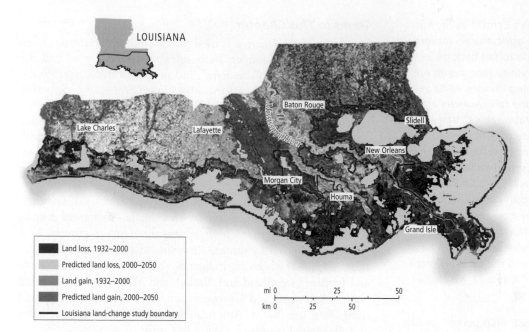

FIGURE 2.3 Wetland loss in Louisiana. The Louisiana coastline and the lower Mississippi River basin are vital to the nation's interests. They are the end point for the vast Mississippi drainage basin and provide coastal wildlife habitats, recreational opportunities, and transportation lanes that connect the vast interior of the country to the ocean and to offshore oil and gas. Most important, the wetlands provide a buffer against damage from hurricanes. Unfortunately, Louisiana has lost one-quarter of its total wetlands over the last century, largely because of human activity. The remaining 3.67 million acres constitute 14 percent of the total wetland area in the lower 48 states. [Source consulted: USGS/National Wetlands Research Center]

The much older and hence more eroded Appalachian Mountains stretch along the eastern edge of North America from New Brunswick and Maine to Georgia (see Figure 2.1A). This range resulted from very ancient collisions between the North American Plate and the African Plate.

Between these two mountain ranges lies the huge central lowland of undulating plains that stretches from the Arctic to the Gulf of Mexico. This landform was created by the deposition of deep layers of material eroded from the mountains and carried to this central North American region by wind and rain and by the rivers flowing east and west into what is now the Mississippi drainage basin.

During periodic ice ages over the last 2 million years, glaciers have covered the northern portion of North America. In the most recent ice age (between 10,000 and 25,000 years ago), the glaciers, sometimes as much as 2 miles (about 3 kilometers) thick, moved south from the Arctic, picking up rocks and soil and scouring depressions in the land surface. When the glaciers later melted, these depressions filled with water, forming the Great Lakes. Thousands of smaller lakes, ponds, and wetlands that stretch from Minnesota and Manitoba to the Atlantic were formed in the same way (see Figure 2.1B). Melting glaciers also dumped huge quantities of soil throughout the central United States. This soil, often many meters deep, provides the basis for large-scale agriculture but remains susceptible to wind and water erosion.

East of the Appalachians, the Atlantic coastal lowland stretches from New Brunswick to Florida (see Figure 2.1C). It then sweeps west to the southern reaches of the central lowland along the Gulf of Mexico. In Louisiana and Mississippi, much of this lowland is filled in by the Mississippi River delta—a low, flat, swampy transition zone between land and sea. The delta was formed by massive loads of silt deposited during floods over the past 150-plus million years by the Mississippi, North America's largest river system. The delta deposit originally began at what is now the junction of the Mississippi and Ohio rivers at Cairo, Illinois; slowly, as ever more sediment was deposited, the delta advanced 1000 miles (1600 kilometers) into the Gulf of Mexico.

Over the centuries, human activities such as deforestation, deep plowing, and heavy grazing have led to erosion and added to the silt load of the rivers. The construction of levees during the last 300 years along riverbanks has drastically reduced flooding. Because of this flood control, much of the silt that used to be spread widely across the lowlands during floods is being carried to the southern part of the Mississippi delta—a low, flat zone characterized by swamps, lagoons, and sandbars. The intrusion of silt is destroying wetlands and, as the silt load extends further out into deep waters, the extra weight is causing the delta to sink into the Gulf of Mexico, a process called *subsidence* **(FIGURE 2.3)**.

CLIMATE

The landforms across this continental expanse influence the movement and interaction of air masses and contribute to its enormous climate variety **(FIGURE 2.4)**. Along the southern west coast of North America, the climate is generally mild (Mediterranean)—dry and warm in summer, cool and moist in winter. North of San Francisco, the coast receives moderate to heavy rainfall. East of the Pacific coastal mountains, climates are much drier because as the moist air sinks into the warmer interior lowlands, it tends to hold its moisture. This interior region becomes increasingly arid moving eastward across the Great Basin (see Figure 2.4C) and Rocky Mountains. Many dams and reservoirs for irrigation projects have been built to make agriculture possible. Because of the low level of rainfall, however, the amount of water that is being extracted exceeds the capacity of ancient underground water basins **(aquifers)** to replenish themselves.

aquifers ancient natural underground reservoirs of water

FIGURE 2.4 PHOTO ESSAY: Climates of North America

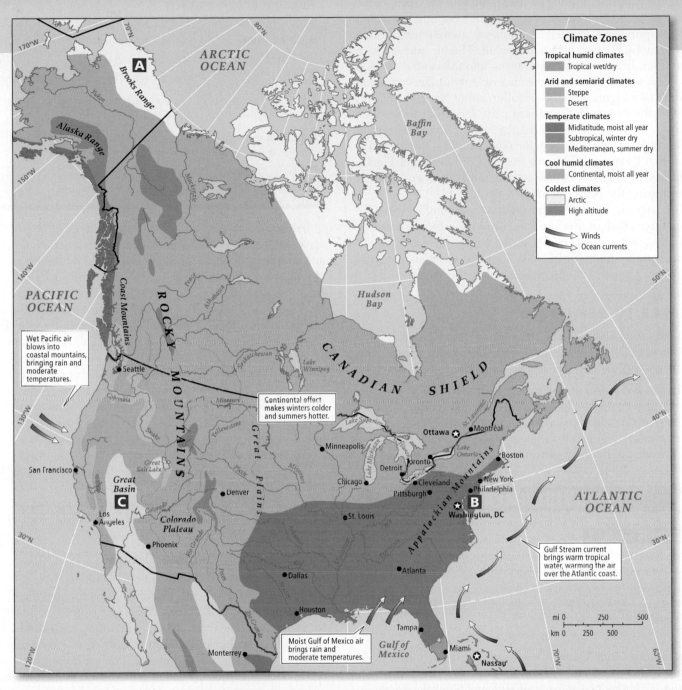

ARCTIC OCEAN

Brooks Range

Alaska Range

PACIFIC OCEAN

Coast Mountains

ROCKY MOUNTAINS

Baffin Bay

Hudson Bay

CANADIAN SHIELD

Great Plains

Great Basin

Colorado Plateau

Appalachian Mountains

ATLANTIC OCEAN

Gulf of Mexico

Climate Zones

Tropical humid climates
  Tropical wet/dry

Arid and semiarid climates
  Steppe
  Desert

Temperate climates
  Midlatitude, moist all year
  Subtropical, winter dry
  Mediterranean, summer dry

Cool humid climates
  Continental, moist all year

Coldest climates
  Arctic
  High altitude

  → Winds
  → Ocean currents

Wet Pacific air blows into coastal mountains, bringing rain and moderate temperatures.

Continental effect makes winters colder and summers hotter.

Gulf Stream current brings warm tropical water, warming the air over the Atlantic coast.

Moist Gulf of Mexico air brings rain and moderate temperatures.

Seattle, San Francisco, Los Angeles, Phoenix, Denver, Minneapolis, Chicago, Detroit, Toronto, Cleveland, Pittsburgh, St. Louis, Ottawa, Montréal, Boston, New York, Philadelphia, Washington, DC, Atlanta, Dallas, Houston, Tampa, Miami, Monterrey, Nassau

mi 0   250   500
km 0   250   500

A Arctic, Alaska

National Geographic Image Collection

B Temperate midlatitude, Maryland

Linda Davidson/The Washington Post via Getty Images

C Desert, Utah

DEA/F. Barbagallo/De Agostini/Getty Images

On the eastern side of the Rocky Mountains, the main source of moisture is the Gulf of Mexico. When the continent is warming in the spring and summer, the air masses above it rise, sucking in warm, moist, buoyant air from the Gulf. This air interacts with cooler, drier, heavier air masses moving into the central lowland from the north and west, often creating violent thunderstorms and tornadoes (see Figure 2.4A). Generally, central North America is wettest in the eastern and southern parts and driest in the west and north (see Figure 2.4B). Along the Atlantic coast, moisture is supplied by warm, wet air above the Gulf Stream—a warm ocean current that flows north from the eastern Caribbean and Florida and follows the coastline of the eastern United States and Canada before crossing the North Atlantic Ocean.

The large size of the North American continent creates wide temperature variations. Because land heats up and cools off more rapidly than water, temperatures in the interior of the continent are hotter in the summer and colder in the winter than in coastal areas, where temperatures are moderated by the oceans.

THINGS TO REMEMBER

- North America has two main mountain ranges, the Rockies and the Appalachians, separated by expansive plains through which run long, winding rivers.

- The size and variety of landforms influence the movement and interaction of air masses, creating enormous climatic variation. Because land heats up and cools off more rapidly than water, temperatures in the interior of the continent are higher in the summer and colder in the winter than in coastal areas.

ENVIRONMENT

GEOGRAPHIC THEME 1

Environment: North America's use of resources has enormous environmental impacts. Home to only 5 percent of the world's population, the region produces 17 percent of humanity's greenhouse gas emissions. Other impacts include the depletion and pollution of water resources and fisheries and the destruction of natural habitats.

North America's wide range of resources, and its seemingly limitless stretches of forest and grasslands, sometimes divert attention from the environmental impacts of settlement and development. However, it is impossible to ignore the many environmental consequences of the North American lifestyle. This section focuses on a few of those consequences: climate change and air pollution, depletion and pollution of water resources and fisheries, and habitat loss.

CLIMATE CHANGE AND AIR POLLUTION

On a per capita basis, North Americans contribute among the largest amounts of greenhouse gases (GHGs) to Earth's atmosphere. No regions have higher per capita emissions. This is largely a result of North America's high consumption of fossil fuels, which is related to several factors. One of these is North America's dominant pattern of urbanization, characterized by vast and still-growing suburbs where automobiles are the main mode

Thinking Geographically

After you have read about the vulnerability to climate change in North America, you will be able to answer the following questions.

A Describe the coastal erosion illustrated by this photo.

B What elements of this community's emergency response system may be compromised by this flooding?

C How might irrigation reduce vulnerability to climate change but also result in more GHG emissions?

D Why are stronger hurricanes more likely as the climate warms?

of transportation. The mostly single-story freestanding buildings across the region's urban landscapes require more energy to heat and cool than do the densely packed, high-rise buildings typical of cities in other world regions. North American industrial and agricultural production also depends very heavily on fossil fuels.

Canada's government was one of the first to commit to reducing the consumption of fossil fuels. Until recently, the United States resisted such moves, fearing damage to its economy. Both countries are now exploring alternative sources of energy, such as solar, wind, geothermal, and nuclear power. GHG emissions fell in both countries beginning in 2008, although this appears mainly to be due to the economic recession. Some reduction in emissions may be related to increased energy conservation. Many utility companies are also switching from coal to natural gas, which produces less CO_2 when burned.

Since 2012, when both economies started to grow again, GHG emissions have increased. Both Canada and the United States possess the technological capabilities needed to lead the world in shifting over to cleaner sources of energy.

Vulnerability to Climate Change

Both Canada and the United States are vulnerable to the effects of climate change. Dense population centers on the Gulf of Mexico and on the Atlantic coast are very exposed to hurricanes, which may become more violent as oceans warm **(FIGURE 2.5B, D)**. Sea level rise and coastal erosion, caused by the thermal expansion of the oceans, are already affecting coastal areas along the Arctic coast of North America (see Figure 2.5A). Here, at least 26 coastal villages are being forced to relocate inland, at an estimated cost of U.S.$130 million per village.

Some studies suggest that regardless of how much humans decrease their GHG output, at least half the populated areas of 400 towns and cities in North America will eventually be directly impacted by sea level rise. Many arid farming zones will dry further as higher temperatures reduce soil moisture, making irrigation crucial (see Figure 2.5C). In all of these areas, resilience to climate change is bolstered over the short term by the region's relatively strong emergency response and recovery systems. Long-term resilience is boosted by careful planning as well as by North America's large and diverse economy, which can provide alternative livelihoods to people who face significant exposure to climate impacts.

FIGURE 2.5 PHOTO ESSAY: Vulnerability to Climate Change in North America

North America's wealth and its well-developed emergency response systems make it very resilient and reduce its overall vulnerability to climate change. However, certain regions are highly exposed to temperature increases, drought, hurricanes, and sea level rise.

A The location of Shishmaref, Alaska, on the Arctic Sea leaves it exposed to coastal erosion, which may be increasing in the area because of warmer air and sea temperatures.

B Florida's low elevations expose it to sea level rise and flooding during and after hurricanes.

Vulnerability to Climate Change

- Extreme
- High
- Medium
- Low

The stark contrast in vulnerability between the United States and Mexico results from the countries' very different sensitivity and resilience to water scarcity. As temperatures rise, this already dry borderland area will have less water. The United States has a much better water infrastructure, reducing its sensitivity to drought, and its emergency response systems make it more resilient to water shortages than Mexico is. With such drastic differences, the U.S.–Mexico border could become an even more contentious zone as the climate changes.

C Higher temperatures raise the rate at which plants lose water, increasing the need for irrigation. Farms that rely on the shrinking aquifers for water have access to less than what they need and so are highly sensitive to the rising temperatures that climate change is bringing.

D Hurricanes gain strength with warmer temperatures. Many low-lying coastal cities are very exposed and sensitive to hurricanes, though their resilience varies. Shown here is a fire that destroyed several homes in New Orleans when the flooding caused by Hurricane Katrina made it impossible for fire trucks to reach the fire.

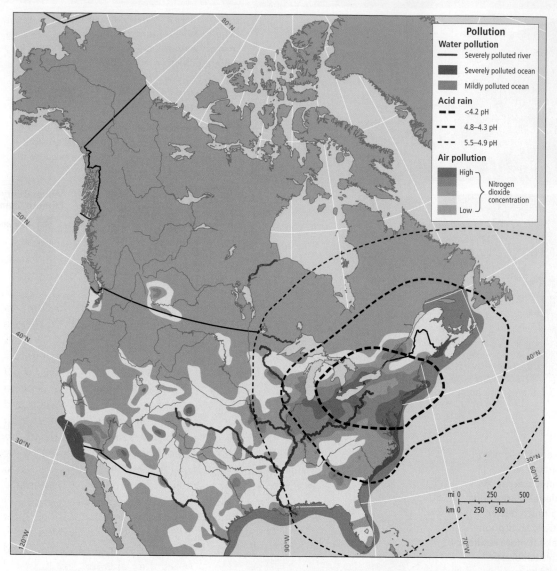

FIGURE 2.6 Air and water pollution in North America. This map shows two aspects of air pollution, as well as polluted rivers and coastal areas. Red and yellow indicate concentrations of nitrogen dioxide (NO_2), a toxic gas that comes primarily from the combustion of fossil fuels by motor vehicles and power plants. This gas interacts with rain to produce nitric acid, a major component of acid rain, as well as toxic organic nitrates that contribute to urban smog. The map also shows polluted coastlines (including all of the coastline from Texas to New Brunswick) as well as severely polluted rivers, which include much of the Mississippi River and its tributaries.

Air Pollution

In addition to climate change, most GHGs contribute to various forms of air pollution, such as smog and acid rain. **Smog** is a combination of industrial emissions, car exhaust, and water vapor that frequently hovers as a yellow-brown haze over cities, including many in North America, and causes a variety of health problems. These same emissions also result in **acid rain,** which is created when pollutants dissolve in falling precipitation and make the rain acidic. Acid rain can kill trees and, when concentrated in lakes and streams, poison fish and wildlife.

The United States, with its large population and extensive range of industries, is responsible for the vast majority of acid rain in North America. Because of continental weather and wind patterns, however, the area most affected by acid rain encompasses a wide swath on both sides of the eastern U.S.–Canada border **(FIGURE 2.6)**. The eastern half of the continent, which includes the entire Eastern Seaboard, from the Gulf Coast to Newfoundland, also is significantly impacted by acid rain.

smog a combination of industrial emissions, car exhaust, and water vapor that frequently hovers as a yellow-brown haze over many North American cities, causing a variety of health problems

acid rain precipitation that has formed through the interaction of rainwater or moisture in the air with sulfur dioxide and nitrogen oxides emitted during the burning of fossil fuels, making it acidic

WATER RESOURCE DEPLETION, POLLUTION, AND MARKETIZATION

People who live in the humid eastern part of North America find it difficult to believe that water is becoming scarce even there. Consider the case of Ipswich, Massachusetts, where the watershed is drying up as a result of overuse. There, innovators are saving precious water through conservation strategies in their homes and businesses. Elsewhere, as populations and per capita water usage grow, conflicts over water are becoming more and more common.

Water Depletion

In North America, water becomes more precious the farther west one goes. On the North American Great Plains, rainfall varies considerably from year to year. To make farming more secure and predictable, taxpayers across the continent have subsidized the building of pumps and stock tanks for farm animals, and aqueducts and reservoirs for crop irrigation. However, more and more water for irrigation is being drawn from fossil water that has been stored over the millennia in aquifers. The *Ogallala aquifer* **(FIGURE 2.7)** underlying the Great Plains is the largest in North America. In parts of the Ogallala, water is being pumped out at rates that exceed natural replenishment by 10 to 40 times.

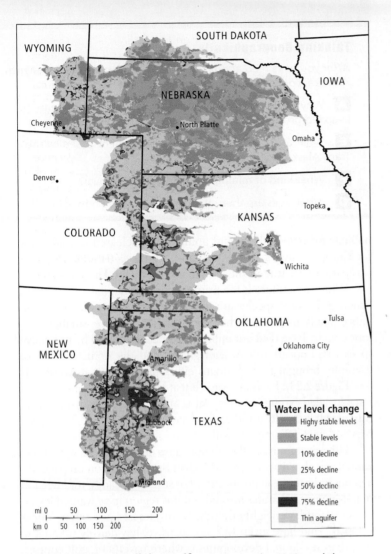

FIGURE 2.7 The Ogallala aquifer. Between the 1940s and the 1980s, the aquifer lost an average of 10 feet (3 meters) of water overall, and more than 100 feet (30 meters) of water in some parts of Texas. During the 1980s, though, there was a period of abundant rain and snow, which meant that water levels in the aquifer did not decline as much. However, in the Ogallala area, the climate fluctuates from moderately moist to very dry, and the dry periods are lengthening. A drought began in mid-1992 and has returned every few years, causing large agribusiness firms to pump Ogallala water to supplement scarce precipitation. Since 1992, water levels in the aquifer declined an average of 1.35 feet per year and now exceed replenishment rates many times over. [Sources consulted: *National Geographic* (March 1993): 84–85, with supplemental information from High Plains Underground Water Conservation District 1, Lubbock, Texas, at http://www.hpwd.com; Erin O'Brien, Biological and Agricultural Engineering, National Science Foundation Research Experience for Undergraduates, Kansas State University, 2001]

Fruit and vegetable crops in California are routinely irrigated with water from surrounding states. Such irrigation, involving expensive and massive engineering projects, accounts for some of the water that goes into the *virtual water footprint* of U.S. consumers, as discussed in Chapter 1 (see also Table 1.1). This water also supplies the cities of Southern California. Water is pumped from hundreds of miles away and over mountain ranges. California uses more energy to move water than some states use for all purposes. Moreover, irrigation in Southern California, especially with the water drawn from the Colorado River,

deprives Mexico of this much-needed resource. The mouth of the Colorado (which is in Mexico) was once navigable. Because of massive diversions, it is now often dry and sandy; only a mere trickle of water gets to Mexico.

Citizens in western North America are recognizing that the use of scarce water for irrigating agriculture, raising livestock, and keeping lawns and golf courses green in desert environments is unsustainable. Conflicts over transporting water from wet regions to dry ones, or from sparsely inhabited to urban areas, are ongoing and have halted some new water projects. However, government subsidies have kept water artificially cheap, and in the past, new water supplies have always been found and harnessed, creating little incentive to change.

Water depletion is an issue even in much wetter eastern North America, as urban populations expand beyond the water resources available to them. The Great Lakes (see the map in Figure 2.1) hold 21 percent of the world's surface fresh water, enough to submerge the entire United States to a depth of 9 feet. Not surprisingly, many cities in the area have long relied on the lakes for their water. As early as 1900, the city of Chicago, in order to clean up sewage it had been dumping into Lake Michigan, gained permission to reverse the flow of the Chicago River. The river now flows into the Mississippi, ultimately transferring Chicago's wastewater to the Gulf of Mexico. This diversion, which moves water at 3200 cubic feet per second from the Great Lakes, opened the doors to requests for large amounts of lake water from cities around and near the lakes, in both Canada and the United States, and even plans to divert huge amounts of lake water to arid western regions of the United States. Concern that the lake water could fall to critical levels, damaging fisheries and inhibiting navigation, led to regulation of water withdrawals. Chicago's diversion was limited by the U.S. Supreme Court in 1967, and in 1985 the Great Lakes Compact (formally known as the Great Lakes–St. Lawrence River Basin Sustainable Water Resources Agreement) was born, though it didn't become law until 2008. Since then, eight U.S. states and two Canadian provinces that border the lakes are legally bound to limit water withdrawals to cities located within the Great Lakes drainage basin, thus limiting the most extreme threats of depletion.

Another area in the eastern United States that is having conflicts about water is the city of Atlanta and the watersheds that it depends on, which stretch into Alabama and Florida. Despite being located in a part of North America with abundant rainfall, Atlanta has outgrown its water resources, prompting desperate efforts to acquire more water. In 2013, the Georgia legislature voted to redraw the boundary between Georgia and Tennessee, correcting an error in a 1818 survey, in order to gain access to the waters of the Tennessee River. Tennessee has so far refused to hand over any land to Georgia, so Atlanta is adopting stricter regulations on urban water use.

Water Pollution

In the United States, 40 percent of rivers are too polluted for fishing and swimming, and more than 90 percent of the *riparian areas* (the interface between land and flowing surface water) have been lost or degraded. Pollution in the rivers of North America comes mainly as storm-water runoff from agricultural areas, urban and suburban developments, and industrial sites.

In the 1970s, scientists studying coastal areas began noticing *dead zones* where water is so polluted that it supports almost no life. Dead zones occur near the mouths of major river systems that have been polluted by fertilizers and pesticides washed from farms and lawns when it rains. A large dead zone is in the Gulf of Mexico near the mouth of the Mississippi, and similar zones have been found in all U.S. coastal areas. Even Canada, where much lower population density means that rivers are generally cleaner, has dead zones on its western coast.

A recently discovered type of water pollution involves pharmaceuticals, such as antibiotics, that are excreted by humans and are not removed during water purification processes. These chemicals then make their way into rivers and lakes, where they enter the food system in drinking water or through fish.

ON THE BRIGHT SIDE: Green Living at a Regional and Global Scale

North Americans are trying to figure out what they can do in their daily lives to ameliorate looming environmental crises. Solutions, such as *greener living*—which involves recycling, driving less, growing a food garden, and improving home energy efficiency—can collectively make a big difference. One major trend is the use of renewable energy, especially solar and wind power, the use of which has grown by 50 percent and 33 percent respectively, per year over the last decade. According to a 2015 World Watch Institute report, renewable energy will account for about one-third of new energy added to the U.S. grid by 2018. [Source: "U.S. Renewable Energy Growth Accelerates," Ben Block, World Watch Institute, 2015, http://www.worldwatch.org/node/5855] ■

Water Marketization

North Americans are used to paying for water, but the cost has usually been just high enough to cover extraction, purification, and delivery in pipes. Threats of polluted drinking water are beginning to change the way water is viewed. For example, in the last few years the public, in response to abundant advertising, has been buying bottled water even when tap water is usually safe. The United States is one of the world's largest national markets for bottled water, which is now a more popular commercial beverage than soft drinks.

A number of North American communities with abundant fresh water have agreed to sell water to beverage companies, without anticipating the effects of massive water withdrawals from local aquifers or rivers. These can include geologic subsidence, loss of aquatic habitats, and the depletion and pollution of natural wells and springs. Some communities, such as Coachella Valley in Southern California, have found that while local wells and streams are drying up because of drought, water is being pumped from local aquifers and bottled at an unrestrained pace by a private company. In the Coachella instance, the Swiss company Nestlé is leasing wells from the Morongo Indian Reservation, which is not subject to California's water conservation regulations.

LOSS OF NATURAL HABITAT AND URBANIZATION

Before the European colonization of North America, which began soon after 1500, the environmental impact of humans in the region was relatively low. Though North America was by no means a pristine paradise when Europeans arrived, millions of acres of forests and grasslands that had served as

urban sprawl the encroachment of suburbs on agricultural land

Thinking Geographically

After you have read about the impacts on the biosphere in North America, you will be able to answer the following questions.

A Why might it make economic and environmental sense to keep old-growth trees, such as the one shown here, standing?

E What are some threats to the environment posed by the Trans-Alaska Pipeline?

F What kind of mining is mountaintop removal?

G What clues are there that this water may be toxic?

habitats for native plants and animals were cleared to make way for European-style farms, cities, and industries **(FIGURE 2.8)**. This was particularly true in the area that became the United States.

While agriculture is the primary driver of habitat loss in North America, invasive species are also a concern. As North American native plants and animals have been forced into ever-smaller territories, many have died out entirely and been replaced by nonnative species (European and African grasses and the domestic cat, for example) brought in by humans either purposely or inadvertently (see Figure 2.23). Estimates vary, but at least 4000 nonnative species have invaded North America. An example is the Asian snakehead fish, which is rapidly invading the Potomac River, where it eats baby bass and fiercely competes with native fish for food.

Urbanization is another major cause of habitat loss, with **urban sprawl** (see the "Farmland and Urban Sprawl" section on page 98) a concern in many communities. For several decades, middle- and upper-income urbanites throughout this region have wanted lower-density suburban neighborhoods. Farms, forests, and other "undeveloped" land have given way to expansive, low-density urban and suburban residential developments where pavement, golf courses, office complexes, and shopping centers cover the landscape. In the process, natural habitats are being degraded even more intensely than they were by farming. The loss of farmland and natural habitat in the urban fringe affects recreational land and the ability to produce local, affordable food for urban populations.

OIL EXTRACTION

In many coastal and interior areas of North America, oil extraction is a large and potentially environmentally devastating industry. This often-overlooked problem was made clear in the spring and summer of 2010, when U.S. waters in the Gulf of Mexico became the site of the largest accidental marine oil spill in world history. In April of 2010, an explosion aboard the Deepwater Horizon, an offshore oil-drilling rig leased by British Petroleum (BP), caused it to sink. One result was a massive leak of oil from the rig's wellhead. Due to its location more than a mile beneath the sea surface, the wellhead could not be capped for almost 4 months, during which time it spewed out at least 200 million gallons of oil. While some of this oil made its way to the surface—damaging shorelines in all the Gulf Coast states—most of the oil remains beneath the surface because of BP's use of chemical dispersants. Studies done in subsequent years have shown that both the dispersants and the oil treated by the dispersants were more toxic to fish than untreated oil. One study suggests that the oil and dispersants

FIGURE 2.8 PHOTO ESSAY: Human Impacts on the Biosphere in North America

While parts of North America are relatively unaffected by humans, much of the region has had low-to-medium impacts from people, and the parts where most people live are very impacted.

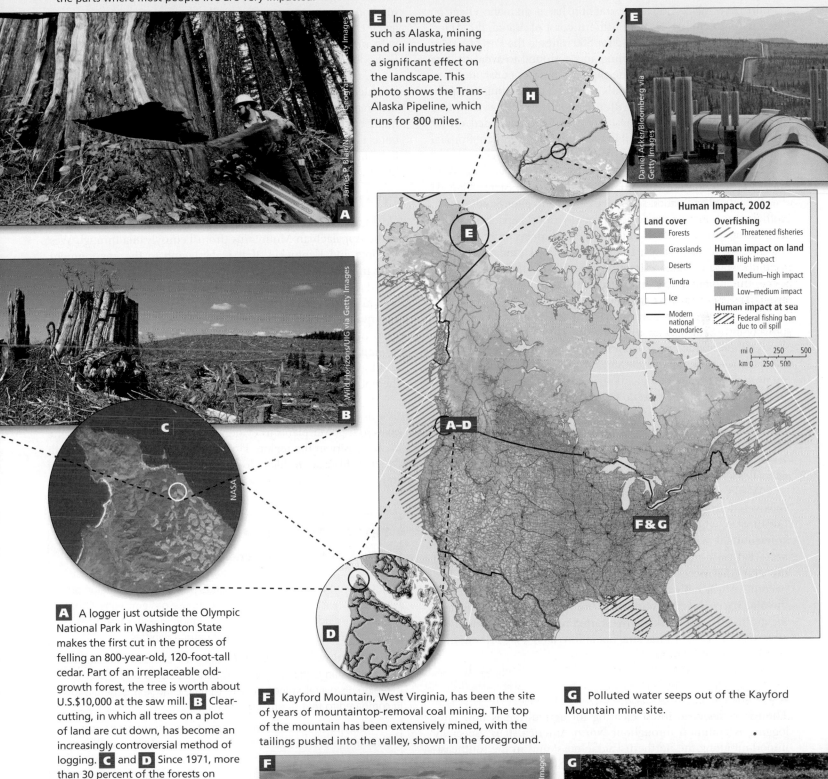

E In remote areas such as Alaska, mining and oil industries have a significant effect on the landscape. This photo shows the Trans-Alaska Pipeline, which runs for 800 miles.

Human Impact, 2002

Land cover
- Forests
- Grasslands
- Deserts
- Tundra
- Ice
- Modern national boundaries

Overfishing
- Threatened fisheries

Human impact on land
- High impact
- Medium–high impact
- Low–medium impact

Human impact at sea
- Federal fishing ban due to oil spill

mi 0   250   500
km 0   250   500

A A logger just outside the Olympic National Park in Washington State makes the first cut in the process of felling an 800-year-old, 120-foot-tall cedar. Part of an irreplaceable old-growth forest, the tree is worth about U.S.$10,000 at the saw mill. **B** Clear-cutting, in which all trees on a plot of land are cut down, has become an increasingly controversial method of logging. **C** and **D** Since 1971, more than 30 percent of the forests on Washington State's Olympic Peninsula have been clear-cut. The clear-cutting of old-growth forests that have never been cut is a practice that has been declining in recent years, but it still continues in some places.

F Kayford Mountain, West Virginia, has been the site of years of mountaintop-removal coal mining. The top of the mountain has been extensively mined, with the tailings pushed into the valley, shown in the foreground.

G Polluted water seeps out of the Kayford Mountain mine site.

73

now covering the sea floor are not being degraded by natural processes and could remain in the food chain for decades, polluting Gulf fish, shellfish, and shrimp catches.

In other places too, oil extraction has a dramatic effect on the environment. Along the northern coast of Alaska, the Trans-Alaska Pipeline runs southward for 800 miles to the Port of Valdez (see Figure 2.8E). Often running above ground to avoid shifting as Earth freezes and thaws, the pipeline poses a constant risk of rupture, which could potentially result in devastating oil spills. The pipeline also interferes with migrations of caribou and other animals that Alaska's indigenous people have depended on for food in the past. Protests about threats to the environment from oil extraction in Alaska tend to be quieted by the yearly rebate of several thousand dollars from oil revenues received by each Alaskan.

With one-tenth the population of the United States, Canada has the largest proven oil reserves in the world after Saudi Arabia, sufficient to meet its needs plus provide export capacity. It is the largest foreign supplier of oil to the United States, but more than 90 percent of this oil is in hard-to-access oil sands in western Canada. The costs of extracting this oil are high in terms of environmental impacts, energy expended, and water resources used. Transport of the extracted oil across the North American continent through the Keystone Pipeline system (part of which is already in use) also poses environmental threats.

While Canada may have more proven oil reserves, the United States is now the largest producer of oil and gas, thanks largely to new drilling technologies known as **fracking** that use high-pressure injections of water to fracture rock formations, releasing natural gas and oil that the formations contain. First developed in the 1940s, fracking became widely used only in the past 10 years, after some important drilling innovations were made. Since then, large portions of the central United States have been opened up to fracking, especially in North Dakota, Texas, and Oklahoma.

While fracking has produced large amounts of fuel at a relatively low cost, it is controversial because of its environmental impacts, which include much higher GHGs emissions than conventional drilling, significant local air pollution, water pollution, and earthquakes. In 2015, scientific studies concluded that the dramatic increase in earthquakes in Oklahoma, which had become a near daily occurrence in some areas, was triggered by the practice of injecting wastewater produced during fracking into the ground. In 2016, Oklahoma started regulating the amount of wastewater injected by fracking operations after a lawsuit was filed by environmental groups.

LOGGING

Though widespread forest clearing for agriculture is now rare, logging is common throughout North America. It is especially important along the northern Pacific coast and in the southeastern United States. Logging in these areas provides most of the construction lumber and much of the paper used in Canada, the United States, and in parts of Asia.

Although the logging industry provides jobs and an exportable commodity, it has been depleting the continent's forests. Environmentalists have focused on the damage created by the logging industry, especially via **clear-cutting**, the main logging method used throughout North America. In this method, all trees on a given plot of land are cut down, regardless of age, health, or species (see Figure 2.8A–D). Clear-cutting destroys wild animal and plant habitats, thereby reducing species diversity. It also leaves forest soils uncovered and very susceptible to erosion.

Concerns about the environmental impacts of logging are heightened because of the dominance of service-sector jobs in the major logging states and provinces. For example, even in many remote areas of the Pacific Northwest where logging was once the backbone of the economy, residents now depend on tourism and other occupations that rely on the beauty of intact forest ecosystems.

COAL MINING

Coal mining is a major industry, especially in Wyoming and parts of the Appalachian Mountains (from Pennsylvania through West Virginia and Kentucky), that is damaging to the environment. Strip mining, in which vast quantities of soil and rock are removed in order to extract underlying coal, can result in visual wastelands and in huge piles of mining waste called *tailings* that pollute waterways and threaten communities that depend on well water. Particularly damaging is a form of strip mining known as *mountaintop removal*, in which the whole top of a mountain is leveled and the tailings are pushed into surrounding valleys, resulting in the pollution of entire watersheds (see Figure 2.8F, G).

In recent years, many have pointed to lower coal consumption in North America and the cancellation of plans to build new coal-fired power plants as evidence of the general decline of the coal industry in the region. However, exports of coal from North America to Asia are likely to keep the industry afloat, and most mines open, for the foreseeable future.

---

fracking a drilling technology that uses high pressure injection of water to fracture rock formations, releasing natural gas and oil that they contain

clear-cutting a method of logging that involves cutting down all trees on a given plot of land regardless of age, health, or species

---

THINGS TO REMEMBER

GEOGRAPHIC THEME 1 • **Environment:** North America's use of resources has enormous environmental impacts. Home to only 5 percent of the world's population, the region produces 17 percent of humanity's greenhouse gas emissions. Other impacts include the depletion and pollution of water resources and fisheries and the destruction of natural habitats.

• While many parts of North America are exposed to multiple climate-change impacts, North America is also a very resilient region. Its overall vulnerability to climate change is generally low compared to that of other world regions.

• Water pollution in North America is a major problem, especially in the United States, where 40 percent of the rivers are too polluted for fishing or swimming.

• Though widespread forest clearing for agriculture is now rare, logging is common throughout North America and remains especially important along the northern Pacific coast and in the southeastern United States.

• In many remote interior areas of North America, coal mining is a major industry that is damaging to the environment.

HUMAN PATTERNS OVER TIME

In prehistoric times, humans came from Eurasia via Alaska, dispersing to the south and east. Beginning in the 1600s, waves of European immigrants and enslaved Africans spread over the continent, primarily from east to west. Today, immigrants are coming mostly from Asia and Middle and South America, arriving mainly in the Southwest and West, where immigrant populations are at their most concentrated. In addition, internal migration is still a defining characteristic of life for most North Americans, who are among the world's most mobile people. The average North American moves nearly 12 times in a lifetime.

THE PEOPLING OF NORTH AMERICA

Recent evidence suggests that humans first came to North America from northeastern Asia at least 25,000 years ago and perhaps earlier, most arriving during an ice age. At that time, the global climate was cooler, polar ice caps were thicker, and sea levels were lower. The Bering land bridge, a huge, low landmass more than 1000 miles (1600 kilometers) wide, connected Siberia to Alaska. Bands of hunters crossed by foot or small boats into Alaska and traveled down the west coast of North America.

The Original Settling of North America

By 15,000 years ago, humans had reached nearly to the tip of South America and had moved deep into that continent. By 10,000 years ago, global temperatures began to rise. As the ice caps melted, sea levels rose and the Bering land bridge was submerged beneath the sea.

Over thousands of years, the people settling in the Americas domesticated plants, created paths and roads, cleared forests, built permanent shelters, and sometimes created elaborate social systems. About 3000 years ago, corn was introduced from Mexico (into what is now the southwestern U.S. desert), as were other Mexican domesticated crops, particularly squash and beans. Such food crops are thought to have been closely linked to settled life and to North America's prehistoric population growth.

These foods provided surpluses that allowed some community members to engage in trade and other specialized activities other than agriculture, hunting, and gathering, making possible large, city-like regional settlements. For example, by 1000 years ago, the urban settlement of Cahokia, including suburban settlements (in what is now central Illinois, across the Mississippi from St. Louis), covered 5 square miles (12 square kilometers) and was home to an estimated 30,000 people (FIGURE 2.9A). Here people specialized in crafts, trade, and other activities beyond the production of basic necessities.

The Arrival of the Europeans

North America was completely transformed by the sweeping occupation of the continent by Europeans. In the sixteenth century, French, Italian, Portuguese, and English explorers came ashore along the Eastern Seaboard of North America, and the Spanish explorer Hernando De Soto made his way from Florida deep into the heartland of the continent in the 1540s. In the early seventeenth century, the British established colonies along the Atlantic coast in what is now Virginia (1607) and Massachusetts (1620). The Dutch explored the Atlantic Seaboard looking for trading opportunities, founding the city of New Amsterdam (now New York) at the mouth of the Hudson River and Beverwick (now Albany) further upstream. The French explored the northern interior of the continent, founding Québec City at the mouth of the St. Lawrence River and Montréal further upstream. Assisted by enslaved Africans, colonists and settlers from northern Europe built villages, towns, port cities, and plantations along the eastern coast over the next two centuries. By the mid-1800s, they had occupied most lands of Native American and Aboriginal peoples into the central part of the continent.

Disease, Technology, and Native Americans

The rapid expansion of European settlement was facilitated by the vulnerability of Native American and Aboriginal populations to European diseases. Having been isolated from the rest of the world for many years, Native American and Aboriginal populations had no immunity to diseases such as measles and smallpox. Transmitted by Europeans and Africans who had built up immunity to them, these diseases killed up to 90 percent of Native American and Aboriginal populations within the first 100 years of contact. It is now thought that diseases spread by early expeditions, such as De Soto's into Florida, Georgia, Tennessee, and Arkansas, so decimated populations in the North American interior that fields and villages were abandoned, the forest grew back, and later explorers erroneously assumed the land had never been occupied.

Technologically advanced European weapons, trained dogs, and horses also took a large toll on Native American and Aboriginal peoples, who often had only bows and arrows. Some Native Americans in the Southwest acquired horses and guns from the Spanish and learned to use them in warfare against the Europeans, but their other technologies could not compete. Numbers reveal the devastating effect of European settlement on Native American and Aboriginal populations: There were roughly 18 million Native Americans and Aboriginal peoples in North America in 1492. By 1542, after just a few Spanish expeditions, there were only half that number. By 1907, slightly more than 400,000, or a mere 2 percent of the original population, remained.

THE EUROPEAN TRANSFORMATION

European settlement erased many of the landscapes familiar to Native American and Aboriginal peoples and imposed new ones that fit the varied physical and cultural desires of the new occupants.

The Southern Settlements

European settlement of eastern North America began with the Spanish in Florida in the mid-1500s and the establishment of the British colony of Jamestown in Virginia in 1607. By the late 1600s, large plantations in the colonies of Virginia, the Carolinas, and Georgia were cultivating crops such as tobacco, rice, and cotton, which became valuable exports.

To secure a large, stable labor force, Europeans brought enslaved Africans into North America beginning in 1619. Within 50 years, enslaved Africans were the dominant labor force on some of the larger Southern plantations (see Figure 2.9B). By the start of the U.S. Civil War in 1861, enslaved people made up about one-third of the population in the Southern states and were often a majority in the plantation regions. Working and living conditions on these plantations were often brutal and hazardous, with

A Cahokia, Illinois, a community of 30,000 that lasted from 700 to 1400 C.E.

B Enslaved people and workers in North America pick leaves and operate machines at a tobacco factory in 1750.

C A log raft being floated down Oregon's Columbia River in 1902.

10,000 B.C.E.	0 C.E.	700 C.E.		1400 C.E.	1700 C.E.

25,000–10,000 B.C.E.
Bering land bridge

700–1400 C.E.
Cahokia flourishes

1492
Arrival of Europeans

1600–1900
Plantations in Southern colonies; industries in New England

NOTE: Timeline range is not to scale

FIGURE 2.9 VISUAL HISTORY OF NORTH AMERICA

Thinking Geographically

After you have read about the human history of North America, you will be able to answer the following questions.

A How did food crops such as corn, beans, and squash influence the development of settlements like Cahokia?

B Why did the plantation system inhibit the establishment of roads, communication networks, and small enterprises that could have boosted other forms of economic activity in the Southern colonies?

infrastructure road, rail, communication networks, and other facilities necessary for economic activity

execution, beatings, and rape regular occurrences. Enslaved people were usually denied formal education and gatherings of any kind were often banned in order to make slave rebellions more difficult to organize. Even so, there were more than 250 documented slave rebellions in North America. The largest concentrations of African Americans in North America are still in the southeastern states (**FIGURE 2.10**).

The plantation system consolidated wealth into the hands of a small class of landowners who made up just 12 percent of Southerners in 1860. These elite members of the planter class kept taxes low and invested money from their exported crops in Europe or the more prosperous Northern colonies instead of in **infrastructure** at home. As a result, the road, rail, communication networks, and other facilities necessary for further economic growth in the South were rarely built.

More than half of Southerners were poor white farmers. Both they and the enslaved population lived simply and their meager consumption did not provide much demand for goods. Because of this, there were few market towns and almost no industries. Plantations tended to import from Europe and the northern United States whatever they couldn't produce for themselves. The result was much slower economic growth in the South relative to areas in the North, where a more diverse range of agriculture and industries supported a larger service economy of small shops, garment making, restaurants and bars, and transportation services.

The inability of the federal government to reconcile the different needs of the agricultural export–oriented Southern economy and the industrializing Northern economies was one of the causes of the Civil War (1861–1865) in the United States, along with the abolition movement to free enslaved Africans and their descendants. After the war, while the victorious North returned to promoting its own industrial development, the plantation economy declined, and the South sank deeply into poverty. The South remained economically and socially underdeveloped well into the 1970s, and even today it remains the poorest subregion of North America.

The Northern Settlements

Throughout the seventeenth century, relatively poor subsistence farming communities dominated the colonies of New England and southeastern Canada. There were no plantations and few enslaved people, and not many cash crops were exported. What exports there were consisted of raw materials like timber, animal pelts, and fish from the Grand Banks off Newfoundland and the coast of Maine. Generally, farmers lived in interdependent communities that prized education, ingenuity, self-sufficiency, and thrift.

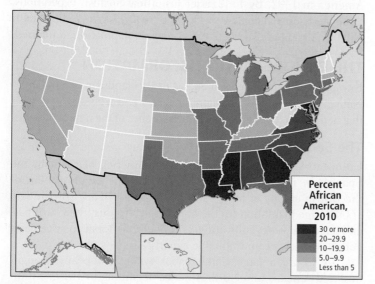

FIGURE 2.10 Percent of African American population in each state, 2010. The percent refers only to those persons who selected "Black, African Am., Negro" as their only race in the 2010 census. It does not include those who selected more than one race that included black. [Source consulted: http://www.census.gov/quickfacts/map/RHI225214/00]

Percent African American, 2010
- 30 or more
- 20–29.9
- 10–19.9
- 5.0–9.9
- Less than 5

D Steel workers in Pittsburgh, Pennsylvania, in 1905.

E Oil wells encroach on agricultural land in Long Beach, California, in 1923.

F A farmer and his children walk through a dust storm in Oklahoma in 1936.

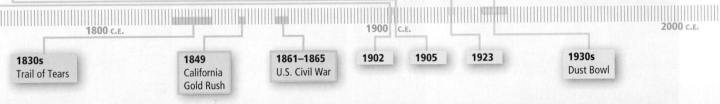

1800 C.E.   1900 C.E.   2000 C.E.

1830s	1849	1861–1865	1902	1905	1923	1930s
Trail of Tears	California Gold Rush	U.S. Civil War				Dust Bowl

C What economic activities now dominate the coastal Pacific Northwest?

D How did the steel industry stimulate development in the "economic core" of North America?

E What activities have become economically important in California other than agriculture and the oil industry?

F What were the causes of the Dust Bowl?

By the late 1600s, New England was implementing ideas and technology from Europe that led to the first industries. By the 1700s, diverse industries were supplying markets in North America and the Caribbean with metal products, pottery, glass, and textiles. By the early 1800s, southern New England, especially the region around Boston, had become the center of manufacturing in North America. It drew largely on young male and female immigrant labor from French Canada and Europe.

> ### THINGS TO REMEMBER
>
> • Recent evidence suggests that humans first came to North America from northeastern Asia at least 25,000 years ago and perhaps earlier, most arriving during an ice age.
>
> • To secure a large, stable labor force, Europeans brought enslaved Africans into North America beginning in 1619.
>
> • By the late 1600s, New England was implementing ideas and technology from Europe and developing some of the first industries in North America.

The Mid-Atlantic Economic Core

The colonies of New York, New Jersey, Pennsylvania, and Maryland eventually surpassed New England and southeastern Canada in population and in wealth. This mid-Atlantic region benefited from more fertile soils, a slightly warmer climate, multiple deep water harbors, and better access to the resources of the interior. By the end of the Revolutionary War in 1783, the mid-Atlantic region was on its way to becoming the **economic core,** or the dominant economic region, of North America. Port cities such as New York, Philadelphia, and Baltimore prospered as the intermediaries for trade between Europe and the vast American continental interior.

In the early nineteenth century, both agriculture and manufacturing grew and diversified, drawing immigrants from much of northwestern Europe. As farmers became more successful, they bought more and more mechanized equipment, appliances, and consumer goods made in nearby cities. By the mid-nineteenth century, the economy of the core was increasingly based on the steel industry, which spread westward to Pittsburgh and the Great Lakes industrial cities of Cleveland, Detroit, and Chicago. The steel industry relied on the mining of coal and iron ore deposits throughout the region and beyond. Steel became the basis for mechanization, and the region was soon producing heavy farm and railroad equipment (see Figure 2.9D).

By the early twentieth century, the economic core stretched from the Atlantic to St. Louis on the Mississippi (including many small industrial cities along the river, plus Chicago and Milwaukee), and from Ottawa to Washington, DC. It dominated North America economically and politically well into the middle of the twentieth century. Most other areas produced food and raw materials for the core's markets and depended on the core's factories for manufactured goods.

EXPANSION WEST OF THE MISSISSIPPI AND GREAT LAKES

The east-to-west trend of settlement continued as land in the densely settled eastern parts of the continent became too expensive for new immigrants. By the 1840s, immigrant farmers from central and northern Europe, as well as European descendants born in eastern North America, were pushing their way beyond the Great Lakes and across the Mississippi River, north and west into the Great Plains of Canada and the United States **(FIGURE 2.11)**.

The Great Plains

Much of the land west of the Great Lakes and the Mississippi River was dry grassland or prairie. The soil usually proved very productive in wet years, and the area became known as North America's *breadbasket.* But the naturally arid character

economic core the dominant economic region within a larger region

FIGURE 2.11 Nineteenth-century transportation. [Sources consulted: James A. Henretta, W. Elliot Brownlee, David Brody, and Susan Ware, *America's History*, 2nd ed. (New York: Worth, 1993), pp. 400–401; James L. Roark, Michael P. Johnson, Patricia Cline Cohen, Sarah Stage, Alan Lawson, and Susan M. Hartmann, *The American Promise: A History of the United States*, 3rd ed. (Boston: Bedford/St. Martin's, 2005), p. 601]

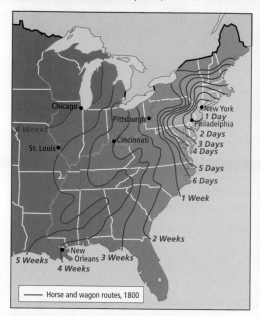

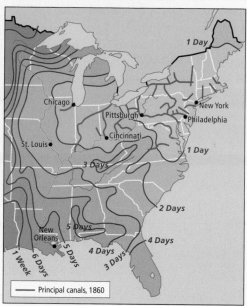

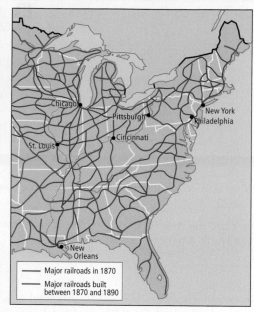

(A) Travel times from New York City, 1800. It took a day to travel by wagon from New York City to Philadelphia and a week to go to Pittsburgh.

(B) Travel times from New York City, 1860. The travel time from New York to Philadelphia was now only 2 or 3 hours and to Pittsburgh less than a day because people could go part of the way via canals (dark blue). Using the canals, the Great Lakes, and rivers, they could easily reach principal cities along the Mississippi River, and travel was less expensive and onerous than it had been.

(C) Railroad expansion by 1890. With the building of railroads, which began in the decade before the Civil War, the mobility of people and goods increased dramatically. By 1890, railroads crossed the continent, though the network was densest in the eastern half.

of this land eventually created an ecological disaster for Great Plains farmers. In the 1930s, after 10 especially dry years, a series of devastating dust storms blew away topsoil by the ton. This hardship was made worse by the widespread economic depression of the 1930s. Many Great Plains farm families packed up what they could and left what became known as the Dust Bowl (see Figure 2.9F), heading west to California and other states on the Pacific Coast.

The Mountain West and Pacific Coast

Some immigrants and settlers from east of the Mississippi, alerted to the possibilities farther west, skipped over the Great Plains entirely. By the 1840s, they were coming to the valleys of the Rocky Mountains, to the Great Basin, and to the well-watered and fertile coastal zones of what was then known as the Oregon Territory and California. In 1848, only one year after California was forcibly annexed from Mexico to the United States, news of the discovery of gold in California drew thousands of people with the prospect of getting rich quickly. The vast majority of gold seekers were unsuccessful, however, and by 1852 they had to look for employment elsewhere. Farther north, logging eventually became a major industry (see Figure 2.9C).

The extension of railroads across the continent in the nineteenth century facilitated the transportation of manufactured goods to the West as well as raw materials and eventually fresh produce to the East. Today, the coastal areas of this region, often called the Pacific Northwest, have thriving, diverse, high-tech economies and flourishing populations. Partially in response to the long history of

resource extraction in the Pacific Northwest, residents are on the forefront of so many efforts to reduce human impacts on the environment that the region has been nicknamed "Ecotopia."

The Southwest

People from the Spanish colony of Mexico first colonized the Southwest in the late 1500s. Their settlements were sparse. As immigrants from the United States moved into the region, drawn by the cattle-raising industry, Mexico found it progressively more difficult to maintain control, and by 1850, nearly the entire Southwest was under U.S. control.

By the twentieth century, a vibrant agricultural economy had developed in central and southern California, supported by massive government-sponsored water-movement and irrigation projects. The mild Mediterranean climate made it possible to grow vegetables almost year-round. With the advent of refrigerated railroad cars, fresh California vegetables could be sent to the major population centers of the East. Southern California's economy rapidly diversified to include oil (see Figure 2.9E), entertainment, and a variety of engineering- and technology-based industries.

EUROPEAN SETTLEMENT AND AMERICAN INDIANS AND ABORIGINAL PEOPLES

As settlement relentlessly expanded west, Native American and Aboriginal peoples who had survived early encounters with

Europeans and were living in the eastern part of the continent occupied land that European newcomers wished to use. During the 1800s, almost all the surviving Native American and Aboriginal peoples were killed in innumerable skirmishes with European newcomers, absorbed into the societies of the Europeans (through intermarriage and acculturation), or forcibly relocated west to relatively small reservations with few resources. The largest relocation, in the 1830s, involved the Choctaw, Seminole, Creek, Chickasaw, and Cherokee of the southeastern states. These people had already adopted many European methods of farming, building, education, government, and religion. Nevertheless, they were rounded up by the U.S. Army and marched to Oklahoma along a route that became known as the Trail of Tears because of the more than 4000 Native Americans who died along the way.

As Europeans occupied the Great Plains and prairies, many of the reservations were further shrunk or relocated onto even less desirable land. Today, reservations cover just over 2 percent of the land area of the United States.

In Canada the picture is somewhat different. Reservations now cover 20 percent of Canada, mostly because of the creation of the Nunavut Territory (now known simply as Nunavut) in 1999 in the far north and the ceding of Northwest Territory land to the Tłı̨chǫ Aboriginal people (also known as the Dogrib) in 2003 (see the Figure 2.1 map). These Canadian Aboriginal peoples stand out as having won the right to legal control of their lands. In contrast to the United States, it had been unusual for native groups in Canada to have legal control of their territories.

After centuries of mistreatment, many Native American and Aboriginal peoples still live in poverty and, as in all communities under severe stress, rates of alcohol and drug addiction and violence are high. Until the 1980s many children from Native American and Aboriginal families were forced to attend boarding schools where their languages were made illegal, their cultures denigrated, and where death rates from disease were far above average. However, in recent decades some tribes have developed more affluence on the reservations and territories by establishing manufacturing industries; extracting fossil fuel, uranium, and other mineral deposits under their lands; or opening casinos. One measure of this economic resurgence is population growth. Expanding from a low of 400,000 in 1907, the Native American population, at less than 2 percent of the total population, stood at approximately 6 million in 2010 in the United States. In Canada, Aboriginal peoples number over 1.2 million, or 3.8 percent of the population in 2011.

THE CHANGING REGIONAL COMPOSITION OF NORTH AMERICA

The regions of European-led settlement still remain in North America, but they are now less distinctive. The economic core region is less dominant in industry, which has spread to other parts of the continent. Some regions that were once dependent on agriculture, logging, or mineral extraction now have high-tech industries as well. The West Coast, in particular, has blossomed with a high-tech economy and a rapidly expanding population that includes many immigrants from Asia and Middle and South America. The West Coast also benefits from North America's trade with Asia, which now surpasses trade with Europe in volume and value.

THINGS TO REMEMBER

• By the early twentieth century, North America's economic core was well established. Stretching from the Atlantic Ocean to St. Louis, Chicago, and Milwaukee, and from Ottawa to Washington, DC, it dominated North America economically and politically well into the middle of the twentieth century.

• The push to settle the Great Plains, the Mountain West, and the Pacific Coast attracted many immigrants, from both Mexico and Europe, interested in farming, mining, and cattle raising.

• In 1492, roughly 18 million Native Americans and Aboriginal peoples lived in North America. By 1542, after only a few Spanish expeditions, there were half that many. By 1907, only about 2 percent of the original population remained; however, by 2010, the Native American and Aboriginal populations had partially rebounded to approximately 7 million.

GLOBALIZATION AND DEVELOPMENT

GEOGRAPHIC THEME 2

Globalization and Development: Globalization has transformed North America, reorienting it toward knowledge-intensive jobs that require education and training. Income inequality has risen as most manufacturing jobs have been moved to countries with cheaper labor or have been replaced by technology. North America's size and wealth have made it the center of the global economy, and its demand for imported goods and its export of manufacturing jobs make it a major engine of globalization.

Like their political systems, the economic systems of Canada and the United States have much in common. Both countries evolved from societies based mainly on family farms. Both then had an era of industrialization followed by a transformation to a primarily service-based economy, and both have important technology sectors and an economic influence that reaches worldwide.

THE DECLINE IN MANUFACTURING EMPLOYMENT

By the 1960s, the geography of manufacturing in North America was changing as corporations tried to boost profits with cheaper labor and more reliance on technology. In the Old Economic Core, higher pay and benefits and better working conditions won by labor unions led to increased production costs. This lowered the high profits demanded by the owners and shareholders of manufacturing corporations. A number of companies began moving their factories to the southeastern United States, where wages were lower and corporate profits higher because of the absence of labor unions. Half a century into this process of deindustrialization, many cities of the Old Economic Core have large abandoned industrial districts.

Since the late 1980s, the United States and Canada have entered into free trade agreements (see Chapter 1) that have enabled many manufacturing industries (such as clothing, electronic assembly, and auto parts manufacturing), to move south to Mexico or overseas to China and other places where labor is significantly cheaper than it is in North America. Companies have also saved on production costs because laws that in the United States and Canada regulate environmental protection and safe and healthy workplaces are absent or less strictly enforced in those other countries.

Technology and automation have also contributed to the decline of manufacturing employment. The steel industry provides an illustration. In 1980, huge steel plants, most of them in the economic core, employed more than 500,000 workers. At that time, it took about 10 person-hours and cost about U.S.$1000 to produce 1 ton of steel. Spurred by more efficient foreign competitors, the North American steel industry applied new technology to lower production costs, improve efficiency, and increase production. By 2014, steel was being produced at the rate of 1.9 person-hours per ton and at a cost of about U.S.$200 per ton. The reorganized steel industry in the United States now produced much of the steel in small, highly efficient mini-mills distributed throughout the United States. In total, the steel industry now employs fewer than half the workers it did in 1980. Throughout North America, this trend towards higher efficiency has resulted in fewer people producing more of a given product at a far lower cost than was the case 40 years ago. Remarkably, even as employment in manufacturing has declined over the last three decades, the actual amount of industrial production has steadily increased.

GROWTH OF THE SERVICE SECTOR

The economic base of North America is now a broad *tertiary sector* in which people are engaged in various services such as transportation, utilities, wholesale and retail trade, health, leisure, maintenance, finance, government, information, and education.

As of 2016, in both Canada and the United States about 80 percent of jobs and a similar percentage of the GNI were in the tertiary, or service, sector. There are high-paying jobs in all the service categories, but low-paying jobs are far more common. The largest private employer in the United States is the discount retail chain Walmart (1.4 million employees), where the average wage is U.S.$12 an hour, or U.S.$24,000 a year, full time. This is just barely above the poverty level for a family of four in the United States. Moreover, roughly half of Walmart employees are part time, earning roughly U.S.$10 an hour on average and receiving no company benefits or health care. Because of low wages, many Walmart workers are forced to turn to government programs to meet their basic needs, costing taxpayers as much as U.S.$6 billion a year. Walmart creates mostly retail jobs because, for the most part, its wares are manufactured abroad.

The Knowledge Economy

An important subcategory of the service sector involves the creation, processing, and communication of information—what is often called the *knowledge economy*, or the *quaternary sector*. The knowledge economy includes workers who manage information, such as those employed in finance, journalism, higher education, research and development, and many aspects of health care. It also includes the *information technology* or *IT* sector, which deals with computer software and hardware and the management of digital data.

digital divide the discrepancy in access to information technology between small, rural, and poor areas and large, wealthy cities that contain major governmental research laboratories and universities

Industries that rely on the use of computers and the Internet to process and transport information are freer to locate where they wish than were the manufacturing industries of the Old Economic Core, which needed locally available material resources such as steel and coal. These newer industries are more dependent on skilled managers, communicators, thinkers, and technicians, and are often located near major universities and research institutions.

Crucial to the knowledge economy is the Internet, which was first widely available in North America and has emerged as an economic force more rapidly there than in any other region in the world. With only 5 percent of the world's population, North America accounted for 9.5 percent of the world's Internet users in 2015. Roughly 87 percent of the population of both the United States and Canada use the Internet, compared to 73 percent of the European Union and 39 percent of the world as a whole. The total economic impact of the Internet in North America is hard to assess, but retail Internet sales increase every year. Indeed, although overall purchases were down from 2008 to 2011, online purchases in the United States and Canada have steadily increased since.

The Internet has entered many aspects of life in North America. In the political sphere, social networking (Facebook, Twitter, YouTube, and others) plays a large role in recruiting volunteers and eliciting cash contributions, especially in recent U.S. election cycles. Since 2009, social networking through Facebook, Twitter, and YouTube has become an integral part of life in North America.

The growth of Internet-based activity makes access to the Internet crucial in North America. Unfortunately, a **digital divide**—a discrepancy in access to information technology between small, rural, and poor areas and large, wealthy cities—has developed, because about a quarter of the North American population is not yet able to afford computers and home Internet connections.

GLOBALIZATION AND FREE TRADE

The United States and Canada are major engines of globalization that impact the world through the size and technological sophistication of their economies. Their combined economy is almost as large as that of the entire European Union. North America's advantageous position in the global economy is also a reflection of its geopolitical influence—its ability to mold the pro-globalization free trade policies that suit the major corporations and the governments of North America.

Free trade has not always been emphasized the way it is now. Before North America's rise to prosperity and global dominance, trade barriers were important aids to the region's development. For example, when it became independent of Britain in 1776, the new U.S. government imposed tariffs and quotas on imports and gave subsidies to domestic producers. This protected fledgling domestic industries and commercial agriculture, allowing its economic core region to flourish.

Now, because both are wealthy and globally competitive exporters, Canada and the United States see tariffs and quotas in other countries as obstacles to North America's economic expansion abroad. Thus, they usually advocate heavily for trade barriers to be reduced worldwide. Critics of these free trade policies point out a number of inconsistencies in the current North American position on free trade. First, North America once needed tariffs and quotas to protect its firms, much like many currently poorer countries still need. Furthermore, contrary to their own free trade precepts, both the United States and Canada still give significant subsidies to their farmers. These subsidies make it possible for North American farmers to sell their crops on the world market at such low prices that farmers elsewhere are hurt or even driven out of business (see the vignette on page 64). For example, many Mexican farmers have lost their small farms because of competition from large U.S. corporate farms, which receive subsidies from the U.S. government. The critics add that beyond agriculture, in North America, the benefits of free trade go mostly to large manufacturers and businesses and their managers, while many workers end up losing their jobs to cheaper labor overseas, or see their incomes stagnate.

Trade between the United States and Canada has been relatively unrestricted for many years. The process of reducing trade barriers began formally with the Canada–U.S. Free Trade Agreement of 1989. Mexico was included when the **North American Free Trade Agreement (NAFTA)** was created in 1994. The major long-term goal of NAFTA is to increase the amount of trade between Canada, the United States, and Mexico. Today, it is the world's largest trading bloc in terms of the GDP of its member states. NAFTA is only 1 of 14 free trade agreements that the United States and Canada have with countries around the world. Both Canada and the United States are members of the WTO (see Chapter 1).

The impacts of free trade agreements are hard to assess because it is difficult to tell whether the many observable changes in the North American economy that have occurred over the past three decades, when most free trade agreements were entered into, have been caused by the agreements or by other changes in regional and global economies. However, a few things are clear. Trade has increased and many companies are making higher profits because they now have larger markets. Since 1990, exports among the NAFTA countries have increased in value by more than 300 percent. By value, NAFTA's exports to the world economy have increased by about 300 percent for the United States and Canada and by 600 percent for Mexico. Some U.S. companies, such as Walmart, expanded aggressively into Mexico after NAFTA was passed. Mexico now has more Walmarts (2306 retail stores) than any country except the United States, which has 5249 **(FIGURE 2.12)**. Canada has 395 Walmarts.

Free trade seems to have worsened the tendency of the United States to spend more money on imports than it earns from exports. This imbalance is called a **trade deficit.** Before NAFTA, the United States usually had much smaller trade deficits with Mexico and Canada. After the agreement was signed, these deficits rose dramatically, especially with Mexico. For example, between 1994 and 2015, the value of U.S. exports to Mexico increased from roughly U.S.$50 billion to $238 billion, while the value of imports increased from U.S.$49 billion to $295 billion. This amounts to a shift from a slight trade surplus in 1994 of U.S.$1 billion to a deficit in 2015 of $57 billion.

North American Free Trade Agreement (NAFTA) a free trade agreement made in 1994 that added Mexico to the 1989 economic arrangement between the United States and Canada

trade deficit the extent to which the money earned by exports is exceeded by the money spent on imports

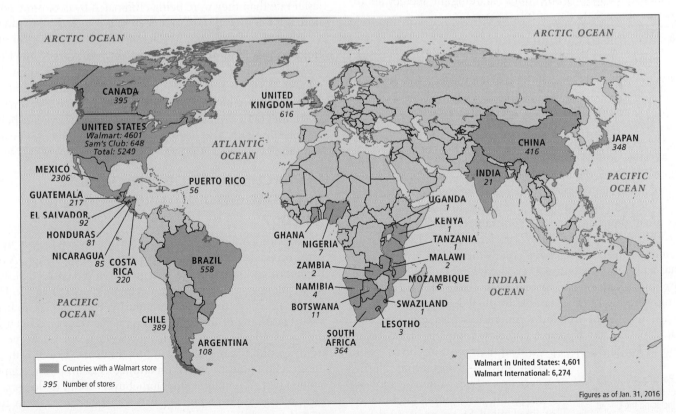

FIGURE 2.12 Walmart on the global scale. Today, Walmart has more than 11,000 retail stores in 28 countries, and employs 2.2 million people around the world (about 1.3 million in the United States). [Source: http://corporate.walmart.com/our-story/our-locations#/tions]

Free trade may have resulted in a net loss of jobs in the United States. Since 1994, increased imports from Mexico and Canada have displaced about 2 million U.S. jobs, while increased exports to these countries have created only about 1 million jobs. However, the export-related jobs that free trade agreements facilitate pay up to 18 percent more than the average North American wage. However, those jobs are usually in different locations than the ones that were lost and the people who take them tend to be younger and more skilled than those who lost jobs. Former manufacturing workers often end up with short-term contract jobs or low-skill, often service sector, jobs that pay the minimum wage and include no benefits. In Canada, these impacts of free trade have been moderated by Canada's more extensive assistance to low-income people.

As the drawbacks and benefits of free trade are being assessed, talk of an agreement spanning the entire Western Hemisphere has stalled, while one including much of the Pacific Rim is moving forward. The proposed *Free Trade Area of the Americas* (*FTAA*) would include all of North America and Middle and South America, but Brazil, Bolivia, Ecuador, and Venezuela are wary of their economies being overwhelmed by that of the United States. However, even in the absence of such an agreement, trade between North America and Middle and South America is growing faster than trade with Asia and Europe (see Chapter 3).

The **Trans-Pacific Partnership (TPP),** a new trade agreement signed by 12 Pacific Rim countries, including the United States, Japan, Mexico, Canada, Australia, Malaysia, Vietnam, Singapore, Chile, and Peru, is poised to create the world's largest free trade area. It was signed despite objections regarding the secrecy about the agreement's specific contents.

The Asian Link to Globalization

The reduction of trade barriers with Asia is a powerful force for globalization in North America. The seemingly endless variety of goods imported from China—everything from underwear to the chemicals used to make prescription drugs—is made possible in large part by China's lower wages. Indeed, many factories that first relocated to Mexico from the southern United States have now moved to China to take advantage of its enormous supply of cheap labor. Despite adjustments in the Chinese economy that may raise prices, trade with China promises to remain quite robust for some time.

U.S. and Canadian companies also want to take advantage of China's broad domestic markets. For example, the U.S. fast-food chain KFC now has more than 2100 locations in 450 cities in China, and its business is growing rapidly there. The U.S. government has a powerful incentive to encourage such overseas expansion by companies like KFC that are headquartered in the United States and whose profits are taxable by the federal government.

Asian investment in North America is also growing. For example, Japanese and Korean automotive companies have located plants in North America to be near their most important pool of car buyers—commuting North Americans. They often establish their plants in the rural mid-South of the United States or in southern Canada. Here they can benefit from an educated and highly productive

Trans-Pacific Partnership (TPP) a trade agreement signed by 12 Pacific Rim countries, including the United States, Japan, Mexico, Canada, Australia, Malaysia, Vietnam, Singapore, Chile, and Peru

labor force that is cheap relative to many parts of the region, and an excellent transportation infrastructure via the Interstate Highway System. Import restrictions that the United States, Canada, and Mexico place on Japanese- and Korean-made cars have provided an incentive for automakers from these countries to manufacture within North America. However, reductions in these restrictions as part of the TPP may remove incentives to manufacture cars within the United States and Canada.

Japanese and Korean carmakers succeeded in North America even as U.S. auto manufacturers such as General Motors were struggling. This was primarily because more advanced Japanese automated production systems—requiring fewer but better-educated workers—produce higher-quality cars, which sell better both in North America and around the world. Some analysts predict that foreign carmakers will eventually take over the entire North American market, while others note that some American car companies are expanding their market share by turning out better-built and more fuel-efficient cars.

IT Jobs Face New Competition from Developing Countries

By the early 2000s, globalization was resulting in the *offshore outsourcing* of information technology (IT) jobs. A range of jobs—from software programming to telephone-based, customer-support services—shifted to lower-cost areas outside North America. By the middle of 2003, an estimated 500,000 IT jobs had been outsourced, and another 3.3 million are forecasted to follow by 2020. During the recent recession, IT jobs were still being created at a faster rate than they were being eliminated by economic contraction, but the overall trend has been that fewer of the world's total IT jobs are staying in North America. There are new IT centers in India, China, Southeast Asia, the Baltic states in North Europe, Central Europe, and Russia. In these areas, large pools of highly trained, English-speaking young people work for wages that are 20 to 40 percent of their American counterparts' pay. Some argue that rather than depleting jobs, outsourcing will actually help create jobs in North America by saving corporations money, which will then be reinvested in new ventures.

REPERCUSSIONS OF THE GLOBAL ECONOMIC DOWNTURN BEGINNING IN 2007

The severe worldwide economic downturn that began in 2007 came on the heels of a long global economic expansion. A booming housing industry and related growth in the banks that finance home mortgages fueled general economic growth in the United States. In 2007, the housing industry collapsed as it became clear that much of the growth in previous years had been based on banks allowing millions of buyers to purchase homes with mortgages that were well beyond their means. The growth in the construction industry, based on erroneous assumptions, came to a sudden halt. When too many homebuyers could no longer afford their mortgage payments, the banks that had lent them money started to fail. This produced worldwide ripple effects because many foreign banks were involved in the U.S. housing market. Between September 2008 and March 2009, the U.S. stock market fell by nearly half, wiping out the savings and pensions of millions of Americans. Similar plunges followed in foreign stock markets, ultimately resulting in a worldwide economic downturn because businesses could no longer find money to fund expansion. As the recession intensified, job losses in the United States caused a sharp drop in consumption, which further affected world markets.

Thanks to its strong regulatory controls, Canada did not have bank failures. However, because so much of the Canadian economy is linked to exports and imports from the United States (see Figure 2.13), Canada underwent a slowdown and many Canadians lost their jobs. However, Canada's recovery was quicker than that of the United States, possibly because household consumption was buttressed by Canada's stronger social safety net (see page 93).

Efforts to deal with the difficulties that caused the recession, as well as the long-term problems made worse by it, haven't yet been very successful. The financial industry has funded extensive lobbying to counter attempts to better regulate the banks that, through their lending practices, sparked the recession.

The recession worsened the trend toward larger wealth disparities that has been underway since the 1970s. Disparities are most extreme in the United States, where in 2015 the wealthiest 1 percent of households owned 35 percent of the country's total wealth, and the next wealthiest 19 percent owned roughly another 52 percent, leaving the remaining 80 percent of the population with only 13 percent of the wealth. Most North Americans are not in favor of this kind of wealth distribution but there are many political barriers that make it difficult to reduce such inequality.

ECONOMIC INTERDEPENDENCIES

Canada and the United States are perhaps most intimately connected by their long-standing economic relationship. The two countries engage in mutual tourism, direct investment, migration, and most of all, trade. The equivalent of nearly U.S.$1.5 billion is traded daily between the two countries. Canada is a larger market for U.S. goods than are all 28 countries in the European Union. By 2015, that trade relationship had evolved into a two-way flow of U.S.$660 billion annually (FIGURE 2.13). In 2015, thirteen percent of U.S. imports came from Canada and 17 percent of its exports went to Canada. Canada sells 67 percent of its exports to the United States and buys 76 percent of its imports from the United States.

The asymmetry of the economic relationship became more apparent in 2015 when, after many years of being the United States' largest trading partner, Canada was surpassed by China. China became the United States' largest source of imports, but not of exports. And while trade with Canada supports more jobs in the United States (1.7 million) than does trade with China (940,000), the number of jobs connected to trade with China is growing much faster.

WOMEN IN THE ECONOMY

While women have made steady gains in terms of pay equity and overall participation in the labor force, there are still important ways in which they lag behind their male counterparts. On average, U.S. and Canadian female workers earn about 80 cents for every dollar that male workers earn for doing the same job (FIGURE 2.14). For example, a female architect earns approximately 80 percent of what a male architect earns for performing comparable work. This is actually an improvement over previous decades. During World War II, when large numbers of women first started working in male-dominated jobs, North American female workers earned, on average, only 57 percent of what male workers earned. The advances made by this older generation of women and the ones that followed have transformed North American workplaces. For the first time in history, women now represent more than half of the North American labor force, though most still work for male managers.

ON THE BRIGHT SIDE: Women in Business and Education
Throughout North America, the number of women entrepreneurs is on the rise, and women start nearly half of all new businesses. While women-owned businesses tend to be small and less financially secure than those in which men have most of the control, credit opportunities for businesswomen have been improving as more upper-level jobs in banking are being held by women. Women are also facing less discrimination in hiring; a 2012 study shows that many male executives now prefer to hire qualified women because they are particularly ambitious and willing to gain advanced qualifications.

In secondary and higher education, North American women have equaled or exceeded the level of men in most categories. In 2015 in the United States, 39 percent of women between the ages of 25 and 34 held an undergraduate degree, compared to just 32 percent of men. Nevertheless, women are still paid less than men for equivalent work. ∎

NORTH AMERICA'S CHANGING FOOD-PRODUCTION SYSTEMS

Agriculture is somewhat paradoxical in this region, occupying huge areas of land and producing enormous amounts of food, but employing relatively few people and contributing relatively little to each country's GDP. Agriculture is more predominant in the United States, where 44 percent of the total land area is under cultivation. Only 7 percent of land is cultivated in Canada, due to much harsher winters (FIGURE 2.15). North America has long been a major food exporter, and at one time, exports of agricultural products were the backbone of the North American economy. Because of growth in other sectors, agriculture now accounts for less than 1.2 percent of the United States' GDP and less than 2 percent of Canada's. Less than 2 percent of North Americans are employed directly in agriculture.

The shift to mechanized agriculture in North America brought about sweeping changes in employment and farm management. In 1790, agriculture employed 90 percent of the American workforce; in 1890, it employed 50 percent. Until 1910, thousands of very productive family-owned farms, located over much of the United States and southern Canada, provided for most domestic consumption and the majority of all exports. Today, the vast majority of these family farms have been replaced by farms owned by corporations.

Family Farms Give Way to Agribusiness
Family farms began to be mechanized and use chemical fertilizers in the late nineteenth century. Mechanical corn-seed planters

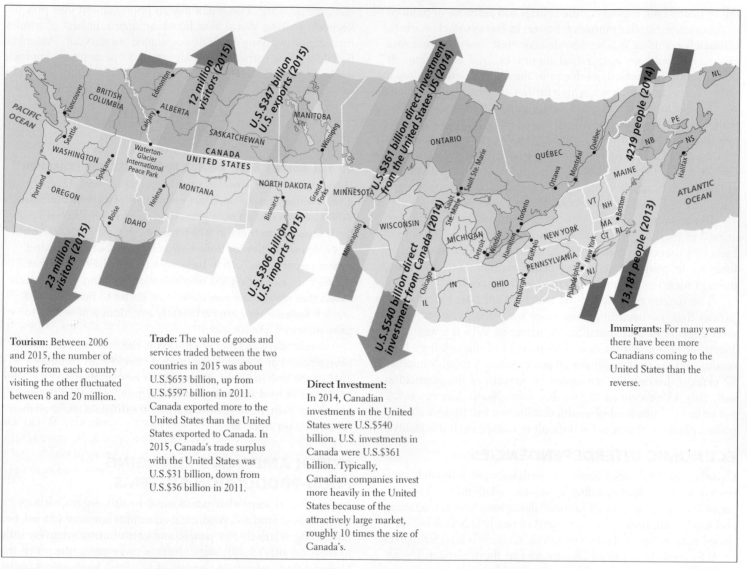

FIGURE 2.13 Transfers of tourists, goods, investment, pollution, and immigrants between the United States and Canada. Canada and the United States have the world's largest trading relationship. The flows of goods, money, and people across the long Canada–U.S. border are essential to both countries. However, because of its relatively small population and economy, Canada is more reliant on the United States than the United States is on Canada. All amounts shown are in U.S. dollars. [Sources consulted: *National Geographic*, February 1990: 106–107, and augmented with data from http://travel.trade.gov/view/m-2015-I-001/index.html, http://en.destinationcanada.com/research/statistics-figures/international-visitor-arrivals, https://ustr.gov/countries-regions/americas/canada, http://www.statcan.gc.ca/daily-quotidien/150424/dq150424a-eng.htm, http://www.migrationpolicy.org/article/canadian-immigrants-united-states, http://www.dhs.gov/sites/default/files/publications/immigration-statistics/yearbook/2011/ois_yb_2011.pdf, http://www.cic.gc.ca/english/resources/statistics/facts2014/permanent/10.asp, and http://www.cic.gc.ca/english/resources/statistics/facts2014/permanent/10.asp#figure7]

and steam-powered threshing machines, along with methods of supplying farmers with large amounts of plant nutrients such as nitrogen, phosphate, and potassium, reduced the need for labor on farms. Because of the cost of this machinery, farmers needed to make ever-larger investments in land in order for their farms to remain profitable. By the 1940s, the use of pesticides (chemicals that kill insects and other pests) and herbicides (chemicals that kill weeds) also became widespread, adding further to both the productivity and cost of farming. By the mid-twentieth century, the number of farms began to decline

agribusiness the business of farming conducted by large-scale operations that purchase, produce, finance, package, and distribute agricultural products

rapidly, as only wealthier farmers could invest in these green revolution methods (see Chapter 1). Some farmers prospered (see the vignette on page 86), while many with fewer resources sold their land, hoping to make a profit sufficient for retirement. Indebtedness and bankruptcies have become increasingly common for both large and small farmers.

A major part of the transformation of North American agriculture has been the growth of large **agribusiness** corporations that sell machinery, seeds, and chemicals. These corporations may also produce and process crops themselves on land they own, or they may contract to purchase crops from independent farms. The financial resources of agribusiness corporations have facilitated

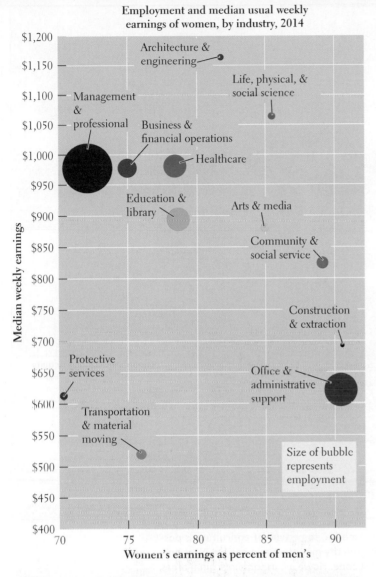

Employment and median usual weekly earnings of women, by industry, 2014

- Architecture & engineering
- Life, physical, & social science
- Management & professional
- Business & financial operations
- Healthcare
- Education & library
- Arts & media
- Community & social service
- Construction & extraction
- Protective services
- Office & administrative support
- Transportation & material moving

Size of bubble represents employment

Median weekly earnings: $1,200, $1,150, $1,100, $1,050, $1,000, $950, $900, $850, $800, $750, $700, $650, $600, $550, $500, $450, $400

Women's earnings as percent of men's: 70, 75, 80, 85, 90

FIGURE 2.14 Women's earnings and employment by industry, 2014. On average, women earned 83 percent of men's median income in 2014 if they worked full time in a wage or salary job: U.S.$719 a week compared to men's $871. [Source consulted: "Highlights of Women's Earnings in 2014," U.S. Department of Labor, Bureau of Labor Statistics, November 2015, at http://www.bls.gov/opub/reports/cps/highlights-of-womens-earnings-In-2014.pdf]

the transition to green revolution methods, enabling large investments in the research and development of new products. The corporations can also provide loans and cash to individual farmers, often as a part of contracts that leave farmers with little actual control over which crops are grown and what methods are used.

While green revolution agriculture provides a wide variety of food at low prices for North Americans, the shift to these production methods has depressed local economies and created social problems in many rural areas. Communities in places such as the Great Plains of both Canada and the United States were once made up of farming families with similar middle-class incomes, social standing, and commitment to the region. Today, farm communities are often composed of a few wealthy farmer-managers amid a majority of poor, often migrant Latino or Asian laborers who work on large farms and in food-processing plants for wages that are too low to

provide a decent standard of living. These workers also struggle to be accepted into the communities where they live.

Food Production and Sustainability

Can green revolution agriculture, like that practiced by successful North American corn farmers, persist over time? Many modern strategies to increase yields, including the use of chemical fertilizers, pesticides, and herbicides, and the large-scale production of meat on "factory farms," can have negative effects. These methods can threaten the health of farmworkers and nearby residents, pollute nearby streams and lakes, and even affect distant coastal areas.

Some irrigation methods can deplete scarce water resources and reduce soil fertility over time. In addition, many North American farming areas have lost as much as one-third of their topsoil because of deep plowing and other farming methods that create soil erosion. A major health issue is that many crops are designed to be processed into high-sugar, high-carbohydrate foods that contribute to obesity and diabetes.

Recently, researchers have been studying the long-term impacts of genetically modified (GMO) crops on human health and the environment. While most scientific studies suggest that GMO plants have little impact on human health, many of these studies have been funded or even conducted by the agribusiness corporations that create GMO crops, making it possible that the studies might be biased. The biggest documented impact that GMO plants have had on the environment is in the way they have influenced the use of herbicides and pesticides. While some varieties of GMO crops have been developed to be more resistant to insects and other pests, and so require fewer pesticides, others are designed to tolerate and even require more intensive use of herbicides.

There is much concern about the raising of animals in "factory farms" in which cows, pigs, or poultry are raised in crowded conditions and fed chemicals to make them gain weight. As with other forms of chemically intensive agriculture, the most well-documented threats to human health are to farmworkers and nearby residents who are exposed daily to many hazardous chemicals. The large quantities of animal waste these farms produce can also severely pollute the air and nearby streams.

Critics of these food production systems often argue that the government subsidies currently being given to farmers should be used as a tool for reform. They advocate directing subsidies away from factory farms and large-scale, chemically intensive crop production and toward small farmers who are willing to use methods that have fewer negative impacts.

ON THE BRIGHT SIDE: Organically Grown

Throughout North America, there is a burgeoning revival of small family farms that supply **organically grown** (produced without chemical fertilizers, herbicides, or pesticides) vegetables and fruits and grass-fed meat directly to consumers. Farmers' markets have popped up across the country (FIGURE 2.16), and more and more people are buying locally grown organic foods, paying higher prices than those in traditional grocery stores. This movement is gaining such favor that large corporate farms are seeing the potential for high profits in sustainable (that is, organic), if not locally grown, food production.

organically grown products produced without chemical fertilizers and pesticides

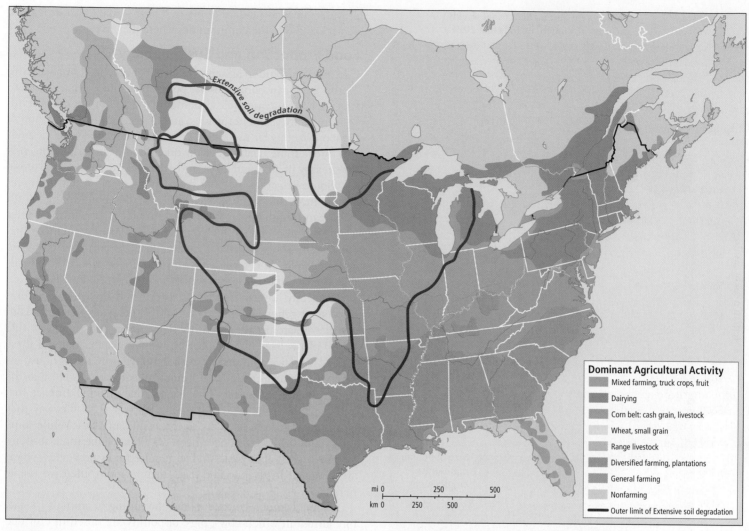

FIGURE 2.15 Agriculture in North America. Throughout much of North America, some type of agriculture is possible. The major exceptions are the northern parts of Canada and Alaska and the dry mountain and basin region (the continental interior) that lies between the Great Plains and the Pacific coastal zone. However, in some marginal areas, such as Southern California, southern Arizona, and the Utah Valley, irrigation is needed for cultivation. (Hawaii is not included here because it is part of Oceania, which is covered in Chapter 11.) [Source consulted: Arthur Getis and Judith Getis, eds., *The United States and Canada: The Land and the People* (Dubuque, IA: William C. Brown, 1995), p. 165]

Dominant Agricultural Activity
- Mixed farming, truck crops, fruit
- Dairying
- Corn belt: cash grain, livestock
- Wheat, small grain
- Range livestock
- Diversified farming, plantations
- General farming
- Nonfarming
- Outer limit of Extensive soil degradation

FIGURE 2.16 A Place of the Heart. A Tennessee wife-and-husband team sell produce from their farm, A Place of the Heart, at a local farmers' market. They also deliver a basketful of vegetables each week to a set of customers who pay a seasonal fee of U.S.$700 for the produce.

Organic food production can reduce harmful impacts on the environment because it uses less-polluting methods. Consumers also gain from higher-quality, toxin-free vegetables and fruits. ■

VIGNETTE The flip side of this story of the decline in North America of the family farm way of life is represented by the burgeoning prosperity of farmers like 37-year-old Brandon Hunnicutt, who runs his family's 3600-acre (1457-hectare) farm above the vast Ogallala Aquifer in Nebraska. He uses the aquifer water to irrigate in dry times and his computer to garner the latest information on nearly every aspect of his operation—from the diagnosis of a crop disease to the right moment for marketing his crops. In the midst of the global recession, he was selling corn at twice the price he received a few years before.

How can this be? In 2007, the rising price of petroleum products spurred the U.S. government to require that ethanol made from corn be added to gasoline in order to lower dependency on foreign energy sources. More than one-third of the corn produced in the United States is now converted to ethanol in order to fuel cars and trucks. By 2008, the demand for corn had radically increased, as had its price. As the recession deepened, the value of the U.S. dollar was lowered in order to spur exports. This gave U.S. corn producers an advantage in the world market. Meanwhile, as the economies of Asian countries continued to grow, Asia began buying more corn-based food products, especially sweeteners. Bad weather in Russia and Ukraine, which also produce grains, further increased the demand for corn. Corn prices rose globally, to the great advantage of large U.S. producers like Brandon Hunnicutt but to the great anguish of the poor across the world, who had formerly relied on cheap corn as a mainstay in their diets. As shown in the discussion of food security in Chapter 1, the global economy can bring very different outcomes to different parts of the world. Financial security for farmers like the Hunnicutts can mean *food insecurity* for the poor in Africa, Asia, and Middle and South America. [*Source: NPR staff, All Things Considered. For detailed source information, see Text Sources and Credits.*] ■

CHANGING TRANSPORTATION NETWORKS AND THE NORTH AMERICAN ECONOMY

North America depends on an extensive network of road and air transportation that enables the high-speed movement of people and goods. The road system originated with the first European settlement in the 1500s in Florida, but until the development of the first railways in the 1820s, most goods were moved on rivers, canals, and the coastal seas. By the 1870s, the railroads had formed a nationwide network that dominated transportation until inexpensive, automobiles were mass-produced in the 1920s. Beginning in the 1950s, the growth of automobile- and truck-based transportation was helped by the U.S. Interstate Highway System and the Trans-Canada Highway System—a huge network of high-speed, multilane roads that continues to dominate transportation in North America. Because this network is connected to the vast system of local roads, it can be used to deliver manufactured products faster and with more flexibility than can be done using the rail system. The highways have thus made it possible to disperse industry and related services into suburban and rural locales across the country, where labor, land, and living costs are lower. An often-noted aspect of the North American system is the absence of effective inter-city passenger rail for most of the region.

After World War II, air transportation became a central part of transportation in North America. The primary niche of air transportation is business travel, because face-to-face contact remains essential to American business culture despite the growth of telecommunications and the Internet. With many industries widely distributed across numerous medium-size cities, air service is organized as a *hub-and-spoke network*. Hubs are strategically located airports, such as those in Atlanta, Chicago, Dallas, and Los Angeles. These airports serve as collection and transfer points for passengers and cargo continuing on to smaller cities and towns. Most airports are also located near major highways, which provide an essential link for high-speed travel and cargo shipping.

THINGS TO REMEMBER

GEOGRAPHIC THEME 2 • **Globalization and Development:** Globalization has transformed North America, reorienting it toward knowledge-intensive jobs that require education and training. Income inequality has risen as most manufacturing jobs have been moved to countries with cheaper labor or have been replaced by technology. North America's size and wealth have made it the center of the global economy, and its demand for imported goods and its export of manufacturing jobs make it a major engine of globalization.

• North America's advantageous position in the global economy is a reflection of its size, technological sophistication, and geopolitical influence, all of which enable it to mold the pro-globalization free trade policies that suit the major corporations of North America.

• The major long-term goal of NAFTA is to increase the amount of trade between Canada, the United States, and Mexico. NAFTA is currently the world's largest trading bloc in terms of the GDP of its member states.

• Flows of trade and investment between North America and Asia have increased dramatically in recent decades.

• Globalization has resulted in the offshore outsourcing of hundreds of thousands of information technology (IT) jobs.

• North American farms have become highly mechanized, chemically intensive operations that need few workers but require huge amounts of land to be profitable.

• In the twentieth century, the mass-production of inexpensive automobiles and trucks, as well as the Interstate Highway System, fundamentally changed how people and goods move across the continent.

POWER AND POLITICS

GEOGRAPHIC THEME 3

Power and Politics: Compared to many world regions, North America has relatively high levels of political freedom. However, there is also widespread dissatisfaction with the political process here. Internationally, Canada plays a modest role while the United States has enormous global political influence, although its status as the world's predominant "superpower" is declining.

THE EXPANSION OF POLITICAL FREEDOMS IN THE UNITED STATES AND CANADA

Relatively high levels of political freedom in this region are part of a long-term trend toward more openness in the political process. However, many in this region are concerned about the role that money, or the lack thereof, plays in shaping the public's influence on governmental decision making.

Voting rights have steadily expanded over the history of both countries. They have been extended from including only white male property owners (United States, mid-1700s; Canada, 1758) to including all adult white males (United States, 1856; Canada,

1898), to including men of African descent (United States, 1870; Canada, 1837), women (United States, 1920; Canada, 1918), and indigenous peoples (United States, 1924; Canada, 1960).

The ways in which candidates for elected office are selected has also opened up in both countries. In the past, a few political insiders selected the candidates that they wanted to run for a particular office. Starting in the 1920s in the United States, candidates for office began to be selected via primaries, which are elections that determine who a political party will nominate to run as their candidate for a particular office. For example, in 2016, Hillary Clinton was chosen to be the Democratic Party's candidate for president of the United States only after she won primaries in enough states to defeat democratic socialist Bernie Sanders and move on to the general election. Primaries have only recently been adopted in Canada, in part because they are a much more expensive way to choose candidates.

Political Disillusionment and the Role of Money in Politics

While there has been a trend toward more openness in the political process, there is widespread disillusionment about politics in both the United States and Canada. Evidence of this can be seen in *voter turnout*—the percentage of people who decide to cast a vote in an election—which is low relative to many other counties with similarly high levels of economic development, especially those of Europe. The United States has a generally lower voter turnout than Canada: 57.5 percent of voting-age U.S. citizens voted in the presidential election of 2012, and 61.1 percent of voting-age Canadians voted in 2015.

A major reason for the lower turnout in the United States is the frustration created by the role of money in politics. A potential candidate for a major office in the United States must now spend millions or, in the case of the office of the U.S. president, more than a billion dollars, to win an election. This cost is a major barrier to people who want to enter into politics, as only those able to raise large amounts of money have any chance to win. Expensive elections also mean that successful candidates have to spend more of their time raising money in order to be reelected and less time doing their job representing the people who elected them. As a result, politicians tend to focus on wealthy people and corporations who can make the donations that will help them get reelected. Recent political science research comparing how public policy in the United States reflects the preferences of the voting public suggests that the policies that the federal government adopts are generally those supported by high-income "economic elites" and their associated interest groups. The research also suggests that the preferences of lower-income people have little to no impact on federal policy.

DEBT AND POLITICS IN THE UNITED STATES

There is much focus in the United States on the national debt. The debt consists of the money the United States borrows by issuing Treasury securities to cover the expenses it has that exceed its income from taxes and other revenues.

To whom does the United States owe its debt? About 66 percent of the total debt is owed to U.S. citizens who own Treasury securities, and to various federal government entities, like the Social Security Trust Fund. About 34 percent is owed to foreign governments (8 percent to China, 5 percent to Japan, 2 percent to the United Kingdom, and smaller amounts to Brazil, Taiwan, and Hong Kong).

Responses to the national debt highlight the differences between the two major political parties in the United States. Most Republicans believe that the debt is dangerous and needs to be reduced immediately by drastically reducing government spending, especially on social services such as the Affordable Health Care Act, education, and various kinds of aid to the poor. Many Democrats are less troubled by the debt and believe that some government spending, especially military spending and subsidies to large corporations, should be cut in order to fund social services, and that further revenues should be raised by taxing the very rich.

These differences between the two major political parties show up repeatedly in the United States, as the government's ability to raise revenue via taxation and the ways it spends this revenue are central to almost every major political issue.

INFLUENCE OF THE U.S. AND CANADA ABROAD

The United States and Canada have dramatically different roles in the global geopolitical order. The United States is recognized as the most powerful country in the world, with the world's largest economy and a military budget that is larger than those of the next largest 12 or 13 countries combined. While the promotion of democracy and political freedoms is an official goal of U.S. foreign policy, in practice, the United States tends to focus its activities abroad more on its own economic and strategic military interests, which are often linked. This can be seen, for example, in the recent U.S. war in Iraq (2003), which numerous planners of the war within the Bush Administration (2000–2008) now acknowledge was motivated more by a desire to control access to Iraq's lucrative oil reserves than to bring democracy to Iraq (FIGURE 2.17).

U.S. economic and strategic interests are reflected in the global distribution of its military bases and its spending on aid to foreign governments. Four main concentrations of bases and spending, where the United States has for years had strong strategic and economic interests, can be seen on the map in Figure 2.17 in Europe, North Africa and Southwest Asia (especially Egypt, Israel, Iraq, Jordan), Afghanistan and Pakistan in South Asia, and island and peninsular East Asia. The first and third concentrations of bases and spending (those in Europe and East Asia) relate to strategic and economic interests dating from World War II that have remained relevant due to the Cold War (see Chapter 1), the subsequent collapse of the Soviet Union (see Chapter 5), the continued large volume of trade between the United States and Europe, and the rapid growth of U.S. trade with East Asia since the opening up of China (see Chapter 9; see also Figure 2.17C). The second concentration relates directly to U.S. interests in Iraq, Iran, Afghanistan, Pakistan, Israel, and Egypt, and the oil and mineral resources of South and Southwest Asia in general.

The map in Figure 2.17 also shows that there are many places where the United States does not have many bases and where spending on foreign aid is at low or moderate levels. These are generally places where the United States has fewer strategic and economic interests. If U.S. policies to support democracy abroad were a strong factor in its allocation of assistance, one might expect that the focus of U.S. spending on military assistance and foreign aid would be in the parts of sub-Saharan Africa and elsewhere that have low levels of democratization. However, due largely to the poverty and political instability of sub-Saharan Africa, the United States has few economic or strategic interests in the region (see Figure 2.17D). (The same could be said of Haiti in the Caribbean; see Figure 2.17E). And yet the United States has few bases and spends little on military assistance and foreign aid in these parts of the world.

Canada takes a more "live and let live" approach on the world stage. While trade is a similarly strong motivator for Canada's foreign policies and foreign aid projects, Canada has far fewer strategic military interests abroad. Arguably, Canada's greatest geopolitical impact comes from its ability to influence the United States. As the second-largest trading partner of the United States, and one with considerable energy resources, Canada has helped to reduce U.S. dependence on overseas oil. This relationship was highlighted by the 9/11 attacks and the subsequent Iraq War. At that time (the early 2000s), about 25 percent of the oil imported into the United States came from the Organization of the Petroleum Exporting Countries (OPEC), which was mostly made up of the countries along the Persian Gulf, where the terrorists had come from. Recent data show that 69 percent of oil imported into the United States comes from countries that are not part of OPEC. Of this, Canada supplies 40 percent; Venezuela, 9 percent; Mexico, 8 percent; and 20 other countries supply the remaining 12 percent **(FIGURE 2.18A)**. Overall U.S. imports of oil have declined sharply over the past few years thanks to new oil extraction techniques, from 60 percent of the total amount of crude oil used in 2012 to 25 percent in 2016.

The extraction of fuel from the oil sands of western Canada is motivated in part by U.S. uneasiness about being dependent on oil from OPEC countries. The oil sands could potentially double the amount of oil that Canada exports to the United States, substantially reducing the need for imports from OPEC countries. The United States has had antagonistic relationships with some Persian Gulf states for a number of years, which to some justifies the many environmental costs of developing Canada's oil sands.

CHALLENGES TO THE UNITED STATES' GLOBAL POWER

A number of challenges to U.S. global power have emerged in recent decades. The wars in Iraq and Afghanistan exposed the limits of the ability of the United States to bring about change in other countries, and a new alliance is challenging the United States and its allies in Europe.

Immediately after the attacks on New York City and Washington, DC, on September 11, 2001 (9/11), the international community extended warm sympathy to the United States. Many governments, despite the protests of large numbers of their citizens, supported then-president George W. Bush in his launching of the *War on Terror*, defined as a defense of the American way of life and of democratic principles. The first target was Afghanistan, which was then thought to be host to Osama bin Laden and the elusive Al Qaeda network that claimed credit for masterminding the 9/11 attacks. The aim was to capture bin Laden—accomplished a decade later in 2011 when bin Laden was killed by U.S. Army Special Forces in Abbottabad, Pakistan—and remake Afghanistan into a stable ally, which still had not been done by 2016.

Although NATO (North Atlantic Treaty Organization) forces (including Canadian troops) joined U.S. forces in Afghanistan, the war proved difficult to resolve because of heavy resistance from tribal leaders within the country and from insurgents in adjacent Pakistan. Also, after the spring of 2003, attention and troop support were diverted as President Bush brought the War on Terror to Iraq, which was later shown to have had no role in the 9/11 attacks.

The wars in Iraq and Afghanistan eventually ended with few concrete gains, despite heavy losses of civilian life. U.S. combat troops withdrew from Iraq in 2011 and from Afghanistan in 2014, only to return shortly thereafter as both countries have remained plagued by violence. In the case of Iraq, international public opinion polls showed that the U.S. image abroad was badly damaged by what was seen as an unjust war aimed mainly at controlling Iraq's oil resources (see Figure 2.17A, B, and the figure map).

Since 2001, an alliance of five countries has emerged as a symbol of resistance to the global power of the United States and its allies. Together, these countries—Brazil, Russia, India, China, and South Africa (known collectively as BRICS)—account for over a quarter of the world's land area and more than 40 percent of its population, and by 2027 they will together form a larger group of economies than the United States and its major allies (Canada, France, Germany, Italy, Japan, and the United Kingdom), which are collectively known as the G7. Meeting at yearly summits, the BRICS countries are committed to establishing a "more multipolar world order" in which the United States and its allies are less dominant. The BRICS countries are the most prominent of several groups of rapidly developing countries that are publicly challenging the current global geopolitical order.

RELATIONSHIPS BETWEEN CANADA AND THE UNITED STATES

Citizens of Canada and the United States share many characteristics and concerns. Indeed, in the minds of many people—especially those in the United States—the two countries are like one. Yet that is hardly the case. Three key factors characterize the interaction between Canada and the United States: *asymmetries*, *similarities*, and *interdependencies*.

Asymmetries

Asymmetry means "lack of balance." Although the United States and Canada occupy about the same amount of space **(FIGURE 2.19)**, much of Canada's territory is cold and sparsely inhabited. The U.S. population is about ten times the Canadian population. While Canada's economy is one of the largest and most productive in the world, producing U.S.$1.6 trillion (PPP) in goods and services in 2016, it is dwarfed by the U.S. economy, which is more than ten times larger, at U.S.$18 trillion (PPP) in 2016.

FIGURE 2.17 PHOTO ESSAY: Power and Politics in North America

Thinking Geographically

After you have read about the power and politics in North America, you will be able to answer the following questions.

A What economic interests does the United States have in Europe?

B What strategic and economic interests make it unlikely that the United States will withdraw completely from Afghanistan and Iraq in the near future?

D What has kept U.S. economic interests in Africa at a low level relative to those in other regions?

E What has kept U.S. economic interests in Haiti at a low level relative to those in other countries?

Political freedoms are generally well protected and democratization is at a relatively high level in North America. The United States is often thought of as using its power to promote political freedoms and democratization on a global scale. There is some truth to this. However, strategic and economic interests often play a greater role in shaping U.S. actions abroad. For example, some U.S. officials who planned the Iraq War point out that the desire to control Iraq's oil resources and those of its neighbors influenced U.S. actions there more than did the promotion of democracy. If the main U.S. interest abroad were the promotion of democracy, then sub-Saharan Africa would be a major focus of U.S. foreign aid and military installations, especially given the many violent conflicts that plague this region. As the map below shows, sub-Saharan Africa receives relatively little aid and has few U.S. military bases. Nevertheless, the United States has increased its presence there in recent years.

In international affairs, Canada quietly supports civil society efforts abroad, while the United States is an economic, military, and political superpower preoccupied with maintaining its role as a world leader. In framing foreign affairs policy, the United States regards Canada's position only as an afterthought, in part because the country is such a secure ally. But managing its relationship with the United States is a top foreign policy priority for Canada. As former Canadian Prime Minister Pierre Trudeau once told the U.S. Congress, "Living next to you is in some ways like sleeping with an elephant: No matter how friendly and even-tempered the beast, one is affected by every twitch and grunt."

Similarities

Notwithstanding the asymmetries, the United States and Canada have much in common. Both are former British colonies and have retained English as the dominant language. Both experienced settlement and exploration by the French. From their common British colonial heritage, they developed comparable democratic political traditions. Both are federations (of states or provinces), and both are representative democracies, with similar legal systems.

Not the least of the features they share is a 4200-mile (6720-kilometer) border, which until 2009 contained the longest sections of unfortified political boundary in the world. For years, the Canadian border had just 1000 U.S. guards, while the Mexican border, which is half as long, had nearly 10,000 agents. In 2009, the Obama administration decided to equalize surveillance of the two borders for national security reasons. Where some rural residents once passed in and out of Canada and the United States unobserved many times in the course of a routine day, now there are drone aircraft with night-vision cameras and cloud-piercing radar scanning the landscape for smugglers, undocumented immigrants, and terrorists.

Canada and the United States share many other landscape similarities. Their cities and suburbs look much the same. The billboards that line their highways and freeways advertise the same brand names. Shopping malls and satellite business districts have followed suburbia into the countryside, encouraging comparable types of mass consumption and urban sprawl. The

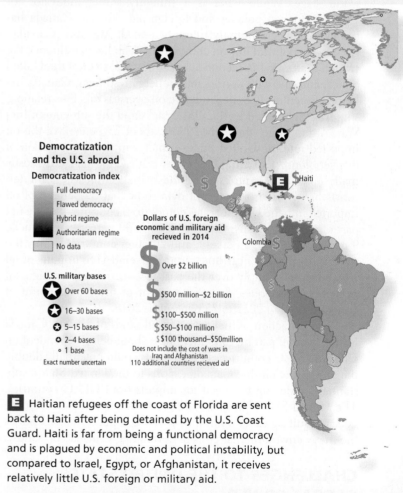

Democratization and the U.S. abroad

Democratization index
- Full democracy
- Flawed democracy
- Hybrid regime
- Authoritarian regime
- No data

U.S. military bases
- ★ Over 60 bases
- ★ 16–30 bases
- ✪ 5–15 bases
- ✪ 2–4 bases
- ○ 1 base
- Exact number uncertain

Dollars of U.S. foreign economic and military aid recieved in 2014
- $ Over $2 billion
- $ $500 million–$2 billion
- $ $100–$500 million
- $ $50–$100 million
- $ $100 thousand–$50million
- Does not include the cost of wars in Iraq and Afghanistan
- 110 additional countries recieved aid

E Haitian refugees off the coast of Florida are sent back to Haiti after being detained by the U.S. Coast Guard. Haiti is far from being a functional democracy and is plagued by economic and political instability, but compared to Israel, Egypt, or Afghanistan, it receives relatively little U.S. foreign or military aid.

Thony Belizaire/AFP/Getty Images

A A Canadian soldier wounded in Afghanistan is lifted off a plane at Ramstein Air Base, one of 260 U.S. bases in Germany. The large U.S. military presence in Europe is a legacy of World War II and the Cold War era that followed. The purpose of the military presence was to discourage potential aggression from the (now-defunct) Soviet Union against U.S. allies in Western Europe.

B Afghan women line up to vote in the country's first legitimate elections in decades. A large part of U.S. foreign aid in Afghanistan is used to promote democratic practices. Afghanistan became a major recipient of U.S. aid and military intervention only after the attacks of September 11, 2001 (masterminded by the Al Qaeda terrorist network, based in part in Afghanistan). Before 9/11, U.S. foreign aid to Afghanistan was much lower and U.S. military bases there were nonexistent.

C A U.S. military base in Japan, one of 130 in the country, is surrounded by dense urban development. These bases are first and foremost a projection of U.S. power designed to counter any future aggression by China, North Korea, the former Soviet Union, South Korea, or by Japan itself.

D A Kenyan police officer investigates a burning roadblock set up in the aftermath of disputed presidential elections in 2013. Democracy is fragile throughout much of sub-Saharan Africa, and elections are often plagued by violence. While U.S. financial support for democracy in sub-Saharan Africa is growing, U.S. foreign aid to the area is relatively small and historically the United States has done little to promote democracy in the region. The United States has few strategic or economic interests, and few military bases, in the region.

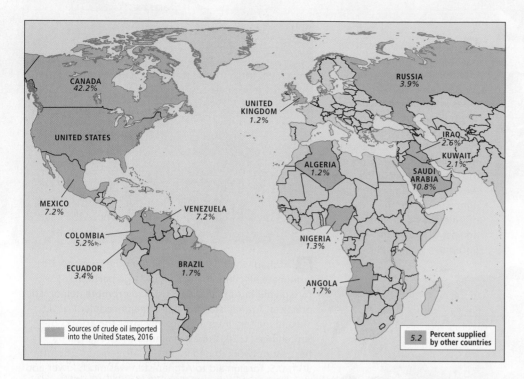

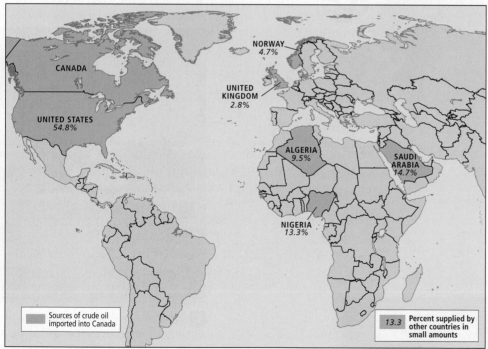

FIGURE 2.18 **(A) Sources of average daily crude oil imports into the United States, 2016.** U.S. dependence on imported crude oil has declined dramatically in recent years, dropping from nearly 60 percent of the total amount of crude oil used in the United States in 2012 to 25 percent in 2016. This map shows the average percentage of crude oil (in millions of barrels) imported daily into the United States from the top 15 exporting nations in January 2016. Canada's share of crude oil imports into the United States rose from 21 percent in September 2009 to nearly 42 percent by January 2016. [Sources consulted: "U.S. Imports by Country of Origin," U.S. Energy Information Administration, at https://www.eia .gov/dnav/pet/pet_move_impcus_a2_nus_ep00_im0_ mbbl_m.htm. For more detail, see http://www.eia .gov/petroleum/imports/samples/ and http://www .eia.gov/energyexplained/index.cfm?page=us_ energy_home]

FIGURE 2.18 **(B) Sources of average daily crude oil imports into Canada, 2016.** In 2016, Canada produced more than twice as much oil as it used. Because of transportation and refining bottlenecks, though, much of the oil produced in western Canada, which accounts for most of Canada's oil, has to be refined in the United States. Some of this oil is sold back to Canada, making the United States Canada's largest source of imported oil. [Sources consulted: *Statistical Handbook,* Canadian Association of Petroleum Producers, at http://www5.statcan.gc.ca /cimt-cicm/topNCountries-pays?lang=eng&sectionId=5& dataTransformation=0&refYr=2016&refMonth=1&freq =6&countryId=0&usaState=0&provId=1&retrieve= Retrieve&save=null&country=null&tradeType=3& topNDefault=25&monthStr=null&chapterId=27& arrayId=0&sectionLabel=V%20-%20Mineral%20 products&scaleValue=0&scaleQuantity=0&commodity Id=270900]

two countries also share similar patterns of ethnic diversity that developed in nearly identical stages of immigration from abroad.

Democratic Systems of Government: Shared Ideals, Different Trajectories

Canada and the United States have similar democratic systems of government, but there are differences in the way power is divided between the federal government and provincial or state governments. There are also differences in the way the division of power has changed since each country became independent.

Both countries have a federal government in which a union of states (United States) or provinces (Canada) recognizes the sovereignty of a central authority, while states/provinces and local governments retain many governing powers. In both Canada and the United States, the federal government has an elected executive branch, elected legislatures, and an appointed judiciary. In Canada, the executive branch is more closely bound to follow the will of the legislature. The Canadian federal government has more and stronger powers (at least constitutionally) than does the U.S. federal government.

FIGURE 2.19 Political map of North America.

Over the years, both the Canadian and U.S. federal governments have moved away from the original intentions of their constitutions. Canada's initially strong federal government has become somewhat weaker, largely in response to demands by provinces, such as the French-speaking province of Québec (see the map in Figure 2.1), for more autonomy over local affairs.

The more limited federal government that the United States had at its outset has expanded its powers. The U.S. federal government's original source of power was its mandate to regulate trade between states. Over time, this mandate has been interpreted ever more broadly. Now the U.S. federal government has a powerful effect on life even at the local level, primarily through its ability to dispense federal tax monies via such means as grants for school systems, federally assisted housing, military bases, drug and food regulation and enforcement, urban renewal, and the rebuilding of interstate highways. Money for these programs is withheld if state and local governments do not conform to federal standards. This practice has made some poorer states dependent on the federal government. However, it has also encouraged some state and local governments to enact more enlightened laws than they might have done otherwise. For example, in the 1960s the federal government promoted civil rights for African American citizens by requiring states to end racial segregation in schools in order to receive federal support for their school systems.

The Social Safety Net: Canadian and U.S. Approaches

The Canadian and U.S. governments have responded differently to workers who have been displaced by economic change.

Ultimately, these differences derive from prevailing political positions and widely held notions each country has about what the government's role in society should be. In Canada there is broad political support for a robust **social safety net,** the services provided by the government—such as welfare, unemployment benefits, and health care—that prevent people from falling into extreme poverty. In the United States there is much less support for these programs and a great deal of contention over nearly all efforts to strengthen the U.S. social safety net.

For many decades, Canada has spent more per capita than the United States on social programs. These programs, especially unemployment benefits, have generally made the financial lives of Canadians more secure. These policies also reflect the workings of Canada's democracy; voters have supported tax hikes to fund several major expansions of the social safety net during the twentieth century.

Compared to Canada, the United States provides much less of a social safety net. The prevailing political argument against the expansion of the U.S. social safety net has been that the system offers lower taxes for businesses, corporations, and the wealthy, which in turn invest their profits in expansions or in new businesses, thus creating jobs and new taxpayers. These benefits of low taxes are assumed to *trickle down* to those most in need. The accuracy of this theory has often been challenged, especially during times of economic recession.

Two Health-Care Systems The contrasts between health-care systems in the two countries are striking. Reformed in the 1970s, the Canadian system is heavily subsidized and covers 100 percent of the population. In the United States, health care has generally been private, relatively expensive, and tied to employment, which made changing jobs difficult for those with preexisting conditions. Also, many small businesses did not provide health insurance coverage for employees, who had to pay for it themselves or go without. The Affordable Care Act (ACA), which has been dubbed "Obamacare" after President Obama, who pushed for the legislation, reduced the percentage of people who have no health-care coverage from 18 percent to 11.4 percent.

The financial viability and medical outcomes of the ACA are not yet known, but the most recent statistics show why change was sought. In 2013, the United States spent more per capita on health care than any other industrialized country— U.S.$8713 per capita, for a total of 16.4 percent of its GDP per capita. Canada spent U.S.$4351 per capita, or just 10.2 percent of its GDP per capita (GNI figures are not available)—yet Canada had better health outcomes, outranking the United States on most indicators of overall health, such as infant mortality, maternal mortality, and life expectancy **(TABLE 2.1)**.

Gender in National Politics

North America has some powerful political contradictions with regard to gender. While women voters are a potent political force, successful female politicians are not as numerous as one might expect. Women cast the deciding votes in the U.S. presidential election of 1996, voting overwhelmingly for

social safety net the services provided by the government—such as welfare, unemployment benefits, and health care—that prevent people from falling into extreme poverty

TABLE 2.1 Health-related indexes for Canada and the United States

Country	Public health expenditures as a percent of GDP in 2013[2]	Percentage of population with no insurance, 2015[5]	Deaths per 1000 live births in 2013[6]	Infant mortality per 1000 live births, 2013[2]	Maternal mortality per 100,000 births, 2013[1]	Life expectancy at birth (years), 2015[1]	Public health expenditures per capita (PPP U.S.$), 2014[4]
Canada	17.1	0	7	5.9	11	81	6105
United States	10.9	11.9	8	4.6	28	79	9523

[1] Population Reference Bureau, *2015 World Population Data Sheet*, at http://www.prb.org/Publications/Datasheets
[2] United Nations Development Program, *2015 UN Human Development Report* (New York: United Nations Development Programme), Table 9, at http://hdr.undp.org /en/2015-report
[3] United Nations Development Program, *2015 UN Human Development Report* (New York: United Nations Development Programme), Table 16, at http://hdr.undp.org /en/2015-report
[4] Centers for Medicare & Medicaid Services, *2014 NHE Fact Sheet*, at https://www.cms.gov/Research-Statistics-Data-and-Systems/Statistics-Trends-and-Reports /NationalHealthExpendData/NHE-Fact-Sheet.html and Canadian Institute for Health Information, Spending, at https://www.cihi.ca/en/spending-and-health-workforce /spending
[5] "Nearly 90 percent of Americans have health coverage," by Chris Isidore, CNNMoney, April 13, 2015, at http://money.cnn.com/2015/04/13/news/economy /obamacare-uninsured-gallup/
[6] The World Bank, Death rate, crude (per 1,000 people), at http://data.worldbank.org/indicator/SP.DYN.CDRT.IN

Bill Clinton. In the 2006 interim elections, 55 percent of women voted for the Democratic Party candidates, registering strong anti-war sentiment and making it possible for Democrats to gain many seats in Congress. In 2008 and 2012, Barack Obama won 55 percent of the female vote. Deciding factors in the election included

issues that are often priorities for women, such as reproductive rights, health care, family leave, equal pay, day care, and ending wars.

And yet in terms of electing female political leaders to office, women have not had the success warranted by their powerful role as voters. Of the 535 people in the U.S. Congress as of 2016, only 104 members, or 19.4 percent, were women. Gender equity was somewhat closer at the state level, where 24.4 percent of legislators were women. In Canada, women have had a little more success. As of 2016, Canadian women constituted 26 percent of the House of Commons in Parliament and held 27 percent of provincial legislative seats. Neither country compares well to the world at large in that, as **FIGURE 2.20** shows, over the last decade other countries have added substantially more women to legislatures.

Things are somewhat more equitable for Canadian women holding gubernatorial and territory executive offices. In 2016, Canada had 3 women as provincial governors and another as executive of the Nunavut Territory. (There are ten provinces and three territories.) In 2016, the United States had 6 out of a possible 50 female governors. Canada briefly had a female prime minister in 1993. In 2016, Hillary Clinton was the first woman to have a serious chance at becoming the U.S. president.

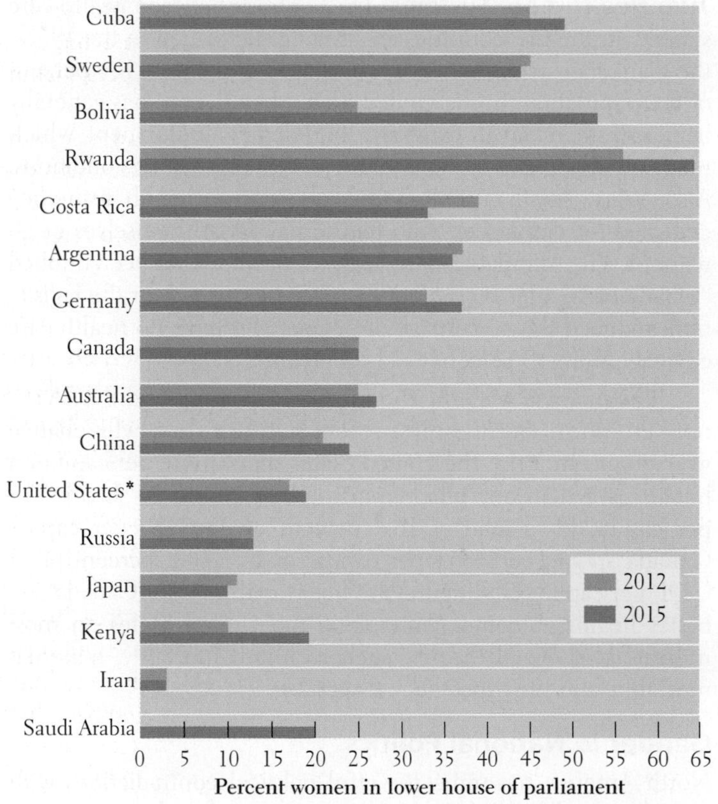

Drugs and Politics in North America

Drugs have had a significant influence on society in North America, as drug use is relatively high in this region on a per capita basis. The United States is the world's largest importer of illegal drugs. Its trade in illegal drugs is worth U.S.$400 billion to $500 billion per year, equal to roughly 2.6 percent of the U.S. economy or roughly twice the amount of the entire agriculture sector.

Governmental responses to the illegal drug trade highlight inequalities in wealth and power, with hundreds of thousands of poor people sentenced to years in jail, often for possessing small amounts of illegal drugs, while wealthier drug users are usually placed in drug treatment programs and sentenced to

* Percentage for the entire U.S. Congress is used, which has 100 senators (20 are women) and 435 congresspeople (77 are women)

FIGURE 2.20 Country comparison of women in national legislatures, 2012 and 2015. [Source consulted: http://unstats.un.org /unsd/gender/chapter5/chapter5.html]

community service. Meanwhile, the four largest banks in North America—JPMorgan Chase, Bank of America, Wells Fargo, and Citigroup—have all recently been caught helping Mexican and Colombian drug cartels to conceal hundreds of billions of dollars from U.S. law enforcement, highly illegal activities for which the banks have merely been fined. In the 1980s the U.S. government was itself involved in the illegal drug trade, with U.S. Senate investigations revealing that the CIA cooperated with cocaine traffickers as part of its effort to fund a rebellion against the Nicaraguan government.

Illegal drugs are available almost everywhere and drug overdose is the leading cause of accidental death in the United States, topping even motor vehicle accidents, at more than 50,000 deaths per year. Mexico has been plagued by decades of violence associated with the vast trade catering to the huge demand for illegal drugs in North America.

In recent years, there has been an escalation in prescription drug abuse, primarily of painkillers, which are responsible for more than half of all deaths from overdose in the United States. Heroin has surged in popularity as policies that reduce the availability of prescription painkillers have taken effect, and about a third of overdose deaths are now related to heroin. Methamphetamine (meth) is also a growing concern, as it is highly addictive, extremely toxic and can be made from household chemicals. Drug cartels in Mexico have recently shifted over to heroin and meth after decades of declining use of cocaine, and heroin and meth addiction is often severe along the major transportation routes that connect northern Mexico with North American cities.

The Failures of the War on Drugs

A *war on drugs* was declared in the United States under the Nixon Administration in 1971, with the goal of reducing the drug trade via three tactics discussed below: *eradication, interdiction,* and *incarceration.* Eradication involves the destruction of drug crops, interdiction intercepts shipments of drugs before they reach users, and incarceration imprisons drug offenders. These policies have been implemented in Canada as well, though with a generally less strict approach. The war on drugs has cost over a trillion dollars and yet compared to 1971, the drug trade is more widespread and drugs are cheaper and more easily obtained. Currently the United States has higher rates of drug use relative to countries such as Mexico, the Czech Republic, the Netherlands, Portugal, and Uruguay that have either legalized most types of drugs or decriminalized them, meaning that an offense is treated like a parking violation.

Eradication is often accomplished by burning, as is often done with drugs found in North America, or else aerial spraying of live drug crops with herbicides, as is often done in U.S.-backed antidrug operations in source countries, such as Mexico or Colombia. Herbicide spraying has not reduced the amount of land under drug cultivation or reduced harvests because the profits from growing drug crops are simply too high. However, spraying has damaged fragile environments, hurt many nondrug crops, and endangered the health of farmers.

Interdiction can yield seemingly spectacular drug busts, but the amounts of drugs that are successfully intercepted are too small to discourage smugglers, given the enormous profits they make from the many shipments that do get through.

Incarceration has done relatively little to reduce drug use or decrease the flow of drugs into the United States, but it has played a major role in giving the United States the largest prison population in the world. With 4 percent of the world's population, the United States has almost a quarter of the world's prisoners, more than China and India combined (which together account for over a third of the world's population). Half of federal prisoners and around 16 percent of state prisoners are drug offenders.

Racism influences incarceration rates in the United States, where minorities are jailed more often and for longer terms than whites. While roughly 68 percent of U.S. drug users are white, 14 percent are African American, and 16 percent are Latino, African Americans make up 39 percent of those incarcerated for drugs, and Latinos, 29 percent. Studies suggest several important sources of racial bias in drug-related incarceration. Federal drug laws impose far longer prison terms for crack cocaine, which is more common in low-income minority communities, than for powder cocaine, which is more common in higher-income white communities.

Local police departments also target minority neighborhoods disproportionately. In part this is because their antidrug funding from the federal government is based on the numbers of people they arrest on drug offenses, and the quickest and cheapest way for them to raise those numbers is to focus on poor and generally minority neighborhoods, where drug dealing and use is often more visible on the street relative to higher-income majority-white neighborhoods, where drug use usually happens behind closed doors.

Incarceration has had a particularly profound effect on African American communities, where 1 in 9 children has a parent in prison. In 2014 California passed a law aimed at reducing racial discrimination in drug laws by equalizing prison terms for crack cocaine and powder cocaine.

New Approaches to Managing Drugs

Because of the ineffectiveness and high cost of the war on drugs, new approaches are being tried. Drug courts, which supervise treatment for nonviolent drug addicts instead of sending them to jail, are a big component of this. Incarcerating a drug user costs around U.S.$25,000 per year, and most users return to drugs upon their release. In contrast, drug courts generally get users off drugs 60 percent of the time for a cost of $900–$3500 per person. Begun in the late 1980s in Miami in the midst of a crack cocaine epidemic, drug courts have since spread to every state, saving billions as a result. For example, Texas, a strong proponent of incarceration and home to the largest prison population in the United States (larger than the prison population of the United Kingdom, France, and Germany combined) has been able to close three of its bigger prisons since implementing drug courts. While significantly more funding has been given to drug treatment in recent years, the United States still spends more money on eradication, interdiction, and incarceration than it does on treatment.

The legalization of drugs is another tactic. In North America most of the focus has been on legalizing marijuana, though other countries, such as neighboring Mexico, have decriminalized cocaine, heroin, and marijuana. Roughly half of North America's population currently has access to legal marijuana for medicinal use. The use of marijuana for medical reasons in North America was documented as early as 1764, and recreational and medicinal use spread throughout much of urban North America during the nineteenth century. Efforts to regulate marijuana were initiated only in the early twentieth century, when campaigns to outlaw its use in the United States associated marijuana with "dangerous" Mexican immigrants. Marijuana was increasingly regulated during the 1930s, but arrests and incarceration for possessing and selling marijuana did not become common until the 1970s. Use surged during the Vietnam War, when many soldiers first encountered it in Southeast Asia, not far from South Asia, where marijuana was first domesticated.

Marijuana use is on the rise in North America. The highest rates of use are in the northern parts of Canada, the West Coast of the United States and Canada, and New England. In these areas marijuana is either legal, decriminalized, or essentially legal because of lax enforcement of drug laws. In 1996, California legalized the medical use of marijuana, and since then Canada and 24 U.S. states have done so as well. Recreational use of marijuana is now allowed on much of the U.S. West Coast, and in 2015 Justin Trudeau was elected prime minister of Canada on a platform that included the nationwide legalization of marijuana.

In the United States, a majority of the population supports the legalization of marijuana for recreational use. Some proponents of legalization focus on the financial benefits they see of legalizing marijuana and on issues of justice. They note the billions of dollars that could be saved in law enforcement costs and the billions more that could be raised by taxing marijuana at rates similar to those on alcohol and tobacco. They also discuss the need to avoid racially biased drug-related incarceration. Opponents of marijuana legalization point to research showing that marijuana has some adverse effects, such as an increased risk of cardiovascular disease and that it can impede concentration and reduce motivation to reach or set goals. Marijuana remains illegal under federal U.S. law, and around one-third of incarcerated federal drug offenders are serving prison sentences for marijuana-related offenses.

THINGS TO REMEMBER

GEOGRAPHIC THEME 3 • **Power and Politics:** Compared to many world regions, North America has relatively high levels of political freedom. However, there is also widespread dissatisfaction with the political process here. Internationally, Canada plays a modest role while the United States has enormous global political influence, although its status as the world's predominant "superpower" is declining.

• While an oft-stated ideal of U.S. official foreign policy (it is less of one for Canada) is to promote democracy abroad, the actual global distribution of U.S. spending on foreign aid and military assistance indicates that U.S. strategic and economic interests take precedence over promoting democracy.

• Canada continues to have a more robust social safety net, including health care, than does the United States.

Thinking Geographically

After you have read about urbanization in North America, you will be able to answer the following questions.

A Is Vancouver's growth pattern typical for North America?

B What factors reduce livability in New York?

C What factors reduce livability in Las Vegas?

D What factors reduce livability in Atlanta?

• While women voters are a potent political force in the United States and Canada, the percentage of elected politicians who are female is much less than the percentage of women in the general population of each country.

• More than a trillion dollars has been spent on the war on drugs, with generally little impact on the drug trade but profound impacts on North American society.

URBANIZATION

GEOGRAPHIC THEME 4

Urbanization: Since World War II, North America's urban areas have become less densely populated, expanding spatially much faster than their populations have increased numerically. This is primarily because of suburbanization and urban sprawl, which have transformed this highly urbanized region.

Close to 80 percent of North Americans live in **metropolitan areas**—cities of 50,000 or more, plus their surrounding suburbs and towns. Most of these people live in the car-dependent suburbs that were built after World War II, where urban life bears little resemblance to that in the central cities of the past.

In the nineteenth and early twentieth centuries, cities in Canada and the United States consisted of dense inner cores and less-dense urban peripheries that graded quickly into farmland. Starting in the early 1900s, central cities began losing population and investment, while urban peripheries—the **suburbs**—began growing (**FIGURE 2.21**). Workers were drawn by the opportunity to raise their families in single-family homes in secure and pleasant surroundings on lots large enough for recreation. Most continued to work in the city, traveling to and from on streetcars.

After World War II, suburban growth dramatically accelerated as cars became affordable to more people (**FIGURE 2.22**). In the United States, suburban growth was also encouraged by the federally funded Interstate Highway System, a network of free, high-speed roads that passed through most cities, providing easy access to surrounding land along the interstate. Canada, with a less extensive interstate highway system and slightly fewer people who could afford cars, had less suburban growth. The U.S. 2010 census data indicates that 51 percent of people live in suburbs, 32 percent are urbanites, and 17 percent live in rural areas. The 2011 census data for Canada indicate

metropolitan areas cities of 50,000 or more and their surrounding suburbs and towns

suburbs populated areas along the peripheries of cities

FIGURE 2.21 PHOTO ESSAY: Urbanization In North America

North America is very urbanized, with 79 percent of the population living in cities. Some of the wealthiest cities in the world are located here, though only a few have very high levels of "livability" compared to cities in Europe or to Oceania. Many cities, especially those in the United States, are characterized by sprawling development patterns that make people dependent on their automobiles. [Sources consulted: http://www.demographia.com/db-worldua.pdf; and *2015 World Population Data Sheet,* Population Reference Bureau, at http://www.prb.org/pdf15/2015-world-population-data-sheet_eng.pdf]

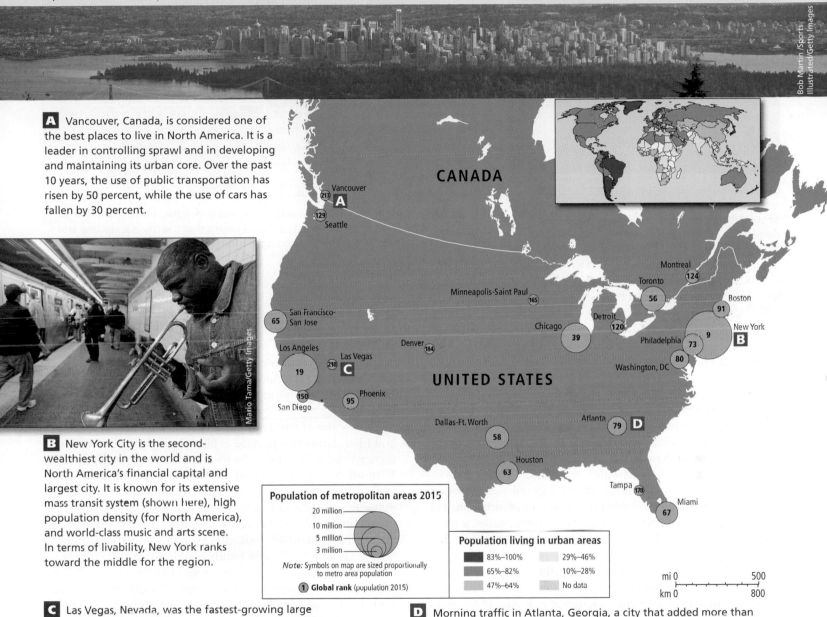

A Vancouver, Canada, is considered one of the best places to live in North America. It is a leader in controlling sprawl and in developing and maintaining its urban core. Over the past 10 years, the use of public transportation has risen by 50 percent, while the use of cars has fallen by 30 percent.

B New York City is the second-wealthiest city in the world and is North America's financial capital and largest city. It is known for its extensive mass transit system (shown here), high population density (for North America), and world-class music and arts scene. In terms of livability, New York ranks toward the middle for the region.

Population of metropolitan areas 2015

20 million
10 million
5 million
3 million

Note: Symbols on map are sized proportionally to metro area population

① **Global rank** (population 2015)

Population living in urban areas

83%–100%	29%–46%
65%–82%	10%–28%
47%–64%	No data

mi 0 — 500
km 0 — 800

C Las Vegas, Nevada, was the fastest-growing large city in North America between 2000 and 2007, with 31.8 percent growth. The recession that began in 2007 led to a population decline by 2010 as jobs in tourism shrank. For many, leaving was complicated by the inability to sell homes that had lost value. Las Vegas generally ranks toward the lower end of livability for North America.

D Morning traffic in Atlanta, Georgia, a city that added more than a million people between 2000 and 2010, the third largest amount of any city in North America (after Houston and Dallas). Atlanta is among the least dense and most sprawling cities in the region and depends almost entirely on cars for transportation. Atlanta is generally in the lower-middle rankings of livability for North America.

FIGURE 2.22 Urban sprawl in Phoenix, Arizona.

(A) Phoenix grew rapidly between 1950 and 2000, resulting in a demand for housing. The photo is of Sun City, a new, car-oriented suburb designed for senior citizens, located about 30 miles from Phoenix's center.

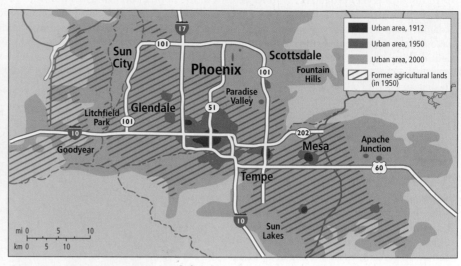

(B) This map shows the original urban settlement and how it has grown from 1912 to the present. A preference for single-family homes means that the city is sprawling into the surrounding desert rather than expanding vertically into high-rise apartments.

conurbation an area formed when several cities expand so that their edges meet and coalesce

that 39 percent of Canadians live in suburbs, 41 percent are urban-ites, and 20 percent are rural. Both urban and suburban populations are considered urban in Figure 2.21 and in similar maps in other chapters.

With the rise in popularity of the family car, interest in publicly funded, non-car-based forms of transportation for these expanding urban areas dwindled. This process was hastened by a consortium of automobile manufacturers, tire companies, and oil companies that purchased, dismantled, and replaced many urban streetcars and light-rail trains with bus systems.

As North American suburbs grew and spread out, nearby cities often coalesced into a single urban mass, or **conurbation.** The largest conurbation in North America, sometimes called a *megalopolis*, is the 500-mile (800-kilometer) band of urbanized area stretching from Boston through New York City, Philadelphia, and Baltimore, to south of Washington, DC. Other large conurbations in North America include the San Francisco Bay Area, Los Angeles and its environs, the region around Chicago, and the stretch of urban development from Seattle–Tacoma to Vancouver, British Columbia.

This pattern of *urban sprawl* requires residents to drive automobiles to complete most daily activities such as grocery shopping and commuting to work. Two major side effects of the dependence on vehicles that comes with urban sprawl are air pollution and emission of the greenhouse gases that contribute to climate change. Another important environmental consequence is the habitat loss that results when suburban development expands into farmland, forests, grasslands, and deserts.

Farmland and Urban Sprawl

Urban sprawl drives farmers from land that is located close to urban areas, because farmland on the urban fringe is very

attractive to real estate developers. The land is cheap compared to urban land, and it is easy to build roads and houses on since farms are generally flat and already cleared. As farmland is turned into suburban housing, property taxes go up and surrounding farmers who can no longer afford to keep their land sell it to housing developers. In North America each year, 2 million acres of agricultural and forest lands make way for urban sprawl.

Advocates of farmland preservation argue that beyond food and fiber, farms also provide economic diversity, soul-soothing scenery, and habitat for some wildlife. For example, the town of Pittsford, New York (a suburb of Rochester), decided that farms were a positive influence on the community. Mark Greene's 400-acre, 200-year-old farm lay at the edge of town. As the population grew, the chances of the farm remaining in business for another generation looked dim. Land prices and property taxes rose, and the Greene family could not meet their tax payments. Residents in the new suburban homes sprouting up on what had been neighboring farms pushed local officials to halt normal farm practices, such as noisy nighttime harvesting or planting, spreading smelly manure, and importing bees to pollinate fruit trees. Pittsford, however, decided to stand by the farmers by issuing U.S.$10 million in bonds so that it could pay Greene and six other farmers for promises that they would not sell their 1200 acres to developers but would instead continue to farm them.

VIGNETTE For Canada's Aboriginal peoples, the chance to indulge in the extravagance of urban sprawl is viewed positively. The Tsawwassen are one of 630 native groups that are negotiating with the Canadian government to gain full control over their ancestral territories. The land they recently received by a treaty lies 20 miles south of Vancouver, near the Strait of St. George. Eager to finally enjoy some of the fruits of development, they are planning a 2 million square foot "big box" and indoor shopping mall in the midst of what has been rural farmland. The surrounding communities, which had previously been able to shield their village-like settlements against urban sprawl, feel that there are enough shopping opportunities already. The Tsawwassen see the mall as an overdue chance for the prosperity that their neighbors already enjoy. ■

Smart Growth and Livability

In recent years the pace of urban sprawl has slowed as center cities have become more livable and drawn in more people. The term *smart growth* has been coined for a range of policies that shift urban growth away from car-dependent suburbs toward compact, walkable urban centers with mixed commercial and residential development and access to transit. *Livability* relates to factors that lead to a higher quality of life in urban settings, such as safety, good schools, affordable housing, quality health care, numerous and well-maintained parks that offer recreational opportunities for people and their pets (**FIGURE 2.23**), and well-developed public transportation systems. Some indexes of livability for major cities worldwide are put out every year by the *Economist* magazine and the Mercer Quality of Living Survey.

In smart growth planning, environmental benefits are envisioned as well. Creating areas where life can be lived locally, in which mass transit and walking replace the family car, and gardening for food is a form of community entertainment, results in less energy use, less air pollution, and fewer CO_2 emissions (**FIGURE 2.24**). While smart growth and the shift toward livability are just starting to take hold throughout North America, a few cities, such as Vancouver, British Columbia, and Portland, Oregon, are much further along in implementing these principles (see Figure 2.21A).

Some older cities are enjoying a renaissance as people move back to them in search of more livability. Places like New York City, which developed long before the term "smart growth" became popular, are arguably the furthest along in crucial aspects of smart growth, given their high density and widespread use of mass transit (see Figure 2.21B). However, the high cost of living in these places can reduce their livability.

It may be a while before smart growth and livability make much of an impact in some of North America's fastest-growing large cities, such as Dallas, Atlanta, and Phoenix, where in recent years car-dependent suburbs have spread out over enormous areas. Not surprisingly, none of these cities ranks high on livability indexes. Car dependence and rapid growth have brought Atlanta massive traffic jams and some of the worst air quality in the United States (see Figure 2.21D). Livability in the Las Vegas metro area is reduced by poor air and water quality as well as by high crime rates (see Figure 2.21C).

Inner-city decay is another impact of sprawl on livability. This is especially true in the United States, where many inner cities are dotted with large tracts of abandoned former industrial land and neighborhoods debilitated by persistent poverty and loss of jobs. Old industrial sites that once held factories or rail yards are called **brownfields.** Because they are often contaminated with chemicals and covered with obsolete structures, they can be very expensive to redevelop for other uses.

ON THE BRIGHT SIDE: The Greening of Detroit

Detroit, once the center of American auto industries, began to decline in the 1950s as manufacturing facilities were moved to the suburbs around the city, and to other parts of the United States, in part to avoid unionized labor.

> **brownfields** old industrial sites whose degraded conditions pose obstacles to redevelopment

FIGURE 2.23 LOCAL LIVES: People and Animals in North America

A Labrador retrievers are the most popular breed of dog in North America. Fishers on the island of Newfoundland, part of Canada's province of Newfoundland and Labrador, bred these dogs to retrieve fishing nets, which accounts for the breed's webbed paws and love of water.

B The ancestors of the Maine coon cat, one of the largest of the popular domestic cat breeds, were developed to control rodent populations aboard ships trading along the shore of Maine and other parts of the Atlantic coast of North America.

C Alaskan Malamutes pull a sled, which is the job that they were bred for by the Kuuvangmiut (formerly known as the Mahlemut) Inuit tribe of northwestern Alaska. Their large size and heavy build makes them ideal for pulling heavily loaded sleds.

Education Images/UIG via Getty Images

Kasia Wandycz/Paris Match via Getty Images

Yuri Yuriev/AFP/Getty Images

Jordan Siemens/Getty Images

FIGURE 2.24 Smart growth planning leads to increases in livability.

Faced with stiff competition from car makers in Japan, companies like Ford and General Motors cut thousands of jobs and the city's population shrank.

Forty years later, the depressing decay of a now much smaller Detroit is beginning to sprout with greenery and flowers as inner-city dwellers grow gardens on urban plots once occupied by houses that have since burned down. These green oases draw neighbors together and reconnect people with natural cycles. Children learn that food is something one can grow, not just buy. ■

Inner City Decay, Housing Segregation, and Redlining

Also left behind in the inner cities are people, many of whom were drawn in generations ago by the promise of jobs that have since moved out to the suburbs or overseas. Often the majority of the population in these inner cities is a mixture of African Americans, Asian Americans, Latinos, and new immigrants. Many are struggling to revitalize their communities, where a lot of the people are poor and in great need of jobs and social services, such as health care and education, that have moved to the suburbs.

A major feature of many U.S. cities is **housing segregation,** in which impoverished minority groups and the neighborhoods they live in have for years been denied fair access to housing by a variety of practices instituted by banks, landlords, and local governments.

housing segregation a common phenomenon in cities, in which ethnic or racial or economic classes live in separate neighborhoods and the segregation is maintained by discriminatory customs, not law

redlining The practice used by banks of drawing red lines on urban maps to show where they will not approve loans for home buyers regardless of qualifications

gentrification the renovation of old urban districts by affluent investors, a process that often displaces poorer residents

Banks in many U.S. cities have a history of denying mortgage loans on homes in the suburbs to minority borrowers even when they can afford them, and also of **redlining**—a practice that involves denying mortgages in minority neighborhoods in inner cities. The effect has been that many minorities have been shut out of home ownership and can only rent homes from landlords in poor, inner-city neighborhoods. Here rental units are often in a state of disrepair because landlords can't get loans to finance renovation. Landlords in suburbs will often refuse to rent to minorities, which allows landlords in inner cities to increase their rental prices. Local governments often concentrate low-income rental housing, subsidized by the federal government, in poor, inner-city neighborhoods where minorities are in the majority.

Legislation prohibiting redlining and other practices that create housing segregation has helped some inner-city minority communities get more access to mortgages and enabled more minority households to move to more affluent suburbs. However, many U.S. cities remain highly segregated, with whites dominating affluent suburban areas and minorities dominating poorer inner cities and their immediate suburbs. Canadian cities have had much less housing segregation, as redlining there focused more on unregulated suburban development.

The new emphasis on smart growth and livability has often perpetuated housing segregation in the United States through the **gentrification** of old inner-city neighborhoods. When affluent, usually white, people invest substantial sums of money in renovating old houses and buildings, poor, often minority, inner-city residents can be displaced in the process. The effect of gentrification on the displaced poor appears to be somewhat less harsh in Canada than in the United States, primarily because Canada's stronger social safety net better ensures social services, housing, and help with housing maintenance.

> ## THINGS TO REMEMBER
>
> **GEOGRAPHIC THEME 4** • **Urbanization** Since World War II, North America's urban areas have become less densely populated, expanding spatially much faster than their populations have increased numerically. This is primarily because of suburbanization and urban sprawl, which have transformed this highly urbanized region.
>
> • After World War II, suburban growth dramatically accelerated as cars became affordable to more people.
>
> • Urban sprawl drives farmers from land that is located close to urban areas, because farmland on the urban fringe is very attractive to real estate developers.
>
> • The term *smart growth* has been coined for a range of policies aimed at shifting urban growth away from car-dependent suburbs and toward compact and livable urban centers.
>
> • Urban sprawl is linked, especially in the United States, to inner-city decay. Many inner cities are dotted with large tracts of abandoned former industrial land and have neighborhoods that are debilitated by persistent poverty and job loss.

POPULATION AND GENDER

GEOGRAPHIC THEME 5

Population and Gender: For more than two centuries, North American birth rates have been in decline. A major factor in this has been women delaying childbearing to pursue education and careers. Declining childbirth rates play a role in the aging of North American populations, which may slow economic growth.

GENDER AND FERTILITY

North America is well along the demographic transition, with the number of children the average woman will have in her lifetime (also known as the *fertility rate*) declining since the early 1800s. This has happened in response to a number of factors: increasing economic development, improved health care, urbanization, and more women participating in the workforce.

All of these factors are connected in how they influence fertility. For example, social scientists have documented that increasing economic development is associated with decreased fertility. It is also usually associated with more women participating in the workforce, higher rates of urbanization, and improved health care. Similarly, urbanization is also associated with more women in the workforce, as families that move to the city are more likely to need the cash income that an adult woman who works outside the home can provide. This in turn is associated with lower fertility because women who have careers tend to delay childbearing. With regard to health care, improvements in the quality of health care and better access to medical professionals and medicines tend to lower the fertility rate as families choose to have fewer children because more will survive into adulthood. This choice is made easier with better access to birth control. All of these factors have come together to influence both gender and fertility in North America.

By the early 1800s, small numbers of North American women were starting to work in urban factories, where they had better access to health care. As more women chose this path, the fertility rate fell steadily from a high of 7 births per woman in 1800 to 3 births per woman in the 1920s. After the rapid rise in fertility following World War II, there was a sharp fertility decline as North American women worked more and gained access to birth control via "the pill," which became widespread in the 1960s. Since a low in the 1970s, North American fertility rates have remained more or less flat, at a level just below replacement level, contributing to an overall aging of the population.

AGING IN NORTH AMERICA

The aging of the region's population relates both to longer life expectancy, which increases the number of older people, and to a declining fertility rate, which decreases the number of younger people (FIGURE 2.25). During the twentieth century, the number of older North Americans grew rapidly. In 1900, one in 25 individuals was over the age of 65; by 2014, the number was 1 in 7. By 2050, when most of the current readers of this book will be over 50, it is likely that 1 in 5 North Americans will be elderly.

Dilemmas of Aging

Aging populations in developed countries like Canada and the United States present us with unfamiliar and as-yet unresolved dilemmas. On the one hand, it is widely agreed that globally, population growth should be reduced to lessen the environmental impact of human life on Earth, especially in those societies (such as in North America) that consume the most. On the other hand, slower population growth means that there will be fewer working-age people to keep the economy going and to provide the physical care the increasing number of elderly people will require.

Given that the population is aging, should retirement ages be extended from 65 to perhaps 70? Might this deprive young

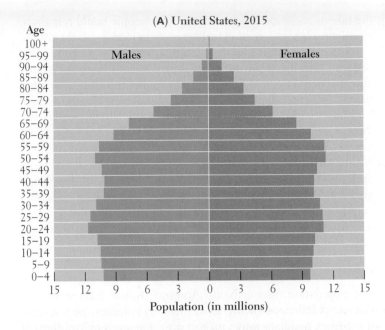

(A) United States, 2015

Population (in millions)

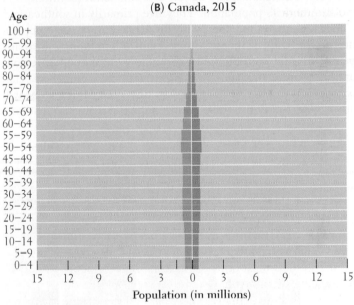

(B) Canada, 2015

Population (in millions)

FIGURE 2.25 Population pyramids for the United States and Canada, 2015. The "baby boomers" (those born between 1947 and 1964), constitute the largest age group in North America, as indicated by the wider middle portion of these population pyramids. [Sources consulted: International Data Base, U.S. Census Bureau, at https://www.census.gov/population/international/data/idb/region.php?N=%20Results%20&T=12&A=separate&RT=0&Y=2015&R=-1&C=US and https://www.census.gov/population/international/data/idb/region.php?N=%20Results%20&T=12&A=separate&RT=0&Y=2015&R=-1&C=CA]

people of access to jobs? These worries have greatly preoccupied Europeans, for whom retirement ages can be as low as 58 and whose populations are aging faster than those of North America. Economists who have studied this issue say that it clearly makes sense to extend the retirement age. For one thing if those over 65 remain self-supporting longer, this could result in higher youth employment, because there would be more working people spending their incomes and creating jobs by so doing. Moreover, the talents and experience of these older people could remain in circulation longer, creating further economic and social benefits.

Eventually most elderly people will need special attention of some sort. To avoid the social isolation that so many elderly already encounter, it will be necessary to develop affordable alternative living arrangements. **FIGURE 2.26** shows how the elderly of various ethnic groups in the United States are living now. The fact that women in all groups are the ones most likely to live alone may relate to women's generally longer life expectancy. Although North Americans of all ages are choosing to live alone more than they did in the past, this can be problematic for the elderly as physical frailty increases and incomes shrink. Day care and cohousing for the elderly, where residents look after each other with the aid of a small staff, are two affordable strategies to provide care and companionship.

POPULATION DISTRIBUTION

The population map of North America **(FIGURE 2.27)** shows the uneven distribution of the more than 358 million people who live here. Canadians make up just over one-tenth (37 million) of North America's population. They live primarily in southeastern Canada, close to the border with the United States. The population of the United States is more than 321 million, with many people moving from the Old Economic Core into other regions of the country that are now growing much faster. The U.S. Census Bureau predicts that the Northeast and Middle West will grow more slowly than the South and West. Canada's national

statistics agency predicts a similar pattern of western and southward movement of population.

The Geography of Population Change in North America

Every year, almost one-fifth of the U.S. population and two-fifths of Canada's population relocate. Some people are changing jobs and moving to cities; others are attending school, retiring to a warmer climate or a smaller city or town; and others are merely moving across town or to the suburbs or the countryside. Still other people are arriving from outside the region as immigrants.

In many farm towns and rural areas in the *Middle West*, or *Midwest* (the large central farming region of North America), populations are shrinking. As family farms are consolidated under corporate ownership, labor needs are decreasing and young people are choosing better-paying careers in cities. Midwestern cities are growing only modestly but are becoming more ethnically diverse, with rising populations of Latinos and Asians in places such as Indianapolis, St. Louis, and Chicago.

In the western mountainous interior, traditionally settlement has been light (see Figure 2.27) due to the rugged topography, lack of rain, and in northern or high-altitude zones, a growing season too short to sustain agriculture. There are some population clusters in irrigated agricultural areas, such as in the Utah Valley, near rich mineral deposits or resort areas. The gambling economy and frenetic construction activity generated by real estate speculation account for several knots of dense population at the southern end of the region.

Along the Pacific coast, a band of population centers stretches north from San Diego to Vancouver (see Figure 2.21A) and includes Los Angeles, San Francisco, Portland, and Seattle. These are all port cities engaged in trade around the **Pacific Rim** (all the countries that border the Pacific Ocean). Over the past several decades, these North American cities have become centers of technological innovation.

The rate of natural increase in North America (0.4 percent per year) is low, less than half the rate of the rest of the Americas (1.2 percent). Still, North Americans are adding to their numbers fast enough through births and immigration that the population could reach 401 million by 2030 and 445 million by 2050.

Pacific Rim a term that refers to all the countries that border the Pacific Ocean

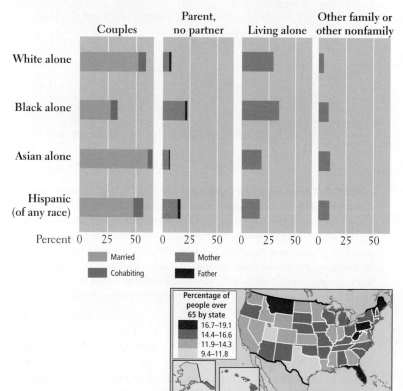

FIGURE 2.26 Living arrangements of U.S. people age 65 and over by sex, race, and ethnicity, 2015. [Sources consulted: https://www.census.gov/hhes/families/files/graphics/HH-7a.pdf and http://www.census.gov/quickfacts/map/AGE775214/00]

THINGS TO REMEMBER

GEOGRAPHIC THEME 5 • **Population and Gender:** For more than two centuries, North American birth rates have been in decline. A major factor in this has been women delaying childbearing to pursue education and careers. Declining childbirth rates play a role in the aging of North American populations, which may slow economic growth.

• North America's population is rapidly aging, and by 2050, one in five people will be over the age of 65. There will be fewer young people to work, pay taxes, and take care of the elderly.

• The population of North America is changing in distribution and becoming more diverse.

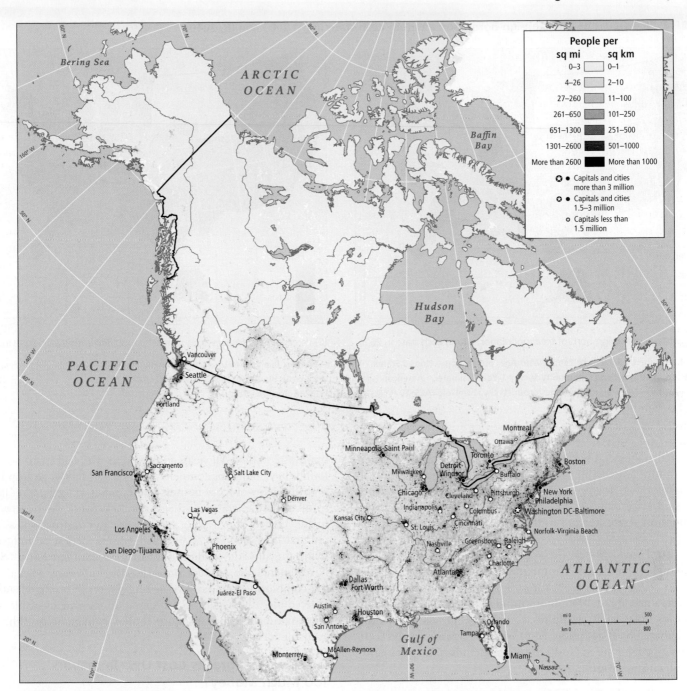

FIGURE 2.27 Population density in North America.

- North Americans are highly mobile. Every year, for a variety of reasons, almost one-fifth of the U.S. population and two-fifths of the Canadian population move to a different location.

- The population of North America is changing in geographic distribution, becoming more diverse, and slowly growing.

SOCIOCULTURAL ISSUES

North America is increasingly diverse, with new immigrants from across the globe adding to a dominant culture that has European roots.

IMMIGRATION AND DIVERSITY

Immigration has played a central role in populating both the United States and Canada. Most people in North America descend from European immigrants, but there are many who have roots in Africa, Asia, Middle and South America, and Oceania. New waves of migration from Middle and South America and parts of Asia promise to make North America a region where people of non-European or mixed descent will become the majority **(FIGURE 2.28)**. Most major North American cities are already characterized by ethnic diversity, with the once numerically dominant Caucasian population now a minority in many big cities. However, most large cities are still highly

FIGURE 2.28 Percent of total foreign-born people within each state in (A) 2000 and (B) 2010.

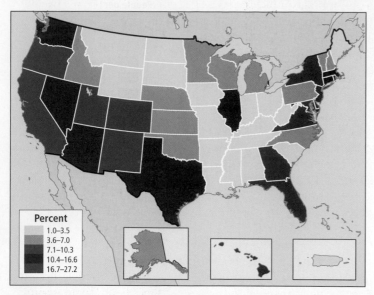

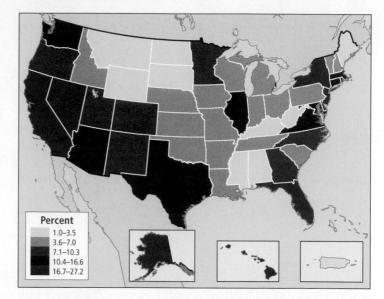

(A) Percentage of foreign-born of total population within each state in 2000.

(B) Percentage of foreign-born of total population within each state in 2010.

[Sources consulted: "Percent of People Who Are Foreign Born—United States—Places by State; for Puerto Rico Universe: Total Population, 2010 American Community Survey 1-Year Estimates," American FactFinder, U.S. Census Bureau, at http://factfinder2 .census.gov/faces/tableservices/jsf/pages/productview.xhtml?pid=ACS_10_1YR_GCT0501.US13PR&prodType=table]

segregated at the neighborhood level, with different ethnic groups living mostly among themselves. There are many methods of measuring urban segregation, but generally speaking the least segregated large cities are on the west coast of North America and the most segregated are in the Old Economic Core and the South.

In the United States, the spatial pattern of immigration is also changing. For decades, immigrants settled mainly in coastal or border states such as New York, Florida, Texas, or California. However, since about 1990, more immigrants have been settling in interior states such as Illinois, Colorado, Nevada, and Utah (see Figure 2.28). The influence of immigrants can also be seen in many aspects of life in North America, including the popularity of ethnic cuisines (**FIGURE 2.29**).

Why Do People Decide to Immigrate to North America?

Given the diverse economic opportunities that North America offers, immigration is often considered in terms of positive **pull factors** that draw in migrants. However, the decision to leave one's homeland is usually not an easy one. There are almost always negative **push factors** that cause people to leave home, family, and friends, and strike out into the unknown with what are usually limited resources. These push factors can be civil or political unrest or some kind of discrimination, but most often they involve a lack of economic opportunity. Many push factors relate to the forces of globalization. For example, when NAFTA enabled subsidized U.S. agribusinesses to sell their corn in Mexico on a massive scale, as

pull factors positive factors that draw in migrants

push factors negative factors that cause people to leave home, family, and friends

many as 2 million Mexican corn farmers who had small holdings found they could not compete on price and had to give up farming. But many of these farmers had families to feed and educate, so often they migrated north in search of work. Many now live in the United States, where they work in construction, as agricultural field laborers, or in a wide variety of service industries, living frugally and sending most of their earnings home. Some have entered the United States legally, but many—because of the complexities and costs of getting papers to immigrate—have entered illegally. Legal U.S. immigration visas are available in far fewer numbers than both the potential immigrants and the jobs available for them to fill.

Do New Immigrants Cost U.S. Taxpayers Too Much Money?

Many North Americans are concerned that immigrants burden schools, hospitals, and government services. And yet numerous studies have shown that, over the long run, immigrants contribute more to the U.S. economy than they cost. Legal immigrants have passed an exhaustive screening process that assures they will be self-supporting. As a result, most start to work and pay taxes within a week or two of their arrival in the country. Most immigrants who draw on taxpayer-funded services tend to be legal refugees fleeing a major crisis in their homeland and are dependent only in the first few years after they arrive. More than one-third of immigrant families are firmly within the middle class, with incomes of U.S.$45,000 or more. Even undocumented (illegal) immigrants play important roles as payers of payroll taxes, sales taxes, and indirect property taxes through rent. Because they fear deportation, undocumented immigrants are also the least likely to take advantage of taxpayer-funded social services.

On average, immigrants are healthier and live longer than native U.S. residents, according to studies by the National Institutes of Health (NIH). They therefore create less drain on the health-care and social service systems than do native residents. The NIH attributes this difference to a stronger work ethic, a healthier lifestyle that includes more daily physical activity, and the more nutritious eating patterns of new residents compared to those of U.S. society at large. Unfortunately, these healthy practices tend to diminish the longer immigrants are in the country, and the healthier status does not carry over to immigrants' children, who are nearly as likely to suffer from obesity as native-born children.

Do Immigrants Take Jobs Away from U.S. Citizens?

The least educated, least skilled American workers are the most likely to end up competing with immigrants for jobs. In a local area, a large pool of immigrant labor can drive down wages in fields like roofing, landscaping, and general construction. Immigrants with little education now fill many of the very lowest-paid service, construction, and agricultural jobs.

It is often argued that U.S. citizens have rejected these jobs because of their low pay, which results in immigrants being needed to fill the jobs. Others say that these jobs might pay more and thus be more attractive to U.S. citizens if there were not a large pool of immigrants ready to do the work for less pay. Research has failed to clarify the issue. Some studies show that immigrants have driven down wages by 7.4 percent for those U.S. natives who do not have a high school diploma. However, other studies show no drop in wages at all. Employers across the country speak of the superior work ethic of immigrants, especially in the construction industry, which makes them happy to pay these workers decent wages.

Professionals in the United States occasionally compete with highly trained immigrants for jobs, but such competition is usually in occupations where there is a scarcity of native-born people who are trained to fill these positions. The computer engineering industry, for example, regularly recruits abroad in such places as India, where there is a surplus of highly trained workers. In this case, it is unclear whether these skilled immigrants are driving down wages. Until the recession caused jobs to be eliminated, it was quite clear that there were not enough sufficiently trained Americans to fill the available positions. In fact, India's growth in IT jobs and related services has been so impressive that many who came to North America to work in these jobs have elected to go back to India to work in the burgeoning IT industry in India, a development that has not been welcomed by their American employers.

Are Too Many Immigrants Being Admitted to the United States?

Many people are concerned that immigrants are coming in such large numbers that they will strain resources here. There is some validity to this. Immigrants and their children accounted for 78 percent of the U.S. population growth in the 1990s. At the current growth rates, by the year 2050, the U.S. population could reach 445 million, with immigration accounting for the majority of the increase. But one of the problems with making these projections is that no one really knows how many undocumented immigrants there are. Most research indicates that illegal immigration possibly exceeded legal immigration rates during the mid-1990s. While estimates of the undocumented immigrant population currently in the United States range from 7 to 20 million, reports

---

FIGURE 2.29 LOCAL LIVES: Foodways in North America

A Pork ribs are smoked and slow cooked in the "pit" at Spoon's Barbecue in Charlotte, North Carolina. Originally, Native Americans would first dig a pit or trench and fill it with burning coals or hot rocks. They would place meat on top and either slow cook it in the open at a low temperature for several hours or cover the entire pit and let it cook for a day or more. Most barbeque is now cooked in a metal smoker, still called a pit, which is often attached to a trailer so that it can be brought to special events.

B Poutine, a French Canadian dish, now hugely popular throughout Canada, consists of French fries served with cheese curds (a tasty by-product of cheese making) and covered in gravy. One story holds that in the 1950s, a take-out customer of Fernand LaChance's restaurant in Warwick, Québec, asked for the combination of fries and cheese curds, to which the owner responded "Ça va faire une maudite poutine!" ("That's going to make a cursed mess!"), thus giving the dish its name. The gravy was later added to keep the fries warm.

C New England–style clam chowder served in a sourdough bread bowl is a classic San Francisco dish that springs from two American traditions. New England clam chowder was developed by fishers and consists of clams and other seafood, potatoes, and seasonings, cooked in a broth to which hard "sea biscuits" were mixed to thicken the chowder. In San Francisco, sourdough bread came via pioneer families and was later popularized by French bakers. It has a particularly sour, tangy flavor that may be the result of the climate of the San Francisco Bay Area.

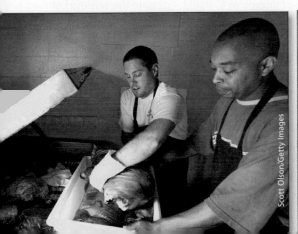
Scott Olson/Getty Images

Christinne Muschi/Toronto Star via Getty Images

Andrew McKinney/Getty Images

since 2012 indicate that migration—legal and illegal—is down to less than half that of previous years, and that more undocumented immigrants are now leaving the United States than entering.

Undocumented immigrants tend to lack skills, and they are not screened for criminal background, as are all legal immigrants. However, research also shows that undocumented immigrants are not only less likely to partake of social services, but are also less likely than the general population to participate in criminal behavior, with only tiny percentages of them having committed offenses.

Why Are So Many Unaccompanied Children Coming as Illegal Immigrants?

In recent years, thousands of unaccompanied children have arrived in the United States. Most are coming from Central America, where economic instability and violence have torn many families apart. Several types of U.S. policies also play a role in this migration. The *Central America Free Trade Agreement* (CAFTA) is an expansion of NAFTA that has created many of the same problems for poor farmers in Central America as NAFTA did in Mexico. U.S. policies that grant special rights to children also created the belief among some Central American families that their children would be granted asylum in the United States if they arrived alone. These policies have been suspended and thousands of children have been deported. Many more are being held in prison-like detention facilities run by U.S. corporations. Relatively little action has been taken by the U.S. government or governments in Central America to address the root causes of the migration.

ethnicity the quality of belonging to a particular culture group

VIGNETTE Every youth who arrives at the border detention shelter in Arizona writes a life story in a creative writing class. A 14-year-old indigenous girl from Honduras started hers: "At 10 years old, my papa started to tell me the good things and the bad things." The next day the teacher asked what this meant. After sitting in silence for 30 minutes, the girl wrote: "While my father is a good man, he did not always do good things. Some of these things created a bad situation for me." Afraid to tell her mom, she called her brothers to help her escape.

Three years before, at the peak of the global recession, her then-teenage brothers, unable to find work in their rural area, decided to take the month-long journey to the United States by jumping trains through Mexico and then walking for days across the mountainous and arid terrain of Texas. Once out of Texas, they found migrant farm work in Southern states. The brothers were working in Florida when they spoke to their sister on the phone.

Within days, the brothers sent their sister all of their savings—$8000—and arranged for a *coyote* to accompany the girl to the United States. As an unaccompanied and attractive female, she was more vulnerable to exploitation, abuse, and sex trafficking. Like most detained in the border shelter, she is unwilling to discuss what happened on her journey, describing it as "the necessary suffering to become someone." Luckily, she made it without falling victim to the drug trafficking and fatal gang violence along the border.

In the 2 months it took her to arrive, she lost contact with her brothers. Lacking family contacts and money, her deportation is likely. Her biggest desire is "to work hard and give the money back to my brothers." [*Source: Material for this vignette was pieced together by Lydia Pulsipher in April 2012 from personal correspondence with Elizabeth Kennedy and Stuart Aitken, geographers who participate in the ISYS Unaccompanied Minors project.*] ∎

RACE AND ETHNICITY IN NORTH AMERICA

Despite strong scientific evidence to the contrary, people across the world still perceive skin color and other visible anatomical features to be significant markers of intelligence and ability. This racialized view of skin color is partially a remnant of European colonialism, which was based on assigning some people to a perpetually low status while giving others privilege. As discussed previously (see Chapter 1), the science of biology tells us there is no scientific validity to such racialized assumptions and practices. There is no valid way to separate groups of people according to skin color or hair or any other feature. The same is true for **ethnicity,** which is the cultural counterpart to race, in that people may ascribe overwhelming (but scientifically unwarranted) significance to cultural characteristics such as religion, family structure, or gender customs. Thus, race and ethnicity are very important sociocultural factors not because they *have* to be, but because people *make* them so.

Extending Equal Opportunity

Numerous surveys show that a large majority of Americans of all backgrounds favor equal opportunities for minority groups. Nonetheless, in both the United States and Canada, many middle-class people of African descent, Latinos, Native Americans and Aboriginal peoples, and people of Asian descent report experiencing both overt and covert discrimination that affects them economically as well as socially and psychologically. And indeed, even a cursory examination of statistics on access to health care, education, and financial services shows that, on average, North Americans have quite unequal experiences based on their racial and ethnic characteristics.

Diversity is increasing across North America, as is the acceptance of diversity. Take, for example, the enthusiasm with which various cultural festivals are attended by Americans regardless of their ethnic heritage (FIGURE 2.30). Still, the issue of race remains important in both the United States and Canada, where whiteness is considered the norm and people of non-European descent are often made to feel out of place. Prejudice has clearly hampered the ability of these groups to reach social and economic equality with Americans of other ethnic backgrounds. In the United States—and to a considerably lesser degree in Canada—despite the removal of legal barriers to equality, these groups (with the notable exception of Asians) still have higher poverty rates and thus lower life expectancies, higher infant mortality rates, lower levels of academic achievement, and more unemployment than other groups.

FIGURE 2.31 shows the changes in the ethnic composition of the North American population from 1950 to 2010 and the projected changes for 2050. In 2001, Latinos overtook African Americans as the largest minority group in the United States. Because of a higher birth rate and a high immigration rate, the Latino population increased by 58 percent in the 1990s, to 16.4 percent of the total population by 2010. By 2010, Asian Americans

FIGURE 2.30 LOCAL LIVES: Festivals in North America

A Mardi Gras Indian Chief Golden Comanche leads a procession through a neighborhood in New Orleans, Louisiana. Mardi Gras, which means "Fat Tuesday," marks the last day before the start of Lent, a 40-day period of fasting and abstinence in Christianity that precedes Easter. Mardi Gras Indians originated as a tribute to Native American communities that had harbored runaway enslaved African Americans.

B A temporary sculpture at Burning Man, a weeklong "art event and transitory community based on radical self-expression and self-reliance." Begun in 1986 as a bonfire on the beach in San Francisco, Burning Man is guided by principles of inclusive participation, anticommercialism, and the goal of not leaving any physical trace of the festival after it happens each year.

C Bull riding is an event at the Calgary Stampede, a 10-day rodeo, exhibition, and festival that was started in 1886 to help draw settlers from the east. Billed as the "Greatest Outdoor Show on Earth," the Stampede draws more than a million people each year. The highlight of today's Stampede is its rodeo, which offers more prize money than any other.

made up just 5 percent of the U.S. population, but their numbers increased by 80 percent between 1990 and 2010. Of the foreign-born population in Canada in 1981, Asians made up 14.1 percent and those of European birth, 66.7 percent. Asians appear to be the fastest-growing nonnative-born group in both countries.

Income Discrepancies

Over the past few decades, many non–Euro-Americans and those of mixed heritage have joined the middle class, finding success in the highest ranks of government and business. In particular, African Americans have completed advanced degrees in large numbers, and more than one-third now live in the suburbs. Yet, as overall groups, African Americans, Latinos, and Native Americans remain the country's poorest and least educated people **(FIGURE 2.32)**. Similar patterns exist in Canada, though income differences are less, and some Asian groups have lower incomes than African Canadians and Latino Canadians.

U.S. Population by Race and Ethnicity, 1950, 2014, and 2050 (projected)

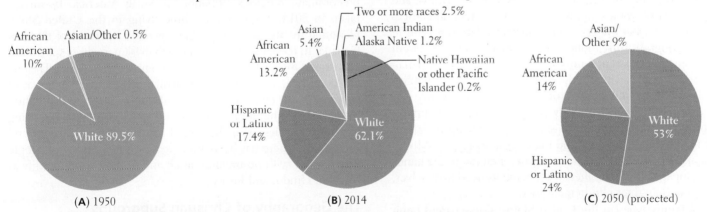

(A) 1950

African American 10%
Asian/Other 0.5%
White 89.5%

(B) 2014

Asian 5.4%
African American 13.2%
Two or more races 2.5%
American Indian / Alaska Native 1.2%
Native Hawaiian or other Pacific Islander 0.2%
Hispanic or Latino 17.4%
White 62.1%

(C) 2050 (projected)

Asian/Other 9%
African American 14%
White 53%
Hispanic or Latino 24%

FIGURE 2.31 The changing ethnic composition in the United States, 1950, 2014, and 2050 (projected). As the percentage of ethnic minorities increases in the United States, the percentage of whites decreases. By 2050, if present reproductive trends continue, whites will constitute only slightly more than half of the population. [Sources consulted: Jorge del Pinal and Audrey Singer, "Generations of Diversity: Latinos in the United States," Population Bulletin 52 (October 1997): 14; U.S. Census Bureau, "Race by Sex, for the United States, Urban and Rural, 1950, and for the United States, 1850 to 1940," *Census of Population, 1950*, Vol. 2, Part 1, United States Summary, Table 36, 1953, at http://www2.census.gov/prod2/decennial/documents/21983999v2p1ch3.pdf; http://2010 .census.gov/news/releases/operations/cb11-cn125.html; http://quickfacts.census.gov/qfd/states/00000.htm]

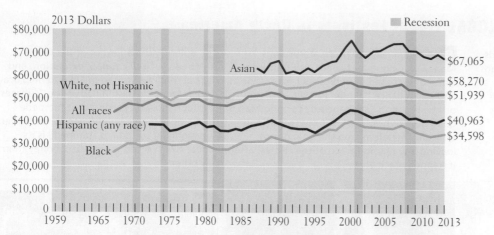

FIGURE 2.32 **Median household income by race and ethnicity, in U.S. dollars, 1967–2013.** [Source consulted: http://www.census.gov/content/dam/Census/library/publications/2015/demo/p60-252.pdf]

The Culture of Poverty

Is anything other than prejudice holding back some in these ethnic minority groups? Some social scientists suggest that persistently disadvantaged Americans of all ethnic backgrounds may suffer from a *culture of poverty*, meaning that poverty itself forces coping strategies that are counterproductive to social advancement. Plans to get higher education, for example, must be abandoned because of lack of funds. This and low social status have bred the perception among the poor that there is no hope for those trapped in the culture of poverty and therefore no point in trying to succeed. The reality is that most people living in poverty have few opportunities to complete a college education. They work at jobs with low wages, and because of housing segregation few can find decent affordable housing.

One component in the culture of poverty is the growing numbers of low-income, single-parent families. In 2013, only 25 percent of Euro-American children and 16 percent of Asian American children lived in single-parent families, while 67 percent of African Americans, 52 percent of Native Americans, and 42 percent of Latinos did. The reasons for these stark differences are often systemic. For example, many social welfare programs in the U.S. give significantly more income support to single-parent families than to two-parent families. This has the effect of penalizing married couples with children and rewarding those that live separately. In these situations, children usually stay with their mothers, and fathers are less likely to be active in their support and upbringing. Moreover, the assistance packages always fall far short of meeting actual needs, and the enormous responsibilities of both child rearing and breadwinning are often left in the hands of undereducated young mothers who, being in need themselves, are unable to help their children advance.

Of all family types in the United States, single-parent families headed by women have the lowest median income. They also have the lowest levels of education, which is linked to low income. One aspect of the proliferation of single-parent households with lower education levels and incomes is that children in the United States are disproportionately poor. In 2010 in the United States, 22 percent of children aged 0 to 17 (16.4 million) lived in poverty, whereas only 11.4 percent of adults did. In

Canada, 14.2 percent of children were poor. By comparison, in Sweden, just 2.4 percent of children lived in poverty; in Ireland, 12.4 percent; and in Poland, 12.7 percent.

ON THE BRIGHT SIDE: More Discussion of Race and Racism

In 2008 and again in 2012, the United States made history by electing its first African American president. Some commentators mistakenly heralded this as the end of racism and of the need to even talk about race. As if to prove them wrong, an avalanche of racially charged stories entered the news cycle of the mainstream media. For example, in 2014 the national news focused on the shooting of Michael Brown, an unarmed black teenager, by Darren Wilson, a white police officer in Ferguson, Missouri. The shooting produced an outpouring of discussion, protests, and other actions that were extensively covered by the media, raising awareness of race in the United States.

Lost in much of the discussion was the fact that shootings like this, even though they are quite common, have historically rarely made even the local news, much less the national news. Currently, most major U.S. national news outlets feature an entire section on race and ethnicity, at least in their online offerings, which they didn't do even just a few years ago. It may be a long time before any person's or organization's claims to be "beyond race" ring true in the United States, but at least the subject is receiving more attention and being more openly discussed. ■

RELIGION

Because so many early immigrants to North America were Christian in their home countries, Christianity is currently the predominant religious affiliation in North America. In surveys taken in 2014, 70 percent of those living in the United States identified themselves as Christian, as did 67 percent of those in Canada. In the United States, more than 67 percent of Christians are Protestants; in Canada, 57 percent are Catholic. Despite the numerical dominance of Christianity, about 23 percent of people in both countries reported having no religious affiliation **(FIGURE 2.33)**. The number who identify as Christian is declining in both countries, with the largest losses among Catholics and Protestants. There has been an increase in the number of people who identify with no tradition at all and of those who identify as Muslim, Hindu, and Jewish.

The Geography of Christian Subgroups

There are many versions of Christianity in North America, and their geographic distributions are closely linked to the settlement patterns of the immigrants who brought them here (see Figure 2.33). Roman Catholicism dominates in regions where Latino, French, Irish, and Italian people have settled—in southern Louisiana, the Southwest, and the far Northeast in the United States, and in Québec and other parts of Canada. Lutheranism is

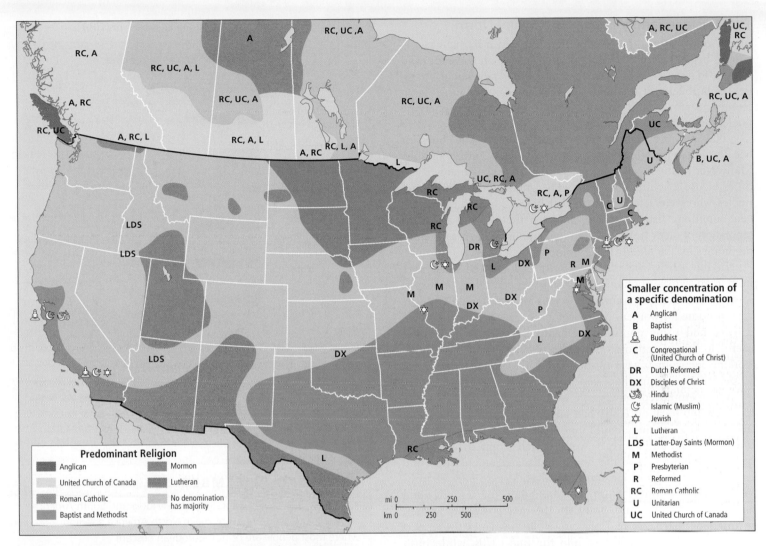

FIGURE 2.33 Religious affiliations across North America. [Sources consulted: Jerome Fellmann, Arthur Getis, and Judith Getis, *Human Geography* (Dubuque, IA: Brown & Benchmark, 1997), p. 164; and http://www.pewresearch.org/topics /religious-affiliation/2015/pages/3/]

dominant where Scandinavian people have settled, primarily in Minnesota and the eastern Dakotas. Mormons dominate in Utah.

Baptists, particularly Southern Baptists and other evangelical Christians, are prominent in the *Bible Belt*, which stretches across the Southeast from Texas and Oklahoma to the Eastern Seaboard. Evangelical Christianity is such an important part of community life in the South that frequently the first question newcomers to the region are asked is what church they attend. Among all the subregions in North America, rates of weekly attendance at a religious service are highest in the Southern United States.

The Relationship Between Religion and Politics

Just how interactive religion and politics should be in North American life has long been a controversial issue in the United States, more so than in Canada. This is true in large part because the framers of the U.S. Constitution, to help ensure religious freedom, declared that church and state should remain separate. However, religion has had a powerful influence on the politics of this region, mostly in the form of Christian political movements surrounding such issues as the abolition of slavery (1770s

to 1865); the prohibition of the sale, production, and distribution of alcohol (1840s to 1933); the movement to ban abortion (1970s to the present) and the promotion of prayer in the public schools (1960s to the present), just to name a few.

National surveys have consistently indicated that a substantial majority of North Americans favor the separation of church and state and support personal choice in belief and behavior.

THE AMERICAN FAMILY

The family is often seen as being endangered by North America's fast-changing culture. A century ago, most North Americans lived in large, extended families of several generations. Families pooled their incomes and shared chores. Aunts, uncles, cousins, siblings, and grandparents were almost as likely to provide daily care for a child as were the mother and father. Though the **nuclear family,** consisting of a married father and mother and their children, has been a part of society going back through history, it became especially widespread in the post-1900 industrial age.

nuclear family a family consisting of a married father and mother and their children

The Nuclear Family Becomes a Shaky Norm

After World War I, and especially after World War II, many young people began to leave their large kin groups on the farm and migrate to distant cities, where they established new nuclear families. Suburbia, with its many similar single-family homes, soon seemed to provide the perfect domestic space for the emerging nuclear family.

This compact family type suited industry and business because it had no firm ties to other relatives and so was portable. Many North Americans born since 1950 moved as many as ten times before reaching adulthood. The grandparents, aunts, and uncles who were left behind missed helping raise the younger generation and had no one to look after them in old age. This separation of older from younger generations is one reason nursing homes for the elderly have proliferated.

In the 1970s, the nuclear family encountered challenges from many areas, including changing gender roles. Suburban sprawl meant onerous commutes to jobs for men and long, lonely days at home for women. More women began to want their own careers, and rising consumption patterns made their incomes useful to family economies. By the 1980s, seventy percent of working-age women were in the paid workforce, compared with 30 percent of their mothers' generation.

Once employed, however, a woman could not easily move to a new job location if she had an upwardly mobile husband. Also, working women could not manage all of the family's housework and child care, as well as a job. Some married men began to handle part of the household management and child care, but the demand for commercial child care grew sharply. With family no longer around to strengthen the marital bond and help with child care, and with the new possibility women in unhappy marriages had of being able to financially support themselves, divorce rates rose into the mid-1970s, while birth rates fell. Those who married after 1975, however, have had a slowly declining rate of divorce. This may be related to the growing number of couples with college educations, as this group is less likely to divorce.

The Current Diversity of North American Family Types

There is no longer a typical American household, only a diversity of household forms and ways of family life **(FIGURE 2.34)**. In 1960, the nuclear family—households consisting of married couples with children—comprised 74.3 percent of U.S. households. This number declined to 56 percent by 1990 and to 49.3 percent by 2010. Family households headed by a single person (female or male) rose from 10.7 percent in 1960, to 15 percent in 1990, to 18 percent in 2010. Nonfamily households (unrelated by blood or marriage) rose from 15 percent of the total in 1960, to 29 percent in 1990, to nearly 34 percent in 2010. Canada has similar patterns of change: nuclear family households shrank from 91.6 percent of all Canadian households in 1961 to 67 percent in 2011; family households headed by a single parent rose from 8.4 percent in 1961 to 16.3 percent in 2011; and nonfamily households rose from 8.6 percent in 1961 to 17.1 percent in 2011.

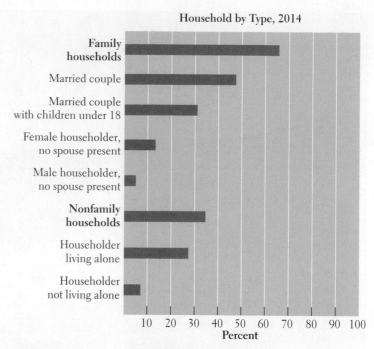

FIGURE 2.34 U.S. households by type, 2014.

[Sources consulted: "Selected Population Profile in the United States: 2007–2009 American Community Survey 3-Year Estimates," *American FactFinder*, U.S. Census Bureau, at http://factfinder2.census.gov/faces/tableservices/jsf/pages/productview.xhtml?pid=ACS_09_3YR_S0201&prodType=table; and http://www.census.gov/content/dam/Census/library/publications/2014/demo/p60-249.pdf?cssp=SERP]

THINGS TO REMEMBER

• A steady increase in migration from Middle and South America and parts of Asia promises to make North America a region where most people are of non-European descent.

• Statistics on access to health care, education, and financial services by race and ethnicity show that, on average, Americans have quite uneven experiences based on their racial and ethnic characteristics.

• A large percentage of children in the United States live in poverty, compared to other developed countries.

• There is no longer a typical American household, only a diversity of household forms.

GEOGRAPHIC PATTERNS OF HUMAN WELL-BEING

The maps of human well-being for North America tell a quick and somewhat misleading story **(FIGURE 2.35)**. First of all, we learn from Map A that both countries in this region have very high average per capita GNI (PPP) figures and that in the global context (see the global inset for Map A), these two countries are among the wealthiest. But this is a classic example of why average GNI figures are not a sufficient measure of well-being, because we now know that there is much disparity of wealth in North America that is not revealed by country-level GNI figures. This map does not show how the wealth is distributed geographically or between

FIGURE 2.35 Maps of human well-being.

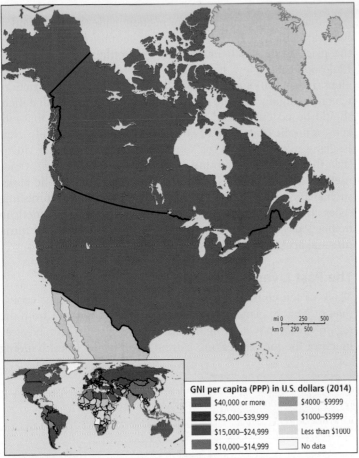

(A) Gross national income (GNI) per capita, adjusted for purchasing power parity (PPP).

GNI per capita (PPP) in U.S. dollars (2014)

$40,000 or more	$4000–$9999
$25,000–$39,999	$1000–$3999
$15,000–$24,999	Less than $1000
$10,000–$14,999	No data

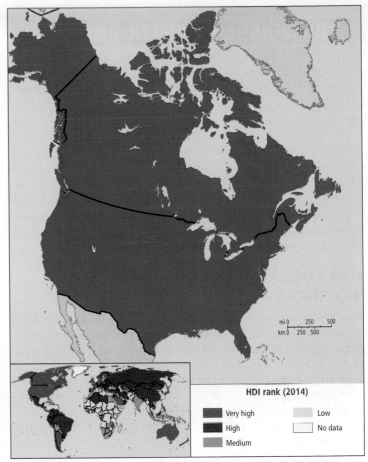

(B) Human Development Index (HDI) rank.

HDI rank (2014)

Very high	Low
High	No data
Medium	

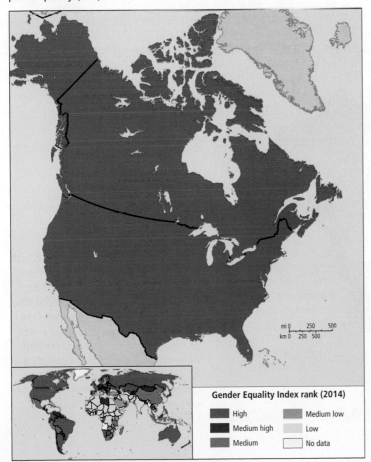

(C) Gender Development Index (GDI) rank.

Gender Equality Index rank (2014)

High	Medium low
Medium high	Low
Medium	No data

social classes. It gives no hint that 22 percent of children in the United States and 14 percent in Canada are classed as poor despite having at least one working parent, that the situation has deteriorated over the last decade, or that in these countries the richest 10 percent have more than 14 times the wealth of the poorest 10 percent. The truth is that at the local scale the poor are geographically separated from the middle class so completely that poor children often experience life only in very deprived communities.

Virtually the same problem is presented by Map B, which shows the ranks of these two countries on the United Nations Human Development Index (HDI). Because the data are presented for the entire countries, not at the province or state level, the fact that many parts of both countries provide very well for their citizens in terms of income, health care, and education boosts the average figures for all. Glossed over are the significant pockets of poverty, ill health, and illiteracy that are present in virtually every state and province (they are less devastating in Canada because of its more robust safety net).

Map C, the Gender Development Index (GDI), measures disparities in HDI by gender. HDI values are estimated separately for women and men, and the ratio of these two values is the GDI. The closer the ratio is to 1, the smaller the gap between women and men. Map C shows that the United States and Canada have fairly equitable human development across genders, relative to other places on Earth. However, it does not illustrate how Australia and North and West Europe are doing better at equalizing pay and opportunities for female and male citizens.

SUBREGIONS OF NORTH AMERICA

When geographers try to understand patterns of human geography in a region as large and varied as North America, they usually impose some sort of subregional order on the whole (FIGURE 2.36). The order used in this book divides North America into eight subregions. Each section on a subregion sketches in the features that give the subregion its distinct "character of place." As we noted in Chapter 1, however, geographers rarely reach consensus on just where regional boundaries should be drawn or specifically how to define a given region. There are many schemes for dividing North America into subregions. The scheme used here is partly based on a book by Joel Garreau, *The Nine Nations of North America* (1981), in which he proposed a set of regions that cut across not only state boundaries but also national boundaries.

NEW ENGLAND AND THE ATLANTIC PROVINCES

New England and Canada's Atlantic Provinces (FIGURE 2.37) were among the earliest parts of North America to be settled by Europeans. Of all the regions of North America, this one

may maintain the strongest connection with the past, and it holds a reputation as North America's *cultural hearth*, meaning the place from which much American culture has emanated. Philosophically, New Englanders laid the foundation for religious freedom in North America through their strong conviction, which eventually became part of the U.S. Constitution, that there should be no established church and no requirement that citizens or public officials hold any particular religious beliefs. New England in particular is the source of a classic American village style that is arranged around a town square, or *common*. Once used for grazing animals, many commons are now public parks, such as Boston Common (FIGURE 2.38). Many interior furnishing styles also originated in New England; in all classes of American homes today, there are copies of New England–designed furniture and accessories.

The Past Lives in the Present

Many of the continuities between the present and past economies of New England and the Atlantic Provinces derive from the region's geography. During the last ice age, glaciers scraped away much of the topsoil, leaving behind some of the finest building

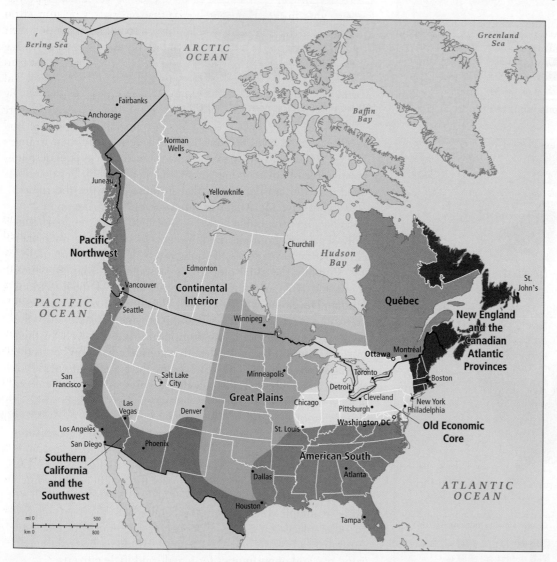

FIGURE 2.36 Subregions of North America. The division of North America into subregions, as shown here, is partly based on the book *The Nine Nations of North America* (1981) by Joel Garreau. He proposed a set of regions that cut across state and national boundaries. Though Garreau includes parts of Mexico and the Caribbean, our map includes territory only within the United States and Canada.

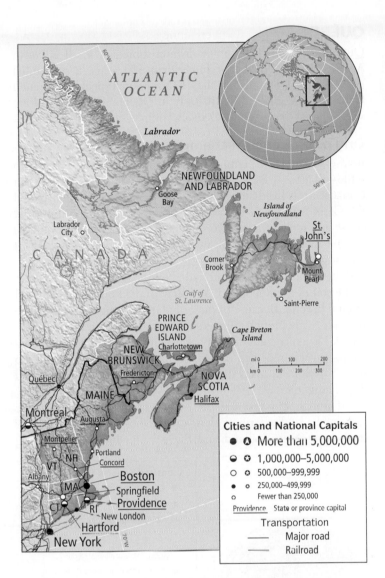

FIGURE 2.37 New England and the Atlantic Provinces subregion.

FIGURE 2.38 A New England common. Most towns in New England are centered around an open space called a common. Once used for grazing animals, most commons are now parks used for recreation and community events. This common is in Boston, Massachusetts.

industrial fishing practices (some of these large, factory-type vessels came from Japan and Russia) that have a severe impact on marine ecosystems. Canada and the United States extended their legal boundaries to 200 miles offshore, but unsustainable fishing practices were retained until, in the 1980s, there was a crash in cod and other fish species. Tens of thousands of New England and Atlantic Province fishers had to find alternative employment. By 2015 science-based catch limits, set by the New England Fishery Management Council, had failed to curtail the crash of cod fisheries. New research suggests that higher ocean temperatures related to climate change are now preventing cod populations from rebounding.

Strong and Diverse Human Resources

Many parts of southern New England now thrive on service- and knowledge-based industries that require skilled and educated workers: insurance, banking, high-tech engineering, and genetic and medical research, to name but a few. New England's considerable human resources derive in large part from the strong emphasis placed on education, hard work, and philanthropy by the earliest Puritan settlers, who established many high-quality schools, colleges, and universities. The cities of Boston and Cambridge have some of the nation's foremost institutions of higher learning (Harvard, the Massachusetts Institute of Technology, and Boston University, for example). The Boston area has capitalized on its supply of university graduates to become North America's second-most-important high-technology center, after California's Silicon Valley.

At present, New Englanders are confronted by some startling discontinuities with their past. Although cities (such as Boston, Providence, and Hartford) have historically been important in the region, New England is becoming even more urban and ethnically diverse. The remaining agricultural enterprises and rural industries have become more mechanized, so rural New Englanders are flocking to the cities for work. There they discover that New

stone in North America but only marginally productive land. Although farmers settled here, they struggled to survive. Although many areas in the region now try to capitalize on their rural and village ambiance and historic heritage by enticing tourists and retirees, economically, northern New England and the Atlantic Provinces of Canada remain relatively poor.

After being cleared in the days of early settlement, New England's evergreen and hardwood deciduous forests have slowly returned to fill in the fields abandoned by farmers. Some of these second-growth forests are now being clear-cut by logging companies and in rural areas, paper milling from wood is still supplying jobs as other blue-collar occupations die out.

Abundant fish was the major attraction that drew the first wave of Europeans to New England. In the 1500s, hundreds of fishers from Europe's Atlantic Coast came to the Grand Banks, offshore of Newfoundland and Maine, to take huge catches of cod and other fish. The fishing lasted for more than 500 years, but eventually the Grand Banks fish stocks were depleted by modern

England is not the Anglo-American stronghold they expected. Anglo-Americans have for years shared New England with Native Americans, African Americans, and immigrants from Portugal and southern Europe. But today, in the suburbs of Boston, corner groceries may be owned by Koreans or Mexicans; Jamaicans may be law students or street vendors selling meat patties and hot wings; restaurants serve food from Thailand; and schoolteachers may be Filipino, Brazilian, or West Indian. The blossoming cultural diversity of New England and the entrepreneurial skills of new migrants are helping New England keep pace with change across the continent. These trends are less obvious in the Atlantic Provinces of Canada, but there too, immigrants from Asia and elsewhere are influencing ways of life.

THINGS TO REMEMBER

• Religious freedom was one of the primary motivations for early settlement in this subregion and became the foundation for the separation of church and state in the U.S. Constitution.

• New England house and furniture styles and village/town spatial organization have been highly influential in North America.

• The blossoming cultural diversity of New England and the entrepreneurial skills of new migrants are helping New England keep pace with change across the continent.

QUÉBEC

Québec is the most culturally distinct subregion in all of North America (FIGURE 2.39). For more than 300 years, a substantial portion of the population has been French-speaking. French Canadians are now in the majority in Québec Province and are struggling to resolve Québec's relationship with the rest of Canada.

Origins of French Settlement

In the seventeenth century, France encouraged its citizens to settle in Canada. By 1760, there were 65,000 French settlers in Canada. Most lived along the St. Lawrence River, which linked the Great Lakes (and the interior of the continent) to the *world ocean* and global trade. The settlers lived on long, narrow strips of land that stretched back from the river's edge. This *long-lot system* gave the settlers access to resources from both the river and the land: they fished and traded on the river, they farmed the fertile soil of the floodplain, and they hunted on higher forested ground beyond. Because of the orientation of the long lots to the riverside, early French colonists joked that one could travel along the St. Lawrence and see every house in Canada. Later, the long-lot system was repeated inland, so that today narrow farms also stretch back from roads that parallel the river, forming a second tier of long lots.

Through the first half of the twentieth century, Québec remained a land of farmers eking out a living on poor soils similar to those of New England and the Atlantic Provinces, growing

---

FIGURE 2.39   The Québec subregion.

A Québec City was founded in 1608, making it one of the oldest cities in North America. The cityscape is dominated by the Château Frontenac hotel, which opened in 1893. Québec City is the provincial capital and the seat of Québec's parliament, and has become a biomedical high-tech center featuring leading-edge software development, especially in medicine.

Cities and National Capitals
● ✪ More than 5,000,000
◒ ✪ 1,000,000–5,000,000
○ ✪ 500,000–999,999
• ○ 250,000–499,999
○ Fewer than 250,000
Québec   State or province capital
▲ Native American reservation
Transportation
—— Major road
—— Railroad

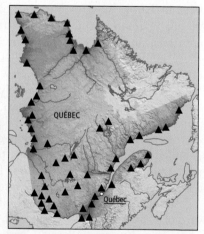

B Aboriginal peoples from various ethnic groups live throughout Québec, but one-fourth of them live in far northern and eastern rural Québec.

only enough food to support their families. After World War II, Québec's economy grew steadily, propelled by increasing demand for the natural resources of northern Québec, such as timber, iron ore, and hydroelectric power. In the St. Lawrence River valley, cities prospered from the river transport and the processing of resources (see Figure 2.39). Most of these enterprises were in the hands of Anglo-Canadians (those with ancestry in the British Isles), who were interested in exporting the natural resources of Québec. Québec's prosperity was most visible in the rapid growth of Montréal, located near the confluence of several rivers.

The Quiet Revolution

In the mid-twentieth century, as rural French Canadians moved into Québec's cities, the so-called Quiet Revolution began. Because their access to education and training was growing, the Québécois were for the first time able to compete with English-speaking residents of Québec for higher-paying jobs and political power. Gradually, their conservative Catholic, rural-based culture gave way to a more cosmopolitan one that began to challenge discrimination at the hands of English speakers. The Québécois resisted the disparagement of French culture and efforts to block access to education and economic participation. In the 1970s, the Québécois pushed for more autonomy from Canada and then for outright national sovereignty (separation from Canada). The province passed laws that heavily favored the French language in education, government, and business. In response, many English-speaking natives of Québec left the province. A referendum on Québec sovereignty failed narrowly in 1996. By 2009, a new, more articulate and less radical movement had emerged among young adults, seeking Québécois control over all manner of social policies and even over foreign policy. Specifically, they sought foreign policy control because of strong opposition to Canada's participation with the United States in the war in Afghanistan. The issue of sovereignty for Québec has remained on the back burner.

Québec extends north into areas around Hudson Bay that are rich in timber and mineral deposits (iron ore, copper, and oil) and are also the homelands of many Aboriginal peoples (see Figure 2.39B). The resources are hard to reach in the remote, difficult, glaciated terrain of the Canadian Shield. Although these are the native lands of the Cree and Inuit Aboriginal peoples, the Québec provincial government has legal control of resources and since the 1970s has pushed through the development of a series of hydroelectric power projects in the vicinity of James Bay (part of Hudson Bay) in order to run mineral-processing plants, sawmills, and paper mills. There is significant public Québécois support for the project, which is seen as serving the popular goals of eliminating oil imports and shifting to hydroelectric power and wind power. However, since the 1970s, the Cree and Inuit have protested the clear-cutting of their forests for paper production, the diversion of pristine rivers for hydropower dams, and the loss of natural ecosystems. By 2015, the James Bay project had been largely completed, with protests quieted by large cash payments to the Cree and Inuit. Some electric power generated in northern Québec is sold to the northeastern United States.

> **THINGS TO REMEMBER**
>
> • Québec is the most culturally distinct subregion in all of North America.
>
> • The Québécois are sensitive to discrimination in Canada and periodically lobby actively for sovereignty.
>
> • In Québec there has been disagreement between the Cree and Inuit and the Québécois over development of the subregion's natural resources via the James Bay project.

THE OLD ECONOMIC CORE

The Old Economic Core—an area that includes southern Ontario and the north-central part of the United States, from Illinois to New York (**FIGURE 2.40**)—represents less than 5 percent of the total land area of the United States and Canada, but it was once the economic core of North America. As recently as 1975, its industries produced more than 70 percent of the continent's steel and a similar percentage of its motor vehicles and parts, an output made possible by the availability of energy and mineral resources in the region or just beyond its boundaries. The industrial economy of this region has gone into severe decline, and large cities dependent on manufacturing, such as Detroit and Cleveland, and hundreds of smaller cities and towns, have suffered near economic collapse. Meanwhile, some of the region's largest cities, such as New York, Toronto, and Chicago, continue to prosper, largely because of their strong service industries, which connect them to regional and global economies.

Loss of Economic Dominance

Plants started closing in the 1970s, and by the 1990s, most manufacturing jobs had moved outside the old industrial heartland to the South, Middle West, Pacific Northwest, and abroad. Industrial resources for factories now come from elsewhere: coal is mined in Wyoming, British Columbia, Alberta, and Appalachia; steel, the mainstay of the automotive and construction industries, can be more cheaply obtained from scrap metal or purchased from China, India, or Brazil.

The trend of industrial shutdowns worsened with the closing of auto plants in the Detroit suburbs during the 2000s, a time when Detroit's population plunged 25 percent. Large, outdated factories and some public buildings (**FIGURE 2.41**) now sit empty or underused throughout much of the Old Economic Core, which some people refer to as the *Rust Belt*.

The Impact of the Economic Decline on People

After World War II, industrial jobs drew millions of rural men and women, both black and white, from the South to the industrial core. Many of their sons and daughters, now without work and without funds to retrain or relocate, constitute some of the more than 47 million U.S. citizens living in poverty as of 2014. Consider the case of one inner-city neighborhood on the west side of Chicago, called North Lawndale, which has had repeated massive job losses.

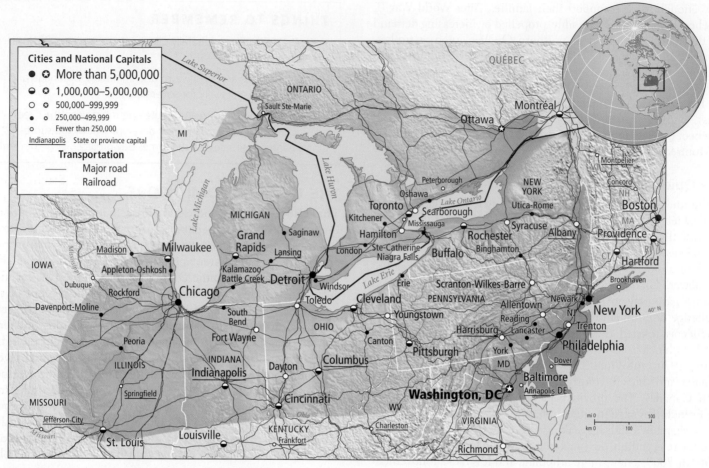

FIGURE 2.40 The Old Economic Core subregion.

FIGURE 2.41 Michigan Central Railroad Station in Detroit.
When it was built in 1913, Michigan Central Station was the tallest train station in the world. With the rise of automobile-based transportation, though, by 1956 the building was virtually unused. It finally closed in 1988 and has been deteriorating ever since. Now the question is: Will concerns about fuel costs and CO_2 emissions decrease automobile use and reintroduce an era of rail transportation?

J.D. Pooley/Getty Images

In 1960, there were 125,000 people living in North Lawndale. It had two large factories employing 57,000 workers, and a large retail chain's corporate headquarters that had thousands of secretaries and office workers. One by one, the large employers closed their doors, and by the 1980s North Lawndale was disintegrating. The housing stock deteriorated as families and businesses left to look for opportunities elsewhere. By 2010, North Lawndale had shrunk to just 36,000 people, and the median household income was roughly half of that for Chicago as a whole. The North Lawndale Industrial Development Team, with the help of a family foundation and local civic groups, still actively recruits industry and offers to provide a custom-trained labor force.

Despite the importance of industry in this region, it would be wrong to think of the economic core as solely based on factories and mines. In between the great industrial cities of this region are thousands of acres of some of the best temperate-climate farmland in North America. And as during its industrial heyday, the thousands of miles of shoreline around the Great Lakes and oceanfront along the Eastern Seaboard provide both winter and summer recreational landscapes, as do the mountains of New York State and Pennsylvania. These ancillary economic activities constitute an important source of supplementary employment for families devastated by factory closings.

THINGS TO REMEMBER

• The Old Economic Core was once the heart of industrial production in North America.

• Changes in the production of steel and in energy costs, plus the loss of industry to foreign sites, brought on a major decline in the Old Economic Core.

• The loss of millions of manufacturing jobs in this subregion has had serious social consequences.

THE AMERICAN SOUTH (THE SOUTHEAST)

The regional boundaries of the American South are perhaps less distinct and based more on a perceived state of mind and way of life than are those of most other U.S. regions. This region, in fact, covers only the southeastern part of the country, not the whole of the southern United States **(FIGURE 2.42)**. Somewhere west of Houston, Texas, the American South grades into the Southwest, a region with noticeably different environmental and cultural features.

Multiple Images Characterize the South

The American South is dominated by a complex of features that many people would identify as *Southern*: a wide range of dialects; barbecue and other comfort foods from both white and African American traditions (see Figure 2.29A); stock car racing; horse farms; bluegrass, jazz, country, and blues music; the Baptist version of Christianity; the open friendliness of the people; conservative stances on politics, gun ownership, patriotism; the prominent role of religion in public life; the early onset of spring; and rural settlement patterns, such as those that keep large kin groups together in clusters of old wooden cabins, mobile homes, or finely built new houses.

Some places in the South have few recognizable Southern qualities **(FIGURE 2.43)**. Parts of Missouri, Oklahoma, and Texas have strong ties to the Great Plains; Miami, on the far southern tip of Florida, with its cosmopolitan Latino culture and its trade and immigration ties to the Caribbean and South America, hardly seems Southern at all. New Orleans, also in the heart of the South, has a unique set of cultural roots—French, African, Cajun, Creole—that set it apart from other Southern cities.

North Americans and foreign visitors to the South tend to hold many outdated images from the Civil Rights era of the 1960s and even from the Civil War era. It is true that significant racial segregation still persists in that black people and white people tend to live and worship separately, but in fact, many more whites and blacks share neighborhoods in the South than in the Old Economic Core. In rural and urban settings, schools are becoming more integrated as a result of residential patterns as well as busing plans and magnet schools. The workplace is now integrated, and it is not uncommon to find African Americans and other minorities in supervisory and administrative positions, especially in business, government, and educational institutions. Biracial families are common.

Since the 1990s, Latino immigrants have come by the thousands into Southern cities. Of the 11 U.S. cities with the largest Hispanic populations, 3 are in the Southeast. Much of the restaurant service, road maintenance, landscaping, roofing, and construction work is now done by Spanish-speaking workers from Mexico and Central America, many of whom are sinking roots in the region and saving to establish businesses and buy homes.

Poverty Persists for Some, But Return Black Migration Is a Boon

The South has the nation's highest concentration of families living below the poverty line, and most of them are white. On the other hand, most Southerners—black, white, or Hispanic—are able to maintain a substantially better standard of living than their parents did.

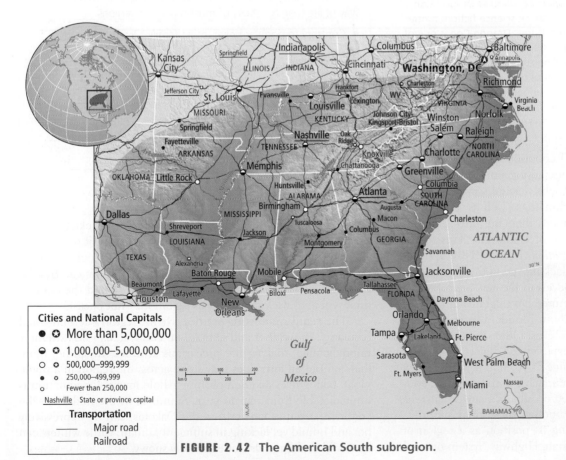

FIGURE 2.42 The American South subregion.

FIGURE 2.43 Atlanta's Dragon-Con. Cosplayers (short for costume players) dressed as Star Wars stormtroopers at Atlanta's Dragon-Con, a 4-day multi-genre convention that focuses on science fiction, fantasy, comics, cartoons, and associated fan cultures. Dragon-Con is one of the city's largest conventions, drawing 50,000 participants each year and contributing more than U.S.$25 million to the metropolitan economy. Atlanta is widely considered the capital of the American South subregion, given its large population (just under 6 million people), its central location within the region, and its historical importance.

Many African Americans left the South in the 1950s and 1960s for factory jobs in the Old Economic Core in what was called the *Great Migration*. Now increasing numbers are coming back to the region because of jobs, business opportunities, the lower cost of living, a milder climate, as well as safer, more spacious, and friendlier neighborhoods than the ones they left in places such as New York, Illinois, or California. This return is being called the *New Great Migration*. In the 1990s, 3.5 million people who self-identified as black moved to the South from other parts of the United States. Seven of the ten metropolitan areas that gained the most African American migrants during the 1990s were in the South, principally in Florida, Georgia, North Carolina, Virginia, Tennessee, and Texas. Other Southern states had no increase, and Louisiana experienced black emigration even before Hurricane Katrina. Since 2000, half a million new black residents have moved to Atlanta, many of them young, educated professionals. Another 500,000 of Atlanta's newest citizens are Hispanic, and Asians constitute another substantial minority.

The South is steadily improving its position as a region of growth. The federally funded Interstate Highway System opened the South to auto and truck transportation. Inexpensive industrial locations, close to arterial highways, have drawn many businesses to the South, including those involved in automobile and modular-home manufacturing, light-metals processing, high-tech electronics assembly, and high-end crafts production. For several decades, tourists and retirees by the hundreds of thousands have been driving south on the interstate highways, many attracted by the bucolic rural landscapes, the historic sites, the warm temperatures, the natural environments of the national parks, the outlet mall shopping, and the recreational theme parks.

Southern agriculture is now mechanized, and it is at once more diversified and more specialized than ever before. Strawberries, blueberries, peaches, tomatoes, mushrooms, and wines from local vineyards are replacing the traditional cash crops of tobacco, cotton, and rice. Those who produce these often organically grown luxury crops for urban consumers often do so on smallholdings as part-time farmers who may also have jobs in nearby factories or service agencies. Most of the country's broiler chickens are now raised by large, nonorganic operations throughout the South. Interestingly, the laborers willing to take low-wage jobs on these factory-like chicken farms are often immigrants from places such as Russia, Ukraine, Vietnam, Haiti, and Honduras.

THINGS TO REMEMBER

• The regional boundaries of the American South are perhaps less distinct and based more on a perceived state of mind and way of life than are those of most other U.S. regions.

• Contrary to popular belief, most families living below the poverty line in the South are white.

• While the South still remains the subregion with the nation's highest poverty rate and lowest educational attainment, it is growing in population, number of businesses, and high-tech industries; it is improving with regard to its overall human well-being; and it is increasingly becoming the choice of residence for immigrants from all over the world.

THE GREAT PLAINS BREADBASKET

The Great Plains **(FIGURE 2.44)** received its nickname, the *Breadbasket*, from the immense quantities of grain it produces—wheat, corn, sorghum, barley, and oats. Other agricultural products include soybeans, sugar beets, sunflowers, and meat. But the Great Plains has many agricultural challenges, and the agricultural system as we know it in this region may not persist.

The gently undulating prairies give the region a certain visual regularity; its weather and climate, in contrast, can be extremely unpredictable, making life precarious at times. In spring, more than a hundred tornadoes may strike across the region in a single night, taking lives and demolishing whole towns. Precipitation is unpredictable from year to year. Summers in the middle of the continent, even as far north as the Dakotas, can be oppressively hot and humid yet lacking in sufficient rainfall, while winters can be terribly cold and dry, or extremely snowy.

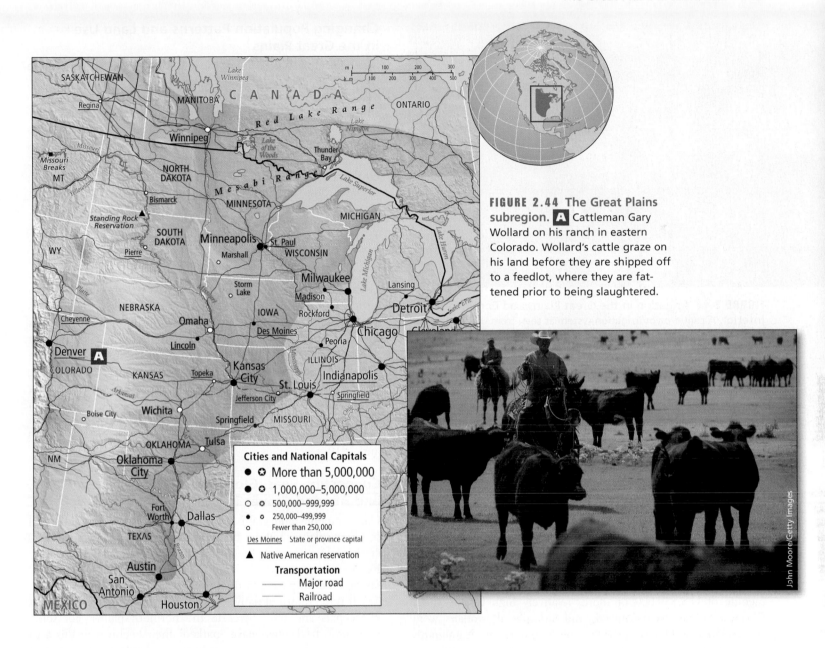

FIGURE 2.44 The Great Plains subregion. **A** Cattleman Gary Wollard on his ranch in eastern Colorado. Wollard's cattle graze on his land before they are shipped off to a feedlot, where they are fattened prior to being slaughtered.

Adapting to the Challenges of the Great Plains

The people of the plains learned to adapt to these challenges long before European settlers began to stake claims there in the 1860s. As people do in even drier regions in the Continental Interior, Great Plains farmers often irrigate their crops with water pumped from deep aquifers and delivered with central-pivot sprinklers that irrigate circular-shaped areas **(FIGURE 2.45)**. However, to produce enough grain for one slice of bread in irrigated fields takes 10.4 gallons of water, a rate of water usage that is not sustainable (see Table 1.2 in Chapter 1). Much of the primary crop, wheat, is genetically selected to resist frost damage. Most wheat is harvested by an energy-intensive system that involves traveling teams of combines that are used to start harvesting in the south in June and move north over the course of the summer. To select the most profitable crops and find the best time to sell them, plains farmers keep close tabs on the global commodities markets, usually with computers installed in barns, homes, or even in the cabs of the huge (and very costly) farm machines they use to work their land.

Animal raising, the other important activity in the Great Plains, has become problematic from environmental and animal-rights perspectives (see Figure 2.45A). Rather than being herded on the open range, as in the past, cattle are now raised in fenced pastures and then shipped to feedlots, where they are fattened for market on a diet rich in sorghum and corn, but often in muddy, inhospitable pens. And like wheat, meat production uses a great deal of water in an area prone to drought. For example, to grow and process the amount of beef in one hamburger requires 640 gallons of water.

Erosion is a serious problem on the plains: soil is disappearing more than 16 times faster than it can form. Grain fields and pasture grasses do not hold the soil as well as the original dense,

David Boyer/National Geographic/Getty Images

FIGURE 2.45 Irrigation in the Great Plains and Continental Interior. A center-pivot irrigation system as seen from the air in eastern Oregon. Often a well at the center of each irrigator taps an underground aquifer at an unsustainable rate.

undisturbed prairie grasses once did. Plowing and the sharp hooves of the cattle loosen the soil so that it is more easily carried away by wind and water erosion. Experts estimate that each pound of steak produced in a feedlot results in 35 pounds of eroded soil. Given the costs in soil and water, raising cattle in this region is by no means a sustainable economic activity.

Globalization Impacts in the Great Plains

Cattle and other livestock, such as hogs and turkeys, are slaughtered and processed for market in small plants across the plains by low-wage, often immigrant, labor. This is new. In the 1970s, a meatpacking job in the Great Plains provided a stable annual income of U.S.$30,000 or more, relatively high for the time. The workforce was unionized, and virtually all workers were descendants of German, Slavic, or Scandinavian immigrants who immigrated in the nineteenth century. But in the 1980s, a number of unionized meatpacking companies closed their doors. Other nonunion plants opened, often in isolated small towns in Iowa, Nebraska, and Minnesota. The labor is no longer local residents but immigrants from Mexico, Central America, Laos, and Vietnam. Pay is often minimum wage, and with union work rules gone, safety is often haphazard. Working hours are long or short, at the convenience of the packing-house manager, overtime pay is rare and those who protest may lose their jobs.

Many of the immigrant workers are refugees from war in their home countries, and their path to assimilation into American society is blocked by wages that are not sufficient to provide a decent life for their children. The Reverend Tom Lo Van, a Laotian Lutheran pastor in Storm Lake, Iowa, says of the Laotian families in his church that the youth are unlikely to prosper from their parents' toil. "This new generation is worse off," he says. "Our kids have no self-identity, no sense of belonging…no role models. Eighty percent of [them] drop out of high school."

Changing Population Patterns and Land Use in the Great Plains

Mechanization has reduced the number of jobs in agriculture and encouraged the consolidation of ownership. Increasingly, corporations own the farms of the Great Plains. Individuals who still farm may own several large farms in different locales. They often choose to live in cities, traveling to their farms seasonally. As a result of these trends, rural depopulation is severe. Thousands of small prairie towns are dying out. Between 2000 and 2007, 60 percent of the counties in the Great Plains lost population; the area now has fewer than 6 people per square mile (2 per square kilometer). The few cities around the periphery of the region (Denver, Minneapolis, St. Louis, Dallas, Kansas City) are growing because young people are leaving the small towns on the prairies.

Some suggest that European agricultural settlement in the Great Plains may be an experiment that should end. Modern agriculture, with its high need for water and for fertile soil, has proven too stressful for the plains environments, and these days the only Great Plains rural counties with increasing populations are those with significant numbers of Native Americans. In the western Great Plains, some Native Americans are returning to work in newly established gambling casinos, new ranching establishments, and other enterprises. Although still only a fraction of the total plains population, the Native American population in this region grew by 40 percent between 1990 and 2010.

Connecting Bees, Global Corn, the Northern Plains, and Almond Groves in California

Cattle ranches and grain farms still prevail in the middle and southern plains, but in the northwestern plains, many farmers have been putting their land into the Federal Conservation Reserve program, which is aimed at stopping soil erosion and saving water by paying farmers to take land out of production.

It is the return of native grasses and wildflowers that brought beekeepers and their hives to the northern plains. For some years now, beekeepers have "parked" their beehives in this area so that their bees can produce honey by feeding on the nectar and pollen of prairie wildflowers. During a few lucrative weeks in February, when the almond groves near Fresno in the Central Valley of California are in bloom, the beekeepers truck their thousands of beehives (containing as many as a billion bees) to pollinate California's Central Valley almond flowers **(FIGURE 2.46)**. California produces two-thirds of the world's almonds; without the bees to pollinate, there would be no almonds. When the almond bloom is over, the bees need to return to their Great Plains wildflower sanctuary.

Here is the problem: Since 2008, when corn became a commodity in high demand for making ethanol, producing corn-based sweeteners, and feeding the world's hungry, Great Plains farmers who had their land in the Federal Conservation Reserve suddenly could make more money putting their land back into corn. They plowed up the prairie and the bees lost at least half of the wildflower cover on which they had previously thrived. This then threatened almond production in California.

Phil Hawkins/Bloomberg, via Getty Images

FIGURE 2.46 Bees, almonds, and ethanol. Flowering almond trees in California are pollinated by high plains honeybees, which live in hives (the white boxes shown in the picture) that are usually trucked in from South Dakota, where they flourish among the fields of wildflowers that have sprung up on abandoned farmland. However, California's almond industry has recently been struggling to find enough bees because the fields in South Dakota are being brought back into corn production to satisfy the demand for ethanol fuel made from corn, reducing the amount of land available for wildflowers for the bees. With the wildflowers gone, bee populations will decrease rapidly.

ON THE BRIGHT SIDE: Lead a Revival of Sustainability in the Great Plains

Mike Faith, a Sioux trained in animal science, manages a large bison herd at Standing Rock Reservation in South Dakota, not far from where Sitting Bull was killed in 1890. Bison meat, which is low in cholesterol, may help control obesity, a recognized problem among American Indians. Faith says, "Just having these animals around, knowing what they meant to our ancestors, and bringing kids out to connect with them has been a big plus." ∎

THINGS TO REMEMBER

- Plains farmers keep close tabs on the global commodities markets, usually with computers installed in barns, homes, or even in the cabs of their huge (and costly) farm machines.

- Population patterns are changing throughout the Great Plains as rural counties lose population to urban areas and local farmworkers are replaced by immigrants working at lower wages.

- Soil erosion and water depletion are serious problems and ultimately will cause significant changes in agricultural practices in this subregion.

- A combination of old and new agricultural strategies are being tried in the Great Plains.

- The landscapes of the Great Plains change according to changing farming methods and to rising and falling global demands for certain crops, such as corn. These landscape changes are linked to factors elsewhere, such as the global demand for food and the production of crops—for example, almonds in the Central Valley of California.

THE CONTINENTAL INTERIOR

Among the most striking features of the Continental Interior are its huge size, its physical diversity, and its very low population density **(FIGURE 2.47)**. This is a land of extreme physical environments, characterized by rugged terrain, extreme temperatures, and lack of water (to gain appreciation for these characteristics, compare the map in Figure 2.1 and Figure 2.4C and the related map to Figure 2.47). These extreme physical features restrict many economic enterprises and account for the low population density (see Figure 2.27) of fewer than 2 people per square mile (1 person per square kilometer) in most parts of the region.

Four Physical Zones

Physically, there are four distinct zones within the Continental Interior: the Canadian Shield, the frigid and rugged lands of Alaska, the Rocky Mountains, and the Great Basin. The *Canadian Shield* is a vast, glaciated territory north of the Great Plains that today is characterized by thin or nonexistent soils, innumerable lakes, and large meandering rivers. The rugged lands of Alaska lie to the northwest of the shield. The shield and Alaska have northern coniferous (*boreal*) and subarctic (*taiga*) forests along their southern portions. Farther north, the forests give way to the **tundra,** a region of winters so long and cold that the ground is permanently frozen several feet below the surface. Shallow-rooted, ground-hugging plants such as mosses, lichens, dwarf trees, and some grasses are the only vegetation.

The Rocky Mountains stretch in a wide belt from southeastern Alaska to New Mexico. The highest areas are generally treeless, with glaciers or tundra-like vegetation, while forests line the rock-strewn slopes on the lower elevations. Between the Rockies and the Pacific coastal zone is the *Great Basin*, which was formed by an ancient volcanic mega crater. It is a dry region of widely spaced mountains, covered mainly by desert scrub and occasional woodlands.

Native Americans and Aboriginal Peoples

The Continental Interior has the greatest concentration of native people in North America, most of them living on reservations, as native lands are called in the United States. Because of the removal of Native Americans and Aboriginal peoples from their ancestral lands in the nineteenth century, only those who live in the tundra and northern forests of the Canadian Shield still occupy their

tundra a region of winters so long and cold that the ground is permanently frozen several feet below the surface

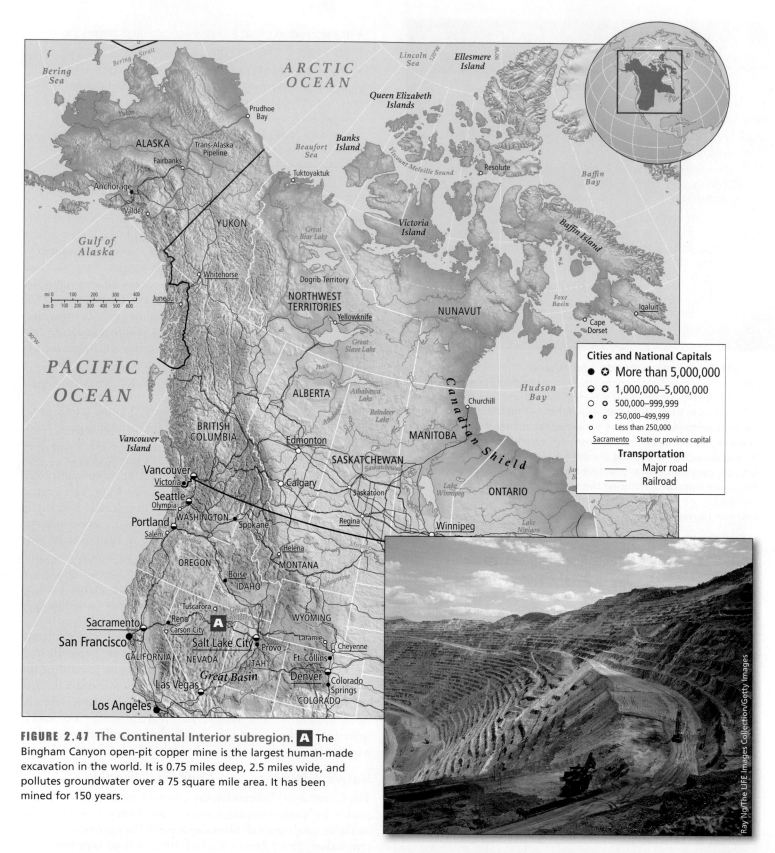

FIGURE 2.47 The Continental Interior subregion. **A** The Bingham Canyon open-pit copper mine is the largest human-made excavation in the world. It is 0.75 miles deep, 2.5 miles wide, and pollutes groundwater over a 75 square mile area. It has been mined for 150 years.

original territories. The Nunavut, for example, recently won rights to their territory (part of the former Northwest Territory) after 30 years of negotiation. They are now able to hunt and fish and generally maintain the ways of their ancestors, although most of them now use snowmobiles and modern rifles. In September 2003, the Tłı̨chǫ Aboriginal people (formerly known as the Dogrib), a group of 3500, reclaimed 15,000 square miles of their land just south of the Arctic Circle. The Nunavut and the Tłı̨chǫ, who have a strong sense of cultural heritage and an entrepreneurial streak as well, have websites and manage their lives and the resources of their ancestral lands—oil, gas, gold, and diamonds—with the aid of Internet communication.

Settlements

The Continental Interior is a fragile environment. It is one of the most intense battlegrounds in North America between environmentalists and resource developers.

Although much of the Continental Interior remains sparsely inhabited, nonindigenous people have settled in considerable density in a few places. Where irrigation is possible, agriculture has been expanding—in Utah and Nevada; in the lowlands along the Snake and Columbia Rivers of Idaho, Oregon, and Washington; and as far north as the Peace River district in the Canadian province of Alberta (see Figure 2.47). There are many cattle and sheep ranches throughout much of the Great Basin. Overgrazing and erosion are problems, but more serious concerns include unsustainable groundwater extraction and pollution from chemical fertilizers. An additional pollutant is the malodorous effluent from feedlots where large numbers of beef cattle are fattened for market. Efforts by the U.S. government to curb abuses on federal land, which is often leased to ranchers in extensive holdings, have not been very successful.

Increasing numbers of people are migrating to the Continental Interior. The many national parks in the region attract both seasonal and permanent workers and millions of tourists. Other people come to exploit the area's natural resources. Towns such as Laramie, Wyoming, and Calgary, Alberta, have swelled with workers in the mining and fossil fuel industries. These groups often find themselves in conflict with each other and with the Native American and Aboriginal peoples of the Continental Interior.

In the United States, environmental groups have pressured the federal government to set aside more land for parks and wilderness preserves and to stop building pipelines for oil and gas transport from Alberta, Canada. Environmentalists also want to limit or eliminate activities such as mining and logging, which they see as damaging to the environment and the scenery. A switch to more recreational and preservation-oriented uses throughout the region would change, and probably lessen, employment opportunities, while the growing numbers of visitors would place stress on natural areas, particularly on water resources.

Federal Lands

More than half the land in the Continental Interior is federally owned. Mining and oil drilling, often on leased federal land, are by far the largest industries (see Figure 2.47A). Since the mid-nineteenth century, the region's wide range of mineral resources has supported the major cities and towns. These mineral resources link the settlements to the global economy, but fluctuating world market prices for minerals have resulted in alternating periods of booms and busts. In recent times, the most stable mineral enterprises have been oil-drilling operations along the northern coast of Alaska. From here, the Trans-Alaska Pipeline runs southward for 800 miles (1300 kilometers) to the Port of Valdez.

> **THINGS TO REMEMBER**
>
> • The Continental Interior subregion is the largest and most physically diverse in North America, with four distinct zones.
>
> • Large numbers of Native Americans from many different tribal groups in this subregion are often in conflict with mining and logging interests as well as with the increasing number of tourists and tourism developments.
>
> • Mining and oil drilling, often on leased federal land, are by far the largest industries of the Continental Interior.

THE PACIFIC NORTHWEST

Once a fairly isolated region, the Pacific Northwest **(FIGURE 2.48)** is now at the center of debates about how North America should deal with environmental and development issues. The economy in much of this region is shifting from logging, fishing, and farming to information technology industries. As this is happening, its residents' attitudes about their environment are also changing. Forests that were once valued primarily for their timber are now also prized for their recreational value, especially their natural beauty and their wildlife, both of which are threatened by heavy logging. Energy, needed to run technology and other industries, is now more than ever a crucial resource for the region.

The physical geography of this long coastal strip consists of fjords and islands, mountains and valleys. Most of the agriculture, as well as the largest cities, are located in the southern part, in a series of lowlands lying east of the long, rugged coastal mountain ranges that extend north and south. Throughout the region, the climate is wet. Winds blowing in from the Pacific Ocean bring moist and relatively warm air inland, where it is pushed up over the mountains, resulting in usually copious orographic rainfall throughout the area and seasonal snowfall in the north. The close proximity of the ocean gives this region a milder climate than is found at similar latitudes farther inland. In this rainy, temperate zone, enormous trees and huge forests have flourished for eons (see Figure 2.5A). The balmy climate in much of the region, along with the spectacular scenery, attracts many vacationing or relocating Canadians and U.S. citizens.

Older Extractive Industries

Pacific Northwest logging provides most of the construction lumber and an increasing amount of the paper used in North America. Lumber and wood products are also important exports to Asia, and the lumber industry is responsible directly and indirectly for hundreds of thousands of jobs in the region. As the forests shrink, environmentalists are harshly critical of the logging industry for such practices as clear-cutting (see Figure 2.8B–D), which destroys wild animal and plant habitats, thereby reducing species diversity, and leaves the land susceptible to erosion and the remaining adjacent forests susceptible to diseases and pests. The criticisms of clear-cutting are obvious and legitimate; the more expensive alternative logging method of selective cutting is still disruptive to habitats and does not preserve forest diversity, but it is healthier for soils and maintains some diversity of fauna and flora.

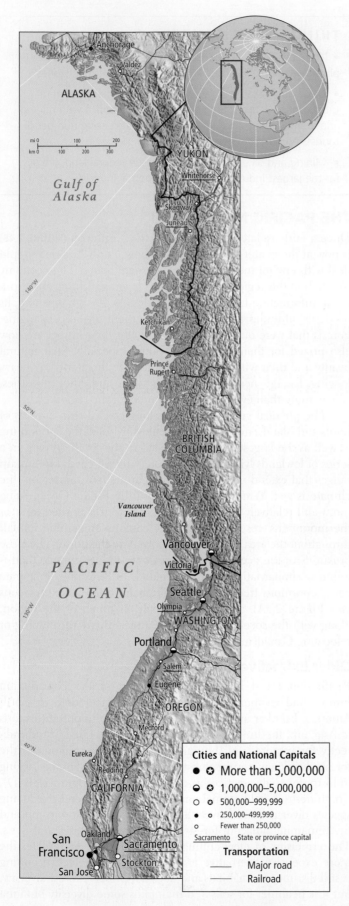

FIGURE 2.48 The Pacific Northwest subregion.

Disputes also rage over how to acquire sufficient energy for the region. Large hydroelectric dams in the Pacific Northwest have attracted industries that need cheap electricity, particularly aluminum smelting and associated manufacturing industries in aerospace and defense. These industries are major employers, but their demand for labor is erratic and involves frequent periodic layoffs. In addition, another major employer in the region, the fishing industry, finds fault with hydroelectric dams because they block the seasonal migrations of salmon to and from their spawning grounds, so some of the region's most valuable fish cannot produce young. Around 40 dams have been removed to improve ecological conditions in rivers in the Pacific Northwest.

Newer IT Industries

Given the environmental impacts of the logging and hydroelectric industries, it is not surprising that many people in the region are enthusiastic about the role of information technology within the major urban areas of the Pacific Northwest: San Francisco, Portland, Seattle, and Vancouver. With Silicon Valley just outside San Francisco and with a growing number of IT companies around Seattle, the Pacific Northwest is a world leader in information technology. However, IT firms produce hazardous waste such as lead and other heavy metals used in circuit boards, and they are particularly dependent on secure and steady sources of electricity—whether from water power, gas, oil, coal, or nuclear fuels—that have significant negative environmental impacts.

The Pacific Northwest, and particularly its major cities, often leads the rest of the United States and Canada in social and economic innovation. This role is currently illustrated by the region's growing connections to Asia. The adjustment of a predominantly Euro-American society to large numbers of Asian immigrants has been underway along the Pacific Coast for many decades. This transition has been accompanied by the rise in importance of the countries around the Pacific Rim as trading partners. Vancouver, Seattle, Portland, and San Francisco are orienting toward Asia, not just in the trade of forest products, but in banking and tourism and in the design and manufacture of IT equipment, some of which is being outsourced to Asia.

THINGS TO REMEMBER

• IT firms have become a major source of the subregion's economy and are challenging the traditional logging, fishing, farming, and hydroelectric industries for primacy, especially given the latter's detrimental impacts on the environment.

• The Pacific Northwest, and particularly its major cities, often lead the rest of the United States and Canada in social and economic innovation.

• The Pacific Northwest continues to grow and adjust its economic and cultural ties with countries across the Pacific as large numbers of Asian immigrants move to the subregion.

• The growth of the global economy has boosted IT industries in the Pacific Northwest at the same time that Asian markets for raw materials from the region have opened up.

SOUTHERN CALIFORNIA AND THE SOUTHWEST

Southern California and the Southwest (FIGURE 2.49) are united primarily by their attractive warm, dry climate; by their increasing dependence on water from outside the region; by their long and deep cultural ties to Mexico; by the hope that NAFTA will make the subregion a nexus of development; and by worries over the problems posed by the border they share with Mexico.

The varied landscapes of the Southwest include the Pacific coastal zone, the hills and interior valleys of Southern California, the mountains and dramatic mesas and canyons of southern Arizona and New Mexico, and the gentle coastal plain of south Texas. This region has warm average year-round temperatures and an arid climate, usually receiving less than 20 inches of rainfall annually. The natural vegetation is scrub, bunchgrass, and widely spaced trees. Ranching is the most widespread form of land use, but other forms are more important economically and in numbers of employees.

Agriculture, Energy, Transportation, and Information Service Industries

Where irrigation is possible, farmers take advantage of the nearly year-round growing season to cultivate high-quality fruits, nuts, and vegetables. As discussed in the opening vignette of this chapter (see also Figure 2.49), California's Central Valley is a leading agricultural producer in the United States and is one of the most valuable agricultural districts in the world. Many of the crops are produced on massive, plantation-like farms. Migrant Mexican workers, often undocumented immigrants and sometimes mere children, supply most of the labor and lower the cost of North American food by working for very low wages.

Although irrigated agriculture is important, the bulk of the region's economy is nonagricultural. Information and service industries—high-tech, software, financial, and the entertainment industries—are the most profitable and employ the most people, but the coastal zones of Southern California and Texas are also home to oil drilling, refining, and associated chemical industries, as well as other industries that need the cheap transport provided by the ocean. Los Angeles—with its major trade, transportation, media, entertainment, finance industries, and research facilities—is now the second most populous city in the United States. Like San Francisco, Seattle, and other Pacific coastal cities, Los Angeles is strategically located for trade with Pacific Rim countries; it recently replaced New York as the largest port in the United States.

On the eastern flank of this region, Austin, Texas, has attracted IT industries, lured by research activities connected

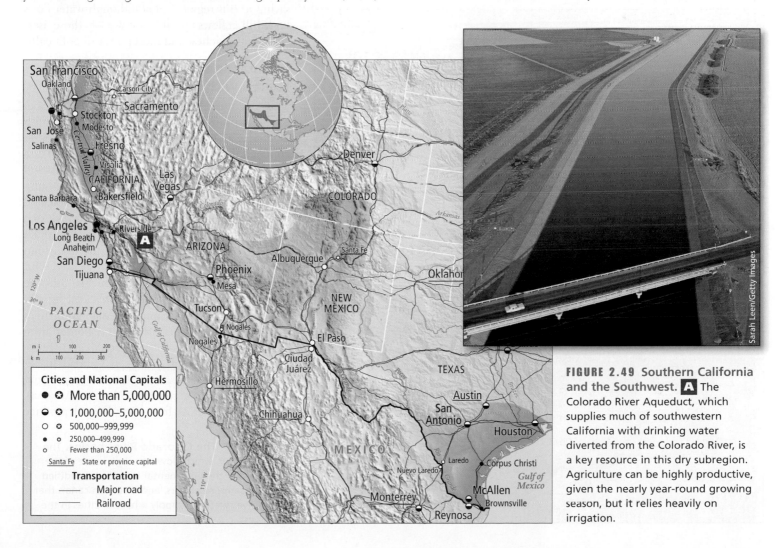

FIGURE 2.49 Southern California and the Southwest. **A** The Colorado River Aqueduct, which supplies much of southwestern California with drinking water diverted from the Colorado River, is a key resource in this dry subregion. Agriculture can be highly productive, given the nearly year-round growing season, but it relies heavily on irrigation.

to the University of Texas, the warm climate, and the laid-back, folksy lifestyle of central Texas. Austin has more than 2 million people and doubled in population between 1990 and 2010. The region as a whole draws large numbers of retirees; entire planned communities of 10,000 and 20,000 people each have sprung up in the deserts of New Mexico and Arizona within just a few years.

From the start of the first decade of the 2000s, residents of Southern California and the Southwest have had to worry about energy costs, congestion, smog, and water scarcity. In the case of energy, a number of factors have provoked questions about the region's environmental sustainability. Energy usage has been high, in part because of historically low energy prices. Deregulation of energy providers leads to escalating prices and corruption among some private generators and distributors, resulting in decreased energy supplies. In the short term, the electricity supply problem has been alleviated by controls on prices and reduction of nonessential use. But like the rest of the United States, Southern California and the Southwest have been consuming energy at rates that are not sustainable, affordable, or safe for the environment.

Water: Scarcities and Disparities of Access

Increasingly scarce water is the most worrisome environmental problem in this region. Because rainfall is light and becoming more irregular, water must be imported to serve the dense urban settlements and industries, of which agriculture is the most demanding water user. Most water comes from the Colorado River, which flows naturally southwest from near Denver, through portions of Utah, Arizona, Nevada, and along the California–Arizona border, through a small part of Mexico, and on to the Gulf of California (FIGURE 2.50). Since the middle of the twentieth century, the water has been captured by such dams as Glen Canyon, Hoover, Davis, and Parker, and held in reservoirs, where it is used first to generate electricity.

The water is also sent through aqueducts to irrigate agriculture and supply domestic and industrial users in Phoenix and Tucson, Arizona; Las Vegas, Nevada; and the Imperial Valley and cities in Southern California (see Figure 2.49A). Recently there has been more awareness that at current rates of development, the environment of the Colorado River basin will not be able to sustain even present rates of water diversion. Control of *wastage*—water lost through evaporation, percolation, and leaks in canals, as well as through nonessential uses such as lawn watering—would help significantly.

Raising the cost of water to users would control its use; now it is so cheap that few take the trouble to conserve because it seems almost a free commodity. But higher user costs could seriously impact those who use the least and are the poorest. A popular saying in this region is that although water flows downhill, it flows uphill to money—to those users who are the richest and most powerful politically.

Cultural and Economic Interests Shared with Mexico

Much of this region was originally a colony of Spain, and the area has maintained the Spanish language, a distinctive Latino culture, and other connections with Mexico. Today, the Latino culture is gaining prominence in the Southwest and spreading beyond it. As we have seen, large numbers of immigrants are arriving from Mexico, and the economies of the United States and Mexico are becoming more interdependent as NAFTA fosters greatly increased trade (see the discussion on pages 81–82).

Among these interdependencies are the factories, known as *maquiladoras*, set up by U.S., Canadian, European, and Asian companies in Mexican towns just across the border from U.S. towns in California, Arizona, and Texas. Maquiladoras in such places as Ciudad Juárez (across from El Paso), Nuevo Laredo (across from Laredo), Nogales (across from Nogales, Arizona), and Tijuana (across from San Diego) produce

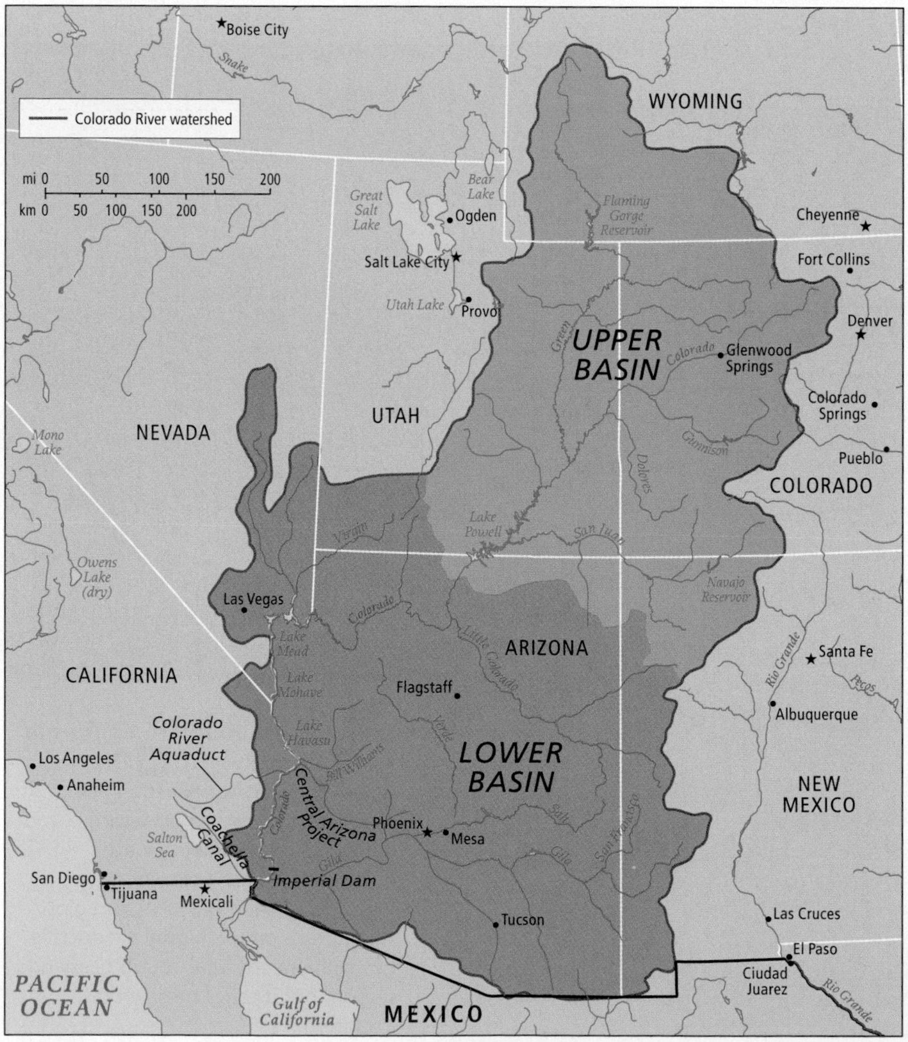

FIGURE 2.50 The Colorado River basins. The Colorado River, which flows through some of the driest land in the Continental Interior, is modified by many dams, reservoirs, and diversion canals that enable river water to supply a host of cities in the Southwest.

manufactured goods for sale primarily in the United States and Canada. They reduce costs by taking advantage of lower wages in Mexico, cheaper land and resources, lower taxes, and weaker environmental regulations. (The maquiladora phenomenon is discussed further in Chapter 3.) These factories are a key part of a larger transborder economic network that stretches across North America.

Until recently, the U.S.–Mexico border was been one of the world's most *permeable national borders*, meaning that people and goods flowed across it easily (though not without controversy). This permeability, greatly increased by NAFTA, meant that for years there were as many as 230 million legal border crossings each year through 35 points of entry. Since the terrorist attacks of 2001, the border has been made much less permeable—a high wall now runs along much of the border—and for undocumented travelers the trip has become much more dangerous. Smugglers who move workers north and illegal drugs and weapons both north and south across the more heavily guarded border charge high fees. They often cheat their passengers, trade the women and girls to sex traffickers, and in a number of cases have left whole truckloads of people to die in the desert heat without food or water.

Contentious Border Issues

As discussed earlier, several contentious issues surround the estimated 11 million undocumented immigrants currently residing in the United States, of whom perhaps 6 million are Mexican. One conflict is about language. Although English and bilingual speakers in the Southwest far outnumber those who speak only Spanish, some fear that English could be challenged in much the same way it has been challenged by French Canadians in Québec. Accordingly, Arizona in 2006 and California in 1986 made English the official language. Some bilingual programs, initiated in the 1970s to help migrant children make a smooth transition from Spanish to English, have been abolished by those who feel it is better that children learn English as quickly as possible. Scientists have found, however, that migrant children do better in math and science and have higher self-esteem if they can study for a period in their native language.

THINGS TO REMEMBER

• North America gets nearly half of its fresh vegetables, fruit, and nuts from the varied agricultural economy of Southern California and the Southwest, which itself relies heavily on access to increasingly scarce water.

• Information technology and other service businesses are the most profitable and employ the most people in the subregion, but agriculture, oil, entertainment industries, and Pacific Rim trade remain significant elements of the economy.

• A fast-growing Latino population, both citizens and recent immigrants, live in and contribute much to the subregion. Some English speakers do not support moves toward bilingual education.

GEOGRAPHIC THEMES: North America Review and Self-Test

1. Environment: North America's use of resources has enormous environmental impacts. Home to only 5 percent of the world's population, the region produces 17 percent of humanity's greenhouse gas emissions. Other impacts include the depletion and pollution of water resources and fisheries and the destruction of natural habitats.

• What aspects of North American lifestyles have increased the emissions of GHGs? What efforts have been made to control these emissions?

• How does urban sprawl influence the average North American's consumption of gasoline?

• How might the livable city movement and smart growth affect GHG emissions? What are some responses to reducing GHGs that you would be willing to take part in yourself?

• What activities in North America are major contributors to water pollution and water scarcity, and what are the resulting impacts for simple water use and beyond?

• What effect do green revolution agriculture techniques have on aquatic ecosystems?

2. Globalization and Development: Globalization has transformed North America, reorienting it toward knowledge-intensive jobs that require education and training. Income inequality has risen as most manufacturing jobs have been moved to countries with cheaper labor or have been replaced by technology. North America's size and wealth have made it the center of the global economy, and its demand for imported goods and its export of manufacturing jobs make it a major engine of globalization.

• How has globalization transformed the North American job market, including in the steel industry?

• How did free trade principles contribute to NAFTA, and what has been the effect of NAFTA on manufacturing jobs and immigration in North America?

• What are some exceptions to U.S. support of free trade principles?

3. Power and Politics: Compared to many world regions, North America has relatively high levels of political freedom. However, there is also widespread dissatisfaction with the political process here. Internationally, Canada plays a modest role while the United States has enormous global political influence, although its status as the world's predominant "superpower" is declining.

• What role does the urge to spread democracy play in the foreign policy of North America?

• How does Canada's approach to foreign relations differ from that of the United States?

• What role did the ideal of democracy play in justifying the Iraq and Afghanistan wars?

• To what extent is the allocation of U.S. foreign aid linked to U.S. strategic military interests?

• How do Canada and the United States differ in their support for government programs that help low-income people?

4. Urbanization: Since World War II, North America's urban areas have become less densely populated, expanding spatially much faster than their populations have increased numerically. This is primarily because of suburbanization and urban sprawl, which have transformed this highly urbanized region.

• What role has car ownership played in suburban growth?

• How does urban sprawl impact farmers?

• What is smart growth?

5. Population and Gender: For more than two centuries, North American birth rates have been in decline. A major factor in this has been women delaying childbearing to pursue education and careers. Declining childbirth rates play a role in the aging of North American populations, which may slow economic growth.

• How do education and work opportunities for contemporary women affect the North American population growth rates?

• What factors have created an aging North American population, and how do those factors connect with immigration and household types in Canada and the United States?

• What are three ways in which you personally are likely to be affected by the aging of North America's population?

• What is the connection between single-parent households and the fact that 22 percent of U.S. children and 14 percent of Canadian children live in poverty?

Critical Thinking Questions

1. Discuss the ways in which North American culture is adapted to high levels of spatial (geographic) mobility.

2. North American family types are changing. Explain the general patterns and discuss the reasons for these changes.

3. The influence of globalization is now felt in small, even isolated, places in North America. Pick an example from the text or from your own experience and explain at least four ways in which this place is now connected to the global economy or global political patterns.

4. Some argue that North Americans profit from having undocumented workers produce goods and services. Explain why this may or may not be the case and discuss how the situation should be changed or left as is.

5. How do you think life in North America would differ if the United States was no longer the world's superpower? How might U.S. attitudes toward the rest of the world be modified? Will these modifications be useful or destructive in the long run?

6. When you compare the economies of the United States and Canada and the ways they are related, what are three important factors to mention?

7. When the U.S. federal government gives subsidies to farmers or producers, what is intended? Why do farmers and other food producers in poor countries say that this practice hurts them?

8. The United States and Canada have different approaches in their health-care and social welfare policies. Explain those different approaches.

Chapter Key Terms

acid rain 70
agribusiness 84
aquifers 66
brownfields 99
clear-cutting 74
conurbation 98
digital divide 80
economic core 77
ethnicity 106
fracking 74

gentrification 100
housing segregation 100
infrastructure 76
Latino 65
metropolitan areas 96
North American Free Trade Agreement
  (NAFTA) 81
nuclear family 109
organically grown 85
Pacific Rim 102

pull factors 104
push factors 104
redlining 100
smog 70
social safety net 93
suburbs 96
trade deficit 81
Trans-Pacific Partnership (TPP) 82
tundra 121
urban sprawl 72

Map Labels

U.S.

Tijuana
El Paso
Ciudad Juárez
San Antonio
Houston
New Orleans

ATLANTIC
Bermuda (U.K.)

Jacksonville
Orlando
Tampa
Miami

OCEAN

30°N
Hermosillo
Mulegé
Baja California
Gulf of California
Nuevo Laredo
Monterrey
Matamoros
Gulf of Mexico
BAHAMAS

A
MEXICO
Sierra Madre Occidental
Durango
Sierra Madre Oriental
Iquitos
Tampico
Mazatlán
San Luis Potosí

Tropic of Cancer

Havana
CUBA
GREATER ANTILLES
DOMINICAN REPUBLIC

20°N
Puerto Vallarta
Guadalajara
Bay of Campeche
Mérida
Yucatán Channel
Cancún
San Juan
Puerto Rico (U.S.A.)
D

Mexico City
Puebla
Veracruz
H
Yucatán Peninsula
Belize City
Port-au-Prince
Kingston
HAITI
Santo Domingo
Montserrat (U.K.)
LESSER ANTILLES

PACIFIC
Acapulco
Oaxaca
BELIZE
Belmopan
JAMAICA
ST. LUCIA
ST. VINCENT
BARBADOS

Guatemala City
GUATEMALA
HONDURAS
Tegucigalpa
Caribbean Sea
Gulf of Venezuela
GRENADA

OCEAN
Gulf of Tehuantepec
San Salvador
EL SALVADOR
NICARAGUA
Managua
Lake Nicaragua
TRINIDAD & TOBAGO
Port of Spain

San José
COSTA RICA
Barranquilla
Maracaibo
Valencia
Caracas

Panama Canal
Colón
Panamá
Lake Maracaibo
VENEZUELA

10°N
PANAMA
Gulf of Panama
Medellín
Llanos
Orinoco
Georgetown
Paramaribo
GUYANA
SURINAME

Bogotá
Guiana Highlands
FRENCH GUIANA (France)

Cali
COLOMBIA

Quito
ECUADOR
Guayaquil

0° Equator
Galápagos Islands (Ecuador)
Gulf of Guayaquil
Iquitos
Marañón
Manaus
Amazon Basin
E F

Piura
Solimões
Madeira

PERU
BRA

10°S
Callao
Lima
Cusco
Andes
BOLIVIA
La Paz
Cochabamba
Mato Grosso
Xingu

Lake Titicaca
Altiplano
Sucre
Santa Cruz
Potosí
B

Tropic of Capricorn
Antofagasta
Atacama Desert
PARAGUAY
Paraná

20°S
Easter Island (Chile)
CHILE
Mts.
Asunción

Córdoba
Pôrto Alegre

Valparaíso
Rosario
URUGUAY

30°S
Santiago
Buenos Aires
Montevideo
La Plata
Río de la Plata
I

ARGENTINA
Pampas

Puerto Montt
Monte Verde

40°S
Patagonia
Comodoro Rivadavia

Falkland Islands (U.K.)

Strait of Magellan
Punta Arenas
Stanley
C
Tierra del Fuego

Photo Captions

A Sierra Madre, Mexico
Michel Setboun/Gamma-Rapho via Getty Images

B Andes, Chile
Hoberman Collection/Universal Images Group/Getty Images

C Tierra del Fuego, Argentina
Travel Ink/Gallo Images/Universal Images Group/Getty Images

D Soufrière Hills Volcano, Montserrat

Canary
Islands

Western
Sahara

MAURITANIA

CAPE
VERDE

SENEGAL
Praia
Dakar

ATLANTIC OCEAN

★ Cayenne

Belém

Fortaleza

Z I L

Recife

São Francisco

Tocantins

Brazilian Highlands

Salvador

★ Brasília

Belo
Horizonte

G

São Paulo   Rio de Janeiro

Land Elevations

meters	feet
4877	16,000
3353	11,000
2134	7000
914	3000
305	1000
152	500
0	0

mi 0   200   400   600
km 0   200   400   600   800   1000

1:37,000,000
Azimuthal Equidistant Projection

Middle and South America

G Rio de Janeiro, Brazil

H Yucatán Lowlands, Mexico (with Mayan temple)

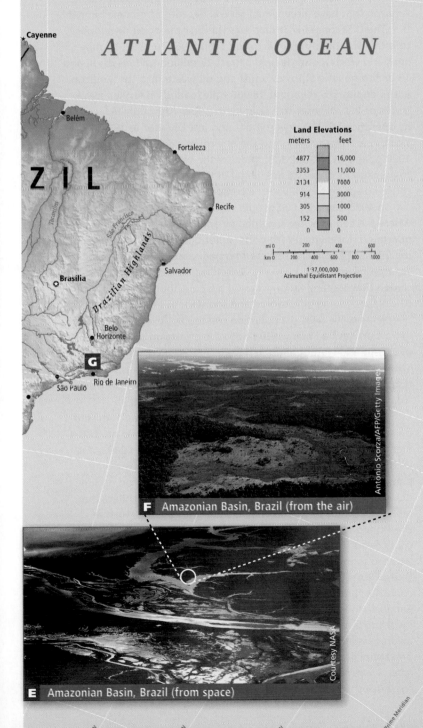
F Amazonian Basin, Brazil (from the air)

E Amazonian Basin, Brazil (from space)

I Pampas, Argentina

FIGURE 3.1 Regional map of Middle and South America.

GEOGRAPHIC THEMES

After you read this chapter, you will be able to discuss the following issues as they relate to the five thematic concepts:

1. Environment: Deforestation in this region contributes significantly to global climate change. In addition, some areas are experiencing a water crisis related to climate change, inadequate water infrastructure, and the intensified use of water, despite the region's overall abundant water resources.

2. Globalization and Development: This region's integration into the global economy has left it with the widest gap between rich and poor in the world. Poverty, economic instability, and the flow of resources and money out of the region have resulted in many conflicts and inspired numerous efforts at reform. Recently, though, several countries in this region have emerged as global economic leaders.

3. Power and Politics: The region was ruled for decades by elites and the military, and was periodically subject to disruptive foreign military interventions. Almost all countries now have multiparty political systems and elected governments, and political freedoms are expanding. The international illegal drug trade continues to be a source of violence and corruption in the region.

4. Urbanization: Since the 1950s, cities have grown rapidly in this region as rural people have migrated to cities and towns. A lack of urban planning has created densely occupied urban landscapes that often lack adequate support services and infrastructure.

5. Population and Gender: During the early twentieth century, the combination of cultural and economic factors and improvements in health care created a population explosion. By the late twentieth century, improved living conditions; better access to education and medical care; urbanization; and changing gender roles were all working together to reduce population growth.

The Middle and South American Region

Middle and South America, like North America, is a region of great physical and cultural diversity; but here the cultural mix is richer and there are wider disparities of wealth than in any other world region **(FIGURE 3.1)**. In spite of the many similarities between the two regions, modern history in Middle and South America has been very different from that of North America.

The five thematic concepts in this book are explored as they arise in the discussion of regional issues. Vignettes—like the one that follows about the Secoya—illustrate one or more of the themes as they are experienced in individual lives.

GLOBAL PATTERNS, LOCAL LIVES The boat trip down the Aguarico River in Ecuador took me into a world of magnificent trees, river canoes, and houses built high up on stilts to avoid floods. I was there to visit the Secoya, a group of 350 indigenous people locked in negotiations with the U.S. oil company Occidental Petroleum over Occidental's plans to drill for oil on Secoya lands. Oil revenues supply 40 percent of the Ecuadorian government's budget and are essential to paying off its national debt. The government had threatened to use military force to compel the Secoya to allow drilling.

The Secoya wanted to protect themselves from pollution and cultural disruption. As Colon Piaguaje, chief of the Secoya, put it to me, "A slow death will occur. Water will be poorer. Trees will be cut. We will lose our culture and our language, alcoholism will increase, as will marriages to outsiders, and eventually we will disperse to other areas." Given all the impending changes, Chief Piaguaje asked Occidental to use the highest environmental standards in the industry. He also asked the company to establish a fund to pay for the educational and health needs of the Secoya people.

Like the Secoya, indigenous peoples around the world are facing environmental and cultural disruption arising from economic development efforts. Chief Piaguaje based his predictions for the future on what has happened in other parts of the Ecuadorian Amazon that have already had several decades of oil development.

The U.S. company Texaco was the first major oil developer to establish operations in Ecuador. From 1964 to 1992, its pipelines and waste ponds leaked almost 17 million gallons of oil and 16 billion gallons of toxic runoff and oil waste into the Amazon Basin, enough to fill about 18,000 fully loaded oil tanker semi-trucks or 250 Olympic-size swimming pools. Although Texaco sold its operations to the government and left Ecuador in 1992, its oil wastes continue to leak into the environment from approximately 1000 open pits **(FIGURE 3.2)**. Many of the techniques Texaco used to manage its oil waste are illegal in the United States but were chosen because they saved the company money.

In 1993, some 30,000 people sued Texaco in New York State, where the company (now owned by and called Chevron) is head-quartered, for damages from the pollution. Those suing were both indigenous people and settlers who had established farms along Texaco/Chevron's service roads. Several epidemiological studies concluded that contamination from oil has contributed to higher rates of childhood leukemia, cancer, and spontaneous abortions among people who live near the pollution created by Texaco/Chevron. Oil extraction has had many other negative effects on the environment. Air and water pollution have increased the rates of illness. The wildlife that the Secoya used to depend on, such as tapirs, has disappeared almost entirely because of overhunting by new settlers from the highlands who are working in the oil industry.

In 2002, the Ecuadorian suit against Chevron was dismissed by the U.S. Court of Appeals, which argued that it had no jurisdiction in Ecuador. The case was refiled in Ecuador in 2003, and in 2011 an Ecuadorian court ruled against Chevron, assessing damages of U.S.$9.5 billion. However, because Chevron no longer has any assets in Ecuador, the plaintiffs tried to collect the money in countries where Chevron does have assets, such as the United States, Canada, Argentina, and Brazil. In 2014, a U.S. court ruled that the 2011 Ecuadorian judgment against Chevron was too tainted by bribery and corruption to be honored in the United States. The villagers are still pursuing Chevron in other countries, and in 2015 they won the right to sue Chevron in Canada. Now one of the longest-running legal battles in history involving a major multinational corporation, the case has raised awareness about how industries often fail to take environmental precautions in developing countries. *[Sources: Alex Pulsipher's field notes; Amazon Watch, 2006; BBC, 2014. For detailed source information, see Text Sources and Credits.]* ∎

FIGURE 3.2 Pollution from oil development in Ecuador. A local resident samples one of the several hundred open waste pits that Texaco left behind in the Ecuadorian Amazon. Wildlife and livestock trying to drink from these pits are often poisoned or drowned. After heavy rains, the pits overflow, polluting nearby streams and wells.

The rich resources of Middle and South America have attracted outsiders since the first voyage of Christopher Columbus in 1492. Europe's encounter with this region marked a major expansion of the global economy. However, during most of the period of expansion, Middle and South America occupied a disadvantaged position in global trade, supplying cheap raw materials that aided the industrial revolution in Europe and then North America but reaping few of the profits. These extractive industries did little to advance economic development within the region, as most profits went to foreign investors, and the negative environmental effects were largely ignored. Recently, countries such as Ecuador, Mexico, Bolivia, Brazil, and Venezuela have worked to control their own resources, develop local manufacturing and service-based industries, keep profits at home, and limit environmental pollution. Trade blocs within the region are creating conditions in which these countries can prosper from trade with each other.

The Ecuadorians' efforts to secure a damage settlement against a powerful multinational corporation is indicative of changing attitudes in this region and others toward outside developers **(FIGURE 3.3)**. Governments are now somewhat warier when

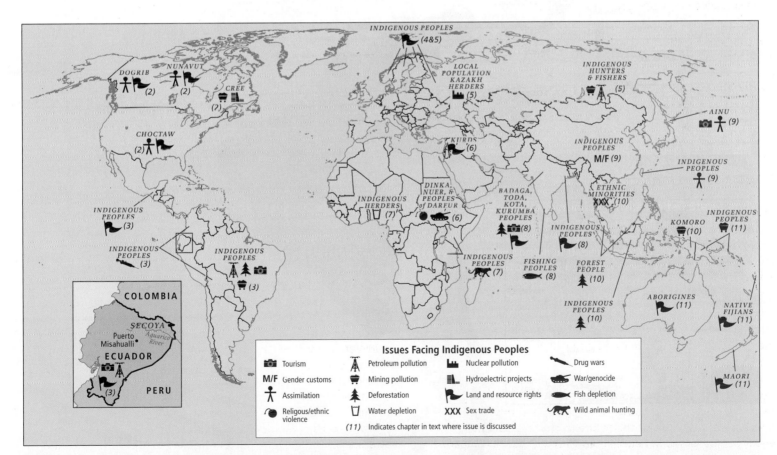

FIGURE 3.3 Indigenous peoples and environmental issues in this text. Issues relating to indigenous peoples are mentioned in many places in this book. In all cases, the issues are in some way related to interactions with the outside world and usually to uses of indigenous peoples' resources by the outside world.

they try to attract investors to develop new extractive, manufacturing, and service industries. Local people are more aware that they must be vigilant to ensure that development serves their interests.

What Makes Middle and South America a Region?

Physically, the Middle and South America region consists of Mexico, which geologically is part of the North American continent; the **isthmus** (land bridge) of Central America; and the continent of South America. For the last 500 years, the region of Middle and South America has been defined by a colonial past very different from that of Canada and the United States. Most of the countries in Middle and South America were at one time colonies of Spain. The exceptions are Brazil, which was a colony of Portugal, and a few small countries that were possessions of Britain, France, the Netherlands, Germany, or Denmark (see Figure 3.12).

Today, Middle and South America is a region of contrasts and disparities. Culturally, this region has large **indigenous** populations that have contributed to every aspect of life, blending with and changing the European, African, and Asian cultures introduced by the colonists. Social stratification based on income, class, race, and gender is notable. Politically, the region's more than three dozen countries have a range of governing ideologies, from the socialism of Cuba to the capitalism of Chile. Yet despite these contrasts and disparities, there are significant commonalities across the region, such as the Spanish language, Catholicism, and development trajectories that are connected to the global economy.

isthmus a narrow strip of land that joins two larger land areas

indigenous native to a particular place or region

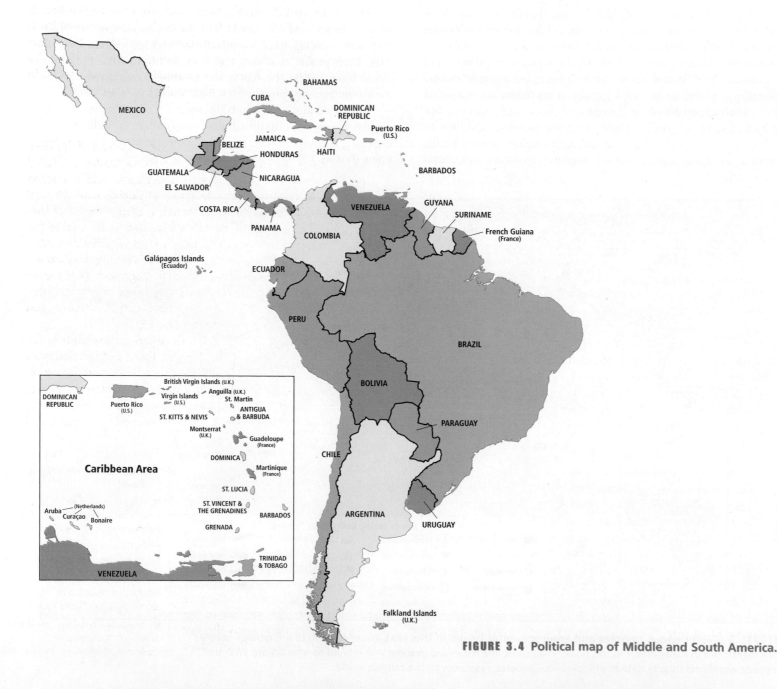

FIGURE 3.4 Political map of Middle and South America.

Terms in This Chapter

In this book, **Middle America** refers to Mexico, Central America (the narrow ribbon of land, or isthmus, that extends south of Mexico to South America), and the islands of the Caribbean (FIGURE 3.4). **South America** refers to the continent south of Central America. The term *Latin America* is not used in this book because it describes the region only in terms of the Roman (Latin-speaking) origins of the former colonial powers of Spain and Portugal. It ignores the region's large indigenous groups, its African, Asian, and Northern European populations, as well as the many mixed cultures, often called *mestizo* cultures, which have emerged. In this chapter, we use the term *indigenous groups* or *peoples* rather than *Native Americans* to refer to the native inhabitants of the region.

THINGS TO REMEMBER

- In this book, Middle America refers to Mexico, Central America (the narrow ribbon of land, or isthmus, that extends south of Mexico to South America), and the islands of the Caribbean. South America refers to the continent south of Central America.

- As people in this region gain more control over their own resources, some are searching for more sustainable ways to develop economically.

PHYSICAL GEOGRAPHY

Middle and South America extend south from the midlatitudes of the Northern Hemisphere across the equator through the Southern Hemisphere, nearly to Antarctica (see Figure 3.1). This long north-south expanse combines with variations in altitude to create the wide range of climates in the region. Tectonic forces have shaped the primary landforms of this huge territory to form an overall pattern of highlands to the west and lowlands to the east.

LANDFORMS

There are a wide variety of landforms in Middle and South America, and this variety accounts for the many different climatic zones in the region. But for ease in learning, landforms are here divided into just two categories: highlands and lowlands.

Highlands

A nearly continuous chain of mountains stretches along the western edge of the American continents for more than 10,000 miles (16,000 kilometers) from Alaska in the north to Tierra del Fuego at the southern tip of South America (see Figure 3.1C). The middle part of this long mountain chain is known as the Sierra Madre in Mexico (see Figure 3.1A), by various names in Central America, and as the Andes in South America (see Figure 3.1B, C). It was formed by a lengthy **subduction zone,** which runs thousands of miles along the western coast of the continents (see Figure 1.8). Here, two oceanic plates—the Cocos Plate and the Nazca Plate—plunge beneath three continental plates—the North American Plate, the Caribbean Plate, and the South American Plate.

In a process that continues today, the leading edge of the overriding plates crumples to create mountain chains. In addition, molten rock from beneath Earth's crust ascends to the surface through fissures in the overriding plate to form volcanoes. Such volcanoes are the backbone of the highlands that run through Middle America and the Andes of South America (see Figure 3.1). Although these volcanic and earthquake-prone highlands have been a major barrier to transportation, communication, and settlement, people now live close to quiescent volcanoes and in earthquake-prone zones, which can pose deadly hazards (for example, the 2010 and 2015 earthquakes in Chile).

The chain of high and low mountainous islands in the eastern Caribbean is also volcanic in origin, created as the Atlantic Plate thrusts under the eastern edge of the Caribbean Plate. It is not unusual for volcanoes to erupt in this active tectonic zone. On the island of Montserrat, for example, people have been living with an active, and sometimes deadly, volcano for more than a decade (see Figure 3.1D). Eruptions have taken the form of violent blasts of superheated rock, ash, and gas (known as *pyroclastic flows*) that move down the volcano's slopes at speeds upward of 450 miles (700 kilometers) per hour. The unusually strong earthquake in Haiti in January 2010 was also the result of plate tectonics.

Lowlands

Vast lowlands extend over most of the land to the east of the western mountains. In Mexico, east of the Sierra Madre, a coastal plain borders the Gulf of Mexico (see Figure 3.1H). Farther south, in Central America, wide aprons of sloping land descend to the Caribbean coast. In South America, a huge wedge of lowlands, widest in the north, stretches from the Andes east to the Atlantic Ocean. These South American lowlands are interrupted in the northeast and the southeast by two modest highland zones: the Guiana Highlands and the Brazilian Highlands (see Figure 3.1G). Elsewhere in the lowlands, grasslands cover extensive flat expanses, including the *llanos* of Venezuela, Colombia, and Brazil, and the *pampas* of Argentina (see Figure 3.1I).

The largest feature of the South American lowlands is the Amazon Basin, drained by the Amazon River and its tributaries (see Figure 3.1E, F). This basin lies within Brazil and the neighboring countries to its west. Earth's largest remaining expanse of tropical rain forest gives the Amazon Basin global significance as a reservoir of **biodiversity.** Hundreds of thousands of plant and animal species live here.

The basin's water resources are also astounding. Twenty percent of Earth's flowing surface waters exist here, running in rivers so deep that ocean liners can steam 2300 miles (3700 kilometers) upriver from the Atlantic Ocean all the way to Iquitos, jokingly referred to as Peru's "Atlantic seaport." The Amazon River system starts as streams high in the Andes. These streams eventually unite to become rivers that flow eastward toward the Atlantic. Once they reach the flat land of the Amazon Plain, their

Middle America in this book, a region that includes Mexico, Central America, and the islands of the Caribbean

South America the continent south of Central America

subduction zone a zone where one tectonic plate slides under another

biodiversity the variety of life forms to be found in a given area

velocity slows abruptly; fine soil particles, or **silt,** then sink to the riverbed. When the rivers flood, silt and organic material transported by the floodwaters renew the soil of the surrounding areas, nourishing millions of acres of tropical forest. Not all of the Amazon Basin is rain forest, however. Variations in weather and soil types, as well as human activity, have created grasslands and seasonally dry deciduous tropical forests in some areas.

CLIMATE

From the jungles of the Caribbean and the Amazon to the high, glacier-capped peaks of the Andes to the parched moonscape of the Atacama Desert and the frigid fjords of Tierra del Fuego, the range of climates in Middle and South America is enormous **(FIGURE 3.5).** In this region, the wide range of temperatures reflects both the great distance the landmass spans on either side of the equator and the tremendous variations in altitude across the region's landmass (the highest point in the Americas is Aconcagua in Argentina, at 22,841 feet [6962 meters]). Patterns of precipitation are affected both by the local shape of the land and by global patterns of wind and ocean currents that bring moisture in varying amounts.

Temperature-Altitude Zones

Four main **temperature-altitude zones,** shown in **FIGURE 3.6,** are commonly recognized in the region. As altitude increases, the temperature of the air decreases by about 1°F per 300 feet (1°C per 165 meters) of elevation. Thus temperatures are highest in the lowlands, which are known in Spanish as the *tierra caliente,* or "hot land." The *tierra caliente* extends up to about 3000 feet (1000 meters), and in some parts of the region these lowlands cover large areas. Where moisture is adequate, tropical rainforests thrive, as does a wide range of tropical crops, notably bananas, sugarcane, cacao, and pineapples. Many coastal areas of the *tierra caliente,* such as northeastern Brazil, have become zones of plantation agriculture that support populations of considerable size.

Between 3000 and 6500 feet (1000 to 2000 meters) is the cooler *tierra templada* ("temperate land"). The year-round, spring-like climate of this zone drew large numbers of indigenous people in the distant past and, more recently, has drawn Europeans. Here, crops such as roses, corn, beans, squash, various green vegetables, wheat, and coffee are grown.

Between 6500 and 12,000 feet (2000 to 3600 meters) is the *tierra fría* ("cool land"). A variety of crops, such as wheat, fruit trees, potatoes, and cool-weather vegetables—cabbage and broccoli, for example—do very well at this altitude. Many animals, including dogs, llamas, sheep, and guinea pigs, are raised for pets, food, and fiber **(FIGURE 3.7).** Several modern population centers are in this zone, including Mexico City, Mexico, and Quito, Ecuador.

Above 12,000 feet (3600 meters) is the *tierra helada* ("frozen land"). In the highest reaches of this zone, vegetation is almost absent, and mountaintops emerge from under snow and glaciers. A remarkable feature of such tropical mountain zones is that in a single day of strenuous hiking, one can encounter many of the climate types found on Earth.

silt fine soil particles

temperature-altitude zones regions of the same latitude that vary in climate according to altitude

trade winds winds that blow from the northeast and the southeast toward the equator

El Niño periodic climate-altering changes, especially in the circulation of the Pacific Ocean, now understood to operate on a global scale

Precipitation The pattern of precipitation throughout the region is influenced by the interaction of global wind patterns with mountains and ocean currents (see Figure 1.11). The **trade winds** sweep off the Atlantic, bringing heavy seasonal rains to places roughly 23° north and south of the equator (see the Figure 3.5 map). Winds from the Pacific bring seasonal rains to the west coast of Central America, but mountains block those rains from reaching the Caribbean side, which receives heavy rainfall from the northeast trade winds.

The Andes are a major influence on precipitation in South America. They block the rains borne by the trade winds off the Atlantic into the Amazon Basin and farther south, creating a rain shadow on the western side of the Andes in northern Chile and southwestern Peru (see Figure 3.5). Southern Chile is in the path of eastward-trending winds that sweep off of the Southern Ocean, bringing steady, cold rains that support forests similar to those of the Pacific Northwest in North America. The Andes block this flow of wet, cool air and divert it to the north. They thereby create another extensive rain shadow on the eastern side of the mountains along the southeastern coast of Argentina (Patagonia).

Adjacent oceans and their currents also influence the pattern of precipitation. Along the west coasts of Peru and Chile, the cold surface waters of the Peru Current bring cold air that cannot carry much moisture. The combined effects of the Peru Current and the central Andes rain shadow have created what is possibly the world's driest desert, the Atacama of northern Chile (see Figure 3.5B).

El Niño

One aspect of the Peru Current, that is only partly understood, is its tendency to change direction every few years (on an irregular cycle, possibly linked to sunspot activity). When this happens, warm water flows eastward from the western Pacific, bringing warm water and torrential rains, instead of cold water and dry weather, to parts of the west coast of South America. The phenomenon was named **El Niño,** or "the Christ Child," by Peruvian fishermen, who noticed that when it does occur, it reaches its peak around Christmastime. Peru's major banks plan for slower economic growth during El Niño years because of the increased flooding and damage to roads and bridges, increased waterborne disease, and reduced fish catches during an El Niño year.

El Niño also has global effects, bringing cold air and drought to normally warm and humid western Oceania and unpredictable weather patterns to Mexico and the southwestern United States. The El Niño phenomenon in the western Pacific is discussed further in Chapter 11, where Figure 11.6 illustrates its trans-Pacific effects.

Hurricanes

In this region, many coastal areas are threatened by powerful storms that can create extensive damage and loss of life. These form annually, primarily in the Atlantic Ocean north of the equator and close to Africa. A tropical storm begins as a group of thunderstorms. A few hurricanes also form in the southeastern Pacific and can affect the western coasts of Middle America before turning west toward Hawaii. When enough warming wet air comes together, the individual storms organize themselves into a swirling spiral of wind that moves across Earth's surface. The

FIGURE 3.5 PHOTO ESSAY: Climates of Middle and South America

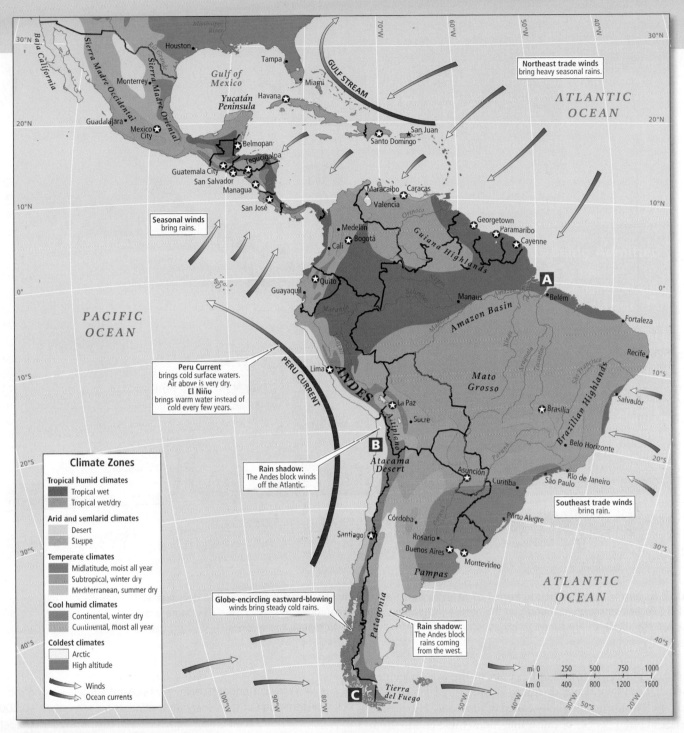

Climate Zones

Tropical humid climates
- Tropical wet
- Tropical wet/dry

Arid and semiarid climates
- Desert
- Steppe

Temperate climates
- Midlatitude, moist all year
- Subtropical, winter dry
- Mediterranean, summer dry

Cool humid climates
- Continental, winter dry
- Continental, moist all year

Coldest climates
- Arctic
- High altitude

- Winds
- Ocean currents

Northeast trade winds bring heavy seasonal rains.

Seasonal winds bring rains.

Peru Current brings cold surface waters. Air above is very dry.
El Niño brings warm water instead of cold every few years.

Rain shadow: The Andes block winds off the Atlantic.

Southeast trade winds bring rain.

Globe-encircling eastward-blowing winds bring steady cold rains.

Rain shadow: The Andes block rains coming from the west.

A Tropical wet, Belém, Brazil

B Desert, Atacama, Chile

C Continental, moist all year, Tierra del Fuego

Holger Leue/Lonely Planet Images/Getty Images

Veronique Durruty/Gamma-Rapho/Getty Images

DEA/A. Garozzo/Getty Images

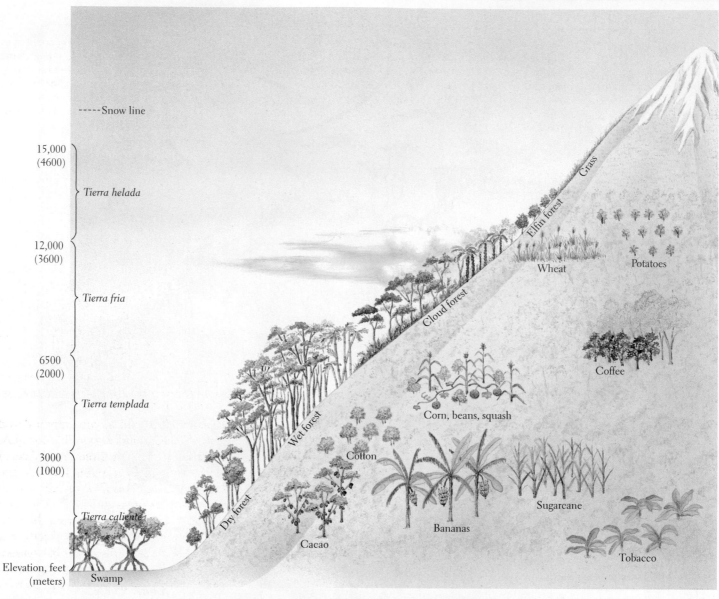

15,000 (4600)

Tierra helada

12,000 (3600)

Tierra fria

6500 (2000)

Tierra templada

3000 (1000)

Tierra caliente

Elevation, feet (meters)

-----Snow line

Snow line

Grass

Elfin forest

Wheat

Potatoes

Cloud forest

Coffee

Corn, beans, squash

Wet forest

Cotton

Sugarcane

Bananas

Tobacco

Dry forest

Cacao

Swamp

FIGURE 3.6 **Temperature-altitude zones of Middle and South America.** Temperatures tend to decrease as altitude increases, resulting in changes in the natural vegetation on mountainsides, as shown here. The same is true for crops, some of which are suited to lower, warmer elevations and some to higher, cooler ones.

highest wind speeds are found at the edge of the eye, or center, of the storm. Once wind speeds reach 74 miles (121 kilometers) per hour, such a storm is officially called a *hurricane*. Hurricanes usually last about 1 week; because they draw their energy from warm surface waters, they slow down and eventually dissipate as they move over cooler water or land. As the populations of coastal areas grow, more people are being exposed to hurricanes. Some scientists also think that climate change is leading to an increase in the number and intensity of hurricanes (see Figure 3.10B).

THINGS TO REMEMBER

• Middle and South America extend south from the midlatitudes of the Northern Hemisphere across the equator through the Southern Hemisphere, nearly to Antarctica.

• A nearly continuous chain of mountains stretches along the western edge of the American continents for more than 10,000 miles (16,000 kilometers), from Alaska in the north to Tierra del Fuego at the southern tip of South America.

• The rain forests of the Amazon Basin are planetary treasures of biodiversity that also play a key role in regulating Earth's climate.

• There are four temperature-altitude zones in the region that influence where and how people live and the crops they can grow.

• Significant environmental hazards in the region include earthquakes, volcanic eruptions, and hurricanes.

FIGURE 3.7 LOCAL LIVES: People and Animals in Middle and South America

A The Mexican hairless dog, or *xolo*, was bred 3000 years ago as a hunting dog and companion, and for food. The xolo was considered sacred by the Aztecs, Maya, and other indigenous groups, all of whom believed the dogs helped their masters' souls pass safely through the dangers of the underworld realm of Mictlan.

B *Cuy*, or guinea pigs, have been raised throughout the Andes region for at least 5000 years. Primarily a source of meat, they are also considered spiritual mediums by traditional Andean healers who use them to diagnose illnesses in people.

C A llama being led by a girl in traditional Quechua dress in Cusco, Peru. Domesticated by pre-Incan peoples thousands of years ago, llamas were raised for their fur, their meat, and their labor as pack animals, and later they started to be used as guard animals for sheep, which were introduced by the Spanish.

ENVIRONMENT

GEOGRAPHIC THEME 1

Environment: Deforestation in this region contributes significantly to global climate change. In addition, some areas are experiencing a water crisis related to climate change, inadequate water infrastructure, and the intensified use of water, despite the region's overall abundant water resources.

Environments in Middle and South America have long inspired concern about the use and misuse of Earth's resources. Millennia before Europeans arrived in this region, human settlements in the Americas had major environmental impacts (see the "Human Patterns over Time" section). However, modern impacts are particularly severe because both population density and per capita consumption have increased so dramatically, and because local environments now supply global demands.

TROPICAL FORESTS, CLIMATE CHANGE, AND GLOBALIZATION

A period of rapid deforestation that has had global repercussions began in the 1970s in Brazil with the construction of the Trans-Amazon Highway. Migrant farmers followed the new road into the rain forest. They began clearing terrain to grow crops, prompting a few observers to warn of an impending crisis of deforestation. Initially, concern focused on the loss of plant and animal species, as the rain forests of the Amazon Basin are some of the most biodiverse on the planet. (Note that the Amazon includes Colombia, Peru, Bolivia, and Brazil; see Figure 3.1.) However, climate change now dominates concerns about deforestation in the Amazon and the rest of this region. As explained

in Chapter 1, forests release oxygen and absorb carbon dioxide (CO_2), the greenhouse gas (GHG) most responsible for global warming. The loss of large forests, such as those in the Amazon Basin, contributes to global warming by releasing the CO_2 once locked in the woody bodies of trees. This CO_2 release happens as trees are burned to make way for crops and roads. Also, when there are fewer trees, less CO_2 can be absorbed from the atmosphere. Together, all of Earth's tropical rain forests absorb about 18 percent of the CO_2 added to the atmosphere yearly by human activity. Because 50 percent of Earth's remaining tropical rain forests are in South America, keeping these forests intact is crucial to minimizing climate change.

Middle and South American rain forests are being diminished by multiple human impacts (FIGURE 3.8). One of the biggest of these is caused by the clearing of land to raise cattle and grow crops such as soybeans (for animal feed), sugarcane (for ethanol), and African oil palm (for cooking oil) (see Figure 3.8C–E). Brazil has in recent years ranked between the world's fourth and seventh largest emitter of greenhouse gases, after the United States, China, and Indonesia. In the Amazon, Middle America, and elsewhere in the region, forests are cleared to create pastures for beef cattle, many of which are destined for the U.S. fast-food industry. If deforestation continues in Middle America at the current rate, the natural forest cover will be entirely gone in 20 years.

Hardwood logging and the extraction of underlying minerals, including oil, gas, and precious stones, also contribute to deforestation. Investment capital in logging industries is coming from Asian multinational companies that have turned to the Amazon forests after having logged as much as 50 percent of the tropical forests in Southeast Asia. The access roads that have been built to support these activities help accelerate deforestation by opening

FIGURE 3.8 PHOTO ESSAY: Human Impacts on the Biosphere in Middle and South America

A A newly built road in Suriname.

B A family on a river raft in the Peruvian Amazon.

C African oil palm is planted on land recently cleared for agriculture.

While there are a wide variety of human impacts on the environments of this region, here we focus on the processes of land cover and land use change that lead to the conversion of forests to grazing land or farmland. The forces guiding this process are complex, driven by poor people's need for livelihoods, governments' desire to assert control over lightly populated areas, and the demands for wood, meat, and food in distant urban centers and the global market. Together, these forces have led to a rapid loss of forest cover throughout much of this region. Brazil loses more forest cover each year than does any other country on the planet.

Human Impact, 2002

Land cover
- Forests
- Grasslands
- Deserts
- Tundra
- Ice
- National boundaries

Overfishing
- Threatened fisheries

Human impact on land
- High impact
- Medium–high impact
- Low–medium impact

Acid rain
- 4.8–4.3 pH
- 5.5–4.9 pH

D Cattle on land that has just been burned in Rondônia, Brazil.

E Soy fields recently cleared of forest in Mato Grosso, Brazil.

Thinking Geographically

After you have read about the impacts on the biosphere in Middle and South America, you will be able to answer the following questions.

A What environmental impact is clearly visible in this photo?

D What land use likely preceded cattle grazing on the land shown in this photograph?

E Where are the soybeans grown on this land likely to end up?

new forest areas to migrants. The governments of Peru, Ecuador, and Brazil encourage impoverished urban people to occupy cheap land along the newly built roads (see Figure 3.8A), and provide chainsaws to the settlers to help remove the trees. However, the settlers have had a difficult time learning to cultivate the poor soils of the Amazon. After a few years of farming, they often abandon the land, now eroded and depleted of nutrients, and move on to new plots. Ranchers sometimes then buy the worn-out land from these failed small farmers to use as cattle pastures (see Figure 3.8D).

Many people in this region are particularly vulnerable to the increasing threats of climate change. Figure 3.10 illustrates several such cases: shortages of clean water brought on by intensifying droughts and glacial melting, vulnerability to rising sea levels, and the effects of increasingly violent storms.

There is now an international effort to combat climate change by preserving the world's remaining forests—especially the crucial tropical rain forests found in this region. Brazil made major progress from 2005 to 2010, when it introduced new regulations that discouraged the cultivation of soybeans on once-forested lands. These policies reduced the rate of deforestation in the Brazilian Amazon by 75 percent. Because so many trees were being saved and not burnt, Brazil's GHG emissions plunged by 39 percent, more than any other country in the world. However, Brazil's economy has slowed in recent years, and there is now pressure to spur economic development with new road and hydroelectric projects in the Amazon. As a result, there has been an increase in deforestation in Brazil's portion of the Amazon.

INTEGRATING ENVIRONMENTAL PROTECTION WITH ECONOMIC DEVELOPMENT

In the past, governments in the region argued that economic development was so desperately needed that environmental regulations were an unaffordable luxury. Now, though, these governments are beginning to embrace the integration of both economic development and environmental protection. One example of this new approach is ecotourism.

Ecotourism

Many countries are now trying to earn money from the beauty of still-intact natural environments through **ecotourism**—in which nature-oriented vacations, often taken in endangered areas, are offered to travelers usually hailing from affluent cities or foreign nations. Ecotourists travel to these places so they can see and appreciate ecosystems and wildlife that do not exist where they live (**FIGURE 3.9**; see also "On the Bright Side: Alternatives to Deforestation").

Ecotourism has its downsides, however. Mismanaged, it can be similar to other kinds of tourism that damage the environment and return little to the surrounding community. While the profits of ecotourism can potentially be used to benefit local communities and environments, the profit margins may be small.

> **ecotourism** nature-oriented vacations, often taken in endangered and remote areas, usually by travelers from affluent nations

ON THE BRIGHT SIDE: Alternatives to Deforestation

Ecotourism is now the most rapidly growing segment of the global tourism and travel industry, which by some measures is the world's largest industry, accounting for U.S.$3.5 trillion in annual expenditures. Many Middle and South American nations have spectacular national parks that can provide a basis for ecotourism. As an alternative to deforestation, development based on ecotourism has the potential to preserve this region's biodiversity, reduce emissions of greenhouse gases, and provide less affluent and indigenous people with a chance to use their skills to teach tourists. ■

FIGURE 3.9 Ecotourism in the Amazon.

(A) A tourist poses at the bottom of a giant ceiba tree in the Ecuadorian Amazon.

(B) An Amazon river dolphin being fed by an ecotour guide.

(C) A tourist explores a walkway suspended in the canopy of tall rainforest trees in the Ecuadorian Amazon.

VIGNETTE Puerto Misahualli, a small river boomtown in the Ecuadorian Amazon, is currently enjoying significant economic growth. Its prosperity is due to the many European, North American, and other foreign travelers who come for experiences that will bring them closer to the now-legendary rain forests of the Amazon.

The array of ecotourism offerings can be perplexing. One indigenous man offers to be a visitor's guide for as long as desired, traveling by boat and on foot, camping out in "untouched forest teeming with wildlife." His guarantee that they will eat monkeys and birds does not seem to promise the nonintrusive, sustainable experience the visitor might be seeking. At a well-known *eco-lodge*, visitors are offered a plush room with a river view, a chlorinated swimming pool, and a fancy restaurant serving "international cuisine." All of this is on a private, 740-acre nature reserve separated from the surrounding community by a wall topped with broken glass. It seems more like a fortified resort than an eco-lodge.

By contrast, the solar-powered Yachana Lodge has simple rooms and local cuisine. Its knowledgeable resident naturalist is a veteran of many campaigns to preserve Ecuador's wilderness. Profits from the lodge fund a local clinic, a high school, and various programs that teach sustainable agricultural methods that protect the fragile Amazon soils while increasing farmers' earnings from surplus produce. The nonprofit group running the Yachana Lodge—the Foundation for Integrated Education and Development—earns just barely enough to sustain the clinic, the school, and the agricultural programs. [*Source: Alex Pulsipher's field notes in Ecuador.*] ■

THE WATER CRISIS

Although Middle and South America receive more rainfall than any other world region and have three of the world's six largest rivers (in volume), parts of the region are experiencing water crises. Most of the factors causing the water crises are induced by humans, who in turn are exposed to a variety of water-related stresses. These stresses are most severe in Haiti; the high Andean mountain zones of Peru and Bolivia; Central America; and northern Mexico. Of these areas, only northern Mexico and the high Andes are dry year round.

Haiti and Central America's problems with freshwater storage and distribution relate mostly to poverty and the mismanagement of resources. These are the poorest areas in the region and are plagued by corruption, poorly trained civil servants, and huge disparities in wealth and power that skew resources toward the rich and away from poor people. For example, only 64 percent of the population in Haiti has access to clean water, and even this access is prone to lengthy interruptions, especially during times of drought. Haiti receives enough precipitation that simple rainfall collection and storage systems would provide low-cost access to safe water for many, but corruption and poor planning make these systems rare. Natural disasters, such as hurricanes (shown in **FIGURE 3.10 B, C**) often vividly expose the weaknesses of water systems in these countries, which may take decades to recover.

In northern Mexico, south of the U.S. states of Arizona and New Mexico, water scarcity is caused not only by a dry climate,

Thinking Geographically

After you have read about the vulnerability to climate change in Middle and South America, you will be able to answer the following questions.

A What clues can be seen in this photo that the neighborhood lacks a centralized water distribution system?

B From which direction are hurricanes most likely to hit Honduras?

C In addition to exposure to tropical storms and hurricanes, what else contributes to Haiti's vulnerability to climate change?

D Why is glacial melting of particular concern to cities in Bolivia?

but also by inadequate planning and lax environmental policies. Cities have grown rapidly due to the expansion of numerous factories, or *maquiladoras*, that have been set up to take advantage of NAFTA-related trade with the United States and Canada (see the discussion of maquiladoras on page 152). The factories have been allowed to pollute water resources with few restraints, causing many waterways along the Mexican border with the United States to be polluted.

Many *colonias*, or communities on the urban fringes of large cities such as Nogales, Tijuana, and Ciudad Juárez, have become host to large numbers of migrants who work in the maquiladoras. The cities have not built enough new water and sanitation facilities to keep pace with the rapid growth in the colonias (see Figure 3.10A). This lack of infrastructure has caused waterborne illnesses to become widespread. Similar problems are found in many cities throughout the region. Even in Mexico City, the region's largest and wealthiest city, it is estimated that as much as 90 percent of urban wastewater goes untreated.

In some of the largest cities, the water infrastructure is so inadequate that as much as 50 percent of the fresh water is lost because of leaky pipes. A substantial part of the population does not have access to toilets, leaving many people to relieve themselves on the city streets. This poses a major health hazard, and as local water resources become too polluted to be drinkable, many cities must bring in potable water from distant areas.

Some sources of water in this region are threatened by climate change. In the Andean mountain zones of Peru and Bolivia, glaciers feed rivers that are the main source of water for millions of people (see Figure 3.10D). Should the glaciers actually disappear, many rivers will run much lower for parts of the year, straining communities, farms, and industries that depend on them.

Behind these immediate problems lie systemic policy failures and inadequate planning, such as that in Cochabamba, Bolivia (discussed in Chapter 1). There, efforts to improve the city's inadequate water supply system focused on "marketizing" the water system. The water supply, long thought of as a public resource, was sold to a group of foreign corporations led by Bechtel of San Francisco, California. The hope was that Bechtel, in return for profits, would make investments in infrastructure that Cochabamba's notoriously corrupt water utility would not make. Unfortunately,

FIGURE 3.10 PHOTO ESSAY: Vulnerability to Climate Change in Middle and South America

Adapting to the multiple stresses that climate change is bringing to this region is proving to be quite a challenge. Water-related troubles, such as drought, hurricanes, flooding, and glacial melting, are combining with growing populations and persistent poverty to create a complex landscape of vulnerability to climate change.

Phys geog of more vulnerable?

A A poor neighborhood in Nogales, Mexico, where there is no centralized water infrastructure and where people depend on water gathered off of the roofs of their homes or brought in by truck. The higher temperatures that climate change is bringing could make this area, where water is already scarce, even drier. Millions of Mexicans who have moved to work in factories along the U.S.–Mexico border are thus highly vulnerable to climate change.

B A major bridge in Honduras that was washed out by Hurricane Mitch, which killed 18,000 people. Hurricanes are likely to intensify as temperatures rise with climate change. Poor countries like Honduras are particularly vulnerable to the damage these storms bring.

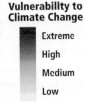

Vulnerability to Climate Change

- Extreme
- High
- Medium
- Low

C Haitians examine damage to crops from rain and flooding associated with Hurricane Sandy. So much of Haiti's forest cover has been removed that even mild tropical storms can cause catastrophic flooding. Much of Haiti's impoverished population is already dependent on foreign aid, a situation likely to worsen with climate change.

D La Paz and many smaller cities and towns in Bolivia's drought-prone highlands receive much of their drinking water from glaciers in the Andes. Higher temperatures are causing these glaciers to melt rapidly. Many have already disappeared and the rest could be gone in 15 years.

A Teotihuacan, Mexico, once home to an estimated 150,000 to 250,000 people.

B Machu Picchu, an estate built for the Incan emperor of the fifteenth century.

| 10,000 B.C.E | 5000 B.C.E | 0 C.E. | 1300 C.E. | 1400 C.E. |

23,000 B.C.E.–12,000 B.C.E
Bering land bridge

200 B.C.E.–800 C.E.
Teotihuacan flourishes

1325 C.E.
Aztec capital of Tenochtitlán founded

FIGURE 3.11 A VISUAL HISTORY OF MIDDLE AND SOUTH AMERICA

Thinking Geographically

After you have read about the human history of Middle and South America, you will be able to answer the following questions.

A How does this image of the ruins of Teotihuacan lend credence to the assertion that some indigenous groups may have been better off than their contemporaries in Europe?

B What about Machu Picchu in the Andean highlands indicates that it was more than a mere summer residence?

Bechtel, unfamiliar with the actual living conditions of the majority of Cochabamba's citizens, immediately increased water prices to levels that few urban residents could afford, while doing little to improve water supply or delivery systems. At one point Bechtel even charged urban residents for water taken from their own wells and rainwater harvested off their roofs! Popular protests forced Bechtel to abandon Cochabamba's water utility, which remains plagued by corruption and an inadequate infrastructure.

THINGS TO REMEMBER

GEOGRAPHIC THEME 1

• **Environment** Deforestation in this region contributes significantly to global climate change. In addition, some areas are experiencing a water crisis related to climate change, inadequate water infrastructure, and the intensified use of water, despite the region's overall abundant water resources.

• One of the biggest impacts on rain forests is the clearing of land to raise cattle and grow crops.

• Hardwood logging and the extraction of underlying minerals, including oil, gas, and precious stones, are also contributing to deforestation.

• Governments in the region are beginning to embrace both economic development and environmental protection. One example of this new approach is ecotourism.

• Lax environmental policies in much of Middle and South America have allowed industries and cities to pollute both air and water with few restraints.

HUMAN PATTERNS OVER TIME

The conquest of Middle and South America by Europeans set in motion a series of changes that continue to shape this region today. The conquest wiped out most of the indigenous civilizations and the conquerors set up new societies in their place. Many cultural features from the time of European colonialism endure to this day, as do some vestiges of the precolonial era.

THE PEOPLING OF MIDDLE AND SOUTH AMERICA

Recent evidence suggests that between 14,000 and 25,000 years ago, groups of hunters and gatherers from northeastern Asia spread throughout North America after crossing the Bering land bridge on foot, or moving along shorelines in small boats, or both. Some of these groups ventured south across the Central American isthmus, reaching the tip of South America by about 13,000 years ago.

By 1492, there were 50 to 100 million indigenous people in Middle and South America. In some places, population densities were high enough to threaten sustainability. People altered the landscape in many ways. They modified drainage to irrigate crops, they terraced hillsides, and they built paved walkways across swamps and mountains. They constructed cities with sewer systems, freshwater aqueducts, and huge earthen and stone ceremonial structures that rivaled the pyramids of Egypt **(FIGURE 3.11A)**.

The indigenous people also practiced the system of **shifting cultivation** that is still common in wet, hot regions in Central America and the Amazon Basin. In this system, also known as *slash and burn*, small plots are cleared in forestlands, the brush is dried and

C A mosaic in Lima, Peru, depicting Francisco Pizarro, conqueror of the Inca Empire.

D Chile wins independence from Spain in 1818 with help from Argentina.

The Battle of Maipu on the 5th April, 1818, printed by Rafael/Index/Bridgeman Images

E Former slaves cultivate sugar cane in Puerto Rico in 1899.

Library of Congress Prints and Photographs Division

Danita Delimont/Gallo Images/Getty Images

| 1500 C.E. | 1600 C.E. | 1700 C.E. | 1800 C.E. | 1900 C.E. | 2000 C.E. |

1492 Arrival of Europeans

1519–1521 Aztec Empire conquered

1533 Inca Empire conquered

1750 Andean potato fuels population explosion in Europe

1791–1822 Wars of independence from Spain and Portugal

1821–1888 Slavery abolished throughout mainland Middle and South America

C Describe the mood of this depiction of the Spanish conquest of the Incas.

D In what way does this painting of Creole Argentinians and Chileans celebrating independence suggest that they were unlikely to found egalitarian societies?

E Even after the end of slavery, who constituted the labor force in sugar cultivation?

burned to release nutrients in the soil, and the clearings are planted with multiple crop species. Each plot is used for only 2 or 3 years and then abandoned for several decades—long enough to allow the forest to regrow. If there is sufficient land, this system is highly productive per unit of land and labor, and is sustainable for long periods of time. However, if population pressure increases to the point that a plot must be used before it has fully regrown and its fertility has been restored, its yields decrease drastically. Recent discoveries of extensive geometric earthworks in the Amazon suggest widespread human inhabitance there within the past several thousand years that may have been dependent on shifting cultivation or more intensive agriculture.

The **Aztecs** of the high central valley of Mexico had some technologies and social systems that rivaled or surpassed those of Asian and European civilizations of the time. Particularly well developed were urban water supplies, sewage systems, and elaborate marketing systems. Historians have concluded that by 1500 C.E., Aztecs probably lived more comfortably on the whole than their contemporaries in Europe.

In 1492, the largest state in the region was that of the **Incas,** stretching from what is now southern Colombia to northern Chile and Argentina. The main population clusters were in the Andean highlands, where the cooler temperatures at these high altitudes eliminated the diseases of the tropical lowlands, while proximity to the equator guaranteed mild winters and long growing seasons. For several hundred years, the Inca empire was one of the most efficiently managed empires in the history of the world. Most remarkably, the Inca did not use money, conducting trade instead through a barter system. Taxes were collected in the form of labor, which was used in highly organized public works projects, such

as paved roads, elaborate terraces, irrigation systems, and great stone cities in the Andean highlands (see Figure 3.11B). Incan agriculture was very advanced, particularly in the development of crops, which included numerous varieties of potatoes and grains.

EUROPEAN CONQUEST

The European conquest of Middle and South America was one of the most significant events in human history (see Figure 3.11C). It rapidly altered landscapes and cultures and, through disease and slavery, ended the lives of millions of indigenous people.

Columbus established the first Spanish colony in 1492 on the Caribbean island of Hispaniola (which is now occupied by Haiti and the Dominican Republic). After learning of Columbus's exploits, other Europeans, mainly from Spain and Portugal on Europe's Iberian Peninsula, conquered the rest of Middle and South America.

The first part of the mainland to be invaded was Mexico, home to several advanced indigenous civilizations, most notably the Aztecs. The Spanish were unsuccessful in their first attempt to capture the Aztec capital of Tenochtitlán, but they succeeded a few months later after a smallpox epidemic decimated the native population. The Spanish demolished the grand Aztec capital in 1521 and built Mexico City on its ruins.

shifting cultivation a productive system of agriculture in which small plots are cleared in forestlands, the dried brush is burned to release nutrients, and the clearings are planted with multiple species; each plot is used for only 2 or 3 years and then abandoned for many years of regrowth

Aztecs indigenous people of high-central Mexico noted for their advanced civilization before the Spanish conquest

Incas indigenous people who ruled the largest pre-Columbian state in the Americas, with a domain stretching from what is now southern Colombia to northern Chile and Argentina

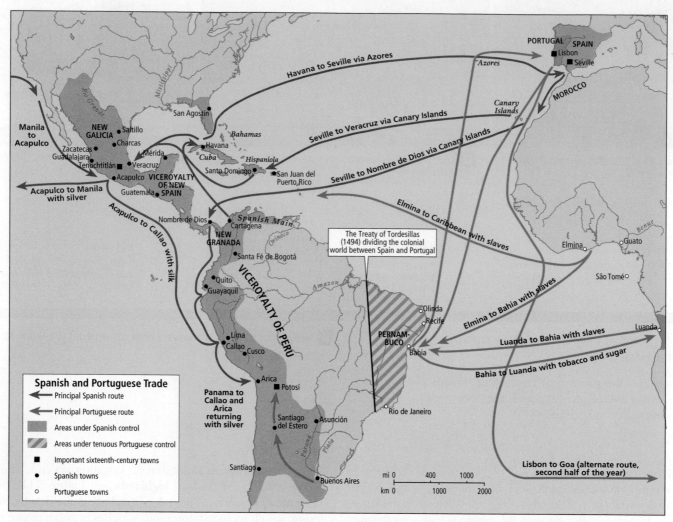

FIGURE 3.12 Spanish and Portuguese trade routes and territories in the Americas, circa 1600. The major trade routes from Spain to its colonies led to the two main centers of its empire, Mexico and Peru. The Spanish colonies could trade only with Spain, not directly with one another. By contrast, there were direct trade routes from Portuguese colonies in Brazil to Portuguese outposts in Africa. Between 4 and 5 million Africans were enslaved and traded to Brazilian plantation and mine owners (as well as a million to the Spanish empire, 2 million to the British Caribbean, and 1.3 million to the French Caribbean and Dutch colonies in Middle and South America). [Source consulted: *Hammond Times Concise Atlas of World History* (Maplewood, NJ: Hammond, 1994), pp. 66–67]

A tiny band of Spaniards, again aided by a smallpox epidemic, conquered the Incas in South America. Out of the ruins of the Inca empire, the Spanish created the Viceroyalty of Peru, which originally encompassed all of South America except Portuguese Brazil. The newly constructed capital of Lima flourished, in large part as a transshipment point for enormous quantities of silver extracted from mines in the highlands of what is now Bolivia.

Diplomacy by the Roman Catholic Church prevented conflict between Spain and Portugal over the lands of the region. The Treaty of Tordesillas of 1494 divided Middle and South America at approximately 46° W longitude **(FIGURE 3.12)**. Portugal took all lands to the east and eventually acquired much of what is today Brazil (where Portuguese is still the primary language); Spain took all lands to the west.

The superior military technology of the Spanish and Portuguese sped the conquest of Middle and South America. A larger factor, however, was the vulnerability of the indigenous people to diseases carried by the Europeans. In the 150 years

following 1492, the total population of Middle and South America was reduced by more than 90 percent, to just 5.6 million. To obtain a new supply of labor to replace the dying indigenous people, the Spanish initiated the first shipments of enslaved Africans to the region in the early 1500s. Between 10 and 12 million enslaved people were brought to this region, compared to the less than 400,000 brought to North America.

By the 1530s, a mere 40 years after Columbus's arrival, all major population centers of Middle and South America had been conquered and were rapidly being transformed by Iberian colonial policies. The colonies soon became part of extensive regional and global trade networks, the latter with Europe, Africa, and Asia.

A GLOBAL EXCHANGE OF CROPS AND ANIMALS

From the earliest days of the conquest, plants and animals were exchanged between Middle and South America, Europe, Africa,

TABLE 3.1 Globally important domesticated plants that originated in the Americas.

Type	Names and places of origin
Seeds Quinoa	Amaranth—*Amaranthus cruentus*, S. Mexico, Guatemala Beans—*Phaseolus* (four species), S. Mexico Maize (com)—*Zea mays*, valleys of Mexico Peanut—*Arachis hypogaea*, central lowlands of S. America Quinoa—*Chenopodium quinoa*, Andes of Chile and Peru Sunflower—*Helianthus annuus*, southwest and southeast N. America
Tubers Potato	Manioc (cassava)—*Manihot esculenta*, lowland of Middle and S. America Potato (numerous varieties)—*Solanum tuberosum*, Lake Titicaca region of Andes Sweet potato—*Ipomoea batatas*, S. America Tannia—*Xanthosoma sagittifolium*, lowland tropical America
Vegetables Tomato	Chayote (christophene)—*Sechium edule*, S. Mexico, Guatemala Peppers (sweet and hot)—*Capsicum* (various species), many parts of Middle and S. America Squash (including pumpkin)—*Cucurbita* (four species), tropical and subtropical America Tomatillo (husk tomato)—*Physalis ixocarpa*, Mexico, Guatemala Tomato (numerous varieties)—*Lycopersicon esculentum*, highland S. America
Fruit Pineapple	Avocado—*Persea americana*, S. Mexico, Guatemala Cacao (chocolate)—*Theobroma cacao*, S. Mexico, Guatemala Papaya—*Carica papaya*, S. Mexico, Guatemala Passion fruit—*Passiflora edulis*, central S. America Pineapple—*Ananas comosus*, central S. America Prickly pear cactus (tuna)—*Opuntia* (several species), tropical and subtropical America Strawberry (commercial berry)—*Fragaria* (various species), genetic cross of Chilean berry and wild berry from N. America Vanilla—*Vanilla planifolia*, S. Mexico, Guatemala, perhaps Caribbean
Ceremonial and drug plants Coca	Coca (cocaine)—*Erythroxylon coca*, eastern Andes of Ecuador, Peru, and Bolivia Tobacco—*Nicotiana tabacum*, tropical America

and Asia via the trade routes illustrated in Figure 3.12, in what is often called the *Columbian exchange*. Many plants essential to agriculture in Middle and South America today—rice, sugarcane, bananas, citrus, melons, onions, apples, wheat, barley, and oats, for example—were all originally imports from Europe, Africa, or Asia. When disease decimated the native populations of the region, the colonists turned much of the abandoned land into pasture for herd animals imported from Europe, including sheep, goats, oxen, cattle, donkeys, horses, and mules.

Just as consequential were plants first domesticated by indigenous people of Middle and South America. These plants have changed diets everywhere and have become essential components of agricultural economies around the globe. The

potato, for example, had so improved the diet of the European poor by 1750 that it fueled a population explosion. Manioc (cassava) played a similar role in West Africa. Corn, peanuts, vanilla, and cacao (the source of chocolate) are globally important crops to this day, as are peppers, pineapples, and tomatoes (Table 3.1).

THE LEGACY OF UNDERDEVELOPMENT

In the early nineteenth century, wars of independence left Spain with only a few colonies in the Caribbean (see Figure 3.11D). **FIGURE 3.13** shows the European colonizing countries and the dates of independence for the various Middle and South American countries. The supporters of the nineteenth-century revolutions

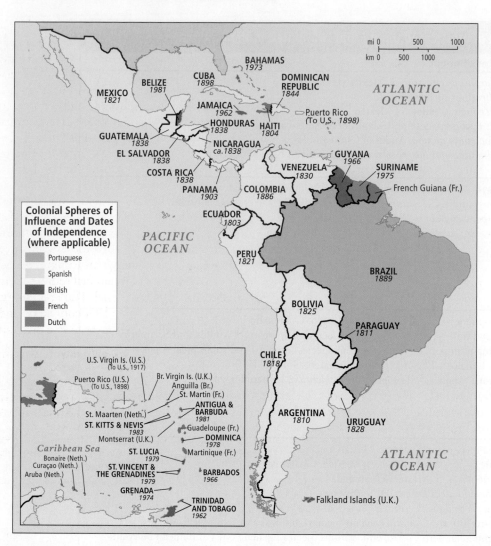

FIGURE 3.13 The colonial heritage of Middle and South America. Most of Middle and South America was colonized by Spain and Portugal, but important and influential small colonies were held by Britain, France, and the Netherlands. Nearly all the colonies had gained independence by the late twentieth century. Those for which no date appears on the map are still linked in some way to the colonizing country. [Source consulted: *Hammond Times Concise Atlas of World History* (Maplewood, NJ: Hammond, 1994), p. 69]

THINGS TO REMEMBER

• Recent evidence suggests that between 14,000 and 25,000 years ago, groups of hunters and gatherers from northeastern Asia spread throughout North America after crossing the Bering land bridge on foot, or moving along shorelines in small boats, or both.

• Shifting cultivation of small plots is a traditional and potentially sustainable agricultural strategy still used by some throughout the region.

• Spain and Portugal were the primary colonizing countries of Middle and South America, with smaller colonies held by Britain, France, Denmark, and the Netherlands. By the mid-nineteenth century, most of the larger colonies had gained independence.

• Exploitative development, incomplete revolutions, a dependence on raw materials exports, and the failure of elites to reinvest profits all led to underdevelopment in the region.

were primarily **Creoles** (people of mostly European descent born in the Americas) and relatively wealthy **mestizos** (people of mixed European, African, and indigenous descent). The Creoles' access to the profits of the colonial system had been restricted by policies favoring the "mother" country, and the mestizos were excluded by racist colonial policies. Once these groups gained power, however, they became a new elite who controlled the state and monopolized economic opportunity. Because they did little to expand economic development or access to political power for the majority of their populations (see Figure 3.11E), the revolutions were incomplete.

The economies of Middle and South America are much more complex and technologically sophisticated than they once were. Nevertheless, disparities persist: about 30 percent of the population is poor and lacks access to land, adequate food, shelter, water and sanitation, and basic education. Meanwhile, a small upper class has levels of affluence equivalent to those of the very wealthy in the United States. These conditions are in part the lingering result of colonial economic policies that favored the export of raw materials and fostered privileges for wealthy elites and outside investors who often spent their profits elsewhere rather than reinvesting them within the region. These policies are further discussed in the "Globalization and Development" section below.

Creoles people mostly of European descent born in the Americas

mestizos people of mixed European, African, and indigenous descent

GLOBALIZATION AND DEVELOPMENT

GEOGRAPHIC THEME 2

Globalization and Development: This region's integration into the global economy has left it with the widest gap between rich and poor in the world. Poverty, economic instability, and the flow of resources and money out of the region have resulted in many conflicts and inspired numerous efforts at reform. Recently, though, several countries in this region have emerged as global economic leaders.

For decades, Middle and South America has been one of the poorer world regions, though not as poor, on average, as sub-Saharan Africa, South Asia, or Southeast Asia. Following centuries of inequality and widespread poverty, Middle and South America experienced many reform efforts aimed at spreading wealth and opportunity more widely. Real change has been rare, though several countries have had significant economic growth in recent decades.

ECONOMIC INEQUALITY AND INCOME DISPARITY

With the exception of a few small countries, **income disparity**—the gap between rich and poor—in this region is one of the biggest in the world. One way of measuring income disparity is with the *Gini index*, a ratio that describes the distribution of income within a country, with 1 being perfect inequality and 0, perfect equality. The **FIGURE 3.14** map of the Gini index for countries around the world shows that Middle and South America have relatively high levels of income disparity. In recent years, disparities have been shrinking in Argentina, Colombia, Brazil, Chile, Mexico, and Venezuela, primarily because of new economic and social policies aimed at reducing disparities. The poverty rate for the region as a whole is around 33 percent, but in the poorer countries (Haiti, Guatemala, Honduras, Nicaragua, Peru, Bolivia, and Paraguay), more than half the population lives in poverty.

Phases of Economic Development

The current economic and political situation in Middle and South America derives from the region's history, which can be divided into three major phases: the early extractive phase, the import substitution industrialization (ISI) phase, and the current structural adjustment and marketization phase, which includes sharply rising investment from abroad. All three phases have helped entrench wide income disparities.

Early Extractive Phase From the time European conquerors first arrived, economic development was guided by a policy of **mercantilism,** in which Europeans extracted resources and controlled much of the economic activity in the American colonies in order to increase the power and wealth of their mother countries. All aspects of production, transport, and commerce within colonies, as well as trade between colonies and other countries, were shaped by mercantilism. Even after the colonies became independent countries, many extractive enterprises were owned by foreign investors who had few incentives to build stable local economies.

As a modest flow of foreign investment and manufactured goods entered the region, a vast flow of raw materials left for Europe and beyond. The money to fund the farms, plantations, mines, and transportation systems that enabled the extraction of resources for export came from abroad, first from Europeans and later from North Americans and other international sources. One example is the still very lucrative Panama Canal. The French first attempted to build the canal in 1880. It was later completed and run by the United States and finally turned over to Panamanian control only in 1999. The profits from these ventures were usually banked abroad, depriving the region of investment funds and tax revenues that could have made it more economically independent. Industries were slow to develop in the region, so even essential items, such as farm equipment, had to be

income disparity the gap in income between rich and poor

mercantilism the policy by which Europeans sought to increase the power and wealth of their mother countries by managing all aspects of production, transport, and commerce in their colonies, as well as trade between colonies and the mother countries

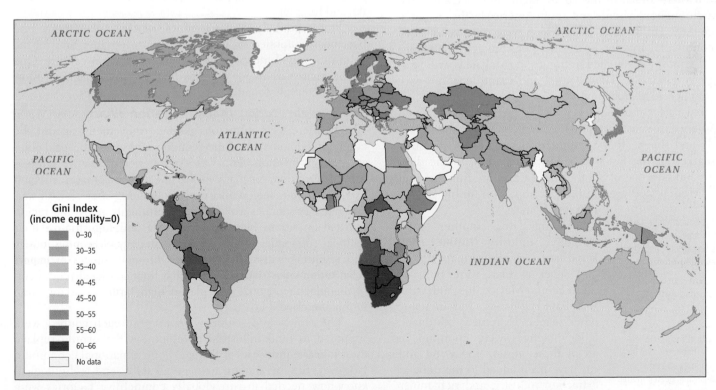

FIGURE 3.14 Gini index map of income distribution by country. This map, created by the World Bank, shows the level of income inequality within each country. Red countries have higher inequality while green countries have lower inequality.

purchased from Europe and North America at relatively high prices. Many people simply did without.

Several economic institutions arose in the early colonial era of Middle and South America to supply food and raw materials to Europe and North America. Large rural estates called **haciendas** were granted to colonists as a reward for conquering territory and people for Spain. For generations these essentially feudal estates were then passed down through the families of those colonists. Over time, the owners, who often lived in a distant city or in Europe, lost interest in the day-to-day operations of the haciendas, productivity stagnated, and hacienda laborers remained extremely poor. Nevertheless, haciendas produced a diverse array of products (cattle, cotton, rum, sugar) for local consumption and export.

Plantations are large factory farms. In addition to growing crops such as sugar, coffee, cotton, or (more recently) bananas, crops are often processed on site for shipment. Plantations were established early on in the colonial era and persist to this day, in part because they are more efficient and profitable than haciendas due to more intensive management and bigger investments in equipment. However, as was the case in the southern United States, plantations had relatively little local economic impact. Instead of employing local populations, plantation owners imported enslaved people from Africa. The equipment that the plantations used was usually imported from Europe, which was also where plantation owners preferred to invest their profits. As a result, little money was available to support the development of local industries that could have grown up around the plantations.

First developed by the European colonizers of the Caribbean and northeastern Brazil in the 1600s, plantations became more common throughout Middle and South America by the late nineteenth century. Unlike haciendas, which were often established in the continental interior in a variety of climates, plantations were for the most part situated in tropical coastal areas with year-round growing seasons. Their coastal and island locations gave them easier access to global markets via ocean transport.

As markets for meat, hides, and wool grew in Europe and North America, the livestock ranch emerged, specializing in raising cattle and sheep. Today, commercial ranches serving such global markets as the fast-food industry are found in the drier grasslands and savannas of South America, Central America, and northern Mexico, and even in the wet tropics on freshly cleared rain forest lands.

Mining was another early extractive industry **(FIGURE 3.15)**. Important mines (primarily gold and silver at first) were built in north-central Mexico, the Andes, the Brazilian Highlands, and in many other places. Extremely inhumane labor practices were common in all of these mines. Today, oil and gas have been added to the mineral extraction industry, and rich mines throughout the region continue to

hacienda a large agricultural estate in Middle or South America, more common in the past; usually not specialized by crop and not focused on market production

plantation a large factory farm that grows and partially processes a single cash crop

import substitution industrialization (ISI) policies that encourage local production of machinery and other items that previously had been imported at great expense from abroad

FIGURE 3.15 A copper mine in northern Chile, where some of the largest copper mines in the world are located. Copper accounts for 13 percent of Chile's GDP, and Chile produces one-third of the world's copper, more than any other country. Chile's copper mines are dependent on imported machinery, such as large dump trucks that are made in the United States.

produce gold, silver, copper, tin, precious gems, titanium, bauxite, and tungsten.

Profits from the region's mines, ranches, plantations, and haciendas continued to leave the region even after most of the countries gained independence in the nineteenth century. One of the main reasons for this was that wealthy foreign investors retained control of many of the extractive enterprises.

Import Substitution Industrialization Phase After World War II, there were numerous political movements against the continuing domination of the economy and society by local elites and foreign investors. This was also a period of relative prosperity because prices for many of the raw materials that this region exported were rising. With Europe rebuilding after years of war, and other economies expanding, governments across the region pursued ambitious plans to modernize and industrialize their national economies. Many adopted a strategy of keeping money and resources within their borders through policies of **import substitution industrialization (ISI)** that financed local production of manufactured goods and placed high tariffs on many commonly imported items.

The money and resources kept within each country were supposed to fund industrial development that would replace the extractive industries as the backbone of national economies. However, the investment, managerial skills, and technological know-how needed to run globally competitive factories were often lacking. Local consumers were numerous enough to

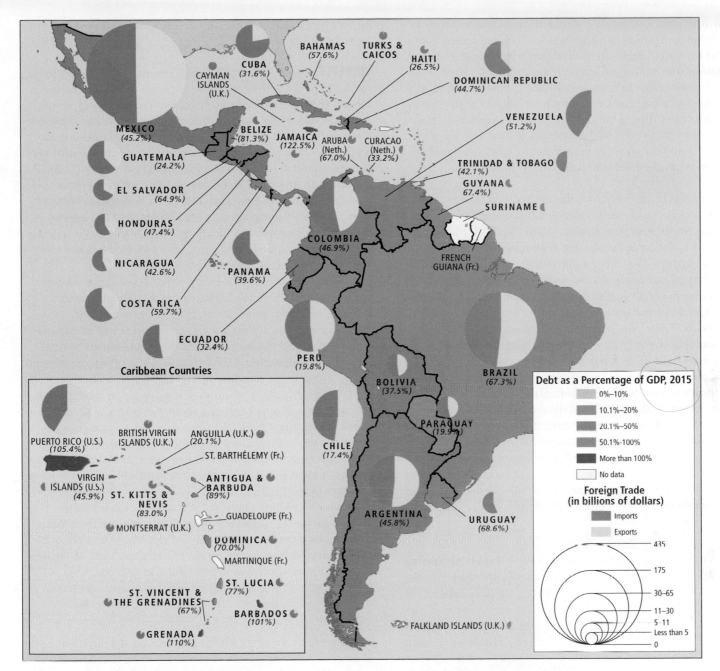

FIGURE 3.16 Economic issues: Public debt, imports, and exports. By the early 1970s, global prices of raw materials were falling; but instead of scaling back their ISI programs, governments borrowed millions of dollars from major international banks, most of which were in North America or Europe. When a worldwide economic recession hit in the 1980s, many governments were unable to repay their loans.

fund these industries with their purchases only in the larger countries, such as Brazil and Mexico. Brazil and Mexico both had success with some ISI programs; Brazil because of its strong aircraft, armament, oil export, and auto industries, and Mexico because of its oil and gas industries. However, most countries did not have successful ISI programs and stayed dependent on the export of raw materials. By the early 1970s, global prices of raw materials were falling, but instead of scaling back their ISI programs, governments borrowed millions of dollars from major international banks, most of which were in North America or Europe. When a worldwide economic recession hit in the

1980s, many governments were unable to repay their loans (**FIGURE 3.16**).

Structural Adjustment and the Marketization Phase Alarmed over the mounting debt of their clients, the international banks took action in the early 1980s through the International Monetary Fund (IMF, see Chapter 1) which developed and enforced **structural adjustment programs (SAPs)** that

structural adjustment programs (SAPs) policies that require economic reorganization toward less government involvement in industry, agriculture, and social services; sometimes imposed by the World Bank and the International Monetary Fund as conditions for receiving loans

ensured that sufficient money would be available to repay the international banks. Often described as "belt-tightening" measures, SAPs were based on several interdependent concepts: **privatization** (the selling of formerly government-owned industries and enterprises to private investors), **marketization** (the development of a free market economy in support of free trade), and *globalization* (the opening of national economies to global investors; see Chapter 1). At the time, these concepts were considered the soundest ways for countries to expand economically and thus to repay their debts to banks in North America, Europe, and Asia.

In the case of privatization, the investors to whom government firms were sold were often multinational corporations located in North America, Europe, and Asia. Thus the SAP era resulted in industries across the region being turned over to foreign ownership. The other main SAP policy, marketization, required that governments remove tariffs on imported goods of all types in order to obtain further loans. The removal of tariffs then caused many local industries to fail.

Crucially, SAPs also reversed the ISI-era trend of expanding government social programs and building infrastructure. To free up funds in order to repay debt, governments were required to fire many civil servants and drastically reduce spending on public health, education, job training, day care, water systems, sanitation, and infrastructure building and maintenance. In other words, SAPs took government services away from the poor and middle-class workers. SAPs encouraged the expansion of industries that were already earning profits by lowering taxes on their activities, thus further shrinking government revenues to pay for services. The most profitable industries remained those based on the extraction of raw materials for export.

Export Processing Zones A major component of SAPs was the expansion of manufacturing industries in **export processing zones (EPZs)**, also known as *free trade zones*—specially created areas within a country where, in order to attract foreign-owned factories, taxes on imports and exports are not charged. The main benefit to the host country is the employment of local people, which eases unemployment and brings money into the economy. Products are often assembled strictly for export to foreign markets.

There are EPZs in nearly all countries on the Middle and South American mainland and on some Caribbean islands (and in many other world regions). However, the largest of the EPZs is the conglomeration of assembly factories, called **maquiladoras,** that are located along the Mexican side of the U.S.–Mexico border. Although these factories do provide employment, jobs are often not secure, as the following vignette illustrates.

privatization the selling of formerly government-owned industries and firms to private companies or individuals

marketization the development of a free market economy in support of free trade

export processing zones (EPZs) specially created legal spaces or industrial parks within a country where, to attract foreign-owned factories, duties and taxes are not charged

maquiladoras foreign-owned, tax-exempt factories, often located in Mexican towns just across the border from U.S. towns, that hire workers at low wages to assemble manufactured goods which are then exported for sale

nationalize to seize private property and place it under government ownership, with some compensation

VIGNETTE In August of 2003, Orbalin Hernandez returned to his self-built shelter in the town of Mexicali on the border between Mexico and the United States. Recently fired for taking off his safety goggles while loading TV screens onto trucks at Thomson Electronics, he had just gone to the personnel office to ask for his job back. Because there were no previous problems with him, he was rehired at his old salary of U.S.$300 per month ($1.88 per hour). Thomson is a French-owned electronics firm that took advantage of NAFTA (see page 153) when it moved to Mexicali from Scranton, Pennsylvania, in 2001. There, its 1100 workers had been paid an average of $20 per hour. Many of the Scranton workers were unable to find work after Thomson left and still feel bitter toward the Mexicali workers.

In 2006, Orbalin and his fellow Mexicali workers were told that their wages were too high to allow their employers to compete with companies located in China, where in 2005, workers with the same skills as Orbalin earned just U.S.$0.35 an hour. Indeed, that year, 14 Mexicali plants closed and moved to Asia. Those firms remaining in Mexicali cut wages and reduced benefits.

By 2009, it appeared that the global recession was affecting Chinese–Mexican relations in new ways. Rising costs for fuel and storage and some questions about the quality of Chinese goods meant that it made less sense to move a Mexican factory to China. In fact, in 2012, China began to invest in Mexico in order to produce such things as motorcycles and trucks for the Mexican market. It is not yet clear whether this trend will be short-lived or the wave of the future.

Though Orbalin's salary at Thomson-Mexicali is barely enough for him to support his wife Mariestelle and their four children, he is grateful for the job. He and Mariestelle are originally from a farming community in the Mexican state of Tabasco, where the minimum wage in 2015 is U.S.$4.68 a day. *[Sources: NPR, ZNet, Bloomberg Businessweek, and Maquiladora Portal. For detailed source information, see Text Sources and Credits.]* ∎

The Backlash Against SAPs The SAP era did not produce the sustained economic growth that was expected to relieve the debt crisis and create broader prosperity. SAPs failed to stimulate economic growth in large part because they encouraged more dependence on exports of raw materials just when prices for these items were falling in the global market.

Since the late 1990s, millions of voters in Middle and South America have registered their opposition to SAPs. Starting in 1999, presidents who explicitly opposed SAPs have been elected in eight countries in the region: Argentina, Bolivia, Brazil, Chile, Ecuador, Nicaragua, Uruguay, and Venezuela. Both Brazil and Argentina made considerable sacrifices to pay off their debts and liberate themselves from the restrictions of SAPs. Venezuela, under President Hugo Chávez, emerged as a leader of the SAP backlash. In 2005, the Chávez administration **nationalized** foreign oil companies operating in Venezuela. Chávez also used Venezuela's oil wealth to help other countries, such as Argentina, Bolivia, and Nicaragua, pay off their debts. Bolivia, led by Evo Morales (often an ally of Chávez), reversed the SAP policy of privatization by nationalizing that country's natural gas industry.

The past decade has seen many changes in the region's relationship with the IMF and SAPs. China's demand for the resources of this region, and its willingness to lend directly to governments as a means of securing access to these resources, has dramatically reduced the need for loans from the IMF. Loans from the IMF dropped from U.S.$48 billion in 2003 to just U.S.$1 billion in 2008 and have remained low since. In contrast to the years after the global recession of the early 1980s, when almost every country in this region turned to the IMF for loans, only a few countries in Central America and the Caribbean currently have significant loans from the IMF.

Beginning several years ago, the IMF reversed its policies from promoting SAPs to promoting *Poverty Reduction Strategies Papers* (PRSPs), which attempt to combine marketization with strong investment in social programs. The IMF is now implementing PRSPs around the world.

THE CURRENT ERA OF INTERNATIONAL TRADE AND INVESTMENT

Growth in international trade and foreign investment has transformed this region over the past decade but has also brought back familiar patterns. Several countries—most notably Brazil, Mexico, and Chile—have become more influential in the global economy as their trade with the rest of the world (especially China and the United States) has grown. These countries have also had massive inflows of *foreign direct investment* (FDI)—investment funds that come in to enterprises from outside the country (see Figure 3.16). However, of these three, only Mexico has growth that is driven by the production of manufactured goods (most of which is destined for the United States). Chile and Brazil's more spectacular growth has been based mainly on its exports of raw materials to China, which is now their largest trading partner. China is the second- or third-largest trading partner for every other major country in the region, including Mexico, Peru, Colombia, Argentina, Venezuela, Ecuador, and the Dominican Republic.

When China's growth slowed after the global recession that began in 2007, China's once-insatiable demand for raw materials also fell. China will likely remain a major trade partner for this region, but it may reduce its imports in coming years as it shifts toward a model of more moderate economic growth that is based on serving the needs of China's huge and increasingly wealthy urban populations.

Regardless, a decade of trade with China has brought many changes to this region. Brazil is now part of the BRICS trade consortium with Russia, India, China, and South Africa (see Chapter 1). The BRICS countries are helping each other counter the influence of the United States and Europe on the global economy. Rich in resources and busily transforming themselves

into large middle-income, global economic powerhouses, the BRICS countries recently formed their own bank as an alternative to the IMF and World Bank. Meanwhile, in 2014 the IMF asked for help from Brazil, Mexico, and Peru in designing and funding a bailout package for overly indebted countries in the European Union—the first time countries in Middle and South America have been asked to play such a role.

The Role of Remittances

Migration for the purpose of supporting family members back home with remittances is an extremely important economic activity throughout this region (see further discussion on page 161). Each year, remittances amount to roughly double the U.S. foreign aid to the region. Most such remittance migrations are within countries. While the amounts of cash sent home are small, these remittances are crucial for families that may not have much other income.

One report (by geographer Dennis Conway and anthropologist Jeffrey H. Cohen) suggests that couples who migrate to the United States from indigenous villages in Mexico typically work at menial jobs and live frugally in order to save a substantial nest egg. They may then return home for several years to build a house (usually a family self-help project) and buy furnishings. When the money runs out, the couple, or just one member of the couple, may migrate again to save up another nest egg.

The Informal Economy

For centuries, low-profile businesspeople throughout Middle and South America have operated in the informal economy—they have supported their families through inventive entrepreneurship but their work is not officially recognized in the statistics and they do not pay business, sales, or income taxes. Most are small-scale operators involved with street vending or recycling used items such as clothing, glass, or waste materials. Often people stay in the informal economy despite the job insecurity and lower pay because the work gives them more freedom to work when and where they want. Avoiding government oversight can also be a motivation, as some countries have elaborate regulations for even small businesses.

Regional Trade and Trade Agreements

The growth in international trade and foreign investment in the region, in part the result of SAPs, has been joined by growth in regional free trade agreements. Such agreements reduce tariffs and other trade barriers among a group of neighboring countries. The two largest free trade agreements within the region are NAFTA and UNASUR.

The 1994 **North American Free Trade Agreement (NAFTA)** established a free trade bloc consisting of the United States, Mexico, and Canada, and containing more than 450 million people. The main goal of NAFTA is to reduce barriers to trade, thereby creating expanded markets for the goods and services produced in the three countries. Since 1994, the value of the economies of these

North American Free Trade Agreement (NAFTA) a free trade agreement made in 1994 that added Mexico to the 1989 economic arrangement between the United States and Canada

three countries has grown steadily; by 2016, it was worth more than U.S.$20 trillion, making it the world's largest trade bloc. An attempt by the United States to create a NAFTA-like trade bloc for all of the Americas (FTAA) has stalled in recent years, largely because of dissatisfaction with the effects of globalization.

UNASUR, a union of South American nations, was organized in May of 2008; it supersedes *Mercosur* and the *Andean Community of Nations*, two previous customs unions. The 12 member countries (Argentina, Bolivia, Brazil, Chile, Colombia, Ecuador, Guyana, Paraguay, Peru, Suriname, Uruguay, and Venezuela) have more than 415 million people, with a combined economy worth nearly U.S.$8 trillion. UNASUR appears to want to emulate the European Union by having a common passport, a parliament, and a single currency by 2019.

The overall record of regional free trade agreements so far is mixed. While they have increased the amount of trade, the benefits of that trade usually are not spread evenly among regions or among all sectors of society. In Mexico, for instance, the benefits of NAFTA have gone mainly to the wealthy investors who are concentrated in the northern states that border the United States. There, the agreement facilitated the growth of maquiladoras by easing cross-border finances and transit. But significantly, as many as one-third of small-scale farmers throughout Mexico have lost their jobs because of the increased competition from subsidized corporate farms in the United States, which, through NAFTA, now have unrestricted access to Mexican markets. In particular, corn imports from the United States have driven down the price of corn in local markets, pushing small farmers out of business.

Rapid Regional Expansion of Internet Use Overall, Middle and South America are more advanced in terms of information technology than are many other developing regions of the world—a fact that could open up many new economic opportunities for the region in the near future. Brazil ranks fifth in the world in terms of total number of Internet users and has two world-class technology hubs in the environs of São Paulo.

Mexico, with 49 percent of its population connected, recently launched a program to give its citizens access to training and higher education via the Internet **(FIGURE 3.17)**. Similar efforts throughout the region are part of a long-term shift, especially in urban areas, toward more technologically sophisticated and better-paid service sector employment. In 2000, only 3 percent of the region's population used the Internet, but that figure had risen to nearly 60 percent by 2015, an increase of nearly 2000 percent.

Food Production and Development

Food production throughout Middle and South America is shifting away from small-scale, often subsistence-level production toward large-scale, green revolution agriculture that is aimed at earning cash. This shift has forced many small-scale farmers who cannot afford the investment in machinery or chemicals off their land and into cities or to wealthier countries where

UNASUR a union of South American nations that was organized in May of 2008; it emulates the European Union, linking the economies of Argentina, Bolivia, Brazil, Chile, Colombia, Ecuador, Guyana, Paraguay, Peru, Suriname, Uruguay, and Venezuela

they often work illegally. In significant numbers, ordinary people have begun to effectively advocate for policies that could help lift them out of landlessness and poverty.

For centuries, while people labored for very low wages on haciendas, the hacienda owners would at least allow them a bit of land to grow a small garden or some cash crops. SAPs encouraged a shift to green revolution agriculture, which was based on crops that could be exported for cash. This type of production, it was hoped, could help countries pay off debts faster. SAPs also made it easier for foreign multinational corporations, such as Del Monte, to use their financial resources to buy up many haciendas and other farms, converting them to plantations. Many rural people have been forced off lands they once cultivated and onto plantations, where they now work as migrant laborers for low wages, as the following story illustrates.

VIGNETTE Aguilar Busto Rosalino used to work on a Costa Rican hacienda. He had a plot on which to grow his own food and in return worked 3 days per week for the hacienda. Since a banana plantation took over the hacienda, he rises well before dawn and works 5 days a week from 5:00 A.M. to 6:00 P.M., stopping only for a half-hour lunch break. Because he now lacks the time and land to farm, he must buy most of his food.

Aguilar places plastic bags containing pesticide around bunches of young bananas. He prefers this work to his last assignment of spraying a more powerful pesticide, which left him and 10,000 other plantation workers sterile. He works very hard because he is paid according to how many bananas he treats. Usually he earns between U.S.$5.00 and $14.50 a day.

It is common practice for these banana operations to fire their workers every 3 months so that they can avoid paying the employee benefits that Costa Rican law mandates. Although Aguilar makes barely enough to live on, he has no plans to press for higher wages because if he did he would be put on a "blacklist" of people that the plantations agree not to hire. [*Source: Andrew Wheat. For detailed source information, see Text Sources and Credits.*] ∎

Because of the moneymaking potential and political power of large-scale agriculture, conflicts are only rarely resolved in favor of small farmers and agricultural workers. Moreover, in rapidly urbanizing countries, green revolution agriculture is seen as the only way to supply cheap food for the millions of city dwellers. Ironically, many of these new urbanites were once farmers capable of feeding themselves who only moved to the cities because they were unable to compete with the new green revolution systems.

Many areas have had large farms, such as haciendas, for centuries. These are now being converted to green revolution agriculture, similar to farms that can be found in the Midwestern United States. There is a wide belt of large-scale farming and ranching from the Argentine pampas into Brazil **(FIGURE 3.18)**. And, as we saw in the earlier discussion of environmental issues (see page 139), rice, corn, soybeans, and meat animals, which are often produced by large green revolution farms on once-forested lands in and around the Amazon Basin, are all major exports for countries like Brazil and Argentina.

FIGURE 3.17 Internet use in Middle and South America, 2015. The figure shows the number of Internet users in each country and the percentage of the country's population that uses the Internet. By November 2015, more than 339 million people (54 percent of the region's population) were using the Internet, a 44 percent increase over 2011. Argentina has the highest percentage of Internet users (80 percent) in South America; Costa Rica has the highest percentage (88 percent) in Middle America; and Curaçao has the highest percentage (94 percent) in the Caribbean.

At a medium scale are zones of modern mixed farming (those that produce meat, vegetables, and specialty foods for sale in urban centers), located on large and small plots around most major urban centers. **FIGURE 3.19** illustrates how the foodways of the area are primarily based on indigenous plants (corn, potatoes, tomatoes) and methods of cooking (baking, grilling), and involve beef, pork, rice, seasonings, and cooking methods (frying) imported from Europe, Asia, and Africa.

THINGS TO REMEMBER

GEOGRAPHIC THEME 2 • **Globalization and Development** This region's integration into the global economy has left it with the widest gap between rich and poor in the world. Poverty, economic instability, and the flow of resources and money out of the region have resulted in many conflicts and inspired numerous efforts at reform. Recently, though, several countries in this region have emerged as global economic leaders.

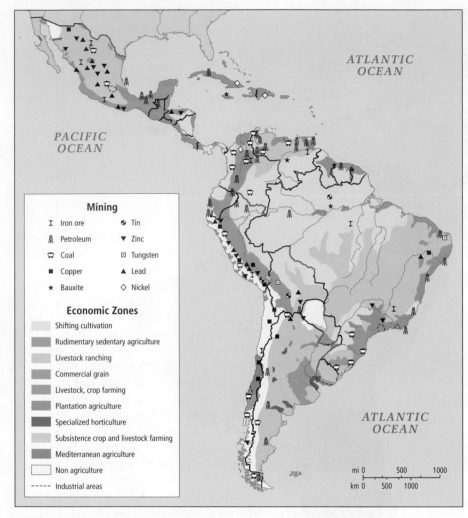

FIGURE 3.18 Agricultural and mineral zones in Middle and South America.
[Source consulted: Adapted from *Goode's World Atlas,* 21st ed. (Chicago: Rand McNally, 2010), pp. 160–161, Minerals and Economic map]

- Three phases of economic development—extractive, import substitution industrialization (ISI), and structural adjustment programs (SAPs)—dominated the region into the twenty-first century.

- Since 1999, presidents who explicitly opposed SAPs have been elected in eight countries in the region: Argentina, Bolivia, Brazil, Chile, Ecuador, Nicaragua, Uruguay, and Venezuela.

- Brazil is part of the BRICS trade consortium with Russia, India, China, and South Africa. These are emerging economies, rich in resources and busily transforming themselves into highly developed, highly productive powerhouses.

- Since the 1990s, some countries have adopted free trade policies, and regional trade blocs have been formed and re-formed, with mixed results.

- Food production throughout this region is undergoing a shift away from small-scale, often subsistence-level production toward large-scale, green revolution agriculture that is aimed at earning cash.

FIGURE 3.19 LOCAL LIVES: Foodways in Middle and South America

A　Corn tortillas are prepared in Mexico City. Corn has been a pillar of diets in Mexico and the countries of Central America for thousands of years. Now cultivated throughout the world, corn was first domesticated from a wild grass in southern Mexico.

B　An Argentine *asado,* or barbeque. Here beef is roasted over an open bed of coals, using a special *parilla,* or grill. This method of cooking is a variant of methods first developed by the indigenous peoples of the grasslands of Argentina. Asado is the national dish of Argentina; it is prepared widely throughout Uruguay, Paraguay, and southern Brazil.

C　*Lomo saltado,* a Peruvian dish with Asian influences. First developed in restaurants started by Chinese immigrants, it is a mix of Chinese and Peruvian ingredients and culinary traditions. It consists of strips of beef or pork marinated in vinegar and stir-fried with onions, parsley, tomatoes, and other vegetables, and then served over rice and french fries.

POWER AND POLITICS

GEOGRAPHIC THEME 3

Power and Politics: The region was ruled for decades by elites and the military, and was periodically subject to disruptive foreign military interventions. Almost all countries now have multiparty political systems and elected governments, and political freedoms are expanding. The international illegal drug trade continues to be a source of violence and corruption in the region.

In the last 30 years, there have been repeated peaceful and democratic transfers of power in countries once dominated by rulers who seized power by force. These **dictators** often claimed absolute authority, governing with little respect for the law or the rights of their citizens. Their authority was based on alliances between the military, wealthy rural landowners, wealthy urban entrepreneurs, foreign corporations, and even foreign governments such as the United States.

Although a slow expansion of political freedoms has transformed the politics of the region, problems remain (as seen in **FIGURE 3.20**). Elections are sometimes poorly or unfairly run and their results are frequently contested. Elected governments are sometimes challenged by mass citizen protests or threatened with a **coup d'état,** in which the military takes control of the government by force. Such coups are usually a response to policies that are unpopular with large segments of the population, powerful elites, the military, or the United States. In the last decade, coups have been threatened in Venezuela, Colombia, Ecuador, and Bolivia and occurred in Honduras. But peaceful democratic elections are increasingly the norm in this region.

Populist Political Movements There is a long tradition of *populist political movements,* which appeal to the interests of the majority of people, who in this region of extreme inequalities tend to be poor. Such movements have emerged intermittently for centuries, but most have been kept from gaining much power by alliances between the wealthy, the military, and foreign investors and governments that have often dominated politics in this region. However, the expansion of political freedoms over the past several decades has opened the way for a wave of populist movements that gained control of national governments. Disgusted by the failures of SAPs and empowered by the revenues of trade with China, these governments embarked on ambitious programs to address the problems of extreme inequality. Although financial strains created recently by China's declining imports have unseated some of these governments and constrained the ambitions of others, this wave of populist political movements has powerfully transformed the region.

The current wave of populist political movements gained considerable momentum in Venezuela with the landslide election of Hugo Chávez as president in 1998. Elites had long dominated Venezuelan politics, and profits from the country's rich oil deposits had not reached the poor. One-third of the population lived on U.S.$2 (PPP) a day, and per capita GNI (PPP) was lower in 1997 than it had been in 1977. Chávez campaigned as a champion of the poor, calling for the government to subsidize job creation, community health care, and food for the large underclass.

He also redirected the profits of the largely foreign-managed oil industry to support government social programs. These policies, all of which were enacted soon after he was elected, eroded Chávez's original support from the small middle class and elites, who began to fear a turn toward broad government control of the economy. They also earned him intense criticism from the U.S. media, as well as the U.S. government, which he responded to with colorful rhetoric criticizing U.S. leaders and their policies. When addressing the UN in 2006, a day after President George Bush spoke, he said, "The devil came here yesterday, and it stinks of sulfur still today." Chávez helped lead the region-wide backlash against SAPs and condemned the United States' history of intervention in the internal affairs of countries across the region.

Chávez was reelected in 2000, and then briefly deposed in a coup d'état in 2001 in which the United States participated covertly. He was quickly reinstated, however, and his presidency was sustained in the landslide election of 2006. Soon after that voters approved a constitutional amendment removing presidential term limits, thereby allowing Chávez to run for president indefinitely, a move that some saw as a triumph of populism, others as a loss for democracy and a drift back toward dictatorship (see the Figure 3.20 map). Hugo Chávez died in March of 2013 from cancer after having been treated numerous times by specialists in Cuba. However, Chávez's policies lived on under the protection of his hand-picked successor, Nicolas Maduro.

Since 2000, populist political movements have taken control of the national governments in Brazil, Argentina, Bolivia, Ecuador, and Nicaragua. Many of these populist governments have since fallen in democratic elections, often due to corruption scandals but also because of difficult economic circumstances related to declining revenues from trade with China. The investments in education, health care, and poverty reduction these populist governments made have only begun to pay off, and the vast inequalities between rich and poor have declined only slightly in recent years. In today's more open political climate, inequalities are still large enough to fuel powerful populist movements for decades to come.

THE DRUG TRADE, CONFLICT, AND LEGALIZATION

The international illegal drug trade is a major source of violence and corruption throughout the region. The primary drugs being traded are cocaine, heroin, and marijuana. Most drugs are produced in or pass through northwestern South America, Central America, and Mexico. **FIGURE 3.21** illustrates some geographic aspects of the cocaine trade.

Production of cocaine and heroin is illegal in all of Middle and South America. However, public figures, from the local police on up to high officials, are paid to turn a blind eye to the industry. Most coca growers are small-scale farmers of indigenous or mestizo origin, working in remote locations where they can make a better income for their families from

dictator a ruler who claims absolute authority, governing with little respect for the law or the rights of citizens

coup d'état a military- or civilian-led forceful takeover of a government

FIGURE 3.20 PHOTO ESSAY: Power and Politics in Middle and South America

Political freedoms are expanding in Middle and South America, although many barriers persist. Decades of elite and military rule, combined with rampant corruption, drug trafficking, and political repression, have at times inspired popular revolutionary movements, often resulting in civil wars that have cost many lives. Nevertheless, in the region as a whole, there has been a decline in political violence in recent decades. [Source: *Economist* Intelligence Unit, Democracy Index 2015, Table 2, at http://www.yabiladi.com/img/content /EIU-Democracy-Index-2015.pdf]

A Federal agents search cars at a checkpoint in Juárez in Chihuahua, Mexico. Mexico's drug war has led to between 100,000 to 160,000 deaths over the past decade and corruption that has eroded local political institutions, especially near the U.S. border. Most of Mexico's drug production and distribution supplies the United States.

Democratization and Conflict

Democratization index

- Full democracy
- Flawed democracy
- Hybrid regime
- Authoritarian regime
- No data

Armed conflicts and genocides with high death tolls since 1990

- Ongoing conflict
- 1000–20,000 deaths
- 20,000–50,000 deaths
- 50,000–100,000 deaths
- 100,000–200,000 deaths

B Colombian police officers, crippled in combat with a rebel group, undertake a protest march destined for the capital in Bogotá. Popular revolutionary movements and a drug war drive this complex conflict, the worst in the region. Political institutions and the rule of law have been badly damaged, with many high-ranking officials linked to extreme human rights and civil liberties abuses.

C Cuban revolutionaries Che Guevara (left), Camilo Cienfuegos (center), and Juan Antonio Mella are immortalized in an aging mural in Havana. Since coming to power in 1959, Cuba's communist government has been criticized for jailing and sometimes executing leaders of the political opposition.

Thinking Geographically

After you have read about power and politics in Middle and South America, you will be able to answer the following questions.

A What evidence can you find in the photo of a violent situation in Mexico?

B What are the indications that these men were once involved in violent combat?

C Why would the Cuban government lionize these men in a mural?

these plants than from other cash crops. Many rent land from the *drug cartels* that get the drugs to global markets. The cartels are monopolistic and violent, and those who oppose them—whether farmers, journalists, or law enforcement officials—often end up kidnapped, tortured, and brutally murdered.

In Colombia, the illegal drug trade has financed all sides of a continuing civil war that has threatened the country's democratic traditions and displaced more than 1.5 million people over the past several decades (see Figure 3.20B). Mexico has endured a decade of armed conflict between powerful drug cartels that control a hugely profitable U.S.-oriented drug trade as well as the Mexican police and military (see Figure 3.20A). As many as 160,000 people have died throughout Mexico during this "drug war."

The U.S. "war on drugs" in this region has been ineffective. Since its inception in 1971, the U.S. military has dominated the efforts, often in collaboration with national militaries. The focus has been on the eradication of drug crops through spraying of herbicides or burning and on the interdiction of drug shipments before they reach the United States. Almost half a century later, illegal drug production levels in Middle and South America are higher than ever, exceeding demand in the United States, where street prices for many drugs, such as cocaine and heroin, have fallen in recent years. One result of this war on drugs is that U.S. military aid to this region through antidrug programs is now about equal to U.S. aid for education and other social programs in the region.

Many who have studied the drug trade between the United States and this region call for radical changes in international drug policies, including the legalization of drug use in the main (U.S.) market, which they assert would cause criminals to lose interest in the drug trade. Some countries have taken matters into their own hands. In 2009, Mexico decriminalized possession of small amounts of most drugs, including marijuana, cocaine, heroin, methamphetamine, and LSD. Argentina, Brazil, Ecuador, and Costa Rica have made similar moves. Argentina, Chile, and Peru have legalized medical marijuana and in 2013 Uruguay legalized and regulated marijuana use, cultivation, sale, and distribution to help reduce drug-related violence and deprive criminals of a major source of income. Throughout the region there has been a shift toward treating drug use as a public health issue instead of a crime, though drug dealers are still often incarcerated when caught.

While slow implementation makes the impacts of these laws hard to assess, a recent decline in incarceration among women in Uruguay, whose crimes are usually drug related, is being celebrated by supporters of drug legalization. Marijuana has been legalized for recreational and medical use in much of the United States, which has had an impact on Mexico, where the drug cartels' profits from marijuana exports to the north have declined. Most cartels have adjusted by increasing their exports of methamphetamine and heroin.

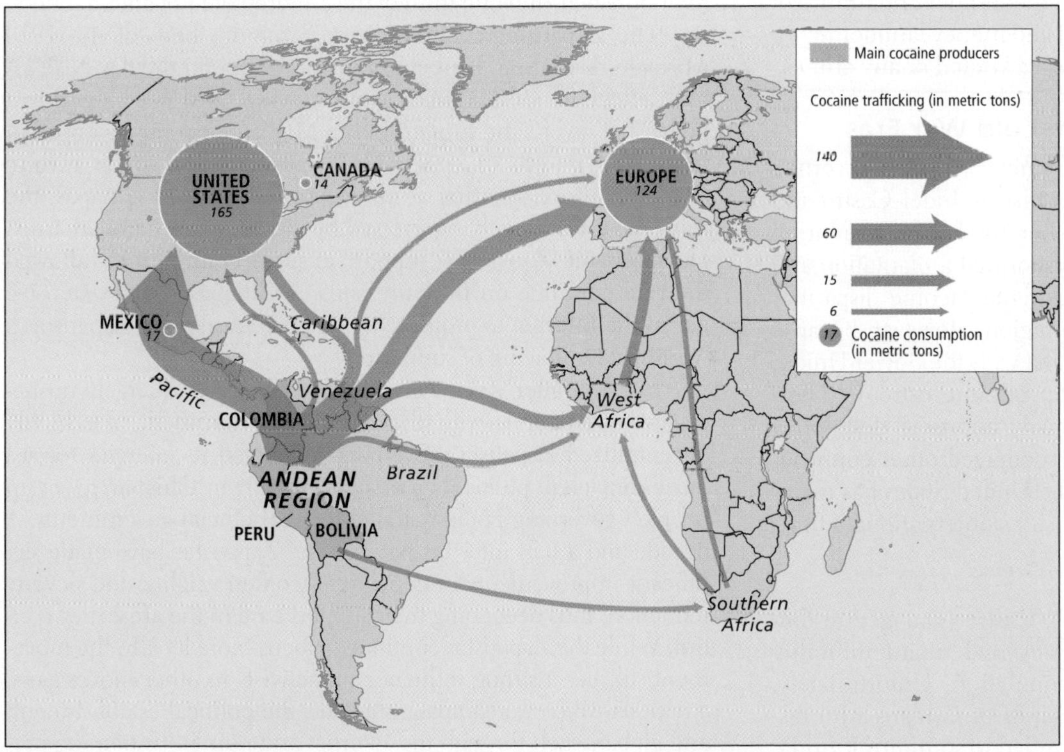

FOREIGN INVOLVEMENT IN THE REGION'S POLITICS

Interventions in the region's politics by outside powers have frequently compromised political freedoms and human rights. Although the former Soviet Union, Britain, France, and other European countries have wielded much influence, by far

FIGURE 3.21 Linkages: Cocaine sources, trafficking routes, and seizures worldwide, 2008. Colombia, Peru, and Bolivia are the most important sources of cocaine cultivation and production in the world. The big cocaine markets are in the Americas; use has declined by 36 percent in the United States since 1998 and increased in South America. Cocaine use in Western Europe and West Africa is on the rise, due in part to new trade routes through West Africa to Europe. [Source consulted: World Drug Report 2010, United Nations Office on Drugs and Crime, at http://www.unodc.org/unodc/en/drug-trafficking/index.html]

the most active foreign power in this region has been the United States.

In 1823, the United States introduced the Monroe Doctrine to warn Europeans that no further colonization would be tolerated in the Americas. Subsequent U.S. administrations interpreted this policy more broadly to mean that the United States itself had the sole right to intervene in the affairs of the countries of Middle and South America, and it has done so many times. The official goal for such interventions was usually to make countries safe for democracy, but in most cases the driving motive was to protect U.S. political and economic interests.

At various times during the past 150 years, U.S.-backed unelected political leaders, many of them military dictators, have been installed in many countries in the region. After World War II, worried that Communism would infiltrate Middle and South America, the United States focused special military scrutiny on a number of countries. Since the 1960s, the United States has funded armed interventions in Cuba (1961), the Dominican Republic (1965), Nicaragua (1980s), Grenada (1983), and Panama (1989). Perhaps the most infamous intervention took place in Chile in 1973. With U.S. aid, the elected socialist-oriented government of Salvador Allende was overthrown. Allende was killed, and a military dictator, General Augusto Pinochet, was installed in his place. Over the next 17 years, the Pinochet regime imprisoned and killed thousands of Chileans who protested the loss of their political freedoms. Voted out of the presidency in 1990, Pinochet remained head of the Chilean armed forces until 1998. In 2000, Pinochet was put on trial for his role in the murder and torture of several Chileans, as well as for tax fraud and money laundering. In 2006, Pinochet died before he could be convicted of any crimes.

Cuba in the Cold War and Post–Cold War Eras

Since 1959, the Caribbean island of Cuba has been governed by a radically socialist government led first by Fidel Castro and then, starting in 2006, by Fidel's brother Raúl. The revolution that brought Fidel Castro to power transformed a plantation and tourist economy once known for its extreme income disparities into one of the most egalitarian in the region. However, because Castro adopted socialism and allied Cuba with the Soviet Union, Cuba's relations with the United States became extremely hostile. The United States has funded many efforts to destabilize Cuba's government, and it has long discouraged other countries from trading with Cuba. Since 1960, the United States has maintained an embargo against Cuba, and so far international efforts to lift the embargo have failed.

With help from the Soviet Union, Castro managed to dramatically improve the country's literacy and infant mortality rates and the life expectancy of its population. Unfortunately, he also imprisoned or executed thousands of Cubans who disagreed with his policies, many of whom fled to southern Florida. Following the demise of the Soviet Union, Castro opened the country to foreign investment. Many countries responded, and Cuba is now a major European tourist destination. Formal diplomatic relations were restored between Cuba and the United States in 2014, and President Obama visited Cuba in 2016. Travel and mail service between the two countries was restored in 2016. However, political repression in Cuba persists and the economic embargo looks likely to remain in place for some time.

contested space any area that two or more groups claim or want to use in different and often conflicting ways, such as the Amazon or Palestine

RECENT REVOLUTIONARY MOVEMENTS

In response to continued poverty and inequality, and building on Middle and South America's long history of revolutionary politics, two major broad-based rural political movements have arisen in recent years. While neither poses any direct threat to national governments, and both may at times ally with national governments, they are building a strong network of support for political and economic change at the local level. The areas in which these movements take place are examples of what geographers call **contested space,** where various groups are in conflict over the right to use a specific territory as each sees fit.

The Zapatista Rebellion

In the southern Mexican state of Chiapas, indigenous farmers have mobilized against the economic and political systems that have left them poor and powerless. The Mexican government redistributed some hacienda lands to poor farmers early in the twentieth century, but most fertile land in Chiapas is still held by a wealthy few who use green revolution agriculture to grow cash crops for export. The poor majority farm tiny plots on infertile hillsides. In 2000, about three-fourths of the rural population was malnourished, and one-third of the children did not attend school.

The Zapatista rebellion (named for the hero of the 1910 Mexican revolution, Emiliano Zapata) began on the day NAFTA took effect in 1994. The Zapatistas view NAFTA as a threat because it diverts the support of the Mexican federal government from land reform to large-scale, export-oriented green revolution agriculture. Starting as an armed rebellion in Chiapas, the movement was largely suppressed by the Mexican national army. However, the Zapatistas' support in local communities allowed them to continue on in many parts of Chiapas. Their extensive use of the Internet to promote their cause earned the Zapatistas a worldwide following of supporters.

In 2003, after 9 years of armed resistance and largely unsuccessful negotiations with the Mexican government, the highly decentralized Zapatista movement redirected its energies toward local nonviolent political campaigns, mainly in Chiapas, to set up people's governing bodies parallel to local official governments. A decade and a half into this process, the Zapatistas have made significant improvements in education, women's rights, and poverty reduction, thus deepening their support base in the areas they control. While the Zapatistas continue to focus more locally, the movement still has a strong influence on activists in other movements, a national voice via campaigns against the political establishment, and global reach through the Internet and numerous training programs for supporters of the movement around the world.

Brazil's Landless Movement

Sixty-five percent of Brazil's arable and pasture land is owned by wealthy farmers who make up just 2 percent of the population.

Since 1985, more than 2 million small-scale farmers have been forced to sell their land to larger farms that practice green revolution agriculture. Because the larger farms specialize in major export items, such as cattle and soybeans, they have been favored by governments wishing to increase exports. As a result, many poor farmers have been forced to migrate to urban areas.

To help these people, organizations such as the Movement of Landless Rural Workers (MST) began taking over unused portions of some large farms. Since the mid-1980s, the MST has coordinated the occupation of more than 51 million acres of Brazilian land (an area about the size of Kansas). Some 250,000 families have gained land titles, while the elite owners have been paid off by the Brazilian government and have moved elsewhere. There are now 1.5 million members of the MST spread across 23 of Brazil's 26 states, and movements with similar goals have been launched in Ecuador, Venezuela, Colombia, Peru, Paraguay, Mexico, and Bolivia.

ON THE BRIGHT SIDE: Children and the Landless Movement

In Brazil, the children of those in the MST movement generally find living on the land that the movement occupies a dramatic improvement over their previous homes, which were often crowded slums in towns or cities. Better access to nature provides the children with opportunities to learn about plants and animals, and community libraries help expose them to scientific, cultural, and political literature that was usually lacking in their old communities. Many of the children take pride in their family's role within the landless movement, as the video by Michalis Kontopodis, *Landless Children/Sem Terrinha*, at http://tinyurl.com/9pxeclw, attests. ∎

THINGS TO REMEMBER

GEOGRAPHIC THEME 3 • **Politics and Power** The region was ruled for decades by elites and the military, and was periodically subject to disruptive foreign military interventions. Almost all countries now have multiparty political systems and elected governments, and political freedoms are expanding. The international illegal drug trade continues to be a source of violence and corruption in the region.

• Elected governments are at times challenged by citizen protests or with a coup d'état because policies are unpopular with large segments of the population, powerful elites, or the military.

• The drug trade has emerged as a major obstacle to democracy, with huge amounts of illicit cash being used to pay off elected officials, civil servants, police forces, and the military.

• Interventions in the region's politics by outside powers have frequently compromised political freedoms and human rights.

• In response to continued poverty and inequality, and building on this region's long history of revolutionary politics, two major broad-based rural political movements have arisen in recent years.

URBANIZATION

GEOGRAPHIC THEME 4

Urbanization: Since the 1950s, cities have grown rapidly in this region as rural people have migrated to cities and towns. A lack of urban planning has created densely occupied urban landscapes that often lack adequate support services and infrastructure.

More than 80 percent of the people in the region live in settlements of at least 2000 people. One city, known as a **primate city,** is vastly larger than all the others, accounting for a large percentage of the country's total population (FIGURE 3.22). Examples in Middle and South America are Mexico City (with 21 million people, a bit more than 17 percent of Mexico's total population); Santiago, Chile (7 million, 39 percent of that country's population); Lima, Peru (9.7 million, 31 percent); Buenos Aires, Argentina (13 million, 30 percent); and Managua, Nicaragua (2.5 million, more than 40 percent).

The concentration of people into just one or two large cities in a country leads to uneven spatial development and to government policies and social values that favor the largest urban areas. Wealth and power are concentrated in one place, while distant rural areas and even other towns and cities have difficulty competing for talent, investment, industries, and government services. Many provincial cities languish as their most educated youth leave for the primate city.

RURAL-TO-URBAN MIGRATION

It always takes some resourcefulness to move from one place to another. Those people who already have some years of education and strong ambition are the ones who migrate to cities. The loss of these resourceful young adults in which the community has invested years of nurturing and education is referred to as **brain drain.** Brain drain happens at several scales in Middle and South America: through rural-to-urban migration from villages to regional towns and from towns and small cities to primate cities. There is also an international brain drain when migrants move to North America and Europe; for example, one in four U.S. doctors are foreign born. Migrants often want better access to jobs and education and to help their families with money they send back home.

Urban Landscapes and Planning

A lack of planning to accommodate the massive rush to the cities has created dense urban landscapes with often inadequate infrastructure. The arrangement of urban landscapes is also very different from the common U.S. pattern. In the United States, a poor, older inner city is usually surrounded by affluent suburbs, with clear and planned spatial separation of residences by income as well as separation of residential, industrial, and commercial areas. In Middle and South America, by contrast, both

primate city a city, plus its suburbs, that is vastly larger than all others in a country and in which economic and political activity is centered

brain drain the migration of educated and ambitious young adults to cities or foreign countries, depriving the communities from which the young people come of talented youth in whom they have invested years of nurturing and education

FIGURE 3.22 PHOTO ESSAY: Urbanization in Middle and South America

Urbanization is at the heart of many transformations in this region. In the last several decades, cities have grown extremely quickly, and 77 percent of the region's population now lives in cities. New opportunities have opened for migrants from rural areas, but new stresses have emerged as well. The low-skilled jobs they can get often don't pay well, and many people are forced to live in unplanned neighborhoods on the outskirts of huge cities.

A Santiago is one of the region's primate cities, home to 39 percent of Chile's population. It dominates Chile's economy, generating 40 percent of the country's GDP.

B A child works as a street vendor in Salvador, Brazil. Many new urban migrants work in similar informal sector jobs that generally pay less and are much less secure than formal sector jobs.

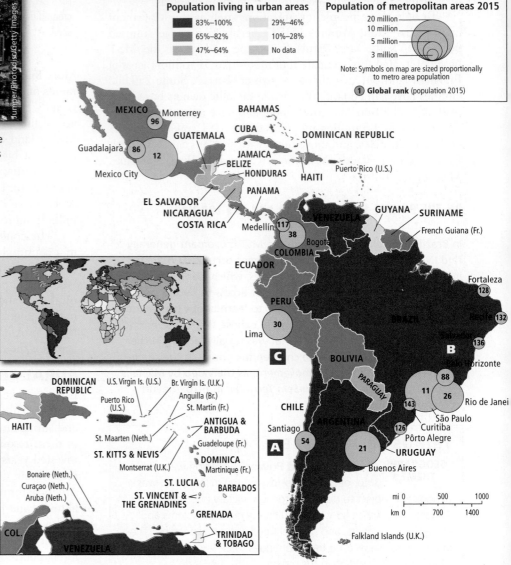

Population living in urban areas

83%–100%	29%–46%
65%–82%	10%–28%
47%–64%	No data

Population of metropolitan areas 2015

20 million
10 million
5 million
3 million

Note: Symbols on map are sized proportionally to metro area population

① **Global rank** (population 2015)

C Police evict 2500 people from an informal squatter settlement recently formed outside Lima, Peru. Structures made out of woven grass mats are the first to be built, followed by wooden shacks and eventually cement or brick houses. Water and electricity may eventually be extended to such settlements, but only after years or decades of occupation.

Thinking Geographically

After you have read about urbanization in Middle and South America, you will be able to answer the following questions.

A What about this picture suggests that there may be considerable wealth in Santiago?

B Why might work in the informal economy be considered insecure?

C What are some defining characteristics of informal squatter settlements?

affluent and working-class areas have become unwilling neighbors to unplanned slums filled with poor migrants. Too destitute even to rent housing, these "squatters" occupy parks and small patches of land wherever they can be found, as depicted in the diagram in **FIGURE 3.23**. Unplanned communities have a history in this region; many were founded first by freed slaves in the late nineteenth century. Most date from the period of urban growth that took place after World War II, when many rural people moved to the cities to work in new industries.

The best known of these unplanned communities are Brazil's **favelas (FIGURE 3.24)**. In other countries they are known as slums,

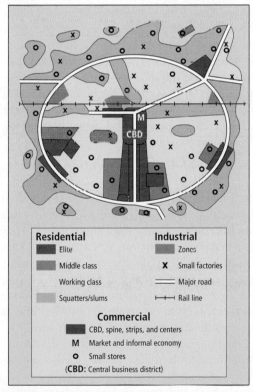

Residential
- Elite
- Middle class
- Working class
- Squatters/slums

Industrial
- Zones
- **X** Small factories
- Major road
- |—•—| Rail line

Commercial
- CBD, spine, strips, and centers
- **M** Market and informal economy
- **o** Small stores
- (**CBD:** Central business district)

FIGURE 3.23 Crowley's model of urban land use in mainland Middle and South America. William Crowley, an urban geographer who specializes in Middle and South America, developed this model to depict how residential, industrial, and commercial uses are mixed together, with people of widely varying incomes living in close proximity to one another and to industries. Squatters and slum-dwellers ring the city in an irregular pattern. [Source consulted: *Yearbook of the Association of Pacific Coast Geographers,* 57(28): 1995; printed with permission]

FIGURE 3.24 A favela in Rio de Janeiro. A favela in Rio de Janeiro clings to a once-vacant hillside that was considered too steep for apartment buildings. Now home to thousands of people, it is still served mainly by footpaths and narrow alleyways. Water is only available sporadically, so many residents store water in tanks on their roof.

shantytowns, *colonias*, barrios, or *barriadas*. The settlements often spring up overnight on vacant land after a coordinated collective occupation by poor homeless families. They rarely occur with the landowner's permission, and are without city-supplied water, electricity, or sewer service. Housing is hastily constructed with whatever materials are available. Once the settlements are established, efforts to eject squatters usually fail, though police are often called in to attempt to evict them (see Figure 3.22C). The impoverished are such a huge portion of the urban population that even those in positions of power will not challenge them directly. Nearby wealthy neighborhoods simply barricade themselves with walls and security guards.

The squatters are frequently enterprising people who work hard to improve their communities. They often organize to press governments for social services. Some cities, such as Fortaleza in northeastern Brazil, even contribute building materials so that favela residents can build more permanent structures with basic indoor plumbing. Over time, as shacks and

> **favelas** Brazilian urban slums and shantytowns built by the poor; called colonias, barrios, or barriadas in other countries

lean-tos are transformed through self-help initiatives into crude but livable suburbs, the economy of favelas can become quite vibrant, with much activity in the informal sector. Housing may be intermingled with shops, factories, warehouses, and other commercial enterprises. Favelas and their counterparts can become centers of pride and support for their residents, where community work, folk belief systems, crafts, and music (for example, Favela Funk) flourish. Many of the best steel bands of Port of Spain, Trinidad, have their homes in the city's shantytowns.

VIGNETTE Favelas are everywhere in Fortaleza, Brazil, where they provide a home to about 12 percent of the city's population. Fortaleza grew from about 1 million in the late 1970s to over 2 million by the year 2000. In 2014, there were more than 3.6 million residents, many of whom had fled drought and rural poverty in the interior of the country. During a period of rapid growth in the 1980s, city parks of just a square block or two in residential areas were invaded by squatters. In one upscale neighborhood, 10,000 people occupied a single park. Because of the lack of water and sanitation in the early days of the migration, migrants often had to relieve themselves on the street.

One day, while strolling on the Fortaleza waterfront, Lydia Pulsipher chanced to meet a resident of a beachfront favela who invited her to join him on his porch. There, he explained how he and his wife had come to the city 5 years before, after being forced to leave the drought-plagued interior when a newly built irrigation reservoir flooded the rented land their families had cultivated for generations. With no way to make a living, they set out on foot for the city. In Fortaleza, they constructed the building they used for home and work from objects they collected along the beach. Eventually, they were able to purchase roofing tiles, which gave the building an air of permanency. At the time of the visit, he maintained a small refreshment stand and his wife a beauty parlor that catered to women from the beach settlements. [*Source: Lydia Pulsipher's field notes. For detailed source information, see Text Sources and Credits.*] ∎

Elite Landscapes

In contrast to colorful but problematic favelas are districts in nearly every city that are modern, planned, and similar to elite urban districts across the world. Examples of such urban landscapes can be found in Rio de Janeiro, São Paulo, and other cities in Brazil's southeastern industrial heartland. These cities have districts that are elegant and futuristic showplaces, resplendent with the very latest technology and graced with buildings and high-end shops that rival those in New York, Singapore, and even Tokyo. Known as **urban growth poles,** these areas attract investment, trade, educated immigrants, and overall economic development—all of which make them stand out from surrounding landscapes.

Planners throughout the region now acknowledge, however, that in the rush to develop modern urban landscapes, municipal governments and developers neglected to underwrite the parallel development of a sufficient urban infrastructure (for an exception, see the discussion of the southern city of Curitiba below).

urban growth poles locations within cities that are attractive to investment, innovative immigrants, and trade, and thus attract economic development like a magnet

In short supply are sanitation and water systems, an up-to-date electrical grid, transportation facilities, schools, adequate housing, and medical facilities—all necessary to sustain modern business, industry, and a socially healthy urban population. In recent decades, there have been massive investments in these systems, especially in the larger and wealthier countries, such as Brazil, Chile, Argentina, and Mexico, that can afford the improvements. These investments are done in recognition that planned development of human capital through education, health services, and community development, whether privately or publicly funded, is a primary part of building strong urban economies.

Urban Transportation

In large, rapidly expanding, partially unplanned urban areas with millions of poor migrants, transportation can be a special challenge. Favelas are often on the urban fringes, far from available low-skill jobs and ill-served by roads and public transportation (see Figure 3.23). Workers must make lengthy, time-consuming, and expensive commutes to jobs that pay very little. The entire urban population gets caught up in endless traffic jams that cripple the economy and pollute the environment.

Yet the southern Brazilian city of Curitiba has carefully oriented its expansion around a master plan (dating from 1968) that includes an integrated transportation system funded by a public-private collaboration. Eleven hundred minibuses making 12,000 trips a day bring 1.3 million passengers from often remote neighborhoods to terminals, where they meet express buses to all parts of the city. The large, pedestrian-only inner-city area is a boon for merchants because minibus commuters spend time shopping rather than looking for parking. Being able to get to work quickly and cheaply has helped the poor find and keep jobs. The reduced use of cars means that emissions and congestion are lowered and urban living is more pleasant.

Gender and Rural–Urban Migration Interestingly, rural women are just as likely as rural men to migrate to the city. This is especially true when employment is available in foreign-owned factories that produce goods for export. Companies prefer women for such jobs because they are a low-cost and usually passive labor force. Factors that push people out of rural areas can affect women disproportionately. For example, the shift toward green revolution agriculture has had a particularly hard impact on rural women because they are rarely considered for such jobs as farm-equipment operators or mechanics. In urban areas, unskilled migrant women usually find work as street vendors or domestic servants.

Male urban migrants tend to depend on short-term, low-skill day work in construction, maintenance, small-scale manufacturing, and petty commerce. Many work in the informal economy as street vendors, errand runners, car washers, and trash recyclers, and some turn to crime. In an urban context, both men and women feel sorely the loss of family ties and village life; the chances for recreating the family life they once knew are extremely low.

Once relocated to a city, migrant families may disintegrate because of extreme poverty and malnutrition, long commutes for both working parents, and poor quality of day care for children. At too young an age, children are left alone or sent into the streets to scavenge for food or to earn money for the family as street vendors (see Figure 3.22B).

THINGS TO REMEMBER

GEOGRAPHIC THEME 4

• **Urbanization** Since the 1950s, cities have grown rapidly in this region as rural people have migrated to cities and towns. A lack of urban planning has created densely occupied urban landscapes that often lack adequate support services and infrastructure.

• One city, known as a primate city, is far bigger than all the others and accounts for a large percentage of the country's total population.

• Settlements known as slums, shantytowns, *colonias,* barrios, or *barriadas* often spring up overnight on vacant land after a coordinated collective occupation by poor homeless families.

• The southern Brazilian city of Curitiba has carefully oriented its expansion around a master plan that includes an integrated transportation system funded by a public-private collaboration.

• Rural women are just as likely as rural men to migrate to the city.

POPULATION AND GENDER

GEOGRAPHIC THEME 5

Population and Gender: During the early twentieth century, the combination of cultural and economic factors and improvements in health care created a population explosion. By the late twentieth century, improved living conditions; better access to education and medical care; urbanization; and changing gender roles were all working together to reduce population growth.

By 2016, about 630 million people were living in Middle and South America, close to ten times the highest estimated population of the region in 1492. Today, populations in Middle and South America continue to grow but are doing so at an ever-slower pace because of diminishing birth rates.

LOWER BIRTH RATES

Birth rates stayed high for much of the twentieth century for a variety of reasons. In agricultural areas, children were seen as sources of wealth because they could do useful farm and household work at a young age and eventually would care for their aging elders. Worry over infant death rates persisted, so some parents had four or more children to be sure of raising at least a few to adulthood. Another factor was the Roman Catholic Church's opposition to systematic family planning.

By the 1970s, the region was undergoing a demographic transition (see Figure 1.28). Between 1975 and 2014, the annual rate of natural increase (the birth rate minus the death rate) for the entire region fell from about 1.9 percent to 1.2 percent—a rate of growth equal to the world average. Between 1990 and 2014, the rate of natural increase fell in every country in the region, as shown in **FIGURE 3.25**.

The population pyramids shown in **FIGURE 3.26** illustrate the contrast between countries that are well along the demographic transition, such as Brazil, and others that are just starting, such as Guatemala. Brazil's pyramid is narrower at the bottom than in the middle, indicating an aging population and a sharply declining birthrate. Guatemala's pyramid is widest at the bottom, indicating a younger population with birth rates that are declining more gradually.

The changes outlined in Figures 3.25 and 3.26 happened for a variety of reasons. Better access to food, shelter, sanitation, and medical care reduced infant mortality, encouraging couples to have only two or three children instead of five or more. As populations urbanized and economies grew, women gained better access to education and employment outside the home, and so delayed childbearing and had smaller families. This trend is continuing as women move into high-level jobs, such as management positions in corporations and government, that traditionally have gone to men. Perhaps the most visible example of this is the women who have become presidents of Chile, Argentina, and Brazil in the last decade.

However, because 28 percent of the region's population is under the age of 15, even if couples have only one or two children, the population will continue to grow as this large group

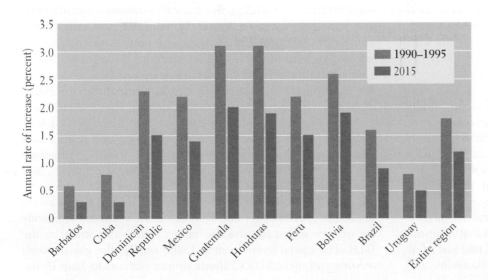

FIGURE 3.25 Trends in natural population increases, 1990–2015. The orange and blue columns show that rates of natural increase have declined steadily throughout the region. Although they are projected to continue to do so into the future, in many countries, natural population increase remains high enough to outstrip efforts to improve standards of living. Note that the rates of natural increase are for a few selected countries and for the entire region.

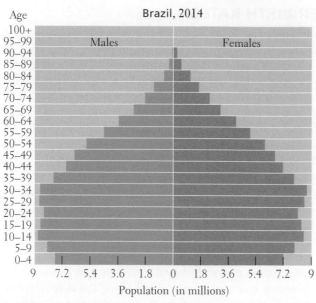

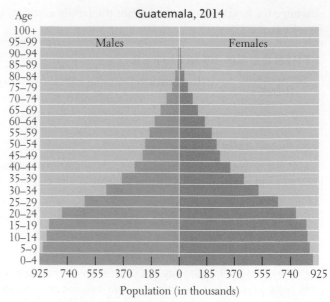

FIGURE 3.26 Population pyramids for Brazil and Guatemala, 2015. Note that the pyramids for these countries are in two different scales. Brazil's is in millions, while Guatemala's is in thousands.

reaches the age of reproduction. Population projections for the year 2050 are for 780 million people, and supplying this extra 180 million people with food, water, homes, schools, and hospitals will be a challenge.

POPULATION DISTRIBUTION

The population density map of this region reveals a very unequal distribution of people. Comparing the population density map (**FIGURE 3.27**) with the landforms map (see Figure 3.1), you will find areas of high population density in a variety of environments. Some of the places with the highest densities, such as those around Mexico City and in Colombia and Ecuador, are in highland areas. But high concentrations are also found in lowland zones along the Pacific coast of Central America, and especially along the Atlantic coast of South America. The cool uplands (tierra templada) were densely occupied even before the European conquest. Most coastal lowland concentrations, in the tierra caliente, are near seaports with vibrant economies and a cosmopolitan social life that attracts people.

HUMAN WELL-BEING

The wide disparities in wealth that are a feature of life in Middle and South America have been discussed above. Too often, development efforts linked to urbanization and globalization have only increased the gap between rich and poor. Each of the three maps in **FIGURE 3.28** shows one of the various indicators of well-being used by the United Nations to show how countries are doing in providing a decent life for their citizens.

Map A in Figure 3.28 is of average gross national income (GNI) per capita (PPP). It is immediately apparent that no country in this region falls in the two highest categories, and just a few islands in the Caribbean (Trinidad and Tobago, Barbados, and

Antigua and Barbuda) fall into the third-highest category. Most of the Caribbean is in the global middle class, with one major exception, Haiti. Across the mainland, only Mexico, Panama, Venezuela, Uruguay, Argentina, Brazil, and Chile have an average per capita GNI between U.S.\$10,000 and U.S.\$14,999. All other countries in the region have average per capita GNIs that are less than \$9999, many of them significantly less. Also, most of these countries have rather wide disparities in wealth that are masked on a map like this. In Brazil, for example, the most affluent 10 percent have incomes that are 41 times greater than the bottom 10 percent, and in actuality, some Brazilians survive on less than \$1000 a year while others have incomes of well over \$100 million per year.

Map B in Figure 3.28, which charts the Human Development Index (HDI) rank for Middle and South America, shows that most countries rank in the medium and high ranges in providing the basics (education, health care, and income) for their citizens. Most of Africa, Central Asia, and South Asia rank lower, as can be seen in the inset map, while North America, Europe, Australia, Japan, and South Korea rank higher. Much of East Asia, Russia, and some of the states of the former Soviet Union are in the same range as Middle and South America. Of course, the UN HDI is a countrywide index and does not give any indication of how well-being is distributed within a given country. On this map and the GNI map, the countries of the Central American isthmus stand out as poorer than Mexico, the Caribbean, and South America.

Map C, the Gender Development Index (GDI), measures disparities in HDI by gender. HDI values are estimated separately for women and men, and the ratio of these two values is the GDI. The closer the ratio is to 1, the smaller the gap between women and men. Map C shows similar patterns to Map B, the

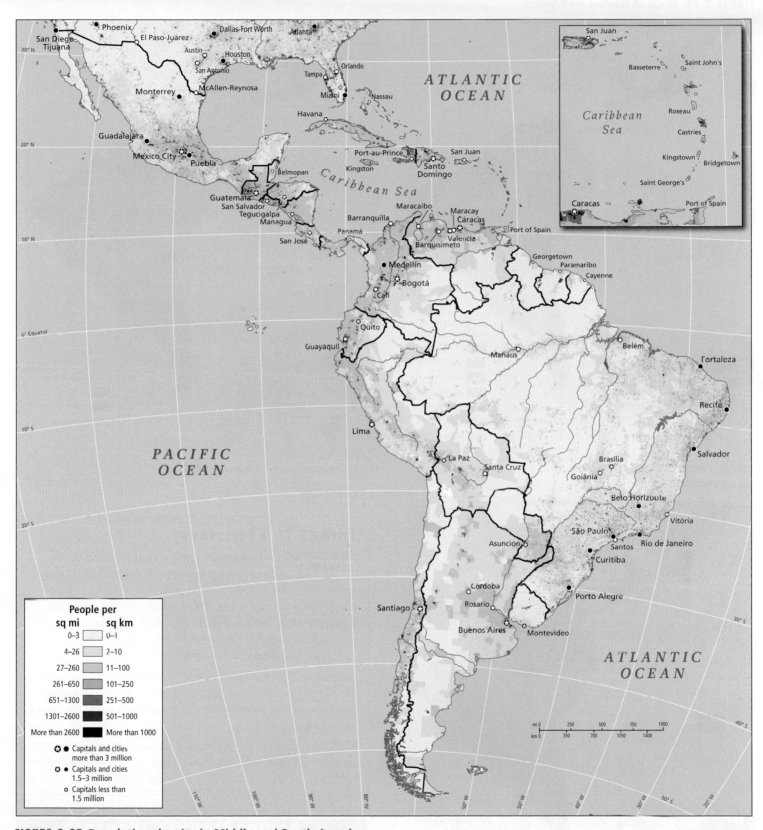

FIGURE 3.27 Population density in Middle and South America.

Human Development Index, though most countries rank only in the medium ranges in providing equitable human development across genders. Most of Africa, South Asia, and Southeast Asia rank lower, as can be seen in the inset map, while North America, Europe, Australia, Japan, and South Korea rank higher. The countries of the Central American isthmus stand out as the least equitable across genders, with Chile, Argentina, and Uruguay the most equitable.

FIGURE 3.28 Maps of human well-being.

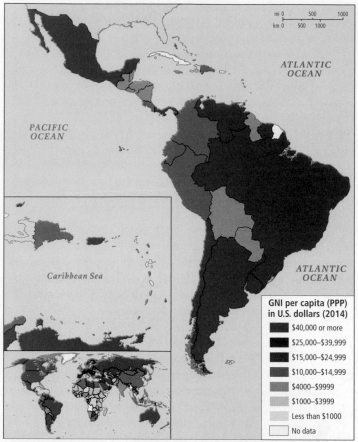

(A) Gross national income (GNI) per capita, adjusted for purchasing power parity (PPP).

GNI per capita (PPP) in U.S. dollars (2014)
- $40,000 or more
- $25,000–$39,999
- $15,000–$24,999
- $10,000–$14,999
- $4000–$9999
- $1000–$3999
- Less than $1000
- No data

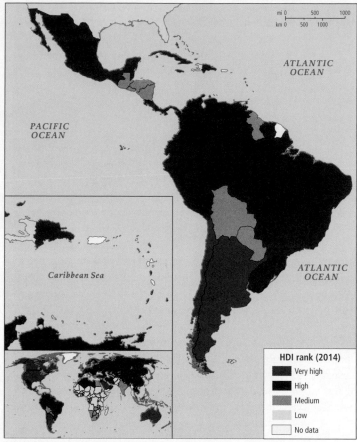

(B) Human Development Index (HDI).

HDI rank (2014)
- Very high
- High
- Medium
- Low
- No data

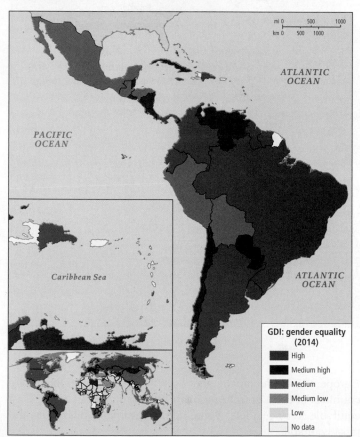

(C) Gender Equality Index (GEI).

GDI: gender equality (2014)
- High
- Medium high
- Medium
- Medium low
- Low
- No data

THINGS TO REMEMBER

GEOGRAPHIC THEME 5

• **Population and Gender** During the early twentieth century, the combination of cultural and economic factors and improvements in health care created a population explosion. By the late twentieth century, improved living conditions; better access to education and medical care; urbanization; and changing gender roles were all working together to reduce population growth.

• Because 27 percent of the region's population is under age 15, even if couples have only one or two children, the population will continue to grow as this large group reaches the age of reproduction.

• Most countries in the region rank in the medium to high range of the Human Development Index (HDI), and the medium ranges of the Gender Development Index (GDI).

SOCIOCULTURAL ISSUES

Under colonialism, a series of social structures evolved that guided daily life—standard ways of organizing the family, community, and economy. They included rules for gender roles, race relations, and religious observance. While these traditional social structures are now changing in response to a variety of forces discussed earlier in this chapter, much continuity with the past is also apparent. Here we look at the region's cultural diversity, family and gender roles, the impact of the Internet, and religion.

CULTURAL DIVERSITY

The region of Middle and South America is culturally complex because many distinct indigenous groups were already present when the Europeans arrived, and then many cultures were introduced during and after the colonial period. From 1500 to the early 1800s, some 10 million people from many different parts of Africa were brought to plantations on the islands and in the coastal zones of Middle and South America. After the emancipation of enslaved peoples of African descent in the British-controlled Caribbean islands in the 1830s, more than half a million Asians were brought there from India, Pakistan, and China as indentured agricultural workers. Their cultural impact remains most visible in Trinidad and Tobago, Jamaica, and the Guianas. In some parts of Mexico, Central America, the Amazon Basin, and the Andean Highlands, indigenous people have remained numerous. To the unpracticed eye, they may appear little affected by colonization, but this is not the case.

In the Caribbean and along the east coast of Central America as well as the Atlantic coast of Brazil, mestizos (people who have a mixture of African, European, and some indigenous ancestry) make up the majority of the population. Some of the most colorful festivals in the Americas, such as versions of Mardi Gras, known here as Carnival, are based on a melding of cultural practices from these three heritages **(FIGURE 3.29A)**. In some areas, such as Argentina, Chile, and southern Brazil, people of Central European descent are also numerous. The Japanese, though a tiny minority everywhere in the region, increasingly influence agriculture and industry, especially in Brazil, the Caribbean, and Peru. Alberto Fujimori, a Peruvian of Japanese descent, campaigned on his ethnic "outsider" status to become president of Peru during a time of political upheaval.

In some ways, diversity is increasing as the media and trade introduce new influences from abroad. At the same time, the processes of **acculturation** and **assimilation** are also accelerating, thus erasing diversity to some extent as people adopt new ways. This is especially true in the biggest cities, where people of widely different backgrounds live in close proximity to one another.

Race and Skin Color

People from Middle and South America, especially those from Brazil, often proudly claim that race and color are of less consequence in this region than in North America. They are only partially right. In all of the Americas, skin color is now less associated with status than in the past. By acquiring an education, a good job, a substantial income, the right accent, and a high-status mate, a person of any skin color may become recognized as upper class.

> **acculturation** adaptation of a minority culture to the host culture enough to function effectively and be self-supporting; cultural borrowing
>
> **assimilation** the loss of old ways of life and the adoption of the lifestyle of another culture

FIGURE 3.29 LOCAL LIVES: Festivals in Middle and South America

A A member of a *bateria*, or percussion band, is parading during Rio de Janeiro's Carnival, the largest in the world. Carnival, which is celebrated in much of the United States as Mardi Gras, takes place in the days and weeks before Lent, the 40-day period in Christian traditions of fasting and abstinence that precedes Easter. Elaborate celebrations take place in Barranquilla in Colombia, Port-au-Prince in Haiti, Port of Spain in Trinidad, and Montevideo in Uruguay.

B A procession commemorating Jesus walks on an *alfombra*, or carpet, made of colored sawdust, painstakingly drawn by hand on the streets of Antigua, Guatemala, during *Semana Santa*, or Holy Week, the last week of Lent. The processions will destroy the alfombras. This tradition is an adaptation of the Mayan practice of making elaborate ground designs with flowers and feathers for Mayan royalty to walk on as they made their way to important ceremonies.

C A richly decorated grave in Tijuana, Mexico, during Día de los Muertos—Day of the Dead—a holiday celebrated throughout Mexico on November 1. Families gather in cemeteries, where they build private altars and other decorations dedicated to deceased loved ones. The holiday is a "Christianized" version of an Aztec festival dedicated to the goddess Mictecacihuatl, Queen of the Underworld, whose job is to watch over the bones of the dead.

Nevertheless, the ability to reduce the significance of skin color through one's actions is not quite the same as race having no significance at all. Overall, those who are poor, less educated, and of lower social standing tend to have darker skin than those who are educated and wealthy. And while there are poor people of European descent throughout the region, most light-skinned people are middle and upper class. In Brazil, where it is popular to downplay the significance of race, the average black worker earns 60 percent of what the average white worker earns. By contrast, in the United States, the average black worker earns 65 percent of what the average white worker earns. Indeed, race and skin color have not disappeared as social factors in this region any more than in the United States. In some countries—Cuba, for example, where overt racist comments are socially unacceptable—it is common for a speaker to use a gesture (tapping his or her forearm with two fingers) to indicate that the person referred to in the conversation is of African descent.

THE FAMILY AND GENDER ROLES

The basic social institution in the region is the **extended family,** which may include cousins, aunts, uncles, grandparents, and more distant relatives. It is generally accepted that the individual should sacrifice many of his or her personal interests to those of the extended family and community and that individual well-being is best secured by doing so.

The arrangement of domestic spaces and patterns of socializing illustrate these strong family ties. Families of adult siblings, their mates and children, and their elderly parents frequently live together in domestic compounds of several houses surrounded by walls or hedges. Social groups in public spaces are most likely to be family members of several generations rather than unrelated groups of single young adults or married couples, as would be the case in Europe or the United States. A woman's best friends are likely to be her female relatives. A man's social or business circles will include male family members or long-standing family friends.

Gender roles in the region have been strongly influenced by the Roman Catholic Church. The Virgin Mary is held up as the model for women to follow through a set of values known as **marianismo,** which emphasizes chastity, motherhood, and service to the family. In this still-widespread tradition, the ideal woman is the day-to-day manager of the house and of the family's well-being. She trains her sons to enter the wider world and her daughters to serve within the home. Over the course of her life, a woman's power increases as her skills and sacrifices for the good of all are recognized and enshrined in family lore.

extended family a family that consists of related individuals beyond the nuclear family of parents and children

marianismo a set of values based on the life of the Virgin Mary, the mother of Jesus, that defines the proper social roles for women in Middle and South America

machismo a set of values that defines manliness in Middle and South America

Her husband, the official head of the family, is expected to work and to give most of his income to his family. Still, men have much more autonomy and freedom to shape their lives than women because they are expected to move about the larger community and establish relationships, both economic and personal. A man's social network is considered just as essential to the family's prosperity and status in the community as is his work.

There are very different sexual standards for males and females. While all expect strict fidelity from a wife to a husband in mind and body, a man is much freer to associate with the opposite sex. Males measure themselves by the model of **machismo,** in which manliness is considered to consist of honor, respectability, fatherhood, household leadership, attractiveness to women, and the ability to be a charming storyteller. Traditionally, the ability to acquire money was secondary to other symbols of maleness. Increasingly, however, a new, market-oriented culture prizes visible affluence as a desirable male attribute.

LGBT Issues

While a general inclination to keep non-heterosexual relationships "in the closet" persists, over the last decade attitudes have begun to change. Respect for the rights of gays and lesbians has grown significantly in recent years: Argentina legalized same-sex marriage in 2010, as did Brazil and Uruguay in 2013, and Mexico in 2015, but homosexuality is still illegal in Guyana and Belize. Even in more accepting countries, social attitudes toward LGBT people are often complex, contradictory, and ultimately disapproving. In Brazil, state legislator and pastor Sargento Isidório, who describes himself as an "ex-homosexual," was reelected in a landslide and gained national notoriety in 2014 when he blamed the recent drought in São Paolo on the city's LGBT "Pride Parade," which, not incidentally, is also the world's largest such event, drawing more than 3 million people annually. Throughout the region, violence against LGBT people is common and is rarely addressed by law enforcement. Transgender people are seen as running particularly afoul of the ideals of *machismo*. Officials of the Catholic church are often among the most outspoken in their disapproval of LGBT people. For example, Chile's Cardinal Jorge Medina stated in 2010 that "if a person has a homosexual tendency it is a defect, like missing an eye, a hand, a foot."

RELIGION IN CONTEMPORARY LIFE

Judaism, Islam, Hinduism, and indigenous beliefs are found across the region, but the majority of the people are at least nominal Christians, most of them Roman Catholic. While the Roman Catholic Church remains highly influential and relevant to the lives of believers, it has had to contend with popular efforts to reform it, as well as with rising competition from other religious movements. From the beginning of the colonial era, the church was the major partner of the Spanish and Portuguese colonial governments. It received extensive lands and resources from colonial governments, and in return built massive cathedrals and churches throughout the region, sending thousands of missionary priests to convert indigenous people. The Roman Catholic Church encouraged these people to accept their low status in colonial society, to obey authority, and to postpone rewards for their hard work until heaven.

People throughout the region converted to the faith, some willingly and some only after the threats, intimidation, and torture of the Spanish Inquisition were brought to the Americas. Many indigenous people put their own spin on Catholicism, creating multiple folk versions of the Mass with folk music, more participation by women in worship services, and interpretations of Scripture that vary greatly from European versions. A range of African-based belief systems (Candomblé, Umbanda, Santería, Obeah, and Voodoo) combined with Catholic beliefs can be found in Brazil, northern South America, Middle America, and the Caribbean—wherever the descendants of Africans have settled. These African-based religions have attracted adherents of European or indigenous backgrounds as well, especially in urban areas.

The power of the Roman Catholic Church began to erode in the nineteenth century in places such as Mexico. Social movements aimed at addressing the needs of the poor masses seized and redistributed some church lands. They also canceled the high fees the clergy had been charging for simple rites of passage such as baptisms, weddings, and funerals. Over the years, the Catholic Church became less obviously connected to the elite and more attentive to the needs of poor and non-European people. By the mid-twentieth century, the church was abandoning many of its earlier racist policies and ordaining clergy of African and indigenous descent. Women were also given more of a role in religious ceremonies.

In the 1970s, a Catholic movement known as **liberation theology** was begun by a small group of priests and activists. They were working to reform the church into an institution that could combat the extreme inequalities in wealth and power common in the region. The movement portrayed Jesus Christ as a social revolutionary who symbolically spoke out for the redistribution of wealth when he divided the loaves and fish among the multitude. The perpetuation of gross economic inequality and political repression was viewed as sinful, and social reform as liberation from evil.

The legacy of liberation theology is now fading. At its height in the 1970s and early 1980s, liberation theology was the most articulate movement for region-wide social change. It had more than 3 million adherents in Brazil alone. But the Vatican objected to this popularized version of Catholicism, and its influence diminished. In countries such as Guatemala and El Salvador, governments targeted liberation theology participants, vilifying them as communist collaborators; hundreds were tortured and assassinated. Liberation theology has also had to compete with newly emerging evangelical Protestant movements.

The Americas' First Pope

The election of Argentina's Cardinal Jorge Bergoglio to become the leader of the Catholic Church in 2013 reflected realities both in this region and outside of it. Pope Francis, as Bergoglio is now known, is the first non-European pope in almost 13 centuries, and the first ever from the Americas. His election reflects the dramatic decline of Europe's share of the world's Catholics, down from 65 percent in 1910 to only 24 percent in 2010, and the rise of Middle and South America's share of the world's Catholics, up from 24 percent in 1910 to 39 percent in 2010. Pope Francis was also elected by a group of Catholic cardinals who want to reform the church and the Vatican bureaucracy in light of the recent financial corruption and sexual abuse scandals that have damaged the Church's image around the world.

While Francis's election was widely celebrated in Argentina, he was also criticized by some for his complicity in Argentina's *dirty war*, a period from 1976 to 1983 when a military dictatorship engaged in widespread torture and carried out more than 30,000 executions and "disappearances" of people it considered terrorists or otherwise politically dangerous. As the head of the Jesuit order in Argentina, Francis—then known as Father Bergoglio—was accused in 1985, 2005, and again in 2010 of handing over to the military priests and Catholic laypeople, some of whom were never seen again, and of doing comparatively little to shield others from the dictatorship's death squads.

Bergoglio denied these accusations, but such behavior would not have been out of the ordinary for someone in his position, as most of Argentina's Catholic clergy, especially the upper levels of the Church's hierarchy, publicly supported the dictatorship and willingly supplied it with information about supposed terrorists and political activists who opposed the regime. In this they were joined by most of Argentina's elite, including wealthy landowners, industrialists, and the managers of most newspapers. The Argentine Catholic clergy's collaboration with the military, which it formally apologized for in 1996, stands in contrast to the actions of Brazil and Chile's Catholic clergy, who were often highly critical of military governments that perpetrated violence similar to Argentina's dirty war during the 1960s, 1970s, and 1980s.

Evangelical Protestantism has spread from North America into Middle and South America, and is now the region's fastest-growing religious movement. About 10 percent of the population, or at least 50 million people, are adherents. The movement is growing rapidly in Brazil, Chile, the Caribbean, Mexico, and Middle America, especially among poor and middle classes in both rural and urban settings. It does not, however, share liberation theology's emphasis on combating the region's extreme inequalities in wealth and power. Some evangelical Protestants teach a *gospel of success*, stressing that those true believers who give themselves to a new life of hard work and clean living will experience prosperity of the body (wealth) as well as of the soul.

The movement is *charismatic*, meaning that it focuses on personal salvation and empowerment of the individual through miraculous healing and psychological transformation. Evangelical Protestantism is not hierarchical in the same way as the Roman Catholic Church, and there is usually no central authority; rather, there are a host of small, independent congregations led by entrepreneurial individuals who may be either male or female.

Perhaps two of the most important contributions of evangelical Protestantism are the focus (as in Umbanda and other African-based religions) on helping people, especially urban migrants, cope with the strains of modern life; and the emphasis placed on involving women in leadership roles in church activities and services.

liberation theology a movement within the Roman Catholic Church that uses the teachings of Jesus to encourage the poor to organize to change their own lives and to encourage the rich to promote social and economic equity

evangelical Protestantism a Christian movement that focuses on personal salvation and empowerment of the individual through miraculous healing and transformation; some practitioners preach to the poor the gospel of success—that a life dedicated to Christ will result in prosperity for the believer

THINGS TO REMEMBER

• The region of Middle and South America is culturally complex because many distinct indigenous groups were already present when the Europeans arrived, and then many cultures were introduced during and after the colonial period.

• The basic social institution in the region is the extended family, which may include cousins, aunts, uncles, grandparents, and more distant relatives.

• Gender roles in the region have been strongly influenced by the Catholic Church through the ideals of *marianismo* and *machismo*.

• Evangelical Protestantism came to the region from North America and is now the fastest-growing religious movement in the region, though Catholicism is still the dominant religion.

SUBREGIONS OF MIDDLE AND SOUTH AMERICA

This tour of the subregions of Middle and South America focuses primarily on the themes of cultural diversity, economic disparity,

and environmental deterioration. These themes are reflected somewhat differently from place to place. For each subregion, examples of connections to the global economy are given.

THE CARIBBEAN

Much of the Caribbean has a strong record of fostering human well-being, but visitors are often unaware of this focus and tend to be struck by the failure of the ramshackle, lived-in landscapes of the islands to match the glamorous expectations inspired by the tourist brochures and television ads. Because Caribbean tourists are usually (and unnecessarily) isolated in hotel enclaves or on cruise ships, glimpsing only tiny swatches of island settlements through green foliage, they rarely learn that the humble houses, garden plots, and quaint, rutted, narrow streets mask social and economic conditions that are supportive of a modest but healthy and productive way of life.

Political, Social, and Economic Change

Over the past half-century, most of the island societies have emerged from colonial status to become independent, self-governing states **(FIGURE 3.30)**. With the exception of Haiti and parts of the Dominican Republic, these islands are no longer the poverty-stricken places they were 40 years ago. Rather, they

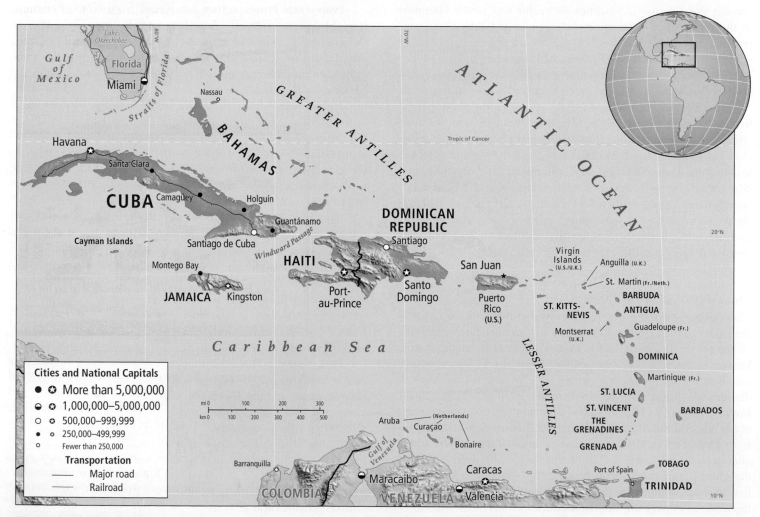

FIGURE 3.30 The Caribbean subregion.

are managing to provide a reasonably high quality of life for their people. Children go to school, and literacy rates for all but the elderly average close to 95 percent. There is basic health care: mothers receive prenatal care, nearly all babies are born in hospitals, infant mortality rates are low, and diseases of aging are competently treated. Life expectancy is in the seventies, and people are choosing to have fewer children; the overall rate of population increase for the Caribbean is the lowest in Middle and South America. A number of Caribbean islands rank high on the HDI, and some (the Bahamas, Barbados, Cuba, and Trinidad and Tobago) do particularly well in making opportunities available to women. Returned emigrants often say that the quality of life on their home Caribbean islands actually exceeds that of societies that are far more materially endowed because life is enhanced by strong community and family support. And, of course, there is the healthful and beautiful natural environment.

Island governments continually search for ways to turn former plantation economies, once managed from Europe and North America, into more self-directed, self-sufficient, and flexible entities that can adapt quickly to the perpetually changing markets of the global economy. To guide their islands toward a specialized activity such as tourism is tempting, but island economists (such as Sir Arthur Lewis, who won the Nobel Prize in Economics in 1979) are wary of the dependence and vulnerability that too much specialization can bring. When plantation cultivation of sugar, cotton, and *copra* (dried coconut meat) died out in the 1950s and 1960s, some islands turned to producing "dessert" crops, such as bananas and coffee, which they sold for high prices under special agreements with the countries that once held them as colonies. Now these protections are disappearing as free market agreements in the EU make such special arrangements illegal. Other island countries turned to the processing of their special resources (petroleum in Trinidad and Tobago, bauxite in Jamaica), the assembly of such high-tech products as computer chips and pharmaceuticals (in St. Kitts and Nevis), or the processing of computerized data (in Barbados). Most islands combine a changing array of one or more of these strategies with tourism development.

In 2016, tourism and related activities contributed at least 60 percent of the gross national product in island countries such as Antigua and Barbuda, Barbados, the Bahamas, and St. Martin. In most years, Antigua and Barbuda (population 91,000) hosts more than four times its population in tourists, and St. Martin (population 71,000) hosts 47 times its population. Heavy borrowing to import food as local agriculture declines and to build hotels, airports, water systems, and shopping centers for tourists has left some islands with debts many times their income. Furthermore, there is a special stress that comes from dealing perpetually with hordes of strangers, especially when they act in ways that violate local mores and make erroneous assumptions about the state of well-being in the islands (see Figure 3.28B).

Host countries consider tourism difficult to regulate, in part because it is a complex multinational industry controlled by external interests in North America and Europe, such as travel agencies, airlines, and the investors needed to fund the construction of hotels, resorts, restaurants, and golf courses. Caribbean cruise ships now routinely bring several thousand passengers into small port cities such as Castries or Soufrière in St. Lucia. Large ships are hard to accommodate because of their enormous size and the numbers of day visitors (as many as 6000) they discharge into small ports. The environmental pollution these ships create, such as garbage and sewage dumping, is hard for tiny countries to regulate or punish. To exercise more control over the industry, some islands are looking for tourists who will stay on shore for a week or more and show a more informed interest in island societies. Ecotourism, sport tourism (such as small-boat sailing), and special-interest seminar tourism (such as cooking, art, yoga, and physical fitness retreats) are three variations that are being pursued as low-density alternatives to the "sand, sea, and sun" mass tourism that brings the majority of visitors to the region.

Cuba and Puerto Rico Compared

Cuba and Puerto Rico are interesting to compare because they shared a common history until the 1950s, then diverged starkly. North American interests dominated both islands after the end of Spanish rule around 1900. U.S. investors owned plantations, resort hotels, and other businesses on both islands. To protect their interests, the U.S. government influenced local dictatorial regimes to keep labor organization at a minimum and social unrest under control. By the 1950s, poverty was widespread in both Cuba and Puerto Rico. Infant mortality was more than 50 per 1000 births, and most people worked as agricultural laborers. Then each island took a different course toward social transformation. In 1959, Cuba went through a Communist revolution under the leadership of Fidel Castro; Puerto Rico underwent a more gradual capitalist metamorphosis into a high-tech manufacturing center.

Cuba Since Castro seized control (see Figure 3.20C), Cuba has dramatically improved the well-being of its general population in all physical categories of measurement (life expectancy, literacy, and infant mortality). By 1990, it had a solid human well-being record despite its persistently rather low GNI per capita (see Figure 3.28). Cuba thus proved that, once the requisite investment in human capital is made, poverty and social problems are not as intractable as some people had thought. Unfortunately, Cuba's successes are not replicable elsewhere in the region, first, because Cuba has been extremely politically repressive, forcing out the aristocracy and jailing and executing dissidents. Second, Cuba has been inordinately dependent on outside help to achieve its social revolution. Until 1991, the Soviet Union was Cuba's chief sponsor. It provided cheap fuel and generous foreign aid and bought Cuba's main export crop, sugar, at artificially high prices.

With the demise of the Soviet Union, Cuba's economy declined sharply during what became known as the *Special Period*, when all Cubans were asked to make major sacrifices and slash the use of food and fuel. To survive this economic crisis, Cuba opened its economy to outside, especially European, investment capital, to help redevelop its tourism industry. Medical tourism and pharmaceuticals are special emphases because early in the revolution, Castro invested in high-quality medical education and encouraged poor, young people to become scientists and doctors. More than 5000 patients, mostly from South America, come per year for cosmetic and other procedures. Canada has invested in joint biotech research on hepatitis and cancer drugs with Cuban medical scientists. In exchange for Cuban pharmaceuticals, high-tech medical equipment, and the services of medical social workers and several thousand medical doctors, for years Venezuela furnished Cuba with around 50,000 barrels of oil a day (2015) in a similar exchange. Oil was discovered off Cuba's north shore in 2012, and oil and gas production could become a major component of Cuba's economy in the near future, as numerous oil exploration firms from around the world are already vying for development rights.

Leisure tourism remains Cuba's biggest source of income. Cuban tourism grossed an estimated U.S.$2 billion in 2015, hosting 3.1 million visitors, mostly Canadians and Europeans, who spent a week or more at luxury beach resorts. More than 140,000 tourists came from the United States in 2014 under special visa agreements. While a growing number are making a decent living working in the tourism industry (nearly 500,000 jobs are connected in some way to tourism), incomes remain low for most Cubans. For them, consumer goods and even food remain in short supply.

Political relations between the Castro regime and successive U.S. administrations remained on Cold War terms from 1961 to 2009, when President Obama loosened the rules on travel to Cuba and proposed wider changes in U.S. relations with Cuba. These changes began to happen more rapidly in 2015, when President Obama opened diplomatic relations with Cuba. So far, the U.S. Congress has refused to cancel the 1996 Helms-Burton Act, which was intended to stop all trade with Cuba. Meanwhile, governments in Europe and the Americas continue to trade with and invest in Cuba.

Another contentious issue between Cuba and the United States is the naval base at Guantánamo in far southern Cuba, on territory held by the United States since the Spanish-American War (1898). This base gained international notoriety after 2002, when the United States, under the Bush administration, began using it to indefinitely detain prisoners—counter to the Geneva Convention on prisoners of war—captured in Afghanistan and elsewhere who were suspected of having links to terrorism. President Obama announced plans to close Guantánamo, but as of 2016 he had not done so.

Puerto Rico In the 1950s, Puerto Rico began Operation Bootstrap, a government-aided program to transform the island's economy from its traditional sugar plantation base to modern industrialism. Many international industries took advantage of generous tax-free guarantees and subsidies to locate plants on the island. By 1965, this industrial sector was shifting from light to heavy manufacturing, from assembly plants to petroleum processing and pharmaceutical manufacturing. The Puerto Rican seaboard became heavily polluted, partly as a consequence of the chemicals released by these more recent industries.

In the 1990s, the U.S. Congress began to eliminate the tax advantages of locating in Puerto Rico and by 2006, many industries had relocated, some to Asia. The resulting recession caused the economy to shrink; many islanders left for the mainland. Government expenses remain high for social services and pension obligations and so the island sinks ever deeper into debt— U.S.$72 billion as of December 2015.

Because Puerto Rico is a commonwealth within the United States and its people are U.S. citizens, many Puerto Ricans migrate to work in the United States. There are already one-and-a-half times as many Puerto Ricans in the United States as there are in Puerto Rico. In the past, the remittances, as well as the manufacturing jobs in Puerto Rico, greatly improved living conditions on the island. Many Puerto Ricans also receive some sort of support from the federal government, including retirement benefits and health care. All of these factors have improved living standards, especially health care. For example, the infant mortality rate in 2015 was just 7.2 infant deaths per 1000 live births, among the lowest in the region, and life expectancy was relatively high, at 79 years. These figures are roughly comparable to Cuba, where the infant mortality rate was 4.2, and life expectancy 78 years). The connection with the United States has helped Puerto Rico's tourism and light-industry economy, but outside of San Juan, with its skyscraper tourist hotels, Puerto Rico's landscape reflects stagnation: a shortage of well-paying jobs, an inadequate transportation system, mediocre schools, and few opportunities for advanced training, all of which encourage migration to the mainland.

Largely as a result of globalizing forces, Puerto Ricans have few options to improve their situation. For now their fate lies with the U.S. Congress. Some support independence from the United States, arguing that not only would this open some financial possibilities (like the right to declare bankruptcy), it would bring closer ties to Middle and South America and help Puerto Rico retain the island's Spanish linguistic and cultural heritage. Others advocate for statehood for Puerto Rico, which would mean higher status in the United States and perhaps some advantages to its tourism industry. It is already too late to keep its previous assembly and industrial base from moving to cheaper labor markets in Asia.

Haiti and Barbados Compared

Haiti and Barbados present another study in contrasts **(FIGURE 3.31)**. During the colonial era, both were European possessions with plantation economies—Haiti a colony of France and Barbados a colony of Britain. Yet they have had very different experiences, and today they are far apart in economic and social well-being. Haiti, though not without useful resources, is the poorest nation in the Americas, with a UN HDI rank of 163 out of 188. Barbados, by contrast, has one of the highest HDI rankings in the entire region (57), even though it has much less space, a higher population density than Haiti, and few resources other than its limestone soil, beaches, and people.

Haiti At the end of the eighteenth century, Haiti had the richest plantation economy in the Caribbean. When Haitian slaves revolted against the brutality of the French planters in 1804, Haiti became the first colony in Middle and South America to become independent. Haiti's early promise was lost, however, when the former slave-reformist leaders were overthrown by other former slaves who were violent and corrupt militarists. They neither

FIGURE 3.31 Contrast in the Caribbean: Haiti and Barbados. While both islands had similar beginnings as European colonies, their present situations are very different.

(A) Haiti is the poorest country in the region and suffers political instability that has brought repeated military intervention by outsiders. Shown here are UN troops from Brazil distributing water in a slum in Port-au-Prince.

(B) Barbados, on the other hand, is politically stable and has a much more prosperous economy based on services (especially tourism) and export-oriented light manufacturing. Barbadians are generally highly educated and often have time to enjoy spectator sports like horse racing.

reformed the exploitative plantation economy nor sought a new economic base. Under a long series of incompetent and avaricious authoritarian governments, the people sank into abject poverty, while the land was badly damaged by particularly wasteful, unprofitable plantation cultivation. In the middle of the twentieth century, a class-based reign of terror under François "Papa Doc" Duvalier (1957–1971) and his son Jean-Claude "Baby Doc" Duvalier (1971–1986), pitted the mulatto elite against the black poor. They were followed by a series of weak leaders who promised social reform but delivered little.

Today, Haiti remains overwhelmingly rural, with widespread illiteracy, an infant mortality rate of 42 per 1000 births (Barbados's rate is 19), and painfully few jobs. Multinational corporations opened maquiladora-like assembly plants, employing primarily young women, but the plants have not flourished and they drew thousands in from the countryside for whom there were no jobs. Although minerals such as bauxite, copper, and tin exist in Haiti, cost-effective development of these resources is not yet possible. Haiti's lands are deforested and eroded and subject to disastrous flooding (see Figure 3.10C). Traditional agriculture, potentially quite productive, has been defeated by extreme environmental deterioration. Efforts to establish democracy have repeatedly devolved into violence (see the Figure 3.20 map). Since the early 1990s the UN has maintained peacekeeping troops in Haiti, and several humanitarian aid organizations, in addition to the U.S. government, run programs there. The United States and other foreign powers increased their presence after four hurricanes struck the island in 2008 and a massively devastating earthquake killed 200,000 to 250,000 people in January 2010 and left 1.5 million homeless. A cholera epidemic then struck. Earthquake and cholera aid supplied by an unprecedented number of countries from across the globe—amounting to U.S.$9.5 billion—began to make a difference, in part because the scope of the disaster weakened sources of resistance to change. But because of the endemic

corruption of local officials—the government demanded large fees to allow in medicine and other relief aid—most of this aid was administered by outsiders and much aid money went to pay outside consultants.

ON THE BRIGHT SIDE: Life in the Caribbean Is Good

With the exception of Haiti, close examination of Caribbean landscapes and ways of daily life reveal a particularly healthful and rewarding, if not affluent, standard of living. Caribbean women hold region-wide leadership positions. Caribbean intellectuals have won several Nobel Prizes. At the 2012 Olympics in London, the Caribbean as a whole won more gold medals than any single country other than the United States, China, the UK, and Russia, and this with a combined population of only 20 million and an overall GDP lower than that of Portugal. ∎

Barbados Tiny but far more prosperous, Barbados has fewer natural resources and is more than twice as crowded as Haiti. With just 166 square miles (430 square kilometers), Barbados has 1684 people per square mile (650 per square kilometer); by contrast, Haiti, with 10,700 square miles (27,800 square kilometers) has only 850 people per square mile (328 per square kilometer). Although both Haiti and Barbados entered the twentieth century with large, illiterate, agricultural populations, their development paths have diverged sharply. In 2012, two-thirds of Haitians were still agricultural workers, and only half were able to read and write. Barbados, on the other hand, now has 100 percent literacy and a diversified economy that includes tourism, sugar production, remittances from migrants, information processing, offshore financial services, and modern industries that sell products throughout the Caribbean. Barbados's present prosperity is explained by the fact that its citizens successfully pressured the British government to invest in the people and infrastructure of its colony before giving it independence in 1966. Barbadians hold jobs requiring sophisticated skills; they are well educated and well fed, and most are homeowners. Furthermore, the Barbadian government and private businesspeople constantly seek new employment options for the citizens and occupy a central role in Caribbean economic and social development. In 2015 Barbados had a per capita GNI of U.S.$14,750; Haiti's was U.S.$1092.

THINGS TO REMEMBER

• Sun, sand, and sea tourism is the main source of economic well-being in most of the Caribbean subregion, but these island nations are working to adopt more flexible and diversified economic strategies in the globalizing world.

• Over the past half-century, many of the island societies have emerged from colonial status to become independent, self-governing states, some more democratic than others. Several Caribbean islands are not yet independent, self-governing states; a few remain colonies, and others are states or entities of France, Great Britain, the Netherlands, or the United States.

• Cuba and Puerto Rico now have quite divergent fates despite colonial similarities. The same could be said for Haiti and Barbados.

• The poverty and social dysfunction of Haiti is atypical of the Caribbean and is the result of a long series of circumstances and disasters, only some of which could have been prevented.

MEXICO

Mexico is a country in transition, working toward becoming a reasonably well-managed, middle-income democracy **(FIGURE 3.32)**, but the situation is fragile; the economic recession, corruption, and drug-cartel violence threaten Mexico's future (see Figure 3.20A, Figure 3.21, and the discussion on pages 157–159). Most of the efforts to achieve economic success and political stability are linked in some way to Mexico's relationship with the United States, its main trade partner. The integration of the Mexican, U.S., and Canadian economies has proceeded rapidly since NAFTA became official in 1994, and the results have been

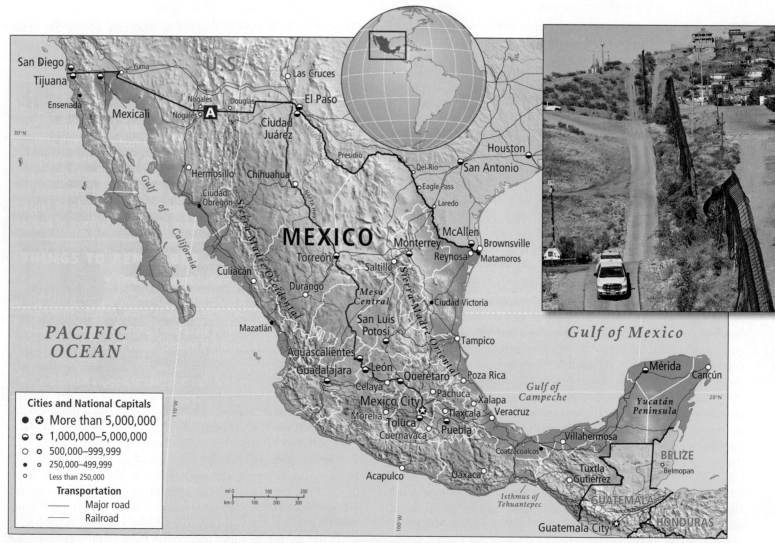

FIGURE 3.32 The Mexico subregion. **A** The U.S.–Mexico border barrier that divides Nogales, Arizona, from Nogales, Mexico.

mixed. Geographic markers of the successes and failures of this integration abound: the stream of legal and undocumented migrants into the United States—many of whom lost their livelihoods in Mexico due to the effects of NAFTA; the towering fences along the border with the United States (see Figure 3.32A); the interconnecting highway systems crowded with trucks carrying products between the three countries; and the ever-changing maquiladora phenomenon that continues to attract recession-plagued U.S. manufacturers, but is also shrinking due to the recession.

Formal and Informal Economies

Mexico's formal economy has modernized rapidly in recent decades. Its main components are mechanized, export-oriented agriculture, a vastly expanded manufacturing sector, petroleum extraction and refining, tourism, a service and information sector, and the remittances of millions of migrants working abroad. Mexico's large and varied informal economy employs about half of the total working population of 52 million. Informal economy workers contribute at least 13 percent of the GDP through such activities as subsistence agriculture, fine craft making, and many types of services, ranging from street vending and tutoring to trash recycling. The informal economy also includes any remittances smuggled home, a flourishing black market, and the crime sector.

Geographic Distribution of Economic Sectors

Mexico's economic sectors have a distinct geographic distribution pattern. The agricultural (or primary) sector employs about 14 percent of Mexico's population. However, agriculture produces only about 4 percent of Mexico's GDP, and only 12 percent of Mexico's land is good for agriculture, much of that comprised of hillsides. In the northwest and northeast, large corporate farms, often run by American companies, have replaced the old, inefficient haciendas and ranches that formerly hired scores of workers at very low wages. The corporate farms employ only a few laborers and technicians. With the aid of government marketing, research, mechanization, and irrigation programs, their products—high-quality meat and produce such as asparagus, peppers, winter vegetables, and raspberries, as well as corn and sorghum—go primarily to the North American market. In the tropical coastal lowlands farther south, subsistence cultivation on small plots is much more common, but here too, tropical fruits and products such as jute, sisal, and tequila are produced for export as well as for internal consumption. In Mexico's convoluted mountains, many small family farms produce a wide variety

of crops, including coffee, corn, tomatoes, sesame seeds, hot peppers, and flowers.

The service (or tertiary) sector is the largest, employing 62 percent of the population and producing 62 percent of the GDP. Service jobs are most varied and noticeable in the big cities. Work in services includes restaurants, hotels, tourism facilities of all sorts, financial and educational institutions, consulting firms, and museums. The service sector is dominated by small operations that cater to everyday needs: shoe repair, Internet cafés, cleaning services, beauty parlors—many in the informal economy.

The Maquiladora Phenomenon

The industrial (or secondary) sector, which includes Mexico's profitable petroleum refining industry that is located along the Gulf coastal plain, employs about 24 percent of Mexico's registered workforce and accounts for 34 percent of the GDP. Industries are found in many geographical niches, rural and urban, but a particular kind is concentrated along the U.S. border; these are the maquiladoras, in cities that include Tijuana, Mexicali, Ciudad Juárez, and Reynosa, that hire people to assemble manufactured goods, such as clothing or electronic devices, usually from duty-free imported parts.

Location of Maquiladoras The Mexican side of the border is a desirable location for the maquiladoras because it provides an inexpensive labor pool within easy reach of the U.S. market **(FIGURE 3.33)**. In 2016, Mexican factory workers earned an average of about U.S.$17 per day ($2.20 per hour). Compared to the United States, Mexico has far fewer regulations covering worker safety, fringe benefits, and environmental protection, and it charges much lower taxes on industries. These conditions attracted a considerable number of maquiladoras even before NAFTA took effect in 1994; after NAFTA took effect, the number of maquiladoras increased dramatically.

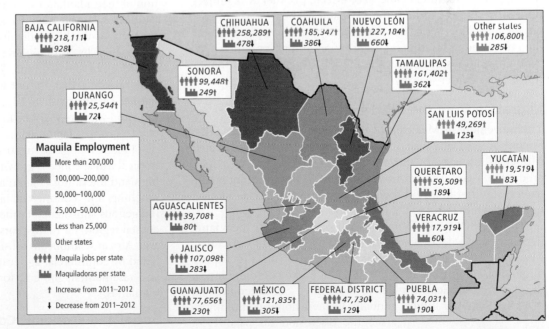

FIGURE 3.33 Location of maquiladoras in Mexico, 2011. The number of facilities and jobs in each state is shown on the map.

Gender in Maquiladoras For many years women have been the employees of choice in maquiladoras because they are more compliant than men, they work on short-term contracts for less pay and, if fired, they leave quietly. Thousands of young women previously sheltered by close-knit families now live in precarious, crowded circumstances where they are preyed upon by employers and by violent mates, gangs, and drug cartels. For example, the murders of hundreds of young women in Juárez remain unsolved. Nonetheless, young women continue to migrate to the maquiladoras because the remittances they send home are crucial to their families' welfare, and these financial contributions give female workers a high status in the family that earlier generations of women did not have. Men, recently displaced from agriculture and local factory work, are increasingly employed in maquiladoras, filling traditionally female jobs such as sewing or assembling electronic devices. They are viewed as more competent than female workers. Recent studies have shown that the majority of women employees do not use maquiladora work as a precursor to migration to the United States, as often has been claimed; if these migrants fail to find maquiladora work, they tend to return to their home villages.

The Downside of Maquiladoras for Mexico Although many Mexicans are pleased to have the maquiladoras and the jobs they bring, the country is also facing negative side effects. The border area has developed serious groundwater and air pollution problems because of the high concentration of poorly regulated factories and the hundreds of thousands of people living in unplanned worker shantytowns. Social problems are also developing among those crowded into unsanitary living conditions, and family violence is high. Governmental authority is fragmented along the long border zone where there are 14 paired U.S.–Mexican cities. The multiplicity of governments makes it difficult to address common problems. To address this issue of overlapping governments, the U.S.–Mexico Border XXI Project, under the auspices of both the U.S. and Mexican governments, has organized joint task forces that focus on infant immunization, the procurement of drinking water, water and air pollution mitigation, hazardous waste prevention and cleanup, environmental studies, and border security.

Migration from the Mexican Perspective

The movement of Mexicans to the United States to find work is as controversial in Mexico as it is in the United States because Mexico is losing many citizens (and parents) during their most productive years. Parents leave children to be raised by grandparents or other kin. And there is worry over the safety of family members while crossing the border and when on the job.

Unlike the many Europeans who cut ties with their families and native land when they migrated to North America, Mexicans often undertake their migrations with the express purpose of helping out their families and home communities. In recent years, Mexican workers in the United States remitted an estimated U.S.$20 billion annually to their home communities (see the discussion on page 153). Most Mexican households receiving remittances are located in relatively

ladino a local term for mestizo, used in Central America

better-off states such as Michoacán and Zacatecas, because residents of the poorest states find it difficult to finance a migration to the United States.

Does migration have to be the main route to advancement for Mexico's youth? Maybe not. Mexicans with more than a high school education are much less likely to migrate than those with just 9 to 12 years of schooling. Since about 49 percent of the Mexican population already uses the Internet (see Figure 3.17), government programs are trying to expand use of the Internet as a means of getting training and higher education to young adults in order to give them a potentially brighter future at home than they are likely to find as migrants in North America. Nonetheless, migration is still an important way in which Mexican families improve their well-being.

THINGS TO REMEMBER

• Mexico has a fragile democracy that is threatened by the violence and civil disorder of the traffic in illegal drugs by powerful drug cartels.

• Mexico's diversifying economy is in part based on the maquiladoras that have resulted from NAFTA, as well as petroleum revenues, remittances, and tourism.

• Mexico must deal with the problems posed by emigration to the United States.

CENTRAL AMERICA

It has long been said that Central America's wealth is in its soil. While this may no longer be true, as industry and services have come to account for a greater proportion of the gross domestic product of all countries, fully one-third of the people of the seven countries of Central America remain dependent on the production of their plantations, ranches, and small farms or *minifundios* **(FIGURES 3.34 AND 3.35)**. Most of the land is controlled by a tiny minority of wealthy individuals and companies, which contributes to high levels of income disparity.

Most of the Central American isthmus consists of three physical zones that are not well connected with each other: the narrow Pacific coast; the highland interior; and the long, sloping, rain-washed Caribbean coastal region. Along the Pacific coast, mestizo (**ladino** is the local term) laborers work on large plantations that grow sugarcane, cotton, and bananas and other tropical fruits; coffee is grown in the hills behind the coast. In the highland interior of Guatemala, Honduras, and Nicaragua, cattle ranching and commercial agriculture have recently displaced indigenous subsistence farmers. Similarly, the humid Caribbean coastal region, for many years sparsely populated with indigenous and African-Caribbean subsistence farmers, has become dominated by commercial agriculture, forestry, tourism development, and resettlement projects for small farmers displaced from the highlands.

Social and Economic Conditions

In Central America, the majority of people are either indigenous or ladino, and about a quarter still live in small rural villages

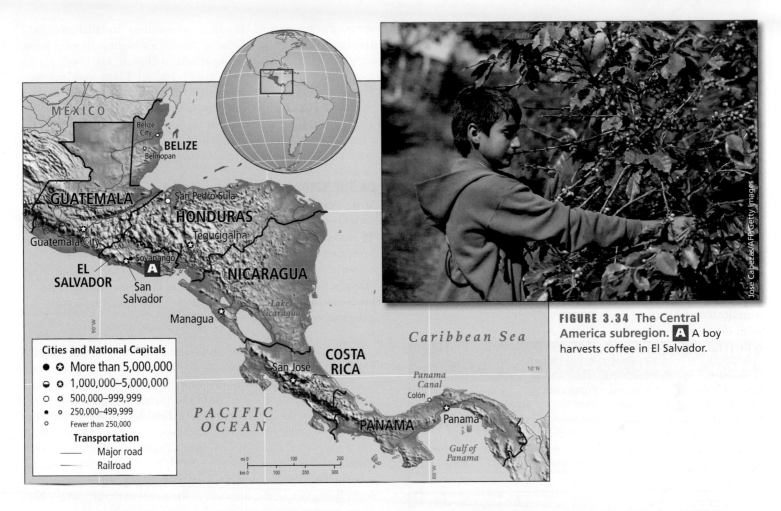

FIGURE 3.34 The Central America subregion. **A** A boy harvests coffee in El Salvador.

FIGURE 3.35 Minifundios. The pattern of tiny farm plots (*minifundios*) surrounding a town is visible in this photo of Costa Rica. Some minifundios are the result of government efforts to break up large land-holdings and redistribute them among poor farmers. Most mini-fundios are too small to provide more than supplemental income to their owners.

surrounded by tiny *minifundios* (see Figure 3.35). In these villages most people make a sparse living by cultivating their own food and cash crops on the minifundios; by working land they rent as sharecroppers; or by working as seasonal laborers on large farms and plantations (see the story of Aguilar Busto Rosalino in the vignette on page 154). The people of this region have experienced centuries of hardship, including long hours of labor at low wages and the loss of most of their farmlands to large landholders. In both rural and urban areas, infrastructure development has lagged. Roads are primitive and few. Most people lack clean water, sanitation, health care, protection from poisoning by agricultural chemicals, and basic education. Often they do not have access to enough arable land to meet their basic needs. The majority of the land is held in huge tracts—ranches, plantations, and haciendas—owned by a few families. On the small bits of farmland available to the poor for growing their own food and cash crops, local densities may be 1000 people or more per square mile. These circumstances all affect the well-being of the people, as reflected in Figure 3.28 and **TABLE 3.2**.

Table 3.2 Population data on Mexico and Central America, 2014–2015

Country (ordered by size), 2015	Population, in millions, 2015	Population density per square kilometer of arable land, 2015	Rate of natural increase (percent), 2015	Literacy rate (percent), 2014
Mexico	127.0	526	1.4	93.5
Guatemala	16.2	1056	2.6	No data
Honduras	8.3	819	2.0	85.1
El Salvador	6.4	904	1.4	84.5
Nicaragua	6.3	416	1.9	78
Costa Rica	4.8	1972	1.1	96.3
Panama	4.0	744	1.5	94.1
Belize	0.4	475	1.9	No data

Sources: Population Reference Bureau, 2015 *World Population Data Sheet*; and *United Nations Human Development Report 2014* (New York: United Nations Development Program), Table 9.

The Exceptions: Costa Rica and Panama

Two exceptions to these extreme patterns of elite monopoly, mass poverty, and rapid population growth are Costa Rica, and to some extent, Panama (see Figure 3.28 and Table 3.2). The huge disparities in wealth between colonists and laborers did not develop in Costa Rica, chiefly because there were no precious metals to extract and the fairly small native population died out soon after the conquest. Without a captive labor supply, the European immigrants to Costa Rica set up small but productive family farms that they worked themselves, not unlike early North American family farms. Costa Rica has democratic traditions that began in the nineteenth century, and it has unusually enlightened elected officials as well as one of the region's soundest economies. Population growth is low for the region, at 1.3 percent per year, and investment in human capital—in schools, health care, social services, and infrastructure—has been high. With no standing army, the country spends little on military installations. As a result, Costa Rica has Central America's highest literacy rates, and on many scales of comparison, including GDP per capita and HDI, the country stands out for its high living standards (see Figure 3.28). Costa Rica has often been hailed as a beacon for the more troubled nations of Central America. Nevertheless, while average incomes have risen in recent decades, income inequality, which was already high, has also grown, and for the past two decades poverty has hovered at around 20 to 25 percent.

ON THE BRIGHT SIDE: A Role Model

Costa Rica is a relatively affluent, healthy, and well-educated society—and an innovator in environmental activism, with 25 percent of its land area set aside in national parks and protected areas. Its economy has moved from dependence on agricultural exports to a more diverse basis in specialized manufacturing, pharmaceuticals, and ecotourism. As a successful democracy, it serves as an important role model for future development in Central America. ∎

Panama is known primarily for the canal that joins the Atlantic Ocean and Caribbean Sea with the Pacific, precluding the need for a long sea voyage around the tip of South America. Opened in 1914, the Panama Canal was built primarily with money from the United States and labor from the British Caribbean. Initially, the United States managed the canal and maintained a large military presence in Panama; as a result, Panama remained a virtual colony of the United States. In 1999, the United States turned over the canal to Panama and removed itself as a dominant presence. Interestingly, the turnover of the canal came at a time when it was increasingly obsolete and facing competition from nearby oil and gas pipelines and potentially from another canal route through Nicaragua. For decades the canal was a bottleneck for world trade as it was not large enough to accommodate the huge cargo, tanker, and cruise ships of the modern era. However, in 2016, after a decade of renovation, the canal was able to accommodate all but the very largest of ships in current use.

Environmental Concerns in Central America

Farmers often use unwise practices to wrench from their tiny plots of land enough food to feed their families. The news media often then blame them for environmental degradation. However, most environmental problems in Central America are caused not by the practices of small farmers but rather by large-scale agriculture and cattle ranching. In Honduras, the reservoir for a large hydroelectric dam built only a few years ago has been nearly filled with silt eroded from surrounding cleared land. Its electricity output must now be supplemented with generators run on imported oil.

Costa Rica has been a leader in the environmental movement in Central America. In the 1980s, the country established wetland parks along the Caribbean coast and encouraged ecotourism at a number of nature preserves, while at the same time acknowledging the potentially negative environmental side effects of tourism. Costa Rica supports scientific research through several international study centers in its central highlands and lowland rain forests, where students from throughout the hemisphere study tropical environments. Elsewhere in Central America, support for national parks is growing. With the help of the U.S. National Park Service and international NGOs such as the Nature Conservancy and the Audubon Society, 175 small parks have been established in the subregion.

Decades of Civil Conflict

Frustrated with elite-dominated governments unresponsive to the widespread poverty, many people in Guatemala, Honduras, Nicaragua, and El Salvador organized movements for political change, both armed and peaceful, for centuries. Despite requests for only moderate reforms, these movements were met with stiff resistance from wealthy elites assisted by national military forces and at times by the U.S. military and intelligence services. Conflicts came to a head in a series of civil wars that flared up intermittently from the 1960s to the 1990s, though the most recent coup d'état occurred in Honduras in 2009. In some cases, Central American political movements had the help of liberation theology advocates and Marxist revolutionaries. Violence was particularly intense in Guatemala, where 200,000 civilians were killed or "disappeared" by the military. The United States often backed and armed military dictatorships (that of General Ríos Montt, in the case of Guatemala) because it was convinced that the revolutionaries posed a Communist-inspired threat to the United States.

Rigoberta Menchú, an indigenous Guatemalan woman, won the 1992 Nobel Peace Prize for her efforts to stop government violence against her people. Her autobiographical account attracted public attention to the carnage and was important in awakening worldwide concern. Eventually, after a number of regional peacemakers joined Menchú in bringing international pressure to bear on the Guatemalan government, the Guatemalan Peace Accord was signed in September 1996.

The Central America Free Trade Association (CAFTA)

Many rural Central American grassroots organizations have fought the globalization of markets across the region. From their perspective, the Central America Free Trade Association (CAFTA), which has extended many of the provisions of NAFTA to Central America, is a threat to their survival. Before CAFTA was implemented in 2009, Oxfam, an international NGO concerned with issues of hunger and human rights, estimated that under the free market rules of CAFTA, U.S. exports of subsidized, mass-produced corn to Central America would increase dramatically, resulting in a lower market price that would force many farmers to sell their land and migrate to find work, essentially recreating the effects of NAFTA in Central America. In the years since CAFTA's implementation, some farmers have been able to adapt to new opportunities, for example by switching crops from corn to fresh vegetables which they can sell to new international buyers that have appeared in recent years. However, as Oxfam predicted, many poorer and less educated farmers have not been able to make the switch and have sold their land to large *agribusiness* (commercial agriculture) corporations, migrating to cities to look for work in new factories that pay below minimum wage. Others have become part of a huge flow of undocumented immigrants to the United States.

A CASE STUDY OF CIVIL CONFLICT: Nicaragua

Until the late twentieth century, Nicaragua had landownership patterns characteristic of the region: A tiny elite held the usual monopoly on land, while the mass of laborers lived in poverty. By 1910, the American company United Fruit had coffee and fruit plantations in the Nicaraguan Pacific uplands and on the Pacific coastal plain. Between 1912 and 1933, the United States kept marines in Nicaragua to quell labor protests that threatened U.S. interests in Nicaragua's food export economy. This U.S. military support helped the wealthy Somoza family to establish its members as brutal dictators in Nicaragua in the 1930s.

The Marxist-leaning Sandinista revolution of 1979 finally ousted the Somoza regime. The Sandinistas, who eventually won several national elections, embarked on a program of land and agricultural reform and improved basic education and health services. However, the country was soon mired in a debilitating war with the *Contras,* right-wing counterinsurgents backed by local elites and the United States. In what became known as the *Iran–Contra affair,* the Communist-wary Reagan administration secretly supported the Contras (for which Congress had explicitly denied funds) with money from covert arms sales to Iran. A trade embargo imposed by the United States further contributed to the ruin of the Nicaraguan economy. By the end of the 1980s, Nicaragua was one of the poorest nations in the Western Hemisphere.

In national elections in 1990, an electorate weary of violence voted the Sandinistas out. Starting in 1997, several free elections, in which 75 percent of the eligible citizens voted, resulted in moderate governments that continued to find it difficult to bring Nicaragua any measure of prosperity. In 2006, Daniel Ortega, a Sandinista leader and president in the late 1980s, was elected president. Now in his third term, he takes populist and anti-U.S. positions similar to those once taken by Chávez in Venezuela and the Castros in Cuba, but he also makes overtures to foreign investors. Strong and stable economic growth combined with effective antipoverty policies have made Ortega hugely popular. ■

THINGS TO REMEMBER

• Behind much of the civil disorder and violence in Central American countries is persistent disparity of wealth and a long-standing unwillingness to invest in human capital by providing education and social services.

• Costa Rica is an exception to the pattern of civil disorder and control by elites in Central America.

• Although participatory democracy is growing, a tiny minority of elites and the military continue to control political and economic power, often in collusion with foreign businesses and governments (including that of the United States). The result has been bloody conflicts that have killed thousands and seriously inhibited development.

THE NORTHERN ANDES AND CARIBBEAN COAST OF SOUTH AMERICA

The five countries in the northernmost part of South America share a Caribbean coastline and extend south into a remote interior of wide river basins and humid uplands **(FIGURE 3.36)**. The Guianas resemble Caribbean countries in that they were once traditional plantation colonies worked by slaves and indentured labor, and today their multicultural societies are made up of descendants of African, East Indian, Pakistani, Southeast Asian, Dutch, French, and English settlers. Venezuela and Colombia share a Spanish colonial past; their populations are largely mestizo, but there is also a small upper class of primarily European heritage and a small population of African derivation in the western and Caribbean lowlands. In all the countries of this subregion, small indigenous populations survive, mainly in the interior lowland Orinoco and Amazon basins, where they hunt, gather, and grow subsistence crops.

The Guianas

To the north of Brazil lie three small countries known collectively as the Guianas: Guyana, Suriname, and French Guiana. Guyana gained independence from Britain in 1966 and Suriname from the Netherlands in 1975. French Guiana, on the other hand, is not independent; it is still part of France. Today, the common colonial heritage of these three countries remains visible in both their economies and their culture. Sugar, rice, and banana plantations established by the Europeans in the coastal areas continue to be economically important, but logging and gold, diamond, and bauxite mining in the resource-rich highlands are now the leading economic activities.

The population descends mainly from laborers who once worked the plantations. These laborers formed two major cultural groups: Africans, brought in as slaves from 1620 to the early 1800s, and South and Southeast Asians, brought in as indentured servants after the abolition of slavery. The descendants of Asian indentured servants are mostly Hindus and Muslims. Many of them became small-plot rice farmers or owners of small businesses. Those of African descent are primarily Christian; they are both agricultural and urban workers. Politics in the Guianas is complicated by the social and cultural differences between citizens of Asian and African descent. These differences have also slowed economic and human development to the point that these three countries lag a bit behind several Caribbean island countries that have similar colonial histories, such as Trinidad and Tobago and Barbados.

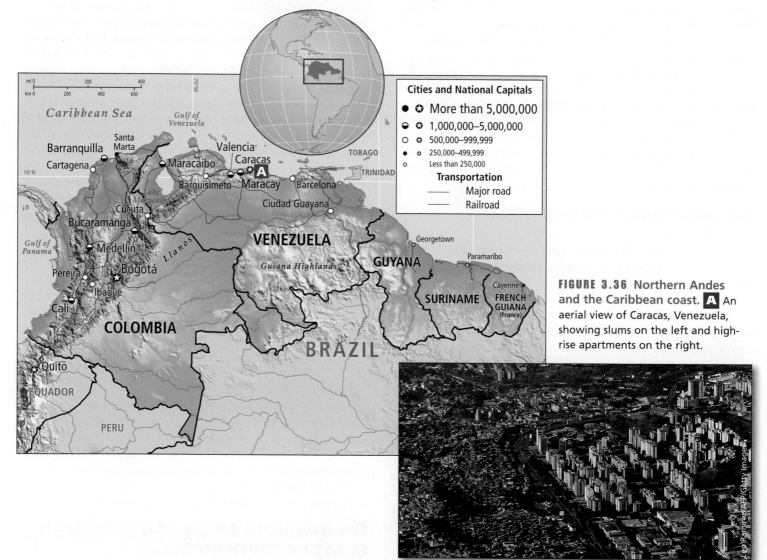

FIGURE 3.36 Northern Andes and the Caribbean coast. A An aerial view of Caracas, Venezuela, showing slums on the left and high-rise apartments on the right.

Venezuela

Venezuela has long had the potential to become one of the wealthier countries in South America, primarily because it holds large oil deposits and is an active member of the Organization of Petroleum Exporting Countries (OPEC). Oil has been the backbone of the country's economy since the mid-twentieth century. Venezuela not only is among the top suppliers of oil to the United States, but also supplies oil to Cuba, Canada, Central and South America, and China, and is seeking agreements with Japan and Europe. The U.S. invasion of Iraq in 2003 illustrates how geopolitics can link distant regions: Venezuelans feared that the invasion would result in the takeover of Iraqi oil management by the United States and Britain. This could have caused Venezuela to lose market share in the United States, and the result in Venezuela, already economically strapped, could have been civil disorder. That scenario did not materialize, however. Instead, a drawn-out war in Iraq allowed Venezuela to gain power globally as oil prices rose sharply.

Although Venezuela's oil resource was nationalized in the 1970s, with the idea that the profits would fund public programs to help the poor majority, pre-Chávez market-oriented governments relaxed this policy, and for a while foreign firms such as Chevron exercised considerable control over production. As a result, oil profits stayed with the elite and the small middle class. By the late 1990s, 63 percent of the country's wealth was controlled by 10 percent of the population, mainly those of European descent, leaving those of mixed native and African descent at the bottom of the income pyramid. Taxes on the wealthy and on foreign investors were kept low, so oil and other assets did not generate enough government revenue to fund badly needed improvements in education, health care, transportation, and communications. This failure to invest in its own people meant Venezuela did not build a base for general economic and social advancement. The landscape reflects this failure: aerial views of the capital city of Caracas show rows of high-rise buildings and modern freeways (see Figure 3.36A), but a closer look reveals expanses of poor shantytowns (discussed on page 163). The shantytowns are home to millions of people who live in deep poverty and lack access to clean drinking water, sanitation, and adequate education and transportation.

The global economic downturn that began in 2007 had a noticeable effect in Venezuela primarily because the resultant fall in oil prices severely curtailed personal and national income. Oil wealth had allowed Venezuela to invest in its own infrastructure and to address both the long-unmet needs of its giant underclass and to push for a greater regional leadership role. The fall in oil prices diminished, at least temporarily, Venezuela's expanding role in Middle and South American geopolitics.

Colombia

A civil war in Colombia has raged on and off for more than 50 years, fueled by economic inequalities and political repression. The conflict has produced so many refugees that Colombia has one of the highest totals of internally displaced people in the world. Although Colombia is the world's second-largest exporter of coffee and a major exporter of oil and coal, a small proportion of the population, mostly of European descent, has monopolized most of the income by keeping wages low and resisting paying taxes. In addition, the wealthy have pushed back against efforts to redistribute some of their extensive landholdings to landless rural people.

On one side of the civil war are revolutionary guerrillas who want government-sponsored reforms for the poor. One such group, the Revolutionary Armed Forces of Colombia (FARC), controls a large part of the Colombian interior south of Bogotá. On the other side are the private armies of the wealthy, who have little faith in the will and ability of the ill-equipped Colombian military to defeat the guerrillas.

The hostilities are complicated by the participation of all the warring parties in the cocaine trade. This trade, as we have seen, is derived from the leaves of the ancient Andean coca plant, traditionally chewed as a fatigue and hunger suppressant, but now processed into the much stronger cocaine and sold internationally (see Figure 3.21). Although coca growers make a somewhat better income from coca than they would from other crops, competing drug-smuggling rings reap most of the profits, intimidating and murdering Colombian government officials who try to reduce or eradicate the drug trade.

Per capita drug use by Colombians is quite low (about one-quarter of the U.S. rate), and there are signs that in-country drug-related violence is beginning to decrease. The Council on Hemispheric Affairs, a nonprofit research organization, suggests that the low rate of drug use is related to the strong family ties of most Colombian youth and to local community initiatives such as the *No Mas* (No More) movement, a civic effort to stop the drug trade violence. Millions have demonstrated against the drug cartels, and some success in abating the violence has come from rehabilitating drug trade operatives.

Much work is being done to find sources of income for rural people other than coca. For example, in the mountains of southwestern Colombia, 1200 farm families have formed a cooperative that produces a line of 20 products—preserved fruits, sauces, and candies—marketed especially to Latinos in North America and Europe. The cooperative also focuses on child and adult education and on enhancing people's marketable skills. Elsewhere in the country, rural people working with the food scientists at the International Center for Tropical Agriculture have developed new varieties of corn that will be more productive and nutritious.

Repeated attempts to end the conflicts via peace talks have failed, but negotiations between the FARC and the Colombian government have been ongoing since 2013, many partial accords have been reached, and both sides appear willing to make crucial concessions to achieve peace. FARC leaders admit that they are unlikely to ever take power by force and now see that they may be able to win through elections, once they disarm. Both sides have agreed to eventually stop their drug trafficking.

Despite all the news about drug violence, Colombia has maintained a vibrant tourist economy. Cartagena, for example, is known for its elegant colonial buildings, beautiful beaches, and lively entertainment for affluent and budget-minded tourists alike **(FIGURE 3.37).**

FIGURE 3.37 Tourist hotels on the water in Cartagena, Colombia. High-rise hotels for tourists and apartment buildings for wealthy Colombians line the beach in Cartagena.

THINGS TO REMEMBER

• All five countries of the Northern Andes and Caribbean Coast depend on extractive industries—logging and mining, petroleum products, and/or drugs—as major income sources.

• The five countries of this subregion have quite different political systems, with varying outcomes for civil society.

• A civil war in Colombia has simmered for more than 50 years, producing so many refugees that Colombia has one of the highest totals of internally displaced people in the world.

THE CENTRAL ANDES

The central Andes, which includes the countries of Ecuador, Peru, and Bolivia, consists of a high and wide system of parallel mountain ranges and plateaus that carries 70 percent of the world's tropical glaciers, a narrow coastal lowland in Ecuador and Peru (Bolivia has been landlocked since its mineral-rich Atacama Desert coastline was annexed by Chile in 1884), and a wide interior apron of lowlands in all three countries that drains east into the Amazon **(FIGURE 3.38)**. On the eve of the Spanish conquest (1532), the coast and mountainous zone was home to the Inca Empire. The legacy of the Incas is still prominently reflected in the landscape: in the thousands of miles of paved trade footpaths; in the numerous massive stone ruins, such as those at Machu Picchu in Peru (see Figure 3.11B); and especially in the roughly one-half of the population that is indigenous—the largest proportion in South America.

Altiplano an area of high plains in the central Andes of South America

After the fall of this area to the Spanish, a tiny group of landowners who were descended from Europeans prospered, while the vast majority of indigenous people and mestizos lived in poverty and servitude, working on large haciendas and in rich mines (copper, lead, and zinc in Peru; tin, bauxite, lead, and zinc in Bolivia). Most of the twentieth century was marked by failed efforts at social change and the violence resulting from those failures. Now the growing political involvement of the large indigenous population may help create greater social equity.

Settlement Patterns

Settlement has a distinct lowland/highland pattern in this region. The Pacific coast is home to large and modern cosmopolitan cities, including Peru's capital and commercial center, Lima, and Ecuador's leading industrial center, Guayaquil. The lowland people are mainly mestizo, some with African heritage. The majority of indigenous people live in the **Altiplano** (highlands). There are several large cities in the Altiplano, including Cusco, Peru (the former capital of the Incan empire); La Paz, the capital of Bolivia; and El Alto, Bolivia, which, at 13,615 ft (4150 m) and home to more than 1.5 million people, is the highest large city in the world. The interior lowland is less densely settled, mostly with indigenous lowlanders, but that is changing with the development in the western Amazon lowlands of mining and of oil and gas for export.

The coast is also a zone of productive agricultural land where, despite the often-dry climate, plantations and other agricultural enterprises produce crops for export. The irrigated production of export crops such as bananas, cotton, tobacco, grapes, citrus, apples, and sugarcane has increased dramatically in recent decades, with most of the profits going to large agribusiness firms. Irrigation and other aspects of export-oriented agriculture are often funded with IMF loans that must be repaid.

The nutrient-rich ocean off Peru currently nourishes a highly productive fishery. The vibrant export-oriented fishing industry is funded with international loans and aimed at European and North American markets. Corporate shrimp farms supply this market as well. Because corporate farms and fishing industries producing for the export market are displacing the small, traditional farmers and fishers who once served the local market, local working people who once fed themselves now face food insecurity.

Climate Change, Water, and Development Issues

A major environmental issue in the central Andean region is climate change. The glaciers of the Andes are melting rapidly, having lost 22 percent of their surface area over the last 35 years. Though general public awareness of this approaching hazard is still low, the loss of Andean glaciers threatens most of the region's rivers because they are fed by the normally slow seasonal glacial melt. The annual melt is now faster than winter replacement; thus the rivers have less volume and their ability to irrigate and generate hydropower is diminishing. The water supply of 30 million people is also at risk because highland cities and rural villages derive most of their water from glaciers. Water shortage is already affecting highland crops; the deterioration of pastureland is curtailing sheep and llamas' production of wool.

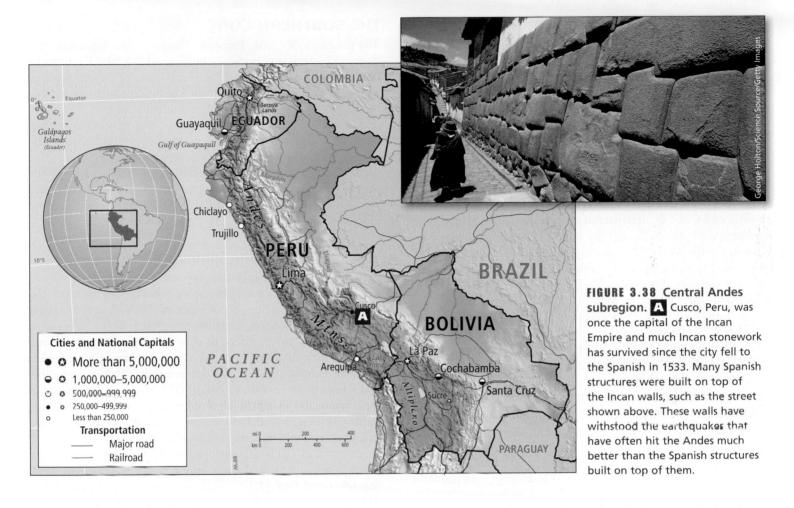

FIGURE 3.38 Central Andes subregion. A Cusco, Peru, was once the capital of the Incan Empire and much Incan stonework has survived since the city fell to the Spanish in 1533. Many Spanish structures were built on top of the Incan walls, such as the street shown above. These walls have withstood the earthquakes that have often hit the Andes much better than the Spanish structures built on top of them.

Agricultural Restoration in the Peruvian Highlands

In the mountainous interior, the indigenous Aymara and Quechua people are using traditional crops to help create sustainable agricultural change. Support for this movement comes in part from the United States, which would like coca farmers to find a profitable alternative crop. In 2004, the then-oil-rich Chávez government in Venezuela provided funds to indigenous farmers in the Altiplano to help them develop their traditional crops for North American organic and gourmet markets. The Venezuelan grants encouraged the market production of quinoa, a 5000-year-old domesticated plant that produces a grain-like kernel extraordinarily nutritious and palatable for those with wheat gluten allergies. Quinoa has since become quite popular with European and North American health-conscious consumers.

Social Inequalities, Exports, and Social Unrest

Improving living standards for the poorest people in the central Andean region is important for humanitarian and environmental reasons. Structural adjustment policies once mandated by the IMF forced the privatization of state-run industries and the streamlining of government. The result was job loss and social turmoil, with only modest economic growth. Governments dramatically increased prices for gasoline, electricity, and transportation to help raise funds to pay off debts. Such policies hurt the urban poor, whose low wages could not cover the increases. In rural areas, SAPs removed government assistance to small-scale farmers, encouraging the trend of small farms producing local food being replaced by export-oriented corporate farms and fisheries and other high-tech operations. Most of these receive government assistance even though they cause environmental degradation and the loss of livelihood for poor people.

One result of the growth of export-oriented agriculture is that highland people who could not get access to land or jobs have moved east into Amazon lowland regions. These lowlands, long the home of scattered though occasionally large and dense groups of indigenous people, have recently undergone rapid and often destructive exploitation of natural resources. National governments, eager to fulfill their SAP obligations, have encouraged export-oriented extraction, first of trees and then of minerals and agricultural products, and these activities have severely damaged the home territories of indigenous people (see the discussion that opens this chapter). Roads built to exploit timber and minerals have also opened this region to waves of landless and jobless migrants from the highlands and coastal zones.

Ecuador's Correa Various strategies for lessening gross inequalities have been tried in the Andes, with varying success. In 2006, Ecuadorians elected a U.S.-educated economist, Rafael Correa, who favors higher taxes on the wealthy and substantial government spending to alleviate social ills. He appears to share with Hugo Chávez and Evo Morales, the president of Bolivia, an anti-U.S. sentiment, but he has a reputation for being incorruptible and was reelected in 2009 and again in 2013. In the wake of weak oil prices and protests against some of his policies, Correa announced that he would not seek reelection in 2017. Shortly thereafter, Ecuador's National Assembly (its congress) passed a constitutional amendment that would let Correa run for the presidency indefinitely beginning in 2021.

Bolivia's Morales Gains in political participation by indigenous people are perhaps most impressive in Bolivia. In the 1980s, Bolivia embraced the free market model and sold off state-owned industries, mostly to foreign interests. These policies, urged on Bolivia by the World Bank and IMF to reduce its debt, enriched a few and brought job losses to the majority and an increase in income disparity. Protests in 2003 were spurred by government plans to export natural gas to the United States under conditions unfavorable to Bolivia. Protesters interviewed by *New York Times* reporter Larry Rohter said they saw an "unbroken line" between the rapacious Spanish colonial policies of the past and modern movements linked to globalization and free trade. In one weekend, the Bolivian military shot dead 50 unarmed protesters who were demanding President Lozada's resignation. Lozada eventually resigned in October 2003, at a time when income disparity in Bolivia was escalating alarmingly. In 2005, Bolivians elected in a landslide the first indigenous head of state in the Americas: Evo Morales, a socialist, former coca farmer, and head of the coca producers' union, who has proven to be innovative, independent, and controversial.

Morales began his presidency advocating against the privatization of water (see the discussion on page 152) and for coca farmers, saying that they should not lose the right to grow an age-old crop that produces only a mild high in its natural state just because an outsider had figured out how to make the powerfully addictive drug cocaine from coca. President Morales, who seems intent on changing how and to whose benefit Bolivia's resources are used, continued to alarm outside investors and Bolivian elites when he nationalized the oil and gas industries. Most recently, the Bolivian government has entered into production of the mineral lithium, an essential component in batteries for electric cars and other electronic devices. Bolivia holds one-half of the world's lithium, mostly under desert salt flats occupied by salt gatherers and quinoa farmers. In 2015, the Bolivian government created its first lithium production plant, and plans to use the revenues to support its ambitious social programs.

THINGS TO REMEMBER

• With the glaciers melting, climate change is a main issue for the central Andes.

• The five countries of this subregion have quite different political systems, with varying outcomes for civil society.

• Each country's large indigenous populations are flexing their political muscle to overcome past inequalities.

THE SOUTHERN CONE

The countries of Chile, Paraguay, Uruguay, and Argentina have diverse physical environments but remarkably similar histories (FIGURE 3.39). The so-called *Southern Cone* of South America had little European settlement during the Spanish empire, but in the late nineteenth and early twentieth centuries, European immigrants—mainly Germans, Italians, and Irish—were drawn there by temperate climates, economic opportunities, and the prospect of landownership. These immigrants soon outnumbered the surviving indigenous populations throughout most of the region. Paraguay is the only one of these countries that has a predominantly mestizo population. It is the poorest country of the four (see Figure 3.28).

The Economies of the Southern Cone

Agriculture was once the leading economic sector in the Southern Cone, but it is no longer the main source of income or the main employer, having been replaced by the service sector. Nonetheless, agriculture remains prominent in the identity of the region and in its ability to earn foreign exchange through exports. The primary agricultural zone, the *pampas*, is an area of extensive grasslands and highly fertile soils in northern Argentina and Uruguay (and southern Brazil). The region is famous for its grain and cattle. Sheep raising dominates in Argentina's drier, less fertile southern zone, Patagonia. On the Pacific side of the Andes, in Chile's central zone, the Mediterranean climates of southern Argentina and Chile support large-scale fruit production that caters to the winter markets of the Northern Hemisphere. Both Argentina and Chile have greatly expanded their production of wine (see Figure 3.39B). Chile also benefits from considerable mineral wealth, especially copper.

Although the agricultural and mineral exports of the Southern Cone created considerable wealth in the past, fluctuating prices for raw materials on the global market have periodically sent the economies of this subregion into a sudden downturn. The desire for economic stability was a major impetus for industrialization and urbanization in the mid-twentieth century. At first, the new industries were based on processing agricultural and mineral raw materials. Later, state policies supported diversification into import substitution manufacturing industries (see page 152). However, inefficiencies, corruption, low quality, and the small size of local markets prevented manufacturing from becoming a leading sector for the region. Although the service sector has now surpassed agriculture and industry in the Southern Cone countries, it too is lackluster, providing only modest contributions to the GDP.

Economic policy has been a source of conflict within the subregion for decades. Despite their considerable resources, these countries have always had substantial impoverished populations that have created political pressure for more jobs and better living standards. In response, each country developed limited mechanisms for redistributing wealth—government subsidies for jobs, food, housing, basic health care, and transportation, for example—that did alleviate poverty to some degree. When global prices for raw materials fell in the 1970s and produced an

SambaPhoto/Cassio Vasconcellos/Getty Images

A Itaipu Dam is one of the largest power stations in the world, supplying 90 percent of Paraguay's electricity.

Alfredo Máiquez/Lonely Planet Images/Getty Images

B Food processing is a leading industry throughout the region. Shown here are three workers in a Chilean winery.

FIGURE 3.39 The Southern Cone subregion.

economic downturn that increased poverty, people clamored for more fundamental changes.

Operation Condor and the "Dirty Wars" In response to political agitation for antipoverty programs in the 1950s, 1960s, and 1970s, military leaders, supported by economic elites and foreign intelligence agencies, took control of governments in Paraguay (1954), Chile (1973), Uruguay (1973), and Argentina (1976). These governments were supplied with communications, weapons, and other support by the intelligence agencies of France and the U.S. Central Intelligence Agency, in what was called *Operation Condor.* The result was the dirty war, in which bloody campaigns of repression were waged against virtually anyone who

investment. An IMF package of structural reforms and loans to the Argentine government held off crisis but did not resolve it. In 2006, Argentina, like Chile, elected a moderate socialist president, Néstor Kirchner, who increased government spending to quell social unrest and promote more equitable economic growth. Cristina Fernández de Kirchner, elected Kirchner's successor after his death, continued these policies. In 2015, Argentina's debt was 46 percent of GDP, which is just over half of the U.S. debt and far below that of many countries in the debt-plagued EU.

FIGURE 3.40 Buenos Aires. Children in a poor part of Buenos Aires simulate snow with polystyrene from a stuffed animal. Behind them looms an unfinished hospital intended to serve their neighborhood; it has instead been left vacant for more than 50 years. Approximately 25 percent of Argentina's population lives in poverty.

Buenos Aires: A Primate City

The primary urban center in the Southern Cone is Buenos Aires, the capital of Argentina and one of the world's largest cities **(FIGURE 3.40)**. Forty percent of the country's people live in this primate city, which boasts premier shopping streets, elegant urban landscapes, and dozens of international banks. Yet six decades of decline have left it with empty factories, social conflict, severe poverty, pollution, and declining human well-being. Some people argue that the downward slide in quality of life in Buenos Aires is an unavoidable result of the restructuring required to create an economy that is competitive in the global arena. Past Argentine leaders contended that better integration with the global economy would help to reverse decades of malaise. They wanted Buenos Aires to be seen as a world city—a center with pools of skilled labor that attracted major investment capital, a sophisticated city with a beautiful skyline and a powerful sense of place. The reality is that many of the city's residents live in run-down apartments and on wages too low to afford basic nutrition. It is likely that the city's poor residents wish to give priority to such basics as decent housing, better food, and modernized transportation—goals that will not necessarily draw profit-seeking free market investors.

supported socialist or communist political ideologies, including labor union members, student activists, academics, journalists, and the clergy. About 2000 people in Paraguay, between 3000 and 15,000 people in Chile, and more than 30,000 people in Argentina were killed. Tens of thousands more were jailed and tortured for their political beliefs (see page 160 for a discussion of the overthrow of President Salvador Allende in Chile in 1973). The military regimes were removed from power for different reasons from 1983 to 1990, though new laws granted blanket immunity to military personnel involved in kidnappings, torture, and murder. These laws have more recently been repealed in Argentina and Uruguay.

Debt and the Legacies of SAPs The IMF has pressured governments to repay their massive debts via SAPs that shift countries away from poverty-alleviation policies and toward free market economic reform. The effects of these policies have been mixed in the Southern Cone. Chile's export-oriented industrial, mining, and agricultural sectors have all grown, but not sufficiently to alleviate the middle-class social discontent that in 2006 led to the election of a moderate socialist, Michelle Bachelet, as president. Bachelet chose moderate free market reforms, saving the profits from rich, government-owned copper mines to stimulate the economy with government spending during the recession that began in 2008. A more conservative president was elected in 2010, though Bachelet was reelected in 2014.

Argentina, after suffering a recession of its own, years of job losses, and a crushing debt burden, defaulted on its debt payments in December 2001, causing a precipitous drop in foreign

> **THINGS TO REMEMBER**
>
> • The Southern Cone countries are environmentally diverse and have a fairly large immigrant population from Europe.
>
> • While the Southern Cone has relatively high GDP and HDI rankings, its economies are weakened by the struggle to pay off past debts, build more modern infrastructures, and improve living standards.

BRAZIL

The observant visitor to Brazil is quickly caught up in the country's physical complexity and in the richly exuberant, multicultural quality of its society **(FIGURE 3.41)**. But its landscapes also

FIGURE 3.41 The Brazil subregion.

plainly show the environmental effects of both colonialism and recent underplanned economic development. Brazil's 205 million people live in a highly stratified society made up of a small, very wealthy elite; a modest but rising middle class; and a majority that lives below, or just barely above, the poverty line. In Brazil's megacities of São Paulo and Rio de Janeiro, elegant high-rise buildings are surrounded by vast favelas (see Figure 3.24). High crime rates and homeless street children are signs of the gross inequities in opportunity and well-being. According to the latest figures (2013), the richest 10 percent of Brazil's population has

41 times the wealth of the poorest 10 percent. This remains one of the widest disparities in Middle and South America and the world, though it is declining. Travelers find themselves delighted by the flamboyant creativity and elegance of the Brazilian people, yet sobered by the obvious hardships under which so many labor.

Brazil's Size and Varied Topography

Brazil has about the same area as the continental United States. Its three distinctive physical features are the Amazon Basin, the Mato Grosso, and the Brazilian Highlands. The Amazon Basin, which covers the northern two-thirds of the country, is described on pages 135–136 (see also Figures 3.1 and 3.4). The Mato Grosso is a seasonally wet/dry interior lowland south of the Amazon with a convoluted surface. Once covered with grasses and scrubby trees adapted to long dry periods, in the twentieth century it was extensively cleared for subsistence and commercial agriculture. The southern third of Brazil is occupied mostly by the Brazilian Highlands, a variegated plateau that rises abruptly just behind the Atlantic seaboard 500 miles (800 kilometers) south of the mouth of the Amazon. The northern portion of the plateau is arid; the southern part receives considerably more rainfall. Settlement in northeastern Brazil is concentrated in a narrow band along the Atlantic seaboard and south through the city of Salvador (see Figure 3.22B). In Brazil's temperate zone, near Rio de Janeiro, São Paulo, Curitiba, and Pôrto Alegre, settlement extends deeper inland (see Figure 3.27).

Managing Brazil's Large and Varied Economy

The Brazilian economy is the largest in Middle and South America and the seventh largest in the world. The resources available for development in Brazil are the envy of most nations, but management of those resources and the development of infrastructure to serve Brazil's best interests has been a challenge. Brazil long looked to the Amazon for its future development but successive governments failed to recognize the fragility of Amazonian ecosystems (see the discussion on page 139) and the hidden costs of development. As a result, one of the world's largest tropical forests has been degraded and many poor people have been misled.

Gold, silver, and precious gems have been important resources since colonial days, but it is industrial minerals—chromite, manganese, rare earths, titanium, tungsten, and especially iron ore—that are most valuable today. These minerals are found in many parts of the Brazilian Highlands, where they are mined with insufficient attention to human and environmental consequences.

Brazil's energy resources are diverse and abundant, making it the 8th largest producer in the world, larger than Venezuela. Renewable energy supplies about 40 percent of Brazil's needs, well above the global average of 10 percent renewable energy. Hydroelectric power is widely available, supplying 15 percent of the total energy supply and over 80 percent of electricity, due to the many rivers and natural waterfalls that descend from the highlands. Ethanol derived from sugar cane supplies around 15 percent of the country's energy needs. Brazil is the world's second largest producer of ethanol after the United States, and has used this fuel since the 1920s, though the current level of ethanol use was not reached until the oil crisis of the 1970s. Oil now provides for 39 percent of Brazil's energy needs, and most of it comes from very large offshore reserves in the Atlantic. A reflection of Brazil's unique energy mix, and its ability to innovate is the fact that it now makes cars that can instantly switch from gas to ethanol and get 40 miles to the gallon.

Agriculture Brazil's agricultural economy is also large and varied. In the more temperate far south, rice, poultry, corn, beans, and tobacco dominate, often on large mechanized farms. Further north near São Paulo and Rio de Janeiro, sugar, coffee, cattle, and oranges dominate. In the tropical northeast and in the Amazon basin bananas, cassava, and other warm climate fruits are widely grown, while on the southern borders of the Amazon in Mato Grosso, Mato Grosso do Sul, Goias, and Distrito Federal large-scale soybean cultivation is expanding, along with corn, rice, beans, and livestock.

Much of Brazil's remaining cultivable land is in the tropical Amazon, where soils are fragile. There, agriculture and timber extraction are expanding rapidly at the expense of tropical rain forests. While the rate of deforestation fell from 2005 to 2010, thousands of square miles of forest still fall yearly for the cultivation of such crops as soybeans and pasture grass. Overall, agricultural exports from Brazil are increasing and exceed imports, but as a proportion of Brazil's GDP, agriculture (5.5 percent) is rapidly losing out to industry (27.5 percent) and services (67 percent).

Industry Brazil is the most highly industrialized country in South America, and its global role is expanding. Most of its industries—steel, motor vehicles, aeronautics, appliances, chemicals, textiles, and shoes—are concentrated in a triangle formed by the huge southeastern cities of São Paulo, Rio de Janeiro, and Belo Horizonte. Yet the country's transportation system (modernized roads, airports, and rail connections) is still far from adequate.

Until the 1990s, the vast majority of Brazil's mining and industrial operations were developed using government funding. Many are still owned and run by the government, but some are successful private enterprises. In the 1980s and 1990s, elected governments adopted structural adjustment policies and privatized many formerly government-held industries and businesses in an effort to make them more efficient. Many of these firms were sold to foreign investors at bargain-basement prices. In 2000 alone, direct foreign investors spent U.S.$32.8 billion on Brazilian properties. Privatizing industry often results in greater productivity, and indeed, beginning in 2001, Brazil's exports began to grow dramatically. Moreover, exports regularly exceeded imports, which boosted the country's finances. By 2014, exports were U.S.$256 billion and imports $247 billion, up 450 percent from 2000 when exports were U.S.$57.6 billion and imports $57.7 billion. However, much of this growth was related to China's demand for Brazilian minerals and soybeans, most of which might have occurred without privatization.

Urbanization

Brazil has a number of large and well-known cities: Rio de Janeiro, Curitiba, São Paulo, Salvador, Recife, Fortaleza, Belém, Manaus, and Brasília. All except Manaus and Brasília are located on the Atlantic perimeter of the country. During the global economic depression of the 1930s, farmworkers throughout the country began migrating into urban areas as world prices for Brazil's agricultural exports fell; agricultural changes in the 1960s pushed even more people off the land and the chance for employment in the factories that were being built with government money pulled them into the cities. Brazil's competitive edge in the global market was its cheap labor. The military governments of the time, thinking that the mostly government-owned industries should continue to pay very low wages because there was such a surplus of labor, found that they had to quell many protests by workers who could not live decently on their wages.

By 2016, eighty-six percent of Brazil's population was urban, and at least one-third of the urban dwellers, many of them underemployed, were living in favelas, the Brazilian urban shantytowns (see Figure 3.24). The poverty in the cities of the northeast rivals that of Haiti, the poorest country in the Americas. However, favela dwellers are famous for their strong community life and support for those in distress. Many turn to music, dance, performance, and religion of one sort or another as a source of strength.

VIGNETTE It's Friday afternoon, and a crowd of white-clad women is gathering outside a house in the Felicidad favela in Fortaleza. Like the surrounding houses, this one is modest, but it gleams white; all its surfaces are swathed in marble. Potted palms decorate the porch, and beside them, welcoming the women, is a tall, elegant, middle-aged man also dressed in white, a religious leader in the movement known as Umbanda. Umbanda, Batuque, and related belief systems thrive in all the Atlantic coastal cities of Brazil from Belém to São Paulo (FIGURE 3.42). Each group is led by a male or female spiritual leader who invokes the spirits to help people cope with health problems and the ordinary stresses and strains of a life of urban poverty. During the ceremonies, which can last as long as 8 hours, the central focus is a combination of drumming, dancing, spirit possession, and friendly psychological support. Umbanda grew out of an older African-Brazilian belief system called Candomblé, and similar movements (Voodoo, Santería, Obeah) are found elsewhere in the Americas, including the Caribbean, the United States, and Canada. In Brazil, Umbanda appeals to an increasingly wide spectrum of the populace from African, European, Asian, and indigenous backgrounds. [*Source: Lydia Pulsipher's field notes*] ∎

Brasília Brasília, the modern capital of Brazil, is an intriguing example of the effort to lead development with urban growth poles (discussed on page 164). Built in the state of Goiás in just 3 years beginning in 1957, Brasília lies about 600 miles (1000 kilometers) inland from the glamorous old administrative center of Rio de Janeiro. The official rationale for building a new capital in the remote interior was that the city would serve as a **forward capital,** helping the development of the western highland territories, the Mato Grosso, and eventually, the Amazon Basin. The scholar of Brazilian development, William Schurz, has suggested an alternative explanation for Brasília's location: moving the capital so far away from the centers of Brazilian society was an efficient way to trim the badly swollen and highly inefficient government bureaucracy.

Symbolism figured more prominently than practicality in the design of Brasília, which was laid out to look from the air like a jet plane. There was to be no central business district, but rather shopping zones in each residential area and one large mall. Pedestrian traffic was limited to a few grand promenades; people were expected to move even short distances in cars and taxis. Public buildings were designed for maximum visual and ceremonial drama, but safety was an afterthought.

Over five decades of actually using this urban landscape, people have made all sorts of interesting changes to the formal

FIGURE 3.42 African-derived religions in Brazil. Practitioners of Candomblé carry flowers out to a boat during a ritual to honor Yemanja, goddess of the sea, in Amoreiras, Brazil. Candomblé is one of several African-derived religions that is gaining popularity, especially in the urban areas of South America's Atlantic coast. Ceremonies can last up to 8 hours and involve drumming, dancing, spirit possession, healing, and psychological support.

design. At the Parliament, legislative staff and messengers created footpaths where they needed them: through flowerbeds and—with little steps notched in the dirt—up and over landscaped banks. Thus they efficiently connected the administration buildings, bypassing the sweeping promenades. Little hints of the informal economy that characterizes life in the old cities of Brazil began to show up—a fruit vendor here, a sidewalk manicurist there. And the shantytowns that the planners had tried hard to eliminate began to rise relentlessly around the perimeter. Overall, in the 50 years since its construction, Brasília's success as a forward capital and growth pole has been limited. Although the city has drawn poor laborers, the entire province of Goiás still has only 4.5 million people, just 6 percent of the country's total, and their average income is only half that of the country as a whole. Population and investment remain centered in the Atlantic coastal cities.

Rio and Mega Events

In 2014, Rio de Janeiro was the primary host city for the most watched sporting event on Earth, the FIFA World Cup, and in 2016 the city hosted the world's best-attended sporting event, the Olympics. The preparations for these events were designed to bring in new investment to the city; build and renovate the transportation, tourism, and athletics infrastructure; and to project an attractive image of Brazil to the world.

> **forward capital** a capital city built in the hinterland to draw migrants and investment for economic development and sometimes for political/strategic reasons

Among the most lasting legacies of the World Cup and the Olympics, which together brought millions of people and billions of dollars of new investment to Rio de Janeiro, is the hardship they place on poorer residents of the city. According to Rio de Janeiro's city government, 4100 families were removed from the center city and other areas to the urban periphery to make way for new infrastructure serving the mega events. By contrast, local NGOs claim that 22,000 families comprising 77,000 people were removed between 2009 and 2015 in preparation for the mega events. Most of those removed had been living near roads, airports, sports facilities, and the city's harbor area, all of which were significantly expanded and renovated. Some of the displaced had been living in informal favela settlements, but many legally owned their homes and land. Most were offered some compensation, but those who refused to move have had to battle with riot police and bulldozers to stay in their homes. For those moved to the urban periphery and thus facing lengthy commutes, rapidly increasing public transportation costs are an additional burden especially for Rio's poor, who regularly spend between 15 and 30 percent of their income on transportation.

The scale of these mega events also provided many opportunities for Brazil's notoriously corrupt public officials to steal public funds. In 2015, investigations into corruption at Brazil's state-owned oil company, Petrobras, uncovered millions of dollars in bribes by major Brazilian construction companies to government officials in Rio whose job it was to award construction contracts for the Olympics. Anger at corruption fueled many protests of the overall effect of the mega events, which many see as "sanitizing" Rio into a playground for tourists and the rich.

THINGS TO REMEMBER

• Brazil, about the size of the United States and home to the Amazon Basin, has suffered rather severe environmental degradation due to large-scale logging, industrial farms, and mining.

• Brazil has a highly urbanized population (more than 85 percent of the people live in urban areas). Many live in difficult circumstances in favelas.

GEOGRAPHIC THEMES: Middle and South America Review and Self-Test

1. Environment: Deforestation in this region contributes significantly to global climate change. In addition, some areas are experiencing a water crisis related to climate change, inadequate water infrastructure, and the intensified use of water, despite the region's overall abundant water resources.

• What factors are driving the cycle of deforestation in this region?

• How are the processes of global warming and deforestation linked?

• What are some possibly sustainable alternatives to deforestation?

• Why are parts of this region experiencing a water crisis?

2. Globalization and Development: This region's integration into the global economy has left it with the widest gap between rich and poor in the world. Poverty, economic instability, and the flow of resources and money out of the region have resulted in many conflicts and inspired numerous efforts at reform. Recently, though, several countries in this region have emerged as global economic leaders.

• What are some factors that are hampering the equitable distribution of wealth?

• Why did import substitution largely fail as a strategy to develop manufacturing and service-based industries?

• Why has there been so much dissatisfaction with SAPs?

• How have regional trading blocs influenced livelihoods in this region?

• What changes are occurring in this region with regard to food production?

3. Power and Politics: The region was ruled for decades by elites and the military, and was periodically subject to disruptive foreign military interventions. Almost all countries now have multiparty political systems and elected governments, and political freedoms are expanding. The international illegal drug trade continues to be a source of violence and corruption in the region.

• What kinds of challenges do elected governments in this region face?

• How has the drug trade influenced the political process in this region?

• How have foreign powers impacted politics in this region?

• What factors have shaped the two rural political movements discussed in this chapter?

4. Urbanization: Since the 1950s, cities have grown rapidly in this region as rural people have migrated to cities and towns. A lack of urban planning has created densely occupied urban landscapes that often lack adequate support services and infrastructure.

• How have primate cities impacted countries in this region?

• What kinds of problems do this region's slums, shantytowns, *colonias,* barrios, and *barriadas* have?

• Why is the Brazilian city of Curitiba unique in this region?

• Why might a rural woman decide to migrate to a city?

5. Population and Gender: During the early twentieth century, the combination of cultural and economic factors and improvements in health care created a population explosion. By the late twentieth century, improved living conditions; better access to education and medical care; urbanization; and changing gender roles were all working together to reduce population growth.

• What accounts for the population explosion of the last century?

• How has urbanization resulted in smaller families and more options for women?

• How does the opportunity to become educated affect a woman's fertility?

• Why are populations likely to grow in this region in the near future?

Critical Thinking Questions

1. If the European colonists had come to Middle and South America in a different frame of mind—if, say, they were simply looking for a new place to settle and live quietly—how do you think the human and physical geography of the region would be different today?

2. Explain how tectonic processes have shaped landforms in Middle and South America.

3. Reflecting on the whole chapter, pick some locations that impressed you with regard to the ways in which people are dealing with either environmental issues or with issues of income and wealth disparity or human well-being. Explain your selections.

4. Discuss the ways in which you see the historical circumstances of colonization affecting modern approaches to economic problems in Mexico, Bolivia, Brazil, Venezuela, or Cuba.

5. Describe the main patterns of migration in this region and discuss the effects of migration on both the sending and receiving societies.

6. Name three factors that could increase political freedoms in this region.

7. Explain how the Amazon Basin and its resources constitute an example of contested space.

8. Argue for or against the proposition that free trade blocs such as NAFTA and UNASUR (previously Mercosur) help the lowest-paid workers have some upward mobility.

9. How would you respond to someone who suggested that Middle and South America were helped toward development and modernization by the experience of European colonization?

Chapter Key Terms

acculturation 169
Altiplano 184
assimilation 169
Aztecs 145
biodiversity 135
brain drain 161
contested space 160
coup d'état 157
Creoles 148
dictator 157
ecotourism 141
El Niño 136
evangelical Protestantism 171
export processing zones (EPZs) 152
extended family 170
favelas 163

forward capital 191
hacienda 150
import substitution industrialization (ISI) 150
Incas 145
income disparity 149
indigenous 134
isthmus 134
ladino 178
liberation theology 171
machismo 170
maquiladoras 152
marianismo 170
marketization 152
mercantilism 149
mestizos 148

Middle America 135
nationalize 152
North American Free Trade Agreement (NAFTA) 153
plantation 150
primate city 161
privatization 152
shifting cultivation 144
silt 136
South America 135
structural adjustment programs (SAPs) 151
subduction zone 135
temperature-altitude zones 136
trade winds 136
UNASUR 154
urban growth poles 164

4

Europe

A Saar River, Saarland, Germany

B Rhine River, Basel, Switzerland

C North European Plain, Schleswig-Holstein, Germany

D Alps, Kühtai, Austria

E Danube River, Budapest, Hungary

FIGURE 4.1 Regional map of Europe.

GEOGRAPHIC THEMES

After you read this chapter, you will be able to discuss the following issues as they relate to the five thematic concepts:

1. Environment: Europe is a world leader in responding to climate change. The European Union, a group of 28 countries within Europe, has set goals for cutting greenhouse gas emissions that are complemented by many other strategies for saving energy and resources. Europe's stringent climate amelioration efforts are in response to the fact that much of Europe's air and many of its seas and streams are quite polluted, and that through its high level of consumption, it negatively impacts environments across the globe.

2. Globalization and Development: Following World War II, Europe took steps to create an economic union now known as the European Union, which features open borders, the free movement of people and goods, progress toward a common currency, and efforts to promote human well-being and social cohesion. To improve its global competitiveness, the EU shifted industries from high-wage western Europe to the lower-wage member states of Central and South Europe. Service jobs replaced many industrial jobs, but unemployment increased. The European Union struggles to integrate the economies of its member states and to reduce regional disparities.

3. Power and Politics: By the 1960s Europe was shedding its colonial dependencies, participating with the United States as a leader in global affairs, and implementing economic unification. After the fall of the Soviet Union in 1991, the European Union expanded into Central Europe, emphasizing political freedoms and economic reform. Migration to West and North Europe from Central and South Europe and from former colonies as well as from several conflict areas increased steadily. By 2015 the intractable Syrian revolution and social unrest elsewhere in Asia and Africa instigated the massive migration of asylum seekers and economic migrants into Europe. Islamic jihadism resulted in several terrorist attacks, including two in Paris, France, which garnered international attention.

4. Urbanization: Europe's cities are both ancient and modern, with old town centers now surrounded by modern high-rise suburbs and supported by world-class urban infrastructures. Cities are the heart of Europe's economy, politics, and culture, and more than 70 percent of the EU population lives in urban areas. Yet much of Europe remains idyllic and productive rural countryside, important symbolically, economically, and as a place for regular recreation.

5. Population and Gender: Europe's population is aging as fertility rates decline. Fertility rates are down due to the high cost of raising a child and because career-oriented women are choosing to have only one or two children. Fearing that aging populations will slow economic growth, European countries try to boost their fertility rates with a variety of policies—the most successful of which address the needs of working mothers. Encouraging immigration may be one way to boost population numbers, but eventually migrants may also opt for smaller families. However, accommodating and assimilating migrants, either refugees or those only seeking jobs, has proved a difficult social issue.

The European Region

Over the last 500 years, the region of Europe (shown in **FIGURE 4.1A–E**) has had a central role in world history and geography as a colonizer of distant territories and as a major player in the development of capitalism. In the twentieth century, Europe brought the world to the edge of disaster with two massive wars. Since then it has revived, prospered, and designed a model for regional economic and political integration (the **European Union,** or **EU**) that has been surprisingly successful and has been copied in other world regions. But unfortunately the EU now faces several grave challenges. First, there are worries that the union might dissolve in the face of growing financial and social disparities. Then in June 2016 came the shocking vote in the United Kingdom to leave the EU entirely. Popularly known as BREXIT, the exit of the UK is expected to take years because many unforeseen adjustments must be made. Added to this is the unexpected flood, starting in 2015, of more than 1 million immigrants—many of them refugees from failing countries in North Africa and Southwest Asia, Sub-Saharan Africa, and South Asia. This influx now threatens to swamp the EU's emerging efforts at multicultural democracy and region-wide freedom of movement. The various terrorist episodes in 2015 and 2016 linked to Muslim countries, from which most of the new migrants came, threatens to poison the possibilities for amicable resettlement and jettison the dream of a Europe with no internal borders.

Because the BREXIT is of undetermined dimensions and will take years to negotiate, in this 7th edition the UK will continue to be counted as a member of the European Union.

The five Geographic Themes used in this book, listed and explained above, are explored as they arise in the discussion of issues in this region. Scattered throughout the chapter, vignettes illustrate one or more of the themes as they are experienced in individual lives.

European Union a supranational organization that unites most of the countries of West, South, North, and Central Europe

VIGNETTE Marina's sewing machine hums to a stop as she reaches for her ringing cell phone. It's a call from Catholic Charities, where she works as a volunteer every Monday. There is an emergency! The panicked voice on the phone recounts how thousands of bedraggled refugees from wars in Syria and elsewhere have reached the southern border of her small Central European country of Slovenia (**FIGURE 4.2**). They are on foot, walking in a steady stream, carrying children, the elderly, and a few mementos from home. After a rough crossing in wooden boats and rubber rafts from Turkey to Greece, most have been on the road for weeks, walking north through the early winter rain and cold of the Balkans. The situation is heartbreaking and chaotic and Marina is needed to help sort and distribute donated clothing, blankets, and food.

As Marina prepares to leave she is struck by the thought that after years of steady but slow improvements in her way of life since Slovenia declared its independence from Communist Yugoslavia in 1991, she may be suddenly faced with disruptive changes. Who are these outsiders, desperate to escape war and find a new life in Europe? Will they try to stay in Slovenia? She is sympathetic and wanting to help, but she is also frightened. She

FIGURE 4.2

(A) A stream of refugees walking north from Croatia through the lush green countryside of Slovenia. Most had walked from Greece through the Balkans, carrying children, the infirm, and belongings through rain and mud, often sleeping unprotected in heavy weather.

(B) Discarded items left along the route of the walking refugees. A sympathetic Slovene public brought food, clothing, tents, and even baby carriages to the exhausted refugees as they walked through the countryside. These gifts were useful for a time, but tired walkers could hardly carry all this with them. Dirty clothes and worn-out shoes were discarded; but much more had to be jettisoned in preparation for the long unknown journey ahead.

Thinking Geographically

(A) What do you think would be the best way to help exhausted walking refugees? (B) Why might refugees discard gifts of food, camping equipment, and other goods along the roads they traversed?

knows about war. Her parents and grandparents were full of tales of hunger and hardship during World War II. She can still see the machine gun pockmarks on many of the older buildings in her town. Now, she is frightened that the new people will bring back that chaos. She hears that most are Muslims, about whom she knows little; and there are so many of them. . . . This vignette continues on page 234. ∎

There are now 28 countries in the European Union, and a few more hope to join in the next several years. But in June 2016, the citizens of the United Kingdom (England, Wales, Scotland and Northern Ireland) unexpectedly voted to leave the EU. Popularly known as BREXIT, this decision has far-reaching, but as yet poorly understood, implications for the whole of Europe.

The newest members of the EU (with the exceptions of Malta and Cyprus) are all formerly communist countries of Central Europe: the Czech Republic, Estonia, Hungary, Latvia, Lithuania, Poland, Slovakia, Slovenia (all in 2004), and Croatia (in 2013). They generally have lower standards of living than older EU members. During the transition into democratic societies and capitalist economies, the emphasis on efficiency eliminated jobs in archaic industries and agriculture, forcing hundreds of thousands of workers from Central Europe to look in other EU states for jobs. These jobs are often precarious because the entire EU is adjusting to the globalizing economy.

Since 2008, the European Union has been under financial stress, as a number of the less industrialized and less prosperous countries (Greece, for example) have found it so hard to make loan payments that they face the possibility of national bankruptcy. Residents of the original and more prosperous EU members—such as France and especially Germany—resent having to support these failing economies and also fear that the migration of workers from the EU's south and east is bringing very different people into close contact with each other, resulting in political tensions and higher costs for tax-supported social and educational services. But in mid-2015, worries about the negative effect of migrants from within the EU were eclipsed by worries about migrants from outside Europe (principally from Syria), who are thought to have non-European values and traditions. Economic planners and employers across Europe who previously argued that allowing workers to migrate is crucial to Europe's economic growth and global competitiveness began to have grave reservations about the scale and character of the migrations beginning in 2015. The migrants themselves, whether from inside or outside Europe, fear that forced **cultural homogenization**—the tendency toward uniformity—awaits them.

What Makes Europe a Region?

Physically, Europe is a peninsula extending off the western end of the huge Eurasian continent. There are many peninsular appendages, large and small, on this giant peninsula of Europe. Norway and Sweden share one of the larger appendages. Other large peninsulas

cultural homogenization the tendency toward uniformity of ideas, values, technologies, and institutions among associated culture groups; the loss of ethnic distinctiveness

are the Iberian Peninsula (occupied by Portugal and Spain) and those of Italy and Greece. All extend into either the Atlantic Ocean or into the various seas that are adjacent to the Atlantic (see the Figure 4.1 map).

The perimeter of the region of Europe to the north, west, and south is primarily oceanic coastline. This unique access to the world ocean facilitated European exploration and colonization of distant territories. The eastern physical and cultural border of Europe has been difficult to define. Just where Europe ends and Asia begins has been a source of contention for millennia. In this book, at this time, the eastern limit of the European region is taken to be the eastern border of the European Union. Other potential parts of Europe—Ukraine, Belarus, western Russia, Moldova, and the Caucasus—are covered in Chapter 5, and Turkey in Chapter 6.

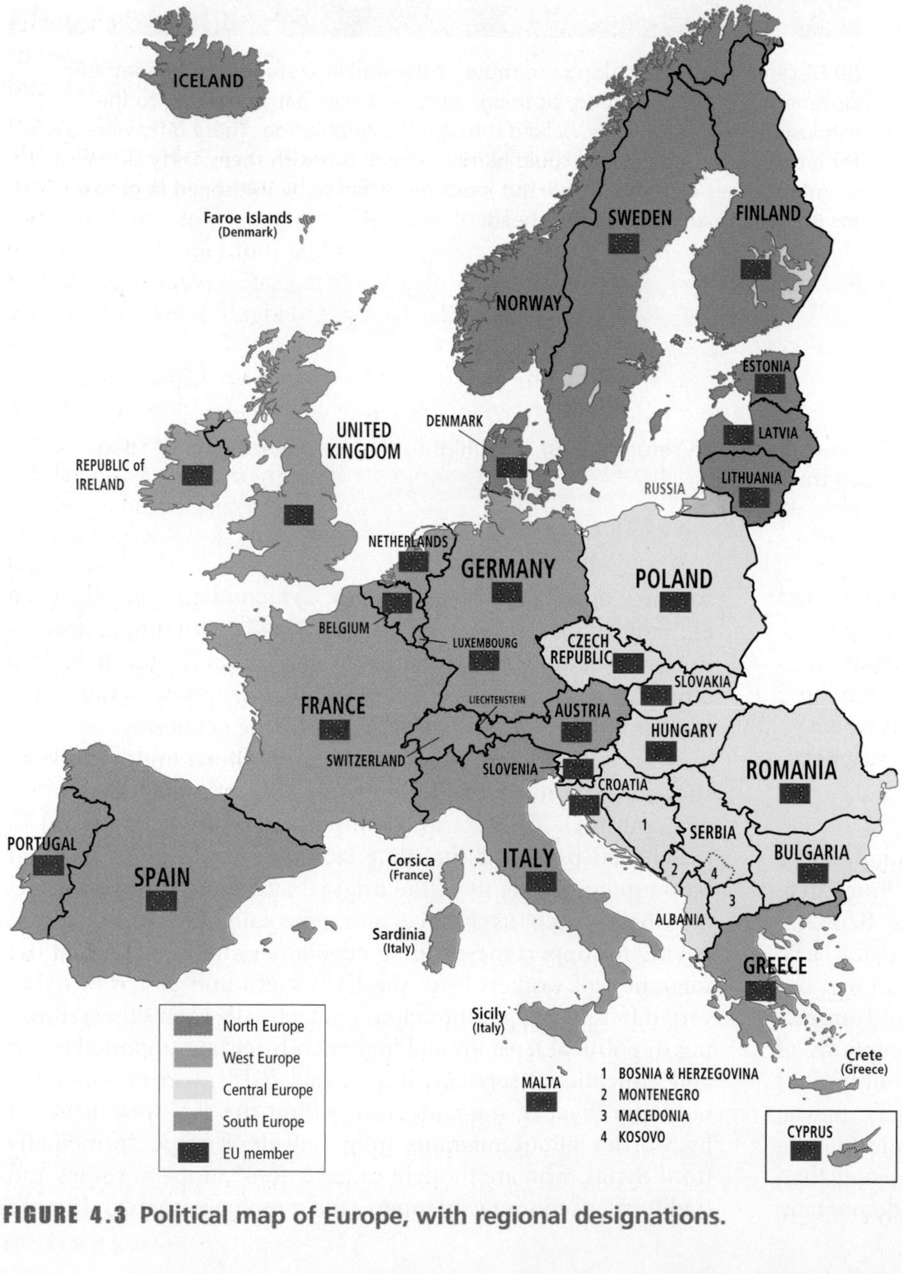

Terms in This Chapter

This book divides Europe into four subregions—*North*, *West*, *South*, and *Central Europe* **(FIGURE 4.3)**. *Central Europe* refers to all those countries formerly in the Soviet sphere (USSR) that are now in the European Union, as well as the countries that were formerly in Yugoslavia, plus Albania.

For convenience, we occasionally use the term *western Europe* to refer to all the countries that were *not* part of the experiment with communism in the Soviet sphere and in Yugoslavia. That is, *western Europe* comprises the combined subregions of North Europe (except Estonia, Latvia, and Lithuania), West Europe (except the former East Germany), and South Europe. When we refer to the countries that were part of the Soviet sphere up to 1989, we use the pre-1989 label *eastern Europe*. *Central Europe* is now the preferred term for what in the Soviet era was called *Eastern Europe*. EU-28 is the term for just the 28 member countries of the European Union. When we refer to the group of countries collectively known to some as the *Balkans* (Albania, Bosnia and Herzegovina, Bulgaria, Croatia, Macedonia, Montenegro, Romania, Serbia, and Slovenia), we prefer the term *southeastern Europe*.

PHYSICAL GEOGRAPHY

Europe's physical geography is shaped by its many peninsulas that reach into surrounding oceans and seas, and by its variable landforms, all of which affect the region's climates and vegetation.

LANDFORMS

Although European landforms are fairly complex, the basic pattern is mountains, uplands, and lowlands, all stretching roughly west to east in wide bands. As you can see in Figure 4.1 (in the

FIGURE 4.3 Political map of Europe, with regional designations.

map and in photo D), Europe's largest and highest mountain chain, the *Alps*, runs west to east through the middle of the continent, from southern France through Switzerland and Austria. The alpine mountain formation extends into the Czech Republic and Slovakia, and curves southeast as the Carpathian Mountains into Romania. This network of mountains is mainly the result of pressure from the collision of the northward-moving African Plate with the southeasterly moving Eurasian Plate (see Figure 1.8). Europe lies on the westernmost part of the Eurasian Plate. South of the main Alps formation, lower mountains extend into the peninsulas of Iberia and Italy and along the Adriatic Sea through Greece to the southeast. The northernmost mountainous formation is shared by Scotland, Norway, and Sweden. These northern mountains are old (about the age of the Appalachians in North America) and have been worn down by glaciers and millions of years of erosion.

Extending northward from the central Alpine zone is a band of low-lying hills and plateaus curving from Dijon (France) through Frankfurt (Germany) to Krakow (Poland). These uplands (see Figure 4.1A) form a transitional zone between the high mountains and lowlands of the *North European Plain*, the most extensive landform in Europe (see Figure 4.1C). The plain begins along the Atlantic coast in western France and covers a wide band around the northern flank of the main European peninsula, reaching across the English Channel and the North Sea to take in southern England, southern Sweden, and most of Finland. The plain continues east through Poland, then broadens to the south and north to include all the land east to the Ural Mountains in Russia.

Crossed by many rivers and holding considerable mineral deposits, the coastal lowland of the North European Plain is an area of large industrial cities and densely occupied rural areas. Over the past thousand years, people have transformed the natural seaside marshes and vast river deltas into farmland, pastures, and urban areas by building dikes and draining the land with wind-powered pumps. This is especially true in the low-lying Netherlands, where concern over climate change and sea level rise is considerable.

The rivers of Europe link its interior to the surrounding seas. Several of these rivers are navigable well into the upland zone, and Europeans have built industrial cities on their banks. The Rhine carries more traffic than any other European river, and the course it has cut through the Alps and uplands to the North Sea also serves as a route for railways and motorways (see Figure 4.1B). The area where the Rhine flows into the North Sea is considered the economic core of Europe. Rotterdam, Europe's largest port, is located here. The Danube River, larger and much longer than the Rhine, flows southeast from Germany, connecting the center of Europe with the Black Sea. As the European Union expands to the east, the economic and environmental roles of the Danube River basin, including the Black Sea, are getting more attention (see Figure 4.1E).

VEGETATION AND CLIMATE

Nearly all of Europe's original forests are gone, some for more than a thousand years. Today, forests with very large and old trees exist only in scattered areas, especially on the more rugged mountain slopes (see Figure 4.1D, A) and in the northernmost parts of Norway, Sweden, and Finland. The dominant vegetation in Europe is crops and pasture grass. Forests are intensively managed and are regenerating on abandoned farmland where agriculture is no longer competitive, so now regrowth forests cover about one-third of Europe.

Europe has three main climate types: temperate midlatitude, Mediterranean, and continental **(FIGURE 4.4)**. The **midlatitude temperate climate,** characterized by year-round moisture, relatively mild winters, and long, mild-to-hot summers, dominates in western Europe, where the Atlantic Ocean moderates temperatures (see Figure 4.4A). To minimize the effects of heavy precipitation runoff, people in these areas have developed elaborate drainage systems for their houses and communities. Forests are both evergreen and deciduous.

A broad, warm-water ocean current called the **North Atlantic Drift** brings large amounts of warm water to the coasts of Europe. It is really just the easternmost end of the Gulf Stream, which carries warm waters from the Gulf of Mexico north along the eastern coast of North America, then curves across the North Atlantic to Europe (see Figure 2.4). The air above the North Atlantic Drift is relatively warm and wet, and the eastward-blowing winds that push this air over North and West Europe and the North European Plain bring moderate temperatures and rain deep into the Eurasian continent, creating a climate that, although still fairly cool, is much warmer than elsewhere in the world at similar latitudes. There is some concern that global climate change could eventually weaken the North Atlantic Drift, leading to a significantly cooler, and perhaps drier, Europe.

Farther to the south, the **Mediterranean climate** prevails— warm, dry summers and mild, rainy winters (see Figure 4.4B). In the summer, warm, dry air from North Africa shifts north over the Mediterranean Sea as far as the Alps, bringing high temperatures and clear skies. Crops grown in this climate, such as olives, grapes, citrus, wheat, apples, and other fruits, must be drought-resistant or irrigated. In the fall, this warm, dry air shifts to the south and is replaced by cooler temperatures and rainstorms sweeping in off the Atlantic. Overall, the climate here is mild, and houses along the Mediterranean coast are often open and airy to afford comfort in the hot, sunny summers. During the short, mild winters, life moves indoors, where wood fires are used to heat just one or two small rooms even in large houses.

In Central Europe and the far northern Scandinavian peninsulas, all of which are situated away from the moderating influences of the North Atlantic Drift and the Mediterranean Sea, the climate is more extreme. In this region of **continental climate,** summers are fairly hot and the winters become longer and colder the farther north or deeper into the interior of the continent one goes (see Figure 4.4C). Here, houses tend to be well

midlatitude temperate climate as in south-central North America, China, and Europe, a climate that is moist all year, with relatively mild winters and long, mild-to-hot summers

North Atlantic Drift the easternmost end of the Gulf Stream, a broad, warm-water current that brings large amounts of warm water to the coasts of Europe

Mediterranean climate a climate pattern of warm, dry summers and mild, rainy winters

continental climate a midlatitude climate pattern in which summers are fairly hot and moist and winters become longer and colder the deeper into the interior of the continent one goes

FIGURE 4.4 PHOTO ESSAY: Climates of Europe

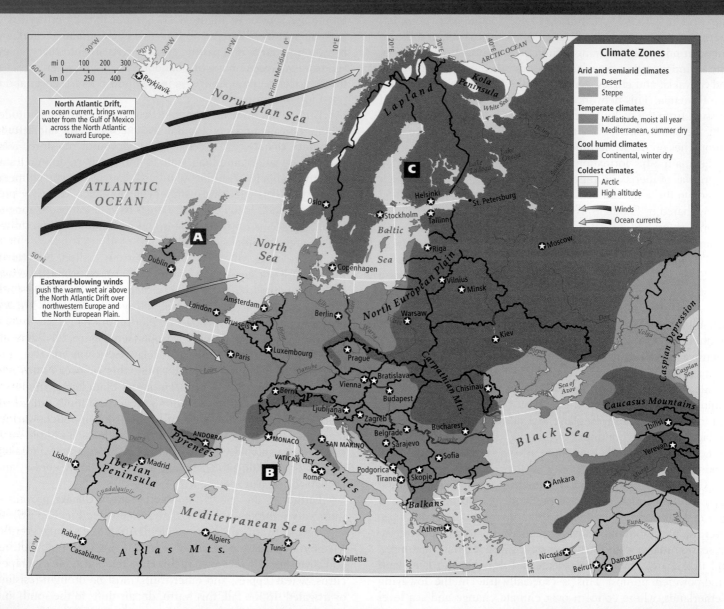

North Atlantic Drift, an ocean current, brings warm water from the Gulf of Mexico across the North Atlantic toward Europe.

Eastward-blowing winds push the warm, wet air above the North Atlantic Drift over northwestern Europe and the North European Plain.

Climate Zones

Arid and semiarid climates
- Desert
- Steppe

Temperate climates
- Midlatitude, moist all year
- Mediterranean, summer dry

Cool humid climates
- Continental, winter dry

Coldest climates
- Arctic
- High altitude

→ Winds
→ Ocean currents

A Midlatitude, moist all year, Scotland

David Henderson/Getty Images

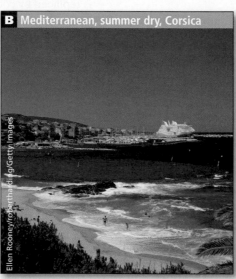

B Mediterranean, summer dry, Corsica

Ellen Rooney/robertharding/Getty Images

C Continental, winter dry, Finland

Raimund Linke/Getty Images

insulated, with small windows, low ceilings, and steep roofs that can shed snow. Crops must be adapted for much shorter growing seasons, and include corn and other grains plus fruit trees and a wide variety of vegetables adapted to the cold, especially root crops and cabbages.

THINGS TO REMEMBER

- Europe is a region of peninsulas upon peninsulas.

- Europe has three main landforms: mountain chains, uplands, and the vast North European Plain.

- Europe has three principal climates: the temperate midlatitude climate, the Mediterranean climate, and the continental climate.

ENVIRONMENT

GEOGRAPHIC THEME 1

Environment: Europe is a world leader in responding to climate change. The European Union, a group of 28 countries within Europe, has set goals for cutting greenhouse gas emissions that are complemented by many other strategies for saving energy and resources. Europe's stringent climate amelioration efforts are in response to the fact that much of Europe's air and many of its seas and streams are quite polluted, and that through its high level of consumption, it negatively impacts environments across the globe.

Europeans are aware of just how dramatically they have transformed their environments over the past several thousand years and are now taking action on many fronts. Nevertheless, the European Union still has a long way to go to meet its goals regarding clean air and water; sustainable development in agriculture, industry, and energy use; and maintaining its biodiversity.

EUROPEAN LEADERSHIP IN RESPONSE TO GLOBAL CLIMATE CHANGE

Europe leads the world in responsiveness to global climate change; at the 2015 Paris climate change conference, EU governments agreed to cut greenhouse gas (GHG) emissions by 40 percent (from a 1990 baseline) by 2030. Although its response can be described as moderate, Europe has been more willing than any other region to address climate change, largely because its governments and corporations see economic advantages to doing so. Recent research suggests that investments in energy conservation, alternative energy, and other measures would cost EU economies 1 percent of their GDP. By contrast, doing nothing about climate change could *shrink* GDP by 20 percent.

Europe's concern about global warming may also be influenced by public alarm at recent extreme weather. The summers of 2003 and 2012 broke high-temperature records across Europe. Crops failed, freshwater levels sank, forests burned, and deaths soared. In 2003, three thousand people died in France alone. On the other hand, in 2006, 2012, and 2013, rainfall and snowfall in Central Europe reached record levels. In 2013, the rivers of Central Europe—the Elbe, the Danube, and the Morava—flooded for weeks. Vacillations in weather that exceed previous records continue: the summer of 2015 brought the most severe drought to central Germany since 2003.

Europe's Vulnerability to Climate Change

There is considerable variation in vulnerability to climate change across Europe. Europe's wealth, technological sophistication, and well-developed emergency response systems make it more resilient to the consequences of climate change than are most regions of the world. Still, some areas are much more vulnerable than others because of their location, dwindling water resources, and rising sea levels, or because of the effects of poverty. **FIGURE 4.5** illustrates some of these vulnerabilities. The European Environment Agency (2015), has created a map that makes it easy to see that climate change is likely to have far-reaching and contradictory effects in different parts of the region. This map, *Key observed and projected climate change and impacts for the main regions in Europe,* can be accessed at http://www.eea.europa.eu/data-and-maps/Figures/key-past-and-projected-impacts-and-effects-on-sectors-for-the-main-biogeographic-regions-of-europe-4.

The Mediterranean is likely to have the most consequential changes, including the risk of desertification, loss of biodiversity, water scarcity, more subtropical diseases, and an impact on tourism. Central and eastern Europe will face decreasing precipitation and risk of forest fires, while northwestern and northern Europe will experience increasing precipitation and flooding, rising seas, and possibly better crop yields.

Europe's Green Behavior

By global standards, Europeans use large amounts of resources and contribute about one-quarter of the world's GHG emissions. However, one European resident averages only one-half the energy consumption of the average North American resident. This is possible because Europeans live in smaller dwellings that need less energy to heat or cool. They drive smaller, more fuel-efficient cars and have easy access to public transportation, which is widely used. Because communities are more spatially compact, many people walk or bicycle wherever they need to go.

These energy-saving practices are related in part to the population densities and social customs of the region, and also to widespread explicit popular support for ecological principles. **Green** political parties influence national policies in all European countries, and Green policies are central to the agenda of the European Union. Results include strong regional advocacy for emission controls, well-entrenched community recycling programs, and grassroots work on local environmental concerns.

Green an adjective indicating a person or group that is environmentally conscious

ON THE BRIGHT SIDE: Guerrilla Gardeners

Under cover of darkness, gardeners—who in the daytime are bureaucrats, stock traders, and computer programmers—sneak

FIGURE 4.5 PHOTO ESSAY: Vulnerability to Climate Change in Europe

Despite Europe's wealth, technological sophistication, and well-developed emergency response systems, parts of the region are vulnerable to the effects of climate change. Many locations are exposed to drought, flooding, or sea level rise, and in some of these areas, people are too poor to afford appropriate precautions.

A As the water level in a reservoir in Catalonia, Spain, sinks, a once-submerged eleventh-century church reemerges. Rising temperatures will make Spain's climate drier, its evaporation rate increase, and its scarce water resources shrink further, threatening agriculture and drinking water, especially along the Mediterranean coast.

B Part of the massive Delta Works that protect the Netherlands from rising sea levels during storms. A response to massive flooding in 1953, the Delta Works significantly reduces the Netherlands' sensitivity to sea level rise, making the Netherlands only slightly more vulnerable to climate change than the rest of Europe, even though 60 percent of its population lives below sea level.

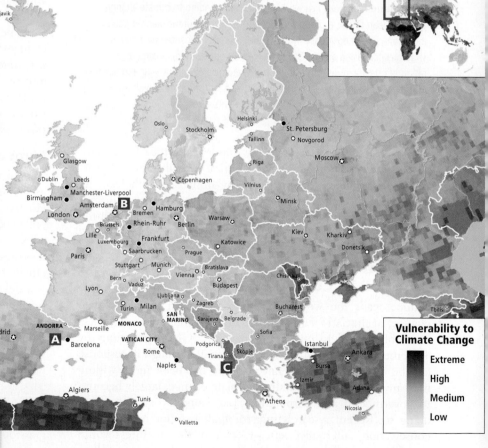

Vulnerability to Climate Change

- Extreme
- High
- Medium
- Low

C Albania's vulnerability to climate change is related to its low incomes and inadequate infrastructure of many kinds, including transportation. Albania's sensitivity to many potential disturbances is thus high and its resilience low. Many problems stem from misguided government policies. For example, until 1992 it was illegal for Albanians to own private automobiles. Even today, many poorer Albanians depend on horses and small, slow, horse-drawn carts as their main mode of transportation.

Thinking Geographically

After you have read about vulnerability to climate change in Europe, you will be able to answer the following questions:

A What clues in this photo, other than the caption, indicate that the reservoir level is very low?

B These structures are the modern equivalent of the windmills that remain a symbol of the Netherlands. What function did they both perform?

C How might owning a car decrease a family's sensitivity to flooding?

into Central London to plant colorful flowers and foliage in traffic islands and roundabouts **(FIGURE 4.6)**. These Green activists are part of a movement called *Guerrilla Gardeners*, which has thousands of activists in more than 30 countries, and a website, http://guerrillagardening.org/.

Bypassing the town councils (which tend to impose crippling rules), the Guerrilla Gardeners make quick assaults late at night, armed with trowels, spades, mulch, and watering cans. Authorities seem unable to stop the guerrillas from covering neglected urban land with blooming hyacinths, tulips, marigolds, shrubs, and even trees. Sneakier yet are the Seed Bombers, a group that packs flower seeds, soil, and water into compact parcels and tosses them into derelict patches of public land, where they shatter on impact, spewing forth seeds that produce plants capable of outcompeting the weeds. *[Source: NPR. For detailed source information, see Text Sources and Credits; see also Philip Booth's "Seed Bombers in Stroud," at http://www.youtube.com/watch?v=vY02FKd1Uco.]* ■

Changes in Transportation

Europe has an extraordinarily dense and advanced transportation network. Although Europeans have for many years favored fast rail networks for both passengers and cargo (see Figure 4.23D) rather than private cars, trucks, and multilane highways, they have

FIGURE 4.6 Guerilla gardeners.

recently been drifting closer to the American model of private cars and trucks driven on sweeping freeways. This change is now happening in even the poorest parts of Europe, where cars have been scarce in the past (see Figure 4.5C). However, in response to rising fuel costs and CO_2 emissions, the European Union has developed long-term plans that reduce the emphasis on cars and trucks and involve designing *multimodal transport* to link high-speed rail to road, air, and water transportation **(FIGURE 4.7)**. This is because highway transport, while convenient, is inefficient and polluting relative to rail or water transport. One gallon (3.7 liters) of gasoline can move one ton of cargo a distance of 59 miles (94 kilometers) by truck, 202 miles (323 kilometers) by rail, and 514 miles (822 kilometers) by waterway. Private cars used for passenger transportation produce three times more CO_2 emissions than does rail-based public transportation.

Europe's long, irregular coastline and the low cost of water transportation have been a boon for the development of links to global trade. Europe has numerous modern ocean ports that cater to container ships: Helsinki, Riga, Hamburg, Copenhagen, Antwerp, Rotterdam, Plymouth, Southampton, Le Havre, Marseille, Barcelona, and Koper. The EU transportation plan, expected to be completed in 2020 (see Figure 4.7), includes "Motorways of the Sea" (upgraded shipping lanes) through the Baltic, the North Sea, the English Channel, and along the eastern Atlantic. In the Mediterranean, shipping lanes are being improved from the western Mediterranean to the northern Adriatic, and several ports across North Africa are being added. All of these changes will improve trade not only for European markets, but also for markets of Morocco, Algeria, Tunisia, Malta, Libya, Egypt, Israel, Palestine, Lebanon, and Turkey. One-third of the world's container traffic now goes through the Mediterranean. The container ship industry is keenly attuned to the concerns about CO_2 and climate change, and is now using optimal (often slower) speeds that are carefully calculated to minimize fuel consumption and emissions while maximizing profits.

EUROPE'S IMPACT ON THE BIOSPHERE

There is a geographic pattern to the ways in which human activities have transformed Europe's landscapes over time **(FIGURE 4.8)**. Western Europe shows the effects of dense population and heavy mining and industrialization (see Figure 4.8A). Even innovative new energy sources like wind power can have negative environmental effects (see Figure 4.8B). Central Europe still shows the old disregard for the environment under communism and central planning and continues to have a major impact on the biosphere through the air and water pollution it generates. In South Europe, successful export agriculture produces health problems for workers, water shortages, and visual pollution of the landscape (see Figure 4.8C).

ON THE BRIGHT SIDE: Energy Solutions

The European Union wants to increase its use of renewable energy in order to reduce fuel imports and thereby strengthen its energy security, stimulate the economy with new energy-related jobs, and combat climate change. The EU's goals are to reduce its CO_2 emissions by 40 percent by 2030, and by 2020, increase use of renewable energy by 20 percent and power 60 percent of EU homes with renewably generated electricity. A wide array of alternative energy projects is now attracting significant investment. Wind power is generally the favored technology: Europe had 34 percent of the world's installed wind capacity in 2015, enough to supply about 13 percent of the EU's electricity. The use of solar power, while currently supplying only 4 percent of the EU's electricity, is surging, in part due to rapidly declining prices for Chinese-made solar panels. Europe has about 75 percent of the world's installed solar power capacity. In Germany, more than 70 percent of the country's electricity demand can be met by

Thinking Geographically

After you have read about the human impacts on the biosphere in Europe, you will be able to answer the following questions:

A How might this large, open pit threaten groundwater resources?

B Why might Danish people object to wind farms on beaches, fields, and meadows?

C How might the large areas covered by plastic contribute to a water shortage?

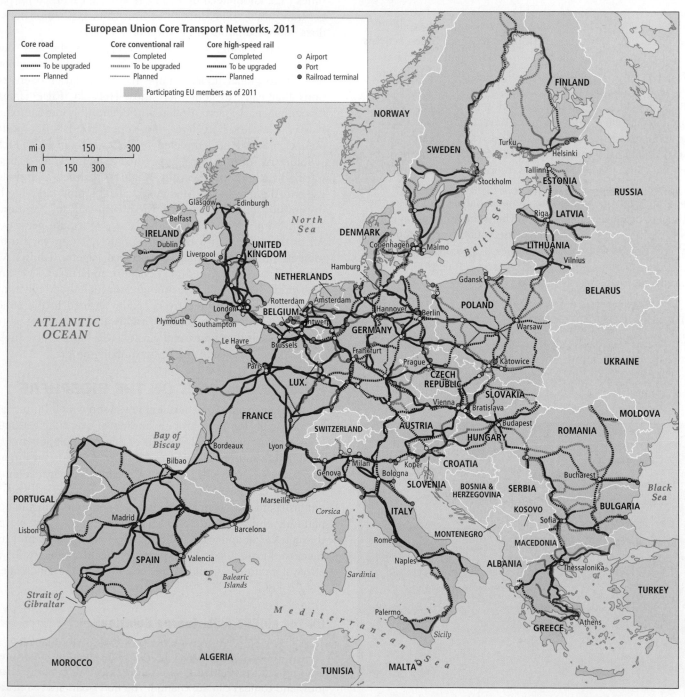

FIGURE 4.7 The EU core transport network components. [Source consulted: European Commission Mobility and Transport]

FIGURE 4.8 PHOTO ESSAY: Human Impacts on the Biosphere in Europe

Most of Europe has been transformed by human activity. Western Europe has some of the most heavily impacted landscapes and ecosystems, but new efforts to address high GHG emissions can impact landscapes as well. In South Europe, intensive, modernized agriculture is creating new problems. The map shows some of the impacts on Europe's land, sea, and air.

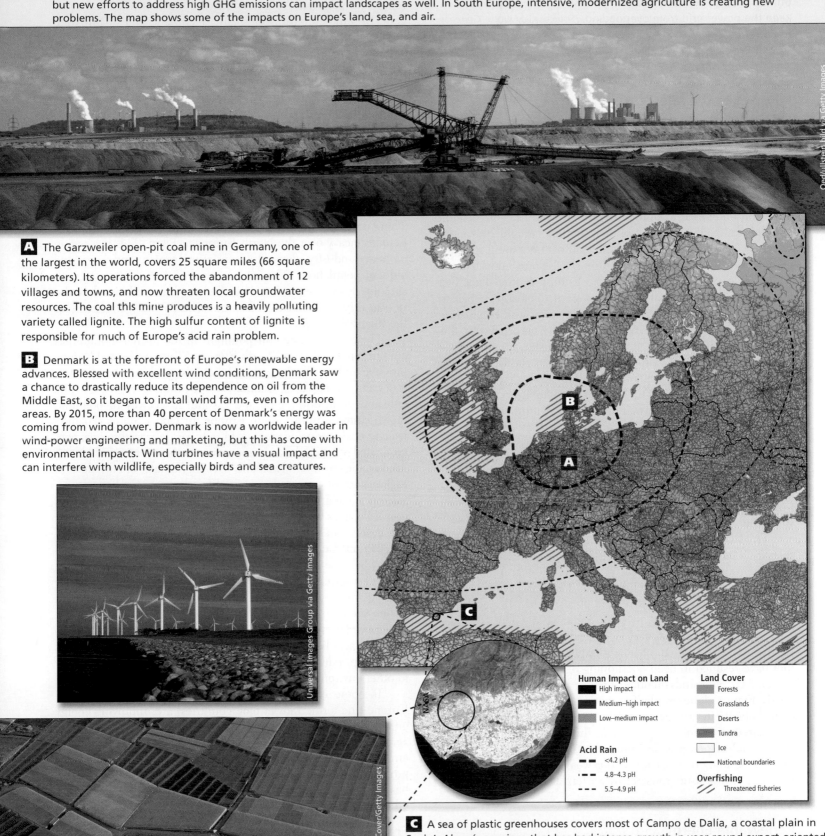

A The Garzweiler open-pit coal mine in Germany, one of the largest in the world, covers 25 square miles (66 square kilometers). Its operations forced the abandonment of 12 villages and towns, and now threaten local groundwater resources. The coal this mine produces is a heavily polluting variety called lignite. The high sulfur content of lignite is responsible for much of Europe's acid rain problem.

B Denmark is at the forefront of Europe's renewable energy advances. Blessed with excellent wind conditions, Denmark saw a chance to drastically reduce its dependence on oil from the Middle East, so it began to install wind farms, even in offshore areas. By 2015, more than 40 percent of Denmark's energy was coming from wind power. Denmark is now a worldwide leader in wind-power engineering and marketing, but this has come with environmental impacts. Wind turbines have a visual impact and can interfere with wildlife, especially birds and sea creatures.

Human Impact on Land
- High impact
- Medium–high impact
- Low–medium impact

Acid Rain
- `– –` <4.2 pH
- `˙ ˙ ˙` 4.8–4.3 pH
- `- - -` 5.5–4.9 pH

Land Cover
- Forests
- Grasslands
- Deserts
- Tundra
- Ice
- — National boundaries

Overfishing
- ⫫ Threatened fisheries

C A sea of plastic greenhouses covers most of Campo de Dalía, a coastal plain in Spain's Almería province that has had intense growth in year-round export-oriented vegetable production over the past several decades. Inside the greenhouses, pesticides and fertilizers are used so extensively that health problems are arising among the mainly Moroccan migrant workers and their families who work and live among the greenhouses. Meanwhile, water shortages are severe enough that a desalination plant is being built nearby.

solar and wind power on a sunny summer day, which means that utilities must periodically shut down fossil fuel power plants to keep the power grid from overloading. [*Source: Kiley Kroh, "Germany sets new record, generating 74 percent of power needs from renewable energy," Climate Progress, May 13, 2014, http://thinkprogress .org/climate/2014/05/13/3436923/germany-energy-records/]* ■

Europe's Energy Resources

Europe's main energy sources have shifted over the years from wood to coal and, more recently, to petroleum and natural gas and in some countries to nuclear power. However, in response to rising energy costs and commitments to cut GHG emissions, renewable energy sources are increasingly emphasized.

The 28 members of the European Union (the EU-28) get a large portion of their fossil fuel supplies from Russia—32 percent of their crude oil, 31 percent of their natural gas, and 26 percent of their coal, as of 2015. Just about half of the natural gas now comes from Russia via pipelines through Belarus and Ukraine, while Turkey supplies Russian gas to Europe via the Black Sea and hosts a pipeline that carries gas from the Caspian Sea to Europe. Russia is negotiating for another trans-Turkey pipeline to carry oil and gas to the Mediterranean. Europeans fear this dependency will be used against them by Russia, which periodically withholds flows to Europe through Belarus and Ukraine. But actually, Russia is itself dependent on the EU fuel trade because 70 percent of Russia's fossil fuel exports go to the European Union and these sales account for half of Russia's government budget.

Another 30 percent of the gas and oil that the European Union consumes comes from various Middle Eastern producers. Large oil and gas deposits in the North Sea, most controlled by Norway (not an EU member), have alleviated Europe's dependence on "foreign" sources of energy, but the production of oil from the North Sea has already peaked and is expected to run out sometime in the next decade.

The EU is increasingly promoting energy independence and security. The use of nuclear power to generate electricity has been more common in Europe than in North America; Finland, the Czech Republic, Slovakia, Hungary, and Bulgaria have Russian-built nuclear power stations. Russia also supplies 18 percent of the mined uranium used in such plants. But concerns about safety—especially after the 2011 Fukushima reactor disaster in Japan—and the problem of how to safely manage spent nuclear fuel have resulted in deep questioning of the viability of nuclear power. EU-28 depends on nuclear power for 25 percent of its total needs; France alone uses half of that total. In France, 77 percent of the electricity was generated by nuclear power during 2015 (compared with only 20 percent in the United States), but the use of nuclear power is declining even in France. Germany has stopped building new reactors and is weaning itself off of nuclear power as enthusiasm grows across western Europe for renewable energy.

Renewable Energy

Europe has the highest per capita use of renewable energy on Earth and, as mentioned, the EU-28 are on track to meet a new target: by 2020, renewable energy will account for at least 20 percent of total energy used. Wood is the largest source of renewable power, at 40 percent; hydropower is next, at 20 percent. In some countries (Germany, Denmark, and Spain) wind and solar power provide a rapidly growing share of the renewable power. Importantly, both wind and solar power are creating *green* jobs. Denmark, with its long seacoast, leads in the use of wind power and has profited greatly from wind power industries. Note that even renewable energy has environmental impacts: wind power's impact has been discussed (see Figure 4.8B); solar panels affect the ecology of the land they occupy; wood removed from forests changes habitats and reduces CO_2 absorption; and all these sources of energy produce pollution, especially during materials manufacturing stages.

Air Pollution

There is significant air pollution in much of Europe, but it is particularly heavy over the North European Plain. This is a region of heavy industry, dense transportation routes, and large and affluent populations. The intense fossil fuel use associated with such lifestyles results not only in the usual air pollution but also in *acid rain*, which can fall far from where it was generated (see the Figure 4.8 map).

There is also much air pollution in the former Communist countries in Central and North Europe. Mines in Central Europe produce highly polluting soft coal that is burned in out-of-date factories and power plants. Central Europe produces high per capita emissions from burning oil and gas and it receives air pollution blown eastward from western Europe. In Upper Silesia (Poland's leading coal-producing area), acid rain has destroyed forests, contaminated soils and the crops grown on them, and raised water pollution to deadly levels. Residents have higher rates of birth defects, higher rates of cancer, and lower life expectancies than comparable populations in the rest of Europe. Industrial pollution was one of Poland's biggest obstacles to entry into the European Union and it only barely met the EU's environmental requirements.

Central Europe's severe environmental problems developed in part because the theories and policies promoted by the Soviet Union portrayed nature as existing only to serve human needs. During the Soviet era, little pollution data was collected and public protests against pollution were prohibited. However, the shift toward more open societies after 1991 has made activism possible in places like Hungary and Bulgaria, and popular protest has resulted in reductions in air pollution as well as public attention to other environmental issues.

To some extent, the new EU member states in Central Europe (the former Soviet bloc countries) are improving energy efficiency and reducing emissions. Power plants, factories, and agriculture are polluting less, and the countries with the worst emissions records, such as Poland, have been making the most progress. Nonetheless, the emphasis on market economies also has brought many more ways to generate air pollution, along with other forms of pollution.

Freshwater and Seawater Pollution

Sources of water pollution in Europe include insufficiently treated sewage, chemicals and silt in the runoff from agricultural plots and urban areas, consumer packaging litter, petroleum residues, and industrial effluent **(FIGURE 4.9)**. Most inland waters contain a variety of such pollutants. Any pollutants that enter Europe's inland wetlands, rivers, streams, and canals eventually reach Europe's surrounding coastal waters. The Atlantic Ocean, the Arctic Ocean, and the northern reaches of the North Sea are better able to disperse chemical pollutants dumped into them because they are part of, or closely connected to, the circulating flow of the world ocean. In contrast, the Baltic, Mediterranean, and Black seas, along with the southern North Sea, are nearly landlocked bodies of water that do not have the capacity to flush themselves out quickly and are prone to accumulating chemical pollution and plastic litter.

In the Mediterranean, the effect of the pollution that pours in from rivers, adjacent cities, industries, hotel resorts, and farms is exacerbated by the fact that the sea has just one tiny opening to the world ocean (see Figure 4.1). At the surface, seawater flows in from the Atlantic through the narrow Strait of Gibraltar and moves eastward. At the bottom of the sea, water exits through the same narrow opening, but only after it has been in the Mediterranean for 80 years or more. The natural ecology of the Mediterranean is attuned to this lengthy cycle, but the nearly 460 million people now living in the countries surrounding the sea have upset the balance. Their pollution stays in the Mediterranean for decades. As a result, fish catches have declined, beloved seaside resorts have become unsafe for swimmers, and agricultural workers have become sick.

There are 34 countries with coastlines on Europe's many seas, all with different economies, politics, and cultural traditions. Such diversity makes it difficult to cooperate to minimize pollution or even reduce the risk of severe pollution. Although most of the Mediterranean's pollution is generated by Europe, rapidly growing populations and economic development on the North African and eastern Mediterranean coasts of the sea also pose environmental threats because these countries still lack adequate urban sewage treatment and environmental regulations to control agricultural and industrial wastes. Over the past decades, wars and armed conflict throughout the eastern Mediterranean (Libya, Egypt, Israel, Lebanon, Syria, and Turkey) have added to the pollution and disrupted efforts to stop long-standing polluting customs such as dumping household sewage and industrial waste into the sea and the streams that feed into it.

The Wide Reach of Europe's Environmental Impact

Europeans are global consumers: their food, clothing, building materials, and industrial products are all based on at least some resources from abroad. One way to grasp the enormity of Europe's global environmental impact is to look at virtual water. Introduced in Chapter 1, *virtual water* is the volume of water required to produce, process, and deliver a good or service that a person consumes. Table 1.1 shows the virtual water content of a variety of products. The average amount of water required to

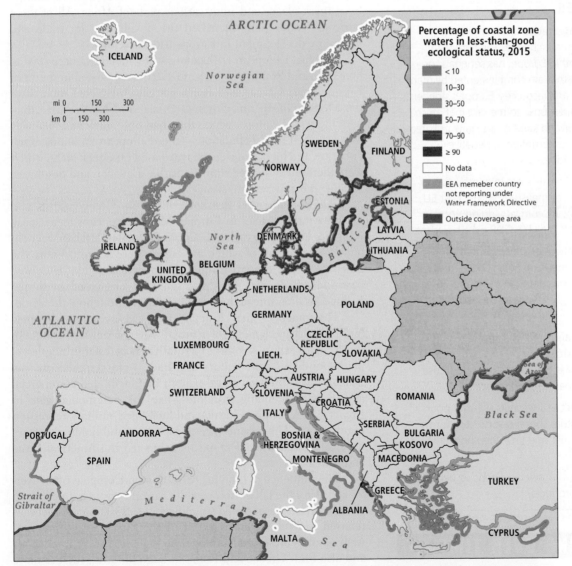

FIGURE 4.9 Percentage of coastal zone waters in less than good ecological status, 2015. The waters along Europe's coastal zones are to varying extents polluted by agricultural, industrial, and fossil fuel toxins. The nearly landlocked Baltic Sea has the worst pollution, but pollution along the edges of the North Sea is also severe. The white shading around Norway merely means that the government did not provide data (Norway is not in the EU).

meet a person's basic needs is about 5 to 13 gallons (20 to 50 liters) per day. To arrive at a person's annual total water footprint, one must add the water required for basic needs to a person's virtual water usage. Although Europeans do not import as many of their consumer goods as do Americans, they still consume one-fifth of the world's imports; many of these goods have a high *virtual water component* (the water consumed in the production process). Nearly all EU countries import more virtual water than they export, and Europe as a region imports more virtual water than any other. The countries that produce Europe's imported virtual water often have very little water themselves to begin with, and the costs of this water loss are not being adequately figured into the price of the products exported to Europe. In all fairness, then, the environmental impacts of Europe's virtual water consumption should be counted against Europe's total impact on the biosphere.

THINGS TO REMEMBER

GEOGRAPHIC THEME 1 • **Environment:** Europe is a world leader in responding to climate change. The European Union, a group of 28 countries within Europe, has set goals for cutting greenhouse gas emissions that are complemented by many other strategies for saving energy and resources. Europe's stringent climate amelioration efforts are in response to the fact that much of Europe's air and many of its seas and streams are quite polluted, and that through its high level of consumption, it negatively impacts environments across the globe.

• Recent research suggests that investments in energy conservation, alternative energy, and other measures would cost EU economies 1 percent of their GDP. By contrast, doing nothing about climate change could shrink the EU's GDP by 20 percent.

• European residents average only one-half the energy consumption of the average North American. Fuel costs and CO_2 emissions are increasingly being considered in the design of multimodal transport that links high-speed rail to road, air, and water transportation.

• Europe's main energy sources are shifting from petroleum and natural gas to renewable sources that do not add carbon emissions to the atmosphere. Some countries continue to use nuclear power but most are decreasing such use.

• In the Mediterranean, the effect of the pollution that pours in from rivers, adjacent cities, industries, hotel resorts, and farms is exacerbated by the sea having just one tiny opening to the world ocean.

• Europe's impact on global environments is extensive because of its consumption of imported products.

HUMAN PATTERNS OVER TIME

The range of explanations for Europe's powerful and sustained impact on the world vary widely. One argument, for which there is no evidence, is that Europeans are somehow a superior breed of humans. Another is that Europe's many bays, peninsulas, and navigable rivers have promoted commerce to a larger extent there than elsewhere, a deterministic argument that is inadequate. In fact, much of Europe's success is based on technologies and ideas it borrowed from elsewhere. For example, the concept of the peace treaty, so vital to current European and global stability, was first documented not in Europe but in ancient Egypt. In order to try to understand how Europe gained the leading role around the globe that it still struggles to retain, despite considerable challenges, it is helpful to look at the history of this area.

SOURCES OF EUROPEAN CULTURE

Starting about 10,000 years ago, the practice of agriculture and animal husbandry gradually spread into Europe from the uplands and plains associated with the Tigris and Euphrates rivers in Southwest Asia and from farther east in Central Asia and beyond. Mining, metalworking, and mathematics also came to Europe from these places and from various parts of Africa. All of these borrowed innovations increased the possibilities for trade and economic development in Europe (FIGURE 4.10).

The first European civilizations were ancient Greece (800 to 86 B.C.E.) and Rome (753 B.C.E. to 476 C.E.). Located in southern Europe, both Greece and Rome initially interacted more with the Mediterranean rim, Southwest Asia, and North Africa than with the rest of Europe, which then had only small and relatively impoverished rural populations. Later European traditions of science, art, and literature were heavily based on Greek ideas, which were themselves derived from yet earlier Egyptian and Southwest Asian (Arab and Persian) sources.

The Romans, after first borrowing from Greek culture, also left important legacies in Europe. Many Europeans today speak *Romance* languages, such as Spanish, Portuguese, Italian, French, and Romanian, all of which are largely derived from Latin, the language of the Roman Empire. European laws that determine how individuals own, buy, and sell land originated in Rome. Europeans then spread these legal customs throughout the world.

The practices that Romans used when colonizing new lands also shaped much of Europe. After a military conquest, the Romans secured control in rural areas by establishing large, plantation-like farms. Politics and trade were centered on new Roman towns built on a grid pattern that facilitated commercial activity. Commanding structures, like the ceremonial gate in Figure 4.10A, inspired loyalty and facilitated military repression of rebellions. These same systems for taking and holding territory were later used when Europeans colonized the Americas, Asia, and Africa.

The influence of Islamic civilization on Europe is often overlooked. After the fall of Rome, while Europe was in a time known as the *early medieval period* (roughly 450 to 1300 C.E.; sometimes referred to as the *Dark Ages*), Turkish, Persian, Egyptian, and Arab scholars preserved learning from their own ancient traditions and also from Greece and Rome in large libraries, such as those in Alexandria, Egypt, and Constantinople (now Istanbul), Turkey. Also, Muslim Arabs originally from North Africa ruled Spain from 711 to 1492 (see Figure 4.10B), and from the 1400s through the early 1900s, the Ottoman Empire (based in what is now

Turkey) dominated much of southeastern Europe and Greece. The Arabs, Persians, and Turks all brought ideas, new technologies, food crops, architectural principles, and textiles to Europe from Arabia, Persia, Anatolia, China, India, and Africa. Arabs also brought Europe its numbering system, mathematics, and significant advances in medicine and engineering, building on ideas they picked up in South Asia.

Beginning 500 years ago, Europe also began to draw on a host of cultural features from the various colonies that Europe established in the Americas, Asia, and Africa. For example, many food crops now popular in Europe came from the Americas (potatoes, corn, peppers, tomatoes, beans, squash, pineapples, and cacao; see Table 3.1) and Southwest and Central Asia (wheat, leafy greens, garlic, onions, and apples).

THE INEQUALITIES OF FEUDALISM

As the Roman Empire declined, a social system known as *feudalism* evolved during the *medieval period* (450–1500 C.E.). This system originated from the need to defend rural areas against local bandits and raiders from Scandinavia and the Eurasian interior. The objective of feudalism was to have a sufficient number of heavily armed, professional fighting men, or *knights*, to defend a much larger group of *serfs*, who were legally bound to live on and cultivate plots of land for the knights. Over time, some of these knights became a wealthy class of warrior-aristocrats, called the *nobility*, who controlled certain territories. Some nobles gained power over other knights, amassing vast kingdoms.

The often-lavish lifestyles and elaborate castles of the wealthier nobility were supported by the labors of the serfs **(FIGURE 4.11)**. Most serfs lived in poverty outside castle walls and, much like enslaved people, were legally barred from leaving the lands they cultivated for their protectors.

THE ROLE OF EARLY URBANIZATION IN THE TRANSFORMATION OF EUROPE

While rural life followed established feudal patterns, new political and economic institutions were developing in Europe's towns and cities. Here, thick walls provided defense against raiders, and commerce and crafts supplied livelihoods, allowing the people more independence from feudal knights and kings.

Located along trade routes, people in Europe's urban areas were exposed to new ideas, technologies, and institutions from Southwest Asia, India, and China. Some institutions, such as banks, insurance companies, and corporations, provided the foundations for Europe's modern economy. Over time, citizens of Europe's urban areas established a pace of social and technological change that left the feudal rural areas far behind.

Urban Europe flourished in part because of laws that granted basic rights to urban residents. With adequate knowledge of these laws set forth in legal documents called *town charters*, people with few resources could protect their rights even if challenged by those who were more wealthy and powerful. Town charters provided a basis for European notions of *civil rights*, which have proved hugely influential throughout the world. With strong protections for their civil rights, some of Europe's townsfolk grew into a small middle class whose prosperity moderated the feudal system's extreme divisions of status and wealth (see Figure 4.10C). A related outgrowth of urban Europe was a philosophy known as **humanism,** which emphasizes the dignity and worth of the individual, regardless of wealth or social status.

The liberating influences of European urban life transformed the practice of religion. Since late Roman times, the Catholic Church had dominated not just religion but also politics and daily life throughout much of Europe. Roman Catholic churches were the center of community life and remain prominent in European rural and urban landscapes. In the 1500s, however, a movement known as the *Protestant Reformation* arose in the urban centers of the North European Plain. Reformers, among them Martin Luther, challenged Catholic practices—such as selling *indulgences*, which diminished the time a sinner was to suffer in purgatory, and holding church services in Latin, a language only a tiny educated minority understood. The inability to understand Latin services stifled public participation in religious discussions. Protestants also promoted individual responsibility and more open public debate of social issues, altering the perception of the relationship between the individual and society. These ideas spread faster with the invention of the European version of the printing press, which enabled widespread literacy.

EUROPEAN COLONIALISM: THE FOUNDING AND ACCELERATION OF GLOBALIZATION

A direct outgrowth of the greater openness and connectivity of urban Europe was the exploration and subsequent colonization of much of the world by Europeans. Spain, Portugal, the United Kingdom (UK), and the Netherlands conquered vast overseas territories and created worldwide trade relationships that transformed Europe economically and culturally and laid the foundation for the modern global economy. This commerce and cultural exchange began a period of accelerated *globalization* that persists today (see the discussion in Chapter 1).

In the fifteenth and sixteenth centuries, Portugal took advantage of advances in navigation, shipbuilding, and commerce to set up a trading empire in Asia (see Figure 4.10D) and a colony in Brazil. Spain soon followed, founding a vast and profitable empire in the Americas and the Philippines. By the seventeenth century, however, England, the Netherlands, and France had seized the initiative from Spain and Portugal. All European powers implemented **mercantilism,** a strategy for increasing a country's power and wealth by acquiring colonies and managing all aspects of their production, transport, and trade for the colonizer's benefit **(FIGURE 4.12)**.

Mercantilism brought wealth that supported the Industrial Revolution in Europe (page 212) by supplying cheap resources from around the globe (see Figure 4.12) for Europe's new factories. The colonies also became markets for European manufactured goods.

humanism a philosophy and value system that emphasizes the dignity and worth of the individual, regardless of wealth or social status

mercantilism a strategy for increasing a country's power and wealth by acquiring colonies and managing all aspects of their production, transport, and trade for the colonizer's benefit

A A street in the Roman colonial town of Timgad, Algeria, founded around 100 C.E.

B Alhambra, a palace built by Spain's North African Islamic rulers in the thirteenth century.

C A holiday festival in a Belgian town, painted by Pieter Bruegel (the elder) in 1559.

| 10,000 B.C.E. | 1000 B.C.E. | 500 B.C.E. | 0 C.E. | 500 C.E. | 1000 C.E. | 1400 C.E. | 1500 C.E. |

8000 B.C.E.
Agriculture and animal husbandry introduced via Southwest Asia

753 B.C.E.–476 C.E.
Roman civilization

1400
North African Islamic rule of Spain

711–1492

1500s
Protestant Reformation

FIGURE 4.10 A VISUAL HISTORY OF EUROPE

Thinking Geographically

After you have read about the human history of Europe, you will be able to answer the following questions:

A What about the photo of the ruins of Timgad reflects a particular Roman strategy for organizing space?

B What about the architecture of Alhambra suggests technological sophistication and an astute understanding of climate?

FIGURE 4.11 Predjama Castle in Slovenia. Predjama Castle, perched in the mouth of a cave in southwestern Slovenia, exemplifies the feudalism from which modern Europe eventually emerged. First mentioned in the historical record in the thirteenth century, Predjama became legendary as an impregnable fortress in the fifteenth century, when the Austrian Imperial Army laid siege to it for over a year, but for months the Austrians were ignorant of a passage through the cave that kept the castle supplied. The Austrians finally succeeded in taking the castle and then the region when Predjama's owner, Erazem Leuger, was betrayed by one of his own men, who informed the Austrians of the castle's greatest vulnerability. A cannonball killed Erazem while he sat in Predjama's unprotected outhouse!

By the mid-1700s, wealthy merchants in London, Amsterdam, Paris, Berlin, and other western European cities were investing in new industries. Workers from rural areas poured into urban centers in England, the Netherlands, Belgium, France, and Germany to work in the manufacturing industries (see Figure 4.10F) and mining. Wealth and raw materials flowed in from colonial ports in the Americas, Asia, and Africa. Some cities, such as Paris and London, were elaborately rebuilt in the 1800s to reflect their roles as centers of global empires.

By 1800, London and Paris, each of which had a million inhabitants, were Europe's largest cities, and eventually became *world cities* (cities of worldwide economic or cultural influence). London is a global center of finance; Paris is a cultural center that has influence over global consumption patterns, from food to fashion to tourism.

By the middle of the eighteenth century, the English, French, and Dutch empires were more profitable and powerful than those of Spain and Portugal. These empires extended European influence into India, Southeast Asia, and Africa. By the twentieth century, European colonial systems had strongly influenced nearly every part of the world. Within Europe, the overseas empires of England, the Netherlands, and eventually France shifted wealth, investment, and general economic development toward western Europe and away from southern Europe and the Mediterranean.

URBAN-BASED REVOLUTIONS IN INDUSTRY AND POLITICS

The wealth derived from Europe's colonialism helped fund two of the most dramatic transformations in a region already characterized by rebirth and innovation: the industrial and democratic revolutions. Both first took place in urban Europe.

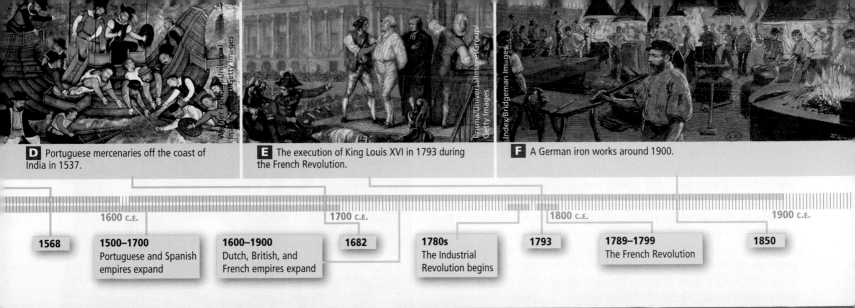

D Portuguese mercenaries off the coast of India in 1537.

E The execution of King Louis XVI in 1793 during the French Revolution.

F A German iron works around 1900.

| 1568 | 1500–1700 Portuguese and Spanish empires expand | 1600–1900 Dutch, British, and French empires expand | 1682 | 1780s The Industrial Revolution begins | 1793 | 1789–1799 The French Revolution | 1850 |

1600 C.E.   1700 C.E.   1800 C.E.   1900 C.E.

C What about this picture suggests relative prosperity in sixteenth-century Belgium?

D What about this picture suggests how the Portuguese felt about the value of trade with India?

E How is the execution of Louis XIV related to the evolution of democracy in Europe?

F What does this photo indicate about the impact of the Industrial Revolution on employment in Europe?

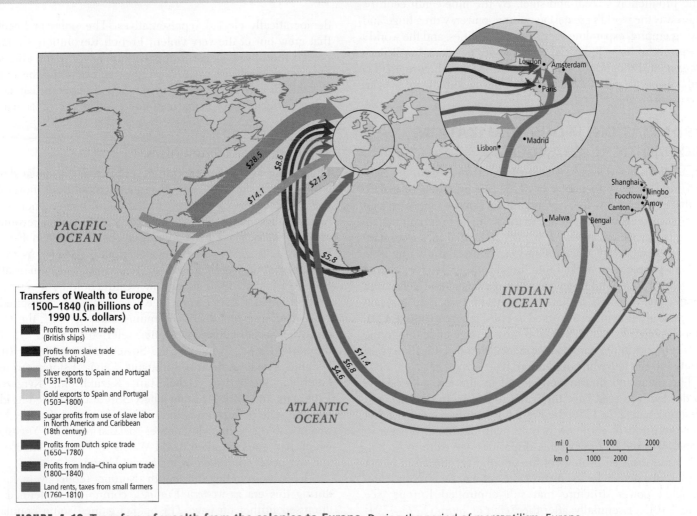

Transfers of Wealth to Europe, 1500–1840 (in billions of 1990 U.S. dollars)

- Profits from slave trade (British ships)
- Profits from slave trade (French ships)
- Silver exports to Spain and Portugal (1531–1810)
- Gold exports to Spain and Portugal (1503–1800)
- Sugar profits from use of slave labor in North America and Caribbean (18th century)
- Profits from Dutch spice trade (1650–1780)
- Profits from India–China opium trade (1800–1840)
- Land rents, taxes from small farmers (1760–1810)

FIGURE 4.12 Transfers of wealth from the colonies to Europe. During the period of mercantilism, Europe received billions of dollars of income from its overseas colonies. This one-way flow of minerals, agricultural products, and profits of the slave trade fueled the industrial revolution in Europe, the construction of magnificent urban structures in Europe's capital cities, and subsequent infrastructure development. [Source consulted: Alan Thomas, *Third World Atlas* (Washington, DC: Taylor & Francis, 1994), p. 29]

211

The Industrial Revolution

Europe's Industrial Revolution—particularly Britain's ascendancy as the leading industrial power of the nineteenth century—was intimately connected with colonial expansion and sugar production. In the seventeenth century, Britain developed a small but influential trading empire in the Caribbean, North America, and South Asia, which provided it with access to a wide range of raw materials and to markets for British goods.

Sugar, produced by British colonies in the Caribbean, was an especially important trade crop (see Figure 4.12). Sugar production is a complex process that requires major investments in equipment, and for which a great deal of labor was once needed. Enslaved people were forcibly brought from Africa to help grow and process sugar. The skilled management and large-scale organization needed for the production and distribution of sugar later provided a production model for the Industrial Revolution. The mass production of sugar also generated enormous wealth that helped fund industrialization.

By the late eighteenth century, Britain was introducing mechanization into its industries, first in textile weaving and then in the production of coal and steel. By the nineteenth century, Britain was the world's greatest economic power, with a huge and growing empire, expanding industrial capabilities, and the world's most powerful navy. Industrial technologies developed in Britain were spread throughout continental Europe (see Figure 4.10F), North America, and elsewhere, transforming millions of lives in the process.

URBANIZATION, INDUSTRIALIZATION, AND POLITICAL CHANGE

Industrialization led to massive growth in urban areas in the eighteenth and nineteenth centuries. The concentration of people and the extremely low living standards in Europe's cities created tremendous pressures for change in the political order, ultimately leading to democratization. Most early industrial jobs were dangerous and unhealthy, demanding long hours and offering little pay. Poor people were packed into tiny dwellings that they shared with many others. Children were often weakened and ill because of inadequate nutrition, sanitation, and health care; industrial pollution; and from being overworked as child laborers **(FIGURE 4.13)**. Water was often contaminated, and most sewage ended up in the streets. Opportunities for advancement through education were restricted to the few who could afford it.

The ideas and information gained by the few poor people who did learn to read gave them the incentive to organize and protest for change. In 1789, the French Revolution led to the first major inclusion of common people in the political process in Europe. Angered by the extreme disparities of wealth in French society, and inspired by news of the popular revolution in North America, the poor rebelled against the monarchy and the elite-dominated power structure that still controlled Europe (see Figure 4.10E). Eventually, after the *Reign of Terror* (1793–1794) in France—when more than 40,000 people were executed as rival political factions battled for control—the general populace, especially in urban areas, became involved in governing through

FIGURE 4.13 An engraving from 1871 of barefoot children working in a brickyard in England. Notice that both boys and girls of about age 11 to 13 are carrying heavy loads of clay on their heads. There are no apparent efforts to keep them safe from hazards.

democratically elected representatives. The political freedoms that grew out of the very violent French Revolution ultimately proved short-lived, as governments dominated by the elite soon regained control in France. Nevertheless, over time the French Revolution provided crucial inspiration to later urban democratic political movements in France and many other parts of the world.

The Impact of Communism

During the struggles of the nineteenth century, popular discontent took the form of new revolutionary political movements that proposed radical changes to the established economic and political order. The political philosopher and social revolutionary Karl Marx framed the mounting social unrest in Europe's cities as a struggle between socioeconomic classes. His treatise *The Communist Manifesto* (1848) helped social reformers across Europe articulate ideas about how wealth could be more equitably distributed. East of Europe in Russia, Marx's ideas inspired the creation of a revolutionary communist state in 1917, the Union of Soviet Socialist Republics, often referred to as the *USSR* or the *Soviet Union*. Eventually, the Soviet Union included Russia, Ukraine, Moldova, Belarus, Lithuania, Latvia, Estonia, Armenia, Georgia, Azerbaijan, Uzbekistan, Kazakhstan, Kyrgyzstan, Tajikistan, and Turkmenistan. The USSR also extended its ideology and state power throughout much of Central Europe, but it was not able to dominate another socialist republic, Yugoslavia, which had emerged after World War II in southeastern Europe and was never controlled by the Soviet Union.

In West, North, and South Europe (collectively known during this era as *western Europe*), communism (defined and discussed further on page 214 below) gained some popularity, but this tended to lead to the formation of Communist political parties that operated peacefully within the context of democratic political systems. In countries like France, Italy, and Spain,

Communist political parties are still influential and often form governing coalitions with other parties.

After lengthy and often violent struggles against the rich and politically powerful, Europe's huge working class won greater political freedoms, such as the right to vote in elections. By the early twentieth century, these struggles also gained urban industrial workers the right to form unions that could bargain with employers and the government for higher wages and better working conditions. Eventually, political power, wealth, and opportunity were distributed more evenly throughout society. However, it is important to note that the road to these sorts of changes in Europe was rocky and often violent, just as are democratizing processes now in many other parts of the world.

Popular Democracy and Nationalism

During the nineteenth and twentieth centuries, the development of democracy was also linked to the idea of **nationalism,** or allegiance to the state. The notion spread that all the people who lived in a certain area formed a nation and that loyalty to that nation should supersede loyalties to family, clan, or individual monarchs. Eventually, the whole map of Europe was reconfigured, and the mosaic of kingdoms gave way to a collection of *nation-states*. All of these new nations were, at varying paces, transformed into democracies by the political movements arising in Europe's industrial cities. Nationalism was a major component of both World War I and II, so in the period after the wars—called the *post-war era*—as the European Union was constructed, nationalism was recognized as a problem and deemphasized. Nonetheless, the urge for a strong national identity lives on in European countries, especially in times of political or economic troubles.

Democracy and the Welfare State

Channeled through the democratic process, by the mid-twentieth century, public pressure for improved living standards moved most European governments toward becoming **welfare states.** In welfare states, the government accepts responsibility for the well-being of citizens, guaranteeing basic necessities such as education, affordable housing and food, unemployment insurance, and health care for all. In time, government regulations of wages, hours, workplace safety, and vacations established more harmonious relations between workers and employers. The gap between rich and poor declined, and overall civic peace increased. And although welfare states were funded primarily through taxes, industrial productivity and overall economic activity did not decline but rather increased, as did general prosperity.

nationalism devotion to the interests or culture of a particular country, nation, or cultural group; the idea that a group of people living in a specific territory and sharing cultural traits should be united in a single country to which they are loyal and obedient

welfare state a government that accepts responsibility for the well-being of its people, guaranteeing basic necessities such as education, affordable food, employment, housing, and health care for all citizens

FIGURE 4.14 LOCAL LIVES: People and Animals in Europe

A The bloodhound, a scent-tracking dog, may have been developed by the Romans in the third century c.e. and later brought to England from France before the fourteenth century. Originally bred to track wild game, bloodhounds are renowned for their ability to track individual people by their body scent. Sometimes called *sleuth hounds*, they are still used today by law enforcement.

B A rose-ringed parakeet of the type that has been kept as a pet in Europe also since the days of the Romans. Prized for their sociable nature and ability to mimic human speech, parakeets and parrots are popular pets throughout Europe. Some have escaped captivity to form wild populations in many European cities.

C Originally bred by nomads along the Dalmatian (Croatian) coast as a guard dog, Dalmatians became popular throughout Europe and North America during the nineteenth century. In the days of horse-drawn fire carriages, Dalmatian "firehouse dogs" cleared the path for oncoming fire carriages and made the horses run faster by nipping at their legs.

Rudi Von Briel/Getty Images

Manfred Pfefferle/Getty Images

davespilbrow/Getty Images

Modern Europe's welfare states have yielded generally adequate to high levels of well-being for all citizens. However, just how much support the welfare state should provide is still a subject of intense debate—one that is currently being addressed in different ways across Europe. A discussion of *social protection systems* (the preferred term to *welfare*), begins on page 226, and maps of the various human well-being outcomes of social protection systems in Europe begins on page 239.

TWO WORLD WARS AND THEIR AFTERMATH

Despite Europe's many advances in industry and politics, by the beginning of the twentieth century, the region still lacked a system of collective security that could prevent war among its rival nations. Between 1914 and 1945, two extremely destructive world wars left Europe in ruins, no longer the dominant region of the world. At least 20 million people died in World War I (1914–1918) and 70 million in World War II (1939–1945). During World War II, Germany's Nazi government killed 15 million civilians in its failed attempt to conquer the Soviet Union. Eleven million civilians died at the hands of the Nazis during the **Holocaust,** a massive execution of 6 million Jews and 5 million gentiles (non-Jews), including ethnic Poles and other Slavs, **Roma** (Gypsies), disabled and mentally ill people, gays, lesbians, transgendered people, and political dissidents.

World War II divided Europe and much of the world into two battling camps. While a long list of countries participated, the main combatants were the *Axis* powers of Germany, Italy, and Japan; and the *Allies*—Russia, the United Kingdom, the United States, and France (except during the German occupation of 1940–1944). After World War II ended in 1945 with an Allied victory, finding a lasting peace became a primary goal that eventually led to the formation of the European Union. But before that, a number of changes took place that still influence the relationships of European countries today. Germany was divided into two parts. West Germany became an independent democracy allied with the rest of western Europe—especially Britain and France—and the United States.

Russia controlled East Germany and the rest of Central Europe (Latvia, Lithuania, Estonia, Poland, Czechoslovakia, Hungary, Romania, Bulgaria, Ukraine, Moldova, and Belarus). After the war, under Russia's dominant leadership, these war-devastated countries solidified into an ideologically communist alliance known as the *Soviet bloc*. The line between East and West Germany was part of what was called the **Iron Curtain,** a long, fortified border zone that separated western Europe from what was then called *eastern Europe*. Yugoslavia in southeastern Europe did not become part of the Soviet bloc but also kept itself insulated from the West, partly to protect its socialized economy and to validate its leadership of countries not aligned with either side of the Cold War.

The Cold War

The division of Europe created a period of conflict, tension, and competition between the United States and the Soviet Union known as the **Cold War,** which lasted from 1945 to 1991. During this time, once-dominant Europe, and indeed the entire world, became a stage on which the United States and the Soviet Union competed for supremacy. The central issue was the competition between **capitalism**—characterized by privately owned businesses and industrial firms that adjust prices and output to match the demands of the market—and **communism** (actually a version of **socialism**), in which the state owns all farms, industry, land, and buildings (the so-called *means of production*).

After 1945, in most of what we now call Central Europe, the Soviet Union forcibly implemented a Communist-inspired economic model known as **central planning,** in which a central bureaucracy dictated prices and output, with the stated aim of allocating goods and services equitably across society according to need. This system was successful in alleviating long-standing poverty and illiteracy among the working classes and for women, but it was rife with waste, corruption, and bureaucratic bungling. It ultimately collapsed in the 1990s due to inefficiency, insolvency, high levels of environmental pollution, and public demands for more political freedoms.

During the post-war period, the rest of Europe, especially western Europe, developed capitalist economies with financial support from the United States under the *Marshall Plan* and from the governments of the involved countries. Basic infrastructure, such as roads, housing, and schools, were rebuilt. Economic reconstruction proceeded rapidly in the decades after World War II and included special attention to the social welfare needs of the general public.

Decolonization

Europe's decline during the two world wars also led to the loss of its colonial empires. Many European colonies had participated in the wars, with some (India, the Dutch East Indies, Burma, and Algeria) suffering extensive casualties, and almost all emerging economically devastated. After World War II, Europe was less able to assert control over its colonies as their demands for independence grew. By the 1960s, most former European colonies had gained independence, often after bloody wars fought against European powers and their local allies (see chapters 3, 7, 8, and 10).

Holocaust during World War II, a massive execution by the Nazis of 6 million Jews and 5 million gentiles (non-Jews), including ethnic Poles and other Slavs, Roma (Gypsies), disabled and mentally ill people, gays, lesbians, transgendered people, and political dissidents

Roma the now-preferred term in Europe for Gypsies; some prefer to be called Gypsies as a point of ethnic identity

Iron Curtain a long, fortified border zone that separated western Europe from (then) eastern Europe during the Cold War

Cold War a period of conflict, tension, and competition between the United States and the Soviet Union that lasted from 1945 to 1991

capitalism an economic system characterized by privately owned businesses and industrial firms that adjust prices and output to match the demands of the market

communism a political ideology and economic system, based largely on the writings of the German revolutionary Karl Marx, in which, on behalf of the people, the state owns all farms, industry, land, and buildings (a version of socialism)

socialism primarily an economic system in which all people have access to basic goods and services and all large-scale industries are collectively owned, with any profits benefitting society as a whole

central planning a communist economic model in which a central bureaucracy dictates prices and output, with the stated aim of allocating goods equitably across society according to need

THE BIRTH OF THE EUROPEAN UNION

At the end of World War II, European leaders concluded that the best way to prevent the kinds of hostilities that had led to two world wars would be to forge closer economic ties among the European nation-states. The first step toward developing an economic union in Europe began in 1951 with the creation of a common market for coal and steel (two resources essential for war industries) that, in the rebuilding of Europe, enabled the monitoring of Germany's industrial rehabilitation. The next major step took place in 1958, when Belgium, Luxembourg, the Netherlands, France, Italy, and West Germany formed the *European Economic Community* (*EEC*). The members of the EEC agreed to eliminate certain tariffs against one another and to promote mutual trade and cooperation. Denmark, Ireland, and the United Kingdom joined in 1973; Greece, Spain, and Portugal in the 1980s; and Austria, Finland, and Sweden in 1995, bringing the total number of member countries to 15. In 1992, the EEC became the European Union, with the intent to work toward a higher level of economic, social, and political integration that would make possible the free flow of goods and people across national borders.

THINGS TO REMEMBER

- During the medieval period, political and economic transformations in Europe's towns and cities challenged the feudal system that dominated rural areas.

- The many overseas colonies of various European states created trade relationships that continue in the modern global economy. Europe was transformed and continues to benefit.

- Most of the countries in the world have been ruled by a European colonial power at some point in their history.

- Industrialization and the growth of cities set the stage for increasing the political power of ordinary people.

- The expansion of political freedoms in Europe has proceeded at different rates across the region and was interrupted by two world wars.

- The European Union was created with the intent to work toward a higher level of economic, social, and political integration that would make possible the free flow of goods and people across national borders.

GLOBALIZATION AND DEVELOPMENT

GEOGRAPHIC THEME 2

Globalization and Development: Following World War II, Europe took steps to create an economic union now known as the European Union, which features open borders, the free movement of people and goods, progress toward a common currency, and efforts to promote human well-being and social cohesion. To improve its global competitiveness, the EU shifted industries from high-wage western Europe to the lower-wage member states of Central and South Europe. Service jobs replaced many industrial jobs, but unemployment increased. The European Union struggles to integrate the economies of its member states and to reduce regional disparities.

The European Union uses a number of economic development strategies designed to ensure its ability to compete with the United States, Japan, and the developing economies of Asia, Africa, and South America. To increase Europe's export potential, the EU has focused on lowering the cost of producing goods in Europe. One strategy it uses to do this has been to shift labor-intensive industries from the wealthiest EU countries in western Europe, where wages are high, to the relatively poorer, lower-wage member states of Central Europe. This has generally worked well, helping poorer European countries prosper while keeping the costs of doing business low enough to reduce the incentive for European companies to move away from Europe to places where labor and resource costs are lower still. However, while the economies of Central Europe have grown, in Europe's wealthiest countries, the resultant reduction of industrial capacity (*deindustrialization*) has led to higher unemployment rates. And despite these efforts, some EU firms have moved abroad to cut costs. **FIGURE 4.15** shows the 28 members of the EU and those countries that anticipate joining the EU or that have declined to become members. The GNI data for each country shows the economic disparities across the EU.

One way that the EU attempts to address economic and development disparities across the region is to redistribute wealth via the EU budget. One third of the EU-28 budget is devoted to what is called the **Cohesion Policy** (discussed further on page 226) aimed at removing disparities between countries and improving overall social integration. The entire budget is made up of funds that EU members have agreed to pool for the good of Europe as a whole. The money is used to improve transportation, energy use, and communications between members; to support the use of sustainable growth to protect the environment; to make Europe competitive globally; and to encourage cooperation in science and learning. The idea is that the EU will be more economically healthy and better able to compete in the global economy if stark disparities in standards of living and well-being are addressed. Therefore, contributions to the EU budget and allocations from the budget are not equitably distributed. In the interest of the common good, some countries contribute much more than they receive and some countries receive much more than they contribute. **FIGURE 4.16** illustrates some of the budget contributions and distributions by member countries.

EUROPE'S GROWING SERVICE ECONOMIES

As industrial jobs have declined across the region, most Europeans (about 70 percent) have found jobs in the service economy. Services, such as the provision of health care, education, finance, tourism, and information technology, are now the base of Europe's integrated economy, drawing hundreds of thousands of employees to the main European cities. For example, financial corporations that are located in London and serve the entire world play a huge role in the British economy. Many multinational companies are headquartered in London in part to take advantage of the city's financial sector. For the 30 percent of workers who have not been able to adjust to the new job market, in

decolonization the process of dissolving or ending colonial relationships, institutions, and mind-sets

Cohesion Policy EU guidelines for programs aimed at removing social and economic disparities across Europe

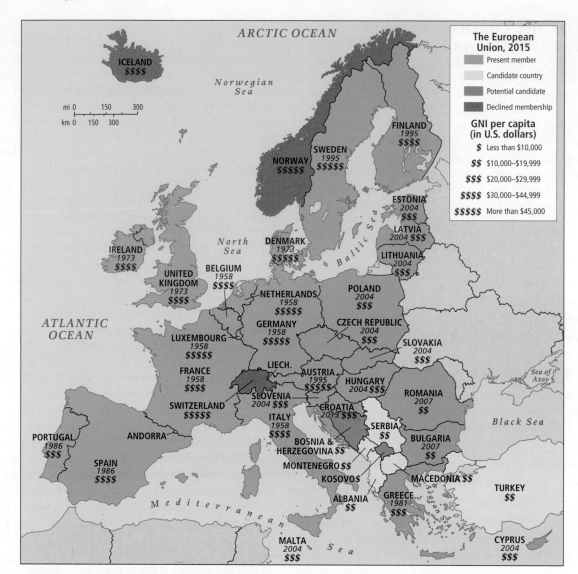

FIGURE 4.15 The current members of the European Union. The current (2015) members of the EU, their dates of joining, and their GNI per capita data. The European Union, formed as the EEC in 1958 with the initial goal of economic integration, became the EU in 1992, and has led the worldwide movement toward greater regional cooperation. It is older and more deeply integrated than its closest competitor, the North American Free Trade Agreement (NAFTA). [Sources consulted: Official website of the European Union, http://europa.eu/about-eu/countries/member-countries/index_en.htm; and http://www.prb.org/pdf15/2015-world-population-data-sheet_eng.pdf, accessed June 7, 2016]

which many positions require special training and skills, underemployment has become chronic.

Europe's highly educated population is transforming the service sector, with almost 40 percent of all employment in the EU now in *knowledge-intensive services*. These jobs require considerable formal education, such as high-end technical training or a college or graduate degree, and usually extensive training on the job. In recent years, though, there has been concern within the EU about the declining profitability of European high-tech companies, such as the cell phone manufacturer Nokia (bought by the U.S. company Microsoft in 2014). Many of these companies have been unable to compete either with Asian companies that can build products more cheaply or with more innovative U.S. companies. While the EU is making little effort to

compete with Asian companies on labor costs, it is now investing more in research and development to better compete with U.S. companies.

A major component of Europe's service economy is tourism. Europe is the most popular tourist destination in the world and Paris is the most popular city. One job in eight in the European Union is related to tourism. Tourism generates a bit more than 10 percent of the EU's GDP and 15 percent of its taxes, although this varies with global economic conditions. Europeans are themselves enthusiastic travelers, frequently visiting one another's countries to attend local festivals as well as traveling to many distant locations throughout the world **(FIGURE 4.17)**. This travel is made possible by the long paid annual vacations—usually 4 to 6 weeks—that Europeans are granted by employers. Vacation days

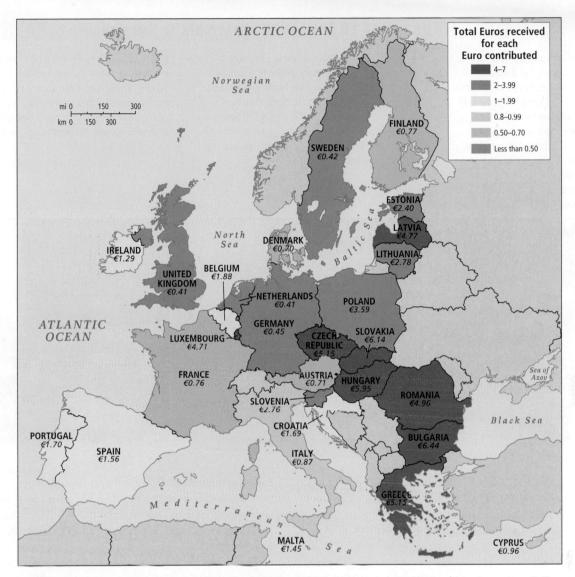

FIGURE 4.16 The EU budget. The budget of the European Union is remarkable in that some countries, in the interest of the common good, pay far more than they receive in benefits, while others receive much more in benefits than they contribute. The intention is that eventually the less affluent countries will be brought up to the same level as the other countries and will be able to contribute equally or perhaps more generously than others. These strategies are meant to promote cohesion among the many varied EU member states. [Source consulted: Official website of the European Union, http://europa.eu/about-eu/basic-information/money/expenditure/index_en.htm, accessed June 7, 2016.]

can be used a few at a time so that people can take numerous short trips. The most popular holiday destinations among EU members in 2014 were Spain, France, Italy, Germany, and Austria; the first three were favored for their sunny weather and beaches.

ECONOMIC COMPARISONS OF THE EU AND THE UNITED STATES

The EU economy now encompasses more than 510 million people (out of a total of 742 million in the whole of Europe)—roughly 200 million more than live in the United States. Collectively, the EU countries are wealthy, with a joint economy slightly larger than that of the United States, making the EU the largest economy in the world (China is the second largest; the United States, third) and one that exerts a powerful influence on

the global trading system **(FIGURE 4.18)**. Even so, trade between countries within the EU amounts to about twice the monetary value of trade between the EU as a whole and the outside world.

In contrast to the United States, which usually imports far more than it exports—resulting in a trade deficit—the EU usually maintains a trade balance in which the values of imports and exports are roughly equal, but as Figure 4.18D shows, the balance is now tipped a little toward more imports than exports. This shift is one of the indicators that EU industries may be losing some of their global competitiveness; for example, Europe's clothing stores, once full of rather expensive garments made in Europe, now sell many less expensive products made in Asia.

According to the World Bank the average GNI per capita (PPP) for the European Union (U.S.$36,280) in 2015 was

FIGURE 4.17 LOCAL LIVES: Festivals in Europe

A Every year in Buñol, Spain, participants in *La Tomatina* throw tomatoes at each other. Supposedly, this festival began in 1945 when a group of young men, not allowed in a parade, grabbed tomatoes from a nearby vegetable stand and hurled them. The young men returned the next year and repeated the brawl, though they brought their own tomatoes. Today, truckloads of tomatoes are brought in, as shown here.

B Riders in *Il Palio*, a horse race held each year in Siena, Italy. Jockeys from each of the city's 17 major neighborhoods circle the main plaza; they ride bareback and frequently get thrown from their horses. A colorful pageant precedes the race, which has been run since 1656, though similar races were run even earlier.

C Glastonbury Festival of Contemporary Performing Arts (known as *Glasto*) is held each year in Somerset, England. First organized in 1970 by a farmer who was inspired by a Led Zeppelin performance, Glastonbury has grown to become one of Europe's largest music festivals.

significantly less than that of the United States (U.S.$55,860; China's was U.S.$13,130). Notably, despite the lower GNI per capita (PPP), poverty levels are lower in the European Union because the disparity of wealth there is about one-third that of the United States.

Careful regulation of EU economies contributes to generally slower economic growth than in the United States. However, these regulations also make the European Union more resilient to global financial crises, such as the crisis beginning in 2008. This is largely because EU banking industries and financial markets are more strongly controlled than in the United States.

ECONOMIC INTEGRATION AND A COMMON CURRENCY

Economic integration has addressed a number of problems in Europe. Individual European countries have relatively small populations, which means that they have smaller internal markets for their products. As a result, their companies earn lower profits than those in large countries. The European Union tackled this problem by joining European national economies into a common market and by removing many of the controls on the movement of people and products posed by country borders. The **Schengen Agreement** (signed in 1985 and implemented in 1995, and

Schengen Agreement an agreement first signed in 1985 and implemented in 1995 that allows for free movement across common borders in the European Union

economies of scale reductions in the unit cost of production that occur when goods or services are efficiently mass-produced, resulting in increased profits per unit

euro (€) the official (but not required) currency of the European Union as of January 1999

repeatedly modified) gradually removed border checks, making it possible to travel throughout the EU with few if any border stops. Open borders gave companies in any EU country access to a much larger market and allowed them to potentially make larger profits through **economies of scale**—reductions in the unit costs of production that occur when goods or services are efficiently mass-produced, resulting in increased profits per unit. Before the European Union, when businesses sold their products to neighboring countries, their earnings were diminished by tariffs and other regulations, as well as by fees for currency exchanges.

The Schengen Agreement also meant that countries relinquished national sovereignty—long a sensitive issue in the region. Since the recession of 2008, which still lingers in some countries, the EU economy has been a bit fragile. This and the recent influx of immigrants has weakened the conviction to maintain open borders and the Schengen Agreement will likely be renegotiated.

The **euro (€)**, the official currency of the European Union is now used in 19 countries: Austria, Belgium, Cyprus, Estonia, Finland, France, Germany, Greece, Ireland, Italy, Latvia, Lithuania, Luxembourg, Malta, the Netherlands, Portugal, Slovakia, Slovenia, and Spain. Countries that use the euro have a stronger voice in the creation of EU economic policies, and the use of a common currency greatly facilitates trade, travel, and migration within the European Union. All non-euro member states, except Sweden and the United Kingdom, have currencies whose value is determined by that of the euro, which has considerable international standing. Depending on global financial conditions, either the euro or the U.S. dollar is the preferred currency of international trade and finance.

(A) The Top 10 EU-28 trading partners in goods, 2014
(% share of world exports)

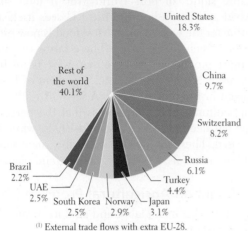

(B) The Top 10 EU-28 trading partners in goods, 2014
(% share of world imports)

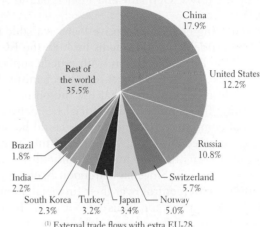

(C) The Top 10 EU-28 trading partners in services, 2014
(% share of service exports)

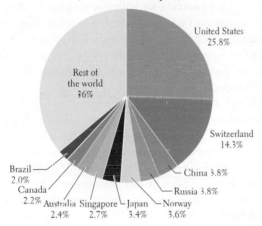

(D) The Top 10 EU-28 trading partners in services, 2014
(% share of service imports)

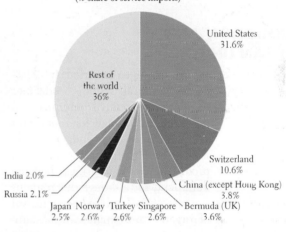

The percent share in the world trade in
goods and services, 2014

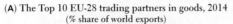

FIGURE 4.18 Share of world trade in goods and services, 2014.

THE EURO AND DEBT CRISES

In recent years, there have been major challenges to the euro, most significantly in the form of a debt crisis that has tested the mechanisms that governments use to maintain economic stability. The debt crisis first emerged in Greece, where government spending was significantly outpacing tax revenues. This was compounded by the growth of Greece's trade deficit during the global economic slowdown starting (in Europe) in 2008, which affected tourism and reduced demand for Greece's exports. Similar conditions in other countries led to similar debt crises in other euro zone countries, especially Ireland, Spain, and Portugal.

Normally, governments respond to high levels of debt by borrowing from other countries and then reducing the value of their currency, relative to that of the lenders, which makes these debts easier to pay. This was not an option for any member of the euro zone because no country could by itself manipulate the value of the euro. The EU has the European Central Bank (ECB), which controls the value of the euro, but it is strongly influenced by the larger euro zone economies, such as those of Germany and France. Banks in these countries had loaned large amounts of money to Greece and the other indebted governments of the euro zone, and they would lose money if the value of the euro declined. However, because all the economies of the euro zone were threatened by the crises, Germany and France had to act; hence the loans.

In addition to making large, inexpensive loans available to Greece and the other indebted governments through the ECB over the short term, the European Union, with crucial support by Germany and France, developed longer-term mechanisms to limit government debt in all countries that use the euro. This involved strengthening the political and financial ties between EU countries so that common economic policies could be better developed and enforced by EU institutions. While many euro zone countries—even those in financial difficulty, such as Greece, Spain, Cyprus, and Slovenia—strenuously object to giving EU institutions more control over their financial affairs, nonetheless, their urgent need for loans from the ECB has made them reconsider these objections. Although short- and long-term solutions have been implemented and countries like Spain and Slovenia are more stable, the euro zone crisis is still ongoing.

FOOD PRODUCTION AND THE EUROPEAN UNION

Concerns about food security in Europe have led to heavily subsidized and regulated agricultural systems. **Subsidies** to agriculture-related enterprises in the 1980s accounted for more than 70 percent of the budget of the European Union. Large-scale food production by agribusiness became dominant, and although that remains the case, there has also been a revival of organic and small-scale sustainable techniques by European farmers and there is a growing market for organically raised produce.

subsidies monetary assistance granted by a government to an individual or group in support of an activity, such as farming, that is viewed as being in the public interest

Common Agricultural Program an EU program meant to guarantee secure and safe food supplies at affordable prices, that are produced sustainably by farmers who earn a fair income

Food security is especially important to Europeans, partly because stories of post–World War II food shortages are still so vivid and partly because the EU sees itself as ensuring food security for the wider world. Most food is now produced on large, efficient mechanized farms. These farms require less labor and are more productive per acre than were farms before the 1970s. One result is that only about 2.3 percent of Europeans are now engaged in full-time farming. A second result of the efficiencies of agricultural mechanization is that the percentage of land in crops has declined since the mid-1990s, while forestlands have increased. These forests are sustainably managed to provide fuel, building materials, wilderness habitat, and places for recreation.

The Common Agricultural Program (CAP)

The drastic decline of labor and land in farming has had an emotional effect on Europeans, who see it as endangering their cultural heritage and their goal of food self-sufficiency **(FIGURE 4.19A–C)**. To address these worries, the European Union established its wide-ranging **Common Agricultural Program (CAP),** meant to guarantee secure and safe food supplies at affordable prices, produced sustainably by farmers who earn a fair income. The CAP is also meant to balance territorial development across the EU and to preserve the quaint rural landscapes nearly everyone associates with Europe.

Until it was drastically revised in 2013, the CAP was criticized for undermining its own goals with its use of tariffs on imported agricultural goods and for giving subsidies (in the form of payments to farmers) to underwrite the costs of food production. Tariffs kept out competition and effectively raised food costs for millions of EU consumers. The subsidies tended to favor large, often corporate-owned farms because payments were based on the amount of land under cultivation, and also tended to encourage overproduction. CAP policies often resulted in too much food, which lowered market prices and discouraged small family farmers.

FIGURE 4.19 LOCAL LIVES: Foodways in Europe

A *Tortellini*, stuffed ring- or navel-shaped pasta from the region around Modena and Bologna, Italy. One legend claims that the gods Venus and Jupiter arrived at a tavern in Bologna one night, weary from an ongoing battle between Bologna and Modena. When they retired to their room, the innkeeper peeked through the keyhole and saw only Venus's navel. Inspired, he rushed off to the kitchen to create the first tortellini.

B Swedish *gravlax*—raw salmon cured in salt and dill—is cut into thin slices. Gravlax was invented by fishermen who preserved salmon by salting it and then burying it in the sand on a beach above the high-tide line. This process gave the dish its name: *grav* means grave and *lax* means salmon.

C In French *confit* of duck legs, the meat is salted, seasoned, and slowly cooked in its own fat. The meat is then stored in the fat, which can preserve it for several weeks. Confit is considered a regional specialty of southwestern France.

On the global scale, the CAP actually undermined worldwide food security because protective agricultural policies like tariffs and subsidies—also used in North America, Japan, and elsewhere—hurt farmers in the developing world. Tariffs lock farmers from poorer countries out of major markets. And subsidies encourage overproduction in rich countries; the resulting occasional gluts of farm products such as butter, eggs, and corn are then dumped—that is, sold cheaply on the world market. Dumping by big producers can lower global prices to the point that farmers in developing countries are driven out of business.

In 2013 the CAP was revised. Subsidies to produce certain crops were eliminated and instead needy farmers were given payments to raise their standard of living to an acceptable level. The EU instituted market reforms to increase efficiency and competitiveness, and put more emphasis on improving environmental sustainability and equalizing agricultural development across the EU, including the maintenance of family farms. Part of the 2013 reforms included shrinking the CAP budget which, at a bit less than 40 percent of the total EU budget, is still substantial.

The Growth of Corporate Agriculture and Food Marketing

Despite the fact that family farms are valued in the European Union, they continue to decline in number—just as they did several decades ago in the United States. Large farms are created by consolidating small family plots into more profitable operations run by European and foreign corporations. These farms tend to employ very few laborers and use more machinery and chemical inputs.

The move toward corporate agriculture is strongest in Central Europe. When Communist governments gained power in the mid-twentieth century, they consolidated many small, privately owned farms into large collectives (except in Slovenia; see the case study). After the breakup of the Soviet Union, these farms were rented to large corporations, which further mechanized the farms and laid off all but a few laborers. Rural poverty rose. Small towns shrank as farmworkers and young people left for the cities. With EU expansion, the CAP has provided even more incentives for large-scale mechanized agriculture in Central Europe. Efforts to counter this trend include green food production and farm tourism, both of which have received CAP subsidies. In Slovenia, some families have retained their small farms by having one or more family members commute to urban factories. Slovenes have found green agriculture an attractive alternative to large-scale commercial agriculture, as illustrated by the following case study.

CASE STUDY: Green Food Production in Slovenia

During the Communist era, Slovenia was unlike most of the rest of Central Europe in that the farms were not collectivized. As a result, the average farm size is just 8.75 acres (3.5 hectares). Although Slovenia has plenty of rich farmland, as standards of living have risen it has become a net importer of food. Nonetheless, Slovenia's new emphasis on private entrepreneurship, combined with a demand throughout Europe for organic foods, has encouraged some Slovene farmers to carve out a niche for themselves in organic farming, first in local markets and eventually as exporters. The case of Vera Kuzmic is illustrative (FIGURE 4.20). ■

FIGURE 4.20 Vera Kuzmic in her market stall in Ljubljana.

Lydia Pulsipher

Thinking Geographically
How can you tell that Vera is involved in small-scale agriculture?

VIGNETTE Vera Kuzmic (a pseudonym) lives 2 hours by car south of Ljubljana, Slovenia's capital. For generations, her family has farmed 12.5 acres (5 hectares) of fruit trees near the Croatian border. In the economic restructuring that took place after Slovenia became independent in 1991, Vera and her husband lost their government jobs. The Kuzmic family decided to try earning its living in vegetable-market gardening because vegetable farming could be more responsive to market changes than fruit tree cultivation. By 2000, the adult children and Mr. Kuzmic were working on the land, and Vera was in charge of marketing their produce and that of neighbors whom she had also convinced to grow vegetables.

Vera secured market space in a suburban shopping center in Ljubljana, where she and one employee maintained a small vegetable and fruit stall (see Figure 4.20). Her produce had to compete with less expensive, Italian-grown produce sold in the same shopping center—all of it produced on large corporate farms in northeastern Italy and trucked in daily. But Vera gained market share by bringing her customers special orders and by guaranteeing that only animal manure and no pesticides or herbicides were used on the fields. For a while, her special customer services and her organically grown produce kept her in business. But when Slovenia joined the European Union in 2004, she had to do more to compete with produce growers and marketers from across Europe who now had access to Slovene customers.

Anticipating the challenges to come, the Kuzmics' daughter Lili completed a marketing degree at the University of Ljubljana. The family incorporated their business and Lili is now its Ljubljana-based director, while Vera manages the farm. Lili's market research showed the wisdom of diversification, so the Kuzmics continue to focus on selling their produce to those in Ljubljana's expanding professional population, who are willing to pay extra for fine organic vegetables and fruits. But now, in a banquet facility on the farm built with CAP funds, Vera also prepares special dinners for bus-excursion groups from across Europe interested in witnessing traditional farm life and in tasting Slovene ethnic dishes made from homegrown organic crops. *[Source: Lydia Pulsipher. For detailed source information, see Text Sources and Credits.]* ∎

THINGS TO REMEMBER

GEOGRAPHIC THEME 2 • **Globalization and Development:** Following World War II, Europe took steps to create an economic union now known as the European Union, which features open borders, the free movement of people and goods, progress toward a common currency, and efforts to promote human well-being and social cohesion. To improve its global competitiveness, the EU shifted industries from high-wage western Europe to the lower-wage member states of Central and South Europe. Service jobs replaced many industrial jobs, but unemployment increased. The European Union struggles to integrate the economies of its member states and to reduce regional disparities.

• Collectively, the EU countries are wealthy, with a joint economy slightly larger than that of the United States, making the European Union the largest economy in the world.

• In recent years, there have been major challenges to the euro, most significantly in the form of a debt crisis that has tested the mechanisms that governments use to maintain economic stability.

• Concerns about food security in Europe led to a heavily subsidized and regulated agricultural system known as the Common Agricultural Policy (CAP).

• The CAP has been revised to help support economic viability, environmental sustainability, and balanced territorial development.

• Agriculture, though a small part of the European Union GDP, remains socially important across the EU and now accounts for about 40 percent of the budget of the European Union.

POWER AND POLITICS

GEOGRAPHIC THEME 3

Power and Politics: By the 1960s Europe was shedding its colonial dependencies, participating with the United States as a leader in global affairs, and implementing economic unification. After the fall of the Soviet Union in 1991, the European Union expanded into Central Europe, emphasizing political freedoms and economic reform. Migration to West and North Europe from Central and South Europe and from former colonies as well as from several conflict areas increased steadily. By 2015 the intractable Syrian revolution and social unrest elsewhere in Asia and Africa instigated the massive migration of asylum seekers and economic migrants into Europe. Islamic jihadism resulted in several terrorist attacks, including two in Paris, France, which garnered international attention.

After World War II, most of West, South, and North Europe strengthened their commitment to the expansion of political freedoms. This transition was most pronounced in Germany, where a new post-war constitution strongly protected political freedoms in an effort to make the authoritarianism of Nazi Germany impossible in the future. A similar post-war constitution was drawn up in Italy.

In several other parts of Europe, the expansion of political freedoms came later or was compromised by violence. Spain remained a dictatorship until 1975, and Portugal formally democratized only after a revolution related to decolonization in Africa in 1974 **(FIGURE 4.21D)**. Until the late 1990s, Northern Ireland had sectarian violence between Catholics and Protestants so severe that free and fair elections were not possible (see Figure 4.21C). Central Europe did not significantly expand political freedoms until the Soviet Union dissolved in 1991 as its former satellites declared independence, one by one. Yugoslavia, a large Communist republic in southern Central Europe, always firmly outside the Soviet bloc, also dissolved beginning in the early 1990s. There the growth of political freedoms was hampered by a powerful wave of ethnic xenophobia and violence (see Figure 4.21A).

THE POLITICS OF EU EXPANSION

After the fall of the Soviet Union, the European Union expanded into Central Europe, transforming the region and setting the stage for future expansions that could reach well beyond the borders of Europe as it is currently defined (see Figure 4.15).

As early as the 1980s, Soviet control over Central Europe began to falter in the face of a workers' rebellion in Poland known as *Solidarity*. Discontent spread even into Russia, where rebellion against authoritarianism led to a softening of control and an opening known as *Glasnost*. By 1990, East Germany had reunited with West Germany. The dissolution of the Soviet Union in 1991 brought the collapse of many economic and political relationships in Central Europe. Thousands of workers lost their jobs in state-owned factories. In some of the poorest countries, including Romania, Bulgaria, and Hungary, social turmoil and organized crime threatened stability. Membership in the European Union became especially attractive to Central European political leaders and citizens who thought, overly optimistically, it would spur economic development for them and would bring investment by the wealthier EU member countries in West and North Europe.

Standards for EU membership, however, are rather demanding and specific. A country must have both political stability and a democratically elected government. Each country has to adjust its constitution to EU standards that guarantee the rule of law, human rights, and respect for minorities. Each must also have a functioning market economy that is open to investment by foreign-owned companies and that has well-controlled banks. Finally, farms and industries must comply with strict regulations governing the finest details of their products and the health of environments. Meeting these requirements has been a challenge for members such as Romania, Bulgaria, and Hungary.

Two very wealthy countries chose not to join the European Union: Switzerland and Norway. They, plus Iceland, have long treasured their neutral role in world politics, and they were also concerned about losing control over their domestic affairs. After the 2008 recession, Iceland lost enormous wealth because of banking speculation. Long a candidate, Iceland has now chosen not to join the European Union, but it maintains a close relationship with the EU through the European Economic Area (EEA) and border agreements in the Schengen Agreement. Iceland participates in the EU open market, has adopted many EU laws, and contributes to the EU budget (see Figure 4.16).

Several countries on the perimeter of Europe are candidate EU countries. They include Turkey, Macedonia, Montenegro, and Serbia. The prospect of Turkey joining the EU has always been in doubt and now seems more remote. Turkey has strained relations with the island country of Cyprus (which was admitted to the European Union in 2004) and has a history of human rights violations against minorities (especially against its large Kurdish population). Women in Turkey have experienced a serious contraction of civil liberties in recent years. There are also issues regarding the separation of religion and state. Turkey would be the first majority Muslim country to join the European Union and recent policies indicate that Turkey is abandoning its official secular posture. Turkey itself has some reservations. Until recently, its economy had grown faster than the EU average, leading some Turks to question the need to join the EU at all. Moreover, Turkey appeared to view its geopolitical advantage as being less with Europe and more as a leader in the Middle East. As the attempted revolutions collectively referred to as the *Arab Spring* (beginning in 2010) developed, Turkey subtly changed its behavior toward Europe, going from being a voice of calm and reason regarding political developments in the Middle East to adopting heavy-handed responses to a wave of Arab Spring–type protests arising within Turkey itself. By 2016, as the lengthy crisis in Syria persisted, accompanied by the flight of many Syrian refugees through Turkey and increasing terrorist activity within Turkey, the behavior of Recep Tayyip Erdoğan (first as prime minister and then as president of Turkey) became increasingly autocratic. The trajectory of Turkey's relationships with Europe and the Middle East became ever more difficult to discern.

A few other countries—Ukraine, Moldova, and perhaps even the Caucasian republics (Armenia, Azerbaijan, and Georgia)—may eventually be invited to join the EU. However, there is strong opposition to this within Europe, and Europe's huge and potentially powerful neighbor, Russia, opposes quite actively the expansion of the European Union to the east into what it considers its sphere of influence.

EU GOVERNING INSTITUTIONS

Somewhat similar to the United States, the European Union has one executive branch and two legislative bodies. The *European Commission* acts like an executive branch of government, proposing new laws and implementing decisions. Each of the 28 member states gets one commissioner, who is appointed for a 5-year term, subject to the approval of the European Parliament. Commissioners are expected to uphold common EU-28 interests and not those of their own countries. The head of the commission is elected by the European Parliament and serves a 5-year term in what amounts to being a prime minister or head of state. The

entire commission must resign if censured by Parliament. The European Commission also includes 33,000 civil servants who work in Brussels to administer the European Union on a day-to-day basis.

EU citizens directly elect the *European Parliament*. Each country elects a proportion of seats based on its population, much like the U.S. House of Representatives. Laws must be passed in Parliament by 55 percent of the member states, which must contain 65 percent of the EU total population. In other words, a simple majority does not rule. The European Parliament employs about 6000 civil servants.

The *Council of the European Union* is similar to the U.S. Senate in that it is the more powerful of the two legislative bodies. However, its members are not elected but consist of one minister of government from each EU country. The Council of Europe employs about 3500 people and is responsible for implementing the broad range of EU social cohesion policies.

ON THE BRIGHT SIDE: The Nobel Peace Prize
Europe's long history of institution building has facilitated the healing of wounds. After both world wars, the countries of the region used their social and economic momentum to work cooperatively with each other to create and maintain the European Union. This accomplishment was recognized in October of 2012 with the awarding of the Nobel Peace Prize, a prize that normally goes to an individual, to the entire European Union. ■

NATO AND THE RISE OF THE EUROPEAN UNION AS A GLOBAL PEACEMAKER

The European Union acts as a global peacemaker and peacekeeper through the **North Atlantic Treaty Organization (NATO)**, which is based in Europe. During the Cold War, European and North American countries cooperated militarily through NATO to counter the influence of the Soviet Union. NATO originally included the United States, Canada, the countries of western Europe, and Turkey; it now includes almost all the EU countries as well. Russia views this expansion as a threat. In 2015, NATO invited Macedonia to join, possibly as a signal to Russia as Russia ever more assertively tried to increase its influence.

Since the breakup of the Soviet Union, NATO has focused mainly on providing the international security and cooperation needed to expand the European Union. When the United States invaded Iraq in 2003, most EU members opposed the war. As worldwide opposition to the United States built, the global status of the European Union rose. Thereafter, with the United States preoccupied in Iraq and Afghanistan, NATO and the EU assumed more of a role as a global peacekeeper. For example, NATO forces fought off attempts by pirates to seize merchant ships during the 2009 Somali pirate crisis off the northeast coast of Africa, and undertook a number of operations in the Mediterranean related to the Arab Spring in Algeria, Tunisia, Libya,

North Atlantic Treaty Organization (NATO) a military alliance between European (including Turkey) and North American countries, developed during the Cold War to counter the influence of the Soviet Union; after the breakup of the Soviet Union in 1991, NATO expanded to include much of eastern Europe. NATO now focuses on providing the international security and cooperation needed to expand the European Union

FIGURE 4.21 PHOTO ESSAY: Power and Politics in Europe

Thinking Geographically

After you have read about power and politics in Europe, you will be able to answer the following questions:

A What clues in this photo suggest war and not another calamity?

C What are the political implications of the sign announcing entry into Free Derry?

D Why could it be said that independence movements in Portugal's African colonies brought democracy to Portugal?

Patterns of political power and conflict in Europe today and elsewhere tend to reflect the extent to which people enjoy political freedoms. The most violent recent conflicts in Europe took place in Central Europe (Hungary, Romania, Greece, the Balkans) during the transition away from communism. The denial of political freedoms allowed extreme ethnic conflict to erupt. The map of Europe shows armed conflicts and genocides with high death tolls since 1990.

Outside Europe, long and brutal wars were fought to gain independence from European colonial economic exploitation. The denial of local political freedoms by colonial administrations was a major impetus for independence movements. The map of Africa and Asia shows how low levels of democratization persisted in the countries long after the formal end of colonial rule. European imperialism helped set many of these countries on an authoritarian trajectory that has only recently started to change.

and Syria. After the November 2015 attacks in Paris (see the discussion below), there was talk of NATO coming to the aid of France in retaliating against ISIS in Syria. However, the phrasing of the mutual defense provision (Article V of the North Atlantic Treaty) allows all allies to make their own decisions regarding such aid, so NATO, as such, is unlikely to aid in any retaliation.

POST-COLONIAL TRIBULATIONS IN EUROPE

Despite being recognized for its rapid and peaceful development after World War II, and despite its success in quieting disastrous ethnic hostilities in southeastern Europe during the 1990s, by 2014, Europe was encountering a rash of political problems related to its former colonialist activities around the world. Many trace the persistence of dictatorships in the Middle East, for example, to the manipulation by Europe of the politics and economies of Middle Eastern countries. Europe's interference was especially prominent after World War I and has extended into the present in such countries as Algeria, Morocco, Egypt, and Syria. France is a particular case in point, especially with regard to its former colony of Algeria, which became independent after an 8-year war, in 1962. Since the nineteenth century, hundreds of thousands of French had migrated to Algeria to run farms and industries there, exporting most products back to France. French rule was harsh and Algerians were kept poor and powerless. After independence, political turmoil and killings continued, so both French colonists and many Algerian workers migrated to France.

Over time, the Algerian French became concentrated and socially excluded in suburban Paris housing blocks, far from jobs, training, and other services. They were not admitted into normal French society unless they completely assimilated as French. Most were stigmatized by their appearance, names, and Muslim faith. Some disillusioned youth, even those born and raised in France, were attracted to **Islamic jihadism,** which promised a robust identity and a mission to claim power through terrorism. This is

Islamic jihadism a personal or group struggle to promote Islamic revivalism, often with the threat of or actual use of force

the backstory to the terrorist attacks in France in 2015 and Belgium in 2016, but similar scenarios can be sketched in for the UK's colonial relationships with the Middle East

D A group of Angolan rebels captured during Angola's war of independence from Portugal, which fought three major wars in Africa during the 1960s and 1970s to hold on to its colonies there. The rising human and financial costs of these wars eventually brought about a revolution in Portugal itself in 1974, which at the time was an undemocratic authoritarian dictatorship. The revolution resulted in independence for the colonies and the eventual establishment of democracy in Portugal. None of Portugal's colonies have well-protected political freedoms, in part because of the authoritarian governing structures put in place by the Portuguese and because of the lengthy civil wars that followed decolonization.

Weber/Hulton Archive/Getty Images

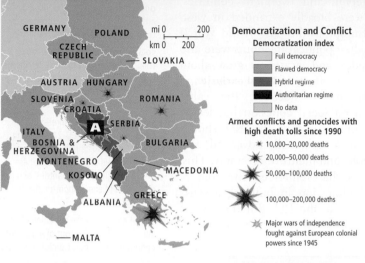

Democratization and Conflict

Democratization index

- Full democracy
- Flawed democracy
- Hybrid regime
- Authoritarian regime
- No data

Armed conflicts and genocides with high death tolls since 1990

- ✷ 10,000–20,000 deaths
- ✷ 20,000–50,000 deaths
- ✷ 50,000–100,000 deaths
- ✷ 100,000–200,000 deaths

- ✷ Major wars of independence fought against European colonial powers since 1945

A The aftermath of Bosnia's civil war in a suburb of Sarajevo in 1996. Attempts by Serbia to limit democratic decision making within the former Yugoslavia sparked several wars during the 1990s. In Bosnia, a civil war that pitted different ethnicities against each other resulted in 175,000 deaths.

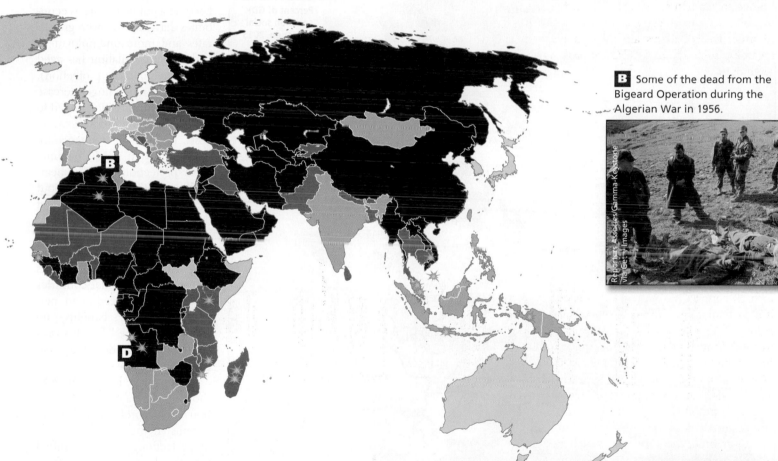

B Some of the dead from the Bigeard Operation during the Algerian War in 1956.

C Two murals in Derry, Northern Ireland, mark an entrance to a "nationalist" neighborhood. The mural on the right depicts a boy wearing a gas mask and holding a homemade petrol bomb. Both murals were created during "the Troubles," a time of violent conflict from 1969 to 1998. For many decades Northern Ireland was the site of repeated terrorist attacks between Protestants and Catholics. Competing claims about Northern Ireland's democratic legitimacy still flourish, though the violence has subsided since peace was agreed to in 1998.

and South Asia, as well as for Portugal's and Belgium's relationships with parts of Africa (see Figure 4.21). The case of Algeria is only one variant of the difficult **post-colonial** struggles of Muslims to find acceptance and to create a life for themselves in their traditional or adopted countries.

SOCIAL PROTECTION SYSTEMS: GOALS AND OUTCOMES

All European countries have tax-supported systems of *social welfare*, or **social protection** (the EU term). The stated goal is to use taxes to fight poverty and promote **social cohesion** by providing all EU citizens with basic health care; assistance with parenting responsibilities, free or low-cost higher education; affordable housing; and pension, survivor, disability, and unemployment benefits. The ideas behind these systems developed during the nineteenth and twentieth centuries, and were broadly expanded in western Europe in the post–World War II era, when political freedoms were also expanded and awareness grew about the serious effects of **social exclusion.**

In recent decades, the value placed on social cohesion has led to the consensus that the EU should become a technically advanced, sustainable, and inclusive economy. This, in turn, has reinforced the role of social protection in building a sense of loyalty to the EU. But the cost of European social protection systems has

post-colonial the conditions and attitudes that persist in a society after colonization is over

social protection in Europe, tax-supported systems that provide citizens with benefits such as health care, long-term care, pensions, and relief from poverty and social exclusion

social cohesion an expression of solidarity within the EU-28, involving balanced and sustainable development, reduced structural disparities between regions and countries, and the promotion of equal opportunities for all people

social exclusion the condition of being systematically left out of important opportunities for participation in society

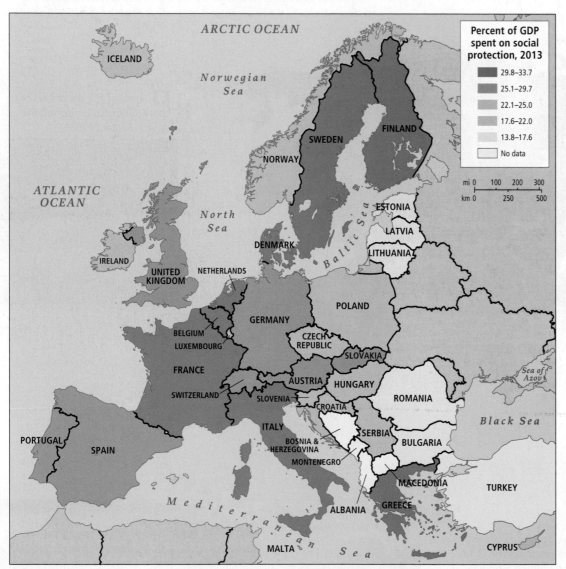

become a political issue as economies have had slower growth rates and more unemployment and as the population has aged (thus needing more protection). Then, the sudden huge increase in refugee immigration added to political contentions regarding costs, because the migrants are so obviously in need of immediate protection. At the heart of the political debates about social protection costs is the fact that there is a relatively wide disparity of income and well-being among EU countries. Since social protection benefits are paid by individual countries based on an agreed-upon percentage of GDP, benefits vary widely across the EU. This variability tends to undermine social cohesion.

The map in **FIGURE 4.22** shows the geographic distribution of social protection (welfare) systems in the EU. The levels of protection can be classified into five basic categories, from least social protection coverage to most. The three countries that have the most well-funded social protection are in North Europe (Finland, Sweden, and Denmark); they attempt to create equality across gender and class lines by providing extensive health care, education, housing, and child- and elder-care

FIGURE 4.22 European expenditures for social protections. In 2013 (the latest data available from the EU), the expenditures for social protection in the EU-28 averaged about 29.5 percent of total EU GDP. The map shows the percent of GDP each country in the EU-28 spent on social protection in 2013. Some non-EU members are included for comparison. [Source consulted: http://ec.europa.eu/eurostat/web/social-protection /data/main-tables]

benefits to all citizens, from cradle to grave. In North Europe, child care is state supported and widely available, in part to help women enter the labor market but also to give all children a certain common set of values. The hope is that this will ensure that in adulthood every citizen will be able to contribute to the best of his or her capability, and that citizens will not develop criminal behavior or abuse drugs. While finding comparable data is very difficult, surveys of crime victims across the European Union show that the crime rate in Scandinavia is generally lower than in other parts of Europe.

The West Europe countries of France, Belgium, and the Netherlands (also dark green on the Figure 4.22 map) have somewhat less lofty goals. France, Belgium, and the Netherlands and those colored medium green (the UK, Germany, Austria, Switzerland, Italy, and Portugal) assist those in need but do not try to promote upward mobility. For example, in France, college education is free or heavily subsidized for all, but strict entrance requirements can be hard for the poor to meet, a common complaint among Muslim and sub-Saharan African migrants. State-supported health-care and retirement pensions are available to all, but for the most part social protection systems still reinforce the traditional "housewife contract" by assuming that women will stay home and take care of children, the sick, and the elderly. In this group, Germany probably has the most generous system and is presently liberalizing its policies to aid working women.

The former Soviet bloc countries—Bulgaria, Romania, Slovakia, and the Baltic Republics—spend the least of their GDP on social protection (14.4 percent to 18.4 percent). And because these countries all have fairly low GDPs, it is clear that citizens in these countries in Central Europe have the least protection. Poland, as well as countries in southeastern Europe, plus Greece, which have not released data, are also low in social protection. Slightly better protected are citizens in the Czech Republic, Hungary, Malta, and Croatia.

During the Communist era, social protection was comprehensive, resembling the current cradle-to-grave social democratic system in Scandinavia, except that women were pressured to work outside the home. Benefits, usually tied to one's job, often extended to nearly free apartments; health care; state-supported pensions; subsidized food; fuel; vacations; and early retirement. However, in the post-Communist era, state funding has collapsed, forcing many people to do without basic necessities. In many post-Communist countries, welfare systems are now being revised, but usually with the goal of reducing benefits drastically and raising the retirement age to over 65.

Europeans generally pay much higher taxes than North Americans (the overall sum of taxes and compulsory contributions for the EU-28 countries was in 2014 about 40 percent of GDP. For the United States it was 25 percent; and for Canada, less than 30 percent). In return, Europeans expect more in terms of services. In some cases, European governments are able to deliver services more cheaply than the so-called free market does elsewhere. For example, the European Union spends on average between U.S.$3000 and U.S.$4000 per person for health care, while the United States spends more than U.S.$8000. Even at this much lower cost, the European Union has more doctors, more acute-care hospital beds per citizen, and better outcomes than the United States in terms of life expectancy and infant mortality.

While some argue that Europe can no longer afford high taxes if it is to remain competitive in the global market, others say that Europe's economic success and high standards of living are the direct result of the social contract to take care of basic human needs for all. The debate has been resolved differently in different parts of Europe, and the resulting regional differences have become a source of concern in the European Union, because, with open borders, unequal benefits can encourage those in need to flock to a country with a generous welfare system and overburden the taxpayers there.

The refugee crisis in 2015–2016 highlighted several intertwined concerns: Already strapped taxpayers couldn't manage to support hundreds of thousands of needy newcomers. The refugees did seem to be well informed about variable levels of social protection in the EU, as they flocked to Germany and Sweden (after first arriving in Italy and Greece). The aging of Europe's population is yet another point of concern because as more and more people retire, the tax burden on a shrinking labor force increases. Of course, the refugees tend to be young, educated, and ambitious, so once they are helped to adjust and find work, they will become new taxpayers. Europeans are increasingly aware that their social protection systems are among the most expansive and well-funded in the world and that basic reforms are necessary to preserve them.

ON THE BRIGHT SIDE: Social Protection, Social Cohesion, and Migrants in Europe

Despite the fact that there is not agreement on just how generous investment in social protection systems should be, the EU has managed to reach consensus on a growth strategy that emphasizes social cohesion. In order to reach the agreed-upon common goals of having technically advanced, sustainable, and inclusive economies, each of the EU-28 countries has adopted its own national targets regarding employment, innovation, education, social inclusion, GHG emissions, and energy usage, to be reached by 2020. When the refugee crisis erupted in the EU in 2015, it looked like the abrupt arrival of well over a million migrants, many of whom were Muslim, might spell the end of social cohesion and the consensus on growth—even, perhaps, the end of the EU open-border policies. Many countries seemed to be vying over which one could be the least welcoming. Then, slowly and spasmodically, EU citizens were awakened to the fact that many parts of the EU could use a boost of young, educated workers who would soon become taxpayers. In early 2016, when quotas on the number of migrants each country would be asked to accept were announced (based on the host country's population), many countries first objected strenuously, and then within a few days changed their minds. At least some began to see that in the medium and longer term, the migrants could have a positive impact on the labor supply, economic growth, and the tax base, provided the right integration and assimilation policies could be developed. ∎

THINGS TO REMEMBER

GEOGRAPHIC THEME 3 • **Power and Politics:** By the 1960s Europe was shedding its colonial dependencies, participating with the United States as a leader in global affairs, and implementing economic unification. After the fall of the Soviet Union in 1991, the European Union expanded into Central Europe, emphasizing political freedoms and economic reform. Migration to West and North Europe from Central and South Europe and from former colonies as well as from several conflict areas increased steadily. By 2015 the intractable Syrian revolution and social unrest elsewhere in Asia and Africa instigated the massive migration of asylum seekers and economic migrants into Europe. Islamic jihadism resulted in several terrorist attacks, including two in Paris, France, which garnered international attention.

• After the fall of the Soviet Union, the European Union expanded into Central Europe in a series of enlargements that have transformed this region and set the stage for future expansions that could eventually reach well beyond the borders of Europe as it is currently defined.

• Somewhat similar to the United States, the European Union has one executive branch and two legislative bodies.

• A new role for the European Union as a global peacemaker and peacekeeper is developing through the North Atlantic Treaty Organization (NATO), which is based in Europe.

• Its long history as a colonizer of other parts of the world is coming back to haunt Europe. Many of Europe's former colonies have had persistent political upheavals that have destabilized their societies and provoked massive migrations and political backlash in the form of terrorism.

• In nearly all European countries, tax-supported systems of social protection provide all citizens with basic health care; free or low-cost higher education; affordable housing; pension, survivor, and disability benefits; and unemployment benefits. Across the EU these benefits are being reevaluated because of their costs.

URBANIZATION

GEOGRAPHIC THEME 4

Urbanization: Europe's cities are both ancient and modern, with old town centers now surrounded by modern high-rise suburbs and supported by world-class urban infrastructures. Cities are the heart of Europe's economy, politics, and culture, and more than 70 percent of the EU population lives in urban areas. Yet much of Europe remains idyllic and productive rural countryside, important symbolically, economically, and as a place for regular recreation.

Though historically based on trade and manufacturing, most European cities are now primarily service oriented. Education and research have been important in urban Europe since

Thinking Geographically

After you have read about urbanization in Europe, you will be able to answer the following questions:

A What are the likely reasons why Prague is located along a river?

B What can you see in this picture that illustrates what you have read about urban life in Europe?

C Describe the people you see walking in this picture. How might walking as a mode of transportation affect the environment and urban spaces of Barcelona?

D Earlier you read about how the amount of greenhouse gas produced by rail-based public transportation compares to that of private automobile–based transportation. Apply that knowledge to life in Europe's cities.

medieval times, and now Europe has numerous hubs of innovation in the knowledge-intensive service sector, which is so crucial to the modern European economy.

One of the most important aspects of the service sector is tourism. With thousands of years of history and culture visible in architectural gems and well-funded museums, concert halls, theaters, and other cultural institutions, many cities in Europe draw a disproportionate share of the world's tourists. In 2013, EU-28 countries accounted for 40 percent of global tourist arrivals, a very large share given the region's size and population (see the discussion on page 216).

Many European cities began as trading centers more than a thousand years ago and still bear the architectural marks of medieval life in their historic centers (**FIGURE 4.23A**). These old cities are located either on navigable rivers in the interior or along the coasts because water transportation figured prominently (as it still does) in Europe's trading patterns.

Since World War II, nearly all the cities in Europe have expanded around their perimeters in concentric circles of apartment blocks (see Figure 4.23B). Usually, well-developed rail and bus lines link the blocks to one another, to the old central city, and to the surrounding countryside. Land is scarce and expensive in Europe, so only a small percentage of Europeans live in single-family homes, although the number is growing. Even single-family homes tend to be attached or densely arranged on small lots. Except in public parks, which are common, one rarely sees the sweeping lawns familiar to many North Americans. Because publicly funded transportation is widely available, many people live without cars, in apartments as close to the city center as they can afford (see Figure 4.23C, D). People walk more in Europe and when necessary they commute by bus or train (and increasingly by car) from suburbs or ancestral villages to work in nearby cities.

Although deteriorating housing and slums do exist (especially in some of the poorer cities and towns of Central Europe and in very large cities like Paris, Brussels, and London), substantial public and private investments in housing, public transportation, sanitation, water, and utilities mean that most people maintain a generally high standard of urban living.

FIGURE 4.23 PHOTO ESSAY: Urbanization in Europe

Europe's cities are famous throughout the world for their architecture, economic dynamics, and cultural variety. Many are quite ancient, with quaint city centers that are now surrounded by a ring of 18th- and 19th-century buildings and then large, more modern apartment blocks. The various parts of some cities are connected by public transportation, but in others transportation outside of the center is difficult, time consuming, or expensive. Despite their global draw, many European cities will shrink in the next several decades due to declining national populations (as in Italy and parts of Central Europe) as well as to deliberate efforts (as in the case of London) to shift growth to other urban centers.

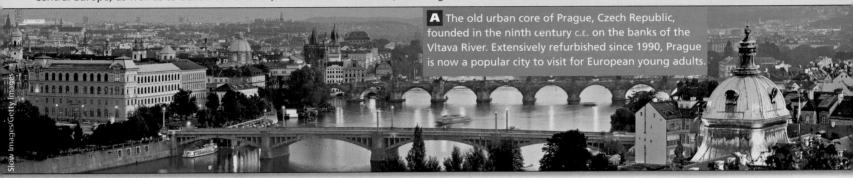

A The old urban core of Prague, Czech Republic, founded in the ninth century c.e. on the banks of the Vltava River. Extensively refurbished since 1990, Prague is now a popular city to visit for European young adults.

B High-rise apartment blocks dot the periphery of Berlin's old urban core. Berlin does a better job of linking its high-rise suburbs to urban life than do Paris or Brussels, where immigrants are often isolated and excluded from urban advantages.

C La Rambla, a tree-lined pedestrian mall in Barcelona, Spain, is a popular strolling place among locals and tourists.

D A tram in Paris, France. The city's metropolitan transit system moves roughly 6 million people per day, and is the second-largest system in Europe, after London's.

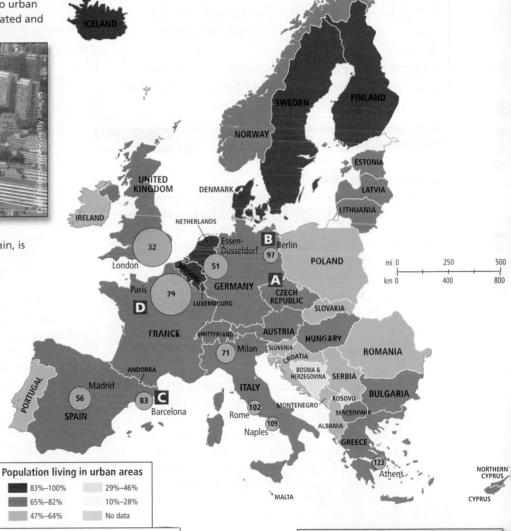

Population living in urban areas

- 83%–100%
- 65%–82%
- 47%–64%
- 29%–46%
- 10%–28%
- No data

Population of metropolitan areas 2015

- 10 million
- 6 million
- 3 million

Note: Symbols on map are sized proportionally to metro area population

① **Global rank** (population 2015)

229

THINGS TO REMEMBER

GEOGRAPHIC THEME 4 • **Urbanization:** Europe's cities are both ancient and modern, with old town centers now surrounded by modern high-rise suburbs and supported by world-class urban infrastructures. Cities are the heart of Europe's economy, politics, and culture, and more than 70 percent of the EU population lives in urban areas. Yet much of Europe remains idyllic and productive rural countryside, important symbolically, economically, and as a place for regular recreation.

• Though once based on employment in industry and manufacturing, most European cities now employ primarily service-oriented workers.

• Many older cities are located either on navigable rivers in the interior or along the coasts because water transportation figured prominently (as it still does) in Europe's trading patterns.

• Since World War II, nearly all the cities in Europe have expanded around their perimeters in concentric circles of apartment blocks.

• Land is scarce and expensive in Europe, so only a small percentage of Europeans live in single-family homes, although the number is growing.

POPULATION AND GENDER

GEOGRAPHIC THEME 5

Population and Gender: Europe's population is aging as fertility rates decline. Fertility rates are down due to the high cost of raising a child and because career-oriented women are choosing to have only one or two children. Fearing that aging populations will slow economic growth, European countries try to boost their fertility rates with a variety of policies—the most successful of which address the needs of working mothers. Encouraging immigration may be one way to boost population numbers, but eventually migrants may also opt for smaller families. However, accommodating and assimilating migrants, either refugees or those only seeking jobs, has proved a difficult social issue.

Europe has gone through the demographic transition (see Chapter 1) and there are several economic consequences: the now low birth rate means fewer consumers are being born. The economies and the tax bases of countries could contract over time. Demand for new workers, especially highly skilled ones, may go unmet. Furthermore, the number of younger people available to provide expensive and time-consuming health care for the elderly, either personally or through tax payments, may decline. Currently, for example, there are just two German workers for every retiree.

EUROPE'S POPULATION IS AGING RAPIDLY

Between 1960 and 2014, the proportion of the European population that is 15 years and under declined from 27 percent to 16 percent (well below the global average of 26 percent), while those over 65 increased from 9 percent to 17 percent (well above the global average of 8 percent). Longer life expectancies have influenced this trend; the average life expectancy of an EU citizen is 78 for men and 83 for women. However, a larger contributor to the aging of the population is the declining fertility rate.

Overall, Europe is now close to a negative rate of natural increase (<0.0), the lowest in the world. In all of the EU-28, fertility rates are below the replacement rate for developed countries (2.1 children per woman aged 15–49). Most EU countries have fertility rates of 1.5 or lower. The one-child family is becoming common throughout Europe. France has a fertility rate of 2.0, mostly because its large immigrant population tends to have slightly bigger families until that population assimilates to French culture. Ireland has a rate of 2.0 because of the strength of Roman Catholic rules against birth control, but other Roman Catholic countries (Spain, Italy, Portugal) have rates below 1.5.

Europe's work force is shrinking. The rates vary, but in rural areas of Spain, for example, more than 1500 villages have been abandoned and if present trends hold, every future generation of Europeans will be noticeably smaller than the previous one. The declining birth rate is illustrated in the population pyramids of European countries, which look more like lumpy towers than pyramids. **FIGURE 4.24** shows the pyramids for Germany, Sweden, and of the whole European Union. The pyramids' narrowing base indicates that by the mid-1970s, the post-war baby boom of the 1950s and 1960s had ended across Europe. By 2000, 25 percent of Europeans were having no children at all.

Why are Europeans choosing to not have children? Urbanization and strong economic development play a role, as does the fact that more and more women want professional careers. In Germany, for example, birth rates are lowest among the most educated women, a group that is growing in size every year. Twenty-five percent of all German women are now choosing to remain unmarried well into their thirties, which means that they will likely have only one or two children, if any.

European governments have tried to boost fertility rates with a variety of policies, the most successful of which have addressed the problems faced by working mothers. Current research on fertility policy suggests that career-minded working mothers are better aided by help with child care than by the generous, full-pay maternity leave of several months to a year common in this region. This may relate to the professional consequences of taking long periods off from work, such as missed opportunities for promotions; also, employers are leery of hiring young women because of the cost of extended maternity leaves. Fertility rates in both France and Sweden increased slightly after each country began providing free child care and preschools. In Germany, where there has historically been little government assistance for day care or preschools, fertility rates remain much lower. There is now a movement within the European Union to provide subsidized child care and more preschools and to lengthen school days so mothers can work full time.

Immigration could provide a partial solution to the dwindling number of young people. Immigrants from outside the region are the major source of population growth today in the EU, and they are providing new workers and taxpayers and are slowing the

FIGURE 4.24 Population pyramids for Germany, Sweden, and the EU, 2014. These population pyramids have quite different shapes, but all have a narrow base, indicating low birth rates. The EU pyramid **(C)** is at a different scale than those of **(A)** Germany and **(B)** Sweden because of the EU's large total population. [Sources consulted: "Population Pyramid of Germany," "Population Pyramid of Sweden," and "Population Pyramid of the European Union," International Data Base, U.S. Census Bureau, 2014, at http://www.census.gov/population/international/data/idb/informationGateway.php?cssp=SERP]

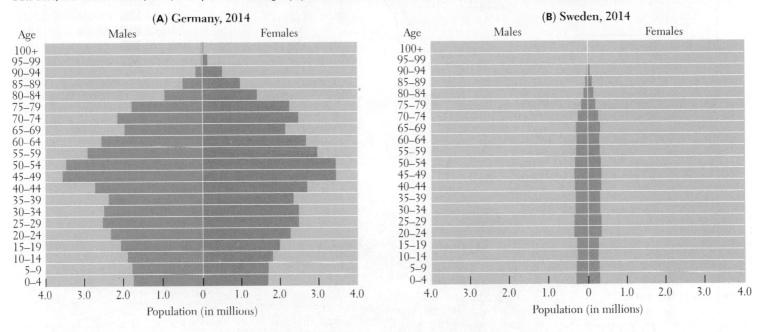

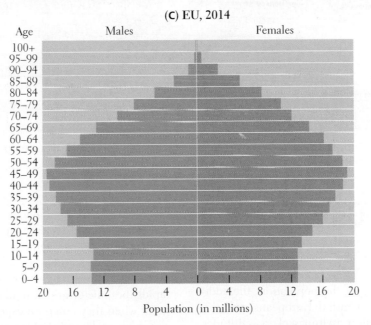

trend toward population aging. However, to completely counter the low fertility rates common throughout this region, millions of new immigrants would be required (many more than the surge of close to a million refugees in 2015). This prospect is politically unpopular in Europe, where many voters are alarmed by the potential cultural changes brought by so many immigrants, who often come from very distant places with different customs and value systems. Moreover, after a generation, most immigrants choose to have smaller families too, making immigration (while still advantageous for many reasons) a poor long-term solution to the aging of the population.

There are currently about 742 million Europeans. Of these, 510 million live within the European Union. The highest population densities stretch in a discontinuous band from the United Kingdom south and east through the Netherlands and central Germany into northern Switzerland **(FIGURE 4.25)**. Northern Italy is another zone of high density, as are pockets in many countries along the Mediterranean coast.

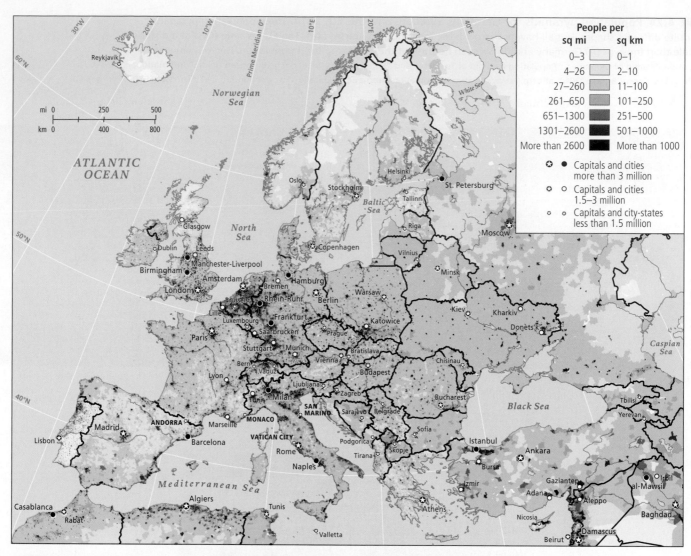

FIGURE 4.25 Population density in Europe. Overall, Europe is one of the more densely occupied regions on Earth, as shown in the world population density map in Figure 1.26.

SOCIOCULTURAL ISSUES

The European Union was conceived primarily to promote economic cooperation and free trade, but its programs have social implications as well. As population patterns change across Europe—lower fertility rates, aging of the population, the addition of more conservative-minded Central Europeans, and the addition of large numbers of Muslim asylum seekers—attitudes toward aging, immigration, and gender roles are also evolving. The European Union needs immigration, but the idea that many of these new people will be of the Muslim faith causes apprehension for some people in Europe. On the other hand, attitudes about gender are becoming more progressive, and these progressive ideas tend to discourage people from having two or more children.

ACTIVE AGING

Concern over the negative effects that might result from the aging of Europe's population has reinforced many stereotypes about

older citizens. They have been described as obsolete, under-achieving, unable to learn new skills, a drag on the economy, and expensive to care for. For a time, these descriptors were used to push people in their late fifties into early retirement. Anyone over 40 was thought to be not worth investment in training. But it rapidly became clear that sending such young workers into retirement when they could be expected to live another 30 years was not only socially rude but terribly expensive, and resulted in large numbers of impoverished and idle elderly.

As part of its emphasis on social cohesion, the EU has begun to examine the issue of aging by developing a range of strategies. **Active Aging** has three broad dimensions that address employment, social participation, and independent living. The Active Aging Index, mapped in **FIGURE 4.26**, measures how individual countries are doing in these areas. The *employment initiative* does not merely emphasize staying longer at work and retiring later, but rather making those added years of work

> **Active Aging** a three-pronged EU policy aimed at changing society's attitudes toward the abilities and needs of people as they age

FIGURE 4.26 Aging: The north-south and east-west divides. One way the EU is addressing issues raised by its aging population is to implement a new program called Active Aging, which has three components: employment, social participation, and independent living. An Active Aging Index has been calculated for the EU-28 and has revealed that there are clear north-south and east-west divides. Countries in the north and west are better prepared in all three categories for the increase of older people than are those in the south and east. These differences are linked to many other factors, such as levels of education and wealth. This map depicts those regional differences. [Source consulted: European Commission EUROSTAT website on Employment, Social Affairs and Inclusion: http://ec.europa.eu/social/main.jsp?langId=en&catId=1063&newsId=2430&further News=yes, accessed June 7, 2016.]

especially productive. Appreciating the potential of older workers is emphasized. The *social participation initiative* recognizes the important contributions to society of older workers and family members who either do paid work or volunteer to help within the family and within the community. The *independent living initiative* places special responsibility on the aging person to remain fit and healthy and independent, so that, barring catastrophic illness, the need for care and support is limited to the last few days of life. As might be expected from previously discussed information, and as shown in Figure 4.26, the countries of North Europe are the best places to age, while Central and South Europe are places where active aging is more difficult to achieve.

MIGRANTS, REFUGEES, AND BORDERS: NEEDS AND FEARS

Until the mid-1950s, the net flow of migrants was out of Europe, to the Americas, Australia, and elsewhere. By the 1990s, the net flow was into Europe from many global locations, many of them former European colonies. In the 1990s, most of the European Union (plus Iceland, Norway, and Switzerland) implemented the Schengen Agreement, an agreement that allows the free movement of people and goods across much of Europe, with the exception of the UK (which opted out of the accord), as well as Ireland, Romania, Bulgaria, Croatia, and non-EU states in southeastern Europe **(FIGURE 4.27)**.

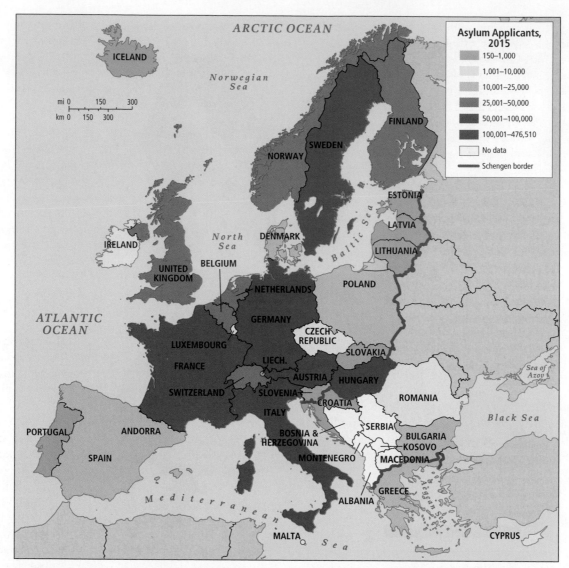

FIGURE 4.27 The EU Schengen Area and asylum applicants into Europe, 2014 and 2015. Countries within the Schengen Border are indicated with the heavy blue line. Migration into Europe continued to increase into the twenty-first century for most countries and remains a crucial issue in EU debates. The number of asylum applicants (migrants) into each country in 2015 (in some cases, 2014) is shown on the map. Many of these people are fleeing conflict from North Africa and Southwest Asia, while economic migrants are escaping from poverty and the absence of jobs in their home country. The distinction between "economic migrants" and "asylum seekers" is flawed, as discussed in the text. [Sources consulted: European Commission EUROSTAT websites http://ec.europa.eu/eurostat/web/asylum-and-managed-migration/data/main-tables; http://www.economist.com/blogs/economist-explains/2015/08/economist-explains-18]

The opening of Europe's borders facilitated trade, employment, tourism, and somewhat controversially, the migration of EU citizens. The Schengen Agreement has also indirectly increased both access for immigrants from outside the European Union and their mobility once they are in the European Union. Greece and Italy, both with long Schengen Area coastlines along the Mediterranean, have been major entry points for migrants, but both are also in serious financial straits and are unable to police their borders effectively. During the rapid influx of immigrants in 2015–2016, a few countries reinstated border controls when national security became an issue; but not until France did so after the November 2015 Paris attacks were borders closed for

any more than a few days. The repeated terrorist attacks in 2015 and 2016 and the simultaneous arrival of more than a million migrants (many of them refugees from Syria seeking asylum) caused much speculation that the Schengen Agreement was dead, an eventuality that would drastically change life and economies in Europe. It is not dead, but it is certainly under threat.

VIGNETTE (continued from page 196) After the call from Catholic Charities, Marina packed a small bag and left to volunteer in assisting the refugees as they came across the Croatian border. There she was shocked to see the pitiful condition of the people and their children. They were impossibly tired and muddy. To her surprise, she could talk with most of them because they were obviously educated and used English as a second language, like she does. They were, indeed, mostly Muslim, but they didn't seem alien at all and were so grateful for kind words and a change of clothes or a carriage for a toddler. After a week on the border, the chaos began to clear. Dozens of buses came to pick up the thousands of travelers to drive them to the Austrian border, from where they would proceed to Germany. Marina asked them why they didn't stay in Slovenia, which they had the right to do. Most refugees had Germany as a goal because there they felt they could find work as the doctors, teachers, and businesspeople they had been back home. Opportunities in Slovenia were completely unknown to them. In fact, no one she spoke with had ever even heard of Slovenia until now. ■

Like some citizens of the United States, some Europeans are ambivalent about migrants from anywhere, even though many are themselves descendants of migrants. Until recently, the internal flow of migration has been mostly from Central Europe into North, West, and South Europe. These Central European migrants were generally treated fairly, although prejudices against the supposed backwardness of Central Europe remain evident. Immigrants from outside Europe, so-called *international immigrants*, have met with varying levels of acceptance. Attitudes reached a new low during the refugee influx of 2015–2016. Perhaps because these days few Europeans identify themselves as belonging to any religious faith, they were left discomforted by shows of religious fervor by Muslim immigrants; but some went way beyond expressing discomfort to outwardly demonstrating hate.

International immigrants often come both legally and illegally from Europe's former colonies and protectorates across the globe. Many Turks and North Africans come legally as **guest workers** who are expected to stay for only a few years, fulfilling Europe's need for temporary workers in certain sectors. Other immigrants are refugees from the world's trouble spots, such as Syria, Afghanistan, Iraq, Haiti, and Central Africa. Many also come in illegally from all of these areas.

CASE STUDY: The Rules for Assimilation: Muslims in Europe

In Europe, culture plays as much of a role in defining differences between people as race and skin color. An immigrant from Asia or Africa may be accepted into the community if he or she has gone through a comprehensive change of lifestyle. **Assimilation** in Europe usually means giving up the home culture and adopting the ways of the new country, including skill in its language and conformity in dress. If minority groups—such as the Roma (see page 214), who have been in Europe for more than a thousand years—maintain their traditional ways, it is nearly impossible for them to blend into mainstream society.

Europe's small but growing Muslim immigrant population (**FIGURE 4.28**) is currently the focus of assimilation issues in the European Union. Muslims come from a wide range of places and cultural traditions, including groups that have lived in Europe for a thousand years or more. Others have come more recently from

Turkey, North Africa, Syria, Lebanon and Jordan, Pakistan, India and Bangladesh. Some of the more recent immigrants maintain traditional dress, gender roles, and religious values, while others assimilate into European culture quickly. Those who do not assimilate, especially in dress and language, run the risk of social exclusion. ■

While some Europeans see international immigrants as important contributors to their economies—providing needed skills and labor and diminishing the impact of low birth rates—many oppose recent increases in immigration. Studies of public attitudes in Europe show that immigration is least tolerated in areas where incomes and education are lowest. Central and South Europe are the least tolerant of new immigrants, possibly because of fears that immigrants may drive down wages that are already relatively low. North and West Europe, with higher incomes and generally more stable economies, are the most tolerant. The European Union is working on curbing illegal immigration from outside Europe while encouraging EU citizens to be more tolerant of legal migrants.

Here should also be mentioned the vocal but increasingly problematic debate about the differences between **economic migrants** and **refugees**. Economic migrants—those who come in search of better job opportunities than they had in their home country—may have immigrated legally or illegally. Refugees (people who have fled from their homes because of war or violent social unrest or discrimination) often become **asylum seekers**—those who seek international protection because they fear war and persecution at home—and hence are classed as legal immigrants. Economic migrants are often thought to be unworthy because they are coming out of ambition rather than necessity. There are several things wrong with this designation of "unworthy." Economic migrants are often themselves fleeing conflicts or discrimination of one sort or another in their home countries and therefore may deserve to be viewed as refugees. Also, Europe badly needs more workers who are ambitious, so why should these economic migrants be unwelcome? It may be because so many of them are black- or brown-skinned and from former colonies of Europe. Furthermore, the refugees are for the most part self-identified and often carry no papers, so confirming their claims to be fleeing terrorism and war is very difficult.

The deepening alienation and anger that has boiled over in recent years among some Muslim immigrants and their children—resulting in protests, riots, and sometimes even terrorism, as in Paris in 2015—relates primarily to the systematic exclusion of these less assimilated Muslims from meaningful employment, from adequate social services, and from higher education. When the French media investigated protests by Muslim residents, they found that the

guest workers legal workers from outside a country who help fulfill the need for temporary workers but who are expected to return home when they are no longer needed

assimilation the loss of old ways of life and the adoption of the lifestyle of another country

economic migrants those who seek employment opportunities better than what they have at home

refugees those who have fled from their homes because of war or violent social unrest or discrimination

asylum seekers refugees who left their homes because of violent persecution and seek new legal status in an adopted country

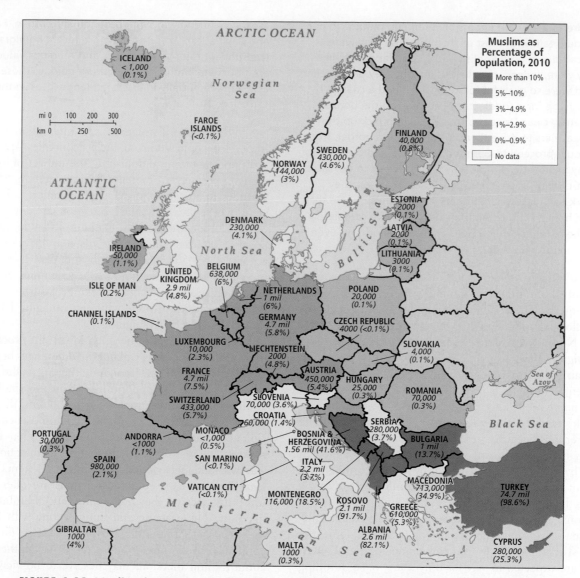

FIGURE 4.28 Muslims in Europe. Muslims are a small minority in most European countries, but their population has grown from a total (in the whole of Europe) of 29.6 million in 1990 to 44.1 million in 2010—amounting to about 6 percent of the total population—and is expected to rise to more than 58 million, or 8 percent of Europe's population, by 2030. In just the EU-28, there were an estimated 20 million Muslims in 2016—including those who came in 2015 and 2016—or about 4 percent of the total EU-28 population. Many intend to make the region their permanent home. Just how, or if, the assimilation of Muslims into generally secular Europe will proceed is a major topic of public debate. See the case study, "The Rules for Assimilation: Muslims in Europe," on page 235. [Sources consulted: New data (2010) from Pew Research Center on Muslims in each country and as a percent of total population of each country; and The Future of the Global Muslim Population. Pew Research Center. Retrieved December 22, 2011. Official data for the recent (2015–2016) influx of Muslim migrants was not available in June 2016.]

protestors' complaints were indeed legitimate. Young Muslims born in Europe who have never known life in any other place and who have been schooled in the lofty ideals of the European Union, harbor the greatest resentment against the constricted opportunities they face. However, many Europeans, unaware of the extent to which their own societies discriminate against and exclude Muslims, have come to view Muslim protesters simply as malcontents.

In some cases, conflicts have also arisen over European perceptions of cultural aspects of Islam, such as the *hijab* (one of

several traditional coverings for women). For example, in 2004, the wearing of the hijab by observant Muslim schoolgirls became the center of a national debate in France about civil liberties, religious freedom, and national identity. French authorities viewed the hijab as a symbol of what they perceived to be the subjugation of women in Islam. They wanted to ban the hijab but were wary of charges of discrimination. Eventually all symbols of religious affiliation were declared illegal in French schools, including Christian crosses and yarmulkes (the Jewish head covering for men). This ban has only reinforced among some Muslims' feelings of being excluded and made the hijab an important symbol of identity and pride.

CHANGING GENDER ROLES

Gender roles in Europe have changed significantly from the days when most women married young and worked in the home or on the family farm. Growing numbers of European women are working outside the home, and the percentage of women in professional and technical fields is rising rapidly. Nevertheless, European public opinion among both women and men largely holds that women are less able than men to perform the types of work typically done by men and that men are less skilled at domestic, caregiving, and nurturing duties. In most of Europe, men still have greater social status, hold more managerial positions, and have more autonomy in daily life than women—more freedom of movement, for example. These advantages for men are stronger in Central and South Europe than they are in West and North Europe.

The Gender Pay Gap

Perhaps most significantly, the gender pay gap in Europe is wide and persistent. **FIGURE 4.29** shows this gap in 2013 and 2014 (the latest data available), in every employment category within the EU. After adjusting for differences in education and experience, men still earn considerably more than women. Pay for both genders is more equal in the education and health professions (except among physicians), where salaries are also lower on average than in manufacturing and finance. Pay is the most discriminatory in those latter two sectors, except among clerical support workers. Of course, this gender pay gap, which is similar to the global gender pay gap, impacts the financial well-being of European women and their families.

The **double day**—the unpaid domestic work that must be done daily in addition to paid employment—continues to fall mostly to women. Certainly, younger European men now assume more cooking, housework, and child-care duties than did their fathers, but women who work outside the home are usually expected to do most of this unpaid domestic work. UN research shows that in most of Europe, women's domestic duties average 4 hours and 20 minutes each day, while men put in 2 hours and 16 minutes.

Women burdened by the double day generally operate with somewhat less efficiency in a paying job than do men. They also tend to choose employment that is closer to home and that offers more flexibility in the hours and skills required. These more flexible jobs (often erroneously classified as part time) almost always offer lower pay and less opportunity for advancement, though not necessarily fewer work hours.

Influence of Women on Policy Formation

Despite the persistent gender inequality, many EU policies officially encourage gender equality, and the number of managerial posts in the EU bureaucracy held by women is growing. Well over half the university graduates in Europe are now women. Membership by women in European national parliaments is increasing, although women constitute less than one-third of

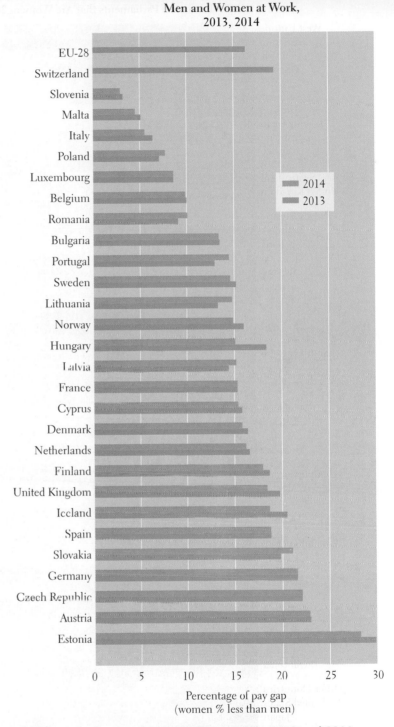

FIGURE 4.29 The gender gap in the EU-28 in 2013 and 2014.

elected representatives in 21 of the 28 countries that make up the EU. Only in Finland (43), Spain (41), and Sweden (44) do women come close to filling 50 percent of the seats in the legislature **(FIGURE 4.30)**. Despite their higher rates of university and professional education, the political influence and economic well-being of European women lag behind those of European men.

double day the longer workday of women with jobs outside the home who also work as caretakers, housekeepers, and/or cooks for their families

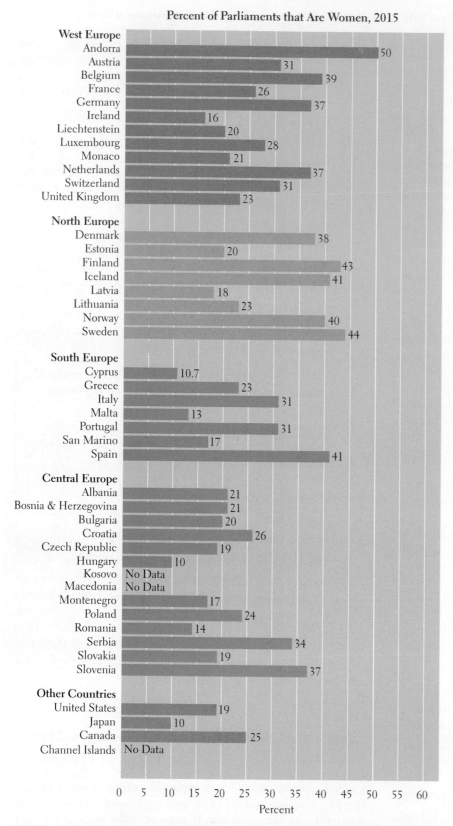

Percent of Parliaments that Are Women, 2015

West Europe
- Andorra — 50
- Austria — 31
- Belgium — 39
- France — 26
- Germany — 37
- Ireland — 16
- Liechtenstein — 20
- Luxembourg — 28
- Monaco — 21
- Netherlands — 37
- Switzerland — 31
- United Kingdom — 23

North Europe
- Denmark — 38
- Estonia — 20
- Finland — 43
- Iceland — 41
- Latvia — 18
- Lithuania — 23
- Norway — 40
- Sweden — 44

South Europe
- Cyprus — 10.7
- Greece — 23
- Italy — 31
- Malta — 13
- Portugal — 31
- San Marino — 17
- Spain — 41

Central Europe
- Albania — 21
- Bosnia & Herzegovina — 21
- Bulgaria — 20
- Croatia — 26
- Czech Republic — 19
- Hungary — 10
- Kosovo — No Data
- Macedonia — No Data
- Montenegro — 17
- Poland — 24
- Romania — 14
- Serbia — 34
- Slovakia — 19
- Slovenia — 37

Other Countries
- United States — 19
- Japan — 10
- Canada — 25
- Channel Islands — No Data

Percent (axis: 0 5 10 15 20 25 30 35 40 45 50 55 60)

FIGURE 4.30 Women in national parliaments of Europe, 2015. Women comprise about half the adult population of the EU, but only in a very few countries do they make up close to half of European legislatures; thus, their influence on policies and legislation is much restricted.

Numerous women have held high offices in the United Kingdom, but elsewhere in West Europe, this trend is only beginning. In 2005, Germany elected Angela Merkel as its first woman chancellor (prime minister) and elected her to a third term in 2014; in France in 2007, Nicolas Sarkozy defeated his female opponent Ségolène Royal but then appointed women to a number of French cabinet-level positions. In 2012, fully one-half of newly elected French President François Hollande's cabinet was female. Slovenia chose Alenka Bratušek as prime minister in 2013. In 2014, Croatia elected a female president, Kolinda Grabar-Kitarović. Still, across Europe, women generally serve only in the lower ranks of government bureaucracies, where they implement policies but have limited power to formulate them.

Although change is clearly underway in the European Union, economic empowerment for women has been slow on many fronts. Norway is a recognized global leader in redefining gender in society. It does this by directing much of its most innovative work on gender equality toward advancing men's rights in traditional women's arenas. For example, men and women are allowed to share the year of paid parental leave with a newborn or adopted child. This policy, recognizing a father's responsibility in child rearing and the need for father and child to bond early in life, is gaining popularity among celebrities in the United States under the name *family leave*. In 2015, Mark Zuckerberg, the Facebook cofounder and chief executive announced that he was taking a 2-month leave to be with his newborn daughter, Max.

THINGS TO REMEMBER

GEOGRAPHIC THEME 5 • **Population and Gender:** Europe's population is aging as fertility rates decline. Fertility rates are down due to the high cost of raising a child and because career-oriented women are choosing to have only one or two children. Fearing that aging populations will slow economic growth, European countries try to boost their fertility rates with a variety of policies—the most successful of which address the needs of working mothers. Encouraging immigration may be one way to boost population numbers, but eventually migrants may also opt for smaller families. However, accommodating and assimilating migrants, either refugees or those only seeking jobs, has proved a difficult social issue.

• Europeans are choosing to have fewer children; as a result, the population as a whole is aging. Small families are in part a result of women pursuing careers that require post-secondary education.

- Because Europe's population is not growing and the number of consumers is staying the same or decreasing, economies may contract over time.

- Until the mid-1950s, the net flow of migrants was out of Europe, to the Americas, Australia, and elsewhere. By the 1990s, the net flow was into Europe.

- While some Europeans see international immigrants as important contributors to their economies—providing needed skills and labor and diminishing the impact of low birth rates—many oppose recent increases in immigration under the assumption (usually erroneous) that the immigrants will not be self-supporting and contributing to society.

- Increasing numbers of European women are working outside the home, and the percentage of women in professional and technical fields is growing rapidly. But the gender pay gap is large.

- Although change is clearly underway in the European Union, economic empowerment for women has been slow on many fronts, especially in influential policy-formulating positions.

HUMAN WELL-BEING IN EUROPE

To a significant extent, the social welfare policies described above affect patterns of well-being which, not surprisingly, have a geographical pattern. As we have observed, although Europe is one of the richest regions on Earth, there is still considerable disparity in wealth and well-being (FIGURE 4.31A–C). Income per capita provides no way to determine if a few are rich and most poor, or if income is more equally apportioned.

Figure 4.31A, gross national income (GNI) per capita (PPP) for all of Europe, illustrates two points. Most of the region has a GNI of at least U.S.$12,000 per capita: only in three small countries in southeastern Europe—Albania, Bosnia and Herzegovina, and Kosovo—is the GNI less than U.S.$12,000 per capita. The map also shows that West and North Europe (except for Estonia, Latvia, and Lithuania) are the wealthiest regions. It is worth noting that Greece, which in the 1970s was still quite poor, joined the EU in 1981 and began to realign its economy (industrializing, reducing its debt and rate of inflation) in order to meet EU specifications. While it is remarkable that in fewer than 30 years Greece appeared to have joined the ranks of Europe's wealthiest countries, the extent to which Greece's prosperity was financed with expensive borrowing was revealed in 2010. These huge debt obligations have lowered the standard of living in Greece, at least for a while. The inset map shows how GNI in Europe ranks in comparison to other parts of the world.

Figure 4.31B, the Human Development Index (HDI) rank, shows that it is possible to have a high GNI rank and a lower HDI rank if social services are inadequately distributed or if there is a wide disparity of wealth. Although the maps in Figure 4.31A and B look very similar, they show that there are only a few countries in Europe that rank both in the highest GNI and highest HDI categories—Norway, Denmark, the Netherlands,

Luxembourg, and Switzerland. Most countries have a GNI ranking that is lower than their HDI rank, which indicates that all of these countries, regardless of their actual wealth, provide well for their people and that wealth disparity is under control. In many cases, this is because the countries have strong social safety nets or were until recently in communist societies that emphasized meeting basic needs for all.

Figure 4.31C, the Gender Development Index (GDI) reveals that while European countries generally treat males and females fairly equally, some countries (Ireland, the UK, Malta, most of Central Europe, plus Greece), have a GDI ranking that is markedly lower than that of the rest of Europe. This raises concerns about the well-being of women in these countries. On the other hand, many countries rank higher on GDI than they do on HDI. There is still cause for concern even in these countries, given what we know about ubiquitous disparities between women's and men's incomes, but in access to health care, education, and empowerment, these countries may come closer to equality between men and women than do most others.

THINGS TO REMEMBER

- Europeans generally value social protection safety nets for all, and standards of human well-being are generally high across Europe.

- Because European social protection systems vary according to how much citizens are willing to pay, the levels of well-being also vary.

- All social protection systems and gender equality policies are now being re-evaluated.

THE SUBREGIONS OF EUROPE

The subregions of Europe take on added significance these days because of the social and financial crises faced by the European Union. The refugee crisis, general immigration, unemployment, aging populations, global competitiveness, and recovery from economic recession are perceived, fairly or not, to fall along subregional divisions. Countries in South and Central Europe are seen as mismanaging the problems, and countries in West and North Europe think of themselves as the more stable and financially responsible entities that must bail out the others.

WEST EUROPE

The countries of West Europe include the United Kingdom, the Republic of Ireland, France, Germany, Belgium, Luxembourg, the Netherlands, Austria, and Switzerland (FIGURE 4.32). Despite their economic success, these countries are contending with several issues that will affect their development and overall well-being:

- Challenges to what had been a continual trend toward economic union with each other and with neighboring countries to the north, east, and south.

FIGURE 4.31 Maps of human well-being.

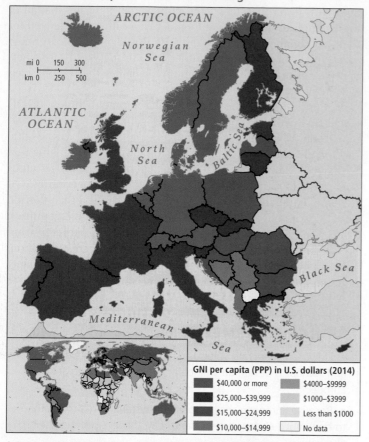

(A) Gross national income (GNI) per capita, adjusted for purchasing power parity (PPP).

GNI per capita (PPP) in U.S. dollars (2014)
- $40,000 or more
- $25,000–$39,999
- $15,000–$24,999
- $10,000–$14,999
- $4000–$9999
- $1000–$3999
- Less than $1000
- No data

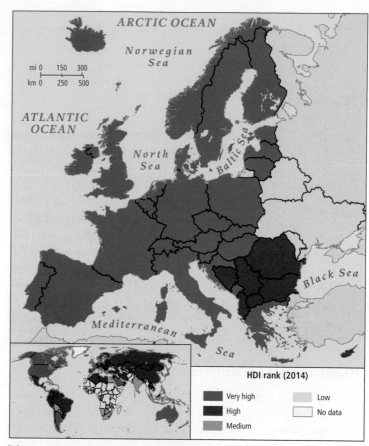

(B) Human Development Index (HDI) rank.

HDI rank (2014)
- Very high
- High
- Medium
- Low
- No data

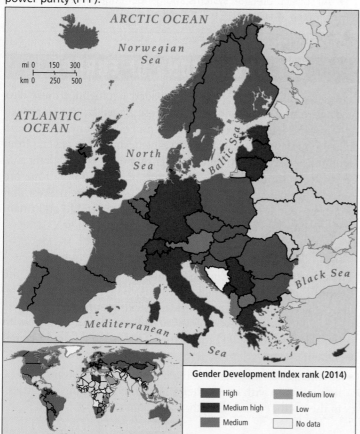

(C) Gender Development Index (GDI) rank.

Gender Development Index rank (2014)
- High
- Medium high
- Medium
- Medium low
- Low
- No data

- Social tensions, such as those generated by the influx of immigrants from inside and outside Europe, many of them refugees, and failures to incorporate migrants into societies.

- Worries about security in light of recent terrorist attacks, and confusion about how to promote the dream of free movement through Europe yet secure the borders against terrorists.

- The persistence of high unemployment despite general economic growth.

- The need to increase the global competitiveness of the subregion's agricultural, industrial, and service sectors while maintaining costly but highly valued social welfare programs.

Benelux

Belgium, the Netherlands, and Luxembourg—often collectively called the *Low Countries* or *Benelux*—are densely populated countries that have very high standards of living (see Figure 4.31A). These three countries are well located for trade: they lie close to North Europe and the British Isles and are adjacent to the commercially active Rhine Delta and the industrial heart of Europe. The coastal location of Benelux and its great port cities of Antwerp, Rotterdam, and Amsterdam give these countries easy access to the global marketplace. In addition, the Benelux countries have played a central role in the European Union for many years: Brussels, Belgium, is considered the EU capital, with most EU headquarters, along with the NATO headquarters, located there, which also draws other business to Brussels.

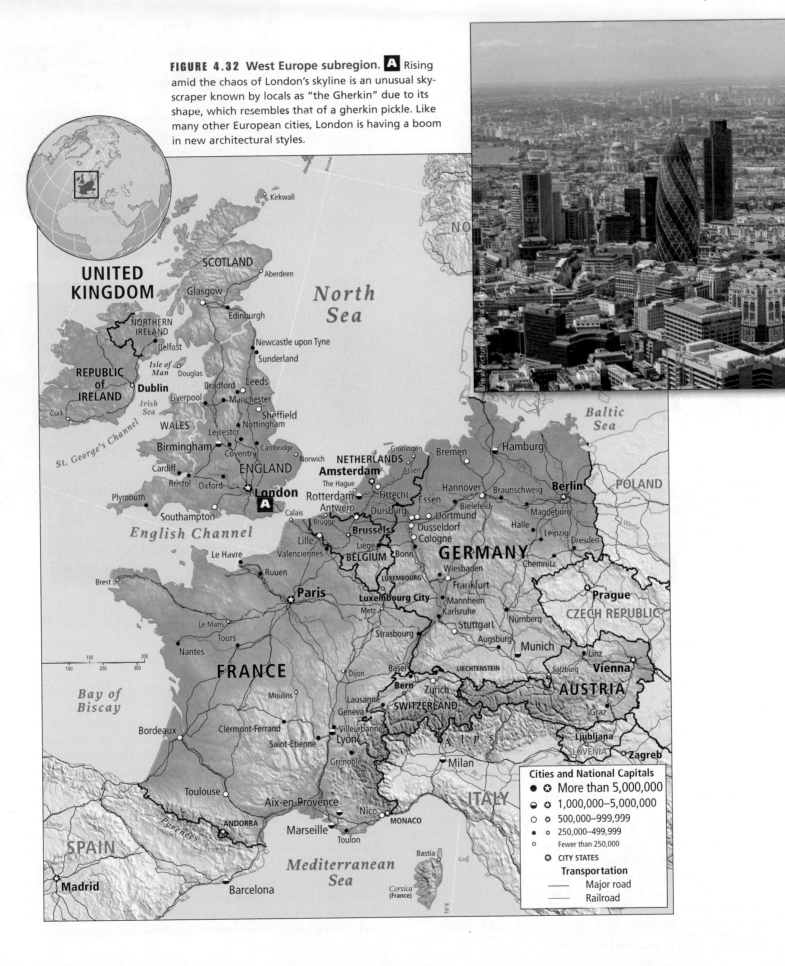

FIGURE 4.32 West Europe subregion. **A** Rising amid the chaos of London's skyline is an unusual skyscraper known by locals as "the Gherkin" due to its shape, which resembles that of a gherkin pickle. Like many other European cities, London is having a boom in new architectural styles.

UNITED KINGDOM

SCOTLAND
Kirkwall
Aberdeen
Glasgow
Edinburgh

North Sea

NORTHERN IRELAND
Belfast
Newcastle upon Tyne
Sunderland
Isle of Man Douglas
Bradford Leeds
REPUBLIC of IRELAND
Dublin
Liverpool Manchester
Irish Sea
Cork
Sheffield
WALES Leicester Nottingham
St. George's Channel
Birmingham
Coventry Cambridge
Cardiff
ENGLAND
Bristol Oxford
Plymouth
London
A
Southampton
Calais

English Channel

Le Havre
Brest
Rouen
Paris
Le Mans
Tours
Nantes

FRANCE

Bay of Biscay

Moulins
Bordeaux
Clermont-Ferrand
Saint-Etienne
Lyon
Villeurbanne
Grenoble
Toulouse
Aix-en-Provence
ANDORRA
Marseille
Toulon
SPAIN
Madrid
Barcelona

Pyrenees

Mediterranean Sea

Groningen
NETHERLANDS
Amsterdam
The Hague
Assen
Rotterdam Utrecht
Antwerp Duisburg
Brugge
Brussels
Lille
Valenciennes Liège
BELGIUM Bonn
LUXEMBOURG
Luxembourg City
Metz

Bremen Hamburg

Baltic Sea

Hannover Braunschweig **Berlin** POLAND
Bielefeld
Essen Magdeburg
Dortmund
Dusseldorf Halle Leipzig
Cologne Chemnitz Dresden

GERMANY

Wiesbaden
Frankfurt
Mannheim **Prague**
Karlsruhe Nürnberg CZECH REPUBLIC
Stuttgart
Strasbourg Augsburg
Basel Munich Linz
Dijon LIECHTENSTEIN Salzburg **Vienna**
Bern Zurich AUSTRIA
Lausanne SWITZERLAND Graz
Geneva

Alps

Ljubljana
SLOVENIA **Zagreb**
Milan
ITALY

Nice
MONACO

Bastia
Corsica (France)

Cities and National Capitals

Symbol		Population
●	✪	More than 5,000,000
◒	☆	1,000,000–5,000,000
○	✪	500,000–999,999
•	○	250,000–499,999
○		Fewer than 250,000
✪	CITY STATES	

Transportation
—— Major road
—— Railroad

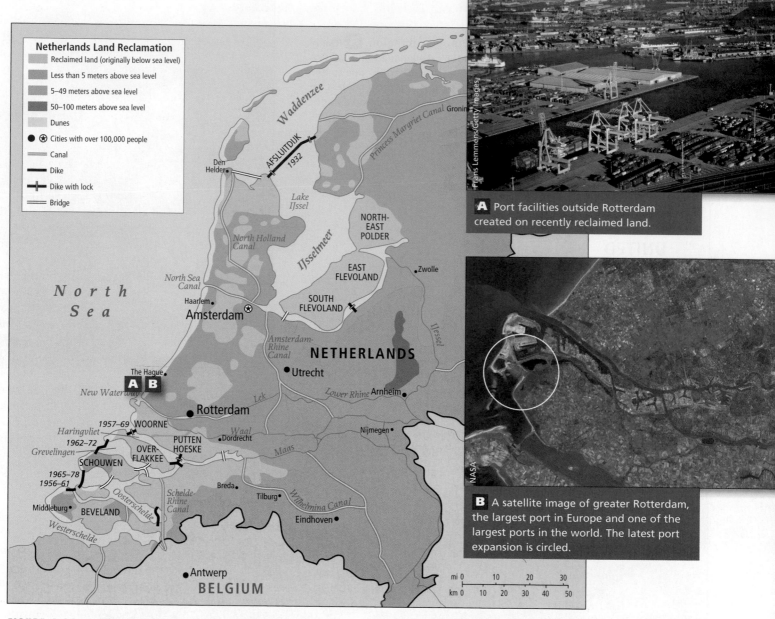

Netherlands Land Reclamation
- Reclaimed land (originally below sea level)
- Less than 5 meters above sea level
- 5–49 meters above sea level
- 50–100 meters above sea level
- Dunes
- ⬤ ⊛ Cities with over 100,000 people
- Canal
- Dike
- Dike with lock
- Bridge

A Port facilities outside Rotterdam created on recently reclaimed land.

B A satellite image of greater Rotterdam, the largest port in Europe and one of the largest ports in the world. The latest port expansion is circled.

FIGURE 4.33 Land reclamation in the Netherlands.

International trade has long been at the heart of the economies of Belgium and the Netherlands. Both were active colonizers of tropical zones: Belgium in Africa and the Netherlands in the Caribbean, Africa, and Southeast Asia. Their economies benefited from the wealth extracted from the colonies and from global trade in tropical products such as sugar, spices, cacao beans, fruit, wood, and minerals. Private companies based in Benelux still maintain advantageous relationships with the former colonies, which supply raw materials for European industries. Cacao is an example of this: It is exported from West Africa in a raw state to Benelux and other European countries, where it is refined into high-end elegant products by the upscale chocolatiers there. Some of these products are now part of the fair trade movement in Europe and America that channels a bigger share of profits to cacao producers in Africa and elsewhere. Getting some of the profits to the actual field laborers is still problematic.

The largest Benelux nation, the Netherlands, is noted particularly for having reclaimed land that was previously under the sea and for creating attractive living spaces in wetlands (**FIGURE 4.33A, B** and the figure map). Today, its landscape is almost entirely a human construct. As its population grew during and after the medieval period, people created more living space by filling in a large, natural coastal wetland. To protect themselves from devastating North Sea surges, they built dikes, dug drainage canals, pumped water with windmills, and constructed artificial dunes along the ocean. Today, a train trip through the Netherlands between Amsterdam and Rotterdam takes one past the port of Amsterdam with lots of ships, fuel storage facilities, and raised rectangular fields crisply edged with narrow drainage ditches and wider transport canals. The fields are filled with commercial flower beds, and one is also likely to see vegetable gardens and grazing cattle, which feed the primarily urban population.

Although the Netherlands has a high standard of living, with 16.9 million people, it is the most densely settled country in Europe (with the exception of Malta), and this density has consequences. There is no land left for the kind of high-quality suburban expansion preferred in the Netherlands that wouldn't intrude on agricultural space. And there is not nearly enough space for recreation; bucolic as they appear, transport canals and carefully controlled raised fields are not venues for picnics and soccer games. People now travel great distances to reach their jobs, and these long commutes sap time and patience and add to air pollution and traffic jams. Even for weekend getaways, people usually leave the Netherlands for neighboring countries that have more natural areas. The choice to maintain agricultural space is largely a psychological and environmental one, as agriculture accounts for only 2.8 percent of GDP and 1.8 percent of employment.

France

France is shaped like an irregular hexagon. It is bound by the Atlantic on the north and west, by mountains on the southwest and southeast, by Mediterranean beaches on the south, and by lowlands on the northeast. The capital city of Paris, in the north of the hexagon, is the heart of the cultural and economic life of France. As one of Europe's two world cities (London is the other), Paris has an economy tied to providing highly specialized services—traditionally, fine food, food processing, fashion, and tourism, but increasingly, accounting, advertising, design, and financial, scientific, and technical services. These service industries draw in a steady stream of workers from Europe and beyond.

Few would quarrel with the idea that French culture has a certain cachet and that France is an arbiter of taste and a model for relaxed urban living. Its magnificent marble buildings, manicured parks, museums, and grand boulevards have made France the leading tourist destination on Earth. In 2014, France attracted about 100 million tourists, well more than the population of the entire country (64.3 million). The countries with the largest increases in visitors to France are Russia, China, and Brazil.

Paris, though elegant, is also a working city. It is the hub of a well-integrated water, rail, and road transportation system, and its central location and transport links attract a disproportionate share of trade and businesses, so it qualifies as a primate city. Together with its suburbs, Paris has almost 11 million inhabitants—roughly one-sixth of the country's total. In the 1970s, French planners decided that Paris was large enough, and they began diverting development to other parts of the country, especially to Toulouse and farther south to the Mediterranean.

To the west and southwest of Paris (toward the Atlantic) are less densely settled lowland basins that are used primarily for agriculture. France has the largest agricultural output in the European Union; globally, it is second only to the United States in value of agricultural exports. Its climate is mild and humid, though drier to the south. Throughout the country, farmers have found profitable specialties, such as wheat (France ranks fifth in world production), grapes (France and Italy compete for first place in world wine production and wine exports), and cheese (France is second in world production). Despite France's

agricultural leadership, agriculture accounts for only 1.7 percent of the French national GDP (2014) and employs only 3 percent of the population. Modernization has made possible ever-higher agricultural production with ever-lower labor inputs. The manufacturing sector, which expanded quickly after World War II, absorbed many former farmworkers; but after 1960, the high-end service sector grew even more rapidly. French industry now accounts for 19.4 percent of France's GDP, and services account for nearly 80 percent.

The Mediterranean coast is the site of the French Riviera and France's leading port, Marseille. Development here is booming. In part because of all of the tourism to the area, so many marinas, condominiums, and parks have been built that they threaten to occupy every inch of waterfront. The downside of this wealthy, densely populated region is that it produces large amounts of the pollutants that contribute to Mediterranean environmental problems (see the discussion on page 207).

France derived considerable benefit and wealth from its large overseas empire, which it ruled until the middle of the twentieth century. Many of the citizens of former French colonies in the Caribbean, North America, North Africa, sub-Saharan Africa, Southeast Asia, and the Pacific (and their descendants) now live in France and bring to it their skills and a multicultural flavor. Paradoxically, while the European French are proud of being cosmopolitan, they are also very protective of their distinctive culture and wish to guard what they consider to be its purity and uniqueness. This has led to a marginalization of some (especially Muslim) immigrant minorities, who tend to live on the fringes of the big cities in high-rise apartment ghettos, where they are poorly served by public transportation, have trouble accessing higher education and technical training, and have unemployment rates that are twice the national average.

Many people in France are deeply concerned about what they perceive to be a decrease in France's world prestige as its industrial competitiveness declines, unemployment rises, social welfare benefits decrease, and traditional French culture is diluted by an influx of migrants from former colonies. Such feelings have resulted in occasional swelling of popular support for the National Front, France's xenophobic right-wing party. So far, these tendencies have been countered by France's deep pride in its sophisticated multicultural identity and contribution to an envisioned egalitarian global community. But during the current influx of Muslim immigrants, xenophobia has increased, strengthening the National Front. From a purely practical point of view, because it has a low birth rate, France needs immigrants to keep its economy humming, its technology on the cutting edge, and its tax coffers full.

France's status as a leading member of the European Union rivaled that of Germany for many years. France's trade volume (imports and exports) remains the sixth largest in the world and it produces at least 25 percent of the EU agricultural products. Recently though, its economic productivity has begun to suffer and its national debt has increased, due in part to expensive social programs and lenient workday rules (the French work only a 35-hour week and have especially generous unemployment benefits). During the EU economic crisis, France paid for an ever-smaller portion of the bailout funds. Some have predicted that France will itself need a bailout because of constrictions in the manufacturing and service sectors.

Germany

The famous image of a happy crowd in 1989 dismantling the Berlin Wall, as the symbolic end of Soviet influence in Central Europe, had particular significance for Germany. For 40 years, the country had been divided into two unequal parts. This is because, at the end of World War II Russian troops, instead of going home, occupied the eastern third of Germany, and by 1949, the Soviets and their East German counterparts had turned it into a Communist state known as East Germany. From the beginning, many East Germans tried to flee to what was then West Germany. To retain the remaining population, the Soviets literally walled off the border in the summer of 1961. East Germany's rich resources of skilled labor, minerals, and industrial capacities were used to buttress the socialist economies of the Soviet bloc and to support the military aims of the Soviet Union.

When the Berlin Wall fell and the two Germanys were reunited, East Germany came home with enormous troubles that proved expensive to fix. Its industries were outdated, inefficient, polluting, and in large part irredeemable. Much of its decaying or substandard infrastructure (bridges, dams, power plants, and housing) did not meet the standards of western Europe and had to be remodeled or dismantled and replaced. Its industrial products were not competitive in world markets; East German workers, though considered highly competent in the Soviet sphere, were undereducated and underskilled by western European standards. Reunified Germany is Europe's most populous country (81 million people), the world's fifth-largest economy, and a global leader in industry and trade. But the costs of absorbing the poor eastern zone dragged Germany down in many rankings for years. Unemployment rose sharply (especially among women) in the 1990s and again during the global recession beginning in Europe in 2008. Not only have there been layoffs in the east (where unemployment is nearly double the national average), but workers throughout Germany, accustomed to high wages and generous benefits, are also losing jobs as firms mechanize or move overseas, some to the United States.

During periodic global economic recessions, Germany's many multinational corporations have an especially difficult time because they are so dependent on global trade in industrial products. Germany long had the largest export economy in the world ($1.28 trillion in 2014), but China—of course a far larger country in area and population—recently overtook Germany, and in 2014 had exports worth U.S.$2.34 trillion. The automobile industry, centered in Stuttgart, is an example of German commercial ventures that are moving operations out of Germany to countries, including the United States, where the markets for German cars lie. Expansion into Asia came not only because factory costs are lower there, but also because the market for luxury brands such as Mercedes-Benz is soaring, especially in China. Daimler Trucks, with home offices in Stuttgart, is the world's largest truck manufacturing company, with plants across Europe and in Africa, Asia, and Brazil. Daimler Trucks North America, now headquartered in Portland, Oregon, is North America's largest manufacturer of heavy-duty trucks.

FIGURE 4.34 Berlin's "bear pit" karaoke. Every Sunday afternoon in Berlin, an event that many visitors describe as magical and heartwarming is held. In the *Mauerpark*, or "wall park"—a green space established after the dismantling of the Berlin Wall in 1989—a crowd of thousands gathers in a stone amphitheater to cheer on total strangers performing karaoke. Singers almost always get a huge round of applause—whether they're good at karaoke or just come across as a nice person. Gareth Lennon, an Irish bike messenger and karaoke enthusiast, started the event in 2009 as he rode around Berlin with a bike-mounted, battery-powered karaoke system and filmed people as they performed. He happened on the amphitheater one Sunday afternoon; within months, hundreds and then thousands of people began to show up every weekend.

Because Germany was regarded as the instigator of two world wars, for years West Germany had to tread a careful path. It had to see to its own economic and social reconstruction and to rebuilding a prosperous industrial base without appearing to become too powerful economically or politically. For the most part, West Germany played this complex role successfully. Since the early 1980s, it has been a leader in building the European Union; with the EU's largest economy, it has borne the greatest financial burden of the European unification process. This burden has only increased during the 2008 economic crisis, the Greek debt crisis, and the 2015 refugee crisis. It is the German people who have taken in most of the refugees (close to 800,000 asylum seekers in 2015) and provided a majority of the funds used to assist Ireland, Greece, Spain, Portugal, and Italy.

Germany's leading role in the European Union has positioned Berlin to be highly attractive to Europe's young adults. It is now considered to be one of the world's most livable cities. **FIGURE 4.34** shows a rollicking Sunday afternoon in 2012 along the old route of the infamous Berlin Wall.

The British Isles

The British Isles, located off the northwestern coast of the main European peninsula, are occupied by two countries: (1) the United Kingdom of Great Britain (England, Scotland, and

Wales) and Northern Ireland—often called simply Britain or the United Kingdom (UK); and (2) the Republic of Ireland (not to be confused with Northern Ireland; see also Figure 4.32).

The Republic of Ireland The main physical resources in the Republic of Ireland (usually called simply *Ireland*) are its soil, abundant rain, and beautiful landscapes. Ireland's considerable human capital remained underdeveloped for centuries during and after the time it served as a test-case colony where Britain could hone the technique of mercantilism that it then used in the Caribbean, North America, Asia, and Africa. The British confiscated land for their own use and for the use of their local sympathizers; thus, large numbers of Irish people were turned into landless and starving paupers. When they protested, they were imprisoned or sent as indentured labor to the new Caribbean colonies of the 1600s. Into the modern era, the Irish depended on small-plot agriculture and eventually on tourism, while industrialization lagged. As a result, until very recently, the Irish people were the poorest in West Europe. Over the last two centuries, some 7 million Irish emigrated to find a better life. In the 1990s, however, a remarkable turnaround began. Ireland attracted foreign manufacturing companies by offering cheap, educated labor; accessible air transport; access to EU markets; low taxes; and other financial incentives. The economy grew so quickly that for a time Irish labor was in short supply and the unemployment rate was 1 percent.

To supply the workers required, Ireland invited back those who had emigrated, and recruited other foreign workers from many locales, especially the former Soviet satellites (for example, Latvia and Poland) now in the European Union. In 2000, approximately 42,000 people immigrated to Ireland, 18,000 of whom were returning emigrants, including some information technology (IT) workers from the United States. Czech workers packed Irish meat; Filipino nurses worked in most Irish hospitals. The cost of living rose sharply, however, and this rise plus the global recession meant that Ireland was no longer able to attract or even retain foreign investment. Ireland originally had not favored the admission of Central European countries to the European Union because it feared those countries would attract investment away from Ireland by offering skilled but even cheaper labor pools. This was beginning to happen by 2007. Nevertheless, until the global recession of 2008, Ireland had one of Europe's fastest-growing (though still small) economies, and industry in Ireland played nearly as important a role as services in the country's GNI. In 2007, Ireland had Europe's second-highest per capita GNI, but by the end of 2011, its unemployment rate was 15 percent, affecting 400,000 workers across the skills spectrum. Thousands of guest workers returned to Central Europe, and tens of thousands of native Irish left to other parts of the European Union, to North America, Australia, and New Zealand, leaving behind mortgages they could no longer afford. A banking crisis and rising government debt left Ireland in need of a bailout that amounted to €85 billion (U.S.$108 billion) from the European Union and the IMF, including more than 12 billion pounds (U.S.$19.3 billion) from the United Kingdom. After a stringent austerity period, foreign investors attracted by low taxes built up Ireland's

export capacity and by 2015 Ireland was beginning to rebound; its belt-tightening was being recommended as a model for Greece, Spain, and others.

Northern Ireland Northern Ireland consists of six counties in the northeastern corner of the island; it is distinct from the Republic of Ireland and is administratively part of the United Kingdom. Northern Ireland began to emerge as a political entity in the 1600s, when Protestant England conquered the whole of Catholic Ireland. England removed or killed many of the indigenous people and settled Lowland Scots and English farmers on the vacated land. The remaining Irish resisted English rule with guerrilla warfare for nearly 300 years, until the Republic of Ireland gained independence from the United Kingdom in 1921. Northern Ireland remained part of the United Kingdom. Protestant majorities held political control in Northern Ireland, and the minority Catholics were subject to severe economic and social discrimination. Catholic nationalists unsuccessfully lobbied using constitutional (peaceful) means for a united Ireland. Other Catholic groups, the most radical among them being the Irish Republican Army (IRA), resorted to violence, including terrorist bombings, often against civilians and British peacekeeping forces, who were seen as supporting the Protestants. The Protestants reciprocated with more violence. By 1995, more than 3000 people had been killed in the Northern Ireland conflict or by violence that spread to the United Kingdom.

In 1998, tired of seeing the development of the entire island—and especially of Northern Ireland—blighted by the persistent violence, the opposing groups finally reached a peace accord known as the *Good Friday Agreement*, which was overwhelmingly approved by voters in both Northern Ireland and the Republic of Ireland. The plan for Northern Ireland provides for more self-government shared between Catholics and Protestants, the creation of human rights commissions, the early release of convicted terrorist prisoners, the decommissioning of paramilitary forces (such as the IRA), and the reform of the criminal justice system. Since 1998, violence has periodically risen and then abated, and the society has slowly begun to accept peaceful coexistence. The city of Belfast remains mostly divided along religious lines, but some neighborhoods are beginning to integrate.

The United Kingdom The United Kingdom consists of England, Scotland, Wales, Northern Ireland, and 14 of what are known as British Overseas Territories, remnants of the British Empire. Scotland, one of the four main parts of the U.K., considered separating from the U.K. administratively, but Scottish voters did not accept this separation.

The United Kingdom has a mild, wet climate (thanks to the North Atlantic Drift) and a robust agricultural sector that is based on grazing animals. Its usable land is extensive, and its mountains contain mineral resources, particularly coal and iron. In contrast to Ireland, the United Kingdom has been operating from a position of power for many hundreds of years. Beginning in the seventeenth century, Britain acquired resources from its colonies, first in Ireland and then in the Americas, Africa, and Asia. Together, these resources were sufficient to make Britain the leader of the Industrial Revolution in the late eighteenth century. By the nineteenth century, the British Empire covered nearly a quarter of the world's land surface. As a result, British culture was diffused far and wide, and English became the *lingua franca* (the language of trade between speakers of different languages) of the world.

Britain's widespread international affiliations, set up during the colonial era, positioned it to become a center of international finance as the global economy evolved. Today, London is the leading global financial center, handling more value in financial transactions than New York City, its closest rival. This remains true despite the UK's rejection of the euro (€). That the UK continues to use the pound (£) may have helped its economy during the EU economic crisis; and in fact, anti-EU sentiment in the UK, fed by the UK's declining political and economic status and by contentions over the influx of immigrants, of which the UK has a great number, resulted in an unexpected vote in June 2016 to leave the European Union. Immediately there were regrets because no real plans existed on how such an important country would leave the EU or what the implications might be. The situation remains unresolved.

The United Kingdom is no longer Europe's industrial leader. At least since World War II, and some say earlier, the United Kingdom has been sliding down from its high rank in the world economy toward the position of an average European nation. The discovery of oil and gas reserves in the North Sea gave the United Kingdom a cheaper and cleaner source of energy for industry than the coal on which it had long depended, but this did not stop the economic decline. Cities such as Liverpool and Manchester in the old industrial heartland and Belfast in Northern Ireland have had long depressions.

Britain has successfully established technology industries in regions that previously were not industrial—for example, near Cambridge and west of London, in a locale called Silicon Vale. The service sector currently dominates the economy (accounting for 80 percent of the GDP and 84 percent of the labor force). In the last decade, new, though not high-paying, jobs have been created in health care, food services, sales, financial management, insurance, communications, tourism, and entertainment. Even the movie industry has discovered that England is a relatively cheap and pleasant place for film editing and production. Britain's past experience with a worldwide colonial empire has left it well positioned to provide superior financial and business advisory services in a globalizing economy. Such service jobs, however, tend to go to young, educated, multilingual city dwellers, many of them from elsewhere in the European Union, not to the middle-aged, unemployed ironworkers and miners left in the old industrial U.K. heartland. Because of the EU's open borders, there are now more than 400,000 foreign nationals working in Greater London alone.

The Jamaican poet and humorist Louise Bennett used to joke that Britain was being "colonized in reverse." She was referring to the many immigrants from the former colonies who now live in the United Kingdom. Indeed, it would be hard to overemphasize the international spirit of the U.K. London, for example, has become an intellectual center for debate in the Muslim world. Salman Rushdie, an Indian Muslim writer, lived under British protection after his controversial book *The Satanic Verses* was published and he received death threats; many bookstores carry Muslim literature and political treatises in a variety of languages; Israelis and Palestinians talk privately about peace; descendants of Indian and Bangladeshi immigrants run popular restaurants; and Saudi Arabians have a strong presence. A leisurely stroll through London's Kensington Gardens reveals thousands of people from all over the Islamic world who now live, work, and raise their families in London. They are joined by many others from the British Commonwealth—Indian, Malaysian, African, and West Indian families—pushing baby carriages and playing games with older children. Impromptu soccer games may have team members from a dozen or more countries.

THINGS TO REMEMBER

• Colonialism transformed Europe economically and culturally; but today, Europe is not only struggling to remain globally competitive, it is struggling to find a way to absorb immigrants from former colonies.

• Despite their long-term economic success, the countries of West Europe face several issues that affect their citizens' well-being, such as the persistence of high unemployment. Social tensions caused in part by the influx of immigrants, and the need to continue to increase the region's global competitiveness while maintaining costly but highly valued social welfare programs are precipitating deep divisions that threaten the stability of the EU.

• The economic crisis in the European Union has resulted in the countries of West Europe being called upon to contribute to the financial support of other countries with failing economies. Germany has borne the greatest burden.

• West Europe has some of the world's most attractive cities, which provide not only jobs for Europe's upwardly mobile young, but endless affordable diversions and opportunities to enjoy the diversity in this region of Europe.

SOUTH EUROPE

The Mediterranean subregion of South Europe **(FIGURE 4.35)** was once unrivaled in wealth and power. The Greek and Roman empires in the ancient period were followed in the medieval period by the Italian trading cities of Venice, Florence, and Genoa, with contacts stretching across the Indian Ocean and by land through Central Asia to eastern China. During the sixteenth century, Spain and Portugal developed large colonial empires, primarily in the Americas but also in Africa and Southeast Asia.

Then times changed, as the rest of Europe's economy expanded to a global scale with England, France, and the Netherlands' development of distant colonies. The Mediterranean ceased to be a center of trade. South Europe declined, and the Industrial Revolution reached this subregion relatively late. Even today, Portugal and Greece and parts of Italy and Spain remain poor in comparison to the rest of the continent; only some of the countries of Central Europe are poorer.

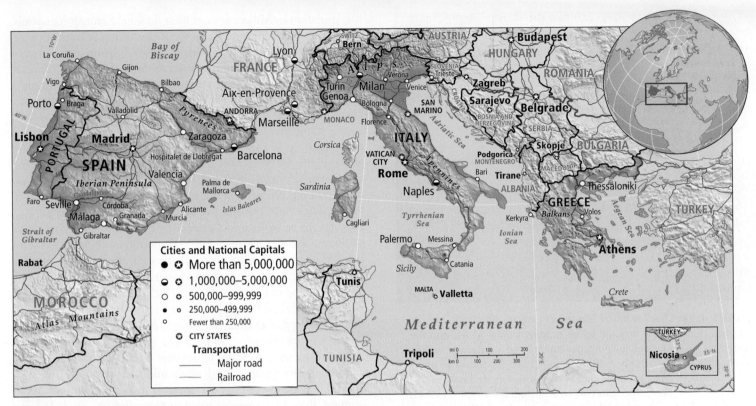

FIGURE 4.35 The South Europe subregion.

This section pays particular attention to Spain and Italy, contrasting the parts of each country that are now prospering with the parts that have remained poor. Then Greece, an early member of the European Union, is discussed in relation to the effect of its recent financial problems on its status as an EU member and how its financial problems have recently been alleviated by the fact that it is a main entry point for refugees/asylum seekers from the war-torn regions of Southwest Asia and North Africa.

In all six countries of South Europe—Portugal, Spain, Italy, Malta, Cyprus, and Greece—agriculture was the predominant occupation through most of the twentieth century. Farmers often lived in poverty, producing crops for local consumption. The lack of industrialization meant emigration was the only route to monetary success. In the 1970s, however, agriculture was modernized and mechanized and focused on irrigating large holdings to grow commercial crops for export to the rest of Europe and beyond (see Figure 4.8C). The abundance of agribusiness produce flowing from South Europe is not without problems. The use of pesticides and chemical fertilizers has negatively affected producers and consumers. The surpluses have driven down prices, hurting smaller family farmers across Europe. Agriculture (measured as a percentage of GDP and as a percentage of the labor force) has declined in importance in South Europe as manufacturing, tourism, and other services have increased. Middle-class tourist money has brought new life to the economy of the region, but it has done little to elevate the lives of poor, rural people. In Greece, for example, income from tourism has been confined to scenic coastal and island locations, and many interior rural areas remain poor and isolated. Ironically, in much of South Europe, the very qualities that attract tourists are being threatened by the sheer number of visitors and the negative effects they have on coastal environments and on rural culture.

Spain and Portugal

In the last quarter of the twentieth century, both Spain and Portugal emerged from centuries of underdevelopment. Just 40 years ago, Spain was a poor and underdeveloped country wracked by years of civil war and then a military dictatorship led by Francisco Franco. Portugal was similarly authoritarian and was still trying to hold on to its African colonies. Even the physical location of these two countries was no longer the advantage it had been during the days of empire, as commercial activity had long ago shifted away from the Mediterranean to West Europe and the wider world **(FIGURE 4.36)**. The Pyrenees kept Spain isolated from the more prosperous parts of Europe, making land travel and commerce difficult (see Figure 4.35).

In 1975, Spain made a cautious transition to democracy; in 1986, it joined the European Union under a cloud of suspicion that it would not measure up. Portugal, after a bloodless coup in 1974, underwent democratic reforms and granted independence to its overseas colonies. Since then, growth has accelerated in both countries with the help of EU funds aimed at bringing these countries into economic and social harmony with the rest of the European Union through investment in infrastructure and in human resources. Spain especially has had success as a result of these improvements. Foreign businesses invested in industry, and factories were modernized. The cities of Barcelona and Madrid became centers of population, wealth, and industry. Both are now

FIGURE 4.36 The Tower of Belem in Lisbon, Portugal. In order to protect the wealth brought by his empire in Asia, Africa, and the Americas, King Manuel I of Portugal constructed a system of defense around the capital of Lisbon. Built from 1513–1519, during an era of great prosperity, the Tower of Belem was both a central part of this defense and a ceremonial gateway to and from the city. For many of the Portuguese expeditions of discovery and colonization, including most of the colonization of Brazil, the tower of Belem was the last architectural symbol of home the travelers saw as they left on their long voyages.

linked to the rest of Europe by a network of roads, airports, and high-speed rail lines (see Figure 4.7).

For some time, segments of Spanish and Portuguese society have not been able to participate in the new prosperity. Many small farmers remain poor, particularly in the northwestern and southwestern provinces of Spain, which have few resources, and throughout Portugal. Recently, even in the relatively prosperous zones of Barcelona, Madrid, and Lisbon, the global recession and a housing bubble similar to that in the United States resulted in many workers being laid off. In 2008, unemployment in Spain was 13.9 percent; by 2014, it was 24.5 percent.

In view of the high unemployment figures, Spanish and Portuguese workers have been willing to work for significantly lower wages than those paid in North and West Europe, and these lower wage scales have helped draw foreign investment from around the world. This is especially true for Spain, where other attractions include its now-solid democratic institutions, which enable the country to withstand crises (such as the terrorist bombing in Madrid in March 2004, and the economic recession beginning in 2008); an educated and skilled workforce (partly the result of EU funds for training); and somewhat lenient environmental and workplace regulations. Spain's economic base includes agribusiness, food and beverage processing, and the production of chemicals, metals, machine tools, textiles, and automobiles and auto parts. Ford, Daimler, Citroën, Opel, Renault, MDI, and Volkswagen all have factories in Spain; but because of the debt crisis in the European Union and a 2015 scandal involving the

manipulation by Volkswagen of their diesel car emission controls, economy car sales have been down recently across Europe.

Regional disparities in wealth within Spain and between Spain and Portugal contribute to persistent levels of social discord. Two of Spain's most industrialized and now wealthiest regions—the Basque country in the central north (adjacent to France) and Catalonia in the east (also bordering France)—have especially strong ethnic identities. Their inhabitants often speak of secession from Spain as a response to years of repression by the Spanish government. In November 2015, the Catalan effort at secession was declared unconstitutional. The Basques, once poor but now quite well off, object to being taxed to support poorer districts in Spain. They have organized a separatist movement that periodically resorts to violence. Galicia, in the far northwestern corner of Spain, also has a strong cultural identity and a proclivity for separatist sentiments. All three of these ethnic enclaves emphasize their distinctive languages as markers of identity and chafe against the use of Castilian Spanish as Spain's official language.

In Portugal, although unemployment rates are lower than in Spain, concern over financial futures is higher than in any other South Europe country, and people have cut expenditures in nearly all categories except education. In an effort to ameliorate various divisive cultural and economic disparities, the Spanish and Portuguese governments borrowed so heavily to boost development that, like Greece, their budget deficits exceeded (by three times) the Euro-zone limit. By 2012, Spain and Portugal were in danger of defaulting on loans and requiring a bailout. EU-enforced belt tightening then brought on social unrest: unemployment rose to 24 percent in Spain (but dropped to 14 percent in Portugal); miners protesting in Madrid were fired upon with rubber bullets; a range of middle-class citizens joined raucous demonstrations; and across Spain, individual provinces sought emergency federal financial aid.

In Spain, migration is another important part of the country's current situation. Spain's connections to North Africa date back more than 1300 years; Moorish Muslim conquerors came in 700 C.E. and stayed until they were expelled in the late 1400s, deeply influencing Spanish culture, language, architecture, cuisine, and attitudes toward gender roles. Today, more than 600,000 immigrants to Spain from North Africa (many not in Spain legally) are working in cleaning and maintenance or are factory workers. They are joined by several million people from sub-Saharan Africa and from Spain's former colonies in Middle and South America. A significant number of these are professionally trained people who despaired of making a decent living in their home countries. It is important to note, however, that Spain can absorb immigrant workers because the net migration rate is *out* of Spain. Since the 1970s, several million Spanish workers have migrated to work in other European countries, creating employment niches for those coming in. The brain drain to EU countries may actually benefit Spain eventually as the emigrant Spaniards return with new skills and ideas.

Italy

Italy has the largest economy in South Europe and a high GNI per capita (PPP); and on the UN HDI, Italy (27) ranks just below Spain (26) (see Figure 4.31A, B). The north of Italy is industrialized and prosperous, while the south is agricultural, is dependent on government payments, and has a high unemployment rate.

Over the millennia, Italian traders have contributed greatly to European culture and prosperity. During the late 1200s, Marco Polo, from a wealthy trading family in Venice, took a trip through Central Asia to China and returned after some 20 years to give medieval Europeans their first impressions of life and commerce in China. The broader view of the world that Polo brought home contributed to the European Renaissance in northern Italy. Wealthy families such as the Medicis were patrons of the arts and literature and invested in beautifying public and private spaces. Italy's enlightened ideas about philosophy, art, and architecture spread throughout Europe.

Italy's prominence faded, however, as first Spain and Portugal and then England, France, and the Netherlands grew rich on their colonies in the Americas, Asia, and Africa. During the 1700s and 1800s, Italy got caught up in a long series of wars between its then-richer and more powerful neighbors, France, Spain, and Austria. During this time, progress was rocky for Italy's ordinary people, buffeted by the political ambitions of politicians and religious leaders (the Pope, the leader of the Catholic Church, remains headquartered at the sovereign Vatican City in Rome). Wide disparities in wealth and power left whole sections of the country lawless and in decline, especially the southern half of the peninsula and the islands of Sardinia and Sicily, which were collectively known as "the South." There, absentee landlords and central government tax collectors from northern Italy hired *mafiosi* to collect fees from those who worked the land but were given no chance for self-government. Economically, the south remains woefully behind the north, due in part to the pervasive corruption that in some places has a stranglehold on civic institutions.

Northern Italy, by contrast, has one of the most vibrant economies in the world, producing products renowned for their quality and design: Ferrari automobiles, Olivetti (now a subsidiary of Telecom Italia Group) office equipment, and high-quality musical instruments. In Milan, one of the largest cities in Italy, fashion designers such as Giorgio Armani have surpassed even their rivals in Paris. However, the mainstay of the Italian economy has traditionally been the exporting power of thousands of small mom-and-pop factories that make everything from machine valves to buttons, shoes, and leather clothing—all of which can now be made more cheaply in China. The small factories are closing, Italy's share of global trade is dropping, and its budget deficits have exceeded the limits that the European Union places on its member countries (Germany and France also have large budget shortfalls but are more solvent). Like all countries in the Euro zone, Italy can no longer simply devalue its money, formerly the *lira*, to make its products less expensive on the world market.

For all of Italy's stylish success as an industrial and services leader in Europe and the world, its ability to continue attracting investment has been damaged by an inefficient bureaucracy, high tax rates, dense tax rules, inadequate infrastructure, and corruption. In 2015, although improvements were instigated by a new, more financially savvy prime minister, Italy had a global competitiveness rank of 43 out of 140, below that of many countries in Asia, Southeast Asia, and Southwest Asia.

Italy has been a democracy since World War II and is a charter member of the European Union. Because of its domestic politics, though, Italy was long known as the "bad boy" of Europe, a designation usually meant to be taken humorously. The label comes in large part from unfair stereotypes of Italians as clever operators on the fringes of legality. But it also derives from the fact that Italians vote governments in and out in rapid succession, and they frequently reelect leaders who have been mixed up in corruption. Italians have devised ways to live with official indiscretions without descending into national crises. During the dozens of governmental emergencies since 1950, a quasi-government based on informal relationships—the *sottogoverno*, or *undergovernment*—has taken over whenever a predicament has developed, allowing daily life to proceed apparently unimpaired.

Italy, like Spain, has for a decade or more dealt with large numbers of immigrants, principally from North and sub-Saharan Africa. As globalization spreads, agricultural and industrial policies that favor development in Europe and other rich areas have had the side effect of jeopardizing farmers, fishers, and small artisans across Africa. Hundreds of thousands of young Africans, usually those with some education, have, in desperation, migrated to Europe to look for work so that they can support families back home. Many have arrived along Italy's southern shores in leaky boats via North Africa, which has few protections for refugees. This poorest part of Italy has not welcomed the Africans, who are stigmatized as public enemies and immediately arrested or victimized by unscrupulous labor and sex worker scouts. During 2015, more than 2000 migrants drowned before reaching shore.

Greece

The Kingdom of Greece became independent from the Ottoman Empire after World War I and for a while was a republic, but in 1935 the Greek monarchy was reestablished. Greece was occupied by both Italy and Germany during World War II; and after the war, it entered a protracted era of Communist/anti-Communist civic unrest. Greece joined NATO in 1952, a move that seemed to strengthen ties with post-war Europe; but in the mid-1960s, a military coup established a dictatorship that lasted until 1974. Democratic elections followed, and the Greek monarchy was abolished.

In 1981, after years of instability, Greece joined the emerging European Community. When Greece signed on to the 1992 EU Treaty of Maastricht, it became obligated to limit deficit spending and debt levels. Always one of the poorer countries in Europe and one that spent more than it took in, Greece entered the Euro zone in 2001, at a time when low interest rates and the strong Euro made excessive borrowing possible. Greece borrowed for infrastructure projects and to fund state jobs and benefits, including pensions. The quality of the infrastructure projects was compromised by a culture of corruption and tax evasion that put public funds in the pockets of officials and other elites. Many of the intended recipients of benefits were cheated, and the Greek deficit increased. Greece did not report these problems to the EU banks and regulators. The interest rate on Greece's debt shot up when the debt—then €360 billion (U.S.$384 billion)—was revealed to be so large. Already short of cash, Greece found it impossible to pay even the interest charges and entered a long period of being near default while the debt mounted quickly. Despite many threats to cut Greece loose, the Euro zone countries and the IMF continued to negotiate repeated bailouts, which by December of 2015 totaled €340 billion (U.S.$363 billion). In return, the EU and the IMF demanded stern austerity measures, the brunt of which are falling on the elderly and the poorest in Greece.

Greece's situation became more difficult as a result of the refugee crisis. Several Greek islands lie close to the Turkish port city of Izmir, from which many thousands of Syrian refugees and other migrants have paid smugglers to take them to the Greek islands. Because Greece is a member of the European Union and within the Schengen Border zone, the migrants assumed they would not be turned back. Several hundred thousand migrants illegally entered Europe through Greece during 2015. The migrant crisis bought Greece some time and money with which to work on easing the debt crisis. By the beginning of 2016, the EU was helping Greece to care for the migrants and defend the Schengen Border against further immigrant arrivals, and had postponed some of the more dire consequences of the debt.

THINGS TO REMEMBER

• Spain and Italy both had periods of growth and prosperity, though regional disparities in wealth and development were common in both countries. More recently, both countries have had financial difficulties related to borrowing too much and to their competitive disadvantages in the global economy. These problems have been exacerbated by large numbers of immigrants seeking opportunities.

• Italy, long known for its thousands of small, family-based manufacturing plants, has been especially hard hit by competition from Asia. High-tech firms remain dominant in northern Italy.

• Manufacturing, tourism, and other services have increased in importance in South Europe as the role of agriculture has declined.

• Spain and Portugal were transformed economically and culturally by their respective eras of colonialism; but today, they are not only struggling to remain globally competitive, they are struggling to find a way to absorb immigrants from their former colonies.

• Greece's financial and political circumstances have long been troubled; by the mid-2000s, the mishandling of public funds had left Greece so indebted to European banks that its persistence as a member of the European Union has been threatened.

• Greece's difficult financial situation was made more severe by the arrival of hundreds of thousands of desperate refugees from Syria and migrants from elsewhere, all seeking to enter the European Union illegally.

• South Europe is less urbanized than other subregions. Rural areas do not have the prosperity of the modernized urban areas, yet they wield more political power than the size of their populations would indicate.

• Democratic institutions in South Europe are stronger than in the past but are still weak and face issues such as corruption, more new immigrants from Africa than they can manage, and mounting debt.

NORTH EUROPE

North Europe **(FIGURE 4.37)** traditionally has been defined as the countries of Scandinavia—Iceland, Denmark, Norway, Sweden, and Finland—and their various dependencies. The Faroe Islands (and Greenland) in the North Atlantic are territories of Denmark; the island of Svalbard in the Arctic Ocean is part of Norway. North Europeans are now including the three Baltic states of Estonia, Latvia, and Lithuania in their region, as we do here. All three Baltic states were part of the Soviet Union until September 1991, and these small countries have many remaining links to Russia, Belarus, and other parts of the former Soviet Union. Although their cultures are not Scandinavian, the three countries are trying to reorient their economies and societies in varying degrees to North Europe and the West.

The countries of North Europe are linked by their locations on the North Atlantic, the North Sea, and the Baltic Sea. Most citizens can drive to a coast within a few hours. The main cities of the region—Copenhagen, Oslo, Stockholm, Riga, and Helsinki—are vibrant ports that have long been centers of shipbuilding, fishing, and the transshipment and warehousing of goods. They are also home to legal and financial institutions related to maritime trade.

Scandinavia

Most people in North Europe live in the southern parts of Scandinavia (see Figure 4.25), where economic activity is concentrated. The landscapes of these warmer lowlands of southern Scandinavia are agricultural but the economies of these countries are based on industries and services. For example, even though the small country of Denmark produces most of North Europe's poultry, pork, dairy products, wheat, sugar beets, barley, and rye, 76 percent of the Danish economy consists of services (finance, education, design, tourism), high-tech manufacturing, construction and building trades, and fisheries. Sweden is the most industrialized nation in North Europe. The Swedes, who also like to underscore their quaint rural roots **(FIGURE 4.38)**, produce most of North Europe's transportation equipment and two highly esteemed brands of automobiles, the Saab and the Volvo (though some Volvo products are now built outside Europe and Saab is owned by a consortium based in China and Japan). Helsinki, Finland, is home to some of the world's most successful information technology companies, such as Tieto, Nokia, and Novo Group. IKEA, founded in Sweden, is the world's largest furniture manufacturer. Through 375 stores in 50 countries, IKEA markets furniture and housewares famed for their affordability, spare, elegant design, and their necessity to be assembled at home. IKEA is a major consumer of woods logged in the world's tropical zones (see Figure 10.8); China is the largest producer of IKEA products.

The northern part of Sweden and almost all of Norway are covered by mountainous terrain, while Finland is a low-lying

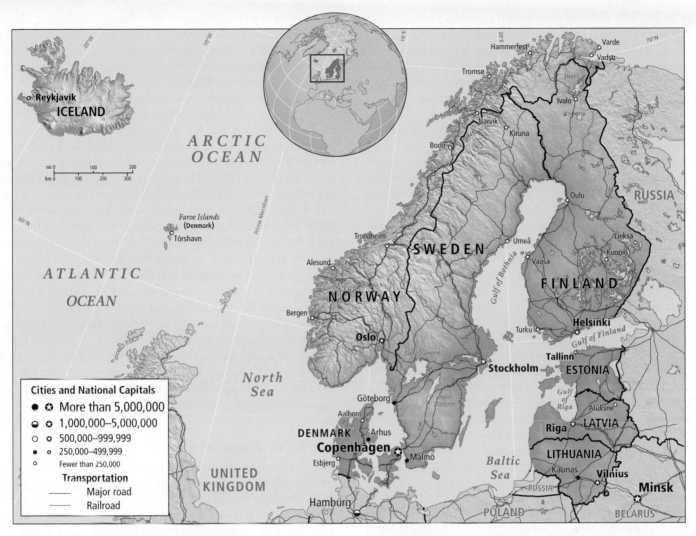

FIGURE 4.37 The North Europe subregion.

land dotted with glacial lakes, much like northern Canada. Sweden and Finland contain most of Europe's remaining forests. These well-managed forests produce timber and wood pulp. Norway, which has less usable forestland, had been considered poorly endowed with natural resources compared with the rest of Scandinavia, but the discovery of gas and oil under the North Sea in the 1960s and 1970s has been a windfall to the country. The exploitation of these resources dwarfs all other sectors of Norway's economy. Norway is able to supply its own energy needs through hydropower, and it exports oil and gas to the European Union. It is now one of Europe's wealthiest countries and ranks highest in the world in human well-being. In anticipation of the North Sea oil running out, Norway has invested a large proportion of its oil profits for use by future generations.

The fishing grounds of the North Sea, the North Atlantic, and the southern Arctic Ocean are also an important resource for the countries of North Europe (as well as for several other countries on the Atlantic). In recent decades, however, overfishing has severely reduced fish stocks and by 2010 most cod eaten in Europe was imported. Still, fishing remains culturally important and rights of access to the fishing grounds are a cause

FIGURE 4.38 The Swedish village of Gamleby. This village in Västervik, Sweden, is known for its colorfully painted wooden houses along the old naval port. Many Swedish towns feature such bright paint jobs. A favorite color is Falun red, a copper-based paint originating in Falun in Dalarna County.

of dispute. In 1994, the European Union created a joint 200-mile (320-kilometer) coastal exclusive economic zone (EEZ) that ensures equal access and sets fishing quotas for member states. In January 2001, in order to give North Sea fish populations a chance to rebound, EU members plus Norway and Iceland agreed to an annual 3-month hiatus in cod fishing, and a ban on using juvenile fish to feed salmon in commercial farms. By 2006, however, it was clear that these short respites from fishing had not increased the cod population; so a hiatus of as long as 12 years was called for over a much bigger area of the North Sea. A small recovery from 2009 to 2014 gave cause for hope, but the situation remained precarious. Current thinking is that global warming may account for some of the drastic slump in cod catches.

Two important characteristics distinguish the Scandinavian countries from other European nations: their strong social protection systems (see pages 226–227) and the extent to which they have equal levels of participation and well-being for men and women (see Figure 4.30C; see also 237–238 on gender). Sweden's comprehensive social protection system provides stability and a safety net for every one of the country's 9.8 million people. This system is founded on three values. First is the idea of security: that all people are entitled to a safe, secure, and predictable way of life with as little discomfort as possible. This concern extends to interior design in subsidized housing, which is modern, elegant, and practical. Second, the appropriate life is ordered, self-sufficient, and quiet, not marred by efforts to stand out above others. Third, when the first two concepts are practiced properly, the ideal society, *folkhem* ("people's home"), is the result.

These three values help explain why Swedes are willing to pay for a social protection system that provides child care,

parental leave, health care, sick leave, elder care, housing subsidies, and other benefits. By 2013, the system amounted to close to one third of the government's annual budget and was not confined only to citizens; immigrants were also covered, though some Swedes expressed their dissatisfaction with this. In the early 2000s Sweden began to experiment with privatization of education and services under a center-right government; but by 2014, as student performance fell, economic disparities increased, and the poor (many of them recent Iraqi refugees) took to the streets, Swedes began to lobby for a return to high levels of tax-supported social protection. The immigrant crisis of 2015 shook Sweden's commitment to social welfare for all because the wave of migrants was so huge; 100,000 newcomers were expected and more than twice that number came in. Because of the open-border rules of the Schengen zone, at first Sweden kept no records. The country's security and its ability to assimilate the newcomers was questioned. Sweden closed its borders for a few days with the intention of reopening them as soon as possible; but this caused worries that the Schengen Agreement was now dead.

For all their emphasis on an orderly peaceful life that is fair for all, Scandinavians are noted for occasional outbursts of exuberant enthusiasm for over-the-top adventures. One such example is the Scandinavian love for heavy metal music. **FIGURE 4.39** shows the Finnish band Apocalyptica, known for its inventive orchestral embellishments of metal music.

The Baltic States

The three small Baltic states of Estonia, Latvia, and Lithuania are situated together on the east coast of the Baltic Sea. Culturally, these countries are distinct from one another, with different languages, myths, histories, and music. After World War II, all three were forced to become Soviet republics. When the Soviet Union collapsed in 1991, they regained their independence. Despite their recent connection to Russia, their cultural ties traditionally lean toward West and North Europe. Ethnic Estonians, who make up 60 percent of the people in Estonia, are nominally Protestant, with strong links to Finland; they view themselves as Scandinavians. Ethnic Latvians, who make up just 50 percent of the population of Latvia, are mostly Lutheran; they see themselves as part of the old German maritime trade tradition. Lithuanians are primarily Roman Catholic and also see themselves as a part of western Europe.

Populations in all three countries are decreasing and aging because birth rates are very low and young people now have the option to migrate to North and West Europe. Ethnic Russians, who today make up about a third of the population in both Estonia and Latvia, are having more children than the indigenous populations are. They could become the dominant culture group within a few decades—a crucial development if these ethnic Russians continue to cultivate strong political ties with Russia. Ironically, although the Russian minority was placed in the Baltic states by Russia to counteract Westernization and maintain the influence of the Soviet Union, under EU rules (all three countries are now EU members), these Russians are minorities whose rights must be protected by the EU.

Of all the countries of North Europe, the Baltic states have the most precarious economic outlook. Under Soviet rule, the

FIGURE 4.39 Scandinavia and heavy metal music. Riding a decades-long Scandinavian obsession with heavy metal music, the Finnish band Apocalyptica, composed of three cellists and a drummer, formed in Helsinki in 1993. Originally a tribute band that played covers of the U.S. band Metallica, Apocalyptica has become a pioneer of a more symphonic- and folk-influenced style of metal. Finland, like much of Scandinavia, became a hotbed of heavy metal in the 1980s; the tradition continues today as metal bands, clubs, and even metal-based church services grow in popularity.

Russians expropriated their agricultural and industrial facilities and used them primarily for Russia's benefit. Until 1991, 90 percent of the Baltic states' trade was with other Soviet republics. Since then, Estonia has undertaken the most radical economic change, moving toward a market economy and increasing its trade with the West. Estonia is the only one of the three to have had real economic growth, and its accomplishments have attracted foreign aid and private investors. Heavily industrialized Latvia and Lithuania now trade more with the UK, Germany, and the West than with Russia. They have had to increase their standards of quality, but many of their factories are out of date and still pollute heavily.

The Baltic states see national security as their major problem because their strategic position along the Baltic Sea is coveted by Russia, which has few easy outlets to the world's oceans. In this regard, the status of the Russian exclave of Kaliningrad is crucial. Kaliningrad is called an **exclave** because, although it is an actual part of Russia, it lies far from the main part of Russia. Kaliningrad is situated along the Baltic Sea, between two EU states, Lithuania and Poland (see Figures 4.1 and 4.40). The Russian Baltic fleet is headquartered in Kaliningrad, which has a relatively ice-free port (see the discussion in Chapter 5). Russia's strategic interest in Lithuania is unlikely to diminish, and Lithuania fears that political instability in Russia could result in Russia reinvading Lithuania, which is not a member of NATO. The countries of western Europe, unwilling to commit to supporting Lithuania militarily, hope to resolve rising differences with the Russians with diplomatic strategies.

THINGS TO REMEMBER

- Five of the northern Europe countries rank very high on the human development index; in fact, they are in the top 16 in the world and all have the most complete, albeit expensive, social protection systems in Europe (and probably the world). Of all world regions, formal policies to achieve gender equality have been most successful here.

- This entire region is linked to the world ocean and environmental concerns are common points of public discussion and action.

- The three post-communist Baltic states all rank in the high range of human development but are struggling to improve their economic competitiveness. All three see security as a major issue because of Russia's maneuvering for control of territory and its desire for access to the Baltic Sea.

CENTRAL EUROPE

Central Europe has undergone a series of profound changes since World War II (**FIGURE 4.40A** and the figure map). First, the region experienced harsh economic and political conditions under communist governments. Then in 1989, in the euphoria at the end of the Cold War, many believed that democracy and market forces would quickly turn around the region's formerly centrally planned, sluggish and highly polluting economies. Instead, the southern parts of Central Europe (traditionally called the Balkans) suffered violent political turmoil, and virtually all had temporary drops in their levels of well-being.

Over time two tiers of countries emerged in Central Europe. In the 1990s, those countries physically closest to western Europe—Poland, the Czech Republic, Slovakia, Hungary, and Slovenia—elected new democratic governments, began more careful environmental policies, made deliberate progress toward market economies, and began to attract foreign investment. Access to consumer goods improved markedly. In 2004, all of these countries joined the European Union, and Croatia joined in 2013. Although conforming to the wide range of economic, environmental, and social requirements set forth by the European Union has not been easy, for many the challenge has been invigorating.

The countries with the greatest economic and social difficulties remain Albania, Bulgaria, and Romania, and some of the countries that were in the former Yugoslavia (Bosnia and Herzegovina, Macedonia, Montenegro, and Serbia—including the region of Kosovo). Romania and Bulgaria joined the EU in 2007; but for all these countries, adjusting to democracy and privatization has been difficult. Plagued by corruption and economic chaos, many governments relaxed the pace of reforms as they struggled merely to maintain civil peace. In the 1990s, all of the former Yugoslavia republics except Slovenia became mired in a series of terribly costly genocidal ethnic conflicts that belatedly drew the military intervention of Europe (including NATO) and the United States.

A Time of Troubles in Southeastern Europe

Shortly after the exhilaration of the breakup of the Soviet Union, the socialist counties of Yugoslavia and Albania went through a rather unexpected but brutal war. This region between Austria and Greece, formerly known as the Balkans, is home to many ethnic groups of differing religious faiths (**FIGURE 4.41**). Too often the assumption has been that it was this diversity that led to the conflict; but over many generations, the various ethnic groups managed to live together peacefully by intermarrying, blending, and realigning into new groups. Often it was outsider conquerors who encouraged ethnic rivalry as a method of control, as was the case with the Hapsburg empire and later the Nazis during World War II.

Marshal Josep Broz Tito, Yugoslavia's founder and leader until his death in 1980, recognized the ethnic complexity of the countries he was governing. He tried through the power of his personality to foster pan-Yugoslav nationalism. There was ethnic peace during Tito's rule, but no national dialogue about the nature and history of the differences or of the benefits of ethnic diversity. When Tito died, Serbs—the most populous ethnic group in the military and the federal government bureaucracy—tried to assert control. In response to the deteriorating economy and burgeoning Serbian nationalism, Slovenia and Croatia, Yugoslavia's northernmost and wealthiest provinces, voted to declare independence in late 1990. Slovenes and Croats themselves became more nationalistic, and a spiral into competing nationalisms threatened.

Given the scale of the interethnic violence that occurred in the following years, it is not surprising that many people, especially outsiders, viewed ethnic hatred as the overriding characteristic of the region and the cause of the conflict. However, more likely causes include an overall lack of ethical leadership, including little guidance by Marshal Tito

exclave a portion of a country that is separated from the main part

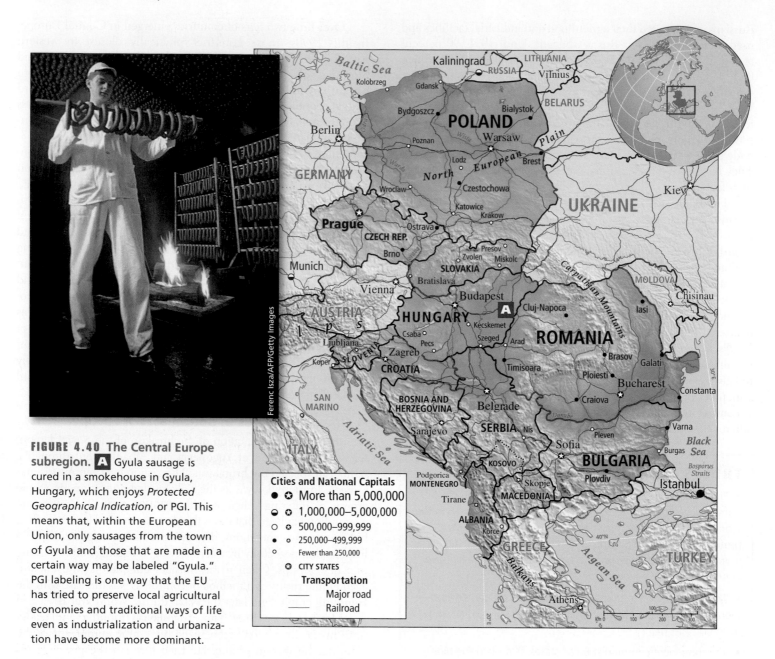

FIGURE 4.40 The Central Europe subregion. A Gyula sausage is cured in a smokehouse in Gyula, Hungary, which enjoys *Protected Geographical Indication*, or PGI. This means that, within the European Union, only sausages from the town of Gyula and those that are made in a certain way may be labeled "Gyula." PGI labeling is one way that the EU has tried to preserve local agricultural economies and traditional ways of life even as industrialization and urbanization have become more dominant.

Ferenc Isza/AFP/Getty Images

on multicultural affairs, invented myths about ethnically pure nation-states, and Serbian geopolitics.

In June of 1991, the Serb-dominated government and military of Yugoslavia allowed the province of Slovenia to separate after just a short skirmish—this happened in part because of Slovenia's location close to the heart of Europe and far from Serbia and in part because there were no significant supportive enclaves of ethnic Serbs in Slovenia. After this loss, the Yugoslav government feared that unless it made a strong show of force in dealing with the Croats and the Bosnians, the rest of the non-Serb populations of Yugoslavia would also move to secede and take with them valuable territory, resources, and skilled people. This fear became a reality when Croatia and Bosnia and Herzegovina attempted to secede from Yugoslavia. The Serbs reacted brutally. While suffering ethnic cleansing, Croats and Bosnians themselves retaliated with brutality and ethnic cleansing, as did ethnic Albanians (also known as *Kosovars*).

Conflict sprang up again in the late 1990s in the province of Kosovo, the southern area of Serbia, dominated by ethnic Albanians. In the late 1980s, in order to divert public attention away from the country's economic deterioration, Yugoslav President Slobodan Milosevic, a Serb, aggressively claimed that Serbs were being threatened by Kosovar Albanians. In response, the guerrilla Kosovo Liberation Army was formed, supported by expatriates of Albanian ethnicity, some from the United States.

The various conflicts in and around Serbia devastated the economies of southeastern Europe. Dozens of bridges were bombed; the Danube River, a major transportation conduit through Europe to the Black Sea, was blocked with wreckage. In 2000, an election in Serbia resoundingly removed Slobodan Milosevic from power, and in 2002 he was brought to trial for war crimes in the International Court of Justice (World Court) in The Hague (where he died of natural causes in 2006). To forestall any future troubles, the European Union and the countries of

FIGURE 4.41 Ethnic groups in southeastern Central Europe. The patchwork of cultural and religious groups in southeastern Central Europe developed over the last several thousand years as different ethnic groups migrated to the area from the west and east. They had to contend with each other's differences and with conquerors from the north and the south. Normally, relations were amicable and intermarriage was common, but occasionally hostilities arose, often as a result of outside pressures.

southeastern Europe had earlier agreed that, once Milosevic was gone, they would enact the Balkan Stability Pact. Funded by the European Union, this pact promotes the recovery of the region through public and private investment as well as through social training to enhance the public acceptance of multicultural societies and the strengthening of informal democratic institutions. In May 2006, the people of Montenegro voted to declare independence from Serbia; in 2008 Kosovo declared independence, but its status remains in contention.

Transitions in Agriculture, Industry, and Society

After all this turmoil, not surprisingly, by 2000 many people in all 13 countries in Central Europe were looking back nostalgically to the Communist era, when there was a strong authoritarian government, jobs were stable, and health care and education were free. Still, even the seven poorest and most politically unstable parts of Central Europe have useful natural resources for both agriculture and industry, and all have large, skilled but inexpensive workforces that have slowly begun to attract foreign investment. The most important trading partners for Central Europe are Germany, France, the Netherlands, Italy, the United Kingdom, Russia, and the United States, but the combination changes from country to country (Serbia and those further south are just developing wider trade networks). Venture capitalists

from Hungary, the Czech Republic, Slovenia, and even Croatia are beginning to invest in their neighbors to the south.

The Czech Republic, now in the middle tier of prosperity in Central Europe, can be taken as an example of transitions in agriculture. There, land was expropriated after World War II and turned into state farms and cooperatives. In the 1990s these large state farms were reorganized into private owner-operated cooperatives or restored to the original owners and their descendants. By this time, though, the original owners had mostly become city dwellers so, rather than farming the land themselves, they usually leased their regained land, either in small parcels to those who wished to become family farmers or in large parcels to the new private cooperatives. Eventually, 1.25 million acres (500,000 hectares), representing 30 percent of the Czech Republic's agricultural land, went to individual private farmers, large and small.

The EU economic advisers who readied the Czech Republic for entry into the European Union preferred larger, commercial farms because overproduction, which could contribute to falling commodity prices across Europe, is easier to control on large, corporately managed farms. EU advisers then suggested that the smaller family farms, which tend to be owned by independent-minded individuals, find more lucrative uses for their land, such as farm tourism. "Dude farms"—similar to dude ranches in the western United States—which provide living history demonstrations of traditional food cultivation and preparation techniques, now attract affluent urban families from all over Europe and

FIGURE 4.42 Organic vineyards in Slovenia. Bozidar Zorijan, an organic wine producer in Slovenia, explains how he prunes his vineyards and keeps pests in check by spraying with natural plant-based teas and by encouraging useful insects to prey on those that do damage. As news of his fine wines spreads, scores of wine enthusiasts visit his elegant tasting room bedecked with trophies he has won.

beyond (see the vignette below and another on page 221, along with Figure 4.20).

VIGNETTE In Slovenia, Marinko Rodica proudly polishes the fine wood table in his small, elegant tasting cellar near the village of Truške, 1000 feet (300 meters) above the Adriatic Sea, near the giant port of Koper. To tourists, he offers up seven different wines (one a blend) made from the six grapes he grows: three red—refošk, merlot, and cabernet sauvignon—and three white varieties—muškat, malvazija, and sivi pinot (pinot gris). Marinko recently won gold medals for his Truške Rdece and Muškat at international fairs.

The climate and the location are ideal for growing grapes, and people have been making wine in this part of the Adriatic for about 2400 years. Over the last decade, Marinko turned his entire farm into an organic vineyard, with the aim of marketing his wine in the local tourist economy and eventually selling it internationally.

Membership in the European Union has brought both benefits and challenges to wine producers (FIGURE 4.42). Their wines now have a wider market—Central Europe, Germany, Austria, Italy, Japan, and the United States—but vintners must adhere to strict EU agricultural policies. For example, Europeans are actually producing too much wine, lowering wine prices for all producers. The European Union wants to reduce the total vineyard acreage across Europe by some 400,000 hectares. Fortunately for Marinko, Slovenia, as one of the tiniest EU members, will not be grossly affected by this reduction. It could even increase demand for Marinko's wine. Currently more than 90 percent of the 100 million liters of wine produced annually in Slovenia is consumed in Slovenia; plans are to increase wine exports over the next several years.

Marinko is ebullient. He has just entered into an association with the University of Primorska in Koper to teach viticulture and has donated some acreage and vines for university students to use to learn the craft. [Source: From field notes and research by Mac Goodwin and Simon Kerma, 2008, 2012.] ∎

Industrial production was a specialty of Central Europe when these countries were part either of the Soviet bloc or Yugoslavia. So it would seem that manufacturing would be the best hope for competing in the EU marketplace. Unfortunately, much of Central Europe's considerable industrial base is still burdened by legacies from the Communist era; it is polluting and energy intensive, with an emphasis on coal and steel production, which can't compete in the global market. In south Central Europe, the turn to producing much needed consumer goods to stimulate the domestic economy is just emerging (see the discussion on page 253).

Despite industry in Central Europe being slow to succeed, high-end product design is improving in places like Slovenia; there, regional trade shows are full of attractive, if still mostly locally sold, goods. Expensive artisanal products—such as handmade designer clothes and shoes, crystal glassware, luxury baby clothes, and super-modern kitchens are sold throughout Europe and the United States. Meticulously constructed prefabricated energy-efficient houses are being exported to Europe and Africa. Pan-Les, in southern Slovenia, produces a charming 500 square foot house with a total annual heating bill of U.S.$300 a year. Beautiful, handcrafted Skrabl pipe organs from the same region are catching the attention of American churches.

Corruption has inhibited post-Communist industrial reform. In 2009, both Romania and Bulgaria narrowly avoided censure by the European Union for failing to rein in corruption. In some countries, the involvement of organized crime has plagued the sale of formerly state-owned industries, especially in military sectors. It has been difficult to develop financial regulations and freedom of the press, among other EU requirements, because of the amount of suspicion and wariness on the part of some political leaders. However, because the countries in the region want to remain in the good graces of the European Union, most have been working recently to control risky banking practices, to limit unscrupulous deals, and to reduce organized crime.

Progress in Market Reform

The governments of Central Europe are taking "free market–friendly" steps to help their economies grow more quickly. One strategy is to encourage foreign investment. But the question is: To what extent will foreign investment help those displaced by the changes? Virtually all of the countries of Central Europe actively try to recruit foreign investors and after some discouragement about the endless paperwork, firms from western Europe, the United States, and China have been coming to the region since 2012 with investment projects. For example, hog farming in Romania first attracted the U.S. Smithfield firm with major European Union CAP subsidies. Then China stepped in, but local hog farmers were bypassed and many quit farming (see the following vignette).

VIGNETTE Grigore Chivu (a pseudonym) wanders through the empty hog pens on his farm near Lugoj in western Romania. For generations, his family made a meager but rewarding living by raising hogs and then processing and selling the meat. A few years ago, just before Romania joined the European Union (EU), he had more than 250 hogs. At Christmas, he and thousands of hog farmers across the country would slaughter a certain number and preserve the meat using time-honored methods. Hog slaughtering was a time of high spirits, celebration, and cooperation as the farm families contemplated the coming feast and their profits. Flavorful sausages, which they smoked and hung high in the rafters of their kitchens for further drying, were slowly sold to select customers, providing a steady income well into summer.

Romania's entrance into the European Union required all farmers to conform to EU standards for processing meat. The old methods were no longer allowed and for some the new standards were prohibitively expensive. At first Chivu and his farmer friends were uncertain just how to respond. Before they could organize a butchering cooperative that would conform to EU standards and for which development funds were available, an American company (based in Virginia) stepped into the breach. Smithfield, the world's largest pork producer, was expanding into Central Europe, where it planned to produce pork products on an industrial scale and market them globally. Eager to enter the EU market, Smithfield enlisted the help of Romanian politicians and got permission (and even EU subsidies) to establish a conglomerate that included feed production, hog breeding, modernized sanitary barns for fattening thousands of hogs in small cages, and slaughterhouses.

Then, in September of 2013, Smithfield was bought by a Chinese company, Shuanghui, which planned to redirect much of the company's global production toward supplying the growing demand for pork in China's new middle class. Shortly after Shuanghui purchased Smithfield, Romanian officials signed agreements with their Chinese counterparts to export more than 3 million hogs to China each year.

As the old, picturesque Romanian agricultural landscape is transformed into huge factory farms, the number of independent hog farmers has been reduced by more than 90 percent (**FIGURE 4.43**). Unable to compete with the lower prices Smithfield can charge, Grigore Chivu, like thousands of his fellow hog farmers, is preparing to migrate to western Europe, where, because of his age and farming skills, he will obtain only a low-wage job to support his family. There is no work for him in Romania and he is too young for a pension. *[Sources: Margareta Lelea; New York Times. For detailed source information, see Text Sources and Credits.]* ∎

ON THE BRIGHT SIDE: The Development of Market Economies in Tourism

As the switch to market-based economies in Central Europe has displaced workers and changed land usage, tourism has been recognized as an important way to employ workers and enhance local economies. The Adriatic attracts tourists from western Europe who are looking for an inexpensive sunny vacation. Medieval castles and wine festivals draw weekend visitors to rural areas. The cities of Central Europe, some of Europe's oldest and most distinguished municipalities, are becoming magnets for

FIGURE 4.43 Hog farms in Romania and the United States.

(A) Hogs on a small animal farm in Romania.

(B) An industrial hog farm in the United States, where thousands of hogs are raised at a time.

Thinking Geographically
What about (A) suggests the small-scale rearing of pigs?

tourists from America, Asia, and western Europe. One example is Budapest, the capital of Hungary (see Figure 4.1E). Situated on the Danube River and built on the remnants of the third-century Roman city of Aquincum, Budapest was one of the seats of the Austro-Hungarian Empire, which ruled central Europe until World War I. By 2000, Budapest had reemerged as one of Europe's finest metropolitan centers, filled with architectural treasures, museums, concert halls, art galleries, boutiques, discos, cabarets, theaters, and several universities of distinction. ■

New Experiences with Democracy

The advent of democracy in Central Europe has reduced some of the tensions associated with the economic transition because people have been able to vote against governments whose reforms created high levels of unemployment and inflation. Most countries have now had several rounds of parliamentary and local elections, and voters are becoming accustomed to Western-style political campaigns. However, the voters' tendency to remove reformers before they can accomplish anything significant has dampened the enthusiasm of potential foreign investors, who are waiting for real and sustained change before they offer their money. The financial crisis in the European Union has brought pressure from voters and experts for revisions that will make the social safety net less costly and more efficient. For example, there

are plans for new national pension systems to replace defunct Communist pension plans. Viable pension systems are important on several counts: they will preserve financial stability for those who labored under Communism and lost their safety net; they will keep active workers productive longer by pushing the retirement age from 58 to 65; retirees will have pensions adequate enough to spend on goods and services, thus creating jobs; and they will relieve younger workers from the burden of supporting impoverished aged parents.

THINGS TO REMEMBER

- The advent of democracy in Central Europe has reduced some of the tensions associated with the economic transition to capitalist market systems because people have been able to vote against governments whose reforms created high levels of unemployment and inflation.

- The states (with the exception of Slovenia) formerly known as the Balkans underwent a long, tragic, interethnic (and partially religious) conflict resulting from the breakup of the former Yugoslavia. These states, most now reorganized as countries, are attempting to recover and meet EU membership requirements, which Croatia did in July 2013.

GEOGRAPHIC THEMES: Europe Review and Self-Test

1. **Environment:** Europe is a world leader in responding to climate change. The European Union, a group of 28 countries within Europe, has set goals for cutting greenhouse gas emissions that are complemented by many other strategies for saving energy and resources. Europe's stringent climate amelioration efforts are in response to the fact that much of Europe's air and many of its seas and streams are quite polluted, and that through its high level of consumption, it negatively impacts environments across the globe.

- How does Europe's physical geography complicate problems of river and sea pollution? What could the European Union do to reduce its contribution to the pollution of Europe's seas?

- Why might it make economic sense for Europe to address climate change now?

- On a more local level, what evidence is there that European ways of life contribute less to global warming than do North American ways?

- Which technological advances are being implemented in Europe to address climate change?

- What are some positive and negative environmental and economic impacts of the strategies being taken to address climate change?

2. **Globalization and Development:** Following World War II, Europe took steps to create an economic union now known as the European Union, which features open borders, the free movement of people and goods, progress toward a common currency, and efforts to promote human well-being and social cohesion. To improve its global competitiveness, the EU shifted industries from high-wage western Europe to the lower-wage member states of

Central and South Europe. Service jobs replaced many industrial jobs, but unemployment increased. The European Union struggles to integrate the economies of its member states and to reduce regional disparities.

- Why have some countries chosen not to adopt the euro?

- What challenges to the euro have emerged in recent years?

- What impact do Europe's heavily subsidized and regulated agricultural systems have on the rest of the world? What worries confront the European Union as it addresses its large investment in regulations of, and subsidies to, agricultural enterprises?

- What are some of the ways that the colonial empires created by European countries affect life in Europe today? How do colonial relationships still affect modern trading patterns?

3. **Power and Politics:** By the 1960s Europe was shedding its colonial dependencies, participating with the United States as a leader in global affairs, and implementing economic unification. After the fall of the Soviet Union in 1991, the European Union expanded into Central Europe, emphasizing political freedoms and economic reform. Migration to West and North Europe from Central and South Europe and from former colonies as well as from several conflict areas increased steadily. By 2015 the intractable Syrian revolution and social unrest elsewhere in Asia and Africa instigated the massive migration of asylum seekers and economic migrants into Europe. Islamic jihadism resulted in several terrorist attacks, including two in Paris, France, which garnered international attention.

- How has the expansion of the EU into Central Europe transformed this region?

• Compare and contrast the EU and the United States politically.

• Compare and contrast the international leadership roles played by the European Union and the United States.

• What new role for the EU is developing through the North Atlantic Treaty Organization (NATO)?

• What are the major types of tax-supported systems of social welfare (or social protection) in Europe?

4. Urbanization: Europe's cities are both ancient and modern, with old town centers now surrounded by modern high-rise suburbs and supported by world-class urban infrastructures. Cities are the heart of Europe's economy, politics, and culture, and more than 70 percent of the EU population lives in urban areas. Yet much of Europe remains idyllic and productive rural countryside, important symbolically, economically, and as a place for regular recreation.

• How did the Industrial Revolution lead to urbanization in Europe?

• What is the connection between urbanization and democratization in Europe?

• Which economic sectors are most European cities now oriented toward? Why?

• Why are so many cities in Europe located either on navigable rivers in the interior or along the coasts?

• Why do only a small percentage of Europeans live in single-family homes?

5. Population and Gender: Europe's population is aging as fertility rates decline. Fertility rates are down due to the high cost of raising a child and because career-oriented women are choosing to have only one or two children. Fearing that aging populations will slow economic growth, European countries try to boost their fertility rates with a variety of policies—the most successful of which address the needs of working mothers. Encouraging immigration may be one way to boost population numbers, but eventually migrants may also opt for smaller families. However, accommodating and assimilating migrants, either refugees or those only seeking jobs, has proved a difficult social issue.

• What are some of the potential consequences to Europe of aging populations and negative population growth rates?

• How do improvements in education and work opportunities for European women affect population growth?

• Which policy interventions have been most successful in countering low birth rates in Europe?

• How do immigration policies influence aging in Europe? What about these policies worries Europeans?

Critical Thinking Questions

1. How do the issues of immigration in Europe compare with those in the United States? If you were a poor, undocumented immigrant searching for a way to support your family, would you choose the United States or Europe as a possible destination? Why?

2. What are the ways in which Europe is still linked to its former colonies?

3. How is the organization of urban space (housing, transportation, shopping) better suited to energy efficiency in Europe than in the United States?

4. What are some of the potential societal consequences of low birth rates in Europe?

5. How might the EU's status as the world's largest economy, and the rising importance of NATO, lead to changes in global power relationships?

6. What does the evolution of democracy in Europe suggest about current efforts to promote democracy in Iraq or Egypt?

7. Why might a pig farmer in West Africa criticize the EU's generous support for European agribusiness farmers?

8. Which European state-supported systems have offered women the best opportunities for employment outside the home?

9. Cite the evidence that Europe is seeking ways of life that contribute less to global warming.

10. Where is Europe's contribution to water-body pollution most severe? What could the European Union do to reduce its contribution to the pollution of Europe's oceans and seas?

Chapter Key Terms

active aging 232
assimilation 235
asylum seekers 235
capitalism 214
central planning 214
Cohesion Policy 215
Cold War 214
Common Agricultural Program (CAP) 220
communism 214
continental climate 199
cultural homogenization 198
decolonization 214
double day 237
economic migrants 235

economies of scale 218
euro (€) 218
European Union (EU) 196
exclave 253
green 201
guest workers 235
Holocaust 214
humanism 209
Iron Curtain 214
Islamic jihadism 224
Mediterranean climate 199
mercantilism 209
midlatitude temperate climate 199
nationalism 213

North Atlantic Drift 199
North Atlantic Treaty Organization (NATO) 223
post-colonial 226
refugees 235
Roma 214
Schengen Agreement 218
social cohesion 226
social exclusion 226
socialism 214
social protection 226
subsidies 220
welfare state 213

A North European Plain, Belarus

Viktor Drachev/AFP/Getty Images

B Caucucus Mountains, Georgia

Amar Grover/Getty Images

C Volga River, Russia

Walter Bibikow/Getty Images

D Ural Mountains, Russia

John Cancalosi/Getty Images

Prague
CZECH
REP.
Vienna
POLAND
Warsaw
Stockholm
Baltic Sea
Kaliningrad
FINLAND
Helsinki
Tallinn
Bratislava
SLOVAKIA
udapest
NGARY
LITHUANIA
Vilnius
Riga
ESTONIA
LATVIA
St. Petersburg
Gulf of Finland
Gulf of Riga
NORWAY
Murmansk
Kirovsk
Kola Peninsula
Barents
Sea
Svalbard
North Land
Novaya Zemlya
Kara
Sea
Laptev
Sea
Vienna
Lviv
Brest
A
Minsk
BELARUS
Pripyat
Chernobyl
Homyel
MOLDOVA
Chisinau
Kiev
UKRAINE
Smolensk
Tver
Moscow
Canal
Bryansk
Moscow
Yaroslavl
Tula
Ivanovo
Kursk
Lipetsk
Nizhniy Novgorod
Kirov
Voronezh
Penza
Kazan
Perm
Lake
Ladoga
Lake
Onega
Northern
Dvina Canal
Divina
Dvina
North European Plain
Arkhangelsk
Kara Strait
White
Sea
Pechora
Salekhard
Vorkuta
D
Pechora
Basin
West Siberian Plain
E
Yamal
Peninsula
North Siberian Lowland
Dikson
Taymyr Peninsula
Lena R.
Norilsk
F Central
Siberian
S i b e
Plateau
Vilyuy
Res.
Olenek
ROMANIA
Odessa
Mykolayiv
Kherson
evastopol
Dnipropetrovsk
Donetsk
Rostov-on-Don
Sea of
Azov
Black Sea
40°E
Caucasus Mts.
GEORGIA
B
Batumi
URKEY
Tbilisi
Groznyy
ARMENIA
Yerevan
AZER.
AZERBAIJAN
Baku
Lake
Urmia
AQ
Zagros Mts.
IRAN
Tehran
Esfahan
Mashhad
Elburz Mts.
Saratov
Volga Don
Canal
Tolyatti
C
Samara
Volgograd
Astrakhan
Caspian Depression
Caspian Sea
Volga
Kama
Don
Volga
Ufa
Orenburg
Magnitogorsk
Orsk
Ural Mountains
U r a l
Nizhniy Tagil
Yekaterinburg
Tyumen
Chelyabinsk
Ob
Irtysh
Surgut
Omsk
Seversk
Tomsk
Novosibirsk
Novokuznetsk
Krasnoyarsk
Gladkaya
Bratsk
Tayshet
Bratsk
Res.
Lake
Baikal
Irkutsk
Basin
Angarsk
Irkutsk
Ulan-Ude
RUSSIAN FEDERATION
(RUSSIA)
Tunguska Basin
Ob
Yenisey
Angara
Kan
S t e p p e s
Aral
Sea
KAZAKHSTAN
Astana
Qaraghandy
Leninsk
Aktogay
Syr Darya
Lake
Balkhash
W. Sayan Mts.
E. Sayan Mts.
Kyzyl
Altai Mts.
Ulan Bator
MONGOLIA
Junggar Basin
Urumqi
TURKMENISTAN
UZBEKISTAN
Ashkhabad
Samarkand
Tashkent
Dushanbe
TAJIKISTAN
Pamirs
Shymkent
Bishkek
Almaty
KYRGYZSTAN
Tien
Amu Darya
Lake
Issyk-Kul
Tarim
Tari
Tak
Hindu
Kush
bul
Islamabad
Faisalabad
Lahore
AFGHANISTAN
PAKISTAN
INDIA
New Delhi

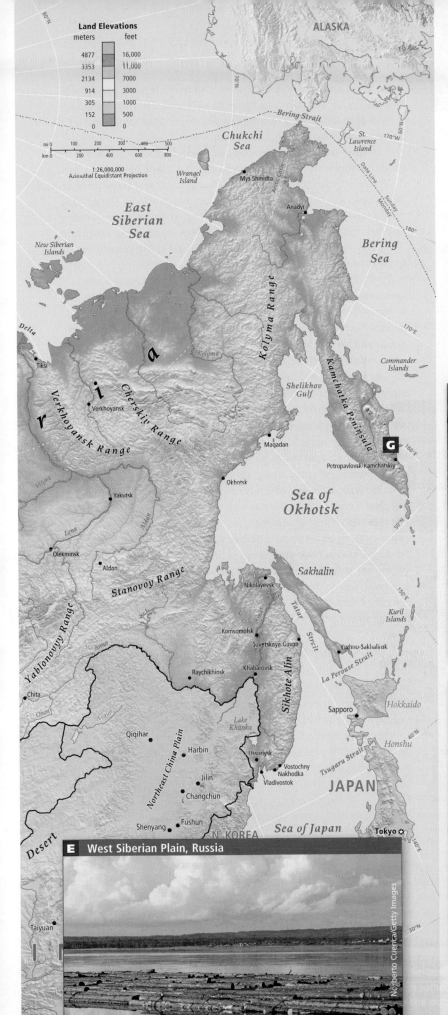

Land Elevations

meters	feet
4877	16,000
3353	11,000
2134	7000
914	3000
305	1000
152	500
0	0

mi 0   100   200   300   400   500
km 0   200   400   600   800

1:26,000,000
Azimuthal Equidistant Projection

5
Russia and the Post-Soviet States

F Central Siberian Plateau, Russia

E West Siberian Plain, Russia

G Pacific Mountain Zone, Russia

FIGURE 5.1 Regional map of Russia and the post-Soviet states.

After you read this chapter, you will be able to discuss the following issues as they relate to the five thematic concepts:

1. Environment: Economic development has taken precedence over environmental concerns in Russia and the post-Soviet states, resulting in numerous environmental problems. Economic activities in the region have expanded since the 1990s economic collapse, and contribute to climate change. Food production systems and water resources have medium to high levels of vulnerability to the impacts of climate change.

2. Globalization and Development: After the fall of the Soviet Union, wealth disparity increased and jobs were lost as many Communist-era industries were closed or sold to the rich and well connected. The region is now largely dependent on its role as a leading global exporter of energy resources.

3. Power and Politics: This region has a long history of authoritarianism. Even if the public now desires expanded political freedoms, there are few opportunities to influence the political process. Elected representative bodies often act as rubber stamps for strong presidents and exercise only limited influence on policy making.

4. Urbanization: A few large cities in Russia and Central Asia, such as Moscow, are growing fast, as they are primate cities or are fueled by the expansion of energy exports. Elsewhere, many cities are suffering from a lack of investment as their economies struggle in the post-Soviet era.

5. Population and Gender: Populations are shrinking in many parts of the region. High levels of participation by women in the workforce and the economic decline in the 1990s that followed the collapse of the Soviet Union resulted in low birth rates. Economic problems have also led to declines in life expectancy.

The Russia and Post-Soviet States Region

Russia and the post-Soviet states **(FIGURE 5.1)** make up a large portion of Eurasia, the world's largest continent. It is a vast region but is sparsely populated and with a climate that, in many places, is unforgiving.

GLOBAL PATTERNS, LOCAL LIVES *It is a cold, snowy day in February 2012 in Moscow, the capital of the Russian Federation. Tens of thousands of protesters hold hands to form an enormous circle around the city along one of the traffic-clogged ring roads that Moscow is known for. National elections are coming up soon, and the demonstrators are trying to defeat the favorite presidential candidate, Vladimir Putin. Marina Segupova, a 28-year-old interior decorator, participates in the demonstration, which she hopes will spur further protest. Using a metaphor apt for the moment, she describes the protest movement: "We are a snowball and we are rolling." At*

Union of Soviet Socialist Republics (USSR) the multinational union formed from the Russian empire in 1922 and dissolved in 1991; commonly known as the *Soviet Union*

Soviet Union see *Union of Soviet Socialist Republics*

the same time, she acknowledges, change will not come easily. Another middle-class professional, the 56-year-old accountant Galina Venediktova, came to express her opinion about corruption and poor public services. During that same month, people held other, more unconventional protests. The feminist punk band Pussy Riot staged an unauthorized "guerilla performance" at a Moscow cathedral, in which they decried the policies of Vladimir Putin and became a worldwide media sensation.

Despite the efforts of Segupova, Venediktova, Pussy Riot, and other demonstrators, Putin became the president of Russia after he decisively won the March 2012 election and will serve until at least 2018. The outcome of the election was hardly in doubt, due to the lack of other serious contenders for the presidency—some had been jailed—and Putin's dominance in state media reporting.

After Putin's election, protests continued in the streets and squares of central Moscow (FIGURE 5.2). However, being a dissenter in Russia became more difficult as the government passed new laws aimed at cracking down on protesters. The three members of Pussy Riot were put on trial and served prison terms for "hooliganism," which has reinforced the notion that political dissent in Russia will be repressed by the authorities. Subsequently, the band has continued to stage protests, often resulting in more jail time.

As Russia slowly becomes wealthier, are people demanding more political freedoms? Some in the middle class, the urbanites in large cities (particularly those in Moscow and St. Petersburg), and the country's youth are. But so far, politicians like Vladimir Putin have the support of the majority of Russians, who perceive them as strong leaders. People in hardscrabble industrial cities, small towns, and rural areas, along with the elderly, view Putin as a force for stability, which is favored by many, considering the political turbulence Russia has had during the last two decades. Putin being reelected also has had larger global geopolitical ramifications, as his annexation of Crimea and aggressive policies toward Ukraine have stirred up nationalist sentiment in Russia, which has solidified his popularity and grip on power, while the relationship between Russia and the West has been frosty.
[Sources: Christian Science Monitor, New York Times, and Sky News. For detailed source information, see Text Sources and Credits.] ■

Russia is the largest country in a region that has entirely changed its political and economic systems in a short period of time. Barely two decades ago, the **Union of Soviet Socialist Republics (USSR),** more commonly known as the **Soviet Union,** was the largest country in the world, stretching from Central Europe to the Pacific Ocean. It covered one-sixth of Earth's land surface. In 1991, the Soviet Union broke apart, ending a 70-year era of nearly complete governmental control of the economy, society, and politics. Over the course of a few years, attempts were made to substitute the Communist Party's economic control with capitalist systems similar to those of Western countries, which are based on competition among private businesses. The transition has proven difficult.

After 70 years of authoritarian rule, elections are becoming the norm throughout this region. Whether these are actually free and fair elections is highly debatable, as opposition candidates are marginalized by lack of access to both print and broadcast media. An explosion of crime and corruption has been a further setback to the democratization of the region.

Politically, the Soviet Union has been replaced by Russia and 11 independent post-Soviet states—the European states of Ukraine, Belarus, and Moldova; the Caucasian states of Georgia, Armenia, and Azerbaijan; and the Central Asian

FIGURE 5.2 Political demonstration in Moscow. Riot police in Moscow detain journalist and civil rights activist Alexandr Podrabinek during a rally protesting the inauguration of Vladimir Putin as president of Russia in 2012. The region of Russia and post-Soviet states continue to have a political culture in which authority is rarely questioned.

Thinking Geographically
Where in Russia are people more likely to support the policies of Vladimir Putin?

FIGURE 5.3 Political map of Russia and the post-Soviet states. Note that we use the borders that are recognized by most of the international community. Specifically, the Crimean Peninsula is mapped as part of Ukraine, but in reality is controlled by Russia.

THINGS TO REMEMBER

• Politically, the Soviet Union has been replaced by Russia and 11 independent post-Soviet states. Most of these countries cooperate in a loose alliance known as the Commonwealth of Independent States.

• A process of geopolitical realignment is taking place, in which some post-Soviet states are pursuing closer relations with Europe, while others are developing ties with nearby Asian countries. However, most maintain strong connections with Russia, which dominates this region.

• Elections have become common in the region, although some are not free and fair. Political systems in the region are also threatened by crime and corruption.

states of Kazakhstan, Kyrgyzstan, Tajikistan, Turkmenistan, and Uzbekistan **(FIGURE 5.3)**. Most of these countries now cooperate in a loose alliance called the Commonwealth of Independent States, although the level of cooperation between them has diminished over time. Three former Soviet republics, the Baltic states of Lithuania, Latvia, and Estonia, are now part of the European Union (see Chapter 4). Russia, which was always the core of the Soviet Union, remains dominant in the region and is influential in the world because of its size (at roughly three-quarters the size of the former Soviet Union, it is still the largest country in the world), population, military, and huge oil and gas reserves.

Geopolitically, this region is still sorting out its relationships. The Cold War between the Soviet Union and the United States and its allies was replaced by a long thaw, although the relationship has grown chilly again. Most former Soviet allies in Central Europe have already joined the European Union, and some other countries in the region may eventually do the same. Trade with Europe, especially in oil and gas, is important, yet future trade is uncertain, as Europe has imposed sanctions on Russia due to its military involvement in Ukraine. In the future, the Central Asian states currently allied with Russia may forge stronger connections with their neighbors in Southwest Asia or South Asia. The far eastern parts of Russia—and Russia as a whole—are already finding common trading ground with East Asia and Oceania **(FIGURE 5.4)**.

What Makes Russia and the Post-Soviet States a Region?

Russia and the post-Soviet states are today associated as a region primarily because of their history from the nineteenth century onward. Imperial Russia underwent a convulsive revolution in 1917 that resulted in a new political and economic experiment with socialism that lasted until 1991. The leaders of this revolution continued with the imperial practice of territorial expansion that ultimately enveloped countries from the Baltic Sea to the Pacific and from the Arctic Sea to the mountains of **Caucasia** and southern Central Asia (see the Figure 5.1 map). Control of the economic and political systems of the Soviet Union was exercised primarily from the Russian capital, Moscow, through authoritarianism reinforced with militarism. After the collapse of the Soviet Union in 1991, Moscow's dominance diminished somewhat as western parts of the region oriented more toward Europe and the Central Asian republics became

Caucasia the mountainous region between the Black Sea and the Caspian Sea

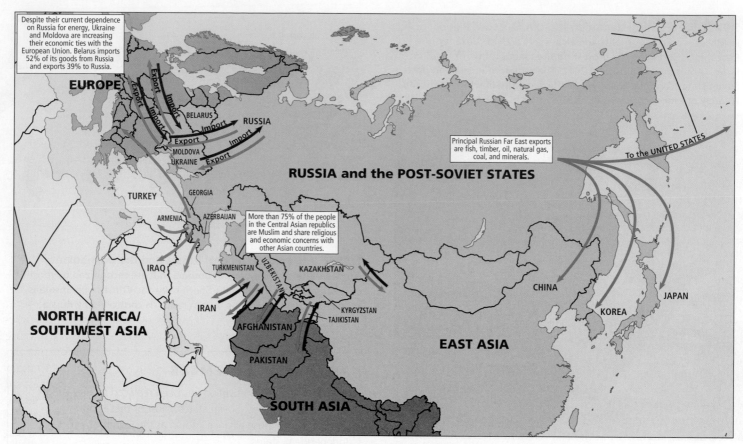

Despite their current dependence on Russia for energy, Ukraine and Moldova are increasing their economic ties with the European Union. Belarus imports 52% of its goods from Russia and exports 39% to Russia.

Principal Russian Far East exports are fish, timber, oil, natural gas, coal, and minerals.

More than 75% of the people in the Central Asian republics are Muslim and share religious and economic concerns with other Asian countries.

RUSSIA and the POST-SOVIET STATES

EUROPE
BELARUS
RUSSIA
MOLDOVA
UKRAINE
TURKEY
GEORGIA
ARMENIA
AZERBAIJAN
IRAQ
TURKMENISTAN
UZBEKISTAN
KAZAKHSTAN
IRAN
AFGHANISTAN
KYRGYZSTAN
TAJIKISTAN
PAKISTAN
NORTH AFRICA/SOUTHWEST ASIA
SOUTH ASIA
EAST ASIA
CHINA
KOREA
JAPAN
To the UNITED STATES

FIGURE 5.4 Russia and the post-Soviet states: Contacts with other world regions. The region of Russia and the post-Soviet states is in flux as all of the component countries rethink their geopolitical positions relative to one another and to adjacent regions. The arrows indicate various types of contact, from economic to religious and cultural.

Russian Federation Russia and its political subunits, which include 21 internal republics

Siberia the territories of Russia that are located east of the Ural Mountains

steppes semiarid grass-covered plains

more autonomous (see Figure 5.4). Nevertheless, Russia remains the dominant power in this region.

Terms in This Chapter

There is no entirely satisfactory name for the former Soviet Union. In this chapter we use *Russia and the post-Soviet states*. One reason we do so is that economic, political, and social developments from the Soviet days still shape the region today. The post-Soviet states continue to be closely associated with Russia economically, but they are independent countries separate from Russia. Russia itself is formally known as the **Russian Federation** because it includes more than 21 mostly ethnic *internal republics*—such as Chechnya, Karelia, and Tatarstan—which all together constitute about one-tenth of the Federation's territory and one-sixth of its population. *Federation* should not be taken to mean that the internal republics share power equally with the central Russian government in Moscow.

PHYSICAL GEOGRAPHY

The physical features of Russia and the post-Soviet states vary greatly over the huge territory they encompass. The physical vastness of the region (especially Russia) is and has always been

a major challenge for effective governance, military activities, transportation, resource extraction, and trade.

LANDFORMS

Because the region is relatively complex physically, a brief summary of its landforms is useful (see the Figure 5.1 map). Moving west to east, there is first the eastern extension of the North European Plain, including the plain of the Volga River (see Figure 5.1A, C), followed by the Ural Mountains (see Figure 5.1D), which are often considered the dividing line between European Russia and **Siberia.** East of the Urals are the West Siberian Plain (see Figure 5.1E), followed by an upland zone called the Central Siberian Plateau (see Figure 5.1F), and finally, in the Far East, a series of mountain ranges bordering the Pacific (see Figure 5.1G). To the south there are mountains and uplands (the Caucasus, shown in Figure 5.1B, and Central Asia, depicted in Figure 5.8A) as well as, in western Central Asia, semiarid grasslands, or **steppes** (see Figure 5.5B).

The eastern extension of the North European Plain stretches 1200 miles (about 2000 kilometers), from Central Europe to the Ural Mountains (see Figure 5.1A, C, D). The part of Russia west of the Urals is often called *European Russia* because the Ural Mountains are traditionally considered part of the indistinct border between Europe and Asia. European Russia is the most densely settled part of the entire region (see

Figure 5.23) and is its agricultural and industrial core. Its most important river is the Volga, which flows into the Caspian Sea. The Volga River and its tributaries form a major transportation route that connects many parts of the North European Plain, including Moscow, St. Petersburg and the Baltic and White seas in the north, and the Black and Caspian seas in the south (see Figure 5.1C).

The Ural Mountains extend in a fairly straight line south from the Arctic Ocean into Kazakhstan (see the Figure 5.1 map). A low-lying range similar in elevation to the Appalachians, the Urals are not much of a barrier to humans and are only partially a barrier to nature (some European tree species do not extend east of the Urals). There are several easy passes across the mountains, and winds carry moisture all the way from the Atlantic into Siberia. Much of the Urals' once-dense forest has been felled to build and fuel new industrial cities.

The West Siberian Plain, east of the Urals, is the largest plain in the world (see Figure 5.1E). A vast, mostly marshy lowland about the size of the eastern United States drains toward the north into the Arctic Ocean. The northward-flowing rivers of Siberia have done little to improve Russia's need for east-west communication. Long, bitter winters mean that in the northern half of this area, a layer of permanently frozen soil **(permafrost)** lies just a few feet beneath the surface. Permafrost is formed when the ground warms up during the short summer but the surface layer of organic material and soil insulates against this warming effect, leaving the subsurface always frozen. In the far north, the permafrost comes to within a few inches of the surface. Because water does not percolate down through this frozen layer, surface water accumulates to create wetlands that form above the permafrost. In the summer months these wetlands provide habitats for many migratory birds. Due to climate change, some permafrost in Siberia is melting, which releases carbon from the thawing vegetation into the atmosphere.

In the far north lies a treeless area called the **tundra,** where mostly only mosses and lichens can grow because of the extreme cold, the shallow soils, and the permafrost. The West Siberian Plain is one of the world's largest oil and natural gas sources, although the harsh climate and permafrost make extraction difficult.

The Central Siberian Plateau and the Pacific Mountain Zone, farther to the east, together equal the size of the United States (see Figure 5.1F, G). Permafrost prevails except along the Pacific coast. There, the ocean moderates temperatures; the many active volcanoes created as the Pacific Plate sinks under the Eurasian Plate supply additional heat. Lightly populated places like the Kamchatka Peninsula, Sakhalin Island, and Sikhote-Alin on Russia's Pacific coast are havens for wildlife.

To the south of the West Siberian Plain, steppes and deserts stretch from the Caspian Sea to the Chinese border. To the west of these grasslands are the Caucasus Mountains (see Figure 5.1B), and to the southeast, facing China and Mongolia, are a series of other mountain chains, including the Pamir and Tien Shan. The rugged terrain has not deterred people from crossing these mountains. For tens of thousands of years, from the Caucasus to the Pamir and Tien Shan, people have exchanged plants (apples, onions, citrus, rhubarb, wheat),

animals (horses, sheep, goats, cattle), technologies (cultivation, animal breeding, portable shelter construction, rug and tapestry weaving), and religious belief systems (principally Islam and Buddhism, but also Christianity, Hinduism, and Judaism).

CLIMATE AND VEGETATION

The climates and associated vegetation in this large region are varied but less so than in other regions because so much territory here is taken up by expanses of midlatitude grasslands and by northern forest and tundra that are cold much of the year. No inhabited place on Earth has as harsh a climate as the northern part of the Eurasian landmass occupied by Siberian Russia (shown in **FIGURE 5.5C** and the figure map). Winters are long and cold, with only brief hours of daylight. Summers are short and range from cool to hot, with long days. Precipitation is moderate, coming primarily from the west. In the northernmost areas, the natural vegetation is tundra grasslands and other hardy ground cover plants. The major economic activities here are the extraction of oil, gas, and some minerals, as well as reindeer herding by the local indigenous population. Just south of the tundra is a vast, cold-adapted coniferous forest known as **taiga** that ranges from northern European Russia to the Pacific (and from a global perspective, this coniferous belt also includes much of Alaska, Canada, and Scandinavia). The largest portion of taiga lies east of the Urals, and here forestry—often unrestrained by ecological concerns—is a dominant economic activity. The short growing season and the large areas of permafrost generally limit crop agriculture, except in the southern West Siberian Plain, where people grow grain.

Because massive mountain ranges to the south block access to warm, wet air from the Indian Ocean, most rainfall in the entire region comes from storms that blow in from the Atlantic Ocean far to the west (see the Figure 5.5 map). But by the time these air masses arrive, most of their moisture has been squeezed out over Europe. A fair amount of rain does reach Ukraine, Belarus, European Russia (see Figure 5.5A), and the Caucasian republics; these regions are especially important areas for food production (vegetables, fruits, and grain). The natural vegetation in these western zones is open woodlands and grassland, though in ancient times, forests were common.

East of the Caucasus Mountains, the lands of Central Asia have semiarid to arid climates (see Figure 5.5B), influenced by their location in the middle of a very large continent. The summers are scorching and short and the winters are intense. In the desert zones, daytime-to-nighttime temperatures can vary by 50°F (28°C) or more. Northern Kazakhstan produces grain and grazing animals. The more southern areas (southern Kazakhstan, Uzbekistan, and Turkmenistan) have grasslands (steppes), which are also used for herding. Some of the land has been converted to irrigated commercial agriculture using water from glacially fed rivers, but most of it is not useful for farming. The climates in the more mountainous area where Kazakhstan, Uzbekistan, Tajikistan,

permafrost permanently frozen soil that lies just beneath the surface

tundra a treeless area, between the ice cap and the tree line of arctic regions, where the subsoil is permanently frozen

taiga subarctic coniferous forests

FIGURE 5.5 PHOTO ESSAY: Climates of Russia and the Post-Soviet States

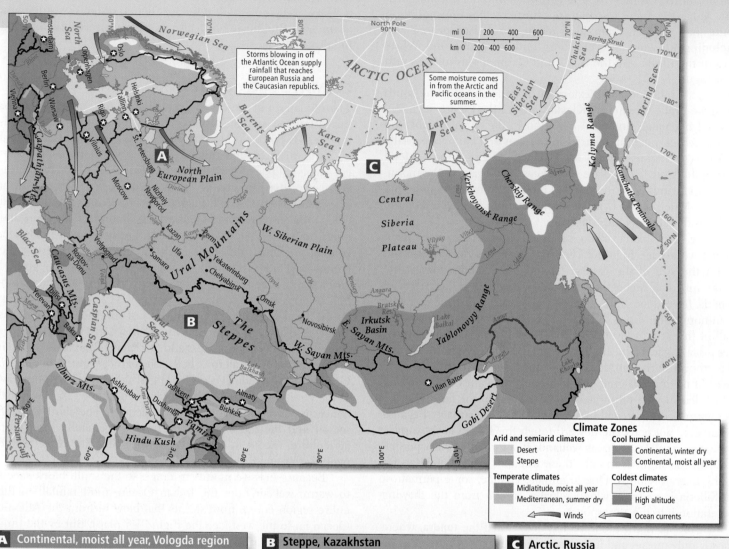

Storms blowing in off the Atlantic Ocean supply rainfall that reaches European Russia and the Caucasian republics.

Some moisture comes in from the Arctic and Pacific oceans in the summer.

Climate Zones

Arid and semiarid climates
- Desert
- Steppe

Temperate climates
- Midlatitude, moist all year
- Mediterranean, summer dry

Cool humid climates
- Continental, winter dry
- Continental, moist all year

Coldest climates
- Arctic
- High altitude

← Winds   ← Ocean currents

A Continental, moist all year, Vologda region

vicsa/Getty Images/iStockphoto

B Steppe, Kazakhstan

Fotosearch/Getty Images

C Arctic, Russia

Daisy Gilardini/Getty Images

and Kyrgyzstan meet are varied and support a number of small-scale agricultural activities, some of them commercial.

The construction of land transportation systems has been held back by the climate of the region. Especially to the north, it is difficult to build during the long, harsh winters, which eventually give way to a spring period called the *rasputitsa*, or the "quagmire season," when melting snow and ice (or, to a lesser extent, autumn rain) turn many roads and construction sites into impassable mud pits. Huge distances between populated places, as well as the complex topography—especially in Siberia—also complicate the situation. As a result, few roads or railroads have ever been built beyond western Russia. One exception is the Trans-Siberian Railway, which crosses the continent. The railway was a major infrastructural achievement in the early twentieth century.

THINGS TO REMEMBER

- The part of Russia west of the Urals is often called European Russia because the Ural Mountains are traditionally considered part of the indistinct border between Europe and Asia.

- In the far north lies a treeless area called the tundra, where mostly mosses and lichens grow because of the extreme cold, the shallow soils, and the permafrost.

- Just south of the tundra is a vast, cold-adapted coniferous forest known as taiga that ranges from northern European Russia to the Pacific.

- Because massive mountain ranges to the south block access to warm, wet air from the Indian Ocean, most rainfall in the entire region comes from storms that blow in from the Atlantic Ocean far to the west.

- The construction of land transportation systems has been held back by the climate of the region. Especially to the north, it is difficult to build during the long, harsh winters, which eventually give way to a spring period called the rasputitsa, or the "quagmire season," when melting snow turns many roads and construction sites into impassable mud pits.

ENVIRONMENT

GEOGRAPHIC THEME 1

Environment: Economic development has taken precedence over environmental concerns in Russia and the post-Soviet states, resulting in numerous environmental problems. Economic activities in the region have expanded since the 1990s economic collapse, and contribute to climate change. Food production systems and water resources have medium to high levels of vulnerability to the impacts of climate change.

In Soviet ideology, nature was considered the servant of industrial and agricultural progress, and humans were meant to dominate nature on a grand scale. While this sentiment was common throughout much of the world at the time of the formation of the USSR, it does seem to have been taken further here than elsewhere. Joseph Stalin, leader of the USSR from 1922 to 1953 and a major architect of Soviet policy, is famous for having said,

"We cannot expect charity from Nature. We must tear it from her." During the Soviet years, huge dams, factories, and other industrial facilities were built without regard for their effect on the environment or on public health. Russia and the post-Soviet states now have some of the worst environmental problems in the world.

Since the collapse of the Soviet Union, the region's governments, beset with myriad problems, have been both reluctant to address environmental issues and incapable of doing so. As one Russian environmentalist put it, "When people become more involved with their stomachs, they forget about ecology." Pollution controls are complicated by a lack of funds and by an official unwillingness to correct past environmental abuses. **FIGURE 5.6** shows examples of human impacts on the region's environment, most of which are related to ongoing industrial pollution (as in Norilsk, in Figure 5.6A, B), nuclear contamination (as in Pripyat, Ukraine, in Figure 5.6C), or the extraction of fossil fuels. The sale of oil and gas on the global market is, during most years, a major source of income for Russia and several of the post-Soviet states, but relatively little attention is being paid even today to the environmental impact of the extraction of fossil fuels (see Figure 5.6D).

URBAN AND INDUSTRIAL POLLUTION

Urban and industrial pollution was often ignored during Soviet times as cities expanded quickly—with workers flooding in from the countryside—to accommodate the new industries. Citizens' concerns about pollution were suppressed and environmental movements such as those in North America and western Europe never developed.

It is often difficult to link pollution directly to health problems because the sources of contamination are multiple and difficult to trace. Such **nonpoint sources of pollution** include untreated automobile exhaust, raw sewage, and agricultural chemicals that drain from fields into water supplies. In all urban areas of the region, air pollution resulting from the burning of fossil fuels is skyrocketing as more people purchase cars and as the industrial and transport sectors of the economy continue to grow.

Some cities were built around industries that produce harmful by-products. The former chemical weapons-manufacturing center of Dzerzhinsk is listed by the nonprofit Blacksmith Institute as one of the ten most polluted cities in the world. It is competing with the city of Norilsk, where much of the vegetation has been killed off around the city's metal-smelting complex—the largest facility of its kind in the world (see Figure 5.6B).

NUCLEAR POLLUTION

Russia and the post-Soviet states are also home to extensive nuclear pollution, the effects of which have spread globally. The world's worst nuclear disaster occurred in Ukraine in 1986, when the Chernobyl nuclear power plant exploded. The explosion severely contaminated a vast area in northern Ukraine, southern Belarus, and Russia. It spread a cloud of radiation over much of Central Europe, Scandinavia, and eventually the entire planet. In the area surrounding Chernobyl, more than 300,000

nonpoint sources of pollution diffuse sources of environmental contamination, such as untreated automobile exhaust, raw sewage, and agricultural chemicals that drain from fields into water supplies

FIGURE 5.6 PHOTO ESSAY: Human Impacts on the Biosphere in Russia and the Post-Soviet States

This region is home to some of the worst pollution in the world. More than 70 years of industrial development with few environmental safeguards have wreaked havoc on ecosystems. There are some attempts at becoming cleaner and less polluting, but decades may pass before significant improvements are realized.

A A cloud of pollution from a nearby nickel smelter hangs over the city of Norilsk. The smelter emits about 1 percent of all global emissions of sulfur dioxide, which helps form acid rain. Enough heavy metals have accumulated in the soils around Norilsk that they can now be mined commercially. The company that runs the smelter has promised improvements by 2020.

B This false-color satellite image shows Norilsk. Purple indicates areas where vegetation has been killed by pollution. The city is located just below the main lake shown in the upper part of the image.

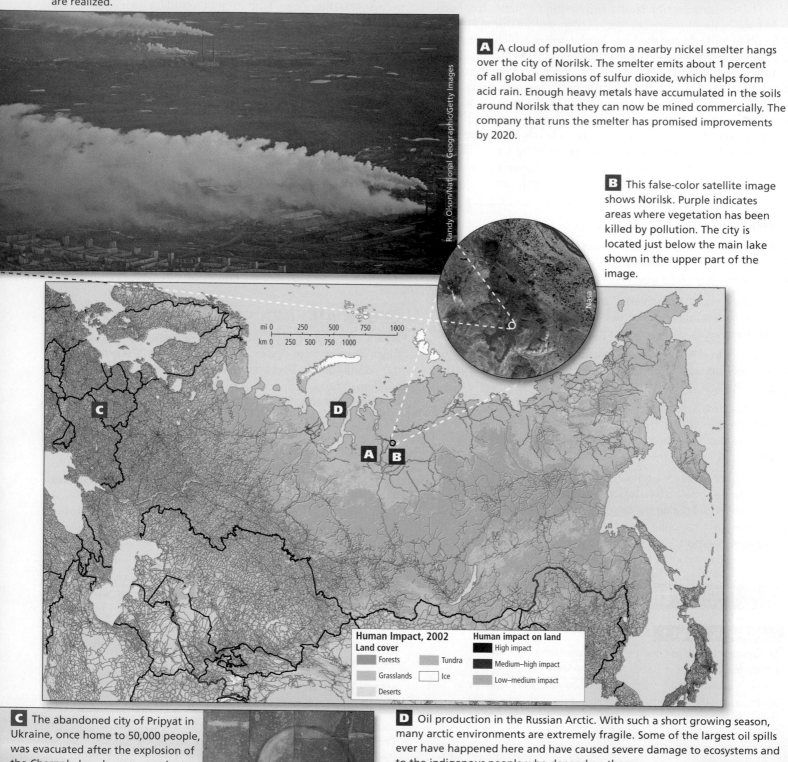

Randy Olson/National Geographic/Getty Images

Nasa

Human Impact, 2002
Land cover

- Forests
- Grasslands
- Deserts
- Tundra
- Ice

Human impact on land

- High impact
- Medium–high impact
- Low–medium impact

mi 0   250   500   750   1000
km 0   250   500   750   1000

C The abandoned city of Pripyat in Ukraine, once home to 50,000 people, was evacuated after the explosion of the Chernobyl nuclear power plant, visible on the horizon.

Chernobyl nuclear power plant

Sergei Supinsky/Getty Images

D Oil production in the Russian Arctic. With such a short growing season, many arctic environments are extremely fragile. Some of the largest oil spills ever have happened here and have caused severe damage to ecosystems and to the indigenous people who depend on them.

Alexander Zemlianichenko Jr./ Bloomberg via Getty Images

Thinking Geographically

After you have read about the human impact on the biosphere in Russia and the post-Soviet states, you will be able to answer the following questions:

A How is the global economy related to environmental problems in Norilsk?

B How extensive is local environmental degradation in Norilsk?

C Why was Pripyat evacuated?

D From what form of pollution, in addition to oil pollution, has the Arctic portion of this region suffered?

people were evacuated from their homes, and Pripyat, which used to have a population of 50,000 people, has been an abandoned ghost town since the disaster (see Figure 5.6C). The ultimate health effects are impossible to assess exactly. The United Nations estimates that 6000 people developed cancer due to the accident, although many believe the number to be much higher.

When the Soviet Union collapsed, many of the post-Soviet states inherited nuclear facilities and even nuclear weapons. One such country is Kazakhstan (see Figure 5.3). Remote areas of eastern Kazakhstan were used as a testing ground for Soviet nuclear devices. The residents were sparsely distributed and no one took the trouble to protect them from nuclear radiation. A museum in Kazakhstan now displays the preserved remains of hundreds of deformed human fetuses and newborns.

On the bright side, Kazakhstan's government has disarmed the nuclear warheads that it inherited from the Soviet Union and has worked with U.S. authorities to safeguard its remaining nuclear weapons material. Kazakhstan plans to use its nuclear technology to become a global leader in the nuclear energy sector. It is already the single biggest source of uranium in the world, generating 38 percent of the world's total.

In addition to mining and exporting the raw material, Kazakhstan's government wants to process uranium to nuclear fuel both for global markets and domestic nuclear power plants and to develop commercial repositories for radioactive waste from other countries. However, global nuclear expansion remains uncertain following the 2011 Fukushima accident in Japan (see Chapter 9), and no reliably safe system has yet been found for storing nuclear waste until it is no longer radioactive. Environmental and human rights NGOs have raised strong opposition to Kazakhstan's nuclear plans, pointing to its poor environmental track record. Kazakhstan's population may also be suspicious of nuclear power because they have lived for so long with land contaminated by radioactive waste.

ON THE BRIGHT SIDE: Controlling Nuclear Material

In the post–9/11 world, people became concerned that military corruption in this region could put nuclear weapons from the former Soviet arsenal into the hands of terrorists. There had been cuts to military funding, and weapons, uniforms, and even military rations were being sold on the black market. Russian smugglers were even trying to sell nuclear materials.

Recent developments have been more encouraging. In an effort to fight corruption and the temptation to sell nuclear materials on the black market, the Russian government has given impoverished military personnel long-overdue pay raises or compensation for being terminated from their jobs. In addition, all countries in the region now cooperate with the International Atomic Energy Agency in controlling nuclear material. ∎

THE GLOBALIZATION OF RESOURCE EXTRACTION AND ENVIRONMENTAL DEGRADATION

Russia and the post-Soviet states have considerable natural and mineral resources (see Figure 5.14). Russia itself has the world's largest natural gas reserves, major oil and coal deposits, and forests that stretch across the northern reaches of the continent. Russia also has major deposits of industrial minerals. The Central Asian states share substantial deposits of oil and gas, which are centered on the Caspian Sea and extend east toward China.

Despite obvious environmental problems, cities like Norilsk (see Figure 5.6A), which sits on huge mineral deposits, continue to attract investment crucial to the new Russian economy. The area is rich in nickel—used in steel and other industrial products—and other minerals, such as copper. Norilsk Nickel, a company that was privatized at the end of communism and is now the largest producer of nickel in the world, dominates the city. With an extensive base of mineral resources, and unconstrained by meaningful environmental regulations, Russian-controlled Norilsk Nickel has used its enormous profits to become a global player in the mining industry. Despite its poor environmental record, it owns mining operations in Australia, Finland, the United States, and southern Africa and employs almost 100,000 people worldwide.

Rivers, Irrigation, and the Loss of the Aral Sea

In the Central Asian states, the Aral Sea, once the fourth-largest lake in the world, was fed by the Syr Darya and Amu Darya rivers for millions of years. These rivers brought water from melting glaciers and snow in the mountains of Kyrgyzstan, Tajikistan, and Afghanistan.

In the 1960s, the Soviet leadership ordered that the two rivers be diverted to irrigate millions of acres of cotton in naturally arid Kazakhstan and Uzbekistan. So much water was consumed by these projects that within a few years, the Aral Sea had shrunk measurably **(FIGURE 5.7A)**. By the early 1980s, no water at all from the two rivers was reaching the Aral Sea, and eventually it lost most of its volume and shrank into three smaller lakes. The region's huge fishing industry, which accounted for one-sixth of the fish catch in the entire USSR, died out because of increased salinity in the water. Once-active ports were marooned many kilometers from the water. The decline and disappearance of the Aral Sea has been described as the largest human-made ecological disaster in history.

The shrinkage of the Aral Sea has caused many human health problems. Winds that sweep across the newly exposed seabed pick up salt and chemical residues, creating poisonous dust storms.

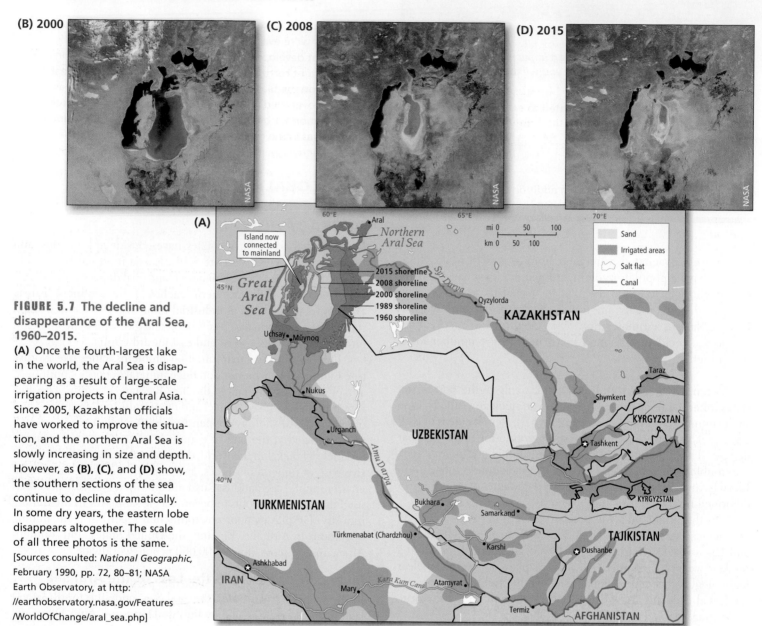

(B) 2000

(C) 2008

(D) 2015

FIGURE 5.7 The decline and disappearance of the Aral Sea, 1960–2015.

(A) Once the fourth-largest lake in the world, the Aral Sea is disappearing as a result of large-scale irrigation projects in Central Asia. Since 2005, Kazakhstan officials have worked to improve the situation, and the northern Aral Sea is slowly increasing in size and depth. However, as **(B)**, **(C)**, and **(D)** show, the southern sections of the sea continue to decline dramatically. In some dry years, the eastern lobe disappears altogether. The scale of all three photos is the same. [Sources consulted: *National Geographic*, February 1990, pp. 72, 80–81; NASA Earth Observatory, at http://earthobservatory.nasa.gov/Features/WorldOfChange/aral_sea.php]

People around the sea report chronic illnesses caused by air pollution, a lack of clean water, and underdeveloped sanitation.

Efforts to increase water flows to the Aral Sea have been hampered by politics because the now-independent countries of Uzbekistan and Kazakhstan share the sea. In particular, Uzbekistan wants to continue its massive irrigation programs, which have made it the world's sixth-largest cotton grower. Nevertheless, some restorative actions have been taken. Kazakhstan, which relies on oil more than cotton for income, built a dam in 2005 to keep fresh water in the northern Aral Sea. By the following year, the water had risen 10 feet, and the fish catch was improving. In the southern Aral Sea, however, water levels remain very low (see Figure 5.7B, C). In Uzbekistan, for instance, where 26 percent of the workforce is employed in agriculture, the limitation of irrigation has forced economic diversification, and cotton production now makes up just 10 percent of the country's exports, far less than the 45 percent it represented in 1990.

CLIMATE CHANGE

Climate change is an issue in this region for many reasons. Although high levels of CO_2 and other greenhouse gases (GHGs) are produced here, Russia's forests play a major role in absorbing global CO_2. And while much of the region is only somewhat vulnerable to the negative effects of a changing climate (shown on the map in **FIGURE 5.8**), Central Asia, which is prone to drought and water scarcity, has high to extreme levels of vulnerability.

Even though its CO_2 and other GHG emissions have declined since the Soviet era, Russia still has the fourth-highest rate of GHG emissions in the world. Its emissions rate is high in part because of its use of energy-inefficient technologies and practices such as burning off or "flaring" the natural gas that comes to the surface when oil wells are drilled. While Russia has higher GHG emissions per capita than European countries, its levels are not as high as those of the United States.

Russia has shown some willingness to limit its GHG emissions by, for example, signing international treaties such as the Kyoto Protocol and the Copenhagen Accord. Such treaties, however, use 1990 as a baseline for emissions. Russia's ability to meet international targets to reduce GHG emissions is largely related to the country's post-1990 economic decline after the end of communism. Lower levels of industrial output have meant lower emissions. More recently, however, its emissions have increased.

Central Asia is particularly vulnerable to the impacts of climate change because of its dependence on water from rivers that are fed by glaciers in the Pamirs and other mountains in Kyrgyzstan and Tajikistan (see Figure 5.8). In recent decades, the glaciers have shrunk because of more extreme temperature ranges and decreases in rain and snowfall. In this area, most precipitation falls in the mountains in the winter and spring. The glaciers act as water-storage systems, supplying melted ice flow to the rivers during the summer, when irrigation is most needed for growing crops. If current trends continue, the glaciers will melt earlier in the year, resulting in lower river flows in the summer. Central Asia's agricultural systems would either have to adapt to a spring growing season (winters are too cold), or farmers would have to store water for use in the summer. Either situation would demand complex and expensive changes on a massive scale.

ON THE BRIGHT SIDE: Better Times Ahead?

While environmental pollution remains serious throughout this region, recent surveys in Russia suggest that people are becoming more aware of environmental issues and more willing to act to protect the environment. The types of environmental activism that people are engaged in often addresses tangible environmental problems that affect ordinary Russians—the types of air, water, and soil pollution for which there is an obvious source of pollution as well as a remedy.

People's awareness about more indirect environmental issues, such as global warming, is still lagging. One reason why environmentalism is only slowly gaining ground is related to the vastness of the land, which reinforces the deeply held public attitude that resources are unlimited. If Russia progresses economically over time, it is possible that Russians' environmental consciousness will continue to grow, mirroring the historical trend in, for example, the large and resource-rich North American continent. ■

THINGS TO REMEMBER

GEOGRAPHIC THEME 1 • **Environment:** Economic development has taken precedence over environmental concerns in Russia and the post-Soviet states, resulting in numerous environmental problems. Economic activities in the region have expanded since the 1990s economic collapse, and contribute to climate change. Food production systems and water resources have medium to high levels of vulnerability to the impacts of climate change.

• Urban and industrial pollution was often ignored during Soviet times as cities expanded quickly—with workers flooding in from the countryside—to accommodate the new industries that often generated lethal levels of pollution.

• The world's worst nuclear disaster occurred in Ukraine in 1986, when the Chernobyl nuclear power plant exploded, severely contaminating portions of northern Ukraine and southern Belarus.

• The Aral Sea has shrunk to a fraction of its former size because the rivers that fed it have been redirected since the 1960s to irrigate millions of acres of cotton crops in naturally arid Kazakhstan and Uzbekistan.

• Central Asia is especially vulnerable to climate change, given its dependence on irrigated agriculture that uses water from rivers fed by glacial melt.

HUMAN PATTERNS OVER TIME

The goals of the Soviet experiment begun in 1917 were unique in human history: to quickly and totally reform an entire human society. The Soviet Union's collapse brought an equally rapid shift from centrally planned economies to market economies. The transition remains incomplete and for many, life is proceeding amid economic, political, and social uncertainty.

The core of the entire region has long been European Russia, the large and densely populated homeland of the ethnic Russians. Expanding gradually from this center, the Russians conquered a large area inhabited by a variety of other ethnic groups. These conquered territories remained under Russian control as part of the Soviet Union (1917–1991), which attempted to create an integrated social and economic unit out of the disparate territories. The breakup of the Soviet Union has diminished Russian domination throughout this region, though Russia's influence is still considerable.

THE RISE OF THE RUSSIAN EMPIRE

For thousands of years, the militarily and politically dominant people in the region were **nomadic pastoralists** who lived on the meat and milk provided by their herds of sheep, horses, and other grazing animals, and used animal fiber to make yurts, rugs, and clothing. As possibly the first people to domesticate horses, their movements followed the changing seasons across the wide grasslands that stretch from the Black Sea to the Central Siberian Plateau. Pastoralists, possibly from the steppes north of the Black Sea, migrated over long distances and have left their genetic mark on people today living anywhere from Europe deep into Asia, as can be seen by DNA analysis. Studies of linguistics also show the reach of this migration: the Indo-European language family connects most of the European languages with languages such as Hindi in India.

Towns arose in two main areas: the dry lands of greater Central Asia and the forests of Caucasia, Ukraine, and Russia. As early as 5000 years ago, Central Asia had settled communities that were supported by irrigated croplands (see Figure 5.10A). These communities were enriched by their locations, which were central for trade along what became known as the Silk Road, that vast, ancient, interwoven ribbon of trading routes

nomadic pastoralists people whose way of life and economy are centered on tending grazing animals who are moved seasonally to gain access to the best pastures

FIGURE 5.8 PHOTO ESSAY: Vulnerability to Climate Change in Russia and the Post-Soviet States

Much of the region has a medium level of vulnerability to climate change, but Central Asia has high and extreme levels of vulnerability. In Central Asia, vulnerability will increase as temperatures rise and glaciers in the Pamirs and other mountains melt. The dependence of so many Central Asians on agriculture makes them very sensitive to drought and water scarcity. Widespread poverty, combined with poorly developed disaster response systems, reduces the overall resilience to these and other disturbances.

A A river fed by glacial melting in Tajikistan. Most of Central Asia depends on rivers for household water use and irrigation water. If these rivers receive less water from glaciers, there will not be enough water to supply people and their crops.

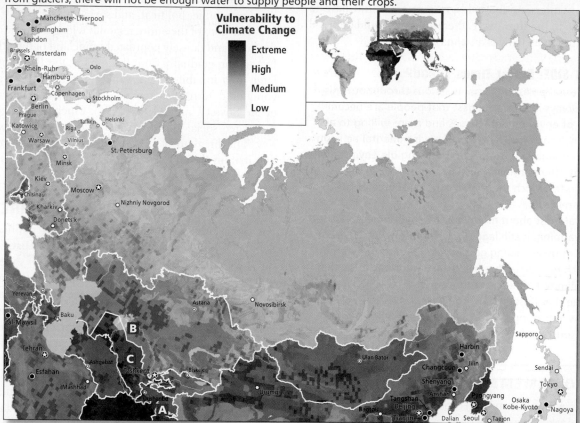

Vulnerability to Climate Change

- Extreme
- High
- Medium
- Low

B A fishing boat stranded in what used to be part of the Aral Sea, which has shrunk dramatically because the river waters that feed the sea have been diverted to irrigate cotton and food crops. Even less water will reach the Aral Sea as glaciers continue melting.

C An irrigated cotton field in Uzbekistan. Twenty-six percent of Uzbeks work in agriculture, and cotton is the country's leading export. If irrigation water were to become less available, the livelihoods of millions of Uzbeks would be threatened.

Thinking Geographically

After you have read about vulnerability to climate change in Russia and the post-Soviet states, you will be able to answer the following questions.

A Why would climate change cause rivers like the one in this photo to receive less water over the long term?

B What major resource did the Aral Sea once provide in abundance?

C How would a decline in Uzbekistan's cotton industry affect the country's economy?

In the twelfth century, the Mongol armies of Genghis Khan conquered the forested lands of Ukraine and Russia. The **Mongols** were a loose confederation of nomadic pastoral people centered in East and Central Asia. Moscow's rulers became tax gatherers for the Mongols, dominating neighboring kingdoms and eventually growing powerful enough to challenge local Mongol rule. In 1552, the Slavic ruler Ivan IV ("Ivan the Terrible") conquered the Mongols, marking the beginning of the Russian empire. St. Basil's Cathedral, a major landmark in Moscow, commemorates the victory (**FIGURE 5.11**).

By 1600, Russians centered in Moscow had conquered many former Mongol territories, integrating them into the growing empire and extending it to the east, as shown in **FIGURE 5.12**. The first major non-Russian area to be annexed was western Siberia (1598–1689). Russian expansion into Siberia (and even into North America, along its Pacific coast) resembled the spread of European colonialism around the world, as well as the future westward spread of the United States into contiguous but sparsely populated areas inhabited by indigenous people. Russian colonists took land and resources from the Siberian populations and treated the people of those cultures as inferior. So many laborers migrated from Russia to Siberia that they outnumbered the indigenous Siberians by the eighteenth century. By the mid-nineteenth century, Russia had also expanded its reach to the south in order to gain control of Central Asia's cotton crop, its major export. The building of railroads played an important role in the movement of people and goods.

To the west, the Russian empire was intertwined with the politics of

between China and the Mediterranean (**FIGURE 5.9**). Many commodities were traded along the route, but Chinese silk fabric became highly valued in the West, which is why the term *Silk Road* was used by later historians. This was a golden age for Central Asian countries because they benefitted greatly from their position in the middle portions of the trade route.

About 1500 years ago, the **Slavs,** a group of farmers—among them those known as the *Rus* (probably of Scandinavian origin)—emerged in what are now Poland, Ukraine, and Belarus. They moved east, founding numerous settlements, including the towns of Kiev and Moscow. The Slavs prospered from a lucrative trade route over land and along the Volga River that connected North Europe and Asia. Powerful kingdoms developed in Ukraine and European Russia. In the mid-800s, Greek missionaries introduced both Christianity (now known as "Orthodox Christianity") and the Cyrillic alphabet that we now associate with the Russian language (**FIGURE 5.10B**). The Cyrillic alphabet is still used in most of the region's countries.

> **Slavs** a group of people who originated in what are now Poland, Ukraine, and Belarus
>
> **Mongols** a loose confederation of nomadic pastoral people centered in East and Central Asia, who by the thirteenth century had established by conquest an empire that stretched from Europe to the Pacific

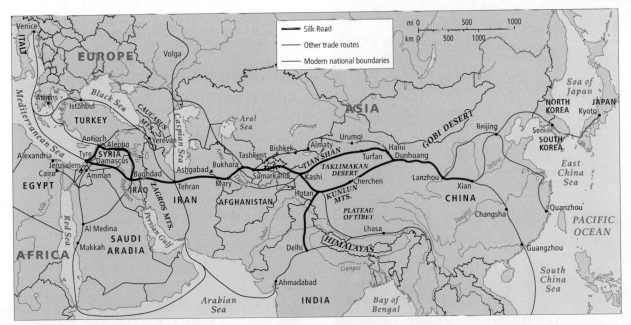

FIGURE 5.9 The ancient Silk Road and related trade routes. Merchants who worked the Silk Road rarely traversed the entire distance. Instead, they moved back and forth along part of the road, trading with merchants to the east and west. [Source consulted: "Maps of the Silk Road," the Silk Road Project and the Stanford Program on International and Cross-Cultural Education, at http://virtuallabs.stanford.edu/silkroad/SilkRoad.html]

A The fort at Bukhara, Uzbekistan, founded in 500 B.C.E. along the Silk Road.

B An Orthodox Christian monastery in Russia, founded around 1500 C.E.

C The summer palace of Czar Peter the Great in St. Petersburg, built from 1714 to 1755.

| 4000 B.C.E. | 500 B.C.E. | 500 C.E. | 1600 C.E. | 1700 C.E. |

3000 B.C.E.
Settled communities in Central Asia, supported by irrigated agriculture and trade on the Silk Road

500 B.C.E.
Fort at Bukhara founded along the Silk Road

500 C.E.
Slavs emerge

1500 C.E.

1598–1689
Russian annexation of western Siberia

FIGURE 5.10 VISUAL HISTORY OF RUSSIA AND THE POST-SOVIET STATES

Thinking Geographically

After you have read about the human history of Russia and the post-Soviet states, you will be able to answer the following questions:

A In addition to trade along the Silk Road, what supported Central Asian cities?

B From where was Orthodox Christianity introduced into this region?

European royalties and empires. In the early nineteenth century, the French emperor Napoleon attempted to conquer Russia but underestimated its size and harsh winter climate. The failed attempt by Napoleon to conquer Russia actually enabled Russia to expand its empire into Poland and Finland (see Figure 5.12).

The Russian empire was ruled by a powerful monarch, the **czar**, who—along with a tiny aristocracy—lived in splendor, while the vast majority of the people lived short, brutal lives in poverty (see Figure 5.10C, D). Many Russians were *serfs* who were legally bound to live on and farm land owned by an aristocrat. If the land was sold, the serfs were transferred with it. Serfdom was legally ended in the mid-nineteenth century, but the inequities of Russian society persisted into the twentieth century, fueling opposition to the czar.

THE COMMUNIST REVOLUTION AND ITS AFTERMATH

Periodic rebellions against the czars and elite classes took place over the centuries. By the mid-nineteenth century, these rebellions were being led not only by serfs, but also by members of the tiny educated and urban middle class. In 1917, Czar Nicholas II was overthrown in a revolution. What followed was a civil war between rival factions with different ideas about how the revolution should proceed. Eventually the revolution brought a complete restructuring of the Russian economy and society, according to what at first promised to be a more egalitarian model.

czar the ruler of the Russian empire

FIGURE 5.11 St. Basil's Cathedral, Moscow. Now the most recognized building in Russia, St. Basil's Cathedral was built between 1555 and 1561 to commemorate the defeat of the Mongols by Ivan the Terrible, the first Russian czar. A central chapel, capped by a pyramidal tower, stands amid eight smaller chapels, each with a colorful "onion dome," that commemorate an Orthodox saint.

D Poor men and women, known as Burlaks, hauling freight along the Volga River in 1873, as depicted by the Russian painter Ilya Repin.

E A World War II–era poster of a Soviet soldier.

F A journey to the International Space Station begins at Baikonur Cosmodrome in 2012.

1800 C.E.		1900 C.E.		2000 C.E.

1714–1755

1796–1914
Russian annexation of Central Asia and Caucasia

1873

1917
Bolshevik Revolution

1922–1953
Stalin era

1941–1945

1945–1991
Cold War

2012

D How does this painting convey the poverty in nineteenth-century Russia?

E How many Soviets were killed during World War II?

F How did the Cold War strain the USSR?

Of the disparate groups that coalesced to launch the Russian Revolution, the **Bolsheviks** succeeded in gaining control during the post-revolution civil war. The Bolsheviks were inspired by the German revolutionary philosopher Karl Marx, whose writings explained the principles of **communism.** Marx criticized the societies of Europe as inherently flawed because they were dominated by **capitalists**—the wealthy minority who owned the factories, farms, businesses, and other means of production. Marx pointed out that the wealth of capitalists was actually created in large part by workers, who nonetheless remained poor because the capitalists undervalued their work. Under communism, workers were called upon to unite to overthrow governments that supported the capitalist system, take over the means of production (farms, factories, services), and establish an egalitarian society. The philosophy was in part based on the idea that people would work out of a commitment to the

Bolsheviks a faction of Communists who came to power during the Russian Revolution

communism an ideology, based largely on the writings of the German revolutionary Karl Marx, that calls on workers to unite to overthrow capitalists and establish an egalitarian society where workers share what they produce

capitalists a wealthy minority that owns the majority of factories, farms, businesses, and other means of production

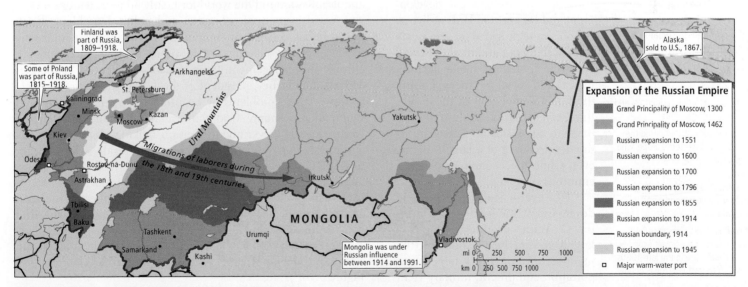

FIGURE 5.12 Russian imperial expansion, 1300–1945. A series of powerful rulers expanded Russia's holdings across Eurasia to the west and east. Expansion was particularly vigorous after 1700, when Russia acquired Siberia. Note that the centuries-long imperial expansion was reversed in 1991, when the Soviet Union was disbanded and the new post-Soviet states were created. [Source consulted: Robin Milner-Gulland with Nikolai Dejevsky, *Cultural Atlas of Russia and the Former Soviet Union,* rev. ed. (New York: Checkmark Books, 1998), pp. 56, 74, 128–129, 177]

common good, sharing whatever they produced so that each had their basic needs met.

The Bolshevik leader Vladimir Lenin argued that the people of the former Russian empire needed a transition period in which to realize the ideals of communism. Accordingly, Lenin's Bolsheviks formed the **Communist Party,** which set up a powerful government based in Moscow. Lenin believed that the government should run the economy, taking land and resources from the wealthy and using them to benefit the poor majority.

The Stalin Era

In 1922, Lenin's successor Joseph Stalin reorganized the society and economy in a highly authoritarian style. His 31-year rule of the Soviet Union brought a mixture of brutality and revolutionary change that set the course for the rest of the Soviet Union's history. He tried to increase general prosperity through rapid industrialization made possible by a **centrally planned,** or **socialist, economy.** All economic activity was run by government officials in Moscow, and the state owned all real estate and means of production. The state determined where all factories, apartment blocks, and transportation infrastructure would be located and managed all production, distribution, and pricing of products throughout the country. The idea was that under a socialist system, the economy would grow more quickly, which would hasten the transition to the idealized communist state in which everyone shared equally.

The centerpiece of Stalin's vision was massive government investment in gigantic development projects, such as factories, dams, and chemical plants, some of which are still the largest of their kind in the world. To increase agricultural production, he forced farmers to join large, government-run collectives.

The Soviet strategy of government control met many expectations. It brought rapid economic development and higher standards of living. Schools were provided for previously uneducated poor and rural children, and contributed to major technological and social advances. During the Great Depression of the 1930s, the Soviet Union's industrial productivity grew steadily, allowing for the development of a large military, even while the economies of other countries stagnated.

However, Stalin's economic development model also had deep flaws. With production geared largely toward heavy industry (the manufacture of machines, transportation equipment, military vehicles, and armaments), the Soviets paid less attention to consumer goods and services that could have further improved living standards.

Communist Party the political organization that ruled the USSR from 1917 to 1991; other communist countries, such as China, Mongolia, North Korea, and Cuba, also have ruling communist parties

centrally planned, or **socialist, economy** an economic system in which the state owns all land and means of production, while government officials direct all economic activity, including the locating of factories, residences, and transportation infrastructure

Cold War the contest that pitted the United States and western Europe, who were espousing free market capitalism and democracy, against the USSR and its allies, who were promoting a centrally planned economy and a socialist state

perestroika literally, "restructuring"; the restructuring of the Soviet economic system that was done in the late 1980s in an attempt to revitalize the economy

glasnost literally, "openness"; the policies instituted in the late 1980s under Mikhail Gorbachev that encouraged more transparency and openness in Soviet government and society

Ultimately, the Soviet Union proved to be less innovative than non-communist countries in western Europe and less able to develop advanced technologies to sustain economic growth over time.

The most destructive aspect of Stalin's rule, however, was his ruthless use of the secret police and mass executions to silence anyone who opposed him. Those who resisted were forced to work in prison labor camps, the so-called *gulags*, many of which were located in Siberia. Prisoners there mined for minerals and also built and worked in newly constructed industrial cities. Millions were executed or died from overwork under harsh conditions. Estimates vary greatly, but between 20 and 60 million people were killed as a result of Stalin's policies.

WORLD WAR II AND THE COLD WAR

During World War II, the Soviet Union did more than any other country to defeat the armies of Nazi Germany (see Figure 5.10E). Hitler exhausted his war machine in his failed attempt to conquer the Soviet Union. In the process of defeating Germany on the Eastern Front, 23 million Soviets were killed, more than all the other European casualties combined. After the war, the Soviet Union was determined to erect a buffer of allied communist states in Central Europe that would be the battleground of any future war with Europe. Because they were unwilling to risk another war, the leaders of the United States and the United Kingdom ceded control of much of Central Europe to the Soviet Union while they busied themselves with reconstructing western Europe. However, when Stalin's intention to utterly dominate Central Europe became clear, the United States and its allies organized to stop the Soviets from extending their power even further into Europe and elsewhere. The result was a politically and economically divided Europe—separated by what was called the "Iron Curtain"—and the **Cold War,** a global geopolitical rivalry that pitted the Soviet Union and its allies against the United States and its allies around the world for nearly 50 years **(FIGURE 5.13)**.

A wide variety of economic and political problems led to the eventual collapse of the Soviet Union. Fundamental flaws in Soviet central planning led to innumerable inefficiencies and chronic mismanagement compared to most market economies. Mikhail Gorbachev, president of the USSR from 1985 to 1991, addressed these problems by beginning to permit a limited private market economy. Under an overall policy of **perestroika,** or "restructuring," government decision making was decentralized and private entrepreneurs were allowed to legally sell goods and services. Gorbachev also introduced policies collectively known as **glasnost,** or "openness," which encouraged more transparency in the way government operated and more openness in society in general, which allowed for criticism of the government. However, many problems remained, such as the diversion of scarce resources to the military, including the Soviet space program, and away from much-needed economic and social development (see Figure 5.10F).

Soviet efforts to promote communism in less developed countries were also expensive, as was extensive political, economic, and military coercion in Central Europe and Central Asia. As problems multiplied, so did resistance to Soviet control, especially

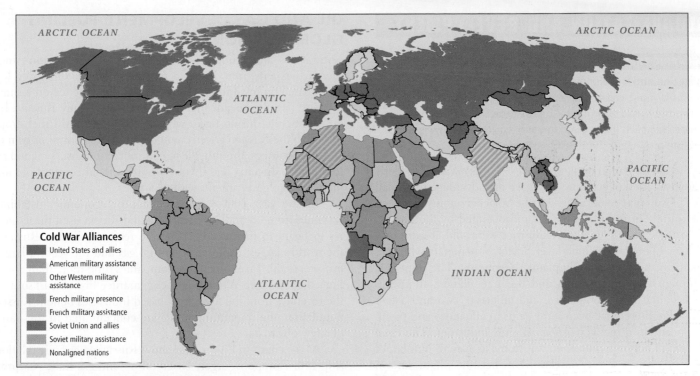

FIGURE 5.13 The Cold War in 1980. In the post–World War II contest between the Soviet Union and the United States, both sides enticed allies through economic and military aid. The group of countries militarily allied with the Soviet Union was known as the *Warsaw Pact*. Some crucial countries remained nonaligned. [Source consulted: Clevelander, at http://en.wikipedia.org/wiki/Image:New_Cold_War_Map_1980.png]

in Central Europe in the late 1980s. The Soviet policies of perestroika and glasnost also meant that Central European countries felt more comfortable enacting their own reforms and distancing themselves from the Soviet Union. Because Gorbachev, unlike his predecessors, did not resort to violence to control the countries of Central Europe, they were able to abandon communism and instead establish democracy. Eventually, this dissension spread to the republics that made up the Soviet Union. Communist hardliners attempted a coup against Gorbachev and his reforms in order to save the Soviet Union. The coup failed during a few tense and confusing days in 1991, which paved the way for the dissolution of the USSR

A war in Afghanistan also helped to stretch the USSR past the breaking point (see Figure 5.17A). In the 1970s in Afghanistan, which lies just to the south of the Central Asian states, there was contention among a small group who favored a Western-style democratic government, those who favored communism, and those who favored a theocratic state based on Islamic law. The Soviets feared that the strongly anti-communist Muslim movement in Afghanistan would influence nearby Central Asian Soviet republics to unite under Islam and rebel against the USSR. In response, the USSR backed an unpopular communist government that allowed Soviet military bases in Afghanistan. This brought strong resistance from the Muslim fundamentalist mujahedeen movement, and in 1979, Russia launched a war to support the Afghan government. A bloody, decade-long conflict ensued, concluding with the Soviets being defeated by Afghan guerillas who were financed, armed, and trained by the

United States. Within two years of the Soviet withdrawal from Afghanistan, the USSR disintegrated. The Soviet–Afghan conflict and its aftermath are discussed further in Chapter 8.

THINGS TO REMEMBER

• As early as 5000 years ago, Central Asia had settled communities that were supported by irrigated croplands and enriched by trade along what became known as the Silk Road

• The Russian empire (1598–1917) was ruled by a powerful monarch, the czar, who (along with a tiny aristocracy) lived in splendor, while the vast majority of the people lived in poverty. Many Russians were serfs who were legally bound to live on and farm land owned by an aristocrat.

• The Bolsheviks, a group inspired by the principles of communism, were the dominant leaders of the Russian Revolution of 1917. Their goal, which was never achieved, was an egalitarian society where people would work out of a commitment to the common good, sharing whatever they produced.

• Taking control in 1922, Joseph Stalin brought a mixture of brutality and revolutionary change that set the course for the rest of the Soviet Union's history. He established a centrally planned, or socialist, economy in which the state owned all property and means of production, while government officials in Moscow directed all economic activity.

• After World War II, a global geopolitical rivalry known as the Cold War situated the Soviet Union and its allies against the United States and its allies for nearly 50 years.

GLOBALIZATION AND DEVELOPMENT

GEOGRAPHIC THEME 2

Globalization and Development: After the fall of the Soviet Union, wealth disparity increased and jobs were lost as many Communist-era industries were closed or sold to the rich and well connected. The region is now largely dependent on its role as a leading global exporter of energy resources.

When the Soviet Union disbanded in 1991, there was a disorderly transition to a market economy that gave the advantage to a small number of well-placed bureaucrats. The Soviet economy consisted almost entirely of industries owned and operated by the government. In the early 1990s, these industries were sold at extremely low prices, often to the high-level Soviet bureaucrats who had run them in the old centrally planned economy. The hope was that through this **privatization** and the free market, industries would become more competitive and efficient. Two and a half decades later, there have been some gains in efficiency and profitability, mostly in a few key export-oriented industries. The most widespread effect of privatization was to make a small number of people fabulously rich very quickly. Many became what is known as **oligarchs**—wealthy and politically connected people who wield enormous, often clandestine, power. These oligarchs continue to exercise power in government and private enterprise.

The Russian government instituted a wide range of policies to try to revitalize the sluggish economy, moving away from the old centrally planned economy as quickly as possible. This transition to a market economy took place so rapidly that it was known as *shock therapy*. Today, nearly 65 percent of Russia's economy is held privately, a momentous change from 1991, when virtually 100 percent of the economy was run by the state.

Unfortunately, shock therapy came at a steep cost for many Russians. A major part of the reforms was the abandonment of government price controls that once kept goods affordable. Instead, in the new market economy, prices skyrocketed for the many goods that were in high demand but short supply. While a few people became rich, many lost their jobs and what little savings they had were lost because of inflation and they could barely pay for basic necessities such as food. Toward the end of the 1990s, as opportunities opened and competition developed, the supply of goods increased and prices fell. But even today, the effects of shock therapy are still being felt. For example, average incomes have grown by 45 percent since 1991, but almost all of this gain has gone to the wealthiest 20 percent of the population; the bottom 20 percent of the population is earning about half what they did in 1991. The concentration of wealth among the richest is extreme. Thirty-five percent of the country's wealth is in the hands of only 110 billionaires, according to a report by the Swiss bank Credit Suisse. Such concentration of economic power in the hands of a very few people confirms that Russia has evolved into an oligarchical society.

privatization the sale of industries that were formerly owned and operated by the government to private companies or individuals

oligarchs in Russia, those who acquired great wealth during the privatization of Russia's resources and who use that wealth to exercise power

Gazprom in Russia, the state-owned energy company; it is the largest gas company in the world

OIL AND GAS DEVELOPMENT: FUELING GLOBALIZATION

Natural gas and crude oil have emerged as the region's most lucrative exports, introducing a new wave of economic development based on fossil fuels. However, this dependence on a single industry has also meant that the region—especially Russia—has great difficulty during times when the price of fossil fuel drops. Although the Russian government continues to control its own oil and gas resources, the struggle to control Central Asia's oil and gas resources is ongoing and involves multinational corporations and many foreign governments **(FIGURE 5.14)**.

By the year 2000, after a decade of economic instability, revenues from oil and gas had risen, and Russia used those monies to begin to finance its economic recovery. Today, Russia is the world's largest exporter of natural gas and the second largest exporter of oil. Taxes on energy companies fund more than half of the federal budget. While this does guarantee that at least some of the revenues from the fossil fuel–based industries benefit Russia's population, the remaining profits are concentrated in the hands of a few oligarchs.

As the charts in Figure 5.14 show, Russia's economic relationship with Europe is largely based on oil and gas trade. **Gazprom** is Russia's largest company as well as the world's largest gas company, and because Gazprom is state controlled, Russia can use the company to promote its geopolitical interests. Russia occasionally tries to use its gas exports to influence the politics and economies of Ukraine, Belarus, and countries in Central Europe. EU member states receive about 25 percent of their natural gas from Gazprom, with Central Europe much more dependent on Gazprom than other parts of the EU. Russia is also very reliant on this trade; approximately 60 percent of Gazprom's profits come from sales to the European Union.

Even before the war in Ukraine erupted, Russia curtailed Ukraine's access to Gazprom's gas in a dispute over prices and as a response to Ukraine moving closer to the EU rather than being a Russian ally. Because pipelines that supply Europe with gas run through Ukrainian territory, this caused gas shortages in parts of Europe. The EU became concerned about its reliance on Russian energy and turned rapidly to other sources of fossil fuels as well as renewable energy sources.

Central Asia has yet to stabilize because a tug-of-war has evolved between Russia and foreign multinational energy corporations over the right to develop, transport, and sell Central Asian oil and gas resources. The new Central Asian states do not have enough capital to develop the resources themselves, but they nevertheless benefit economically from selling the rights to exploit oil and gas to foreign companies. Due to corruption, some of this income ends up in the hands of government officials and well-connected people.

Because Central Asia is landlocked, the only way to get oil and gas from there to world markets is via pipelines (see Figure 5.14), the routes of which have become a major source of dispute. Russia, the United States, the EU, Turkey, China, and India have all proposed or developed various routes. The oil pipelines from Kazakhstan and the gas pipelines from Turkmenistan are especially important, as they are part of the infrastructure that supplies

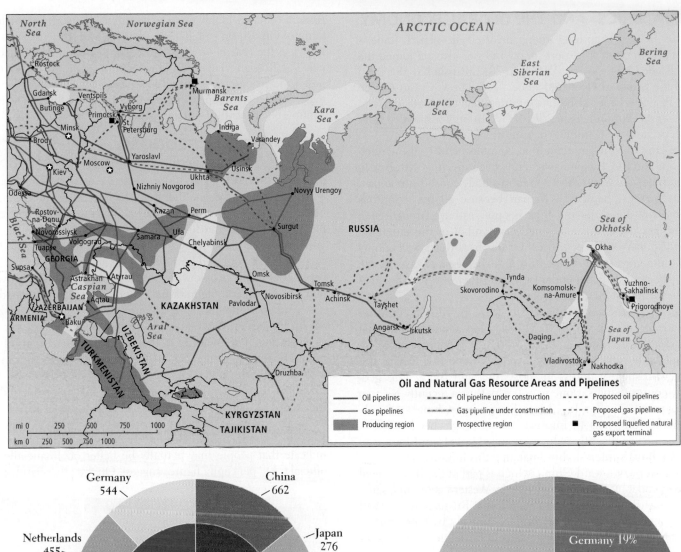

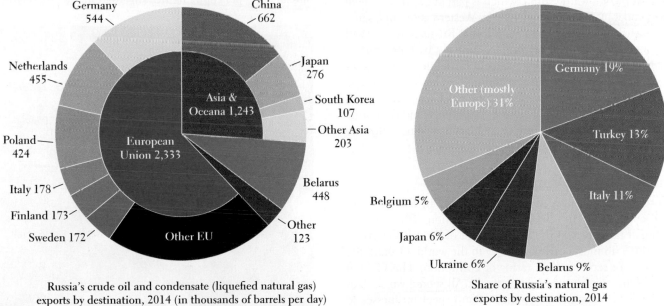

Russia's crude oil and condensate (liquefied natural gas) exports by destination, 2014 (in thousands of barrels per day)

Share of Russia's natural gas exports by destination, 2014

FIGURE 5.14 Oil and natural gas in Russia and the post-Soviet states. [Sources consulted: U.S. Department of Energy, Energy Information Administration, "Russia Country Analysis Brief," May 2008, and "Russia: International Energy Data and Analysis," July 28, 2015, at https://www.eia.gov/beta/international/analysis_includes/countries_long/Russia/russia.pdf]

energy to China, whose energy usage is rising. The United States maintains a military presence to protect its interests in countries such as Georgia and Uzbekistan, both of which, along with Turkey, have pipelines to Europe. Russia has built pipelines that traverse its own territory and reach into Europe. Each of these parties has made numerous efforts to encourage Central Asian countries to use their pipelines instead of those of their competitors.

RUSSIA, BRICS, AND THE GLOBAL ECONOMY

Because it was becoming more involved in global trade—particularly through its exports of oil and gas—Russia joined the World Trade Organization (WTO) in 2012. Russia has also been part of the **Group of Eight (G8)** since 1997. The G8 is an organization of eight large, affluent countries that meets regularly to discuss world economic issues. Russia is currently suspended from the G8 because of its military actions in Ukraine.

Looking to form an economic counterweight to the United States and the EU, Russia has recently been meeting with Brazil, India, China, and South Africa to form a trade consortium known as BRICS (see Chapter 1). In part because all five countries are large and populous and their economies have strengthened over the last decade, the alliance is a symbol of the emerging power of non-Western countries. Of the five, Russia's economy may be among the most sluggish and the least diverse in the group—because it is dependent on fossil fuel exports—but it is also the wealthiest on a per capita basis.

The economic and political diversity of the BRICS countries has made it difficult for BRICS to develop a united bloc that would challenge Western power. Some BRICS countries are also geographically isolated from each other, except for Russia and China, who share a border, as do China and India. This geographic connectedness is one reason why Russia recently became the largest oil supplier to China. Russia and China are working together on plans to improve Russia's high-speed railways, which could significantly reduce the travel time from Moscow to Beijing. China is interested in financing this infrastructure.

Russia's pivot toward China, which is part of the BRICS goal of further collaboration, is also driven by Western economic sanctions against Russia and the realization that Russia and the West are no longer post–Cold War military strategic partners, but in competition with each other over the old Soviet-dominated territories (see the Power and Politics section). This has forced Russia to look for new markets and alliances, such as China. Russia has increased its military budget to support its political and economic objectives after years of reducing its military spending following the collapse of the Soviet Union.

Institutionalized Corruption

The cost of doing business in Russia is affected by Russia's high level of corruption. In 2014 the organization Transparency International rated Russia the most corrupt of the BRICS countries and the 35th most corrupt country in the world. Overall, the region does not score well in terms of corruption. The Central Asian states are considered highly corrupt. All scored worse than Russia. Civil servants and the police often expect businesses to offer bribes to receive licenses and to avoid other regulatory hurdles. This affects both large corporations and small entrepreneurs.

The propensity for corruption is intertwined with government bureaucracies. Even those who work their way through the labyrinth of setting up a business face protection racketeers, as organized crime has become commonplace since the end of the Communist era. Many oligarchs have become closely connected to the

Group of Eight (G8) an organization of eight countries with large economies: France, the United States, Britain, Germany, Japan, Italy, Canada, and Russia

Russian Mafia, a highly organized criminal network dominated by former *KGB* (Russia's intelligence agency during the Cold War, now known as the *Federal Security Service*, or FSB) and military personnel. The Russian Mafia has influence in nearly every corner of the post-Soviet economy, especially in illegal activities and the arms trade, and has developed international networks through which it controls illicit activities abroad, particularly in Europe.

THE INFORMAL ECONOMY

Much of the economic activity in the region takes place in the informal sector **(FIGURE 5.15)**. To some extent, the new informal economy is an extension of the one that flourished under communism. The black market of that time was based on currency exchange and the sale of hard-to-find luxuries. In the 1970s, for example, savvy Western tourists could enjoy a vacation on the Black Sea paid for by a pair or two of smuggled Levi's blue jeans and some Swiss chocolate bars. Today, many people who have lost stable jobs due to privatization now depend on the informal economy for their livelihood.

Workers in the informal economy tend to be young, unskilled adults; retirees on tiny pensions; and those with only a low level of education. The majority of these workers operate out of their homes by selling, among other things, cooked foods, vodka made in their bathtubs, and clothing and electronics that have been smuggled into the country. The World Bank has estimated that the informal sector makes up about 40 percent of the Russian economy. In the Caucasian countries, the number is as high as 50 to 60 percent. On the one hand, such large percentages indicate that people may actually be better off financially than official GDP per capita figures suggest. On the other hand, a large

FIGURE 5.15 Street vendors in Moscow, Russia.

Natalia Kolesnikova/Getty Images

informal economy is commonly considered a sign of economic weakness and underdevelopment.

Despite the fact that the informal economy helps people survive, it is not popular with governments. Unregistered enterprises do not pay business and sales taxes and usually are so underfinanced that they tend not to grow into job-creating formal sector businesses. In many cases, informal businesspeople must pay protection money to local gangsters (the so-called *Mafia tax*) to keep from being reported to the authorities.

VIGNETTE Natasha is an engineer in Moscow. She has managed to keep her job and the benefits it carries, but in order to better provide for her family, she sells secondhand clothes in a street bazaar on the weekends. Asked about her customers, Natasha says, "Many are former officials and high-level bureaucrats who just can't afford the basics for their families any longer." Some are older retired people whose pensions are so low that they resort to begging on Moscow's elegant shopping streets close to the parked Rolls-Royces of Moscow's rich. These often highly educated and only recently poor people buy used sweaters from Natasha and eat in nearby soup kitchens. *[Source: This composite story is based on work by Alessandra Stanley, David Remnick, David Lempert, and Gregory Feifer.]* ∎

UNEMPLOYMENT AND THE LOSS OF SOCIAL SERVICES

Since becoming privatized, most formerly state-owned industries have cut jobs in an attempt to compete with foreign companies. Losing a job is especially devastating in a former Soviet country because one also loses the subsidized housing, health care, and other social services that were often provided along with the job. The new companies that have emerged rarely offer full benefits to employees. There is little job security because most small private firms appear quickly and often fail. Discrimination is also a problem, given the absence of equal opportunity laws. Job ads often contain wording such as "only attractive, skilled people under 30 need apply."

Unemployment figures for the region vary widely. Official unemployment rates for 2014 were 6 percent in Russia and 10 percent in Ukraine. In Belarus, unemployment is officially less than 1 percent; however, the old Soviet-style system of employing workers whether they are needed or not is still practiced in Belarus. Also, many people do not bother to register as unemployed because of the absence of unemployment benefits. The rate of **underemployment,** which measures the number of people who are working too few hours to make a decent living or who are highly trained but are working at low-paying jobs, is even higher in all of these countries.

FOOD PRODUCTION IN THE POST-SOVIET ERA

Across most of the region, agriculture is precarious at best, either hampered by a short growing season and boggy soils or requiring expensive inputs of labor, water, and fertilizer. Because of Russia's harsh climates and rugged landforms, only 10 percent of its vast expanse is suitable for agriculture. In the Caucasus mountain

zones, rainfall adequate for agriculture coincides with a relatively warm climate and long growing seasons. Together with Ukraine and European Russia, this area is the agricultural backbone of the region (FIGURE 5.16). The best soils are in an area stretching from Moscow south toward the Black and Caspian seas, and extending west to include much of Ukraine and Moldova. In Central Asia, irrigated agriculture is extensive, especially where long growing seasons support cotton, fruit, and vegetables.

Changing Agricultural Production

During the 1990s, agriculture went into a general decline across the entire region. In most of the former Soviet Union, yields dropped by 30 to 40 percent compared to previous production levels. This was due mostly to the collapse of the subsidies and internal trade arrangements of the Soviet Union. Many large and highly mechanized collective farms were suddenly without access to equipment, fuel, or fertilizers. Since that time, agricultural output has rebounded but even today it remains below Soviet levels. Russia's commercial agricultural sector is still characterized by many large and inefficient farms, most of which are now run by corporations. At the same time, Russians are heavily dependent on "homegrown" food—small household plots, gardens, and family farms that produce more than half of the total agricultural output on only 20 percent of the arable land. In the wake of its conflict with Ukraine (see page 283), Russia has stopped importing food from EU and other Western countries to retaliate for sanctions that those countries imposed on Russia. This could redirect investment into the agricultural sector in Russia and increase domestic commercial production.

In Central Asia, agriculture has been reorganized with more emphasis on smaller, food-producing family farms. Grain and livestock farming (for meat, eggs, and wool) are increasing. Production levels per acre and per worker also are increasing. China, through the Asian Development Bank, has provided assistance in reducing the use of agricultural chemicals, which were used extensively during Soviet times.

Farmers in Georgia, favored with warm temperatures and abundant moisture from the Black and Caspian seas, can grow citrus fruits and even bananas; they do so primarily on family, not collective, farms. Before 1991, most of the Soviet Union's citrus and tea came from Georgia, as did most of its grapes and wine. Georgia still exports some food to Russia, but because of political tensions between the two countries, Georgia is increasing the amount of food and wine it sells to Europe and other markets.

> **underemployment** the condition in which people are working too few hours to make a decent living or are highly trained but working at menial jobs

THINGS TO REMEMBER

GEOGRAPHIC THEME 2 • **Globalization and Development:** After the fall of the Soviet Union, wealth disparity increased and jobs were lost as many Communist-era industries were closed or sold to the rich and well connected. The region is now largely dependent on its role as a leading global exporter of energy resources.

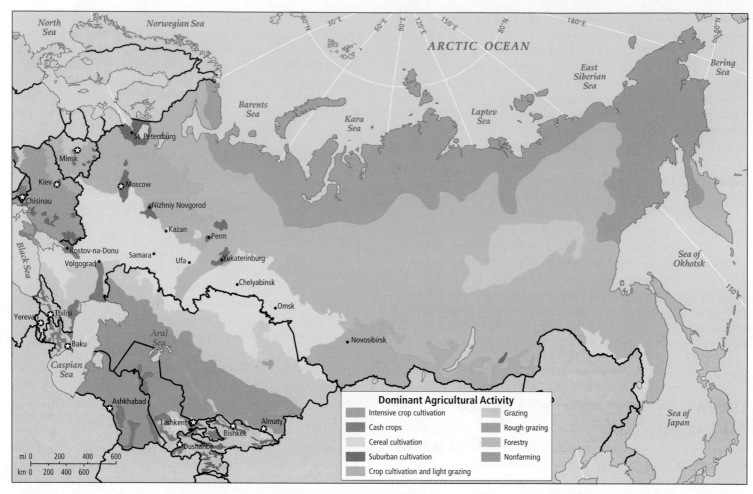

FIGURE 5.16 Agriculture in Russia and the post-Soviet states. Agriculture in this part of the world has always been difficult, due to the cold climate, the short growing seasons, low soil fertility, and the lack of rainfall in all but a few places (Ukraine, Moldova, and Caucasia). [Source consulted: Robin Milner-Gulland with Nikolai Dejevsky, *Cultural Atlas of Russia and the Former Soviet Union,* rev. ed. (New York: Checkmark Books, 1998), pp. 186–187, 198–199, 204–205, 216–217]

- Even though average incomes in Russia have grown by 45 percent since 1991, almost all of this gain has gone to the wealthiest 20 percent of the population. Income inequality is high and much of Russia's wealth is in the hands of a few oligarchs.

- Natural gas and crude oil have emerged as the region's most lucrative exports, introducing a new wave of globalization and fossil fuel–dependent economic development.

- The World Bank has estimated that the informal sector makes up about 40 percent of the Russian economy and 50 to 60 percent of the economies of the Caucasian countries.

- Russians are heavily dependent on "homegrown" food—small household plots, gardens, and family farms that produce more than half of the total agricultural output on only 20 percent of the arable land.

POWER AND POLITICS

GEOGRAPHIC THEME 3

Power and Politics: This region has a long history of authoritarianism. Even if the public now desires expanded political freedoms, there are few opportunities to influence the political process. Elected representative bodies often act as rubber stamps for strong presidents and exercise only limited influence on policy making.

Since 1991, the politics in this region have remained very authoritarian and largely dominated by Russia, which sees itself as the successor to the Soviet Union. For almost two decades, Russia's political scene has been controlled by one person, Vladimir Putin, a former high official in the KGB. Putin was elected president in 2000 and has been the country's de facto leader as either president or prime minister ever since. During this time, he has exercised tight control over political and economic policy and consolidated political power in Moscow. He has been popular for bringing more peace and prosperity to Russia and for restoring Russia's image of itself as a world power. Putin has also been criticized for preventing meaningful reforms at the local, state, and national levels, and for extending government control over the press and the media (see page 287). While criticism of the government is formally allowed, most people (correctly) fear that such criticism would be met with retribution from the *siloviki* (literally "men of force," many of whom have military or intelligence backgrounds) who control many governing institutions. Political analysts agree that the siloviki have become more prominent within the Russian political structure since the start of the Ukrainian war in 2014. Journalists and others who document corruption or speak out against the government have been killed and jailed.

Putin is the most dominant political person in the region, and is likely to remain in power into the near future. Long after that, authoritarianism will likely hold sway in this region. One reason is that Russia's current constitution, written in 1993, gives sweeping powers to the president, which will mean that a strong, and likely authoritarian, leader will succeed Putin. Many scholars of this region have also noted that Russia's geography, featuring a vast territory and few mountain or other barriers to prevent invasion by outsiders, favors a single powerful authoritarian ruler with a strong military. Others argue that the presence of authoritarian rulers throughout Russian history has created a political culture in which strong and sometimes brutal leaders such as Vladimir Putin thrive.

Elsewhere in the region, authoritarianism and limited political freedoms are also the norm. Belarus, Ukraine, Moldova, the Caucasian republics (including Chechnya, discussed on pages 285–287), and Central Asia are still most often run by authoritarian leaders who are unwilling to share power or allow criticism (see the map in **FIGURE 5.17**). Elections often involve systematically intimidating voters and arresting the political opposition. Not only is much of the region authoritarian, but its ratings on the democratization index have worsened over the last decade.

THE COLOR REVOLUTIONS AND POLITICAL FREEDOMS

A few of the post-Soviet states have transitioned to somewhat greater political freedoms. A series of so-called *color revolutions* took place, led by coalitions of educated young adults who rallied under various symbols, usually colors or flowers, meant to unite them despite their differences. The color revolutions include the Rose Revolution in Georgia (2003–2004), the Orange Revolution in Ukraine (January 2005), and the Tulip Revolution in Kyrgyzstan (April 2005). Funded sometimes by foreign NGOs and governments, including the United States, that wanted to hasten democratization, it appeared for a time that these movements would result in significant reforms. However, while these countries have now held elections and tolerate some freedom of expression, they retain many forms of authoritarian control and practices that constrain political freedoms. Much like the outcome of the Arab Spring (see Chapter 6), the transition toward greater democracy here has proven to be hard.

CRISIS IN UKRAINE

In early 2014 in Ukraine, one of the most challenging crises in this region since the fall of the Soviet Union emerged. The crisis has positioned Russia against the EU and the United States in a seeming throwback to the days of the Cold War.

Ukraine is divided in terms of its geographic orientation. Areas to the west, which border Central Europe, tend to favor closer ties with Europe, while areas to the east, where many of Ukraine's Russian-speaking minority reside, lean toward Russia. Then-president Viktor Yanukovych backed away from previous commitments to strengthen the economic ties between Ukraine and the EU, resulting in a wave of protests in Ukraine's capital of Kiev. Like the color revolutions, these protests were supported by U.S. and European governments and pro-democracy NGOs. After the police tried to quell the protests, which resulted in more than 80 deaths, the Yanukovych government itself was toppled by an alliance of demonstrators and political groups in favor of closer integration with the EU. This prompted a reaction in eastern Ukraine. First, Crimea, an autonomous republic within Ukraine that is dominated by ethnic Russians, voted in a referendum to secede from Ukraine. The referendum was declared illegal by Ukraine, which only allows referendums at the national scale. Russian military forces, already present in large numbers in Crimea as part of post-Soviet agreements with Ukraine, supported local pro-Russian militias who took control of the province, and the Russian government then annexed Crimea (**FIGURE 5.18**). However, this de facto Russian control of Crimea is not internationally recognized.

Russia's claim to Crimea has historical roots. For almost 200 years Crimea was a Russian province. During the Soviet days, Crimea was administratively transferred to Ukraine, which is why it became part of Ukraine in the post-Communist era. While 58 percent of the population is ethnic Russian, Crimea has a complex ethnic mix; Ukrainians and Tartars make up significant minorities. Crimean Tartars were persecuted and forcibly removed from their homeland during the Stalin-era Soviet Union and are today wary of Crimea's return to Russian control. Russia annexed Crimea in part to control the large naval base in the city of Sevastopol, which gives Russia better access to the Black Sea.

Parallel to the Crimea events, the Russian army began massing on the borders of Ukraine's eastern provinces, several of which have large ethnic Russian populations that support closer ties with Russia. Throughout eastern Ukraine, administrative and police buildings were taken over by local pro-Russian forces. Although Russia denies having done so, most observers agree that the Russian military covertly assisted the rebels. Figure 5.18 shows the territories in eastern Ukraine that are held by the rebels as self-proclaimed "people's republics" independent from Ukraine (as of March 2016). Russia may attempt to annex eastern Ukraine in the same way that it did Crimea. While the rebel-held territories, called the Donetsk Basin by geographers, make up only a small part of Ukraine, it is a populous and heavily industrialized area. Its economy is oriented toward Russia and most people's first language is Russian. (Although to complicate the picture, many Russian speakers consider themselves ethnically Ukrainian!) Figure 5.18 also shows how other regions in eastern and southern Ukraine are

Since World War II, most conflicts in this region have been in Caucasia and Central Asia. Most took place just before or shortly after the breakup of the Soviet Union. Since then, popular movements pushing for more political freedoms have emerged in several countries. However, violent conflict continues to be a major source of political instability in this region.

A Children in Afghanistan play on a tank dating from the Soviet Union's 10-year war in that country. The Soviets intervened in 1979 to prop up an authoritarian and unpopular Afghan government that had initiated radical reforms and murdered much of the political opposition. The war resulted in at least 1.5 million deaths and a humiliating defeat for the USSR. Fourteen thousand Soviet soldiers died, billions of dollars were spent, and practically nothing was gained. Public outrage over the Afghan war played a major role in the breakup of the Soviet Union. Since the fall of the USSR, the expansion of political freedoms has proceeded unevenly throughout the region.

Democratization and Conflict

Armed conflicts and genocides with high death tolls since 1945

- Ongoing conflict
- 1000–5000 deaths
- 5000–50,000 deaths
- 50,000–300,000 deaths
- 300,000–1,000,000 deaths

Democratization index
- Full democracy
- Flawed democracy
- Hybrid regime
- Authoritarian regime
- No data

RUSSIA
BELARUS
MOLDOVA
UKRAINE
GEORGIA
ARMENIA
AZERBAIJAN
TURKMENISTAN
UZBEKISTAN
KAZAKHSTAN
RUSSIA
KYRGYZSTAN
TAJIKISTAN
AFGHANISTAN

mi 0   500   1000
km 0   500   1000

B Pro-Russian separatist rebels on a road near Donetsk in eastern Ukraine. Pro-Russian rebels and the Ukrainian army have been engaged in conflict since 2014. The rebels fly a flag that represents the Russian Orthodox religion, which indicates that ethnic and religious identity are closely intertwined.

C Chechen police in a show of force in Grozny, Chechnya, where rebels have fought two unsuccessful wars for independence from Russia and have conducted ongoing terror campaigns against Russia since the fall of the Soviet Union. Political instability in Caucasia is one of the major obstacles to the expansion of political freedoms in this region.

Thinking Geographically

After you have read about power and politics in Russia and the post-Soviet states, you will be able to answer the following questions.

A How did the USSR's war in Afghanistan ultimately contribute to the dissolution of the Soviet Union?

B How do proximity to Russia and the geography of ethnicity and language in Ukraine correlate with conflict?

C How does Russia's unwillingness to let Chechnya secede relate to fossil fuel resources?

a blend of Ukrainian and Russian cultures, which makes them potential flashpoints for a geographically expanded conflict.

Regardless of the eventual outcome of the ongoing confrontation between Russia and Ukraine, underlying tensions will remain. Ukraine's government will be challenged to balance the demands of citizens in the western parts of the country, many of whom want closer ties with the EU, with those of ethnic Russians in the east who want closer ties and possibly even political union with Russia. Russia will have to weigh the economic benefits of its gas trade with Europe and the cost of economic sanctions that Western governments have imposed, against its long-term opposition to the expansion of the EU into Ukraine. Additionally, the EU and the United States will have to decide whether supporting political movements such as the color revolutions is worth risking much more tense relations with Russia.

CULTURAL DIVERSITY AND RUSSIAN DOMINATION

Russia's long history of expansion into neighboring lands has left it and neighboring countries with exceptionally complex internal political geographies. As the Russian czars and then the Soviets

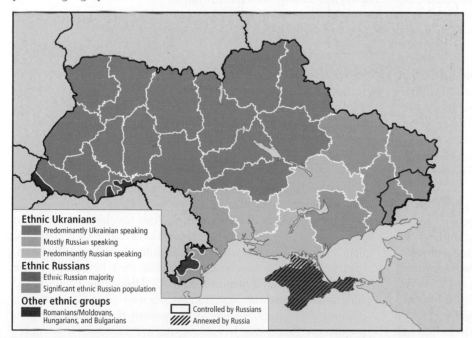

Ethnic Ukranians
- Predominantly Ukrainian speaking
- Mostly Russian speaking
- Predominantly Russian speaking

Ethnic Russians
- Ethnic Russian majority
- Significant ethnic Russian population

Other ethnic groups
- Romanians/Moldovans, Hungarians, and Bulgarians

- [ ] Controlled by Russians
- [/] Annexed by Russia

FIGURE 5.18 The geography of conflict and ethnicity in Ukraine. Pro-Russian forces currently control the eastern portion of Ukraine and call it the Luhansk and Donetsk People's Republic. Along with the Crimean Peninsula, now annexed by Russia, this is an area where the Russian-speaking population of Ukraine is most numerous.

pushed the borders of Russia eastward toward the Pacific Ocean over the past 500 years, they conquered a number of non-Russian areas. Russians are now the main inhabitants of many of those areas, some of which are now within countries neighboring Russia, such as Ukraine and Kazakhstan. Other territories have been designated *semiautonomous internal republics* (FIGURE 5.19) within Russia and have significant ethnic minority populations that are descendant from indigenous people or trace their origins to long-ago migrations from Germany, Turkey, or Persia.

Both the czars and the USSR had a policy of **Russification,** whereby large numbers of ethnic Russian migrants were settled in non-Russian ethnic areas and given the best jobs and most powerful positions in regional governments. The goal was to force potentially rebellious regional minorities to conform to the state's goals. However, even during the Soviet period, but especially after, minorities organized to resist Russification and to enhance local ethnic identities. Today, the reversed process—de-Russification—is happening especially in Central Asia, where the Russian language and general Russian influence is discouraged (see, for example, the vignette on page 289).

THE CONFLICT IN CHECHNYA

Shortly after the breakup of the Soviet Union, several internal republics demanded more autonomy, and two of them, Tatarstan and Chechnya, declared outright independence. Tatarstan has since been placated with more economic and political autonomy, and as an enclave surrounded by other parts of the Russian Federation, its independence is probably unrealistic (see Figure 5.19). However, Chechnya's stronger resistance to Moscow's authority has led to the worst bloodshed of the post-Soviet era.

Chechnya, located on the fertile northern flanks of the Caucasus Mountains (see the inset map in Figure 5.19), is home to 800,000 people. Partially in response to Russian oppression, the Chechens converted from Orthodox Christianity to the Sunni branch of Islam in the 1700s. Since then, Islam has served as an important symbol of Chechen identity and as an emblem of resistance to the Orthodox Christian Russians, who annexed Chechnya in the nineteenth century.

In 1942, during World War II, as the Germans invaded Russia, a group of Chechen rebels simultaneously waged a guerilla war against the Soviets. Near the end of the war, Stalin exacted his revenge by deporting the majority of the Chechen population (as many as 500,000 people) to Kazakhstan and Siberia. Here they were held in concentration camps, and many died of starvation. The Chechens were finally allowed to return to their villages in 1957, but a heavy propaganda campaign portrayed them as traitors to the Soviet Union, a designation that greatly affected their daily lives.

In 1991, as the Soviet Union was dissolving, Chechnya declared itself an

Russification the czarist and Soviet policy of encouraging ethnic Russians to settle in non-Russian areas as a way to assert political control

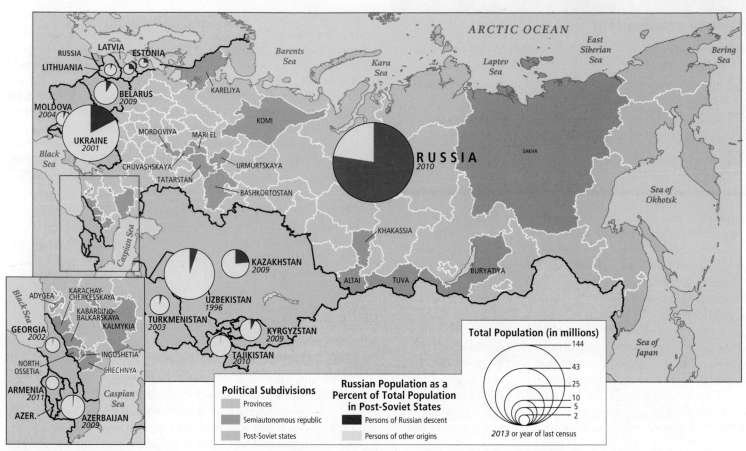

FIGURE 5.19 **Ethnic character of Russia and percentage of Russians in the post-Soviet states.** Russia, with all of its internal republics and administrative regions, is formally called the Russian Federation. The ethnic character of many internal republics was changed during the Soviet era, so Russians now form significant minorities in the republics. The pie charts for each country indicate the percentages of Russians in the post-Soviet states. In 2010, the ethnic makeup of Russia was 77.7 percent Russian, 3.7 percent Tatar, 1.4 percent Ukrainian, 1.0 percent Chuvash, 1.0 percent Chechen, and 10.2 percent other. [Sources consulted: James H. Bater, *Russia and the Post-Soviet Scene* (London: Arnold, 1996), pp. 280–281; Graham Smith, *The Post-Soviet States* (London: Arnold, 2000), p. 75; "Ethnic Groups," *The World Factbook 2015*, Central Intelligence Agency, at https://www.cia.gov/library/publications/the-world-factbook/geos/rs.html]

independent state. Russia saw this as a dangerous precedent that could spark similar demands by other cultural enclaves throughout its territory. Russia also wished to retain the agricultural and oil resources of the Caucasus; it had planned to build pipelines across Chechnya to move oil and gas to Europe from Central Asia. Russia responded to the Chechen insurgency with bombing raids and other military operations that killed tens of thousands of people and created 250,000 refugees.

Most Chechen guerrillas have since given up, most Russian combat troops have been pulled out of Chechnya, and Russia has begun making substantial investments in rebuilding the capital city of Grozny. While this shift is a welcome change, some Chechen rebels have continued their struggle by carrying out brutal terrorist attacks, many of which now take place in Moscow and other areas outside Chechnya. Chechnya remains highly militarized (see Figure 5.16C), and the ongoing conflict there highlights Russia's inability to peacefully address internal political dissent. The terrorist attacks have fueled Russian xenophobia against people from the Caucasus region, which has made a compromise about the future status of Chechnya hard to achieve. The conflict has even been linked to global terrorism,

as Muslims from the Caucasus have joined extremists elsewhere in the world. For example, the two brothers responsible for the Boston Marathon bombings in 2013 came from a Chechen background and had been influenced by Islamic radicalism.

THE CONFLICT IN GEORGIA

In the post-Soviet era, conflicts between Russia and Georgia over the ethnic republics of South Ossetia and Abkhazia, which lie south of Chechnya, have worsened. Both republics became part of Georgia at the behest of Joseph Stalin (a native Georgian). He then moved Georgians and Russians into Ossetia and Abkhazia, making the native people a minority in their homeland. More recently, Russia has strategically supported the ethnic Ossetian and Abkhazian populations' agitation for secession from Georgia, even granting Russian citizenship to 90 percent of the non-Georgian population. Russia may have done this in retaliation for Georgia's increasingly close relations with the United States and the EU, as evidenced by its candidacy for NATO membership. A pipeline that links the oil fields of Azerbaijan with the Black Sea via Georgia is another source of contention (see Figure 5.14). The Black Sea route is important because it is an expedient way to ship oil to world markets, as the Black Sea connects to the Mediterranean Sea and, ultimately, to the Atlantic Ocean. Russia would like to enhance its control over the oil resources of the Caspian Sea region by routing all pipelines through Russian-controlled territory.

In 2008, Georgia's military, attempting to gain control over rebelling parts of South Ossetia, engaged with the Russian army, which supported the rebels, saying that they had the right to protect (recent) Russian citizens. Today, the Georgian government does not control Abkhazia and South Ossetia and it is not clear whether the two will break away and become independent countries, become provinces within the Russian Federation, or be satisfied by offers of more autonomy within Georgia's federal structure. This ambiguous geopolitical situation is referred to as a *frozen conflict* and is similar to the Ukrainian/Crimean conflict, as Russia directly or indirectly controls territories just beyond its internationally recognized borders.

THE MEDIA AND POLITICAL REFORM

During the Soviet era, all media were under government control. There was no free press, and public criticism of the government was a punishable offense. Between 1991 and the early 2000s, the communications industry was a center of privatization, and several media tycoons emerged to challenge the authorities. Privately owned newspapers and television stations regularly criticized the policies of various leaders of Russia and the other states. It appeared that a free press was developing.

During Vladimir Putin's rise to power, however, the most outspoken newspapers and TV stations in Russia were shut down. Since then, critical analysis of the government has become rare throughout Russia. Journalists openly critical of government policies have been treated to various forms of punishment: censorship, exile, or violence. Since 2000, about 200 journalists have died under suspicious circumstances.

THINGS TO REMEMBER

GEOGRAPHIC THEME 3   • **Power and Politics:** This region has a long history of authoritarianism. Even if the public now desires expanded political freedoms, there are few opportunities to influence the political process. Elected representative bodies often act as rubber stamps for strong presidents and exercise only limited influence on policy making.

• A series of so-called *color revolutions* took place throughout this region. The revolutions were aimed at hastening democratization but had limited success.

• In early 2014 in Ukraine, one of the most challenging crises in this region since the fall of the Soviet Union emerged, pitting Russia against the EU and the United States in a seeming throwback to the days of the Cold War.

• In 2014, Russia annexed Crimea. While Russia de facto controls Crimea, it is internationally recognized as Ukrainian territory.

• Russia's long history of expansion into neighboring lands has left it and surrounding countries with exceptionally complex internal political geographies.

URBANIZATION

GEOGRAPHIC THEME 4

Urbanization: A few large cities in Russia and Central Asia, such as Moscow, are growing fast, as they are primate cities or are fueled by the expansion of energy exports. Elsewhere, many cities are suffering from a lack of investment as their economies struggle in the post-Soviet era.

Even though Russia and the post-Soviet states have the largest land area of any world region, and immense areas of lightly populated or uninhabited land, this is still largely a region of cities. Especially Ukraine, Belarus, and Russia are highly urbanized. There, 70 percent or more of the population lives in urban areas. The Caucasian republics of Georgia, Armenia, and Azerbaijan are somewhat less urbanized (around 60 percent), and Central Asia is only 47 percent urban, which is below the world average of 53 percent.

Russia's capital city of Moscow, with approximately 12 million people, is a primate city (see Chapter 3 for the definition of primate city), as are most capital cities in this region. The dominance of primate cities reflects the importance of politics and government in determining where economic development took place during the Soviet and post-Soviet eras.

Urban life for most people remains shaped by the legacy of Communist-era central planning. Giant apartment blocks built for industrial workers dominate most cities, especially on the fringes of the older, pre-Soviet urban cores (**FIGURE 5.20B**). Cramped, drab apartments, badly in need of repair, with shared kitchens and bathrooms, are common. At the community scale, inadequate sewage, garbage, and industrial waste management pose serious long-term health and environmental threats.

In Moscow, housing shortages are particularly severe because the city has grown so rapidly since 1991, when it began receiving much of the new investment in the region. Due to high demand, the cost of housing has increased so dramatically that Moscow is now on par with famously expensive cities like New York and Tokyo. Shortages are also exacerbated by the ability of wealthy people to buy up multiple apartments and refurbish them into one luxurious dwelling. This is a form of gentrification similar to the process taking place in North American cities (Chapter 2).

Outside Moscow, St. Petersburg, and a few other cities that are growing due to energy revenues (see Figure 5.20C), many urban areas are deteriorating and suffering from the absence of funds for maintenance because economic and population growth has been slow or nonexistent. In some cases, population decline is a cause of the deteriorating quality of housing (see Figure 5.20B).

URBAN LEGACIES OF SOVIET REGIONAL ECONOMIC DEVELOPMENT

One of the more difficult legacies of the Soviet era involves the huge industrial cities in the lightly inhabited expanses of Siberia and the Pacific coast. Founded originally as trading posts and fortress towns during the pre-Soviet Russian empire, these towns became the focus of Soviet planners, who were eager to solidify their control of the vast interior east of Moscow. They became home to major extractive industries and defense-related

FIGURE 5.20 PHOTO ESSAY: Urbanization in Russia and the Post-Soviet States

An uneven pattern of urbanization is developing in this region. In several countries, the largest cities are growing rapidly, as they are centers of new development and globalization. Elsewhere, however, many cities are struggling economically and even shrinking in terms of population because they are less able to attract new investment.

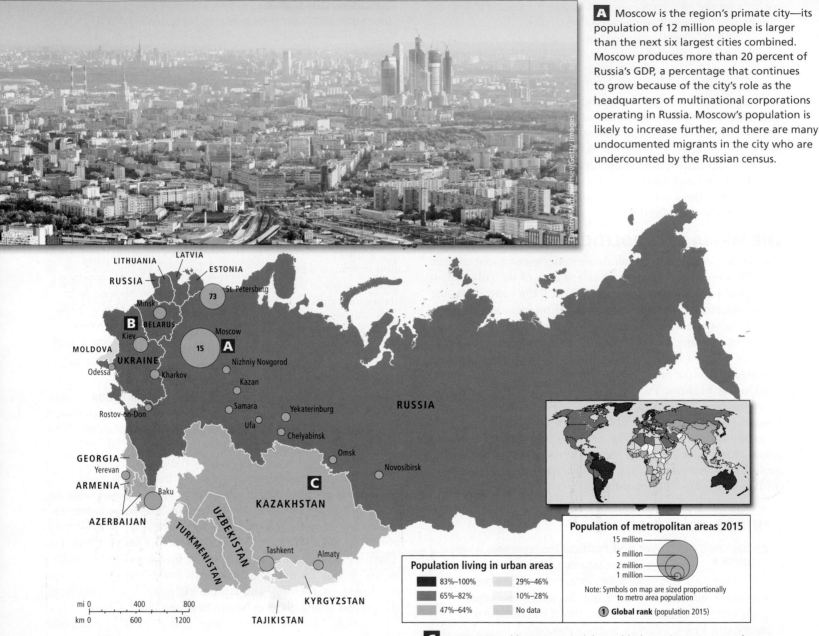

A Moscow is the region's primate city—its population of 12 million people is larger than the next six largest cities combined. Moscow produces more than 20 percent of Russia's GDP, a percentage that continues to grow because of the city's role as the headquarters of multinational corporations operating in Russia. Moscow's population is likely to increase further, and there are many undocumented migrants in the city who are undercounted by the Russian census.

Population living in urban areas

- ■ 83%–100%
- ■ 65%–82%
- ■ 47%–64%
- 29%–46%
- 10%–28%
- No data

Population of metropolitan areas 2015

- 15 million
- 5 million
- 2 million
- 1 million

Note: Symbols on map are sized proportionally to metro area population

① **Global rank** (population 2015)

mi 0   400   800
km 0   600   1200

C Astana, Kazakhstan's capital, has added massive amounts of new development, which has largely been supported by the increased exploitation of Kazakhstan's oil resources by local and international companies. The population of Astana doubled in the 10 years after Astana was designated the new capital in 1997 (the old capital was Almaty).

B An apartment building in Kiev, Ukraine. Large-scale homogeneous apartment blocks were built during the Soviet era. They are often run down today and are not considered desirable properties.

Thinking Geographically

After you have read about urbanization in Russia and the post-Soviet states, you will be able to answer the following questions.

A Why is Moscow growing so rapidly?

B Where in the post-Soviet city would you find large-scale apartment blocks?

C What is responsible for the growth of Astana?

infrastructure, but further growth was always hampered by the vast distances and challenging physical geography that separate them from the main population centers in the west. Even so, nearly 90 percent of Siberia's people are concentrated in the few large urban areas. Life in these cities is very different from the more traditional rural lifestyles that characterized this region only a few generations ago **(FIGURE 5.21)**.

Because the region's rivers run mainly north-south, there is a need for land transportation systems, such as railroads and highways, that run east to west to connect these cities with the west. The harsh climate of the region makes this infrastructure exceedingly expensive to build and maintain, and much has fallen into disrepair in the post-Soviet era. Only one poorly paved road and one main rail line—the Trans-Siberian Railway **(FIGURE 5.22)**—runs the full east-west length,

connecting Moscow with Vladivostok, the main port city on the Russian Pacific coast.

URBAN CENTRAL ASIA AND CAUCASIA

While Central Asia has historically had a relatively low rate of urbanization, the recent expansion of the energy industries has fueled new urban growth. Cities here have also grown because of the region's relatively high overall population growth. Tashkent, the capital of Uzbekistan, has 2.3 million people, making it the fourth-largest city in the region, after Moscow, St. Petersburg, and Kiev. Almaty, the former capital of Kazakhstan, is the next-largest city in Central Asia, at 1.3 million people, followed by Astana, the new capital of Kazakhstan, which has 750,000 people. The largest city of Caucasia is Baku, the capital of Azerbaijan, which is located on the shores of the Caspian Sea. With 2 million residents, and growing, it combines an ancient urban heritage with energy-driven modernization.

VIGNETTE Yernar Zharkeshov is a 24-year-old university-educated man who is moving back home to Kazakhstan after having spent time in Britain and Singapore. He is back in his homeland seeking new opportunities, which are increasing because of its extensive oil exports. The natural choice for him is Astana, a city that was designated the capital of Kazakhstan in 1997. Yernar quickly found a job as a government economist; he is just one of

FIGURE 5.21 LOCAL LIVES: People and Animals in Russia and the Post-Soviet States

A A trained hunting eagle is released by its master in Kazakhstan. Training an eagle requires great devotion over a long period of time. Raised by hand by their master, eagles are kept blindfolded from birth to make them depend on and trust their human companions. A Kazakh proverb says that "as the man trains his eagle, so does the eagle train his man." A golden eagle is on the flag of Kazakhstan.

B Siberian huskies take a break from training for a long-distance sled race. One of the oldest dog breeds, huskies were bred several thousand years ago by the Chukchi people in far eastern Siberia, near Alaska (see Figure 5.1) to pull heavy loads on sleds over long distances. With their legendary strength and stamina, huskies helped the Chukchi and other peoples explore and populate new territories. They were also central to U.S. penetration into Alaska in the early twentieth century.

C Reindeer are tended by a group of the indigenous Nenets people in western Siberia. Reindeer have been raised throughout northern Russia for thousands of years for their meat, hides, antlers, milk, and to pull sleds. Having weathered the collapse of Soviet state-run herding and meat-production economies, reindeer herders in some areas now find themselves in conflict with Russia's oil and gas producers, whose facilities often block important migration routes.

Vyacheslav Oseledko/AFP/Getty Images

Dmitry Kostyukov/AFP/Getty Images

Hans Neleman/Getty Images

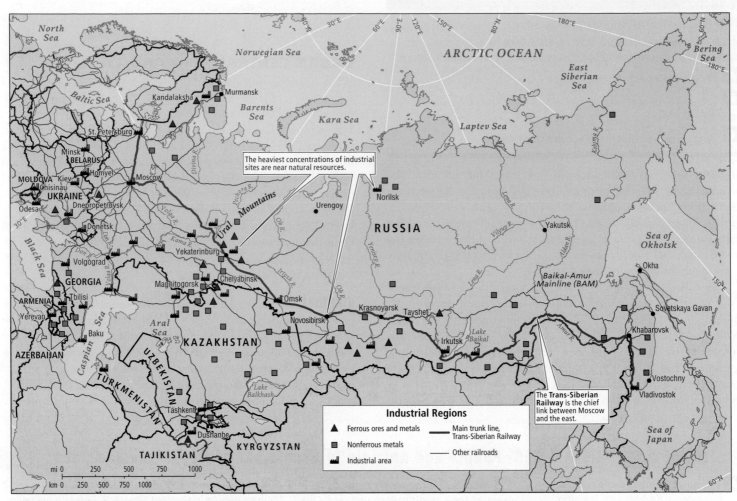

FIGURE 5.22 Principal industrial areas and land transport routes of Russia and the post-Soviet states.
The industrial, mining, and transportation infrastructure is concentrated in European Russia and adjacent areas. The main trunk of the Trans-Siberian Railway and its spurs link industrial and mining centers all the way to the Pacific, but the frequency of these centers decreases with distance from the borders of European Russia. [Sources consulted: Robin Milner-Gulland with Nikolai Dejevsky, *Cultural Atlas of Russia and the Former Soviet Union,* rev. ed. (New York: Checkmark Books, 1998), pp. 186–187, 198–199, 204–205, 216–217; http://www.travelcenter.com.au/russia/images/trans-sib-map-v3.jpg]

thousands of young people who have migrated to the growing capital.

The idea of Astana came from Nursultan Nazarbayev, who has been president of Kazakhstan since its independence from the Soviet Union in 1991. Without much public debate, he moved the capital from Almaty in the south to the remote and sparsely populated steppe in the north that is known for its harsh climate. Here he realized his vision for a grand and impressive new capital, built more or less from scratch. Astana is meant to symbolize a new, forward-looking Kazakhstan, and Nazarbayev has spared no expense. The city is full of stunning buildings, either designed by world-renowned architects or by Nazarbayev himself (see Figure 5.19C).

Like many other migrants to Astana, Yernar Zharkeshov is an ethnic Kazakh. President Nazarbayev's unspoken motive for moving the capital is to consolidate the country's northern territories, which are currently dominated by ethnic Russians. The presence of these Russians is the result of the practice of Russification during the days of the Soviet Union. Astana has been designed to

promote the opposite—the "Kazakhification" of the country's north. [Source: National Geographic. For detailed source information, see Text Sources and Credits.] ▪

THINGS TO REMEMBER

GEOGRAPHIC THEME 4
• **Urbanization:** A few large cities in Russia and Central Asia, such as Moscow, are growing fast, as they are primate cities or are fueled by the expansion of energy exports. Elsewhere, many cities are suffering from a lack of investment as their economies struggle in the post-Soviet era.

• Russia's capital city of Moscow, with 12 million people, is a primate city, as are most capital cities of the countries in the region.

• Urban life for most people remains shaped by the legacy of Communist-era central planning. Giant apartment blocks built for industrial workers dominate most cities, especially on the fringes of the older, pre-Soviet urban cores.

- One of the more difficult legacies of the Soviet era involves the huge industrial cities in the lightly inhabited expanses of Siberia and the Pacific coast, where urban growth has long been hampered by spatial isolation from the main population centers of the region.

- While Central Asia and Caucasia have historically had a relatively low rate of urbanization, population growth and recent expansion in the energy industries has fueled new urban growth.

POPULATION AND GENDER

GEOGRAPHIC THEME 5

Population and Gender: Populations are shrinking in many parts of the region. High levels of participation by women in the workforce and the economic decline in the 1990s that followed the collapse of the Soviet Union resulted in low birth rates. Economic problems have also led to declines in life expectancy.

With high death rates and low birth rates, this region is undergoing a unique, late-stage variant of the demographic transition (see Chapter 1). In all but the Caucasian and Central Asian states, populations are shrinking faster than in any world region. During the Soviet era, women's opportunities to become educated and work outside the home curtailed population growth. Free health care and adequate retirement pensions also helped lower incentives for large families, as people didn't have to depend on their children for support in old age. Severe housing shortages were an additional disincentive to having children, and many families chose to have only one or two children. All of these factors were leading to slower population growth during the last years of the Soviet Union, but the population didn't actually decline until the economic collapse and the "shock therapy" of the 1990s, which had major social repercussions.

By the mid-2000s, the rate of shrinkage had slowed down considerably across the region, and by 2015 Russia's population was holding steady. The death rate is the same as the birth rate, resulting in a natural increase rate of zero. Russia is attracting some migrants, mostly from neighboring countries such as the Central Asian states, and so currently has a small population increase. But this growth may not last; Russia is predicted to lose about 10 million people by 2050. The population decline of the 1990s means that fewer people will be in their reproductive years in the coming decades. These population trends are geographically uneven. Rural areas are suffering from severe population decline and many villages are emptying out; only the elderly remain.

Russia has made considerable efforts to fight population decline. Couples are now offered a bonus, equivalent to more than 2 years' worth of wages for the average Russian, for having a second child. Monetary incentives increase further for having more than two children. However, similar policies in Europe have been less successful than efforts to address the concerns of career-minded working mothers, who often value access to inexpensive, high-quality day care and longer school days more than they do money. Also, the limited financial resources of the Russian state means that these child subsidies may be phased out in the near

future. Much like in Europe, immigration could help offset potential population decline, especially among the working-age population, but anti-immigrant sentiment is an obstacle towards such a solution. According to a 2013 UN report, Russia is home to 11 million migrants, predominantly low-wage workers from other post-Soviet states. That is the largest immigrant population in the world after that of the United States, although it has remained roughly the same since the end of the Soviet Union in 1991.

GENDER AND LIFE EXPECTANCY

Much of the reason for the population decline, as well as the recent slowing of this decline, relates to life expectancy. Life expectancy differs dramatically according to gender, with women living 10 to 11 years longer than men in Russia, Ukraine, and Belarus. By comparison, around the world, women live on average only 5 years longer than men. The countries in Central Asia and Caucasia conform to this international pattern. Life expectancy for women declined only marginally after the collapse of the Soviet Union, but life expectancy for men dropped considerably in many places. In Russia, for example, between 1987 and 1994, male life expectancy dropped from 64.9 to 57.6 years. Since then, male life expectancy has recovered to Soviet-era levels, but at 65 years, it is still the shortest in any developed country. Female life expectancy has held steady at 75 to 76 years throughout this period.

Major causes of low life expectancies in the region are the loss of health care, which was usually tied to employment, and the physical and mental distress caused by lost jobs and social disruption. These factors, though, might not explain much of the gender difference in life expectancy since the fall of the Soviet Union, given that women worked outside the home almost as much as men during the Soviet era and seem to have been more likely to have lost jobs during the 1990s. Other possible causes for low male life expectancy are alcohol abuse and alcohol-related accidents and suicides, which are much more common among men. Russia has by far the world's highest rate of alcohol-related health disorders (1277 per 100,000 people in Russia, compared to 600 per 100,000 people in the United States), and more than half of all deaths among the working-age population are alcohol-related. On the positive side, recent increases in male life expectancy may be a result of declining rates of alcohol abuse. This change is likely an outcome of government and private campaigns to reduce alcohol consumption and improve treatment of alcoholism, as well as general economic improvements. On the other hand, during the post-Soviet era, heroin started to flow from Afghanistan, through porous borders of Central Asia, and into Russia. The UN reports that Russia is the world's largest heroin consumer.

Population pyramids for several countries **(FIGURE 5.23)** show the overall population trends in the region and reflect geographic differences in life expectancy and fertility. The pyramids for Belarus and Russia resemble those of European countries (for example, Germany, as shown in Figure 4.24), but there are also differences. First, they are significantly narrower at the bottom, indicating that birth rates in the last several decades have declined sharply. Because so few children are born, in 20 years there will be few prospective parents, which means that low birth rates and population stagnation or decline will probably continue in the near future. Also, the narrower point at the top

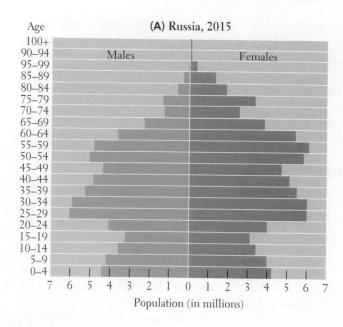

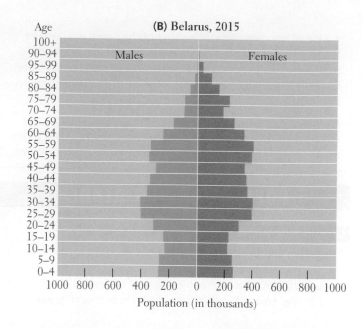

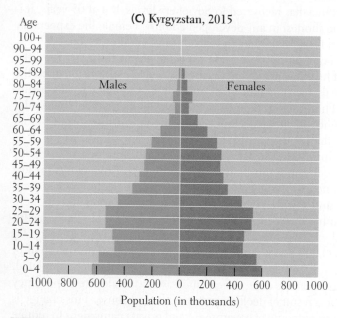

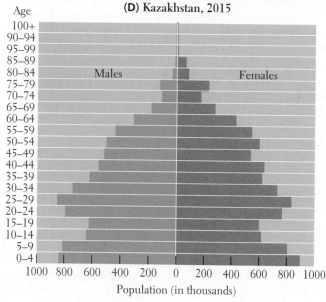

FIGURE 5.23 Population pyramids for Russia, Belarus, Kyrgyzstan, and Kazakhstan. Note that the pyramid for Russia is at a different scale (millions) than the other three pyramids (thousands) because of Russia's larger total population. This difference does not significantly affect the pyramid's shape. [Sources consulted: "Population Pyramids of Russia," "Population Pyramids of Belarus," "Population Pyramids of Kyrgyzstan," and "Population Pyramids of Kazakhstan," *International Data Base*, U.S. Census Bureau, 2015, at https://www.census.gov/population/international/data/idb/informationGateway.php]

for males shows their much shorter average life span compared to women.

In all countries, there has been a recent rebound in birth rates in the youngest age group, ranging from a slight increase in Russia to a significant rebound in Kazakhstan. In Russia, the birth rate rebounded mostly because those who are 25 to 35 years old make up a large age cohort, and as they are the parents of today's young children, the birth rate has gone up. Kyrgyzstan's pyramid indicates a younger population structure, which is common in less affluent

economies. In the Central Asian countries, the population is currently growing and will likely expand during the next decades.

POPULATION DISTRIBUTION

A broad area of moderately dense population forms an irregular triangle that stretches from Ukraine on the Black Sea north to St. Petersburg on the Baltic Sea and east to Novosibirsk, the largest city in Siberia. In this triangle, settlement is highly urbanized, but the cities are widely dispersed **(FIGURE 5.24)**.

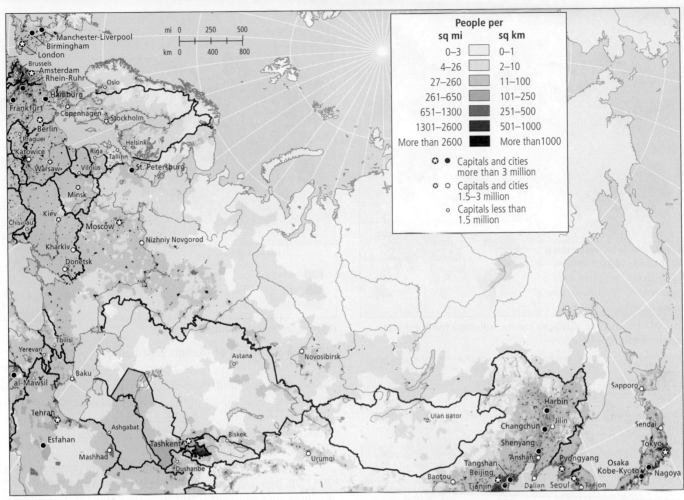

FIGURE 5.24 Population density in Russia and the post-Soviet states. Population trends in this region are quite uneven. While Central Asia and some countries in Caucasia are growing, populations in the rest of the region are shrinking. Some of this unevenness may be related to cultural differences or to varying levels of dependence on social welfare institutions that collapsed with the end of the Soviet Union.

A secondary spur of dense settlement extends south from Russia into Caucasia, the mountainous region between the Black Sea and the Caspian Sea, where there are several primate cities of well over 1 million people each. In Central Asia, another patch of relatively dense settlement is centered on the city of Tashkent. During the Soviet period, the development of irrigated cotton farming and mineral extraction along major Central Asian rivers resulted in patches of high rural density, fueled partly by ethnic Russian immigration.

Geographic Patterns of Human Well-Being The recent increases in life expectancy and birth rates in some countries may indicate improvements in the general well-being of the population after the difficult period immediately following the Communist era. In **FIGURE 5.25**, the color maps show that Russia and the post-Soviet states mostly fall in the middle range of the three well-being indicators that are presented.

As we have observed elsewhere, GNI per capita (PPP adjusted) can be used as a general measure of well-being because it shows in a broad way how much people in a country have to spend on necessities (see Figure 5.25A). However, because GNI per capita is an average, it does not indicate how income is distributed or how accessible health care, education, and other social services are to most people. In this particular region, the intermediate HDI ranks are in large part a holdover from the Soviet era, when socialism kept wages relatively equal (though they were usually lower outside European Russia) and the strong social safety net provided basic health care, food, and housing for all. Since 1991, disparities in wealth have steadily increased within the Russian Federation and each of the post-Soviet states. As we have observed, in Russia a few people have become fabulously wealthy; their wealth has undoubtedly skewed the average figure. For many Russians, income is actually well below the country's $24,700 GNI per capita (PPP) because Russia has changed from a society based on economic equality to one where the gap between poverty and wealth is similar to that of the United States. The difference between the lives of the economic elite in Moscow and those of people left behind in rural villages is extreme. Significant differences in income among the new states has become more apparent in the post-Soviet era. GNI per capita (PPP) in Belarus is $17,600 and in Kazakhstan is $21,580, while in Kyrgyzstan and Tajikistan it hovers around $3000.

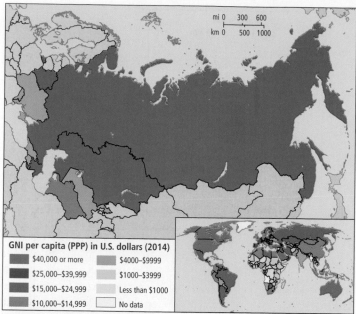

(A) Gross national income (GNI) per capita, adjusted for purchasing power parity (PPP).

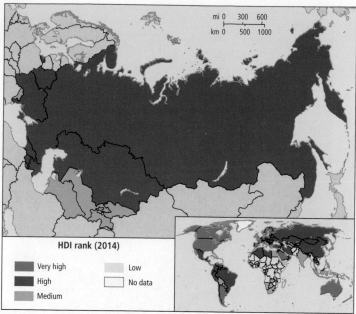

(B) Human Development Index (HDI).

FIGURE 5.25 Maps of human well-being.

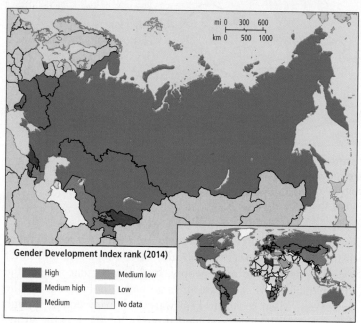

(C) Gender Development Index (GDI).

Figure 5.25B depicts the ranks on the HDI, which is a calculation (based on adjusted real income, life expectancy, and educational attainment) of how adequately a country provides for the well-being of its citizens. It is possible to have a low GNI rank and a higher HDI rank if social services are sufficient or if there are no wide disparities of wealth. Belarus, the Russian Federation, and Kazakhstan stand out as ranking high on the HDI scale. Georgia, Azerbaijan, Ukraine, and Armenia rank a bit lower, but still in the high category despite their modest incomes. In the case of Ukraine, the ongoing conflict is likely to be negatively affecting human development. From a global perspective, Russia and the post-Soviet states rank slightly higher on HDI relative to their GDP, which may be a leftover result of the Soviet emphasis on

broad education and basic social services that improved human well-being during most of the twentieth century. The lower HDI rankings (medium) of Moldova, Uzbekistan, Kyrgyzstan, and Tajikistan are matched by their GNI per capita (PPP), which hovers between $3000 and $6000.

Figure 5.25C ranks countries into five categories (high to low) on how equitable they are between the genders in terms of longevity, education, and income. HDI values are estimated separately for women and men and the ratio of the two is noted in the Gender Development Index (GDI). First, as is true across the world, nowhere in the region are women close to being equal to men. However, Russia and the post-Soviet states perform a little bit better with regard to gender equality than they do in terms of their GNI and HDI. All countries are in the medium or medium-high GDI categories. This means that, unlike for GNI and HDI, there is little GDI variation across the region. Again, the Soviet system, which encouraged education and careers for women, is partly responsible for this record. The country with the lowest gender equality (Tajikistan, in category 3) is not much more unequal than the country with the highest gender equality rank (Russia, in category 1). Unlike for HDI, the geographic gender equality differences that do exist among the countries in the region seem unrelated to economic differences; that is, some countries with less gender equality have a low GNI, while other such countries have a medium GNI.

THINGS TO REMEMBER

GEOGRAPHIC THEME 5

• **Population and Gender:** Populations are shrinking in many parts of the region. High levels of participation by women in the workforce and the economic decline in the 1990s that followed the collapse of the Soviet Union resulted in low birth rates. Economic problems have also led to declines in life expectancy.

- By the mid-2000s, the rate of population decline had slowed down considerably, and since 2013 Russia's population has increased slightly, although the prediction is that the population will again decline during the next few decades.

- Life expectancy in the region differs dramatically according to gender, with women living 10 to 11 years longer than men in Russia, Ukraine, Moldova, and Belarus. By comparison, around the world, women live on average only 5 years longer than men.

- Recent increases in male life expectancy may be a result of declining rates of alcohol abuse since the mid-2000s. This change is likely an outcome of campaigns to reduce alcohol consumption and improve treatment of alcoholism, as well as an improving economy.

- This region does fairly well in terms of GNI, HDI, and GDI, primarily because of the lingering effects of the Soviet system, which encouraged education and health care for all and a strong safety net despite people's fairly low incomes. As capitalism has gained footage, these benefits have decreased and for some, well-being has decreased.

SOCIOCULTURAL ISSUES

When the winds of change began to blow through the Soviet Union in the 1980s, few people anticipated the speed and depth of the transformations, or the social instability that resulted. On the one hand, new freedoms have encouraged self-expression, individual initiative, and cultural and religious revivals **(FIGURE 5.26)**; on the other hand, the post-Soviet era has brought very hard times to many, as access to jobs and the social safety net has become more difficult. Men and women have been differently affected by the massive changes in this region.

GENDER: CHALLENGES AND OPPORTUNITIES IN THE POST-SOVIET ERA

Soviet policies that encouraged all women to work for wages outside the home have transformed this region. By the 1970s, 90 percent of able-bodied women of working age had full-time jobs, giving the Soviet Union the highest rate of female paid employment in the world. However, the traditional attitude that women are the keepers of the home persisted. The result was the *double day* for women. Unlike men, most women worked in a factory or office or on a farm for 8 or more hours and then returned home to cook, care for children, shop daily for food, and do the housework (without the aid of household appliances). Because of shortages (the result of central-planning miscalculations), they also had to stand in long lines to procure necessities for their families.

These special pressures on working women affected diets because there was so little time or space for cooking and there were so few helpful appliances. The everyday cuisine in most of the region was limited during the Communist era; only recently have

FIGURE 5.26 LOCAL LIVES: Festivals in Russia and the Post-Soviet States

A A man performs stunts on horseback during the festival of Nauryz in Kyrgyzstan. Nauryz is a Zoroastrian (an ancient religion with roots in Iran) festival that marks the New Year and the first day of spring on March 21. Forbidden during the Soviet era, Nauryz celebrations are now a source of pride throughout Central Asia.

B Revelers at KaZantip, a multiweek electronic dance music festival held on Ukraine's Crimean Peninsula during the summer. DJs entertain more than 100,000 revelers at the festival each year. A treasured notion is that KaZantip is an independent republic, complete with its own constitution prohibiting a wide variety of "chauvinistic" behavior and promoting love and fun. Due to the conflict in Ukraine, the festival has recently been held at other sites, but is expected to make a return to Crimea.

C Women eat pancakes during Maslenitsa. Celebrated seven weeks before the Eastern Orthodox Christian Easter, Maslenitsa is a week-long festival that marks the end of winter and the last chance to revel before the onset of Lent. Maslenitsa has counterparts in Catholic countries, such as the carnivals in Venice and Rio, and Mardi Gras in New Orleans (see Chapter 2). During Maslenitsa, eating meat is forbidden, but dairy products and pancakes are consumed in large quantities.

market forces made a wider range of food available. Traditional recipes are now being revived. **FIGURE 5.27** shows several distinct dishes of the region that are prepared at home or in restaurants. The popularity of hearty soups and stews and the common use of root vegetables and grains (bread is an essential part of every meal) reflect the region's climate, soil, and agricultural systems.

When the first market reforms in the 1980s reduced the number of jobs available to all citizens, President Gorbachev encouraged women to go home and leave the increasingly scarce jobs to men. Many women lost their jobs involuntarily. By the late 1990s, women made up 70 percent of the registered unemployed, despite the fact that because of illness, death, or divorce, many if not most were the sole supporters of their families. Consequently, many had to find new jobs. Today, unemployment for women is slightly lower than for men, although 3 in 10 women don't participate in the workforce.

On average, the female labor force in Russia is now more educated than the male labor force. The same pattern is emerging in Belarus, Ukraine, Moldova, and parts of Muslim Central Asia. In fact, the same is true in the United States and western countries as well. In Russia, the best-educated women commonly hold professional jobs, but they are unlikely to hold senior supervisory positions and are paid less than their male counterparts. In 2014, the wages of women—both professional and low-skilled workers—averaged 33 percent less than those of men. Still, this region ranks higher in gender income equity than many others.

The Trade in Women

During the post-Communist era, the "marketing" of women became one of the less savory entrepreneurial activities. One part of this has been the Internet-based mail-order bride services aimed at men in Western countries. A woman in her late teens or early twenties,

trafficking the recruiting, transporting, and harboring of people through coercion for the purpose of exploiting them

usually trying to escape economic hardship, pays a small fee to be included in an agency's catalog of pictures and descriptions, which may feature tens of thousands of women. She is then assessed via email or Facebook by the prospective groom, who then travels—usually to Russia or Ukraine—to meet women he has selected from the catalog and to potentially have one of them accompany him to the West to marry.

Sex work has also increased in recent years. Precise numbers are hard to come by, but a 2010 UN report estimates that 140,000 women have been smuggled into Europe, where they become part of a sex trade that generates $3 billion annually. The post-Soviet states comprise the primary source region for Europe's sex workers. Most of the women—coming from countries like Ukraine, Moldova, and Russia—are supplied by members of the Russian Mafia, who have been known to kidnap schoolgirls or deceive women who are looking for jobs as maids or waitresses in Europe, and then force them to work as prostitutes. This is a form of **trafficking,** which is defined by the UN as the recruiting, transporting, and harboring of people through coercion for the purpose of exploiting them.

The Political Status of Women

One way for women to address institutionalized discrimination is to increase their political power. Although women were granted equal rights in the Soviet constitution, they never held much power. In 1990, women accounted for 30 percent of Communist Party membership but just 6 percent of the governing Central Committee. The very few in party leadership often held these positions at the behest of male relatives.

Since the fall of the Soviet Union, the political empowerment of women has grown somewhat, but long-standing cultural biases against women in positions of power remain. In many countries, the number of women in legislatures has risen **(FIGURE 5.28)**, but often this is because male leaders want to appear more progressive in the eyes of international donors. Frequently

FIGURE 5.27 LOCAL LIVES: Foodways in Russia and the Post-Soviet States

A A bowl of borscht, or beet soup, a Ukrainian dish with many variants, is popular throughout this region. Beets are an extremely nutritious root vegetable that can tolerate cold weather well, enabling farmers to make the most out of the short growing season found throughout much of this region. Borscht is served warm in the winter and chilled in the summertime.

B Pirozhki are baked yeast buns stuffed with meat, vegetables, or fruit, and glazed with egg to give them a golden color. Shown here are cabbage, egg, and dill pirozhki. They are a popular fast food throughout this region, and even in neighboring Mongolia. They are often made at home and, depending on the stuffing, can be an appetizer, a main dish, or a dessert.

C A girl prepares *non*—a kind of bread cooked throughout Central, South, and Southwest Asia (where it is also known as *naan*). Non is considered sacred by many Central Asians and is traditionally served with great reverence. In Uzbekistan, for example, it is never cut with a knife, but rather is broken by hand and put near each place setting at a table, always with the flat side down (serving it flat side up is considered insulting). One tradition holds that when an Uzbek person leaves a house, he or she should bite off a piece of bread, which will be kept for that person to eat upon return.

Alexandra Grablewski/Getty Images

Lara Hata/Getty Images

Victor Drachev/AFP/Getty Images

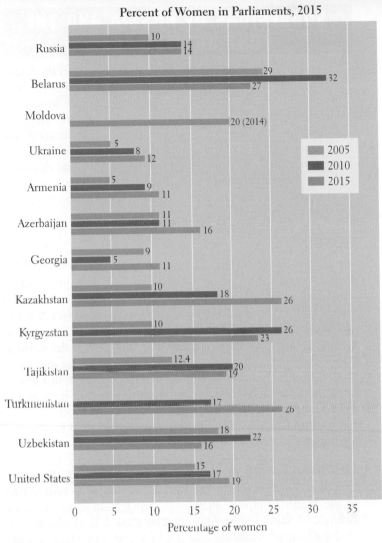

FIGURE 5.28 Women legislators, 2005, 2010, and 2015. This graph shows the percentage of legislators in Russia and the post-Soviet states (with the United States for comparison) who are female. The general trend has been an increase in the number of women in parliament. However, the countries with the largest percentage of female legislators are not necessarily those with the most open democratic participation. [Sources consulted: *The World's Women 2015: Trends and Statistics*. New York: United Nations, at http://unstats.un.org/unsd/gender /chapter5/chapter5.html; and United Nations Development Program, *Women in Politics in Moldova*, http://www.md.undp.org/content/moldova/en/home /operations/projects/effective_governance/women-in-politics-in-moldova.html]

they promote women who are the least likely to work for change. Support among women themselves for women's political movements is not widespread, as many fear being seen as anti-male or against traditional feminine roles. To be a feminist is to occupy a tenuous position in society.

RELIGIOUS AND NATIONALIST RESURGENCE IN THE POST-SOVIET ERA

The official Soviet ideology was atheism, and religious practice and beliefs were seen as obstacles to revolutionary change. Orthodox Christianity, which was the official religion of the Russian empire, was tolerated, but few people went to church, in part because the open practice of religion could be harmful to one's career. Now, religion is a major component of the general cultural resurgence across the former Soviet Union.

The overall level of religiosity is moderate in Russia by global standards but a number of people, especially those from indigenous ethnic minority groups, are turning back to ancient religious traditions. In Russia, Ukraine, Moldova, Belarus, Georgia, and Armenia, most people have some ancestral connection to Orthodox Christianity. Those with Jewish heritage form an ancient minority, mostly in the western parts of Russia and Caucasia, where they trace their heritage back to 600 B.C.E. Religious observances by both groups increased markedly in the 1990s, and many sanctuaries that had been destroyed or used for nonreligious purposes by the Soviets were rebuilt and restored.

Even President Putin in Russia commonly appears in public with Orthodox priests and attends religious services. This signals a support for the traditional values that the church represents, as well as an emphasis on Russian culture in general. Putin is tapping into growing nationalist sentiment. Russia has been defining itself in opposition to Western culture, which is seen by Russian traditionalists as too secular, multicultural, and accepting of homosexuality. In fact, the Russian parliament passed a law in 2013 against homosexual "propaganda." This vague law makes it possible for the government to target and persecute the country's LGBT (lesbian, gay, bisexual, and transgender) community. Linking the LGBT community with the West also conveniently positions Russia as having a heterosexual, Slavic, and Orthodox culture that needs to remain "pure" of outside influences. By appealing to the sense of national identity among many Russians, it is easier for the government to rally support for policies that are intended to advance Russia's influence around its borders.

In Central Asia, Islam was repressed by the Soviets, who feared Islamic fundamentalist movements would cause rebellion against the dominance of Russia. Today, Islam is openly practiced and is increasingly important politically across Central Asia, Azerbaijan, and some of the Russian Federation's internal ethnic republics, such as Chechnya and Tatarstan. Especially in the Central Asian states, however, the return to religious practices is often contentious. Some local leaders still view traditional Muslim religious practices as obstacles to social and economic reform.

Militant Islamic movements have often been violently repressed by Central Asian governments, in part out of fear of influences from nearby countries, where Islam plays a greater role in politics. In some cases, these influences are clear—as in Tajikistan's civil war (1992–1997), which involved Taliban fighters from across the border in Afghanistan; and the conflict in Chechnya in which Saudi militants fought. At times, government repression has backfired, as when Uzbekistan and Kyrgyzstan joined forces to eliminate an extremist Islamic movement in 2000. In their fervor to eliminate radicals, these governments persecuted ordinary devout Muslims, especially men, which actually helped recruit more Islamic militants instead. Human Rights Watch, an organization that monitors human rights abuses worldwide, reports that Uzbekistan's government jails and tortures believers who worship independently outside the strict supervision of the state. These measures are done to try to prevent the political instability unleashed by the Arab Spring (see Chapter 6) from spreading to Uzbekistan.

VIGNETTE After Russia's bad experience with market reforms during the 1990s, the country searched for a new identity that was neither a return to communism nor affirmation of Western-style democracy. The answer has been a rebirth of Russian nationalism. Nowhere is the newfangled celebration of the nation more overt than in the stands of soccer stadiums. Especially for young Russian men, supporting a club team or the national team has been an outlet for nationalist sentiment.

Thirty-two-year-old Ivan Katanaiev says that when the United States and Europe wanted to push Russia towards their "decadent" societal model, soccer became a source of national resistance. The messages displayed on banners in the stadiums, he continues, hark back to our history and heritage. Katanaiev believes that the form of ultranationalism displayed by the soccer fans is endorsed on the highest political level. Even President Putin paid tribute to the leader of a soccer fan club who had been killed in a confrontation with a Muslim from the Caucasus. The ethnocentrist nationalism in the soccer community has also meant that it has been fertile ground for recruiters who are looking for people to volunteer to fight on the side of ethnic Russians in the Ukraine conflict.

However, not all soccer fans are comfortable with the mixing of nationalist politics and soccer. Robert Ustian, who hails from the province of Abkhazia in Georgia, is a fan of the club CSKA Moscow, but he has been dismayed by racist chants among the fans. He founded the organization "CSKA Fans Against Racism," which sends a very different message about what it means to be a soccer fan in Russia today. ∎

THINGS TO REMEMBER

• By the 1970s, 90 percent of able-bodied women of working age had full-time jobs, giving the Soviet Union the highest rate of female paid employment in the world.

• Today, female participation in the workforce is somewhat lower than during the Soviet era, and while women are better educated, their pay is not commensurate with men's pay.

• The official Soviet ideology was atheism, and religious practice and beliefs were seen as obstacles to revolutionary change. Now, religion is a major component of the general cultural resurgence across the former Soviet Union. Nationalist sentiments are also on the rise.

• In Central Asia, Islam was repressed by the Soviets, who feared Islamic fundamentalist movements would cause rebellion against the dominance of Russia. Today, Islam is more openly practiced and increasingly important politically across Central Asia, Azerbaijan, and some of the Russian Federation's internal ethnic republics, such as Chechnya and Tatarstan.

SUBREGIONS OF RUSSIA AND THE POST-SOVIET STATES

This section begins with Russia, the largest country in the region and the one that has dominated the entire region for many hundreds of years in what most scholars now recognize as a quasi-colonial manner. The post-Soviet states of Belarus, Moldova, and Ukraine are discussed next because of their location and their close social, cultural, and economic associations with European Russia. The Caucasian states of Georgia, Armenia, and Azerbaijan are then covered. Finally, the countries of Central Asia—Kazakhstan, Kyrgyzstan, Tajikistan, Turkmenistan, and Uzbekistan—are discussed.

RUSSIA

In the post-Soviet era, Russian influence in the region remains strong, even as the post-Soviet states have begun to construct their own economies and political identities. Russia is still the largest country on Earth in terms of area—nearly twice the size of Canada, the United States, or China—and is rich in natural resources. With 144 million people, it ranks eighth in the world in population. For convenience, we have divided Russia into three parts: European Russia; Central Siberia, which is composed of about half the landmass east of the Urals; and the Russian Far East. There is no clear boundary between Central Siberia and the Far East; in fact, the term *Siberia* is commonly used for Russian territories all the way to the Pacific.

European Russia

European Russia is the area of Russia that shares the eastern part of the North European Plain with Latvia, Lithuania, Estonia, Belarus, Ukraine, and Moldova (**FIGURE 5.29**). It is usually considered the heart of Russia because it is here that early Slavic peoples established what became the Russian empire, with its center in Moscow. Although it occupies only about one-fifth of the total territory of present-day Russia, European Russia has most of the industry, the best agricultural land, and about 70 percent (100 million people) of the population of the Russian Federation. A tiny exclave of European Russia is Kaliningrad, located on the Baltic Sea between Lithuania and Poland. The Kaliningrad port is important to Russia because it is free of ice for more of the year than St. Petersburg and serves as an outlet to the Atlantic Ocean. The port of Murmansk, which faces the Arctic Ocean, also connects to the Atlantic. Because of warm ocean currents, it is actually ice-free year-round despite its location north of the Arctic. However, Murmansk is located far from main commercial routes, which means that it is mostly important for military purposes. After the Cold War, Russia reduced its investment in Murmansk, but is now investing there and in other military bases that face the Arctic Ocean. Russia's interest in the Arctic is driven in part by climate change. Fossil fuel gas resources can be exploited and maritime trade routes that are opening up because of the melting ice need to be defended.

The vast majority of European Russians live in cities, their parents and grandparents having left rural areas when Stalin collectivized agriculture and established heavy industry. Most of

Cities and National Capitals

● ✪ More than 5,000,000
◖ ✪ 1,000,000–5,000,000
○ ✪ 500,000–999,999
• ◦ 250,000–499,999
○ Less than 250,000

Transportation

—— Major road
—— Trans-Siberian Railway
—— Other railroad

Internal Republics

1 Adygea
2 Karachay-Cherkesskaya
3 Kabardino-Balkarskaya
4 North Ossetia
5 Chechnya & Ingushetia
6 Dagestan
7 Kalmykiya

FIGURE 5.29 European Russia subregion. **A** The Kotelnicheskaya was completed in 1952 as an apartment building for workers and bureaucrats in Moscow, with multiple families sharing a kitchen and bathroom. It is one of seven such grand buildings—nicknamed the Seven Sisters—all of which are famous Moscow landmarks. Today, the building has been extensively remodeled and apartments are owned or rented by an affluent clientele.

these people went to work in one of four major industrial regions that were chosen by central planners for their accessibility and their location near crucial mineral resources. One industrial region is centered on Moscow; another in the Ural Mountains and foothills (part of this zone lies in Siberian Russia); a third along the Volga River from Kazan to Volgograd; and the fourth just north of the Black Sea, extending into Ukraine around Donetsk (see Figure 5.22).

The most densely occupied part of European Russia—stretching from St. Petersburg on the Baltic, south to Caucasia, and east to the Urals (see Figure 5.24)—coincides with that part of the continental interior that has the most favorable climate (see Figure 5.5). Even so, because the region lies so far to the north—Moscow is at a latitude 100 miles (160 kilometers) north of Edmonton, Canada—and in a continental interior, the winters are long and harsh and the summers short and mild.

Urban Life and Patterns in European Russia Moscow remains at the heart of Russian life. The city has grown in the post-Soviet era; about 12 million people live within the city limits and another 4 million live in the metropolitan area, making Moscow the largest city in greater Europe. Always a center of Russian culture, Moscow has so changed since 1991 that a common saying is "Moscow and Russia are two different countries." The availability of goods—food, electronics, furniture, appliances, clothing, automobiles, and especially luxury products and entertainment of all types—has exploded. Prices are much higher here than in most Russian cities. The criminals are more concentrated, more innovative, and more violent in Moscow. Entrepreneurs are especially active. While the city is full of apparently affluent, educated young people, the level of inequality is dramatic. Even many educated individuals are struggling to make ends meet in this high-cost city.

The privatization of real estate in Moscow gives an interesting insight into pre- and post-Soviet circumstances. During the Communist era, all housing was state owned and there was a general housing shortage; thousands of families were always waiting for housing, some for more than 10 years. Most apartments were built after the mid-1950s in large, shoddily constructed gray concrete blocks; the units included one to three rooms that housed an entire family. These monotonous, high-rise neighborhoods are typically located on the periphery of the city. In 1991, the government allowed residents of Moscow to acquire, free, about 250 square feet (23 square meters) of usable living space per person—about the size of a typical American living room. Most families simply took ownership of the apartments in which they had been living, so that by 1995, a majority of Muscovite families owned their apartments. A lively real estate market quickly emerged, with a rising wealthy elite willing to pay for an apartment in the city center. Purchase prices in excess of U.S.$1,000,000 are not unusual today. Rents also skyrocketed. Those who could find alternative housing for themselves made tidy sums simply by renting out or selling their city-center apartments (see Figure 5.29A). The demand for commercial space for private businesses reduced the number of dwelling units, exacerbating the housing shortage.

St. Petersburg, renamed Leningrad during most of the Communist period (1924–1991), is Russia's second-largest city,

with about 5 million people. It is a shipbuilding and industrial city located at the eastern end of the Gulf of Finland, on the Baltic Sea. Czar Peter the Great ordered the city built in 1703 as part of his effort to make Russia more European. From 1713 to 1918, it served as the Russian capital, and it has been one of Russia's cultural centers ever since. The city has a rich architectural heritage—including many palaces, public parks, and canals—which has drawn comparisons with the ambience of Venice. Since 1991, it has been undergoing a renaissance; the transportation system is being refurbished, urban malls are proliferating, and historic sites are being renovated. The czars' Winter Palace, now called the Hermitage Museum, houses one of the world's most important art collections (see Figure 5.10C).

Central Siberia

Central Siberia encompasses the West Siberian Plain beyond the Urals and about half the Central Siberian Plateau (FIGURE 5.30). It is so cold that 60 percent of the land's subsurface is *permafrost*, some of which appears to be melting due to climate change (see page 270). This vast central portion of Russia is home to just 31 million people, many of whom moved into the region from European Russia. Settlement is concentrated in cities in the somewhat warmer southern tier.

For millennia, Siberia has been a land of wetlands and quiet, majestic forests. The Siberian taiga, south of the permafrost range, is one of the largest forests in the world and sequesters a great deal of atmospheric carbon. Like all economic activities, forest harvesting declined during the post-Communist period, but as the economy has improved recently, more clear-cutting is taking place, which lowers the rate of *carbon sequestration*—a process whereby plants absorb carbon though photosynthesis and store it in their biomass. In the northern zones closer to the Arctic, climate change appears to be melting the permafrost, and water that used to be retained in lakes is escaping downstream, as melting permafrost also alters the topography of the land. Another effect is that organic material that used to be frozen is now exposed to the air, which results in the release of methane and other GHGs. Basically, global warming generates more GHG emissions, which further heats the climate through this ecological feedback mechanism.

Although the entire area of Russia east of the Urals continues to be occupied by a scattering of indigenous groups—some still practicing nomadic herding, hunting, and gathering—Central Siberia is dotted with bleak urban landscapes and industrial squalor (see the discussion of Norilsk on page 267). This is the legacy of the Soviet effort to claim territory from the indigenous peoples by settling it with Russians and to establish industries that would exploit Siberia's valuable mineral, fish, game, and timber resources. Novosibirsk, Siberia's twentieth-century capital and financial center, is an example of the result of this policy.

Novosibirsk is the largest Russian city east of the Urals (1.5 million people). It is some 1600 miles (2500 kilometers) east of Moscow and is a major stop along the Trans-Siberian Railway (see Figure 5.30A). This isolated commercial center has the highest concentration of industry between the Urals and the Pacific. During the Cold War, it was a center for strategic industries such

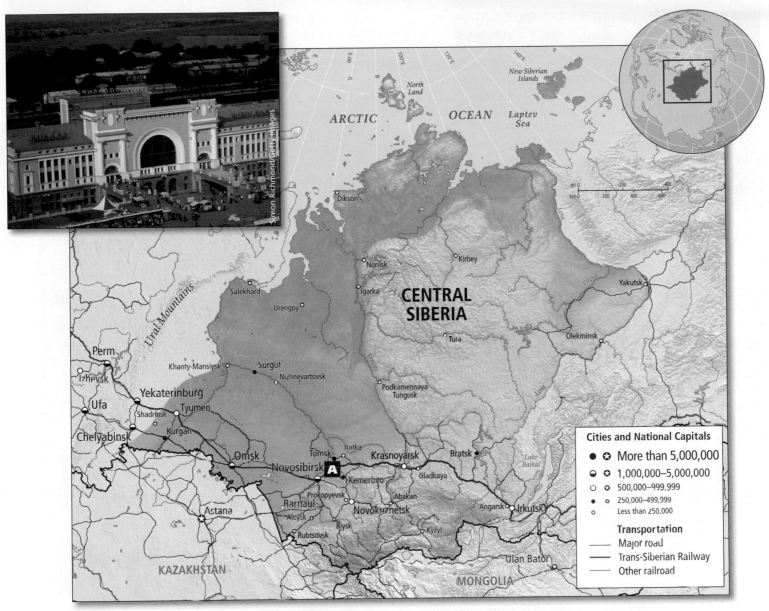

FIGURE 5.30 Central Siberia subregion. A The Trans-Siberian Railway is the main connection between Siberia and western Russia. Hence, Novosibirsk's main railway station is a hugely important fixture in the city's downtown area. With 1.5 million people, Novosibirsk is Russia's third-largest city, after Moscow and St. Petersburg, yet it is extremely isolated from other large cities. It takes about 50 hours by train to get from Novosibirsk to Moscow, which lies more than 1600 miles to the west.

as optics and weaponry; it contained more than 200 heavy industry plants. These factories are now infamous for being obsolete and having high rates of pollution, but they can also potentially be profitable in the future if they are privatized and reorganized. Novosibirsk's banks and financial services are drawing outside investors interested in acquiring factories in the vicinity.

The Russian Far East

Here in this extensive territory of mountain plateaus with long coastlines on the Pacific and Arctic oceans **(FIGURE 5.31)**, almost 90 percent of the land is permafrost. The coastal volcanic mountains stop the warmer Pacific air from moderating the Arctic cold that spreads deep into the continent. Making up over one-third of the entire country of Russia, this area is nearly two-thirds the size of the United States. If the population of 6.3 million were evenly distributed across the land, there would be just 2.6 persons per square mile (1 per square kilometer), but 75 percent of the people live in just a few cities along the Trans-Siberian Railway and on the Pacific coast.

The residents of the Russian Far East are indigenous people or immigrants and exiles from Russia. Indigenous groups include the Yakuts, which is the largest ethnicity of the vast Sakha internal republic to the north. The Yakuts historically hunt, fish, and keep reindeer. In southern Siberia near Mongolia, the Buryats share language and customs with other Mongol groups, such as nomadic herding. Some of the ethnic Russians are descendants

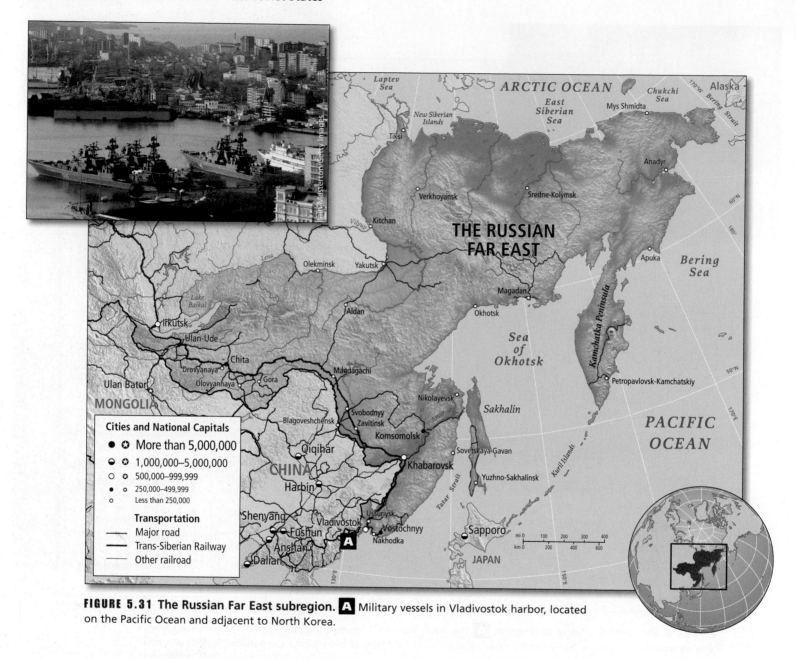

FIGURE 5.31 The Russian Far East subregion. **A** Military vessels in Vladivostok harbor, located on the Pacific Ocean and adjacent to North Korea.

of those sentenced to hard labor in the region's labor camps for political crimes in the Soviet era (some made famous in Aleksandr Solzhenitsyn's book *The Gulag Archipelago* were along the Kolyma River). These Far Easterners worked in timber and mineral extraction enterprises and in isolated industrial enclaves. Since the 1980s, many immigrants have headed for cities along the Pacific coast, especially Nakhodka and Vladivostok, near North Korea. Vladivostok is strategically important as a naval base and is also Russia's largest Pacific port (see Figure 5.31A). In this frontier culture, there is a spirit of independence and entrepreneurship that is rare elsewhere in Russia.

The interior of the Russian Far East has many resources, but it has not been developed because of its distance from European Russia and its difficult physical environments. Only 1 percent of the territory is suitable for agriculture, mostly along the Pacific coast and in the Amur River basin. Relatively unexploited stands

of timber cover 45 percent of the area. The wilderness supports many endangered species, such as the Siberian tiger, the Far Eastern leopard, and the Kamchatka snow sheep. The region's timber, coal, natural gas, oil, and minerals first attracted Soviet central planners and now attract private investors. The proximity to Chinese markets has increased timber harvesting, some of it illegally logged. The American multinational company Exxon has started oil production off Sakhalin Island that is destined for Asian markets. Since 2009, liquefied natural gas has been shipped to Japan, which is just 30 miles south of Sakhalin. With no pipelines in place and with few domestic energy sources, Japan is willing to pay the extra price for natural gas that has to be liquefied before it can be transported on oceangoing tankers.

The region has further potential for linkages with the greater Pacific Rim. In recent decades, port facilities on the Pacific coast have made the region more accessible to Asia. An oil pipeline

FIGURE 5.32 The Belarus, Moldova, and Ukraine subregion. **A** Skateboarders in downtown Minsk, Belarus.

from Siberia is now expected to terminate at Nakhodka on the Russian Pacific coast, which will give Russia access to petroleum markets in Japan and the western Pacific, and a spur to China will supply that country as well. Some geographers and economists speculate that, because of its rich resource reserves and its location so far from the Russian core, the Russian Far East is likely to integrate economically with other Pacific Rim nations, and might eventually seek further independence from European Russia. The Far East has a strategic position connecting the BRICS partners Russia and China, which means that Russia is likely to pay close attention to the development of the region. So far, there has been limited economic development and areas in Siberia and the Far East that are not part of the growing natural resource economy have suffered from economic and population decline.

THINGS TO REMEMBER

- Russia, the largest country in the world in area, has three principal geographic divisions, and although linked politically, each has distinct economies and cultural traditions.

- Moscow is Russia's fast-growing primate city; it attracts an affluent, young, and educated population.

- Central Siberia and the Far East are sparsely populated, with only a handful of isolated commercial and industrial centers between the Urals and the Pacific. The natural resource economy of these areas is increasingly oriented towards Asia.

BELARUS, UKRAINE, AND MOLDOVA

Sandwiched between Russia and the Central European countries that were once part of the Soviet Union's sphere of influence are the post-Soviet states of Belarus, Ukraine, and Moldova (**FIGURE 5.32**). Each of these countries is the home of a distinct ethnic group, but Russian residents form significant and influential minorities in all three. Due to their location and history, these countries continue to maintain close ties with Russia, though Moldova and Ukraine are considering closer associations with Europe. This has caused serious geopolitical tension with Russia, most notably the conflict in Ukraine.

Belarus

In size and terrain, Belarus resembles Minnesota; its flat, glaciated landscape is strewn with forests and dotted with thousands of small lakes, streams, and marshes that are replenished by abundant rainfall. Much of the land has been cleared and drained for collective farm agriculture. The stony soils are not particularly rich; nor, other than a little oil, are there many known useful resources or minerals beneath their surface. Belarus absorbed 70 percent of radiation contamination after the Chernobyl nuclear accident in Ukraine in 1986 because of its proximity to the accident site. Twenty percent of Belarus's agricultural land and 23 percent of its forestland were contaminated. Scientific and agricultural methods have been developed to reduce the level of radiation in the soil, but many people still rely on contaminated food.

During the twentieth century, Belarus was rather thoroughly Russified by a relatively small but influential group of Russian workers and bureaucrats. Although 84 percent of the population of 9.5 million is ethnic Belarusian (a Slavic group) and only 8 percent is Russian, the Russian language predominates, and Belarusian culture survives primarily in museums and historical festivals. The drab urban concrete landscape, such as in the capital Minsk, looks and feels like Soviet Russia (see Figure 5.32A). The Belarusian economy remains dominated by state firms that sell goods to Russia; only a few retail shops and industries are privatized. Belarus's GDP per capita is significantly higher than in Ukraine and Moldova, in part because Belarus receives subsidies from Russia, such as oil at below market price, which is economically beneficial. The economy, however, has struggled with slow growth for a number of years.

Belarus became an independent state when the Soviet Union collapsed in 1991, but it remains tightly controlled by its autocratic leader, Alexander Lukashenko, who has been in power since 1994 and allows little meaningful democracy and freedom of speech. Lukashenko no longer accedes to all Russian demands, occasionally spars with Russia over trade agreements, and appears to be making overtures to the European Union, but because Belarus relies heavily on Russia—especially for cheap energy—it is unlikely to seriously rebel. Russia is also able to manipulate the flow of oil and gas to Europe that is transported via pipelines across Belarus (see Figure 5.14).

Russia sees Belarus as a useful **buffer state** against the growing influence of the EU and NATO. A buffer state is situated between two rival and more powerful political entities, in this case Russia and the European Union, and serves as neutral territory that limits direct contact and potential conflict between the two powers.

Moldova and Ukraine

Moldova has a mix of cultures—Romanian in the west and Slavic in the east—while Ukraine is primarily Slavic in its traditions, including language, cuisine, religious customs, and architectural forms. The distinctive ethnic artistry of these two countries (elaborate holiday breads and pastries, intricate embroidery and lace, and in Ukraine, ornately painted Easter eggs) is prominently displayed in homes and marketed at home and abroad. Both countries seem to be tending toward closer association with the EU, but in Ukraine, long-standing ties to Russia complicate that propensity. Moldova is also affected by internal divisions that have developed as Slavic minorities in the eastern section of the country have formed a breakaway state called Transnistria, whose future political status is uncertain, although it currently receives political and economic support from Russia. Transnistria has the potential to be another political and military flashpoint, just like Ukraine.

Moldova and Ukraine have warmer climates than do Belarus and Russia, and with their agricultural resources, they have the potential to increase their agricultural production significantly. Figure 5.16 shows that intensive crop cultivation is far more typical here than elsewhere in the post-Soviet states. Not only is the climate warmer, but the area's open farming landscape is endowed with rich, black-colored soil (called *chernozem*). Moldova, which borders Romania to the west, is much smaller (3.5 million people) than the more economically diverse Ukraine (44 million people) but produces many of the same products—grain, sugar, milk, and vegetables, for example—as well as manufactured goods related to agriculture. Moldova is pursuing the possibility of increasing the amount of higher-value commodities, such as wine, nuts, and fruits, that it markets to Europe. Ukraine and Moldova will probably continue to trade their agricultural products primarily to Russia, but as farms are privatized and modernized, possibly through investment from Europe, both countries will likely increase their participation in the global agricultural market. The future of Ukraine's industrial production is uncertain because the industrial heartland is located in the eastern part of the country, which is where the military struggle between government and separatist forces is taking place (see page 283).

The European Union has a strong interest in Ukraine and Moldova. The two countries are potential future members of the EU, but their membership is dependent on economic and political progress toward greater democracy, and is affected by the ebb and flow of EU relations with Russia, which does not want either country to be reoriented toward Europe. The relationship is highly strained at the moment as Russia supports separatists in eastern Ukraine and has annexed Crimea. Another issue is border security. The EU's long eastern border faces Ukraine and Moldova. The EU cooperates with Ukraine and Moldova to tighten border security in order to reduce the number of undocumented immigrants and the illegal drugs that follow a route from Asia, through Ukraine and Moldova, and on to western Europe. Finally, oil and gas pipelines from Russia cross Moldova and Ukraine on their way to Europe (see Figure 5.14), and the region is also a potential route for future pipelines from Caucasia to EU countries.

buffer state a country that is situated between two rival and more powerful political entities and serves as a neutral territory, limiting direct contact and potential conflict between the two powers

---

THINGS TO REMEMBER

• Belarus was thoroughly Russified in the past and tends to be closely associated with Russia today. Belarus has not implemented as many market reforms as other post-Soviet states have and its economy remains dominated by state firms that sell to Russia. Politically, the country is tightly controlled by its Soviet-style leader.

- Moldova and Ukraine have significant agricultural resources, including relatively warm climates and rich soil. Their agricultural trade is mainly with Russia, but as farms are privatized and modernized, both countries' participation in the global agricultural market probably will increase. Because Ukraine and Moldova make up a significant portion of the EU's border to the east, the European Union has taken a strong interest in the economic and political development of these countries. Mixed ethnicities within both countries have resulted in internal tensions over political power and, in the case of Ukraine, outright war.

CAUCASIA: GEORGIA, ARMENIA, AND AZERBAIJAN

Caucasia is located around the rugged spine of the Caucasus Mountains, which stretch from the Black Sea to the Caspian Sea. (The language map in **FIGURE 5.33** serves as the map for this

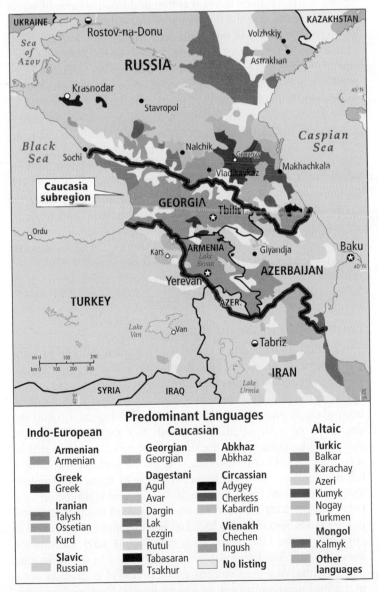

Predominant Languages

Indo-European

Armenian
Armenian

Greek
Greek

Iranian
Talysh
Ossetian
Kurd

Slavic
Russian

Caucasian

Georgian
Georgian

Dagestani
Agul
Avar
Dargin
Lak
Lezgin
Rutul
Tabasaran
Tsakhur

Abkhaz
Abkhaz

Circassian
Adygey
Cherkess
Kabardin

Vienakh
Chechen
Ingush

No listing

Altaic

Turkic
Balkar
Karachay
Azeri
Kumyk
Nogay
Turkmen

Mongol
Kalmyk

Other languages

FIGURE 5.33 The Caucasia subregion. This map illustrates the diversity of ethnolinguistic groups in Caucasia and its adjacent, culturally related neighbors.

subregion.) Culturally and physically, this region includes a piece of the Russian Federation on the northern flank of the mountains as well as the three independent states of Georgia, Armenia, and Azerbaijan, which occupy the southern flank of the mountains in what is known as *Transcaucasia*. Transcaucasia is a band of subtropical intermountain valleys and high volcanic plateaus that drop to low coastal plains near the Black and Caspian seas (see Figure 5.1).

In this mountainous space the size of California, live more than 50 ethnic groups—Armenians, Chechens, Ossetians, Karachays, Abkhazians, Georgians, and Tatars, to name but a few—most speaking separate languages. The groups vary widely in size, from a few hundred (for example, the Ginukh in Dagestan) to more than 6 million (the Turkic Azerbaijanis). Some are Orthodox Christians, many are Muslims, a few are Jews, and some retain ancient elements of local animistic religions. All these groups, including Chechens and others who live in Russian Caucasia (see the discussion on political conflict in Caucasia), are remnants of ancient migrations. For thousands of years, Caucasia was a stopping point for nomadic peoples moving between the Central Asian steppes, the Mediterranean, and Europe. Other ethnic enclaves were created more recently by the Soviet-instigated relocation and then return of minorities (for example, the Chechens). The region's history as a global meeting place, together with the mountainous topography that has fostered isolated self-reliance, has resulted in the ethnolinguistic patchwork that is Caucasia. Today, many Caucasians maintain ties to Europe, Russia, and North America, where hundreds of thousands of emigrants from the region live. The trend of emigration intensified after the Communist era, which has led to the depletion of an able-bodied workforce and brain drain; but it has also had some positive effects, as those living abroad have sent remittances back to people in the Caucasian countries.

As a result of external political maneuvering and boundary making, an ethnic group may now have members in several different Caucasian states or in the internal republics of Russia. Because of their strong ethnic loyalties, this pattern can result in persistent conflict **(FIGURE 5.34)**. After the Soviet collapse, the three culturally distinct states of Georgia, Azerbaijan, and Armenia each gained independence. Unlike in most other parts of the former Soviet Union, the move to independent status resulted in ethnic warfare. Very quickly, several ethnic groups within these already tiny states took up arms to obtain their own independent territories. The Abkhazians and the South Ossetians took up arms against Georgia. The Christian Armenians in Nagorno Karabakh (an exclave of Armenia located inside Azerbaijan) fought against the Muslim majority of Azerbaijan; 15,000 people died before a truce was signed in 1994. On the Russian side of the border, separatists in Chechnya took up arms for political independence. Although the ethnic strife across Caucasia has some local causes, it has often been instigated by the greater powers that surround Caucasia—Turkey, Russia, Iran—who are focused on their own national agendas: access to strategic military installations or to agricultural and mineral resources. The historic animosity between the nations of Armenia and Turkey is particularly prominent; it dates back to the mass killings of Armenians in Turkey during World War I, a

FIGURE 5.34 Areas of contention in Caucasia.

tragic event that Armenia wants Turkey to accept, unsuccessfully so far, as genocide.

Access to the region's significant oil and gas reserves has recently been another source of conflict in Caucasia. Especially getting the oil and gas safely into the world market is a major problem (see Figure 5.14). Pipelines, trucks, and ocean tankers all have to pass through contested territory or across difficult terrain. The routing of necessary infrastructure involves considering which countries are at odds with each other and which are allies. Regional governments and Western oil companies have built a pipeline from Azerbaijan (where most of the oil is located) to the Black Sea, where the oil can easily be shipped to Europe. Transit through Chechnya is problematic because of Chechnya's conflict with Russia. There are now pipelines through Caucasian Russia, and new ones are under construction. A pipeline also carries oil from Azerbaijan through Georgia to Turkey's Mediterranean coast.

THINGS TO REMEMBER

• Caucasia, a mountainous subregion about the size of California, is home to more than 50 ethnic groups, several belief systems, three independent countries, and seven Russian internal republics in the northern reaches of the Caucasus Mountains.

• A number of political conflicts emerged in the post-Soviet era. Various ethnic groups and governments have laid claim to territories across the region; some of these conflicts remain unresolved.

• The ownership of the Caspian Sea petroleum riches is contested, and transporting the petroleum is also challenging because the routes both to internal and foreign markets—especially Western markets—involve passing the petroleum through contested space.

THE CENTRAL ASIAN STATES

Since 1991, following nearly 12 decades of Russian colonialism, five independent nations have emerged in Central Asia: Kazakhstan, Uzbekistan, Turkmenistan, Kyrgyzstan, and Tajikistan **(FIGURE 5.35)**. Each of these new nations has traditions that are recognizably different from the others, yet all draw on the deep, common traditions of this ancient region.

Physical Setting

Central Asia lies in the center of the Eurasian continent, in the rain shadow of the lofty mountains that lie to the south in Iran, Afghanistan, and Pakistan (see Figures 5.1 and 5.5). Its dry continental climate is a reflection of this location. What rain there is falls mainly in the north, on the Kazakh steppes (see Figure 5.5B), where wide grasslands now support limited grain agriculture and large herds of sheep, goats, and horses. In the south are deserts crossed by rivers that carry glacial meltwater from the high peaks still farther to the southeast. In Soviet times, these rivers were tapped to irrigate huge fields of cotton and smaller fields of wheat and other grains (see the discussion of irrigation and the Aral Sea on pages 268–270). In southern settlements, houses have food gardens walled to protect against drying winds. These gardens nurture many types of melons, herbs, onions and garlic, and tree crops—plums, pistachios, apricots, and apples—all of which were first domesticated in this region in ancient times.

Uzbekistan and Turkmenistan are largely low-lying plains. Kazakhstan has plains in the north and uplands in the southeast that grade into high mountains. Kyrgyzstan and Tajikistan lie high in the Hindu Kush, the Pamir, and Tien Shan mountains to the southeast of Kazakhstan. These two countries have exceedingly rugged landscapes: Tajikistan's elevations go from near sea level to more than 22,000 feet (6700 meters), and those in Kyrgyzstan are similar. Both offer little in economic potential; Tajikistan is the poorest country in the entire region and has undergone numerous changes in government and a 1990s civil war involving many political and religious factions, from which the country has not yet recovered.

Central Asia in World History

Civilization flourished in Central Asia long before it did in lands to the north. The ancient Silk Road, a continuously shifting ribbon of trade routes connecting China with the fringes of Europe, operated for thousands of years, diffusing ideas and technology from place to place (see Figure 5.9; see also Figure 1.6 and the discussion of a hypothetical Central Asia region). Vestiges of nomadic life, common in the days of the Silk Road, can still be seen from Uzbekistan to Siberia. The *yurt*, a portable dwelling of felt on a collapsible wood frame, is a type of domestic structure that has been used for more than 2000 years and is still being used today **(FIGURE 5.36)**. Modern versions of ancient trading cities still dot the land; Samarkand, which has been dubbed a "crossroad of cultures" by the UN, is one such city **(FIGURE 5.37)**. Commerce along the Silk Road diminished as trade shifted to sea lanes after 1500. Central Asia then entered a long period of stagnation until, in the mid-nineteenth century, czarist Russia developed an interest in the region's major export crop, cotton.

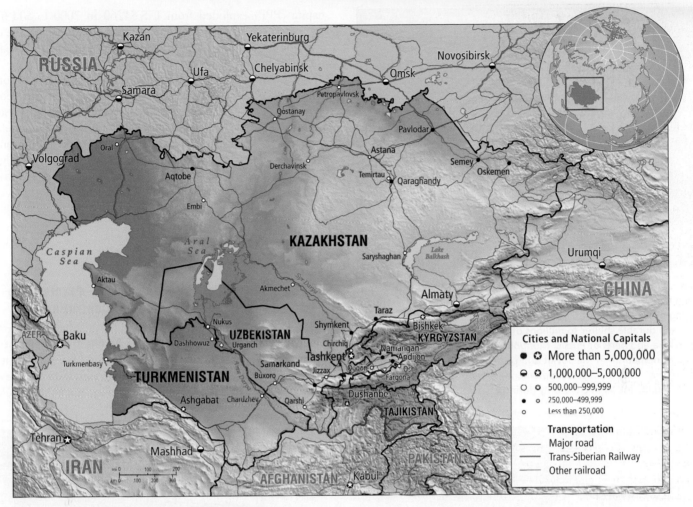

FIGURE 5.35 The Central Asian States subregion.

Russian imperial agents built modern mechanized textile mills to replace small-scale textile firms that had employed people to do traditional cotton, wool, and silk weaving **(FIGURE 5.38)**. As part of the Russification process, the new mills employed imported Russians and were often located in newly built Russian towns alongside railroad lines. Consequently, these industries benefited few Central Asians. In rural areas, cotton fields often replaced the wheat fields that had fed local people. Occasional famines struck whenever people could not afford imported food or when food supplies were interrupted.

Russian domination intensified in the Soviet era, after 1918. Only a few Central Asian political leaders actually joined the Communist Party; those who did were thoroughly Russified. Kazakhstan, with a long border with Russia, is the most Russified of the Central Asian states today; almost a quarter of the population is ethnically Russian, the Russian language has official status in the country, and Orthodox Christianity is the second most popular religion after Islam. Islam—dominant in the other Central Asian states—was officially tolerated during the Soviet era but was undermined by the Communist government's promotion of atheism. Fearing that the open practice of Islam would encourage pan-Islamism and weaken Russia's hold on the area, Russia restricted Islamic practices, such as access to mosques, pilgrimages to Makkah (Mecca), and Muslim-influenced dress.

For Central Asia, the transition since independence in 1991 has been rocky. Some conditions of daily life have deteriorated significantly. Trade patterns were interrupted; factories closed; transportation and services, never well developed, declined; and poverty spread as health standards fell. Although elections are held and free markets are opening up opportunities for entrepreneurship, the government remains authoritarian and patriarchal, and the heads of state tend to be autocratic. Elections are routinely hijacked by fraudulent vote counts. Most Central Asian governments, though nominally Muslim themselves, fear Islamic fundamentalism and repress even moderate Muslims. Three of the countries—Turkmenistan, Uzbekistan, and Tajikistan—border volatile Afghanistan. The conflict in Afghanistan has the potential to create a "spillover" effect of religious strife into Central Asia.

Russia has tried to maintain a strong influence on Central Asia, but international interest in the region's oil and gas resources is challenging that influence. In Kazakhstan especially, large oil and gas reserves are driving rapid economic growth. GDP per

capita (PPP) exploded from U.S.$1260 in 2000 to $11,670 in 2014, and growth may continue in the future, based on the development of new oilfields in western Kazakhstan near the Caspian Sea and the construction of new pipelines (see Figure 5.14). Since the Soviet days, Kazakhstan's existing pipeline network has connected with Russia, but a new pipeline also links the country to China. Part of Kazakhstan's new wealth is being invested in constructing a lavish new capital, Astana (see the vignette on page 289). Oil wealth is also beginning to make a difference in human well-being, as is reflected in Kazakhstan's relatively high HDI rank, which is similar to that of Russia and significantly above those of the other Central Asian states.

FIGURE 5.36 The yurt: A celebrated but fading way of life. A Kazakh herding family compound, centered on the yurt, the traditional dwelling of the steppes of Central Asia. Yurts once provided mobile, warm, and durable shelter to millions of nomadic people throughout Central Asia and parts of Russia, Mongolia, and western China. The decline of nomadism in the Soviet and post-Soviet era has meant that yurts are now used mainly as supplementary living space in rural areas, and are usually found in cities only during festivals.

ON THE BRIGHT SIDE: The Free Market—Persistence of an Ancient Central Asian Invention

The Central Asian states appear to be finding their capitalist "legs" fairly quickly. As communism fades and the Russians leave, Central Asians are taking up old, market-based skills that have been part of their heritage since the ancient Silk Road days, when

FIGURE 5.37 Samarkand, Uzbekistan. Samarkand has been designated a World Heritage Site by UNESCO. In the center of the city is the Registan complex, which is a public plaza flanked by three Islamic schools (madrassas) that were built between 1400 and 1600. Samarkand was an important point along the Silk Road.

FIGURE 5.38 **Women weaving silk carpets in Margilan, Uzbekistan.** Carpets have been produced across Central Asia for thousands of years, usually from the wool of sheep. They carry a great deal of symbolism and ethnic distinctiveness and vary in color and design from place to place. Carpets probably originated as portable furnishings for nomadic peoples and are used as lining for yurts, for sleeping and sitting on, and keeping warm in general.

FIGURE 5.39 **A market in Central Asia.** The marketplace in Osh, Kyrgyzstan, bustles with the activity of shoppers and entrepreneurs, much as have similar places throughout Central Asia for thousands of years.

trade and long-distance travel were the heart of the economy. The Chinese government is pursuing a new "Silk Road strategy" that is intended to reignite these economic connections. A largely contraband trade in consumer goods (clothes, electronics, cars) and illegal substances (from drugs to guns) with Iran, Afghanistan, and western China is already blossoming. Through microcredit (small, low-interest loans to the poor—see Chapter 8), women are becoming involved in entrepreneurism. Opportunities to render services such as food vending, auto repair, transportation, and accommodations to tourists and traveling business-people are opening up for workers displaced in the transition (FIGURE 5.39). Meanwhile, all five countries are negotiating with competing multinational oil companies to market Central Asian oil and gas to their best advantage. ■

THINGS TO REMEMBER

• The five newly independent (since 1991) countries in predominantly Muslim Central Asia remain relatively poor and are governed by autocratic leaders, although a fossil fuel–based economy, especially in Kazakhstan, has generated economic growth and development.

• While Russia tries to maintain some control over these former territories, primarily through economic pressure, the countries are reconnecting with nearby Asian nations and are also taking up old, market-based skills that have been part of their heritage since the ancient Silk Road days, when trade and long-distance travel were the heart of the economy.

GEOGRAPHIC THEMES: Russia and the Post-Soviet States: Review and Self-Test

1. Environment: Economic development has taken precedence over environmental concerns in Russia and the post-Soviet states, resulting in numerous environmental problems. Economic activities in the region have expanded since the 1990s economic collapse, and contribute to climate change. Food production systems and water resources have medium to high levels of vulnerability to the impacts of climate change.

• How did Soviet authorities regard the problem of urban and industrial pollution?

• Describe how cities in this region are contaminated by pollution, and describe the sources of pollution.

• What factors led to the shrinking of the Aral Sea?

• Why is Central Asia especially vulnerable to climate change?

2. Globalization and Development: After the fall of the Soviet Union, wealth disparity increased and jobs were lost as many Communist-era industries were closed or sold to the rich and well connected. The region is now largely dependent on its role as a leading global exporter of energy resources.

• Who has gained the most from recent economic growth in Russia?

• Discuss how Russia's natural resources are transforming the role of Russia in the global economy and in regional politics.

• How does Central Asia's geographic situation affect the export of its energy resources?

• Why is the informal sector of the economy important in this region?

• Where does most of the agricultural output of the region come from? Compare that to the population distribution of the region. What are the resulting patterns of food exports and distribution?

3. Power and Politics: This region has a long history of authoritarianism. Even if the public now desires expanded political freedoms, there are few opportunities to influence the political process. Elected representative bodies often act as rubber stamps for strong presidents and exercise only limited influence on policy making.

• Why might the conflict in Ukraine be seen as a revival of Cold War tensions? In what ways is it different from Cold War

conflicts? How is the conflict related to the internal ethnic geography of Ukraine?

• How has Russia's history of expansion into neighboring lands influenced the political geography of the region? What are the social, political, and economic impacts of Russification policies?

• What impact did Vladimir Putin's rise to power have on the media in Russia?

4. Urbanization: A few large cities in Russia and Central Asia, such as Moscow, are growing fast, as they are primate cities or are fueled by the expansion of energy exports. Elsewhere, many cities are suffering from a lack of investment as their economies struggle in the post-Soviet era.

• How are the reasons for housing shortages today different from those of the Soviet era?

• Which capitals of this region are primate cities? How does the primate city phenomenon affect these countries?

• How is Communist-era central planning still evident in cities in the region?

• What factors hold back economic growth in the industrial cities located in Siberia and along the Pacific coast?

• What has fueled recent urbanization in Central Asia?

5. Population and Gender: Populations are shrinking in many parts of the region. High levels of participation by women in the workforce and the economic decline in the 1990s that followed the collapse of the Soviet Union resulted in low birth rates. Economic problems have also led to declines in life expectancy.

• What Soviet-era developments helped curtail population growth? What is being done in the post-Soviet era to encourage population growth?

• How does life expectancy vary according to gender in this region?

• How does alcohol use relate to male life expectancy in Russia?

• How and why has access to health care and other social services changed for so many in this region since the fall of the USSR?

• How have women participated in the workforce, from the Soviet era until today?

Critical Thinking Questions

1. Considering the importance on fossil fuel extraction for some countries in the region, discuss the pros and cons of fossil fuel from an economic development perspective. What are the economic and geopolitical implications of fluctuating oil prices?

2. In general terms, discuss how the change from a centrally planned economy to a more market-based economy has affected career options and standards of living for the elderly, for young professionals, for unskilled laborers, for members of the military, for women, and for former government bureaucrats.

3. Discuss the changing birth rate in Russia and other countries in the region. What are the major causes of change? What do you think the

impact of this decline can have on neighboring countries and regions? Where are similar declines happening elsewhere?

4. Given Russia's crucial role in World War II and the defeat of Germany, discuss what led to the Cold War between the Soviet Union and the West.

5. Discuss the religious affiliations of the different states and how those have impacted geopolitics in the region.

6. Ethnic and national identities are affecting how the internal republics of Russia envision their futures. How might the geography of Russia change if these feelings were to intensify? In what ways might the global economy be affected?

Chapter Key Terms

Bolsheviks 275
buffer state 304
capitalists 275
Caucasia 263
centrally planned, or socialist, economy 276
Cold War 276
communism 275
Communist Party 276
czar 274
Gazprom 278
glasnost 276

Group of Eight (G8) 280
Mongols 273
nomadic pastoralists 271
nonpoint sources of pollution 267
oligarchs 278
perestroika 276
permafrost 265
privatization 278
Russian Federation 264
Russification 285
Siberia 264

Slavs 273
Soviet Union 262
steppes 264
taiga 265
trafficking 296
tundra 265
underemployment 281
Union of Soviet Socialist Republics (USSR) 262

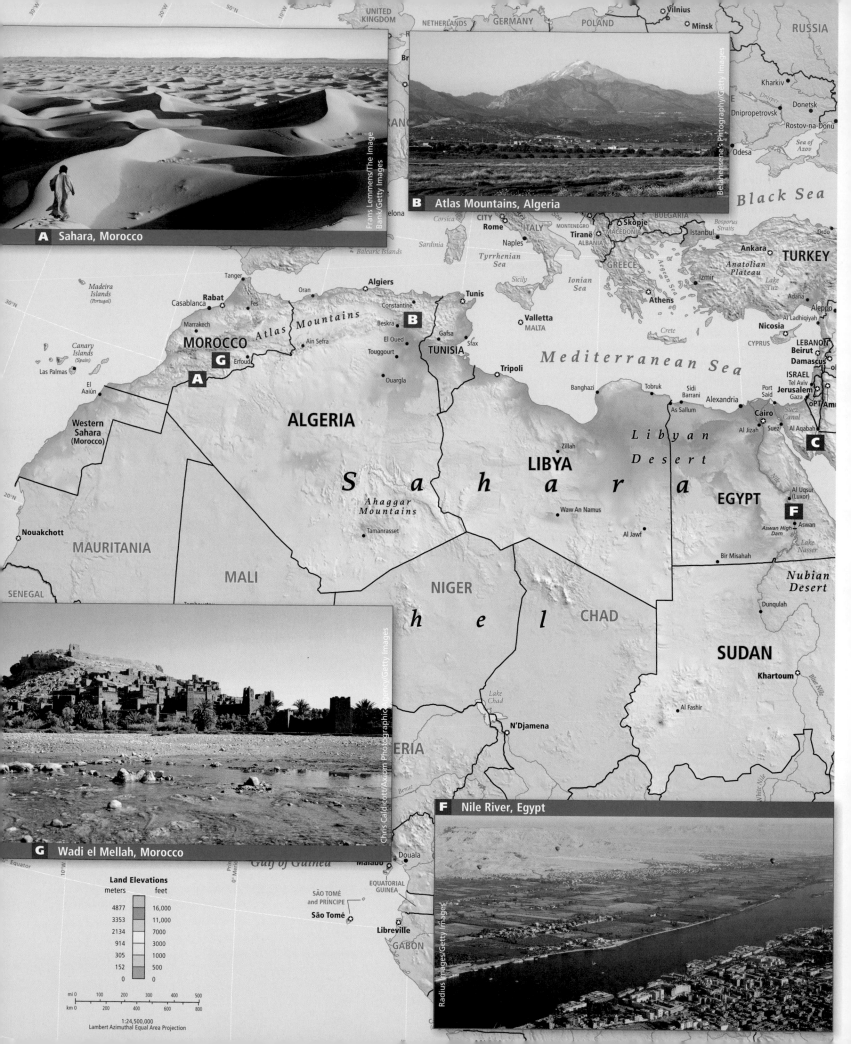

A Sahara, Morocco

B Atlas Mountains, Algeria

F Nile River, Egypt

G Wadi el Mellah, Morocco

6
North Africa and Southwest Asia

Map labels:

Volgograd

Caucasus

GEORGIA

ARMENIA · **Yerevan**

AZERBAIJAN · **Baku**

FranZ Aberham/Photographer's Choice RF/Getty Images

E Mount Ararat, Turkey

TURKMENISTAN · **Ashgabat**

Mt. Ararat elev. 16,946' ▲ **E**

Caspian Sea

Mashhad

Tabriz

Van

Lake Urmia

Elburz Mts.

AFGHANISTAN

SYRIA

Mosul

Tehran

Tabas

Kermanshah

IRAN

Esfahan

Kerman

Baghdad

Zagros Mts.

IRAQ

PAKISTAN

JORDAN

Abadan

Al Basrah

Shiraz

Bushehr

Bandar-e Abbas

Persian

Rafha

Kuwait

KUWAIT

Tabuk

Ad Dammam

BAHRAIN

Gulf

Strait of Hormuz

Bandar Beheshti

QATAR

Dubai

Gulf of Oman

Manamah

Abu Dhabi

Muscat

Doha

UNITED ARAB EMIRATES

Arabian

Riyadh

Peninsula

OMAN

20°N

Al Madinah (Medina)

SAUDI ARABIA

D

Al Jawarah

Jiddah

Makkah (Mecca)

Rub' al Khali

Red

Port Sudan

Abha

Arabian Sea

Sea

YEMEN

ERITREA

Sana'a

Asmera

Taizz

Aden

Gulf of Aden

Lake Tana

DJIBOUTI · **Djibouti**

Somali Peninsula

Horn of Africa

D Rub'al Khali dunes, Saudi Arabia

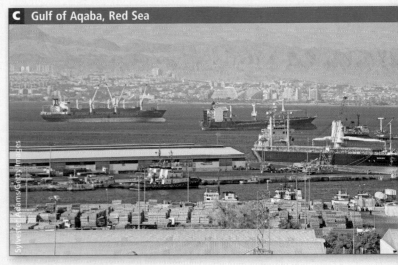

C Gulf of Aqaba, Red Sea

Sylvester Adams/Getty Images

FIGURE 6.1 Regional map of North Africa and Southwest Asia.

Malcolm MacGregor/AWL Images/Getty Images

0° Equator

Nairobi

INDIAN

OCEAN

Victoria

SEYCHELLES

ANIA

Dodoma

313

GEOGRAPHIC THEMES: North Africa and Southwest Asia

After you read this chapter, you will be able to discuss the following issues as they relate to the five thematic concepts:

1. Environment: The population of this predominantly dry region is growing fast, while access to cultivable land and water for agriculture and other human uses is limited. Therefore, many countries here are dependent on imported food. Climate change could reduce food output and make water, already a scarce commodity, a source of conflict. New technologies offer solutions to water scarcity but are expensive and unsustainable.

2. Globalization and Development: The vast fossil fuel resources of a few countries in this region have transformed economic development. In these countries, economies have become powerfully linked to global flows of money, resources, and people. Politics have also become globalized, with Europe and the United States strongly influencing the region.

3. Power and Politics: Authoritarian power structures prevail throughout much of North Africa and Southwest Asia. Beginning with the Arab Spring of 2010, waves of protests swept the region. The outcome of the Arab Spring has varied from reform to repression and civil war.

4. Urbanization Two patterns of urbanization have emerged in the region. In the oil-rich countries, there are spectacular new luxury-oriented urban areas. In the oil-poor countries, many people live in overcrowded urban slums with few services, and jobs for rural migrants are scarce.

5. Population and Gender: This region has the second-highest population growth rate in the world. Part of the reason for this is that women are generally poorly educated and tend not to work outside the home. Childbearing thus remains crucial to a woman's status, a situation that encourages large families.

The North Africa and Southwest Asia Region

The region of North Africa and Southwest Asia (shown in **FIGURES 6.1** and **6.2**) contains 21 countries plus the occupied Palestinian Territories. Physically, the region is part of two continents: Africa and Eurasia. The North African countries of this region stretch from Western Sahara and Morocco on the Atlantic Ocean, through Algeria, Tunisia, Libya, and Egypt along

> **Islam** a monotheistic religion that emerged in the seventh century C.E. when, according to tradition, God revealed the tenets of the religion to the Prophet Muhammad
>
> **Islamism** a grassroots religious revival in Islam that seeks political power to curb what are seen as dangerous secular influences; also seeks to replace secular governments and civil laws with governments and laws guided by Islamic principles

the Mediterranean Sea, plus Sudan, which lies south of Egypt. Eurasia begins on the eastern side of the Red Sea with the Arabian Peninsula, which consists of Saudi Arabia, Yemen, Oman, the United Arab Emirates, Qatar, Bahrain, and Kuwait (see Figure 6.1C); the countries of the Eastern Mediterranean *littoral* (shoreline), including Jordan, Israel, the occupied Palestinian Territories, Lebanon, and Syria; and Turkey,

Iraq, and Iran. This region is often referred to as the *Middle East*, a term we do not use in this book for two reasons. First, it arose during the colonial era and describes the region from a limited Euro-American geographic perspective. To someone in China or Japan, the region lies to the far west, and to someone in Russia, it lies to the south. Second, the term Middle East, while geographically imprecise, typically refers to the area east of the Mediterranean but does not include all the countries in North Africa and Southwest Asia.

The five thematic concepts in this book are explored as they arise in the discussion of regional issues; the interactions between two or more themes are often featured. Vignettes, like the one on page 316 about the changing role of women in this region, illustrate one or more of the themes as they are experienced in individual lives.

The Arab Spring, a movement that will likely continue evolving for decades, was brought on by deep and structural problems. Large numbers of the poor, the uneducated, the educated unemployed, minorities, and women have been pushed further and further into poverty and powerlessness by political and economic systems that have privileged the wealthy and politically connected. Movements like the Arab Spring arise when enough people decide they can no longer tolerate these systems.

The issue of gender equity in North Africa and Southwest Asia is controversial because some traditionalists in the region are convinced that only Westerners raise this issue, and that they do so in order to belittle Muslim culture for having a set of values that places females in a special, and subservient, category. But Mona Eltahawy's article (see page 316) and the response it has received across the region suggest that gender equity is becoming a central theme in the Arab Spring movement and that women are leading a profound transformation of this region.

What Makes North Africa and Southwest Asia a Region?

To most outsiders, North Africa and Southwest Asia is a region characterized by five qualities: It is the center of the religion of Islam; it holds a great deal of Earth's petroleum reserves; water is generally scarce here; Arab culture dominates most countries; and women are discriminated against more intensely than in most of the rest of the world. While these five features are all present, they alone provide a far too simplistic picture of the region.

The vast majority of people practice **Islam,** a monotheistic religion that emerged between 601 and 632 C.E., which is when the founder of the religion, the Prophet Muhammad, received revelations from God, according to the believers. Islam is a faith that is interpreted in many different ways. Most Muslims are moderate in their thinking and accept the validity of other beliefs, especially that of Christianity; for example, Jesus (a central figure in Christianity) plays an important role in Islam. Some Muslims are drawn to ultra-fundamentalist versions of Islam known as **Islamism.** The term Islamism refers to a Muslim religious activist movement that seeks to curb secular influences that are spreading as a result of globalization. Some Islamists think that Western culture has a corrupting influence on Islamic countries, while others are more open to a diversity of ideas. The global public opinion institute Pew Research Center reports that people in many

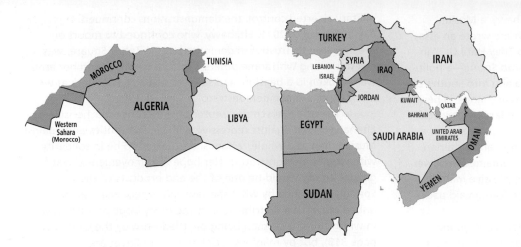

FIGURE 6.2 Political map of North Africa and Southwest Asia.

Muslim countries strongly support religious pluralism and the freedom to worship, while also embracing Islamic principles as a basis for legislation, somewhat contradictory findings. However, when it comes to violent, radical groups such as Al Qaeda, the Taliban, and the Islamic State, only 13 to 15 percent of Muslims surveyed around the world support them.

Fossil fuel reserves—made up of oil, coal, and natural gas formed over millions of years from the fossilized remains of dead plants and animals—are highly uneven in their global distribution (see Figure 6.18). Fossil fuel resources include all existing fossil fuels, but only those that can be extracted under current technological and economic conditions are considered reserves. In North Africa and Southwest Asia, oil and natural gas reserves are mainly located and extracted around the Persian Gulf. These fossil fuels are extracted and exported throughout the world at tremendous profit, but the profits are not equitably distributed to the people in the countries from which they are generated. For example, in the oil-rich areas around the Persian Gulf, profits have traditionally gone to the members of a few large, privileged families, and only modest amounts are spread to the rest of the citizens. More recently, however, a number of these **Gulf states** have invested in public services, education, and infrastructure to promote a higher level of well-being among most people. Countries without significant fossil resources remain poor, which has resulted in a region characterized by uneven economic development.

This region is generally arid, but the degree of aridity and its impact on development vary widely. Newly revealed water resources that are found deep underground may change development options.

Arab culture and language is widespread, but many people in the region are not Arab. The population of Turkey and Iran, the second and third most populous countries in the region, are of non-Arab ethnicities, as are many minority populations, such as the Kurds, Berbers, and Jews. Some Christians in the region are ethnically Arabs, while others are not.

Finally, the role and status of women, a point of contention for many years, are in transition. The experiences of women vary widely from country to country. In Turkey, women have many more options for education and work than in Saudi Arabia or Yemen. Gender issues also vary between more rural and urban areas. Urban women tend to be more educated, outspoken, and active in commerce, public life, and government, while rural women tend to lead more secluded domestic lives with few educational opportunities. The movement toward greater gender equity is generally strongest in urban areas.

Terms in This Chapter

In this book, as mentioned earlier, we choose to not use the common term Middle East. The term *Arab world* is used only where it applies, since many people in the region are not of Arab ethnicity.

We use the term **occupied Palestinian Territories (oPT)** to refer to Gaza and the West Bank, those areas where Israel still exerts control despite treaty agreements; the word "occupied" is lowercase to show the supposed temporary quality of the occupation. The U.S. Department of State uses the term *Palestinian Territories*. The United Nations, upon recently agreeing to give the territory observer status, now refers to it as the *State of Palestine*.

The Arabic language uses a different alphabet than English, which means that Arabic words and place names have to be represented with English letters. The goal is to use a spelling that is understandable to English-speakers and as close as possible to the original pronunciation. However, such transliteration—the representation of the sounds of another language—is not a straightforward process. In fact, different spellings of the same word are not unusual. For example, in this book we use Qur'an (instead of Koran) and Makkah (instead of Mecca).

One related problem is what to call the Islamic State, the extremist political movement that operates in Iraq and Syria. The group's full Arabic name is long and complex, which means that various shorter names and abbreviations have appeared in English, including ISIS, ISIL, and Daesh. We use the term *Islamic State* because it is probably the most common, but even that name is problematic. By calling itself the Islamic State, the group claims to represent Islam and "State" implies that it has aspirations to control and govern a larger territory. Using the term Islamic State in the text is obviously not an endorsement of the group's agenda.

fossil fuel a source of energy formed from the remains of dead plants and animals; fossil fuel includes oil, coal, and natural gas

Gulf states countries that border the Persian Gulf: Saudi Arabia, Kuwait, Bahrain, Oman, Qatar, and the United Arab Emirates

occupied Palestinian Territories Palestinian lands occupied by Israel since 1967

GLOBAL PATTERNS, LOCAL LIVES Mona Eltahawy, a Muslim Egyptian-American journalist, was angry when she wrote an essay for *Foreign Policy* magazine entitled "Why Do They Hate Us?" in the spring of 2012. It is easy to see why. The *Arab Spring*—a political movement that washed across North Africa and into Southwest Asia, beginning in 2010—held the promise of real change in her native land. For a time, the tens of thousands of largely peaceful male and female demonstrators in Cairo's Tahrir Square shared a common purpose: the resignation of the country's authoritarian president, Hosni Mubarak (FIGURE 6.3). He had remained in power for nearly 30 years, aided by a powerful and oppressive military, and supported by internal corruption and large foreign aid packages from the United States.

For the many thousands of women—students and young professionals—who participated in the Cairo demonstrations, it was exhilarating to be welcomed by male compatriots, many of whom were members of the Muslim Brotherhood, a large and amorphous organization of men opposed to the corrupt Mubarak regime and its affiliations with the West. The Muslim Brotherhood, however, is also known for its stance against liberalizing women's rights. For a time, the common purpose of ousting Mubarak united everyone.

Eltahawy, who grew up in Saudi Arabia, the United Kingdom, and Egypt, and who has focused her career on social justice and women's rights in Egypt, joined the throng of demonstrators. As the protests went on, not only did retaliation by the military and police toward demonstrators become increasingly brutal, but antagonism toward women by the Muslim Brotherhood demonstrators and even from young, more liberal male demonstrators grew, leading to random violence against women. Mubarak's resignation in February 2011 left undetermined the way in which a new government would be established; so as the

military asserted control, the demonstrations continued. One day in the fall of 2011, Eltahawy, who continued to report on the increasingly frustrated demonstrators in Tahrir Square, was arrested along with some others. While in custody, both her arms were broken in a beating, and she was sexually assaulted, as were many other women demonstrators.

In her essay, which she wrote some months after her release, Eltahawy raged against repressive political authorities and against Muslim men as a whole, whom she expected to be in solidarity with women demonstrators. Her hope that movement toward gender equity would be one of the end products of the Arab Spring was dashed by what she saw and experienced, so she lashed out with a diatribe against the many large and small ways in which she saw women being belittled—not by the Qur'an (see page 318), but by religious social practice in Egypt and elsewhere. In thus expressing her rage, Eltahawy seems to have invigorated a cross-regional discussion on gender equity that is attracting serious male and female participants. They are expressing a surprising level of agreement that the Arab Spring must address the human rights of women. *[Source: Foreign Policy, Al Jazeera. For more detailed information, see Text Sources and Credits.]* ∎

FIGURE 6.3 Women and the Arab Spring. Women have been very active in protests in Egypt and elsewhere in the region in recent years, such as the protester shown here during a 2014 rally in Istanbul. Certain groups, including members of Egypt's government, have felt threatened by this. Some female protesters have become a target for sexual assaults designed to make others perceive them as dangerous, "dirty," or immoral.

Ozan Kose/AFP/Getty Images

> **THINGS TO REMEMBER**
>
> • The Arab Spring, a broad decentralized movement against corruption and toward more political freedom, has spread across the region since 2010, but with limited success.
>
> • The vast majority of the people in the region practice Islam. Islamism, an activist movement to curb secular influences in society, has gained strength.
>
> • The extraction of fossil fuel has increased wealth in some countries, especially those around the Persian Gulf, while other countries remain poor. The result is a region characterized by uneven economic development.
>
> • Women are increasingly participating in political movements, and issues of gender equity are important in the public debate.

PHYSICAL GEOGRAPHY

Landforms and climates are particularly closely related in this region. The climate is dry and hot in the vast stretches of the relatively low, flat land; it is somewhat moister where mountains capture orographic rainfall. The lack of vegetation perpetuates aridity. Without plants to absorb and hold moisture, the rare but occasionally copious rainfall simply runs off, evaporates, or sinks rapidly into underground aquifers, of which there are many across the region.

CLIMATE

No other region in the world is as dry as North Africa and Southwest Asia (FIGURE 6.4A). A belt of dry air that circles the planet between roughly 20° N and 30° N creates desert climates in the Sahara of North Africa, the Eastern Mediterranean, the Arabian Peninsula, Kuwait, Iraq, and Iran.

FIGURE 6.4 PHOTO ESSAY: Climates of North Africa and Southwest Asia

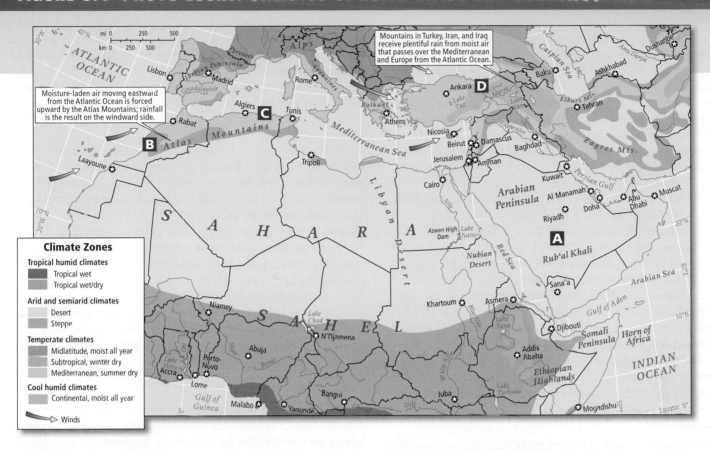

Mountains in Turkey, Iran, and Iraq receive plentiful rain from moist air that passes over the Mediterranean and Europe from the Atlantic Ocean.

Moisture-laden air moving eastward from the Atlantic Ocean is forced upward by the Atlas Mountains; rainfall is the result on the windward side.

Climate Zones

Tropical humid climates
- Tropical wet
- Tropical wet/dry

Arid and semiarid climates
- Desert
- Steppe

Temperate climates
- Midlatitude, moist all year
- Subtropical, winter dry
- Mediterranean, summer dry

Cool humid climates
- Continental, moist all year

→ Winds

A Desert, Saudi Arabia

B Steppe, Morocco

C Mediterranean, summer dry, Algeria

D Continental, moist all year, Turkey

The Sahara's size and location under this high-pressure belt of dry air make it a particularly hot desert region. In some places, temperatures can reach 130°F (54°C) in the shade at midday (see Figure 6.1A). With little cloud cover, water, or moisture-holding vegetation to retain heat, nighttime temperatures can drop quickly to below freezing. Nevertheless, in even the driest zones, humans survive as traders and nomadic herders at scattered oases, where they maintain groves of drought-resistant plants such as date palms. Desert inhabitants often wear light-colored, loose, flowing robes that reflect the sunlight, retain body moisture during the day, and provide warmth at night.

In the uplands and at the desert margins, enough rain falls to nurture grass, some trees, and limited agriculture. Such is the case in northwestern Morocco; along the Mediterranean coast in Algeria, Tunisia, and Libya; in the highlands of Sudan, Yemen, and Turkey; and in the northern parts of Iraq and Iran (see Figure 6.4B–D). The rest of the region, generally too dry for cultivation, has for generations been the prime herding lands for nomads, such as the Kurds of Southwest Asia, the Berbers and Tuareg in North Africa, and the Bedouin of the steppes and deserts on the Arabian Peninsula. Recently, most nomads and their descendants have settled in farming communities or urban places. Some of their lands are now irrigated for commercial agriculture, but the general aridity of the region means that sources of irrigation water (rivers and aquifers) are scarce.

LANDFORMS AND VEGETATION

The rolling landscapes of rocky and gravelly deserts and steppes cover most of North Africa and Southwest Asia (see the Figure 6.1 map and Figure 6.4). In a few places, mountains capture sufficient moisture to allow plants, animals, and humans to flourish. In northwestern Africa, the Atlas Mountains, which stretch from Morocco on the Atlantic coast to Tunisia on the Mediterranean coast, lift damp winds from the Atlantic Ocean, creating orographic rainfall of more than 50 inches (127 centimeters) per year on windward slopes (see Figure 6.1B and C). In some Atlas Mountain locations, there is enough snowfall to support a ski industry.

A rift that formed between two tectonic plates—the African Plate and the Arabian Plate—separates Africa from Southwest Asia (see Figure 1.8). The rift, which began to form about 12 million years ago, is now filled by the Red Sea. The Arabian Peninsula lies to the east of this rift. Mountains bordering the rift in Yemen rise to 12,000 feet (3658 meters). They capture enough rainfall to sustain agriculture, which has been practiced there for thousands of years. Irrigation has been used for millennia to increase yields.

Behind these mountains to the northeast lies the great desert region of the Rub'al Khali (which means "the empty quarter"). Like the Sahara, it has virtually no vegetation. The desert is generally flat although sand dunes of the Rub'al Khali, which are constantly moved by strong winds, are among the world's largest, some as high as 2000 feet (610 meters; see Figure 6.1D).

The landforms of Southwest Asia are more complex than those of North Africa. The Arabian Plate is colliding with the Eurasian Plate

Qur'an (or **Koran**) the holy book of Islam, believed by Muslims to contain the words God revealed to Muhammad

and pushing up the mountains and plateaus of Turkey and Iran (see Figure 1.8). Turkey's mountains lift damp air that passes over Europe and the Mediterranean from the Atlantic, resulting in considerable rainfall (see Figure 6.1E). The rain makes it to the mountains of western Iran, but the farther east one goes, the drier it gets. The tectonic movements that create mountains also create earthquakes, a common hazard in Southwest Asia, especially Turkey and Iran.

There are only three major river systems in the entire region, and all have attracted human settlement for thousands of years. The Nile flows north from the moist central East African highlands (see Figure 6.1F). It crosses arid Sudan and desert Egypt and forms a large delta on the Mediterranean. The Euphrates and Tigris rivers both begin with the rain that falls in the mountains of Turkey. The rivers then flow southeast to the Persian Gulf. A much smaller river, the Jordan, starts as snowmelt in the uplands of southern Lebanon and flows through the Sea of Galilee to the Dead Sea. Most other streams are for most of the year dry riverbeds, or *wadis*, carrying water only after the generally light rains that fall between November and April (Figure 6.1G shows a wadi in Morocco).

North Africa and Southwest Asia were home to some of the very earliest agricultural societies (see page 325). Today, rain-fed agriculture is practiced primarily in the highlands and along parts of the Mediterranean coast, where there is enough precipitation to grow citrus fruits, grapes, olives, and many vegetables, though supplemental irrigation is often needed. While irrigated agriculture is an age-old type of farming, it has become increasingly significant in modern times, both in the valleys of the major rivers, where seasonal flooding fills irrigation channels, and where aquifers are tapped by wells to provide water for cotton, wheat, barley, vegetables, and fruit trees.

ENVIRONMENT

GEOGRAPHIC THEME 1

Environment: The population of this predominantly dry region is growing fast, while access to cultivable land and water for agriculture and other human uses is limited. Therefore, many countries here are dependent on imported food. Climate change could reduce food output and make water, already a scarce commodity, a source of conflict. New technologies offer solutions to water scarcity but are expensive and unsustainable.

Environmental concerns are only beginning to become an overt focus in this region. In part, this is because for thousands of years people here have confronted the challenges of a naturally arid environment and have been quite attuned to its scarcities, which exceed those in most other places on Earth (FIGURE 6.5).

AN ANCIENT HERITAGE OF WATER CONSERVATION

The **Qur'an** (or **Koran**), the holy book of Islam, guides believers to avoid spoiling or degrading human and natural environments and to share resources, especially water, with all forms of life. In actual practice, the residents of this region conserve water better than most people in the world. Daily bathing is a religious requirement and is often done in public baths, where water use is minimized.

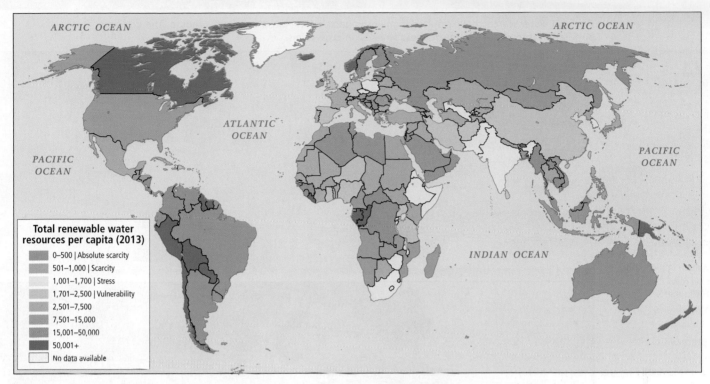

FIGURE 6.5 Total renewable water resources per capita in cubic meters, 2013. Most of the region experiences water stress or scarcity, which occur when so much water is withdrawn from rivers, lakes, and aquifers that not enough water remains to meet growing human and ecosystem requirements for sustainability.
[Source: *UN World Water Development Report 2015: Water for a Sustainable World*, p. 12.]

For millennia, mountain snowmelt has been captured and moved to dry fields and villages via constructed underground water conduits, called *qanats*, that minimize evaporation. Likewise, traditional architectural designs that maximize shade and airflow are used to create buildings that stay cool. This includes thick stone walls and small windows that keep out intense daytime heat, and wind towers that funnel breezes above the ground level into the first floor of buildings.

Despite their history of water-conserving technologies and practices, however, this region's 540 million residents now have such limited water resources that even clever combinations of ancient and modern measures are no longer sufficient to ensure an adequate supply of water. Growing populations, water pollution, and unwise modern usages of water virtually guarantee that water shortages will be more extreme in the future **(FIGURE 6.6A–C)**, especially with the likelihood of climate change–related drier conditions. As shown in the map in Figure 6.5, Turkey has the most plentiful water resources, followed by Iran, Iraq, and Syria. However, all other countries in the region suffer from physical *freshwater scarcity*, meaning that they have less water than the minimum the United Nations considers necessary to support basic human development—1000 cubic meters per person per year. The poorer countries in the region also suffer from economic water scarcity, which is when people or countries do not have the financial means to utilize existing water resources.

A recent reassessment of groundwater storage in North Africa by a consortium of British scientists revealed that deep aquifers in Algeria, Libya, Egypt, and Sudan may hold up to 100 times the water available through renewable freshwater resources

(FIGURE 6.7). This *fossil water*, deposited thousands of years ago, is unevenly distributed and so deep that special pumps are required to bring it to the surface. There probably isn't enough to meet the expected increases in demand for agricultural irrigation and drinking water for rapidly growing urban populations, but this stored groundwater may help alleviate scarcities in the short term.

WATER AND FOOD PRODUCTION

The predominant use of water in North Africa and Southwest Asia is for irrigated agriculture, even though agriculture does not contribute significantly to national economies. In Tunisia, for example, agriculture accounts for just 9.5 percent of GDP but 83 percent of all the water used. Only a small amount of available water is used for household and industrial purposes. Most national governments actually subsidize irrigated agriculture because they are worried about depending completely on imported food. Irrigated agriculture also supports jobs and family economies that are crucial to rural communities. Some of this region's food specialties appear in **FIGURE 6.8**.

Until the twentieth century, agriculture was confined to a few coastal and upland zones where rain could support cultivation, and to river valleys (such as the Nile, Tigris, and Euphrates valleys) where farms could be irrigated with simple gravity-flow technology. Some agriculture has also been possible in *oases*—small areas in the desert where vegetation can survive because groundwater is found near the land surface. However, to accommodate population growth and development, Libya, Egypt, Saudi Arabia, Tunisia, Syria, Turkey, Israel, and Iraq all now have ambitious mechanized irrigation schemes that have expanded

FIGURE 6.6 PHOTO ESSAY: Human Impacts on the Biosphere in North Africa and Southwest Asia

War and water scarcity have resulted in major impacts on the biosphere in this region. The growing population threatens to intensify these stresses on ecosystems.

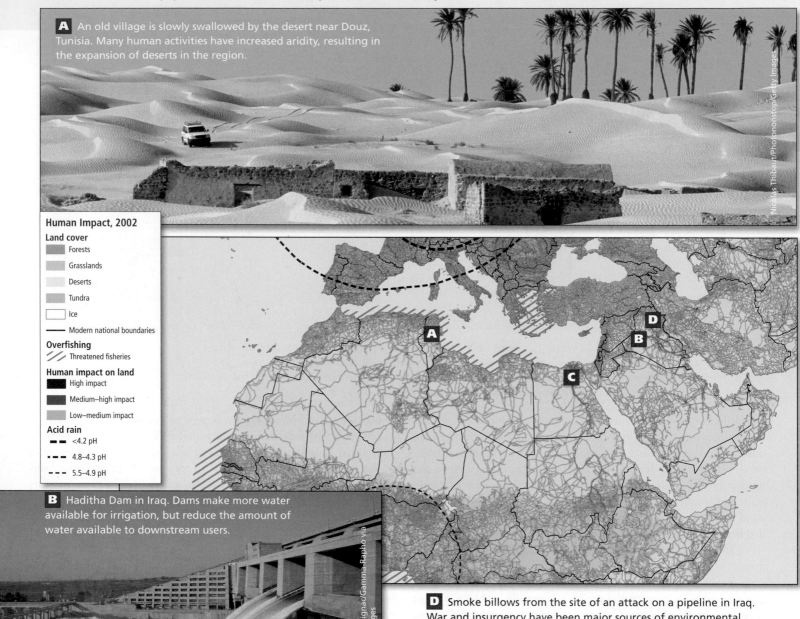

A An old village is slowly swallowed by the desert near Douz, Tunisia. Many human activities have increased aridity, resulting in the expansion of deserts in the region.

Human Impact, 2002

Land cover
- Forests
- Grasslands
- Deserts
- Tundra
- Ice
- —— Modern national boundaries

Overfishing
- /// Threatened fisheries

Human impact on land
- High impact
- Medium–high impact
- Low–medium impact

Acid rain
- ▬ ▬ <4.2 pH
- ▬·▬ 4.8–4.3 pH
- ▬ ▬ 5.5–4.9 pH

B Haditha Dam in Iraq. Dams make more water available for irrigation, but reduce the amount of water available to downstream users.

C Outside Alexandria, Egypt, an automated irrigation system moves through a potato field located on what would otherwise be desert land. The irrigation water arrives via canals from the nearby Nile Delta.

D Smoke billows from the site of an attack on a pipeline in Iraq. War and insurgency have been major sources of environmental degradation in this region. The largest oil spill in history occurred during the first Gulf War (1991) when the Iraqi government spilled 300 million gallons of oil into the Persian Gulf in order to thwart a land invasion by the United States. During the Gulf War, there were also oil fires, like the one in the picture, that burned for months and they had measurable impact on greenhouse gas emissions.

Thinking Geographically

After you have read about the impacts on the biosphere in North Africa and Southwest Asia, you will be able to answer the following questions.

B What about a dam creates soil moisture problems for downstream users of river water once the dam is operational?

C How might a dam like the Aswan on the upper part of the Nile River in Egypt contribute to the salinity and lowered fertility in lowland regions that were once naturally flooded but are now irrigated by pumped water?

agriculture deep into formerly uncultivable desert environments (see Figure 6.6 B, C).

Over time, irrigation projects damage soil fertility through **salinization.** When irrigation is used in hot, dry environments, pure water evaporates, leaving behind a salty residue of the minerals that exist in the water. When too much residue accumulates, the plants are unable to grow or even survive. This human-induced loss of fertility is one of the largest environmental water issues that the world faces today in arid and semiarid regions. Note that the process of salinization occurs naturally as well. Some of the deserts in the region are salt flats where, at some point in time, surface water has evaporated, resulting in a top layer of salty minerals.

Israel has developed relatively efficient techniques of *drip irrigation* that uses hoses and pipes that deliver water, drop by drop, to each plant with minimal runoff. This dramatically reduces the amount of water used, limits salinization, and frees up water for other uses. Until very recently, however, poorer states have been unable to afford this somewhat complex technology.

salinization a process where evaporation of water leaves salty minerals behind in the soil; may occur naturally or as a result of irrigation

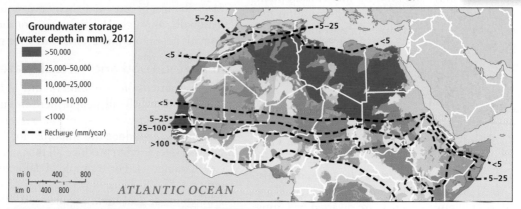

FIGURE 6.7 Water for the future? A map of newly discovered fossil groundwater deep below the deserts in North Africa. This ancient water deposit, already tapped in Libya, is expensive to extract and is unlikely to meet future needs because it is not being replenished. [*Source consulted:* A. M. MacDonald, H. C. Bonsor, B. E. O. Dochartaigh, and R. G. Taylor, "Quantitative Maps of Groundwater Resources in Africa," *Environmental Research Letters*, April 19, 2012, http://iopscience.iop.org /1748-9326/7/2/024009/pdf/1748-9326_7_2_024009.pdf.]

FIGURE 6.8 LOCAL LIVES: Foodways in North Africa and Southwest Asia

A A Moroccan *tagine* stew is prepared in a heavy clay pot (also called a tagine) with two parts: a round bottom and a cone-shaped lid with a hole in the top. The bottom holds the food, while the hole in the lid allows excess steam to escape, preventing the food from overcooking. The cone shape of the lid encourages condensation of moisture, thus reducing water loss and keeping flavors in the stew. Tagines are placed directly on hot coals or a grill and left to cook for 3 or more hours.

B Falafel is fried in Gaza, Palestine. Either fava beans or chickpeas are mixed with parsley, scallions, garlic, and spices and then formed into a ball or patty that is deep-fried. Falafel is often made into a flatbread sandwich and served with tomatoes, cucumbers, lettuce, and sauces based on tahini, a paste made from ground sesame seeds. The origin of this popular vegetarian street food, which has spread throughout this region and much of the world, is unclear. Egyptian Coptic Christians say they invented it as a substitute for meat during Lent. Others claim it originated in ancient Egypt, or even in South Asia.

C *Khubz*, or flatbread, is made in a *tannur*, a wood- or charcoal-fired oven. Possibly originating in ancient Sumer (today's Iraq), khubz consists of a dough of flour, water, and salt that is rolled flat and then pressed onto the interior side walls of the tannur, where the radiant heat from the fire rapidly bakes it. Khubz is so central to diets in this region that in recent decades most governments have subsidized it, keeping prices artificially low. Policies to remove subsidies have resulted in food riots among the urban poor, which can escalate to political uprisings.

VULNERABILITY TO CLIMATE CHANGE

North Africa and Southwest Asia are especially vulnerable to the multiple effects of climate change. Higher temperatures are already increasing evaporation rates where water scarcity has been a problem for many years (FIGURE 6.9B, C). These climatic changes, along with people's intensified use of water, are transforming more nondesert lands into deserts. Also, as the climate warms, a sea level rise of a few feet could severely impact the Mediterranean coast, especially the Nile Delta, one of the poorest and most densely populated lowland areas in the world (see Figure 6.9A), and the Persian Gulf coasts, where vast new high-tech cities that lie at sea level would likely be flooded (see Figure 6.24A). Conversely, under some climate-change scenarios, periods of unusually intense rainfall could increase flooding.

Independent of any environmental effects of climate change, global efforts to reduce dependencies on fossil fuel consumption could devastate oil- and gas-based economies and transform the region's geopolitics, leaving countries vulnerable to losing some of their global power and to having lower overall income levels. Despite the enormous wealth from fossil fuel sales that has flowed into some of these countries over the past 40 years, few countries have undertaken the significant economic diversification needed to prepare for the reduced global consumption of the fossil fuels they sell. The countries in the region also contribute to climate change by emitting greenhouse gases. In fact, the affluent Gulf states have the highest emissions in the world per capita. Most recent data showed that Qatar had the world's highest emissions, followed by the United Arab Emirates and Kuwait. These are small countries with overall emissions that are dwarfed by China and the United States, but their development is nevertheless based on wasteful energy consumption.

Climate Change and Desertification

Climate change could accelerate **desertification,** the conversion of nondesert lands into deserts (see Figure 6.6A). As temperatures increase, soil moisture decreases because of evaporation. As a result, plant cover is reduced, which causes less protection for arid soil. Bare patches of soil can become badly eroded by wind, and eventually sand dunes can blow onto formerly vegetated land.

In the grasslands (steppes) that often border deserts, a wide array of land use changes contributes to the general drying. For example, as groundwater levels fall because water is being pumped out for the needs of increasing populations in urban areas and irrigated agriculture, some plant roots no longer reach sources of moisture, causing the plants to die. In other cases, international development agencies have inadvertently contributed to desertification by encouraging nomadic herders to take up settled cattle ranching. Agencies encourage settlement because the mobility of nomads is viewed as complicating management for modern states where records must be kept, taxes collected, and children sent to school with regularity. However, ranching on fragile grasslands may lead to overgrazing, which can kill

desertification a set of ecological changes where a dry area with some vegetation is turned into desert

seawater desalination the removal of salt from seawater—usually accomplished through the use of expensive and energy-intensive technologies—to make the water suitable for drinking or irrigating

Thinking Geographically

After you have read about the vulnerability to climate change in North Africa and Southwest Asia, you will be able to answer the following questions.

A What about Alexandria's location makes it particularly vulnerable to sea level rise?

B What are some of the factors in the region that makes it vulnerable to climate change and complicate emergency responses to climate change, thus limiting resilience?

the grasses and increase evaporation. If irrigation is used to support ranching, such projects can also deplete groundwater resources, resulting in long-term desertification. Incidentally, encouraging nomads to settle may actually increase their vulnerability to climate change, because it is their very mobility and skills at cycling seasonally through multiple environments that have helped nomads adapt to climate variability over many generations.

Imported Food and Virtual Water

Because agriculture is so difficult due to the aridity of this region, the diets of nearly all people here include imported food. The total per capita water use of this region must thus include the water used to produce this imported food. *Virtual water* is the volume of water used to produce all that a person consumes in a year (see Chapter 1). For example, 1 kilogram (2.2 pounds) of beef requires 15,500 liters (4094 gallons) of water to produce, while 1 kilogram of goat meat requires just 4000 liters (1056 gallons). Goat meat, which is popular in this region and produced locally, has a much less significant water component than beef, but beef consumption is on the rise. One kilogram of corn requires 900 liters (238 gallons) of water, and 1 kilogram of wheat, 1350 liters (357 gallons). Beef, corn, and wheat are common imports; in fact, North Africa and Southwest Asia import more wheat than any other world region, which is ironic because this is where the wheat plant was first domesticated.

Strategies for Increasing Access to Water

Some strategies have been developed for increasing supplies of fresh water, but each presents a set of difficulties. All of them are expensive and some have enormous potential to cause more wasting of water.

Seawater Desalination

The fossil fuel–rich countries of the Persian Gulf have invested heavily in **seawater desalination** technologies that remove the salt from seawater, making it suitable for drinking or irrigation. For example, desalination plants supply 70 percent of Saudi Arabia's drinking water and some of its wheat field irrigation. Desalination is an industrial process that separates salt from fresh water; however, it uses huge amounts of energy, burns fossil fuels, and results in emissions that contribute to global warming. If all the costs of producing food with desalinated water were counted—the fuel burned, the cost of irrigation and other equipment, and the fact that irrigated soil inevitably loses productivity due to soil salinization—the wheat produced by this method would be far too

FIGURE 6.9 PHOTO ESSAY: Vulnerability to Climate Change in North Africa and Southwest Asia

Water scarcity, sea level rise, desertification, food scarcity, and political instability are some of the factors that make parts of this region highly vulnerable to climate change.

A Alexandria, Egypt, located on the Mediterranean coast of the Nile Delta region, is a low-lying city that is exposed to rising sea levels. Efforts to improve the sea wall around the city, visible at the right of the photo, may be outpaced by rising sea levels. The entire Nile Delta is struggling to adapt to salt water that is moving up rivers and permeating soils, making agriculture extremely difficult and polluting fresh groundwater resources that cities depend on. Half of Egypt's population lives in the Nile Delta, and 80 percent of the country's imports and exports run through Alexandria.

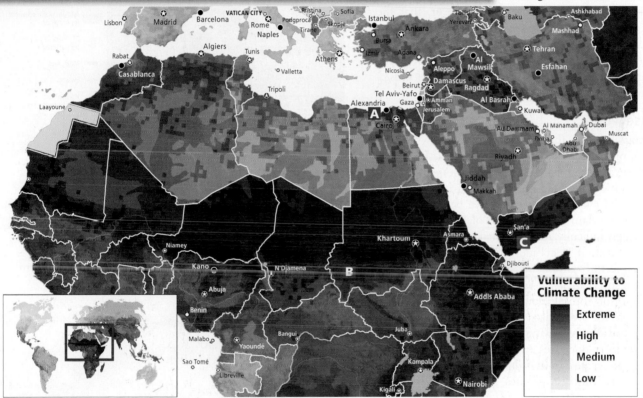

Vulnerability to Climate Change

- Extreme
- High
- Medium
- Low

B Refugees from Darfur, Sudan, line up for food and water in neighboring Chad. Sudan is significantly exposed to drought, and widespread poverty has left the population very sensitive to any disruption in food or water supplies. Meanwhile, political instability is uprooting people from their homes and complicating emergency response and longer-term planning efforts that might lead to greater resilience.

C Residents of Wadi Mur, Yemen, ride out a sandstorm. The same factors that make Sudan so vulnerable to climate change also affect Yemen. Additionally, Yemen is a transit country for refugees from Somalia and other parts of Africa, who eventually seek to travel elsewhere. A conflict within the country has resulted in even more internally displaced people. These populations further stretch Yemen's resources and frustrate planning efforts.

expensive for anyone to buy. Governments have not charged agricultural users the true production cost of water, which has kept the wheat artificially cheap, though there are now plans to remove these subsidies and divert more water to urban populations.

Groundwater Pumping

Many countries pump groundwater from underground aquifers to the surface for irrigation or drinking water. Starting in the 1980s, Libya invested some of its earnings from fossil fuels into one of the world's largest groundwater pumping projects, known as the *Great Man-Made River.* This project includes a series of underground pipelines that draws on the ancient fossil water deposit under the Saharan desert mentioned earlier to supply billions of gallons of water per day to Libya's coastal cities and agricultural fields. With this irrigated agriculture, Libya can reduce its dependency on food imports and even export some foods to the European Union. It should be recognized, however, that the rate of natural replenishment of the aquifer is far slower than the rate of extraction, making this use of groundwater ultimately unsustainable. Hydrologists also report fissures in the land surface above the aquifer that they associate with land sinking, or *subsidence*, as the water is withdrawn.

Dams and Reservoirs

Dams and reservoirs built on the regions' major river systems to increase water supplies have created new problems. In Egypt, for example, the Aswan dams on the Nile River generate much of the country's electricity, reduce flood risk, and improve navigation, but the natural cycles of the river have been altered by the construction of the dams. Downstream of these two dams, water flows have been radically reduced, making it necessary to use supplemental irrigation (see Figure 6.6C), and the lack of flooding means that fertility-enhancing silt is no longer deposited on the land. As a result, expensive fertilizers are now necessary. Moreover, with less silt coming downstream, parts of the Nile Delta are sinking into the sea or are experiencing saltwater intrusion because the seawater is flowing into aquifers and wells in coastal zones, rendering once-fresh water unusable. Upstream, the artificial reservoir created by the dams has flooded villages, fields, wildlife habitats, and historic sites, and created still-water pools that harbor parasites.

Dams Strain Cross-Border Relations

The need to share the water of rivers that flow across or close to national boundaries often complicates the political relationships among the countries involved. The Aswan High Dam sits on Egypt's border with Sudan, an area that has been a source of contention for many years. The colonial-era Nile Waters Agreement created by the British gives most of the river's flow to Egypt, some to Sudan, while 9 other upstream countries in sub-Saharan Africa were not included in the treaty. Therefore, when Ethiopia began to build a hydroelectric dam on the Blue Nile tributary in 2011, Egypt expressed serious concerns. In the short term, the dam will be able to hold back one year's flow of the Blue Nile that won't reach Egypt. Second, the Ethiopian dam may allow Sudan to increase the amount of water it uses for irrigation because the dam will send water year-round downstream to Sudan rather than

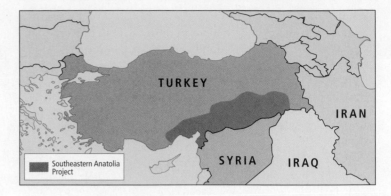

FIGURE 6.10 Dams on the Tigris and Euphrates drainage basins. Turkey's projects to manage the Tigris and Euphrates river basins through dam construction have international implications. Water that is retained in Turkey will not reach its neighbors. The main map shows Turkey's dams on the headwaters of the two rivers, as well as dams built in Syria, Iraq, and Iran, all of which will have environmental effects, especially on the lower reaches of both rivers in Iraq. The smaller map shows the full extent of Turkey's Southeastern Anatolia Project. [Source consulted: United Nations Environmental Programme, Vital Water Graphics: Problems Related to Freshwater Resources, "Turning the Tides" map, http://www.unep.org/dewa/assessments/ecosystems /water/vitalwater/22.htm.]

the current uneven seasonal flow. The three countries are now negotiating a treaty to share water in the context of the new dam.

In the case of Turkey's Southeastern Anatolia Project, which involves the construction of several large dams on the upper reaches of the Euphrates River for hydropower and irrigation purposes, the dams have reduced the flow of water to the downstream countries of Syria and Iraq (FIGURE 6.10). In negotiations over who should get Euphrates water, for example, Turkey argues that it should be allowed to keep more water behind its dams because the river starts in Turkey and most of its water originates there as mountain rainfall. Meanwhile, Iraq's claim to the water is linked to the fact that the Euphrates travels the longest distance in Iraq.

THINGS TO REMEMBER

• Landforms affect climates in this driest region in the world. Rolling deserts and steppes cover most of the land. In a few places, mountains capture moisture, allowing plants, animals, and humans to flourish.

GEOGRAPHIC THEME 1 • **Environment:** The population of this predominantly dry region is growing fast, while access to cultivable land and water for agriculture and other human uses is limited. Therefore, many countries here are dependent on imported food. Climate change could reduce food output and make water, already a scarce commodity, a source of conflict. New technologies offer solutions to water scarcity but are expensive and unsustainable.

HUMAN PATTERNS OVER TIME

Often heralded as a "cradle of civilization" and the birthplace of three major world religions, North Africa and Southwest Asia have more recently struggled with the impacts of outsiders. In the current era, social and political change has lagged behind economic development, which itself varies widely around the region. Only recently has the wealth generated by oil resulted in a spreading of opportunities, such as broad public education for both men and women. As indicated in the opening vignette, change is now under way across this region and is gaining momentum, but there are countervailing forces that inhibit political reform, political opportunities for women, and economic development.

Important developments in agriculture, societal organization, and urbanization took place long ago in this part of the world. Also, three of the world's great religions were born here: Judaism, Christianity, and Islam.

AGRICULTURE AND THE DEVELOPMENT OF CIVILIZATION

Between 8000 and 10,000 years ago, formerly nomadic peoples founded some of the earliest known agricultural communities in the world. These communities were located in an arc formed by the uplands of the Tigris and Euphrates river systems (in modern Turkey and Iraq) and the Zagros Mountains of modern Iran (see Figure 6.15A). This zone, which stretches to the eastern shore of

the Mediterranean, is often called the **Fertile Crescent** because of its once-plentiful fresh water; its fertile soil that is seasonally replenished by flooding; its open forests and grasslands; its abundant wild grains; and its fish, goats, sheep, wild cattle, camels, horses, and other large animals (**FIGURES 6.11** and **6.12**).

The skills of these early people in domesticating plants and animals allowed them to build ever more elaborate settlements. The settlements eventually grew into societies based on widespread irrigated agriculture along the foot of the mountains and in river basins, especially along the Tigris and Euphrates. Nomadic herders living in adjacent grasslands traded animal products for the grain and other goods produced in the settled areas. Over several thousand years, agriculture spread to the Nile Valley and across North Africa, north and west into Europe, and east to the mountains of Persia (modern Iran). Ultimately, other cultivation systems across the world were influenced by developments in the Fertile Crescent.

The Fertile Crescent is also considered the birthplace of urbanization. Eventually, small settlements associated with agriculture took on urban qualities: dense populations, specialized occupations, concentrations of wealth, and centralized government and bureaucracies. City dwellers, who did not engage in agriculture, thrived because surrounding rural areas were capable of producing a surplus of food. The center of early urbanization was Sumer (in modern southern Iraq), where a number of cities existed 5000 years ago. By contemporary standards Sumerian cities were not very large— perhaps 50,000 residents—but these city states extended their influence over the surrounding territory. Today, these places only exist as archeological sites, but cities in the western Fertile Crescent that emerged a little bit later are still inhabited. Jericho (in modern occupied Palestinian Territories) and Damascus (in modern Syria) are among the cities that claim to be the "oldest in the world."

AGRICULTURE AND GENDER ROLES

Increasing research evidence suggests that the dawning of agriculture may have marked the transition to markedly distinct roles for men and women. Archaeologist Ian Hodder reports that at the 9000-year-old site of Çatalhöyük (see Figure 6.11), near modern Konya in south-central Turkey, where the economy was primarily based on hunting and gathering, there is little evidence of gender differences. Families were small and men and women performed similar chores in daily life. Both had comparable status and power, and both played key roles in social and religious life. Archaeological evidence elsewhere indicates that gender roles were also egalitarian in other preagricultural societies.

Scholars believe that after the development of agriculture, as the accumulation of wealth and property became more important in human society, concerns about family lines of descent and inheritance emerged. This led, in turn, to the idea that women's bodies needed to be controlled so that a woman could not become pregnant by a man other than her mate and thus confuse lines of inheritance. From this core concern about secure lines of descent grew many practices aimed at reinforcing the idea that the mating of women had to be controlled and, by extension, that women's daily spatial freedom, interaction

Fertile Crescent an arc of lush, fertile land, where nomadic peoples began the earliest known agricultural communities and where the world's first cities emerged

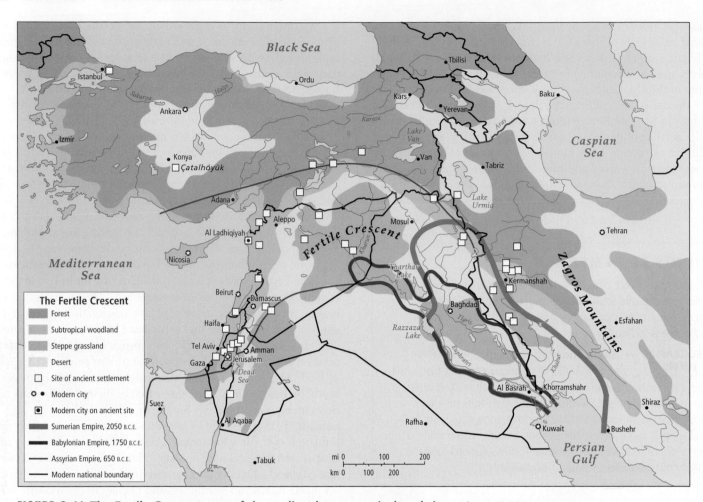

FIGURE 6.11 The Fertile Crescent, one of the earliest known agricultural sites. About 10,000 years ago, people in the Fertile Crescent began domesticating cereal grains, legumes, and animals, especially sheep and goats. The uses of domesticated animals spread to Europe and Africa as agricultural peoples traded their surpluses for other goods or moved into other regions. Three major empires developed successively in the eastern part of the Fertile Crescent: the Sumerian, the Babylonian, and the Assyrian. [Source consulted: Bruce Smith, *The Emergence of Agriculture* (New York: Scientific American Library, 1995), p. 50.]

with men, and sexuality needed to be curtailed. These attitudes still prevail in the region (see page 354).

THE COMING OF MONOTHEISM: JUDAISM, CHRISTIANITY, AND ISLAM

The very early religions of this region were founded on *polytheism*, a belief in many gods who controlled natural phenomena; such was the case through the Greek era and into the Roman period. But starting several thousand years ago, **monotheism**—the belief system based on the idea that there is only one god—began to emerge. The three major monotheistic world religions—Judaism, Christianity, and Islam—all have connections to interrelated sacred texts. For Jews, it is the Torah (the first five books in what Christians refer to as the Old Testament); for Christians, the Old

monotheism the belief system based on the idea that there is only one god

Judaism a monotheistic religion characterized by the belief in one god, a strong ethical code summarized in the Ten Commandments, and an enduring ethnic identity

diaspora the dispersion of Jews around the globe after they were expelled from the Eastern Mediterranean by the Roman Empire beginning in 73 C.E.; the term can also refer to other dispersed culture groups

and New Testaments of the Bible; and for Muslims, both the Bible and the Qu'ran. Moreover, the three religions all consider the city of Jerusalem to be a sacred site. Muslims also revere Makkah (Mecca) and Al Madinah (Medina) in Saudi Arabia.

Judaism was founded approximately 4000 years ago. According to tradition, the patriarch Abraham led his followers from Mesopotamia (modern Iraq) to the shores of the Eastern Mediterranean (modern Israel and the occupied Palestinian Territories), where he founded Judaism. Judaism is characterized by the belief in one god, a strong ethical code summarized in the Ten Commandments, and an enduring ethnic identity reinforced by dietary and religious laws.

After rebelling against the Roman Empire, which culminated in their expulsion in 73 C.E. from the Eastern Mediterranean, some Jews were enslaved by the Romans and most migrated to other lands in a movement known as the **diaspora.** The term diaspora can mean the dispersion of any group of people from their original homeland, but it is especially associated with the historical movement of Jews from Palestine across North Africa, Europe, and various parts of Asia. After 1500, Jews were among the earliest European settlers in all parts of the Americas.

FIGURE 6.12 LOCAL LIVES: People and Animals in North Africa and Southwest Asia

A A camel race in Jordan. Camels were probably domesticated between 4500 and 5000 years ago in the southern Arabian Peninsula. Extensively adapted to arid climates, camels have long been used to carry people and heavy loads across deserts. They are also prized for their lean meat and highly nutritious milk.

B An Arabian horse and his trainer at the stud farm of a Saudi prince in Saudi Arabia. Bred for war and raids by the nomadic desert-dwelling Bedouin people of Southwest Asia, the Arabian horse is known for its endurance, intelligence, speed, stealth, and ability to survive on relatively little food and water. Arabian horses are one of the oldest horse breeds, with evidence of their domestication going back 4500 years. Now they are a part of most major horse breeds throughout the world.

C A mummified cat from ancient Egypt, where cats were revered and sometimes worshiped as symbols of the cat goddess Bastet. Part of a wider tradition of animal worship in Egypt, cats were often treated with the same respect accorded humans, including mummification before burial. Though cat worship was officially banned in 390 B.C.E., cats have been kept as pets and for pest control.

Christianity is based on the teachings of Jesus of Nazareth, a Jew who, claiming to be the son of God, gathered followers in the area of Palestine about 2000 years ago. Jesus, who became known as Christ (meaning *anointed one* or *Messiah*), taught that there is one God who primarily loves and supports humans, but who will judge those who do evil. This philosophy grew popular, and both Jewish religious authorities and Roman imperial authorities of the time saw Jesus as a dangerous challenge to their power.

After Jesus's execution in Jerusalem in about 32 C.E., his teachings were written down (the Gospels) by those who followed him, and his ideas as interpreted by these writers spread and became known as Christianity. Centuries of persecution ensued,

but by 400 C.E., Christianity had become the official religion of the Roman Empire. However, following the spread of Islam after 622 C.E., only remnants of Christianity remained in Southwest Asia and North Africa.

Islam, now the overwhelmingly dominant religion in the region, emerged in the seventh century C.E., after the Prophet Muhammad wrote down what was conveyed to him by Allah ("God" in Arabic) in the Qur'an and transmitted it to his followers. Born in about 570 C.E., Muhammad was a merchant and caravan manager

Christianity a monotheistic religion based on the belief in the teachings of Jesus of Nazareth, a Jew who described God's relationship to humans as primarily one of love and support, as exemplified by the Ten Commandments

in the small trading town of Makkah (Mecca) on the Arabian Peninsula near the Red Sea (see Figure 6.16B). Followers of Islam, called **Muslims,** believe that Muhammad was the final and most important in a long series of revered prophets, which includes Abraham, Moses, and Jesus.

Unlike many versions of Christianity, Islam has virtually no central administration and only an informal religious hierarchy (this is somewhat less true of the Shi'ite version of Islam; see the discussion on page 329). The world's 1.6 billion Muslims may communicate directly with God. A clerical intermediary is not necessary, though there are numerous clerical leaders who help their followers interpret the Qur'an. An important effect of the lack of a central authority is that the interpretation of Islam varies widely within and among countries, from group to group, and from individual to individual.

Today, 93 percent of the people in the region are followers of Islam; for them, the Five Pillars of Islam embody the central teachings of the religion. Not all Muslims are fully observant, but the Pillars have had an enormous impact on daily life, including on festivals and religious holidays, since the time of Muhammad (FIGURE 6.13).

THE PILLARS OF ISLAM

1. The declaration of faith among the believers—a phrase that states Allah is the only God and Muhammad is his messenger.

2. Daily prayer at five designated times (daybreak, noon, midafternoon, sunset, and evening). Although prayer is an individual activity, Muslims are encouraged to pray in groups and in mosques. The call to prayer five times per day defines the aural character of cities and towns in all parts of the region. The call to prayer typically comes from minarets that tower above the landscape. In the past, the call was done in person; today, it is more likely to be a voice recording. Nevertheless, the chant is quite audible and one of the most distinct features of Muslim countries. As it is proscribed that prayer should take place five times per day, in order for Muslims to have easy access to a nearby mosque, the urban landscape is dense with mosques.

Muslims followers of Islam

hajj the pilgrimage to the city of Makkah (Mecca) that all Muslims are encouraged to undertake at least once in a lifetime

3. Obligatory fasting (no food, drink, or smoking) during the daylight hours of the month of Ramadan, followed by a celebratory meal after sundown (Figure 6.13B). Ramadan falls in the ninth month of the Islamic calendar, which is a lunar calendar meaning that Ramadan will fall during a different time every year. Ramadan celebrates the revelation of the Qur'an to Muhammad.

4. Obligatory almsgiving (*zakat*) in the form of a "tax" of 2.5 percent. The alms are given to Muslims in need. *Zakat* is based on the recognition of the injustice of economic inequity. Although it is usually an individual act, the practice of government-enforced *zakat* is returning in certain Islamic countries. The tradition of almsgiving means that charitable institutions are numerous and important in Islamic life. It is doubtful, however, that this pillar has significantly reduced inequality in the region.

5. Pilgrimage **(hajj)** at least once in a lifetime to the Islamic holy places in or near Makkah (Mecca) during the twelfth month of the Islamic calendar. The most important hajj sites are the Masjid Al-Haram (the largest mosque in the world) and the Kaaba, a holy structure that sits in the middle of the courtyard of the mosque.

Saudi Arabia occupies a prestigious position in Islam, as it is the site of two of Islam's three holy shrines: Makkah, the

FIGURE 6.13 LOCAL LIVES: Festivals of North Africa and Southwest Asia

A Colombian singer Shakira performs at Mawazine, a world music festival held each year in Rabat, the capital of Morocco. Part of many conscious efforts to portray Morocco and Rabat as "open to the world," the festival has featured great traditional and international artists such as Cheb Khaled and Stevie Wonder. It has also come under criticism from Islamist politicians who see it as encouraging immoral behavior.

B A banquet is set in Gaza, Palestine, for Iftar, the sunset meal that Muslims eat to break the fast each night in the holy month of Ramadan. During Ramadan, observant Muslims refrain from eating, drinking, quarreling, or having sex between sunrise and sunset. They offer extra prayers each day as a demonstration of their submission to Allah. Wealthy Muslims often sponsor public banquets, such as the one shown below.

C An Iranian woman celebrates Nauryz, an ancient New Year holiday of Zoroastrian origin celebrated throughout Iran, parts of Turkey, and much of Central Asia (see Figure 5.26 on page 295). People jump over bonfires while singing a verse of purification that is meant to remove sickness and problems, replacing them with warmth and energy.

birthplace of the Prophet Muhammad and of Islam, and Al Madinah (Medina), the site of a mosque that is located on the site of Muhammad's home and also contains his burial place. (The third holy shrine is in Jerusalem.) The fifth pillar of Islam has placed Makkah and Al Madinah at the heart of Muslim religious geography for more than 1400 years. Today, a large private sector service industry, owned and managed by members of the huge Saud family that rules Saudi Arabia, organizes and oversees the 5- to 7-day hajj for more than 3 million devout foreign visitors. The hajj represents the largest temporary movement of people in the world. The number of people descending on Makkah every year has increased as there are more Muslims than ever (by natural increase or religious conversion), and the level of affluence in the Islamic world has increased so that more can afford the trip. This presents numerous problems for the city of Makkah, from providing housing for all pilgrims to managing huge crowds at sacred sites. During the 2015 hajj, as many as 2000 people were crushed to death in a stampede as too many worshippers congregated at the same site outside Makkah. Unfortunately, such events have been recurrent over the years during the hajj.

Islamic Religious Law and the Sunni–Shi'ite Divide

Beyond the Five Pillars, Islamic religious law, called **shari'a,** or "the correct path," guides daily life according to the principles of the Qur'an. But there are many interpretations of the Qur'an, several renderings of shari'a, and a wide variety of versions of observant Muslim life. Some Muslims believe that no other legal code is necessary in an Islamic society, because shari'a provides guidance in all matters of life, including worship, finance, politics, marriage, sex, diet, hygiene, war, and crime. Other Muslims think that secular law is more useful in modern societies that are increasingly multicultural because secular law makes allowances for different religious sensibilities. The debate about whether shari'a or secular law is best has raged for hundreds of years, it continues to divide traditionalists and moderates in Islamic countries, and it recently flared up again as part of the Arab Spring.

Insofar as the interpretation of the Qur'an and shari'a is concerned, the Muslim community is split into two major groups that formed after the death of Muhammad, when divisions arose over who should succeed the Prophet and have the right to interpret the Qur'an for all Muslims: **Sunni** Muslims, who today account for 85 percent of the world community of Islam, and **Shi'ite** (or **Shi'a**) Muslims, who live primarily in Iran. A majority of Muslims in Iraq and Bahrain are also Shi'ite, while large Shi'ite minorities exist in Lebanon. Elsewhere, Sunnis dominate. While Sunnis have a relatively decentralized religious system for interpreting shari'a, Shi'ites recognize an authoritative priestly class, whom they alternately call imams, ayatollahs, or mullahs.

This division continues today with different ceremonies and disagreements about theology and the interpretation of shari'a. These religious differences are exacerbated by countless local disputes over land, resources, and political power. In Saudi Arabia, for example, where the Shi'ites are a minority, the group lives in segregated places that are marginalized and underfunded by the Sunni-dominated government. In another case, Iraq, the long-standing conflict between Sunnis and Shi'ites intensified after the U.S. invasion in 2003, as rivalries arose over which group

should control government affairs and fossil fuel resources. As the main Shi'ite country, Iran plays an influential role in Shi'ite communities elsewhere. One aspect of the Shi'ite tradition is that the clerics also tend to be political leaders. In Sunni countries, there is a clearer divide between religious leadership and the secular affairs of the government, even if government policy is influenced by religion. The political role of Shi'ite clerics may come from the minority status of Shi'ite communities around the region, and as they tended to be shut out of national political power, the mosque became the community center for both spiritual and political matters. As conflict in the region has intensified, the Sunni–Shi'ite divide has also become more pronounced in many countries. What has also emerged is a geopolitical struggle between the regional powers Saudi Arabia and Iran, who represent two opposing visions of Islam.

THE DIFFUSION OF ISLAM

Among the first converts to Islam were the Bedouin—nomads of the Arabian Peninsula. By the time of Muhammad's death in 632 c.e., they were already spreading the faith and creating a vast Islamic sphere of influence. Over the next century, Muslim armies built an Arab–Islamic empire over most of Southwest Asia, North Africa, and the Iberian Peninsula of Europe **(FIGURE 6.14)**.

The spread of Islam is an example of **diffusion**—the process by which ideas, things, and people move across space and time. The diffusion pattern of Islam was dependent on multiple political, cultural, and geographic factors. The Arab world adopted Islam first in a contiguous pattern of expansion farther away from the core region around Makkah and Al Madinah. In other places, there were natural barriers (the massive Saharan desert) or political and cultural barriers (established Christian kingdoms in Europe) that delayed or limited the spread.

Not only is Islam spread by person-to-person interactions, but many other forms of Arab culture as well. While most of Europe was stagnating during the medieval period (450–1500 c.e.), the Arab–Islamic empire nurtured learning and economic development. Muslim scholars traveled throughout Asia and Africa, advancing the fields of architecture, history, mathematics, geography, and medicine. Centers of learning flourished from Baghdad (Iraq) to Toledo (Spain). During the early Arab–Islamic era, the development of banks, trusts, checks, receipts, and bookkeeping fostered vibrant economies and wide-ranging trade. The Islamic world was centrally located between the Mediterranean and Asia, which allowed it to play a dominant role in trade. The traders founded settlements and introduced new forms of urban and living spaces, such as the courtyard design (see page 354). The architectural legacy of Arabs and Muslims not only diffused to, but lives on in, Spain, India, Central Asia, the Americas, and countless places across the world.

By the end of the tenth century, the Arab–Islamic empire had

> **shari'a** Islamic religious law that guides daily life according to the interpretations of the Qur'an
>
> **Sunni** the larger of two major groups of Muslims
>
> **Shi'ite** (or **Shi'a**) the smaller of two major groups of Muslims; Shi'ites are found primarily in Iran and southern Iraq
>
> **diffusion** the process by which ideas, things, and people move across space and time

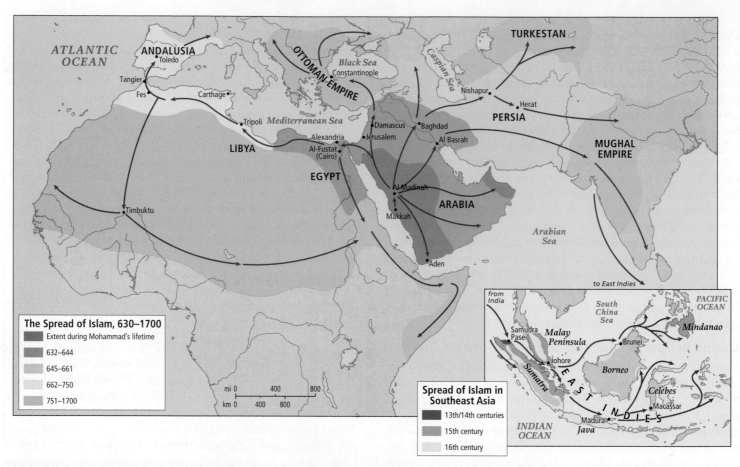

FIGURE 6.14 The diffusion of Islam, 630–1700. In the first 120 years after the death of the Prophet Muhammad in 632, Islam spread, primarily by conquest. Over the next several centuries, Islam was carried to distant lands by both traders and armies. [Source consulted: Richard Overy, ed., *The Times History of the World* (London: Times Books, 1999), pp. 98–99.]

begun to break apart. From the eleventh to the fifteenth centuries, Mongols from eastern Central Asia (eventually converting to Islam by 1330) conquered parts of the Arab-controlled territory, forming the Muslim Mughal Empire, centered in what is now north India. Meanwhile, beginning in the 1200s, nomadic Turkic herders from Central Asia began to converge in western Anatolia (Turkey) where they eventually forged the **Ottoman Empire,** which became the most influential Islamic empire the world has ever known, and lasted until the end of World War I in 1918.

By the 1300s, the Ottomans had become Muslim, and by the 1400s, they had defeated the Christian Byzantine Empire centered in Constantinople, which was the successor to the Roman Empire. The Ottomans took over Constantinople, renamed it Istanbul, and soon controlled most of Egypt, the Eastern Mediterranean, and areas further east in Southwest Asia. By the late 1400s, they also controlled much of southeastern and central Europe and by the 1600s had taken over parts of coastal North Africa from the Arabs. Previously, in the 1490s, the Arab Muslims had lost their control of the Iberian Peninsula to Christian kingdoms. Islam currently still dominates in a huge area that stretches

Ottoman Empire influential Islamic empire centered in today's Turkey that lasted from the 1200s to the early 1900s

from Morocco to western China and includes northern South Asia, as well as Malaysia, Brunei, and Indonesia in Southeast Asia (see the inset of Figure 6.14). The presence of Islam in Southeast Asia is a reminder of the historic Arab-controlled trade around the Indian Ocean and a form of non-contiguous diffusion.

Once a location was completely conquered, the Ottoman Empire, like the Arab–Islamic empire before it, encouraged religious tolerance toward the conquered peoples so long as they adhered to a religion with a sacred text. Jews, Christians, Buddhists, and Zoroastrians (adherents of a religion that predated Islam in Iran and surrounding areas) were allowed to practice their religions, although there were attractive economic and social advantages to converting to Islam. Multicultural urban life in Ottoman cities facilitated vast trading networks spanning the known world, and Istanbul became a cosmopolitan capital with elaborate buildings and lavish public parks that outshined anything in Europe until the nineteenth century (**FIGURE 6.15C**).

WESTERN DOMINATION AND STATE FORMATION

The Ottoman Empire ultimately withered in the face of a Europe made powerful by colonialism and the Industrial Revolution.

Throughout the nineteenth century, North Africa provided raw materials for Europe in a trading relationship dominated by European merchants. By 1830, France was exercising direct control over parts of the North African territory of Algeria (see Figure 6.15D). France took control of Tunisia in 1881 and Morocco in 1912, Spain controlled a bit of the Mediterranean coast and what is now called Western Sahara, Britain gained control of Egypt in 1882 and Sudan in 1898, and Italy took control of Libya in 1912 (FIGURE 6.16A).

The modern borders of North Africa and Southwest Asia are in many cases the product of European colonialism. The two most important colonizers—Great Britain and France—negotiated among themselves how to control the region as the power of the Ottoman Empire waned. This is evident today through the straight lines that were devised by the colonial mapmakers. For example, the borders that separate Syria, Jordan, and Iraq today were originally intended to create two spheres of influence by separating British and French-governed territories.

World War I (1914–1918) brought the fall of the Ottoman Empire, which had allied itself with Germany. At the end of the war, the victorious powers dismantled the Ottoman Empire and all of the former Ottoman territories; only Turkey was recognized as an independent country. The rest of the formerly Ottoman-controlled territory was allotted to France and Britain as protectorates (see Figure 6.16B). On the Arabian Peninsula, Bedouin tribes were consolidated under Sheikh Ibn Saud in 1932, and Saudi Arabia began to emerge as an independent country.

World War II (1939–1945) further affected the political development of North Africa and Southwest Asia. Most significantly, in the aftermath of the Holocaust in Europe, the Jewish state of Israel was created in the Eastern Mediterranean on land inhabited by Arab farmers and nomadic herders as well as by some Jews (see the discussion on page 342; see also Figure 6.15E).

By the 1950s, European and U.S. energy companies played a key role in influencing who ruled Iran and Saudi Arabia—countries where vast oil deposits were to become especially lucrative. In Egypt, a major cotton producer, European textile companies played a similar role. Officials in the governments of these countries received financial benefits from foreign companies and showed their loyalty to these companies with low taxes on oil and cotton exports and easy access to land. While a tiny ruling elite grew fabulously wealthy, even the modest revenues generated by oil and other taxes were not invested in creating opportunities for the vast majority of poor people. Over time, ever more political power accrued, especially to foreign energy companies. The United States and Western Europe supported those autocratic local leaders who, in turn, were sympathetic to the interests of Western companies and Cold War strategies vis-à-vis the Soviet Union. As a result of these concerns, both Cold War adversaries supported undemocratic governments and stood in the way of reforms that would have resulted in a more educated populace able to participate in the full range of democratic institutions. The Assad regime in Syria is such an example, and current conflict there cannot be fully understood without taking into account how international geopolitics have shaped the region for decades (Figure 6.15F).

THINGS TO REMEMBER

- About 10,000 years ago in the Fertile Crescent, formerly nomadic peoples founded some of the world's earliest known agricultural communities. The domestication of plants and animals allowed them to build ever more elaborate settlements that eventually grew into cities and advanced societies that were supported by widespread irrigated agriculture.

- The three major monotheistic world religions—Judaism, Christianity, and Islam—all have their origins in this region in the Eastern Mediterranean. Islam is by far the largest in numbers of adherents in the region (93 percent), and it is the principal faith in all of the region's countries except Israel.

- For Muslims, the Five Pillars of Islamic Practice embody the central teachings of Islam. Some Muslims are fully observant, some are not; but the Pillars have had an impact on daily life for 1400 years.

- Beginning in the nineteenth century and continuing through the end of World War II, European colonial powers ruled or controlled most countries in the region.

- Following World War II, the state of Israel was created, and in countries with oil deposits and other useful assets, the United States and Western Europe supported autocratic local leaders who maintained a friendly attitude toward U.S. and European business and geopolitical strategic interests.

GLOBALIZATION AND DEVELOPMENT

GEOGRAPHIC THEME 2

Globalization and Development: The vast fossil fuel resources of a few countries in this region have transformed economic development. In these countries, economies have become powerfully linked to global flows of money, resources, and people. Politics have also become globalized, with Europe and the United States strongly influencing the region.

There are major economic barriers to peace and prosperity within North Africa and Southwest Asia. In some countries, the wealth from fossil fuel exports remains in the hands of an elite few, while most people work at low-wage urban jobs or are relatively poor farmers or herders. Persian Gulf states have started to make significant investments in education and social services, but even there a large underclass of immigrant workers exists. The economic base is unstable because the main resources are fossil fuels and agricultural commodities, both of which are subject to wide price fluctuations on world markets. Meanwhile, in many poorer countries, limited financial resources and the need to stay competitive in the global market are forcing governments to streamline production and cut jobs and social services.

FOSSIL FUEL EXPORTS AND DEVELOPMENT

This region, which has close to two-thirds of Earth's known reserves of oil, is the major oil supplier to the world. Natural gas reserves and exports are significant as well. Most oil and

A Mud-brick houses in Harran, Turkey, were inhabited for at least 5000 years.

B Masjid Al-Haram, built in Makkah in 630.

C Istanbul's "Blue Mosque," completed in 1616.

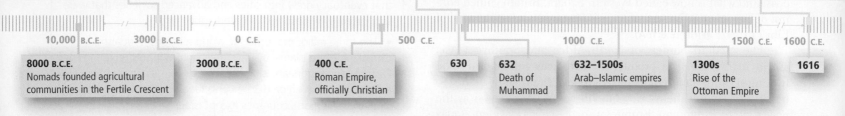

| 10,000 B.C.E. | 3000 B.C.E. | 0 C.E. | 500 C.E. | 1000 C.E. | 1500 C.E. | 1600 C.E. |

8000 B.C.E.
Nomads founded agricultural communities in the Fertile Crescent

3000 B.C.E.

400 C.E.
Roman Empire, officially Christian

630

632
Death of Muhammad

632–1500s
Arab–Islamic empires

1300s
Rise of the Ottoman Empire

1616

FIGURE 6.15 VISUAL HISTORY OF NORTH AFRICA AND SOUTHWEST ASIA

Thinking Geographically

After you have read about the human history of North Africa and Southwest Asia, you will be able to answer the following questions:

A What allowed for urban settlements like Harran to develop?

B What ceremonial purpose is filled by the Masjid Al-Haram and the Kaaba in Makkah?

gas reserves are located around the Persian Gulf (**FIGURES 6.17** and **6.18**) in the countries of Saudi Arabia, Kuwait, Iran, Iraq, Oman, Qatar, and the United Arab Emirates (UAE). The North African countries of Algeria, Libya, and Sudan also have oil and gas reserves. It is important to note, however, that several countries in this region—Morocco, Tunisia, Syria, Lebanon, Jordan, Israel, and Turkey—have minimal oil and gas resources and are net importers of fossil fuels. Meanwhile, Egypt and Sudan produce enough to supply most of their own needs, but export little. Regardless of their own petroleum resources, all countries in the region are affected by the geopolitics of oil and gas.

FIGURE 6.16 Colonial regimes in North Africa and Southwest Asia.

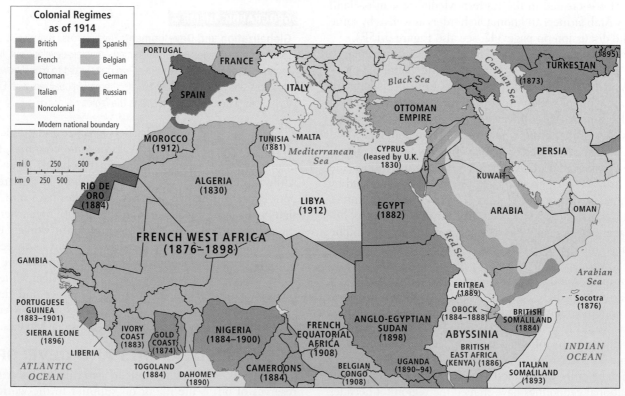

(A) European powers began influencing the affairs of the region in the nineteenth century and expanded their control by 1914 at the beginning of World War I. The dates on the map indicate when the Europeans took control of each country. [Source consulted: *Hammond Times Concise Atlas of World History* (Maplewood, NJ: Hammond, 1994), pp. 100–101.]

D France's wars in Algeria in the 1830s.

E Jewish, Arab, and UN representatives negotiate a cease-fire during the 1948 war.

F People undergo security checks in Baghdad on their way to Karbala to attend a religious festival.

1700 C.E.   1800 C.E.   1900 C.E.   2000 C.E.

1830s

1830–1918
European colonial expansion

1918
Fall of the Ottoman Empire

1948
War resulting in the creation of the state of Israel

1950s–present
Strong European and U.S. influence on politics

2015

C Which Turkish city had lavish buildings and parks that outshined anything in Europe until the nineteenth century?

D When did France first assert control over the territory in what is now Algeria?

E The state of Israel was formally created in the aftermath of what event in Europe?

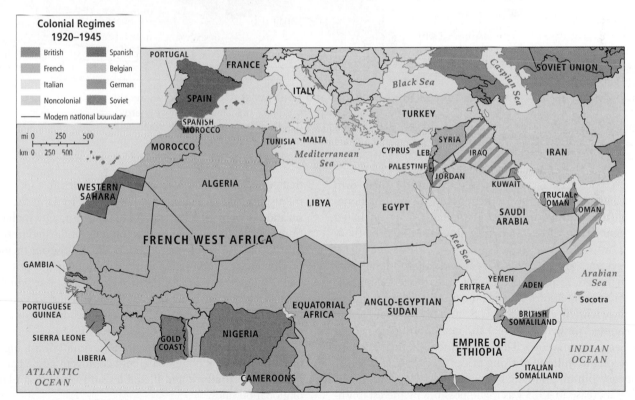

Colonial Regimes 1920–1945

- British
- French
- Italian
- Noncolonial
- Spanish
- Belgian
- German
- Soviet
- Modern national boundary

(B) Between 1920 and 1945, what was left of the Ottoman Empire in the Eastern Mediterranean became *protectorates* (territories that have their security and foreign affairs managed by other countries) administered by the British and French. The striped areas reflect the British colonial practice of allowing local rulers to govern while the British retained control of many of the colonies' policies and actions. [Sources consulted: *Rand McNally Historical Atlas of the World* (Chicago: Rand McNally, 1965), pp. 36–37; *Cultural Atlas of Africa* (New York: Checkmark Books, 1988), p. 59.

European and North American companies were the first to exploit the region's fossil fuel reserves early in the twentieth century. These companies paid governments a small royalty for the right to extract oil (natural gas was not widely exploited in this region until the 1960s, though gas extraction has grown rapidly since then). Oil was processed at on-site refineries owned by the foreign companies and sold at very low prices, primarily to Europe and the United States and eventually to other countries, such as Japan. Since then, most of the countries in the region have established national oil companies that are majority

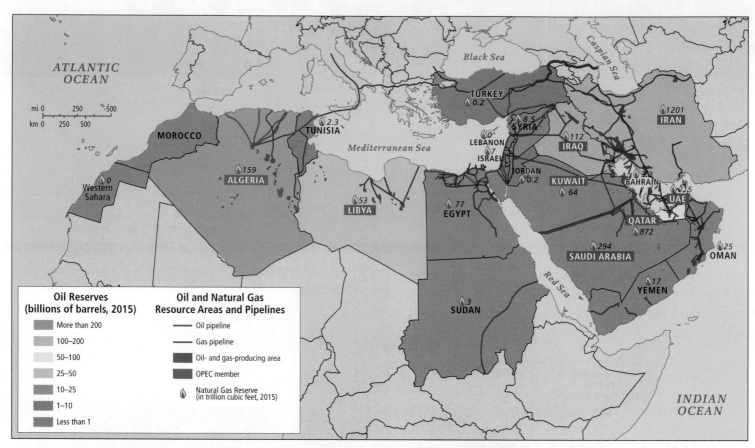

FIGURE 6.17 **Oil and gas reserves and resources in North Africa and Southwest Asia.** The map shows the oil reserves, oil and gas resource areas, and pipelines in the region. Sources consulted: U.S. Department of Energy, Country Analysis Briefs, http://www.eia.doe.gov/emeu/cabs/Region_me/html; U.S. Department of Energy, "Selected Oil and Gas Pipeline Infrastructure in the Middle East," http://www.eia.doe.gov/cabs/Saudi_Arabia/images/Oil%20 and%20Gas%20Infrastructure%20Persian%20Gulf%20(large)%20(2).gif.]

government owned. This enables them to control their own resources rather than being overly dependent on multinational corporations from Europe or the United States. It is estimated that 90 percent of global oil reserves is now controlled by these national oil companies.

Global oil and gas prices have risen and fallen dramatically since 1973, when governments in the Gulf states raised the price of oil and gas for several reasons. For one, the Gulf states were imposing a kind of penalty in response to U.S. support for Israel in the 1973 Yom Kippur War between Israel and neighboring Arab countries (discussed further below). The oil-producing states were also reacting to a long-term trend of rising prices on products exported from European countries and the United States to the Gulf states. The oil price increase in 1973 was made possible by the establishment of national oil companies, and importantly, the founding in 1960 of a **cartel**—a group of producers strong enough to control production and set prices for products—known as **OPEC (Organization of the Petroleum Exporting Countries).** For a few months after the Yom Kippur War, there was a total halt of oil shipments from the Gulf states to the United States and to a few other countries

cartel a group of producers that controls production and set prices for products

OPEC (Organization of the Petroleum Exporting Countries) a cartel of oil-producing countries— currently, Algeria, Angola, Iran, Iraq, Kuwait, Libya, Nigeria, Qatar, Saudi Arabia, the United Arab Emirates, Indonesia, Ecuador, and Venezuela—that was established to regulate the production and price of oil

that supported Israel. When the Gulf states resumed their shipments, they raised the price of oil. OPEC remains the main organization of oil-producing states; membership, which changes from time to time, now includes all of the states indicated in **FIGURE 6.19**. OPEC members cooperate to periodically restrict or increase oil production, thereby significantly influencing the price of oil on world markets. In 2015, the price of oil was low, which lowered revenues for oil producers around the world and triggered cutbacks in production. However, many OPEC countries, especially the Gulf states, have low production cost due to geologically accessible oil. Therefore, OPEC decided on maintaining high levels of production even if doing so perpetuates low prices because it is still profitable to produce and it enables OPEC to gain a larger share of global oil sales.

Many OPEC countries, which were exceedingly poor just 40 years ago, have become much wealthier since 1973, but they have also become more vulnerable to economic downturns. For example, oil income in Saudi Arabia shot up from U.S.$2.7 billion in 1971 to U.S.$110 billion in 1981, and to U.S.$592 billion in 2012. However, Saudi Arabia and the other Gulf states were slow to invest their fossil fuel earnings in basic human resources at home, and they only gradually began to explore other economic strategies in case oil and gas ran out. Like their poorer neighbors, OPEC countries remain very dependent on outside countries for their technology, manufactured goods, skilled labor, and expertise.

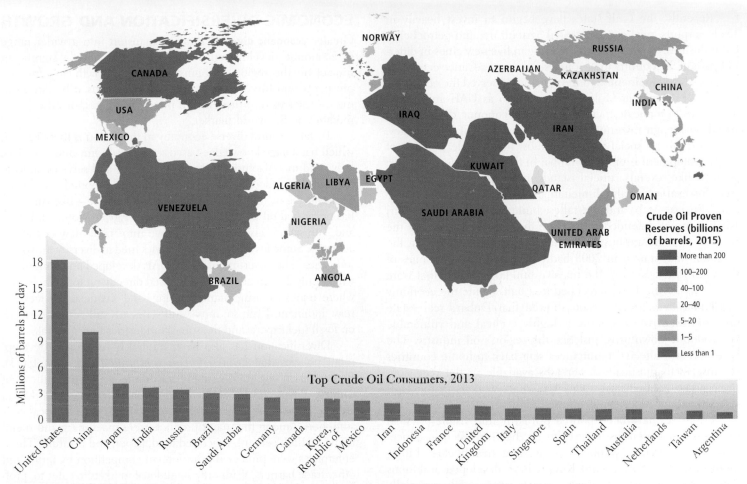

FIGURE 6.18 Oil reserves by country. In the map, the size of countries corresponds to the amount of oil reserves held. This can be contrasted with the consumption of oil in the lower chart. [Sources consulted: U.S. Department of Energy, Energy Information Administration, http://www.eia.gov/beta/international/rankings/#?prodact=57-6&cy=2015, http://www.eia.gov/beta/international/rankings/#?prodact=5-2&cy=2013&pid=5&aid=2&tl_id=2-A&tl_type=a.]

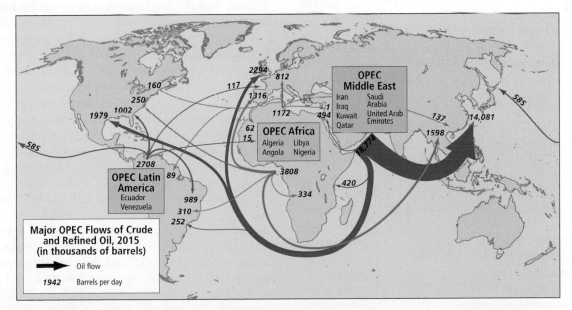

FIGURE 6.19 Major OPEC oil flows in 2015. The map shows the average number of barrels (in thousands) of crude oil and petroleum products distributed per day by OPEC to regions of the world. [Sources consulted: *2015 OPEC Annual Statistical Bulletin*, Graph 5.8, p. 64, http://www.opec.org/opec_web/static_files_project/media/downloads/publications/ASB2015.pdf.]

Recently, the Gulf states have begun to invest heavily in roads, airports, new cities, irrigated agriculture, and petrochemical industries. Saudi Arabia is investing in five new cities in different parts of the country to spur investment and advance economic and social development. Saudi Arabia has observed the economic success of its smaller Gulf neighbors, such as UAE and Qatar, and wants to replicate their approach to economic development based on foreign investment, reduced government bureaucracy, and more liberal social rules. The most high-profile example of new development is probably Dubai in the UAE. There, massive projects like several artificial islands in the shape of palm trees (see, for example, Palm Jumeirah in Figure 6.24A), the iconic hotel Burj al Arab, and the tallest building in the world, Burj Khalifa, are all intended to increase the profile of Dubai as the premier tourism and business center of the region. However, the recession that started in 2008 badly damaged the economies of the Gulf states. Most of the massive building projects that were under way in the UAE screeched to a halt. While the economy has rebounded, the recession showed that Dubai's real estate and tourism-driven economy is highly cyclical and vulnerable to economic downturns, just like the region's oil industry. The recession also affected remittances sent back to home countries because, with drastically fewer jobs available in the recession, those countries that have many temporary workers in the Gulf states experienced a steep decline in remittances.

Resources have been poured primarily into visible physical development; much less oil wealth has been invested in education, social services, public housing, and health care. Libya, before the Arab Spring, and Kuwait have developed ambitious plans for addressing all social service needs, with especially heavy investment in higher education. Increasingly, Gulf states like UAE and Qatar are pouring money into education as well to develop domestic human resources. Iran is another possible exception, although because the Iranian government shares very little of its data with the rest of the world, it is hard to be certain how much Iran is investing in social services. Meanwhile, the non-OPEC countries (those that do not produce fossil fuels) do not share directly in the wealth of the Persian Gulf states and cannot afford comprehensive public services. For the most part, the oil-rich countries have not helped their poorer neighbors develop.

Many factors beyond OPEC's control strongly influence world oil prices. After the 1973 OPEC oil embargo, other countries made successful efforts to increase their domestic supply. More recently, rapid industrialization in China and India has driven up their use of fuel and other petroleum products, creating long-term pressures that are raising the price of oil. Employing a theory known as "peak oil," some experts argue that in an era of diminished conventional oil reserves—which these experts say has already begun—prices will be driven higher. On the other hand, new oil exploration technologies, such as "fracking," have increased supply, which put pressure downward on oil prices. Efforts to combat climate change with renewable energy sources have been increasingly successful and could also reduce the demand for oil. However, so long as gasoline-fueled cars remain a major mode of transportation, demand for oil will remain robust. Current global oil flows are summarized in Figure 6.19.

economic diversification the expansion of an economy to include a wider array of activities

ECONOMIC DIVERSIFICATION AND GROWTH

Greater **economic diversification**—expansion into a wider range of economic development strategies—could have a significant impact on the region, bringing economic growth and broader prosperity and thereby limiting the damage caused by diminishing oil reserves or a drop in the prices of oil, gas, or other commodities on the world market.

By far the most diverse economy in the region is that of Israel, which has a large knowledge-based service economy and a particularly solid manufacturing base. Israel's goods and services and the products of its modern agricultural sector are exported worldwide. Turkey is the next most diversified, in part because—like Israel—it has never had oil income to fall back on. Egypt, Morocco, Libya, and Tunisia are also starting to move into many new economic activities. Some fossil fuel–rich countries tried to diversify into other industries. The government in the UAE developed an economy in which only 25 percent of GDP is based directly on fossil fuels, and where trade, tourism, manufacturing, and financial services are now dominant. Even so, most Gulf states are still very dependent on fossil fuel exports and the spin-off industries they generate.

Diversification has also been limited by economic development policies that favored *import substitution* (see Chapter 3). Beginning in the 1950s, many governments, such as those of Turkey, Egypt, Iraq, Israel, Syria, Jordan, Tunisia, and Libya, established state-owned enterprises to produce goods for local consumption. Among the major products were machinery and metal items, textiles, cement, processed food, paper, and printing. These enterprises were protected from foreign competition by tariffs and other trade barriers. With only small local markets to cater to, profitability was low and the goods were relatively expensive. Without competitors, the products tended to be shoddy and unsuitable for sale in the global marketplace. Meanwhile, the extension of government control into so many parts of the economy nurtured corruption and bribery and squelched entrepreneurialism. Many of the state-owned import substitution enterprises have subsequently either gone bankrupt or been sold to private investors.

Economic diversification and export growth were also limited by a lack of financing, private and public, from within the region. For example, rather than invest locally in mundane but needed consumer products, members of the Saudi royal family generally invest their wealth in more profitable private firms in Europe, North America, and Southeast Asia. Even when private and public investment has stayed in the region, it has gone into lavish projects and may not necessarily prove to be economically profitable. Only recently have governments recognized that they need to invest locally in order to plan wisely for a time when oil and gas run out.

Finally, both international and domestic military conflicts and the ensuing political tensions have stymied economic diversification because they have resulted in some of the highest levels (proportionate to GDP) of military spending in the world. Military spending diverts funds from other types of development, such as health care and education **(FIGURE 6.20)**. The top spenders in the world are Saudi Arabia and Oman, which both dedicate more than 10 percent of their GDP to the military. In fact, seven of the top ten per capita spenders are in North Africa and Southwest Asia. The aftermath of the Arab Spring has made the region even more insecure and the perceived need for a strong military is as marked as ever.

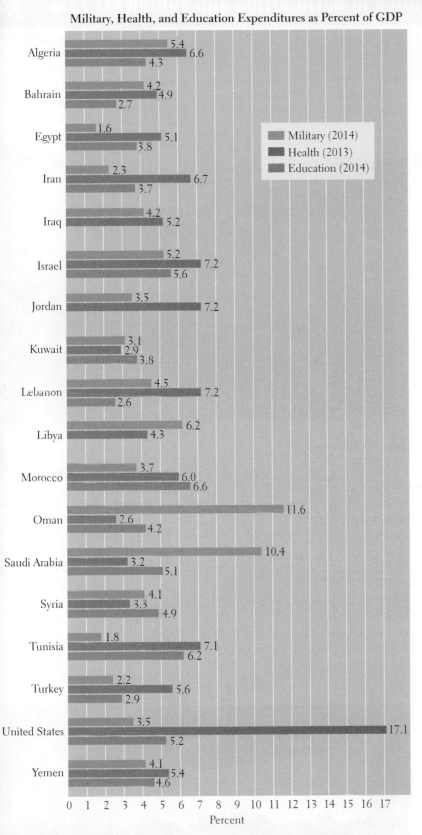

FIGURE 6.20 Military, health, and education expenditures as a percent of GDP, 2013–2014. The graph shows those countries in the region that have data for each variable. The United States is included for comparative purposes. [Sources consulted: SIPRI Military Expenditure Database, http://www.sipri.org/research/armaments/milex/milex_database; *2015 UN Human Development Report*, Tables 9–10; *The World Factbook*, https://www.cia.gov/library/publications/the-world-factbook/rankorder/2034rank.html.]

THINGS TO REMEMBER

GEOGRAPHIC THEME 2 • **Globalization and Development:** The vast fossil fuel resources of a few countries in this region have transformed economic development. In these countries, economies have become powerfully linked to global flows of money, resources, and people. Politics have also become globalized, with Europe and the United States strongly influencing the region.

• Profits in the oil- and gas-producing countries were low until OPEC initiated price increases in 1973. Since then, increasing wealth has led to modernization among oil-rich countries. In many places, wealth accrued mainly to a privileged few, although a broader spectrum of the population now experience some of the benefits of development.

• Until very recently, countries in the region, with the exception of Turkey and Israel, had not diversified their economies. In the Gulf states, huge profits were not invested at home, but rather abroad, thus limiting diversification. When import substitution policies have been tried, they have failed to lift countries out of poverty. Today, more investments in infrastructure and human resources are being made in countries that can afford it and import substitution has been mostly abandoned.

POWER AND POLITICS

GEOGRAPHIC THEME 3

Power and Politics: Authoritarian power structures prevail throughout much of North Africa and Southwest Asia. Beginning with the Arab Spring of 2010, waves of protests swept the region. The outcome of the Arab Spring has varied from reform to repression and civil war.

Political and economic cooperation in the region has been thwarted by a complex tangle of hostilities between neighboring countries and within countries. Some are instigated by antagonism between Sunnis and Shi'ites, but many of these hostilities are the legacy of outside interference by Europe and the United States in regional politics, including colonial intrusions in earlier times, and more recently, by the activities of global oil, gas, industrial, and agricultural corporations. The Israeli–Palestinian conflict, which has profoundly affected politics throughout the region, is rooted in the persecution of the Jews in nineteenth-century Europe, culminating in the Holocaust during World War II. The Iran–Iraq war of 1980–1988 and the Gulf War of 1990–1991 were instigated, in part, by pressures on petroleum resources, again from Europe and the United States. The violence in Iraq, though begun by a homegrown dictator, came to a head when the United States invaded and occupied Iraq, beginning in March 2003. Though the United States has now formally withdrawn, violence continues to plague Iraq and neighboring Syria with the emergence of a radical group, the Islamic State.

THE ARAB SPRING

The political protests that swept across North Africa in 2010 and into Southwest Asia by 2011 eventually resulted in the toppling of

FIGURE 6.21 PHOTO ESSAY: Power and Politics in North Africa and Southwest Asia

Thinking Geographically

After you have read about power and politics in North Africa and Southwest Asia, you will be able to answer the following questions.

A What circumstances made many suspicious of the U.S. claim that the 2003 war with Iraq was undertaken to bring democracy to Iraq?

B Why is the 1973 war between Israel and Egypt cited as one of the causes of political repression within countries of the region?

C What were the issues that motivated Syrians to take to the streets in prolonged demonstrations in 2011?

D How has the geopolitical situation changed for Kurds?

E What accounts for the higher death toll in Libya's Arab Spring than Egypt's?

Political freedoms are relatively few in this region, which is one of the least democratized in the world. Many countries suffer from a violent political culture, as conflicts that might have been resolved peacefully through a free media and fair elections have instead been prolonged, embittering and enraging otherwise reasonable people. Meanwhile, wars and terrorism have given authoritarian regimes an excuse to forcibly repress legitimate political opposition groups.

Frank Rossoto Stocktrek/Getty Images

A U.S. Navy warships patrol the Persian Gulf. U.S. and European militaries have had a large presence in this region for a number of years, especially because of the importance of the region's oil exports to the global economy. The U.S.-led invasion of Iraq in 2003 was justified as an effort to rid Iraq of weapons of mass destruction and to ensure greater political freedoms for the people of Iraq. However, many U.S. officials have openly admitted that safeguarding foreign access to the region's oil reserves was a larger motivation.

E Thousands celebrate in Tripoli, Libya, on the second anniversary of the end of Libya's revolution of 2011. Like the war in Syria, this conflict, in which 25,000–30,000 people were killed, was ignited by the government's brutal repression of Arab Spring protests. Civil strife has not ended either; the central government is weak and local militias, and even competing governments, wield power in Libya.

Mahmud Turkia/AFP/Getty Images

long-standing dictatorships in Tunisia, Libya, and Yemen. A protest also overthrew the military regime in Egypt, but the military is back in power again as of 2013. In Syria, the outcome of political protest has been most problematic where initial opposition to President Assad morphed into a civil war involving multiple groups. The early demonstrations across the region were precipitated by high unemployment and rising food prices caused by the global recession that began in 2008; poor living conditions; government corruption; and the absence of freedom of speech and other political freedoms. An undercurrent of dissent among women was also palpable in every country. Movements for women's rights specifically and human rights generally were rejuvenated (see the opening vignette).

The political disquiet reflected the fact that, for much of their history, most governments in this region gave citizens very little ability to influence how decisions were made. Constitutions were not constructed to facilitate widespread participation in public discourse or to protect the rights of women and minorities. Laws were simply interpreted to suit the factions that held power. Elections were either nonexistent or were rigged to reelect those already in office or their chosen successors. Some Gulf states are absolute monarchies even without the pretense of democracy. Meanwhile, freedom of speech and of the press was strictly limited, especially if it involved criticism of the government. As a result, most people across the region saw their governments as unresponsive and corrupt and viewed themselves as powerless to influence government in any way other than by massive protests. The role that Islam should play in society has been an especially contentious issue, and it relates directly to determining which system of laws should be adopted as well as which protections women should be afforded.

Different Paths Toward Reform

The Arab Spring movement began in Tunisia and spread to Egypt, then to Libya, and eventually to most countries in the region. The size and focus of the protests varied—see the Arab Spring map in **FIGURE 6.21**—but they all raised hopes for real political reforms. However, there were no viable reform models extant within the region to follow. Should constitutions be revised first to guarantee freedoms and equal participation for all, or should elections come first, with the winning majority framing the new constitutions?

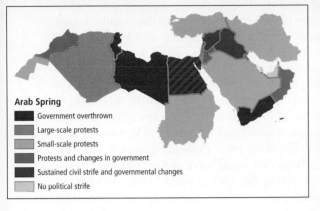

Arab Spring

- Government overthrown
- Large-scale protests
- Small-scale protests
- Protests and changes in government
- Sustained civil strife and governmental changes
- No political strife

B An Egyptian man displays a Star of David painted on his shoe, an act of symbolic desecration, during an anti-Israel rally. Ongoing tensions between Israel and its Arab neighbors have worked against political freedom throughout the region. Many Arab governments use the conflict to justify repressing legitimate political parties. Meanwhile, Israel bans its citizens from visiting "enemy countries" (a number of Arab states and Iran) without special permission.

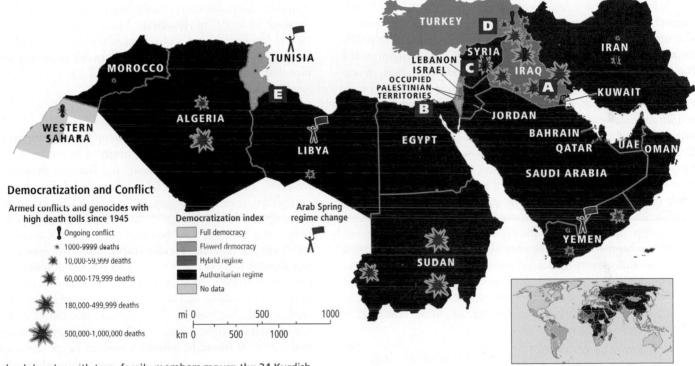

Democratization and Conflict

Armed conflicts and genocides with high death tolls since 1945

- Ongoing conflict
- ✳ 1000-9999 deaths
- ✳ 10,000-59,999 deaths
- ✳ 60,000-179,999 deaths
- ✳ 180,000-499,999 deaths
- ✳ 500,000-1,000,000 deaths

Democratization index

- Full democracy
- Flawed democracy
- Hybrid regime
- Authoritarian regime
- No data

Arab Spring regime change

mi 0   500   1000
km 0   500   1000

TURKEY D
SYRIA IRAN
LEBANON C IRAQ
ISRAEL A
OCCUPIED KUWAIT
PALESTINIAN
TERRITORIES B JORDAN
TUNISIA BAHRAIN
MOROCCO QATAR UAE OMAN
E EGYPT SAUDI ARABIA
ALGERIA
WESTERN SAHARA LIBYA YEMEN
SUDAN

D Near Turkey's border with Iraq, family members mourn the 34 Kurdish civilians killed by the Turkish air force, which mistook them for rebels. Kurds form a majority of the population in southeastern Turkey and in neighboring parts of Syria, Iraq, and Iran. Kurds have long been denied political freedoms and in some cases even the right to speak their own language. The relationship between the Turkish government and the Kurds remains conflictual, while in northern Iraq the Kurds have greater autonomy than they previously had, although they now face a new threat, the Islamic State.

C A rebel sniper in the city of Aleppo, Syria. A civil war was sparked by the Syrian government's harsh crackdown on Arab Spring protests, beginning in 2011. More than 200,000 people have been killed and millions have been displaced by the conflict.

Tunisia and Egypt chose to have elections first, but turnout was low with not even one-third of the eligible voters participating. In Tunisia, a coalition between secular and Islamic parties has led to constitutional compromises, although many women are worried about the influence of conservative Islamists. Another challenge is posed by Islamist militants who have launched terrorist attacks in the country. In Egypt, an Islamist-oriented government was elected that proceeded to adopt a constitution that did not adequately protect the rights of more secular political groups, minorities, or women. Political turmoil and violence resulted as the Islamists governed in ways that outraged their numerous political opponents, women, and minorities. In 2013 the Islamists were removed from power by Egypt's military who tolerate little dissent from either democracy activists or Islamists.

More incremental approaches to reform were chosen by the kings of Jordan and Morocco, resulting in less chaos and violence. In both cases, the respective kings negotiated with reformists well before full-blown Arab Spring mass movements developed, bargaining to keep some of their royal powers while agreeing to give up others. They both supported the idea that lasting political reform should begin with constitutional reforms hammered out not by elected bodies but by constitutional councils made up of a broad range of people. These councils were charged with making the political systems more democratic and inclusive, and in both cases women were significantly represented, as were religious minorities and those with secular points of view. When elections were held, candidates from a wide range of political perspectives won positions. In Morocco, for example, an Islamist party won enough seats in parliament to form a government with allies; but with only a little over a quarter of the total seats, the Islamists had to accept cooperation and compromise as a necessity, an important precedent for the future. Jordan's path toward reform has been somewhat slower. The monarch has appointed reform-minded governments rather than declaring open elections. Freedom of the press is also more limited in Jordan than it is in Tunisia and Morocco, which suggests a more open political culture in North Africa compared to Jordan, where the ruler is rarely criticized publicly.

Islamism in a Globalizing World

Ironically, the Arab Spring, with its emphasis on political freedoms, actually opened up space for Islamist groups who took advantage of the new openness to assert their own agenda. Contemporary radical Islamism has its roots in two interrelated movements dating back to the eighteenth and nineteenth centuries—**Salafism** that originated in Egypt and **Wahhabism** in Saudi Arabia. They both stress a return to what they see as the roots of the religions, which is common among fundamentalists of different beliefs. Salafist scholars from Egypt were

Salafism a religious movement that started in Egypt and that has influenced the ideology of radical Islamism

Wahhabism a religious movement that started in Saudi Arabia and that has influenced the ideology of radical Islamism

jihadists militant Islamists engaging in jihad, especially in opposition to Western influence

theocratic states countries that require all government leaders to subscribe to a state religion and all citizens to follow rules decreed by that religion

secular states countries that have no state religion and in which religion has no direct influence on affairs of state or civil law

welcomed to Saudi Arabia where their emphasis on pan-Islamism (uniting the Islamic world) coalesced with the Wahhabists' rejection of everything modern. This worldview became powerful in schools and mosques in 1970s Saudi Arabia; it thoroughly shaped the belief system of a generation of Saudis, and has spread to other Muslim countries since. Today's radical Islamists believe in an extreme, purist Qur'an-based version of Islam that has little room for adaptation to modern times. Globalization is a particular problem for these conservative Muslims, who lament what they see as the global spread of open sexuality, consumerism, and hedonism, transmitted in part by TV, movies, and popular music. Salafists were once explicitly peaceful, focusing mostly on the much-needed social services they provided to the poor. But since the 1990s, the idea of *jihad* (a struggle) against the West has gained popularity among younger Islamists. The concept of jihad is multifaceted—it can mean a personal struggle to be a good Muslim or to do peaceful missionary work, but has in public discourse increasingly been associated with religiously inspired violence by **jihadists.** Many conservative Muslims—radical Islamists or not—object to what they see as the Western emphasis on the liberalization of women's roles, especially the idea that women should be educated and active outside the home (see Figure 6.30).

Historically, Islamism is rooted in the interaction between religious and governmental authority. Governments in Saudi Arabia, Yemen, the UAE, Oman, and Iran are **theocratic states** in which Islam is the official religion and political leaders are considered to be divinely guided by both Allah and the teachings of the Qur'an. Elsewhere, such as in Turkey as well as in Tunisia, Libya, and Syria (at least before the 2011 rebellion began), governments are officially **secular states,** where religious parties are not allowed and the law is neutral on matters of religion. In practice, even in this region's secular states, Islamic ideas influence government policies. Combined with an authoritarian political culture, this has meant that political freedoms, such as free speech and the right to hold public meetings, were severely limited *outside of Islam.* In countries such as Egypt, for decades the only public spaces in which people were allowed to gather were mosques, and the only public discussions free of censorship by the government were religious discourses. In this context, many political movements became rooted in Islam that might not have done so otherwise. In Egypt, the Muslim Brotherhood, which is an old organization with Salafist roots, won the first post–Arab Spring election because it was highly organized and could mobilize its supporters. The Brotherhood developed for decades as a religious social movement, and even though it was banned on and off by the Egyptian authorities, it emerged as a potent political force during the Arab Spring. In a similar fashion, Shi'ite minorities around the region were shut out of political power; instead, they often rallied around their local religious leaders who then became de facto political leaders.

The militancy often associated with Islamism is characteristic of many popular political movements in this region, where challenges to the authority of governments are frequently met with violent repression (see the Figure 6.21 map). In both secular and theocratic governments, political freedoms are sometimes so weak that minorities have been denied the right to engage in their own cultural practices and to speak their own languages. Journalists and private citizens have been harassed, jailed, or even killed for criticizing governments or exposing corruption. While the Arab Spring protests were, in part, a response to these conditions, the extent to which the new governments and political reforms will protect political freedoms remains to be seen.

Will There Be an Iranian Spring?

When the Arab Spring erupted across North Africa, many wondered if Iran (which is not Arab) would join in the protests. A seeming prelude occurred in June 2009, when many Iranians suspected the presidential elections had been rigged. Hundreds of thousands of people in Tehran took to the streets in protest. The demonstrations first centered on election fraud, but when the Iranian police killed several unarmed demonstrators, the brutality of the autocratic, theocratic state became a major focus. The Iranian street protests persisted into 2010, when the activities of the Arab Spring in North Africa began. However, there were no major protests during the 2013 Iranian elections, partly because the leaders of the 2009 protests were in jail or in exile, but also because a moderate presidential candidate, who eventually ended up winning the election, was backed by reformists. A president in Iran has power over day-to-day affairs of the government, but ultimate power rests with the Supreme Leader, which is a lifetime appointment. Since the 1979 Iranian Revolution, the Supreme Leader, in Shi'ite fashion, has always been a cleric. This system is backed up by the Revolutionary Guard, a militia that cracks down on any threat to Iran's Islamic system of government.

THE ROLE OF MEDIA AND THE INTERNET IN POLITICAL CHANGE

In some countries—Tunisia, Israel, and Lebanon, for example—the press is reasonably free and opposition newspapers can be aggressive in their criticism of the government. But in much of the region, journalism can be a risky career, often leading to imprisonment. In Saudi Arabia, a dozen newspapers are available every morning, but all are owned or controlled by the royal Saud family, and all journalists are constrained by the fact that they may not print anything critical of Islam or of the Saud family, which numbers in the tens of thousands. When accidents happen or some malfeasance by a public official is revealed, the story is blandly reported with little effort to explore the causes of events or their effects, or to hold responsible officials accountable.

An example of journalistic change is the broadcasting network Al Jazeera, founded by the ruling emir of Qatar, but now privately owned and independent of the Qatar government. Al Jazeera is credited with changing not only the climate for public discourse across the region but also public opinion outside of the region through Al Jazeera English, the source of some of the information used in the opening vignette. The various versions of Al Jazeera now openly cover controversy, including reforms proposed by the most radical Iranian and Arab Spring protesters, with no apparent censorship.

The role of the Internet and cell phone technology as a force for political change emerged first during the Iranian protests beginning in 2009, when the fact that so many Iranians had video-capable cell phones meant that via the Internet, the world saw instantly the brutality of the Iranian police and the Revolutionary Guard. Use of the Internet (e-mail, Facebook, crowdsourcing, and Twitter) became commonplace during the Arab Spring, as ordinary people with nothing but an inexpensive cell phone were able to give real-time reports about activities in the streets of Cairo, Tripoli, and Benghazi. Twitter became the chief medium because it is free, mobile, and quick. Crowdsourced data could then be compiled and mapped via software such as *Ushahidi* (invented and made popular by several Kenyan engineers; see the discussion in the opening vignette of Chapter 7), which allows protesters to see how and where a crisis unfolds and coordinate their actions accordingly.

DEMOCRATIZATION AND WOMEN

Most countries in this region now allow women to vote, and two countries—Israel and Turkey—have elected female leaders (prime ministers) in the past. Nevertheless, women who want to actively participate in politics still face many barriers. In the Gulf states, where women's political status is the lowest, circumstances vary from country to country. Oman gave all women the right to vote in 2003. In Kuwait in 2009, four highly educated women were elected to parliament; they quickly energized the pace of law making, often publicly criticizing their male colleagues, who were frequently absent for crucial votes. In Saudi Arabia, by contrast, women were given the right to vote and run in local elections as recently as 2015; twenty-one Saudi women now serve on local councils. (Note that only local elections are allowed in Saudi Arabia and only since 2005.) Across the region, women average only 14 percent of national legislatures, the lowest of any world region. However, female parliamentary representation in parts of North Africa (31.5 percent in Tunisia, 26 percent in Algeria) and Iraq (26.5 percent) is higher than in the United States. Women lost ground in Egypt when the post–Arab Spring elections in 2012 brought fewer women to office than had been the case beforehand. Moreover, there is a tendency—even among women—to not support female candidates.

ON THE BRIGHT SIDE: Reforms for Women Could Address Labor Problems

An important impetus for change in the region is the increasing number of women who are becoming educated and employed outside the home. In a number of countries, women outnumber men in universities; most notably, in Saudi Arabia women make up 70 percent of university students (but only 5 percent of the workforce). Recently, a Saudi prince, known for his interests in reform, suggested that one way to reduce dependence on foreign workers is to allow Saudi women to become drivers of taxis and trucks, as is now allowed in Jordan. More broadly, unleashing the full economic potential of this increasingly well-educated segment of the population would have tremendous positive impacts on the region. ∎

THINGS TO REMEMBER

GEOGRAPHIC THEME 3 • **Power and Politics:** Authoritarian power structures prevail throughout much of North Africa and Southwest Asia. Beginning with the Arab Spring of 2010, waves of protests swept the region. The outcome of the Arab Spring has varied from reform to repression and civil war.

• While the immediate causes of the Arab Spring demonstrations may be economic, activists who have for years promoted increased political freedoms as a path to peace in this region have been reinvigorated by the protests and continue their work to end authoritarian rule. Both male and female participants have thus far been both energized and disappointed by the results.

• Demand for more political freedoms is growing and public spaces for debate, other than mosques, are emerging. The press is reasonably free in some countries and severely curtailed in others. Al Jazeera is credited with changing both the climate for public discourse across the region and public opinion outside the region.

• Most countries now allow women to vote, and the two most developed countries have elected female leaders. Even though women who want to actively participate in politics still face barriers, patterns are changing as more women are becoming educated and are working outside the home.

THREE WORRISOME GEOPOLITICAL SITUATIONS IN THE REGION

The region of North Africa and Southwest Asia is known as a center of especially troublesome political issues that command the attention of major global powers. These issues are related to the politics of the international oil trade and the presence of the state of Israel (see Figure 6.21A), and at least one issue may be related to global climate change. Two of the three situations we discuss here—in Iraq and Syria—are increasingly interrelated as the Islamic State group controls territories that span both countries.

Situation 1: The State of Israel and the "Question of Palestine"

The Israeli–Palestinian conflict has lasted more than 60 years and has included several major wars and innumerable skirmishes. It is a persistent obstacle to political and

Zionism a movement that started in nineteenth-century Europe to create a Jewish homeland in Palestine

economic cooperation within the region, and it complicates the relations between many countries in this region and the rest of the world.

Israel's excellent technical and educational infrastructure, its diverse and prospering economy, and the large aid contributions (public and private) it receives from the United States and elsewhere have made it one of the region's wealthiest, most technologically advanced, and militarily powerful countries.

The Palestinian people, by contrast, after years of conflict, are severely impoverished with inadequate access to education, training, jobs, and the basics of human well-being (see **TABLE 6.1**). Many have lived in refugee camps for decades. Through a series of events over the last 70 plus years, Palestinians have lost most of the land on which they used to live. They now reside on the West Bank (home to 2.8 million Palestinians) and the Gaza Strip (1.9 million), with many also living in neighboring Jordan. The Jordanian census does not distinguish between people of Palestinian origin and Jordanian natives, but the two groups are probably equal in size. The West Bank and Gaza are both highly dependent on Israel's economy, where about 100,000 work in the construction, manufacturing, and service industries (Figure 6.23B). Israel often takes military action in the West Bank and Gaza in retaliation for Palestinian suicide bombings and rocket fire launched primarily from Gaza. Many parts of the West Bank occupied Palestinian Territories continue to be encircled by security walls built by Israel to discourage Palestinian suicide bombers from entering Israel. In effect, the wall also curtails Palestinian access to places they need to go on a day-to-day basis (Figure 6.23A).

The Creation of the State of Israel In the late nineteenth century, as a response to centuries of discrimination and persecution in Europe and Russia, the **Zionist** movement emerged with the aim of establishing a homeland for the Jewish diaspora. Small groups of European Jews began to purchase land from absentee landowners in a part of the Ottoman Empire known at the time as Palestine. Some of the purchased land was uncultivated but landowners had also leased their lands for many years to Arab Palestinian tenant farmers and herders or had granted them the right to use the land freely. Such rights were no longer recognized by the Zionists who acquired the land to be farmed and occupied by Jews. Historians still debate whether or not the Zionists or the former landlords adequately compensated the displaced indigenous inhabitants.

TABLE 6.1 Circumstances and human well-being among Palestinians and Israelis, 2015

| | Population in millions, mid-2015 | GNI PPP in U.S.$, 2014 | Life expectancy at birth | | HDI rank | Infant mortality rate, 2013 | Overall life satisfaction, 2014 index* |
			Males	Females			
Palestinians	4.5	5,080	72	75	113	18.6	4.7
Israelis	8.4	32,550	80	84	18	3.2	7.4

*Ranked on a scale of 1 to 10, with 0 representing least satisfied and 10 representing most satisfied.
Source: 2015 Population Data Sheet and 2015 Human Development Report, Tables 9 and 16.

On their newly purchased lands, the Zionists established communal settlements called *kibbutzim,* into which a small flow of Jews came from Europe and Russia. While Jewish and Palestinian populations intermingled in the early years, Zionist land purchases displaced more and more Palestinians, and violence became increasingly common in the 1920s and 1930s.

In 1917, the British government adopted policies that favored the establishment of a national home for Jews in the Palestine territory. It was at this time that the word *Palestine,* with roots far back in history, began to be used officially. In 1923 after World War I, the United Kingdom received what is called a "mandate" from the League of Nations—the precursor to the United Nations—to administer the territory of Palestine **(FIGURE 6.22A)**. British control of Palestine made it easier for Jewish people to migrate there. By 1946, following the genocide of 6 million Jews in Nazi Germany's death camps during World War II, strong sentiment had grown throughout the world in favor of a Jewish homeland in the space in the Eastern Mediterranean that Jews had shared in ancient times with Palestinians and other Arab groups. Jewish migration to Palestine continued to accelerate.

Decades of Conflict and the Two-State Solution In
November 1947, after an intense debate in the UN General Assembly, the UN adopted the "Plan of Partition with an Economic Union" that ended the earlier mandate and called for the creation of both Arab and Jewish states (the *two-state solution* still under discussion today) with special international status for the city of Jerusalem (see Figure 6.22B).

The Palestinians and neighboring Arab countries fiercely objected to the establishment of a Jewish state and feared that they would continue to lose land and other resources. Then, as now, the conflict between Jews and Palestinian Arabs was less about religion than control of land, settlements, and access to water. The sequence of the changes in allotments of land to Israel and to the Palestinians over the last 90 years can be followed in Figure 6.22.

On the same day that the British reluctantly ended the mandate and withdrew their forces, May 14, 1948, the Jewish Agency for Palestine unilaterally declared the state of Israel on the land designated to them by the Plan of Partition. Warfare between the Jews and Palestinians began immediately. Neighboring Arab countries—Lebanon, Syria, Iraq, Egypt, and Jordan—supporting the Palestinian Arabs invaded Israel the next day. Fighting continued for several months, during which Israel prevailed militarily. An armistice was reached in 1949, and as a result, the Palestinians' land shrank still further, with the remnants (Gaza and the West Bank) incorporated into Egypt and Jordan, respectively (see Figure 6.22C). In essence, the Palestinians ended up as a national entity, but without a state of their own.

In the repeated conflicts over the next decades—such as the Six-Day War in 1967 and the Yom Kippur War in 1973—Israel again defeated its larger Arab neighbors, expanding into territories formerly controlled by Egypt (Gaza and Sinai), Syria (Golan Heights), and Jordan (West Bank of the Jordan River; see Figure 6.22D). Since 1948 hundreds of thousands of Palestinians have fled the war zones, many ending up in refugee camps in nearby countries. Some Palestinians stayed inside Israel, especially to the north near Lebanon, and became Israeli citizens. Approximately 20 percent of Israelis are Palestinians, but they have not been treated by the authorities as equal to Jewish Israelis. De facto discrimination has resulted in fewer social services in Palestinian neighborhoods, less access to government jobs, and restrictive building permits that limit residential choice.

In 1987, the Palestinians mounted the first of two prolonged uprisings, known as the **intifada,** characterized by escalating violence. The first ran until 1993, when the Oslo Accords provided that Israel withdraw from parts of the Gaza Strip and the West Bank, and that the Palestinian Authority be the entity which would enable Palestinians to govern themselves in their own state. For a period during the 1990s, the situation improved as Palestinians seemed to have achieved something close to statehood. However, the second intifada began in 2000 and continued to 2005, fueled by extremists on both sides, failed peace negotiations, and ongoing tensions over the expansion of Israeli settlements in Gaza, the West Bank, and the Golan Heights (see Figure 6.22F). Both sides have suffered substantial trauma and casualties. According to B'Tselem, an Israeli human rights group, far more Palestinians have died in this violence (since 2000, there have been 7704 Palestinian deaths versus 1147 Israeli deaths).

Over the years since 1947, the vision of a two-state solution has not died. In 2012, the Palestinians were awarded nonmember observer state status at the United Nations, and thus the Palestinians gained official recognition that they had never before achieved, even if they don't have an official vote in the UN as an independent country. Israeli allies in the UN, primarily the United States, have blocked the full recognition of Palestine as a state. From the perspective of international law, to consider Palestine as an independent state is also problematic, because it does not have full control over its territories; the Israeli military is, in practice, in charge of much of the West Bank.

Territorial Disputes When Israel occupied Palestinian lands
in 1967, the UN Security Council passed a resolution requiring Israel to return those lands in exchange for peaceful relations between Israel and neighboring Arab states. This *land-for-peace* formula, which set the stage for an independent Palestinian state, has been only partially fulfilled, as noted above.

Despite the land-for-peace agreement, between 1967 and 2013, Israel secured ever more control over the land and water resources of the occupied territories. Israel took what appeared to be a major step toward peace in 2005 when it removed all Jewish settlements and military on-the-ground presence from the Gaza Strip, but this progress was negated by a blockade of Gaza's economy. Israel, and to some extent Egypt, have control over almost all the goods and people that cross in and out of the land and sea borders of Gaza. Israel also controls the supply of water, electricity, and communications infrastructure in Gaza. In response to attacks from Gaza, Israel has additionally engaged in air strikes on Gaza that have devastated the territory. The problem is that the Palestinian Authority is divided between the Islamist Hamas movement, which is in control of Gaza, while the more moderate Fatah controls the West Bank. This

intifada a prolonged Palestinian uprising against Israel

FIGURE 6.22 Israel and Palestine, 1923–1949, and Israel and the Palestinian Territory after 1949.

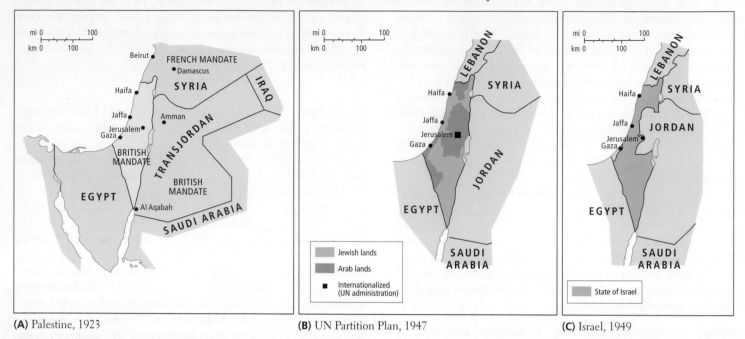

(A) Palestine, 1923

(B) UN Partition Plan, 1947

(C) Israel, 1949

Israel and Palestine, 1923–1949. (A) Palestine, 1923. Following World War I, Britain controlled Palestine and Trans-Jordan (the precursor to Jordan). **(B)** The UN Partition Plan, 1947. After World War II, the United Nations developed a plan for separate Jewish and Palestinian (Arab) states. **(C)** Israel, 1949. The Jewish settlers won a war against Arab forces, creating the state of Israel, which was larger than the original UN-proposed Jewish state. [Sources consulted: Colbert C. Held, *Middle East Patterns—Places, Peoples, and Politics* (Boulder, CO: Westview Press, 1994), p. 184; *The Israeli Settlements in the Occupied Territories*, 2002, Foundation for Middle East Peace, http://www.firstpr.com.au/nations.]

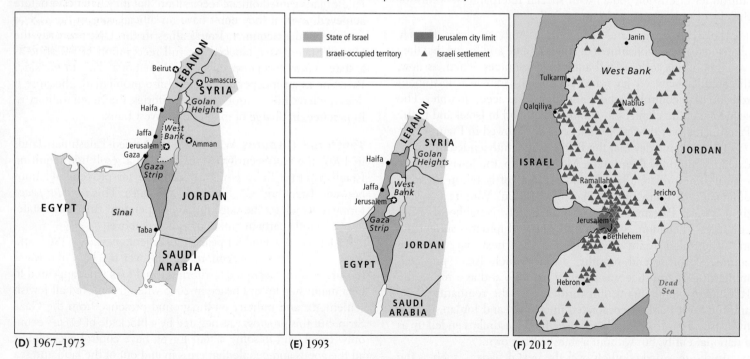

(D) 1967–1973

(E) 1993

(F) 2012

Israel and the Palestinian Territory after 1949. When the state of Israel was created in 1949, its Arab neighbors were opposed to a Jewish state. **(D)** In 1967, Israel soundly defeated combined Arab forces and took control of Sinai, the Gaza Strip, the Golan Heights, and the West Bank. **(E)** In subsequent peace accords, Sinai was returned to Egypt, but Israel maintained control of the Golan Heights, the Gaza Strip, and the West Bank, claiming that they were essential to Israeli security. **(F)** Although the Palestinians were granted some autonomy in the Gaza Strip and West Bank, during the 1990s, the Israelis continued to build Jewish settlements in the West Bank and Golan Heights. This map shows West Bank settlements where 350,000 Jewish settlers live. [Sources consulted: "Map: Golan Heights," http://www.fmep.org/reports/archive/vol.-22/no.-6/map-golan-heights; Geoffrey Aronson, "The Occupation Returns to Center Stage," *Foundation for Middle East Peace Settlement Report*, 22(6), http://www.fmep.org/reports/archive/vol.-22 /no.-6/the-occupation-returns-to-center-stage; http://thinkprogress.org/security/2012/05/14/483860/eu-settlements-threaten-two-states/?mobile=nc; http://thinkprogress .org/wp-content/uploads/2012/05/APN-Settlement-Map1.png.]

situation with two geographically separate territories governed by two different administrations has been a reality since 2007. Hamas gained legitimacy among many Palestinians because, modeled after the Muslim Brotherhood in Egypt (see page 340), it provided much-needed social services. But periodically, as in August 2016, there are violent protests by Palestinians against Hamas' inefficient administration of Gaza. At the same time, because of its continuing attacks on Israel, Hamas is considered a terrorist organization by the United States and the EU.

The territorial dispute is also affected by continuous Israeli settlements in the West Bank. From 2005, some 200,000 new Israeli settlers have colonized the West Bank and thousands of Palestinians have been displaced as a result. The **West Bank barrier,** a high containment wall **(FIGURE 6.23A)** built beginning in 2003, encircles the Jewish settlements on the West Bank and separates approximately 25,000 Palestinian farmers from their fields. It also blocks roads that once were busy with small businesses, effectively annexes 6 to 8 percent of the West Bank to Israel, and severely limits Palestinian access to much of the city of Jerusalem, most of which is now on the Israeli side of the barrier. The barrier, declared illegal by the World Court and the United Nations, and opposed by the United States, is nonetheless very popular among Israelis because it has reduced the number of Palestinian suicide bombings.

The most intractable of territorial disputes may be over the city of Jerusalem. The city is situated right at the boundary between Israel and the West Bank. West Jerusalem is heavily Jewish, while East Jerusalem is predominantly populated by Palestinians, although a large number of Jewish settlers have moved there over time. In the middle sits the old city of Jerusalem. This is where three world religions have converged for centuries, and all three now claim the right to sacred spaces in the city. Since the days of the Crusades, many Christian churches have been built in the old city and pilgrims from around the world visit there. The Western Wall is a remnant of the Jewish temple from biblical times and the most important site of prayer in contemporary Judaism. Only a few hundred feet from there is the Dome of the Rock, a Muslim shrine at the location of Mohammad's ascent to heaven according to Islamic thought. Next to it sits the Al Aqsa Mosque, commonly considered the third holiest site in Islam (after Makkah and Al-Madinah). A potential two-state solution will have to resolve who should control and govern the city, and how different faiths would have access to Jerusalem. No negotiations in terms of how that could be achieved have been attempted.

ON THE BRIGHT SIDE: Economic Interdependence as a Peace Dividend

Though rarely covered in the media, economic cooperation is a fact of life for Israelis and Palestinians and is a crucial component for any two-state solution. Economic ties between Palestine and Israel have been essential to both for many years. Israel is the largest trading partner of the West Bank and the Gaza Strip, and provides Palestinians with a currency (the Israeli shekel), electricity, and most imports. Israel needs the labor of tens of thousands of Palestinian workers in fields, factories, and homes (see Figure 6.23B). During periods of calm, Israelis and Palestinians quietly reach agreements to expand Palestinian access to jobs in, and trade with, Israel. Outbreaks of violence impede the ease of movement of people through roadblocks, but the two sides continue to remain economically interdependent. ■

> **West Bank barrier** an Israeli-built concrete wall or fence that now surrounds much of the West Bank and encompasses many of the Jewish settlements there

FIGURE 6.23 Conflict and cooperation between Israelis and Palestinians.

(A) Palestinians on their way to Jerusalem to celebrate Ramadan pass through a heavily guarded checkpoint at the West Bank barrier in Bethlehem, Palestine. The barrier has hindered the flow of people and goods between Israel and the West Bank, severely damaging the latter's economy.

(B) Palestinian workers at a SodaStream factory in the West Bank. Israeli companies like SodaStream have access to the expertise, equipment, and investment capital that make the growth of manufacturing industries possible, while Palestinians supply cheap labor. SodaStream's operations are controversial and a boycott movement of Israeli companies making products in the occupied Palestinian Territories is gaining traction in Europe and the United States.

Situation 2: Fifty Years of Trouble Between Iraq and the United States

The origins of the U.S. war with Iraq, beginning in 2003, lie in 1963, when the United States backed a coup that installed a pro-U.S. government that evolved into the regime of Saddam Hussein. For decades, the United States had an amicable relationship with the Iraqi government, driven in large part by Iraq's considerable oil reserves, the fifth largest in the world after those of Venezuela, Saudi Arabia, Canada, and Iran. The United States publicly supported Iraq in the 1980–1988 war between Iraq and Iran. Relations with Iraq took a dramatic turn for the worse in 1990 when Saddam Hussein, Iraq's dictator, invaded Kuwait. The United States and its allies forced Iraq's military out of Kuwait in the Gulf War of 1990–1991, and afterward the United Nations placed Iraq under crippling economic sanctions, although they failed to tame Saddam Hussein (see Figure 6.6D).

After the terrorist attacks of September 11, 2001, the United States first launched a war on Afghanistan, but soon shifted its focus to Iraq (see Chapter 2), which resulted in a 2003 invasion of the country. The initial invasion met little resistance; however, the subsequent U.S. occupation encountered insurgent attacks, with violence peaking between 2006 and 2007, after which it gradually decreased, leading to the end of official U.S. combat operations in 2012. By 2014, a total of 4486 U.S. troops had been killed and many more wounded or diagnosed with some form of serious mental disorder, including posttraumatic stress disorder (PTSD). The death toll for Iraqis, including civilians, was much higher, estimated in the hundreds of thousands and possibly over 1 million, when counting Iraqi deaths caused by the harsh conditions created by the war.

The failure to understand the simmering tensions between Iraq's major religious and ethnic groups crippled the U.S. intervention. Sunnis in central Iraq had dominated the country under Saddam, although they constituted only about 35 percent of the population. Shi'ites in the south, with over 60 percent of the population, have dominated politics since the fall of Saddam. This has given more influence to neighboring Iran, whose mostly Shi'ite population has an affinity with Iraqi Shi'as. Most observers agree that the new Shi'ite-dominated government has not been very inclusive; instead, Iraqi Sunnis have increasingly felt alienated from political decision making. Meanwhile, Kurds in the northeast, once brutally suppressed under Saddam Hussein, are allied with Kurdish populations in Turkey, Iran, and Syria. They have resisted cooperation with the Iraqi national government in Baghdad by forming an autonomous and self-governing region.

The military force of Iraq proved to be ineffectual in maintaining the security of the country. Soldiers were poorly trained and prone to desertion, while Sunnis were excluded from serving altogether. In this weak security space, a militia that we know as the Islamic State quickly gained ground in the territories in central Iraq where most of the population was Sunni. The roots of the Islamic State can be traced to Al Qaeda insurgents who fought against the U.S. occupation and later the Shi'ite-led government of Iraq. Now its own organization, the Islamic State took over Mosul, Iraq's second-largest city, in 2014. At current writing, it has expanded its control into parts of central and northern Iraq, as well as into bordering Syria.

The objective of the Islamic State is pan-Islamic rather than just overthrowing the government in Iraq or Syria. It wishes to establish a new *caliphate*—a historic term for extensive Islamic empires during the medieval "golden era" of Muslim power—that would unite Muslims everywhere against Western powers. As unrealistic as this goal is, the Islamic State has been successful in using social media to spread its vision and to disseminate successes on the battlefield, including brutal tactics such as beheadings of opponents. Through this media campaign, it has been able to attract foreign fighters. There are tens of thousands of young men, and some women, that have come from other Muslim countries, and a few from the West, to join the fight. The Islamic State has overshadowed Al Qaeda as the leading radical Islamist group in the world, as well as becoming a source of inspiration for Islamists everywhere to carry out terrorist attacks against perceived enemies.

Situation 3: Failure of the Arab Spring in Syria

As noted above, the conflict in Iraq recently spilled over into Syria. The reason why is that Syria was already embroiled in a civil war and Islamic State forces could relatively easily move across the border and establish a presence in Syria.

The background to this development is the Arab Spring protests that spread around Syria in 2011. Protests escalated into a rebellion, unleashing the same questions about the future as in all the Arab Spring movements. Can autocratic governments continue to block democratic reforms? Will Islamist factions more often than not succeed in taking control? Will women be able to change patriarchal customs? Will gains in economic development be overwhelmed by the material destruction caused by civil war? In Syria, yet another concern has been raised that carries implications for the entire region: Climate change and the water scarcity that accompanies it (discussed later in this section) are now thought to be among the stressors that helped to bring on the Syrian rebellion.

The political causes of rebellion in Syria are many. The existence of Syria is a legacy of French colonialism, which brought together a large number of ethnic and religious groups. Several factions of Muslims, as well as Christians and even a few Jews, lived together. Markets were full of life, women could have careers and go out alone past midnight, old men played board games while children played along community streets, tourists flocked to ancient historic sites, and Syrian TV dramas disseminated the Syrian hospitable way of life across the region. But after the Assad family swept to power in a bloodless coup in 1970, Syria slowly morphed into an authoritarian state. By the time Bashar al-Assad took over from his father Hafez in 2000, Syria had gained a reputation for backing assassinations in Lebanon and bombings in Iraq, and for supporting *Hezbollah*—a Lebanese-based radical Shi'ite group that has close links to Iran, opposes Israel, and uses terrorist tactics.

Although some Syrians enjoyed the way the Assads thumbed their noses at the United States and Europe, more educated and secular Syrians could see that their access to decent lives and participation in civic life was being curtailed by the Assad regime. The trappings of a free society, such as shopping malls, Coca-Cola, Internet cafés, and Facebook, were not enough to pacify people when open debate, elections, and creativity were stifled by an authoritarian government and a society rife with corruption.

Dictatorial rule for forty years and ethno-religious divisions set up Syria for failure during the Arab Spring. The Assad family belongs to the minority Alawite sect of Shi'ites, while the majority of Syrians are Sunni Muslims. To keep political equilibrium, the Assads favored a secular state similar to Turkey's. However, Sunnis were kept out of governing circles. Sunni antagonism against Alawites grew because Alawites led particularly privileged lives, with many advantages coming to them from the Assad family.

Arab Spring protests were met with brutal repression in Syria. In time, it became apparent that the rebel factions came from opposing points of view—some favoring pro-democratic secular changes, others wanting more Islamist conservative reforms. The outcome of the conflict in Syria has been tragic. Different sources estimate the number of deaths from 200,000 to 300,000. Moreover, 4.6 million refugees have left the country, while 6.6 million are internally displaced within Syria (see Figure 6.15F). Such numbers are staggering for a country of 17 million. Most Syrian refugees remain in camps in bordering countries. The impact of this movement of people will be felt for decades in the region in ways that are not yet obvious. The refugee stream to Europe is widely written about in media, but it should be noted that only 10 percent of Syrian refugees have made it to Europe; most remain in Jordan, Turkey, and Lebanon.

Western governments, while expressing sympathy for the secular factions of the war, have been reluctant to intervene for fear of becoming bogged down in a conflict without a clear end point. Russia, on the other hand, supports the Assad regime, which is an old ally. The reasons are twofold: Russia's only military base in the region is in Syria and it seeks to maintain that military presence. Also, there are Islamists along the southern border of Russia in Caucasia and Central Asia; thus, Islamist ideologies are as much of a threat to Russia as they are to the West. Another country that would like to see a resolution to the conflict is Turkey. It wants to limit the refugee movement into its country; ethnic Kurds live on both sides of the border and such connections can be destabilizing; and terrorist attacks have occurred recently in Turkey, some of which may be attributed to the Islamic State.

To conclude, the Islamic State plays an outsized role in Syria as well. The group has established a capital of sorts in the city of Raqqa in eastern Syria. It is financing its war with oil sales, the looting of antiquities, and taxing the local population. What the Islamic State lacks, though, are the knowledge and skills to administer a territory beyond intimidation and violence, so its long-term sustainability is questionable.

THINGS TO REMEMBER

• The decades-old conflict between Israel and the Palestinians is unlikely to be resolved due to competing claims of territorial control. The situation continues to be an obstacle to widespread political and economic cooperation in the region.

• The United States has maintained a long and complicated relationship with Iraq that dates back to the 1960s, when the United States tried to influence Iraq's internal affairs.

• After the 2003 U.S. invasion, sectarian conflict plagued Iraq, which ultimately led to the rise of the Islamic State. This group now also vies for control in Syria and has destabilized much of the region.

• Arab Spring–inspired political protests against the autocratic Assad regime in Syria involved competing groups, such as the Islamic State, other Islamists, secularist, Kurds, and pro-democracy dissident factions. The result has been a civil war that devastated Syria as well as forced people to leave their homes and seek refuge in neighboring countries and Europe.

URBANIZATION

GEOGRAPHIC THEME 4

Urbanization: Two patterns of urbanization have emerged in the region. In the oil-rich countries, there are spectacular new luxury-oriented urban areas. In the oil-poor countries, many people live in overcrowded urban slums with few services, and jobs for rural migrants are scarce.

The rate of urbanization in Southwest Asia and North Africa varies greatly—from 100 percent in the small Gulf city-states Bahrain and Qatar to 33 to 34 percent in Sudan and Yemen. On average, the region is more urban than the world as a whole. In part, that is an outcome of a long historical process whereby people have clustered in cities as agricultural opportunities in the desert landscape were marginal and as cities of the region played an important role in far-flung trade. However, contemporary forms of urbanization are also transforming this region. Two globalized patterns of urbanization with distinct geographic signatures are now apparent: one pattern in the newly oil-rich countries, and another in those countries where development has primarily resulted in rural-to-urban migration, with its associated crowded slum housing.

Determining the size of cities in historical times is not always easy, but estimates indicate that five (six if counting Córdoba, Spain, which was Muslim at the time) of the ten largest cities in the world in 1000 C.E. were in Southwest Asia and North Africa; four of the top ten in the year 1500. This is not surprising considering the importance of the region in economic, cultural, and political affairs at the time. Today, there are still numerous cities with a medieval core that is dominated by narrow, pedestrian-only streets where commerce is taking place. Behind the bustle of these public spaces, the traditional living arrangement was that of a courtyard surrounded by residential spaces, often for extended families. These ancient city centers are referred to as the *medina*, or simply the old city. Under the influence of European colonizers, this form of indigenous urban design was abandoned and new neighborhoods were built with Western-style apartment blocks and street networks.

Most of the population still lived in small rural settlements until the 1970s, when people began to migrate in significant numbers from villages to urban areas in response to economic forces driven by oil wealth and globalization, and by agricultural modernization that displaced small farmers. In locations other than the Gulf states, the shift toward green revolution export-oriented agriculture reduced the amount of labor required to produce crops headed for the world market. Even though the green revolution created needed export income for the countries involved, people who became part of the excess labor force had to leave the countryside for the cities. In North Africa, drought also instigated rural-to-urban migration (FIGURE 6.24B). These poor migrants ended up in emerging slums on the periphery or in the historic old city where dwellings didn't have modern amenities. By 2015, 61 percent of the region's people lived in urban areas (see the Figure 6.24 map) and 48 cities could claim a population greater than 1 million people. The largest metropolitan area in the region, and one of the largest in the world, is Cairo, with almost 19 million residents in 2015 (Figure 6.24C). Istanbul in Turkey is home to 14 million people. Both were centers of historic empires that have continued to grow in modern times.

The petroleum-rich Gulf states are now highly urbanized. Between 70 and 100 percent of the (still small) population now live in urban areas, which are extravagant in design, sprawling and auto-dependent, and very modern in appearance. Some of these cities have grown from almost nothing to over 1 million in population in just a few decades. These modern Gulf cities draw investment in high-tech ventures attracted by a low rate of taxation or no taxes, and high-end tourism. The new ventures and the construction of office and living space require a wide range of skilled workers from all over the world. For example, Dubai in the United Arab Emirates has built elaborate new high-rise condominiums for the rich. Palm Jumeirah is a fanciful palm tree–shaped artificial peninsula off the Gulf coastline with huge homes (see Figure 6.24A). At the same time, poor migrants who provide the necessary construction labor for such high-profile luxury projects are concentrated in overcrowded apartments or temporary labor camps on the outskirts of the city.

Outside the Gulf states, there have been far less planning and financing for urban growth. For example, in 1950, Cairo had about 2.4 million residents, while today it is home to almost 19 million people. Some of these millions of new residents live in huge makeshift slums, both on the outskirts of the city and the medieval interiors of the old city, where ancient houses are crumbling and plumbing and other services are chronically dysfunctional. In Egypt, 17 percent of city dwellers live in what the UN defines as slum conditions. In the poorest countries in the region, conditions are much worse. In Yemen, 67 percent of urban residents live in slums; in Sudan, an astounding 94 percent do.

The planning that has taken place in Egypt to deal with overcrowding has largely failed. Everybody agrees that Cairo, Egypt's primate city, has simply grown too large, and in order to spur decentralization, urban planners in Egypt have for decades laid out satellite developments on undeveloped land at some distance from Cairo. However, these "new towns" have not attracted many new residents, and some homes have remained vacant for years and even decades. The problem is not that they are unattractive places, but that most Egyptians trying to escape overcrowded

Thinking Geographically

After you have read about urbanization in North Africa and Southwest Asia, you will be able to answer the following questions.

A What circumstances dealt a blow to high-tech and tourism development in the Gulf states, such as Dubai?

B By 2015, what percent of the population in this region lived in urban areas?

C What has driven the migration of rural people to the region's old established cities, such as Cairo?

Cairo can't move there because of the lack of employment opportunities and transportation services. There are not enough local jobs or public transit options for those without cars to places where job prospects are better.

Internal and International Migration The prospect of better educational opportunities, jobs, and living conditions draw rural internal migrants to Cairo and other cities outside the Gulf states. However, because stable, well-paying jobs are scarce, many migrants end up working in the informal economy (see Chapter 3), sometimes as street vendors, as casual laborers, or as menial service providers. Unfortunately, education also does little to ensure employment; for decades, this region has been noted for its large numbers of unemployed and underemployed university graduates.

In the Gulf states, on the other hand, there has been a deficit of trained native young people willing and able to work in white-collar jobs, yet surplus trained workers from neighboring countries have not been welcomed. Instead, immigrants come from all over the world to function as temporary guest workers. Some work as laborers on construction sites and as low-wage workers throughout the service economy, but many work in shops and professions. In some countries, such as the UAE and Qatar, guest workers make up 80 to 90 percent of the population. Most employers prefer Muslim guest workers, and over the last two decades, several hundred thousand Muslim workers have arrived from Palestinian refugee camps in Lebanon and Syria, as well as from Egypt, Pakistan, and India. Some female domestic and clerical workers come from Muslim countries in Southeast Asia, including the Philippines. Overwhelmingly, these immigrant workers are temporary residents with no job security and no right to become citizens. Their visas are contingent upon an employer who is willing to hire them and their passports are typically confiscated by their employers; lose your job and you will be deported. The migrants remit most of their income to their families at home and often live in stark conditions alongside the opulence of those enriched by oil and gas. The wealth of a country like Qatar, the country with the highest GDP per capita in the world in 2014, is built on extreme labor exploitation.

Refugees comprise a major category of migrants, as this region has the largest number of refugees in the world. Usually, they are escaping human conflict, but environmental disasters such as earthquakes or long-term drought also displace many

FIGURE 6.24 PHOTO ESSAY: Urbanization in North Africa and Southwest Asia

Globalization has brought two distinct patterns of urbanization to this region. In fossil fuel–rich countries, populations are more urbanized and city governments have undertaken lavish building booms, drawing in laborers and highly skilled workers from across the globe. In countries without fossil fuel wealth, cities are receiving massive flows of poor, rural migrants. This is partially a result of economic reforms (emphasizing export agriculture and industrialization) aimed at making rural and urban economies more globally competitive.

A Palm Jumeirah is an artificial palm-shaped island in Dubai, UAE. Built by more than 40,000 workers, mostly low-wage migrants from South Asia, Palm Jumeirah will eventually house 65,000 people, mostly in villas and condos that cost millions of dollars each. Palm Jumeirah and several similar developments are central to Dubai's efforts to build a globalized tourism-based economy that will prosper long after the region's fossil fuel resources are exhausted.

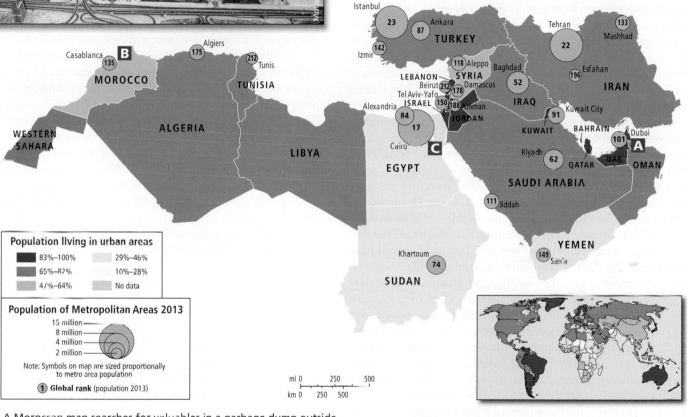

Population living in urban areas

- 83%–100%
- 65%–82%
- 47%–64%
- 29%–46%
- 10%–28%
- No data

Population of Metropolitan Areas 2013

- 15 million
- 8 million
- 4 million
- 2 million

Note: Symbols on map are sized proportionally to metro area population

① **Global rank** (population 2013)

ml 0   250   500
km 0   250   500

B A Moroccan man searches for valuables in a garbage dump outside Casablanca. The city's population is growing quickly due largely to migration from Morocco's drought-stricken interior, but the lack of jobs in the city is resulting in escalating unemployment and crime.

C A market in Cairo. After decades of migration from rural areas, Cairo is the region's largest city; it is severely overcrowded.

people. When Israel was created in 1949, many Palestinians were placed in refugee camps in Lebanon, Syria, Jordan, the West Bank, and the Gaza Strip. Palestinians still constitute the world's oldest and largest refugee population, numbering at least 5 million, although many of them were born in the country where they now reside and exhibit the characteristics of a diaspora more than refugees. Countries like Jordan and Lebanon, which already had a large Palestinian refugee population, have been the recipients of a large number of recent refugees from the conflict in Syria. Jordan has the largest number of refugees, 2.8 million people, in the world. To the north, Turkey is currently the temporary home of 1.6 million refugees, almost all of them from Syria. Elsewhere, Iran is sheltering almost a million Afghans and Iraqis because of continuing violence and instability in their home countries. Across the region, even more people are refugees within their own countries, a category called *internally displaced people*. Almost 4 million Iraqis are internally displaced. This number has skyrocketed since 2014 when the Islamic State unleashed a new round of violence. In Sudan, about 2.3 million internal refugees are living in camps, largely as an outcome of the conflict in Darfur (Figure 6.9B).

Refugee camps often become semipermanent communities, in essence urban slums, of stateless people in which whole generations are born, mature, and die. Although residents of these camps can show enormous ingenuity in creating a community and an informal economy, the cost in terms of social disorder is high. Tension and crowding create health problems. Because birth control is generally unavailable, refugee women have a high fertility rate. Disillusionment is widespread. Years of hopelessness, extreme hardship, lack of education, or well-paid employment take their toll on youth and adults alike. Because of these factors, it is easy for Islamists and jihadists to find new recruits, which they do by providing basic services to camp residents. Moreover, even though international organizations also provide for refugees, the refugees constitute a huge drain on the resources of their host countries. In Jordan, for example, native Jordanians are a minority in their own country because more than 3 million Palestinian refugees and their children, and the new influx of Syrians, account for well over half the total population of the country, and their presence has changed life for all Jordanians.

THINGS TO REMEMBER

GEOGRAPHIC THEME 4 • **Urbanization:** Two patterns of urbanization have emerged in the region. In the oil-rich countries, there are spectacular new luxury-oriented urban areas. In the oil-poor countries, many people live in overcrowded urban slums with few services, and jobs for rural migrants are scarce.

• The region is home to the largest refugee population in the world. Palestinians, Iraqis, and most recently Syrians have escaped conflicts and are refugees in neighboring countries and Europe, or are internally displaced. Many refugee camps have turned into semipermanent urban slums over time.

POPULATION AND GENDER

GEOGRAPHIC THEME 5

Population and Gender: This region has the second-highest population growth rate in the world. Part of the reason for this is that women are generally poorly educated and tend not to work outside the home. Childbearing thus remains crucial to a woman's status, a situation that encourages large families.

Although fertility rates have dropped significantly since the 1960s, to 3.2 children per adult woman in 2014, they are still higher than the world average of 2.5. Only sub-Saharan Africa, at 5.0 children per woman, is growing faster. At current growth rates, the population of the region will be about 570 million by 2030. This growth will severely strain supplies of fresh water and food, and worsen shortages of housing, jobs, medical care, and education.

As noted in many world regions, population growth rates are higher in societies where women are not accorded basic human rights, are less educated, and work primarily inside the home. In places where women have opportunities to work or study outside the home, they usually choose to have fewer children. Figure 6.30 shows that as of 2013, considerably less than 50 percent of women across the region (except in Israel and Qatar) worked outside the home at jobs other than farming. Moreover, on average, only 75 percent of adult females can read—a fact that limits their employability—whereas 89 percent of adult males can read. The literacy gender gap in the region may not be as high as in South Asia and sub-Saharan Africa, but it is nevertheless significant, with Yemen having, in fact, the highest such gap in the world.

For uneducated women who work only at home or in family agricultural plots, children remain the most important source of personal fulfillment, family involvement, and power. This may partially explain why in 2014 only 41 percent of women in this region were using modern methods of contraception, and only 53 percent any method of contraception at all. Both of these numbers are well below the world averages of 56 and 62 percent, respectively. Other factors resulting in low use of contraception are male dominance over reproductive decisions and the unavailability or high cost of effective birth control products.

The deeply entrenched cultural preference for sons in this region is both a cause and result of women's lower social and economic standing. It also contributes to population growth, as families sometimes continue having children until a desired number of sons is reached. At the same time, these effects vary from country to country, which is evident in the different shapes of the population pyramids in **FIGURE 6.25** (see also the discussion in Chapter 1). Iran had high population growth until approximately 20 years ago. Since then birth rates have dropped dramatically, which is represented by the narrowing of the bottom of the pyramid. This is not happening in Israel, where women still have as many as 3.3 children, on average (compared to Iran's 1.8). In Qatar and the UAE, gender imbalance is extreme (see Figure 6.25C). The unusually large numbers of males over the age of 15 are the result of the presence of numerous male guest workers.

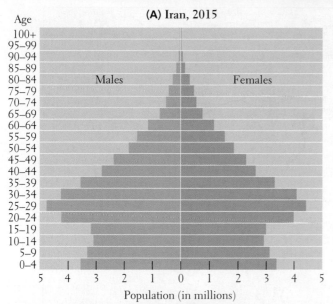

(A) Iran, 2015

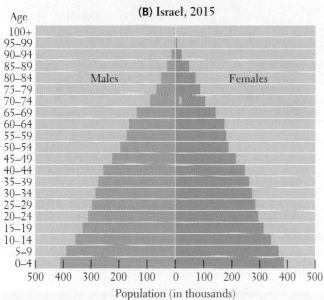

(B) Israel, 2015

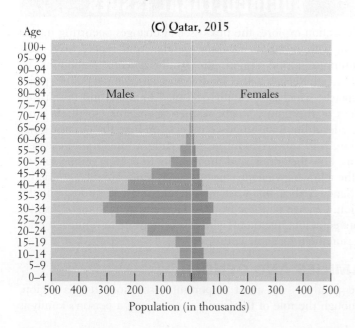

(C) Qatar, 2015

Overall, opportunities for women are opening and this change is likely to reduce women's interest in having large families. Slower population growth will mean that fewer resources will have to be invested in housing, schools, and services.

POPULATION DISTRIBUTION

Although the region as a whole is nearly twice as large as the United States, most of the population is concentrated in the few areas that are useful for agriculture. Vast tracts of desert are virtually uninhabited, while the region's 540 million people are packed into coastal zones, river valleys, and mountainous areas that capture orographic rainfall (compare **FIGURE 6.26** with Figure 6.4). Population densities in these areas can be quite high. For example, the well-watered Nile delta is home to a very dense rural population, while some of Egypt's urban neighborhoods have more than 260,000 people per square mile (100,000 people per square kilometer), a density 4 times higher than that of New York City, the densest city in the United States.

HUMAN WELL-BEING

As we have observed in previous chapters, gross national income (GNI) per capita can be used as a general measure of well-being because it shows broadly how much money people in a country have to spend on necessities. As discussed in earlier chapters, the term GNI is often followed by PPP, which indicates that all figures have been adjusted for cost of living and so are comparable.

FIGURE 6.27A points to two situations regarding GNI in this region. First, from country to country, there is wide variation—less than U.S.$4000 per person per year in Sudan and Yemen and over U.S.$40,000 per person per year in Kuwait, the UAE, Saudi Arabia, and Qatar. Second, oil and gas wealth does not necessarily translate into high GNI per capita (compare Figure 6.27A with Figure 6.18)—Iraq has significant oil wealth but only modest GNI per capita. Libya formerly ranked higher but has fallen to moderate levels in the aftermath of the Arab Spring. Turkey, which has a diversified economy and very little oil, nonetheless has a GNI rank equal to that of Iran (in the middle range), which has significant oil and gas reserves.

GNI per capita is only an average figure and does not indicate how income is actually distributed across the population. Thus, it does not show whether income is equally apportioned or whether a few people are rich and most are poor. Data on income inequality in the region are not always available, but those that do exist suggest that most countries have moderate levels of inequality. Gulf states with great contrasts between low-wage immigrant workers and fabulously wealthy locals should be assumed to have great disparities.

FIGURE 6.25 Population pyramids for Iran, Israel, and Qatar. The population pyramid for Iran **(A)** is at a different scale (millions) from those for Israel **(B)** and Qatar **(C)** (thousands). The imbalance of Qatar's pyramid in the 25–54 age groups is caused by the presence of numerous male guest workers. [Source consulted: "Population Pyramids for Iran," *International Data Base*, U.S. Census Bureau, 2015, http://www.census.gov/population/international/data/idb/informationGateway.php.]

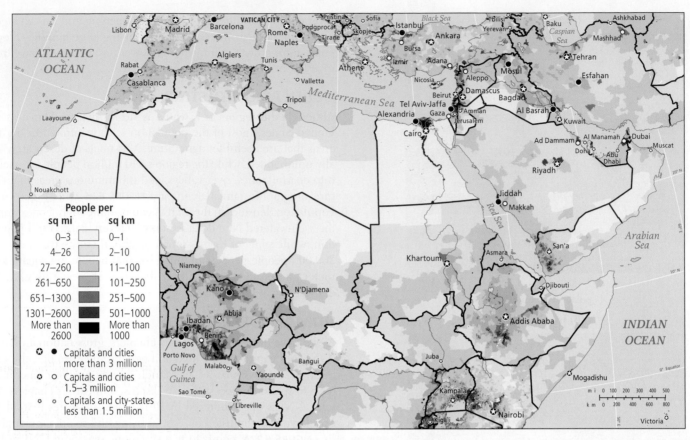

FIGURE 6.26 Population density in North Africa and Southwest Asia.

Figure 6.27B depicts countries' rank on the Human Development Index (HDI), which is a calculation (based on adjusted real income, life expectancy, and educational attainment) of how adequately a country provides for the well-being of its citizens. In this region, HDI varies greatly from the low to the very high range, with Israel ranking the highest (18th in the world). If on the HDI scale a country drops below its approximate position on the GNI scale, one might suspect that the lower HDI ranking is due to poor distribution of wealth and poor social services: A critical mass of people are simply not getting a sufficient share of the national income to receive good health care and an education. Iraq ranks lower on HDI than it does on the GNI scale, indicating that the revenue generated from oil is not reaching many people in the country. Jordan and Tunisia, on the other hand, rank high on the HDI scale but somewhat low on GNI per capita. This probably signifies that the government is investing in social services that will enhance human well-being.

Figure 6.27C shows the regional gender development index (GDI). Countries are ranked according to the degree to which women and men are equal with regard to longevity, education, and income. Of the 21 countries, only Qatar ranks in the highest category. The small but wealthy country has made rapid progress in providing opportunities for women. Two other Gulf states, Kuwait and UAE, rank in the medium high category

together with Israel and Libya. On the other hand, 12 countries rank in the lowest category, which shows that much of the region ranks lower on gender development than do most other parts of the world.

SOCIOCULTURAL ISSUES

This section explores the broad social changes occurring in this region with regard to families, gender, and space. Like most institutions in this region, the family remains predominantly patriarchal. Related to this are divisions between the public spaces of society, which are usually occupied by men, and the private spaces of the home, where women spend most of their time. More legal limits are imposed on women here than in any other region, with various countries restricting women's political freedoms, how they dress, and their ability to occupy public spaces. As the following vignette suggests, such limits are most restrictive in Saudi Arabia, but exist to different degrees throughout the region. Change on these fronts is happening, as women push for more equitable treatment, which is evident in an improved GDI measurement for some countries.

FAMILIES AND GENDER

The family is the most important institution in this region. Although the role of the family is changing, a person's family is

FIGURE 6.27 Global maps of human well-being.

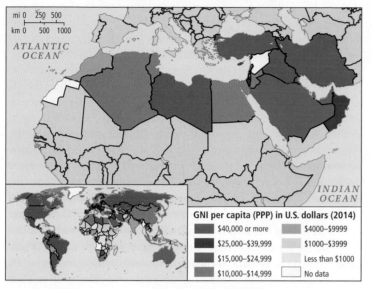

(A) Gross national income (GNI) per capita, adjusted for purchasing power parity (PPP)

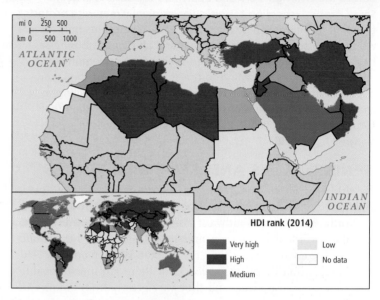

(B) Human Development Index (HDI)

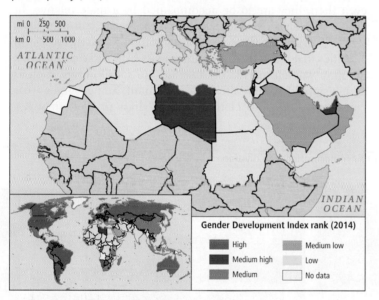

(C) Gender Development Index (GDI)

still such a large component of personal identity that the idea of individuality is almost a foreign concept. Each person is first and foremost part of a family, and the defining parameter of one's role in the family is gender identity (see Figure 6.3).

The head of the family is nearly always a man; even when a woman is widowed or divorced, she is under the tacit supervision of a male, perhaps her father, her adult son, or another male relative. An educated unmarried woman with a career outside the home is likely to live in the home of her parents or a brother, and to defer to them out of respect in decision making. Traditionally, men are considered more stable and capable of making decisions, and thus it is believed that they should be in charge. Women usually agree or at least comply with this viewpoint with minimal argument. Such **patriarchal** ideas are beginning to change as the result of modernization. During the Arab Spring, a few brave women have challenged convention and taken a political stance (see the opening vignette; see also the discussion below).

VIGNETTE Because Saudi Arabia is an absolute monarchy, the legal system consists of a limited set of actual laws and a large number of royal decrees; some of the latter may be inconsistent and contradictory. The legal rights of citizens are therefore not always easy to decipher. Recently, a small number of women started to challenge discriminatory practices. Granted, Islamic shari'a law is the basis for weak women's rights in family law that governs marriage and divorce, as well as courtroom proceedings that stipulate a woman's testimony is worth less than a man's. However, legal practices are also a matter of interpretation by a rigidly patriarchal society. Twenty-two-year-old law student Mohra Ferak, concerned about women's rights, organized a series of popular seminars directed at women to inform them about their legal rights and how to maneuver the existing legal system. Such information is not always publicly available in a country where the desire for privacy in family matters usually trumps public debate. Another young and educated female, the attorney Bayan Mahmoud Zahran, spends time at a shelter for divorced and widowed women—two vulnerable groups in Saudi Arabia—offering them legal advice. On the one hand, the fact that women now study law and can become lawyers is a step forward. On the other hand, King Salman, the ruler of the kingdom, is less interested in social reform than economic development. New technology, however, has the capacity to empower people. Women, who are not allowed to drive in Saudi Arabia, have started to use the app for the car service Uber, which provides them with greater spatial freedom as they no longer have to wait for a male family member to drive them around.

Many aspects of the Saudi legal system are based on gender segregation, which has some eerie parallels with the American South

patriarchal relating to a social organization in which the father is supreme in the clan or family

before the civil rights era and apartheid-era South Africa. In Saudi Arabia, schooling for girls was originally met with protests that had to be quelled by military troops, restaurants only serve women in a separate secluded area, and all types of businesses have to design gender-segregated spaces for both employees and customers according to the specifications of the labor department. Women like Bayan and Mohra are not willing to challenge these conventions yet, but perhaps they will in the future. [Source consulted: K. Zoepf, "Sisters in Law," The New Yorker, January 11, 2016.] ∎

GENDER ROLES AND GENDERED SPACES

Carefully specified gender roles are common in many cultures, and there is often a spatial component to these roles. In the region of North Africa and Southwest Asia, in both rural and urban settings, the ideal is for men and boys to go forth into *public spaces*—the town square, shops, the market. Women are expected to inhabit primarily *private spaces*. But there are many possible exceptions to those ideals and they vary greatly from country to country.

To facilitate this ideal of public/private spaces for the sexes, traditional family compounds included a courtyard that was usually a private, female space within the home (FIGURE 6.28A); the only men who could enter it were relatives. For others, a separate room to entertain guests was maintained. For the urban upper classes, female space was an upstairs set of rooms with latticework or shutters at the windows, which increased the interior ventilation and from which it was possible to look out at street life without being seen. These lattice works also protrude from the façade so that the person looking out can observe in all directions (see Figure 6.28B, C). Today, the majority of people in the region live in urban apartments designed much like elsewhere in the world, yet even in these places, there is a clear demarcation of public and private space. One or two formally furnished reception areas are reserved for nonfamily visitors, and rooms deeper into the dwelling are for family-only activities. When guests visit, women in the family are usually absent or present only briefly. Customs vary not only from country to country but also from rural to urban settings, by social class, and by personal preference. Today, many women as well as men go out into public spaces, but the conditions under which women enter these spaces remain an issue and a woman who challenges convention does so at her own peril and that of her family's reputation.

The requirement that women stay out of public view (also known as **female seclusion**) is most strictly enforced in the more conservative Muslim countries of the Gulf states. Women in these countries are generally expected to remain in private spaces except when engaged in important business; even then, they are to be accompanied by a male relative. In the more secular Islamic countries—Morocco, Tunisia, Libya, Egypt, Turkey, Lebanon, and Iraq—women regularly engage in activities that place them in public spaces. In these countries, a group of women may go out together for social events, such as a high-spirited evening meal at a restaurant. Some women wear conservative religious clothing; others dress in Western styles,

female seclusion the requirement that women stay out of public view

FIGURE 6.28 Domestic spaces. Many of the older cities and buildings in this region reflect the division between public space and private or domestic space.

(A) As seen from above, almost all houses in the old part of Baghdad are arranged in a courtyard pattern. Trees sometimes provide shade in these private spaces.

(B) Projecting windows covered with lattice or louvers, known as a *mashrabiya*, in Jeddah, Saudi Arabia. Mashrabiyas allow women to look out on the street below and catch breezes without being seen.

(C) An interior view of a courtyard in Cairo. Courtyards are places where women can do chores and manage children while remaining secluded from the outside world.

at least in the big cities. Increasingly, female doctors, lawyers, teachers, and businesspeople are found in even the most conservative societies.

Affluent urban women may observe seclusion either rarely (especially if they are highly educated), or, if they are relatively affluent but less educated, even more strictly than do rural women. Although rural women are often more traditional in their outlook, they have many tasks that they must perform outside the home: agricultural work, carrying water, gathering firewood, and buying or selling food in markets. In North Africa, women of the rural Berber ethnicity often travel to sell their goods at markets and have more spatial freedom than most Arab women. Meanwhile, upper-class women who can afford servants to perform daily tasks in public spaces can more easily stay secluded and afford the amenities (TV, DVDs, cell phones) that relieve the boredom and isolation of seclusion. It should also be noted that seclusion customs are in flux, being relaxed in some countries, such as Morocco, Turkey, and Tunisia, but becoming more strictly enforced in other countries, such as post–Arab Spring Egypt and areas where Islamists are newly active.

Many women in this region use clothing as a way to create private space **(FIGURE 6.29)**. This is done with the many varieties of the **veil,** which may be a garment that totally covers the woman's body and face or just a scarf that covers her hair. There are many different names for such clothing depending on its design and the country. A *hijab* is a headscarf, the *niqab* covers much of the face, while a *chador* or *burka* are full-body garments. In some cultures, even prepubescent girls wear the veil; in others, they go unveiled until their transition to adulthood is observed. The veil allows a devout Muslim woman to preserve a measure of seclusion when she enters a public space, thus increasing the territory she may occupy with her honor preserved. A modern young woman may choose to wear a headscarf with makeup, jewelry, jeans, and a

t-shirt to signal to the public that she is both an up-to-date woman and an observant Muslim. The use of some form of veil is dependent on personal choice, family background and expectations, as well as social pressures. In more traditionalist countries like Iran and Saudi Arabia, "morality police" enforce proper clothing in public spaces.

There is considerable debate about the origin and validity of female seclusion and whether veiling and seclusion are specifically Muslim customs. Scholars say that these ideas, as well as the custom of having more than one wife, actually predate Islam by thousands of years and do not derive from the teachings of the Prophet Muhammad. In fact, Muhammad may have been reacting against such customs when he advocated equal treatment of males and females. Muhammad's first wife Khadija did not practice seclusion, and worked as an independent businesswoman whose counsel Muhammad often sought.

THE RIGHTS OF WOMEN

This region has notably more restrictive customary and legal limits on women than any other. In Saudi Arabia, and to some extent other Gulf states, women cannot travel independently or even drive a car or shop without male supervision. Yet even in these contexts, changes have recently occurred. Some women in the Gulf states have become more active in public life, education, and business. In Qatar, Sheikha Moza, the second of the three wives of the emir (Muslim ruler) of Qatar, Sheikh Hamad bin Khalifa al-Thani, has implemented remarkable changes for women. They can now drive, attend a university, be elected to political office, and work side by

> **veil** the custom of covering the body with a loose dress and/or of covering the head—and in some places the face—with a scarf

FIGURE 6.29 Variations on the veil as portable seclusion. There is an almost infinite variety of interpretations of the veil.

(A) An Iraqi woman wears a head scarf.

(B) Schoolgirls in Iran wear a uniform that covers most of their hair and a suit that covers most of their body.

(C) These Tunisian women are covered except for their eyes.

side with men; the sheikha has founded a battered women's shelter and serves as a UNESCO envoy. In the UAE, a few female activists, along with female attorneys and journalists, have begun to question persistent patriarchal attitudes. Some of the Gulf states, together with Israel, have the highest percentage of the region's women in the labor force **(FIGURE 6.30)**. Saudi Arabia remains the most restrictive country, but even there it is now possible for a woman to register a business without first proving that she has hired a male manager. A few young Saudi women even staged mini-demonstrations by posting YouTube videos of themselves driving. Although they were eventually punished by the state for this behavior, they inspired further challenges to convention. The legal restrictions of Saudi women are explored in the prior vignette.

The Arab Spring in North Africa demonstrated just how elusive political equality for women can be. In Tunisia, Libya, and Egypt, some women were active as public protesters against undemocratic regimes, but military and civilian supporters of the various regimes singled them out for particularly harsh retribution. Among their fellow male demonstrators, it was usually the younger men who supported the women's rights and physically defended them when supporters of the various regimes attacked.

A source of contention within this region and abroad is the practice of polygyny (see Chapter 7): when a man takes more than one wife at a time. Although the Qur'an allows a man up to four wives, it generally does not encourage this and imposes financial limits on the practice by requiring that each wife be given separate and equal living quarters and support. While legal in most of the region (although not in Tunisia or Turkey), polygyny is relatively rare, with less than 4 percent of males in North Africa practicing it. In Southwest Asia, polygyny—not covered in any reliable statistical surveys, but only in secondary reports—may be somewhat more common. About 5 percent in Jordan and 8 to 12 percent of marriages in Kuwait are polygamous. The social justifications given for polygyny further illustrate traditional ideas about women's rights in Islam. In common legal practice in many countries, an unmarried woman under 40 must be regarded as a minor requiring protection. Therefore, if a man takes her as a second or third wife, she gains support and safety. Similarly, a widow is saved from disgrace and poverty if her dead husband's brother takes her as an additional wife.

Female genital mutilation (FGM, see Chapter 7) is a practice that intends to control female sexuality. However, it is not widely practiced in this region except in Sudan and Egypt, where although it is illegal, as of 2013 according to the World Health Organization, more than 90 percent of women have undergone the practice, mostly before age 15. The prevalence of FGM in Sudan and Egypt seems to be culturally connected to nearby sub-Saharan countries where the practice is also very common.

Political and social equality with men can only improve with more economic independence for women (Figure 6.30). With

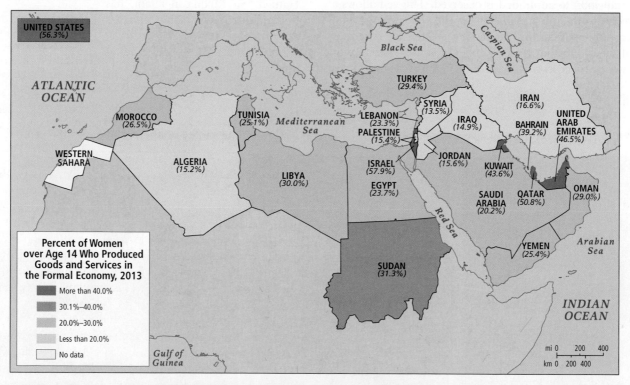

FIGURE 6.30 Percentage of the region's women who are in the labor force (2013). The female participation rate in the labor force has, for the most part, increased over time. However, the percentages in the map should also be compared to higher numbers for working males, which range from 66.4 percent (occupied Palestinian Territories) to 95.5 percent (Qatar) in the region. [Source consulted: *2015 UN Human Development Report*, Table 5.]

the exception of Israel and Qatar, most women in this region do not work outside the home; when they do, they are paid, on average, only about 40 percent of what men earn for comparable work. Only South Asian countries have a similarly large wage gap based on gender.

THE LIVES OF CHILDREN

Three observations can be made about the lives of children in the Islamic cultures of North Africa and Southwest Asia. First, in most families children contribute to the welfare of the family starting at a very young age. In cities, they run errands, clean the family compound, and care for younger siblings. In rural areas, they do all these chores and also tend grazing animals, fetch water, and tend gardens. Second, their daily lives take place overwhelmingly within the family circle. Both girls and boys spend their time within the family compound or in urban areas in adjacent family apartments. Their companions are adult female relatives and siblings and cousins of both sexes. Even teenage boys in most parts of the region identify more with family than with age peers.

In rural areas, prepubescent girls can move around in public space as they go about their chores in the village. The U.S. geographer Cindi Katz found that until puberty, rural Sudanese Muslim girls have considerably more spatial freedom than do girls of similar ages in the United States, who are rarely allowed full access to their own neighborhoods. After puberty, rural girls may be restricted to the family compound and required to wear the veil.

The third observation is that school, television, and the Internet increasingly influence the lives of children and introduce them to a wider world. In the past, many boys went to school for up to a decade, while the schooling of girls only lasted for a few years. Now, a larger percentage of girls than boys go to school in Saudi Arabia, Israel, Jordan, Lebanon, Libya, the occupied Palestinian Territories, Qatar, Tunisia, and the UAE. Like educated women everywhere, these girls will make life choices different from those of their mothers. Overall, educational opportunities have expanded significantly across the region. The "school life expectancy" (the total number of years of education) that a child today can expect to receive is roughly 12 to 14 across the region, and even higher in some countries.

In rural areas, too, it is fairly common for the poorest families to have access to a television, which is often on all day, in part because it provides a window on the world for secluded women. Television can serve either to reinforce traditional cultural values or as a vehicle for secular perspectives, depending on which channels are watched.

THINGS TO REMEMBER

GEOGRAPHIC THEME 5 • **Population and Gender:** This region has the second-highest population growth rate in the world. Part of the reason for this is that women are generally poorly educated and tend not to work outside the home. Childbearing thus remains crucial to a woman's status, a situation that encourages large families.

• The family is the most important societal institution in the region.

• Most Muslim families in this region are patriarchal in structure. The role and status of women, the spaces they occupy, and the clothes they wear are often carefully defined, circumscribed, and in some cases rigidly imposed.

SUBREGIONS OF NORTH AFRICA AND SOUTHWEST ASIA

The subregions of North Africa and Southwest Asia present a mosaic of the issues that have been discussed so far in this chapter. Although all countries in the region except Israel share a strong tradition of Islam, they vary in how Islam is interpreted in national life and the extent to which Westernization is accepted. They also vary in prosperity and in the evenness of the distribution of wealth among the general population. All subregions have a dry landscape, but to different degrees pockets of fertile land exist where there is access to water.

THE MAGHREB

North Africa has often been *eroticized* in old American movies, meaning portrayed as different and distinctly foreign. People may conjure up intriguing images of Berber or Tuareg camel caravans transporting exotic goods across the Sahara from Tombouctou (Timbuktu) to Tripoli; smugglers and Cold War spies in the Moroccan city of Tangier; Barbary Coast pirates; the World War II battles waged by the "Desert Fox," German Field Marshal Erwin Rommel; or the classic lovers played by Humphrey Bogart and Ingrid Bergman in *Casablanca*. These images—some real, some fantasy, some merely exaggerated—are of the western part of North Africa, what Arabs call the Maghreb ("the place of the sunset," or more loosely translated the westernmost part of the Arab world). The reality of the Maghreb, however, is much more complex than popular Western images of it.

The countries of the Maghreb stretch along the North African coast from Western Sahara through Libya (FIGURE 6.31). A low-lying coastal zone is backed by the Atlas Mountains, except in Western Sahara and Libya. Despite the region's overall aridity, these mountains trigger sufficient rainfall in the coastal zone to have a Mediterranean climate that supports export-oriented agriculture and, in winter, even a modest skiing industry. Algeria, Tunisia, and Libya have interiors to the south that reach into the huge expanse of the Sahara.

The cultural landscapes of the Maghreb reflect the long historic relationships all the countries have had with the broader Mediterranean region. Tunisia, situated in a strategic location connecting the eastern and western Mediterranean, is a good example. This is where Carthage emerged as a rival to the Roman Empire. Later Tunisia was influenced by Berber people and cultures (see the discussion below), incorporated into the Ottoman Empire, and eventually it became a French colony. Throughout the subregion, European domination lasted from

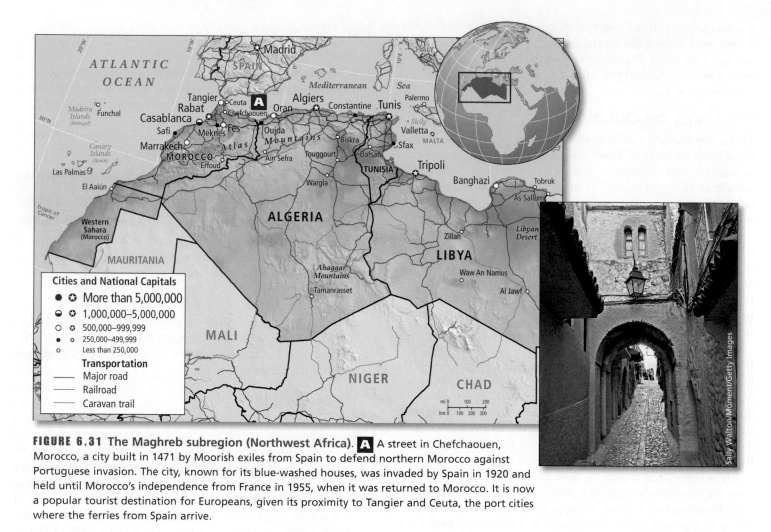

FIGURE 6.31 The Maghreb subregion (Northwest Africa). **A** A street in Chefchaouen, Morocco, a city built in 1471 by Moorish exiles from Spain to defend northern Morocco against Portuguese invasion. The city, known for its blue-washed houses, was invaded by Spain in 1920 and held until Morocco's independence from France in 1955, when it was returned to Morocco. It is now a popular tourist destination for Europeans, given its proximity to Tangier and Ceuta, the port cities where the ferries from Spain arrive.

the mid-nineteenth century well into the twentieth century, during which time the people of North Africa took on many European ways: consumerism; mechanized market agriculture; and manners of dress, language, and popular culture. Especially, French colonialism had a strong impact on the subregion (see Figure 6.16). The cities, beaches, and numerous historic sites of the Maghreb, such as Carthage, continue to attract millions of European tourists every year who come to buy North African products—fine leather goods, textiles, handmade rugs, sheepskins, brass and wood furnishings, and paintings—and to enjoy a culture that, despite Europeanization, seems exotic. The agricultural lands in the coastal zones of the Maghreb are strategically located close to Europe, where there is a strong demand for Mediterranean food crops: olives, olive oil, citrus fruits, melons, tomatoes, peppers, dates, grains, and fish. Large amounts of marijuana are also grown in the Rif Mountains of northern Morocco and smuggled into Europe. Oil and gas production is also very important in two of the countries in the subregion—Algeria and Libya. They supply a large amount of oil and gas to the European Union. Fossil fuel sales account for 30 percent of Algeria's GDP and 95 percent of its exports. With a small population and large reserves, oil and gas are even more important for Libya, at about 65 percent of the GDP. However, the country's fossil fuel sales

collapsed after the Arab Spring as its energy infrastructure—pipelines, refineries, and oil ports—have been seriously disrupted by continuous political instability.

Europe is also a source of jobs for people from the Maghreb. Local firms are supported by European investment and tourism, and millions of guest workers have migrated to Europe, many after losing jobs as a result of agricultural modernization. There are millions of North African migrants particularly in Spain, France, and Italy, many of them undocumented. Libya has also emerged as an important transit country for refugees from countries like Sudan and Somalia who try to make it to Europe. Because of the absence of a functioning government in Libya during the post–Arab Spring period, smugglers can operate with ease in Libya and transport refugees in small vessels across the Mediterranean to the southern shores of Italy, which are located nearby. These voyages are extremely dangerous and thousands of refugees have lost their lives (see Chapter 4).

In North Africa, settlement and economic activities are concentrated along the narrow coastal zone where water is more available (see Figure 6.26). Most people live in the cities that line the Atlantic and Mediterranean shores—Casablanca, Rabat, Tangier, Algiers, Tunis, and Tripoli—and in the towns and villages that link them. The architecture, spatial organization, and lifestyles of these cities mark them as cultural transition zones between African and Arab lands and Europe. The city centers retain an ancient ambience, with narrow walkways and huge heavy doors leading to secluded family compounds, but most people live in modern apartment complexes around the peripheries of these old cities (see Figure 6.31A). Multilane highways, shopping malls, restaurants with international cuisine, and art galleries serve citizens and tourists alike. These developments coexist with a large number of people living in urban slums and hardscrabble rural villages. In Libya, the population cluster of the Maghreb tapers off toward the desert east. The capital and largest city Tripoli is located in the west, while the rest of the country is sparsely populated (with about 6 million people, Libya is the least populous of the Maghreb countries). This geography has meant that the country is organized more around tribal identity rather than a unified national identity. The country was governed by the autocratic leader Muammar el-Qaddafi from 1969 to 2011, and after he was overthrown during the Arab Spring, Libya has fragmented along tribal lines—at the moment there are competing governments in Tripoli and the eastern city of Benghazi. Libya is increasingly becoming a **failed state,** meaning that the government has lost control and is no longer able to fulfill its basic responsibilities as a defender of a sovereign state.

Many North Africans seek an identity that is less European and that retains their own distinctive heritage. Historians note that for those people who were Westernized during the era of European domination and who now make up the educated middle class, independence from Europe meant the right to establish their own secular countries, with constitutions influenced by European models. But by the 1990s, Islamist movements were challenging these independent secular states, linking them negatively with ongoing European and North American influences. The Arab Spring–related shifts toward greater political freedoms have, ironically, tended to strengthen these Islamist political movements, which are opposed to wider political freedoms for all, even though the Islamist movements were themselves previously repressed by secular governments. Now those North Africans who prefer secular governments are fearful of the outcome if Islamists win elections and change constitutions.

The Berber–Arab Cultural Mix

Part of the Maghreb is a mix between **Berber** and Arab cultures. Berbers are the indigenous people of North Africa and they are most numerous in Morocco and Algeria. Census statistics here don't record language and ethnic identity, but there may be around 20 million speakers of Berber languages and many more who are of Berber descent. Ethnic mixing and the lack of a distinct ethnic identity among people of Berber background who live in urban areas also complicate the picture, as does the fact that the Berbers themselves are divided into separate groups with different languages. Among rural Berbers, primarily living in the Atlas Mountains, the ethnic identification is stronger. There, language, traditional clothing, and small-scale agriculture—sometimes subsistence and nomadic in character—have persisted. Berber women often take an active role in the farming household. One typical task is to sell produce at the nearest market, which means that they are more likely to travel independently of men. They are also less likely to be veiled than their Arab counterparts, even if Berbers also are Muslims.

The independence of countries in the Maghreb gave rise to greater Berber awareness and nationalism. The response from the new governments of Morocco and Algeria was often a policy of Arabization—integration of Berber into larger Arabic culture and society. Despite these efforts, Berber society has survived, although emigration from the mountains in search of urban employment has created greater assimilation.

More recently, Berbers have received greater recognition. Their languages are considered "national" or official languages in Morocco and Algeria. Schooling is increasingly taking place in Berber languages as well. The Berber culture itself is also of economic value, both for these indigenous societies as well as for the countries at large, especially in Morocco. Tourists from around the world travel to Berber villages, where they explore the architecture, crafts, and costumes of an exoticized culture.

Violence in Algeria

Algeria, with 40 million people, more than one-quarter of whom are under the age of 15, is just emerging from a long period of turmoil. Inhabited since antiquity by groups that eventually formed the Berber culture, Algeria has also long been home to outsiders who have often dominated the coastal areas. Many Phoenician and Carthaginian coastal settlements flourished during the first millennium B.C.E. Rome controlled much of Algeria following its defeat of the Carthaginians and its control lasted, off and on, until the Arab–Muslim empires swept through the Maghreb in the seventh century C.E. By the 1500s, Algeria was controlled by the Ottoman Empire, during which time the coast became a major center of piracy. This "Barbary Coast" of Algeria was eventually subdued by British and later U.S. forces, and in the 1830s, France invaded Algeria and took control of the country for 130 years.

French rule provoked deep resentment among Algerians that is still felt to this day. Tens of thousands of French colonists immigrated to Algeria, and many were given the most fertile land that had been forcibly taken from Algerians. Resistance and rebellions against French rule were brutally suppressed by the French military, which confiscated land and created food shortages to subdue the Algerian population. Over time, Algeria's economy was reoriented away from local economic development and toward supplying food and raw materials to France's rapidly industrializing economy.

Agitation against French rule increased dramatically after World War II. The violent protests and brutal repression by the French military that resulted paved the way for a war of independence. This brutal war, which lasted from 1954 to 1962, resulted in the deaths of between 400,000 and 1,500,000 people. It brought full independence from France, but also an enduring legacy of violence and authoritarian rule to Algeria.

After independence, Algerian politics were dominated by the military and secular socialists. Such was common around the Maghreb and other countries at this time, and the

failed state a country where the government has lost control and is no longer able to fulfill its basic responsibilities as a defender of a sovereign state

Berber an indigenous group of people in North Africa that are most numerous in Morocco and Algeria

ideology of *Arab nationalism* took hold. That ideology stressed anticolonialism, unity among Arab states, and economic and cultural modernization based on socialism. The long-time leader of neighboring Libya, Muammar el-Qaddafi, was also a forceful proponent of Arab nationalism. The strong role of the military didn't leave much room for democracy, though. Islamists, who wanted an explicitly Islamic government above all else, became the main opposition in Algeria. When elections were allowed, the government, then controlled by the secular socialists, intervened to nullify elections that would have given the Islamists control of the government. The resulting civil war between Islamist militias and the Algerian military resulted in more than 100,000 deaths. A military-backed secularist president has stayed in power from 1999 to the present because of his ability to bring reconciliation between warring parties and stability to the overall political and economic system of Algeria. The protests of the Arab Spring, which (temporarily) brought Islamists to power in neighboring Tunisia and Egypt, reminded many Algerians of the bloody civil war of the 1990s. Because of this, Islamist militants have gained little momentum in current Algerian politics.

Trouble in Western Sahara

Western Sahara is a sparsely populated (570,000 inhabitants) desert country just south of Morocco. The country's importance in the subregion has to do with its status as a disputed territory. Much of Western Sahara is under the control of Morocco, but an independence movement, the Polisario Front, representing the local population, the mainly nomadic Saharawis, has been fighting Morocco's dominance since the 1970s. A long sand wall built by Morocco runs the length of Western Sahara (1,700 miles, or 2,700 kilometers) in an approximately north-to-south direction. The sand wall separates Moroccan-controlled areas to the west from areas to the east controlled by the independence movement, which is headquartered across the border in Algeria. The latter has caused political friction between Morocco and Algeria.

The roots of the conflict, much like in many other places in the region, date back to the colonial era. Spain controlled the territory from the 1800s to the 1970s, when a transition toward independence was planned. Morocco, which borders Western Sahara to the north, then moved hundreds of thousands of Moroccans into Western Sahara as a way to lay claim to the area. One reason is that Western Sahara has phosphate deposits (phosphate is used to produce industrial chemicals and agricultural fertilizers) and there may be oil deposits offshore. United Nations–sponsored peace talks have for decades been unable to resolve the status of Western Sahara. Possible outcomes could be full independence for Western Sahara or some form of semiautonomous status within Morocco.

THINGS TO REMEMBER

• The cultures of the Maghreb reflect the long and changing relationships all of its countries have had with Europe. The local economies are dependent on European investment, markets, and tourism. Millions of guest workers have migrated to Europe, many after losing jobs as a result of agricultural modernization.

• The Maghreb is a blend between Arab and Berber culture. Many Berbers have been assimilated into urban, Arabic life, but traditional Berber lifestyles and language remain in the Atlas Mountains.

• Algerians' experience of a protracted civil war between the secularist military-backed government and Islamists in the 1990s has shaped its response to the protests of the Arab Spring, resulting in little support for Islamist political movements.

• The territory of Western Sahara is occupied by Morocco. UN-sponsored peace talks have yet to resolve the conflict between a local movement and Morocco, and Western Sahara's future status.

THE NILE: SUDAN AND EGYPT

The Nile River begins its trip north to the Mediterranean in the hills of Uganda and Ethiopia in central East Africa. The countries of Sudan and Egypt share the main part of the Nile system **(FIGURE 6.32)**, which is their chief source of water. Although Sudan and Egypt have in common the Nile River, an arid climate, and Islam (over 90 percent of the population are Muslims in both countries), they differ culturally and physically. Egypt, despite troubling environmental problems and major civil disruptions associated with the Arab Spring and government reform, is industrializing and plays an influential role in global affairs, whereas Sudan struggles with civil war, famine, and deep poverty.

Sudan

Sudan is a large country with a relatively small population—just 36 million people, less than half the 88 million in Egypt. The country has two distinct environmental zones. The dry steppes and hills of the south are home to animal herders. The lower, even drier Saharan north stretches to the border with Egypt. Although Sudan mostly consists of dry grassland and desert, it does contain the main stem of the Nile and its two chief tributaries, the White Nile and the Blue Nile. This river system brings water from the upland south that is used to irrigate fields of cotton for export. Most Sudanese live in a narrow strip of rural villages along these rivers (see Figure 6.26; see also Figure 6.32). Most of Sudan's cities are clustered around the capital Khartoum, where the White and Blue Niles join.

South Sudan gained independence from Sudan in 2011 after decades of bloody civil war (discussed further in Chapter 7). The historical roots of animosity between these two states can be found in the politics of oil, religion, and race. The two countries share a large oil field. Islam has been the main religion of Sudan for centuries, but the southern Sudanese people, for the most part, are either Christian or hold various indigenous beliefs, and most are African ethnicities, like Dinka and Nuer. In western Sudan (Darfur), there are also Muslim, Arabic-speaking Africans. Despite the religion and language they share with the north, these southwestern Sudanese, as well as the Christians in what is now South Sudan, are antagonistic toward the Muslim, Arabic-speaking, lighter-skinned Sudanese of the north, who for

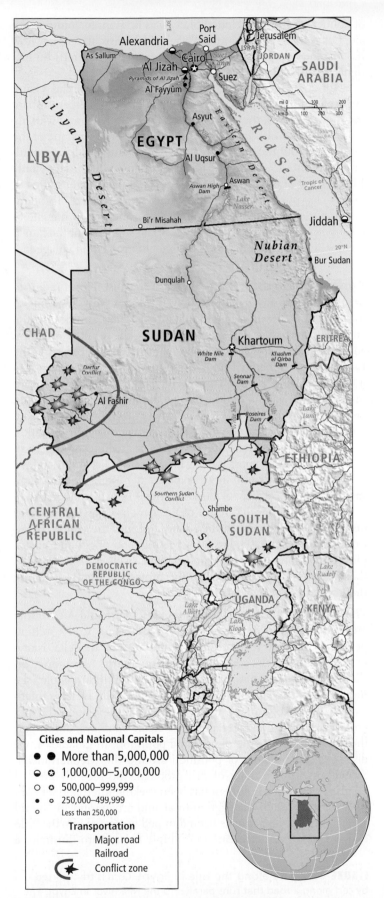

FIGURE 6.32 The Sudan and Egypt subregion.

thousands of years raided the upland south for slaves. In 1983, the Islamist-dominated government in Khartoum decided to make Sudan a completely Islamic state and imposed shari'a on the millions of non-Muslim southerners. As a result, Sudan became home to three civil wars and was classified as a failed state.

The war that led to South Sudan's independence was fought partially over the imposition of shari'a, and also over control of South Sudan's modest oil reserves. From 1983 to the present, this north–south civil war claimed 2 million lives, forced 4 million people to abandon their homes, and orphaned many thousands of children who were then forced to become soldiers for one side or the other. Since a 2005 peace agreement, some 2.2 million people who had fled the southern Sudan conflict have moved back into the region. As a result, the already inadequate resources are being strained. Since then, the fragile and ethnically diverse new country of South Sudan has experienced civil war. Sudan and South Sudan have also struggled over an oil-rich border region.

A second civil war occurred in Darfur, located in western Sudan. This is an area that is part of a cultural and environmental border zone: an ethnically Arab population to the desert north and multiple African ethnicities in the steppe landscape to the south (see discussion of the Sahel in Chapter 7). With limited water resources, persistent droughts, and a population divided between pastoralists (see Chapter 7) and farmers, Darfur is an example of how conflict can be driven by environmental change and geopolitics. Possibly the result of climate change, Darfur had experienced droughts and water shortages for years. Arab herders from north Darfur were forced southward to access water for their farm animals. When doing so, they encroached on farm land used by the local population, mostly of African ethnicities. Some of the herders gave up agriculture altogether and resorted to banditry, forming the Janjaweed militia. To complicate matters, some Darfur groups want greater local autonomy from the central government in Khartoum and a larger share of Darfur's oil reserves. The response from the Khartoum government was to arm the Janjaweed militia. There is strong evidence that the Khartoum government gave the Janjaweed free rein to rape, rob, and kill its opponents. These findings have been documented through the personal accounts of thousands of refugees and by international human rights organizations.

In March 2009, Omar al-Bashir, president of Sudan, became the first sitting head of state to be indicted by the International Criminal Court in the Hague for his role in the killing of thousands of Darfuri villagers by the Janjaweed, although as of this writing, he remains at large. A peacekeeping force under the banner of the African Union reduced some of the violence against civilians, but only after the conflict claimed the lives of more than 400,000 Darfurians and more than 2 million were displaced, many to the neighboring country of Chad. Most of these people remain refugees and dependent on UN humanitarian aid.

Egypt

The Nile flows through Egypt in a somewhat meandering track more than 800 miles (about 1300 kilometers) long, from Egypt's southern border with Sudan north to its massive delta in the southeastern Mediterranean. Egypt is so dry that the Nile Valley and the Nile Delta are virtually the only habitable parts of the country (see Figures 6.1 and 6.32). Ninety-six percent of Egypt is now desert, though archaeological evidence shows that cultivation was once possible far from the Nile basin. In recorded human history, however, it has only been the agriculture just along the banks of the Nile that has fed the country and provided high-quality cotton for textiles.

The Nile's flow is no longer unimpeded. All countries that share the Nile use its waters and there are multiple dams along the river. At the border with Sudan, the river is captured by a 300-mile-long (483-kilometer-long) artificial reservoir, Lake Nasser. The lake stretches back from the Aswan High Dam, completed in 1970 (see Figure 6.32), which controls flooding along the lower Nile and produces hydroelectric power for Egypt's cities and industries. This is so important to Egypt that it has put political pressure on the less powerful upstream countries so that Egypt receives a significant share of the water in the Nile. The downstream river environment, north of the dam, is greatly modified by the dam's effects. Because the floodplain is no longer replenished annually by floodwaters carrying a fresh load of sediment and organic matter from upstream, irrigation and fertilizers are needed to maintain the productivity of the principal crops: cotton, grains, vegetables, sugarcane, and animal feed **(FIGURE 6.33)**. The effects of the Aswan High Dam extend into the Mediterranean. Because the Nile Delta is no longer replenished with sediment, it is eroding away; furthermore, Eastern Mediterranean fisheries are being affected by the lack of freshwater flow and the infusion of chemical fertilizers.

With 88 million people, Egypt is the most populous of the Arab countries, and it has been the most politically influential. Its geographic location, bridging Africa and Asia, gives it strategic importance, and the country plays an influential role in such global issues as world trade and in the peace process between the occupied Palestinian Territories and Israel. Even though Egypt has long been a recipient of U.S. aid (it was the third largest recipient in 2013, after Afghanistan and Israel), which limits its ability to take independent positions, it did oppose the 2003 war in Iraq, saying that the war was likely to increase terrorism. As the Arab Spring failed to bring greater democracy to Egypt after an Islamist-dominated government and subsequent military assertion

of political power, the United States was hesitant to continue the aid, but as Egypt remains an ally against Islamists, the money flow has resumed despite the Egyptian government's dubious human rights record.

Although the cities in Egypt, especially the capital Cairo, are growing most rapidly (in 2015, Egypt was 43 percent urban; see Figure 6.24C), the number of people still living and working on the land has also increased. In fact, Egypt remains one of the most rural countries in the region. Those employed in agriculture constitute 29 percent of Egypt's workforce but produce only 14.5 percent of the GDP. The Nile fields are no longer sufficient to feed Egypt's people, and food must be imported. To afford this imported food for their families, rural men often migrate to neighboring countries, where they work to supplement the family income, leaving the women in charge of farming and village life.

Agricultural Reform In the 1950s, a socialist reform redistributed land from rich owners to poor farmers at very low rents so that the farmers would be able to support their families. In the late 1990s, a new law reversed the process—the result of structural adjustment programs (SAPs) imposed by the World Bank and the International Monetary Fund. The purpose of the new law was to increase the land's productivity by cultivating it in large tracts with machinery, irrigation, and fertilizers. Control of the land would be returned to the original owners, who were allowed to increase rents in order to pay for the new investments. This reorganization of the agricultural system along the lines of green revolution strategies was meant to boost Egypt's food security (its ability to feed its own people) and its ability to export water-intensive crops such as cotton and rice. Few of these intended benefits materialized, and the most noticeable result of the new law was that millions of farmers were forced off their land and into the cities. "Bread riots" occurred in 2008 due to increasing food prices and unemployment among millions of poor, urban dwellers. The plight of those people was also central to the protests of the Arab Spring that rocked Egypt in 2011 through 2013.

The Nile Delta Egypt's two main cities, Cairo and Alexandria, are both in the Nile Delta region. With close to 19 million people in its urban region, Cairo, at the head of the delta, is one of the most densely populated cities on Earth. Alexandria, on the Mediterranean, has a population of 4.8 million people (see the Figure 6.24 map). Outside the cities, the Nile Delta is intensively cultivated farmland with a large rural population. Egypt's crowded delta region faces a crisis of clean water availability and the threat of waterborne disease and pollution. In Chapter 4, we discussed the problem of pollution around the Mediterranean. To address Egypt's share of this problem, the recently created Ministry of the Environment has been seeking international contractors to treat industrial, agricultural, and urban solid and liquid wastes, which for years have been dumped untreated into the Nile and the Mediterranean. But an example from Cairo illustrates a

Education Images/Universal Images Group/Getty Images

FIGURE 6.33 Life along the Nile in Egypt. Corn is transported by cart along a road that runs parallel to the Nile River in Egypt. The fields behind are irrigated with water taken directly from the Nile, as well as with water pumped from underground aquifers.

disturbing pollution linkage between Europe and North Africa. A major industrial complex in Cairo recycles used car batteries sent to Egypt for disposal from all over Europe, because no disposal sites for the batteries are allowed in Europe. However, the recycling complex releases lead into the air in concentrations 30 times higher than those allowed by world health standards. The contaminated air blows across Cairo and into the Eastern Mediterranean. Some children's playgrounds in Cairo are so polluted that they would be considered hazardous waste sites in the United States and Europe; and of course, eventually some of this pollution reenters the air and water of Europe.

The Slow Pace of Economic Development In the 1990s, Egypt's economy, for many years plagued by stagnation, inflation, and unemployment, appeared to be turning around. After years of SAPs (similar to those described above in the section on Egypt's agricultural reform) aimed at fostering private enterprise and reducing the economic role of government, inflation was brought somewhat under control. Induced by tax incentives, multinational companies, such as Microsoft, McDonald's, American Express, Löwenbräu of Germany, and three German automakers, opened subsidiaries in Egypt. For the ordinary working people of Egypt, however, this high-end development did not translate quickly into prosperity; instead, people lost public services as government spending was cut. Egypt's economy, after having improved marginally, was dealt a double blow by the global recession that began in 2008 and by the political and economic turmoil that followed the Arab Spring. The economy actually contracted during the Arab Spring because of precipitous declines in tourism, manufacturing, and construction. Unemployment now stands at 13 percent overall, with roughly 9 percent of men and 25 percent of women unemployed.

One important industry that has been affected by the Arab Spring turmoil is tourism. The Egyptian government and the private sector have invested in a tourism infrastructure since the 1970s. The most obvious tourist destinations are the remnants of the ancient Egyptian civilizations. Many of the famous sites that include pyramids and the famous sphinx are located within comfortable reach from Cairo. Recently, all-inclusive beach resorts in the Sinai Peninsula have also attracted large number of visitors looking for a sun-and-sand vacation at a reasonable price. However, the number of tourists visiting Egypt has dropped from a peak of almost 15 million in 2013 to 9.5 in 2015. Safety concerns are the major reason for this decline. There have been a number of terrorist attacks on tourists and Sinai is a center of Islamist activity in Egypt. In 2015, a Russian airplane full of tourists was brought down by terrorists, which has many wondering whether Egyptian authorities can adequately protect visitors to the country.

A feature of Egypt that benefits its economic development and places the country in a central position of global trade is the Suez Canal. The canal connects the Mediterranean with the Red Sea, and since it was opened in 1869, the canal shortens shipping routes between Europe and Asia enormously compared to the alternative route around Africa. Today, the canal handles 7 percent of global ocean-going trade. The economic benefits of the canal include shipping fees, employment in ports and logistics, and manufacturing based on material that flows through the canal. The canal has been improved and expanded many times; as recently as 2015, Egypt deepened the canal and built new parallel channels to enhance its capacity, which could generate new revenue for the country.

THINGS TO REMEMBER

- Sudan struggles with dissension in the Darfur region and southern Sudan. Violence continues at a lesser rate between the Arab government in the north and non-Arabs in the west and south. The humanitarian needs of millions of people remain large.

- Egypt is the most populous and influential Arab country. This dry country is dominated by the Nile River and its life-giving water; almost all people live in close proximity to the river or in the Nile Delta.

- The Arab Spring has not resulted in greater political and economic stability. Egypt still suffers from poverty, a scarcity of water, high unemployment, food shortages, and environmental deterioration. The political turmoil has also depressed the important tourism industry.

THE ARABIAN PENINSULA

The desert peninsula of Arabia has few natural attributes to encourage human settlement, and today large areas remain virtually uninhabited (see Figure 6.26). The land is persistently dry and has large areas that are barren of vegetation. Streams flow only after sporadic rainstorms that may not come again for years. The peninsula's one significant resource—oil—amounts to about 30 percent of the world's proven reserves. Saudi Arabia has by far the largest portion, with approximately 16 percent of the world's reserves.

Traditionally, control of the land was divided among several ancestral tribal groups led by patriarchal leaders called **sheikhs** (the title is still used today, denoting someone in a leadership position). The sheikhs were based in desert oasis towns and in the uplands and mountains bordering the Red Sea. The tribespeople they ruled were either nomadic herders who followed a seasonal migration path over wide areas in search of pasture and water for their flocks, or poor farmers who settled where rainfall or groundwater supported crops. Until the twentieth century, the sheikhs and those they governed earned additional income by trading with camel caravans that crossed the desert (once the main mode of transportation) and by providing lodging and sustenance to those on religious pilgrimages to Makkah.

Wealth in Saudi Arabia and the Gulf States

Saudi Arabia is not only at the heart of this region culturally because it is the home of Makkah and other Islamic holy sites, but it is politically and economically central as well. Saudi Arabia's leaders, with their stranglehold on the country's vast oil wealth, have sought flashy symbols of modernity **(FIGURE 6.34)** but have managed to resist the political changes sweeping through much of this region. However, many other powerful changes are afoot.

sheikhs traditionally, patriarchal leaders who controlled the use of lands shared among several ancestral tribal groups; today, the title is still used to denote someone in a leadership position

FIGURE 6.34 The Arabian Peninsula subregion. **A** Al-Faisaliah Tower in Riyadh, Saudi Arabia, is named after King Faisal (1903–1975) and is the second-tallest building in the kingdom. It is designed to resemble a ballpoint pen. Inside the golden ball at the top is a revolving restaurant that offers views of the city.

In the early twentieth century, the sheikhs of the Saud family, in cooperation with conservative religious leaders of the Wahhabi sect, united the tribal groups to form an absolutist monarchy called Saudi Arabia, which occupies the central part of the peninsula. The Saud family consolidated its power just as reserves of oil and gas were becoming exportable resources for Arabia. The resulting wealth has added greatly to the power and prestige of the Saud family and their allies, the Wahhabi clerics. Their tight control over Saudi Arabia has inhibited the development of opportunities for young people, despite the considerable material wealth from petroleum resources.

Discontent, especially among young Saudi adults, is rising. Sometimes people express this discontent by embracing fundamentalist Islam—see the discussion of Wahhabism and the rise of Islamism on page 340. A majority of the suicide hijackers involved in the 9/11 attacks in the United States were Saudi, as was the founder of the Al Qaeda movement, Osama bin Laden, a member of a Saudi family that became rich in the construction business. Analysts familiar with Saudi Arabia suggest that Islamist violence has an appeal because there is no civil forum for airing

discontent. The regional press speculates that political power is bound to shift away from the Saud family and others like it as education rates increase and young people find ways to use the Internet as an outlet for political expression.

Despite the overall conservatism of Arabian society, oil money has changed landscapes, populations, material culture, and social relationships across the peninsula. Where there were once mud-brick towns and camel herds, there are now large, modern cities. The population of the peninsula has burgeoned to 80 million people, Saudi Arabia being the largest of those with 32 million. Irrigated agriculture has made the peninsula nearly self-sufficient in food, though only in the short term, since the irrigation is unsustainable (see page 321). Even with all this apparent progress, women are still strictly limited in political power; and many peninsula residents have yet to be included in economic and social development, as the subregion is dependent on immigrant labor.

The six smaller nations on the perimeter of the Arabian Peninsula, now ruled by ancestral clans that resisted the Saudi family expansion in the 1930s, have varying profiles. Kuwait and the United Arab Emirates, each with about 6 percent of the world's oil reserves, are rapidly modernizing and are extremely affluent as the oil incomes only have to be shared among a small population. Kuwait was a forerunner in the Gulf region in terms of providing its citizens with social services like education and health care; now, the UAE and others are following that same path. Qatar also has much oil per capita, and on behalf of the government, the Qatar Investment Authority places the proceeds of oil sales in industries around the world, like Volkswagen, Hollywood production companies, and a major European soccer team. As a way to exhibit its new wealth, Qatar will host the soccer World Cup in 2022, which will cost hundreds of billions of dollars for new stadiums and infrastructure—a staggering expenditure for a country with 2 million residents. Bahrain and Dubai (a semi-independent emirate of the UAE) generate income not so much from oil as from services related to oil production and transport, banking, and by providing entertainment, shopping, and manufactured goods to neighboring wealthy Saudis. In Dubai, where daytime temperatures average 110°F (43°C), a shopping mall features a frigid indoor ski slope where skiers can rent skis and winter clothing (**FIGURE 6.35**). Bahrain was the only one of the small Gulf states that experienced political turmoil during the Arab Spring. There, discontent has been brewing for a long time because the majority Shi'ite population is ruled by a Sunni monarchy. During the Arab Spring, military forces from other Sunni countries in the region intervened on behalf of the Sunni rulers. The situation remains tense.

The final Gulf state country is Oman, which has moderate oil reserves but enough to sustain a high level of well-being. As the gateway to the Persian Gulf, Oman occupied an important position in the historic trade around the Indian Ocean. Contemporary Oman is ruled by a sultan who has directed the country toward a path of modernization. Oman is unique in the region as most people adhere to Ibadism, which is an Islamic sect that is neither Sunni or Shi'ite.

FIGURE 6.35 Ski Dubai in the Mall of the Emirates. Malls in Dubai are noted for their opulence. In a specially insulated section of this mall, visitors can ski on 6000 tons of artificially made snow down a 1300-feet (400-meter) run. Even though the temperature outside may be well above 110°F (43°C), visitors must wear winter clothing to stay warm inside the facility.

ON THE BRIGHT SIDE: Changes in Attitudes Toward Work in the Gulf States

The actual day-to-day work necessary to keep the Arabian Peninsula societies running has been performed by contract laborers from South and Southeast Asia at wages that can be as low as U.S.$1 an hour. Scientific and engineering expertise has been supplied from Europe and the Americas at much higher rates of compensation. Workers and expertise have been imported because a shortage of skills exists in the Gulf states. The causes are multiple: Women tend to not work outside the home even though they may be educated; young men have been discouraged from doing manual labor because of the low status of such work, and they are not motivated to undertake the long study programs necessary to gain professional degrees; institutions of higher learning have not been available within the region until recently; and the motivation to undertake demanding jobs day after day has been squelched by the substantial stipends paid to most citizens from oil revenue.

Times are now changing. When leaders observed that the ideas generated by the Arab Spring in North Africa were spreading to Gulf youth, and that feminist movements were beginning to sprout, they concluded youthful idleness was not healthy. Furthermore, when censuses showed 43 percent of Gulf state populations were expatriate laborers from India and Southeast Asia who were increasingly dissatisfied with their jobs, it became clear that it was time to reorder society. Education, job training for local youth, and the reform of social attitudes toward work are now being emphasized. Expansion of higher education is taking place, although dependent on university lecturers from elsewhere in the world. The dignity and psychic rewards of work are stressed, and former rules against the two genders working together are being relaxed. The goals are to instill pride in personal self-sufficiency and in on-the-job accomplishments and to lessen dependence on workers from outside the region. Meaningful lives for the population will also prove an antidote against political radicalization. ∎

Yemen: Poverty and Conflict

Yemen occupies an important juncture in the world's shipping lanes, at the southern end of the Red Sea on the Gulf of Aden. The country is about one-fourth as large as Saudi Arabia, its neighbor to the north, but the two countries have nearly the same-size population (27 million compared with Saudi Arabia's 32 million). Yemen is not well endowed with oil but it does depend on oil exports. Production, however, has already passed its peak and oil revenues are bound to drop. Yemen's standards of education and of living remain by far the lowest on the peninsula and women's rights are meager.

The Arab Spring brought political change to Yemen in 2012, when a new government was elected, ending a 33-year regime notorious for corruption and authoritarianism. However, the new government struggled to assert its control across the country. In the south, disgruntled sheikhs are defecting from the government, apparently because the patronage system is failing them. Meanwhile, Yemen has attracted several extremist groups that are looking for a safe haven. Most notable among these is Al Qaeda, which periodically trains new recruits in Yemen's uncharted territory that borders the Saudi Arabian desert. Hundreds of young men have fled the failed state of nearby Somalia and come to Yemen to seek some type of work; the $100 or so that Al Qaeda offers new recruits has been enough to buy their loyalty. However, the biggest cause of instability in Yemen has been the Houthi, a Shi'ite ethnic group with roots in the northwest of Yemen. The Houthi's military advances across much of Yemen have prompted air strikes by Saudi Arabia. The Shi'ite Houthi are reputedly funded and armed by Iran, Saudi Arabia's arch enemy. Yemen is both becoming the site of a proxy war between the leading Sunni and Shi'ite countries in the region, as well as drifting toward becoming another failed state.

THINGS TO REMEMBER

• Despite the conservatism of Arabian society in this subregion, oil money has changed landscapes, populations, material culture, and social relationships across the peninsula; and as the number of educated women and men increases, further changes are on the horizon.

• Yemen is a poor and politically unstable state with internal conflicts that has embroiled more politically powerful actors in the region.

THE EASTERN MEDITERRANEAN

Jordan, Lebanon, Syria, Israel, and the occupied Palestinian Territories have all been preoccupied over the last 60 years with political and armed conflict. Much of this arose out of the establishment of the state of Israel, but rivalry over access to land and resources, as well as agitation for greater political freedom, has driven many conflicts **(FIGURE 6.36)**. In the 1970s, the Arab–Israeli conflict (see pages 342–345) spilled over into Lebanon as the country was home to many Palestinian refugees. Lebanon is a diverse but fragmented country with Sunni Muslim, Shi'ite Muslim, and Christian populations about equal in size, with a smattering of other groups, such as the small Druze sect. Lebanon, and especially its capital Beirut, have historically been centers of commerce in the subregion, but the complex internal ethno-religious politics as well as influences from nearby countries—both Israel and Syria have at times occupied parts of Lebanon—has made it politically unstable. Lebanon was actively rebuilding its economy and infrastructure when an armed conflict between Hezbollah (an anti-Israeli Shi'ite militia) and Israel destroyed much of southern Lebanon in 2006. Syria has had conflicts with most of its neighbors for many years and was itself plunged into civil war when its government tried to repress Arab Spring protests, and Jordan has taken in large numbers of Palestinian refugees over many decades.

All Eastern Mediterranean countries have scarce water supplies and must face the likelihood that climate change is already making the situation worse, as indicated by the decade of drought that is likely to continue. The various strategies to acquire more water are discussed on pages 322–324, but there are no easy solutions to the problem. Irrigated agriculture appears to be failing, which may be fortuitous since agriculture is the biggest user of

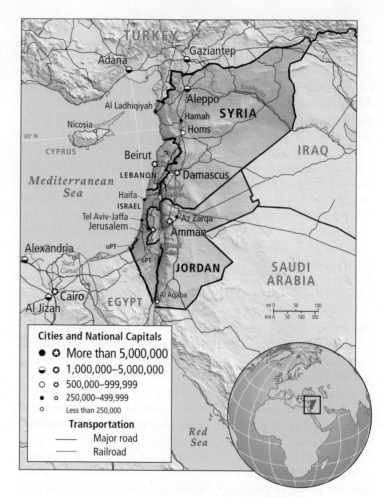

FIGURE 6.36 The Eastern Mediterranean subregion.

water. Population growth and rapid urbanization are increasing per capita water use, in part because piping water into homes removes people from personal familiarity with acquiring water from its sources in nature, leading to more wasteful habits.

If Israeli–Palestinian tensions and Syria's civil war were resolved, the Eastern Mediterranean could become the economic leader in the region. The countries of this subregion are strategically located adjacent to the rich markets of Europe and the potentially lucrative markets of Central Asia and the Persian Gulf states. The strategic location dates back in history. The famous Silk Road that connected Asia with the Mediterranean had its terminus point near Aleppo, Syria, which is also reflected in that city's ancient quarter. Unfortunately, Aleppo has been a flashpoint in the Syrian civil war and much of its heritage was consequently destroyed. Under a peacetime scenario, the potential for tourism in this subregion would be strong. As part of the Fertile Crescent, the cities located here are among the most historic in the world; there are also impressive Roman ruins, and an abundance of important religious sites that would attract Christian, Jewish, and Muslim pilgrims from around the world. At the moment, the subregion is a long way from peace, and in territories controlled by the Islamic State, the group has destroyed historic monuments that reflect the cultural and religious past.

The climate makes the subregion suitable for subtropical export agriculture. Jordan already exports vegetables, citrus fruits, bananas, and olive products. If irrigation systems could be developed and used sustainably (not an easy matter), Syria would also be able to engage in export agriculture. Syria lost Russia as an important trading partner after the breakup of the Soviet Union in 1991, but Russia continues to support the Assad regime with weaponry and other assistance. Until the recent civil unrest, private investors, especially from the Gulf states, helped Syria expand its industrial base to include pharmaceuticals, food processing, and textiles, in addition to gas and oil production. Natural gas discoveries in the waters off Israel, Lebanon, and Syria could make the Eastern Mediterranean subregion more energy independent and interconnected, and even spur exports in the future.

Some argue that Israel could be a model of development for neighboring countries. Despite its rather meager natural resources, Israel has managed to develop economically and now has one of the most educated, prosperous, and healthy populations in the region (see the GNI and HDI information in Figure 6.27). Israeli engineering is world renowned, and Israeli innovations in cultivating arid land have spread to the Americas, Africa, and Asia. A range of sophisticated industries are located along the coast between the large port cities of Tel Aviv and Haifa. Moreover, Israel benefits from qualities that other countries lack. As the homeland for the world's Jews, it has a pool of unusually devoted immigrants and financial resources contributed by the worldwide Jewish community and a number of foreign governments, particularly the United States. Regardless, the reality is that Israel's continual conflicts with its neighbors have turned Israel into an armed fortress and constrained economic growth because it can't trade with surrounding countries in a normal way. In response, young and highly trained Israelis are leaving the country, while ever-smaller numbers of Jews from other parts of the world are choosing to migrate to Israel. However, Israel still has a small positive net migration rate—more immigrants than emigrants.

Israel also has to battle internal divisions constantly; it is not a homogenous Jewish state. Arabs make up close to 20 percent of the population. Moreover, there is a divide in Israeli society between the majority secular-modern Jewish population, many of whom are moderately religious or nonreligious, and the ultra-orthodox community whose members live traditional lives according to their interpretation of religious texts. The latter group makes up only 10 percent of the Israeli population but its birth rates are very high and thus constitutes a growing demographic share. Another division has to do with where one is born; a quarter of the Jewish population comes from somewhere else than Israel. Many are immigrants from the United States, Russia, Africa, or Asia, and they bring with them cultural traditions and attitudes from their place of origin.

THINGS TO REMEMBER

- The ethno-religious makeup of the Eastern Mediterranean is diverse, which is a result of historic patterns as well as contemporary population movements. This complexity, together with limited land and resources, has resulted in volatility and conflict.

- Water scarcity is a rising issue.

- Resolving the hostility between Palestinians and Israelis and ending the Syrian civil war would provide a great stimulus for economic growth and human well-being for the states and peoples of the subregion.

THE NORTHEAST

Turkey, Iran, and Iraq (**FIGURE 6.37**) are culturally and historically distinct from one another. For example, a different language is spoken in each country: Farsi in Iran, Turkish in Turkey, and Arabic in Iraq. Yet the three countries have certain similarities. At various times in the past, each was the seat of a great empire, and each was deeply influenced by Islam. Each occupies the attention of Europe and the United States because of its location, resources, or potential threats. All three countries share common concerns, such as how to allocate scarce water and how to treat the large Kurdish population that occupies a zone overlapping all three countries (see Chapter 1). Moreover, each country has experienced a radical transformation at the hands of idealistic reformist governments, although the paths pursued have varied dramatically.

Turkey

Over the last century, Turkey has been more closely affiliated with Europe and North America than with any country in its home region. Turkey has been a stalwart member of NATO (see Chapter 4) ever since 1952, shortly after this mutual defense organization was created (1949) to address the perceived Cold War threat from the Soviet Union. Greece also joined at that time, and the two countries formed a strategic southern military pivot point for NATO. After this long, successful association with Europe, a strong faction in Turkey wants to join the European Union. Many Turks have spent years in Europe as guest workers, and Europe–Turkey business relations are dense and profitable.

VIGNETTE Veli-Çetin Avci and his wife Elif are an elegant young couple, both mechanical engineers and officers in Mekanik, a family-held firm that imports computer-driven tool-making machines from the European Union (**FIGURE 6.38**). They sell to Turkish firms that produce parts used in the global automobile industry. When asked about Turkey's chances of joining the EU (discussed in Chapter 4) and to what extent membership in the organization would help or hurt Turkey, their answer had two parts. They responded immediately that joining the EU would be good for Turkey because its requirements for membership would force the adoption of higher standards in all aspects of Turkish life and would increase democratic participation. But, they hastened to add, their business would probably fail as a result because, given EU membership requirements for open markets, their middleman position between European and Turkish firms would be quickly taken over by the powerful European firms from which they now bought machines. Nevertheless, these business professionals believe that being part of the EU would give them many opportunities to modify their present business in order to take advantage of freer access to European markets

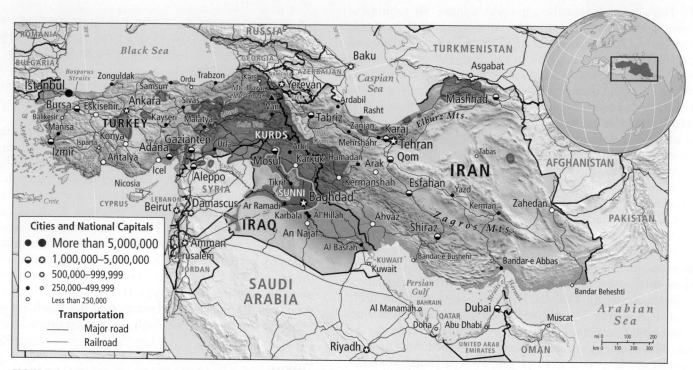

FIGURE 6.37 The Northeast subregion and the Kurds. The area of Kurdish concentration overlaps five countries. The shaded area with a yellow boundary shows the concentration of Arab Sunni Muslims in Iraq.

and to global trade networks. *[Source: From the field notes of Lydia Pulsipher and Mac Goodwin, July 2009, Konya, Turkey.]* ∎

In recent years, enthusiasm for joining the EU has begun to wane as Islamic identity is on the rise in Turkey, the European economy is faltering, and some EU members have put up ever more challenging barriers to Turkish membership. The official reasons the EU has given for not yet accepting Turkey are that

FIGURE 6.38 Elif Avci and her husband Veli-Çetin Avci enjoy a rare evening out with friends.

Turkey has not sufficiently marketized its economy or instituted constitutional human rights guarantees, including freedom of religion and the press, and protections for minorities such as the Kurds. Less publicly stated are European worries about absorbing a predominantly Muslim state into an at least nominally Christian Europe. Turkish membership would also mean that the EU directly borders the conflict zone of Syria and Iraq, which is an undesirable prospect from the European perspective. And, finally, Turkey has an antagonistic relationship with the EU member state of Greece, which would likely try to block potential Turkish membership. Meanwhile, Turks worry that their culture will not be sufficiently respected in Europe, but more importantly, Turkey has assumed an increasingly influential role in its own nearby neighborhood as the implications of the Arab Spring become more discernible.

As discussed earlier, Turkey, once the core of the Ottoman Empire, was dismantled after World War I. After independence in 1923, Turkey undertook a path of radical Europeanization, led by a military officer, Mustafa Kemal Atatürk, who is revered as the father of modern Turkey. Atatürk and his followers declared Turkey a secular state; encouraged women to discard the veil and seclusion while men should do the same with the fez, the traditional Turkish head garment; and modernized the laws and state bureaucracy. Atatürk also promoted state-sponsored industrialization and actively sought to establish connections with Europe. Today, laws that prohibit people from religious head covering in government offices or from students wearing them in schools and universities have been relaxed. While industrialization and modernization have proceeded in western Turkey, much of eastern Turkey remains agricultural and relatively poor. Islam, long deemphasized as a matter of state policy, is nevertheless an overriding influence on daily life, and there is a resurgence of Islamic fundamentalism. A moderate Islamist party has been governing Turkey on and off for many years now. The president's wife is often seen wearing a head scarf in public, which indicates that Turkey is nudging in a conservative direction. The military has traditionally viewed itself as the guardian against Islamism in Turkish society. A military coup was attempted against the sitting government as recently as July 2016, but it failed and has left Turkish politics in turmoil.

Turkey straddles the Bosporus Straits, a narrow passage from the Black Sea to the Mediterranean that often is described (with exaggeration) as the division between Europe and Asia. This location gives Turkey potential advantages as an intermediary if it can successfully increase economic links with Europe and Asia. Istanbul, a booming city of more than 14 million (see the Figure 6.26 map), is located at the Bosporus and is the regional headquarters for hundreds of international companies. Of Turkey's 78 million people, 77 percent are city dwellers. Many of them once lived in rural areas of Turkey, then traveled to Europe to work in factories and on farms; they have returned eager to pursue a standard of urban living similar to what they experienced in Europe. Remittances from Turkish guest workers in Europe have financed innumerable large homes that now dot the landscape of western Turkey. Today, as the Turkish economy has expanded and Turkish migrants have lived in Europe for decades and their connections with the homeland have diminished, the value of remittances has declined.

Turkey's future depends, in part, on its ability to manage its relatively abundant water resources. With its mountainous topography, Turkey receives the most rainfall of any country in the region, and the headwaters of the economically and politically important Tigris and Euphrates rivers are in the mountains of (Kurdish) southeastern Turkey. Agriculture employs 25 percent of Turkey's workforce but accounts for only 8 percent of its GDP (the country exports Mediterranean fruits, nuts, and vegetables). As new irrigation and power generation systems on the Euphrates come on line, agricultural and industrial production should increase dramatically. Despite insufficient electrical power, Turkey's diversified economy manages to produce a wide range of goods, including apparel, food, textiles, transport equipment, and leather goods, all of which are exported to Europe and Asia.

A Future Kurdistan? In addition to the recent civil war in Syria, just to the south, a long-standing obstacle to Turkey's stability is its ongoing conflict with its Kurdish minority (see Chapter 1). The Kurds have lived in the mountainous borderlands of Iran, Iraq, Syria, Turkey, and Caucasia (Figure 6.37) for at least 3000 years. The division of the Ottoman Empire after World War I by France and Britain divided the Kurds among lands held by Turkey, Iran, Iraq, and Syria, leaving the Kurds without a state of their own. All four countries have maintained hostile relations with their Kurdish minorities. The conflict springs from complex disputes over control of territory and resources, but it is exacerbated by cultural and language differences as well as the Kurds' resistance to state control. The total number of Kurds in the region is somewhere around 30 million people, with the highest concentration in Turkey. There is a sizable Kurdish diaspora in Europe as well, especially in

Germany. The Kurds are predominantly Sunni Muslims and their language is related to Farsi, which is spoken in Iran.

The war in Iraq and now the civil conflict in Syria have complicated the Kurds' situation, in part because the loyalties of the transnational Kurdish community to the states in which they are located are now even more in question and also because Kurdish influence in Iraq and Syria has increased in recent years. Kurdish forces have been instrumental in limiting the advances of the Islamic State in both Syria and Iraq. After the U.S. invasion of Iraq in 2003, the Kurds developed great autonomy and self-governance in northern Iraq, which also allowed them to build their military capacities. This is the closest the Kurds have come to an actual state of their own—a Kurdistan. This is a concern to the surrounding countries who do not want to cede territory to a potential future Kurdistan.

Kurdish relations with Turkey have been especially problematic in the past. To diminish the sense of identity among the Kurdish people, it was common practice in Turkey to refer to Kurds as "mountain Turks." Various Kurdish rebel factions fought Turkish forces for decades. The Kurdish struggle within Turkey seemed to be abating as the Turkish government, seeking to stabilize the borders it shares with its unpredictable neighbors, offered to meet many long-denied Kurdish pleas for respect and autonomy. However, the civil war in Syria has reignited the Kurdish conflict in Turkey.

Iran

Iran occupies a transitional geographic position in this subregion. It is predominantly a non-Arab Persian country, but people of Arab, Turkish, Kurdish, and Caucasian heritage occupy its western parts. Its eastern parts are occupied by people with language and ethnic roots in South Asia, especially Afghanistan and Pakistan. Because its northern flank borders the Caspian Sea, Iran shares in the debate over the future of this inland sea and its resources, especially water and oil, with Caucasia and Central Asia. This transitional position of Iran has made it a zone of particular interest for world powers, most recently the United States and the Soviet Union (and now Russia).

Like Turkey, Iran has abundant natural resources and a large, educated population—it has 79 million people and is growing at a modest rate—that could make it a regional economic power. Iran's large petroleum reserves give it influence in the global debate on energy use and pricing, but social and economic turmoil have held the country back for many years.

Iran's recent political history and its status as a theocratic state based largely on a Shi'ite version of Islamism complicate its role in the modern world. The theocracy was formed under the leadership of the Islamic fundamentalist Ayatollah Ruhollah Khomeini, a Shi'ite spiritual leader who led a revolution in 1979 against the last monarch, Shah Reza Pahlavi. Iran had been on a path of secular and economic reforms patterned after those instituted by Kemal Atatürk in Turkey since the 1920s. In 1953, to keep power in the hands of the shah, the United States orchestrated a coup d'état against the winner of the national election. This helped the United States retain access to Iran's oil and allowed the shah to retain power for 26 more years, during which he continued his efforts at Europeanization. But his emphasis on military might, political repression, and royal grandeur paid for with oil money overshadowed any genuine efforts at agricultural reform, industrialization, and social equality. Disparities in wealth and well-being between those who had connections to the shah and those who didn't grew ever wider.

The 1979 Revolution was initially welcomed by many Iranians as an antidote to Western influence. However, those who had hoped for a more democratic and secular Iran were disappointed by the new theocratic state. Resisters were imprisoned, and many were even executed. Among those most affected were women. Many upper-class women had lived emancipated lives under the shah's reforms, studying abroad and returning to serve in important government posts. Now all women past puberty had to wear long black chadors, and they could no longer drive, travel in public alone, or work at most jobs. Despite such restrictions, even some highly educated women supported the return to seclusion, seeing it as a way to counter the unwelcome effects of Western influences. The turmoil of the revolution was characterized by several episodes in which Westerners were taken hostage by the government and held for many months, and by general resentment against the West, which in turn led to decades of isolation and conflict. During the 1980s, Iran engaged in a devastating war with Iraq (see page 371).

Since the 1979 Iranian Revolution, there have been significant social improvements in basic human well-being, and some of the radical restrictions on women's role in public life mentioned above have been liberalized, but only limited economic growth and hesitant moves toward greater political freedom have transpired (see the discussion of the Iranian Spring on page 341). Iran has a GNI per capita commensurate with that of its neighbors; it ranks in the second best category on the UN's Human Development Index, but poorly on gender equality (see Figure 6.27). The Iranian government had begun to work toward economic diversification and industrialization but now increasingly imports products from Asia or Europe, to the detriment of local industry. The Revolutionary Guard, whose main role is as a security force protecting the theocratic state from internal dissent, controls a large share of the Iranian economy, which is also detrimental to the development of a well-functioning market economy. Attempts to improve the country's out-of-date transportation system so that Iran can begin again to participate in trade with its Arab and Central Asian neighbors are moving slowly. Iran's oil refinery capacity also needs to be expanded so that the country can benefit further from its oil wealth. Agriculture additionally remains a weak economic sector. Even though agriculture employs 16 percent of the population, it accounts for only 9 percent of the GDP, and Iran imports more than twice the amount of food it exports.

Moves toward political reform have been constrained by the conservative Islamic clerics who wield immense power through a wide range of governmental bodies. For example, Iran's Council of Guardians, a 12-member body of Islamic clerics, decides who can run for election and how elections are administered; it can veto any law passed by the *Majlis* (the Iranian equivalent of the U.S. Congress). The official head of state is the Supreme Leader, a religious cleric, who wields ultimate power over most political issues. In 1997, the Majlis allowed a reformer named Mohammad

Khatami to be elected president. He set about liberalizing many of Iran's internal policies, including women's rights, and announced an end to its active nuclear armament program. Then in 2004, an Islamist, Mahmud Ahmadi-Nejad, was propelled to the presidency.

His 2009 reelection was highly contentious and street protesters who suspected a rigged election were brutally suppressed. However, this precursor to the Arab Spring in 2010 turned out to be a harbinger of change in Iran as well. It showed that Iran had a well-informed and sophisticated Internet-savvy electorate, skilled at international social networking, using Facebook and Twitter. The election protests also highlighted major internal divisions among the ruling clergy and showcased the Iranian government's willingness to ruthlessly repress its people, many of whom were jailed or even killed during the protests. Young people in Iran have lived under an Islamist regime all their lives and many of them don't particularly like it. Therefore, they are less swayed by the type of jihadism that has found fertile ground among discontented Muslim youth in many other countries. In fact, Iran is a relatively young society with large numbers of people under 30 years of age. Since the 1990s, the Iranian government has promoted family planning measures to limit population growth, and has been successful in doing so.

Iran has maintained frosty relations with the United States since the revolution of 1979, and it supports dictators like Assad in Syria and radical groups like Hezbollah in Lebanon in order to promote an Iran-led Shi'ite agenda across the region. It also has developed a nuclear program that some suspect will be used to create nuclear weapons. Although many Iranians may not be supporters of the cleric-led regime, they view a domestic nuclear program favorably as a matter of national pride and the right to self-determination. On the other hand, there are also geopolitical situations where Iran and the United States have common agendas. Iran wants a stable Afghanistan along its eastern border, and it would like to see an Islamic State-free Iraq along its western border. In 2013, a centrist, Hassan Rouhani, was elected to the presidency by a wide margin. Under his administration, Iran has agreed to permit UN nuclear inspections inside Iran. In return, international sanctions will be lifted so that Iran can participate in global trade once again.

Iraq

Iraq, home to one of the earliest farming societies on Earth and to the Babylonian Empire of biblical times, was carved out of the Ottoman Empire after World War I. Most of Iraq's 37 million people live in the area of productive farmland in the country's eastern half, on the floodplains of the Tigris and Euphrates rivers. Although the present chaos in Iraq makes it difficult to obtain accurate information about population numbers and distributions, generally speaking Sunni Arabs (about 10 million people) reside in central Iraq around the capital of Baghdad and northward from there. About 7 million ethnic Kurds, who are also Sunni Muslims, live mostly in the northern mountains in the regions bordering Turkey, Syria, and Iran. The southern third of the country is occupied by at least 18 million Shi'ite Muslims concentrated around Al Basrah at the head of the Persian Gulf.

Iraq's main oil fields are also in the south. Minority pockets of Shi'ite Muslims and other groups live throughout the country. These groups lived in more ethnically integrated neighborhoods and districts before the Iraq war in 2003.

Foreign activity in Iraq is not new. In the aftermath of World War I, Iraq was taken over by the British until 1932, when it became an independent monarchy. Although approximately half of the Iraqi people are Shi'ite Muslims, since 1932 a powerful minority of Arab Sunni Muslims has controlled the government and most of the wealth. Following independence, the Iraqi monarchy, like many governments in the region, maintained strong alliances with Britain and the United States and gradually lost touch with its people. A tiny minority monopolized the wealth that was generated by increasing oil production, but they did not invest in social or economic development. The Shi'ites living in the south, where the oil is located, and the Kurds in the north benefited little from oil earnings.

In 1958, a group of young military officers, including Saddam Hussein, overthrew the monarchy and created a secular socialist republic—much like the Arab nationalist movement discussed in the Maghreb section—that prospered over the next 20 years despite occasional political disruptions. Saddam founded Iraq's secret police and then assumed leadership of the Baath Party and eventually the presidency of Iraq. Large proven oil reserves (the fifth largest in the world) became the basis of a very profitable state-owned oil industry, and the profits from oil financed a growing industrial base. Agriculture on the ancient farmlands of the Tigris and Euphrates floodplain also prospered, though environmental problems mounted. The proceeds from oil and agriculture produced a decent standard of living for most Iraqis, who also benefited from government-sponsored education and health-care systems, but corruption and political repression were ever present. Friction between Sunnis and Shi'ites was particularly strong under Saddam Hussein, and since the dictator's ousting, strife over political power and access to water and petroleum resources has drastically increased Sunni–Shi'ite hostilities.

Like a number of other countries in the region, Iraq has been in a state of war and crisis for several decades. Two major events—the Iran–Iraq war (1980–1988) and the Gulf War (1990–1991), precipitated by Iraq's invasion of Kuwait—and years of severe economic sanctions that followed crippled the country's economy. In 2003, the United States launched a war against Iraq (discussed further on page 345) that removed Saddam Hussein. Iraq held its first truly democratic elections in 2010, with more than 60 percent of the population casting ballots. Allowing democratic elections has fundamentally changed power relations in Iraq because the Shi'ites are the largest sectarian group in Iraq. During Saddam's long reign, Arab Sunnis dominated the country. The new Shi'ite-led government has ruled in a divisive fashion, favoring Shi'ite interests. Discontent spread among Iraqi Sunnis, spawning the radical rebel movement that the world now knows as the Islamic State (see page 346).

THINGS TO REMEMBER

• Turkey is, in many ways, at a crossroads. A large segment of the country's population wants to continue efforts to Westernize, join the European Union, and remain secular. An equally large, if not larger, group is more conservative and wants to increase the influence of religious Islam in everyday life, engage more with Muslim countries, and curtail ties with the West.

• Due to colonial boundary making, the ethnic Kurds found themselves living in many different countries, all of which discriminated against them. The Kurds became a nation without a state, although they recently gained control of and govern territories in northern Iraq.

• Iran has been through a series of dramatic changes since the overthrow of the shah's regime in 1979. The overthrow led to the establishment of a theocratic state that has been oscillating between reform and a crackdown on domestic political opposition. During the entire post-Revolutionary period, however, it has promoted an Iran-led Shi'ite agenda across the region.

• After the fall of Saddam Hussein, elections in Iraq empowered the country's majority Shi'ites. Some Sunni Iraqis, now shut out of political power, emerged as the Islamic State group.

GEOGRAPHIC THEMES: North Africa and Southwest Asia Review and Self-Test

1. Environment: The population of this predominantly dry region is growing fast, while access to cultivable land and water for agriculture and other human uses is limited. Therefore, many countries here are dependent on imported food. Climate change could reduce food output and make water, already a scarce commodity, a source of conflict. New technologies offer solutions to water scarcity but are expensive and unsustainable.

• How will people get enough water to grow food in the future, especially if climate change makes this dry region even drier?

• What countries do the Euphrates/Tigris and the Nile rivers connect? What are the geopolitical implications of dams on these rivers?

• What factors could make it difficult to obtain enough water for agriculture? What technologies could provide potential solutions? How crucial is imported food for this region?

2. Globalization and Development: The vast fossil fuel resources of a few countries in this region have transformed economic development. In these countries, economies have become powerfully linked to global flows of money, resources, and people. Politics have also become globalized, with Europe and the United States strongly influencing the region.

• How have the huge fossil fuel reserves of some countries transformed economic development and driven globalization in this region?

• What flows of money, resources, and people characterize globalization in the Gulf states? How important are outside political influences?

• How might changes in job opportunities for women in this region relieve the dependency on imported labor? Where is this likely to be most important?

3. Power and Politics: Authoritarian power structures prevail throughout much of North Africa and Southwest Asia. Beginning with the Arab Spring of 2010, waves of protests swept the region. The outcome of the Arab Spring has varied from reform to repression and civil war.

• Is this region democratizing? What are signs for and against democratization?

• How have recent events changed the political landscape of this region?

• How is repression of the political opposition in many countries related to the emergence of the Arab Spring? What has been the effect of integrating political opposition movements, Islamist and others, into the democratic process?

4. Urbanization: Two patterns of urbanization have emerged in the region. In the oil-rich countries, there are spectacular new luxury-oriented urban areas. In the oil-poor countries, many people live in overcrowded urban slums with few services, and jobs for rural migrants are scarce.

• How has globalization shaped different patterns of urbanization throughout the region?

• Why is urbanization currently so different in the Gulf states than in the rest of the region? How have economic reforms aimed at improving global competitiveness, especially in export-oriented agriculture, influenced urbanization outside the Gulf states?

5. Population and Gender: This region has the second-highest population growth rate in the world. Part of the reason for this is that women are generally poorly educated and tend not to work outside the home. Childbearing thus remains crucial to a woman's status, a situation that encourages large families.

• How does the low status of women contribute to this region's high population growth?

• Relative to other regions, how high is this region's population growth rate? How do women's education levels influence how many children they have?

Critical Thinking Questions

1. How is the influence of religion in North Africa and Southwest Asia affected by historical factors, economic conditions, and the ethnic composition of different countries?

2. Describe how the present-day map of North Africa and Southwest Asia is related to the dismantling of the Ottoman Empire after World War I?

3. Discuss the possibilities that scarcity of water is or will become a cause of violence in the region. What is the evidence against this happening?

4. Consider how globalization and immigration have influenced the recent development of urban landscapes in the Gulf states.

5. Compare and contrast the public debate over the proper role of religion in public life in your own country and in one country from this region (for example, Turkey, Morocco, Egypt, or Saudi Arabia). Contrast the roles of religious fundamentalism in the debates in your country and in the country you chose from this region.

6. Gender is a complex subject in this region. Choose a rural, traditional location somewhere in the region and a modern, urban location elsewhere in the region, and make a list of the forces in each that would affect the future of a 20-year-old woman. Describe those hypothetical futures objectively; that is, without using any judgmental terminology.

7. What does it mean if in some countries of this region agriculture may produce only a small amount of the GDP, yet employ 40 percent or more of the people? What public policies would be appropriate in these circumstances—for example, should agriculture be de-emphasized? How might this de-emphasis affect food security?

8. Describe the circumstances that led to support in Europe and the United States for the formation of the state of Israel. Why has the two-state solution (which would include a Palestinian state) not come to fruition?

9. Considering the various factors that encourage relatively high fertility in this region, design themes for a public education program that would effectively encourage lower birth rates. Which population groups would you target? How would you incorporate cultural sensitivity into your project?

Chapter Key Terms

Berber 359
cartel 334
Christianity 327
desertification 322
diaspora 326
diffusion 329
economic diversification 336
failed state 359
female seclusion 354
Fertile Crescent 325
fossil fuel 315
Gulf states 315
hajj 328

intifada 343
Islam 314
Islamism 314
jihadists 340
Judaism 326
monotheism 326
Muslims 328
occupied Palestinian Territories 315
OPEC (Organization of the Petroleum Exporting Countries) 334
Ottoman Empire 330
patriarchal 353
Qur'an (or Koran) 318

Salafism 340
salinization 321
seawater desalination 322
secular states 340
shari'a 329
sheikhs 363
Shi'ite (or Shi'a) 329
Sunni 329
theocratic states 340
veil 355
Wahhabism 340
West Bank barrier 345
Zionism 342

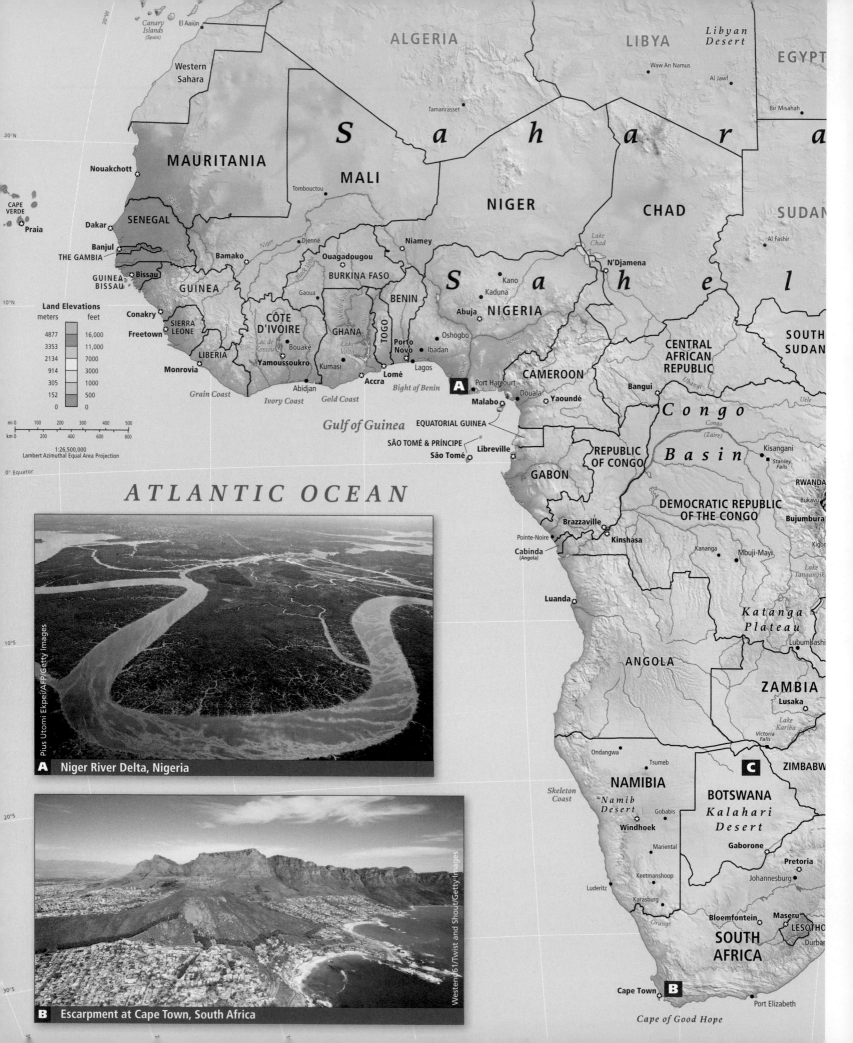

ALGERIA

LIBYA

Libyan Desert

EGYPT

Canary Islands (Spain)

El Aaiún

Western Sahara

Waw An Namus

Al Jawf

Bir Misahah

MAURITANIA

S a h a r a

20°N

Nouakchott

MALI

NIGER

CHAD

SUDAN

CAPE VERDE

Praia

Tombouctou

Lake Chad

N'Djamena

Al Fashir

Dakar

SENEGAL

Djenné

Niamey

Niger

S a h e l

Banjul

THE GAMBIA

Bamako

Ouagadougou

BURKINA FASO

Kano

Kaduna

GUINEA BISSAU

Bissau

GUINEA

Gaoua

BENIN

Abuja

NIGERIA

10°N

Land Elevations

meters	feet
4877	16,000
3353	11,000
2134	7000
914	3000
305	1000
152	500
0	0

Conakry

SIERRA LEONE

Freetown

CÔTE D'IVOIRE

GHANA

TOGO

Oshogbo

CENTRAL AFRICAN REPUBLIC

SOUTH SUDAN

Lac de Kossou

Bouaké

Lake Volta

Porto Novo

Ibadan

Benue

LIBERIA

Yamoussoukro

Kumasi

Lomé

Lagos

Bangui

Ubangi

Uele

Monrovia

Abidjan

Accra

Bight of Benin

A

Port Harcourt

CAMEROON

C o n g o

mi 0   100   200   300   400   500

km 0   200   400   600   800

Grain Coast

Ivory Coast

Gold Coast

Gulf of Guinea

EQUATORIAL GUINEA

Malabo

Douala

Yaoundé

Congo (Zaire)

Kisangani

Stanley Falls

1:26,500,000
Lambert Azimuthal Equal Area Projection

SÃO TOMÉ & PRÍNCIPE

São Tomé

Libreville

B a s i n

RWANDA

Bukavu

0° Equator

GABON

REPUBLIC OF CONGO

DEMOCRATIC REPUBLIC OF THE CONGO

Bujumbura

Kigon

ATLANTIC OCEAN

Brazzaville

Kinshasa

Kananga

Mbuji-Mayi

Lake Tanganyika

Pointe-Noire

Cabinda (Angola)

10°S

Luanda

Katanga Plateau

Lubumbashi

ANGOLA

ZAMBIA

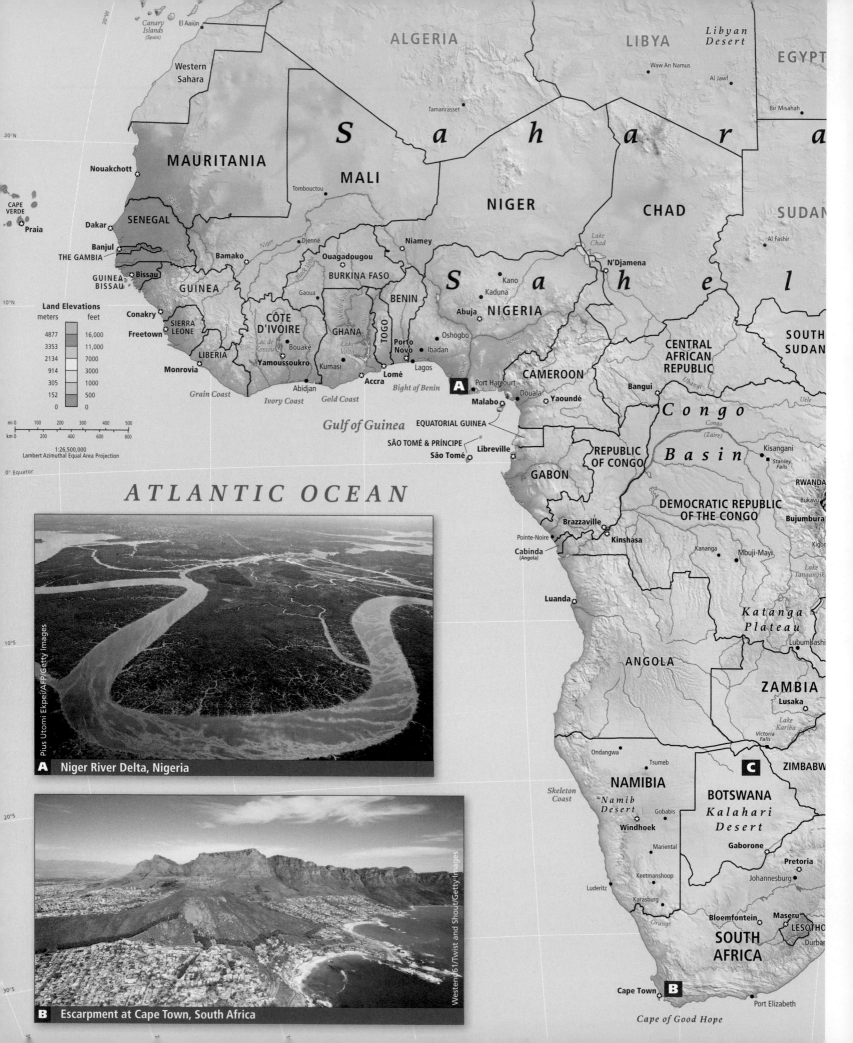

Lusaka

Lake Kariba

Victoria Falls

Ondangwa

C

ZIMBABW

Tsumeb

NAMIBIA

BOTSWANA

Skeleton Coast

Namib Desert

Gobabis

Kalahari Desert

20°S

Windhoek

Gaborone

Mariental

Pretoria

A Niger River Delta, Nigeria

Keetmanshoop

Johannesburg

Luderitz

Karasburg

Bloemfontein

Maseru

LESOTHO

Orange

30°S

B Escarpment at Cape Town, South Africa

SOUTH AFRICA

Durba

Cape Town

B

Port Elizabeth

Cape of Good Hope

Pius Utomi Ekpei/AFP/Getty Images

Westend61/Twist and Shout/Getty Images

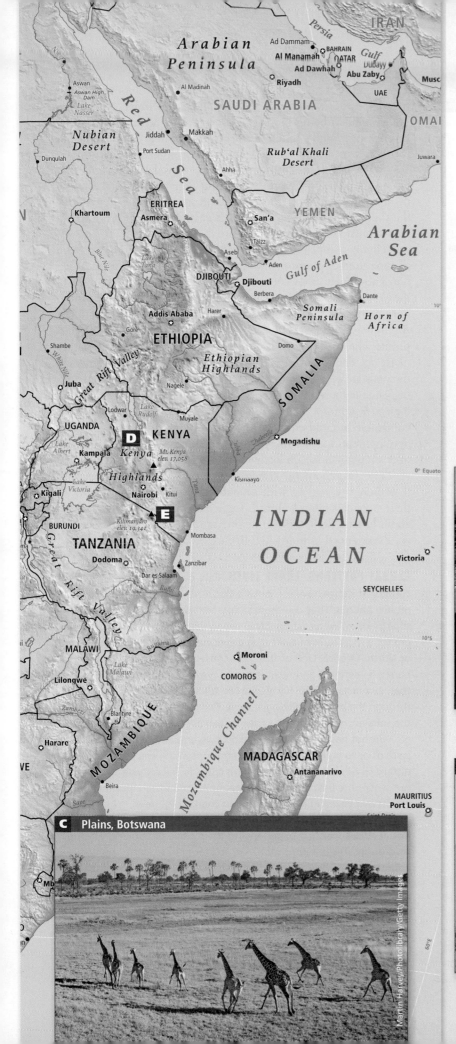

7
Sub-Saharan Africa

D Great Rift Valley, Kenya

Nigel Pavitt/AWL Images/Getty Images

E Mount Kilimanjaro, Tanzania

cversnap/Vetta/Getty Images

FIGURE 7.1 Regional map of Sub-Saharan Africa

C Plains, Botswana

Martin Harvey/Photolibrary/Getty Images

GEOGRAPHIC THEMES: Sub-Saharan Africa

After you read this chapter, you will be able to discuss the following issues as they relate to the five thematic concepts:

1. Environment: Sub-Saharan Africa's poverty and political instability leave its people less able to adapt to climate change than the inhabitants of other world regions. In the past, this region has contributed little to the buildup of greenhouse gases in the atmosphere, but currently deforestation by rural Africans and by multinational logging companies is intensifying global climate change.

2. Globalization and Development: Most sub-Saharan African economies are dependent on the export of raw materials, a legacy of the era of European colonialism. This results in economic instability because prices for raw materials can vary widely from year to year. While a few countries are diversifying and industrializing, most still sell raw materials and all rely on expensive imports of food and manufactured goods.

3. Power and Politics: Despite a shift in sub-Saharan Africa toward more political freedoms, some governments remain authoritarian, nontransparent, and corrupt. Public participation is growing, often due to electronic media, and free and fair elections have brought about dramatic changes in some countries. In other countries, elections and governments tainted by suspicions of fraud have led to surges of violence.

4. Urbanization: Sub-Saharan Africa is undergoing a massive wave of rural-to-urban migration, the fastest rate of urbanization in the world. Much of this growth is unplanned, and 70 percent of the urban population now live in impoverished slums characterized by crowded fire-prone housing, inadequate sanitation, and poor access to jobs, clean water, and food.

5. Population and Gender: Of all the world regions, populations are growing fastest in sub-Saharan Africa, yet growth rates are slowing as the demographic transition takes hold in more prosperous countries where better health care, lower levels of child mortality, and more economic and educational opportunities encourage smaller families. In particular, better educated women are able to pursue careers and to choose to have fewer children.

The Sub-Saharan Region

Sub-Saharan Africa **(FIGURE 7.1)** contains 48 countries and occupies a space bigger than North America and Europe combined. While its population is now growing quickly, the region is not particularly densely populated. Sub-Saharan Africa is where modern humans originated; it has megafauna that most of us will see only in zoos, and it has a broad variety of landscapes and climates. But the region is not well known to most people and is often erroneously thought of as being desperately poverty-stricken and barely capable of helping itself. This chapter will show that Africans across the continent are designing and implementing solutions to Africa's needs.

During the era of European colonialism (the 1500s to the 1950s), this region's massive wealth of human talent and natural resources flowed out of Africa to Europe and the Americas. Even after African countries became politically independent in the 1950s, 1960s, and 1970s, wealth continued to flow out of Africa.

This trend of wealth extraction persisted in development initiatives by Western countries and by China and India. Now, however, wealth is being increasingly created in Africa for Africans, by Africans. While much of the region remains impoverished, with parts in armed conflict and outside interests often taking advantage of this instability, there are many hopeful signs that discord is diminishing and well-being increasing. Corruption is being addressed, and African migrants are returning home with the skills needed to innovate for Africa and to monitor the strategies of outside developers to ensure they are appropriate.

Highly competent women and men use a powerful language of opportunity, optimism, and innovation to characterize the Africa they know today (see Global Patterns, Local Lives below). In 2016, the World Bank described sub-Saharan Africa as one of the fastest-growing regions in the world economically. But caution is in order: A number of studies by the World Bank, the United Nations, and private scholars indicate that although several countries with particularly rich deposits of oil, gold, platinum, copper, and other strategic minerals are beginning to take control of their resources, and adding value to those resources before exporting them, thus increasing their net earnings, reliance on exportable resources has left them vulnerable to rapid changes in global prices. Furthermore, global developments such as the recession in China and overall tightening of credit regulations are likely to constrict African growth.

The five thematic concepts described above will be used to explore this region, with interactions between two or more of these concepts often featured. Vignettes, like the one that follows about Juliana Rotich, illustrate one or more of the themes as they are experienced in individual lives.

GLOBAL PATTERNS, LOCAL LIVES Few people in the East African country of Kenya have computers, but nine out of ten Kenyans have mobile phones. They use these simple and inexpensive devices (not smartphones) with texting (SMS) capabilities and perhaps a camera for a multitude of tasks that are done on computers in Europe or North America. One of the most consequential tasks is mobile banking—sending and receiving money via a mobile phone. More than 20 percent of the Kenyan gross domestic product (GDP) flows through the mobile banking system. This means that owners of even the smallest businesses can order materials and pay bills and employees without having to invest in expensive equipment and accounting systems or to make an expensive time-consuming trip to a bank in a distant city. Small-scale farmers with crops to sell can use SMS to find out where the best market is at the moment and can maximize their profits by bypassing middlemen who charge high fees.

Juliana Rotich **(FIGURE 7.2)** grew up in a small town in Kenya, just as Africa was first experiencing the mobile phone and Internet revolutions in communication. An outstanding student, she was admitted to a U.S. university, where she became an information technology (IT) major. She then collaborated with friends in the online community to found *Ushahidi*, a web-based, open source reporting system with interactive mapping capabilities that is now widely used throughout the world. Ushahidi, which means "testimony" in Swahili, was originally designed for quickly sharing information about the violence that broke out in Kenya during the 2008 elections. A user can connect to Ushahidi through a phone or website to instantly learn specific details about a developing situation.

FIGURE 7.2 **Ushahidi founder.** Juliana Rotich (standing) is a Kenyan technologist and digital activist, a specialist with a computer science degree from the University of Missouri. She founded and is executive program director of Ushahidi, a crowd-sourcing platform that enables mobile phone users to map crisis information in order to be able to quickly respond to threatening situations. Rotich, who also writes a respected blog, *Afromusing*, is a TED Senior Fellow and an MIT Media Lab Director's Fellow. She is part of a growing number of educated Africans who have focused on applying technology to African development needs.

Information gathered from hundreds or even thousands of users (in a process called *crowdsourcing*) can be verified, mapped, and disseminated in real time.

Who benefits most when barriers to the use of technology are lowered and suddenly virtually everyone has a mobile phone? The businesspeople and farmers of Kenya have clearly been beneficiaries. However, with its mapping capabilities, the Ushahidi application has also been used to help rescuers locate people buried by earthquakes in Haiti, Chile, and Japan. Ushahidi enabled crucial communication between dissidents during and after the Arab Spring revolutions in Libya and Egypt; it was used by Syrians, first to protest against the Assad government, then during the Syrian civil war to help victims, and most recently, Syrian refugees fleeing to Europe have depended on it (FIGURE 7.3).

Some of the most exciting innovations that Ushahidi gives Africans access to are in the areas of disease control, health care, political reform, wildlife inventorying, and the control of corruption. Ushahidi thus increases transparency and helps people share their stories. *[Source: "99 Faces." For detailed source information, see Text Sources and Credits.]* ∎

What Makes Sub-Saharan Africa a Region?

Sub-Saharan Africa is a region distinct from North Africa for physical, cultural, and historical reasons. The Sahara Desert and the **Sahel**—the grassy transition zone between the desert and wetter climes to the south—have long presented major physical obstacles to human habitation, though in prehistory the Sahel and Sahara had a moister climate. Trading caravans

> **Sahel** a band of arid grassland, where steppe and savanna grasses grow, that runs east–west along the southern edge of the Sahara

FIGURE 7.3 **Ushahidi platform users worldwide.** The map shows a selection of the locations in which Ushahidi has been deployed since 2008.

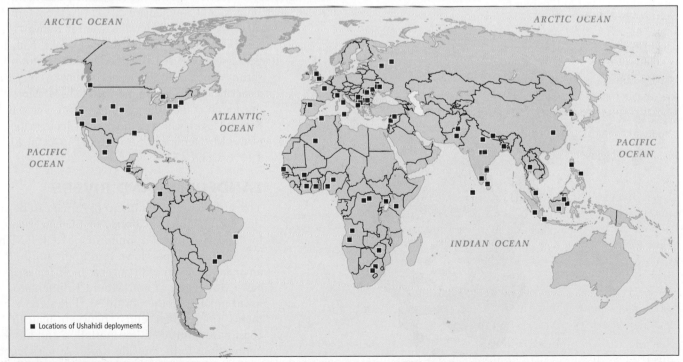

Thinking Geographically

In 2016, the zika virus emerged as an international threat, primarily because of its impact on the brains of developing fetuses. How could Ushahidi contribute to controlling the virus and its ill effects?

have traversed the Sahara and Sahel for thousands of years, but sub-Saharan Africa developed largely separate from North Africa. For millennia, sub-Saharan Africa was best known to the outside world through trade centered on the Indian Ocean. To Europeans, the region was known only by the accounts of a few travelers, such as those of the Arab explorer Ibn Battuta, who traveled widely throughout the region in the 1300s. Then, in the mid-1400s, the Portuguese sent their exploratory fleets down the west coast of Africa, beginning a 600-year-long period of European colonialism that transformed sub-Saharan Africa.

The Europeans noted what for them were exotic qualities: dark-skinned people who had unique and varied ways of life and cultures and possessed valuable trade items, such as precious minerals, exotic plants and animals, and fine textiles. For centuries, however, Europeans discounted the sophistication and complexity of sub-Saharan Africa's cultural and historical heritage of horticulture, weaving, dyeing, mining, and metalwork. Everything south of the Sahara was lumped into "Black Africa," and for some, this term is still used to distinguish the region from Africa as a whole. Now, the reality is that sub-Saharan Africa is coming into an era of growth and global participation.

Terms in This Chapter

In this chapter, we refer only occasionally to the whole continent, and then we refer to it simply as Africa. Sub-Saharan Africa is defined by the countries shown in **FIGURE 7.4**. For cultural and historical reasons, the new country of South Sudan is included in this region. For similar reasons, the country of Sudan is not included because its location on the Nile River and the strong Arab influence in the capital of Khartoum have brought Sudan into closer association with North Africa and Southwest Asia. Sudan is instead discussed in Chapter 6.

The names of African countries can often be confusing. For example, there are two neighboring countries called Congo—the Democratic Republic of the Congo and the Republic of Congo. Because these designations are both lengthy and easily confused, we abbreviate them in this text. The Democratic Republic of the Congo (formerly Zaire) carries the name of its capital in parentheses: Congo (Kinshasa). The Republic of Congo is called Congo (Brazzaville). Check the regional map in Figure 7.1 to see the locations of these countries and their capitals.

THINGS TO REMEMBER

• The many stereotypes of sub-Saharan Africa—that it is a region of conflict, disease, corruption, and poverty—are being countered by young, educated Africans who are constructing new public agencies and private businesses in their home countries, often grounded in modern technology.

• Africans are not only affecting their own countries, they are revising Africa's position vis-à-vis the rest of the world by devising innovative solutions that can be used in many situations across the globe.

• Some sub-Saharan countries now have economies that are among the fastest-growing in the world.

PHYSICAL GEOGRAPHY

The African continent is big—the second largest after Asia and about three times the size of the United States. But Africa's great size is not matched by its surface complexity. Africa has no major mountain ranges, but it does have several high volcanic peaks in the central east, including Mount Kilimanjaro (19,324 feet [5890 meters] high; see Figure 7.1E) and Mount Kenya (17,057 feet [5199 meters] high). At their peaks, both have permanent snow and ice, though these features have shrunk dramatically due to global climate change and associated local environmental changes. More than one-fourth of the African continent is covered by the Sahara Desert, which in the north reaches to the edge of the Mediterranean Sea (see the maps in Figures 1.11, 6.5, and 7.5).

LANDFORMS AND RIVERS

The surface of the continent of Africa can be envisioned as a raised platform, or plateau, bordered by fairly narrow and uniform coastal lowlands. The platform slopes downward to the north; it has an upland region with several high peaks in the east (Mount Kenya and Mount Kilimanjaro) and lower uplands in the northwest. The steep escarpments (long, high cliffs) between plateau and coast have obstructed transportation and hindered connections to the outside world. Africa's lengthy, uniform coastlines provide few natural harbors, Cape Town in South Africa being an exception (see Figure 7.1B).

FIGURE 7.4 Political map of sub-Saharan Africa.

Geologists usually place Africa at the center of the ancient supercontinent Pangaea (see Figure 1.8). As landmasses broke off from Africa and moved away—North America to the northwest, South America to the west, and India to the northeast—Africa readjusted its position only slightly. Because its shift was so small, it did not pile up long, linear mountain ranges, as did the other continents when their plates collided with one another (see Chapter 1).

Africa continues to break apart along its eastern flank. There, the Arabian Plate has already split away and drifted to the northeast, leaving the Red Sea, which separates Africa and Asia. Another split, known as the Great Rift Valley, extends south from the Red Sea more than 2000 miles (3200 kilometers) (see Figure 7.1D). The Great Rift Valley dominates Africa's eastern interior in an irregular series of tectonic fractures that curve inland from the Red Sea and extend southwest through Ethiopia and into the highland Great Lakes region where Lake Victoria forms the headwaters of the White Nile. The Great Rift Valley contains numerous archaeological sites that have shed light on the early evolution of humans and other large mammals.

There are five main rivers in sub-Saharan Africa.

- Lake Tanganyika and Lake Malawi feed watercourses that join the south-flowing Zambezi, which enters the Indian Ocean in Mozambique. The main course of the Zambezi originates in Zambia and flows south and east over Victoria Falls and into Mozambique (see Figure 7.38).

- The Nile River flows north from Lake Victoria toward the Mediterranean, and near Khartoum it is joined by the Blue Nile flowing from the Ethiopian Highlands.

- The Niger River originates in the Guinea Highlands and flows northeast through Mali and Niger into Nigeria, where it flows into the Gulf of Guinea.

- The Congo River originates in the Katanga Plateau in southern Congo (Kinshasa) and flows north along the eastern edge of the central uplands. It then makes a wide arc to the west and flows into the Atlantic at Cabinda. The Congo flows through Africa's high rainfall belt and has a flow second only to the Amazon.

- The Orange River in South Africa is the fifth main river of the continent. It rises in Lesotho and flows west across the country, then it forms the border with Namibia and enters the Atlantic Ocean at Alexandra Bay.

Many rivers in sub-Saharan Africa have been threatened with dams, primarily to aid irrigation and to produce power. There are an estimated 1270 such dams in the region and, like dams everywhere on Earth, they are being re-evaluated because not only have the dams proven to disrupt lives through the ecological changes they cause, there has also been a lack of equity in the distribution of the water and power made possible by them.

CLIMATE AND VEGETATION

Most of sub-Saharan Africa has a tropical climate (see the **FIGURE 7.5** map). Average temperatures generally stay above 64°F (18°C) year-round everywhere, except at the more temperate southern tip of the continent and in the cooler upland zones (hills, mountains, high plateaus). Seasonal climates in Africa differ more by the amount of rainfall than by temperature.

Most rainfall comes to Africa by way of the **intertropical convergence zone (ITCZ),** a band of atmospheric currents that circle the globe roughly around the equator (see photo D and inset map E in Figure 7.5). At the ITCZ, warm winds converge from both the north and the south and push against each other. This causes the air to rise, cool, and release moisture in the form of rain. The rainfall produced by the ITCZ is most abundant in Africa near the equator, where dense tropical rain forests flourish in places such as the Congo Basin (see Figure 7.5A, D).

The ITCZ shifts north and south seasonally, generally following the area of Earth's surface that has the highest average temperature at any given time. Thus, during the height of summer in the Southern Hemisphere in January, the ITCZ might bring rain far enough south to water the dry grasslands, or steppes, of Botswana. During the height of summer in the Northern Hemisphere in August, the ITCZ brings rain as far north as the southern fringes of the Sahara, to the area called the Sahel, a band of arid grassland 200 to 400 miles (320 to 640 kilometers) wide that runs east-west along the southern edge of the Sahara (see the Figure 7.5 map). At roughly 30° N latitude (the Sahara) and 30° S latitude (the Namib and Kalahari deserts), the air, which has dried out during its passage, descends, forming a subtropical high-pressure zone that shuts out lighter, warmer, moister air. As a result of this system (which is in no way precise), deserts tend to be found in Africa (and on other continents) in these zones about 30 degrees north and south of the equator.

The tropical wet climates that support equatorial rain forests are bordered on the north, east, and south by seasonally wet/dry subtropical woodlands (see Figure 7.5B). These give way to moist tropical savannas or steppes, where tall grasses and trees intermingle in a semiarid environment. These tropical wet, wet/dry, and steppe climates have provided suitable land for different types of agriculture for thousands of years, but the amount of moisture available in each of these climates can vary greatly, in cycles that are decades long. Farther to the north and south lie the true desert zones of the Sahara, the Namib, and the Kalahari (see Figure 7.5C). This banded pattern of African ecosystems, as shown on the main map of Figure 7.5, is modified in many areas by elevation and wind patterns.

The volcanic highland parts of equatorial east Africa have an important modifying effect on temperatures and are well watered, with the result that these ecological zones are less prone to insect-borne diseases and are particularly fertile.

Without high mountain ranges to block them, wind patterns can have a strong effect on climates in Africa. For example, winds blowing north along the east coast keep ITCZ-related rainfall away from the **Horn of Africa,** the triangular Somali peninsula that juts out from northeastern Africa below the Red Sea. As a consequence, the Horn of Africa is one of the driest parts of the continent. Along the west coast of the Namib Desert, moist air from the Atlantic is blocked from moving over the desert by cold air above the northward-flowing water of the Benguela Current. Like the cool Peru Current off South America, the Benguela is an oceanic current that is chilled by its passage past Antarctica. Rich in nutrients, it supports a major fishery along the west coast of Africa (see the Figure 7.5 map).

intertropical convergence zone (ITCZ) a band of atmospheric currents that circle the globe roughly at the equator; warm winds from both north and south converge at the ITCZ, pushing air upward and causing copious rainfall

Horn of Africa the triangular Somali peninsula that juts out from northeastern Africa below the Red Sea and wraps around the Arabian Peninsula

FIGURE 7.5 PHOTO ESSAY: Climates of Sub-Saharan Africa

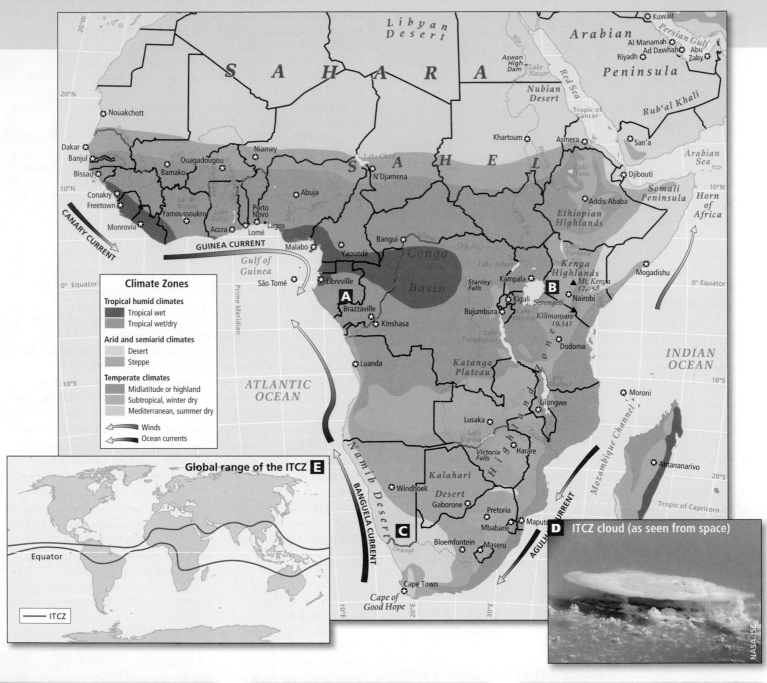

Climate Zones

Tropical humid climates
- Tropical wet
- Tropical wet/dry

Arid and semiarid climates
- Desert
- Steppe

Temperate climates
- Midlatitude or highland
- Subtropical, winter dry
- Mediterranean, summer dry

→ Winds
→ Ocean currents

Global range of the ITCZ **E**

Equator

—— ITCZ

D ITCZ cloud (as seen from space)

NASA-JSC

A Tropical wet, Gabon

Jacques Jangoux/Science Source/Getty Images

B Subtropical, winter dry, Kenya

Nigel Hicks/Dorling Kindersley/Getty Images

C Desert, Namibia

David Yarrow Photography/The Image Bank/Getty Images

THINGS TO REMEMBER

- The surface of the continent of Africa can be envisioned as a raised platform, or plateau, bordered by fairly narrow and uniform coastal lowlands.

- Most of sub-Saharan Africa has a tropical wet or wet/dry climate, except at the southern tip and in the cooler uplands. Seasonal climates in Africa differ more by the amount of rainfall than by temperature.

- Most rainfall comes to Africa by way of the intertropical convergence zone (ITCZ), a band of atmospheric currents that circle the globe roughly around the equator.

ENVIRONMENT

GEOGRAPHIC THEME 1

Environment: Sub-Saharan Africa's poverty and political instability leave its people less able to adapt to climate change than the inhabitants of other world regions. Up to now, this region has contributed little to the buildup of greenhouse gases in the atmosphere, but currently deforestation by rural Africans and by multinational logging companies is adding to the intensification of global climate change.

DEFORESTATION AND CLIMATE CHANGE

Sub-Saharan Africa contributes to an increase in global CO_2 emissions and potential climate change primarily through deforestation. Trees absorb CO_2 as they photosynthesize, thus removing carbon from the air and storing it as biomass, a process known as **carbon sequestration.** When trees are burned or when they decompose after they die, they release the stored CO_2 into the atmosphere. Because of this, deforestation is a major contributor to carbon buildup in the atmosphere and ultimately to climate change.

African countries no longer lead the world in the rate of deforestation—the percentage of total forest area lost. In 2016, countries in Central and South America and in East and Southeast Asia had higher rates of forest removal. But at least six sub-Saharan countries have been severely deforested (Burundi, Togo, Nigeria, Benin, Uganda, and Ghana) and now have less than 16 percent of the forests they did in 1990. This deforestation is responsible for major human environmental impact across the region **(FIGURE 7.6)**. Nigeria and Congo (Kinshasa) have the third and fourth greatest GHG emissions from deforestation, respectively.

VIGNETTE Liberian environmental activist Silas Siakor is an affable and unassuming fellow. But his casual style belies his remarkable sleuthing ability and fierce dedication to his homeland. At great personal risk, Siakor uncovered evidence that 17 international logging companies were bribing Liberia's then-president, Charles Taylor, with cash and guns. Taylor allowed the companies to illegally log Liberia's forests, which are home to many endangered species, including forest elephants and chimpanzees. In return, the companies paid cash and provided weapons that Taylor used to equip his personal armies. Made up largely of kidnapped and enslaved children, Taylor's armies fought those of other Liberian warlords in a 14-year civil war that took the lives of 150,000 civilians. The war also spilled over into neighboring Sierra Leone, where another 75,000 people died. Meanwhile, the logging companies—based in Europe, China, and Southeast Asia—reaped huge fortunes from the tropical forests.

Silas Siakor pulled together publicly available information that had been previously ignored by the international community and information provided by ordinary citizens in ports, villages, and lumber companies. He prepared a clear, well-documented report proving the massive logging fraud, which was delivered to the United Nations. In response to Siakor's report, the UN Security Council voted to impose sanctions that would end the timber trade and to prosecute some of the people involved. Taylor fled to Nigeria, but in 2006 he was extradited to face a war crimes tribunal in The Hague, Netherlands. In April 2012, Taylor received a 50-year sentence.

Eventually, democratic elections in Liberia resulted in the election of Africa's first woman president, Ellen Johnson-Sirleaf, who then boldly moved to address the poverty and political instability of her country and, most importantly, corruption in the resource management sector. She cancelled all of Charles Taylor's former contracts with timber companies, pending a revision of Liberian forestry law. Liberia then became the first African country to join the global anticorruption Extractive Industries Transparency Initiative (EITI), whereby tax revenues are audited and made public. The EITI revealed major flaws in Liberia's taxing of foreign resource extraction firms and the results were presented in well-attended public meetings. The involvement of the public and the transparency of the tax audits ultimately worked to exonerate Johnson-Sirleaf herself from any charges of corruption. In 2011, she was re-elected to a second and final 6-year term. [Sources: Liberia's Johnson-Sirleaf Defends Governance Record, Reuters World, May 17, 2013; Goldman Environmental Prize, National Public Radio, and Silas Siakor. For detailed source information, see Text Credit pages.] ∎

The story of Silas Siakor and Ellen Johnson-Sirleaf illustrates how, like many other countries in this region, Liberia is still in the process of building strong political institutions that guide the extraction and use of the country's resources sustainably and for the benefit of its citizens.

Much of Africa's deforestation is driven by the growing demand for farmland and fuelwood, although logging by international timber companies is also increasing. Africans use wood (or charcoal made from wood) to supply nearly all their domestic energy (see Figure 7.6A, B). Wood remains the cheapest fuel available, in part because of African traditions that consider forests to be a free resource held in common. Even in Nigeria, a major oil producer, most people use fuelwood because they cannot afford petroleum products. Not only are charcoal prices rising across the region

carbon sequestration the removal and storage of carbon taken from the atmosphere

Deforestation, desertification, and increasing water use have had major impacts on sub-Saharan Africa's ecosystems.

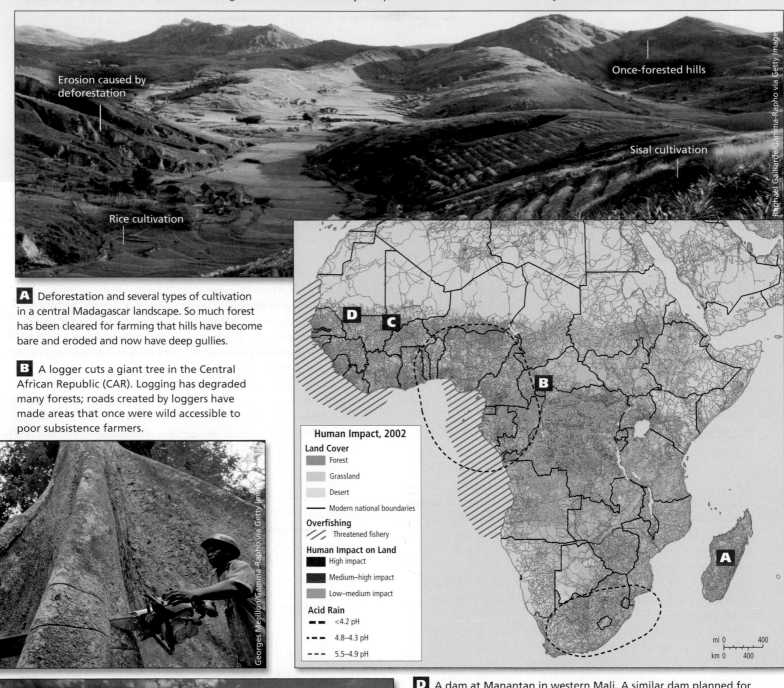

A Deforestation and several types of cultivation in a central Madagascar landscape. So much forest has been cleared for farming that hills have become bare and eroded and now have deep gullies.

B A logger cuts a giant tree in the Central African Republic (CAR). Logging has degraded many forests; roads created by loggers have made areas that once were wild accessible to poor subsistence farmers.

Erosion caused by deforestation

Once-forested hills

Sisal cultivation

Rice cultivation

Human Impact, 2002

Land Cover
- Forest
- Grassland
- Desert
- Modern national boundaries

Overfishing
- Threatened fishery

Human Impact on Land
- High impact
- Medium–high impact
- Low–medium impact

Acid Rain
- — — <4.2 pH
- - - - 4.8–4.3 pH
- - - - 5.5–4.9 pH

mi 0   400

km 0   400

C A fisher casts his net in the Niger River wetlands, where carefully synchronized resource-use patterns have been developed over millennia.

D A dam at Manantan in western Mali. A similar dam planned for upstream of the Niger River wetlands would alter the river's flow in order to supply farmers with irrigation water. The project would increase food security for farmers elsewhere in Mali but threaten the livelihoods of fishers, farmers, and pastoralists in the Niger River wetlands.

Thinking Geographically

After you have read about human impacts on the biosphere in sub-Saharan Africa, you will be able to answer the following questions.

A Other than the caption, do other clues exist to suggest that this landscape has been cleared of forest?

B How do this photo and caption reveal the quandaries faced by those who want to assist the poor in the Central African Republic?

C What aspect of the fishing method depicted in the photo suggests that catches are fairly small in size?

as forests disappear, but the smoke from charcoal is causing asthma and other respiratory problems, as well as creating greenhouse gases.

Deforestation is being increasingly addressed across the continent. In Maputo, the capital of Mozambique, a group of small-scale farmers are engaged in a for-profit project that combines new technology with traditional crops to create a sustainable fuel and food preparation system. The goal is to shift the 1.2 million inhabitants of Maputo from relying on charcoal—made from the fast-disappearing old-growth forests in the north of the country—to using ethanol made from the roots of the cassava plant grown by farmers on their own land. Newly developed small, metal cookstoves run on the cassava-based ethanol.

The multifaceted stove/ethanol/cassava pilot project in Maputo is funded by four American firms that hope to replicate it elsewhere around the world. Their goal is to save the forests, relieve fuel shortages, halt rising fuelwood prices, and alleviate air pollution. Other benefits include improving food security (since cassava is a major food crop in Africa), and creating local small businesses that can convert excess cassava into ethanol and build the innovative cookstoves.

Elsewhere, African governments attempting to slow deforestation are encouraging **agroforestry**—the farming of economically useful trees in conjunction with the usual subsistence and cash crops. Using the farmed trees for fuel and construction helps reduce dependence on old-growth forests and can provide income through the sale of farmed wood. For example, by practicing agroforestry on the fringes of the Sahel, a family in Mali can produce fuelwood, fencing and building materials, medicinal products, and food—all on the same piece of land. This would double what could be earned from a single crop or from animal herding. However, critics caution that agroforestry sometimes introduces invasive species, nonnative plants that threaten many African ecosystems by outcompeting indigenous species.

ON THE BRIGHT SIDE: The Value of Agroforestry

Agroforestry has long been promoted by the Green Belt Movement, founded by Dr. Wangari Maathai of Kenya, which helps rural women plant trees for use as fuelwood and to reduce soil erosion. The movement grew from Maathai's belief that a healthy environment is essential for democracy to flourish. Thirty years and 30 million trees after she commenced her work, Maathai was awarded the 2004 Nobel Peace Prize for her contributions to sustainable development, democracy, and peace. ■

AGRICULTURAL SYSTEMS, FOOD, AND VULNERABILITY TO CLIMATE CHANGE

Agricultural systems in Africa are undergoing a rapid transition from subsistence farming to commercial farming. Unfortunately, this change is happening at a time when the advantages of traditional agriculture are just beginning to be appreciated. The social and environmental effects of turning rapidly to commercial production are not fully understood at this point, especially given the as-yet little-known ecological impacts of climate change.

Traditional Agriculture

Most sub-Saharan Africans practice **subsistence agriculture**—farming that provides food for only the farmer's family—usually on small farms of about 2 to 10 acres (1 to 4 hectares). Most subsistence farmers are female, though males may assist in the initial preparation of a plot. These farmers are also **horticulturists,** meaning that they carefully attend to the needs of individual plants and understand the intricate ecological relationships between plants in a garden. They also practice **mixed agriculture,** which involves raising a diverse array of crops and a few animals as livestock. Family members may also fish, hunt, herd, and gather some of their food from forest or grassland areas. The wide variety of food produced increases food security and also means that the cuisine of Africa is highly varied **(FIGURE 7.7)**.

To maintain soil quality in a tropical wet or wet/dry climate, subsistence farmers have for generations used **shifting cultivation,** in which small patches of forest are cleared, with the detritus either burned or left to decay. The clearings are cultivated with a wide variety of plants for 2 or 3 years and then abandoned for a decade or longer and left to regrow, meanwhile new cleared patches are planted. After a few decades, the soil in the abandoned plots naturally replenishes its organic matter and nutrients and is ready to be cleared and cultivated again. However, when the population increases greatly, demand for farmland means that **fallow periods** are shortened and the soil becomes degraded.

These traditional techniques for growing food—subsistence, mixed, and shifting cultivation—have advantages and disadvantages in the present era of climate change. All these closely related forms of agriculture are quite reliably productive, and because they are closely tended, they provide healthy diets and a diverse array of strategies for coping with the changes in temperature and rainfall that climate change may bring. For example, traditional farmers in Nigeria (usually women) grow complex tropical gardens, often with 50 or more species of plants, as in

agroforestry growing economically useful crops of trees on farms, in conjunction with the usual plants and crops, to reduce dependence on trees from non-farmed forests and to provide income to the farmer

subsistence agriculture farming that provides food for only the farmer's family and is usually done on small farms

horticulture hands-on farming that pays close attention to the specific needs of individual plants

mixed agriculture farming that involves raising a diverse array of crops and animals on a single farm, often to take advantage of several environmental niches

shifting cultivation a productive system of agriculture in which small plots are cleared in forestlands, the dried brush is burned to release nutrients, and the clearings are planted with multiple species; each plot is used for only 2 or 3 years and then abandoned for many years of fallowing

fallow period the time (ideally, a decade or more) during which an agricultural plot is allowed to rest and regrow a natural vegetation cover

FIGURE 7.7 LOCAL LIVES: Foodways in Sub-Saharan Africa

A Ethiopian and Eritrean cuisine features spicy vegetable and meat stews and salads served with *injera,* a type of savory pancake made from fermented grains. Pieces of the injera are torn off and used to scoop up the rest of the meal. Injera is traditionally cooked on a clay plate over an open fire. Because cooking injera on an open fire requires a lot of wood, people also use more efficient stoves, including electric models that are popular in cities.

B Recently harvested cassava root is transported in Nigeria. Originally domesticated in Brazil, cassava was brought to Africa in the seventeenth century as cheap transportable food in the slave trade. Because it grows abundantly in a wide variety of soils and environments and is easy to cultivate, it has become a popular crop, despite its relatively low nutritional value. Sub-Saharan Africa produces more cassava than any world region, with Nigeria being the world's largest producer.

C A *potjiekos,* or "small pot stew," is prepared in South Africa. Cast iron pots for cooking were brought to South Africa by migrants from the Netherlands, who developed the potjiekos during their expansion north from Cape Town into the interior. These "trekkers" made stews of wild game, vegetables, and spices each evening, keeping the leftovers in the pot for the next day's journey. The trekkers also added fresh ingredients as they traveled. Making a potjiekos is a social activity, with fireside conversation throughout the 3 to 6 hours that it takes to cook the stew.

Photostock Israel/Oxford Scientific/Getty Images

Lynn Johnson/National Geographic/Getty Images

Graeme Williams/Gallo Images/Getty Images

many other parts of sub-Saharan Africa. Some of the plants can handle drought, while others can withstand intense rain or heat. If water is needed, it will be brought directly to each plant by the farmer. The presence of so many different species of plants means that some will thrive, no matter what the weather brings.

However, the subsistence nature of most African farming can also leave families without much cash. In years when harvests are too low to provide surplus that can be sold, and hunting and gathering fail to provide supplementary food for the family, there may not be enough money to buy food. While such situations can lead to famine, it is important to note that the most serious famines in Africa have occurred not because of low harvests but rather because of political instability that disrupts economies and the systems that grow and distribute food.

Commercial Agriculture

Much of Africa is now shifting over to **commercial agriculture,** in which crops are grown deliberately for cash rather than solely as food for the farm family. This type of production also has advantages and disadvantages with respect to changing market situations or changing climate. If harvests are good and prices for crops are adequate, farmers can earn enough cash to get

commercial agriculture farming in which crops are grown deliberately for cash rather than solely as food for the farm family

mono-crop agriculture farming in which a single crop species is grown in a large field

them through a year or two of poor harvests. Having some cash income can also allow poor families to invest in other means of earning a living, such as opening a grinding mill to help process other farmers' harvests or building a stone oven to bake neighbors' bread for a fee. In addition, they will have money for school fees so that their children can prepare for careers other than farming.

At whatever scale commercial agriculture is practiced—whether small scale and locally managed, or large scale and owned and operated by international corporations (agribusiness)—problems can develop that may be out of the control of the farmer. Some of the most common commercial crops, such as corn, peanuts, cacao beans (used for making chocolate), rice, tea, and coffee, are often less adapted to environments outside their native range (see Figure 7.8D). If temperatures increase or water becomes scarce or overabundant, these crops are likely to produce poorly. Moreover, to maximize profits, these crops are often **mono-cropped,** that is, grown in large fields containing only a single plant species. This can leave crops vulnerable to pests and diseases. And since only one cash crop is grown, should that crop fail, the farmer may have no other garden foods to rely on. The uncertainties of global climate change can leave farmers of commercial crops vulnerable for another reason: the instability of prices for commercial crops on global markets. Because of overproduction or crop failures in distant places, local prices can rise or fall dramatically from year to year.

Commercial agriculture usually requires permanently cleared fields. Because soil fertility in the tropics declines rapidly once the trees are removed, however, commercial crops are more likely to fail even under ideal climatic conditions. Chemical fertilizers can compensate for soil infertility, but are often too expensive for ordinary farmers, and may lose their effectiveness when used repeatedly. Moreover, rains almost always wash much of the fertilizer into nearby waterways, thus polluting them. This may ultimately hurt farmers and fishers by spoiling drinking water and reducing the quantity of available fish—important sources of protein for rural people.

These aspects of commercial agricultural systems make them a risky choice even under the most ideal climatic conditions, let alone in the hotter and more drought-prone environments that climate change may produce. Foreign agricultural "experts" who often push commercial systems can be woefully ignorant of local conditions and the value of local cultivation techniques. In Nigeria, for example, women grow most of the food for family consumption and are guardians of the knowledge that makes Nigeria's traditional farming systems work. However, despite a growing awareness of women's importance in African agriculture, women are frequently not included or even consulted by the foreign experts who promote commercial agricultural development projects. At best, women are employed as field laborers, rarely as policy planners or project supervisors or even equipment operators. Consequently, diverse subsistence agricultural systems based on numerous plant species and deep knowledge about agriculture have been replaced by less stable commercial systems that are based on a single plant species and do not take into account the skills and knowledge base of women farmers.

In the recent past, agricultural scientists have begun to recognize past mistakes and have been trying to incorporate the traditional knowledge of African farmers, female and male, into commercial agriculture systems. For example, scientists at Nigeria's International Institute for Tropical Agriculture are developing cultivation systems that, like traditional systems, use many species of plants that help each other cope with varying climatic conditions. Most of these systems are designed to include both subsistence and commercial agriculture, and to give families both a stable food supply *and* the cash to help them ride out crop failures and pay school tuition for their children **(FIGURE 7.8C)**.

WATER RESOURCES AND WATER MANAGEMENT ALTERNATIVES WITH CLIMATE CHANGE

Although sub-Saharan Africa has a large wet tropical zone, much of the region is seasonally dry (see the Figure 7.5 map). Freshwater shortages are expected to increase, as population numbers and densities grow, or where people are dislocated due to conflict or other reasons (Figure 7.8A). Groundwater contamination is another cause of shortages, as is increasing demand for irrigation for agricultural systems large and small (see Figure 7.6D). For Africa's millions of smallholder farmers, water is one of the most important factors in production. Access to water and the ability to manage it effectively are key to food production and to environmental sustainability.

New Groundwater Discoveries

Is there sufficient groundwater in Africa? Until recently, the answer was a resounding no. However, a recent comprehensive review and quantitative mapping of all available data on groundwater for the African continent **(FIGURE 7.9)** revealed that the quantity of **groundwater** (water naturally stored in aquifers over periods measuring thousands of years) was nearly 100 times that of previous estimates. In many cases, the water was close enough to the surface to be accessed via inexpensive bore well pumps.

This does not mean that Africa suddenly has an inexhaustible supply of water. A recent global mapping of naturally occurring arsenic contamination of groundwater shows that while Africa has only modest vulnerability to arsenic poisoning of groundwater, a wide band north of the equator from Senegal and Guinea through Nigeria to Ethiopia is likely to become contaminated by arsenic as development proceeds and population increases. Groundwater is also endangered by commercial agriculture because of its reliance on chemical fertilizers and pesticides that eventually work their way into the aquifers.

Nor will it be easy to get the groundwater to the people who need it. The largest reservoirs of groundwater are in lightly populated areas (such as North Africa and Botswana; see Figure 7.1C). The most heavily populated country, Nigeria, has only a relatively small supply of groundwater. Nonetheless, this news on groundwater resources is welcome because climate change is likely to reduce precipitation in irregular and unpredictable patterns across the continent, and because per capita water use is likely to expand exponentially as this region develops economically (see Chapter 1).

New Large- and Small-Scale Irrigation Projects

Government officials continue to favor large-scale irrigation projects to address the need for increased food supplies in the face of current and future water scarcity. Africa has a number of major rivers—the Niger, the Nile, the Congo, the Zambezi, and the Limpopo—and climate change will affect all of them. Here, we focus on the Niger, one of Africa's most important rivers, which flows through several very different ecological zones.

The Niger rises in far western Africa, in the tropical wet Guinea Highlands, and carries summer floodwaters northeast into the normally arid lowlands of first Mali and then Niger (see the Figure 7.5 map and Figure 7.6C, D). There, the waters spread out into lakes and streams that nourish wetlands. For a few months of the year (June through September), the wetlands ensure the livelihoods of millions of fishers, farmers, and pastoralists. These people share the territory in carefully synchronized patterns of land use that have survived for millennia. Wetlands along the Niger produce eight times more plant matter per acre than the average wheat field does. They provide seasonal pasture for millions of domesticated animals, and serve as an important habitat for birds and other wildlife.

The governments of Mali and Niger now want more dams on the Niger River so they can channel its water into irrigated agribusiness projects that they hope will help feed the more than 26 million people in both countries. However, the dams would forever change the seasonal rise and fall of the river. The irrigation systems may also pose a threat to human health. Systems that rely on surface storage not only lose a great deal of water through evaporation and leakage, but

groundwater water naturally stored in aquifers as many as 5000 years ago during wetter climate conditions

FIGURE 7.8 PHOTO ESSAY: Vulnerability to Climate Change in Sub-Saharan Africa

Much of this region is highly exposed to drought, flooding, and other climatic disturbances. Poverty and little access to cash increase sensitivity to climate change, while political instability makes it harder for governments to effectively use disaster management to help boost resilience.

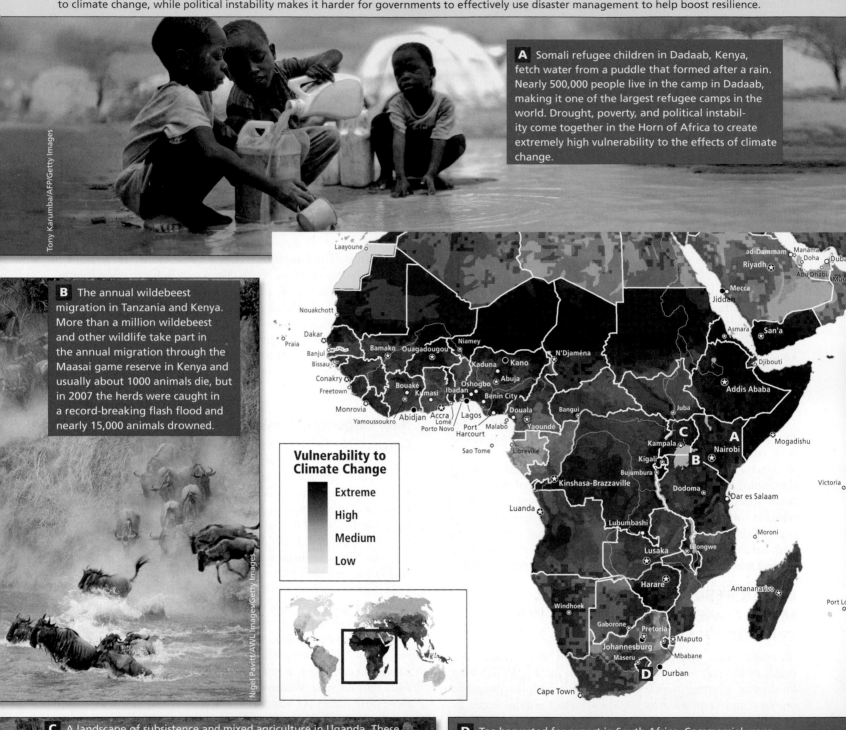

A Somali refugee children in Dadaab, Kenya, fetch water from a puddle that formed after a rain. Nearly 500,000 people live in the camp in Dadaab, making it one of the largest refugee camps in the world. Drought, poverty, and political instability come together in the Horn of Africa to create extremely high vulnerability to the effects of climate change.

B The annual wildebeest migration in Tanzania and Kenya. More than a million wildebeest and other wildlife take part in the annual migration through the Maasai game reserve in Kenya and usually about 1000 animals die, but in 2007 the herds were caught in a record-breaking flash flood and nearly 15,000 animals drowned.

Vulnerability to Climate Change

- Extreme
- High
- Medium
- Low

C A landscape of subsistence and mixed agriculture in Uganda. These diversified systems make farmers somewhat more resilient to climatic disturbances, but provide little cash to help farmers buy other food they might need.

D Tea harvested for export in South Africa. Commercial crops like tea can increase farmers' resilience to climate change by providing them with cash to buy any needed food. But tea, like many commercial crops, requires expensive and polluting fertilizers, and prices for tea are unstable.

Tony Karumba/AFP/Getty Images

Nigel Pavitt/AWL Images/Getty Images

Ivan Vdovin/age fotostock/Getty Images

Hoberman Collection/Universal Images Group/Getty Image

Thinking Geographically

After you have read about sub-Saharan Africa's vulnerability to climate change, you will be able to answer the following questions.

A How is poverty evident in this photo?

B Given what happened to wildebeests in 2007, should they be prevented from migrating?

C How is agricultural diversity illustrated in this photo?

D How might climate change in Africa or elsewhere cause problems for tea producers?

also the standing pools of water they create often breed mosquitoes that spread tropical diseases such as malaria and the newly recognized zika virus, and harbor the snails that host schistosomiasis, a debilitating parasite that enters the skin of humans who spend time standing in still water to fish or do other chores.

Many smaller-scale alternatives are available. In some parts of Senegal, for example, farmers are using hand- or foot-powered pumps to bring water from rivers or ponds directly to the individual plants that need it. This is in some ways a more modern version of traditional African irrigation practices whereby water is delivered directly to the roots of plants by human, often female, water brigades. Smaller-scale projects provide the same protection against drought that larger systems offer, but they conserve water and are much cheaper and simpler for small farmers to operate. They also avoid the social dislocation and ecological disruption (erosion and flooding) of larger projects involving dams and reservoirs. Already used successfully for 15 years, these low-tech pumps will help farmers adjust to the drier conditions that may come with climate change.

Aware of how plans to dam the Niger River could affect ancient ecological systems there, two scholars decided to learn more about water management in cities across the region, especially at the microscale of household use.

CASE STUDY: The Material Culture of Water in Niamey, Niger

In 2013–2014, geographer Sara Beth Keough and anthropologist Scott Youngstedt studied the material aspects of how water, a scarce resource in this part of Africa, is distributed to users in the city of Niamey, located in the Sahel climate zone along the Niger River in the southwest of the country (see Figure 7.5). Water is taken from the Niger River and treated with chemicals and sent through an underground pipe network to homes or standpipes. Only very well-off citizens can afford to have water piped into their homes. Almost everyone else gets water from standpipes either in public spaces or in residential compounds. Water is typically collected in recycled 25-liter yellow plastic jugs that originally held palm oil. Some people collect water themselves, paying for it at the standpipe (at a cost of 3 cents a container). Others, who live far from a standpipe, may have water delivered to their homes (5 to 10 cents a container, depending on the distance traveled). All water vendors are male, but they are migrants from outside Niger. It is also possible to buy water from street vendors in chilled half-liter plastic bags for 5 cents apiece. Looking at the material culture of water distribution helped Keough and Youngstedt understand

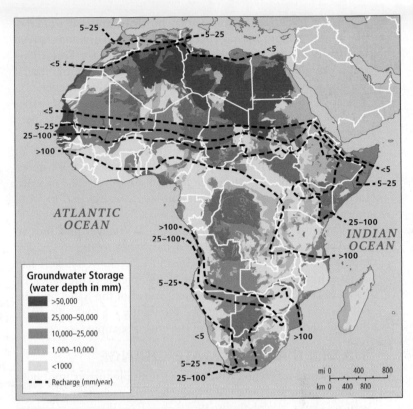

FIGURE 7.9 Newly exposed water resources. It was long thought that Africa had only shallow aquifers in which there were sparse deposits of groundwater. But recent research has revealed that for the African continent, the quantity of water stored in aquifers is nearly 100 times that of previous estimates. However, because much is fossil water deposited as many as 5000 years ago, it cannot be recharged (replenished) quickly under present drier climate conditions. Also, many densely populated areas have only small deposits. Compare this map of the depth of groundwater deposits and number of years needed to recharge with the Figure 7.22 population map. [*Sources consulted:* A. M. MacDonald, H. C. Bonsor, B. E. O. Dochartaigh, and R. G. Taylor, 2012, "Quantitative Maps of Groundwater Resources in Africa," *Environmental Research Letters* 7. doi:10.1088/1748-9326/7/2/024009; boundaries of surficial geology of Africa courtesy of the U.S. Geological Survey; country boundaries sourced from ArcWorld © 1995–2011 ESRI. All rights reserved.]

how community life functioned in Niamey, how careers were built on water distribution and sales, how the system did or did not contribute to water safety and security; they also learned about the role that water plays in defining gender roles and the status of migrants. [*Source:* S. Keough and S. Youngstedt, *Focus on Geography,* Winter 2014, 57(5), 152–163.] ∎

Herding, Desertification, and Climate Change

Herding, or **pastoralism,** is practiced by millions of Africans, primarily in savannas, on desert margins, and in the mixture of grass and shrubs called open bush. Herders live off the milk, meat, and hides of their animals. Pastoralists generally have a finely tuned understanding of environmental geography and seasonal cycles; over the course of a year, they typically circulate through wide areas, taking their animals to available pasturelands and watering holes, meanwhile trading with settled farmers for grain, vegetables, and other necessities.

pastoralism a way of life based on herding; practiced primarily on savannas, on desert margins, and in the mixture of grass and shrubs called open bush

Many traditional herding areas in Africa are now undergoing desertification, the process by which arid conditions spread to areas that were previously moist (see Chapter 6). This drying out is often the result of the loss of native vegetation; traditional herding may be partially to blame, but other causes of vegetation loss are climate change and economic development schemes that encourage commercial cattle raising. Cattle need more water and forage than do traditional herding animals and therefore place greater stress on native grasslands than do goats or camels. Agricultural intensification in the Sahel also contributes to desertification, because scarce water resources are diverted to irrigation, leaving dry, vegetation-less soils exposed to wind erosion.

Over the last century, due to natural changes and human impacts, desertification has shifted the Sahel to the south. For example, the *World Geographic Atlas* in 1953 showed Lake Chad situated in a forest well south of the southern edge of the Sahel. By 1998, Lake Chad sat in the northern fringe of the Sahel and the Sahara itself was encroaching (see the Figure 7.5 map). By 2010, the lake had shrunk to a tenth of the area it occupied in 1953.

WILDLIFE AND CLIMATE CHANGE

Africa's world-renowned wildlife faces multiple threats from both human and natural forces, all of which could become more severe with increases in global climate change. The Intergovernmental Panel on Climate Change (IPCC), a body of scientists tasked by the United Nations with assessing scientific information relevant to understanding climate change, estimates that 25 to 40 percent of the species in Africa's national parks may become endangered as a result of climate change.

Wildlife managers are developing new management techniques to help animals survive. For example, one of the greatest wildlife spectacles on the planet took a tragic turn in 2007. The annual 1800-mile-long (2900-kilometer-long) natural migration of more than a million wildebeests, zebras, and gazelles in Kenya's Maasai Mara game reserve requires animals to traverse the Mara River. In the best of times, this is a difficult migration that usually results in roughly a thousand animals drowning. In the past, the park's managers have taken a hands-off approach to the migration, considering the losses part of the natural order of things. However, in 2007, extremely heavy rains, possibly related to global climate change, swelled the Mara River to record levels. When the animals tried to cross, 15,000 drowned (see Figure 7.8B). Park managers are now taking a more active role in helping the migrating animals cope with unusual climatic conditions that may worsen with climate change. This may involve stopping animals from attempting a river crossing or directing them to a safer crossing.

Sub-Saharan Africa's rich array of animals has long played an economic role in daily life. Three species are discussed in **FIGURE 7.10**. Farmers' dependence on hunting wild game (bushmeat) for part of their food and income is already a major threat to wildlife in much of Africa. For example, farmers who need food or extra income are killing endangered species such as gorillas, chimpanzees, and especially elephants in record numbers (Figure 7.10C). If crop harvests are diminished by global climate change, many farmers will become even more dependent on bushmeat and on income from selling contraband, such as ivory. The threat to wild populations of various species has led to calls to expand and establish new protected areas for wildlife.

FIGURE 7.10 LOCAL LIVES: People and Animals in Sub-Saharan Africa

A The Rhodesian ridgeback, or the African lion hound, helps lion hunters by distracting the lions. Very intelligent and capable hunters themselves (they can kill animals as large as baboons), ridgebacks are now used mostly as guard dogs. The nomadic KhoiKhoi people developed the ridgebacks by interbreeding indigenous dogs of southern Africa with a variety of European dogs.

B African grey parrots, highly intelligent and able to mimic human speech, are popular pets worldwide. One bird famously greeted Dr. Jane Goodall, known for her research with chimpanzees, by saying, "Got a chimp?" (The bird had seen a photo of Goodall with a chimpanzee.) As many as 21 percent of the wild African grey parrots are caught each year, and the International Union for the Conservation of Nature (IUCN) lists the parrots as a vulnerable species.

C Elephants are a major attraction at Africa's many national parks and nature reserves, but park space cannot sustain large numbers of elephants whose natural impulse is to pull down trees to eat leaves. In response, some parks have reduced their elephant populations. The parks, though, may be the only hope for the survival of elephants that are hunted for their ivory tusks. Hunters and poachers have for years tended to kill large-tusked elephants, leading to natural selection for elephants with smaller tusks or with no tusks at all—a once-rare trait that is now found in about 30 percent of the population.

Africa's national parks constitute one-third of the world's preserved national parkland. These parks are struggling to deal with poaching (illegal hunting) within the park boundaries by members of surrounding communities. Poaching is often fueled by local economic hardships and by demand outside Africa for exotic animal parts (tusks, hooves, penises) as medicines and aphrodisiacs, especially in Asia, where efforts to educate consumers about the damage done by their purchases are only beginning. In 2011, the western black rhino was declared extinct, due largely to poaching driven by demand for its horns by practitioners of Chinese medicine.

Some parks are now using profits from ecotourism to fund development in nearby communities that previously depended, in part, on poaching. For example, in 1985, wildlife poaching threatened the animal population in Zambia's Kasanka National Park. Park managers decided to generate employment for the villagers through tourism-related activities. They built tourist lodges and a wildlife-viewing infrastructure and started cottage industries to make products to sell to tourists. Funding also goes to local clinics and schools, and students are included in research projects within the park. Local farmers have expanded into alternative livelihoods, such as beekeeping and agroforestry. Today, poaching in Kasanka is rare, its wildlife populations are booming, tourism is growing, and local communities have an ongoing stake in the park's success.

Nevertheless, without proper public education—both for local tourism policy makers and lawmakers, as well as for visiting tourists—placing tourism ventures in wildlife habitats can backfire. In 2015, for an alleged fee of U.S.$50,000, an American dentist took part in a lion-hunting expedition that he later claimed to have believed was an ecotourism big-game bowhunt. He shot and killed a beloved iconic black-maned lion, a symbol for Hwange National Park in Zimbabwe, who had been lured out of the park. Hwange Park is one of several developed for the purpose of protecting Africa's wildlife and educating local people to value nature. Zimbabwe eventually exonerated the dentist for his actions when it was revealed that Zimbabwean laws were, in fact, inadequate and did not specifically bar such hunts.

THINGS TO REMEMBER

GEOGRAPHIC THEME 1 • **Environment:** Sub-Saharan Africa's poverty and political instability leave its people less able to adapt to climate change than the inhabitants of other world regions. Up to now, this region has contributed little to the buildup of greenhouse gases in the atmosphere, but currently deforestation by rural Africans and by multinational logging companies is intensifying global climate change.

• The primary way in which sub-Saharan Africa contributes to CO_2 emissions and potential climate change is through deforestation. Much of Africa's deforestation is driven by the growing demand for farmland and fuelwood, but corruption on the part of local authorities and international timber companies also plays a role.

• Long-standing physical challenges, such as deforestation, desertification, and increasing water scarcity, have had a major impact on sub-Saharan Africa's ecosystems and difficulties are likely to increase with climate change.

• Food production is crucial to Africa's growing cities. Careful improvements in commercial food production are necessary to avoid ecological drawbacks.

• Traditional small-plot farmers have environmental knowledge that is crucial to food security, but to use this knowledge, they require access to safe water. Challenges to food production are likely to increase with climate change.

• Newly discovered groundwater resources hold promise for alleviating some of Africa's water scarcities.

• Sub-Saharan Africa's wildlife is under human-induced pressure, which has been intensified by climate change.

HUMAN PATTERNS OVER TIME

Africa's rich and ancient history is often overshadowed by the dramatic changes that came with European colonialism, beginning about 600 years ago. These changes were so powerful that even today many countries are still struggling with the lingering effects of the colonial era.

Africa has long been misunderstood and dismissed by people from outside the region. European slave traders and colonizers called Africa the "Dark Continent" and assumed it was a place where little of significance had occurred in all of human history. The substantial and elegantly planned cities of Benin in western Africa, Djenné in the Niger River basin, and Loango in the Congo Basin, which European explorers encountered in the 1500s, never became part of Europe's image of Africa. Even today, most people outside the continent are unaware of Africa's internal history or its contributions to world civilization, let alone its central role in the emergence of humankind as a species.

THE PEOPLING OF AFRICA AND BEYOND

Africa is the original home of humans (**FIGURE 7.11A**). Knowledge of the evolution of humans is advancing so rapidly that it is difficult to keep up with the newest discoveries. We now know that there were a number of extremely ancient hominid species that lived simultaneously in Africa and that moved out of the continent into Asia and later Europe. It was probably in eastern Africa (in what are today the highlands of Ethiopia, Kenya, and Uganda) that one of the first human species (*Homo erectus*) evolved more than 2 million years ago. *Homo erectus* differed anatomically from present-day humans. These early, tool-making humans ventured out of Africa, reaching north of the Caspian Sea and beyond into Asia as early as 1.8 million years ago. Anatomically, modern humans (*Homo sapiens*) evolved about 200,000 years ago from earlier hominoids (probably *Homo erectus*, but perhaps others) in eastern Africa. Like the migrations of earlier human species, those of *Homo sapiens* radiated out of Africa, first toward the eastern Mediterranean (*Homo sapiens* reached the eastern Mediterranean about 90,000 years ago), then spreading across mainland and island Asia, and only later turning west into Europe. After coexisting in Eurasia and Europe with earlier dispersions of *Homo erectus* (such as *Homo neanderthalensis* and other hominid species—all originating in Africa), *Homo sapiens* eventually outcompeted them all.

| **A** A museum diorama of early humans (*Homo erectus*) in southern Africa | **B** The ruins of Great Zimbabwe | **C** Sankore Mosque, Tombouctou, Mali, built in 13... |

| 2,000,000 B.C.E. | 5000 B.C.E. | 2500 B.C.E. | 1000 B.C.E. | 500 C.E. | 1000 C.E. | 1500 C.E. |

2 million years ago
First human species evolve in eastern Africa

5000 B.C.E.
Early agriculture in Sahel

1400 B.C.E.
Iron- and steelmaking in northeastern Africa

700–1500 C.E.

1250–1600 C.E.
Mali Empire

FIGURE 7.11 VISUAL HISTORY OF SUB-SAHARAN AFRICA

Thinking Geographically

After you have read about the human history of sub-Saharan Africa, you will be able to answer the following questions.

A What are the clues in the picture that food may have been a source of conflict between early humans and between humans and other animals?

B Aside from the central fortlike structure, describe any further evidence of human habitation.

EARLY AGRICULTURE, INDUSTRY, AND TRADE IN AFRICA

In Africa, people began to cultivate plants as far back as 7000 years ago in the Sahel and the highlands of present-day Sudan and Ethiopia. Agriculture was brought south to equatorial Africa 2500 years ago and to southern Africa about 1500 years ago. Trade routes spanned the African continent, extending north to Egypt and Rome and east to India and China. Gold, elephant tusks, and timber from tropical Africa were exchanged for salt, textiles, beads, and a wide variety of other goods.

About 3400 years ago, people in the vicinity of Lake Victoria (now adjacent to Uganda, Kenya, and Tanzania) learned how to smelt iron and make steel. By 700 C.E., when Europe was still recovering from the collapse of the Roman Empire, a remarkable civilization with advanced agriculture, iron production, and gold-mining technology had developed in the highlands of southeastern Africa in what is now Zimbabwe. This empire, now known as the Great Zimbabwe Empire, traded with merchants from Arabia, India, Southeast Asia, and China, exchanging the products of its mines and foundries for silk, fine porcelain, and exotic jewelry. The Great Zimbabwe Empire collapsed around 1500 for reasons not yet understood (see Figure 7.11B).

Complex and varied social and economic systems existed in many parts of Africa well before the modern era. Several influential centers, made up of dozens of linked communities, developed in the forest and the savanna of the western Sahel. There, powerful kingdoms and empires rose and fell, such as Djenné in

Mali; Ghana (700–1000 C.E.), centered not in modern Ghana, but in what is now Mauritania and southwestern Mali; and the Mali Empire (1250–1600 C.E.), centered a bit to the east on the Niger River in what became the famous Muslim trading and religious center of Tombouctou (Timbuktu), originally founded about 1100 C.E. by Tuareg nomads as a seasonal camp (see Figure 7.11C). Some rulers periodically sent large trade caravans carrying gold and salt through Tombouctou on to Makkah (Mecca), where their opulence was a source of wonder.

Africans also traded enslaved people. Long-standing customs of enslaving people captured during war fueled this trade, as was the case in many ancient cultures beyond Africa. The treatment of enslaved people within Africa was sometimes brutal and sometimes reasonably humane. Long before the arrival of Islam, a slave trade developed with Arab and Asian lands to the east. After the spread of Islam began, around 700 C.E., the slave trade continued, and Muslim traders exported close to 9 million African slaves to parts of Southwest, South, and Southeast Asia (FIGURE 7.12). Whenever enslaved Africans were traded to non-Africans, the deeply embedded indigenous checks on brutality were often abandoned. For example, to ensure sterility and to help promote passivity, Muslim traders often preferred to buy castrated males.

The course of African history shifted dramatically in the mid-1400s, when Portuguese sailing ships began to appear off Africa's west coast. The names given to stretches of this coast by the Portuguese and other early European maritime powers reflected their interest in Africa's resources: the Gold Coast, the Ivory Coast, the Pepper Coast, and the Slave Coast.

D European slave trading in 1814

E African troops under British control in colonial Nigeria in 1900

F A soccer stadium in Cape Town, South Africa, built for the 2010 World Cup

1600 c.e.   1700 c.e.   1800 c.e.   1900 c.e.   2000 c.e.

1400–1900
European slave trade

1840–1916
European colonial powers, meeting in Europe, establish borders of modern Africa

1948–1994
Official apartheid in South Africa

2010

C What does the existence of a mosque in Mali in 1327 indicate about the geography of trading patterns in this part of Africa?

D What is the evidence of European involvement in the slave trade in this image?

E What about this photo suggests hierarchies of status and power in British colonial Africa?

F What about this stadium conveys the wealth and Westernization of South Africa?

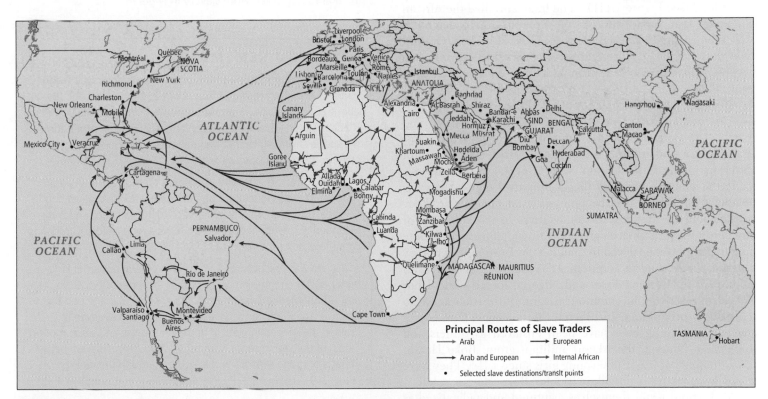

FIGURE 7.12 Map of the African slave trade. The origin of the African slave trade is indigenous and ancient, predating Islam by thousands of years. Outside of Africa, many slave markets were established around the Mediterranean and the Indian Ocean and eventually, after 1500 c.e., millions of enslaved people were brought to the Americas. [*Sources consulted:* Joseph H. Harris, in Monica B. Visona et al., *A History of Art in Africa* (New York: Harry N. Abrams, 2001), pp. 502–503]

By the 1530s, the Portuguese had organized a slave trade with the Americas. The trading of enslaved people by the Portuguese, and then by the British, Dutch, and French, was more widespread and brutal than any trade of African enslaved people that preceded it. Enslaved African people became part of the elaborate production systems supplying the raw materials and money that fueled Europe's Industrial Revolution.

To acquire enslaved people, the Europeans established forts on Africa's west coast and paid nearby African kingdoms with weapons, trade goods, and money to conduct slave raids into the interior. Some enslaved people were taken from enemy kingdoms in battle. Many more were kidnapped from their homes and villages in the forests and savannas. Most enslaved people traded to Europeans were male because they brought the highest prices and the raiding kingdoms preferred to keep captured women for their reproductive capabilities. Between 1600 and 1865, about 12 million captives were packed aboard cramped and filthy ships and sent to the Americas. One-quarter or more of them died at sea. Of those who arrived in the Americas, about 90 percent went to plantations in South America and the Caribbean. Between 6 and 10 percent were sent to North America (see Figure 7.12).

The European slave trade severely drained parts of Africa of human resources and set in motion a host of damaging social responses within Africa that even today are not completely understood (see Figure 7.11D). The trade enriched the African kingdoms that could successfully conquer their neighbors and impoverished or enslaved more peaceful or less powerful kingdoms. It also encouraged the slave-trading kingdoms to become dependent on European trade goods and technologies, especially guns.

Modern slavery, also known as human trafficking, persists in modern Africa, as it does elsewhere on Earth, and it is a growing problem that some argue exceeds the transatlantic slave trade of the past. Today in Africa, slavery is most common in the Sahel region, where several countries have made the practice officially illegal only in the past few years **(FIGURE 7.13)**. People may become enslaved during war. They may be sold by their parents or relatives to pay off debts or forced into slavery when migrating to find a job in a city. Some migrants are even enslaved by gangsters in Europe, where they become street vendors, domestic servants, or prostitutes. In Côte d'Ivoire, Burkina Faso, and Nigeria, forced child labor is used increasingly in commercial agriculture such as cacao (chocolate) production. In Liberia, Sierra Leone, Cameroon, and Congo, young boys have been enslaved as miners and soldiers. In Nigeria, the Islamic militant group Boko Haram has kidnapped at least 2000 young girls since 2014. They were seized apparently to provide wives for Boko Haram soldiers, many of whom were, themselves, captured and radicalized as young boys. It is hard to know exactly how many Africans are currently enslaved, but the Global Slavery Index (http://www.globalslaveryindex.org) estimates there were 5.6 million slaves in sub-Saharan Africa in 2014.

FIGURE 7.13 Modern slavery in Niger. A man born into slavery in Niger is presented with a document that declares his freedom in 2008. Though slavery was "abolished" in Niger in 1960 and made a criminal offense in 2003, it remains deeply embedded in society. At least 43,000 people are enslaved in Niger, most of them born into such status. Enslaved people have no rights and little access to education, and will spend most of their lives herding cattle, tending fields, or working in their "master's" house.

Thinking Geographically
What does this photo suggest about the social relationships of the man being freed from slavery in Niger?

THINGS TO REMEMBER

• Anatomically modern humans (*Homo sapiens*) evolved from earlier hominids in eastern Africa about 250,000 years ago. By about 90,000 years ago, *Homo sapiens* had reached the eastern Mediterranean. They joined and eventually replaced other hominid species from Africa that had migrated throughout Eurasia starting 1.8 million years ago.

• Agriculture, industry, and international trade have an early history in Africa; agriculture began as early as 7000 years ago in northeast sub-Sahara; iron smelting 3400 years ago; and trade extended across the Indian Ocean to Asia long before Europeans came to east Africa.

• Slavery was practiced in Africa within several powerful kingdoms, and then pre-Islam Arab traders extended the custom around the Mediterranean and to Asia. Muslims continued the practice. The European slave trade to the Americas severely drained the African interior of human resources and set in motion a host of damaging social responses within Africa. Slavery persists in parts of modern Africa and elsewhere around the world.

THE SCRAMBLE TO COLONIZE AFRICA

The European slave trade wound down by about the mid-nineteenth century. Significant abolitionist movements in the Americas and Europe helped to raise public consciousness, and

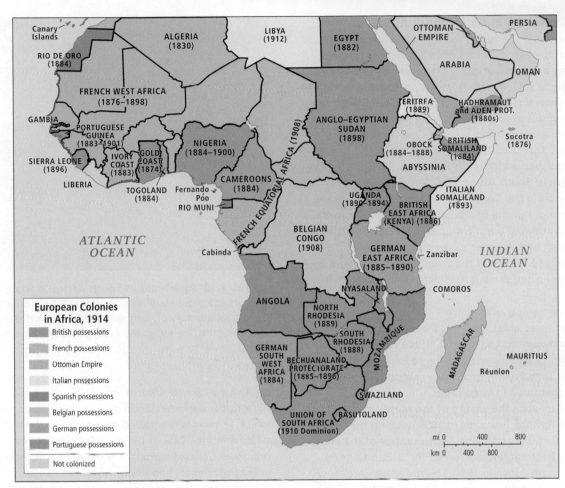

FIGURE 7.14 The European colonies in Africa in 1914. The dates on the map indicate the beginning of officially recognized control by the European colonizing powers. Countries without dates were informally occupied by colonial powers for a few centuries. [*Source consulted:* Alan Thomas, *Third World Atlas* (Washington, DC: Taylor Francis, 1994), p. 43.]

the pressure to enact antislavery legislation increased. The humanitarian movement against slavery did not extend to all Europeans, however. Many European traders and colonial officials found it profitable to use African labor within Africa to extract raw materials for Europe's growing industries, and their treatment of laborers in Africa was little better than that of enslaved people in the West.

European colonial powers competed avidly for territory and resources, and by World War I, only two areas in Africa were still independent (**FIGURE 7.14**): Liberia on the west coast (populated by former enslaved people from the United States) and Ethiopia (then called Abyssinia) in East Africa, which managed to defeat early Italian attempts to colonize it. Otto von Bismarck, the German chancellor who convened the 1884 Berlin Conference at which competing European powers formalized the partitioning of Africa, evinced the common arrogant European attitude when he declared, "My map of Africa lies in Europe." With some notable exceptions, the boundaries of most African countries today derive from the colonial boundaries established between 1840 and 1916 by European administrators and diplomats to suit their purposes (see Figure 7.11E). These territorial divisions lie at the root of many of Africa's current problems.

In some cases, the boundaries were purposely drawn to divide tribal groups and thus weaken them. In other cases, groups who used a wide range of environments over the course of a year were forced into smaller, less diverse lands or forced to settle down entirely, thus losing their traditional livelihoods. As access to resources shrank, hostilities between competing groups developed. Colonial officials often encouraged these hostilities, purposely replacing strong leaders with leaders who could be manipulated. Food production, which was the mainstay of most African economies until the colonial period, was discouraged in favor of activities that would support European industries: cash-crop production (cotton, rubber, palm oil) and mineral and wood extraction. Eventually, many formerly prosperous Africans became hungry and poor.

One of the main objectives of European colonial administrations in Africa, in addition to extracting as many raw materials as possible, was to create markets in Africa for European manufactured goods. The case of South Africa provides insights into the European expropriation of African lands and the subjugation of African peoples. In this case, such aims ultimately led to the infamous system of racial segregation known as **apartheid** (see the Case Study).

apartheid a system of laws mandating racial segregation; South Africa was an apartheid state from 1948 until 1994

CASE STUDY: The Colonization of South Africa

The Cape of Good Hope is a rocky peninsula that shelters a harbor on the southwestern coast of South Africa. Portuguese navigators seeking a sea route to Asia first rounded the Cape in 1488. The Portuguese remained in nominal control of the Cape until the 1650s, when the Dutch took possession with the intention of establishing settlements. Dutch farmers, called Boers, expanded into the interior, bringing with them herding and farming techniques that used large tracts of land and depended on the labor of enslaved Africans. The British were also interested in the wealth of South Africa, and in 1795, they seized control of areas around the Cape of Good Hope. When slavery was outlawed throughout the British Empire in 1834, large numbers of Boers who enslaved people migrated away from British territory to the northeast in order to keep their enslaved people. There, in what became known as the Orange Free State and the Transvaal, the Boers often came into violent conflict with African inhabitants.

In the 1860s, extremely rich deposits of diamonds and gold were unearthed in these areas, securing the Boers' economic future. African men were forced to work in the diamond and gold mines for minimal wages under extreme hardship and unsafe conditions. They lived in unsanitary compounds that travelers of the time compared to large cages.

Britain, eager to claim the wealth of the mines, invaded the Orange Free State and the Transvaal in 1899, waging the Boer War. This brutal war briefly gave the British control of the mines until resistance by Boer nationalists forced the British to grant independence to South Africa in 1910. This independence, however, applied to only a small minority of whites: the Boers (who became known as Afrikaners) and some British settlers who chose to remain. Not until 1994 would full political rights (voting, freedom of expression, freedom of assembly, freedom to live where one chose) be extended to black South Africans—who made up more than 80 percent of the population—and to Asians and mixed-race or "coloured" people (2 percent and 8 percent of the population, respectively).

In 1948, the long-standing customary practice of segregation in South African society was officially reinforced by apartheid, a system of laws that required everyone except whites to carry identification papers at all times, to live in racially segregated areas, and to use segregated transportation and other infrastructure (**FIGURE 7.15**). Eighty percent of the land was reserved for the use of white South Africans, who at that time made up just 10 percent of the population. Black, Asian, and "coloured" people were assigned to ten ethnically based "homelands" (such as Transkei, KwaZulu, and Swazi). The South African government considered the homelands to be independent enclaves within the borders of, but not legally part of, South Africa. Nevertheless, the South African government exerted strong influence in the homelands. Democracy theoretically existed throughout South Africa, but nonwhites were allowed to vote only in the homelands.

The African National Congress (ANC) was the first and most important organization that fought to end racial discrimination in South Africa. Formed in 1912 to work nonviolently for civil

FIGURE 7.15 Apartheid in South Africa. A South African woman protests apartheid by using a "whites only" car on a train in 1952. Africa's long history of European domination officially ended only in 1994 with the first national elections in which black South Africans were allowed to vote.

rights for all South Africans, the ANC grew into a movement with millions of supporters, most of them black, but some of them white. Its members endured decades of brutal repression by the white minority. One of the most famous members to be imprisoned was Nelson Mandela, a prominent ANC leader, who was jailed for 27 years and finally released in 1990.

Violence increased throughout South Africa until the late 1980s, when it threatened to engulf the country in civil war. The difficulties of maintaining order, combined with international political and economic pressure, forced the white-dominated South African government to initiate reforms that would end apartheid. A key reform was the dismantling of the homelands. Finally, in 1994, the first national elections took place in which black South Africans could participate. Nelson Mandela, the long-jailed ANC leader, was elected the country's president and proved to be an extraordinary leader. His efforts at reconciliation between the country's different racial groups earned him a Nobel Peace Prize in 1993. Today in South Africa, finding new leaders with Mandela's unselfish, clear-eyed vision has proven difficult. The thorny process of dismantling systems of racial discrimination and corruption continues, with varying amounts of success. [Sources consulted: Robert Stock, *Africa South of the Sahara: A Geographical Interpretation* (New York: Guilford Press, 2012); Leonard Guelke, *Historical Understanding in Geography: An Idealist Approach* (New York: Cambridge University Press, 1982).] ∎

ABUSE OF POWER IN THE AFTERMATH OF INDEPENDENCE

The era of formal European colonialism in Africa was relatively short. In most places, it lasted for about 80 years, from roughly the 1880s to the 1960s. In 1957, Ghana became the first sub-Saharan African colonial state to become independent. The last sub-Saharan African country to gain independence was South Sudan in 2011, not from a European power, but from its neighbor Sudan, after more than 40 years of civil war.

Africa entered the twenty-first century with a complex mixture of enduring legacies from the past and looming challenges for the future. Although it has been liberated from colonial domination, most old colonial borders remain intact (compare the colonial borders in Figure 7.14 with the modern country borders in Figure 7.4 and with Figure 7.31 of major language groups). Often, these borders exacerbate conflicts between hostile groups by joining them into one resource-poor political entity. Other borders divide potentially powerful ethnic groups, thereby diminishing their influence; or cut off nomadic people from the resources they use on a seasonal basis.

For many years and with few exceptions, leadership has been lacking. Governments continued to mimic the colonial bureaucratic structures and policies that distanced them from their citizens. Corruption and abuse of power by bureaucrats, politicians, and wealthy elites stifled individual initiative, civil society, and entrepreneurialism, creating instead frustration and suspicion. Too often, to change governments, coups d'état were resorted to. Democracy, where it existed, was often weakly connected only to voting and not to true participation in policy making at the community, regional, and national levels.

ON THE BRIGHT SIDE #WhatWouldMagufuliDo

When John Magufuli became president of Tanzania, he did so on a platform of anticorruption that was so austere it began to provoke jokes. To emphasize his intentions, Magufuli cancelled Independence Day celebrations, was seen picking up trash on the streets, and visited government offices to be sure workers were at their desks. Soon, this inspired a flutter of Twitter comments and the hashtag #WhatWouldMagufuliDo. Ludicrous photos demonstrating money-saving ideas filled the airwaves. Magufuli was the butt of the jokes, but his anticorruption campaign was very popular and the public participated with gusto. Soon afterward, the Twitter campaign spread to neighboring Kenya, where corrupt officials were skewered. Because the populations of Tanzania and Kenya are so young—with a median age under 20—and hence familiar with IT culture (see Figure 7.17), the potential for such Twitter conversations to influence public debate and then subsequent elections is huge. [*Source: The Economist,* January 16, 2016, p. 51.] ∎

Even today, African policy makers, like their colonial predecessors, are often insensitive to the widespread poverty and hardship around them. This was demonstrated during the summer of 2010 when South Africa hosted the 2010 World Cup, the world's largest sporting event and a first for this region, which up to that point had never hosted a global sporting event of similar scale. The preparations involved demolishing the homes of tens of thousands of poor, urban residents who were relocated to make way for the elegant new soccer stadiums (see Figure 7.11F). More recently, in an August 2012 crackdown reminiscent of those under apartheid, police shot dead 34 miners during a strike at a platinum mine in Marikana in Northwest Province. (More strikes followed.)

Other relics of the colonial era can be seen in the dependence of African economies on relatively expensive imported food and manufactured goods, as well as on the production of agricultural and mineral raw materials for which profit margins are low and prices on the global market highly unstable. Many sub-Saharan African countries frequently end up competing against each other and remain economically entwined in trade relationships that rarely work to their advantage. South Africa, which for so long was a beacon of hope and inspired by particularly enlightened leaders (including Mandela and Archbishop Desmond Tutu), seems to be slipping into an economic and political decline. Its GNI PPP per capita is no longer the largest in the region, wealth disparity is increasing, and recent leaders have been corrupt and underqualified. Nonetheless, South Africa, with its sophisticated infrastructure, highly profitable manufacturing and service sectors, as well as extractive sectors, remains sub-Saharan Africa's most developed country.

THINGS TO REMEMBER

• The main objectives of European colonial administrations in Africa were to extract as many raw materials as possible; to create markets in Africa for European manufactured goods and food exports; and to divide and conquer by splitting powerful ethnic groups or by placing groups hostile to one another under the same jurisdiction, governed by complicit indigenous leaders.

• During the era of European colonialism, sub-Saharan Africa's massive wealth of human talent and natural resources flowed out of Africa. Even after the countries of the region became politically independent, wealth continued to flow out of Africa because of unfavorable terms of trade and fraud.

• South Africa trod a rocky road through colonial oppression and apartheid to its present relatively precarious position as the most developed country in sub-Saharan Africa.

• Governments have continued to mimic colonial policies, including corruption, in ways that have kept sub-Saharan African countries economically dependent into the twenty-first century. This has stifled individual initiative, civil society, and entrepreneurialism. Political freedoms are often poorly protected.

GLOBALIZATION AND DEVELOPMENT

GEOGRAPHIC THEME 2

Globalization and Development: Most sub-Saharan African economies are dependent on the export of raw materials, a legacy of the era of European colonialism. This results in economic instability because prices for raw materials can vary widely from year to year. While a few countries are diversifying and industrializing, most still sell raw materials and all rely on expensive imports of food and manufactured goods.

Sub-Saharan Africa emerged from the exploitation and dependence of the colonial era just in time to get sucked into the vortex of globalization, where small or weak countries are often left with little control over their own fates.

For many centuries, sub-Saharan Africa has contributed human labor and raw materials to the global economy, but the profits from turning these human resources and raw materials into higher-value manufactured and processed products have gone to wealthier countries. Raw materials that are traded on the world markets for processing or manufacturing elsewhere into more valuable goods are called **commodities.** Sub-Saharan commodities include cotton; cacao beans; coffee; timber; leather; palm oil; unrefined petroleum; gas; precious stones and metals, such as copper, gold, and iron ore. The profits from commodity production are usually too low to lift poor countries out of poverty, because the profit margin on raw materials is usually low and many other poor countries are producing the same commodities. Hence, competition between them on the world market often drives down profits further.

The wide fluctuations of commodity prices on global markets create economic instability for producers. Commodities in their raw state are of more or less uniform quality; therefore, on the world market, they bring about the same price. For example, on the major commodities exchanges, a pound of copper may sell for U.S.$3 to $5, regardless of whether it is mined in South Africa, Chile, or Papua New Guinea. Because of instant communication and the global scale of commodity trading, any change that influences the global supply or demand for a particular commodity can cause immediate instability in the pricing of that commodity. If an earthquake in Chile damages a major copper mine, the temporary restriction in the supply of copper will drive up the price on world markets. In response, governments that make money from sales of raw copper may overestimate their future revenues and commit to expensive infrastructure projects. But the demand for copper can quickly decrease for many reasons; for example, a slowing of housing construction or innovations in building industry materials can reduce the use of copper for wiring and plumbing, resulting in a rapid drop in the world market price of copper. The loss of revenues from copper may suddenly halt infrastructure projects and cut essential services such as electricity, road maintenance, or health care.

Since 2000, there has been a noticeable shift in many African countries away from reliance on just one or two commodities. Nearly all have diversified to some extent, a trend that is discussed in the section "The Current Era of Diverse Globalization" that follows on page 398. Those that rely on petroleum exports are now particularly vulnerable because of recent fluctuations in the price of oil. Expectations are that there will be a general downward trend in oil prices because of two emerging trends: decreasing demand for oil as people turn to alternative energy sources in recognition of the connection between carbon emissions and climate change and increasing supplies of petroleum-based fuels in the United States and elsewhere.

commodities raw materials that are traded, usually to other countries, for processing or manufacturing into more valuable goods

commodity dependence economic dependence on exports of agricultural and mineral raw materials

dual economy an economy in which the population is divided by economic disparities into two groups, one prosperous (usually the smaller of the two) and the other near or below the poverty level

SUCCESSIVE ERAS OF GLOBALIZATION

Successive eras of globalization have transformed Africa over the past several centuries. They include the era of colonial exploitation of commodities; the transition from colonial status to political independence during which large debts were incurred, bringing on the era of structural adjustment and the rise of informal economies; and the current era, in which some African countries are moving away from commodity dependence and toward new types of economic development and regional integration.

The Colonial Exploitation Era

Most early European colonial administrations in Africa (British, French, German, Belgian) evolved directly out of private resource-extracting corporations, such as the German East Africa Company. Colonial economies were based on the export of commodities of agricultural or mineral raw materials. The welfare of Africans and their future development were a low priority for most colonial administrators. Education and health care for Africans were generally neglected by colonial governments, which were guided by mercantilism (see Chapter 3)—the idea that colonies existed to benefit their colonizers. Laborers and farmers were paid poorly and the colonial governments strongly discouraged any economic activities that might compete with those of Europe. African economies were thus hindered from making a transition to the more profitable manufacturing-based industries that were transforming Europe and North America.

For years, **commodity dependence,** widespread poverty, and the lack of internal markets for local products and services characterized all sub-Saharan African economies, with the partial exception of South Africa, the only sub-Saharan African country with a manufacturing base. Early on, profits from its commodity exports (mainly minerals) were reinvested in the manufacturing of mining and railway equipment. In the late twentieth century, South Africa developed a large service sector with particular strengths in finance and communications that support the mining and manufacturing industries. Today, with only 6 percent of sub-Saharan Africa's population, South Africa produces close to 30 percent of the region's economic output.

Unfortunately, the relics of colonialism and apartheid have left most sub-Saharan countries, including South Africa, with a **dual economy**—one part fairly rich, industrialized, and connected; the other below the poverty level and primarily reliant on low-wage labor and small-scale informal enterprises. By the 1940s, even though the labor of black South Africans was essential to the country's prosperity, 84 percent of black South Africans lived at a bare subsistence level. So even in this most industrialized country in sub-Saharan Africa, by 2012, White South African households had six times the income of Black households and nearly 30 percent of the labor force was unemployed. The informal economy was at least 30 percent of the total and was especially important to the income of women.

The Era of Structural Adjustment

By the 1980s, most sub-Saharan countries remained impoverished, dependent as they were on volatile and relatively low-value commodity exports. Attempts at investing in manufacturing industries often failed (discussed below); governments struggled

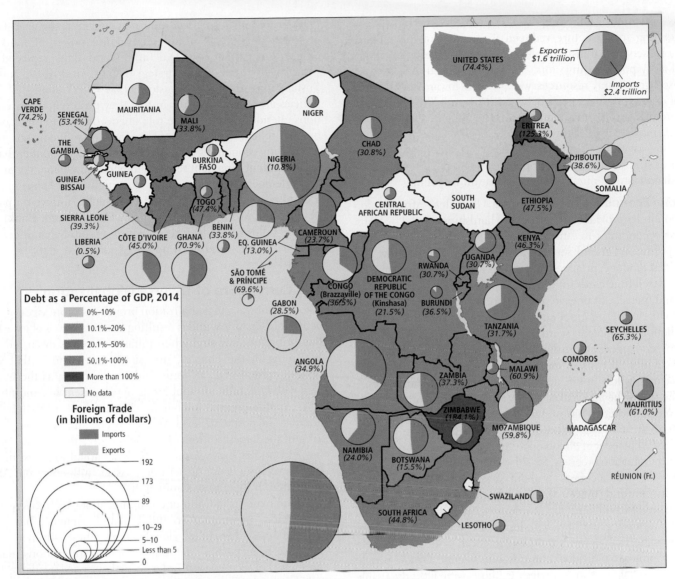

FIGURE 7.16 Economic issues: Public debt, imports, and exports. Public debt is increasing across Africa, partly as a result of borrowing to fund development projects (but the U.S. public debt—see inset—far exceeds that of most sub-Saharan countries). In all but 14 sub-Saharan countries, imports exceed exports. These fourteen are Angola, Botswana, Burundi, Chad, Côte d'Ivoire, Democratic Republic of the Congo (Kinshasa), Republic of Congo (Brazzaville), Equatorial Guinea, Gabon, Nigeria, São Tomé and Príncipe, Sierra Leone, Swaziland, and Zambia. Minerals (copper and cobalt) are the main export earners for Zambia; diamonds for Botswana; cocoa, coffee, tea, and timber for Burundi, Côte d'Ivoire, São Tomé and Príncipe; iron ore and diamonds for Sierra Leone; and the primary export earnings of the other eight derive from oil and petroleum products. [*Sources consulted: World Fact Book 2015,* https://www.cia.gov/library/publications/the-world-factbook/fields/2186.html#ao.]

to make payments on the large loans taken out for these projects. Infrastructure (roads, water, and utilities) was either not built or not maintained. A breaking point came in the early 1980s, when an economic crisis swept through the region and much of the rest of the developing world, leaving most countries unable to repay their debts at all.

In response, the International Monetary Fund (IMF) and the World Bank designed structural adjustment programs (SAPs; see Chapter 3)—belt-tightening imposed to enforce repayment of the loans. SAPs did have some useful results. They tightened bookkeeping procedures and thereby curtailed corruption and waste in bureaucracies. They closed some corrupt state-owned

industrial and service monopolies, opened some sectors of the economy to medium- and small-scale business entrepreneurs, and made tax collection more efficient. But overall, SAPs had many unintended consequences and failed at their primary objective—reducing debt **(FIGURE 7.16)**.

To facilitate loan repayment, SAPs required governments to sell off inefficient government-owned enterprises, often at bargain-basement prices. In addition, government jobs in social services, education, health, and agricultural programs were slashed so that tax revenues could be devoted to loan repayment. If countries refused to implement SAP requirements, the international banks cut off any future lending for economic development.

As unemployment rose, so did political instability. The deteriorating infrastructure reduced the quality of the remaining social services, transportation, and financial services, which scared away potential investors. SAPs also reduced food security because agricultural resources were shifted toward the production of cash crops for export. Between 1961 and 2009, per capita food production in sub-Saharan Africa actually decreased by 14 percent, making it the only region on Earth where people were not eating as well as they had in the past. This nearly 50-year period of structural adjustment policies was a painful period for sub-Saharan Africa, one from which it is still recovering.

Informal Economies to the Rescue Ultimately, Africa's informal economies provided relief from the hardships created by SAPs. Informal economies in Africa are ancient and varied, providing employment and useful services and products. People grow and sell garden produce, prepare food, vend a wide array of products on the street, sell time cards (or credits) for mobile phones, make craft items and utensils, or engage in child and elder care. Others earning a living in the informal economy, however, may do sex work, distill liquor, or smuggle scarce or illegal items, including drugs, weapons, endangered animals, bushmeat, and ivory. Many of these latter activities take place "under the radar," and so may involve criminal acts, wildly unsafe activities, and hazardous substances.

The role of the informal economy is especially important in sub-Saharan cities, where it has grown from one-third of all employment to more than two-thirds. Among women, it is thought to produce more than 90 percent of all job opportunities. Although it is difficult to assess the situation precisely, for sub-Saharan Africa as a whole, the informal economy accounts for about 38 percent of the GDP—the highest in the world **(TABLE 7.1)**. Such jobs are a godsend to the poor, but they create problems for governments because the informal economy typically goes untaxed, so less money is available to pay for government services or to repay debts. Moreover, as the informal sector grows, profits to individual entrepreneurs decline as more people compete to sell goods and services to those with little disposable income. And although women typically dominate informal economies, when large numbers of men lose their jobs in factories or the civil service, they may crowd into the streets and bazaars as vendors, displacing the women and young people.

TABLE 7.1 Regional informal economies as a percentage of GDP

Region	Informal economy as a percentage of GDP
Sub-Saharan Africa	37.6
East Asia and Pacific	17.5
Europe and Central Asia	36.4
Latin America and Caribbean	34.7
Middle East and North Africa	27.3
High-income OECD	13.4
South Asia	25.0
World	17.1

Source: F. Schneider et al. (2012), http://ftp.iza.org/dp6423.pdf.

In 2000, in response to the now widely recognized failures of SAPs, and the overemphasis on the power of markets to guide development, the IMF and World Bank replaced SAPs with Poverty Reduction Strategy Papers, or PRSPs, and with Sustainable Structural Transformation (SST) strategies. These policies are similar to SAPs in that they push market-based solutions intended to reduce the role of government in the economy, but they differ in several ways. They focus on reducing poverty and diversifying economies in terms of manufacturing, rather than just "development" and debt repayment per se, and generally promote more democratic reforms. They also include the possibility that a country may have all or most of its debt "forgiven" (paid off by the IMF, World Bank, or African Development Bank) if the country follows the PRSP rules. Thirty-six sub-Saharan countries had qualified for and been approved to receive debt relief as of July 2015.

The Current Era of Diverse Globalization

The current wave of globalization promises a more vigorous interactive role for Africa, and is resulting in new sources of investment that bring jobs and strengthen infrastructure. As discussed at the beginning of the chapter, one significant trend is that young African professionals at home and abroad (such as those associated with Ushahidi) are supporting development in a multitude of ways. Another trend is foreign direct investment (FDI) at both the country and corporate level. While Europe and the United States are still the largest sources of investment in sub-Saharan Africa, a few African firms are earning sufficient capital to invest in ancillary developments. Furthermore, Asia's influence on African economies is increasing significantly.

The major expansion of communications technology within this region, highlighted in the Ushahidi vignette, has proven to be a major benefit of globalization. These advances would not have been possible without numerous innovations made by Africans. For example, in order to make service available to the very poor, prepayment credits are sold in very small units. In fact, many people now make a living selling such small credits. Other Africans have created ways to charge mobile phones in remote villages where there is no electricity, as the Malawian William Kamkwamba figured out how to do (for more on Kamkwamba's story, see the vignette on page 403).

Foreign private investment is also transforming the communications environment in ways that encourage regional economic development in Africa. The India-based cell phone company Airtel operates a borderless network that now covers 18 countries: Burkina Faso, Chad, Congo (Kinshasa), Congo (Brazzaville), Gabon, Ghana, Kenya, Madagascar, Malawi, Niger, Nigeria, Rwanda, Seychelles, Sierra Leone, Tanzania, South Sudan, Uganda, and Zambia. Airtel's customers can make calls across the network at local rates without incurring surcharges, and recently, the company added Internet access, SMS, international roaming, and portal applications to the local rate. The borderless aspect of Airtel's network, and others like it, is helping to increase trade and interaction between African countries, which have long been goals of African regional economic development efforts. But there are other foreign private investors competing in the African communications market. Hundreds of millions of unserved potential

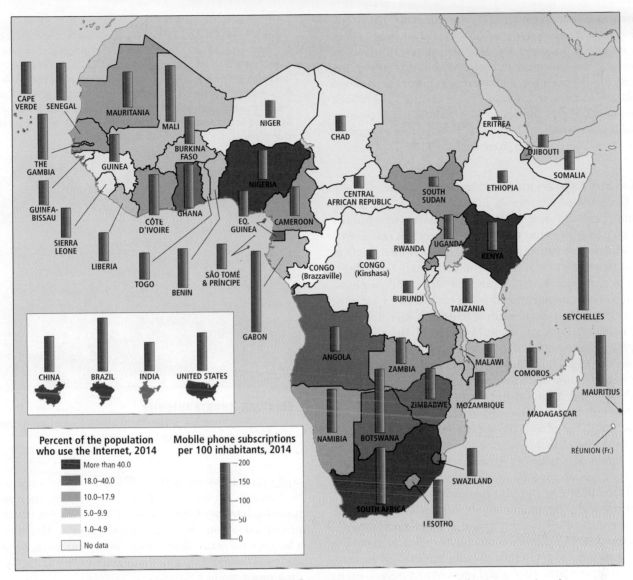

FIGURE 7.17 Mobile phone users in sub-Saharan Africa. Over the past decade, mobile phone use in sub-Saharan Africa has boomed. It is estimated that as of 2016, there were more than a billion mobile phones in the region. Because of innovations that make it possible to buy service access in tiny amounts, many of the new users will be the very poorest citizens. One of the most popular uses of phones in Africa is to move money from one account to another, a primitive but useful form of banking.

subscribers span the continent, and as of 2015, Airtel was losing market share to MTN, Orange Telecom, and Globalcom.

ON THE BRIGHT SIDE: African Migrants Investing in Africa

The role of African professionals in globalization is growing. As was noted in the opening vignette, perhaps the most dramatic sign of change for African economies is that educated Africans who are skilled in IT and other high-tech fields are investing their time and efforts at home instead of migrating to richer regions. Their incomes, along with remittances (money sent home by Africans working and living abroad, primarily in Europe and North America), allow them and those they help support to buy

mobile phones, start small businesses, fund education for children, build houses, and help the needy. Remittances are a more stable source of investment than foreign direct investment, in that they tend to come regularly from committed donors who will continue their support for years. They are also much more likely to reach poorer communities. But payments will stop if the remitters lose their jobs in America or Europe.

Young educated adults are profoundly changing the trajectory of Africa's development. They have been largely responsible for the rapid spread of phone technology. **FIGURE 7.17** is a map of mobile phone subscriptions by country as of 2015. Between 1998 and 2014, mobile phone subscribers in sub-Saharan Africa increased by more than 150 times, from about 4 million to about 608 million. Google estimates that in 2016 mobile phones in sub-Saharan Africa will reach 1 billion, or one for every person, but because few have smartphones (or computers), access to the Internet is limited to about 17 percent. Smartphone connections, however, will number 525 million by 2020 and result in greatly improved access to the Internet. The mobile industry already directly employs several million, creates ancillary opportunities for millions more, and contributes billions of dollars to government coffers through tax payments (in 2013, U.S.$13 billion). [*Source consulted:* Dalberg, "Unlocking 90 Billion Dollars for Development in Sub-Saharan Africa," http://www.gsmamobileeconomyafrica.com/GSMA_ME_SubSaharanAfrica_Web_Singles.pdf.] ∎

Asian Investors Take an Interest Another major trend in globalization has involved China and, to a lesser extent, India, both of whom have begun to view sub-Saharan Africa as a new frontier for their large and growing economies. Both countries now exert a powerful influence on the region not only through their demand for Africa's export commodities, but also through their direct investment in agribusiness, mining, and industry, and through the sale of their low-cost manufactured goods to Africans. Certainly, there are significant possible benefits of trade with Asia. Together, China and India now consume about 15 percent of Africa's exports (especially oil and iron ore), and their share is growing twice as fast as that of any of Africa's other trading partners. All of sub-Saharan Africa is in need of improved infrastructure (roads, ports, utilities, and technology), and Asian investment in such amenities could ultimately not only facilitate the extraction and movement of raw materials to ports, but could also greatly increase intra-African trade, linking countries that have never before been able to trade. China and India's investment in African agriculture, undertaken to meet rising demands for food in China and India, could result in less food for Africa. Or, with sensible management by individual African country governments, Asian investment could result in an increase in food security for Africa: more efficient production through technology transfer could create higher earnings for African farmers and more food supplies for internal urban markets. Consumer goods from China and India like household utensils and appliances could ease life for Africa's homemakers, too, if they are of sufficient quality and durability.

It is important, however, to keep an eye on the extent to which Asian investments in Africa actually benefit Africans rather than simply replicate the injustices of colonialism. For example, Chinese investments in mining and infrastructure development could have boosted local African economies by providing construction jobs for African laborers and professional design and management experience for mid-level educated Africans, but this benefit failed to materialize because many debt-strapped African governments agreed to China's demand that all work be done by Chinese companies using Chinese workers **(FIGURE 7.18)**.

China also gains advantageous access to Africa's oil and minerals by offering low-interest loans to heavily indebted resource-rich countries in return for low prices on these commodities. And, because sub-Saharan African governments have low product safety and environmental standards, China has done such things as "dump" inferior electronics in the markets of Nigeria and befoul wetlands in Gabon with crude oil waste. Most

FIGURE 7.18 China in Angola. The man on the right is one of the estimated 20,000 Chinese workers in Angola. China has given loans and aid to Angola in excess of U.S.$20 billion since 2015, and in return, China has been guaranteed a large portion of Angola's future oil production. In addition, 70 percent of Angola's development projects, such as building and upgrading railways, have been awarded to Chinese companies, most of which import workers from China.

Thinking Geographically
Why might China be particularly interested in investments in Africa's transportation infrastructure?

controversial has been the willingness of Chinese companies to deal with brutal and corrupt local leaders, such as Liberia's now-convicted and imprisoned Charles Taylor (diamonds and timber extraction) and Zimbabwe's Robert Mugabe (mining), who bartered away their countries' resources at bargain prices and used the profits to enrich themselves and to wage war against their own citizens.

Low-cost Chinese goods have made life more affordable for millions, but many Chinese goods are shoddy. Loose threads and crooked stitching on clothes bought in street markets can disappoint, but power strips and electrical equipment that catch fire can destroy whole livelihoods. One gentleman in Lagos, overjoyed to open a computer store, arrived at work one day to find that 30 desktop computers, plus televisions and air-conditioners, had burned up in a fire started by a faulty Chinese power strip.

Perhaps most disturbing is the story of China's displacement of an age-old industry in Nigeria: hand-died fabrics in graphic indigenous patterns, once valued across the world for African authenticity. China has taken these traditional designs home and reproduced them in Chinese factories far more cheaply, thus eliminating 97 percent of Nigeria's textile jobs. Not only have 225,000 Nigerian textile jobs been eliminated, because Nigerians are now buying Chinese-made textiles, Nigeria's trade deficit with China is growing ever larger.

China's arrangements with corrupt governments may become increasingly precarious in light of falling global oil prices. In Angola, for example, oil will be less of a moneymaker, so repaying

8 Main Regional Economic Communities	Abbreviation	Number of countries	Countries
Arab Maghreb Union	AMU	5	Algeria, Libya, Mauritania, Morocco, Tunisia
Common Market for Eastern and Southern Africa	COMESA	19	Burundi, Comoros, Dem. Rep. of the Congo, Djibouti, Egypt, Eritrea, Ethiopia, Kenya, Libya, Madagascar, Malawi, Mauritius, Rwanda, Seychelles, Sudan, Swaziland, Uganda, Zambia, Zimbabwe
Community of Sahel-Saharan States	CEN-SAD	28	Benin, Burkina Faso, Central African Republic, Chad, Comoros, Côte d'Ivoire, Djibouti, Egypt, Eritrea, Gambia, Ghana, Guinea, Guinea-Bissau, Kenya, Liberia, Libya, Mali, Mauritania, Morocco, Niger, Nigeria, São Tomé and Príncipe, Senegal, Sierra Leone, Somalia, Sudan, Togo, Tunisia
Economic Community of Central African States	ECCAS	11	Angola, Burundi, Cameroon, Central African Republic, Chad, Congo, Dem. Rep. of the Congo, Equatorial Guinea, Gabon, Rwanda, São Tomé and Príncipe
Intergovernmental Authority on Development	IGAD	7	Djibouti, Eritrea, Ethiopia, Kenya, Somalia, Sudan, Uganda
Southern African Development Community	SADC	14	Angola, Botswana, Dem. Rep. of the Congo, Lesotho, Malawi, Mauritius, Mozambique, Namibia, Seychelles, South Africa, Swaziland, United Rep. of Tanzania, Zambia, Zimbabwe
East African Community	EAC	5	Burundi, Kenya, Rwanda, Uganda, United Rep. of Tanzania
Economic Community of West African States	ECOWAS	15	Benin, Burkina Faso, Cape Verde, Côte d'Ivoire, Gambia, Ghana, Guinea, Guinea-Bissau, Liberia, Mali, Niger, Nigeria, Senegal, Sierra Leone, Togo

FIGURE 7.19 **Principal trade organizations in sub-Saharan Africa.** The confusing pattern of regional trade organizations now operating in sub-Saharan Africa is indicative of the emerging status of internal African trade. These economic communities in sub-Saharan Africa continually change and some countries belong to two or three such organizations. Usually, one country is dominant in an organization. Notice that the Central African Republic and the Democratic Republic of the Congo belong to several trade groups. Because political stability is essential to trade, some of the strongest trade organizations, such as the Economic Community of West African States (ECOWAS), have become involved in resolving conflicts within and between countries. [*Source consulted:* UN Conference on Trade and Development, *Economic Development in Africa Report 2013*, pp. 9–10, http://unctad.org/en/PublicationsLibrary/aldcafrica2013_en.pdf.]

debts will be harder; and with oil more cheaply available elsewhere, China may lose interest in helping Angola.

REGIONAL AND LOCAL ECONOMIC DEVELOPMENT

As they create alternatives to past development strategies, many African governments have been focusing on regional economic integration similar to that of the European Union (EU). Hoping to utilize existing talent and consumer demand for African-made products, local agencies and public and private donors are pursuing grassroots development designed to foster very basic innovation at the local level that can then be marketed across the region.

The Potential of Regional Integration

According to the World Trade Organization (WTO), less than 20 percent of the total trade of sub-Saharan Africa in 2011 was conducted between African countries. This is true partly because so many countries produce the same raw materials for export. And everywhere except South Africa, industrial capacity is so low that the raw materials cannot be absorbed within the continent, so African countries compete with each other and with all other global producers to sell to the main customers, which are currently in Europe, North America, and industrialized Asia. This failure to trade with each other can be traced to divisive colonial policies, lack of transportation and communications grids, and arcane bureaucratic regulations that impede the flow of goods, people, and ideas.

Over the last several decades, regional trading blocs have been formed to encourage neighboring countries to trade with each other and cooperate in the production of manufactured (value-added) exports. The trade blocs are somewhat fluid; they form and then reorganize as governments or market conditions change **(FIGURE 7.19)**. By combining the markets of several countries, much as the EU does, regional trade blocs can create a market size sufficient to foster industrialization and entrepreneurialism. Africa's many different regional trade blocs share several goals: reducing tariffs between members, forming common currencies, reestablishing peace in war-torn areas, cooperating to upgrade transportation and communications infrastructure, and building regional industrial capacity. Building a full-scale, continent-wide economic union along the lines of the EU is a long-term goal. According to studies by the WTO, if sub-Saharan countries could increase trade with each other by just 1 percent, the region would show a total added income of U.S.$200 billion per year—a substantial internal contribution toward the alleviation of poverty.

The goals of regional and local economic integration are being boosted by the creation of **value chains.** These link parts of the production chain of a final product in order to maximize regional or local economic impact. For example, farmers in Endau, Kenya, were able to increase their profits fivefold by switching from growing corn to producing sorghum after a local nongovernmental organization (NGO) negotiated a deal with a brewery in Nairobi to use the sorghum in its beer. Some farmers have used their boosted income to increase their profits further by buying milling equipment and grinding the sorghum of other local farmers into sorghum flour, which is more valuable than unprocessed sorghum. Countries and regions with strong value chains are better able to keep the profits their industries create from going elsewhere. Such value chains are one of the fastest-growing aspects of regional economic development worldwide.

Local Development

An increasingly common strategy for improving living standards in this region is **grassroots economic development,** often also called **self-reliant development.** Projects using these strategies are designed to provide sustainable livelihoods in rural and urban areas, and often utilize simple technology that requires minimal or no investment in imported materials. Small-scale self-help projects, frequently financed with microcredit loans, use local skills to create products or services for local consumption. One of the most important aspects of this type of development is that control of the projects remains in local hands so that participants retain a sense of ownership and commitment in difficult economic times; the terms of microcredit require that loans be repaid, so new entrepreneurs will have access to the loan money. One district in Kenya has more than 500 such self-reliant groups. Most members are women who build water tanks, plant trees, and terrace land so that it can be farmed for food. They also construct schools and form credit societies. The urban garden movement (see page 411) and the activities of William Kamkwamba in Malawi (page 403) are other examples.

value chains links between various aspects of a production line to maximize efficiency and profits

grassroots economic development economic development projects designed to provide sustainable livelihoods in rural and urban areas; these often use simple technology that requires minimal or no investment in imported materials

self-reliant development small-scale development in rural areas that is focused on developing local skills, creating local jobs, producing products or services for local consumption, and maintaining local control so that participants retain a sense of ownership over the process

Transportation Needs The issue of improving rural transportation illustrates how a focus on local African needs can generate unique solutions. When non-Africans learn that transportation facilities in Africa are in need of development, they usually imagine building and repairing roads for cars and trucks. But a recent study that analyzed village transportation on a local level found that 87 percent of the goods moved are carried via narrow footpaths on the heads of women. Women "head up" (their term) firewood from the forests, crops from the fields, and water from wells. An average adult woman spends about 1.5 hours each day moving the equivalent of 44 pounds (20 kilograms) more than 1.25 miles (2 kilometers).

Unfortunately, the often-treacherous footpaths trodden by Africa's load-bearing women have been virtually ignored by African governments and international development agencies, which tend to focus solely on roads for motorized vehicles (which are also badly needed). Grassroots-oriented NGOs are now making far less expensive but equally necessary improvements to Africa's footpaths. Some women have been provided with bicycles, donkeys, and even motorcycles that can travel on the footpaths. This saves time and energy for the women, who can direct more of their efforts at becoming educated and generating income.

ON THE BRIGHT SIDE: Bamboo Bicycles

As soon as Bernice Dapaah left college in Ghana, she established a little outdoor workshop in her village to build bikes from plants. She wanted to help resolve the transportation problem in her country, and it occurred to her that bike frames could be built from bamboo held together with hemp, epoxy, and lacquer. Bamboo was renewable and it grew right outside the village. She hired young people who helped her perfect the design. The project experienced its share of difficulties at first, but now she has hired and trained 35 people and built hundreds of bikes, some of which are exported to Europe and America. But most are used in Africa, where there is a huge demand for low-cost *green* transport. Business is currently so good that she is able to donate bikes to children who must travel long distances to school. ∎

Energy Needs Africa exports oil but its own energy needs remain largely unmet. Wood is the most common fuel for cooking, and this accounts for a great deal of damaging deforestation. In all of sub-Saharan Africa, between 50 and 75 percent of the homes (620 million people) do not have electricity, and those that do are subject to rolling blackouts on a daily basis. About 60 percent of urbanites have electricity, but only 14 percent of rural people do. A recent World Bank report notes that "more than a century after the light bulb was invented most of the African continent is still in the dark after nightfall." Children can't do their homework; there are no household TVs or computers. Without lighting, businesses cannot grow, and medical facilities can't use modern equipment or refrigerate medicines. Industries operate far below capacity. The unpredictability of power hampers initiative and stifles job growth. Ironically, fixing this deficit would require expending a great deal of fossil fuel energy.

Addressing Africa's energy needs will be costly and will require unusual ingenuity, but as with transportation needs, some of the most basic levels of home and office electricity can be addressed with local solutions, as the following vignettes illustrate.

VIGNETTE In Malawi, 14-year-old William Kamkwamba was forced to drop out of school when a famine struck his country in 2001 and his family could no longer afford the U.S.$80 school fee. Although depressed by the prospect of no future, he visited the local library when he was able. There, he found an English-language book entitled *Using Energy* that described an electricity-generating windmill. With an old bicycle frame, tractor fan blades, PVC pipes, and scraps of wood, he built a windmill that generated enough power (stored in a car battery) to light his home, run a radio, and charge neighborhood cell phones. More elaborate energy projects soon followed.

Now known as "the boy who harnessed the wind," William appeared on Jon Stewart's *Daily Show* in the United States in 2009 and gave a TED talk in 2010. He explained how he planned to start his own windmill company and other ventures that will bring power to remote places across Africa. He graduated from the first pan-African prep school in South Africa and in 2014 from Dartmouth College in the United States. William's web page is http://www.williamkamkwamba.com. ∎

VIGNETTE Myeka High School lies deep in the hinterland of South Africa and far from electric power lines. For years, few students graduated from the destitute school due to lack of educational materials and general demoralization. Then the U.S.-based Solar Electric Light Fund (SELF) invested in solar power for the school, Dell contributed computers, and INFOSAT Telecommunications set up Internet service. Graduation rates soared, and students left with employable skills.

Intrigued by the idea of alternative power sources, a Myeka science teacher and several students devised a biogas system to utilize waste from the school's toilets to create methane gas, which now powers the school's electrical system. Spinoff projects have been impressive: sanitation problems are solved, and manure fertilizes the school's gardens, which feed the many AIDS orphans in attendance. Teachers and students have together mastered science, math, and technical skills that will serve them for years to come. *[Source consulted: N. T. Sibisi and J. M. Green, Journal of Energy in Southern Africa, August 2005, 16(3), http://www.erc.uct.ac.za /jesa/volume16/16-3jesa-sibisi.pdf.]* ∎

THINGS TO REMEMBER

GEOGRAPHIC THEME 2 • **Globalization and Development:** Most sub-Saharan African economies are dependent on the export of raw materials. This pattern, a legacy of the era of European colonialism, results in economic instability because prices for raw materials can vary widely from year to year. While a few countries are diversifying and industrializing, most still sell raw materials and all rely on expensive imports of food and manufactured goods.

• Prospective investors in sub-Saharan Africa have been discouraged by problems that structural adjustment programs (SAPs) either ignored or worsened.

• Recently, Africa has become a new frontier for Asia's large and growing economies, especially those of China and India, which both buy and sell in sub-Saharan Africa. The effects of Chinese investment in West Africa may replicate some aspects of European imperialism.

• Africa's informal economies have provided some relief from the hardships created by SAPs. These economies are ancient, varied, and agile; they provide employment and useful services and products.

• A major and enduring source of investment funds in Africa comes from members of the African diaspora, who send regular remittances to friends and families.

• Many African governments are focusing on regional economic integration along the lines of the EU. Grassroots economic development is also being pursued.

• Local people using local resources and ingenuity are capable of solving many problems.

POWER AND POLITICS

GEOGRAPHIC THEME 3

Power and Politics: Despite a shift in sub-Saharan Africa toward more political freedoms, some governments remain authoritarian, nontransparent, and corrupt. Public participation is growing, often due to electronic media, and free and fair elections have brought about dramatic changes in some countries. In other countries, elections and governments tainted by suspicions of fraud have led to surges of violence.

After years of being ruled by European colonizers and then corrupt elites and the military, Africa is now shifting toward systems with greater **transparency** and political freedoms. Yet change is often blocked by conflict, and even when reforms are enacted and free elections established, violence frequently accompanies these elections.

ETHNIC RIVALRY

Africa has suffered from frequent civil wars that are in many ways the legacy of colonial-era policies of **divide and rule.** Divisions and conflicts between ethnic or religious groups were deliberately exaggerated and aggravated by European colonial powers. To make it difficult for Africans to unite against colonial rule, the borders and administrative units of the African colonies were designed so that large ethnic groups were divided into two or three separate units, while different and sometimes hostile groups were brought together under the same jurisdiction. The situation in Nigeria is illustrated in **FIGURE 7.20**. The language map of sub-Saharan Africa in Figure 7.32 also illustrates how ethnic groups were divided by political borders.

transparency in politics, the state of being open to observation and participation by the public

divide and rule the deliberate intensification of divisions and conflicts by potential rulers; in the case of sub-Saharan Africa, by European colonial powers

After independence, when Africans assumed control of their own countries, governance was complicated because African officials inevitably belonged to one local ethnic group or another and so were not seen as impartial in their attempts to resolve conflicts. Moreover, during the colonial era, older indigenous traditions for ensuring ethical behavior, punishing greed on the part of leaders, and resolving ethnic conflict had been weakened. The political institutions that replaced them were often not open in their discussions of issues and not adept at conflict resolution, resulting in hostilities between ethnic groups that have at times devolved into civil wars **(FIGURE 7.21C)**.

CASE STUDY: Conflict in Nigeria

Nigeria was, and remains in part, a creation of British divide-and-rule imperialism. Many disparate groups—speaking 395 indigenous languages—have been joined into one unusually diverse country (Figure 7.20).

The British reinforced a north–south dichotomy that mirrored the physical and cultural patterns: dry and Muslim in the north, wet and Christian in the south. Among the Hausa

Thinking Geographically

After you have read about the power and politics in sub-Saharan Africa, you will be able to answer the following questions.

A In what ways does this picture show both the strengths and limits of political freedoms in Kenya?

B Why might it be significant that Ellen Johnson-Sirleaf was elected as Liberia's president without the help of quotas that reserve certain elected positions for women?

C What does the photo suggest about the strength of this rebel militia and its ability to acquire uniforms and arms?

D What does this photo convey about the general well-being of the people depicted?

and Fulani ethnic groups in the north, the British ruled via local Muslim leaders who did not encourage public education. In the south, the Yoruba and Igbo ethnic groups, who were primarily

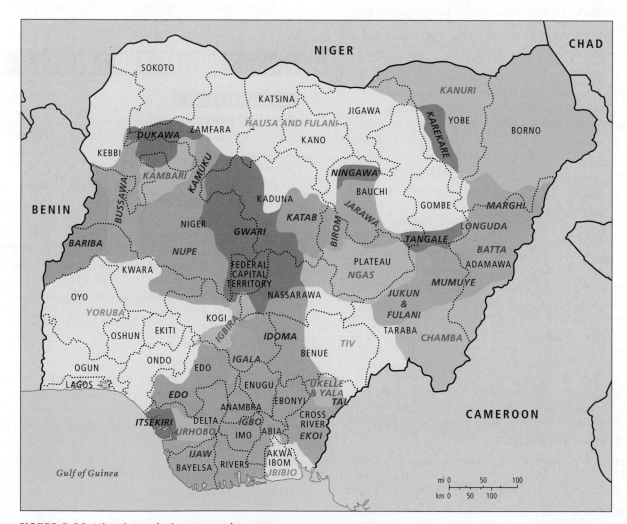

FIGURE 7.20 Nigeria's ethnic geography. There are thousands of ethnic groups distributed across sub-Saharan Africa. Nigeria alone has more than 400. When Europeans set up political units, they either ignored ethnic boundaries or purposely divided large ethnic groups in order to control them more easily (divide and conquer). The black dotted lines are district political boundaries, the colors represent the country's various ethnic groups. [*Source consulted:* Ethnic Map of Nigeria, Online Nigeria: Community Portal of Nigeria, http://www.onlinenigeria.com/mapethnic.asp#.]

FIGURE 7.21 PHOTO ESSAY: Power and Politics in Sub-Saharan Africa

This region has long been plagued by political violence. A long-term shift toward democratization may help reduce bloodshed as governments become more responsive to their citizens. As in other regions, the most democratized countries have had fewer violent conflicts since 1945. Nevertheless, in any country, the transition to democracy can be tumultuous, and elections are often marked by violence.

A A man in Nairobi, Kenya, runs from police at a demonstration where he has been protesting demands by legislators for an increase in their salaries. That this demonstration could take place at all is a step forward, as is the presence of the media. However, the fact that participants in this nonviolent protest were beaten, tear-gassed, and arrested shows how limited the political freedoms are in Kenya.

B Ellen Johnson-Sirleaf of Liberia is the first woman to be elected head of state in this region. Her election in 2005 was considered free and fair by international observers and helped Liberia begin to resolve tensions that had produced two devastating civil wars. She has introduced policies that encourage public participation in government policy making.

Democratization and Conflict

Democratization index

- Full democracy
- Flawed democracy
- Hybrid regime
- Authoritarian regime
- No data

Armed conflicts and genocides with high death tolls since 1945

- ❗ Ongoing conflict
- ✳ 1000–10,000 deaths
- ✳ 10,000–60,000 deaths
- ✳ 60,000–180,000 deaths
- ✳ 180,000–500,000 deaths
- ✳ 500,000–1,000,000 deaths

C Members of a rebel militia in eastern Congo that is battling the government for control of the area's metal resources: gold, tin, and coltan (a mineral used in cell phones).

D Refugees from Sudan relocate to South Sudan after a referendum in 2011 established the latter's independence from the former. South Sudan is plagued by political corruption and is involved in continuing hostilities with Sudan; both of these problems have constrained the export of its main resource, oil.

a combination of animist and Christian people, were ruled more directly by the British, who encouraged attendance at Christian missionary schools that were open to the public. At independence, the south had more than ten times as many primary and secondary school students as the north. The south was more prosperous, and southerners also held most government civil service positions. Yet the northern Hausa dominated the top political posts, in part because of their long association with British colonial administrators in the north. Over the years, bitter and often violent disputes erupted between the southern Yoruba–Igbo and northern Hausa–Fulani regarding the distribution of increasingly scarce clean water, development funds and jobs, and oil revenues, and the severe environmental damage done by oil extraction.

The politics of oil have complicated the troubles in Nigeria. Much of Nigeria's oil is located on lands occupied by the Ogoni people, who live in the south along the edges of the Niger River delta (see Figure 7.1D). The Ogoni are a group of about 300,000 distinct from other ethnicities in Nigeria. Virtually none of the profits from oil production and export and very little of the oil itself go to Ogoniland. While receiving few benefits from oil extraction, Ogoniland has suffered gravely from the resulting pollution (see the photo in Figure 1.4). Oil pipelines crisscross Ogoniland, and spills and blowouts are frequent; between 1985 and 2016, there were hundreds of spills, many larger than that of the *Exxon Valdez* disaster in Alaska. Natural gas, a by-product of oil drilling, is burned off, even though it could be used to generate electricity—something many Ogoni lack. Royal Dutch Shell, a multinational oil company that is very active in the Niger River delta, acknowledges that while historically it has netted U.S.$200 million a year in profits from Nigeria, it has paid a total of just $2 million to the Ogoni community in 40 years.

Nigeria's President Muhammadu Buhari (from the far north), newly elected in 2015, is now backing a plan to reform the entire oil industry, including Shell's mandatory payment of U.S.$1 billion into a fund to clean up the many oil spills—the cleanup is expected to take 30 years—and to compensate the Ogoni. The income from oil exports recently accounted for 70 percent of Nigeria's revenue, but the global drop in oil prices means that those revenues are now cut in half and government employees are being laid off across the country, leading to the threat of civil unrest.

Geographic strategies have been used of late to reduce tensions in Nigeria. One approach has been to create more political states (Nigeria now has 30) and thereby reallocate power to smaller local units with fewer ethnic and religious divisions. However, when large, wealthy states were subdivided, reputedly to reduce conflict and spread oil profits more evenly, an unintended side effect of the subdivision has been to splinter the opposition geographically and mute the voice and power of the public. [*Sources consulted:* Reuters, "Nigerian President to Become Oil Minister in New Cabinet," September 29, 2015, http://www.reuters.com/article/us-un-assembly-nigeria-idUSKCN0RT2T820150929; "Nigerian Government Finally Sets Up Fund to Clean Up Ogoniland Oil Spills," *The Guardian*, August 7, 2015.] ∎

THE ROLE OF GEOPOLITICS

Cold War geopolitics between the United States and the former Soviet Union deepened and prolonged African conflicts that grew out of divide-and-rule policies. After independence, some sub-Saharan African governments turned to socialist models of economic development, often receiving economic and military aid from the Soviet Union. Other governments became allies of the United States, receiving equally dubious aid (see Figure 5.13). Both the United States and the USSR, rather than truly aiding newly independent nations, tried to undermine each other's African allies by arming and financing rebel groups.

In the 1970s and 1980s, southern sub-Saharan Africa became a major area of East–West tension. The United States aided South Africa's apartheid government in military interventions against Soviet-allied governments in Namibia, Angola, and Mozambique. Another area of Cold War tension was the Horn of Africa, where Ethiopia and Somalia fought intermittently throughout the 1960s, 1970s, and 1980s. At different times, the Soviets and Americans funded one side or the other. The failure by both sides of the Cold War to anchor the aid they dispensed to any requirements that it be used for sustainable development became a major stimulus for militarization and corruption. Once this source of cash was removed at the end of the Cold War, corrupt officials turned to selling off natural resources and commodities.

RADICAL ISLAM: A GEOPOLITICAL MANEUVER

Over the last fifteen years, the northern tier of sub-Saharan countries has become the nexus of geopolitical maneuvering, with the focus more and more on radical Islamism. The conflicts in Sudan and South Sudan first made outsiders aware of hostilities between Christian and radical Islamist Africans generated primarily by the desire to capture oil resources. Then further east in the Horn of Africa, in northern Uganda, Ethiopia, and Somalia, Al Shabaab, through the veneer of Islamism (and again with its eyes on valuable resources), has bombed hotels and disrupted civil society for a decade or longer, creating thousands of refugees. In the failed North African state of Libya, leaders proved to be exploiters of unemployed sub-Saharan young Muslim men, enlisting them as mercenaries in the Libyan civil war. Libya also exploited Muslim farmers in Mali and Chad to procure food when civil unrest diminished Libyan food production. In religiously mixed Nigeria, a militant Islamist group known as Boko Haram, that occasionally claims links with ISIS or other Muslim militants in Southwest Asia, has attempted to establish itself in the northern Muslim sections of the country. The suspicion is that Boko Haram is largely made up of those who were captured as children a decade or more ago in Sudan and in Liberia and Sierra Leone and elsewhere throughout West Africa and turned into boy soldiers. Boko Haram, which says it wishes to reinstate an Islamic caliphate, continues to steal children, especially schoolgirls who are then married off to Boko Haram fighters (see Figure 7.33A). The newly elected President Buhari of Nigeria, a Muslim from the north, swore to reign in Boko Haram and in early March 2016, evidence of his success was revealed when dozens of emaciated Boko Haram fighters surrendered to the Nigerian military police.

CONFLICT AND THE PROBLEM OF REFUGEES

Conflicts create refugees and refugees are commonplace across sub-Saharan Africa. Refugees pour back and forth across borders throughout this region and are displaced within their own countries. Often, they are trying to escape **genocide** (the deliberate destruction of an ethnic, racial, religious, or political group). With only 11 percent of the world's population, this region contains about 19 percent of the world's refugees, and if people displaced within their home countries are also counted, the region has about 28 percent of the world's refugee population (see Figure 7.21D). Women and children constitute 75 percent of Africa's refugees because many adult men who would be refugees are either combatants, jailed, or dead.

As difficult as life is for these refugees, they inadvertently place a severe burden on the often poverty-stricken and politically unstable areas that host them. Even with help from international agencies, the host areas find their own development plans deferred with the arrival of so many distressed people who must be fed, sheltered, and given health care. Large portions of economic aid meant to address development needs have been diverted to deal with the emergency needs of refugees.

SUCCESSES AND FAILURES OF DEMOCRATIZATION

In sub-Saharan Africa, efforts to expand political freedoms have produced mixed results. The number of elections held in the region has increased dramatically. In 1970, only 11 states had held elections post-independence. By 2006, 25 out of the then 44 sub-Saharan African states had held open, multiparty, secret ballot elections, with universal suffrage. By 2014, every country had held some form of election, though not all could be considered "free and fair" (see the map in Figure 7.21).

While the implementation of democratic elections has increased (see Figure 7.21B), transparency in everyday government actions is often absent. The growth of political freedoms—such as freedom of speech, the freedom to assemble in public, and the ability to participate in policy formation at the local and national levels—has been irregular and uneven. Public frustration with suspicious election outcomes, authoritarian policies, and violent repression of peaceful protests (see Figure 7.21A) has often led to rebellion (see Figure 7.21C).

At times, flawed elections have brought about massive violence. In 2008, after several peaceful election cycles, Kenya had an election so corrupt that the country erupted in deadly protest riots. More than 1000 people died and 600,000 were displaced by mobs of enraged voters. By 2010, a new Kenyan constitution gave some hope that political freedoms would be better protected. In Nigeria in 2008, local elections sparked similar violence, and in the Congo (Kinshasa) in 2006, the first elections held there in 46 years resulted in violence that left more than 1 million Congolese people refugees within their own country. In the 2012 Congo (Kinshasa) parliamentary elections, the will of the voting public was validated when the ruling party lost more than 40 percent of its legislative seats to opposition parties. Uganda has been the site of repeated elections assessed by outside observers as rigged and fraudulent,

with opposition candidates detained. In 2016, the election that returned Yoweri Museveni to a fourth term was again marred by the arrest of opposition politicians and voter intimidation.

Zimbabwe may represent the worst case of the failed attempts at democratization. In the 1970s, Robert Mugabe became a hero to many for his successful guerilla campaign in what was then called Rhodesia (now Zimbabwe). The defeated white minority government of Rhodesia had been allied with apartheid South Africa but was not formally recognized by any other country because of its extreme racist policies. Mugabe was elected president in 1980, following relatively free multiparty elections. Although he seemed to be a reformist, over the years his authoritarian policies became more and more extreme and corruption flourished yet again. More and more Zimbabweans became impoverished and alienated while Mugabe increased his personal wealth. In the 1990s, he implemented a highly controversial land-redistribution program that resulted in his supporters gaining control of the country's best farmland, much of which had been in the hands of white Zimbabweans. This move decimated agricultural production and contributed to a chronic food shortage and massive economic crisis. The resulting political violence created 3 to 4 million refugees who fled mostly to neighboring South Africa and Botswana. The most talented and best educated have not returned. Mugabe held on to his office through the rigged elections of 2002 and 2008, but was then forced to share power in 2009 with opposition leader Morgan Tsvangirai, who became prime minister. After some of Tsvangirai's policies were implemented, Zimbabwe began to experience some modest economic growth. But, in 2012, Mugabe called for early elections and appeared to be financing his campaign with money from state-owned diamond mines. The elections, which occurred in 2013, returned Mugabe to full power, and were widely condemned as fraudulent. The downward spiral continues and potential investors remain aloof; in 2016, Mugabe turned 92.

In other places, elections (most recently in South Sudan and Rwanda and some years ago in South Africa) helped end civil wars and resettle refugees. The possibility of becoming respected elected leaders induced some former combatants to lay down their arms. In Sierra Leone and Liberia, public outrage against corrupt ruling elites resulted in elections that brought fortuitous changes of leadership. Over the long term, fair and regular elections and transparency can help enable people to have more input at the policy level, which forces governments to be more responsive to the needs of their citizens.

GENDER, POWER, AND POLITICS

The education of African women and their economic and political empowerment are now recognized as essential to development. As girls' education levels rise, population growth rates fall; as the number of female political leaders increases, so does their influence on policies designed to reduce poverty.

One major characteristic of the expansion of political freedoms has been the increase in the number of women across Africa who are assuming positions of political influence. This chapter opened

genocide the deliberate destruction of an ethnic, racial, religious, or political group

with the story of Juliana Rotich, a businesswoman in Kenya whose road to success was the result of policy changes advocated by various female Kenyan political figures over the last decade. In Liberia—a country devastated by logging fraud, child-soldier recruitment, and the general brutality of its dictator Charles Taylor (see the vignette on Silas Siakor on page 381)—President Ellen Johnson-Sirleaf, a former World Bank economist, began to work on democratic and environmental reforms immediately after she took office (see Figure 7.21B). In Nigeria, Ngozi Okonjo-Iweala, the first female finance minister, now fights for the empowerment of women entrepreneurs as a director at the World Bank. In Rwanda, racked by genocide and mass rapes in the 1990s, women currently make up 64 percent of the national legislature, the highest percentage in the world. And Rwandan women are also leaders at the local level, where they represent 40 percent of the mayors. In Mozambique, 40 percent of the parliament is female; in Burundi, 31 percent; in Senegal, 43 percent; and in South Africa, 42 percent. Altogether, there are 23 African countries where the percentage of women in national legislatures is well above the world average of 20 percent (the U.S. figure is 19 percent).

There has been a sea change in attitudes toward women in politics. It might seem that policies which establish quotas for the number of female members of parliament are falsely elevating women who would not be able to compete in the political arena on their own. But, in fact, the establishment of quotas is usually a reflection of a larger, postconflict commitment to the empowerment of women and to involving women in all aspects of political and civil society. And, it is important to note that many female leaders in Africa, including Ellen Johnson-Sirleaf and at least half of Rwanda's female parliamentarians, were elected without the aid of quotas.

THINGS TO REMEMBER

GEOGRAPHIC THEME 3 • **Power and Politics:** Despite a shift in sub-Saharan Africa toward more political freedoms, some governments remain authoritarian, nontransparent, and corrupt. Public participation is growing, often due to electronic media, and free and fair elections have brought about dramatic changes in some countries.

• Africa's frequent civil wars and violent internal strife are in many ways the legacy of colonial-era policies of divide and rule.

• Fair and regular elections and transparent government procedures enable public input into policy formation and better overall government.

• Failure by both sides of the Cold War to require that the aid they dispensed be used for sustainable development became a major stimulus for militarization and corruption.

• With 11 percent of the world's population, this region contains 19 percent of the world's refugees; if people displaced within their home countries are also counted, the region has 28 percent of the world's refugee population.

• Expansion of political freedoms has increased the number of African women in positions of political influence.

Thinking Geographically

After you have read about urbanization in sub-Saharan Africa, you will be able to answer the following questions.

A What do the types of houses, the water, and the boats suggest about the location of this slum?

B What are some factors that contribute to the prevalence of gangs in African slums?

C How might the lack of clean water and sanitation create an incentive for people to have large families?

D How does this project in Nairobi embody the idea of "self-reliant development"?

URBANIZATION

GEOGRAPHIC THEME 4

Urbanization: Sub-Saharan Africa is undergoing a massive wave of rural-to-urban migration, the fastest rate of urbanization in the world. Much of this growth is unplanned, and 70 percent of the urban population now live in impoverished slums characterized by crowded fire-prone housing, inadequate sanitation, and poor access to jobs, clean water, and food.

THE RECENT TREND OF RAPID URBAN GROWTH

Sub-Saharan Africa is still majority rural (62 percent), but it is urbanizing faster than any other world region. This trend is recent. In the 1960s, only 15 percent of sub-Saharan Africans lived in cities; now about 38 percent do (see the **FIGURE 7.22** map). Estimates are that by 2030, half of all Africans will be urban dwellers. In 1960, just one sub-Saharan African city—Johannesburg, South Africa—had more than 1 million people; in 2015, 70 cities did. The largest sub-Saharan African city is Lagos, Nigeria, where in 2015 various estimates put the population at between 11 and 13 million; by 2020, Lagos is projected to have 25 million people. Much of this growth in West Africa is taking place in primate cities (see Chapter 3). For example, Kinshasa, Congo, with 11.6 million people, is almost six times the size of Congo's next-largest city, Lubumbashi; Lagos in Nigeria is nearly four times the size of Kano. In East Africa, the population is more evenly distributed, though primate cities are beginning to sprout up.

What accounts for this rapid growth in African cities? There are several factors. Usually, migration from rural to urban areas accounts for most of the rapid growth of cities in the developing world, but in sub-Saharan Africa, contrary to trends elsewhere, such migration, though extremely significant, accounts for just one-third of urban growth. Rather, it is natural increase that accounts for about half of sub-Saharan cities' growth rate of 5 percent per year, the highest in the world. This fact needs some explanation because normally urban life is thought to decrease fertility rates for a number of reasons. A third factor contributing to urbanization is a change in the way urban areas are defined. More countries are now defining settlements of over 2000 inhabitants as urban, where previously an urban settlement had to have 20,000 inhabitants. Clusters of these settlements often surround large cities, and as they grow, they increasingly interact with the

FIGURE 7.22 PHOTO ESSAY: Urbanization in Sub-Saharan Africa

Urban populations are exploding because of high birth rates in cities, migration from rural areas, and a redefinition of what constitutes urban. By 2030, half of the people in the region will live in cities, most in slums that are plagued by violence and inadequate access to water, food, sanitation, and education. Largely ignored by most governments, people in urban slums survive by helping themselves.

[Sources consulted: 2015 Population Data Sheet and Demographia World Urban Areas 2015:01.]

A A part of the Makoko slum in Lagos, Nigeria, the largest city in this region and one of the fastest growing in the world. Makoko is home to more than 100,000 people. Waterborne diseases such as malaria are common, as is flooding, which kills and displaces residents on a regular basis. Because of these hazards and the fact that the community sits on potentially valuable waterfront property, the municipal government of Lagos periodically evicts some of Makoko's residents and demolishes their homes.

B Opposing gangs battle for control of territory in Kinshasa, Congo. Gangs control most slum neighborhoods in this region. Violence is widespread and includes organized attacks and robberies of passersby, the police, and other gangs.

C A girl drinks from a public water tap in Soweto, South Africa. Slum houses rarely have indoor plumbing, and people obtain water from public spigots. Latrines provide the only sanitation.

Population living in urban areas

- 83%–100%
- 65%–82%
- 47%–64%
- 29%–46%
- 10%–28%
- No data

Population of Metropolitan Areas (2015)

- 15 million
- 8 million
- 4 million
- 2 million

Note: Symbols on map are sized proportionally to metro area population

① Global rank (population 2013)

mi 0   400   800
km 0   600   1,200

D A "sack garden," one of many innovative self-help projects found in the Kibera slum in Nairobi, Kenya. High food costs and high unemployment rates make such urban gardening attractive to many African city dwellers.

urban core; hence, they are now included in the urban population figures.

CHARACTERISTICS OF RAPID URBAN GROWTH

The principal factors that characterize urbanization in sub-Saharan Africa are, first of all, the uncontrolled nature of the growth. Second, as mentioned, rural-to-urban migration plays an important role in rapid urbanization. A puzzling characteristic is that, contrary to expectations, urbanization has not led to a decrease in fertility. And finally, governments, perhaps inadvertently, have retained policies that have an anti-urban bias, which makes improving urban conditions difficult.

The uncontrolled nature and rapidity of urban growth have led to poor housing and a lack of basic services. Governments and private investors have not addressed the need for affordable housing, so most migrants live in extremely substandard crowded spaces that they themselves have often quickly constructed from found materials. Vast, unplanned one-story slums surrounding older urban centers house 72 percent of Africa's urban population. Basic sanitation is absent, as is access to clean water and nourishing food, factors that severely impact urban dwellers. Unplanned growth also means that transportation in these huge and shapeless settlements is a jumble of government buses and private vehicles, and boats in waterfront locales (see Figure 7.22A). Parents frequently have to travel long hours through extremely congested traffic to reach distant jobs, getting much of their sleep while sitting on a crowded bus and leaving their children unsupervised and susceptible to the influence of gangs (see Figure 7.22B).

Uncontrolled urbanization has major implications for the region, economically, politically, demographically, and in relation to gender. Economically, the development of manufacturing industries and the service sector has lagged and jobs needed for the growing urban workforce are in short supply, so much so that women overwhelmingly depend on employment in the informal economy. Politically, angry youth and adults have become a volatile political force. As their numbers and power grow, governments will have to become more responsive to the needs of these urban slum dwellers.

People generally move from the countryside to cities for jobs and educational opportunities. However, the volatility of many African commodity-based economies, combined with the low living standards found in many slums, means that there is a great deal of what geographer Deborah Potts describes as *circular migration*, or back-and-forth migration between rural and urban areas. When urban economies stagnate and jobs become scarce, the higher cost of living in cities combines with the hazards of slum life to convince many migrants to move back to the countryside at least temporarily. This constant movement makes estimating urban or rural populations difficult, and it is possible that urban populations and their growth rates are being overestimated.

In terms of demography, one of the anticipated benefits of urbanization is not materializing: slowing rates of fertility. Normally, urban life strengthens all the factors that influence the *demographic transition*, discussed below on page 412. Certainly, birth rates have declined a bit across sub-Saharan Africa, but there is a lag in the expected effect of urbanization on fertility. Poverty

and disease are still relatively widespread, and clean water and sanitation are generally lacking (see Figure 7.22A, C). Whereas in rural areas, recent local efforts and international aid and investment have improved overall conditions and access to birth control, it has been assumed erroneously that urban life, itself, will act as birth control in Africa's cities. Similarly, there have been lags in providing education and economic opportunities for urban women, both of which are known to significantly lower fertility rates.

The anti-urban bias of governments is demonstrated by the failure to fund and maintain infrastructures, and schools and transportation services, thus contributing to the chaotic nature of urban life. The resistance of governments is partly due to the rural bias of legislatures: cities are drastically underrepresented in national parliaments; hence, enlightened urban policies have little chance of being debated and passed. Finally, locally mobilized community support groups lobbying for human well-being in urban areas are only beginning to confront governments. Nonetheless, the increasing role of women in government, discussed elsewhere, and of young entrepreneurs (for example, those connected to Ushahidi) make it likely that over time life in sub-Saharan Africa's cities will improve.

The Scarcity of Clean Water in Sub-Saharan Cities Public health is a major urban concern, as many water distribution systems are susceptible to contamination by harmful bacteria from untreated sewage and garbage (see Figure 7.22C). Only the largest sub-Saharan African cities have sewage treatment plants, and few of these extend to the slums that surround them. The result is frequent outbreaks of waterborne diseases such as cholera, dysentery, and typhoid. It is estimated that unsafe water and sanitation facilities result in an annual loss of U.S.$28.4 billion from illness and premature death in Africa alone.

Although the safety of city water supplies is improving, there are frequent periodic water shortages in as many as 11 out of 14 sub-Saharan cities (including Niamey, Niger, covered in the Case Study above on page 387). To control usage during a crisis, city officials limit service to a few hours a day, causing people to store water in vessels that may become contaminated. Recent reassessments of the availability of groundwater resources may eventually alleviate these urban water shortages, but recall that the natural distribution of groundwater does not match up well with the locations of the big cities (see page 385; see also Figure 7.9).

The problem of unsafe water for households, whether rural or urban, may have a solution that is simple and affordable, and could create local jobs. A for-profit group, working with grassroots organizations in Uganda but subsidized by American investors, has designed a simple plastic jug, called the TivaWater jug, with an interior sand-and-clay filtering device that purifies water (FIGURE 7.23). The jug is made to be affordable for poor people in Ugandan shantytowns. While initially manufactured in the United States, the jug is now assembled in Uganda, and sold and distributed by small businesses there, the intention being for Ugandans to eventually export the jugs to other African countries.

Food Security in Urban Areas Maintaining adequate food supplies is a long-standing problem for cities everywhere and more so when the customers are poor, refrigerated storage facilities are

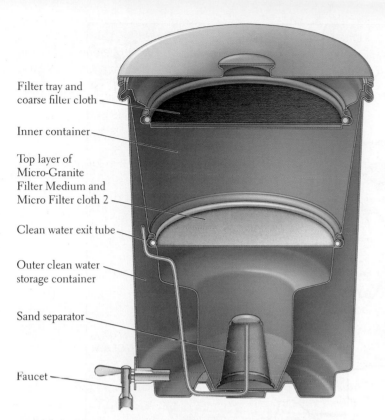

Filter tray and
coarse filter cloth

Inner container

Top layer of
Micro-Granite
Filter Medium and
Micro Filter cloth 2

Clean water exit tube

Outer clean water
storage container

Sand separator

Faucet

FIGURE 7.23 The TivaWater jug. This specially designed water-filtering jug, made of unbreakable food-grade plastic, removes dirt, bacteria, and viruses using a new version of biosand technology. It is easy to maintain, has a faucet that dispenses the water, and can be manufactured locally; because of subsidies, it is also affordable for the very poor.

lacking, transportation is unreliable and expensive, and supplies are unpredictable. Increasingly, food is shipped into sub-Saharan cities from distant parts of the continent or other world regions, but such food is often expensive and waste is common. Not surprisingly, urban Africans are harking back to their agricultural roots and beginning to produce their own food in the tiny spaces between houses and in the wastelands that surround urban shantytowns, as is the case in other world regions.

ON THE BRIGHT SIDE: Urban Food Gardens

A 2012 UN Food and Agricultural Organization (FAO) report, *The Greening of African Cities,* emphasizes that the growing of produce in and around cities, on even the tiniest scraps of land, has an advantage over both rural market gardening and imported food in supplying urban people with safe, nutritious food. Fruits and vegetables can be highly perishable, so if urban consumers can produce their own food, they are likely to eat better, to create less waste, and to experience lower food transport costs. Also, food waste can be immediately recycled as compost, reducing or eliminating the need for fertilizers; the greenbelts created by urban gardens can reduce urban air and water pollution.

If urban gardening is to become a viable food security solution, however, some adjustments will be necessary. At present, urban farmers often use brownfields—land where the soil is contaminated by past industrial activities. Moreover, urban farmers frequently use wastewater that may not be safe to irrigate their crops. But these problems can be corrected. Raised cultivation beds

can be sealed off from the contaminated soil of brownfields, and for irrigation, rainwater can be harvested off nearby roofs, or, alternatively, wastewater can be treated before it is used.

Because few urban gardeners own the land they cultivate, their gardens can be confiscated for development without warning or compensation. One adaptation to this uncertainty is semi-mobile "sack gardens" (see Figure 7.22D) that can be relocated if necessary. Finally, to avoid the overuse of fertilizers and pesticides, urban farmers are learning to compost food waste and use organic techniques to control pests. ■

THINGS TO REMEMBER

GEOGRAPHIC
THEME 4
• **Urbanization:** Sub-Saharan Africa is undergoing a massive wave of rural-to-urban migration, the fastest rate of urbanization in the world. Much of this growth is unplanned, and 70 percent of the urban population now live in impoverished slums characterized by crowded fire-prone housing, inadequate sanitation, and poor access to jobs, clean water, and food.

• In the 1960s, only 15 percent of sub-Saharan Africans lived in cities; now about 37 percent do. Estimates are that by 2030, half of all Africans will be urban dwellers.

• Urbanization has not produced the expected lowering of fertility probably because of poor living conditions, which keep infant mortality rates high; lack of jobs and inadequate basic services impede personal advancement, especially for women.

• When urban economies stagnate and jobs become scarce, the higher cost of living in cities combines with the hazards of slum life to convince many migrants to move back to the countryside at least temporarily, resulting in circular migration, or back-and-forth migration between rural and urban areas.

• In some cities, residents have taken the initiative to produce their own food in the tiny spaces between houses and in the wastelands surrounding urban shantytowns.

• Public health is a major urban concern, as many water distribution systems are susceptible to contamination by harmful bacteria from untreated sewage and garbage.

POPULATION AND GENDER

GEOGRAPHIC THEME 5

Population and Gender: Of all the world regions, populations are growing fastest in sub-Saharan Africa, yet growth rates there are slowing as demographic transition takes hold in more prosperous countries, where better health care, lower levels of child mortality, and more economic and educational opportunities encourage smaller families. In particular, better-educated women are able to pursue careers and to choose to have fewer children.

In a little over 50 years, sub-Saharan Africa's population has more than quadrupled, growing from about 200 million in 1960 to 949 million in 2015. By 2050, the population of this region could reach just over 2 billion. Contrary to the perception outsiders often hold, however, Africa as a whole is much less

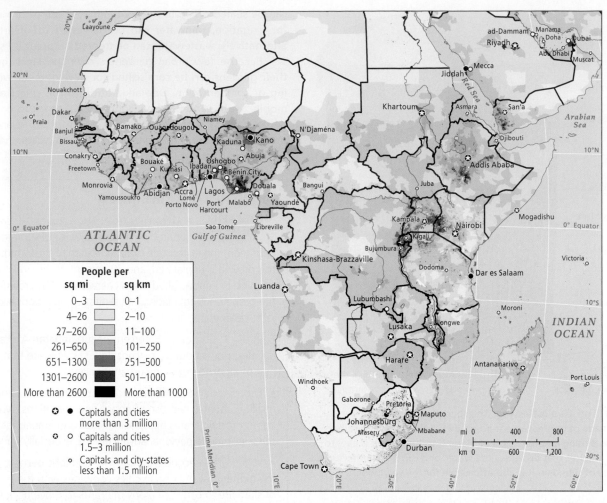

FIGURE 7.24 Population density in sub-Saharan Africa. Of all the world regions, populations are growing fastest in sub-Saharan Africa. The demographic transition is taking hold in a few of the more prosperous countries. In these countries, better health care and more economic and educational opportunities for women are enabling them to pursue careers and to choose to have fewer children. [*Sources consulted:* Deborah Balk, Gregory Yetman, et al., Center for International Earth Science Information Network, Columbia University, http://www.ciesin.columbia.edu.]

densely populated than most of Europe and Asia—36 people per square kilometer, compared to the global average of 51 people per square kilometer (Figure 1.26). If fertility rates remain high, though, places that are relatively uncrowded now may change dramatically over the next few decades **(FIGURE 7.24)**. In some rural areas, the population density may far exceed the carrying capacity of the land, which will lead to widespread impoverishment. Increasingly, Africa's population is becoming concentrated in a few urban areas, such as the cities along the coast of the Gulf of Guinea in West Africa.

POPULATION GROWTH IS SLOWING

At the moment, sub-Saharan Africa's annual fertility rate (5 births per adult woman) is the highest in the world; even so, overall rates are slowing in nearly every country including in cities. Birth rates are as high as they are, in part, because many Africans view children as both an economic advantage and a spiritual link between the past and the future. Childlessness is considered a tragedy, as children ensure a family's genetic and spiritual survival. The result is that sub-Saharan Africa has the world's youngest population: the median age

in 2015 was 19 (half the population was younger than 19 and half the population was older). Large families continue to be viewed as having economic value because children and young adults can perform important work on family farms and in family-scale industries. Moreover, in this region of generally poor health care and resultant high incidence of infant mortality, parents may have extra children in the hope of raising at least a few of them to maturity. In all but a few countries, the *demographic transition*—the sharp decline in births and deaths that usually accompanies economic development (see Figure 1.28)—has barely begun, and more people now survive long enough to reproduce than did in the past. If Africa is on the brink of a development surge, however, fertility rates could decline quite sharply to a replacement level (2.1 births per woman) as women choose roles other than motherhood. Once through the demographic transition, sub-Saharan populations will begin to age.

Five countries have gone through the demographic transition—South Africa, Botswana, Seychelles, Réunion, and Mauritius (the last three being small island countries off Africa's east coast). Circumstances have changed enough to make smaller families desirable and attainable: in all five countries, per capita

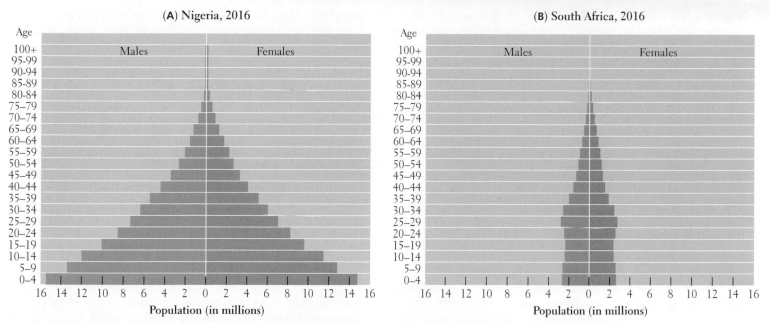

FIGURE 7.25 Population pyramids for Nigeria and South Africa. Note that the two pyramids are at the same scale. (A) Nigeria had a population of 182 million people in 2015 and a growth rate of 2.7 percent, as indicated by its very wide pyramid base. It has the largest population on the continent, nearly twice that of Ethiopia, which has the second-largest population (98 million). (B) South Africa, on the other hand, had a population of 55 million in 2015 and a growth rate of only 1.6 percent; its population has experienced some decline, as shown by the shrinking of the pyramid base. The decline is, in part, due to the severe HIV/AIDS epidemic. [*Source consulted:* International Data Base, U.S. Census Bureau, http://www.census.gov/population/international/data/idb/informationGateway.php.]

incomes (GNI) are five to ten times the sub-Saharan average of U.S.$3480. Advances in health care have also cut the infant mortality rate to less than half the regional average of 64 infant deaths per 1000 live births. Because of this, parents with two or three children can expect them all to live to adulthood. The circumstances of women in these five countries have also improved, as reflected in female literacy rates of about 80 to 90 percent, compared to the regional average of 54 percent. In addition, in these same five countries, women have more opportunities to work outside the home at decent-paying jobs; thus, many women are choosing to use contraception because they have life options beyond motherhood. Indeed, contraception use is up across the continent, but the percentage of married women using contraception in these five countries is double the rate for sub-Saharan Africa as a whole, which is only 30 percent (the world average is 62 percent).

POPULATION PYRAMIDS ILLUSTRATE GROWTH RATES

The population pyramids in **FIGURE 7.25** show the contrast between countries that are growing rapidly, such as Nigeria, Africa's most populous country (with 182 million people and a natural growth rate of 2.7 percent), and countries that have already gone through the demographic transition, such as South Africa (with 55 million people, a natural growth rate of 1.6 percent, and a median age of 26). Nigeria's pyramid is very wide at the bottom because over half its population is under the age of 19. In just 15 years, this entire group will be of reproductive age. Only 15 percent of Nigerian women use any sort of birth control.

In contrast to Nigeria, South Africa's pyramid has contracted at the bottom because its birth rate has dropped from 35 per 1000 to 22 per 1000 over the last 20 years. This decrease is primarily an effect of economic and educational improvements as well as social changes that have come about since the end of apartheid (in 1994). It is likely to persist as the advantages of smaller families become clear, especially to women. The birth rate decrease is partially attributable to the fact that contraception is used by 60 percent of South African women; but part of the birth rate decrease in South Africa is also due to high rates of HIV/AIDS infection among young adults and the lower fertility of those afflicted.

THE ROLE OF MIGRATION WITHIN SUB-SAHARAN AFRICA AND TO OTHER REGIONS

Migration has already been mentioned as a factor in urbanization, but it has several other demographic impacts as well. The migration of sub-Saharan people looking for work in North Africa, Europe, Turkey, or the United States is now a familiar pattern and is discussed briefly below, but the vast majority of African migrants are moving within Africa.

Much of the migration between African cities and rural areas crosses international borders, which makes estimating urban and rural populations even harder. Migrants most often go to rural areas as agricultural laborers, or to the cities to work in various jobs in the informal sector. The five sub-Saharan African countries with the most immigrants in 2005 were: Côte d'Ivoire (2.4 million), Ghana (1.7 million), South Africa (1.1 million),

Nigeria (1.0 million), and Tanzania (0.8 million). Together, these countries have 40 percent of the internal migrants in Africa and a significant number of these migrants are very young, even children. Illustrative of how children become not only internal migrants but also enslaved laborers are a number of lawsuits filed in San Francisco in 2015 against eight well-known chocolate companies for purchasing raw cacao from producers in Côte d'Ivoire, who used the forced labor of children as young as 11 (see the discussion of modern slavery on page 392)

Those sub-Saharan migrants who leave the continent to make their way to North Africa and then Europe number in the hundreds of thousands. These are overwhelmingly young males, who, in addition to seeking opportunities for work and education, plan to send money back to support family members. They tend to be physically fit and already at least partly educated; hence, they represent a loss of human capital to their home communities. Those who do find work usually send remittances home, but they are also subject to danger and exploitation—many hundreds died trying to cross the Mediterranean in 2015–2016, others were exploited and held in bondage by underworld gangs once they reached Europe. Happily, an increasing number of young migrants who do find a measure of success in their destinations eventually return home to start development projects. A number of examples are cited in this chapter. Nonetheless, the fate of thousands of African migrants hoping to reach Europe continues to be thrown into question by the current migrant/refugee crisis discussed in Chapter 4.

POPULATION AND PUBLIC HEALTH

Infectious diseases, including HIV/AIDS, are by far the largest killers in sub-Saharan Africa, responsible for about 50 percent of all deaths. Some diseases are linked to particular ecological zones. For example, people living between the 15th parallels north and south of the equator are most likely to be exposed to sleeping sickness (trypanosomiasis), which is spread among people and cattle through the bites of tsetse flies. The disease attacks the central nervous system and results in death if untreated. Several hundred thousand Africans suffer from sleeping sickness, and most of them are not treated because they cannot afford the expensive drug therapy.

Africa's most common chronic tropical diseases, schistosomiasis and malaria, are linked to standing fresh water. Their incidence has thus increased with the construction of dams and irrigation projects. Schistosomiasis is a debilitating, though rarely fatal, disease that affects about 170 million sub-Saharan Africans. It develops when a parasite carried by a particular freshwater snail enters the skin of a person standing in water. Malaria, spread by the anopheles mosquito (which lays its eggs in standing water), is more deadly. The disease kills at least 1 million sub-Saharan Africans annually, most of them children under the age of 5. Malaria also infects millions of adult Africans who are left feverish, lethargic, and unable to work efficiently because of the disease. The mosquito-borne Zika virus is recently recognized to cause microcephaly in newborns.

Until recently, relatively little funding was devoted to controlling the most common chronic tropical diseases. Now, however, major international donors are funding research in Africa and elsewhere. More than 60 research groups in Africa are working on a vaccine that will prevent malaria in most people. The distribution of simple, low-cost, and effective mosquito nets is also reducing the transmission rates of malaria and Zika.

The 2014 Ebola Epidemic

Ebola is a rare but deadly contagious disease caused by a virus, which is spread only through direct contact with the blood and bodily fluids (including urine and semen) of a person already exhibiting symptoms of the illness. It is not airborne like flu. Symptoms include fever, muscle pain, fatigue, headache followed by vomiting, diarrhea, rash, and internal and external bleeding. Death comes from dehydration, and the only treatment thus far is intravenous rehydration. Vaccines are being tested. Although Ebola originated in Congo (Kinshasa) in 1976, the most recent outbreak occurred in 2014 in West Africa in Sierra Leone, Guinea, and Liberia, where more than 10,000 people died, including hundreds of health workers who had taken extreme precautions to avoid infection. Measures to stop the disease are affecting many African social customs; they include bans on hugs upon greeting a friend, traditional funerals in which custom had required washing of the deceased, and huge wakes at which family and friends are able to say their final goodbyes to the body of the deceased. In November 2015, the end of the Ebola outbreak was officially declared, but then an additional case was identified in January 2016 in Sierra Leone.

HIV/AIDS in Sub-Saharan Africa

A leading cause of death in sub-Saharan Africa, and the leading cause of death for women of reproductive age, is acquired immunodeficiency syndrome (AIDS), caused by the human immunodeficiency virus (HIV) **(FIGURE 7.26)**. Sub-Saharan Africa, with 13 percent of the world's population, has 70 percent of all patients living with HIV. As of 2015, more than 25.8 million sub-Saharan Africans were HIV-infected. In Botswana, one of the richest countries in Southern Africa, nearly 29.2 percent of the adult female population is living with HIV. The Southern Africa subregion alone accounted for 31 percent of global AIDS deaths in 2010. While the epidemic is subsiding a bit, it has significantly constrained economic development.

Worldwide, women account for half of all people living with HIV, but in sub-Saharan Africa, HIV/AIDS affects women more than men: 59 percent of the region's HIV-infected adults are women. The reasons for this pattern are related to the social status of women in sub-Saharan Africa and elsewhere.

The rapid urbanization of Africa has hastened the spread of HIV, which is much more prevalent in urban areas. It is not uncommon for poor, new urban migrants removed from their village support systems to become involved in the sex industry in order to survive economically. In some cities, virtually all sex workers are infected. Meanwhile, many urban men, especially those with families back in rural villages, visit prostitutes on a regular basis. These men often bring HIV back to their rural homes and infect their unwitting wives. As transportation between cities and the countryside has improved, bus and truck drivers have also become major carriers of HIV to rural villages.

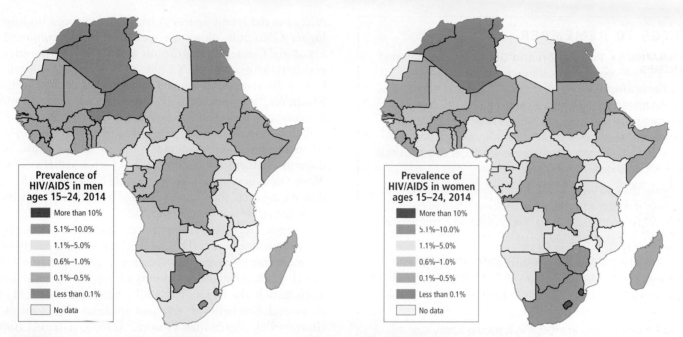

FIGURE 7.26 HIV/AIDS in Africa. New HIV infections fell in sub-Saharan Africa from 2.2 million in 2001 to 1.8 million in 2009. During this period, rates declined more than 25 percent in 22 of the countries. The rate has remained stable elsewhere in the region. The number of deaths from AIDS has declined in sub-Saharan Africa, in large part because of the increased availability of antiretroviral medications. [*Sources consulted:* UNAIDS, *Global Report 2010 Fact Sheet,* http://www.unaids.org/documents/20101123_FS_SSA_em_en.pdf; UNAIDS, *Global Report: UNAIDS Report on the Global AIDS Epidemic 2010,* http://www.unaids.org/globalreport/documents/20101123_GlobalReport_full_en.pdf.]

Myths and social taboos surrounding HIV/AIDS exacerbate the problem of controlling the spread of the disease and make education a key component in combating the epidemic. For example, some men believe that only sex with a mature woman can result in infection, so very young girls are increasingly sought as sex partners (sometimes referred to as "the virgin cure"). Elsewhere, infection is considered such a disgrace that even those who are severely ill refuse to get tested, yet they remain sexually active.

The Social Costs of HIV/AIDS Across the continent, the consequences of the HIV/AIDS epidemic have been enormous. Children contract HIV primarily in utero or from nursing; just over 400 children under the age of 15 die each day from AIDS-related causes. In 2013, there were 210,000 new cases among sub-Saharan children. More than 15 million children have been orphaned, and many have no family left to care for them or to pass on vital knowledge and life skills. Millions of parents, teachers, skilled farmers, craftspeople, and trained professionals have been lost. The disease has severely strained the health-care systems of most countries. Demand for treatment continues to explode, yet drugs are prohibitively expensive, and many health-care workers themselves are infected. Because so many young people are dying of AIDS, decades of progress in improving the life expectancy of Africans have been erased. For example, in 1990, adult life expectancy in South Africa was 63 years, but in 2015, it fell to 61 years and was just 59 years for males.

HIV/AIDS Education and Medication The most effective means of prevention are massive education programs that lower rates of infection among those who can read and understand explanations about how HIV is spread. For example, Senegal started HIV/AIDS education in the 1980s, and the levels of infection there have so far remained low. Major education campaigns in Uganda lowered the incidence of new HIV infections from 15 percent to just 4.1 percent between 1990 and 2004. By contrast, infection rates soared in places such as South Africa, where a few top politicians put forth untenable theories about the causes of HIV infection or denied that HIV/AIDS was a problem at all.

HIV/AIDS medications (known as antiretroviral therapy) are prohibitively expensive for most families in sub-Saharan Africa. The United Nations and a number of private philanthropies cover some of the costs. The medications are not a cure and must be taken daily for the rest of one's life, but they do make it possible to live with the disease. Pharmaceutical firms in India, China, and elsewhere have come out with generic antiretroviral versions that have reduced the cost to about $350 per year, about one-tenth of the average annual income in the region. Approximately five out of every six recipients worldwide reside in sub-Saharan Africa, and most of those are in Southern Africa. As yet, only 60 to 80 percent of those who need the medicines are receiving them; the goal for 100 percent of those with HIV/AIDS to be covered by 2015 was not met. The cost to the world of keeping parents alive with antiretroviral therapy is far less than that of dealing with the social effects of millions of orphans.

THINGS TO REMEMBER

GEOGRAPHIC THEME 5   • **Population and Gender:** Of all the world regions, populations are growing fastest in sub-Saharan Africa, yet growth is slowing as the demographic transition takes hold in more prosperous countries where better health care and lower levels of child mortality, along with more economic and educational opportunities, encourage smaller families. In particular, educated women are better able to pursue careers and to choose to have fewer children.

• In fewer than 50 years, sub-Saharan Africa's population has more than quadrupled, growing from about 200 million in 1960 to 949 million in 2015. By 2050, the population of this region could reach just over 2 billion.

• The overall low population density figures in sub-Saharan Africa may be surprising, but they can be misleading. Densities are extremely high in some places and very low in less habitable areas.

• Women play an important role in reducing family size. As they gain education and job opportunities, women choose to have only one or two children.

• Despite the growing economic importance of women to families, old patriarchal ideas about male sexual freedom result in women falling victim to HIV/AIDS more often than men.

• Migration, both internal and to other continents, plays a significant role in population dynamics.

• Sub-Saharan Africa has higher rates of infectious diseases than any other region in the world: recurring epidemics of malaria, schistosomiasis, sleeping sickness, ebola, Zika, and HIV/AIDS.

GEOGRAPHIC PATTERNS OF HUMAN WELL-BEING

As we have observed in previous chapters, gross national income (GNI) per capita can be used as a general measure of well-being because it shows broadly how much money people in a country have to spend on necessities. However, it is only an average figure and does not indicate how income is actually distributed across the population. Thus, it does not show whether income is equally apportioned or whether a few people are rich and most are poor. (As discussed in earlier chapters, the term GNI is often followed by *PPP*, which indicates that all figures have been adjusted for cost of living and so are comparable.)

The map in **FIGURE 7.27A** (GNI) indicates three things regarding the ability of citizens to have a sufficient share of the national income to have good health care and an education. First, most countries are yellow or light yellow in color, indicating that they fall in the two lowest GNI ranks (with incomes under U.S.$4000). None are in the top two categories of GNI per capita rank (dark red or purple colors), and only five small countries—Botswana, Equatorial Guinea, Gabon, Mauritius, and the Seychelles—are in the medium-high rank (lighter red). A look at the world map inset for GNI shows that of all regions on Earth, sub-Saharan

Africa has the lowest figures. A band of countries with somewhat higher GNIs runs along the west side of the continent from Equatorial Guinea through South Africa. In these countries, GNI per capita ranges from U.S.$4000 to U.S.$24,999, but only little Gabon, Botswana, and the tiny island countries of Mauritius and Seychelles, far to the east in the Indian Ocean, are in the medium category, averaging U.S.$15,000 to U.S.$24,999 per year. Gabon and Botswana depend on extractive industries—selling timber, oil, manganese, and diamonds—and in both nations, income disparity is rather stark. South Africa, which is actually considered Africa's most developed country because of its industry, has a GNI that averages less than U.S.$12,700. There are no available data for Somalia, Réunion, and Djibouti.

Figure 7.27B (HDI) depicts countries' rank on the Human Development Index (HDI), which is a calculation (based on adjusted real income, life expectancy, and educational attainment) of how adequately a country provides for the well-being of its citizens. In this region, HDI ranks stretch from high (Mauritius at 63 and Seychelles at 64) and medium (Botswana, Congo [Brazzaville], Equatorial Guinea, Gabon, Ghana, Namibia, South Africa, and Zambia) to very low. As noted in Chapter 6, if on the HDI scale a country drops below its approximate position on the GNI scale, one might suspect that the lower HDI ranking is due to the poor distribution of wealth and social services: a critical mass of people are simply not getting access to basic necessities.

The Gender Development Index (GDI; Figure 7.27C) shows where sub-Saharan countries rank on equality between the genders on a scale of 1 to 5: 1 represents high equality between males and females, and 5 is low equality. The development indicators used in the calculation of GDI rankings are male and female longevity, male and female educational achievement, and GNI PPP by gender. The map of these rankings shows that only a few countries score high, with most scoring 3, 4, or 5. The thumbnail world map shows that Africa, Southwest Asia, and South Asia have the greatest gender inequality. Other evidence on gender (discussed previously) indicates that women in sub-Saharan Africa are underemployed in cities, are frequently employed in the informal economy, and have difficulty accessing education and training, but despite these drawbacks, they are improving their social and economic status and beginning to assume leadership and policy-making roles.

SOCIOCULTURAL ISSUES

To the casual observer, it may appear that a majority of sub-Saharan Africans live traditional lives in rural villages. It is true that 62 percent of the sub-Saharan population lives in rural areas: however, this region is also the most rapidly urbanizing on Earth. Needless to say, social change is occurring everywhere.

GENDER ISSUES

The fact that the role and status of women can have major consequences for rates of population growth has just been discussed, but gender has enormous influence on many other aspects of daily life as well. Long-standing African traditions dictate a fairly strict division of labor and responsibilities between men and

FIGURE 7.27 Maps of human well-being.

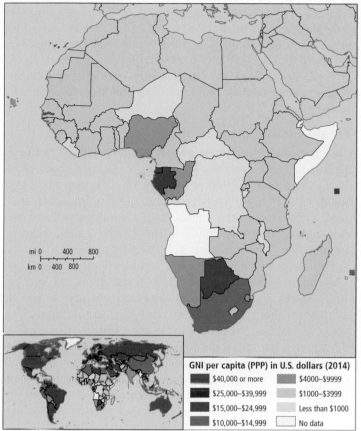

(A) Gross national income (GNI)

GNI per capita (PPP) in U.S. dollars (2014)

- $40,000 or more
- $25,000–$39,999
- $15,000–$24,999
- $10,000–$14,999
- $4000–$9999
- $1000–$3999
- Less than $1000
- No data

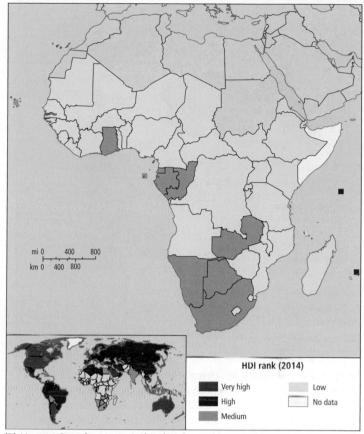

(B) Human Development Index (HDI)

HDI rank (2014)

- Very high
- High
- Medium
- Low
- No data

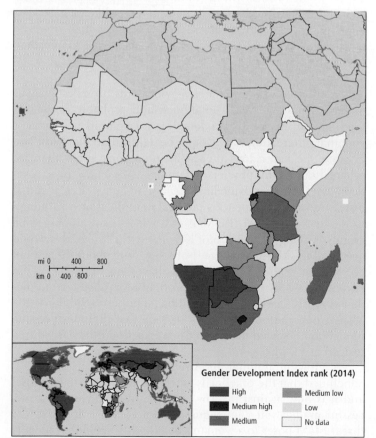

(C) Gender Development Index (GDI)

Gender Development Index rank (2014)

- High
- Medium high
- Medium
- Medium low
- Low
- No data

women. In general, women are responsible for domestic activities, including raising the children, attending to the sick and elderly, and maintaining the house. Women collect water and firewood and prepare nearly all the food. Men are usually responsible for preparing land for cultivation. In the fields intended to produce food for family use, women sow, weed, and tend the crops, as well as process them for storage. In fields where cash crops are grown, men perform most of the work and retain control of the money earned. When husbands in search of cash income for the family migrate to work in mines or in urban jobs, within Africa or beyond, women take over nearly all of the agricultural work, often using simple hand tools, not machines or even draft animals. When there are small agricultural surpluses or handcrafted items to trade, it is women who transport and sell them in the market. Throughout Africa, married couples often keep separate accounts and manage their earnings as individuals, but when a wife sells her husband's produce at the market, she usually gives the proceeds to him.

Ideas in Africa about how males and females should relate to each other come from a variety of sources. Even in the ancient past, social controls tempered gender relationships. Most marriages were social alliances between families; therefore, husbands and wives spent most of their time doing their tasks with family members of their own sex rather than with each other. Then as today, women primarily influenced other women and men influenced men. It is important to note that having multiple

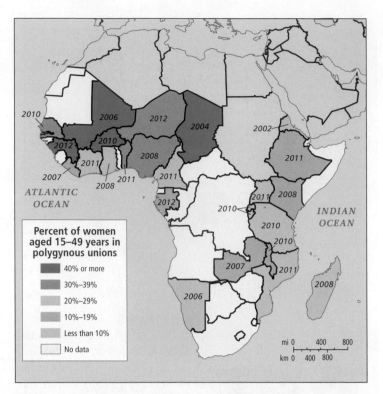

FIGURE 7.28 Polygyny in Africa. The map depicts the spatial distribution of polygyny in sub-Saharan Africa for various years from 2004 to 2012. Region-wide, 72 percent of sub-Saharan women indicated they are their husband's only wife but that their spouses have had relations with other women; 19 percent are in two-wife marriages; 7 percent in three-wife marriages; and the remaining 2 percent are in marriages with four or more wives. Polygynous unions are more common in remote poor areas, where women are less educated and child survival is low. Usually, there is a wide age gap between a polygynous husband and his wives. [*Source consulted:* James Fenske, "African Polygamy: Past and Present," *Editorial Express,* February 15, 2012, https://editorialexpress.com/cgi-bin/conference/download.cgi?db_name=CSAE2012&paper_id=115; data from the Demographic and Health Surveys (DHS) Program STAT compiler (DHS, 2014), published by the UN in *The World's Women 2015,* p. 17, Figure 1.14.]

wives—the practice of **polygyny** (FIGURE 7.28)—is more common in sub-Saharan Africa (where it has ancient pre-Muslim, pre-Christian roots) than in Muslim North Africa (as discussed in Chapter 6). In a 2012 study of the historical causes, distribution, and persistence of polygyny in Africa, economist James Fenske (2013) produced the map in Figure 7.28. He noted that while polygyny is on a marked decline, the reasons for this decline are unclear. Current improvements in women's education don't seem to reduce its incidence, but economic development and declining infant mortality do.

polygyny the practice of having multiple wives

female genital mutilation (FGM) removing the labia and clitoris and sometimes stitching the vulva nearly shut

Women face significant sexual and physical abuse in the home and during civil unrest, when rape and beatings are more common. Many of the female politicians who have come to power in the last decade are now addressing these issues, openly declaring that they themselves have been victims of domestic violence. Most have emphasized the need for more women to be in the position to develop and enforce policies to reduce violence against women. In Liberia, for instance, President Ellen Johnson-Sirleaf—who was, herself, a victim of domestic abuse—has defined women's rights as a national security issue. Liberia has recruited women to serve in both the police and the armed forces.

Female Genital Mutilation

A practice known as **female genital mutilation (FGM)** (formerly called female circumcision) has been documented in 27 countries in the central portion of the African continent plus Egypt (FIGURE 7.29). It also is known in Yemen, India, Indonesia, Iraq, Israel, Malaysia, and the United Arab Emirates (and, in some cases, among recent migrants in the United States, Canada, Australia, and New Zealand). The practice predates Islam and Christianity; today, it is performed among all social classes and in Christian, Muslim, and animist (see the definition on page 419) religious traditions. In the procedure, which is usually performed without anesthesia, a young girl's entire clitoris and parts of her labia are removed. In the most extreme cases (called infibulation), her vulva is stitched nearly shut. This mutilation far exceeds that of male circumcision, eliminating any possibility of sexual stimulation for the female and making urination and menstruation difficult. Intercourse is painful and childbirth is particularly devastating because the flesh scarred by the mutilation is inelastic. A 2006 medical study conducted with the help of 28,000 women in six African countries showed that women who had undergone FGM were 50 percent more likely to die during childbirth, and their babies were at similarly high risk. The practice also leaves women exceptionally susceptible to infection, especially HIV infection.

Traditionally, the practice is viewed as an essential part of a girl's passage into womanhood and probably originated long ago to ensure that a female was a virgin at marriage and that, thereafter, she would be faithful to her husband because she had a low interest in intercourse other than for procreation. Many physicians familiar with FGM have testified that there is no medical benefit to the practice and it instead results in numerous medical complications. Although in decline at present, FGM is still widespread among some groups. Among the Kikuyu of Kenya, nearly all females would have had FGM 40 years ago, but today only about 40 percent do. Kikuyu women's rights leaders have had some success in curbing FGM by replacing it with right-of-passage ceremonies that joyously mark a girl's transition to puberty. Similarly, the Maasai now perform a right of passage that includes dressing in elaborate beaded dresses and participating in ritual bathing instead of FGM.

A growing number of African and world leaders have concluded that the practice is an extreme human rights abuse, and it is now against the law in 16 countries. The most successful eradication campaigns are those that emphasize the threat FGM poses to a woman's health, thus making it socially acceptable to not undergo this ritual. The World Health

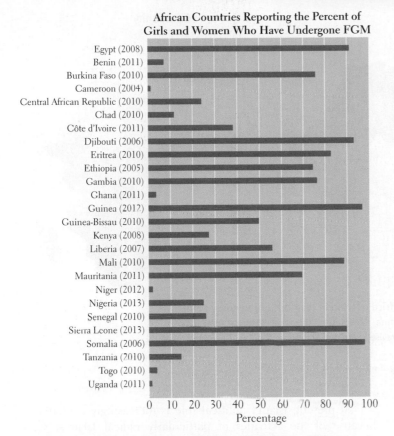

African Countries Reporting the Percent of Girls and Women Who Have Undergone FGM

Egypt (2008)
Benin (2011)
Burkina Faso (2010)
Cameroon (2004)
Central African Republic (2010)
Chad (2010)
Côte d'Ivoire (2011)
Djibouti (2006)
Eritrea (2010)
Ethiopia (2005)
Gambia (2010)
Ghana (2011)
Guinea (2012)
Guinea-Bissau (2010)
Kenya (2008)
Liberia (2007)
Mali (2010)
Mauritania (2011)
Niger (2012)
Nigeria (2013)
Senegal (2010)
Sierra Leone (2013)
Somalia (2006)
Tanzania (2010)
Togo (2010)
Uganda (2011)

0   10   20   30   40   50   60   70   80   90   100
Percentage

FIGURE 7.29 Percent of girls and women living in African countries who have undergone FGM. The World Health Organization estimates that there are 91.5 million girls and women living in Africa today who have undergone some form of FGM. Of those, 12.4 million are girls between 10 and 14 years of age. Close to half of them are in two countries: Egypt (not in sub-Saharan Africa) and Ethiopia. Incidences of FGM have been documented in 27 countries in Africa and in 8 other countries (the above chart only includes countries that provided data). Some cases have also been documented in Europe, North America, Australia, and New Zealand, which are undoubtedly the result of migration rather than local practice. [*Source consulted:* Prevalence of female genital mutilation/cutting (for relevant countries only) at http://genderstats.org/Browse-by-Indicator?ind=50&srid=39.]

Organization (WHO) in 2008 took a strong stand against FGM, saying:

Female genital mutilation has been recognized as discrimination based on sex because it is rooted in gender inequalities and power imbalances between men and women and inhibits women's full and equal enjoyment of their human rights. It is a form of violence against girls and women, with physical and psychological consequences.

[*Source: World Health Organization, Eliminating Female Genital Mutilation—An Interagency Statement (Geneva: World Health Organization Press, 2008), p. 10.]*

Gay, Lesbian, and Transgender Rights

The rights of gay, lesbian, and transgender (LGBT) sub-Saharan Africans are not well protected in any country in this region. Violent homophobia is relatively common and most LGBT people choose to keep their sexual identities hidden. Many countries have criminal sanctions against homosexual behavior or same-sex marriage, and political and religious leaders (some of them evangelical clergy from America active in Kenya and Uganda) play a role in perpetuating the idea that all deviations from heterosexuality are "un-African" and should be punished harshly. Anti-LGBT stances have taken on a geopolitical aspect in that LGBT rights are commonly characterized as a "Western" value that is polluting the non-Western world. Nonetheless, there are a few outspoken supporters of LGBT rights; and South Africa, with its unique experience with anti-apartheid activism, is the one country that has a constitutional ban on discrimination based on sexual orientation.

RELIGION

Africa's rich and complex religious traditions derive from three main sources: traditional or indigenous African belief systems (many of them animist), Islam, and Christianity.

Indigenous Belief Systems

Traditional African religions, found in every part of the continent, may be the most ancient on Earth, since this is where human beings first evolved. **FIGURE 7.30A** highlights the countries in which traditional beliefs remain particularly strong. Traditional beliefs and rituals often are used to bring departed ancestors into contact with living people, who in turn are the connecting links in a timeless spiritual community that stretches into the future. The future is reached only if family members procreate and perpetuate the family heritage through storytelling.

Most traditional African beliefs can be considered **animist** in that spirits, including those of the deceased, are thought to exist everywhere—for example, in trees, streams, hills, and art. In return for respect (expressed through ritual), these spirits offer protection from sickness, accidents, and the ill will of others. African religions tend to be fluid and adaptable to changing circumstances. For example, in West Africa, Osun, the god of water traditionally credited with healing powers, is now also invoked for those suffering economic woes.

Religious beliefs in Africa, as elsewhere, continually evolve as new influences are encountered. If Africans convert to Islam or Christianity, they commonly retain aspects of their indigenous religious heritage. The three maps in Figure 7.30 show a spatial overlap of belief systems, but they do not convey the philosophical blending of two or more faiths, which is widespread. In the Americas, the African diaspora has influenced the creation of new belief systems developed from the fusion of Roman Catholicism and African beliefs. Voodoo in Haiti, Santería in Cuba, and Candomblé in Brazil (see Chapter 3) are examples of this fusion.

Islam and Christianity

Islam began to extend south of the Sahara in the first centuries after Muhammad's death in 632 C.E., and today, according to a recent Pew Research Center study (2010), about 15 percent of sub-Saharan Africans are Muslim (see Figure 7.30B). Islam is now the predominant religion in the north,

animism a belief system in which spirits, including those of the deceased, are thought to exist everywhere and to offer protection to those who pay their respects

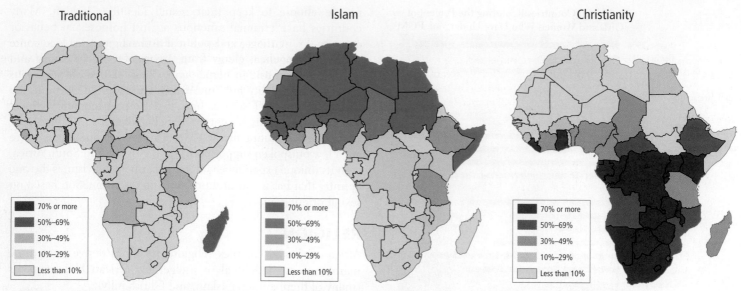

FIGURE 7.30 Religions in Africa. Notice that the various religions in Africa overlap in distribution; in many countries, one is dominant but others are present. [*Sources consulted:* Matthew White, "Religion in Africa," in *Historical Atlas of the Twentieth Century,* October 1998, http://users/erols.com/mwhite28/aforelg.htm; revised with new data from the CIA, *The World Factbook,* 2009, https://www.cia.gov/cia/publications/factbook/index.html.]

throughout the Sahel, and parts of East Africa, where Muslim traders from North Africa and Southwest Asia brought the religion. Powerful Islamic empires have arisen here since the ninth century. The latest of these empires challenged European domination of the region in the late nineteenth century.

Christianity first came to the region through Egypt and Ethiopia shortly after the time of Jesus Christ, well before it spread throughout Europe (see Figure 7.30C). Christianity did not come to the rest of Africa until nineteenth-century missionaries from Europe and North America became active along the west coast. Many Christian missionaries provided the education and health services that colonial administrators had neglected. In the twentieth century, old-line established churches began to gain adherents in Africa. The Anglican Church (Church of England) grew so rapidly that by 2000, Anglicans in Kenya, Uganda, and Nigeria outnumbered those in the United Kingdom. The Anglican Church in Africa attracts the educated urban middle class, and because of their strength in numbers, African Anglicans are able to influence central policy decisions in the wider church, often expressing and enforcing a more conservative point of view than that held by Europeans or American Anglicans. Sub-Saharan Africa is now home to one-fifth of the Christians in the world, and these Christians constitute 57 percent of the sub-Saharan population.

Today, both Muslims and Christians express strong support for democracy and religious freedom, but on the other hand, there is also strong support for government based on the Bible or shari'a law. The very conservative perspectives on LGBT rights discussed above on page 419 are held by both Christians and Muslims. Christians tend to be somewhat more supportive of increasing women's rights than are Muslims.

At present, Islam's political role is increasingly contentious because of the activities of particularly radical Islamist sects (for example, Boko Haram in the West and Al Shabaab in the East; see the discussion on page 406) that have challenged governments in Nigeria, Niger, Mali, South Sudan, Ethiopia, Somalia, Uganda, and Kenya. The appeal, origins, and financing of these radical movements remain mysterious. The media company Al Jazeera investigated Islamist activity in Uganda and Kenya, and discovered a murky connection to the Tabliq sect that gained prominence during the era of the one-time exceedingly violent Muslim dictator in Uganda, Idi Amin (1971–1979), who, when overthrown, fled to Saudi Arabia.

Across the continent, the variety of African religious traditions are celebrated with festivals. As is the case with festivals everywhere, such events offer an opportunity to represent ideal versions of culture, a time to revel in joy and renewed religious dedication, and a chance to make important social contacts, reinforce identity, and perform community service (**FIGURE 7.31**).

Evangelical Christianity and the Gospel of Success In contrast to the Anglican Church, modern evangelical versions of Christianity appeal to the less-educated, more recent urban migrants—the most rapidly growing segment of the population. These independent, evangelical sects interpret Christianity as being aligned with traditional beliefs in the importance they attach to: the need to make sacrificial gifts to spiritual leaders and ancestors, the power of miracles, and the idea that devotion brings wealth. The result is one of the world's fastest-growing Christian movements.

VIGNETTE Preachers at the Miracle Center in Kinshasa describe the gospel of success forcefully: "The Bible says that God will

FIGURE 7.31 LOCAL LIVES: Festivals in Sub-Saharan Africa

A During Timkat, a festival that marks the baptism of Jesus Christ, Ethiopian Orthodox Christian priests in Addis Ababa carry replicas of the Ten Commandments, wrapped in cloth, on their heads.

B South African Muslims spot the new moon that marks the end of Ramadan, a month of fasting during which observant Muslims do not eat, drink, quarrel, or have sex between sunrise and sunset. Throughout Africa, many Muslims mark the end of Ramadan with prayers for peace and unity as well as with drumming, singing, dancing, and general celebration.

C The 8-day festival of the Umhangla (the "Reed Dance" ceremony) protects the virginity of young unmarried women, who gather in the king of Swaziland's village and repair the surrounding fence with tall reeds. Later, in costumes, the women dance for the royal family, foreign dignitaries, tourists, and other spectators. Traditionally, the king also chooses a new wife. The women who wear red feathers in their hair are the daughters of the king.

materially aid those who give to Him. We are not only a church, we are an enterprise. In our traditional culture you have to make a sacrifice to powerful forces if you want to get results. It is the same here."

Generous gifts to churches are promoted as a way to bring divine intervention to alleviate miseries, whether physical or spiritual. Practitioners donate food, television sets, clothing, and money. One woman gave 3 months' salary in the hope that God would find her a new husband.

Like all religious belief systems, the gospel of success is best understood within its cultural context. Many of the believers, new to the city, feel isolated and are looking for a supportive community to replace the one they left behind. In return for material contributions to the church and volunteer services, members receive social acceptance, companionship, and community assistance in times of need. ∎

ETHNICITY AND LANGUAGE

Ethnicity, as we have seen throughout this book, refers to the shared language, cultural traditions, and political and economic institutions of a group. The map of Africa in **FIGURE 7.32** shows a rich and complex mosaic of languages. Yet despite its complexity, this map does not adequately depict Africa's cultural diversity (compare it with Figure 7.20 of Nigeria).

Most ethnic groups have a core territory in which they have traditionally lived, but very rarely do groups occupy discrete and exclusive spaces. Often, several groups share a space, practicing different but complementary ways of life and using different resources. For example, one ethnic group might be subsistence cultivators, another might herd animals on adjacent grasslands, and a third might be craft specialists working as weavers or metalsmiths.

People may also be very similar culturally and occupy overlapping spaces, but identify themselves as being from different ethnic groups. For example, Hutu farmers and Tutsi cattle raisers in Rwanda share similar occupations, languages, and ways of life. However, European colonial policies purposely exaggerated ethnic differences as a method of control, assigning a higher status to the Tutsi. Hutu and Tutsi now think of themselves as having very different ethnicities and abilities. In the 1990s and again in 2004, the Hutu-run government encouraged attacks on the minority Tutsi people. Hutu were also killed, but those killed tended to be people who opposed the genocide. The best estimates indicate that more than 1 million people, most of them Tutsi, died during the two genocide catastrophes.

Some African countries have only a few ethnic groups; others have hundreds. Cameroon, sometimes referred to as a microcosm of Africa because of its ethnic complexity, has 250 different ethnic groups. Different groups often have extremely different values and practices, making the development of cohesive national policies difficult. Nonetheless, the vast majority of African ethnic groups have peaceful and supportive relationships with one another.

To a large extent, language correlates with ethnicity. More than a thousand languages, falling into more than a hundred language groups, are spoken in Africa; Bantu is the largest such group (see Figure 7.32). Most Africans speak their native tongue and a **lingua franca** (language of trade). Some languages are spoken

lingua franca a language of trade

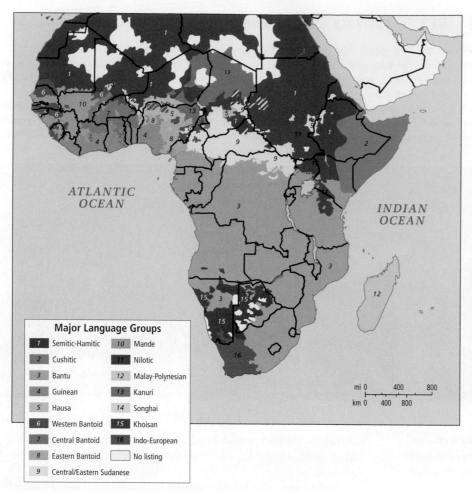

Major Language Groups

1	Semitic-Hamitic	10	Mande
2	Cushitic	11	Nilotic
3	Bantu	12	Malay-Polynesian
4	Guinean	13	Kanuri
5	Hausa	14	Songhai
6	Western Bantoid	15	Khoisan
7	Central Bantoid	16	Indo-European
8	Eastern Bantoid		No listing
9	Central/Eastern Sudanese		

FIGURE 7.32 Major language groups of Africa. Notice the variety of languages in such countries as Nigeria and Kenya; also note how country borders often split language groups into several parts. [*Sources consulted:* Edward F. Bergman and William H. Renwick, *Introduction to Geography—People, Places, and Environment* (Englewood Cliffs, NJ: Prentice-Hall, 1999), p. 256; Jost Gippert, TITUS Didactica, http://titus.uni-frankfurt.de/didact/karten/afr/afrikam.htm.]

by only a few dozen people, while other languages, such as Hausa, are spoken by millions of people from Côte d'Ivoire to Cameroon. Lingua francas—such as Hausa, Arabic, and Swahili—are taking over because they better suit people's needs or have become politically dominant; some African languages are dying out. Former colonial languages such as English, French, and Portuguese (all classed as Indo-European languages in Figure 7.32) are also widely used in commerce, politics, education, and on the Internet.

THINGS TO REMEMBER

• Gender relationships in sub-Saharan Africa are complex and variable, but in this region as in others, women are generally subordinate and are often physically abused. Nevertheless, change is under way.

• Female genital mutilation has terrible consequences for women's health. It has been recognized as sexual discrimination because it is rooted in gender inequalities and power imbalances between men and women.

• Traditional African religions are among the most ancient on Earth and may be found in every part of the continent. Today, however, about one-third of sub-Saharan Africans are Muslim and about half are Christian. Many in the evangelical movement, a subset of Christianity, promote the gospel of success as a way of life.

• Sub-Saharan Africa is a mosaic of more than 1000 languages that are roughly similar to the mosaic of ethnic groups. Hausa, Arabic, and Swahili are now the most commonly used languages of trade.

SUBREGIONS OF SUB-SAHARAN AFRICA

Sub-Saharan Africa can be divided into four subregions that primarily reflect geographic location (rather than other affinities): West Africa, Central Africa, East Africa, and Southern Africa. Factors such as traditions and the current geopolitical situation have also guided the placement of a particular country in a particular subregion.

WEST AFRICA

West Africa, which occupies the bulge on the west side of the African continent, includes 15 countries (**FIGURE 7.33**). West Africa is physically framed on the north by the Sahara, on the west and south by the Atlantic Ocean, and on the east by Lake Chad and the mountains of Cameroon.

Horizontal Zones of Physical and Cultural Difference

West Africa can be thought of as a series of horizontal physical zones, from dry in the north to moist in the south. The north–south division also applies to economic activities linked to the environment—herding in the north, farming in the south—and, to some extent, to religions and cultures—Muslim in the north, Christian in the south. Most of these cultural and physical features do not have distinct boundaries, but rather zones of transition and exchange.

Stretching across West Africa from west to east are horizontal vegetation zones that reflect the horizontal climate zones shown in Figure 7.5. Small remnants of tropical rain forests still exist along the Atlantic coasts of Côte d'Ivoire, Ghana, and Nigeria, although most of this once-great coastal forest has fallen victim to logging, intensifying settlement, and agriculture. To the north, this moist environment changes into drier woodland mixed with savanna. Farther north still, as the environment becomes drier, the trees thin out and the savanna dominates. Only where annual cycles create wetlands, as in the Niger River basin, is moisture sufficient to foster fishing and cultivation as well as grazing. Still further to the north, the savanna blends into the yet drier Sahel (Arabic for "shore" of the desert) region of arid grasslands. In the northern parts of Mali, Niger, and Mauritania, the arid land becomes actual desert, part of the Sahara.

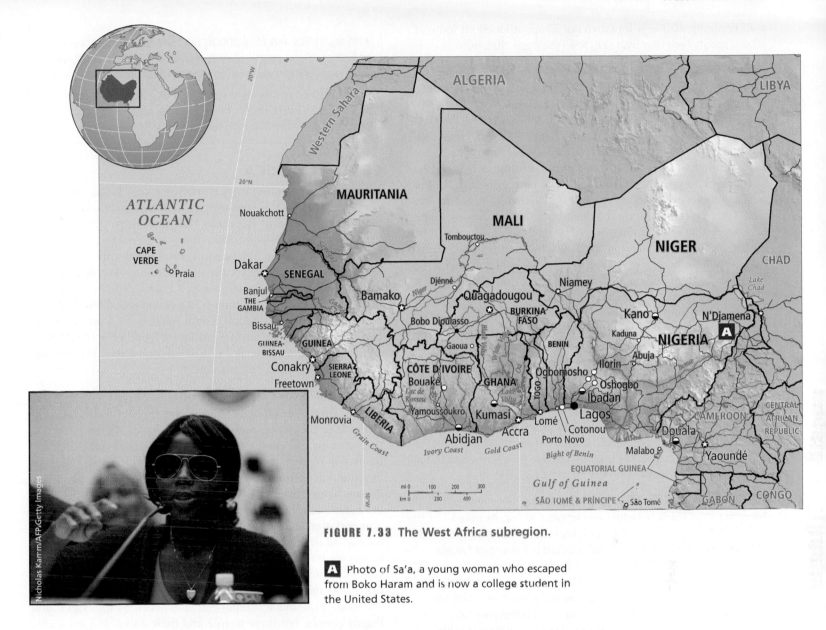

FIGURE 7.33 The West Africa subregion.

A Photo of Sa'a, a young woman who escaped from Boko Haram and is now a college student in the United States.

Along the coast, and stretching inland, abundant rainfall supports crops such as coffee, cacao, yams, palm oil, corn, bananas, sugarcane, and cassava. Farther north and inland, millet, peanuts, cotton, and sesame grow in the drier savanna environments. Cattle are also tended in the savanna and north into the Sahel; farther south, however, the threat of sleeping sickness spread by tsetse flies limits cattle raising. In the Sahel and Sahara, cultivation is possible only at oases and along the Niger River during the summer rains (see page 385). Food, other than that provided by animals and the fields along the riverbanks, must come from trade with the coastal zones and with the global market.

West African environments have dried out as more and more forests and woodlands are cleared by international loggers or by local people for cultivation, grazing, and fuelwood. When the forest cover is gone, the land is more vulnerable to erosion, and sunlight evaporates moisture. Desiccating desert winds now blow south all the way to the coast, and this drying out of the land is exacerbated by the recurrence of natural cycles of drought. The ribbons of differing vegetation running west to east are not as distinct as they once were, and even coastal zones occasionally experience dry, dusty conditions reminiscent of the Sahel.

As mentioned, the geographies of religion and culture in West Africa to some extent reflect the north-south gradation of physical patterns. Generally, the north is Muslim, with Arab and North African influences that are apparent (see Figures 7.30, 7.31, and 7.32). Southern West Africa is populated by people who practice a mix of Christian and traditional (animist) religions, and whose background and culture are Central African. Some large ethnic groups once occupied distinct zones, but as a result of migration linked to colonial and postcolonial influences, members of specific groups are distributed across many different areas and in a number of West African countries. There are hundreds of smaller ethnic groups, each with its own language; the Figure 7.32 map that depicts major language groups in Africa shows the cultural complexity of the entire West Africa subregion. The example of Nigeria, discussed earlier in the chapter (see Figure 7.20), illustrates the potential difficulties faced by a nation attempting to unite a multitude of ethnic groups while dealing with contentions over a lucrative but ill-apportioned oil industry whose profits benefit only a very few (see pages 404–406).

Regional Conflict in Coastal West Africa

Civil conflict in Nigeria and problems caused by destructive resource extraction were discussed on pages 404–406. The 1990s and early 2000s saw continuing bloody conflict in the far West African countries of Guinea, Sierra Leone, Liberia, and Côte d'Ivoire. The wars in these four countries were intertwined, but each conflict was fed by local disputes that ultimately could be traced back to poverty and horrendously bad leadership. In the 1980s, Liberia was a country with a reasonably bright future. That future was lost during the dictatorial rule of Charles Taylor (discussed above) beginning in 1989. Over the years until 2003, Taylor and his enslaved child soldiers, armed with AK-47's, commandeered diamond mines and tropical timbers and murdered hundreds of thousands of civilians. In 2002, the conflict spilled into Côte d'Ivoire, formerly one of the most stable West African countries, but one nonetheless troubled by regional disparities. Abidjan, the former capital on the southern coast, had fine roads, skyscrapers, French restaurants, and a modern seaport, but the northern part of the country was much poorer. In the north of Côte d'Ivoire (still in the tropical wet zone) are cacao plantations that employ immigrant agricultural workers at very low wages and, in some cases, use enslaved children from nearby Burkina Faso and southern Mali. In 2003, UN peacekeepers were belatedly sent to Liberia and Taylor was turned over to authorities, to be tried by a war crimes tribunal in The Hague, Netherlands, for crimes against humanity (see the vignette on page 381).

ON THE BRIGHT SIDE: Efforts at Reform

There are now many hopeful signs in West Africa. As noted earlier in this chapter, in 2006, a highly respected, reform-minded woman, Ellen Johnson-Sirleaf, was elected president of Liberia. Under Johnson-Sirleaf's leadership, Liberia is beginning to rebuild the country physically and socially. Meanwhile, human rights groups and chocolate lovers in Europe and elsewhere (Africans consume very little chocolate or cacao products) successfully pressured international chocolate producers to address a wide range of issues, including the use of enslaved children and adults as laborers on cacao plantations and unsustainable cacao production practices. In May 2006, the first African international cacao summit in Abuja, Nigeria, began the process of creating *value chains* (see page 402) for the benefit of African cacao growers—who in the recent past have produced about 80 percent of the cacao for the U.S.$75 billion annual global chocolate market. Note that in the very recent past (2015), European and American chocolate companies have been sued for continuing to deal with producers who use enslaved children as their labor force. ■

THINGS TO REMEMBER

- West Africa can be thought of as a series of horizontal physical zones, from dry in the north to moist in the south. The north–south division also applies to economic activities—herding in the north, farming in the south—and, to some extent, to religions and cultures—Muslim in the north, Christian in the south.

- In the 1990s and early 2000s, the far West African countries of Guinea, Sierra Leone, Liberia, and Côte d'Ivoire were involved in bloody conflicts traceable to poverty and corruption.

- The election of Ellen Johnson-Sirleaf as president of Liberia and the coalition of chocolate consumers with African cacao producers have been among recent positive developments.

CENTRAL AFRICA

Looking at a map of Africa with no prior knowledge, you might reasonably expect the countries of Central Africa—Cameroon, the Central African Republic, Chad, Congo (Brazzaville), Equatorial Guinea, Gabon, São Tomé and Príncipe, and Congo (Kinshasa)—to be the hub of continental activity, the true center around which life on the continent revolves (see **FIGURE 7.34** map; see also Figure 7.1). Paradoxically, Central Africa plays only a peripheral role on the continent. Its dense tropical forests and its tropical diseases and difficult terrain make it the least accessible part of the continent, and recent armed conflict over resources and political power has left it in a deteriorating state that discourages visitors and potential investors.

Wet Forests and Dry Savannas

Central Africa consists of a core of wet tropical forestlands surrounded on the north, south, and east by bands of drier forest and then savanna (see the Figure 7.5 map). The core is the drainage basin of the Congo River, which flows through this area and is fed by a number of tributaries. Although rich in resources, the forests form a barrier that has isolated Central Africa from the rest of the continent for centuries and has impeded the development of transportation, commercial agriculture, and, until recently, even mining. Moreover, as is true in all tropical zones, the soils of these forests are not suitable for long-term cash-crop agriculture. Most Central Africans are rural subsistence farmers who live along the rivers and the occasional railroad line. The forests themselves are home to nomadic hunter-gatherers, such as the Mbuti and Pygmy people, but these people and their way of life are highly endangered as the scramble for forest and mining resources intensifies (see Figure 7.34A). Although the sixteenth-century city of Loango on the Congo coast was a model of urban development for its time, during the colonial period and until recently, cities in Central Africa have been small in size, with little urban infrastructure. Now, despite crude urban conditions, cities are growing rapidly, and too often, the newcomers are refugees escaping civil violence in the countryside. The new arrivals find a serious lack of housing, clean water, adequate education systems, and urban employment in the formal sector.

A Colonial Heritage of Violence

The region's colonial heritage has contributed to the difficulties described above. In what is now known as the Democratic Republic of the Congo (Kinshasa), for example, the Belgian colonists left a terrible legacy of brutality. In 1885, Belgium's king, Leopold II, with a personal fortune to invest, obtained international approval for his own personal colony, the Congo Free State, with the promise that he would end slavery, protect the native people, and guarantee

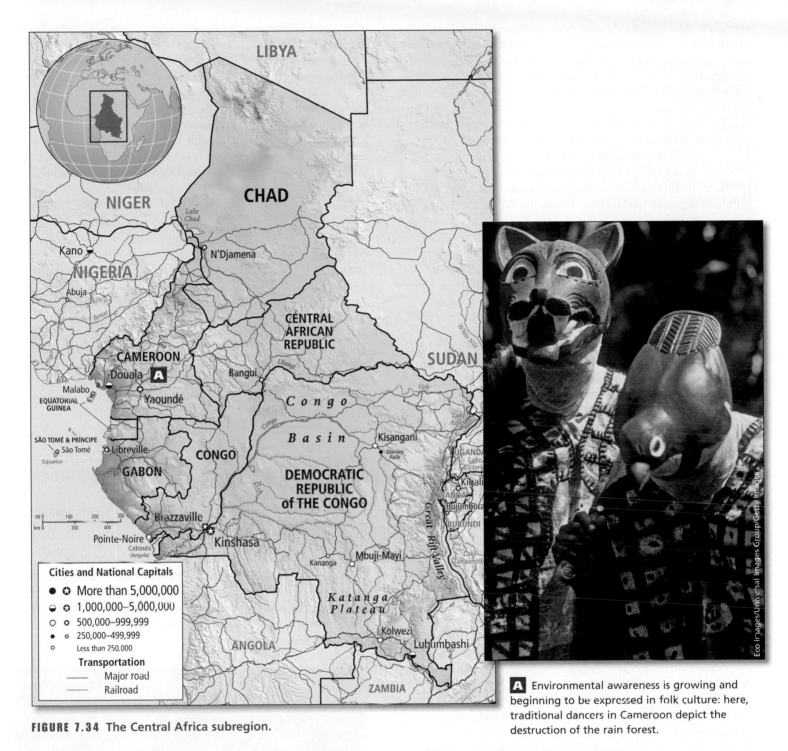

FIGURE 7.34 The Central Africa subregion.

A Environmental awareness is growing and beginning to be expressed in folk culture: here, traditional dancers in Cameroon depict the destruction of the rain forest.

free trade. He did none of these. The Congo Free State consisted of 1 million square miles (2.5 million square kilometers) of tropical forestland rich in wildlife: elephants (who were killed for their ivory), rubber, timber, and copper. King Leopold soon appropriated the labor of the population of 10 million people to extract Congo's resources for his own profit.

The young writer Joseph Conrad witnessed the cruel behavior of Belgian colonial officials when he obtained a job on a steamer headed up the Congo River in 1890. He saw a campaign of terror that featured enslavement, whipping, beheadings, and lopping off the hands of those who didn't meet their production quotas for rubber and other commodities. Conrad's novella *Heart of Darkness* (published in 1902), and a pamphlet by Mark Twain, "King Leopold's Soliloquy" (published in 1905), awakened Europeans and Americans to Leopold's atrocities, and efforts began to end his regime.

Postcolonial Exploitation

When the countries of Central Africa were granted their independence in the 1960s, the Europeans stayed on to continue their economic domination. Resources were stripped and the tax-free profits sent home to France, Belgium, or Portugal, rather than being reinvested in Central Africa. When the European colonizers departed, leaving Africans with little money to develop an infrastructure (including such things as safe water distribution systems), no experience with self-government, and no educated constituency, many Central African countries, such as Congo (Kinshasa), fell into the hands of corrupt leaders. Political upheavals resulted when some ethnic groups were excluded from representation in government and economies went into a downward spiral. As in other parts of sub-Saharan Africa, the average person's purchasing power has decreased over the last few decades. Tax concessions to foreigners, as well as government corruption, have left little revenue to maintain infrastructure. Roads and electricity grids are negligible. Few industries, either government or private, can survive without such facilities. People displaced by civil disruption are so poor that there is little market for any type of consumer product, so even the informal economy is weak.

CASE STUDY: Devolution in Congo (Kinshasa) and Signs of Hope

As Belgian colonial officials withdrew in 1960, they left untrained Congolese in charge of the newly independent government. The new head of government, Patrice Lumumba, tried to correct the inequalities of the colonial era by nationalizing foreign-owned companies in order to keep their profits in the country. But, Lumumba's strategies, adopted during the height of the Cold War, along with the support he received from the Soviet Union, raised the red flag of Communism. The Western allies arranged for Lumumba to be killed. They then supported and armed a politician who opposed him: the notorious and sadistic Mobutu Sese Seko.

Mobutu Sese Seko seized power in 1965 and then he also nationalized the country's institutions, businesses, and industries, but not for the benefit of the Congolese people. Rather, he extracted personal wealth for himself and his supporters from the newly nationalized ventures. As a result, trained technicians and capital left the country, and Congo's economy, physical infrastructure, educational system, food supply, and social structure rapidly declined. By the 1990s, the vast majority of the Congolese people were utterly impoverished. Laurent Kabila, a new dictator, overthrew Mobutu in 1997. Neighbors to the east (Uganda and Rwanda) who coveted Congo's mineral wealth supported Kabila, whose inept regime quickly led to widening strife in Central Africa. By 1999, Angola, Namibia, Zimbabwe, Chad, Rwanda, and Uganda all had troops in Congo and were aggressively competing for parts of its territory and resources. More than 4 million Congolese died as a result of the conflict, and 1.6 million were displaced.

Despite the presence of over 20,000 UN peacekeeping forces, the largest such UN mission in the world, the chaos continued. In 2006, the UN Security Council officially recognized that the ongoing conflicts in the Great Rift Valley region of Africa (which includes the zone where Congo, Sudan, Uganda, Rwanda, and Burundi come together) were all linked to the illegal exploitation and international trade of natural resources, and to the proliferation of arms trafficking. The Security Council urged all countries to disarm and demobilize their militias and armed groups, especially northern Uganda's Lord's Resistance Army, and to promote the lawful and sustainable use of natural resources. In 2006 and again in 2011, Joseph Kabila won the presidency of Congo in the first multiparty elections; he is scheduled to retire in 2016.

By 2016, despite UN efforts, conditions have only deteriorated. The causes of the continuing civil violence (rape, kidnapping, and massacres) remain linked to competition for land, power, and resource wealth, especially along the Great Rift Valley border. Meanwhile, the Chinese government and the American mining firm Freeport-McMoRan are vying to exploit at great profit Congo's considerable mineral resources.

The creativity of returned migrants may break the cycle of devolution in Central Africa. An estimated 30 percent of Central Africans have cell phones and use them daily for such activities as sending money to merchants, talking with relatives in remote regions, accessing medical advice, and conducting routine banking (see Figure 7.17). Cell phones could become a major engine of development, leading to better Internet connectivity, which up to now has been poor—just 16 percent of Central Africans have access to the Internet. *[Sources consulted:* http://www.economist.com/news/middle-east-and-africa/21679750-war-weary-citizens-are-scared-joseph-kabila-may-not-retire-gracefully-will; http://www.huffingtonpost.com/david-tereshchuk/a-giant-leap-in-2016-africa_b_8901556.html; for detailed source information, see Text Credit pages.] ∎

VIGNETTE At a shabby bar in Kinshasa, Sam sings a jazzy solo in the muggy evening air to a sharply dressed crowd (FIGURE 7.35). Dressing stylishly has become a bit of an obsession among Kinshasa's young professionals. The recent spate of peace in Congo has sparked a resurgence in Kinshasa, where luxury apartments now attract talented local entrepreneurs and returning migrants fresh from Paris seeking to support their country, or perhaps make a small fortune in a hoped-for boom-time. Trendy restaurants serve plentiful river shrimp and musicians enliven the nightlife. Money flows at least for the moment. Congo's economic growth is believed to have amounted to 9 percent in 2014.

FIGURE 7.35 Night life in Kinshasa.

But Sam, like most Kinois, lives outside this bubble of prosperity in a crowded slum without clean water or electricity. Almost 90 percent are underemployed. This great divide of wealth leaves Congo politically unstable and hinders continued economic growth. The main sign of progress in the slum is that everyone has a cell phone. That coupled with the return of educated affluent migrants who are tech savvy could mean some hope for an alternative future.
[Source: "Will Kabila Go?," The Economist, December 9, 2015.] ∎

THINGS TO REMEMBER

• Dense tropical forests, tropical diseases, and difficult terrain make Central Africa the least accessible part of the continent. The Central African subregion has an enormous wealth of natural resources: copper, gold, diamonds, oil, and rare tropical plants and animals. Illegal trade in natural resources, ineffective and corrupt leadership, and the proliferation of arms trafficking are seriously impeding progress.

• The roots of conflict in Central Africa date back to the colonial era, Cold War geopolitics, and competition for land and resources in the Great Rift Valley. Armed, outside intervention is a recurring problem, especially in Congo (Kinshasa).

• Finding ways to sustainably exploit Central Africa's resources for the benefit of Central African people is difficult.

• The shift to democratic institutions in Congo (Kinshasa) is not going well. Elections are now more frequent, but corruption is rampant and citizen participation in democratic institutions weak.

• Change could come with the return of migrants and with increasing access to technology, which helps people earn better incomes and gain access to crucial information.

EAST AFRICA

The subregion of East Africa (**FIGURE 7.36**) occupies the Horn of Africa, which reaches along the southern shore of the Red Sea and the Gulf of Aden, and includes the countries of Djibouti, Eritrea, Ethiopia, Somalia, and the newly independent country of South Sudan. To the south of the Horn, the subregion also includes the coastal countries of Kenya and Tanzania; the interior highland countries of Uganda, Rwanda, and Burundi in the Great Rift Valley; and the islands of Madagascar, Comoros, Seychelles, Mauritius, and Réunion in the Indian Ocean across the Mozambique Channel.

Across the subregion, the relatively dry coastal zones contrast with the moister interior uplands; further to the west, the drying effects of the Sahara affect South Sudan, Uganda, and western Tanzania. A vital and productive farming economy has long existed in the interior highlands; across the region today, nearly 80 percent of the population makes a living from farming. A service economy is beginning to grow in urban areas and tourism zones along the coast, in Madagascar, and around the national parks of Kenya, Uganda, and Tanzania (see Figure 7.36B). Industry is only beginning to be developed, and is found primarily in Kenya and Tanzania. On the coast, where port cities have a thousand-year history of trade across the Arabian Sea and Indian Ocean, trade-related services are central to economies.

Conflict in the Horn of Africa

Africa's ability to support its people is being tested in East Africa, where the growing population and related increase in consumption are straining the region's carrying capacity. The entire Horn of Africa has suffered periodic famines since the mid-1980s, caused by governments and political officials who have manipulated innumerable situations for personal gain, resulting in ineffective responses to naturally occurring droughts. The result in Somalia, Ethiopia, and now South Sudan, all often referred to as failing states, has been an environment of chaos and lawlessness for ordinary citizens.

Newly Independent South Sudan South Sudan occupies what was once the southern third of the oil-rich country of Sudan (covered in Chapter 6). The south of Sudan, which has a population mostly made up of darker-skinned Christians who have sub-Saharan origins, has always been culturally quite different from the more Arabic north. South Sudan was often handled roughly by the north and dismissed as backward and out of touch, although the south is where two-thirds of Sudan's oil is located. After a long and bloody struggle over the oil, South Sudan won its independence in 2011, but the border between the two countries remained in contention; South Sudan is landlocked and thus has no easy way to get its oil to the world market. The possession of the oil (and the accompanying right to use the oil) also remains in contention. Private oil companies based in China and Malaysia have brokered a settlement with the intention of acquiring the oil and transporting it to a Red Sea port in pipelines through Sudan.

The Role of Militant Islamists in the Horn Militant Islamists—most prominently, Al Shabaab—have attempted to impose order and conformity in the Horn of Africa, but their rigid rules, especially those regarding women, have been oppressive and met with resistance. Furthermore, persistent violence in Somalia and Kenya in 2015 indicates that Al Shabaab is a rising nexus of Islamist terrorism. The involvement of the United States began in the 1990s with an effort to quell the conflict by siding with corrupt warlords against the militants. This resulted in the debacle in Mogadishu, Somalia, depicted in the 2001 film *Black Hawk Down*. The United States withdrew, and conflict has repeatedly flared. The timeline of Al Shabaab violence for 2015 shows multiple deadly attacks throughout the year, most notably the killing of 142 university students in Garissa University in Kenya in April. Inactive for a few months, Al Shabaab then began a campaign of resurgence in January 2016 with an attack on a Mogadishu hotel where 40 people were killed and another in February 2016 that slaughtered 14.

East African Piracy and Its Root Causes In 2008, the world became aware of the dangers to global shipping posed by a band of pirates from Somalia. During that year, 40 commercial ships from multiple nations (Denmark, Japan, Ukraine, the United States, North Korea, Taiwan, India, China, and others) were attacked or seized. When the privately owned American ship *Alabama* was boarded in April 2009 and the captain taken prisoner, a serious discussion began on the root causes of the piracy phenomenon and how to stop it.

FIGURE 7.36 The East Africa subregion.

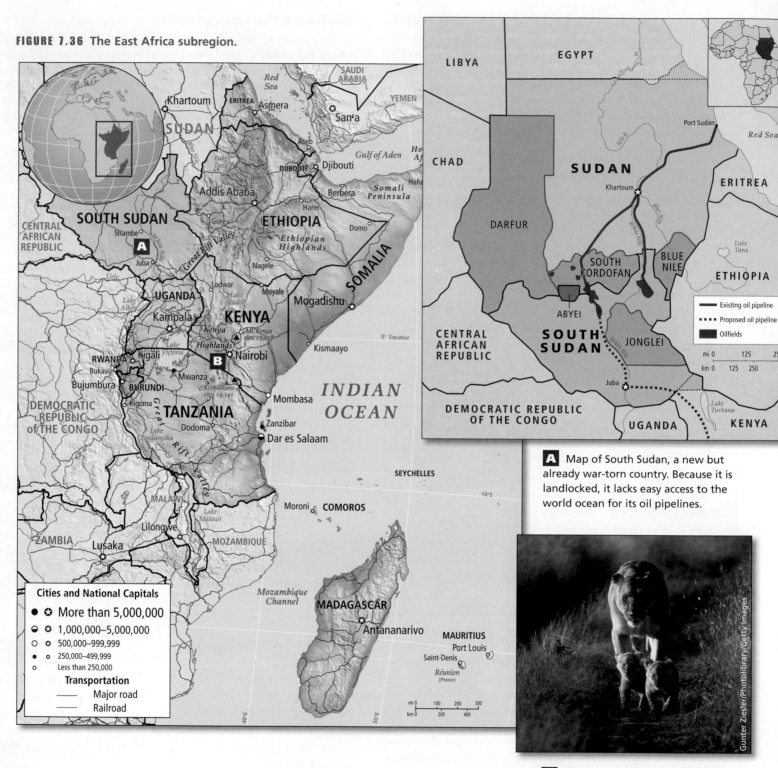

A Map of South Sudan, a new but already war-torn country. Because it is landlocked, it lacks easy access to the world ocean for its oil pipelines.

Existing oil pipeline
Proposed oil pipeline
Oilfields

Cities and National Capitals
- ● ✪ More than 5,000,000
- ◒ ✪ 1,000,000–5,000,000
- ○ ✪ 500,000–999,999
- ● ○ 250,000–499,999
- ○ Less than 250,000

Transportation
—— Major road
—— Railroad

B Lions in Kenya's Maasai Mara game reserve are increasingly being targeted by poachers for their skins.

Gunter Ziesler/Photolibrary/Getty images

In the article "The Two Piracies in Somalia: Why the World Ignores the Other" (2009), Mohamed Abshir Waldo of Mombasa, Kenya, explains that the long conflict in Somalia, which forced many farmers and herders into refugee camps and some into fishing for a living, was a big factor in the turn to piracy. But a more proximate cause was "fishing encroachment" by large commercial fishing fleets (from Italy, France, Spain, Greece, Russia, the Arabian Peninsula, and the Far East), whereby traditional small-boat Somali fishers were displaced and harassed. Coastal villagers in Somalia also accused the foreign fishing fleets of purposefully polluting the seabeds and of violently seizing their boats. Western warships, providing security to the large commercial fishing fleets, further harassed Somali fishers. Eventually, the fishers joined with unidentified criminal groups in East Africa to create this new "entrepreneurial" solution of high-seas piracy. An interactive BBC website, available at http://www.bbc.com/news/world-africa-33822635, provides a map and helpful background information on piracy in East Africa, which, after a lull due to strict enforcement, was by 2015 once again a threat. Plundering of Somali resources by foreign fishing trawlers continued and local fishers were once again desperate.

Interior Valleys and Uplands

The Great Rift Valley, which contains numerous archaeological sites that have shed light on the early evolution of humans and other large mammals, is today covered primarily with temperate and tropical savanna, although patches of forest remain on the high slopes and in southwestern Tanzania. In Ethiopia, deep river valleys, including the one holding the headwaters of the Blue Nile, drain the semiarid highlands. Rain falls during the summer, and conditions vary with elevation. For most rural people who make a living by herding cattle, finding water is always a concern, but in the moister central uplands of Ethiopia and Kenya and in the woodlands of southern Tanzania, herding and cultivation are combined in mixed farming. Animal manure is used to fertilize a wide variety of crops that are closely adapted to particularly fragile environments, and pastures and fields are rotated systematically. These ancient and specialized mixed-agriculture systems were ignored and disdained by European colonizers, and agronomists have only recently begun to appreciate the precision of African upland mixed agriculture, which is now threatened by civil disorder, development schemes, and population pressure.

Coastal Lowlands

To the east and south of the Ethiopian highlands is a broad, arid apron that descends to the Gulf of Aden and to the Indian Ocean. Along the coasts of the Horn of Africa are large stretches of sand and salt deserts, a few rivers, and a handful of port cities. In Somalia, which occupies the low, arid coastal territory in the Horn, people eke out a meager living through fishing, cultivation around oases, and herding in the dry grasslands. Somalia's 8 million people, virtually all Muslim and ethnically Somali, are aligned in six principal clans. Islamic fundamentalists, clan-based warlords, and political adventurers have engaged in a violent rivalry for power over the last several decades. The resulting turmoil has driven many cultivators from their lands and blocked access to some lands that are crucial to the food production system, so the land that is still available is being overused. Thus, it is primarily politics, not rapid population growth, that has increased pressure on arable land in all countries of the Horn of Africa. Nonetheless, at present rates of growth, populations in the Horn are expected to double and even triple by 2050, and conditions will make it difficult to accommodate the increase.

East Africa's shores from southern Somalia southward (known regionally as the Swahili Coast) have long attracted traders from around the Indian Ocean—from China, Indonesia, Malaysia, India, Persia (Iran), Arabia, and elsewhere in Africa. The result has been a grand blending of peoples, cultures, languages, plants, and animals. In the sixth century, Arab traders brought Islam to East Africa, where it remains an important influence in the north and along the coast. Traders, speaking the lingua franca of Swahili, established networks that linked the coast with the interior. These networks penetrated deeply into the farming and herding areas of the savanna and into the semifeudal cultures of the highland lake country. Today, trade across the Indian Ocean continues. Arab countries are East Africa's most important trading partners, but Asia, China, India, and Japan are increasingly important to East African trade; trade with Indonesia and Malaysia is also increasing. East Africa, which needs more industrialization, exports mostly agricultural products (live animals, hides, bananas, coffee, cashews, processed food, tobacco, and cotton) and imports machinery, transportation and communication equipment, industrial raw materials, and consumer goods.

Contested Space in Kenya and Tanzania

Kenya and Tanzania are modestly industrialized countries in East Africa that remain largely dependent on agricultural exports, like cacao, and some finished products: tea, coffee, beer, horticultural products, and processed food, as well as cement and petroleum products. Both countries are regional hubs of trade, yet both are among the poorest countries in the world (see Figure 7.27A).

Beginning in ancient times, traders dealt in slavery, ivory, and rare animals, items that can no longer be legally traded. British colonists took over in the early nineteenth century when the old trading economy started to fail. The British relocated native highland farmers in Kenya, such as the Kikuyu, to make way for coffee plantations; in the savannas of Kenya and Tanzania, they displaced pastoralists, such as the Masai. Some of the lands of the Masai became game reserves designed to protect the savanna wildlife for the use of European hunters and tourists.

National Parks as Protectors of Africa's Wild Heritage

Today, the former colonial game reserves of Kenya and Tanzania are among the finest national parks in Africa, offering protection to wild elephants, giraffes, zebras, and lions, as well as to many less spectacular but no less significant species (see Figure 7.36B). Tourism, mostly in the form of visits to these parks, accounts for 20 percent of foreign exchange earnings; in 2013, it brought both Kenya and Tanzania approximately U.S.$1.13 billion. The future of the parks is precarious, however, because of competing demands for the land by herders and cultivators and the indigenous tendency to view wild animals as a useful, consumable resource (for example, bushmeat and ivory). In Kenya, a living elephant is worth close to $15,000 a year in income from tourists who come to see it. Nonetheless, some still view hunting elephants for their ivory as a profitable activity, even though a hunter earns no more than $1000 per dead animal.

A more sinister, very recent development, reported by Pulitzer Prize–winning journalist Jeffrey Gettleman, is the organized slaughter of elephants in the zone where Congo, South Sudan, Uganda, Kenya, and Tanzania come together. Thousands have been killed, their ivory stolen, and their bodies left to rot by poachers who are apparently using helicopters. Gettleman reported that the ivory trade is fueled largely by Chinese demand, but the perpetrators involve military rogues from all the countries in this zone, as well as the Janjaweed in Darfur in Sudan and Al Shabaab in Somalia.

Agriculture and Parks Vie for Land and Water After Kenya gained its independence from Britain in the early 1960s, government policies resettled African farmers on highland plantations formerly owned by Europeans, giving the country a strong base in export agriculture through the production of coffee, tea, sugarcane, corn, and sisal. Nomadic pastoralists fared less well than farmers because national parks were established on their lands and park managers, unfamiliar with local ecology, inadvertently encouraged the spread of tsetse fly infestations to the local pastoralist herds. Now the pastoralists and cultivators vie for scarce water—the herders need regular sources of water for their animals, while the cultivators are increasing their consumption of water by irrigating their crops. Taking note of income from export agriculture, the government has sided with the farmers, even to the point of policing from armed helicopters the water gathering by nomadic herders.

In both Kenya and Tanzania, dependence on export agriculture has put both livelihoods and valuable national parks at risk. Rapid population growth since the mid-twentieth century is part of the problem. In 1946, Kenya had just 5 million people; by 2015, it had 44.3 million. In the past, this primarily arid land could adequately support relatively small populations when managed according to traditional herding and subsistence cultivation practices. Today, the land's carrying capacity is severely stressed because population pressure is forcing farmers to overplow and herders to overgraze wildland buffer zones surrounding the parks. Hungry people have begun to poach park animals for food, and there is now a large underground trade in bushmeat and ivory.

The Islands

Madagascar, lying across the Mozambique Channel from East Africa (see the Figure 7.36 map), is the fourth-largest island in the world; only Greenland, New Guinea, and Borneo are larger. Its unique plant and animal life is the result of *diffusion* (transmission by natural or cultural processes) from mainland Africa and Asia and subsequent evolution in isolation on the island. Although the island has undergone much change over the last few centuries, it remains an amazing evolutionary time capsule that is highly prized by biologists. They debate the exact features of the natural dispersal and evolutionary processes, and the role played by human occupants who came in ancient times from Africa and other places around the Indian Ocean, including Southeast Asia. Madagascar is increasingly attractive to tourists and developers, and is greatly threatened by environmental degradation **(FIGURE 7.37)**.

The other East African island nations—the Comoros, the Seychelles, and Mauritius—are much smaller and less physically complex than Madagascar. All of the islands have a cosmopolitan ethnic makeup, the result of thousands of years of trade across the Mozambique Channel and the Indian Ocean. Further cultural mixing in the islands took place during European (primarily French and British) colonization. During the colonial era, these islands supported large European-owned plantations worked by laborers brought in from Asia and the African mainland. More recently, tourism has been added to the economic and biological diffusion mix. Most visitors come from the African mainland, from Asian countries surrounding the Indian Ocean, and from Europe.

Madagascar, with 22 million people, half of whom live on less than $1 a day, underwent IMF structural adjustment policies

FIGURE 7.37 Madagascar's unique animal life. Silky sifakas are one of several critically endangered species of lemur, a primate found only in Madagascar. Approximately 250 silky sifakas remain in the wild, all of them in a few protected areas in the rain forests of northeastern Madagascar.

in the 1990s that failed to alleviate poverty. The island's political leaders tend to be wealthy businessmen with little sense of the biological treasure that Madagascar is or of the necessity to address poverty and social welfare issues. In 2009, Andry Rajoelina—a former disc jockey, now a media magnate—staged a coup, with the aid of the military, against the democratically elected government that had begun to make some progress on ecological and development fronts. His coup was not successful, and 7 months of bloody fighting left more than 100 dead and cut tourism in half. The UN and African Union worked to enhance democratic participation, and in 2014 and 2015, peaceful elections were held and winners installed.

THINGS TO REMEMBER

- A physically diverse subregion, East Africa extends south from Eritrea to Tanzania on Africa's east coast and includes the off-coast islands in the Indian Ocean.

- The entire Horn of Africa has suffered periodic bouts of Islamist violence that has produced numerous refugees.

- Piracy on the high seas is a tactic of some coastal fishers who are resisting the poaching of their resources by high-tech fishing vessels from Europe and Asia.

- The former colonial game reserves of Kenya and Tanzania are among the finest national parks in Africa, offering protection to wild elephants, giraffes, zebras, lions, and many less spectacular but no less significant species. High-tech poaching, especially of elephants, is escalating.

- The rapid growth of the human population is resulting in competition over land and resources.

- The islands of East Africa have a cosmopolitan ethnic makeup, the result of thousands of years of trade across the Mozambique Channel and the Indian Ocean.

FIGURE 7.38 The Southern Africa subregion.

A A bridge being built over the Zambezi River in Mozambique, where economic growth has boomed since the end of a long civil war.

B A vineyard in South Africa, where some rural areas have prospered on the basis of high-value, export-oriented agriculture.

SOUTHERN AFRICA

At the beginning of the twenty-first century, hope is often a theme in public discourse across Southern Africa **(FIGURE 7.38)**. Rich deposits of diamonds, gold, chrome, copper, uranium, and coal are cause for optimism, as is the fact that at least in the country of South Africa, apartheid is gone, racial conflict has abated (but not vanished), and relative civil peace has brought substantial economic growth to most of the subregion (see page 394). This justifiable optimism must be balanced with the fact that the region of Southern Africa faces significant agricultural, political,

poverty, health, and development challenges that require enlightened leadership and a concerned participating public.

The region of Southern Africa is a plateau ringed on three sides by mountains and a narrow lowland coastal strip (see Figure 7.38). The Kalahari Desert lies at the center and the Namibian Desert along the southwest Atlantic coast, but for the most part, Southern Africa is a land of savannas and open woodlands. Its population density is low, with the highest densities found in South Africa, Lesotho, Swaziland, and Malawi (see Figure 7.24).

Rich Resources and Qualified Successes

South Africa is the wealthiest and perhaps the most stable country in the whole of sub-Saharan Africa. Some other countries in Southern Africa—Mozambique, Namibia, Botswana, Lesotho, Swaziland, and Angola—have been enjoying a measure of stability and economic growth as well, because of the development of mineral wealth, which is always subject to volatile price changes on the world market. Even predominantly agricultural Malawi improved its economy markedly over the last decade, as did Zambia, which remains overly dependent on copper exports. Only Zimbabwe, once thought of as having great promise, has become a worrisome failed state, largely because of particularly ill-conceived race-based development strategies and bad governance (see page 407).

Angola, on the west coast, and Mozambique, on the east, are two former Portuguese colonies that both endured civil wars in order to gain their independence. Marxist, Soviet-supported governments were on one side and rebels, supported by South Africa and the United States, were on the other. Mozambique ended its civil war in 1994; by 1996, its once-devastated economy had one of the highest rates of growth on the continent (7 percent in 2014; see Figure 7.38A). Nevertheless, nearly 80 percent of the population is still employed in agriculture and half of the citizens live below the poverty line (see Figure 7.27A).

In Angola, oil and mineral wealth have been central to the economy since independence in 1974, but a postindependence civil war raged over these resources until 1998, when the warring parties signed a peace treaty. By 2009, Angola was the largest oil producer in sub-Saharan Africa, and its oil revenues now constitute 50 percent of its GDP. Although Angola has enjoyed a period of rapid growth (and rapid inflation), its position is precarious because of the volatility of oil prices. Wealth disparity remains high and human well-being very low. More than 40 percent of the population live below the poverty line (see Figure 7.27B).

Mineral wealth is also a mixed blessing in Botswana. Diamonds and other minerals brought wealth, but also rapid social change and an increase in income disparity. Botswana, once one of the most successful economies in sub-Saharan Africa, tries to portray itself in the best possible light, but the statistics on well-being that it provides are suspect, and certainly the burgeoning HIV/AIDS epidemic is thwarting Botswana's economic development. Most recently, a precipitous fall in the price of diamonds, due to a contraction of diamond markets in Russia and China, has drastically reduced Botswana's prospects.

In Zimbabwe, chromium, gold, coal, nickel, platinum, and silver mining account for 40 percent of the country's exports. Unfortunately, Zimbabwe's economy has failed because of misguided advice from international lending agencies, the decline of commercial agriculture, and the flagrantly inept and illegitimate regime of Robert Mugabe (see page 407).

Zambia's rich copper resources have drawn investment and management by the Chinese, whose need to bring electricity to China's rapidly industrializing cities has created a large demand for copper wire. The Chinese-run mines are efficiently managed, and production has soared, but very little of the profit reaches Zambian citizens. Despite the low wages of local miners, who earn just U.S.$50 a month, the Chinese mine managers insist on importing Chinese mine workers (see page 400). Working conditions for both Zambian and Chinese laborers can be deadly. Zambian activists note that, as in colonial times, the extraction of Zambia's resources is primarily benefiting foreign investors. Nonetheless, mine development has fueled consumerism, and Zambia now has multiple shopping malls, where a widening range of products are available. Mobile phone use, especially by young adults, is widespread.

HIV/AIDS in Southern Africa A continuing cause of concern in the subregion is the HIV/AIDS epidemic, discussed in some detail above, pages 414–415. Although the epidemic has ebbed a bit since 2005, by 2015, infection rates in Southern Africa remained higher than in any other part of the world. More than 20 percent of adults in Southern Africa have HIV/AIDS, with the highest percentages of infection in Swaziland (23), Lesotho (16), South Africa (20), and Botswana (16). In all cases, women have a higher infection rate than men (see Figure 7.26; see also the discussion on pages 414–415).

Food Insecurity in Southern Africa The three interior countries of Malawi, Zimbabwe, and Zambia remain overwhelmingly rural. Agriculture and related food-processing industries form an important part of their economies and their exports, despite the growing role of mineral mining. Over the last two decades, agricultural productivity has fallen and food has been in seriously short supply. Both the global recession beginning in 2008 and the global commodity price spikes of 2009 share some of the blame for food shortages. The case of Malawi is illustrative.

CASE STUDY: Food Crisis in Malawi

Malawi has one of the world's highest rural population densities and its environment is under extreme stress. Eighty-five percent of Malawians are completely dependent on agriculture for their own wages and food production. In 1980, there were about 6 million Malawians; by 2015, there were 16.3 million. In 1980, food production was still sufficient to feed the people, with surpluses left over for export. Subsequently, land productivity declined sharply because the fragile tropical soils, which need years of rest between plantings, were forced into continuous plantation cultivation of tobacco, maize (corn), tea, and sugar for export. The primary reason for turning to commercial agriculture for export rather than food production for local consumption was a postindependence (1964) government policy copied from the British. Smallholders, the primary local food producers, lost land to large export producers. Their plots became too small to allow for the necessary fallow periods, and their productivity subsequently fell.

By the 1990s, 86 percent of Malawi's rural households had less than 5 acres (2 hectares) of land, yet they were responsible for producing nearly 70 percent of food for internal consumption. HIV/AIDS kept skilled farmers out of the fields because they were sick, or had to tend the sick, so harvests shrank further. Malawi became more and more dependent on imported food. Between 2007 and 2009, the global food crisis and recession raised the cost of imported food. Additionally, in the spring of 2010, drought, which had been a problem for a number of years, grew worse, putting in jeopardy food security for about 300,000 people in the far southern districts. The specter of starvation in a place that had once been food secure was very real. This case of Malawi, and similar situations in Zimbabwe and Zambia, illustrate the point made in Chapter 1: malnutrition and famine are often primarily the result of mismanagement and wrong priorities, rather than the consequence of the simple collapse of food production due to some natural event.

By 2011, partly because the natural environment rebounded, conditions started to improve. Also important were U.S. Agency for International Development (USAID) training in soil and moisture conservation and modest government aid from Australia that emphasized innovative scientific research available to smallholders. Nevertheless, the use of so much of Malawi's land for export agriculture continues to place food security at risk. [*Source: Malawi Food Security Update. For detailed source information, see Text Credit pages.*] ∎

The Precarious Position of South Africa South Africa is a country of beautiful and striking vistas (see Figure 7.38B). Just inland from its long coastline, uplands and mountains afford dramatic views of the coast and the ocean. The interior is a high, rolling plateau reminiscent of the Great Plains in North America. South Africa is not tropical; it lies entirely within the midlatitudes and has consistently cool temperatures that range between 60 and 70°F (16 and 20°C). The extreme southern coastal area around Cape Town has a Mediterranean climate: cool and wet in the winter, dry in the summer.

About twice the areal size of Texas (whose population is 25 million), South Africa has 53 million people. Seventy-five percent of the people are of African descent, 14 percent of European descent, and 2 percent of Indian descent; the rest have mixed backgrounds. When Nelson Mandela was elected president in 1994, after more than 300 years of European colonization and 50 years of apartheid, he and Bishop Desmond Tutu, a prominent leader in the South African Anglican Church (and a Nobel Peace Prize winner himself in 1984) insisted that in order to heal the racial divide, the country needed to face the reality of what had happened under apartheid (see Figure 7.15). Truth and Reconciliation Commissions asked those who had been brutal to step forward and admit what they had done in detail; those who spoke honestly were formally forgiven for their misdeeds.

South Africa's colonial history and economic distinctiveness have been discussed above. After apartheid, South Africa built competitive industries in communications, energy, transportation, and finance (it has a world-class stock market); it supplies goods and services to neighboring countries; and its cities are modern and prosperous **(FIGURE 7.39)**. In 2012, though, its ranking as one of the world's top 10 emerging markets began to slip. Its economy still accounts for one-third of sub-Saharan Africa's production, and other Southern African countries continue to receive many of their imports from South Africa (although China's share is growing rapidly); but South Africa's economy has been suffering since 2008 when trade between South Africa and the EU decreased during that institution's debt crisis starting that year. South Africa's position as the major market for products from all across Africa means that people everywhere in sub-Saharan Africa suffer when South Africa's economy slumps.

Will South Africa be able to retain its position at the helm of economic development as sub-Saharan Africa moves into a more prosperous age? Progress is fragile, as illustrated by the effect of periodic global recessions. During the 2008 to 2012 recession, South Africa suffered multiple blows: a decrease in the amount of aid from the United States and Europe and a hiatus in foreign investment; lower prices for its exports; and skyrocketing food prices. Strikes by miners left dozens dead; riots in the poorest parts of the country broke out over lack of employment and shortages of basic services like electricity and water. Jacob Zuma, a man with a besmirched

FIGURE 7.39 Durban, South Africa. The beachfront and harbor of Durban, South Africa's third-largest city (after Johannesburg and Cape Town). Durban has been a hub of manufacturing, trade, and transportation, and because of its coastal location, it is becoming a center for export-oriented industries. Durban's tropical climate and beautiful beaches make it a popular destination for tourists.

record of corruption, gender discrimination, and failed leadership during the HIV/AIDS epidemic, won a hotly contested presidential election. By 2016, his powers of leadership were slipping further. The country was acquiring debt and Zuma was less and less able to answer his detractors. The severe HIV/AIDS epidemic brought terrible losses to the young and educated, the hope of the future; the gap between rich and poor grew wider than it was at the end of apartheid.

Fortunately, the whole of sub-Saharan Africa may be on the verge of an era of self-guided development. No longer is the region inevitably tied to the success of South Africa, as was the case for so long. There is uncertainty as to whether South Africa will find the creative and courageous leadership necessary to regain its development momentum and move relatively smoothly into the information age of the twenty-first century.

THINGS TO REMEMBER

• Both optimism and worry characterize the voices heard across Southern Africa—optimism because of the resilience and creativity of its citizens and rich resources that are gradually improving well-being for more of the people; worry because of disease, especially HIV/AIDS, and deteriorating leadership and corruption in some countries.

• Foreign access to the resources of countries in the northern parts of Southern Africa could work to the ultimate disadvantage of these countries.

• The food crisis in Malawi, brought on by a complex set of causes—colonial emphasis on cash-crops for exports, changing global markets, and inappropriate cultivation practices—is slowly turning around, with the help of judicious aid from developed countries.

• Racial reconciliation, outstanding post-apartheid leadership, and rich resources helped make South Africa a leader in Southern Africa. Recent misfortunes, some self-inflicted, have brought South Africa to a precarious position. The country's potential to lead sub-Saharan Africa into a more prosperous age is uncertain.

GEOGRAPHIC THEMES Sub-Saharan Africa: Review and Self-Test

1. Environment: Sub-Saharan Africa's poverty and political instability leave its people less able to adapt to climate change than the inhabitants of other world regions. Up to now, this region has contributed little to the buildup of greenhouse gases in the atmosphere, but currently deforestation by rural Africans and by multinational logging companies is intensifying global climate change.

• Why are many African food production systems particularly vulnerable to climate change?

• Why would political instability leave people less able to deal with the fluctuations of climate change?

• How do poverty and having little access to cash influence the vulnerability of many Africans to climate change?

• Describe some of the main forces that result in deforestation in sub-Saharan Africa.

2. Globalization and Development: Most sub-Saharan African economies are dependent on the export of raw materials, a legacy of the era of European colonialism. This results in economic instability because prices for raw materials can vary widely from year to year. While a few countries are diversifying and industrializing, most still sell raw materials and all rely on expensive imports of food and manufactured goods.

• How did European colonization affect economic development and food production in Africa?

• Why does being a supplier of raw materials put a country at a disadvantage in the global economy?

• What are the current positive and negative trends in food production in sub-Saharan Africa?

• Are there any exceptions in the region to the general pattern of economic dependence on exports of cheap raw materials?

• Why is the development of manufacturing industries in sub-Saharan Africa considered such an important step in raising standards of living?

3. Power and Politics: Despite a shift in sub-Saharan Africa toward more political freedoms, some governments remain authoritarian, nontransparent, and corrupt. Public participation is growing, often due to electronic media, and free and fair elections have brought about dramatic changes in some countries. In other countries, elections and governments tainted by suspicions of fraud have led to surges of violence.

• What forces are working for and against democracy in sub-Saharan Africa?

• To what extent can the weakness of democracy in sub-Saharan Africa be traced to institutions put in place during the colonial era?

• What role have elections played in both diffusing and inciting violence?

• What are some examples of authoritarian political structures that have given way to more democratic systems of government?

• How are issues of gender influencing trends toward democratization?

4. Urbanization: Sub-Saharan Africa is undergoing a massive wave of rural-to-urban migration, leading to the fastest rate of urbanization in the world. Much of this growth is unplanned, and 70 percent of the urban population now live in impoverished slums characterized by crowded fire-prone housing, inadequate sanitation, and poor access to jobs, clean water, and food.

• How is urbanization influencing population growth in sub-Saharan Africa? Relate these trends to the demographic transition.

• Discuss the importance of urban gardening as an innovation in sub-Saharan Africa's modernization.

• How does the unplanned nature of sub-Saharan African cities lead to poor access to jobs, clean water, and food?

• What are at least three factors that are contributing to rapid urbanization in sub-Saharan Africa?

5. Population and Gender: Of all the world regions, populations are growing fastest in sub-Saharan Africa, yet growth is slowing as the demographic transition takes hold in more prosperous countries, where better health care and lower levels of child mortality, along with more economic and educational opportunities, encourage smaller families. In particular, educated women are better able to pursue careers and to choose to have fewer children.

• What do population pyramids tell us about a country's population?

• What special spiritual role do children play in family and community life in sub-Saharan Africa?

• What is the relationship between rights and education for women and falling fertility rates?

• What role do infectious diseases play in sub-Saharan African development, life expectancies, and use of resources?

• In what ways do women bear the brunt of the HIV/AIDS epidemic in sub-Saharan Africa?

Critical Thinking Questions

1. If you were to investigate the origins of lumber and lumber products sold where you live, where would you start? What would make you think that you are or are not participating in the African timber trade? How is this trade related to the concept of *neocolonialism*, the revival of colonial-like resource extraction in the modern period?

2. Describe the ways in which the environment is linked to the spread of the chronic communicable diseases of Africa. Why might some scientists argue that malaria is a worse threat to Africa's development than HIV/AIDS?

3. Describe the trajectory of population growth in Africa. What are the factors that contribute to this pattern? How might developed countries help the situation?

4. Reflect on the many passages throughout this chapter that describe gender roles in Africa. Which points do you regard as the most influential in the economic development process?

5. Describe the important ways in which women are exerting an influence on African political processes.

6. Technology is changing life in Africa. What do you see as the most crucial ways in which technology will modify the ability of ordinary people to better their lives and to participate in civil society?

7. What are the various circumstances that make young heterosexual women so susceptible to HIV/AIDS in sub-Saharan Africa? How does the social and economic status of the average woman (compared to the average man) influence the spread of HIV among women?

8. Looking at the various African locations where food is in short supply, explain the main factors that account for these shortages. How would you respond to those who say that drought and climate are the principal causes of famine?

9. If you were asked to deliver a speech to the local Rotary Club about hopeful signs out of Africa, what topics would you include in your talk? Which pictures from this book (or elsewhere) would you show?

Chapter Key Terms

agroforestry 383
animism 419
apartheid 393
carbon sequestration 381
commercial agriculture 384
commodities 396
commodity dependence 396
divide and rule 403
dual economy 396
fallow period 383

female genital mutilation (FGM) 418
genocide 407
grassroots economic development 402
groundwater 385
Horn of Africa 379
horticulture 383
intertropical convergence zone (ITCZ) 379
lingua franca 421

mixed agriculture 383
mono-crop agriculture 384
pastoralism 387
polygyny 418
Sahel 377
self-reliant development 402
shifting cultivation 383
subsistence agriculture 383
transparency 403
value chains 402

A Coastal Lowlands, South India

B Himalaya Mountains, Nepal, as seen from space

SOUTH ASIA

Indus Valley, Northern Pakistan — C

of Tibet

a y a s

Mt. Everest
Elev. 29,028

SIKKIM

Lhasa

Brahmaputra

ARUNACHAL PRADESH

Thimphu
BHUTAN
Paro
Gangtok
Phuntsholing

Gauhati

ASSAM

NAGALAND

Kohima

Khasi Hills
MEGHALAYA

Imphal
MANIPUR

BIHAR

(Ganges)

BANGLADESH

Dhaka

TRIPURA

Aizwal

MIZORAM

Dhanbad

Asansol

Chandpur

Mandalay

WEST BENGAL

Khulna

Kolkata
(Calcutta)

Chittagong

Ganga-Brahmaputra Delta

Arakan Mts.

BURMA
(Myanmar)

Irrawaddy

Mouths of the Ganga

Bhubaneswar

Nay Pyi Taw

THAILAND

Bay of Bengal

Rangoon

Gulf of Martaban

Land Elevations

meters	feet
4877	16,000
3353	11,000
2134	7000
914	3000
305	1000
152	500
0	0

mi 0 50 100 150 200 250
km 0 100 200 300 400

1:13,400,000
Lambert Azimuthal Equal Area Projection

Preparis North Channel

Andaman Islands (India)

Port Blair

Andaman Sea

Ten Degree Channel

OCEAN

Nicobar Islands (India)

Great Channel

Strait of Malacca

Sumatra

Western Ghats, Kerala, India — D

Deccan Plateau, Karnataka, India — E

Indo-Gangetic Plain, North India — F

FIGURE 8.1 Regional map of South Asia.

GEOGRAPHIC THEMES: South Asia

After you read this chapter, you will be able to discuss the following issues as they relate to the five thematic concepts:

1. Environment: Climate change puts more lives at risk in South Asia than in any other region in the world, primarily due to water-related issues. Over the short term, droughts, floods, and the increased severity of storms imperil many urban and agricultural areas. Over the longer term, sea level rise may profoundly affect coastal areas and glacial melting poses a threat to rivers and aquifers.

2. Globalization and Development: Globalization benefits some South Asians more than others. Educated and skilled South Asian workers with jobs in export-connected and technology-based industries and services are paid more and sometimes have better working conditions. Less-skilled workers in both urban and rural areas are left with demanding but very low-paying jobs.

3. Power and Politics: India, South Asia's oldest, largest, and strongest democracy, has shown that the expansion of political freedoms can reduce conflict. Across the region, when people have been able to participate in policy-making decisions and implementation—especially at the local level—seemingly intractable conflict has been defused and combatants have been willing to take part in peaceful political processes.

4. Urbanization: South Asia has two general patterns of urbanization: one for the rich and the middle classes and one for the poor. The areas that the rich and the middle classes occupy are often spacious, clean, orderly, and peaceful, with numerous well-paying jobs. The areas that the urban poor occupy are chaotic, crowded, and violent, with overstressed infrastructures and menial jobs. These two patterns often coexist in very close proximity.

5. Population and Gender: In this most densely populated of world regions, population growth is slowing as the demographic transition takes hold. However, a severe gender imbalance is developing in this region due to age-old beliefs that males are more useful to families than females. As a result, adult males significantly outnumber adult females.

The South Asian Region

South Asia (**FIGURE 8.1**), like so many of today's world regions, began its modern history as the result of European colonization. During that time, economic, political, and social policies rarely put the needs of South Asian people first. Toward the end of the colonial era, major upheavals precipitated by departing colonists left the region with a legacy of distrust and difficult political borders that still lingers.

The five thematic concepts highlighted in this book are explored as they arise in the discussion of regional issues, with interactions between two or more themes featured, as in the geographic themes above. Vignettes, like the one that follows, illustrate one or more of the themes as they are experienced in individual lives.

GLOBAL PATTERNS, LOCAL LIVES In the summer of 2015, Medha Patkar spent several days traveling around the Narmada River Valley, warning people that their villages and farms would soon be flooded. After over 50 years of controversy since the dam's construction began in 1961, and a year after India's Supreme Court gave final approval to raise the height of the Sardar Sarovar Dam to its maximum height of 452 feet (138 meters), the water behind the dam was starting to rise.

This was not the position that Patkar, the leader of the "Save the Narmada River" movement, had hoped to be in back in 1985 when she arrived as a graduate student to survey the villages that would be submerged by the dam (**FIGURE 8.2A**). Since then, and despite decades of hard fought battles to stop the dam's construction, over 320,000 farmers and fishers have been forced to move. Although Indian law requires that these "evacuees" be given land or cash to compensate them for what they have lost, only a fraction have actually received any compensation. Some refuse to move even as the rising waters of the reservoir consume their homes and are eventually forcibly removed by Indian police. Many then relocate to crowded urban slums in nearby cities. Meanwhile, the natural cycles of the remaining unflooded sections of the Narmada River to the west are severely disturbed as poorly managed releases of water from the dam cause massive die-offs of aquatic life, which have put many fisher people out of work.

In approving the raising of the dam to its final height, the Supreme Court of India argued that the project's benefits far outweighed its negative impacts. As many as 30 million people in Gujarat, the state where the dam is located, will have more access to water, primarily for irrigation, via the vast network of canals that has been built to bring water from the dam's reservoir. Gujarat is one of India's major agricultural producers, despite being highly prone to drought, and many support the dam for its irrigation benefits.

For the past 15 years, the dam's greatest proponent has been Narendra Modi, who was chief minister of Gujarat from 2001 to 2014, when he became prime minister of India. Modi's rise to national power has been based largely on his success in managing Gujarat's recent surge of economic growth. Per capita incomes tripled under Modi, which he argues is due to his success in attracting new foreign investments with large infrastructure projects and reduced government interference in the economy. But Modi's critics counter that his large projects, such as the Sardar Sarovar Dam, a new "smart city" for finance and technology companies in Gujarat, and a high-speed "bullet train" from Ahmedabad to Mumbai, benefit the wealthy at the cost of the poor. They argue that Modi's policies made it much easier for companies and the government to acquire land from farmers, forcibly removing them if necessary, as in the case of the Sardar Sarovar Dam and related projects. Modi's policies in New Delhi, India's capital, have mirrored his time in Gujarat. Soon after he became prime minister, his party proposed extending Gujarat's land-acquisition policies nationwide. In this most densely populated of world regions, land acquisition is always a very contentious issue, and Modi's opponents were able to defeat the law, which they branded as "anti-poor" and "anti-farmer." They pointed out that while Gujarat's economy grew dramatically under Modi, important indicators of human well-being, such as child malnutrition and infant mortality, did not improve.

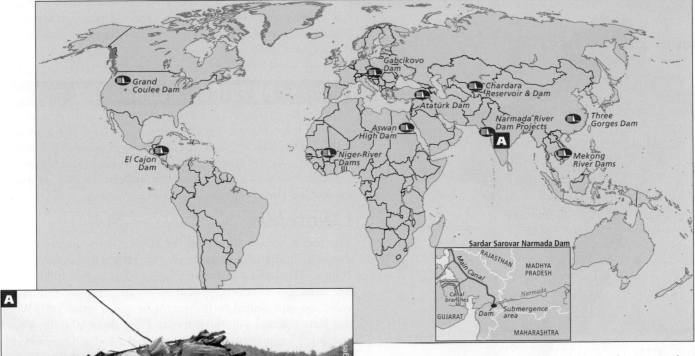

Sardar Sarovar Narmada Dam

FIGURE 8.2 The Sardar Sarovar Dam on the Narmada and other major dams around the world. The Sardar Sarovar Dam on the Narmada River is only one of hundreds of thousands of dam projects throughout the world that together have displaced between 40 and 80 million people. The map shows some of the major dams around the world. **A** Before they were displaced by the reservoir of the Sardar Sarovar Dam, rural people stand outside homes that are now submerged.

Robert Nickelsberg/The LIFE Images Collection/Getty Images

Medha Patkar and the hundreds of thousands of poor farmers displaced by the Sardar Sarovar Dam helped shape the opposition to Modi's land-acquisition law, providing vivid and poignant stories of what happens when poor people "get in the way" of large development projects. Modi has responded to these criticisms with significant investments in sanitation and clean water, on a national scale, designed to benefit the poor. Meanwhile, his ability to govern efficiently and his success in encouraging economic growth have made him the most popular prime minister in decades, and one who will undoubtedly leave his mark on the region. ■

India, the giant of South Asia, has ambitions to become "the next China." Recent years have seen India register some of the highest rates of economic growth in the world, making it currently the third largest economy in terms of purchasing power parity (PPP). Long mired in deep poverty and saddled with a corrupt and inefficient political culture, India now projects a dynamic image to the world with its large, highly skilled English-speaking middle class, a state-of-the-art technology sector, and a government eager to remove obstacles to economic growth. And yet

many "obstacles" to economic growth stem from issues of fairness and human decency in a society where so many people barely survive. Conflicts are common in South Asia, which is a poor and densely populated region, and they may become more common as disparities increase, especially if this region's well-established, democratic decision-making structures are not strengthened and allowed to falter. Several countries in the region with more authoritarian political institutions are struggling with violence and ensuing chaos. India's ability to maintain a hold on democracy is thus more important than ever as it tries to balance the ambitions of leaders like Narendra Modi with the rights of poor people who find themselves "in the way" of economic development and the realities of South Asia's overstretched ecosystems.

What Makes South Asia a Region?

In this book, the countries that make up the South Asia region are Afghanistan and Pakistan in the northwest; the Himalayan states of Nepal and Bhutan; Bangladesh in the northeast; India (including the territories of Lakshadweep, Andaman, and Nicobar Islands); and the island countries of Sri Lanka and the Maldives (**FIGURE 8.3**; see also Figure 8.1). Physically, these countries occupy territory known as the *Indian subcontinent*—the portion of a tectonic plate that joined the Eurasian continent nearly 60 million years ago (discussed below). Historically and culturally, these modern countries have a shared patchwork of religious, social, and political features linked to

FIGURE 8.3 Political map of South Asia.

ancient conquerors from Central Asia, explorers and traders from the Arabian Peninsula and Southeast Asia, and more recent colonizers from Europe. This region can be compared to a patchwork, as its extensive history of internal and external influences has left it with a fragmented and overlapping pattern of religions, languages, ethnicities, economic theories, forms of government, attitudes toward gender and class, and ideas about land and resource use.

Terms in This Chapter

Because its clear physical boundaries set it apart from the rest of the Asian continent, the term **subcontinent** is often used to refer to the entire Indian peninsula, which includes Nepal, Bhutan, India, Pakistan, and Bangladesh (the term usually does not include Afghanistan).

South Asians have recently adopted new place-names to replace the names given to them during British colonial rule. The city of Bombay, for example, is now officially *Mumbai*, Madras is *Chennai*, Calcutta is *Kolkata*, Benares is *Varanasi*, and the Ganges River is the *Ganga River*.

> **subcontinent** a term often used to refer to the entire Indian peninsula, including Nepal, Bhutan, India, Pakistan, and Bangladesh

THINGS TO REMEMBER

• The Sardar Sarovar Dam and similar dams throughout the world are created primarily for irrigation and power generation, but they have side effects that often create more problems than they solve.

• The management of water—where it is apportioned and whether there is too much or too little of it—has been a feature of South Asian life for a long time. Climate change now makes water management more problematic.

• Improving the standard of living of poor populations in many parts of the world nearly always requires increases in water and energy use.

PHYSICAL GEOGRAPHY

Many of the landforms and even climates of South Asia are the result of huge tectonic forces. These forces have positioned the Indian subcontinent along the southern edge of the Eurasian continent, where it is surrounded by the warm Indian Ocean and shielded from cold airflows from the north by the massive mountains of the Himalayas (see Figure 8.1).

LANDFORMS

The Indian subcontinent and the territory surrounding it are a dramatic illustration of what can happen when two tectonic plates collide. The Indian-Australian Plate, which carries India, broke free from the eastern edge of the African continent millions of years ago and drifted to the northeast (see Figure 1.8). When this plate collided with the Eurasian Plate about 60 million years ago, it left a giant peninsula jutting into the Indian Ocean. As the relentless pushing from the south continued, both the leading (northern) edge of what became South Asia and the southern edge of Eurasia crumpled and buckled. The result is the world's highest mountains—the Himalayas, which rise more than 29,000 feet (8800 meters)—as well as other mountain ranges that curve away from the central impact zone—in the west, the Hindu Kush and Karakoram Range, and in the east, the lesser-known ranges that border China and Burma (see Figure 8.1B and the figure map). The continuous compression also lifted the Plateau of Tibet, which rose up behind the Himalayas to an elevation of more than 15,000 feet (4500 meters) in some places. The compression and mountain-building process continue into the present.

The three great rivers of South Asia—the Indus, the Ganga, and the Brahmaputra—all begin within 100 miles (160 kilometers) of one another in the Himalayan highlands near the Tibet, Nepal, and India borders; they are largely fed by glacial meltwater (see the Figure 8.1 map and Figure 8.5). The main river basins of the Indus and Ganga lie to the southwest and south of the Himalayas in what is called the Indo-Gangetic Plain (see Figure 8.1C, F). The Indus flows southwest through Pakistan and ultimately joins the Arabian Sea. The Ganga flows southeast to the Bay of Bengal. The Brahmaputra flows east through Tibet and then turns south through Arunachal Pradesh and Assam into Bangladesh. The mouths of the Ganga and Brahmaputra are blended together in a giant, ever-changing delta.

Lying south of the Indo-Gangetic Plain is the Deccan Plateau, an area of modest uplands (1000–2000 feet [300–600 meters] in elevation) interspersed with river valleys (see Figure 8.1E). This upland region is bounded on the east and west by two moderately high mountain ranges, the Eastern and Western Ghats (see Figure 8.1D). The Ghat mountain ranges descend to long but narrow coastlines interrupted by extensive river deltas and floodplains (see Figure 8.1A). The river valleys and coastal zones are densely occupied; the uplands are only slightly less so.

Because of its continual high degree of tectonic activity and deep crustal fractures, South Asia is prone to devastating earthquakes, such as the magnitude 7.7 quake that shook the state of Gujarat in western India in 2001, which killed over 14,000; the magnitude 7.6 quake that hit the India–Pakistan border region in 2005, which killed over 86,000; and the magnitude 7.8 quake in Nepal in 2015, which killed over 8000. Coastal areas are also vulnerable to tidal waves, or *tsunamis*, that are caused by undersea earthquakes. In 2004, a massive tsunami originating off Sumatra in Southeast Asia wrecked much of coastal Sri Lanka and southern India, killing over 54,000 people in South Asia and another 230,000 in Southeast Asia.

CLIMATE AND VEGETATION

The climate of South Asia is characterized by a seasonal reversal of winds known as monsoons **(FIGURE 8.4)**. These monsoon winds are affected by the *intertropical convergence zone (ITCZ)*, which, as the name suggests, is a zone created when air masses moving south from the Northern Hemisphere and north from the Southern Hemisphere converge near the equator. As the warm air rises and cools, it drops copious amounts of precipitation. As described in Chapter 7, the ITCZ shifts north and south seasonally. The intense rains of South Asia's summer monsoons are likely caused by the ITCZ being sucked onto the land by a vacuum created when huge air masses over the Eurasian continent heat up and rise into the upper troposphere.

The **summer monsoon** (see Figure 8.4A) begins in early June, when the warm, moist ITCZ air first reaches the mountainous Western Ghats. The rising air mass cools as it moves over the mountains, releasing rain that nurtures patches of dense tropical rain forests and tropical crops on the upper slopes of the Western and Eastern Ghats and in the central uplands. Once on the other side of India, the monsoon gathers additional moisture and power in its northward sweep up the Bay of Bengal, sometimes turning into tropical cyclones.

As the monsoon system reaches the hot plains of Bangladesh and the Indian state of West Bengal in late June, columns of warm, rising air create massive, thunderous cumulonimbus clouds that drench the parched countryside. Monsoon rains sweep east to west, parallel to the Himalaya Mountains, moving across northern India to Pakistan and finally petering out over the Kabul Valley in eastern Afghanistan by July. Rainfall is especially intense in the east, north of the Bay of Bengal, where the town of Darjeeling holds the world record for annual rainfall—about 35 feet, even though no rain falls for half the year. These patterns of rainfall are reflected in the varying climate zones **(FIGURE 8.5)** and agricultural zones of South Asia (see Figure 8.17). Although sufficient rain falls in central India to support

> **summer monsoon** rains that begin every June when the warm, moist ITCZ air first reaches the mountainous Western Ghats

FIGURE 8.4 Summer and winter monsoons in South Asia.

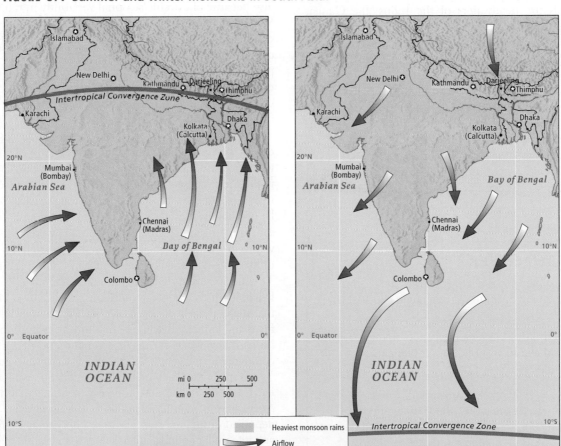

A In the summer, the ITCZ is drawn north, picking up huge amounts of moisture from the ocean, which are then deposited over India, Sri Lanka, and Bangladesh.

B In the winter, cool dry air blows from the Eurasian continent south across India, pushing the ITCZ south.

forests, most land has long been cleared of forest (see the discussion on page 447) and planted in crops. Much of the remaining forests are small and so separated from each other that they no longer provide suitable habitat for India's wildlife. Forest vegetation is more common in the Himalayan highlands and foothills, but even in the mountains there is widespread deforestation, often leaving only patchy areas of formerly richly forested land.

Periodically, the monsoon seasonal pattern is interrupted and serious drought ensues. This happened in July and August of 2009, when the worst drought in 40 years struck much of South Asia (see the discussion below). Then in September 2009, heavy rains came to central south India, causing crop-damaging floods that killed hundreds of people. Increasingly, scientists are concluding that the extreme droughts and floods of recent years are not an anomaly but instead part of global climate change.

The **winter monsoon** (see Figure 8.4B) is underway by November each year, when the cooling Eurasian landmass sends the cooler, drier, heavier air over South Asia. This cool, dry air from the north sinks down to the lower elevations and pushes the warm, wet air back south to the Indian Ocean. Very little rain falls in most of the region during this winter monsoon. However, as the ITCZ retreats southward across the Bay of Bengal, it picks up moisture that is then released as early winter rains over parts of southeastern India and Sri Lanka.

The monsoon rains deposit large amounts of moisture over the Himalayas, much of it in the form of snow and ice that add to the existing mass of glaciers (see Chapter 1). Meltwater from these glaciers feeds the headwaters of the Indus, the Ganga, and the Brahmaputra. These rivers carry enormous loads of sediment, especially during the rainy season. When the rivers reach the lowlands, their velocity slows and much of the sediment settles out as silt, which is then repeatedly picked up and redeposited by successive floods. As illustrated in the diagram of the Brahmaputra River in **FIGURE 8.6**, the movement of silt constantly rearranges the floodplain landscape, complicating human settlement and agricultural efforts. However, the seasonal deposit of silt nourishes much of the agricultural production in the densely occupied plains of Bangladesh. The same is true on the Ganga and Indus plains.

winter monsoon a weather pattern that begins by November, when the cooling Eurasian landmass sends the cooler, drier, heavier air over South Asia

---

THINGS TO REMEMBER

- Because of the high degree of tectonic activity and deep crustal fractures, all the countries of South Asia are prone to devastating earthquakes.

- The three major rivers of this region originate from glacial meltwater high in the northwest Himalaya Mountains.

- Periodic interruptions of the monsoon rainfall patterns can result in devastating droughts.

- The three main rivers—the Indus, the Ganga, and the Brahmaputra—carry enormous loads of sediment and deposit it as silt, a process that is repeated by successive floods. The movement of silt constantly rearranges the floodplain landscape, complicating human settlement and agricultural efforts.

ENVIRONMENT

GEOGRAPHIC THEME 1

Environment: Climate change puts more lives at risk in South Asia than in any other region in the world, primarily due to water-related issues. Over the short term, droughts, floods, and the increased severity of storms imperil many urban and agricultural areas. Over the longer term, sea level rise may profoundly affect coastal areas and glacial melting poses a threat to rivers and aquifers.

Humans have lived in South Asia for at least 50,000 years, but as recently as 1700 c.e. (just before British colonization), population density and human environmental impacts were relatively light compared to the population and impacts of the period that followed; since 1700, they have grown exponentially.

SOUTH ASIA'S VULNERABILITY AND RESILIENCE TO CLIMATE CHANGE

All across South Asia, people face a wide range of challenges made more serious by climate change. Here, we focus on six of these challenges, all of which relate to water in some way, followed by some of the responses to climate change that can increase this region's resilience to climate change.

1. Drought and Flooding

Significant sections of South Asia have lived with water scarcity problems for millennia, mostly due to low average rainfall, and also because of extreme seasonal variability in rainfall. The Indus Valley of Pakistan, for example, has many ancient structures that captured water during wet seasons and floods and held it for use during dry times (see Figure 8.10A). Now, the shifting rainfall patterns and rising temperatures linked to climate change are creating drier conditions in much of South Asia, and occasionally causing abnormally heavy rainfall that can result in widespread flooding, as was the case in Pakistan in 2010.

The primary danger of drought and flooding is their ability to destroy harvests. The most common method of boosting resilience to drought in South Asia is irrigation, with many ancient and new methods in use (see Figure 8.7B). Most irrigation techniques in this region use ditches to spread water through fields, but this requires large amounts of water and can result in the buildup of salt in the soil. Drip irrigation technology is now known to be the most efficient way to irrigate. Using up to 90 percent less water, it is an increasingly utilized method of conserving water in South Asia (see Chapter 6). The Indian government now subsidizes about half of the cost, which is helping the technology spread, as are advances in drip irrigation that make it cheaper. Even so, drip irrigation is still too expensive for many farmers. Because agriculture uses the most water of any human activity (more than 70 percent in some places), drip irrigation could free up enormous amounts of water for other uses.

Some South Asian methods of water conservation improve resilience to both flooding and drought. Since ancient times, many South Asians have used artificial ponds to increase the rate at which water deposited during the summer monsoon percolates through the soil and into underground aquifers rather than evaporating. This practice makes more water available for irrigation during the dry season via wells. More recent methods involve constructing small berms (raised strips of land) on fields and check dams in gullies to hold rainwater for longer so that it can drain into the soil. Because these "artificial

FIGURE 8.5 PHOTO ESSAY: Climates of South Asia

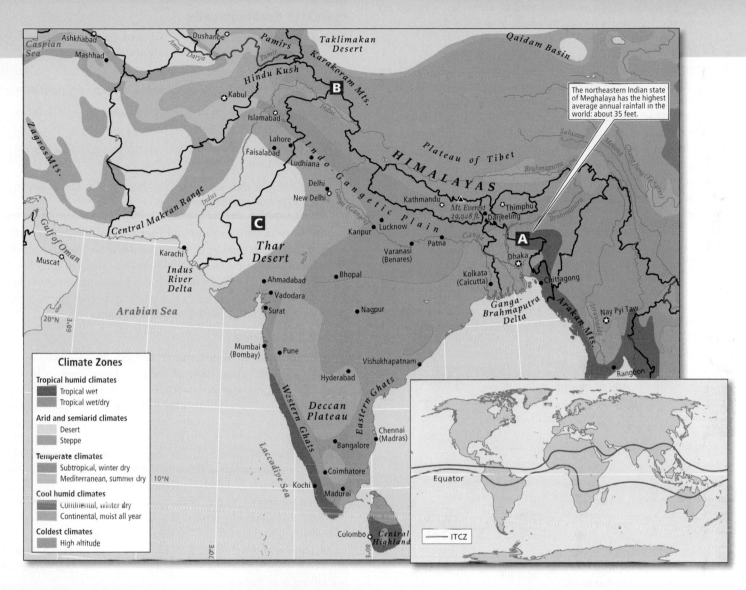

The northeastern Indian state of Meghalaya has the highest average annual rainfall in the world: about 35 feet.

Climate Zones

Tropical humid climates
- Tropical wet
- Tropical wet/dry

Arid and semiarid climates
- Desert
- Steppe

Temperate climates
- Subtropical, winter dry
- Mediterranean, summer dry

Cool humid climates
- Continental, winter dry
- Continental, moist all year

Coldest climates
- High altitude

Equator

—— ITCZ

A Tropical wet, Meghalaya, India

© Tushar S. Chowdhury/Moment Open/Getty Images

B High altitude, Kashmir, India

Tauseef Mustafa/AFP/Getty Images

C Desert, Jaisalmer, India

Shantanu Bedarkar/Moment Open/Getty Images

recharge" techniques reduce the amount of water that remains on the surface during the monsoon, they help limit flood damage.

2. Glacial Melting

Because South Asia's three largest river systems are fed, in part, by glaciers high in the Himalayas, the issue of glacial melting is of particular concern. (Smaller glacial-melt rivers serve Afghanistan and Central Asia; see Figure 5.8A.) Unless effective action against climate change is taken, Himalayan glaciers could eventually disappear, causing many large South Asian rivers to run nearly dry during the winter dry season. As many as 703 million people, almost half of the region's population, depend on these glacially fed rivers for drinking water, domestic and industrial uses, and irrigation.

While the immediate effect of glacial melting may be flooding, the real threat is that as the glaciers shrink, they will provide less water each year to rivers and to recharge the ancient aquifers beneath the heavily populated Indo-Gangetic plains south of the Himalayan Range. The amount of water pumped for multiple uses on these plains already exceeds the rate of recharge; the long-term effect of the pumping, therefore, will be severe water shortages.

Thinking Geographically

After you have read about the vulnerability to climate change in South Asia, you will be able to answer the following questions.

A What suggests that this girl comes from a very poor family?

B Why is the use of simple irrigation technology a sensible strategy in a country like Afghanistan?

C List all the indications in this photo that Mumbai has drainage problems.

D How might such embankments help reduce sensitivity to storm surge and sea level rise?

Both water conservation and increased water storage will be needed for supplies to last through the dry winter monsoon. High Himalayan communities in Ladakh, on the India–China border, are experimenting with various ways of retaining autumn glacial melt, such as in stone catchments where the melt refreezes over the winter and is available for spring irrigation.

3. Sea Level Rise

Tens of millions of impoverished farmers and fishers live near sea level in South Asia, most of them in Bangladesh, which has more people vulnerable to sea level rise (see Chapter 1) than any other country in the world **(FIGURE 8.7A)**. With 165 million people already squeezed into a country the size of Iowa, over 10 percent of Bangladeshis might have to find new homes if sea levels rise by 3 to 5 feet (0.9 to 1.5 meters). The biggest economic impacts of sea level rise would result from the submergence of parts of South Asia's largest cities, such as Dhaka in Bangladesh, Mumbai (see Figure 8.7C and Chapter 1) and Kolkata in India, and Karachi in Pakistan.

Also threatened by continuing sea level rise are the Maldives Islands in the Indian Ocean, 80 percent of which lie 3 feet (or 0.9 meter) or less above the sea. Beach erosion is so severe that homes built only a few years ago in this richest of South Asia's countries are falling into the sea.

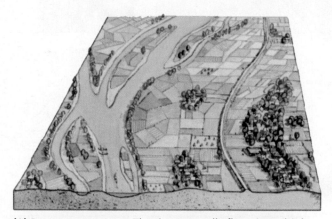

(A) Pre-monsoon stage: The river normally flows in multiple channels across the flat plain.

(B) Peak flood stage: During peak flood times, the great volume of water overflows the banks and spreads across fields, towns, and roads. The force of the water carves new channels, leaving some places cut off from the mainland.

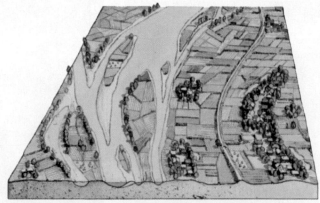

(C) Post-monsoon stage: The river returns to its banks, but some of the new channels persist, changing the lay of the land and access for those who use the land. As the river recedes, it leaves behind silt and algae that nourish the soil. New ponds and lakes form and fill with fish.

FIGURE 8.6 The Brahmaputra River in Bangladesh at various seasonal stages. People who live along the river have learned to adapt their farms to a changing landscape, and along much of the river, farmers are able to produce rice and vegetables nearly year-round. [*Source consulted: National Geographic*, June 1993, p. 125.]

FIGURE 8.7 PHOTO ESSAY: Vulnerability to Climate Change in South Asia

Climate change is putting more lives at risk in South Asia than in any other region. However, many responses to climate change are being developed that may increase resilience to climate-related hazards.

A A girl in coastal Bangladesh washes fish outside her home, which was flooded and partially ruined after a cyclone destroyed the river embankment that once protected her village from high tides. Bangladesh is extremely exposed to sea level rise, increased storm intensity, and flooding related to glacial melting. Because it has a large population of poor people, Bangladesh is very sensitive to these climate hazards.

Munir Uz Zaman/AFP/Getty Images

B In Afghanistan, a man makes improvements to irrigation infrastructure that helps reduce the sensitivity of nearby areas to water shortage. Widespread poverty and political instability contribute to Afghanistan's extremely high vulnerability to climate hazards.

Robert Nickelsberg/The LIFE Images Collection/Getty Images

Vulnerability to Climate Change

- Extreme
- High
- Medium
- Low

C A high tide during monsoon rains floods part of Mumbai's rail system. Low-lying Mumbai (formerly Bombay) is at significant risk for sea level rise; while the city's wealth and well-developed emergency response systems have increased its resilience in the face of climate hazards, much of Mumbai remains quite sensitive.

D Workers add boulders to reinforce an embankment that guards against erosion south of Kolkata (formerly Calcutta) on an island that is very exposed to sea level rise and storm surge. Such embankments can reduce sensitivity to water level changes, but they are expensive to build and require maintenance.

Punit Paranjpe/AFP/Getty Images

Deshakalyan Chowdhury/AFP/Getty Images

Bangladesh v. India, H₂O

All these places are developing responses to climate change, though the resources available to do so are often very limited (see Figure 8.7D). Over the short term, cities that can afford the expense, such as Mumbai or other large coastal cities, are building dikes, storm walls, and other "defensive" structures, or are improving their drainage systems. Smaller cities and towns and rural areas are using cheaper "green" approaches, such as restoring coastal ecosystems, such as mangrove swamps, that can help keep coastlines from eroding. In southern Bangladesh, coastal villagers have developed floating seed beds, where crop seedlings are started during the rainy monsoon season on rafts made of bamboo and water hyacinth. Over the long term, if sea levels rise enough, many people will have to relocate, a scenario that is sure to bring conflict in this crowded region.

4. Water Scarcity and Conflict

With more than 20 percent of the world's population and only 4 percent of its freshwater, South Asia suffers from extreme **water scarcity,** which is a lack of sufficient water resources to meet water needs. This region also has a high potential for conflict over water, as most major rivers travel through multiple countries. The three main river systems all originate partially in Tibet, where the Chinese government is building numerous dams that could reduce flows to South Asia. Within the region, the most water scarce areas are in Afghanistan, Pakistan, and northwest India, but water disputes may be found everywhere. The most well-developed international water treaties in South Asia cover the four rivers that start in India and flow into Pakistan. This is largely because there is so much potential for disputes over water to fuel a war between India and Pakistan. In contrast, of the 54 rivers that flow through India into Bangladesh, only one has a treaty. This reflects both the much larger water resources in eastern South Asia and the generally lower potential for war between India and Bangladesh.

Generally, India uses its size, and political and economic power to get the better of its neighbors when it comes to water. Unlike Pakistan, Bangladesh cannot appeal to internationally recognized treaties with well-developed conflict resolution mechanisms. Instead, Bangladesh often has to wage elaborate diplomatic and public relations campaigns merely to get a response from India. Usually, poor people and river ecosystems end up suffering.

For example, consider the case of the Teesta River, which originates in India and flows into Bangladesh. In the 1970s, there was an agreement between India and Bangladesh that awarded India 39 percent of the Teesta's flow for irrigation and Bangladesh 36 percent; 25 percent remained in the river, to maintain its ecology. As India's demand for water rose, another agreement took shape in 2011 that split the Teesta's flow 50/50 between the two countries, with no provision for the river's ecology. However, the Indian state of West Bengal bargained for more water, so the agreement was delayed. During this period, India constructed structures along the Teesta that divert as much as 85 percent of its flow to irrigation projects in India, leaving Bangladesh with mainly the water that flows into the Teesta channel from within Bangladesh. The losers have been some 9 million Bangladeshis and the wildlife that formerly thrived in the now much diminished Teesta River. New agreements are on the horizon, but they are unlikely to challenge India's claim to the river's water.

The same pattern on a larger scale can be seen in India's diversion of as much as 60 percent of the Ganga River flow to Kolkata to flush out channels where silt is accumulating and hampering river traffic (see the Ganga-Brahmaputra delta on the map in Figure 8.1 and the map in Figure 8.5). These diversions deprive Bangladesh of normal freshwater flow to the Ganga-Brahmaputra delta, allowing salt water from the Bay of Bengal to penetrate inland, ruining agricultural fields. The diversion has also caused major alterations in Bangladesh's coastline, damaging its small-scale fishing industry. Thus, the livelihoods of 40 million rural Bangladeshis have been put at risk to serve the needs of Kolkata's 16 million people, triggering massive protests in Bangladesh. In the late 1990s, India signed a treaty with Bangladesh promising a fairer distribution of water, but Bangladesh still receives a considerably reduced flow. India's latest water policies merely mention the need to resolve the issue.

Without global or regional agreements to help countries adapt to climate change, the potential for conflict over water will increase. Any new global agreements to mitigate its impacts could weaken or strengthen India's ability to dominate water resources in South Asia. If new agreements recognize the rights of nations to access their current water resources at their current levels, Bangladesh would have a powerful new ally in its disputes with India. If such agreements fail to occur, India could even more assertively dominate a Bangladesh weakened by sea level rise, or a Pakistan strained by drought.

5. Rising per Capita Use of Water and Pollution

As living standards rise in South Asia, so do the use of water and its pollution. As we learned in Chapter 1, humans require an average of 5 to 13 gallons (20 to 50 liters) of clean water per day for basic domestic needs: drinking, cooking, and bathing/cleaning. In South Asia's poorest urban and rural areas, per capita domestic water consumption is about 5 to 6 gallons per day. Consumption increases as incomes rise and people acquire conveniences that consume large amounts of water, such as a toilet that can use several gallons in a single flush. The middle and upper classes consume closer to 13 gallons per day, and some much more than that. As South Asian incomes increase and living standards improve, and populations continue to grow, the demand for water is increasing dramatically across this region. A window into exactly how much water would be needed is provided in Delhi (the ancient city of "Old Delhi" and New Delhi are now commonly referred to simply as Delhi). Seventeen five-star hotels, serving a few thousand guests mostly from Europe and North America, use about 210,000 gallons (800,000 liters) of water daily, roughly the same amount as 42,000 people living in Delhi's slums.

As climate change poses ever more serious threats to existing water resources, the most likely way to expand access to these resources is by lessening water pollution. For example, less than 20 percent of this region's wastewater is treated, meaning that many of the most available sources of drinking and irrigation water, such as rivers and streams, are highly polluted. Very few people drink water out of the tap in South Asia, as contamination within urban and rural water supplies means that water must be boiled or otherwise purified before it is consumed. Untreated sewage, and garbage that is dumped into rivers and streams, are major sources of contamination (FIGURE 8.8A), with less than 20 percent of the region's sewage treated at all. Expanding wastewater treatment would be expensive and energy

water scarcity a lack of sufficient water resources to meet water needs

intensive, but with so many people, and so little water, it is the most likely option for growing this region's water resources.

As is often the case in South Asia, religion plays a role in the resolution of serious problems such as water pollution. In 2008, Veer Bhadra Mishra, who was a Brahmin priest and professor of hydraulic engineering at Banaras Hindu University in Varanasi, received approval from India's central government to build a series of processing ponds that use India's heat and monsoon rains to clean Varanasi's sewage at half the cost of other methods. The technology has since become mired in bureaucratic delay, but that didn't stop Mishra from preaching a contemporary religious message to the thousands who visited his temple on the banks of the Ganga River, which is sacred to Hindus: no longer is it valid to believe that the Ganga is a goddess who purifies all she touches while assuming that it is impossible to cause her damage. Rather, Mishra said, because the Ganga is their symbolic mother, it would be a travesty for Hindus to smear her with sewage and other wastes.

Basic water safety is especially crucial in Varanasi, a sacred city where each year millions of Hindus come to die, be cremated, and have their ashes scattered over the Ganga River. The number of such final pilgrimages has increased with population growth and affluence, causing wood for cremation fires to become scarce. As a result, incompletely cremated bodies are being dumped into the river, where they pollute water used for drinking, cooking, and ceremonial bathing (see Figure 8.8B). In an attempt to address this problem, the government recently installed an electric crematorium on the riverbank. It is attracting considerable business, as a cremation in this facility costs 30 times less than a traditional funeral pyre.

6. Virtual Water and Exports

Much of South Asia's water is used in producing things it exports, such as rice, cotton, or clothing made from locally produced cotton. This makes South Asia a major exporter of virtual water (Chapter 1), which is problematic in a region where water is increasingly scarce. Afghanistan, Pakistan, and northwest India are all naturally dry regions that have been made more arid by thousands of years of human use, and all are heavily dependent on agricultural exports that require irrigation. Currently, India is the world's largest exporter of rice, and the second largest exporter of cotton. Pakistan's exports are dominated by cotton clothing, which uses primarily locally grown cotton, and Afghanistan is almost entirely dependent on agricultural exports. The costs of the water resources depleted in the production of these goods are not sufficiently accounted for in their prices.

As climates become hotter and dryer, developing South Asian countries will have to find less water-intensive items to export, or else risk pushing already scarce water resources beyond their limits. This will mean moving more of the population out of the agriculture sector, which currently supplies jobs for close to 50 percent of the region's workforce. In this shift, India is leading the way with its growing high-tech service and manufacturing industries, which contribute much more to the country's gross national income (GNI) than its vast agricultural sector, using a fraction of the water.

DEFORESTATION

Deforestation has been occurring in South Asia since the first agricultural civilizations developed between 5000 and 10,000 years ago. Ecological historians have shown that as the forests vanished, the northwestern regions of the subcontinent (from India to Afghanistan) became increasingly drier. The pace of deforestation has increased dramatically over the past 200 years. By the mid-nineteenth century, perhaps a million trees a year were felled for use in building railroads alone. Such radical deforestation jeopardized the well-being of both people and animals in South Asia (FIGURE 8.9).

South Asia's forests are still shrinking because of commercial logging and expanding village populations that use wood for building and for fuel. Many of South Asia's remaining forests are in mountainous or hilly areas, where forest clearing dramatically increases erosion during the rainy season. In addition to the loss of CO_2-absorbing forests, one result of deforestation is massive landslides that can destroy villages and close roads. With fewer trees and less soil to retain the water, rivers and streams become clogged with runoff, mud, and debris. The effects can reach so far downstream that increased flooding in the plains of Bangladesh is now linked to deforestation in the Himalayas.

ON THE BRIGHT SIDE: Technological Solutions

To address the issues of climate warming and drought, public and private entities are using alternative energy sources to reduce CO_2 emissions in the atmosphere. For example, India, which contributes the seventh-largest amount of greenhouse gases in the world, is also home to the world's largest producer of plug-in electric cars. The Mahindra Reva E20 sells in India for about U.S.$15,000 with a government subsidy. India's small but surging middle class now has the disposable income to afford these economy cars. Even factoring in emissions from the plants that generate the electricity used to charge the cars, electric cars produce substantially lower levels of CO_2 emissions than do gasoline- or diesel-powered automobiles.

Solar and wind energy are the focus of public investment for several South Asian countries because they have the potential to decrease greenhouse gas emissions. In north and west South Asia, where there is less cloud cover and energy is in greatest demand by industries and high-tech firms, the development of solar power is being emphasized. Wind energy is most efficiently produced in the wind-prone state of Tamil Nadu in southern India. Nationally, the use of wind power increased 22 percent per year from 1992 to 2010; by 2011, it constituted 70 percent of India's renewable-energy generation. ∎

Resistance to Deforestation

Unlike China and many other nations facing similar problems, the countries of South Asia have a healthy and vibrant culture of environmental activism that has alerted the public to the consequences of deforestation. In 1973, for example, in the Himalayan district of Uttarakhand (then known as Uttaranchal), India, a sporting-goods manufacturer planned to cut down a grove of ash trees so that the factory, in the distant city of Allahabad, could use the wood to make tennis racquets. The trees were sacred to nearby villagers, however, and when their protests were ignored, a group of local women took dramatic action. When the loggers came, they found the women hugging the trees and refusing to let go until the manufacturer decided to find another grove.

FIGURE 8.8 PHOTO ESSAY: Human Impacts on the Biosphere in South Asia

South Asia's huge population and growing industries have had major impacts on water and air quality, as well as the extent and health of remaining ecosystems, such as forests.

A Students stroll across a bridge over a very polluted and garbage-choked stream in a slum area in Rawalpindi, Pakistan. Nationally, only 8 percent of urban sewage is treated before it is discharged. Meanwhile, 40 percent of premature deaths in Pakistan are attributed to water-related diseases.

B A Hindu woman washes her baby in the polluted Ganga River in Varanasi, India. Many Hindus consider this a sacred act of ritual purification.

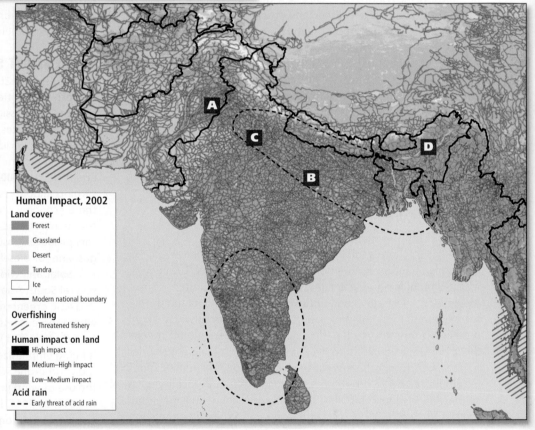

Human Impact, 2002

Land cover
- Forest
- Grassland
- Desert
- Tundra
- Ice
- —— Modern national boundary

Overfishing
- Threatened fishery

Human impact on land
- High impact
- Medium–High impact
- Low–Medium impact

Acid rain
- – – – Early threat of acid rain

C A public bus billows smoke in New Delhi, India, which has the worst air pollution in the world. Sixty-seven percent of air pollution in New Delhi comes from vehicles.

D Women remove firewood from a highland forest in Arunachal Pradesh, India. Deforestation, driven by the needs of poor people as well as by large corporations, has resulted in loss of habitat for many endangered species.

Thinking Geographically

After you have read about the human impacts on the biosphere in South Asia, you will be able to answer the following questions.

A What are the indications that water in this stream is polluted?

B What are some major sources of pollution for the Ganga River?

C What does this photo suggest about sources of air pollution in New Delhi?

D What effects can deforestation have on humans and the environment?

The women's action grew into the *Chipko movement* (literally, "hugging"), which is also known as the *social forestry movement*. The movement has spread to other forest areas and has been responsible for slowing deforestation and increasing ecological awareness. Proponents of the movement argue that the management of forest resources should be turned over to local communities. They say that people living at the edges of forests possess complex local knowledge of those ecosystems that has been gained over generations—knowledge about which plants are useful for food, medicines, and fuel, and as building materials. Those who live in forested areas are more likely to manage the forests carefully because they want their descendants to benefit from them for generations to come.

CASE STUDY: Nature Preserves in the Nilgiri Hills

The Mudumalai Wildlife Sanctuary and neighboring national parks in the Nilgiri Hills (part of the Western Ghats) harbor some of the last remaining forests in southern India. Here, in an area of about 600 square miles(1554 square kilometers), live a few of India's last wild tigers and a dozen or more other rare species, such as sloth bears and barking deer (see Figure 8.9A). Even much smaller forest reserves play an important role in conservation. At 287 acres (116 hectares), Longwood Shola is a tiny remnant of the ancient tropical forests that once covered the Nilgiris.

Phillip Mulley, a naturalist, Christian minister, and leader of the Badaga ethnic group, points out that the indigenous peoples of the Nilgiris must now compete for space with a growing tourist industry (1.7 million visitors in 2005). In addition, huge tea plantations were cut out of forestlands by the Tamil Nadu state government to provide employment for Tamil refugees from the conflict in Sri Lanka (see Figure 8.20B). So while the forestry department and citizen naturalists are trying to preserve forestlands, the social welfare department, faced with a huge refugee population, is cutting them down. [*Source:* Lydia and Alex Pulsipher, and the government of Tamil Nadu. For detailed source information, see Text Sources and Credits.] ■

FIGURE 8.9 LOCAL LIVES: People and Animals in South Asia

A A Bengal tiger (*Panthera tigris tigris*) crosses a stream in Nepal. Fewer than 2500 remain in the wild, many of them in national parks and reserves in India, Bangladesh, Nepal, and Bhutan. Despite ongoing conservation efforts since the 1970s, Bengal tiger populations are in steady decline due to demand for tiger body parts by some practitioners of traditional Chinese medicine. An endangered species, tigers are the national animal of both India and Bangladesh.

B A cow lounges in a fabric store in Varanasi, India. Cattle are allowed to roam free in some cities in India, due to the reverence Hindus widely hold for cows. The killing of cattle for any purpose is banned in 6 of India's 28 states and allowed without restriction in only 4. Scholars have found much evidence to suggest that cows have not always been sacred, and were eaten by Hindus in the distant past.

C Asian elephants at work in Nagarhole National Park in Karnataka, India, carrying fallen branches and brush. Asian elephants (*Elephas maximus*) are a separate species distinct from African elephants (*Loxodonta africana*) and have, among several other unique characteristics, the largest volume of cerebral cortex of any land animal. This portion of the brain is used for memory, attention, perceptual awareness, thought, language, and consciousness. Asian elephants can be trained to follow instructions, a quality that has made them useful to humans for work and warfare for thousands of years. They are particularly prized for their ability to cross difficult terrain while carrying heavy objects such as logs.

Tom Brakefield/Getty Images

Buena Vista Images/The Image Bank/Getty Images

Dibyangshu Sarkar/AFP/Getty Images

A A water storage tank at Mohenjo-daro, in Pakistan.

B The Taj Mahal, a Mughal mausoleum in Agra, India.

| 2000,000 B.C.E. | 5000 B.C.E. | 0 C.E. | 1500 C.E. | 1600 C.E. | 1700 C.E. |

200,000 b.c.e.
Early humans enter South Asia

2500 b.c.e.
Mohenjo-daro

1526–1857 C.E.
Mughal Empire

1643–1653
Taj Mahal

FIGURE 8.10 VISUAL HISTORY OF SOUTH ASIA

Thinking Geographically

After you have read about the human history of South Asia, you will be able to answer the following questions.

A This structure was designed to store water from what source?

B Suggest some principal features of the structure and grounds of the Taj Mahal.

AIR POLLUTION

Of all the world regions, South Asia has the worst air pollution. According to recent studies compiled by the World Health Organization (WHO), of the ten cities with the worst air pollution (as measured by particulate matter suspended in the air), nine are in South Asia (six in India, three in Pakistan). Delhi is widely considered to have the worst air pollution in the world (see Figure 8.8C), and more than half of its children have permanently impaired lung capacities due to the persistent haze of smog and smoke that chokes India's capital and largest city.

In urban areas, much of the blame falls on vehicles, industries, and power plants, but rural South Asia with its massive population may be a larger contributor on a global scale. South Asia burns more fuelwood, agricultural waste, and animal dung than any world region, most of it on small cookstoves found throughout rural areas. For decades this burning has produced a brown cloud of smoke so large that it is visible from space. Known as the *Asian brown cloud*, it influences monsoon precipitation, which is reduced in central India and intensified in northwest India and Pakistan.

Air pollution-related illnesses are the fifth largest killer in India and on the rise throughout South Asia as growing populations and economies result in more pollution. Some pollutants are decreasing in the larger cities due to new laws requiring the use of less polluting fuels. Many regulations on industrial air pollution date from a disastrous event that took place in central India in 1984: an explosion at a pesticide plant in Bhopal produced a gas cloud that killed at least 15,000 people and severely damaged the lungs of 50,000 more. The explosion was largely the result of negligence on the part of the U.S.-based Union Carbide Corporation (which owned the plant) and the local Indian employees who ran the plant. In response to the tragedy, the Indian government launched an ambitious campaign to clean up poorly regulated factories, a project that is far from complete.

ON THE BRIGHT SIDE: M. C. Mehta

M. C. Mehta, a Delhi-based lawyer, became an environmental activist partly in response to the condition of the Taj Mahal, whose white marble was becoming pitted by acid rain. For more than 20 years, he has successfully promoted environmental legislation that has removed hundreds of the most polluting factories from India's river valleys. ■

THINGS TO REMEMBER

GEOGRAPHIC THEME 1
• **Environment:** Climate change puts more lives at risk in South Asia than in any other region in the world, primarily due to water-related issues. Over the short term, droughts, floods, and the increased severity of storms imperil many urban and agricultural areas. Over the longer term, sea level rise may profoundly affect coastal areas and glacial melting poses a threat to rivers and aquifers.

• Unless effective action against climate change is taken, Himalayan glaciers could eventually disappear, causing South Asia's largest rivers to run nearly dry, at least during the winter. As many as 703 million people, almost half of the region's population, depend on these glacially fed rivers for drinking water, domestic and industrial uses, and irrigation.

C The traditional textile-weaving industry in Bengal, India.

D Gandhi leading the Salt March to the sea.

E A shopping mall outside New Delhi, which is a center of employment for India's growing middle class.

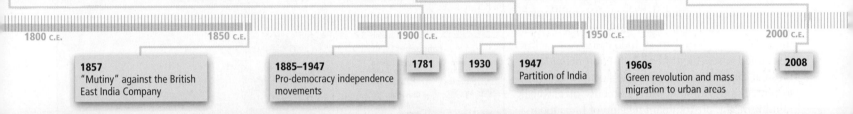

1800 C.E.	1850 C.E.	1900 C.E.		1950 C.E.	2000 C.E.

1857
"Mutiny" against the British East India Company

1885–1947
Pro-democracy independence movements

1781

1930

1947
Partition of India

1960s
Green revolution and mass migration to urban areas

2008

C What is significant about the fact that as of 1750, before the onset of British colonialism, South Asia produced 12 to 14 times more cotton than Britain and the rest of Europe combined?

D What was the immediate result of the Salt March?

E What does this photo suggest about India's modern economy?

- Pakistan, India, and Bangladesh have major air and water pollution problems related to industrial waste and lack of sewage treatment facilities.

- As climates become hotter and dryer, developing South Asian countries will have to find less water-intensive items to export, or else risk pushing already scarce water resources beyond their limits.

- South Asia's forests are shrinking because of commercial logging and expanding village populations that use wood for building and for fuel.

- The Chipko ("tree hugging") social forestry movement in Uttarakhand, India, has spread to other forest areas, slowing deforestation and increasing ecological awareness.

HUMAN PATTERNS OVER TIME

Home to some of Earth's oldest and most influential civilizations, South Asia faces the world's great transitions with an enviable ability to innovate while maintaining ancient traditions. Tens of thousands of years of continuous human occupation, and the integration of numerous influences from outside the region, have given South Asia astounding cultural diversity to draw on in this time of change. In the current era of globalization, South Asia is integrating powerful global economic forces while struggling to achieve an economic transformation that benefits more than a privileged few. The politics of this region are also in transition; people are demanding more political freedom, challenging the authoritarian systems of the past. In this most densely populated of world regions, the demographic transition is under way as population growth slows, while changing gender roles give women more freedom. Throughout these great transitions, the people of South Asia often manage to create a synthesis of tradition and innovation that smooths over the most jarring changes.

A variety of groups have migrated into South Asia, many of them as invaders who conquered peoples already there. Despite much blending over the millennia, the continued coexistence and interaction of many of these groups make South Asia both a richly diverse and an extremely contentious place.

THE INDUS VALLEY CIVILIZATION

There are indications of early humans in South Asia as far back as 200,000 years ago, but the first real evidence of modern humans in the region is about 38,000 years old. The first large agricultural communities, known as the **Indus Valley civilization** (or **Harappa culture**), appeared about 4500 years ago along the Indus River in what is modern-day Pakistan and northwest India. The architecture and urban design of this civilization were quite advanced for the time. Homes featured piped water and sewage disposal. Towns were well planned, with wide, tree-lined boulevards laid out in a grid. Evidence of a trade network that extended to Mesopotamia and eastern Africa has also been found.

Vestiges of the Indus Valley civilization's agricultural system survive to this day in parts of the valley, including infrastructure for storing monsoon rainfall to be used for irrigation in dry times (shown in **FIGURE 8.10A**). Possible cultural and linguistic remnants of the Indus Valley civilization

Indus Valley civilization the first substantial settled agricultural communities, which appeared about 4500 years ago along the Indus River in modern-day Pakistan and northwest India

Harappa culture see Indus Valley civilization

451

survive today among the Dravidian peoples of southern India, who originally migrated from the Indus region beginning about 2000 years ago.

Scholars have long debated the reasons for the decline of the Indus Valley civilization. Some believe that complex geologic (seismic) and ecological changes (drier climate) brought about a gradual demise. Others argue that foreign invaders brought a swift collapse, instigating out-migration.

A SERIES OF INVASIONS

The first recorded invaders to join the indigenous people of South Asia came from Central and Southwest Asia into the rich Indus Valley and Punjab about 3500 years ago. Many scholars believe that these people, referred to as Indo-European (a linguistic term; see Figure 8.11), in conjunction with those of the Harappa and other indigenous cultures, instituted some of the early elements of classical Hinduism, the predominant religion of India today.

Wave after wave of other invaders arrived, including the Persians, the armies of the Greek general Alexander the Great, and numerous Turkic and Mongolian peoples. Defensive structures against these invaders can still be found across northwest South Asia. Jews came to the Malabar Coast of Southwest India more than 2500 years ago, Christians shortly after the time of Jesus. Arab traders came by land and sea to India long before the emergence of Islam; and then starting about 1000 years ago, Arab traders and religious mystics introduced Islam to what are now Afghanistan, Pakistan, and northwest India. By sea, the Arabs brought Islam to the coasts of southwestern India and Sri Lanka. In 1526, the **Mughals,** a group of Turkic Persian people from Central Asia, invaded from the north, intensifying the growth of Islam. The Mughals reached the height of their power and influence in the seventeenth century, controlling the north-central plains of South Asia. The last great Mughal ruler (Aurangzeb) died in 1707, but the cultural legacy of the Mughals remained, even as the power and range of the dynasty declined. One aspect of this legacy is the 520 million Muslims now living in South Asia. The Mughals also left a unique heritage of architecture, art, and literature that includes the Taj Mahal, miniature painting, and a rich tradition of lyric poetry (see Figure 8.10B). The Mughals contributed to the evolution of the Hindi language, which became the language of trade of the northern subcontinent and which is still used by more than 400 million people.

LANGUAGE AND ETHNICITY

One result of the numerous invasions of South Asia over the millennia is that today there are many distinct ethnic groups, each with its own language or dialect. In India alone, 18 languages are officially recognized, but there are actually hundreds of distinct languages. This complexity results partly from strong cultural traditions that allow groups to maintain distinct identities in the midst

Mughals a dynasty of Central Asian origin that ruled India from the sixteenth century to the nineteenth century

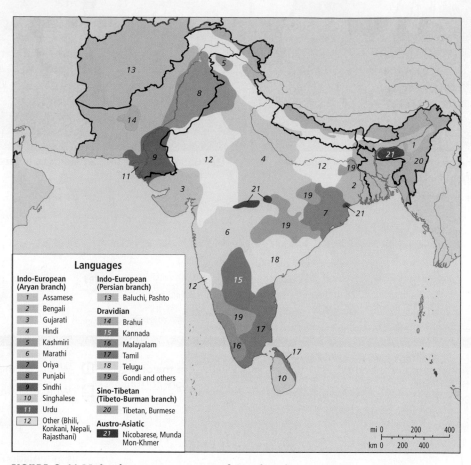

Languages

Indo-European (Aryan branch)
1 Assamese
2 Bengali
3 Gujarati
4 Hindi
5 Kashmiri
6 Marathi
7 Oriya
8 Punjabi
9 Sindhi
10 Singhalese
11 Urdu
12 Other (Bhili, Konkani, Nepali, Rajasthani)

Indo-European (Persian branch)
13 Baluchi, Pashto

Dravidian
14 Brahui
15 Kannada
16 Malayalam
17 Tamil
18 Telugu
19 Gondi and others

Sino-Tibetan (Tibeto-Burman branch)
20 Tibetan, Burmese

Austro-Asiatic
21 Nicobarese, Munda Mon-Khmer

FIGURE 8.11 Major language groups of South Asia. The modern pattern of language distribution in South Asia is testimony to the fact that this region has long been a cultural crossroads. [*Source consulted:* Alisdair Rogers, ed., *Peoples and Cultures* (New York: Oxford University Press, 1992), p. 204.]

of foreign cultures and withstand invasion or even factors that could force an entire group to relocate. As shown in **FIGURE 8.11**, the Dravidian language-culture group, represented by numbers 14 through 19, is an ancient group, thought to have originated in Pakistan, that predates the Indo-European invasions by a thousand years or more. Today, because of ancient migrations from Pakistan, Dravidian languages are found mostly in southern India, but a remnant of the extensive Dravidian past can still be found in the Indus Valley in south-central Pakistan.

By the seventeenth century, Hindi—a blend of Sanskrit-based and Persian-based Indo-European languages—was the dominant language throughout northern India and what is now Pakistan, where it is known as Urdu. Today, variants of Hindi serve as national languages for both India and Pakistan, though it is the first, or native, language of only a minority. English is a common second language throughout the region. As the language of the British colonial bureaucracy, English remains a language used at work by professional people of all categories. Between 10 and 15 percent of South Asians speak, read, and write English. Many others use a version of spoken English.

RELIGIOUS TRADITIONS

The main religious traditions of South Asia are Hinduism, Buddhism, Sikhism, Jainism, Islam, and Christianity (**FIGURE 8.12**). (For a discussion of Islam, Christianity, and Judaism, see Chapter 6.)

on eating beef, may stem from the fact that cattle have been tremendously valuable in rural economies as the primary source of transport, field labor, dairy products, fertilizer, and fuel (animal dung is often burned).

Caste

Hinduism includes the **caste system,** a complex and evolving way of dividing society into hereditary hierarchical categories (see Figure 8.30). In the present form of caste, one is born into a given sub-caste, or community (called a *jati*), that traditionally defined much of one's life experience—where one would live, where and what one could eat and drink, with whom one would associate, one's marriage partner, and often one's livelihood. There are four main divisions or tiers of *jati*, called **varna,** which are organized hierarchically.

Brahmins, members of the priestly caste, are the most advantaged in caste hierarchy. Thus, they must conform to those behaviors that are considered most ritually pure (for example, strict vegetarianism and abstention from alcohol). As is the case with many castes, Brahmins are found in numerous occupations outside their place as priests in the *varna* system (as barbers and hairdressers, for example). Below Brahmins, in descending rank, are *Kshatriyas,* who are warriors and rulers; *Vaishyas,* who are landowning (small-plot) farmers and merchants; and *Sudras,* who are low-status laborers and artisans. A fifth group, the *Dalits*—"the oppressed," or untouchables—is actually considered to be so lowly as to have no caste. Dalits perform those tasks that caste Hindus consider the most despicable and ritually polluting: killing animals, tanning hides, cleaning, and disposing of refuse. A sixth group, also outside the caste system, is the *Adivasis,* who are thought to be descendants of the region's ancient aboriginal inhabitants.

Although *jatis* are associated with specific subcategories of occupations, in modern economies, this aspect of caste is more symbolic than real. Members of a particular *jati* do, however, follow the same social and cultural customs, dress in a similar manner, speak the same dialect, and tend to live in particular neighborhoods or villages. This spatial separation arises from the higher-caste communities' fears of ritual pollution, such as through physical contact or the sharing of water or food with lower castes. When one stays in the familiar space of one's own *jati,* one is enclosed in a comfortable circle

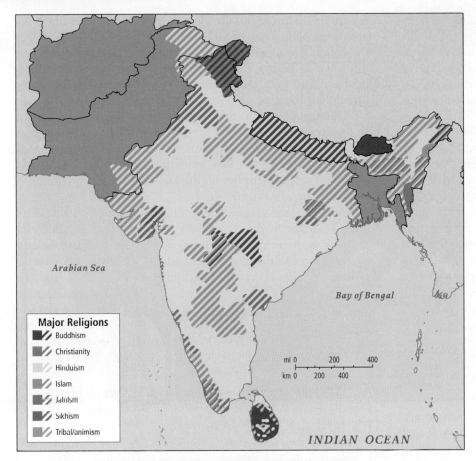

FIGURE 8.12 Major religions in South Asia. Notice that while Hinduism dominates in India, there are variable and overlapping (striped) patterns of other religious traditions in India, Nepal, and Sri Lanka. On the other hand, Bhutan, Bangladesh, Pakistan, and Afghanistan have very little religious diversity. [*Source consulted:* Gordon Johnson, *Cultural Atlas of India* (New York: Facts on File, 1996), p. 56.]

Map legend:

Major Religions
- Buddhism
- Christianity
- Hinduism
- Islam
- Jainism
- Sikhism
- Tribal/animism

Arabian Sea
Bay of Bengal
INDIAN OCEAN

ml 0 — 200 — 400
km 0 — 200 — 400

Hinduism is a major world religion practiced by approximately 900 million people, 800 million of whom live in India. It is a complex belief system, with roots in both ancient literary texts (known as the *Great Tradition*) and in highly localized folk traditions (known as the *Little Tradition*).

The Great Tradition is based on 4000-year-old scriptures called the *Vedas.* Its major tenet is that all gods are merely illusory manifestations of the ultimate divinity, which is formless and infinite. Some devout Hindus worship no gods at all, and instead engage in meditation, yoga, and other spiritual practices designed to bring people to a state described as "infinite consciousness." The average person, however, is thought to need the help of personified divinities in the form of gods and goddesses (the Little Tradition). While all Hindus recognize some deities (such as Vishnu, Shiva, Ganesh, and Krishna), many deities are found only in one region, one village, or even one family.

Some beliefs are held in common by nearly all Hindus. One is the belief in reincarnation, the idea that any living thing that desires the illusory pleasures (and pains) of life will be reborn after it dies. A reverence for cows, which are seen as only slightly less spiritually advanced than humans, also binds all Hindus together (see Figure 8.9B). This attitude, along with the Hindu prohibition

---

Hinduism a major world religion practiced by approximately 900 million people, 800 million of whom live in India; a complex belief system, with roots both in ancient literary texts (known as the Great Tradition) and in highly localized folk traditions (known as the Little Tradition)

caste system a complex, ancient Hindu system for dividing society into hereditary hierarchical classes

jati in Hindu India, the subcaste into which a person is born, which traditionally defines the individual's experience for a lifetime

varna the four hierarchically ordered divisions of society in Hindu India underlying the caste system: Brahmins (priests), Kshatriyas (warriors/kings), Vaishyas (merchants/landowners), and Sudras (laborers/artisans)

of families and friends that becomes a mutual aid society in times of trouble. The social and spatial cohesion of *jatis* helps explain the persistence and respect paid to a system that, to outsiders, seems to put a burden of shame and poverty on the lower ranks. However, it is important to note that caste and class are not the same thing. Class refers to economic status, and there are class differences within caste groups because of differences in wealth.

Most historians, anthropologists, and other social scientists who have examined the origins of caste generally agree that the current form of caste is much stricter and more rigid than what one would have found even three hundred years ago. Rather than placing the origin of caste in ancient Hindu texts, such as the Vedas, these scholars argue that caste may have been shaped more by the needs of rulers, many of them foreigners, to divide society into groups that would be more easily dominated, controlled, and governed. Upper-caste groups (Brahmins and Kshatriyas) owned or controlled most of the land, and lower-caste groups (Sudras) were the laborers, so caste and class tended to coincide. There were many exceptions, however. Today, as a result of legally mandated expanding educational and economic opportunities, caste and class status are less connected. Some Vaishyas and Sudras have become large landowners and extraordinarily wealthy businesspeople, while some Brahmin families struggle to achieve a middle-class standard of living. By and large, however, Dalits remain very poor.

Geographic Patterns of Religious Beliefs

The geographic pattern of religion in South Asia is complex and overlapping. As Figure 8.12 shows, there is a core Hindu area in central India, with other faiths more common on the fringes of the region.

The approximately 575 million Muslims in South Asia form the majority in Afghanistan, Pakistan, Bangladesh, and the Maldives; and the 184 million Muslims in India are a large and important minority, comprising 12 percent of the population. They live mostly in the northwestern and central Ganga River plain.

Buddhism began about 2600 years ago as an effort to reform and reinterpret Hinduism. Its origins are in northern India, where it flourished early in its history before spreading eastward to East and Southeast Asia. About 10 million people—only 1 percent of South Asia's population—are Buddhists. They are a majority in Bhutan and Sri Lanka.

Jainism, like Buddhism, originated as a reformist movement within Hinduism more than 2000 years ago. Jains (about 6 million people, or 0.6 percent of the region's population) are found mainly in cities and western India. They are known for their educational achievements, promotion of nonviolence, and strict vegetarianism.

Sikhism was founded in the fifteenth century as a challenge to both Hindu and Islamic systems. Sikhs believe in one god, hold high ethical standards, and practice meditation. Philosophically, Sikhism accepts the Hindu idea of reincarnation but rejects the idea of caste. (In everyday life, however, caste plays a role in Sikh identity.) The 21 million Sikhs in the region live mainly in Punjab, in northwestern India. Their influence is greater than their numbers because many Sikhs hold positions in the government, the military, and police forces throughout India.

The first Christians in the region are thought to have arrived in the far southern Indian state of Kerala with St. Thomas, the apostle of Jesus, in the first century C.E. Today, Christians and Jews are influential but tiny minorities along the west coast of India. A few Christians live on the Deccan Plateau and in northeastern India near Burma.

Animism, the most ancient religious tradition, is practiced throughout South Asia, especially in central and northeastern India, where there are indigenous people whose occupation of the area is so ancient that they are considered aboriginal inhabitants. (For a discussion of animism, see Chapter 7.)

Buddhism a religion of Asia that originated in India in the sixth century B.C.E. as a reinterpretation of Hinduism; it emphasizes modest living and peaceful self-reflection leading to enlightenment

Jainism a religion of Asia that originated as a reformist movement within Hinduism more than 2000 years ago; Jains are found mainly in western India and large urban centers throughout the region and are known for their educational achievements, nonviolence, and strict vegetarianism

Sikhism a religion of South Asia that combines beliefs of Islam and Hinduism

---

THINGS TO REMEMBER

• The Indus Valley/Harappa civilization, which began 4500 years ago, was remarkable for its innovative developments in water management and agriculture.

• South Asia has been invaded many times, primarily from the north by groups stretching from Greece to Mongolia.

• There are many distinct ethnic groups in South Asia, each with its own language or dialect. Today, variants of Hindi are the principal languages of India and Pakistan, while Bengali is the official language of Bangladesh. English is a common second language throughout South Asia.

• Caste and class are not the same thing: "class" refers to economic status, and "caste" to hereditary hierarchical social categories. There are class differences within caste groups because of differences in wealth. Even though India's constitution bans caste discrimination, caste is still hugely influential in Indian society.

• There is a geographic pattern to religion in South Asia, but it is not absolute and people of different religions often live in close proximity. Relations between Muslims and Hindus can be quite tense, occasionally resulting in violent confrontation. Other religious traditions also play prominent local roles in the life of South Asia.

BRITISH COLONIAL RULE AND ITS LEGACIES

As Mughal rule declined, a number of regional states and kingdoms rose and fought with one another **(FIGURE 8.13)**. The absence of one strong power created an opening for yet another invasion. By the late 1700s, several European trading companies were competing to gain a foothold in the region. Of these, Britain's East India Company was the most successful. By 1857, the East India Company, acting as an extension of the British government, repressed a rebellion against European intrusion and became the dominant power in the region.

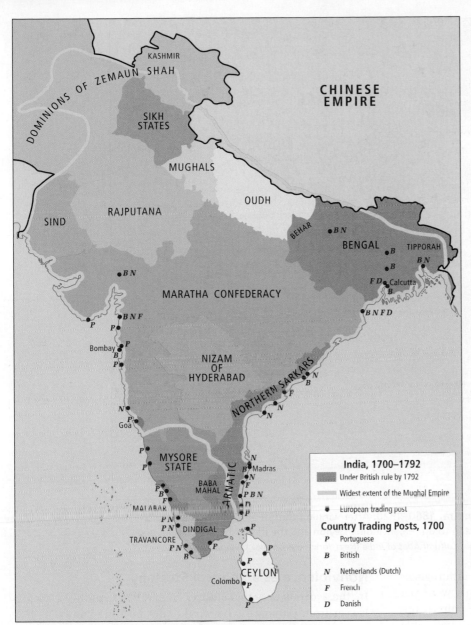

FIGURE 8.13 Precolonial South Asia. By 1700, several European nations had established trading posts along the coast of India and Ceylon (now Sri Lanka). After the death of the Mughal ruler Aurangzeb in 1707, the Mughals' ability to assert strong central rule throughout South Asia declined. A number of emergent regional states competed with one another for territory and power. Among the strongest was the Maratha Confederacy, composed of a number of small states dominated by the Maratha peoples, who were known for their martial skills. Weakness of administrative control at the center and constant rivalries between these South Asian states paved the way for British conquest by the end of the eighteenth century. [*Sources consulted:* William R. Shepherd, *The Historical Atlas* (New York: Henry Holt, 1923–1926), p. 137; Gordon Johnson, *Cultural Atlas of India* (New York: Facts on File, 1996), p. 111.]

The British controlled most of South Asia from the 1830s through 1947 **(FIGURE 8.14)**. By making the region part of the British Empire, the British accelerated the process of globalization in South Asia, transforming the region politically, socially, and economically. Even areas not directly ruled by the British felt the influence of their empire. Afghanistan repelled British attempts at military conquest, but the British continued to intervene there, trying to make Afghanistan a "buffer state" between British India and Russia's expanding empire. Nepal remained only nominally independent during the colonial period, and Bhutan became a protectorate of the British Indian government.

The Deindustrialization of South Asia

As in their other colonies, the British used South Asia's resources primarily for their own benefit, often with disastrous results for South Asians. One example was the fate of the textile industry in Bengal (modern-day Bangladesh and the Indian state of West Bengal).

Bengali weavers, long known for their high-quality muslin cotton cloth, initially benefited from the increased access that British and other traders gave them to overseas markets in Asia, the Americas, and Europe. By 1750, South Asia had an advanced manufacturing economy that produced 12 to 14 times more cotton cloth than Britain alone and more than all of Europe combined. However, during the eighteenth and nineteenth centuries, Britain's own highly mechanized textile industry—based on cotton grown in India, various other colonies, and the American South—developed cheaper cloth that then replaced Bengali muslin. This shift happened first in the British colonies in the Americas, then in Europe, and eventually throughout South Asia. The British East India Company also hindered competition for the British textile industry from weavers in Bengal with economic policies that severely limited the South Asian textile industry, which survived in only a few places (see Figure 8.10C). As one British colonial official put it, "While the mills of Yorkshire prospered, the bones of Bengali weavers bleached on the plains of India."

Many people who were pushed out of their traditional livelihood in textile manufacturing were compelled to find work as landless laborers. But rural South Asia already had an abundance of agricultural labor, so many migrated to emerging urban centers. In the 1830s, a drought worsened an already difficult situation and more than 10 million people starved to death. Throughout the nineteenth century, similar events forced millions of South Asian workers to join a stream of indentured laborers migrating to other British colonies in the Americas, Africa, Asia, and the Pacific, where their descendants can still be found.

Democratic and Authoritarian Legacies

Contemporary South Asian governments retain institutions put in place by the British to administer their vast empire. These governments inherited many of the shortcomings of their colonial forebears, such as authoritarian bureaucratic procedures that tend to resist change. While the governments have proved functional

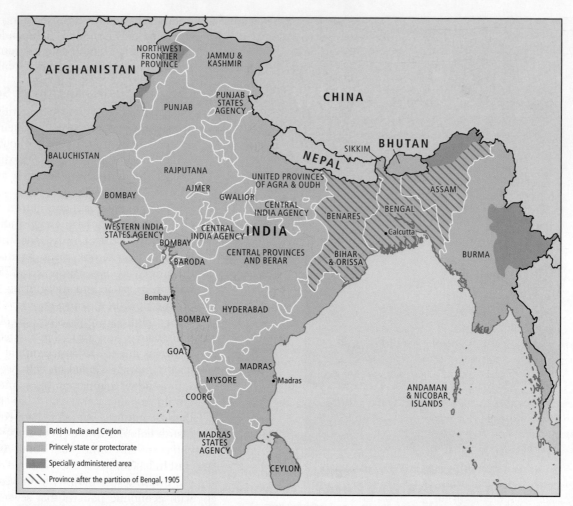

FIGURE 8.14 The British Indian Empire, 1860–1920. After winning control of much of South Asia, Britain controlled lands from Baluchistan to Burma, including Ceylon and the islands between India and Burma. [*Source consulted:* Gordon Johnson, *Cultural Atlas of India* (New York: Facts on File, 1996), p. 158.]

over time, there have been numerous major civil disturbances in virtually every country. Democratic governments were not instituted on a large scale until the final days of the empire, but since independence in 1947 (see below), people have been able to use the political freedoms afforded by democracy to voice their concerns and to make many peaceful transitions of elected governments. Still, the struggle to retain and build democratic institutions continues.

Independence and Partition

The tremendous changes brought about by the British inspired many resistance movements among South Asians. Some of these were militant movements intent on pushing the British out by force, such as the unsuccessful rebellion of South Asian soldiers employed by the British East India Company in 1857. Other movements, such as the Indian National Congress (founded in 1885), used political means to agitate for more democracy, which they saw as the route to South Asian political independence. Although both militant and political actions were brutally repressed, the democracy movements gained worldwide attention and, after decades of struggle, proved successful.

Nonviolence as a Political Strategy

In the early twentieth century, Mohandas Gandhi, a young lawyer from Gujarat—who would later become known by the honorific "Mahatma," meaning "great soul"—emerged as a central political leader in South Asia's independence movement. He used tactics of **civil disobedience** to nonviolently defy laws imposed by the British that discriminated against South Asians. Gathering a large group of peaceful protesters, he would notify the government that the group was about to break a discriminatory law. If the authorities ignored the act, the demonstrators would have made their point and the law consequently weakened. If the government instead used force against the peaceful demonstrators, it would lose the respect of the masses. Throughout the 1930s and 1940s, this technique was used to slowly but surely undermine British authority across South Asia. The most famous example was the Salt March of 1930, when Gandhi led tens of thousands of people on a march to the sea, where they made salt by evaporating seawater, thus breaking the law that made it illegal for South Asians to produce salt (see Figure 8.10D). The purpose of the law had been to facilitate British rule by controlling a vital human nutritional necessity. Breaking the law on such

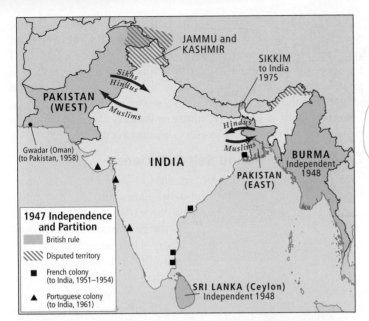

FIGURE 8.15 Independence and Partition. India became independent of Britain in 1947, and by 1948, the old territory of British India was partitioned into the independent states of India and East and West Pakistan. The Jammu and Kashmir region was contested space, and remains so today. Sikkim went to India, and both Burma and Sri Lanka became independent. Following additional civil strife, East Pakistan became the independent country of Bangladesh in 1971. [*Source consulted: National Geographic*, May 1997, p. 18.]

a massive scale created an international media frenzy that catapulted Gandhi to global notoriety; it moved millions of South Asians to support independence from Britain. Gandhi's tactics of civil disobedience became central to South Asian political culture and were used in the United States during the civil rights movement of the 1950s and 1960s.

The Partition of India in 1947

When Britain granted independence to India in 1947, it was divided into two independent countries: India, which was predominantly Hindu; and West and East Pakistan, which was predominantly Muslim **(FIGURE 8.15)**. (Afghanistan, Bhutan, and Nepal were never officially British colonies; Ceylon [now Sri Lanka] became independent in 1948.) This division—called **Partition**—which Gandhi greatly lamented, was perhaps the most enduring and damaging outcome of colonial rule.

Muslim political leaders, concerned about the fate of a minority Muslim population in a united India with a Hindu majority, first suggested the idea of two nations. Though Partition was highly controversial, it became part of the independence agreement between the British and the Indian National Congress (India's principal nationalist party). Northwestern and northeastern India, two very different places historically and culturally, but where the populations were predominantly Muslim, became a single country consisting of two parts known as West and East Pakistan, separated by northern India (see Figure 8.15). Although both India and Pakistan maintained secular constitutions with no official religious affiliation, the general understanding was that Pakistan would have a Muslim majority and India a Hindu

majority. Fearing that they would be persecuted if they did not move, more than 7 million Hindus and Sikhs migrated to India from their ancestral homes in what had become West or East Pakistan. A similar number of Muslims left their homes in India for one of the parts of Pakistan. In the process, civil society broke down: families and communities became divided, looting and rape were widespread, and between 1 and 3.4 million people were killed in numerous local outbreaks of violence. In 1971, after a bloody civil war, Pakistan was divided into Bangladesh (formerly East Pakistan) and Pakistan (formerly West Pakistan).

Partition was the tragic culmination of the divide-and-rule approach the British used throughout their empire (see Chapter 7). This approach heightened tensions between South Asian Muslims and Hindus, thus creating a role for the British as seemingly indispensable and benevolent mediators. Instead of relieving tensions, the partition of India and Pakistan laid the groundwork for the wars, skirmishes, strained relations, and ongoing arms race between India and Pakistan that persist to this day. Conflicts over water among India, Pakistan, and Bangladesh were also complicated by Partition. With religion the deciding factor, borders were created that ignored the difficulties of managing access to water from rivers that cross international borders (as discussed earlier on page 446).

The Post-Independence Period

In the more than 60 years since the departure of the British, South Asians have experienced both progress and setbacks. Democracy has expanded steadily, albeit somewhat slowly. India is now the world's most populous democracy. It is gradually dismantling age-old traditions that hold back women, poor low-caste Hindus, and other disadvantaged groups, and a vibrant, if still small, middle class is emerging (see Figure 8.10E). Pakistan has held a number of elections, and while some of its governments have been effective, more often they have been corrupt, militaristic, authoritarian regimes. The long feud between Pakistan and India about the border between the two countries (discussed on page 464), meanwhile, has sapped resources and ruined innumerable lives. Bangladesh, despite the damage it sustained in its war of independence from Pakistan in 1971, has had a more stable, responsive, and less militaristic (though not trouble-free) government than Pakistan.

Under British rule, agricultural modernization lagged, and during World War II, the Bengal Famine claimed an estimated 4 million lives. After independence, progress in agricultural production remained slow until the late 1960s, when the green revolution brought about a marked increase in harvests (see page 460). The move to modernized farming on large tracts of land with far fewer agricultural workers brought prosperity to some South Asians, relieved food scarcities, and made food exports possible; but it also forced millions to migrate to the cities. This migration intensified the already vast economic disparities in urban South Asia.

Industrialization became a main goal after independence, partly in response to the dismantling of industry during the colonial period. The emphasis on technical training has produced several generations of highly skilled engineers whose

civil disobedience protesting of laws or policies by peaceful direct action

Partition the breakup following Indian independence that resulted in the establishment of Hindu India and Muslim Pakistan

talents are in demand around the world. In most countries in the region, urban-based industrial and service economies now constitute a far larger share of GNI than agriculture (which nonetheless continues to employ almost 50 percent or more of the workforce). The information technology (IT) sector is growing especially rapidly in India, which is now the world's third largest economy in terms of PPP.

> ### THINGS TO REMEMBER
>
> • Following some 300 years of Mughal rule, the British controlled most of South Asia from the 1830s through 1947, profoundly influencing the region politically, socially, and economically.
>
> • British colonial rule in South Asia channeled resources to Europe, and in so doing depressed flourishing industries and inhibited development.
>
> • In the early twentieth century, Mohandas "Mahatma" Gandhi emerged as a central political leader of India's independence movement, using nonviolent civil disobedience that eventually led to independence.
>
> • Instead of relieving tensions, Partition—the 1947 division of British India into two countries, India and Pakistan—laid the groundwork for the repeated wars, skirmishes, strained relations, and ongoing arms race between India and Pakistan.

GLOBALIZATION AND DEVELOPMENT

GEOGRAPHIC THEME 2

Globalization and Development: Globalization benefits some South Asians more than others. Educated and skilled South Asian workers with jobs in export-connected and technology-based industries and services are paid more and sometimes have better working conditions. Less-skilled workers in both urban and rural areas are left with demanding but very low-paying jobs.

South Asia is a region of startling economic contrasts. India is a good example: it is home to hundreds of millions of desperately poor people, but it is also a global leader in the computer software industry and engineering innovation. It is celebrated for being the world's third largest economy in terms of purchasing power, with a large and growing middle class and even a robust space program. And yet its poor often see little improvement in their own lives from this economic development.

FROM SELF-SUFFICIENCY TO GLOBALIZATION

After independence, national self-sufficiency was the main goal for all parts of the region, but after decades of slow economic growth and persistent poverty, most governments began to embrace globalization as a route to greater prosperity. Starting in the 1980s, strong growth occurred in the IT and other high-tech industries, as well as manufacturing in some parts of the region. Meanwhile, changes in agriculture drastically increased the amount of food available while channeling the profits from growing food to an ever-smaller number of farmers.

Agriculture still employs almost 50 percent of the workers in South Asia, but the contribution of agriculture to most national economies is much lower, averaging less than 20 percent of the GNI for the region. The industrial sector employs far fewer people but produces between one-quarter and one-third of GNI in all countries except Nepal and Afghanistan. The service sector has expanded more rapidly than either agriculture or industry and now produces more than half of the GNI in every country except Bhutan.

Central Planning and Self-Sufficiency

After independence from Britain in 1947, the new leaders in India, Pakistan, Bangladesh, and Sri Lanka turned to socialist ideas and industrial success in the Soviet Union, believing that central planning (see Chapter 5) by the government would speed industrialization and reduce poverty. They also wanted their respective countries to become self-sufficient, avoiding dependence on manufactured goods imported from the industrialized world (see Chapter 3 for a discussion of import substitution industrialization). To reach their goal of self-sufficiency, governments took over the industries they believed to be the linchpins of a strong economy: steel, coal, transportation, communications, and a wide range of manufacturing and processing industries.

South Asian central planning in the decades after independence generally failed to meet goals for a variety of reasons. One problem was that policies intended to boost employment often contributed to inefficiency, encouraging industries to employ as many people as possible, even if they were not needed. For example, until recently, it took more than 30 Indian workers to produce the same amount of steel as 1 Japanese worker, making Indian steel uncompetitive in the world market during this period. In addition, as in the former Soviet Union, decisions about which products should be produced were made by ill-informed government bureaucrats and not driven by consumer demand. Until the 1980s, items that would improve daily life for the poor majority, such as cheap cooking pots or simple tools, were produced only in small quantities and were of inferior quality. At the same time, there was a relative abundance of large kitchen appliances and large cars that only a tiny minority could afford.

ECONOMIC REFORMS

During the late 1980s and 1990s, much of South Asia began to undergo economic reforms aimed at increasing competitiveness and creating secure jobs in the private sector. In many other world regions, similar reforms took place as part of structural adjustment programs (SAPs; see Chapter 3) that were mandated by the International Monetary Fund (IMF) and World Bank. In India, by contrast, as a response to an earlier financial crisis in the 1980s, the government itself initiated economic reforms. These consisted of, among other things, the privatization of government-run industries and banks and the relaxation of tariffs on imports. India's economic reforms have been arguably more successful than SAPs and similar reforms imposed by the IMF and World Bank in other countries.

India's economic reforms have freed many private companies from a maze of regulations, enabling both foreign and Indian companies to invest heavily in manufacturing and other industries **(FIGURE 8.16)**. Some Indian companies, such as Reva, which builds economical electric cars for India's middle class, are quite

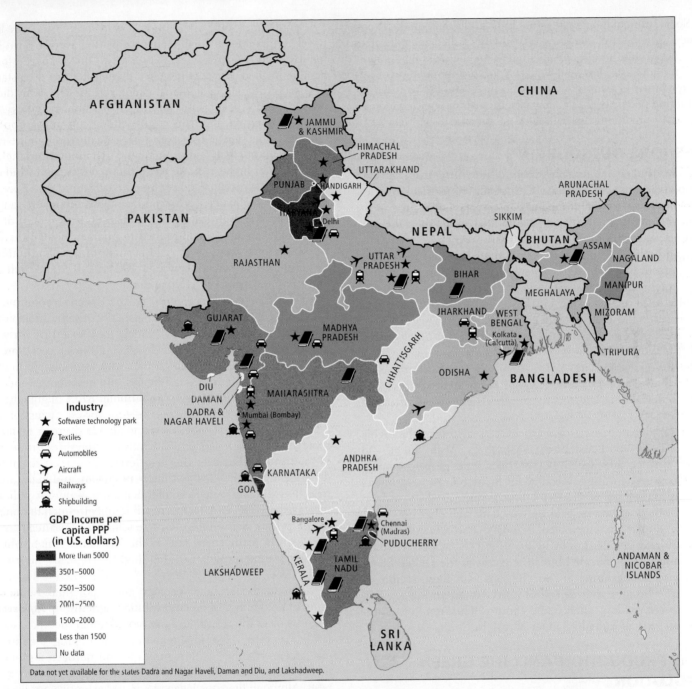

FIGURE 8.16 GDP income per capita (PPP) and industrial and IT centers in India. The presence of industry is often associated with higher incomes, and because of this, planners sometimes try to bring industry to low-income places. In some areas, poverty may be so great that even fairly intensive industrial development is able to raise average incomes above base levels only slowly. (The per capita income PPP information since 2009 for India's states is not available.) [*Sources consulted:* Directorate of Economics & Statistics of each of the respective state governments, 2006; GDP data from the Reserve Bank of India, Table 8, "Per Capita Net State Domestic Product at Factor Cost—State-Wise (at Current Prices)," http://rbidocs.rbi.org.in/rdocs/Publications /PDFs/008T_BST130913.pdf; data from "Comparing Indian States and Territories with Countries: An Indian Summary," *The Economist*, http://www.economist.com/content/indiansummary.]

innovative. Foreign auto companies are also flocking to India, drawn by its large and cheap workforce, its excellent educational infrastructure (for the middle and upper classes), and especially its large demand for manufactured goods of all sorts. Nearly every major global automobile company is currently establishing significant manufacturing facilities somewhere in India—producing everything from economical first cars to luxury brands, such as Mercedes-Benz. So many global manufacturers have flocked to

India in recent years that the country is challenging China as an exporter of manufactured goods.

One possible outcome of all this growth is that as manufacturing jobs increase, many of India's urban poor will see their incomes rise. They will then increase their consumption of Indian-made products, thus fueling further growth. By 2015, GDP per capita (PPP) had risen in all Indian states (see Figure 8.16) and averaged U.S.$6266, nearly 20 times higher than it was in 1992.

So far, most of the benefits from India's self-implemented economic reforms seem to be going to the highly skilled and educated urban upper and middle classes. India's middle class now stands at 267 million—still less than a quarter of India's population of 1.3 billion—but is likely to grow dramatically to perhaps as many as 550 million people by 2025. However, the nation's new economic policies are producing wider urban/rural income disparities.

OFFSHORE OUTSOURCING

India's current manufacturing boom is benefiting from a previous boom in offshore outsourcing that started in the 1990s. In **offshore outsourcing,** a company contracts to have some of its business functions performed in a country other than the one where its products or services are actually developed, manufactured, and sold. Companies in North America and Europe have been outsourcing jobs to cities such as Bangalore, Mumbai, and Ahmadabad to take advantage of India's large, college-educated, and relatively low-cost workforce. Jobs outsourced to India include those in IT, data entry, web design, engineering, telephone support, pharmaceutical research, and "back office" work. Many of the workers in these jobs are women. By holding outsourced jobs with North American firms, South Asian workers have gained experience that positions them perfectly for jobs now being created by South Asian–owned firms in South Asia.

Major global finance firms are increasingly hiring the highly skilled workers on Dalal Street—Mumbai's "Wall Street"—to provide finance and accounting services. Stiff global competition makes cost cutting imperative for finance firms, and India's relatively low salaries provide a solution. While a graduate of a top Indian business school is paid only a quarter of what his or her U.S. counterpart is compensated, in India that worker's salary buys a much higher standard of living than the U.S. employee would have. In fact, well-educated Indian migrants in the United States are now returning home to take the jobs with these seemingly low salaries because they can still live well and join this exciting development phase in their home country. This trend of return-migration could eventually happen across the South Asian region.

FOOD PRODUCTION AND THE GREEN REVOLUTION

Due to changes in food production systems in South Asia, food supplies and the incomes of wealthy farmers have increased. Meanwhile, impoverished agricultural workers, displaced by new agricultural technology, have been forced to seek employment in cities, where they often end up living in crowded, unsanitary conditions, and must purchase food instead of growing their own. Agricultural production per unit of land has increased dramatically over the past 50 years, especially in parts of India; nevertheless, agriculture remains the least efficient economic sector, meaning that it has the lowest return on investments of land, labor, and cash. **FIGURE 8.17** shows the distribution of agricultural zones in South Asia.

offshore outsourcing the contracting of certain business functions or production functions to providers where labor and other costs are lower

agroecology the practice of traditional, nonchemical methods of crop fertilization and the use of natural predators to control pests

Until the 1960s, agriculture across South Asia was based largely on traditional small-scale systems (the average farm holding in South Asia is still less than 2 acres) that managed to feed families in good years, but often left them hungry or even starving in years of drought or flooding. Moreover, these systems did not produce sufficient surpluses for the region's growing cities.

Beginning in the late 1960s, the green revolution (see Chapter 1) boosted grain harvests dramatically through the use of new agricultural tools and techniques. Such innovations included seeds bred for high yield and resistance to disease and wind damage; fertilizers; mechanized equipment; irrigation; pesticides; herbicides; and double-cropping (producing two crops consecutively per year). To a lesser extent, the increase in the amount of land under cultivation also contributed to a larger yield. Where the new techniques were used, yield per unit of farmland improved by more than 30 percent between 1947 and 1979 and both India and Pakistan became food exporters.

Despite the achievements of the green revolution, the benefits have been uneven. The prosperity of some Indian states, such as Punjab and Haryana, which have extensive irrigation networks, has increased tremendously, but the successes are precarious. Many impoverished farmers who were unable to afford the special seeds, fertilizers, pesticides, and new equipment could not compete and had to give up farming. Most migrated to the cities, where their skills matched only the lowest-paying jobs.

South Asia needs alternatives to standard green revolution strategies because it will have difficulty maintaining current levels of food production over the long term. The green revolution's dependency on chemical fertilizers, pesticides, and high levels of irrigation all contribute to aquifer depletion, waterway pollution, increased erosion, and the loss of soil fertility through the buildup of salt in soils. *Soil salinization* (see Chapter 6) is already reducing yields in many areas, such as the Pakistani Punjab, which is Pakistan's most productive—but highly irrigated—agricultural zone.

The methods of agroecology are a potential remedy for some of the failings of green revolution agriculture. **Agroecology** often involves the revival and use of traditional methods, such as fertilizing crops with animal manure, intercropping (planting several species together) with legumes to add nitrogen and organic matter, water conservation, and using natural predators to control pests. Although the knowledge required is extensive, unlike green revolution techniques, the methods of agroecology are advantageous to poor farmers because the necessary resources are readily available in most rural areas and the knowledge can be handed down orally from generation to generation.

While the green revolution has managed to increase food supplies, it has not eliminated hunger and malnutrition. This is because food tends to go to those who have money to buy it, and many of the farm workers who have been pushed off the land and into cities cannot afford enough food to avoid malnutrition. For example, between 1964 and 2015, the amount of food produced per capita in South Asia increased roughly 20 percent and the proportion of undernourished people dropped from 33 percent of the population to 15.7 percent. Nonetheless, this represents 281 million people—more than a third of the world's total undernourished population. Because of corruption and social

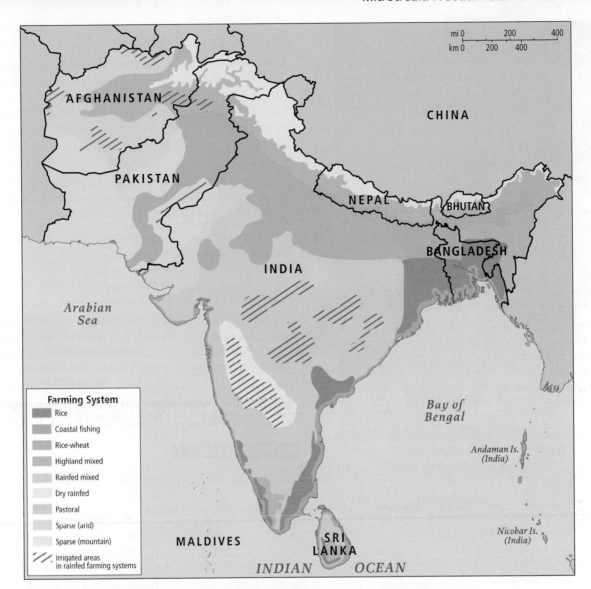

FIGURE 8.17 Major farming systems of South Asia. [*Sources consulted:* John Dixon and Aidan Gulliver with David Gibbon, *Farming Systems and Poverty: Improving Farmers' Livelihoods in a Changing World* (Rome and Washington, DC: FAO and World Bank, 2001), http://www.fao.org/farmingsystems/FarmingMaps/SAS/01/FS/index.html.]

discrimination, government programs that provide food to the poor have generally failed to reach those most in need. For example, despite massive resources devoted to improve child nutrition, 43 percent of children in India show signs of malnutrition. India today is home to almost half of the world's malnourished children.

MICROCREDIT: A SOUTH ASIAN INNOVATION FOR THE POOR

Over the past four decades, a highly effective strategy for lifting people out of extreme poverty and the hunger and malnutrition that so often accompanies it has been pioneered in South Asia. **Microcredit** makes very small loans (generally under U.S.$100) available to very low-income, would-be business owners. Throughout South Asia, as in most of the world, banks are generally not interested in administering the small loans that poor people, especially poor women, need. Instead, the poor must rely on small-scale moneylenders, who often charge interest rates as high

as 30 percent or more *per month*. In 1976, Muhammad Yunus, an economics professor in Bangladesh, started the Grameen Bank, or "Village Bank," which makes small loans, mostly to people in rural villages who wish to start businesses.

The microcredit loans often pay for the start-up costs of small enterprises, which include such wide-ranging ventures as cell phone–based services, chicken raising, small-scale egg production, the construction of pit toilets, and the distribution of home water purification systems. Potential borrowers (more than 90 percent of whom are women) are organized into small groups that are collectively responsible for repaying any loans to group members. If one member fails to repay a loan, then everyone in the group is denied loans until the loan is repaid. This system, reinforced

microcredit a program based on peer support that makes very small loans available to very low-income entrepreneurs

FIGURE 8.18 Microcredit for small-scale entrepreneurs. Borrowers from the Grameen Bank operate a business manufacturing and repairing fishing nets in rural Bangladesh. Following a model established in 1976 by Bangladeshi economist Muhammad Yunus, such groups gather weekly to discuss business and pay their loan installments. For many, this is a treasured social outing.

with weekly meetings, creates incentives for members to repay their loans (**FIGURE 8.18**). The repayment rate on the loans is extremely high, averaging around 98 percent—much higher than the rate most banks achieve. Guided by the notion that poor people know best how to get themselves out of poverty, the Grameen Bank and similar organizations in 40 countries have distributed $25 billion in loans to 100 million people. In 2006, Muhammad Yunus and the Grameen Bank were awarded the Nobel Peace Prize "for their efforts to create economic and social development from below."

Nevertheless, some controversy lingers over the effectiveness of microcredit at reducing poverty, with several studies showing much less success than is commonly claimed. Some research suggests that broader regional or national-scale economic growth raises incomes more noticeably, and that microcredit programs do not contribute enough at these scales to move people out of poverty. Others studies suggest that microcredit programs would have a larger impact if combined with additional financial services such as insurance and savings accounts, as well as legal education and emergency food aid.

ON THE BRIGHT SIDE: MICROCREDIT: A Bangladeshi Innovation

So far, the Grameen Bank has grown dramatically since its founding in 1976 in Bangladesh, where it has loaned over U.S.$11 billion to more than 8 million borrowers. Similar microcredit projects have been established in India and Pakistan and throughout Africa, Middle and South America, North America, and Europe. In 2006, Dr. Yunus and the Grameen Bank were awarded the Nobel Peace Prize for their work in microcredit. In 2009, he received the Presidential Medal of Freedom from U.S. President Barack Obama. ■

THINGS TO REMEMBER

GEOGRAPHIC THEME 2 • **Globalization and Development:** Globalization benefits some South Asians more than others. Educated and skilled South Asian workers with jobs in export-connected and technology-based industries and services are paid more and sometimes have better working conditions. Less-skilled workers in both urban and rural areas are left with demanding but very low-paying jobs.

• During the 1990s, much of South Asia began to undergo economic reforms that increased economic growth, but also produced wider income disparities.

• Due to changes in food production systems in South Asia, food supplies and incomes of wealthy farmers have increased. Meanwhile, poorer farmers and agricultural workers have been displaced by new agricultural technology and forced to seek employment in cities.

• Guided by the notion that poor people know best how to get themselves out of poverty, the Grameen Bank and similar organizations in 40 countries have distributed $25 billion in loans to 100 million people.

POWER AND POLITICS

GEOGRAPHIC THEME 3

Power and Politics: India, South Asia's oldest, largest, and strongest democracy, has shown that the expansion of political freedoms can reduce conflict. Across the region, when people have been able to participate in policy-making decisions and implementation—especially at the local level—seemingly intractable conflict has been defused and combatants have been willing to take part in peaceful political processes.

Conflict is common in South Asia, and the most successful paths to peace enable opposing groups to compete peacefully in democratic elections. Every South Asian country and every foreign country that have intervened in the region's politics have missed opportunities to resolve conflicts through democratic means, resorting instead to the use of force (see Figure 8.20A, B). The underlying causes of conflict in this region are often complex and layered. For example, religious differences are frequently exploited by political parties to win votes by unifying diverse peoples against a perceived common enemy. Meanwhile, underlying forces, such as poverty, economic rivalry between groups, and distrust of governments that are widely regarded as corrupt, intensify these supposedly "religious" conflicts to the point where violence seems inevitable. This and similarly complex and layered conflicts have occurred repeatedly in this region for many decades, with tragic loss of life, unresolved tensions, and damage to struggling economies the results. Fortunately, a number of new developments in the politics of this region may provide novel ways to address conflict and corruption.

The Hindu–Muslim Relationship

Indian independence leaders like Mohandas Gandhi and Jawaharlal Nehru (the first prime minister of India, 1947–1964) emphasized the common cause of throwing off British rule that once united Muslim and Hindu Indians. Since independence, members of the Muslim upper class have been prominent in Indian national government and the military. Muslim generals have served India willingly, even in its wars with Pakistan after Partition. Hindus and Muslims often interact amicably, trade with each other extensively, and the stars of Bollywood (the Mumbai film industry) are idolized by millions of Hindus and Muslims across South Asia regardless of their religion.

But there is a dark side to the Hindu–Muslim relationship that has become tense at the local to international levels. In some Indian villages, Hindus may regard Muslims as members of low castes that they consider unclean. Religious rules about food are often a source of discord, with many Hindus vegetarian and most Muslims occasional meat eaters. There is a long history of political violence between the two religions, dating back to the Mughal conquests, when brutal campaigns killed hundreds of thousands and many important Hindu temples were destroyed and mosques built in their place. British rule did little to pacify tensions, with many policies having the effect of pitting Hindus against Muslims who then required "benign" intervention on the part of British authorities to maintain the peace.

After Partition in 1947, some wealthy Hindu landowners remained in what was then East Pakistan (now Bangladesh). In some Bangladeshi villages today, Hindus are often somewhat wealthier even though Muslims may be a majority. Although the two groups may coexist amicably for many years, they view each other as different, and conflict resulting from religious or economic disputes—euphemistically called **communal conflict**—can erupt over seemingly trivial events, as described in the following vignette.

VIGNETTE The sociologist Beth Roy, who studies communal conflict in South Asia, recounts an incident that she refers to as "some trouble with cows" in the village of Panipur (a pseudonym) in Bangladesh **(FIGURE 8.19)**. The incident started when a Muslim farmer either carelessly or provocatively allowed one of his cows to graze in the lentil field of a Hindu. The Hindu complained, and when the Muslim reacted complacently, the Hindu seized the offending cow. By nightfall, Hindus had allied themselves with the lentil farmer and Muslims with the owner of the cow. More Muslims and Hindus converged from the surrounding area, and soon there were thousands of potential combatants lined up facing each other. Fights broke out. The police were called. In the end, a few people died when the police fired into the crowd of rioters. [*Source: Beth Roy. For detailed source information, see Text Sources and Credits.*] ■

RELIGIOUS NATIONALISM

The association of a particular religion with a particular territory or political unit—be it a neighborhood, a city, or an entire country—to the exclusion of other religions, is commonly called **religious nationalism.** The ultimate goal of such movements is often political control over a given territory.

Although both India and Pakistan were formally created as secular states, religious nationalism has been a reality in both countries, shaping relations between people and their governments.

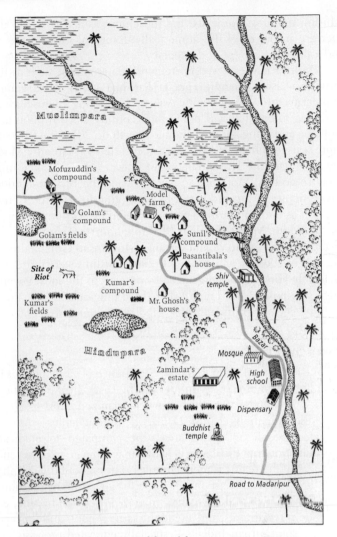

FIGURE 8.19 Some trouble with cows. This map of the village of Panipur (a pseudonym) illustrates how intimately the separate Muslim and Hindu communities were connected. The map shows Muslim and Hindu areas, the area where the riot took place, and numerous other features of village life.

Rejecting the idea of multiculturalism, and referring back to the days of Partition, India is increasingly thought of as a Hindu state, even though it has the second largest Muslim population in the world (after Indonesia), and in Pakistan and Bangladesh the government strongly identifies with Islam. In each country, many people in the dominant religious group associate their religion with their national identity.

In India, urban men from upper-caste groups (often called "forward castes") are the predominant supporters of Hindu nationalism and its associated ideology called *Hindutva*, which conceives of South Asia as the Hindu homeland and the centuries of rule by foreigners and the ongoing influence of foreign ideologies as a source of weakness for Hindus. Proponents of Hindutva often argue that violence against Muslims is justified as self-defense in light of the historical invasions of India by

communal conflict a euphemism for religiously based violence in South Asia

religious nationalism the association of a particular religion with a particular territory or political unit to the exclusion of other religions

Muslim empires. Some see the movement as a reaction to the increase in power of the numerically dominant lower castes in India since the advent of democracy.

Political parties based on religious nationalism, such as the Hindutva-based Bharatiya Janata Parishad (BJP), have gained popularity throughout South Asia by promoting themselves as purifying forces that will purge their country of corruption and violence and bring about economic growth. In 2014, the BJP won more seats in India's Parliament than ever, led by Narendra Modi, the most hardline Hindu nationalist to ever become prime minister. To do this, Modi had to reach beyond the BJP's traditional base among the forward castes by promoting his record as chief minister in Gujarat during a period of booming economic growth. Nevertheless, Modi was shamed throughout the campaign for turning a blind eye to violence against Muslims in Gujarat during the riots of 2002, in which roughly 2000 Muslims were murdered by largely Hindu mobs organized by officials in the local government, the police, and Hindutva-oriented religious organizations. Until he became prime minister, Modi was banned from entering the United States due to his role in the riots.

Caste and Politics

Despite decades of effort to fight the influence of caste, politics in India are still dramatically influenced by caste. In the twentieth century, Mohandas Gandhi began an organized effort to eliminate discrimination against Dalits or "untouchables." Among educated people in urban areas, the campaign to eradicate discrimination on the basis of caste may appear to have succeeded, but the reality is more complex. Throughout the country, there are now Dalits serving in powerful government positions. Members of high and low castes ride city buses side by side, eat together in restaurants, and attend the same schools and universities. For some urban Indians—especially educated professionals who meet in the workplace—caste is deemphasized as the crucial factor in finding a marriage partner. However, studies suggest that fewer than 5 percent of marriages cross *jati* lines, let alone the broader gulf of *varna*. Nearly everyone notices the tiny social clues that reveal whether one is of a higher or lower caste, and in rural areas, where the majority of Indians still reside, the divisions of caste remain prevalent.

One reason caste continues to be relevant is that at the local level, most political parties design their vote-getting strategies to appeal to caste and subcaste loyalties. They often secure the votes of entire *jati* communities with political favors such as new roads, schools, or development projects. These arrangements fly in the face of the official ideologies of the major political parties and of Indian government policies, which actively work to eliminate discrimination on the basis of caste. Ironically, some of these policies also have the effect of encouraging people to identify more strongly with caste.

In the 1920s, after centuries of governing with and promoting the interests of the forward castes, British India instituted policies that reserved a certain number of school admissions and government jobs for three disadvantaged groups: people of lower castes (often referred to as OBC

regional conflict a conflict created by the resistance of a regional ethnic or religious minority to the authority of a national or state government; currently, these are the most intense armed conflicts in South Asia

for "other backward class"), Dalits (untouchables), and Adivasis (the ancient aboriginal peoples of South Asia). After independence, India extended such policies to the point where now 49.5 percent of government jobs are reserved for people from these three groups, who together constitute approximately 64 percent of the Indian population. Critics argue that with so many sought after scholarships and jobs now dependent on one's caste status, people are encouraged to identify with caste now more than ever. Indeed, several new political parties that explicitly support the interests of OBC, Dalits, and Adivasis have gained considerable power in recent decades. The reserving of nearly half of government jobs has also resulted in controversy, with students and faculty at many elite institutions of higher education successfully protesting the established quotas for lower-caste applicants. Meanwhile, many forward castes such as the Patels and a number of Brahmin and Kyshatria castes have agitated for their own special treatment, arguing that some of their less wealthy members are unfairly shut out of opportunities by their caste status.

REGIONAL CONFLICTS

The most intense armed conflicts in South Asia today are **regional conflicts** in which nations dispute territorial boundaries or a minority actively resists the authority of a national or state government. Most of these hostilities arise from the authoritarian tendencies of governments that at times work against the growth of political freedoms that might otherwise defuse political tensions.

Conflict in Kashmir

Since 1947 and the post-independence division of India and Pakistan, between 60,000 and 100,000 people have been killed in violence in Kashmir. At the root of the violence is a struggle for territory between India and Pakistan, neither of which is willing to let the people of Kashmir resolve the dispute democratically (see the **FIGURE 8.20** map).

For many years, Kashmir has been a Muslim-dominated area, and in 1947, some Kashmiris believed that it should be turned over to Pakistan for this reason. Although the Hindu maharaja (king) of Kashmir wished for his country to remain independent at the time, the most popular Kashmiri political leader and significant portions of the populace favored joining India. When Pakistan-sponsored raiders invaded western Kashmir in 1947, the maharaja quickly agreed to his nation's becoming part of India. A brief war between Pakistan and India resulted in a cease-fire line that became a tenuous boundary.

Thinking Geographically

After you have read about the power and politics in South Asia, you will be able to answer the following questions.

A What aspects of this photo suggest that the convoy is part of a government-funded army?

C What direction was the camera facing when this photo was taken?

D How did the Maoists describe the war they waged against King Gyanendra?

FIGURE 8.20 PHOTO ESSAY: Power and Politics in South Asia

Most armed conflicts in South Asia have been worsened or sparked by governmental authoritarianism that has eroded political freedoms. Supporters of political opposition groups have faced disenfranchisement, imprisonment, and sometimes execution. However, in some cases, growing respect for political freedoms has paved the way for peaceful reconciliation between former combatants.

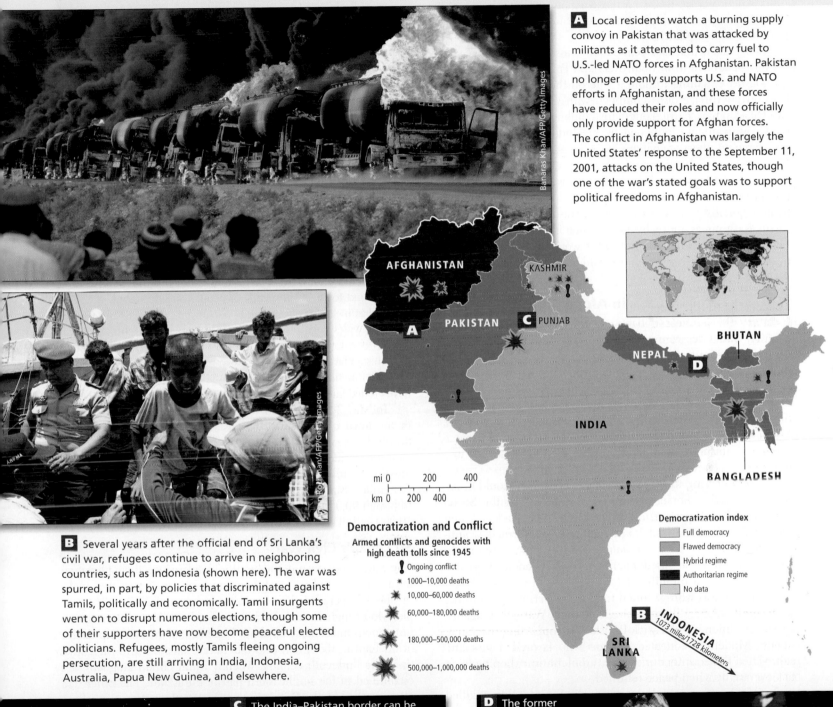

A Local residents watch a burning supply convoy in Pakistan that was attacked by militants as it attempted to carry fuel to U.S.-led NATO forces in Afghanistan. Pakistan no longer openly supports U.S. and NATO efforts in Afghanistan, and these forces have reduced their roles and now officially only provide support for Afghan forces. The conflict in Afghanistan was largely the United States' response to the September 11, 2001, attacks on the United States, though one of the war's stated goals was to support political freedoms in Afghanistan.

B Several years after the official end of Sri Lanka's civil war, refugees continue to arrive in neighboring countries, such as Indonesia (shown here). The war was spurred, in part, by policies that discriminated against Tamils, politically and economically. Tamil insurgents went on to disrupt numerous elections, though some of their supporters have now become peaceful elected politicians. Refugees, mostly Tamils fleeing ongoing persecution, are still arriving in India, Indonesia, Australia, Papua New Guinea, and elsewhere.

Democratization and Conflict

Armed conflicts and genocides with high death tolls since 1945

- ❗ Ongoing conflict
- ✶ 1000–10,000 deaths
- ✶ 10,000–60,000 deaths
- ✳ 60,000–180,000 deaths
- ✱ 180,000–500,000 deaths
- ✱ 500,000–1,000,000 deaths

Democratization index
- Full democracy
- Flawed democracy
- Hybrid regime
- Authoritarian regime
- No data

INDONESIA
1073 miles/1728 kilometers

C The India–Pakistan border can be seen from space, thanks to the floodlights illuminating the fence that stretches for much of its length. The fence was built to discourage arms smuggling related to the region's many conflicts, especially the one in Kashmir.

New Delhi, India

Lahore, Pakistan

India-Pakistan, Border

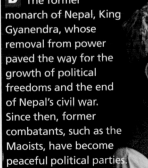

D The former monarch of Nepal, King Gyanendra, whose removal from power paved the way for the growth of political freedoms and the end of Nepal's civil war. Since then, former combatants, such as the Maoists, have become peaceful political parties.

Pakistan attempted to invade Kashmir again in 1965 but was defeated. India and Pakistan are technically still waiting for a United Nations (UN) decision about the final location of the border (Figure 8.20C). The Ladakh region of Kashmir (see the Figure 8.1 map) is the object of a more limited border dispute between India and China.

After years of military occupation, most Kashmiris now support independence from both India and Pakistan. However, neither country is willing to hold a vote on the matter. Anti-Indian Kashmiri guerrilla groups equipped with weapons and training from Pakistan have carried out many bombings and assassinations. Blunt counterattacks launched by the Indian government have killed large numbers of civilians and alienated many Kashmiris.

Another complication in the Kashmir dispute is the fact that both India and Pakistan—which came close to war against each other in 1999 and again in 2002—have nuclear weapons. Because of the nationalistic fervor of the protagonists, many see the conflict in Kashmir as more likely to result in the use of nuclear weapons than any other conflict in the world. Analysts agree that any use at all of nuclear weapons would have severe repercussions for everyone on Earth.

War and Reconstruction in Afghanistan

In the 1970s, political debate in Afghanistan became polarized. On one side were several factions of urban elites, who favored modernization and varying types of democratic reforms. Opposing them were rural, conservative religious leaders, whose positions as landholders and ethnic leaders were threatened by the proposed reforms. Divisions intensified as successive governments, all of which came to power through military coups, became more and more authoritarian. Political opponents were imprisoned, tortured, and killed by the thousands, resulting in a growing insurgency outside the major cities.

In 1979, fearing that a civil war in Afghanistan would destabilize neighboring Soviet republics in Central Asia, the Soviet Union invaded Afghanistan. Rural conservative leaders (often erroneously called "warlords") and their followers formed an anti-Soviet resistance group, the *mujahedeen*. As Afghan resistance to Soviet domination increased, the mujahedeen became ever more strongly influenced by militant Islamist thought and by Persian Gulf Arab activists who provided funding and arms. At the time, the United States, still searching for Cold War allies against the Soviet Union, joined with Pakistan in supporting the mujahedeen. Moderate, educated Afghans who favored democratic reforms fled the country during this turbulent time, hoping to go back eventually when peace returned.

In 1989, after heavy losses—14,000 Soviet soldiers killed and billions of dollars wasted—the Soviets gave up and left Afghanistan. Anarchy prevailed for a time as mujahedeen factions fought one another, adding to the 1.5 million civilians and combatants killed in the war with the Soviets.

In the early 1990s, a radical religious, political, and military movement called the **Taliban** emerged from among the mujahedeen. The Taliban wanted to control corruption and crime and minimize Western ways—especially those related to the

Taliban an archconservative Islamist movement that gained control of the government of Afghanistan in the mid-1990s

role, status, and dress of women—that had been introduced in earlier decades by the urban elites and reinforced or made more extreme by the Russian occupation. The Taliban wanted to strictly enforce *shari'a*, the Islamic social and penal code (see Chapter 6). Efforts by the Taliban to purge Afghan society of non-Muslim influences included greatly restricting women (see page 479), promoting only fundamentalist Islamic education, and publicly banning the production of opium, to which many Afghan men had become addicted, all the while privately promoting its sale to raise funds for their side. By 2001, the Taliban controlled 95 percent of the country, including the capital of Kabul.

Following the events of September 11, 2001, the United States and its allies focused on removing the Taliban, who had given shelter to Osama bin Laden and his international Al Qaeda network. By late 2001, the Taliban were overpowered by an alliance of Afghans supported heavily by the United States, the United Kingdom, and eventually NATO (see the Figure 8.20 map).

In 2003, the United States launched the war in Iraq that diverted national attention, troops, and financial resources away from Afghanistan. Almost immediately, the Taliban resurfaced, effectively thwarting the ability of Afghanistan's new government to ensure security and to meet the needs of people outside Kabul. Based in rural areas in both Pakistan and Afghanistan, the Taliban are now aided by widespread distrust of the government in Kabul, which is seen as corrupt (see Figure 8.20A). It appears that most people in Afghanistan favor a democratic government based on Muslim principles, but functionally flawed elections and corruption in government have eroded faith in democracy since the first elections in 2004.

In May 2011, bin Laden was killed in a raid by U.S. forces in the town of Abbottabad, Pakistan. Another raid resulted in the death of the next-highest Al Qaeda commander, and sporadic drone attacks killed both combatants and civilians. With the threat from Al Qaeda seemingly diminished, calls within the United States for a withdrawal from Afghanistan increased. Although 90,000 U.S. troops have returned home, approximately 10,000 will likely remain indefinitely.

Sri Lanka's Civil War

The Singhalese have dominated Sri Lanka since their migration from Northern India several thousand years ago. Today, they make up about 74 percent of Sri Lanka's population of 20.5 million people. Most Singhalese are Buddhist. Tamils, a Hindu ethnic group from South India, make up about 18 percent of the total population of Sri Lanka. About half of these Tamils have been in Sri Lanka since the thirteenth century, when a Tamil Hindu kingdom was established in the northeastern part of the island. The other half were resettled by the British in the nineteenth century to work on tea, coffee, and rubber plantations. Some Tamils have done well, especially in urban areas, where they dominate the commercial sectors of the economy. However, many others have remained poor laborers isolated on rural plantations.

Upon its independence in 1948, Sri Lanka had a thriving economy led by a vibrant agricultural sector and a government that made significant investments in health care and education. It was poised to become one of Asia's most developed economies. But, driven by nationalism, Singhalese was made the only official language and Tamil plantation workers were denied the right to vote. Efforts were also made to deport hundreds of thousands of Tamils to India. In the 1960s, the government shifted investment away from agricultural development and toward urban manufacturing and textile industries, which were dominated by Singhalese. By 1983, the Tamil minority, lacking political power and influence, chose to use guerilla warfare against the Singhalese, mounting an army known as the Tamil Tigers.

For more than 30 years, the entire island was subjected to repeated terrorist bombings and kidnappings. Peace agreements were attempted several times, but in the end it was an overwhelming military victory by the government, combined with an effective crackdown on international funding for the Tamils, that forced the Tamil surrender in May 2009. However, dissatisfaction with the peace process has resulted in an ongoing flow of Tamil and other refugees from Sri Lanka (see Figure 8.20B). Tamils have since asked the UN to investigate war crimes committed during the war, but the Sri Lankan government has insisted on conducting its own investigations.

Despite many years of violence, economic growth has been surprisingly robust in Sri Lanka. Driven by strong growth in food processing, textiles, and garment making, Sri Lanka is today one of the wealthiest nations in South Asia on a per capita GDP basis, and it provides to a relative degree for the well-being of most of its citizens.

Nepal's Rebels

After a civil war that ended generations of monarchical rule, Nepal has endured years of political turmoil. An elected legislature and multiparty democracy were introduced in Nepal in 1990, but until 2008, a royal family governed with little respect for the political freedoms of the Nepalese people.

In 1996, inspired by the ideals of the late Chinese leader Mao Zedong (but with no apparent support from China), Maoist revolutionaries took advantage of public discontent and waged a "people's war" against the Nepalese monarchy. Following a decade of civil war, during which 13,000 Nepalese died, the Maoists both took military control of much of the countryside and developed strong political support among most Nepalese. Persistent poverty and lack of the most basic development under the dictatorial rule of then-monarch King Gyanendra (see Figure 8.20D) led to massive protests that forced Gyanendra to step down in 2006. Soon thereafter, the Maoists declared a cease-fire with the government.

In 2008, the Maoists won sweeping electoral victories that gave them a majority in parliament and made their former rebel leader prime minister. Then, in May 2009, with his many conditions for reforming Nepalese society still unmet, the Maoist prime minister resigned in a tactical parliamentary move and brought his party into opposition. While the Maoists and other parties collaborated in writing a new constitution, divisive issues have hindered its implementation. These include power sharing between competing political parties, the dominance of forward castes within the government and most political parties, and deciding how the country should be divided into ethnic states. Observers concur, however, that if Nepal's various factions come to see peaceful democratic processes as giving them a voice in their own government, Nepal will probably not return to civil war.

Movements Against Government Inefficiency and Corruption

In recent years, reform movements have gained support from the many South Asians frustrated by government inefficiency, corruption, religious nationalism, caste politics, and the failure of governments to deliver on their promises of broad-based prosperity. Bureaucrats who demand bribes have lately been the focus of such activism. Confronted by a bureaucrat who asked for nearly $200 to issue a legitimate income tax refund, one Indian couple in Bangalore launched the website I Paid a Bribe, aimed at collecting information about crooked officials. The idea caught on quickly and similar sites now exist in more than 17 countries. Kisan Baburao "Anna" Hazare of Maharashtra, a former military man who advocates that those convicted of corruption lose a hand as punishment, is a more militant anticorruption crusader in India. Although most Indians quickly backed off such extreme measures, Hazare has attracted a following in the new middle class, especially among educated women.

Some aspects of the high-tech revolution in South Asia are likely to make it more difficult for bureaucrats and elected officials to demand bribes. For example, cell phones are increasingly being used by governments to pay civil servants and contractors. Because transactions over a cell phone are more traceable, it is now harder for upper-level government officials to demand bribes from the people they supervise in return for issuing their paychecks. Another example is the *Aadhaar Project*, an ambitious effort to provide all 1.2 billion Indians with a unique photo, digitized ID card (UID). By October 2012, a total of 208 million cards had been issued, with complete coverage of India expected before 2020. The UID project is designed to facilitate the distribution of government benefits of all types, much of which currently fail to reach those most in need because there are few formal records that can be used to identify them. Currently, much of the money intended to reach such people ends up in the hands of corrupt officials.

Greater participation by women in the political process may or may not reduce corruption. A recent study by the World Bank showed that Indian village councils which included a greater number of women were slightly less likely to suffer from corruption. One explanation for this is that women leaders are more likely to spend money on smaller projects related to improving schools and clinics, rather than "big ticket" items such as road building that are notorious sources of corruption. However, another study found more corruption in village councils where the position of council leader is reserved for a woman by law. In these localities, officials were more able to engage in corruption because of less effective oversight by less experienced council leaders.

THINGS TO REMEMBER

GEOGRAPHIC THEME 3 • **Power and Politics:** India, South Asia's oldest, largest, and strongest democracy, has shown that the expansion of political freedoms can ameliorate conflict. Across the region, when people have been able to participate in policy-making decisions and implementation—especially at the local level—seemingly intractable conflict has been defused and combatants have been willing to take part in peaceful political processes.

- Religious nationalist movements are increasingly attractive to people frustrated by government inefficiency, corruption, and caste politics, and by the failure of governments to deliver on their promises of broad-based prosperity.

- Despite decades of effort to fight its influence, caste still dramatically affects politics in India.

- The most intense armed conflicts in South Asia today are regional conflicts in which nations dispute territorial boundaries or a minority actively resists the authority of a national or state government.

URBANIZATION

GEOGRAPHIC THEME 4

Urbanization: South Asia has two general patterns of urbanization: one for the rich and the middle classes and one for the poor. The areas that the rich and the middle classes occupy include sleek, modern skyscrapers bearing the logos of powerful global companies, universities, upscale shopping districts, and well-appointed apartment buildings. The areas that the urban poor occupy are chaotic, crowded, and violent, with overstressed infrastructures and menial jobs. These two patterns often coexist in very close proximity.

Although today only 33 percent of the region's population live in urban areas, of the world's top 20 largest metropolitan areas, five are located in South Asia. By 2017, Delhi had the third largest urban population in the world (after Tokyo, Japan, and Jakarta, Indonesia) at 26 million people; Karachi had 22 million; Mumbai 18 million; Dhaka 16 million; and Kolkata 15 million. All these cities have grown quickly and all now have massive slums **(FIGURE 8.21D)**. This growth is likely to continue, and South Asia's current urban population of about 580 million people could expand to as many as 720 million by 2025. Even so, this region's huge rural populations mean that South Asia will not be majority urban until 2045 or later.

Many middle-class South Asians move to cities for education, training, or business opportunities (Figure 8.21C). Because people come to cities where schooling is more readily available, large cities have higher literacy rates than rural areas. Mumbai and Delhi both have literacy rates above 80 percent, approximately 25 percent above the average for the country; Dhaka, in Bangladesh, and Karachi, in Pakistan, both have 63 percent literacy, more than 15 percent above the rate for each country as a whole.

MUMBAI

Bombay is the name by which most Westerners know South Asia's wealthiest, most culturally influential city. It is now called Mumbai, after the Hindu goddess Mumbadevi. Mumbai has the largest deepwater harbor on India's west coast. With more than 22 million people, it is not South Asia's largest metropolitan area but it is the wealthiest, hosting India's largest stock exchange, and paying about a third of the taxes collected in the entire country and bringing in nearly 40 percent of India's trade revenue. While Bangalore is known as India's "Silicon Valley," Mumbai is a close second.

Thinking Geographically

After you have read about urbanization in South Asia, you will be able to answer the following questions.

A and **D** What circumstances have brought South Asians into crowded urban districts, like the two shown in these photos of Mumbai and Dhaka?

C What is notable about the members of this classroom?

Mumbai's wealth also extends into the realm of culture through the city's flourishing creative arts industries, including its internationally known film industry, which produces the most films of any film industry in the world (around four times as many as Hollywood). Known as "Bollywood" (a combination of *Bombay* and *Hollywood*), Mumbai's film industry produces popular Hindi movies mostly focusing on love, betrayal, and family conflicts, all played out on lavish sets and accompanied by popular music and dance. Bollywood sells about twice as many tickets as Hollywood in a given year, but makes about one-sixth the profits.

Mumbai's wealth is most evident when one looks up at the elegant high-rise condominiums built for the city's rapidly growing middle class. But at street level, the urban landscape is dominated by great numbers of people living on the sidewalks, in narrow spaces between buildings, and in large, rambling shantytowns (see Figure 8.21A). The largest of the shantytowns is Dharavi, which houses up to a million people in less than 1 square mile (2.6 square kilometers). Widely considered Asia's most populous slum, Dharavi is rivaled in size only by Orangi Township in Karachi, Pakistan. Dharavi is said to contain 15,000 one-room factories turning out thousands of products that are sold in the global marketplace, many of them from India's recycled plastic and metal.

There is yet a third aspect to urban life in South Asia. Were one to carefully observe places as widely separated and culturally different as Mumbai, Chennai, Kathmandu, and Peshawar, one would discover that beyond the main avenues and shantytowns, these cities also contain thousands of tightly compacted, reconstituted villages where daily life is intimate and familiar, not anonymous as in Western cities. Koli is one such place (see the vignette below).

VIGNETTE Koli is an ancient fishing village that predates the city of Mumbai and is now squeezed between Mumbai's elegant coastal high-rises and the Bay of Mumbai (FIGURE 8.22). Ringed by fishing boats, Koli is a labyrinth of low-slung, tightly packed homes. Some villagers still fish every day. The screeching of taxis and buses is soon lost in quiet calm as one ducks into a narrow covered passageway that winds through the village and branches in multiple directions. At first, Koli appears impoverished, but inside, well-appointed homes, some with marble floors, TVs, and computers, open onto the dimly lit but pleasant alleys. The visitor soon learns that this is no slum or warren of destitute shanties, but rather a community of educated bureaucrats, tradespeople, and artisans who constitute South Asia's rising urban middle class. *[Source: From the field notes of Lydia and Alex Pulsipher, and their colleague Tom Osmand.]* ■

FIGURE 8.21 PHOTO ESSAY: Urbanization in South Asia

A High-rise luxury apartments overlooking the Dhobi Ghat slum in Mumbai, India, illustrate the enormous disparities in wealth to be found in urban South Asia.

10,000 people work in Dhobi Ghat, washing clothes and linens from hotels and hospitals in the concrete enclosures shown here.

Dwellings of people who work as clothes washers.

While only 33 percent of South Asia's population is urban, the region is home to some of the world's largest cities, such as Mumbai, Delhi, Dhaka, Kolkata, and Karachi. Throughout the area, between 50 and 90 percent of the urban population live in slums plagued by shoddy construction and inadequate access to water and sanitation.

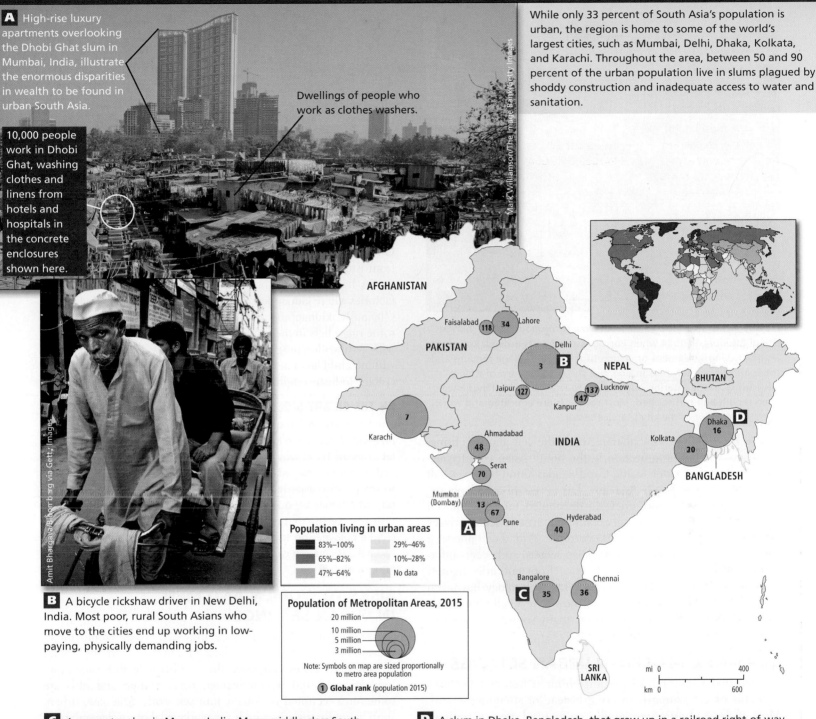

Population living in urban areas

■ 83%–100%	29%–46%
65%–82%	10%–28%
47%–64%	No data

Population of Metropolitan Areas, 2015

20 million
10 million
5 million
3 million

Note: Symbols on map are sized proportionally to metro area population

① **Global rank** (population 2015)

B A bicycle rickshaw driver in New Delhi, India. Most poor, rural South Asians who move to the cities end up working in low-paying, physically demanding jobs.

C A computer class in Mysore, India. Many middle-class South Asians migrate to cities and remain there to take advantage of educational opportunities and the availability of jobs.

D A slum in Dhaka, Bangladesh, that grew up in a railroad right-of-way. Many slums develop in areas that have clear risks to human habitation and few safeguards against potentially deadly hazards.

FIGURE 8.22 The village of Koli. Now surrounded by the vibrant city of Mumbai, in 1534 when Portuguese colonists arrived, Koli was a small fishing settlement on a beautiful bay. The village remains and there are still some fishers in Koli, but this exterior view of Koli is misleading. In the interior, out of sight to strangers, dwellings have been rebuilt and refurbished. Most residents are educated and work as bureaucrats or in the city's service sector.

As it is doing in sub-Saharan Africa, the cell phone is bringing urban-style communication to the South Asian countryside. Mumbai-based reporter Anand Giridharadas found that over half the population of South Asia now has access to a cell phone. Even though this access is unevenly distributed, for many the cell phone has become a means of creating the type of social privacy that the bedroom stands for in Western life. With a cell phone, a person can conduct relationships, keep secrets, and access information, all while avoiding interference from authority figures. One measure of how rapidly cell phone technology has taken over is that many more South Asians now have a cell phone than have the use of a flush toilet. Indeed, many young South Asians now view IT access as essential.

RISKS TO CHILDREN IN URBAN SETTINGS

In South Asia, around 10 million children between the ages of 5 and 14 are primarily working instead of attending school. Most of these children are employed in rural areas as agricultural laborers with their families. In urban areas, children work in much riskier settings, usually outside the home and without supervision by a family member. Many children labor as domestic servants, in small export-oriented factories, as street vendors, as scavengers at dumpsites, or in **sex work** (the selling of sexual acts for a fee).

Clearly, many of these occupations are entirely unsafe and inappropriate for children, who should be protected from such exploitation. But a more complex image emerges when one considers the many urban craft industries, in which children are often trained in an ancient art and

sex work the provision of sexual acts for a fee

supervised by family members, but nonetheless denied full access to education. For example, carpet weaving is an ancient artistic and economic enterprise in South Asia. Traditionally, it has been a family-run enterprise, with women and children serving as weavers and men as merchants. For thousands of years, young children have learned weaving skills from their parents and become proud members of the family's home-based production unit. But in today's rapidly modernizing society, the role of children as family workers must be balanced with the need for children to attend school. One of the chief benefits of urban life is education, where children learn the skills that will enable them to survive and prosper in a modern economy, such as the ability to read, write, and do basic math.

As global trade has increased the demand for fine, hand-woven carpets, most of the profit has gone not to the weaving families but to South Asian middlemen and foreign traders from Europe and America. The possibility of reaping such large profits has led some unscrupulous carpet merchants to set up factories where kidnapped children are forced to produce carpets. Obviously, kidnapping and enslavement must be stopped, and some rug sellers in the United States and Europe have responded with certification programs that ensure their merchandise is made with no child labor at all. The question remains: Is there room for cottage industry employment of children within the family circle?

ON THE BRIGHT SIDE: Can Child Labor Be "Fair Labor"?

The UN, South Asian governments, and nongovernmental organizations (NGOs) like RugMark are addressing the issue of child labor in the hand-woven carpet industry. They have instituted an active program to curb child labor abuses while remaining open to the positive experience of a child learning a skill and becoming part of a family production unit. India has established a national system to certify that exported carpets are made in shops whose child employees go to school, eat an adequate midday meal, and receive basic health care. Such carpets bear the label "Kaleen" or "RugMark." ∎

RISKS TO WOMEN AND MEN IN URBAN SETTINGS

When rural adults move to cities, the need to find employment, the crowded conditions, the lack of housing, and the welter of new experiences can leave them vulnerable to exploitation. Inexperienced and undereducated rural women and girls are sometimes recruited or forced into sex work. The vast, urban, slum-based brothels in which sex workers labor are legendary in the cities of South Asia and have been documented in, among other places, *Half the Sky,* a book and video created by Nicholas Kristof and Sheryl WuDunn.

Brothels are notoriously exploitative of women wherever they are found, but the age of the mobile phone is changing the geography of sex work by allowing sex workers to move out of brothels. As a result, they can develop more control over those who will become their clients, and manage their incomes, which in former times were immediately seized by pimps and madams. However, being spatially autonomous also leaves them less likely to have access to condoms or to learn how to avoid exposure to HIV, for which sex workers remain at high risk.

Rural men new to cities are also at risk: on-the-job accidents, exposure to HIV, extreme pressure to support families on low wages by working long hours at physically demanding jobs (see Figure 8.21B), and the loss of camaraderie with fellow villagers are just a few of the hardships experienced by such men.

THINGS TO REMEMBER

GEOGRAPHIC THEME 4   • **Urbanization:** South Asia has two general patterns of urbanization: one for the rich and the middle classes and one for the poor. The areas that the rich and middle classes occupy include sleek, modern skyscrapers bearing the logos of powerful global companies, universities, upscale shopping districts, and well-appointed apartment buildings. The areas that the urban poor occupy are chaotic, crowded, and violent, with overstressed infrastructures and menial jobs. These two patterns often coexist in very close proximity.

• Mumbai is South Asia's wealthiest city and is a showcase of the economic disparity that characterizes South Asian cities.

• Some South Asian urban neighborhoods share many of the intimate qualities of village life.

• When rural people move to cities, the loss of village constraints on behavior, the need to find employment, the crowded conditions, the lack of housing, and the welter of new experiences can leave them vulnerable to exploitation.

POPULATION AND GENDER

GEOGRAPHIC THEME 5

Population and Gender: In this most populous of world regions, population growth is slowing as the demographic transition takes hold. Birth rates are falling due to rising incomes, urbanization, better access to health care, and the fact that women are finding more opportunities to study and work outside the home, and thus are delaying childbearing and having fewer children. However, a severe gender imbalance is developing in this region due to age-old beliefs that males are more useful to families than are females. As a result, adult males significantly outnumber adult females.

Since overtaking East Asia in 2008, South Asia is now the most populous world region. It is also the smallest world region and the most densely populated (**FIGURE 8.23**; see also Figure 1.26). India alone will have overtaken China by 2030, at which point China's population will likely be shrinking at the rate of about 20 million people per decade, while India's will be growing by over 100 million per decade. However, a notable trend is that current rates of population growth have slowed to approximately 1.15 percent and fertility is declining significantly. Only in Afghanistan and Pakistan are fertility rates still above 2.4 children per woman (**FIGURE 8.24**). Families are choosing to have fewer children for a variety of reasons, such as improved educational and employment opportunities for women, better health care, and urbanization.

Population density in this region is highest along river valleys and coasts. The Ganga-Brahmaputra delta in Bangladesh and India has the highest densities (see Figure 8.23). Densities in the delta's rural agricultural and fishing areas, inhabited for tens of thousands of years, often reach levels associated with urban and suburban areas in most parts of the world. Other major areas of settlement include the entire Ganga River basin, the Indus River basin in Pakistan, and the area centered on Lahore known as the Punjab, named for the five rivers that flow across the foothills of the Himalayas to the Indus River.

GEOGRAPHIC PATTERNS IN THE STATUS OF WOMEN

On average, women's literacy rates, social status, earning power, and welfare are generally lowest in the belt that stretches from the northwest in Afghanistan across Pakistan, western India, Nepal, Bhutan, and the Ganga Plain into Bangladesh. Women fare somewhat better in eastern and central India and considerably better in southern India and in Sri Lanka. In these latter regions, where literacy rates are higher (**FIGURE 8.25**), different marriage, inheritance, and religious practices give women better access to education and resources.

SLOWING POPULATION GROWTH: HEALTH CARE, URBANIZATION, AND GENDER

Efforts to slow population growth through better health care have been under way for more than 50 years. Today, urbanization and the rising status of women are also helping to reduce the birth rate.

With improved health care—such as rehydration to control the effects of diarrhea in infants—far fewer babies are dying in infancy. In 1992, infant mortality rates were 91 per 1000 live births in India, 109 in Pakistan, and 120 in Bangladesh. By 2016, those rates were reduced to 42 in India, 69 in Pakistan, and 38 in Bangladesh. With better assurance that their babies will survive to adulthood and be able to care for their elderly parents, couples now choose more often to have just two children.

At the same time that improved access to health care has helped propel the reduction in family size, the shift in population from rural to urban areas has reduced the economic incentive for having a large family (see the Figure 8.21 map). Whereas children in rural areas (starting at an early age) contribute their labor to farming, thus increasing the family income, the labor of children in urban areas is less likely to boost the family income. Children in urban areas are more likely to attend school, and thus represent a cost to the family in terms of school fees, books, and uniforms.

Improvements in the status of women have also slowed population growth. As in other parts of the world, women in South Asia who have the opportunity to go to school and/or work at a paying job tend to delay childbearing and have fewer children.

A comparison of the population pyramids as well as some other statistics for Sri Lanka and Pakistan helps illustrate the influence of health care and the status of women on birth rates (**FIGURE 8.26**; see also Figures 8.24 and 8.25). As a country that is far along but not completely through a demographic transition (see Figure 1.28), Sri Lanka has a much higher GNI per capita

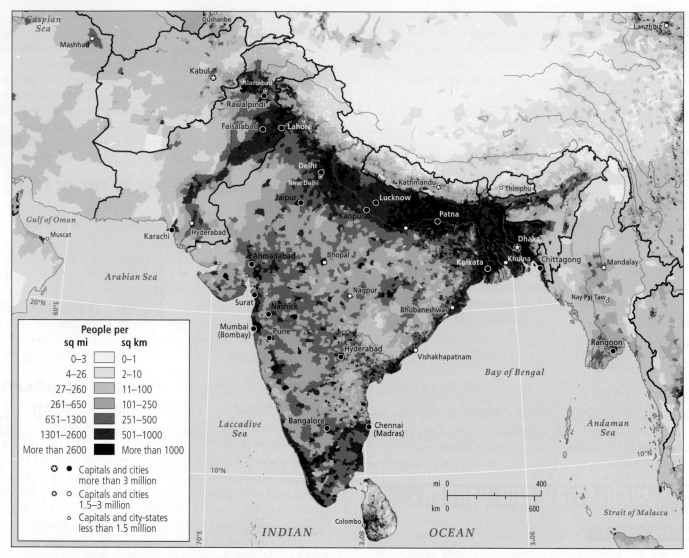

FIGURE 8.23 Population density of South Asia.

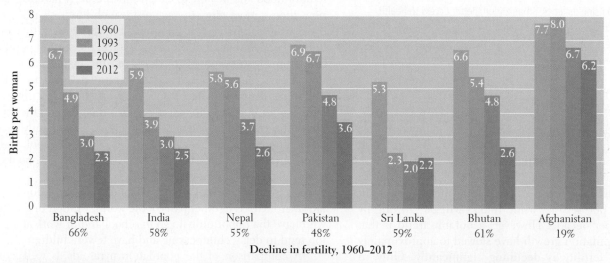

FIGURE 8.24 Total fertility rates in South Asia, 1960–2012. All South Asian countries except Afghanistan have experienced a substantial decline in fertility since 1960. Afghanistan's rate has declined less than 20 percent over the past 50 years, probably because of almost continuous conflicts, the overwhelmingly rural population (77 percent), and the low literacy rates for both men (43.1 percent) and women (12.6 percent). [*Sources consulted: World Population Data Sheet 2012,* Population Reference Bureau, http://www.prb.org/pdf12/2012-population-datasheet_eng.pdf; "Bhutan—Fertility Rate: Fertility Rate, Total (births per woman)," Index Mundi, http://www.indexmundi.com/facts/bhutan/fertility-rate; "Afghanistan—Fertility Rate: Fertility Rate, Total (births per woman)," Index Mundi, http://www.indexmundi.com/facts/afghanistan/fertility-rate; "Afghanistan," *World Fact Book 2012,* Central Intelligence Agency, https://www.cia.gov/library/publications/the-world-factbook/geos/af.html.]

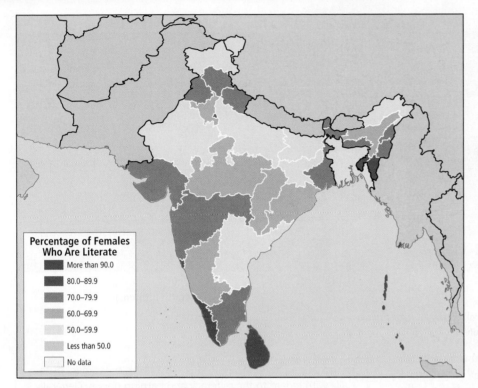

FIGURE 8.25 Female literacy in South Asia. Female literacy lags behind male literacy in all countries in South Asia, and is below 50 percent in four countries. Female literacy is crucial to improving the lives of not only women, but also children: women who can read often seize opportunities to earn an income and nearly always use this income to help their children. In India, overall literacy has risen in the past decade, and in 2011, female literacy there reached 65 percent, up from 54.5 percent in 2005. [Sources consulted: "District Wise Female Literacy Rate of India," Maps of India, http://www.mapsofindia.com/census2001/femaleliteracydistrictwise .htm; *United Nations Human Development Report 2009* (New York: United Nations Development Programme), Table J, "Gender-Related Development Index and Its Components," http://hdr.undp.org/en/reports/global/hdr2009/; "Map of Literacy Rate in India," Maps of India, http://www.mapsofindia .com/census2011/literacy-rate.html; "Literacy," *World Fact Book*, Central Intelligence Agency, 2016, https://www.cia.gov/library/publications /the-world-factbook/ields/2103.html#af.]

(U.S.$10,370) than Pakistan (U.S.$5090). Health care is generally better in Sri Lanka, as indicated by its much lower infant mortality rate (in 2016, Sri Lanka had 9 deaths per 1000 live births versus Pakistan's 69 per 1000). Sri Lanka has also worked harder to educate and economically empower its women. Three key indicators of women's education and empowerment that are associated with reduced fertility rates are much higher in Sri Lanka than in Pakistan: female literacy (92 percent versus 46 percent), the percentage of young women who attend high school (73 percent versus 19 percent), and the percentage of women who work outside the home (36 percent versus 24 percent). Surprisingly, Sri Lanka has achieved all this with an urbanization rate of just 18 percent, far lower than that of Pakistan's 38 percent.

Despite Pakistan's rather dismal statistics compared to those of Sri Lanka, its population pyramid does indicate that birth rates have slowed significantly (notice that the bottom of the pyramid in Figure 8.26B narrows noticeably). If this trend continues, Pakistan, as well as all South Asian countries except Afghanistan,

will enjoy what is called the *demographic dividend*: a surge of young people entering the workforce and choosing to have fewer children, who will thus have the time and energy to study, work, and contribute to the economy and civil society. South Asian countries that have experienced a demographic dividend will have to deal with an aging population for which there will be fewer younger caregivers and taxpayers to support the elderly. Current population growth trends suggest that this transformation may take 50–100 years in the case of Pakistan, and 20–40 years in the case of Sri Lanka.

GENDER IMBALANCE

South Asian populations have a significant gender imbalance because of cultural customs that make sons more likely than daughters to contribute to a family's wealth. A popular toast to a new bride is "May you be the mother of a hundred sons." Many middle-class couples who wish to have sons hire high-tech laboratories that specialize in identifying the sex of a fetus, the intention being to abort female fetuses. This practice is now illegal, but it persists. Poorer South Asians may neglect the health of female children, with some even committing female infanticide (the dowry implications of this are discussed on page 478). India's census of 2011 indicated a gender imbalance of between 30–70 million women, and similar situations are developing in other South Asian countries.

Recent data suggest that India's gender imbalance may be diminishing, but even if the balance were to return to natural levels, the imbalance already created will likely result in as many as 30 percent more men looking to marry than women by 2050. Gender imbalance on this scale could create serious problems, such as surges in crime, drug abuse, and political violence related to the presence of so many young men with no prospect of having a family. In all cultures, the possibility of building a family is a stabilizing influence for young men.

Notably, the one Indian state where women outnumber men is Kerala, where the government has made women's development a priority by funding education for women well beyond basic levels. The results are reflected in its female literacy rate, which is the highest in India at 92 percent (the average for India as a whole is 63 percent; see Figure 8.25 for the female literacy map), and in the number of women who work outside the home. In Kerala, far from being viewed as an economic liability to the family, daughters are seen as an asset. Women can travel the public streets alone and in groups, and commonly work in public places, with many holding high positions in government, education, health care, IT industries, and other professions.

MEASURES OF HUMAN WELL-BEING

Maps using different indicators to measure well-being show that the overall state of human well-being in South Asia is low **(FIGURE 8.27)**. Most of the region has an average annual GNI per capita PPP of

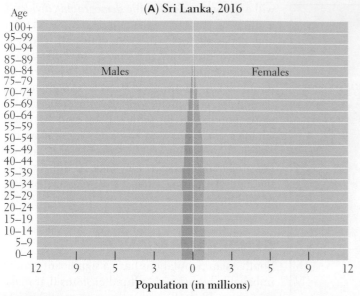

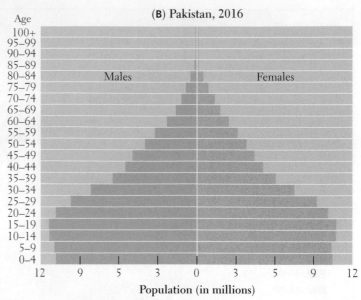

FIGURE 8.26 Population pyramids for Sri Lanka and Pakistan. [*Source consulted:* International Data Base, U.S. Census Bureau, http://www.census.gov/population/international/data/idb/informationGateway.php.]

around U.S.$5000 (see Figure 8.27A) and ranks low on human development and gender development. What these maps fail to reveal is that incomes and human well-being are rising rapidly. India's GNI per capita PPP has doubled since 2005, and Sri Lanka has made the transition from medium to high human development in just the last three years. The maps also fail to show the cities and subnational divisions that have relatively high human development (for example, the state of Kerala in India), but these variances get lost in the countrywide average.

Figure 8.27C illustrates gender inequality, reflecting inequality in achievements between women and men along three dimensions: reproductive health, empowerment, and participation in the labor market. Again, most of South Asia ranks in the low and very low categories, with only Bhutan, Nepal, and Sri Lanka ranking in the medium and medium low categories. Based on the global thumbnail map, it is clear that in terms of gender inequality, South Asia is in a league with only Southwest Asia and sub-Saharan Africa, but virtually nowhere else.

While South Asia's record in human development is not impressive and may even be backsliding a bit because of the global recession, there have nevertheless been economic progress and some policy innovations across the region, as the next section will demonstrate.

THINGS TO REMEMBER

GEOGRAPHIC THEME 5 • **Population and Gender:** In this most densely populated of world regions, population growth is slowing as the demographic transition takes hold. Birth rates are falling due to rising incomes, urbanization, better access to health care, and the fact that women are finding more opportunities to study and work outside the home, and thus are delaying childbearing and having fewer children. However, a severe gender imbalance is developing in this region due to age-old beliefs that males are more useful to families than are females. As a result, adult males significantly outnumber adult females.

• There is a geographic pattern to the status of women, especially with regard to their literacy and earning power. Generally speaking, women are less restricted in the southern and eastern sections of this region.

• While population growth is slowing across the region, momentum provided by a young population means that growth will continue for the foreseeable future.

SOCIOCULTURAL ISSUES

Within the life of one South Asian village or urban neighborhood, there can be considerable cultural variety. Differences based on caste, economic class, ethnic background, gender, religion, and even language are usually accommodated peacefully by long-standing customs, such as religious and ethnic festivals and foodways (**FIGURES 8.28** and **8.29**), that guide cross-cultural interaction. However, South Asia is also undergoing a number of social changes, political shifts, and cultural transitions that challenge many traditional practices, especially those related to gender.

THE TEXTURE OF VILLAGE LIFE

The vast majority—about 70 percent—of South Asians live in the region's hundreds of thousands of villages. Even many of those now living in South Asia's giant cities were born in a village or occasionally visit an ancestral rural community, so for most people village life is a common experience.

VIGNETTE The anthropologist Faith D'Aluisio and her colleague Peter Menzel offer another peek into village life as night falls in Ahraura, a village in the state of Uttar Pradesh in north-central India. In the enclosed women's quarters of a walled compound, Mishri is finishing her day by the dying cooking fire as her 1-year-old son tunnels his way into her sari to nurse himself to sleep. Mishri, who is 27, lives in a tiny world bounded by the walls of the

FIGURE 8.27 Maps of human well-being.

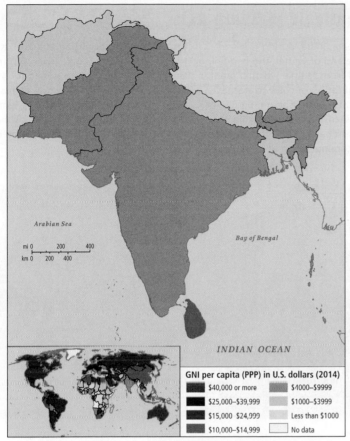

(A) Gross national income (GNI) per capita, adjusted for purchasing power parity (PPP)

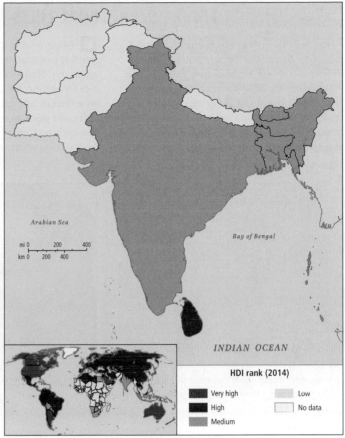

(B) Human Development Index (HDI)

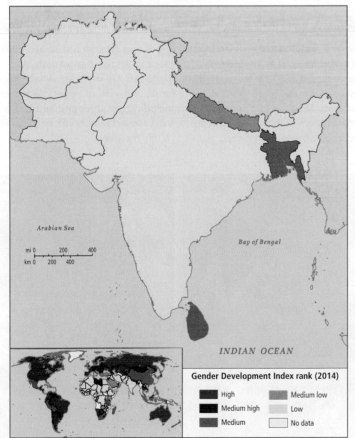

(C) Gender Equality Index (GEI)

courtyard she shares with her husband, five children, and several of her husband's kin. Like many villages in northern India, her village observes the practice of purdah, in which women keep themselves apart from men (see the discussion below).

That Mishri can observe purdah is a mark of status because it shows she need not help her husband in the fields. Within the compound, she works from sunup to sundown, chatting only momentarily with two women who cover their faces and scurry from their own courtyards to hers for the short visit. Mishri is devoted to her husband, who was chosen for her by her family when she was 10; out of respect, she never says his name aloud. [*Source: Adapted from Faith D'Aluisio and Peter Menzel, Women in the Material World (Berkeley, CA: Counterpoint, 1996.)*] ■

VIGNETTE The writer Richard Critchfield, who studied village life in more than a dozen countries, wrote that the village of Joypur (in Bangladesh) in the Ganga-Brahmaputra delta is set in "an unexpectedly beautiful land, with a soft languor and gentle rhythm of its own." In the heat of the day, the village is sleepy: naked children play in the dust, women meet to talk softly in the seclusion of courtyards, and chickens peck for seeds.

In the early evening, mist rises above the rice paddies and hangs there "like steam over a vat." It is then that the village comes to life, at least for the men. The men and boys return from the fields, and after a meal in their home courtyards, the men

FIGURE 8.28 LOCAL LIVES: Festivals in South Asia

A A village festival in Pakistan features *kabaddi,* a popular South Asian sport in which teams take turns sending a "raider" across a field's center line. That person must tag, or in some cases wrestle to the ground, members of the other team and then return to his or her own side without taking a breath. Kabaddi has been played at the Indian National Games since 1939 and at the Asian Games since 1991.

B Celebrants in Kolkata, India, during Holi—a festival celebrating the end of winter and beginning of spring. Holi evolved from temple worship practices involving the application of color to statues. In a riotous and celebratory atmosphere, people of different ages, genders, castes, and economic backgrounds temporarily disregard their differences and hurl the colors of the coming spring at each other.

C Pilgrims participating in a 2010 Kumbh Mela bathe in the Ganga River at Haridwar. During this event, which is held for 45 days every 3 years in different cities across north and central India, Hindus purify themselves by bathing in the sacred waters of the river. Recent Kumbh Melas have attracted over 120 million participants.

FIGURE 8.29 LOCAL LIVES: Foodways in South Asia

A Boys in Hampi, Karnataka, eat a south Indian *thali,* or feast, featuring rice, lentils, and various other vegetarian dishes served on a banana leaf. Most South Asian cuisine is eaten with the hands; in South India and Sri Lanka, banana leaves often serve as plates.

B Chapatis are cooked on a griddle at a wedding in Rajasthan in northwest India. Chapati is a kind of bread made of stone-ground wheat flour and cooked on a griddle or curved iron *tava* located above a fire or oven. The bread is unleavened—it is made of dough that never rises because it doesn't contain yeast. Most popular in northern South Asia, where wheat is more widely grown than rice, chapatis are torn into pieces and used to scoop up vegetable or meat dishes.

C Street food in a crowded marketplace in Dhaka, Bangladesh, during Ramadan. On offer are whole chickens, lamb, and various vegetable dishes, including potatoes and lentils. Usually, the food is taken home and shared with family members.

come "to settle in groups before one of the open pavilions in the village center and talk—rich, warm Bengali talk, argumentative and humorous, fervent and excited in gossip, protest, and indignation" as they discuss their crops, an upcoming marriage, or national politics. [Source: Richard Critchfield, Villages (New York: Anchor Press/Doubleday, 1981).] ■

SOCIAL PATTERNS IN THE STATUS OF WOMEN

The status of women in South Asia varies significantly along rural/urban and religious divides. Generally speaking, rural women have far less freedom than do urban women. In rural India, middle- and upper-caste Hindu women are often more restricted in their movements than are lower-caste women because they have a status to maintain. Meanwhile, lower-caste women who go into public spaces may have to contend with sexual harassment and exploitation from upper-caste men. The socioeconomic status of Muslim women in South Asia is lower than that of their Hindu and Christian counterparts. In India, some of this is related to the generally lower incomes and standard of living for Muslims, which also usually means lower educational levels. Muslim women also work outside the home less than non-Muslim women in India. Low rates of education and workforce participation for women also prevail in Muslim-dominated countries such as Pakistan and Afghanistan, though rates are significantly higher in Bangladesh.

Purdah

The practice of concealing women from the eyes of nonfamily men, especially during women's reproductive years, is known as **purdah.** It is observed in various ways across the region. The practice is strongest in Afghanistan and across the Indo-Gangetic Plain, where within both Muslim and Hindu communities, women are often secluded within structures **(FIGURE 8.30)** and wear veils or head coverings. Purdah is less strict in central and southern India, but even there, separation between unrelated men and women is maintained in public spaces. In general, low-caste Hindus do not observe this custom, but that is changing. In recent decades, as some low-status households have increased their incomes, they have adopted purdah as a sign of their rising wealth.

Purdah practices have influenced the architecture of South Asia. Homes are often situated in walled compounds that seclude kitchens and laundries as women's spaces. In grander homes, windows to the street are usually covered with lattice screens, known as *jalee,* that allow in air and light but shield women from the view of outsiders.

Marriage, Motherhood, and Widowhood

Throughout South Asia, most marriages are arranged by the parents of the prospective bride and groom. Particularly in wealthier, better-educated families, the wishes of the bride and groom are considered, but in some cases they are not **(FIGURE 8.31)**. This is especially true of a child marriage, when a young girl (often as young as 12 and in some places even younger) is married off to a much older man.

Usually, a bride goes to live in her husband's family compound, where she becomes a source of domestic labor for her

FIGURE 8.30 Material culture of purdah. A lattice screen in Fatehpur Sikri, India. Known as *jalee,* lattice screens are often found in parts of South Asia where women are secluded. Like the louvers and latticed windows of Southwest Asia (see Figure 6.27B), *jalee* allow ventilation and let in light but shield women from the view of strangers.

mother-in-law. Most brides work at domestic tasks for many years until they have produced enough children to have their own crew of small helpers, at which point they gain some prestige and a measure of autonomy.

Motherhood in South Asia determines much about a woman's status within her community. A woman's power and mobility increase when she has grown children and becomes a mother-in-law herself. On the other hand, in some communities, the death of a husband, regardless of cause, is a disgrace to a woman and can completely deprive her of all support and even her home, children, and reputation. Widows may be ritually scorned and blamed for their husband's death. Widows of higher caste rarely remarry, and in some areas, they become bound to their in-laws as household labor or may be asked to leave the family home. Most simply become marginalized "aunties" in extended families and help with household duties of all sorts.

purdah the practice of concealing women from the eyes of nonfamily men

(A) A barber, often a member of a "barber caste" or *jati,* practices his trade in Chandigarh, India. Because barbers come into contact with so many people, they are often sought out by parents seeking a spouse for their children.

FIGURE 8.31 Caste and marriage. Caste remains a powerful force with respect to marriage. Fewer than 5 percent of marriages between Hindus cross subcaste or *jati* lines.

(B) A Brahmin priest officiates at a Hindu wedding ceremony in Varanasi, India. The specifics of the ceremony vary considerably, depending on the caste of the bride and groom.

(C) A *hijra* performs at a wedding ceremony in Rawalpindi, Pakistan. *Hijras* are trans women (male to female transgender) who leave their families, often at a young age, to join a "*hijra* family." Here, they undergo a gradual transition toward femininity, sometimes culminating in castration. They also must learn the *hijra* trades, which include singing and dancing for weddings and birth ceremonies, sometimes fortune-telling, and often sex work. Because of their unique identity, *hijras* are allowed to move between different castes with greater ease than most.

Dowry and Violence Against Females

A **dowry** is a sum of money paid by the bride's family to the groom's family at the time of marriage. Dowries originated as an exchange of wealth between Muslim landowners or high-caste families that practiced purdah. With her ability to work reduced by purdah, a woman was considered a liability for the family that took her in. Changing dowry customs appear to be a cause of the growing incidence of various kinds of domestic violence against females in Pakistan, India, and Bangladesh. Until the last several decades, only wealthy families gave the groom a dowry—in this case, a substantial sum that symbolized the family's wealth, meant to give a daughter a measure of security in her new family.

Ironically, the increases in affluence and education have reinforced the custom of dowry and made it much more common for all families. As more men became educated, their families felt that their diplomas increased their worth as husbands and gave them the power to demand larger and larger dowries. Soon, the practice spread through lower-caste families wanting to upgrade their status. Now the dowries they must pay to get their daughters married can cripple poor families. A village proverb captures this dilemma: "When you raise a daughter, you are watering another man's plant." Because it places financial burdens on parents of female children, dowry is thought to contribute to selective abortion and infanticide of females.

dowry a price paid by the family of the bride to the groom (the opposite of bride price); formerly, a custom practiced only by the rich

GENDER, POLITICS, AND POWER

As countries in South Asia have moved toward greater respect for political freedoms, the status of women in the region has risen. India, Pakistan, Bangladesh, and Sri Lanka have all had female heads of state (prime ministers) in the past. However, it is important to note that all of these women were either wives or daughters of previous heads of state. Women have been notably less successful in local elections, and at the parliamentary level Indian and Sri Lankan women remain very poorly represented (Table 8.1).

Table 8.1 Percentages of South Asian women in parliament, 2013

Sri Lanka	4.9
Bhutan	8.5
Pakistan	20.6
Nepal	29.5
Bangladesh	20
India	12.0
Afghanistan	27.7

Source: Women in National Parliaments, Inter-Parliamentary Union website, http://www.ipu.org/wmn-e/classif.htm.

The very low percentage of women in India's parliament inspired a confederation of Muslim and Hindu women's groups to lobby for legislation that would temporarily (for a 15-year trial period) reserve one-third of the seats in the lower house of parliament and in state assemblies for women. Such one-third quotas are already in place in Pakistan (20.6 percent), Nepal (29.5 percent), and Bangladesh (20), but are so far only close to being met in Nepal. If recent voter turn-out and political activism trends continue, as discussed on page 467, there is likely to be a major improvement in female representation in the region's national parliaments and local offices.

Women and the Taliban in Afghanistan

Women in Afghanistan have frequently suffered brutal repression since a conservative Islamist movement, the Taliban, gained control of the government there in the mid-1990s. Prior to that time, rights for women in Afghanistan were slowly but steadily improving, and upper-class women had many freedoms; they could dress in Western styles and they had the right to attend gender-integrated universities. The Taliban support strict and radical interpretations of Islamic law, forcing females, including urban professional women, to live in seclusion. In regions where the Taliban retain control, girls and women are not allowed to work outside the home or attend school. In virtually all parts of the country, despite the decline of Taliban control, women must wear a heavy, completely concealing garment, called a *burqa* (or burka), whenever they leave the house. (Men also must follow a dress code, though a less restrictive one.) Although the Taliban were driven from official power in November 2001, they maintain control of the rural southern provinces and mountainous zones near Pakistan, where cultural and religious conservatism continues to adversely affect Afghan women. Even efforts to provide women and girls with a basic education, such as that described in the following vignette about Radio Sahar, run into hostility, and often violence results.

VIGNETTE From behind her microphone at Radio Sahar ("Dawn"), Nurbegum Sa'idi speaks to a female audience on a wide range of topics. Located in the city of Herat, Radio Sahar is one in a network of independent women's community radio stations that has sprung up in Afghanistan since early 2003. Radio Sahar provides 13 hours of daily programming consisting of educational items that address cultural, social, and humanitarian matters as well as music and entertainment. For example, one recent broadcast aimed at informing women of their legal rights followed the life of a young woman who was physically abused by her husband and his entire family. The woman took the brave step of asking for a divorce. As a result, she was forced into hiding, where she was counseled on the steps she might take next. A reported 600,000 Afghan women and youth listen to Radio Sahar while they do their daily chores.

Girls on the Air, a film by Valentina Monti, reveals the diversity of ideas and hopes for the future shared by the young journalists who founded Radio Sahar. A clip can be seen at http://www.youtube.com/watch?v=c6KxtDHbuuY. [Source: Internews Afghanistan. For detailed source information, see Text Sources and Credits.] ■

ON THE BRIGHT SIDE: Cell Phones and Literacy

In order to reach women held in deep seclusion, the Afghan government is now making available basic reading and writing lessons on special mobile phones distributed free to Afghan women. The reading/writing software was developed by an Afghan IT firm with U.S. Agency for International Development (USAID) assistance. As a woman achieves literacy, she can add other subjects to her phone through free apps. ■

LGBT ISSUES IN SOUTH ASIA

South Asia's stance on LGBT issues is complex and contradictory. Socially, South Asia is not one of the world's more accepting places for LGBT people. Same-sex sexual activity remains illegal in most of South Asia and abuse, violence, and discrimination toward LGBT people are the norm. And yet South Asian countries are global leaders in formally recognizing the existence and rights of transgender people as a "third gender" in constitutions and in government documents, with India now reserving some scholarships and employment opportunities for transgender people. In 2005, India became the first country in the world to allow citizens to self-identify as a third gender on their passports. Nepal followed suit in 2007, Pakistan in 2009, and Bangladesh in 2013, and almost all of South Asia now recognizes a third gender in most types of government documents (Afghanistan and Bhutan are exceptions), with several extending preferential treatment in government hiring, as is offered to other minorities. This formal acceptance is due largely to the deep cultural roots and substantial numbers of transgender people in South Asia.

The largest group of transgender people are the *Hijra,* a trans female identity numbering 6–7 million, found mostly in North India, Nepal, Pakistan, and Bangladesh. *Hijras* are often considered to be the cultural descendants of Mughal era *eunuchs,* who were castrated males employed to guard the harems of rulers and other wealthy and powerful people. However, many references to *hijras* may also be found in pre-Mughal texts, such as the ancient Hindu epics, the Ramayana and Mahabharata. Many *hijras* identify with the most important Hindu gods, such as Shiva and Vishnu, who have transgender forms (called Ardhanari and Mohini, respectively), both of whom play a role in the Mahabharata. The patron goddess of the the *hijras* is Bahuchara Mata, whose temple in Gujarat receives over 1.5 million visitors annually.

The deep cultural roots of the *hijras* have not brought them cultural acceptance. Young boys who display transgender leanings are often rejected by their families and sent to live in all-*hijra* communities centered on a guru, where the predominant occupation is prostitution. One in five *hijras* will contract HIV and most live in relative poverty, subject to regular harassment by gangs and the police. *Hijras* are also seen as having magical powers to facilitate communication between humans and gods and between men and women, making them sought after as fortune-tellers, attendants at the blessings of newly born children and at weddings.

Other LGBT identities generally have fewer rights in South Asian countries, with the notable exception of Nepal, which is the only country in South Asia, and one of only a few in all of Asia, to adopt a constitution that bans all forms of discrimination against LGBT people. In 2007, Nepalese LGBT activists were able to take advantage of political disarray and openness to change after the country's civil war ended to petition for more LGBT friendly policies. Greater acceptance of LGBT people may also derive from Nepal's large tourism industry, which employs nearly a quarter of the non-agricultural labor force, and now makes a concerted effort to attract LGBT tourists. In 2008, Nepal elected its first openly gay member of parliament, who also works in the LGBT tourist industry. However, even in Nepal, LGBT people face severe social discrimination, violence, and daily harassment by police.

Mumbai is widely considered to be South Asia's "LGBT capital" with India's first gay magazine, a celebrated LGBT film festival, and a film industry rumored to employ numerous LGBT people. And yet here, too, discrimination is the norm. Virtually no one in Bollywood has "come out," and the portrayal of LGBT characters in films usually involves ridicule and abuse presented as comedy. As with other major South Asian cities, the anonymity of big city life is a major draw of Mumbai for LGBT people. An increasing number work in the "tech support" call centers serving overseas corporate clients, where one's appearance and ostensible sexual orientation are less important than one's ability to speak English clearly to foreigners. Mumbai's "Pride Parades" are known for their anonymity, with many participants sporting colorful masks that hide their identity. With homosexuality illegal in India, many people fear losing their job or being "outed" to their unknowing family. With anonymity so important, blackmailing of LGBT people has become common, especially when people arrange to meet using the online dating and LGBT chatrooms that have become so popular in recent years.

Throughout South Asia, the recent surge of religious nationalism in politics is opposed by many LGBT people who see religion as an obstacle to social acceptance and political participation. Indeed, many Muslim politicians interpret the Qur'an as forbidding homosexuality, and most proponents of Hindutva argue the same with regard to Hinduism. While some officials of the BJP point out the ambiguous gender of many Hindu gods, none have been willing to advocate for LGBT rights. Throughout South Asia, most of the expansion of political rights for LGBT people has been achieved through the courts, not through legislatures. While a number of *hijras* have run for office, and gay officials—two mayors and one state legislator—have been elected, their status in society remains low. However, their new official status as a third gender has brought some gains. For example, Bangladesh is now recruiting *hijras* as traffic police as a way to offer them a livelihood outside of sex work. Although change may come slowly, few are losing hope yet. As one *hijra* politician running for office in Pakistan put it, "This is the first drop of rain. If we do not have success in this election then next time, next time, next time."

THINGS TO REMEMBER

• Purdah is practiced in both Muslim and Hindu households, but the status of Muslim women is significantly lower than that of their Hindu, Sikh, Jain, Buddhist, and Christian counterparts.

• Changing dowry customs appear to be a cause of the growing incidence of various kinds of domestic violence against females in Pakistan, India, and Bangladesh.

• Motherhood in South Asia determines much about a woman's status within her community.

• In South Asia, a number of women have held very high positions of power, and young women today, on average, have access to more educational and employment opportunities than women of a generation ago. However, the overall status of women in the region is notably lower than the status of men.

• South Asia is a world leader in formally recognizing the rights of transgender people, though other LGBT identities generally have fewer rights in this region than elsewhere.

SUBREGIONS OF SOUTH ASIA

The subregions of South Asia are grouped primarily according to their physical and cultural similarities, so in several cases, parts of India are grouped with adjacent countries.

AFGHANISTAN AND PAKISTAN

Afghanistan and Pakistan **(FIGURE 8.32)** share location, landforms, and history, and they have both been involved in recent political disputes over global terrorism. Historically, many cultural influences have passed through these mountainous countries into the rest of South Asia: Indo-European migrations, Alexander the Great and his soldiers, the continual infusions of Turkic and Persian peoples, and the Turkic-Mongol influences that culminated in the Mughal invasion of the subcontinent at the beginning of the sixteenth century. Today, both countries are primarily Muslim and rural; 75 percent of Afghanistan's people and 65 percent of Pakistan's live in villages and hamlets. Nonetheless, each country has some large cities. In Afghanistan, Kabul has several million more inhabitants than its official population estimate of 3 million people; Kandahar has a half million people; and Herat, 400,000. In Pakistan, there are nine cities with more than a million inhabitants, including Karachi (13 million), Lahore (7 million), Faisalabad (2 million), Rawalpindi (2 million), and Peshawar (1.4 million). The capital of Islamabad has 700,000 residents. Both countries must cope with arid environments, scarce resources, and the need to find ways to provide rapidly growing urban and rural populations with higher standards of living. Both countries are home to conservative Islamist movements. Pakistan has a weak elected government that, to stay in power, is dependent on a strong military whose loyalties—to the government or to Islamic fundamentalists—are constantly in question (see page 483). Afghanistan's more extreme poverty is made worse by the armed strife under which it has suffered for more than 30 years (see below).

FIGURE 8.32 The Afghanistan and Pakistan subregion.  The Hunza Valley in northern Pakistan is framed by the windows of the 700-year-old Baltit Fort in Karimabad. The people of Hunza Valley are known for their relatively high literacy rate (90 percent), which is due in part to educational programs funded by a wealthy hereditary imam known as the Aga Khan.

The landscapes of this subregion are best considered in the context of the ongoing tectonic collision between the landmasses of India and Eurasia. At the western end of the Himalayas, the collision uplifted the lofty Hindu Kush, Pamir, and Karakoram mountains of Afghanistan and Pakistan. This system of high mountains and intervening valleys swoops away from the Himalayas and bends down to the southwest and toward the Arabian Sea. Landlocked Afghanistan, bounded by Pakistan on the east and south, Iran on the west, and Central Asia on the north, lies entirely within this mountainous system. Pakistan has two contrasting landscapes: the north, west, and southwest are in the mountain and upland zone just described; the central and southeastern sections are arid lowlands watered by the Indus River and its tributaries. In both countries, earthquakes regularly occur due to tectonic compression and active mountain building.

Afghanistan

The Hindu Kush and the Pamirs in the northeast of Afghanistan are rugged and steep-sided (see Figure 8.32). The mountains extend west, then fan out into lower mountains and hills, and eventually into plains to the north, west, and south. In these gentler but still arid landscapes, characterized by rough, sparsely vegetated meadows and pasturelands, most in the country struggle to earn a subsistence living from cultivation and the keeping of grazing animals. Rural people remain semi-nomadic, moving in order to locate forage and water resources for their animals (**FIGURE 8.33**). The main food crops are wheat, fruit, and nuts. Opium poppies are native to this region, but they were not an important modern cash crop until the Soviet invasion in 1979; the ensuing wars created the environment for opium production and profits.

FIGURE 8.33 Rural Afghanistan. A young Afghan moves a caravan through Balkh Province, Afghanistan. Camels and donkeys are a common mode of transportation throughout the drier parts of South Asia.

The less mountainous regions in the north, west, and south are associated with Afghanistan's main ethnic divisions. Very few of these groups could properly be called tribes. Rather, they are ancient ethnic groups. In the north along the Turkmenistan, Uzbekistan, and Tajikistan borders are the Afghan Turkmen, Uzbek, and Tajik ethnic groups, who share cultural and language traditions with the people of these three countries and are intermingled with them **(FIGURE 8.34)**. The Hazara, who are concentrated in the middle of the country, trace their biological ancestry to the Mongolian invasions of the past, yet today are closely aligned with Iranian culture and languages. In the south, the Pashto-speaking Pashtuns and Baluchis are culturally akin to groups farther south across the Pakistan border. There is significant variation within each group, especially regarding views on religion, education, modernization, and gender roles. For example, the Taliban are primarily illiterate Pashtuns. On the other hand, educated Pashtuns, long important in Afghan governance, played an important role in the resistance against the Taliban, and many are strong advocates of women's rights and a democratic and corruption-free Afghan government.

As of 2012, Afghanistan had just 33.4 million people, but its rate of increase is the fastest in the region (2.8 percent per year). Women have 6.2 children, on average (see Figure 8.24); as mentioned above, literacy rates are exceedingly low for men and women. The ethnic diversity of Afghanistan and its neighbors has thwarted many efforts to unite the country under one government (see Figure 8.34). Although the various ethnic groups have remained separate and competitive, contrary to Western media reports, they have not been at continuous war with one another. However, the events of the last several decades in the aftermath of the Russian invasion of 1979 (see the discussion on page 466) have disturbed long-standing relationships and feelings of trust. More important, the sheer devastation of almost three decades of war has left Afghanistan crippled. Although the traditional rural subsistence economy proved remarkably resilient in the face of ongoing civil strife, the self-sufficiency of that system has been compromised by an ongoing drought.

The cities of Afghanistan contrast with the countryside and with one another. Back in the 1970s, the capital Kabul was a modernizing city with utilities and urban services, tree-lined avenues and parks, and lifestyles similar to those in cities elsewhere in South Asia. Most women were not veiled; girls attended schools and wore the latest fashions. Then the Russians invaded and brought to Kabul a version of Western culture that quickly alienated the more conservative countryside. Alcohol consumption and partying were common, Russian women dressed scantily and provocatively, respect for religious piety was nil, and all this became negatively associated in Afghan minds with modernization. After the mujahedeen defeated the Russians and the Taliban came to power, modernization became anathema. Kabul suffered destruction first as the Russians were forced to retreat and then as the Taliban were themselves driven out in 2001. Today, Kabul has an official population of 3 million people but the actual number is probably several million greater. The built environment has

Afghanistan Provinces

1. Badakhshan	18. Kunar
2. Badghis	19. Kunduz
3. Baghlan	20. Laghman
4. Balkh	21. Logar
5. Bamiyan	22. Nangarhar
6. Daykundi	23. Nimruz
7. Farah	24. Nurestan
8. Faryab	25. Oruzgan
9. Ghazni	26. Paktia
10. Ghor	27. Paktika
11. Helmand	28. Panjshir
12. Herat	29. Parwan
13. Jowzjan	30. Samangan
14. Kabul	31. Sar-e Pol
15. Kandahar	32. Takhar
16. Kapisa	33. Wardak
17. Khost	34. Zabul

Ethnolinguistic Groups

- Baluch
- Pashtun
- Haraza
- Nuristani
- Ismaelien
- Turkmen
- Uzbek
- Tajik
- Kyrgyz
- Other

mi 0   100   200
km 0 100   200

FIGURE 8.34 Language and ethnicity in Afghanistan. The overlapping of ethnolinguistic groups illustrates how, over the millennia, particular groups have become divided and migrated to different places.

been much diminished by war. The cities of Herat in the west and Kandahar in the south were always more traditional and culturally conservative—more in tune with the hinterlands. Both were devastated during the late-twentieth-century conflicts, and little rebuilding has taken place in either city in recent years.

Pakistan

Although not much larger in area than Afghanistan, Pakistan has more than five times as many people (180.4 million). Pakistanis, too, live primarily in villages (65 percent), though the country has many large cities. Some villages sprinkled throughout the arid mountain districts are associated with herding and subsistence agriculture, but it is the lowlands that have attracted the most settlement. Here, the ebb and flow of the Indus River and its tributaries during the wet and dry seasons form the rhythm of agricultural life. The river brings fertilizing silt during floods and provides water to irrigate millions of cultivated acres during the dry season.

From the 1980s to 2007, Pakistan had a favorable annual economic growth rate of 6 percent or better. In 2009, that rate fell to just 2.7 percent because of the global recession and persistent civil disruptions linked to Al Qaeda and the war in Afghanistan. Agriculture in the Indus River basin, especially in the Pakistani Punjab, was formerly a primary economic sector. Cash crops such as cotton, wheat, rice, and sugarcane were grown in large irrigated fields, as they had been for many thousands of years. Recent irrigation projects, however, have overstressed the system, resulting in waterlogged and salinized soil. As a result, the contribution of agriculture to GDP has shrunk in recent years, with industry and services taking over, in part. Although textile, yarn-making, and embroidery industries are growing around the cities of Lahore and Karachi, providing jobs for some rural people, the wealth generated has not resulted in higher wages or local investment, but rather has gone into the pockets of the wealthy, who invest outside of Pakistan. There are exceptions: Pakistan has a host of well-educated writers who inform the English-speaking world about sociopolitical relations in their country. Some landowning elites invest selflessly in their constituents. An example is the Aga Khan, a member of the hereditary upper class in the Hunza Valley (see Figure 8.32A); his efforts to make public education available for the population have increased literacy to 90 percent.

Pakistan's northern mountainous hinterlands, where government authority is difficult to establish, have become a major home base for Al Qaeda. The country has a particularly problematic and ineffective military, a fragile and dysfunctional political system that stifles dissent, and a nuclear arsenal. These realities have made Pakistan even more worrisome to the West than is Afghanistan. Public opinion among the masses (who have an overall literacy rate of just 54 percent, with the rate for females only 39 percent) is starkly anti-Western, but it also does not favor rule by Islamic militants.

In many ways, Pakistan's difficulties of today are the outcome of Cold War geopolitics, when antidemocratic and militaristic regimes were propped up and heavily armed by the West to gain strategic advantage over the USSR. During this time, Pakistan was able to surreptitiously acquire nuclear capabilities and sell

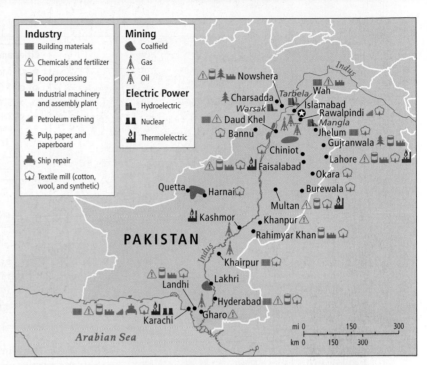

FIGURE 8.35 Industrial, mining, and power centers in Pakistan.

nuclear secrets to Iran, Libya, and North Korea. Billions of dollars of Western aid went to support the Pakistani military, and while too little went to the people for education and social reforms or to strengthen democratic institutions, there was an effort to create an economy that could provide jobs for young adults in government and commerce. **FIGURE 8.35** is a 2012 map that shows a variety of centers of economic activity in Pakistan (compare this with the topographic map of Pakistan in Figure 8.32).

Frequent terrorist bombings show the power of Islamic militants to disrupt Pakistan's economy and society (and that of Pakistan's neighbors). In an effort to calm the situation, in 2012, the U.S. Congress approved an assistance bill to provide U.S.$1.5 billion to strengthen Pakistan's legislative and judicial systems; buttress health care; and improve the public education system, with a special emphasis on classes for women and girls. Many Pakistanis worry that this social aid is merely intended to Westernize them. Meanwhile, educated elites struggle to come to terms with their own privilege, Westernized ways, and impotence in the face of terrorism.

THINGS TO REMEMBER

• Both Afghanistan and Pakistan have high mountains in the north that are the result of the ongoing tectonic collision between the landmasses of India and Eurasia. Both countries have arid lowlands to the south and west; the Indus River and its tributaries water those in Pakistan, but Afghanistan has fewer sources of water.

• Inefficient, corrupt governments plague both countries, and both have predominately conservative, fiercely independent rural populations with diverse ethnic backgrounds. Islamic fundamentalism appeals to many people in both countries, especially in rural areas.

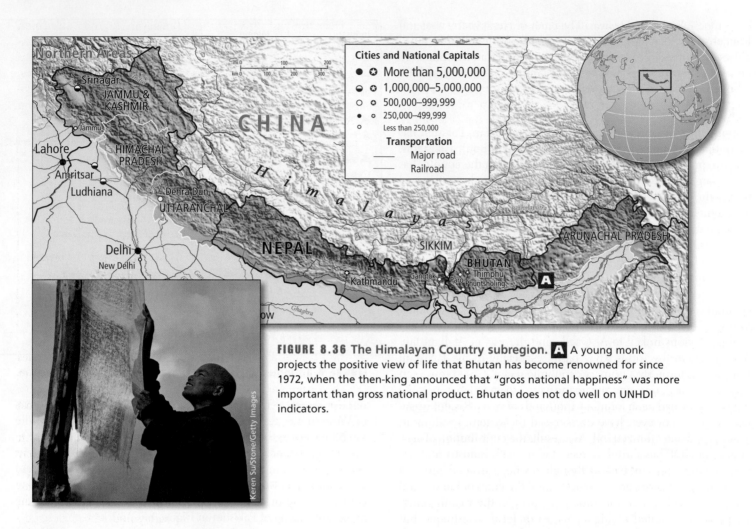

FIGURE 8.36 The Himalayan Country subregion. **A** A young monk projects the positive view of life that Bhutan has become renowned for since 1972, when the then-king announced that "gross national happiness" was more important than gross national product. Bhutan does not do well on UNHDI indicators.

HIMALAYAN COUNTRY

The Himalayas form the northern border of South Asia (FIGURE 8.36). This subregion is constructed of the mountainous portions of the Indian states of Jammu and Kashmir, Himachal Pradesh, Uttaranchal, and Arunachal Pradesh, and the countries of Nepal and Bhutan. Notice that the country of India has a far eastern lobe that borders Burma (Myanmar). Physically, this mostly mountainous subregion grades from relatively wet in the east to dry in the west because the main monsoons move up the Bay of Bengal and deposit the heaviest rainfall there before moving north and west (see Figure 8.4 and Figure 8.1).

This strip of Himalayan territory can be viewed as having three zones: (1) the high Himalayas, (2) the foothills and lower mountains to the south, and (3) a narrow strip of southern lowlands along the base of the mountains. This band of lowlands forms the northern fringe of the Indo-Gangetic Plain in the west and the narrow Assam Valley of the Brahmaputra River in the east. Although some people manage to live in the high Himalayas, most people live in the foothills, where they cultivate strips and terraces of land in the valleys or herd sheep and cattle on the hills. Some of this area is extremely rural in character—for example, Bhutan's capital city, Thimphu, with just 98,000 inhabitants, is the largest town in that country; many towns have 2000 or fewer

people. The capital of Nepal is Kathmandu, a rapidly urbanizing city that (with its suburbs) numbers about 1.5 million people.

Culturally, the Himalayan Country subregion is Muslim in the west and Hindu and Buddhist in the middle; animist beliefs are important throughout but are especially strong in the far eastern portion. Throughout the subregion, but especially in valleys in the high mountains and foothills, indigenous people continue to live in traditional ways, isolated from daily contact with the broader culture. In the Indian state of Arunachal Pradesh, for instance, at the eastern end of the subregion, a population of less than 1 million speaks more than 50 languages.

Despite cultural differences, the Himalayan people have learned to survive in their difficult mountain habitat by relying on one another. An example of this cross-cultural reciprocity comes from central Nepal, where two indigenous groups engage in a complicated seasonal cycle of trade that links them with both Tibet to the north and India to the south.

VIGNETTE The Dolpo-pa people are yak herders and caravan traders who live in the high, arid part of Nepal where they can produce only enough barley and corn to feed themselves for half the year. Through trade, they parley that half-year supply of grain into enough food for a whole year.

At the end of the summer harvest, they load some of their grain onto yaks and head north to Tibet, where they trade grain to Tibetan nomads in return for salt, a commodity in short supply in Nepal. They keep some salt for their own use and carry the rest to the Rong-pa people in the central foothills of Nepal, where they exchange the salt for sufficient grain to last through the winter. The Rong-pa live in a zone where they cultivate wheat and beans but also herd sheep and goats, and salt is a necessary nutrient for them and their animals. If the Dolpo-pa do not bring enough salt to meet all the Rong-pa needs, the Rong-pa load some goats and sheep with bags of red beans and set out for Bhotechaur in western Nepal, where they meet Indian traders at a bazaar. They trade their beans for iodized Indian salt, sell a sheep or two, and buy some cloth and perhaps a copper pan to take home. [*Source: Eric Valli and the movie Himalaya. For detailed source information, see Text Credit pages.*] ∎

Most people in the Himalayan subregion are very poor. Statistics for the Indian states in this subregion hover near those for Nepal, which has an annual per capita GNI of just U.S.$1200, adjusted for PPP. In Nepal, the average life expectancy is 68, literacy is about 60 percent (48 percent for women), and nearly half of the children under 5 years of age are malnourished. At present, only 18 percent of the land is cultivated, and the country's high altitude and convoluted topography make agricultural expansion unlikely. The government's strategy, therefore, is to improve crop productivity and lower the population growth rate (now 1.8 percent per year).

Bhutan, with a GNI per capita that is roughly four times that of Nepal, is still far from rich. The country is reliant on agriculture in difficult mountainous environments and on trade of mostly cottage industry products with India, which also extends various types of aid to Bhutan for infrastructure development. Bhutan has become famous for its positive view of life and was the first country to promote the idea of measuring human well-being with a "gross national happiness" quotient (see Figure 8.36A). It now promotes this concept to environmentally and spiritually conscious tourists who pay a fee of $250 a day to visit the country. Bhutan is also known for being the last place on Earth to acquire television and the Internet, which occurred only over the past decade. A 10-minute video that shows the effect TV has had on Bhutan society can be found at http://www.pbs.org/frontlineworld/stories/bhutan/.

The Indian state of Arunachal Pradesh, at the far eastern end of the Himalayan subregion, is one of the more lightly populated and environmentally rich areas of the subregion. Moist air flowing north from the Bay of Bengal brings plentiful rain as it lifts over the mountains. This is one of the most pristine regions in India. Forest cover is abundant, and a dazzling sequence of flora and fauna occupies habitats at descending elevations: glacial terrain, alpine meadows, subtropical mountain forests, and fertile floodplains. Conditions at different altitudes are so good for cultivation that lemons, oranges, cherries, peaches, and a variety of crops native to South America—pineapples, papayas, guavas, beans, corn (maize), and potatoes—are now grown commercially for shipment to upscale specialty stores, primarily in India.

Over the past 20 years, the Himalayan Country subregion has undergone numerous changes brought about by the increasing numbers of tourists trekking and climbing the mountains and seeking spiritual enlightenment at its numerous holy sites. Tourism, in particular, has been a mixed blessing, creating economic opportunities for some but also having a transforming effect on the culture and the landscape.

THINGS TO REMEMBER

• In this mountainous subregion, there are a variety of cultures and belief systems among relatively sparse populations.

• Many people use multiple livelihood strategies and reciprocal trade to survive, generally making do with little.

• With the possible exception of Arunachal Pradesh, the physical environment of this subregion precludes much expansion into agriculture or industry.

NORTHWEST INDIA

Northwest India stretches almost a thousand miles from the states of Punjab and Rajasthan at the border with Pakistan, eastward to encompass the famous Hindu holy city of Varanasi (FIGURE 8.37) It is dry country, yet it contains some of the wealthiest and most fertile areas in India.

In the western part of this subregion, so little rain falls that houses can safely be made of mud and have flat roofs. Widely spaced cedars and oaks are the only trees, and the landscape has a dusty khaki color. Villagers plow the landscape between the trees using oxen and a humpbacked breed of cattle called Brahman, and then plant crops of barley and wheat, potatoes, and sugarcane. Along the northern reaches of the subregion, the rivers descending from the Himalayas compensate for the lack of rainfall. The most important is the Ganga, which, with its many tributaries, flows east and waters the state of Uttar Pradesh, bringing not only moisture but also fresh silt from the mountains.

The western half of the Northwest India subregion contains one of India's poorest states—Rajasthan—as well as one of its wealthiest—the Indian part of Punjab, where the Sikhs have their holiest site (see Figure 8.37A). Rajasthan, with only a few fertile valleys, is dominated by the Thar Desert in the west, which covers more than a third of the state. Historically, small kingdoms (Rajasthan means "land of kings") were established wherever water could be contained. Seminomadic herders of goats and camels still cross the desert with their animals, selling animal dung or trading it for grazing rights as they travel, thus keeping farmers' fields fertile and their fires burning. Perhaps the best-known seminomads are the Rabari, about 250,000 strong today. Since the 1990s, however, more and more small farmers have occupied former pasturelands. Although these farmers still want dung for their fields, they cannot afford to lose one bit of greenery to passing herds, so the increasing population pressure means that the benefits of reciprocity between herders and farmers are being lost.

In arid Rajasthan, less than 1 percent of the land is arable, but agriculture, poor as it is, produces 50 percent of the state's

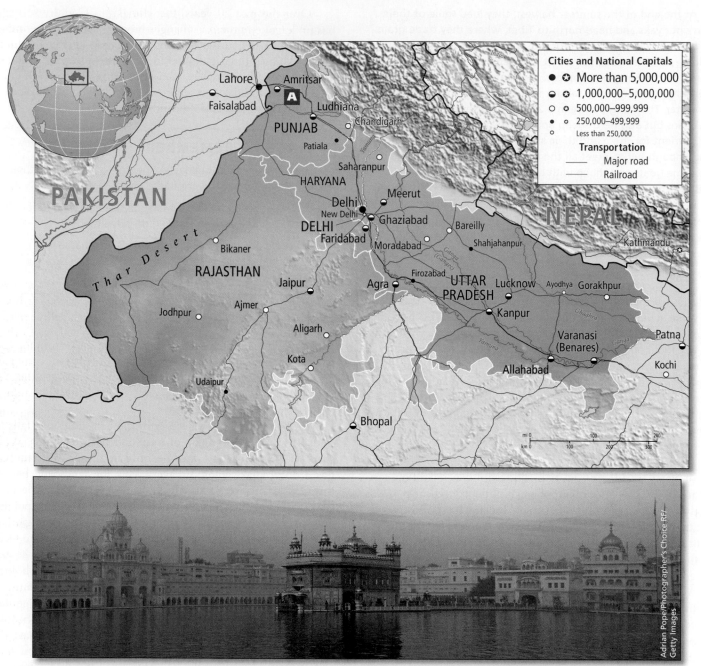

FIGURE 8.37 The Northwest India subregion. **A** The Golden Temple in Amritsar, Punjab, is the most sacred place for Sikhs.

GDP. It is not surprising that the average annual per capita GDP (PPP) is just U.S.$2093 (see Figure 8.27). The crops include rice, barley, wheat, oilseeds, peas and beans, cotton, and tobacco. A thriving tourist industry, focused on the palaces and fortresses constructed by Hindu warrior princes of the past who fought off invaders from Central Asia, accounts for much of the other half of the GDP.

Unlike Rajasthan, Punjab and Uttar Pradesh receive water and fresh silt carried down from the mountains by rivers. Nearly 85 percent of Punjab's land is cultivated, and in some years Punjab alone provides nearly two-thirds of India's food reserves. The typical crops are corn, potatoes, sugarcane, peas and beans, onions, and mustard. Punjab's productive agriculture contributes to its

high average annual per capita GDP (PPP) of about U.S.$4300, more than double that of Rajasthan. These differences in wealth result partly from differences in physical geography, the introduction of *green revolution* technology, and the growing influence of the global economy and associated urbanization.

Although agriculture is the most important economic activity throughout the subregion and employs about 75 percent of the people, well-paying jobs are generated by industry in and around the city of Delhi in Uttar Pradesh and on the Indo-Gangetic Plain. Here, as elsewhere in the subregion, typical industries are sugar refining and food processing, as well as the manufacture of cotton cloth and yarn, cement, glass, and steel. Residents also make craft items and hand-knotted wool carpets.

The city of New Delhi, India's capital, is located approximately in the center of Northwest India. The British built the city center in 1931 just south of the old city of Delhi—an important Mughal city—amid the remains of seven ancient cities. It has all the monumental hallmarks of an imperial capital, and all the problems one might expect of a big city where so many are poor. The Delhi metropolitan area (including the old and new cities and outlying suburbs) counts more than 19.5 million people and attracts a continuing stream of migrants. Most have left nearby states to escape conflict or poverty, or both, such as those who have fled Tibet to escape Chinese oppression and others who have fled the conflicts with Pakistan since the 1980s. Annual per capita GDP (PPP) in Delhi (U.S.$5887) is above the average for India (U.S.$3468), but actual incomes for most people are far lower because Delhi's tiny minority of the extremely wealthy pulls up the average. Delhi's rather high literacy rate (82 percent), while already 15 percent higher than the country's overall rate, is held down by the continual arrival of migrants from poor, rural areas. In fact, Delhi has far fewer schools than it needs for its population because jobs increasingly require a high level of skills, not mere literacy. Because of its rapid growth, the city has difficulty providing even the most basic services; water, power, and sewer facilities are insufficient, and 75 percent of the city's structures violate local building standards. Many people have no buildings to inhabit at all; they live in shanties constructed from found materials.

In 2011, Delhi was found to have worse pollution than Beijing, China. Delhi is landlocked and frequently experiences cold air inversions, which hold the polluted air low over the city. The metropolitan area has an annual pollution-related death toll of 7500. Most of the pollution comes from the more than 3 million unregulated motor vehicles: taxis, trucks, buses, motorized rickshaws, and scooters—most without pollution-control devices, and all competing for space, cargo, and passengers. Because the external air pollution is so intractable, firms dealing in devices to clean interior air abound in New Delhi.

THINGS TO REMEMBER

• Agriculture is the economic base of the region, employing some 75 percent of the population; however, industrial enterprises are increasing around Delhi and along the Ganga Plain.

• The Northeast India subregion is home to New Delhi, India's capital that is host to a never-ending migration of people seeking jobs and education. Pollution is one of the many side effects of rapid growth and modernization.

NORTHEASTERN SOUTH ASIA

Northeastern South Asia, in strong contrast to Northwest India, has a wet, tropical climate. This subregion bridges national and state boundaries, encompassing the states of Bihar, Jharkhand, and West Bengal in India; the country of Bangladesh; and the far eastern Indian states of Meghalaya, Assam, Nagaland, Manipur, Mizoram, and Tripura—all clustered at the north end of the Bay of Bengal (FIGURE 8.38). This area is viewed as a subregion because of its dominant physical and human features. The Ganga and Brahmaputra

rivers and the giant drainage basin and delta region created by those rivers, as well as the wet climate and fertile land, have nourished a population that is now among the most densely populated in the world (see the Figure 8.24 map). Here, we discuss several parts of the Indian portion of this subregion before moving on to Bangladesh.

The Ganga-Brahmaputra Delta

The largest delta on Earth is that formed by the Ganga and Brahmaputra rivers. Every year, the two rivers deposit enormous quantities of silt, building up the delta so that it extends farther and farther out into the Bay of Bengal. The rivulets of the delta change course repeatedly, and then periodically the bay is flushed out by a huge tropical cyclone (see Figure 8.6). The people of the delta have learned never to regard their land as permanent, and they have adapted their dwellings, means of transportation (see Figure 8.38A), and livelihoods to the drastic seasonal changes in water level and shifting deposits of silt. Villages sit on river terraces or, in the lowlands, are raised on stilts above the high-water line. Moving about in small boats, people fish during the wet season, but when the land emerges from the floods, they return to farming. Called *nodi bhanga lok* ("people of the broken river"), those who occupy the constantly shifting silt of the delta region are looked down upon by more permanent settlers on slightly higher ground. Because they must often flee rising floodwaters, they are less secure financially and are thought by their neighbors to lack the qualities of thrift and good citizenship that come from living in one place for a lifetime. Currently, their livelihoods (and at times their very lives) are threatened by periodic cyclones and climate change–related sea level rise, compounded by the loss of nourishing sediment caused by river diversions upstream.

West Bengal

With a population of 80 million packed into an area slightly larger than Maine, West Bengal is India's most densely occupied state (see Figure 8.24). Refugees have dramatically increased the population of West Bengal, which has taken in people from eastern Bengal (now part of Bangladesh) after Partition in 1947; migrants fleeing the Pakistani civil war (1971) that gave Bangladesh its independence; and the Tibetan refugees of the Chinese takeover of Tibet (1979). Today, better employment and farming opportunities draw a continual flow of Bangladeshi migrants (most of them migrating illegally) across the border into India. The result is a mix of cultures that are frequently at odds; often, Hindus and Muslims demonstrate and dispute with indigenous people over land occupied by new migrants.

Nearly 75 percent of the people in this crowded state earn a living in agriculture, but agriculture accounts for only 35 percent of West Bengal's GDP. In addition to growing food for their own consumption, many people work as rice and jute cultivators or as tea pickers—all labor-intensive but low-paying jobs. Twenty-five percent of India's tea comes from West Bengal; the plant is grown in the far north of the state around Darjeeling, a name well known to tea drinkers. One consequence of the dense population's dependence on agriculture is that the land has become overstressed and soil fertility has declined. In addition, impoverished farm laborers who must gather firewood because they cannot afford other fuels deplete the surrounding woodlands.

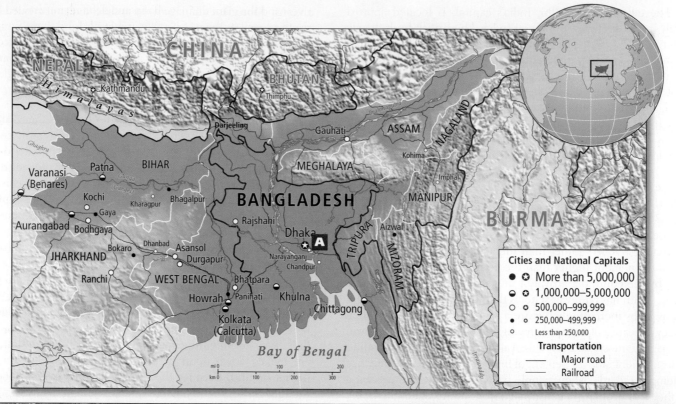

FIGURE 8.38 The Northeastern South Asia subregion. **A** Men row in a traditional boat race on the Buriganga River near the center of Dhaka, Bangladesh. Boat races are an ancient sport in this land of rivers, and boats almost always have a leader who energizes his crew with songs, rhymes, and gestures.

Kolkata (formerly known as Calcutta), a giant, vibrant city of 16 million people, is famous in the West for Mother Teresa's ministrations to its poor at Nirmal Hriday (Home for Dying Destitutes). Bengalis, however, are also proud of their two native Nobel laureates—Rabindranath Tagore (literature) and Amartya Sen (economics)—and of their Academy Award–winning filmmaker Satyajit Ray, who received a lifetime achievement Oscar.

Kolkata was built on a swampy riverbank in 1690 and served as the first capital of British India. But its sumptuous colonial environment has become lost in the squatters' settlements that have invaded the city's parks and boulevards and periphery. The city's decline began in the late 1940s, when Partition and agricultural collapse sent millions of people pouring into Kolkata and other cities. Soon the crowds overwhelmed the city's housing and transportation facilities. Outmoded regulations limited incentives to start businesses and to improve private property. In the late 1990s, the physical decline, economic inertia, and lack of opportunity in Kolkata sent young professionals fleeing to other parts of the world.

By 2005, however, those who had fled Kolkata were returning to partake in an economic revival, based primarily on technology industries, and new graduates began to stay. Unfortunately, old Kolkata is not being refurbished; rather, developers are building new, exclusive gated suburbs on the wetlands south of the city (in areas exposed to future sea level rise). At least 50 large construction projects are under way (though progress slowed during the recession which began there in 2007); they are expected to eventually attract another 7 or 8 million people to this lowland metropolitan area. Thus far, worries about the fate of old Kolkata, or about the environmental ramifications of such massive building on wetlands within the Ganga-Brahmaputra delta, are not getting much coverage in the press.

Far Eastern India

Far Eastern India has two physically distinct regions: the river valley of the Brahmaputra where the river descends from the Himalayas, and the mountainous uplands stretching south and east of the river between Burma on the east and Bangladesh on the west (see the Figure 8.38 map). Although migrants have recently arrived here from across South Asia, ancient indigenous groups that are related to the hill people of Burma, Tibet, and China have traditionally occupied Far Eastern India.

The Indian state of Assam encompasses the Brahmaputra River valley, eastern India's most populated and most productive area. Hindu Assamese make up two-thirds of the population of 29 million, and indigenous Tibeto-Burmese ethnic groups make up another 16 percent; the rest of the people are recent migrants. The Indian government has attempted to reduce the proportion and influence of the Assamese people (who have continually objected to being under Indian control) by making large tracts of land available to outsiders, such as Bengali Muslim refugees, Nepalese dairy herders, and Sikh merchants. Since the late 1970s, mortal disputes have occurred between the Assamese and the new settlers and between Assam and India's national government.

More than half the people in Assam work in agriculture; another 10 percent are employed in the tea industry and in forestry (forests cover about 25 percent of the land area, and bamboo, a new flooring product in America and Europe, is a major forest export). Tea is the main cash crop—Assam produces half of India's tea. Assam also has oil; by the 1990s, Assam's oil and natural gas accounted for more than half that produced in all of India. Given the country's shortage of energy, this alone could explain India's efforts to dominate the Assamese politically.

Colorful names such as "Land of Jewels" and "Abode of the Clouds" convey the exotic beauty of the emerald valleys, blue lakes, dense forests, carpets of flowers, and undulating azure hills in the mountainous sections of eastern India that surround Assam. In these uplands, occupied largely by indigenous ethnic groups, people produce primarily for their own consumption and devise ingenious ways to make use of local natural resources. The state of Mizoram ranks second in India in literacy (88 percent) due to the influence of Christian missionary schools.

ON THE BRIGHT SIDE: The Logistics of Development

Bangladesh's march toward a better life has not been based on luck, but rather on carrying out interlocking, methodical steps. Publicly funded family planning (birth control) has empowered women to have far fewer babies and to turn their energy toward gaining education for themselves and their children. The boom in textile manufacturing has provided jobs for the literate, and microcredit has put entrepreneurial activities within reach of the ambitious poor. Improved rice varieties have protected against crop failures and famine, and the many Bangladeshi working abroad regularly send remittances to their families. Additionally, the government has maintained social safety net spending for the poorest. ∎

Bangladesh

Bangladesh is one of South Asia's poorest countries and one of the world's most densely populated, predominantly agricultural nations. More than 153 million people live in an area slightly smaller than Alabama, with 75 percent of them trying to manage as farmers on severely overcrowded land; meanwhile, urban residents struggle with low standards of living in wet environments. The caloric intake for close to one-third of Bangladeshis is insufficient to meet the minimum daily energy requirements, and 40 percent of its children are underweight.

Nevertheless, Bangladesh is making improvements, bit by bit and on many fronts. The percentage of rural people living in poverty is steadily dropping, according to USAID. Adult literacy went from 34 percent in 1990 to 57 percent in 2011, and the enrollment of eligible children in secondary school increased from 30 percent to 45 percent during the same period. Contraceptives have become more available, and 61 percent of women now use some form of birth control, bringing fertility rates down from 7 children per woman in 1974 to 2.3 in 2012. Infant mortality has also dropped, from 128 per 1000 in 1986 to 43 per 1000 in 2012.

Economically, the textile industry (dismantled by the British) is reviving; Bangladesh is now the second-largest exporter of garments after China and has consistently increased its shipments to the U.S. market. Bangladesh bases its competition for market share with India and China by using the high quality of their textiles as a selling point.

Unfortunately, to keep competitive with cheap labor markets throughout Asia, Bangladesh has allowed very relaxed job safety standards, as the disastrous November 2012 factory fire outside Dhaka attests. One hundred and seventeen workers at Tazreen Fashions, reportedly a supplier to Walmart, Sears, the Gap, Disney, Tommy Hilfiger, and other vendors in North America, lost their lives. Tazreen was known to flout safety regulations, and the American importers did not sufficiently oversee the supply chain.

The real hope for economic growth in Bangladesh lies in development from within. The plan is for Bangladeshis to be increasingly able to purchase domestically produced food, textiles, and technology with income from manufacturing jobs and from their own small businesses funded through microcredit. Microcredit is the remarkable innovation of the Grameen Bank that is now jumpstarting entrepreneurial initiatives by the poor worldwide (see page 461).

THINGS TO REMEMBER

- Northeastern South Asia clusters around the upper reaches of the Bay of Bengal, home to the world's largest delta, formed by the Ganga and Brahmaputra rivers.

- Climate change puts many lives at risk in Northeastern South Asia, primarily because of water-related impacts. Sea level rise, flooding, and the increased severity of storms imperil many urban and rural residents in the lowland parts of the delta region.

- West Bengal, India's most crowded state, is in this subregion; its city of Kolkata has a population of more than 13 million.

- India's far eastern states manage their diverse cultures and many indigenous peoples, despite meddling from Delhi.

- Bangladesh, once noted for its poverty and dependence, is using multiple strategies to become economically self-sustaining.

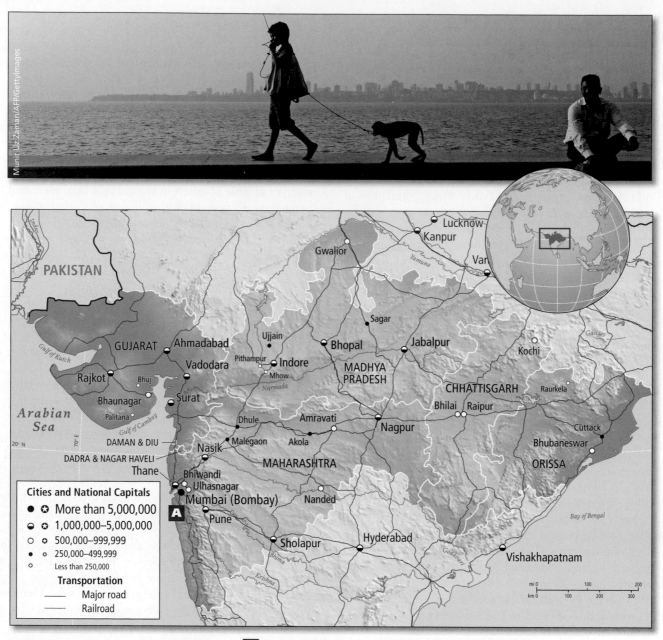

FIGURE 8.39 The Central India subregion. **A** A boy and his trained monkey walk along Mumbai's Marine Drive. In the background is Malabar Hill, one of the most expensive neighborhoods in the world, with costs per square foot higher than those of Manhattan in New York City.

CENTRAL INDIA

The Central India subregion stretches across the middle of India, from Gujarat in the west to Orissa in the east (**FIGURE 8.39**). It contains India's last untouched natural areas as well as much of its industry. The Narmada River, site of many planned hydroelectric and multipurpose dams, flows across the subregion and empties into the Gulf of Cambay.

Some of India's most significant environmental battles are being fought in the central highlands of this subregion; the protests over the damming of the Narmada River, described at the beginning of this chapter, are only one example. Central India has most of India's remaining forest cover and is home to a concentration of national parks and sanctuaries (most notably, tiger reserves; see Figure 8.9A). However, because the forest cover is patchy, not continuous, many wild plants and animals are isolated in such small populations that their extinction is highly likely, even if it is somewhat delayed by the parks' protection. In recognition of this problem, there are now plans to reconstitute forest corridors between parks. Estimates of future human population growth, however, do not bode well for the future of wildlife anywhere in the subregion.

Central India is notable for its several industrial areas, most of which are now rebounding from the global recession. In the state of Gujarat, on the Arabian Sea, the service and industrial sectors account for 75 percent of GDP; the rate is even higher, at 83 percent, in the neighboring state of Maharashtra, where education for females has increased **(FIGURE 8.40)**. Even the central plateau and eastern areas of Central India, although more rural in character, have pockets of industrial activity that are spreading and connecting. An example is the city of Indore (see the Figure 8.39 map), a commercial and industrial hub whose residents think of it as a mini Mumbai. Nearby, the town of Pithampur, known as India's Detroit and now home to 1.2 million people, houses several automobile plants, a steel plant, a container plant, and small appliance factories. Production in Pithampur's automotive industries is climbing back after a slump in 2009, but its continued growth is dependent on the rail connections to Ahmadabad, Mumbai, Indore, and Delhi that use the old British tracks, with their mismatched gauges, being upgraded to the now-standard broad gauge—all of which will require government subsidies. Industrialization, of course, leads to urbanization; in fact, the western part of this subregion is exceptionally urbanized for India. In Gujarat, about 38 percent of the people live in urban areas; in Maharashtra, the home of the megacity Mumbai (with 22.9 million inhabitants; see Figure 8.39A), about 43 percent do. Mumbai exerts enormous influence on all development in Maharashtra. Its many aspects are covered starting on page 458.

FIGURE 8.40 Education and technology access for females. Girls at a primary school in Maharashtra, India, work on Internet-connected computers. Females' access to education and technology is increasing throughout South Asia but still lags behind that of males. For example, in 2011 in India, the percentage of females enrolled in secondary school was only 79 percent of the number of males enrolled.

Eric Vandeville/Gamma-Rapho via Getty Images

THINGS TO REMEMBER

• The Central India subregion contains India's last untouched natural areas as well as much of its industry.

• Mumbai, the principal city in Central India and the country's financial capital and largest city, is joined by other cities that are booming industrial centers.

SOUTHERN SOUTH ASIA

Southern South Asia encompasses the southernmost part of India and the country of Sri Lanka **(FIGURE 8.41)**. It resembles the rest of the region in that the majority of people—from about 70 percent in the west to somewhat more than 50 percent in the east—work in agriculture. But this subregion is set apart by its relatively high proportion of well-educated people; its advanced technology sector in Bangalore; the strong tradition of elected communist state governments in Kerala, where social services are particularly strong; its focus on environmental rehabilitation and preservation; and the overall higher status and well-being of women. The cultural mix here also sets it apart; it is the center of Dravidian cultures and languages that previously extended into north-central India.

This part of India normally receives consistent rainfall and is well suited for growing rice, peanuts, chili peppers, limes, cotton, cinnamon and cloves, and castor oil plants (used in medicines). The southwestern coast of India (known as the Malabar Coast) has a narrow coastal plain backed by the Western Ghats. The sea-facing slopes of these mountains are some of the wettest in India; they support forests containing teak, rosewood, and sandalwood, all highly valued furniture woods. Small parts of the Deccan Plateau, a series of uplands to the east, are also forested. Here, dry deciduous forests yield teak, tree-farmed eucalyptus, cashews, and bamboo. Several large rivers and numerous tributaries flow eastward across this plateau and form rich deltas along the lengthy and fertile coastal plain facing the Bay of Bengal.

Anomalies in weather patterns are always a possibility. In 2009, this region suffered first an unusual drought during the monsoon season, particularly in Andhra Pradesh and Karnataka, and then a period of sudden rains. The dried-out soil could not absorb the rain quickly enough, leading to flooding that killed hundreds and ruined crops that had managed to weather the drought. India's central government was able to provide emergency assistance and food. State governors also planned to import foodstuffs to keep food prices from rising drastically.

Bangalore and Chennai

Bangalore, long known for its strong IT activity, which has drawn thousands of well-trained technical employees from all over India, continues to evolve. It now has many IT rivals throughout South Asia, but remains competitive.

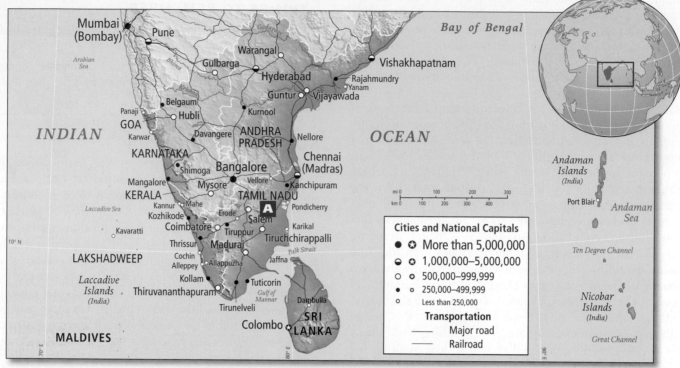

FIGURE 8.41 The Southern South Asia subregion. **A** A Hindu temple in Kanchipuram, Tamil Nadu, India. South India is known for its distinctive temple style, which features ornately decorated towers (called *gopurams*) and a water reservoir (called a *kalyani*).

Meanwhile, manufacturing has gained markedly; the products produced in and around the city range from appliances and home furnishings to solar-powered lighting and pharmaceuticals. Chennai, 200 miles (320 kilometers) to the east on the Bay of Bengal, is known for its innovative engineering and sustainable development research institutes and is cultivating a role as a prime destination for jobs outsourced from Europe. It now also emphasizes manufacturing and is becoming a center for Italian-designed high-fashion shoe manufacturing. Because styles for these high-end shoes change so rapidly, the Chennai factories are assured of the continuous production of about 2 million pairs per year.

Kerala

Kerala, a primarily coastal state in far southwestern India, is the exception that proves the rule. It is often cited as an aberration in South Asia because gender discrimination is far less prevalent and its people enjoy a higher standard of well-being than the rest of the region. Just why Kerala stands out in this way is not entirely understood, but the state has a unique history. Since the 1950s, it has had a series of elected communist governments that have strongly supported broad-based social services, especially education for both males and females (see Figure 8.21). In addition, it has long had a variation on the

FIGURE 8.42 A woman prays at the Golden Temple in Dambulla, Sri Lanka. A gilded statue of the Buddha sits atop a 250-year-old monastery. The statue was added in 1998, using funds donated by Sri Lankans living abroad.

traditional family structure that gives considerable power to women; for example, a husband often resides with his wife's family (a pattern also found in Southeast Asia) instead of the reverse. Female seclusion is less stringently practiced; women are often seen in public, and it is not unusual to see a woman going about her shopping duties alone. Agriculture employs less than half the population of 33.3 million people. Remittances from migrants to the Middle East figure prominently in the state's economy, and fishing is an important economic sector as well.

Another feature that may have added to the openness of society and greater freedom for women is Kerala's significant contact with the outside world. Beginning thousands of years ago, traders from around the Indian Ocean, and especially Southeast Asia, made calls along the Malabar Coast, possibly bringing new ideas about gender roles. In the first century C.E., the apostle Thomas is said to have come to spread Christianity (20 percent of Kerala is Christian). Over time, Muslim and Jewish traders brought their faiths by sea as well (25 percent of Kerala is Muslim). Then, between 1405 and 1433, the Chinese explorer General Zheng visited Cochin and the Malabar Coast repeatedly. This history of cultural and religious variety seems to have led to more open-mindedness toward diversity than exists in neighboring states.

Sri Lanka

Sri Lanka, known as Ceylon until 1972, is an island country off India's southeastern coast known for its beauty. From the coastal plains, covered with rice paddies and nut and spice trees, the land rises to hills where tea and coconut plantations are common. At the center is a mountain massif that reaches to nearly 8200 feet (2500 meters) in elevation. The summer monsoons bring abundant moisture to the mountains, giving rise to lush forests and several unnavigable rivers that produce hydroelectric power. Close to 30 percent of the land is cultivated, and just over 25 percent remains forested.

The original hunter-gatherers and rice cultivators of Sri Lanka, today known as Veddas, now number fewer than 5000. Several thousand years ago, settlers from northern India, who built numerous city-kingdoms, joined them. Known as Singhalese, these descendants of northern Indians now constitute about 74 percent of the population of 20.5 million. The Singhalese brought Buddhism to Sri Lanka, and today 70 percent of Sri Lankans (primarily Singhalese) are Buddhists **(FIGURE 8.42)**.

About a thousand years ago, Dravidian people from southern India, known as Tamils, began migrating to Sri Lanka; by the thirteenth century, they had established a Hindu kingdom in the northern part of the island. Later, in the nineteenth century, the British imported large numbers of poor Tamils from the state of Tamil Nadu in southeastern India to work on British-managed tea, coffee, and rubber plantations. Known as "Indian Tamils," the plantation-based Tamils share linguistic and religious traditions with the "Sri Lankan Tamils" of the north and east, but each community considers itself distinct because of its divergent historical, political, economic, and social experiences. While some Sri Lankan Tamils dominate the commercial sectors of the economy, the Indian Tamils have remained isolated and poverty-stricken plantation workers.

The Sri Lankan civil war between the Singhalese urban majority and the Tamil minority was discussed earlier (on page 466 and throughout the chapter), as has Sri Lanka's strong performance in providing education and social services. Now that peace has prevailed since May 2009, there is widespread hope that the Sri Lankan diversified economy, which has proven remarkably resilient through all of the nation's troubles, will be able to perform well enough to quell with prosperity any further ethnic or religious strife.

THINGS TO REMEMBER

• Although 50 to 70 percent of the people in Southern South Asia work in agriculture, this subregion is set apart by its relatively high proportion of well-educated people, its advanced technology sectors, its strong tradition of elected communist state governments (primarily in Kerala), its focus on environmental rehabilitation and preservation, and the higher status and well-being of women.

• Sri Lanka is emerging from a protracted war between the Singhalese urban majority, who reside in the southern two-thirds of the country, and the Tamil minority, who live in the north.

GEOGRAPHIC THEMES: South Asia Review and Self-Test

1. Environment: Climate change puts more lives at risk in South Asia than in any other region in the world, primarily due to water-related issues. Over the short term, droughts, floods, and the increased severity of storms imperil many urban and agricultural areas. Over the longer term, sea level rise may profoundly affect coastal areas and glacial melting poses a threat to rivers and aquifers.

• Why is the issue of melting glaciers in the Himalayas so important in this region? Compare the short- and long-term effects of this change.

• What additional threats do coastal areas of South Asia face?

2. Globalization and Development: Globalization benefits some South Asians more than others. Educated and skilled South Asian workers with jobs in export-connected and technology-based industries and services are paid more and sometimes have better working conditions. Less-skilled workers in both urban and rural areas are left with demanding but very low-paying jobs.

• What kinds of workers in South Asia have gained from the recent boom in foreign investment? What about South Asian workers is especially attractive to foreign investors in high-tech industries?

• How are jobs in South Asia linked to the global economy?

• What kinds of workers in South Asia are benefiting the least from these new foreign investments?

3. Power and Politics: India, South Asia's oldest, largest, and strongest democracy, has shown that the expansion of political freedoms can reduce conflict. Across the region, when people have been able to participate in policy-making decisions and implementation—especially at the local level—seemingly intractable conflict has been defused and combatants have been willing to take part in peaceful political processes.

• Which conflicts in South Asia have been made worse by an unwillingness on the part of governments and warring parties to allow free and fair elections?

• What democratic strategy has been used to defuse some conflicts in the region, at least for the short term?

• What are some signs that political participation is catching on in South Asia?

4. Urbanization: South Asia has two general patterns of urbanization: one for the rich and the middle classes and one for the poor. The areas that the rich and the middle classes occupy are often spacious, clean, orderly, and peaceful, with numerous well-paying jobs. The areas that the urban poor occupy are chaotic, crowded, and violent, with overstressed infrastructures and menial jobs. These two patterns often coexist in very close proximity.

• What has been the impact on food production of the introduction of new seeds, fertilizers, pesticides, and equipment in South Asia? How has this shift resulted in the growth of cities?

• What kinds of jobs and housing do people in South Asia usually find in urban areas?

5. Population and Gender: In this most densely populated of world regions, population growth is slowing as the demographic transition takes hold. However, a severe gender imbalance is developing in this region due to age-old beliefs that males are more useful to families than are females. As a result, adult males significantly outnumber adult females.

• What has created the strong preference for sons among many South Asian families? Where in South Asia is the preference for sons the weakest?

• What is the link between fertility and education for women of South Asia?

Critical Thinking Questions

1. Dams and the reservoirs they create are sources of irrigation water and generators of electricity. Name some of the factors that have rendered so many places in South Asia, as well as elsewhere around the world, in need of markedly more water and electricity.

2. Explain why some would say that India has been part of globalization for thousands of years. Tie this history to what is happening in the present.

3. What are some of the lingering features of the British colonial era in South Asia? To what extent is globalization reinforcing or erasing these features?

4. How are changes in agriculture resulting in urbanization?

5. Describe microcredit and its impact on extreme poverty in South Asia.

6. Describe how South Asia is vulnerable to climate change. Identify some responses to the climate crisis that are emerging from this region.

7. Describe the main challenges to democracy in South Asia.

8. Identify the factors that have led to a recent boom in manufacturing industries in South Asia.

9. What factors have encouraged high population growth in South Asia in the past? What factors may encourage slower population growth in the future?

10. Describe the main factors that are creating South Asia's gender imbalance.

Chapter Key Terms

agroecology 460
Buddhism 454
caste system 453
civil disobedience 456
communal conflict 463
dowry 478
Harappa culture 451
Hinduism 453
Indus Valley civilization 451

Jainism 454
jati 453
microcredit 461
Mughals 452
offshore outsourcing 460
Partition 457
purdah 477
regional conflict 464
religious nationalism 463

sex work 470
Sikhism 454
subcontinent 440
summer monsoon 441
Taliban 466
varna 453
water scarcity 446
winter monsoon 442

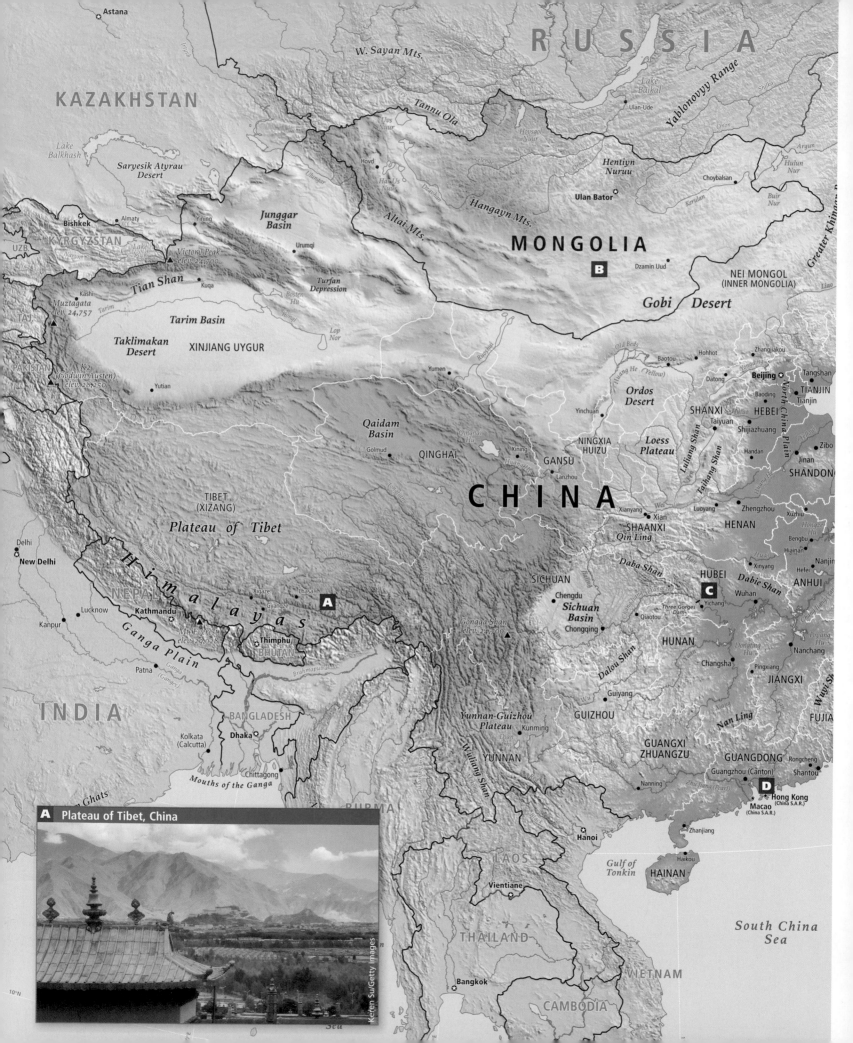

RUSSIA

KAZAKHSTAN

Astana

W. Sayan Mts.

Lake
Balkhash

Saryesik Atyrau
Desert

Tannu Ola

Yablonovyy Range

Lake Baikal

Ulan-Ude

Shilka

Hovd

Uvs
Nuur

Hangayn Mts.

Hentiyn
Nuruu

Choybalsan

Hulun
Nur

Greater Khingan

Bishkek
Almaty

KYRGYZSTAN

UZB

Yining

Junggar
Basin

Urumqi

Altai Mts.

MONGOLIA

Ulan Bator

B

Dzamin Uud

NEI MONGOL
(INNER MONGOLIA)

Victory Peak
elev. 24,700

Tian Shan

Kuqa

Turfan
Depression

Gobi   Desert

TAJ

Muztagata
elev. 24,757

Tarim

Taklimakan
Desert

Tarim Basin

XINJIANG UYGUR

Bosten
Hu

Lop
Nor

Old Beds

Baotou

Hohhot

Zhangjiakou

Huang He (Yellow)

Ordos
Desert

Datong

Beijing

Tangshan

TIANJIN

PAKISTAN

K2
(Godwin Austen)
elev. 28,250

Yutian

Yumen

Yinchuan

NINGXIA
HUIZU

Loess
Plateau

SHANXI

Luliang Shan

Baoding

Tianjin

Taiyuan

Shijiazhuang

Handan

North China Plain

Zibo

Jinan

SHANDONG

Delhi

New Delhi

INDIA

Kanpur

Lucknow

NEPAL

Kathmandu

Himalayas

Ganga Plain

Patna

Ganga
(Ganges)

Qaidam
Basin

Golmud

QINGHAI

Qinghai
Hu

Xining

Tongtian

GANSU

Lanzhou

Huang He

TIBET
(XIZANG)

Plateau of Tibet

Nu (Salween)

Xigaze

Lhasa

Gyangze

A

Mt. Everest
elev. 29,028

Thimphu

BHUTAN

Brahmaputra

CHINA

Wei

Xianyang
Xian

SHAANXI

Qin Ling

Luoyang

Zhengzhou

Xuzhu

HENAN

Daba Shan

HUBEI

C

Yichang

Three Gorges
Dam

Qiaotou

Yangtze

Dabie Shan

Xinyang

Hefei

Bengbu

Hainan

ANHUI

Nanjing

SICHUAN

Chengdu

Gongga Shan
elev. 24,700

Sichuan
Basin

Chongqing

HUNAN

Dongting
Hu

Poyang
Hu

Nanchang

Changsha

Pingxiang

JIANGXI

BANGLADESH

Kolkata
(Calcutta)

Dhaka

Chittagong

Mouths of the Ganga

Ghats

BURMA

Yunnan-Guizhou
Plateau

Dian
Chi

Kunming

YUNNAN

Wuliang Shan

Dalou Shan

Guiyang

GUIZHOU

Nan Ling

GUANGXI
ZHUANGZU

Nanning

Zhu Jiang (Pearl)

GUANGDONG

Guangzhou (Canton)

Rongcheng

Shantou

FUJIA

Wuyi Shan

Xiang

D

Hong Kong
(China S.A.R.)

Macao
(China S.A.R.)

LAOS

Hanoi

Zhanjiang

Gulf of
Tonkin

HAINAN

Haikou

South China
Sea

THAILAND

Vientiane

Bangkok

VIETNAM

CAMBODIA

10°N

A Plateau of Tibet, China

Keren Su/Getty Images

9
East Asia

C Three Gorges Dam, Chang Jiang, Hubei Province, China

ChinaFotoPress/VCG/VCG via Getty Images

D Hong Kong Island and South China Sea

Bruce Dale/National Geographic/Getty Images

E Mount Fuji, Japan

Koichi Kamoshida/Getty Images

FIGURE 9.1 Regional map of East Asia.

Sakhalin

Sea of Okhotsk

Kuril Islands

La Perouse Strait

Tatar Strait

Lesser Khingan Range

HEILONGJIANG

Qiqihar

Ranghulu

Harbin

Sikhote Alin Range

Hokkaido

Sapporo

Changchun   Jilin

JILIN

Northeast China Plain

Changbai Shandi

Fushun

Shenyang

Benxi

Anshan

Pai T'ou Shan

Tsugaru Strait

LIAONING

Dandong-Sinuiju

Sungari Res.

NORTH KOREA

P'yongyang ✧

Sea of Japan

Honshu

Sendai

J A P A N

Korea Bay

Dalian

Bo Hai

Seoul ✧

Taebaek Sanmaek

Taejon

SOUTH KOREA

Taegu

Ulsan

Changwon

Pusan

Kwangju

Korean Archipelago

Korea Strait

Qingdao

Yellow Sea

Fukuoka

Cheju

Kitakyushu

Hiroshima

Okayama Kobe Osaka

Nagoya

Mikuni Sammyaku

Kiso Sammyaku

Mount Fuji

Tokyo ✧

Yokohama

E Shizuoka

Hamamatsu

Shikoku

Kyushu

Osumi Islands

JIANGSU

(Grand Canal)

Zhenjiang

Wuxi

Shanghai

SHANGHAI

Tai Hu

Hangzhou

Ningbo

ZHEJIANG

East China Sea

Ryukyu Islands (Japan)

Amimi Islands

Okinawa Islands

Wenzhou

Senkaku Islands

Fuzhou

Taipei ✧

Taiwan Strait

Taichung

TAIWAN

Kaohsiung

Bashi Channel

Sakishima Islands

P A C I F I C   O C E A N

50°N

40°N

30°N

20°N

140°E

130°E

B Mongolia

robertharding/Getty Images

GEOGRAPHIC THEMES

After you read this chapter, you will be able to discuss the following geographic themes as they relate to the five thematic concepts:

1. Environment: East Asia's most serious environmental problems result from its high population density combined with its rapid urbanization and environmentally unsustainable economic development. Climate change is intensifying the droughts and floods that have long plagued this region.

2. Globalization and Development: East Asia pioneered a spectacularly successful economic development strategy that has transformed economies across the globe. Governments in the region intervened strategically to encourage the production of manufactured goods destined for sale abroad, primarily to the large economies of North America and Europe.

3. Power and Politics: As the countries of East Asia develop economically, people have been pushing for more political freedom. Japan, South Korea, and Taiwan all have high levels of political freedom, and demands for political change in China are increasing, especially in urban areas.

4. Urbanization: Across East Asia, cities have grown rapidly over the last century, fueled by export-oriented manufacturing industries. China has recently undergone the most massive and rapid urbanization in the history of the world. Its urban population, now 755 million people, has more than tripled since China initiated economic reforms in the 1980s.

5. Population and Gender: Although East Asia remains a highly populous world region, families here are having far fewer children than in the past, resulting in slow population growth and populations that are aging. The legacy of China's now-abandoned one-child policy, combined with an enduring cultural preference in China for male children, has created a shortage of females.

The East Asia Region

East Asia, with 1.6 billion people, is the second most populous world region after South Asia **(FIGURE 9.1)**. The region is dominated by China, which has a population of 1.37 billion people. China's dominance is due not only to its population, but also to its physical size and its economic role in the global economy. Just as the striking economic rise of Japan, South Korea, and Taiwan over the past 50 years has been the model for developing countries, the opening of China to the global economy will continue to transform East Asia and the world in the future. The changes underway in this region are immense. Incomes have risen and cities have boomed, but so have air and water pollution and greenhouse gas (GHG) emissions (although there are signs that new, cleaner technologies may mitigate these pollution problems).

The five thematic concepts in this book are explored as they arise in the discussion of regional issues, with interactions between two or more themes featured. Vignettes, like the

hukou system the system in China where each person's permanent residence is registered and any person who wants to migrate must obtain permission from authorities to do so; the system has recently been liberalized

floating population the Chinese term for people who live and work in a place other than their household registration location; many are people who have left economically depressed rural areas for the cities

one that follows about environmental concerns, illustrate one or more of the themes as they are experienced in individual lives.

GLOBAL PATTERNS, LOCAL LIVES When the Chinese journalist Chai Jing was pregnant, she found out that her soon-to-be-born daughter had a tumor. It turned out to be benign, but Jing was convinced that her baby girl's condition was caused by air pollution. Therefore, she embarked on a year-long journey detailing China's horrendous pollution problem. Spending her own money, U.S.$159,000, she turned her investigation into a documentary film called *Under the Dome* (it is available on YouTube). It turned out to be a success—it had more than 100 million views in a 2-day period when it was released in 2015. Surveys of attitudes in China show, not surprisingly, that pollution is one of the leading concerns among people in China. Perhaps inspired by Al Gore's film about climate change, *An Inconvenient Truth,* Jing's film is partly lecture, and partly clips that document children who have never seen white, fluffy clouds against a blue sky; officials who bluntly say that economic development is more important than environmental protection; and polluters who callously ignore the fines that they accrue when they exceed pollution limits.

When she was making the film, Jing had the support from state-supported television, and government censors who reviewed the manuscript gave her the go-ahead to make the film. At first the government was apparently willing to allow some public exposure of the pollution problem, but when the film became hugely successful it banned its distribution. *[Sources: Celia Hatton, "Under the Dome: The Smog Film Taking China by Storm," BBC News, http://www.bbc.com/news/blogs-china-blog-31689232; "China Takes Under the Dome Anti-Pollution Film Offline," BBC News, http://www.bbc .com/news/world-asia-31778115]* ∎

The experience of Chai Jing in the vignette illustrates some of the negative consequences of China's rapid economic development. **FIGURE 9.2** also shows how urbanization and globalization are transforming East Asia as millions of people, mostly rural young adults, flock to East Asia's coastal cities to work in factories. There they produce goods for sale on global markets, learn new skills, and gain new confidence.

Until 2014, most rural-to-urban migration in China was illegal because the ***hukou*** (household registration) **system,** an ancient practice reinforced in the Communist era, effectively tied rural people to the place of their birth. The objective was to maintain social order by limiting migration so cities wouldn't be overwhelmed by people for whom there was inadequate housing and employment. Today the system is being reformed (see page 529) and migrants are being allowed to move to approved cities as long as they obtain permission from the authorities. This is a major change for the more than 160 million people who have been ignoring the hukou system for decades, becoming part of what is called the **floating population,** a term used in China to describe those who have no rights to subsidized housing, schools, or health care in the place to which they have migrated. These migrants generally work in menial, low-wage jobs and make agonizing sacrifices to send money home to children and spouses living in rural areas. The number of people migrating from rural to urban areas in China has increased year by year. China's migrant workers now number 250 million people, approximately half of the working people in China's cities (see Figure 9.2 and Figure 9.20C).

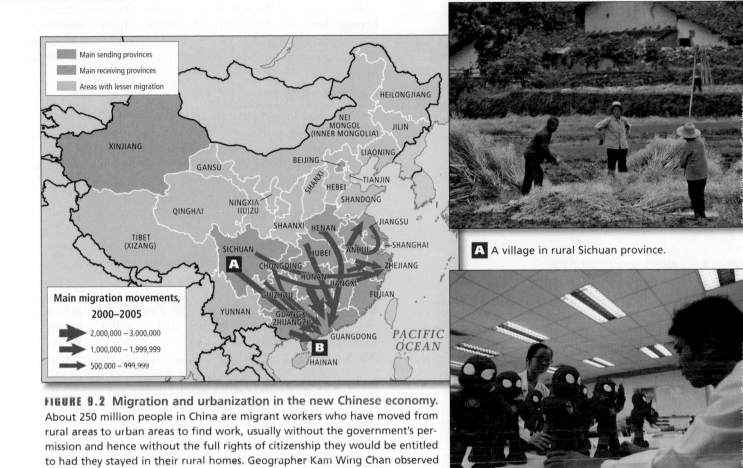

A A village in rural Sichuan province.

B Workers in Dongguan, Guangdong Province, check stuffed toys for defects.

FIGURE 9.2 Migration and urbanization in the new Chinese economy. About 250 million people in China are migrant workers who have moved from rural areas to urban areas to find work, usually without the government's permission and hence without the full rights of citizenship they would be entitled to had they stayed in their rural homes. Geographer Kam Wing Chan observed that the number of workers moving long distances has steadily increased since the 1990s, with the majority moving to Guangdong Province, which some now call the "world factory," and into the lower Chang Jiang region around Shanghai. In the future, however, the pool of cheap, rural labor is predicted to dry up. As migrations slow, the price of labor goes up, and China's coastal manufacturing economy will have to change. [Source consulted: Kam Wing Chan, "Internal Migration in China: Trends, Geography and Policies," *Population Distribution, Urbanization, Internal Migration and Development: An International Perspective* (New York: United Nations Department of Economic and Social Affairs, Population Division, 2011), pp. 81–109]

For decades, the Chinese government only weakly enforced the hukou system because urban industries were in need of a growing workforce. The lax enforcement enabled the mobility of the labor force while keeping that labor vulnerable and exploitable. The current reforms to the hukou system are designed both to encourage more migration to cities, which have had labor shortages in recent years, and to control the places that migrants can relocate to. The goal is to freely allow migration to small and medium-sized cities while still limiting migration to the biggest cities so they are not overwhelmed by new residents. A major benefit for migrants is that they now have better access to housing, schools, and health care. Another new policy reform is that urban migrants no longer have to give up their right to farmland in the village that they left behind, which provides some sense of security if they would one day go back.

What Makes East Asia a Region?

Part of the region of East Asia is China, home to more than one-fifth of humanity. The vast East Asian territory stretches from the Taklimakan Desert in far western China to Japan's rainy Pacific coastline, and from the frigid mountains of Mongolia in the north to the tropical landscapes of Hainan, China's southernmost province, in the south (see Figure 9.1; see also Figure 9.4B). East Asia is comprised of the countries of China, Mongolia, North Korea, South Korea, Japan, and Taiwan (**FIGURE 9.3**; the last has operated as an independent country since World War II but is claimed by China as a province). These countries are grouped together because of their cultural and historical roots, many of which are in China. Because of China's great size, historical influence, enormous population, and huge economy, it is given particular emphasis in this chapter. Japan, whose large and prosperous economy makes it a major player on the world stage, is also emphasized.

FIGURE 9.3 Political map of East Asia.

Terms in This Chapter

East Asian place-names can be very confusing to those unfamiliar with East Asian languages. We give place-names in English transliterations of the appropriate Asian language, taking care to avoid redundancies. For example, *he* and *jiang* are both Chinese words for river. Thus the Yellow River is the Huang He, and the Long River, also called the Yangtze in English, is the Chang Jiang in Chinese; it is redundant to add the term *river* to either name. The word *shan* appears in many place-names and usually means mountain.

Pinyin (a spelling system based on Chinese sounds) versions of Chinese place-names are now commonplace. For example, the city once called Peking in English is now Beijing, and Canton is Guangzhou.

Although China refers to Tibet as Xizang, people around the world who support the idea of Tibetan self-government avoid using that name. This text uses Tibet for the region (with Xizang in parentheses).

tsunami a large sea wave caused by an earthquake

THINGS TO REMEMBER

• East Asia has experienced great economic development, which has ushered in an era of urbanization but has also resulted in environmental problems.

• China's migrant workers now number 250 million people, approximately half of the working people in China's cities; many have moved without legal papers.

• The current reforms to the hukou system are designed to give migrants better access to services while still leaving the government in control over which places migrants can relocate to, thus keeping the biggest cities from being overwhelmed by new residents.

PHYSICAL GEOGRAPHY

A quick look at the regional map of East Asia (see Figure 9.1) reveals that the topography here is perhaps the most rugged of any world region. East Asia's varied climates result from the meeting of huge warm and cool air masses and the dynamic interaction between land and oceans. The region's large human population has affected the variety of ecosystems that have evolved there over the millennia and that still contain many important and unique habitats.

LANDFORMS

The complex topography of East Asia is partially the result of the slow-motion collision of the Indian subcontinent with the southern edge of Eurasia over the past 60 million years. This tremendous force created the Himalayas and lifted up the Plateau of Tibet (see Figure 9.1A; also depicted in gray and gold in the Figure 9.1 map), which can be considered the highest of four descending steps that define the landforms of mainland East Asia, moving roughly west to east.

The second step down from the Himalayas and the Plateau of Tibet is a broad arc of basins, plateaus, and low mountain ranges (depicted in yellowish tan in the Figure 9.1 map). These landforms include the broad, rolling highland grasslands and deep, dry basins and deserts of western China (such as the Taklimakan Desert) and Mongolia (see Figure 9.1B), as well as the Sichuan Basin and the rugged Yunnan–Guizhou Plateau to the south, which is dominated by a system of deeply folded mountains and valleys that bend south through the Southeast Asian peninsula.

The third step, directly east of this upland zone, consists mainly of broad coastal plains and the deltas of China's great rivers (shown in shades of green in the Figure 9.1 map). Starting from the south, this step is defined by three large lowland river basins: the Zhu Jiang (Pearl River) basin, the massive Chang Jiang basin (see Figure 9.1C), and the lowland basin of the Huang He on the North China Plain. Each of these rivers has a large delta. Despite the deltas being subject to periodic flooding, they have historically been used for agriculture. However, coastal cities have now spread into many of these deltas, filling in wetlands and farms with zones of dense population and industrialization (see Figure 9.20). Low mountains and hills (shown in light brown in the Figure 9.1 map) separate these river basins. China's Far Northeast and the Korean Peninsula are also part of this third step.

The fourth step consists of the continental shelf, covered by the waters of the Yellow Sea, the East China Sea, and the South China Sea. Numerous islands—including Hong Kong, Hainan, and Taiwan—are anchored on this continental shelf; all are part of the Asian landmass (see Figure 9.1D).

The islands of Japan have a different geological origin: they are volcanic, not part of the continental shelf. They rise out of the waters of the northwestern Pacific in the tectonically active zone where the Pacific and Philippine plates dive beneath the Eurasian plate. Lying along a portion of the Pacific Ring of Fire (see Figure 1.9), the entire Japanese island chain is particularly vulnerable to disastrous volcanic eruptions, earthquakes, and **tsunamis** (seismic sea waves). Volcanic Mount Fuji, the highest peak in the country and a recognizable national symbol (see Figure 9.1E), last erupted in 1707. However, the mountain is still classed as active, and deep internal rumblings have been detected since 2001. In March of 2011, the largest earthquake in recorded Japanese history (registering 9.2 on the Richter scale) hit off the coast of Honshu, Japan, near the city of Sendai. The quake and subsequent tsunami killed about 18,000 people and damaged several nuclear reactors located on the coast, resulting in the second-worst nuclear accident ever in the world (discussed further on page 505).

The East Asian landmass has few flat portions, and the flattest land is either very dry or very cold. Consequently, the large numbers of people who occupy the region have had to be particularly inventive in creating spaces for agriculture. They have cleared and terraced entire mountain ranges, until recently using only simple hand tools (see Figure 9.5A). They have irrigated drylands with water from melted snow, drained wetlands using elaborate levees and dams, and applied their complex knowledge of horticulture and animal husbandry to help plants and animals flourish in difficult conditions.

CLIMATE

East Asia has two principal contrasting climate zones, shown in the map in **FIGURE 9.4**: the dry interior west and the wet (monsoon) east. Recall from Chapter 8 that the term *monsoon* refers to the seasonal reversal of surface winds that flow from the Eurasian continent to the surrounding oceans during winter and from the oceans inland during summer.

The Dry Interior

Because land heats up and cools off more rapidly than water does, the interiors of large landmasses in the midlatitudes tend to be intensely cold in winter and extremely hot in summer. Western East Asia, roughly corresponding to the first two topographic steps described earlier, is an extreme example of such a midlatitude continental climate because it is very dry. In fact, this area is farther away from an ocean than any other place on Earth's surface (this *pole of inaccessibility* is located in the Junggar Basin in northwestern China). With little vegetation or cloud cover to retain the warmth of the sun after nightfall, summer daytime and nighttime temperatures may vary by as much as 100°F (55°C).

Grasslands and deserts of several types cover most of the land in this dry region (see Figure 9.4C). Only scattered forests grow on the few relatively well-watered mountain slopes and in protected valleys supplied with water by snowmelt. In all of East Asia, humans and their impacts are least conspicuous in the large, uninhabited portions of the deserts of Tibet (Xizang), the Tarim Basin in Xinjiang, and the Mongolian Plateau.

The Monsoon East

The monsoon climates of the east are influenced by the extremely cold conditions of the huge Eurasian landmass in the winter and the warm temperatures of the surrounding seas and oceans in the summer. During the dry winter monsoon, descending frigid air sweeps south and east through East Asia, producing long, bitter winters on the Mongolian Plateau, on the North China Plain, and in China's Far Northeast (see Figure 9.4D). While occasional freezes may reach as far as southern China, winters there are shorter and less severe. The east–west mountain ranges of the Qin Ling in central China partially protect southern China from cold northern air, and the warm waters of the South China Sea moderate land temperatures.

As the continent warms during the summer monsoon, the air over land rises, which creates low air pressure that pulls in high-pressure wet, tropical air from the adjacent seas. The warm, wet air from the ocean deposits moisture on the land in the form of seasonal rains. This effect is reinforced by the Intertropical Convergence Zone (ITCZ)—where air masses from north and south converge, are forced upward into the atmosphere, and generate rainfall—which is located over southern China in the summer (see Chapter 7). As the summer monsoon moves northwest, it must cross numerous mountain ranges and displace cooler air. Consequently, its effect is weakened toward the northwest. Thus, the far southeast is drenched with rain and has warm weather for most of the year (see Figure 9.4A), whereas central China has about 5 months of summer monsoon weather. The North China Plain, north of the Qin Ling and Dabie Shan ranges, receives even less monsoon rain—about 3 months of monsoon each year. Very little monsoon rain reaches the dry interior.

Korea and Japan have wet climates year-round, similar to those found along the Atlantic Coast of the United States, because of their proximity to the sea. They still have hot summers and cold winters because of their northerly location and exposure to the continental effects of the huge Eurasian landmass. Japan and Taiwan actually receive monsoon rains twice: once in spring, when the main monsoon moves toward the land, and again in autumn, as the winter monsoon forces warm air off the continent. This retreating warm air picks up moisture over the coastal seas, which is then deposited on the islands. Much of Japan's autumn precipitation falls as snow, in particular in northern latitudes and at higher altitudes.

Natural Hazards

The entire coastal zone of East Asia is intermittently subject to **typhoons** (which is the same weather phenomenon that is called tropical cyclones in the southern hemisphere or hurricanes off the coasts of the United States). Japan's location along the northwestern edge of the Pacific Ring of Fire (see Figure 1.9) results in volcanic eruptions, earthquakes, and tsunamis. These natural hazards are a constant threat in Japan; the heavily populated zone from Tokyo southwest through the Inland Sea (between Shikoku and southern Honshu) is particularly endangered. Earthquakes are also a serious natural hazard in Taiwan and in China's mountainous interior.

typhoon a tropical storm in the western Pacific Ocean; called a cyclone or hurricane elsewhere

---

THINGS TO REMEMBER

- There are four main topographical zones, or "steps," that form the East Asian continent.

- Japan was created by volcanic activity along the Pacific Ring of Fire.

- East Asia has two principal contrasting climates: the dry continental interior (west) and the monsoon east.

- East Asia faces a wide range of natural hazards, including earthquakes, tsunamis, volcanic eruptions, and tropical storms (typhoons).

FIGURE 9.4 PHOTO ESSAY: Climates of East Asia

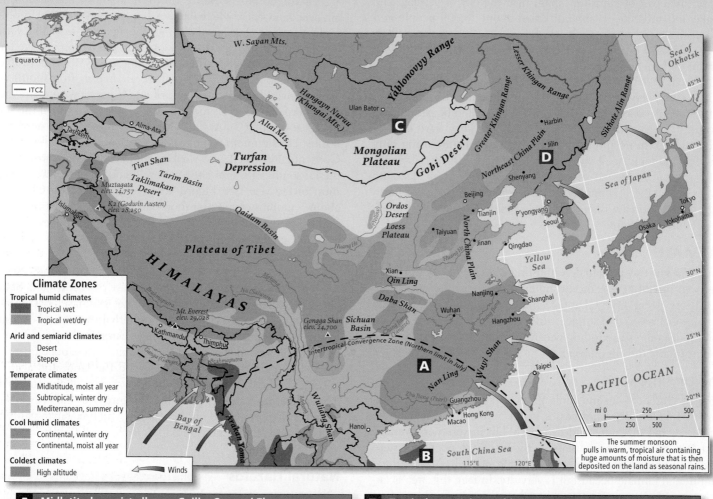

Equator
— ITCZ

W. Sayan Mts.

Yablonovyy Range

Lesser Khingan Range

Sea of Okhotsk

Hangayn Nuruu (Khangai Mts.)

Ulan Bator

C

Altai Mts.

Mongolian Plateau

Gobi Desert

Greater Khingan Range

Sikhote Alin Range

Harbin

Northeast China Plain

Jilin

D

Sea of Japan

Tian Shan

Turfan Depression

Alma-Ata

Tashkent

Tarim Basin

Taklimakan Desert

Muztagata elev. 24,757

K2 (Godwin Austen) elev. 28,250

Islamabad

Shenyang

Beijing

Tianjin

P'yongyang

Seoul

Tokyo

Osaka

Yokohama

Qaidam Basin

Ordos Desert

Loess Plateau

Huang He (Yellow)

Taiyuan

North China Plain

Jinan

Qingdao

Yellow Sea

Plateau of Tibet

Huang He

Xian

Qin Ling

Nanjing

Shanghai

HIMALAYAS

Mekong

Daba Shan

Chang Jiang

Wuhan

Hangzhou

Nu (Salween)

Mt. Everest elev. 29,028

Brahmaputra

Kathmandu

Thimphu

Gonaga Shan elev. 24,700

Sichuan Basin

Chang Jiang (Yangtze)

A

Nan Ling

Wuyi Shan

Taipei

PACIFIC OCEAN

Ganga (Ganges)

Brahmaputra

Intertropical Convergence Zone (Northern limit in July)

Zhu Jiang (Pearl)

Guangzhou

Hong Kong

Macao

mi 0 ... 250 ... 500

km 0 ... 250 ... 500

20°N

Climate Zones

Tropical humid climates
- Tropical wet
- Tropical wet/dry

Arid and semiarid climates
- Desert
- Steppe

Temperate climates
- Midlatitude, moist all year
- Subtropical, winter dry
- Mediterranean, summer dry

Cool humid climates
- Continental, winter dry
- Continental, moist all year

Coldest climates
- High altitude

→ Winds

Bay of Bengal

Arakan Yoma

Wuliang Shan

Hanoi

B

South China Sea

115°E

120°E

The summer monsoon pulls in warm, tropical air containing huge amounts of moisture that is then deposited on the land as seasonal rains.

A Midlatitude, moist all year, Guilin, Guangxi Zhuang Autonomous Region, China

Haibo Bi/Getty Images

B Tropical wet, Hainan, China

TravelAsia/Asia Images/Getty Images

C Steppe, Mongolia

Tuul and Bruno Morandi/The Image Bank/Getty Images

D Continental, winter dry, Jilin, China

Fotosearch/Getty Images

ENVIRONMENT

Environment: East Asia's most serious environmental problems result from its high population density combined with its rapid urbanization and environmentally unsustainable economic development. Climate change is intensifying the droughts and floods that have long plagued this region.

Climate change, especially global warming, is a growing concern in East Asia. China is the world's largest overall producer of greenhouse gases (but still produces less than half of U.S. emissions on a per capita basis; see Chapter 1). China has recently been working to limit its emissions for a variety of reasons, but even though it has significantly increased its efficiency through the use of new technologies, its emissions still make up almost 30 percent of the world's total. This amount will increase as China's urban households use more cars, air conditioning, appliances, and computers. Japan, Korea, and Taiwan are also major greenhouse gas emitters.

China's Vulnerability to Glacial Melting

Glaciers on the Plateau of Tibet capture monsoon moisture and store it as frozen ice, which is then slowly released as meltwater. China's largest rivers, the Huang He and the Chang Jiang, are partially fed by these glaciers, which are now melting so rapidly that scientists predict they will eventually disappear. One effect could be significantly lower flows in these two great rivers during the winter, when little rain falls. Both have already begun to run low during winter, and trade has been affected because riverboats and barges have become stranded on sandbars. Another effect could be a reduction in the amount of irrigation water available for dry-season farming.

Water Shortages

Nearly every year, abnormally low rainfall or abnormally high temperatures create a drought somewhere in China. These droughts often cause more suffering and damage than any other natural hazard.

Droughts can be worsened by human activity on a local or regional scale. When people begin to live or farm in dry environments, as many millions have done in China and Mongolia during the twentieth century, *desertification* (see Chapter 6) can result. People clear vegetation to grow crops, which require more water than the natural vegetation, so the crops must be irrigated with surface water or water pumped from underground aquifers **(FIGURE 9.5A)**. Many dry areas in China are subject to strong winds that can blow away topsoil once the vegetation is removed. Furthermore, irrigated crops are often less able than natural vegetation to hold the soil. This has resulted in huge dust storms much like those that plagued the central United States in the 1930s Dust Bowl (see Chapter 2). High dunes of dirt and sand have appeared almost overnight in some parts of China that border the desert, threatening crops, roads, and homes. Dust storms can also move over long distances from western China and affect large cities, such as Beijing, near the coast. Particulate matter from these dust storms even circles the entire globe in the upper atmosphere.

VIGNETTE In Ningxia Huizu Autonomous Region on the Loess Plateau in north-central China, Wang Youde squints out at what are now sand-colored low hills barren of vegetation, thinking about how these vast tracts of former farmland have been transformed into deserts by a combination of human error and climate change.

The Loess Plateau was already prone to dust storms during times of drought (*loess* means "wind-deposited soil"). Agricultural expansion into loess areas led to the removal of thick, natural, deep-rooted grasses which once helped hold down the soil. One can still see the agricultural terraces on the arid slopes where now not even grass grows. Wang Youde's family and 30,000 others fled the area when Youde was 10 years old because one day a sand dune covered their village. From his early childhood, he remembers flowers, birdsong, and occasional snowfalls; all have vanished. Now Youde is back, heading up a project to revegetate thousands of hectares with drought-resistant plants that will hold the soil. By hand, squares of braided straw or stones are laid down to keep water from running off the land and to protect planted seedlings. The seedling survival rate is only 20 to 30 percent. Yet Youde says, "Every time we see an oasis that we have created we are very satisfied. . . [b]ecause we have poured sweat and blood into our work." His adult children are helping him, hoping to remain with the family and escape the hardships of migrating to find urban factory work. [Source: Asia Pacific News. For detailed source information, see Text Sources and Credits.] ∎

Water shortages are particularly intense in the North China Plain, which produces 60 percent of China's wheat, 45 percent of its corn, and 35 percent of its cotton. Here the water table is falling more than 3 feet per year, mainly because of the increased use of groundwater for wheat irrigation. Groundwater must be used for agriculture because much of the available surface water is dedicated to urban needs. (The North China Plain is one of the most densely populated and urban regions of China.) Withdrawals of water from the Huang He often make the river's lower sections run completely dry during the winter and spring. A gigantic development called the South–North Water Transfer Project is under way, designed to attempt to rectify this problem (see Figure 9.5B). The goal is to divert water from the Chang Jiang basin in central China, which has a relative surplus of water, to the Huang He basin in the north, using dams, canals, and pipelines. The upper reaches of the Huang He receive water from the Plateau of Tibet, while more water is transferred to the lower Huang He in the coastal provinces in the east. The first segment of the project started to bring water to Beijing in 2014, but the project is not expected to be completed until 2050.

China is also trying to avoid water shortages, focusing on water conservation to stretch existing supplies. Already, 30 percent of China's urban water is recycled, and many cities are trying to raise this percentage. China is also making a major effort to remove pollutants from wastewater discharged by industry and farming, which together account for 85 percent of water use.

Japan, the Koreas, and Taiwan have monsoon rainfall patterns, which give them a generally wetter climate and make them less vulnerable to drought than China and Mongolia.

Flooding in China

Flooding has been a hazard in China for hundreds of years. In fact, the six deadliest floods in human history took place in China. The same shifting patterns of rainfall that may worsen droughts can also worsen flooding. Under usual conditions, the huge amounts

FIGURE 9.5 PHOTO ESSAY: Vulnerability to Climate Change in East Asia

China is particularly vulnerable to drought, desertification, flooding, and other hazards that climate change may intensify.

Simon Yu/Getty Images

Goh Chai Hin/AFP Creative/Getty Images

A The terraced hillsides of China's eastern Gansu Province are in a dry upland zone, where agriculture depends on rainfall or irrigation water taken from rivers or underground aquifers. Rainfall is already unpredictable and could become more so with climate change. Melting glaciers on the Plateau of Tibet threaten to reduce river flows, and the overuse of groundwater for irrigation is depleting aquifers. Large impoverished rural populations are increasingly being forced to relocate to other parts of China.

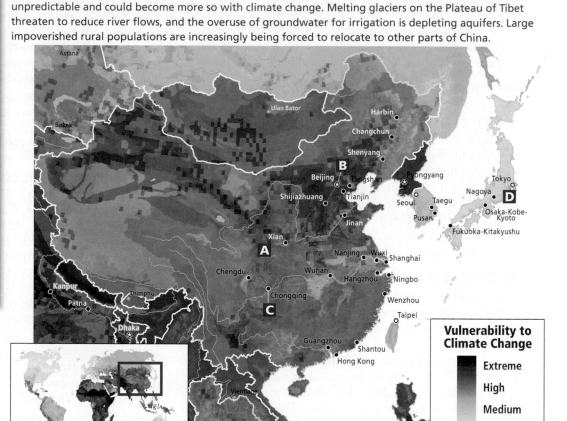

Vulnerability to Climate Change

- Extreme
- High
- Medium
- Low

B A canal outside of Beijing, China, that is subject to severe water shortages and the effects of massive increases in population. China's national government is creating huge canals and other infrastructure to divert water from distant rivers in southern and western China to Beijing and surrounding areas.

C Flooding hits the Chang Jiang at Chongqing, China, carrying off boats and barges. Climate change could result in more flooding. This is especially likely in eastern China, where there is much rainfall during the summer monsoon.

D A solar power station in Kawasaki, Japan. East Asia has made large investments in solar power. China and Japan have the second- and third-largest installed solar power–generating capacity in the world (behind Germany). More renewable power is needed to offset emissions from coal-fired power plants.

ChinaFotoPress/VCG/VCG via Getty Images

Jgotaki/E+/Getty Images

Thinking Geographically

Now that you have read about vulnerability to climate change in East Asia, you should be able to answer the following questions about the photos in this chapter.

A What is one process that can result from people beginning to live or farm in dry environments?

B What is causing the water table to fall each year on the North China Plain?

C Flooding is especially likely in what part of China?

of rain deposited on eastern China during the summer monsoon periodically can cause catastrophic floods along the major rivers. If global warming leads to even slight changes in rainfall patterns, flooding could be much more severe (see Figure 9.5C). Engineers have constructed elaborate systems of dikes, dams, reservoirs, and artificial lakes to help control flooding. However, these systems failed in 1998, when heavy rains in the Chang Jiang basin caused some of the worst flooding in decades, resulting in the death of approximately 4000 people. The pattern is now repeating frequently on the Chiang Jiang and other rivers in central and southern China. During the rainy summer of 2010, the new Three Gorges Dam (see page 508) could absorb much of the floodwaters, but the amount of water was nevertheless too much to handle and the floodgates of the dam had to be opened, leading to mudslides and a large number of casualties from drowning.

Recently, urban areas have suffered the most from flooding. The number of Chinese cities affected by annual floods more than doubled from 2008 to 2015. The problem is that urban sprawl has expanded but the drainage infrastructure has not been sufficiently improved. Since the big flood of 1998, the amount of land in China that has been urbanized has more than doubled. Most of this urbanization has taken place on floodplains, creating high amounts of impervious land cover (asphalt and concrete) that leads to serious drainage problems.

Natural Hazards and Energy Vulnerability

Until the Sendai earthquake and tsunami of 2011, Japan, along with many other countries, planned to increase its use of nuclear power as a way to reduce GHG emissions. However, the tsunami washed over the Fukushima nuclear plant and destroyed its cooling systems, which caused a meltdown of several of the plant's nuclear reactors. The severity of the nuclear disaster that followed in the wake of the tsunami halted or curtailed many plans for expanding nuclear power in Japan and across the world. The accident forced the evacuation of more than 300,000 people and temporarily contaminated the water supply of Tokyo. For miles around the reactor, the ground, food, and livestock were contaminated and thousands of gallons of water containing 10,000 times the normal level of iodine-131 (a radioactive substance) were released into the Pacific Ocean. Some places remain ghost towns; nobody has been allowed to resettle them.

Another unfortunate outcome of the Fukushima disaster is that Japan has since increased its GHG emissions from coal energy because of the suspension of the nuclear power plants. On the positive side, the accident is expected to boost reliance

on renewable resources, including solar power (see Figure 9.5D). Japan restarted its first nuclear reactor in 2015 but whether nuclear power will make a full comeback in Japan is questionable.

FOOD SECURITY AND SUSTAINABILITY

East Asia's **food security**—the capacity of the people in a geographic area to consistently provide themselves with adequate food—is increasingly linked to the global economy. The wealthiest countries in the region can buy food on the global market, but they are less able to produce sufficient food domestically for two main reasons. Much agricultural land has been lost to urban and industrial expansion; and rising levels of affluence have led to a demand for more meat, which is either imported or requires the importation of animal feed.

About 75 percent of the food consumed in Japan, South Korea, and Taiwan is imported. Japan is the third-largest food importer in the world, and with limited amount of grazing land available, it imports more meat than any other country. In China, self-sufficiency in grain production is important to national identity because devastating famines were a recurring problem before and during the Communist Revolution. China is now nearly self-sufficient with regard to basic necessities, but it relies on imports of grain and soybeans for animal feed and luxury food items. China is the second-largest food importer in the world (after the United States).

Recent dramatic increases in global food prices illustrate the perils of dependence on food imports, especially for the poor. From 2007 to 2008, the world market price for grain and other basic food commodities shot up, and it did so again from 2010 to 2011. The reasons for the increases in food prices include high oil prices (because producing food is often energy intensive), the use of corn to produce ethanol fuel (ethanol is used as a replacement for gasoline in the Americas and elsewhere) rather than as a food source, and the rise in demand from China and other economically expanding countries for agricultural commodities. A recent weakening of oil prices has also provided some relief in the cost of food, although the price remains relatively high from a historical perspective.

Food Production

Just over half of East Asia's vast territory can support agriculture (FIGURE 9.6). In much of this area, food production has been pushed well beyond what can be sustained over the long term. As a result, East Asia's fertile zones are shrinking. In China, roughly one-fifth of the agricultural land has been lost since the Communist Revolution of 1949, largely because of urban and industrial expansion and agricultural mismanagement that created soil erosion and desertification (see the vignette on page 503). This is not surprising; when Japan, South Korea, and Taiwan developed, which they did before China developed, their available farmland declined too.

As urban populations become more affluent, they consume more meat and other animal products that require more land and resources than the plant-based national diet of the past. Soybeans are one example of this change in diet. Soybeans are no longer used mainly for human food but for animal feed, for farmed fish food, and

food security when people consistently have access to sufficient amounts of food to maintain a healthy life

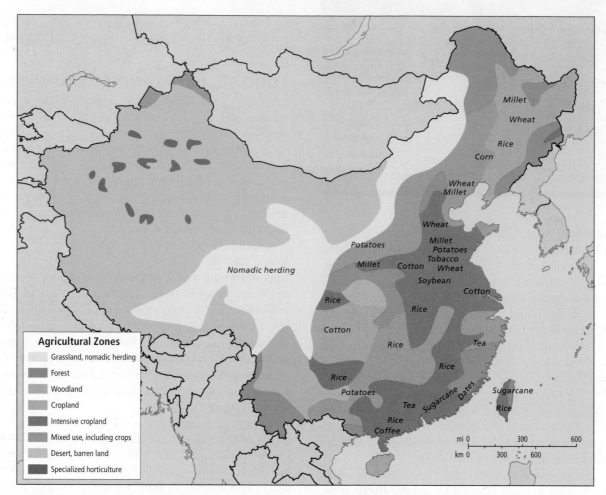

FIGURE 9.6 China's agricultural zones. The economic reforms instituted in recent years in China include more regional specialization in agricultural products. [Sources consulted: "World Agriculture," *National Geographic Atlas of the World,* 8th ed. (Washington, DC: National Geographic Society, 2005), p. 19; "China: Economic, Minerals" map, *Goode's World Atlas,* 21st ed. (New York: Rand McNally, 2005), pp. 39, 207]

especially for high-grade cooking oil. More than 80 percent of the soybeans used in China are now imported, primarily from the United States, Brazil, and Argentina.

Even with the increase in meat consumption, vegetable dishes with an ancient history remain popular throughout East Asia **(FIGURE 9.7A)**, and meat and dairy dishes are still prepared in traditional ways in places where raising farm animals has been the norm (see Figure 9.7C).

ON THE BRIGHT SIDE: Responses to the Climate Crisis

East Asia is responding to the warming aspect of global climate change. Japan has led these efforts for decades. In 1997, the Japanese city of Kyoto hosted the meeting in which countries first committed to reducing their GHG emissions. Japanese automakers such as Toyota were among the first to develop and sell hybrid gas-electric vehicles. China and Japan are now second and third only to Germany in installed solar power–generating capacity (see Figure 9.5D). China is also the world leader in manufacturing the photovoltaic cells—more and more of which are being used domestically—used to generate electricity from solar energy. Large on-the-ground solar panel plants are being installed in China's sunny, dry interior, where the level of incoming solar radiation is high. The electricity from these is transported via the grid to energy-hungry cities in eastern China. However, mitigating climate change is not the sole reason for the growth of green energy; the high level of air pollution from coal-fired power plants must be reduced by turning to cleaner power sources. ∎

Rice Cultivation

The region's most important grain is rice, and over the millennia its cultivation has transformed landscapes throughout central and southern China, Japan, Korea, and Taiwan. In these areas, rainfall is sufficient to sustain **wet rice cultivation,** which can be highly productive.

Wet rice cultivation requires elaborate systems of water management, as the roots of the plants must be submerged in water early in the growing season. Centuries of painstaking human effort have channeled rivers into intricate irrigation systems, and whole mountainsides have been transformed into descending terraces that evenly distribute the water. Writing about wet rice cultivation in Sichuan Province, geographer Chiao-Min Hsieh describes how "[e]verywhere one can hear water gurgling like music as it brings life and growth to the farms." However, these same cultivation techniques, combined with extensive forestry and mining, have led

wet rice cultivation a prolific type of rice production that requires the plant roots to be submerged in water for part of the growing season

to the loss of most natural habitats in all but the most mountainous, dry, or remote areas. With so little suitable agricultural land left, further expansion of the area under wet rice cultivation is unlikely.

Fisheries and Globalization

Many East Asians depend heavily on ocean-caught fish for protein (see Figure 9.7B). People in Japan and China eat more than four times as much fish and shellfish as do Americans. The Japanese in particular have had a huge impact on the seas of not only this region but also of the entire world. There are some 4000 coastal fishing villages in Japan, sending out tens of thousands of small crafts to work nearby waters each day.

There are also many large Japanese fishing vessels, complete with onboard canneries and freezers, that harvest oceans around the world. They can do so because waters at a certain distance away from land, typically 200 nautical miles (230 miles, or 370 kilometers), are considered international waters and outside the jurisdiction of any individual country. Environmentalists have criticized Japan for overfishing. For example, Japanese fishing off western Africa has reduced the catches of local fishers so much that many have been forced to migrate to Europe for work. There is little room for Japan to expand its fish imports and exports because the global wild fish catch has been static or in decline since 1990 due to overfishing. Moreover, consumption in Japan has waned. Japan's per capita fish consumption in 2014 was 59.5 pounds (27 kilograms), compared to 89 pounds (40.2 kilograms) in 2002. In fact, the consumption of meat exceeded fish for the first time in 2006.

Instead, China is today both the largest producer and consumer of fish in the world. Unlike Japan, the world's second-largest consumer of fish, most of China's production is based on aquaculture rather than on wild caught fish. It is likely that aquaculture will expand in the future, in China and elsewhere, as an important source of protein in an increasingly populous world.

People and Animals

Animals are used in East Asia not only for food but also for medicinal, aesthetic, and cultural purposes. For example, traditional Chinese medicine involves, among other things, the capturing of wild animals. Such animals are found in markets around China because they are believed to have healing properties **(FIGURE 9.8A)**. Unfortunately, because of this practice, some species are now endangered.

In Japan, aquaculture provides living ornamentation to gardens and other carefully tended landscapes. Carp have been domesticated and bred to have a variety of colors and patterns (see Figure 9.8C) and are a central part of many Japanese gardens. As an adaptive species that does well under a wide variety of aquatic conditions, ornamental carp are now found around the world. Animals are often central to a culture, as is the case in Mongolia, where many species of animals are fundamental to the nomadic culture that historically dominated there. Sheep, goats, yaks (a long-haired Central Asian bovine), horses, and cattle are the most commonly raised animals. A way of life once common in areas of dry grasslands around the world (see Figure 9.8B), nomadic living is becoming less common than it was in the past as places like Mongolia become more urbanized.

> ### THINGS TO REMEMBER
>
> **GEOGRAPHIC THEME 1**
> • **Environment** East Asia's most serious environmental problems result from its high population density combined with its rapid urbanization and environmentally unsustainable economic development. Climate change is intensifying the droughts and floods that have long plagued this region.

FIGURE 9.7 LOCAL LIVES: Foodways in East Asia

A Kimchi, a Korean dish made of fermented and often highly seasoned vegetables, was developed at least 3000 years ago as a food storage and preservation technique. Specific kinds of kimchi are made at different times of the year because fermentation happens at variable rates, depending on the particular vegetables available for fermenting at that point in the year and on the temperature at which each vegetable ferments.

B A wealthy restaurant owner in Japan poses with a 448-pound (220-kilogram) tuna that he bought for U.S.$1.8 million at the Tsukiji fish market in Tokyo, Japan. All of this fish will be served as sushi, which is a combination of cooked, vinegared rice and other ingredients, usually raw seafood.

C A woman in rural Mongolia prepares *byaslag,* a cheese made in Mongolia of yak, cow, sheep, or goat milk. Unlike most cheeses originating in Europe, byaslag is not aged but rather is desiccated in the dry air of Mongolia. It is one of many unique products made with animal milk in Mongolia.

• China's main river systems are prone to flooding as well as being affected by the melting of glaciers. The melting glaciers will ultimately lead to water shortages.

• The region's most important grain is rice, and over the millennia, rice cultivation has transformed landscapes throughout central and southern China, and Japan, the Koreas, and Taiwan.

• About three-quarters of the food consumed in Japan, South Korea, and Taiwan is imported. So far, China is self-sufficient with regard to basic necessities such as rice, but it is highly dependent on imports for other foods.

• People in East Asia consume much fish and seafood. Some of this food is harvested in waters far away from East Asia.

THREE GORGES DAM: THE POWER OF WATER

The Three Gorges Dam (FIGURE 9.9A) is at this time the largest dam in the world, at 600 feet (183 meters) high and 1.4 miles (2.3 kilometers) wide. It was designed to improve navigation on the Chang Jiang and control flooding, but it is most lauded for its role in generating hydroelectricity for China, a country that uses considerable amounts of energy and is working to reduce its GHG emissions. But dams also come with a social cost. The Chinese environmental activist Dai Qing notes that China has 22,000 large dams, all of which have displaced people—perhaps as many as 60 million—without any attention to their rights as stakeholders in the projects.

Many experts involved with the Three Gorges project see serious design and planning flaws. Of greatest concern is the

Thinking Geographically

After you have read about human impacts on the biosphere in East Asia, you should be able to answer the following questions about the photos in this chapter.

A Which design and planning flaws concern experts involved with the Three Gorges Dam project?

B Where is the worst air pollution in Taiwan?

C Of all the coal burned in the world each year, how much is burned in China?

D What allows particulates from China's coal burning to be transported globally?

dam's position above a seismic fault. The dam was built at the east end of the Three Gorges because the deep canyons provided a prodigious reservoir for water to power turbines, providing electricity for all of central China, from the sea to the Plateau of Tibet. Unfortunately, the enormous weight of the water and its percolation through geological fissures in the 370-mile-long (600-kilometer-long) reservoir behind the dam could trigger earthquakes. Already at issue are huge landslides along the gorges, lubricated by the rising reservoir water. Meanwhile, cracks in the dam raise doubts about its structural integrity. Similar defects led to the failure of China's much smaller Banqiao Dam during a 1975 typhoon, which caused 150,000 deaths from the flooding and an ensuing famine. Even if the Three Gorges Dam holds, its potential to generate power will probably be reduced by the buildup of eroded silt behind the dam.

FIGURE 9.8 LOCAL LIVES: People and Animals in East Asia

A A stall in a night market in Beijing sells dried seahorses, lizards, and starfish, all of which are used in traditional Chinese medicine (TCM). Millions of animals are captured each year for use in TCM, and some are being pushed dangerously close to extinction.

B Sheep shearing in Mongolia, where livestock raising occupies 75 percent of the land. It was the backbone of the economy for thousands of years, until the 1970s, when mining and services took the lead. Many animals are still raised by nomadic and seminomadic herders; however, many former herders have now moved to the city because of economic downturns and an increasingly variable climate.

C *Nishikigoi* (also known as koi), a variety of carp, in a garden in Himeji, Japan. Carp were domesticated in both ancient China and Rome as a food source. Nishikigoi, which means "brocaded carp," are an ornamental variety developed in Japan in the 1820s. Currently, the trade in ornamental nishikigoi is larger than the trade of carp raised for food.

FIGURE 9.9 PHOTO ESSAY: Human Impacts on the Biosphere in East Asia

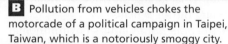

Economic development in East Asia has often proceeded without adequate environmental protection or safeguards for human health.

A China's Three Gorges Dam is the largest dam in the world. It is capable of supplying about 3 percent of China's electricity needs, resulting in significantly less air pollution than the several large coal-fired power plants that it replaced. However, the dam has also caused significant environmental damage related to its 370-mile-long (600-kilometer-long) reservoir, which has altered the habitats of endangered species; submerged the dwellings of 1.3 million people; and become a cesspool of untreated sewage, industrial waste, and garbage that once flowed downstream.

B Pollution from vehicles chokes the motorcade of a political campaign in Taipei, Taiwan, which is a notoriously smoggy city.

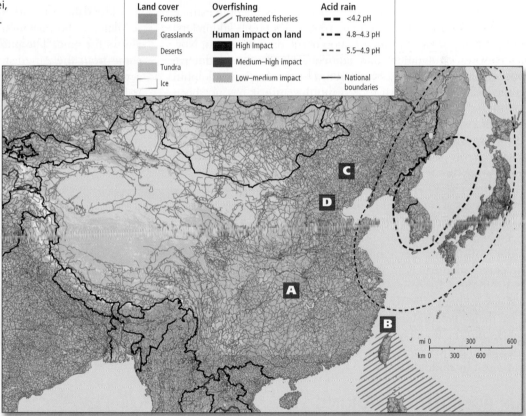

Human Impact, 2002

Land cover		Overfishing		Acid rain	
	Forests	Threatened fisheries		<4.2 pH	
	Grasslands	**Human impact on land**		4.8–4.3 pH	
	Deserts	High Impact		5.5–4.9 pH	
	Tundra	Medium–high impact		National boundaries	
	Ice	Low–medium impact			

C An open-pit coal mine in China's province of Inner Mongolia. China is already the world's biggest producer and consumer of coal, and its demand for coal is likely to increase in the future. China is the largest contributor of GHGs in the world.

D A toxic haze of air pollution engulfs Beijing and the North China Plain, often blowing eastward to reach Korea, Japan, and even the United States and Europe. China's coal-fired power plants, industries, and vehicles are emerging as major global environmental concerns.

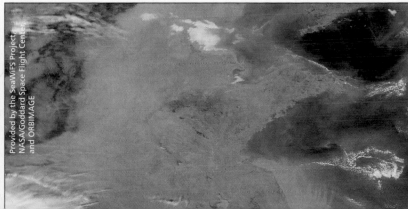

Any failure of the dam would be a financial as well as human disaster. The construction of the dam cost approximately U.S.$25 billion, but the ultimate costs may end up being three times this figure, due in part to unforeseen negative environmental and social impacts and to theft from the project by corrupt officials.

Also of concern is the incalculable cost associated with relocating the 1.3 million people who once lived where the dam now forms a reservoir. Thirteen major cities have been submerged, along with 140 large towns, hundreds of small villages, 1600 factories, and 62,000 acres (25,000 hectares) of farmland. In some cases, communities were rebuilt immediately uphill from the submerged location and people only had to relocate a short distance; others had to migrate to distant parts of the country. The reservoir has also destroyed important archaeological sites as well as some of China's most spectacular natural scenery. There are significant environmental costs as well. The giant sturgeon, for example, a fish that can weigh as much as three-quarters of a ton and is as rare as China's giant panda, may become extinct. Sturgeon used to swim more than 1000 miles (1600 kilometers) up the Chang Jiang, past the location of the dam, to spawn. Now the sturgeon's migration, and consequently their reproductive process, has been irretrievably altered. The rare baiji river dolphin was, in fact, declared extinct recently because it has not been sighted for years.

The plan to build the Three Gorges Dam came just as UN development specialists were deciding that the benefits dams could bring do not sufficiently outweigh the many problems they cause. Decades ago, international funding sources such as the World Bank withdrew their support for the Three Gorges Dam because of concerns over the social and environmental costs and other shortcomings of the project. However, Chinese industrialists who need the energy, construction companies that have prospered from building the dam and its many ancillary projects, and government officials eager to impress the world and leave their mark on China continue not only to support the *Da Ba* (the Big Dam), but also to look for other locations around the world where China can gain influence and profit by building dams.

AIR POLLUTION: CHOKING ON SUCCESS

Air pollution is often severe throughout East Asia, but the air quality in cities is particularly poor. After South Asia, the pollution in East Asian cities ranks among the worst in the world. China's pollution problems are particularly important because its demand for energy is growing so fast. Globally, coal burning is a major source of air pollution, and China is the world's largest consumer and producer of coal, accounting for 47 percent of all the coal burned in the world each year (see Figure 9.9C, D). Between 1980 and 2012, China's coal consumption grew almost sixfold. While China is mining and burning more coal every year, the pace of expansion has slowed somewhat recently in favor of other energy sources.

Nevertheless, at any given point in time, there are still hundreds of new coal-fired plants in the planning stages in China.

The combustion of coal releases high levels of pollutants—suspended particulates, smog-causing ozone, and sulfur dioxide—all of which can cause respiratory ailments. In Chinese cities, these emissions can be many times higher than the World Health Organization (WHO) defines as safe. **FIGURE 9.10** shows the geography of urban air pollution in China. The worst pollution is often in cities, where homes are in close proximity to industries that depend heavily on coal for fuel. The use of coal to heat homes is still common in China and contributes further to pollution, especially during the winter heating season. The cities of northern China suffer from high levels of air pollution, especially in the Hebei province, where a combination of heavy industry and atmospheric conditions that trap pollution are problematic. In southern areas such as the Guangdong province, manufacturing is less energy intensive and the air is better.

The data in Figure 9.10 come from China's environmental regulators, who now monitor air pollution properly. That wasn't always the case. The push toward accurate monitoring actually came from the American embassy in Beijing, which installed a rooftop air-quality monitor device in 2008 and started tweeting real-time data to inform the public when pollution levels were dangerously high. At first, Chinese authorities tried to shut down the monitor, but eventually modeled their own pollution monitoring after the embassy's system, and it has been in place since 2013.

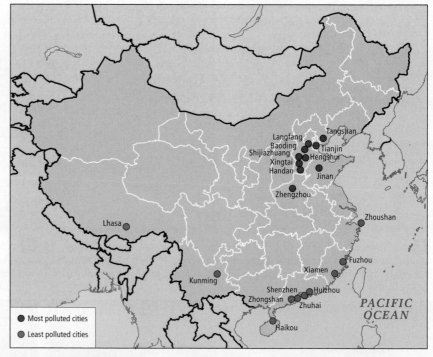

FIGURE 9.10 China's most and least polluted cities in terms of air quality. Based on a 2014 report from China's Ministry of Environmental Protection, the map shows the 10 cities in China that have the most air pollution and the 10 cities with the least air pollution (out of 74 monitored cities). Pollution is measured by the levels of particles (soot), carbon monoxide, and ozone found in the air. [Source consulted: "China Names 10 Most Polluted Cities," *China Daily*, February 2, 2015, at http://www.chinadaily.com.cn/china/2015-02/02/content_19466412.htm]

Sulfur dioxide from coal burning also contributes to acid rain, which is displaced to the northeast by prevailing winds that reach coastal East Asia. Acid rain is particularly damaging to freshwater ecosystems and forests. Figure 9.9B, D and the accompanying map show the zones of heavy pollution. South Korea is particularly afflicted, as the vignette below illustrates. Particulates from China's coal burning are transported globally by high-altitude, west-to-east flowing jet streams, thus affecting air quality in North America and Europe. Locally, toxic mercury from coal plants accumulates in aquatic life and, ultimately, in humans.

VIGNETTE On January 20, 2014, the weather announcer Kim Bo-gyung on the South Korean Arirang News warned her viewers that roughly 2 inches of snow were expected overnight. That in itself was not very unusual; South Korea commonly has chilly winter air that comes from continental Asia and picks up moisture when it moves over the Yellow Sea. This day, however, the snow contained more than just water. It was, in fact, acid snow. Pollutants such as nitrogen oxide and sulfur oxide travel from China and combine with water in the atmosphere, and in winter form snow that is deposited in South Korea. The pH during this weather event was as low as 3.8, which is the same level of acidity as wine or orange juice. The weather announcer cautioned people with sensitive skin to stay away from the snow. Pollution that originates in China sometimes moves beyond South Korea across the entire Pacific Ocean. New research indicates that 10 percent of sulfate in the air over the western United States actually comes from China. [Sources consulted: Trent Knoss, "'Acid Snow' Fell in South Korea This Week," January 24, 2014, at https://theraptorlab. wordpress.com/2014/01/24/acid-snow-fell-in-south-korea-this-week; "Acid Snow Through Tomorrow Morning," Arirang News, January 20, 2014, at http://www.arirang.co.kr/News/News_View.asp?nseq=156238] ∎

Air pollution from vehicles is also severe. While per capita automobile ownership is much lower than in the United States, the streets of large Chinese cities are clogged with traffic. For years, China's vehicles have had very high rates of lead and carbon dioxide emissions. The government has begun to work on the problem. Emission standards are tightened every few years, based on the EU's standards. One major instance of air-quality improvement was during the 2008 Summer Olympics. China removed half of Beijing's cars from the road and a number of polluting industries were temporarily shut down as well. The actions had the intended effect: Beijing's air was noticeably better during the Olympic summer. Since then, though, the city's continuously poor air quality (although Beijing is not on the list of the 10 most polluted cities in Figure 9.10) has been considered a problem for the city's international competitiveness. Multinational corporations are wary of conducting business in Beijing and employees don't want to live there because of the air quality.

Air Pollution Elsewhere in East Asia

Public health risks related to air pollution are also serious in the largest cities of Japan (Tokyo and Osaka), Taiwan (Taipei), Mongolia (Ulan Bator), and South Korea (Seoul), and in adjacent industrial zones. Even with antipollution legislation and increased enforcement, high population densities and rising expectations about better living standards make it difficult to improve environmental quality. Taiwan is a case in point.

Taiwan has some of the dirtiest air in the world. Some of the main causes are the island's extreme population density of 1664 people per square mile (642 per square kilometer), its high rate of industrialization, and its close proximity to industrialized south China. There are now 91 motor vehicles (cars or motorcycles) in Taiwan for every 100 residents—more than 21.4 million exhaust-producing vehicles on this small island. In addition, there are nearly eight registered factories per square mile (three per square kilometer), all emitting waste gases. The government of Taiwan acknowledges that the air is six times dirtier than that of the United States or Europe.

Both North and South Korea depend on hydro- or nuclear-generated electricity to run factories and heat buildings, unlike elsewhere in East Asia. North Korea has relatively few industries and uses very few cars, so its air pollution levels are thought to be low for the region (although few reliable data are available for North Korea). South Korea uses fossil fuels in its numerous industries and has many gasoline-powered cars, which are the sources of most of its internally generated air pollution. Mongolia generally has the region's cleanest air, but in Ulan Bator, the pollution from coal heaters has inspired some imaginative projects. For example, whole sections of Ulan Bator are now heated by centrally located boilers, which supply hot water to apartment buildings and individual dwellings.

> **THINGS TO REMEMBER**
>
> • In an effort to address droughts, floods, and its burgeoning need for electricity, China has built many dams, the largest of which is the Three Gorges Dam, which now poses many potential problems itself.
>
> • Rapid urban economic development in East Asia, combined with weak environmental protection, has resulted in some of the most polluted cities on the planet.

HUMAN PATTERNS OVER TIME

East Asia is home to some of the most ancient civilizations on Earth. Settled agricultural societies have flourished in China for more than 7000 years, which makes China among the oldest continuous civilizations in the world. After centuries of leading the world technologically and economically, East Asia was gradually eclipsed by European powers beginning in the seventeenth century. Outright domination by Europeans occurred only in certain places and not until the nineteenth century. The twentieth century was violent and tumultuous, and after 1945 when World War II ended, the countries of East Asia adopted entirely new economic systems. Three of the countries—China, Mongolia, and North Korea—became communist, and stayed so until the 1980s, when all but North Korea began to allow more capitalist practices. The other three countries—Japan, South Korea, and Taiwan—took a more capitalist route after World War II and today have among the highest standards of living in the world. China, while having a moderate GDP per capita, is now a global economic superpower that rivals the United States and the European Union.

Chinese civilization evolved from several hearths, including the North China Plain, the Sichuan Basin, and the lands of interior Asia that were inhabited by Mongolian nomadic pastoralists. On East Asia's eastern fringe, the Korean Peninsula and the islands of Japan and Taiwan were profoundly influenced by the culture of China, but they were isolated enough that each developed a distinctive culture and maintained political independence most of the time. In the early twentieth century, Japan industrialized rapidly by integrating the European influences that China disdained. As FIGURE 9.11 shows, both ancient and contemporary traditions are celebrated in East Asia today.

BUREAUCRACY AND IMPERIAL CHINA

Although humans have lived in East Asia for tens of thousands of years, the region's earliest complex civilizations appeared in China about 4000 years ago. Written records exist only from the civilization that was located in north-central China. There, a small, militarized, feudal aristocracy (see Chapter 4 for a discussion of feudalism) controlled vast estates on which the majority of the population lived and worked as semi-enslaved farmers and laborers. The landowners usually owed allegiance to one of the petty kingdoms that dotted northern China. These kingdoms were relatively self-sufficient and well defended with private armies.

An important move away from feudalism came with the Qin empire (beginning in 221 B.C.E.), which instituted a trained and salaried bureaucracy in combination with a strong military to extend the monarch's authority into the countryside. One part of the legacy of the Qin empire is shown in FIGURE 9.12B.

The Qin system proved more efficient than the old feudal allegiance system it replaced. The estates of the aristocracy were divided into small units and sold to the previously semi-enslaved farmers. The empire's agricultural output increased because the people more efficiently utilized the land they now owned. In addition, the salaried bureaucrats were more responsible than the aristocrats they replaced, especially about building and maintaining levees, reservoirs, and other tax-supported public works that reduced the threat of flood, drought, and other natural disasters. Although the Qin empire was short-lived, subsequent empires maintained Qin bureaucratic ruling methods, which have proved essential in governing a united China.

FIGURE 9.11 LOCAL LIVES: Festivals in East Asia

A A man dressed as a samurai—a member of Japan's pre-industrial-era warrior class—for a festival in Fukushima, Japan. Thousands of parades, processions, and street festivals are held each year in Japan, often celebrating events in local history or legend.

B Participants in the Boryeong Mud Festival, an annual event created in 1998 to celebrate the medicinal properties of the mud found near the town of Boryeong, South Korea. The 2-week-long festival draws more than 2 million people each year and features a mud pool, mud slides, a mud prison, and cosmetics and treatments that are based on the medicinal qualities of the mud.

C A man participates in *Naadam*, or "Games," which form the basis of festivals held throughout Mongolia in early July to celebrate the country's nomadic history. There are three games practiced: archery, horseracing, and wrestling. Females compete in the archery and horseracing events.

Toru Yamanaka/AFP/GettyImages

Jung Yeon-Je/AFP/Getty Images

Bruno Morandi/robertharding/Getty Images

CONFUCIANISM MOLDS EAST ASIA'S CULTURAL ATTITUDES

The philosophy of **Confucianism** is closely related to China's bureaucratic ruling tradition. Confucius, who lived from 551 to 479 B.C.E., was an idealist who was interested in reforming government and eliminating violence from society. He thought human relationships should involve a set of defined roles and mutual obligations. Confucian values include courtesy; knowledge; integrity; and respect for and loyalty to family lineage, parents, local community, and government officials. These values diffused across the region and are still widely shared throughout East Asia (see Figure 9.12A).

The Confucian Bias Toward Males

The model for Confucian philosophy was the patriarchal extended family. The oldest male held the seat of authority and was responsible for the well-being of everyone in the family. All other family members were aligned under the patriarch according to age and gender. Beyond the family, the Confucian patriarchal order held that the emperor was the grand patriarch of all China, charged with ensuring the welfare of society. Imperial bureaucrats were to do his bidding and commoners were to obey the bureaucrats.

Over the centuries, Confucian philosophy penetrated all aspects of East Asian society. Concerning the ideal woman, for example, a student of Confucius wrote: "A woman's duties are to cook the five grains, heat the wine, look after her parents-in-law, make clothes, and that is all! When she is young, she must submit to her parents. After her marriage, she must submit to her husband. When she is widowed, she must submit to her son." These concepts about limited roles for women affected society at large, where the idea developed that sons were the more valuable offspring, with public roles, while daughters were primarily servants within the home.

The Bias Against Merchants

For thousands of years, Confucian ideals were used to maintain the power and position of emperors and their bureaucratic administrators at the expense of merchants. In parable and folklore, merchants were characterized as a necessary evil, greedy and disruptive to the social order. At the same time, the services of merchants were sorely needed. Such conflicting ideas meant the status of merchants waxed and waned. At times, high taxes left merchants unable to invest in new industries or trade networks. At other times, however, the anti-merchant aspect of Confucianism was less influential, and trade and entrepreneurship flourished. Under communism, merchants again acquired a negative image. Later, when a market economy was encouraged, the social status of merchants rose once again.

Cycles of Expansion, Decline, and Recovery

Although the Confucian bureaucracy at times facilitated the expansion of imperial China, its resistance to change also led to periods of decline (**FIGURE 9.13**). Heavy taxes were periodically levied on farmers, bringing about farmer revolts that weakened imperial control. Threats from outside, particularly invasions by nomadic people from what are today Mongolia and western China, inspired the creation of massive defenses such as the Great Wall (see Figure 9.12C), built along China's northern border. Nevertheless, the Confucian bureaucracy always recovered from invasions. After a few generations, the nomads were indistinguishable from the Chinese. Chinese culture and civilization have absorbed a tremendous mixture of different influences and, as a result, have themselves changed.

One nomadic invasion did result in important links between China and the rest of the world. In the 1200s, the Mongolian military leader Genghis Khan and his descendants were able to conquer all of China. They also pushed west across Asia as far as what are now Hungary and Poland (see Chapter 5). It was during the time of this Mongol empire that traders such as the Venetian Marco Polo made the first direct contacts between China and Europe. These connections proved much more significant for Europe, which was dazzled by China's wealth and technologies, than for China, which saw Europe as backward and barbaric.

Indeed, from 1100 to 1600, China remained the world's most developed region, despite enduring several cycles of imperial expansion, decline, and recovery. It had the largest economy, the highest living standards, and the most magnificent cities. Improved strains of rice allowed dense farming populations to expand throughout southern China and supported large urban industrial populations. Nor was innovation lacking: Chinese inventions included paper making, printing, paper currency, gunpowder, and improved shipbuilding techniques.

WHY DID CHINA NOT COLONIZE AN OVERSEAS EMPIRE?

During the well-organized Ming dynasty, 1368–1644, Zheng He, a Chinese Muslim admiral in the emperor's navy, directed an expedition that could have led to China conquering a vast overseas empire. From 1405 to 1433, Zheng He sailed 250 ships—the biggest and most technologically advanced fleet that the world had ever seen. Zheng He took his fleet throughout established Chinese trade routes to Southeast Asia, across the Indian Ocean, and all the way to the east coast of Africa.

The lavish voyages of Zheng He were funded for almost 30 years, but they never resulted in an overseas empire like those established by European countries a century or two later. These newly explored regions simply lacked much that China needed or wanted. Moreover, back home the empire was continually threatened by the armies of nomads from Mongolia, so any surplus resources were needed for upgrading the Great Wall. Eventually the emperor decided that Zheng He's explorations were not worth the effort. In the years following Zheng He's voyages, China reduced its contacts with the rest of the world as the emperor focused on repelling the Mongols. The stance of isolationism was not only defense driven, but also an outcome of a long-standing attitude in China characterized by suspicion of outside cultures. As a result, the pace of technological change slowed, leaving China ill prepared to respond to growing challenges from Europe after 1600. China did, however, become a regional colonizing power by extending control to include territories in Central Asia and Southeast Asia (see Figure 9.13).

Confucianism a Chinese philosophy that teaches the importance of stability and social order based on traditional institutions such as the patriarchal family, community, and the state

A The oldest Confucian temple, built in Confucius's hometown of Qufu, China, in 478 B.C.E.

B Part of a terra cotta army that has been excavated near Xian. The army was buried with the first Qin emperor in 210 B.C.E.

C A section of the Great Wall built in 1570.

| 50 00 B.C.E. | 4000 B.C.E. | 3000 B.C.E. | 2000 B.C.E. | 1000 B.C.E. | 0 C.E. | 1000 C.E. | 1500 C.E. |

5000 B.C.E.
Agricultural societies develop in China

551–478 B.C.E.
Confucian philosophy begins

221–206 B.C.E.
Qin empire develops alternatives to feudalism

3000 B.C.E.–1644 C.E.
Great Wall of China built (in sections)

FIGURE 9.12 VISUAL HISTORY OF EAST ASIA

Thinking Geographically

Now that you have read about the history of East Asia, you should be able to answer the following questions about the photos in this chapter.

A What is the model for Confucian philosophy?

B How did the Qin empire influence subsequent Chinese empires?

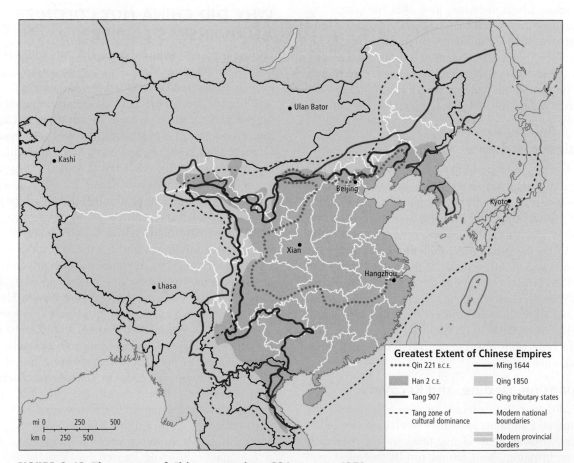

Greatest Extent of Chinese Empires

- ••••• Qin 221 B.C.E.
- ▬ Ming 1644
- ▓ Han 2 C.E.
- ░ Qing 1850
- ▬ Tang 907
- — Qing tributary states
- - - - Tang zone of cultural dominance
- ▬ Modern national boundaries
- ▒ Modern provincial borders

FIGURE 9.13 The extent of Chinese empires, 221 B.C.E. to 1850 C.E. The Chinese state has expanded and contracted throughout its history. [Source consulted: *Hammond Times Concise Atlas of World History* (Maplewood, NJ: Hammond, 1994)]

D British warships attack near Guangzhou in southern China during the First Opium War in 1840.

E A poster made during the Cultural Revolution in 1969, showing workers reading a book of the thoughts of Mao Zedong.

F A high-speed train passes Mt. Fuji in Japan in 2013.

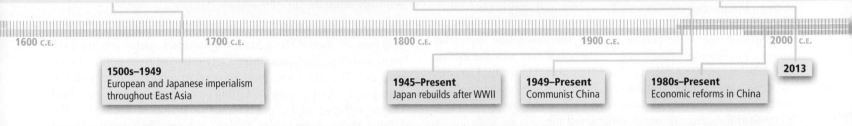

| 1600 C.E. | 1700 C.E. | 1800 C.E. | 1900 C.E. | 2000 C.E. |

2013

1500s–1949
European and Japanese imperialism throughout East Asia

1945–Present
Japan rebuilds after WWII

1949–Present
Communist China

1980s–Present
Economic reforms in China

C The Great Wall was built in response to what?

D Which major Chinese city and port was a British possession from the time of the Opium Wars until recently?

E After the Communist Revolution in China, how did the Chinese government control all aspects of economic and social life?

F How did Japan's political and economic development proceed after World War II?

EUROPEAN AND JAPANESE IMPERIALISM

By the mid-1500s, during Europe's Age of Exploration, Spanish and Portuguese traders interested in acquiring China's silks, spices, and ceramics found their way to East Asian ports. They brought a number of new food crops from the Americas such as corn, peppers, peanuts, and potatoes. These new sources of nourishment contributed to a spurt of economic expansion and population growth during the Qing, or Manchu, dynasty (1644–1912), and by the mid-1800s, China's population was more than 400 million; at that same time, Europe had 270 million people.

By the nineteenth century, European merchants gained access to Chinese markets and European influence increased markedly. In exchange for Chinese silks and ceramics, British merchants supplied opium from India, which was one of the few things that Chinese merchants would trade for. The emperor attempted to crack down on this drug trade because of its debilitating effects on Chinese society. The result was the Opium Wars (1839–1860), in which Britain badly defeated China. Hong Kong became a British possession, and British trade, including its opium trade, expanded throughout China well into the twentieth century (see Figure 9.12D).

The final blow to China's long preeminence in East Asia came in 1895, when a rapidly modernizing Japan won a spectacular naval victory over China in the Sino-Japanese War (and to solidify its emerging regional dominance, Japan also defeated Russia). After this first defeat by the Japanese, the Qing dynasty made only half-hearted attempts at modernization, and in 1912, it was overthrown by an internal revolt and collapsed. From the time of the decline of the Qing empire (1895) until China's Communist Party took control in 1949, much of the country was governed by provincial rulers in rural areas and by a mixture of Chinese, Japanese, and European administrative agencies in the major cities.

CHINA'S TURBULENT TWENTIETH CENTURY

In response to the absence of a central state authority, two rival political groups arose in China in the early twentieth century. The Nationalist Party, known as the Kuomintang (KMT), was an urban-based movement that appealed to a wide variety of social classes. The Chinese Communist Party, on the other hand, found its base among the rural poor. At first the KMT gained the upper hand, uniting the country in 1924. However, Japan's invasion of China in 1931 changed the dynamic.

By 1937, Japan had control of most major Chinese cities. The KMT did not resist the Japanese effectively and were confined to the few deep interior cities not under Japanese control. The Communists, however, waged a constant guerilla war against the Japanese throughout rural China, where they gained widespread support. Japan's brutal occupation caused 10 million Chinese deaths, including those of 250,000 to 300,000 civilians in the city of Nanjing, then China's capital. Similar acts of wartime brutality, as well as the keeping of local "comfort women" (women forced into prostitution) by the Japanese army in the occupied territories during World War II, still complicate the relationship between Japan and its East Asian neighbors.

When Japan finally withdrew in 1945, defeated at the end of World War II by the United States, the Soviet Union, and other Allied forces, the vastly more popular Communists pushed the KMT into exile in Taiwan. In 1949, the Communist Party, led by Mao Zedong, proclaimed the country the *People's Republic of China*, with Mao as president. This is the China we know today.

Mao's Communist Revolution

Mao Zedong's revolutionary government became extremely powerful, dominating all the outlying areas of China—the Far Northeast, Inner Mongolia, and western China (Xinjiang)—and launched a brutal occupation of Tibet (Xizang). The People's Republic of China was in many ways similar to past Chinese empires. The Chinese Communist Party replaced the Confucian bureaucracy and Mao Zedong became a sort of emperor with unquestioned authority. China received support from the Soviet Union, but the two communist states remained wary of each other for decades.

Among the early beneficiaries of the revolution were the masses of Chinese farmers and landless laborers. On the eve of the revolution, huge numbers lived in abject poverty. Famines were frequent, infant mortality was high, and life expectancy was low. The vast majority of women and girls held low social status and spent their lives in unrelenting servitude.

The revolution drastically changed this. All aspects of economic and social life became subject to central planning by the Communist Party. Land and wealth were reallocated, often resulting in an improved standard of living for those who needed it most. Major efforts were made to improve agricultural production and to reduce the severity of floods and droughts. Everyone, regardless of age, class, or gender, was mobilized to construct almost entirely by hand huge public works projects—roads, dams, canals, whole mountains terraced into fields for rice and other crops. "Barefoot doctors" with rudimentary medical training dispensed basic medical care, midwife services, and nutritional advice to people in even the most remote locations. Schools were built in the smallest of villages. Opportunities for women became available, and some of the worst abuses against them, such as the crippling binding of feet to make women's feet small and child-like, were stopped. Most Chinese people who are old enough to have witnessed these changes say that the revolution did a great deal to improve overall living standards for the majority.

Mao's Misdeeds Progress, however, came at enormous human and environmental costs. During the **Great Leap Forward** (a government-sponsored program of massive economic reform initiated in the 1950s), 30 million people died, many from famine brought on by poorly planned development; others because they were persecuted for opposing the reforms. Meanwhile, deforestation, soil degradation, and agricultural mismanagement became widespread. In the aftermath of the Great Leap Forward, some Communist Party leaders tried to correct the inefficiencies of the centrally planned economy, only to be demoted or jailed as Mao Zedong remained in power.

In 1966, partially in response to the failures of the Great Leap Forward, a political movement known as the **Cultural Revolution** enforced support for Mao and punished dissenters. Everyone was required to study the "Little Red Book" of Mao's sayings (see Figure 9.12E). Educated people and intellectuals were a main target of the Cultural Revolution because they

Great Leap Forward a failed economic reform program under Mao Zedong intended to quickly raise China's industrial level

Cultural Revolution a political movement launched in 1966 to force the entire population of China to support the continuing revolution

were thought to instigate dangerously critical evaluations of Mao and the central planning of the Communist Party. Tens of millions of Chinese scientists, scholars, and students were sent out of the cities to labor on farms, in mines and industries, or to jail, where as many as 1 million died. Children were encouraged to turn in their parents. Petty traders were punished for being capitalists, as were those who adhered to any type of organized religion. The Cultural Revolution so disrupted Chinese society that by Mao's death in 1976, the Communist regime had been seriously discredited.

Changes After Mao Two years after Mao's death, a new leadership formed around Deng Xiaoping. Limited market reforms were instituted, but the Communist Party retained tight political control. In 2009, after more than 30 years of reform and remarkable levels of economic growth, China's economy became the third largest in the world, behind those of the European Union and the United States. At that point, China's economy became the largest in East Asia, surpassing its rival Japan (although Japan's per capita wealth is still three times that of China's). However, the disparity of wealth in China has been increasing, human rights are still often abused, and political activity remains tightly controlled even as discontent occasionally boils over into open protests against the government.

THINGS TO REMEMBER

• Settled agricultural societies have flourished in China for more than 7000 years, which makes China among the oldest continuous civilizations in the world.

• In China, an important move away from feudalism came with the Qin empire (beginning in 221 B.C.E.), which instituted a trained and salaried bureaucracy in combination with a strong military to extend the monarch's authority into the countryside.

• For five centuries, from 1100 to 1600 C.E., China was the world's most developed region, despite its cycles of imperial expansion, decline, and recovery.

• Following the Opium Wars (1839–1860)—in which Britain badly defeated China—Hong Kong became a British possession, and British trade, including its opium trade, expanded throughout China, well into the twentieth century.

• After World War II, Mao Zedong's revolutionary communist government took over China; since then, economic reforms have taken place but the Communist Party remains firmly in control.

JAPAN BECOMES A WORLD LEADER

Although China's influence was preeminent in East Asia for thousands of years, Japan, with only one-tenth the population and 5 percent of the land area of China, dominated East Asia economically and politically for much of the twentieth century. Japan's rise as a modern global power resulted largely from its response to challenges from Europe and North America.

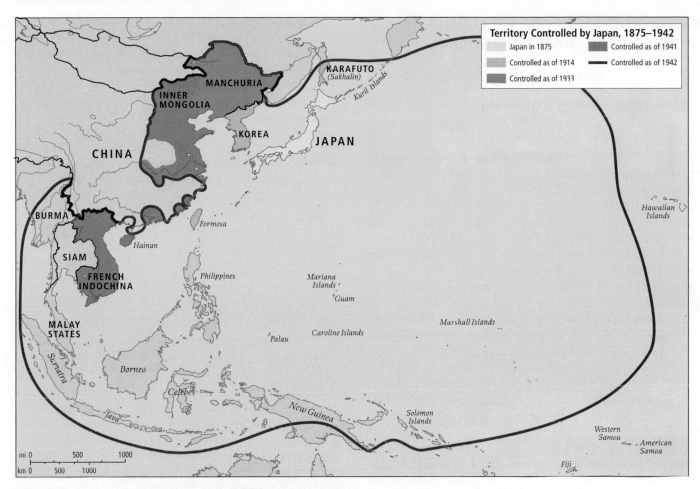

FIGURE 9.14 Japan's expansions, 1875–1942. Japan colonized Korea, Taiwan (then known as Formosa), Manchuria, China, parts of Southeast Asia, and several Pacific islands to further its program of economic modernization and to fend off European imperialism in the early twentieth century. [Source consulted: *Hammond Times Concise Atlas of World History* (Maplewood, NJ: Hammond, 1994)]

Beginning in the mid-sixteenth century, active trade with Portuguese colonists brought new ideas and military technology that strengthened Japan's wealthier feudal lords (*shoguns*), allowing them to unify the country under a military bureaucracy. However, the shoguns monopolized contact with the European outsiders, refusing to allow Japanese people to leave the islands, on penalty of death.

A second period of radical change began when a small fleet of U.S. naval vessels arrived in Tokyo Bay in 1853. The foreigners, carrying military technology far more advanced than Japan's, forced the Japanese government to open the economy to international trade and political relations. In response, a group of reformers (the *Meiji*, which means "enlightened rule") seized control of the Japanese government, setting the country on a crash course of modernization and industrial development that became known as the *Meiji Restoration*. During this time, Japanese students were sent abroad and experts were recruited from around the world, especially from Western nations, to teach everything from foreign languages to modern military technology. Investments emphasized infrastructure development, especially in transportation and communication. The improvements that resulted enabled Japan's economy to grow rapidly, surpassing China's size in the early twentieth century.

Between 1895 and 1945, Japan, which was relatively poor in resources, fueled its economy with resources from a vast colonial empire. Equipped with imported European and North American military technology, its armies occupied first Korea, then Taiwan, then coastal and eastern China, and eventually Indonesia and much of Southeast Asia, as shown in **FIGURE 9.14**. Many people in these areas still harbor resentment about the brutality they suffered at Japanese hands. Japan's imperial ambitions ended with its defeat in World War II and its subsequent occupation by U.S. military forces until 1952.

Immediately following World War II, the U.S. government imposed many social and economic reforms on Japan. Japan was required to create a democratic constitution and reduce the emperor's role to a symbolic one. Its military was reduced dramatically, making Japan reliant on U.S. forces to protect it from attack. Politically, it was advantageous for the United States to transform Japan into a Cold War ally.

With U.S. support, Japan rebuilt rapidly after World War II, and it eventually became a giant in industry and global business, exporting automobiles, electronic goods, and many other products (see Figure 9.12F). Japan's economy is still among the world's largest, wealthiest, and most technologically advanced, even if growth has been sluggish the last two decades (see Figure 9.20D).

CHINESE AND JAPANESE INFLUENCES ON KOREA, TAIWAN, AND MONGOLIA

The history of the remaining East Asia region has largely been shaped by what transpired in China and Japan.

The Korean War and Its Aftermath

Korea was a unified but poverty-stricken country until 1945. At the end of World War II, the United States and the Soviet Union, as victorious allies, agreed to divide the former Japanese colony. The Soviet Union occupied and established a communist regime to the north, while the United States took control of the southern half of the Korean peninsula, where it instituted reforms similar to those in Japan. After the United States withdrew its troops in the late 1940s, North Korea attacked South Korea. The United States returned to defend the south, leading a 3-year war against North Korea and its allies, the Soviet Union and Communist China.

After great loss of life on both sides and the devastation of the peninsula's infrastructure, the Korean War ended in 1953 in a truce. A **demilitarized zone (DMZ)** was established near the 38th parallel; it serves as the de facto border between North and South Korea. A DMZ could be any border area between rival states where military activities are prohibited, but the term usually refers to Korea. This border, which is 2.5 miles (4 kilometers) wide, is called "demilitarized" because it is a buffer zone not claimed by either side, although in the surrounding area, both sides are heavily armed.

North Korea closed itself off from the rest of the world, and to this day it remains isolated and impoverished, occasionally gaining international attention by hinting at its nuclear potential (see Figure 9.18D). South Korea, on the other hand, now has a prosperous and technologically advanced market economy. Relations between the two countries remain tense, with occasional skirmishes breaking out along the border.

Taiwan's Uncertain Status

For thousands of years, Taiwan was a poor, agricultural island on the periphery of China. Then, between 1895 and 1945, it became part of Japan's regional empire. In 1949, when the Chinese nationalists (the Kuomintang) were pushed out of mainland China by the Chinese Communist Party, they set up an anticommunist government in Taiwan, officially naming it the Republic of China. For the next 50 years, the United States aided Taiwan just like it did Japan and South Korea. As a result, Taiwan became modern and industrialized. Its economy quickly dwarfed that of China and remained dominant until the 1990s, serving as a prosperous icon of capitalism right next door to massive Communist China. Today, Taiwan remains an economic powerhouse. Taiwanese investors have been especially active in Shanghai and the cities of China's southeast coast.

Mainland China has never relinquished its claim to Taiwan. As China's economic and military power has increased, the United Nations, the World Bank, and the United States, along with most other countries, agencies, and institutions, have judiciously tiptoed around the issue of whether Taiwan should continue to be treated as an independent country or as a rebellious province of China. Even if the overwhelming number of people in Taiwan are ethnically Han Chinese, the self-perceived identity among most people is Taiwanese rather than Chinese-only. Most Taiwanese also feel that their country should hold on to its sovereignty (see Figure 9.18C).

Mongolia Seeks Its Own Way

For millennia, Mongolia's nomadic horsemen posed periodic threats to China, so much so that the Great Wall was built, reinforced, and extended to combat them (see Figure 9.12C). China has long been obsessed with both deflecting and controlling its northern neighbor; and China did control Mongolia from 1691 until the 1920s. Revolutionary communism spread to Mongolia soon thereafter, and Mongolia continued as an independent communist country under Soviet, not Chinese, guidance until the breakup of the Soviet Union in 1989. Communism brought education and basic services. Literacy for both men and women rose above 95 percent. Situated between two enormous countries, Mongolia has been wary of both Russia and China during the difficult road to a market economy that it has been on since 1989. In need of cash to participate in the modern world, many families have elected to abandon nomadic herding and permanently locate their portable *ger* homes near Ulan Bator and search for paid employment. While the economy has developed, the rapid and drastic change in lifestyle has also resulted in the deterioration of the traditional family structure, a rise in personal debt, and poverty, in a society that formerly took pride in an egalitarian if not prosperous standard of living.

---

demilitarized zone a border area between rival states where military activities are prohibited; usually refers to the border between North and South Korea

---

THINGS TO REMEMBER

• For much of the twentieth century, Japan dominated East Asia economically and politically. Japan's rise as a modern global power resulted largely from its response to challenges from Europe and North America.

• North Korea remains a communist country and is impoverished and cut off from the wider world.

• Taiwan and South Korea emerged after World War II as rapidly industrializing countries.

• Until recently, Mongolia retained communist connections to the Soviet Union.

GLOBALIZATION AND DEVELOPMENT

GEOGRAPHIC THEME 2

Globalization and Development: East Asia pioneered a spectacularly successful economic development strategy that has transformed economies across the globe. Governments in the region intervened strategically to encourage the production of manufactured goods destined for sale abroad, primarily to the large economies of North America and Europe.

After World War II, the countries of East Asia established two basic types of economic systems. The Communist regimes of China, Mongolia, and North Korea relied on central planning by the government to set production goals and to distribute goods among their citizens. In contrast, the governments of first Japan and then Taiwan and South Korea established **state-aided market economies** with the assistance and support of the United States and Europe. In this type of economic system, market forces, such as supply and demand and competition for customers, determine many economic decisions. However, the government intervenes strategically, especially in the financial sector, to make sure that certain economic sectors develop in a healthy fashion. Investment in the country by foreigners is also limited so that the government can retain more control over the direction of the economy and so that economic development benefits domestic interests. In the cases of Japan, South Korea, and Taiwan, government intervention was designed to enable **export-led growth.** This economic development strategy relies heavily on the production of manufactured goods destined for sale abroad—in this case primarily to North America and Europe—while limiting imports for local consumers.

More recently, the differences among East Asian countries have diminished as China and Mongolia have set aside strict central planning and adopted many aspects of state-aided market economies and export-led growth. China in particular relies heavily on exports of its manufactured goods to North America and Europe. Japan, South Korea, and Taiwan became less export-oriented over time as their own affluent domestic markets emerged.

THE JAPANESE MIRACLE

Throughout the nineteenth century, the economies of Japan, Korea, and Taiwan were minuscule compared with China's. During the twentieth century, though, all three grew tremendously, in part because of ideas that originated in Japan.

Japan Rises from the Ashes

Japan's recovery after its crippling defeat at the end of World War II is one of the most remarkable tales in modern history. Except for the imperial capital of Kyoto, which was spared because of its historical and architectural significance, all of Japan's major cities were destroyed by the United States. Most notably, the United States bombed Hiroshima and Nagasaki with nuclear weapons, and leveled Tokyo with incendiary bombs.

Key to Japan's rapid recovery was its state-aided market economy, in which the government guided private investors in creating new manufacturing industries. The overall strategy was one of export-led economic growth, with the Japanese government negotiating trade agreements with the United States and Europe. These and other deals ensured that large and wealthy foreign markets would be willing to import Japanese manufactured goods. South Korea, Taiwan, countries in Southeast Asia, and eventually China and many other parts of the developing world have imitated Japan's model of a state-aided and export-oriented economy. Also central to Japan's recovery, but less imitated abroad, were arrangements between the government, major corporations, and

labor unions that guaranteed lifetime employment by a single company for most workers in return for relatively modest pay.

The government-engineered trade and labor arrangements produced explosive economic growth of 10 percent or more annually between 1950 and the 1970s. The leading sectors were export-oriented automobile and electronics manufacturing. Japanese brand names such as Sony, Panasonic, Nikon, and Toyota became household words in North America and Europe. Products made by these companies sold at much higher volumes than would have been possible in the then relatively small Japanese and nearby Asian economies. However, growth started to slow considerably around 1990, followed by a stagnation phase called the lost decades. Despite this stagnation, Japan's postwar "economic miracle" continues to have an immense worldwide impact as a model, and Japan remains a significant actor in the world economy. Japan purchases resources from all parts of the world for its industries and domestic use. These purchases and investments in various local economies create jobs for millions of people around the globe.

Productivity Innovations in Japan

Over the years, Japan has made major innovations in manufacturing that have boosted its productivity and have been diffused to other industrial economies, changing the spatial arrangements of industries. The **just-in-time system** clusters together companies that are part of the same production process so that they can deliver parts to each other when they are needed **(FIGURE 9.15)**. For example, factories that make automobile parts are clustered around the final assembly plant, delivering parts literally minutes before they will be used. This saves money by making production more efficient and reducing the need for warehouses.

A related innovation is the **kaizen system** of continuous improvement in manufacturing and business management. In this system, production lines are constantly surveyed for errors, which helps ensure that fewer defective parts are produced. Production lines are also constantly adjusted and improved to save time and energy. Both the just-in-time and the kaizen systems have been imitated by companies around the world and have been taken overseas by Japanese companies that invest abroad. For example, Toyota uses the kaizen and just-in-time systems in its U.S. plants.

MAINLAND ECONOMIES: COMMUNISTS IN COMMAND

After World War II, economic development on East Asia's mainland proceeded on a dramatically different course than it did in Japan. Communist economic systems transformed poverty-ridden China, Mongolia, and North Korea. Private

state-aided market economy an economic system based on market principles but with strong government guidance; in contrast to the limited government, free market economic system of the United States and, to a lesser degree, Europe

export-led growth an economic development strategy that relies heavily on the production of manufactured goods destined for sale abroad

just-in-time system the system pioneered in Japanese manufacturing that clusters companies that are part of the same production system close together so that they can deliver parts to each other precisely when they are needed

kaizen system a system of continuous manufacturing improvement, pioneered in Japan, in which production lines are constantly adjusted, improved, and surveyed for errors to save time and money and ensure that fewer defective parts are produced

FIGURE 9.15 Japan's just-in-time system. In this example of the just-in-time system, related industries are clustered together around a Toyota plant in Toyota City, Japan. The facilities supply each other with various automobile parts and other inputs needed for the production process.

property was abolished and the state took full control of the economy, loosely following the example of the Soviet centrally planned economy (see Chapter 5). These sweeping changes transformed life for the poor majority but ultimately proved less resilient and successful than was hoped.

By design, most people in the communist economies could not consume more than the bare necessities. On the other hand, the policy—called the *iron rice bowl* in China—of guaranteeing nearly everyone a job for life, sufficient food, basic health care, and housing was better than what they had before. One drawback was that overall productivity remained low.

The Commune System

When the Communist Party first came to power in China in 1949, its top priority was to make monumental improvements in both agricultural and industrial production. The Communist governments in North Korea and Mongolia held similar goals, though these countries had much smaller populations and resource bases to work with.

In the years following World War II, an aggressive agricultural reform program joined small landholders together into cooperatives so that they could pool their labor and resources to increase production. In time, the cooperatives became full-scale communes, with an average of 1600 households each. The communes, at least in theory, took care of all aspects of life. They provided health care and education, and built rural industries to supply simple items such as clothing, fertilizers, small machinery, and eventually even tractors. The rural communes also had to fulfill the ambitious expectations that the leaders in Beijing had for better flood control, expanded irrigation systems, and especially, more food production.

The Chinese commune system met with several difficulties. Rural food shortages developed because farmers had too little time to farm. They were required to spend much of their time building roads, levees, terraces, and drainage ditches, or working in the new rural industries. Local Communist Party administrators often compounded the problem by overstating harvests in their communes to impress their superiors in Beijing. The leaders in Beijing responded by requiring larger food shipments to the cities, which caused even greater food shortages in the countryside.

Although the Chinese agricultural communes were inefficient and created devastating food scarcities during the Great Leap Forward, they did eventually result in a stable food supply that kept Chinese people well fed.

On the Korean Peninsula, North Korea and South Korea have taken very different economic paths. In North Korea, the Communists have invested in military strength at the expense of broader rural economic development. The country not only lacks good farmland, but the government has not sufficiently funded farming inputs, such as seeds, fuel, and fertilizer, to produce enough food for the population. Commune-style agriculture has been so neglected that production is precarious, with famine a constant threat. In recent years, North Korea has used rocket launchings, nuclear tests, and even the threat of war to intimidate its neighbors, and it is selling its military technology abroad to pay for its food imports. According to the United Nations, roughly 70 percent of North Koreans suffer from food insecurity, which is why the world has allowed food aid into North Korea. But the intransigence of the North Korean regime can also backfire, as cutbacks in food aid from South Korea, the United States, and the UN suggest. The black market is an important source of food imports into North Korea from across the Chinese border.

In Mongolia, Communist policy followed the Soviet model of collectivization, though it was tailored to Mongolia's economy, which at the time was largely one of herding and agriculture. There were minimal changes to the nomadic pastoralist way of life, but the role of mining and industry was emphasized. Today, 29 percent of the population works in agriculture, but other sectors of the economy are more productive, as agriculture's contribution to GDP is less than 17 percent.

Focus on Heavy Industry

The Communist leadership in China, North Korea, and Mongolia believed that investment in heavy industry would dramatically raise living standards. Massive investments were made in the mining of coal and other minerals and in the production of iron and steel. Especially in China, heavy machinery was produced to build roads, railways, dams, and other infrastructure improvements that leaders hoped would increase overall economic productivity. However, much as in India (see Chapter 8), the vast majority of the population remained impoverished agricultural laborers who received little benefit from industries that created jobs mainly in urban areas. Not enough attention was paid to producing consumer goods and household items that would have driven modest internal economic growth and improved living standards for the rural poor. Even in the urban areas, growth remained sluggish because, as also was the case in the Soviet command economy, small miscalculations by bureaucrats resulted in massive shortages and production bottlenecks that constrained economic growth.

Struggles with Regional Disparity

For centuries, China's interior west has been poorer and more rural than its coastal east. The interior west has been locked into agricultural and herding economies, while the economies of the east have benefited from trade and industry in even the most restricted times. The first effort to address regional disparities, right after the revolution, was an economic policy that focused on a combination of investment from the central government and the development of **regional self-sufficiency.** Each region was encouraged to develop as an independent entity with both agricultural and industrial sectors that would create jobs and produce food and basic necessities. This policy did little to lessen regional disparities and in fact constrained economic growth by inhibiting internal trade between regions.

ECONOMIC REFORMS IN CHINA

In the 1980s, China's leaders enacted economic reforms that changed the country's economy in five ways. First, building on Communist-era methods of financial control but also integrating strategies pursued in Japan, China used the state-owned banking system to direct investment into export-oriented manufacturing industries. Second, at the local level, economic decision making was decentralized; this was named the *responsibility system*. Under the responsibility system, farmers now could make household-level decisions, subject to the approval of the commune, about how to use land and which crops to grow. Third, the reforms allowed existing and newly established businesses to sell their produce and goods in competitive markets. Fourth, **regional specialization**, rather than regional self-sufficiency, was encouraged in order to take advantage of regional variations in climate, natural resources, and location, and thereby encourage national economic integration. In practice, this has meant that coastal urban areas focus on export-oriented manufacturing and advanced service production, while resource-based industries and production for domestic consumption are fostered in the interior provinces. Finally, the government allowed foreign direct investment in Chinese export-oriented coastal enterprises and the sale of foreign products in China.

While this five-part shift dramatically improved the efficiency with which food and goods were produced and distributed, regional disparities have widened and on the provincial level, rural–urban disparities are sometimes extreme. For example, rural households in some interior provinces have an income that is less than a third that of their urban counterparts (**FIGURE 9.16**). Internal migration and the floating population discussed at the beginning of this chapter are direct consequences of this spatially uneven development.

China's economic reforms have transformed not only China's economy but also the economies of wider East Asia and indeed the whole world. Today China is the world's largest producer of manufactured goods, supplying consumers across the globe. Of the other Communist-led countries, Mongolia has participated in this transformation only moderately, mainly as a raw material exporter, and North Korea has not participated at all.

East Asia's Role as Mega-Financier As U.S. government debt has mounted in recent decades (up to U.S.$13.6 trillion by the end of 2015), so has concern about the fact that China and Japan now own much of this debt. China owns about 21 percent of U.S. debt, and Japan 19 percent (**FIGURE 9.17**); many people are worried that either country could destabilize the U.S. economy by selling or refusing to buy more U.S. debt. However, such an event is highly unlikely. Both countries' ownership of U.S. debt is a side effect of their huge trade surplus with the United States. These amounts are large enough that, worldwide, the only relatively stable investment available that China and Japan can buy in sufficient quantities with all their dollars is U.S. debt. Also, in order to finance U.S. consumption of foreign goods and to keep the value of the dollar high, the export economies of China and Japan are investing in U.S. dollars. Dumping their vast holdings of U.S. debt and therefore destabilizing the U.S. economy would mean that both China and Japan would not be able to sell as many of their goods to consumers in the United States, which would destabilize their own economies. Hence China and Japan have strong incentives to continue buying U.S. government debt.

East Asians are able to use their reserves to finance development elsewhere in the world, thereby securing privileged trade deals. In recent years, China's lending to developing countries has outpaced even that of the World Bank. For example, China has financially supported many Latin American countries that have an adversarial relationship with Western-dominated lending institutions. Often these loans are made in return for guarantees that the recipient country will supply China with needed industrial materials, resources, and fossil fuels. While China has short-term, concrete objectives involving promoting its own economic interests, some of these investments are meant to project *soft power*—a country's attempt to persuade others to do what it wants without resorting to force. The goal of Chinese soft power is to shape long-term attitudes about China around the world and to ensure friendly allies for decades to come. China is also using investments and soft power in Africa and Southwest Asia but without making any demands on these countries to engage in political reforms, such as improving their human rights records.

regional self-sufficiency an economic policy in Communist China that encouraged each region to develop independently in the hope of evening out the wide disparities in the national distribution of production and income

regional specialization specialization (rather than self-sufficiency) that takes advantage of regional variations in climate, natural resources, and location

THINGS TO REMEMBER

GEOGRAPHIC THEME 2 • **Globalization and Development** East Asia pioneered a spectacularly successful economic development strategy that has transformed economies across the globe. Governments in the region intervened strategically to encourage the production of manufactured goods destined for sale abroad, primarily to the large economies of North America and Europe.

• Japanese management innovations known as the just-in-time and kaizen systems were so successful that they diffused and created clusters of economic activity throughout the world.

• After World War II, Communist economic systems transformed poverty-ridden China, Mongolia, and North Korea. Private property was abolished and the state took full control of the economy, loosely following the example of the Soviet centrally planned economy.

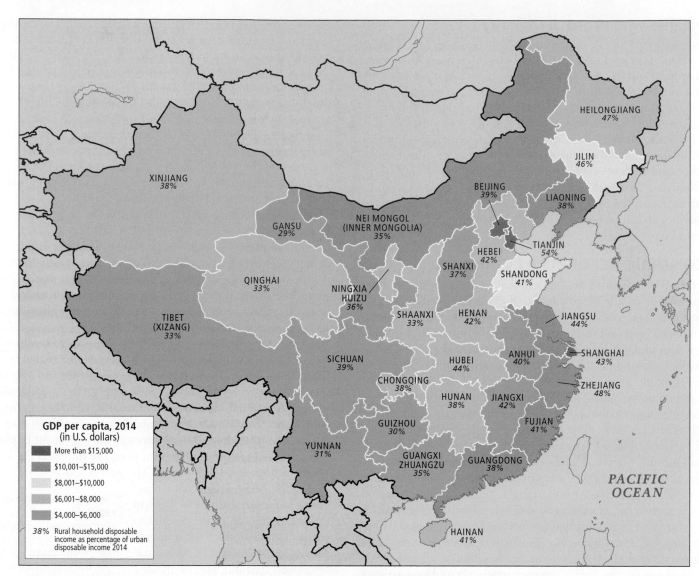

FIGURE 9.16 China's regional GDP and rural–urban income per capita disparities, 2014. Notice the disparity in GDP per capita across China, as indicated by the colors of the provinces, as well as the rural–urban income disparity in each province, as represented by a percentage (the per capita disposable income of rural households as a percentage of per capita disposable income of urban households). [Source consulted: National Bureau of Statistics of China, http://data.stats.gov.cn/english/easyquery.htm?cn=E0103]

• By design, most people in the Communist economies could not consume more than the bare necessities, but the policy—called the *iron rice bowl* in China—of guaranteeing nearly everyone a job for life, sufficient food, basic health care, and housing was better than what they had before.

• In the 1980s, China's leaders enacted reforms that changed the country's economy, enabling China to become the world's largest producer of manufactured goods, supplying consumers across the globe.

• Because China and Japan now own about 40 percent of U.S. public foreign debt, many people are worried that either country could destabilize the U.S. economy by selling or refusing to buy more U.S. debt. However, such an event is highly unlikely.

POWER AND POLITICS

GEOGRAPHIC THEME 3

Power and Politics: As the countries of East Asia develop economically, people have been pushing for more political freedom. Japan, South Korea, and Taiwan all have high levels of political freedom, and demands for political change in China are increasing, especially in urban areas.

Demands for more political freedom are growing throughout East Asia. Japan's current democratic political structure was established after World War II, South Korea's in the late 1980s, and Taiwan's in the mid-1990s. Mongolia has dramatically expanded its political freedom since abandoning communism in 1992. China, however, remains under the tight control of an

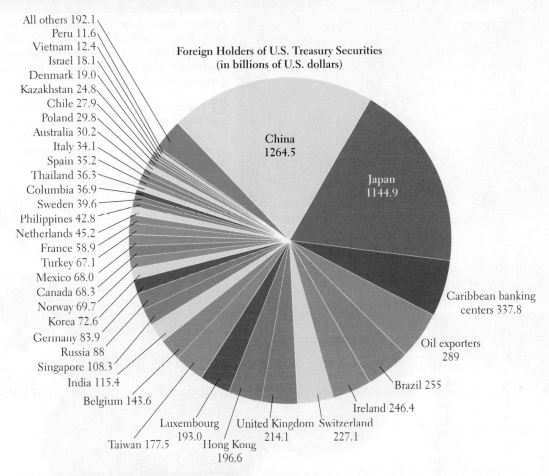

All others 192.1
Peru 11.6
Vietnam 12.4
Israel 18.1
Denmark 19.0
Kazakhstan 24.8
Chile 27.9
Poland 29.8
Australia 30.2
Italy 34.1
Spain 35.2
Thailand 36.3
Columbia 36.9
Sweden 39.6
Philippines 42.8
Netherlands 45.2
France 58.9
Turkey 67.1
Mexico 68.0
Canada 68.3
Norway 69.7
Korea 72.6
Germany 83.9
Russia 88
Singapore 108.3
India 115.4
Belgium 143.6
Taiwan 177.5
Luxembourg 193.0
Hong Kong 196.6
United Kingdom 214.1
Switzerland 227.1
Ireland 246.4
Brazil 255
Oil exporters 289
Caribbean banking centers 337.8
Japan 1144.9
China 1264.5

Foreign Holders of U.S. Treasury Securities
(in billions of U.S. dollars)

FIGURE 9.17 Foreign holders of U.S. Treasury securities, November 2015. The total dollar amount of U.S. securities held by foreign countries is more than $6 trillion; note, though, that the United States also holds large amounts of securities from foreign countries. [Source consulted: U.S. Treasury, "Major Foreign Holders of Treasury Securities (in billions of dollars)," at http://ticdata.treasury.gov/Publish/mfh.txt]

authoritarian regime, and North Korea is even more tightly held. The map in **FIGURE 9.18** shows each country's level of democracy. The red starbursts indicate places where civil unrest has broken out since 1945.

PRESSURES FOR POLITICAL CHANGE IN CHINA

With China now a globalized economy, many wonder how much longer the Communist Party can remain in control without allowing a significant expansion of political freedom throughout the entire country. The Communist Party officially says that China is a democracy, and indeed for many years some elections have been held at the village level and within the Communist Party. However, representatives of the National People's Congress, the country's highest legislative body, are appointed by the Communist Party elite, who maintain tight control throughout all levels of government. Most experts on China agree that while radical change is unlikely in the near future, a slow but steady shift toward more political freedom is underway. Change might be inevitable as the population becomes more prosperous, educated, and widely traveled, thereby becoming exposed to places that have more political freedoms. However, the Communist

Party remains determined to repress any major reform movements that arise from outside the party's senior leadership.

Less than a decade after market reforms began, demands for political change that were inspired by similar events in communist central Europe, culminated in a series of pro-democracy protests that drew hundreds of thousands of people to Beijing's Tiananmen Square in 1989 (see Figure 9.18A). These protests were brutally repressed, with thousands (the precise number is uncertain) of students and labor leaders massacred by the military. The Tiananmen Square event made clear that China's integration into the global economy would take place on political terms dictated by the state.

Informed consumers and environmentalists in developed countries have criticized China for its "no holds barred" pursuit of economic growth. Much of China's growth has been dependent on environmentally destructive activities and harsh conditions for workers, both of which effectively reduce production costs so that the prices of Chinese goods are low on world markets.

Information Technology and Political Freedom

Because of the spread of information via the Internet, people's push for political freedom from within China has increased. Since the revolution in 1949, China's central government has

FIGURE 9.18 PHOTO ESSAY: Power and Politics in East Asia

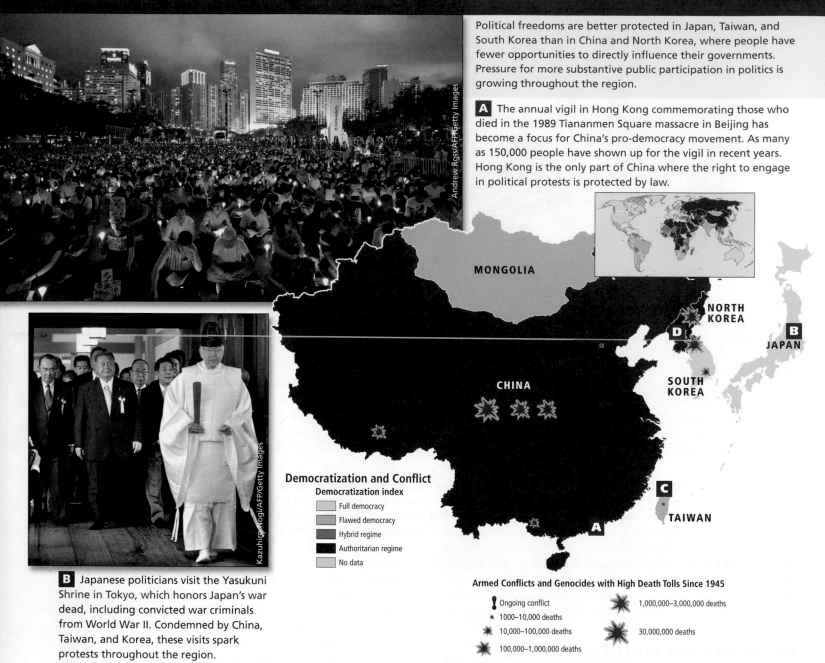

Political freedoms are better protected in Japan, Taiwan, and South Korea than in China and North Korea, where people have fewer opportunities to directly influence their governments. Pressure for more substantive public participation in politics is growing throughout the region.

A The annual vigil in Hong Kong commemorating those who died in the 1989 Tiananmen Square massacre in Beijing has become a focus for China's pro-democracy movement. As many as 150,000 people have shown up for the vigil in recent years. Hong Kong is the only part of China where the right to engage in political protests is protected by law.

Andrew Ross/AFP/Getty Images

MONGOLIA

NORTH KOREA

D

B

JAPAN

CHINA

SOUTH KOREA

C

TAIWAN

A

Kazuhiro Nogi/AFP/Getty Images

Democratization and Conflict
Democratization index

- Full democracy
- Flawed democracy
- Hybrid regime
- Authoritarian regime
- No data

B Japanese politicians visit the Yasukuni Shrine in Tokyo, which honors Japan's war dead, including convicted war criminals from World War II. Condemned by China, Taiwan, and Korea, these visits spark protests throughout the region.

C A Taiwanese activist puts her country's flag on a model of one of the Senkaku/Diaoyu islands, one of several disputed islands in the East China Sea that is claimed by Japan, China, and Taiwan. Many Taiwanese see this and other disputes as ways to reaffirm the country's continued independence from China, which considers Taiwan a rebellious province.

Armed Conflicts and Genocides with High Death Tolls Since 1945

- Ongoing conflict
- 1000–10,000 deaths
- 10,000–100,000 deaths
- 100,000–1,000,000 deaths
- 1,000,000–3,000,000 deaths
- 30,000,000 deaths

D North Korea's longest-range ballistic missile is paraded in P'yongyang. North Korea is testing such missiles and other nuclear weapons. China, Japan, South Korea, Russia, and the United States all express varying levels of concern about this.

STR/AFP/GettyImages

Pedro Ugarte/AFP/Getty Images

Thinking Geographically

Now that you have read about power and politics in East Asia, you should be able to answer the following questions about the photos in this chapter.

A What happened in Tiananmen Square in 1989?

C When was democracy in Taiwan established?

D What is North Korea purchasing with money earned from sales of its military technology abroad?

controlled the news media in the country. By the late 1990s, though, the expanding use of electronic communication devices was loosening central control over information. By 2001, 23 million people were connected to the Internet in China, although that represented just 2 percent of the population. The expansion of Internet access has risen dramatically since then; 674 million people (or 49 percent of the population) were connected by 2015 **(FIGURE 9.19)**. However, Internet access is not evenly distributed. A digital divide has emerged between major cities like Beijing and Shanghai, where 70 to 75 percent of the people have Internet access, and the rural interior, where only about 35 percent do. As economic development and communication infrastructure spreads over time, it is likely that the divide will diminish. In fact, access to the Internet in rural China is significantly higher than it was just a few years ago.

Not too long ago, telephones were very rare and people had to obtain permission to use them, but now millions of Chinese people have access to an international network of information. It is much more difficult for the government to give inaccurate explanations for problems caused by inefficiency and corruption. Analysts, both inside and outside China, see the availability of the Internet to ordinary Chinese citizens as a watershed development that supports the expansion of political freedom.

Social networking technologies have emerged as a powerful tool for collective action in China. In recent years, the use of a cell phone–based Twitter-like service known as *Weibo* has grown exponentially. It has more than 500 million registered users in China. This type of microblogging has been instrumental in forcing major media coverage and public discussion of incidents that expose government corruption and mismanagement.

Nevertheless, the Internet in China is not the open forum that it is in most Western countries. Chinese Internet users are required to register their real names when using an online account. Postings on social media sites like Weibo are constantly screened by the government to monitor anonymous criticisms about party officials. And if people writing blogs in China use the words "democracy," "freedom," or "human rights," they may receive the following reminder: "The title must not contain prohibited language, such as profanity. Please type a different title." Such censorship, informally dubbed the "Great Firewall," has been aided by global technology firms, which have been accused of blocking access to certain websites for users in China. Google, in a dispute with the Chinese government over state-sponsored censorship, moved its operations to Hong Kong, where more freedom is allowed (see Figure 9.19B). Meanwhile, a web browser from the Chinese company Baidu dominates the search engine market, with more than 70 percent of users. It is easier for Chinese authorities to censor a domestic company than to censor other companies whose information may flow from servers around the world.

Protests and Political Freedom

Protests by workers for better pay and living conditions, and by farmers and urban dwellers displaced by new real estate developments, are becoming more common in China. The government has stopped issuing complete statistics on protests, but the nonprofit group Human Rights Watch estimates that 100,000 to 200,000 such protests are being held each year.

Most protests across China are small-scale, local protests that only involve a few dozen people. They are held with some frequency; virtually every day some form of resistance against those in power takes place. The protesters are usually not directly motivated by broader political concerns. The triggering mechanism is often when an individual or small group of people sense that they have been ignored and mistreated by authorities. The international Pew Research Center conducted attitude surveys in 2015, which indicated that ordinary Chinese people think corrupt officials are the most important problem that is facing the country. Corruption is followed by environmental pollution and the growing gap between the rich and poor. It is often these concerns that are the underlying factors of dissent, and when combined with individual grievances, will cause people to engage in public protests.

Since confidence in public officials is weak, as the Pew survey indicates, people don't expect that the legal system will treat them fairly; thus they engage in direct street protests. Most of the time, news about protests doesn't travel far. Internet censorship, as discussed above, has the capacity to spatially contain dissent. For example, for protests that take place in rural areas or small towns far away from the coastal urban centers, attention in Chinese media is limited or nonexistent. People have started to distribute videos of disturbances on social media such as Weibo, but before they "go viral," censors often take note and remove them. At times, though, many people have joined interest groups that pressure the government to take action on particular problems.

More systemic threats to the power monopoly of the Communist Party are dealt with more harshly. The university professor and human rights activist Liu Xiaobo called for a decentralized federal system of government (that is, more power to the provinces) and democratic elections in 2008. He was subsequently arrested, and sentenced to 11 years in prison. In 2010, Liu Xiaobo was awarded the Nobel Peace Prize, though neither he nor his family was allowed to attend the ceremony in Norway. He is still imprisoned in China, even as prominent international human rights activists lobby for his release. Another group that is perceived as a threat by the Chinese government is Falun Gong, a Buddhist-inspired movement. Falun Gong represents an emergence of new religious groups that operate independent from the state. Many of its members have been arrested, but Falun Gong has also been able to mount a worldwide media campaign for its cause, which is another reason it is disliked by the government.

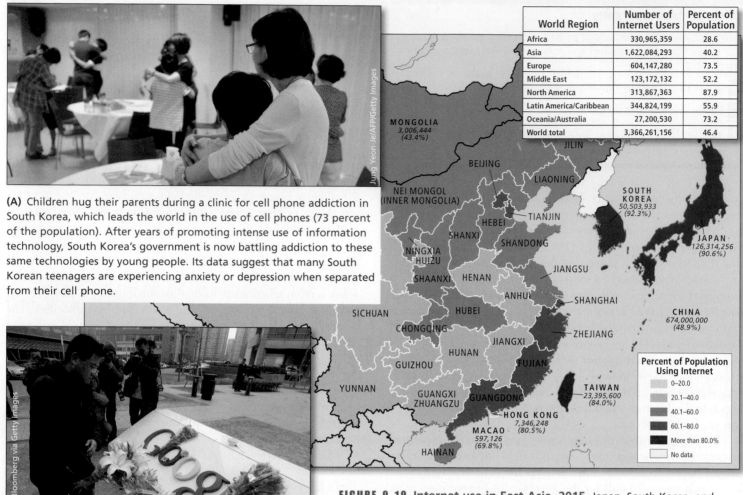

(A) Children hug their parents during a clinic for cell phone addiction in South Korea, which leads the world in the use of cell phones (73 percent of the population). After years of promoting intense use of information technology, South Korea's government is now battling addiction to these same technologies by young people. Its data suggest that many South Korean teenagers are experiencing anxiety or depression when separated from their cell phone.

World Region	Number of Internet Users	Percent of Population
Africa	330,965,359	28.6
Asia	1,622,084,293	40.2
Europe	604,147,280	73.5
Middle East	123,172,132	52.2
North America	313,867,363	87.9
Latin America/Caribbean	344,824,199	55.9
Oceania/Australia	27,200,530	73.2
World total	3,366,261,156	46.4

(B) A Chinese man makes an offering of flowers at Google's Beijing headquarters in support of the company's efforts to keep the government from censoring content delivered by Google's search engine in China. Internet use is growing rapidly in China, as are the government's efforts to control what citizens have access to.

FIGURE 9.19 **Internet use in East Asia, 2015.** Japan, South Korea, and Taiwan have very high Internet use rates, while China has the most Internet users of any country in the world (an estimated 674 million people). Usage in China varies greatly among provinces. [Sources consulted: *China Statistical Yearbook*, http://www.stats.gov.cn/tjsj/ndsj/2014/indexeh.htm; Statistica, "Number of Internet Users in China in 2015, By Region," http://www.statista.com/statistics/277259/number-of-internet-users-in-china-by-province]

In an effort to maintain control over China's increasingly articulate protestors, the government has allowed elections to be held for what are called "urban residents' committees." These elections are supposed to serve as an outlet for people to be able to peaceably voice their frustrations and create a limited amount of change at the local level. However, many of these elections are hardly democratic, with candidates selected by the Communist Party. Although in cities where unemployment is high and where protests have been particularly intense, elections tend to be more free, open, and truly democratic.

JAPAN'S RECENT POLITICAL SHIFTS

Significantly, the government that played such a central role during the post–World War II rise of the Japanese economy was controlled from 1955 to 2009 by one political party, the Liberal Democratic Party (LDP). This may have been an outcome of the homogenous and consensus-oriented culture of Japan. Facing a prolonged long economic slump, the Japanese people temporarily elected a government led by the opposition party, the Democratic Party of Japan (DPJ), from 2009 to 2012. The LDP remains dominant in Japanese politics, although it is likely to face more challenges in the future. One of the main differences between the two parties is that the more nationalistic LDP takes a hawkish stance against China and all of Japan's other neighboring countries (see Figure 9.18B). Having closer relations with China may be advantageous for Japan, though, as China has the potential to be a major market for Japanese exports in the future.

GEOPOLITICS OF THE SEA

Political tensions have been rising between the countries of East Asia even as their economies have become more linked. One especially important issue concerns the territorial control of the waters off the coast of East Asia. While these disagreements are rooted in contemporary economic conditions and political calculations by governments, they are also fueled by the painful legacies of World War II and its aftermath. Japan and Russia are in a long-standing dispute over the Kuril Islands in the Pacific Ocean. Recently, two major disputes between East and Southeast Asian countries have arisen over uninhabited islands that lie over oil and natural gas reserves.

In the East China Sea, the Senkaku Islands (in China, these are known as the Diaoyu Islands) are claimed by China, Japan, and Taiwan. The conflict over the Senkaku/Diaoyu Islands has escalated lately, mainly because of competing claims on them by China and Japan. Because China and Japan are the world's first- and second-largest importers of fossil fuels, respectively, they each have an interest in controlling nearby reserves. The waters around the islands are also rich fishing grounds, in the midst of vital shipping lanes, and militarily strategic. Within China, the conflict often inflames widespread resentment about Japan's brutal occupation of China before and during World War II. Japanese nationalists, on the other hand, eager to counter China's rising power and armed with several treaties that explicitly give it control over the islands, have staged protests and planted Japanese flags on the islands. Chinese ships sail around the islands in what Japan claims are its territorial waters. China also says it should have control over the surrounding air space. Many now see this as East Asia's most risky potential military conflict of the post–World War II period.

The control of the Senkaku/Diaoyu Islands is also important for legal reasons. International law, based on the United Nations Convention on the Law of the Sea (UNCLOS), stipulates that waters extending 12 nautical miles (1 nautical mile is approximately 1.15 miles, or 1.85 kilometers) out from a country's shoreline are its *territorial waters*. Nonmilitary foreign ships have the right to pass through those waters, but otherwise a country can govern and regulate its territorial waters much like it does its land. In addition, an *exclusive economic zone* extends much farther, 200 nautical miles. While this zone is not a country's territory, the country has the exclusive rights to mineral exploitation on the seabed. By claiming the Senkaku/Diaoyu Islands, China and Japan therefore also assert the rights to resources in the East China Sea in the vicinity of the islands. In reality, the exact boundaries of territorial waters and, especially, exclusive economic zones are hard to determine. If no agreement is reached between disputing countries, the UN's International Court of Justice can issue a ruling. That has not been done in this case.

The second dispute concerns the Spratly Islands of the South China Sea (see also Chapter 10). Both China and Taiwan are pursuing claims to these islands, as are several Southeast Asian countries, largely for the same reasons as with the Senkaku/Diaoyu Islands. China's strategy here has been to build an artificial island that now also has an airfield. By actually populating these islands, it is easier for China to invoke the international Law of the Sea in order to claim the rights to oceanic resources. The other countries are worried that China is militarizing its claims over the South China Sea and that it intends to move air force and navy vessels into these waters as a way to protect its interests. None of the other countries in the region are as militarily powerful as China.

THINGS TO REMEMBER

GEOGRAPHIC THEME 3 • **Power and Politics** As the countries of East Asia develop economically, people have been pushing for more political freedom. Japan, South Korea, and Taiwan all have high levels of political freedom, and demands for political change in China are increasing, especially in urban areas.

• By 2015, there were 674 million people in China using the Internet, representing 49% of the country's population. China has strict controls on Internet access and use, but social networking is being used to promote the expansion of political freedoms.

• Chinese citizens commonly engage in public protests against authorities. The government uses media censorship to spatially contain such dissent.

• In Japan, one political party has controlled the government for all but 3 years since 1955.

• Recently, two major disputes between East and Southeast Asian countries have arisen over islands that may lie over oil and natural gas reserves.

URBANIZATION

GEOGRAPHIC THEME 4

Urbanization: Across East Asia, cities have grown rapidly over the last century, fueled by export-oriented manufacturing industries. China has recently undergone the most massive and rapid urbanization in the history of the world. Its urban population, now 755 million people, has more than tripled since China initiated economic reforms in the 1980s.

Throughout most of the twentieth century, the most dynamic East Asian cities were in Japan, Korea, and Taiwan. Tokyo, Japan, is the world's largest urban area, with 38 million inhabitants. The reason why it is so big is that it is a **conurbation**—a number of cities located so close together that they have grown together over time to form one large urban area. Some of the other urban centers in the area, located around Tokyo Bay, include Yokohama and Kawasaki. As the capital and economic center of one of the largest economies in the world, Tokyo has also attracted huge amounts of investment. The vast Kobe–Osaka–Kyoto area in Japan, which is also a conurbation, has 17 million people (see the map in **FIGURE 9.20**). Korea is home to Seoul, East Asia's second-largest urban area, at 23 million people, and the fourth-largest urban economy in the world (after Tokyo, New York, and Los Angeles). The urban area of Taiwan's capital of Taipei is home to 7 million people and has the third-highest per capita income in East Asia (U.S.$46,000), after Hong Kong and Macao. Seoul and Taipei are especially good examples of primate cities that dominate their respective countries (see Chapter 3).

However, for the last 30 years, the strongest urban growth in East Asia, and even in the entire world, has been in China. In 2014, all 40 of the fastest-growing large cities in the world, as measured by increase in GDP per capita, were in China. Despite the rapid growth, urbanites

conurbation an extended urban area consisting of several nearby cities that have grown together over time

FIGURE 9.20 PHOTO ESSAY: Urbanization in East Asia

Some of the largest, wealthiest, and fastest-growing urban areas in the world are in East Asia. China's cities are undergoing particularly rapid growth.

Hong Wu/Getty Images

A A man searches for lumber in the remains of an old neighborhood in Shanghai that is being demolished to make way for high-rise apartments for the city's burgeoning population of wealthy residents. The people of this neighborhood were forcibly evicted by the city government and resettled in cheaper apartments on the urban fringe. In order to modernize, few attempts at renovating historic neighborhoods, called *hutongs,* are being made in urban China.

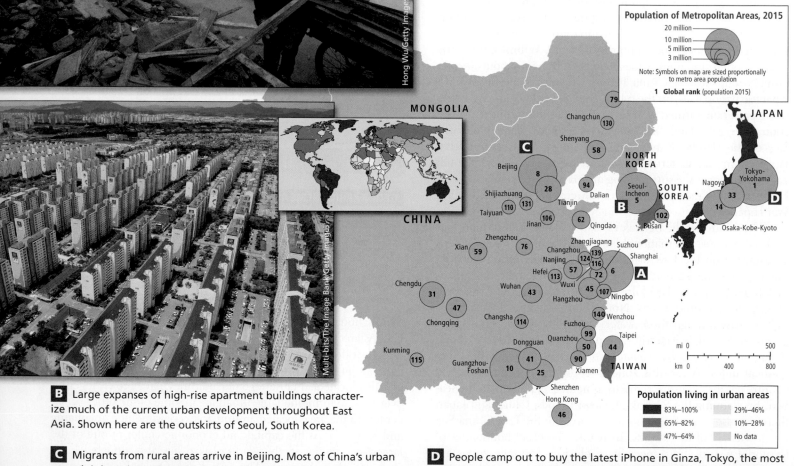

Multi-bits/The Image Bank/Getty Images

B Large expanses of high-rise apartment buildings character-ize much of the current urban development throughout East Asia. Shown here are the outskirts of Seoul, South Korea.

C Migrants from rural areas arrive in Beijing. Most of China's urban growth is based on migration from rural areas.

D People camp out to buy the latest iPhone in Ginza, Tokyo, the most lavish shopping district in the world's largest city.

Map labels

Population of Metropolitan Areas, 2015
- 20 million
- 10 million
- 5 million
- 3 million

Note: Symbols on map are sized proportionally to metro area population

1 Global rank (population 2015)

Population living in urban areas
- 83%–100%
- 65%–82%
- 47%–64%
- 29%–46%
- 10%–28%
- No data

MONGOLIA

CHINA

NORTH KOREA

SOUTH KOREA

JAPAN

TAIWAN

Cities:
- Changchun 130
- Shenyang 58
- 79
- Beijing 8, 28
- Dalian 94
- Shijiazhuang 131
- Taiyuan 110
- Tianjin 106
- Jinan 62
- Qingdao
- Seoul-Incheon 5
- Busan 102
- Nagoya 33
- Tokyo-Yokohama 1
- 14
- Osaka-Kobe-Kyoto
- Zhengzhou 76
- Xian 59
- Zhangjiagang 139
- Changzhou 124
- Suzhou 116
- Nanjing 57
- Shanghai 6
- Hefei 113
- Wuxi 72
- Chengdu 31
- Wuhan 43
- Wuxi 45
- 107
- Ningbo
- Chongqing 47
- Hangzhou
- Changsha 114
- Wenzhou 140
- Fuzhou
- Quanzhou 99
- 50
- Taipei 44
- Kunming 115
- Dongguan 41
- Guangzhou-Foshan 10
- 25
- 90
- Xiamen
- Shenzhen
- Hong Kong 46

mi 0 — 500
km 0 — 400 — 800

Peter Parks/AFP/Getty Images

Tomohiro Ohsumi/Bloomberg via Getty Images

Thinking Geographically

Now that you have read about urbanization in East Asia, you should be able to answer the following questions about the photos in this chapter.

A What happened to many of the former residents of Pudong, the new financial center of Shanghai?

C Migrants to urban areas who ignore the hukou system are considered part of what population?

D The urbanized region that stretches from the cities of Tokyo and Yokohama on Honshu south through the coastal zones of the Inland Sea to the islands of Shikoku and Kyushu is home to what percent of Japan's population?

still represent only 55 percent of China's total population, so more urban growth is likely, whereas in every other country in the region—even North Korea—the majority of the population has been urban for at least two decades. Most East Asian cities are extremely crowded, and future growth poses a challenge for planners and residents (see Figure 9.20A, B). Especially in China, developers are targeting rural land on the outskirts of cities. There, villagers and farmers are rarely given fair compensation for the land that they occupy or farm. The reason is that the individual Chinese farmer does not own farmland; he or she only leases it from the local government, which may be inclined to sell the land to high-bidding developers.

SPATIAL DISPARITIES

China's focus on export-oriented manufacturing based in urban areas has led to rural areas lagging behind cities in access to jobs and income, education, and medical care. The map in Figure 9.16 depicts the problem. GDP per capita is significantly lower in China's interior provinces than it is in coastal provinces, and within each province there is a disparity between rural and urban places. Notice that rural–urban GDP disparities, while still significant, are less extreme in northeastern and coastal provinces than in interior and western provinces. Similar rural-to-urban disparity patterns are found in all other countries in East Asia.

INTERNATIONAL TRADE AND SPECIAL ECONOMIC ZONES

When China initiated economic reforms in the 1980s, many people were wary of the disruption that could result from abruptly opening the economy to international trade. To ease the transition, China first selected five coastal cities as sites where foreign technology and management could be imported to China and free trade established. These **special economic zones (SEZs)** now operate like EPZs found in other world regions (**FIGURE 9.21**; see Chapter 3). Since then, the program has expanded to a large number of cities, many of them in the interior of the country. There are a wide variety of zones with different names and they are administered in different ways, but all provide footholds for international investors and multinational companies eager to establish operations in the country.

This program was successful. Today the SEZs and other zones are China's greatest **growth poles,** meaning that their development, like a magnet, is drawing yet more investment and migration. The first coastal SEZs were spectacularly successful. In just 25 years, many coastal cities grew from medium-sized towns or even villages into some of the largest urban areas in the world. Foreign direct investment remains concentrated on the eastern coast, which accounts for 85 percent of China's exports. Areas along the southern coast near Hong Kong and the central coast near Shanghai have especially well-developed export economies. As shown in Figure 9.16, GDP per capita (PPP) rates remain noticeably higher on the eastern coast relative to China's interior. The government is mindful of these geographic differences and is trying to develop the inland areas. Generating massive amounts of electricity at the Three Gorges Dam is one way of addressing the divide, as it can attract new industrial development to China's interior and act as a counterweight to the booming coast.

Transportation Improvements

Recently, growth poles in the interior have had higher rates of growth than those on the coast. One reason for this is that China spends 9 percent of its GDP on transportation improvements (by comparison, the United States spends 2 to 3 percent), which has made central and western China more accessible. New highways and railroads may eventually enable the interior to catch up with the economic development of the coast. The world's longest high-speed rail now takes a traveler from Beijing to southern China in less than 10 hours, rather than the 22 it used to take, and is routed through central China, not along the coast.

ECONOMIC TRANSITIONS

In 2013, China reached a major milestone that every rapidly developing country eventually reaches: its booming service sector began to create more jobs than did its industrial sector. Many of these jobs offer much better pay and working conditions than industrial jobs do, which is creating a serious shortage of workers in the industrial sector.

Until recently, China's spectacular urban growth was based on hundreds of millions of new urban migrants who were willing to put up with adverse conditions to earn a little cash. Fewer migrants are now willing to work in such jobs, creating periodic labor shortages. Some factory owners have been forced to offer higher pay, better working conditions, and shorter workdays or more time off. The government is also trying to offset labor shortages by making changes to the hukou system, beginning to give formal urban residency status to many people who have migrated to urban areas illegally.

The extra costs these changes impose—especially the higher wages and the improved access to subsidized housing, education, and health care for millions of urban migrants—mean that China is no longer the cheapest place to manufacture products. Some factories have already moved to Vietnam, Bangladesh, and countries in Africa with even cheaper labor.

Some industrial employers hope that enough new migrants will be drawn to the cities by the hukou

special economic zones (SEZs) free trade zones within China, which are commonly called export processing zones (EPZs) elsewhere

growth poles zones of development whose success draws more investment and migration to a region

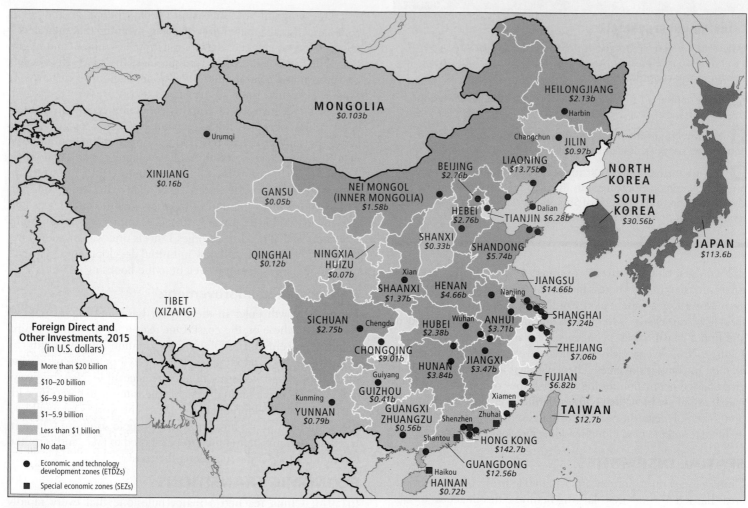

FIGURE 9.21 Foreign investment in East Asia. The map shows China's original special economic zones (SEZs) and one additional important category, the economic and technology development zones (ETDZs). The colors on the map reflect levels of foreign investment in each of the countries of the region and in each of China's provinces. [Source consulted: "Foreign Investment into China by Province, 2011–2015," at http://www .eiu.com/public/thankyou_download.aspx?activity=download&campaignid=chinafdi2012]

reforms that the labor shortages will disappear. The experiences of other countries, though, suggests that China's urban service sector will continue to attract more and more migrants. The rapid growth of the service sector is being fueled by the purchasing power and consumption of China's increasingly wealthy urban populations. Moreover, China's population growth is slowing down significantly, which also means that industries cannot rely on the labor surplus that has so far characterized the Chinese economy.

As noted previously, during Japan's economic miracle from the 1950s to 1970s, its economy grew annually at a rate of 10 percent. After that, the growth slowed down as the economy matured. The same transition towards an affluent economy happened in South Korea and Taiwan as well. China appears to be following the same path. For a long time, China's economic growth was around 10 percent, but it has been nudging downward recently—it was 7.3 percent in 2014 and 6.9 percent in 2015. As the country grows wealthier, maintaining the breakneck pace of development becomes unrealistic. As the vignette below shows, the

relative slowdown has impacted many people who only recently have been exposed to capitalist institutions such as stock markets.

VIGNETTE Jin is a recently retired woman in Shanghai who has found a new hobby: investing in the stock market. As the economy expands, the stock market typically follows along. Why not, Jin reasons, grow your retirement savings by investing in the stock market? She was inspired by watching talk shows on television that promoted easy ways to get rich. Jin had her nephew install trading software called Big Wisdom on her new computer. She also has the Big Wisdom app on her cell phone so she can trade whenever she feels like it.

The emerging culture of trading is also creating new social spaces. In order to tap into the latest news about which stocks are hot, people like Jin go to brokerage houses, which are places that are open to the public where large screens with real-time changes in stock price can be observed and computer stations are available for buying and selling stocks. Over numerous cups of tea, small-time investors socialize and exchange tips all day long.

But here's the problem. China's stock markets have been extremely volatile lately. Trading in China is dominated by individual investors much like Jin. Economic experts say that the demographics of Chinese investing—the investors are also known as the "pyjama traders"—makes the market unpredictable. Small investors who trade frequently and are looking for short-term gains exhibit herd behavior. This creates sudden swings in the market; during 2015–2016, the stock market took sudden dives on multiple occasions. The problem has been exacerbated by a recent decline in GDP growth, which has made investors nervous, and the lack of reliable information to base decisions on. The institutions of capitalism are still quite immature in China. *[Source: Vincent Ni, "China's Stock Market and the Rise of the 'Pyjama Traders'" BBC News, January 13, 2016, at http://www.bbc.com/news /blogs-china-blog-35291536]* ■

HONG KONG'S SPECIAL ROLE

Hong Kong, which was a British possession from the end of the first Opium War in 1842 until 1997, has long had a special role as a link to the global economy for China. Before 1997, some 60 percent of foreign investment in China was funneled through Hong Kong, and since then Hong Kong has remained the financial hub for China's booming southeastern coast **(FIGURE 9.22)**.

Hong Kong is one of the most densely populated cities in the world; its 7.3 million people are packed into only 23 square miles (60 square kilometers), an area roughly the same size as Manhattan in New York City but with four times as many people. Hong Kong is composed of Hong Kong Island, which is the city's center, and a rocky peninsula near the delta of the Zhu Jiang (Pearl River), the most important river in southern China. Skyscraper-dominated Hong Kong is also one of the richest cities in the world: its per capita GNI (adjusted for PPP) in 2014 was slightly above that of the United States, at U.S.$56,570.

In July 1997, Britain's 99-year lease of Hong Kong ran out and Hong Kong became a special administrative region (SAR) of China. Hong Kong still operates under a political and legal system established by the British that allows for more political freedom. Even so, more than a million residents, many of them quite wealthy, left before China took over in 1997, worried that their fortunes and freedoms would diminish. China so far has allowed more political freedom in Hong Kong, in part because of the island's important role as a link to global trade.

However, in 2014, in response to perceived interference from Beijing in Hong Kong's local elections, hundreds of thousands of protesters gathered in the streets of Hong Kong for a period of 3 months. They used umbrellas as shields against the use of pepper spray by the police. These so-called *umbrella protesters* may not have achieved their goals but they are, unlike in the rest of China, allowed to voice their opinions. At least for now.

SHANGHAI'S LATEST TRANSFORMATION

Shanghai has historically been a trendsetting city. Its opening to Western trade in the early nineteenth century spawned a period of phenomenal economic growth and cultural development that led to it being called the "Paris of the East." As a result of China's

recent reentry into the global economy, Shanghai is undergoing another boom, which has enriched some people and dislocated others (see Figure 9.20A). The Shanghai urban area is the largest in China, with more than 23 million residents. On average, these residents of Shanghai are now as affluent as people in Central European countries like Poland and Slovakia, and if current growth trends continue, their wealth will soon be on par with that of residents of Portugal or Slovenia. Shanghai today has the world's single busiest cargo port, and the region around the city is now responsible for as much as a quarter of China's GDP.

In less than a decade, the city's urban landscape has been remade by the construction of more than a thousand business and residential skyscrapers; subway lines and stations; highway overpasses; bridges; and tunnels. Shanghai's shopping district on Nanjing Road is as imposing as any such district around the world. For hundreds of miles into the countryside, suburban development linked to Shanghai's economic boom is gobbling up farmland, and displaced farmers and villagers have rioted.

ON THE BRIGHT SIDE: Shanghai's Pajama Culture

Shanghai's long-standing role as a window to the outside world has meant that it often is host to behavior that is less common elsewhere in the country. For example, for years the people of Shanghai have relaxed in public in their pajamas—light, loose cotton tops and bottoms stamped with images of puppies or butterflies. The city government has tried to squelch the custom, but the citizens have proved recalcitrant. The police seem to understand that images of them arresting old and young alike for wearing pajamas would be ridiculous, especially in the international media, so for now the issue is unresolved. ■

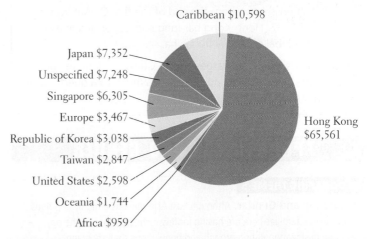

FIGURE 9.22 Net foreign direct investment in China, 2012. Foreign investment in China reached nearly $112 billion in 2012. Note that investment from Hong Kong, even though it is part of China, is considered "foreign" here. Many investments that actually originate in various countries across the globe are funneled through Hong Kong banks, as was the case when Hong Kong was a British territory. The same holds true for small banking centers in the Caribbean. [Source consulted: UNCTAD, "Bilateral FDI Statistics," at http://unctad.org/Sections /dite_fdistat/docs/webdiaeia2014d3_CHN.pdf]

Pudong, the city's new financial center, sits across the Huangpu River from the Bund—Shanghai's famous, elegant row of big colonial-era brownstone buildings that served as the financial capital of China until half a century ago. Previously, Pudong was a maze of dirt paths and sprawling neighborhoods of simple, tile-roofed houses, but its former residents were pushed out to make way for soaring high-rises. New construction is restricted in the historic section of the Bund, which is another reason why Pudong's central location has been attractive as a new business center. Pudong is home to Shanghai's stock exchange and the tallest skyscrapers in China. Buildings like the Shanghai Tower and the Oriental Pearl Tower have become a visual symbol for the economic vitality of Pudong, the city of Shanghai, and all of China.

THINGS TO REMEMBER

GEOGRAPHIC THEME 4

• **Urbanization** Across East Asia, cities have grown rapidly over the last century, fueled by export-oriented manufacturing industries. China has recently undergone the most massive and rapid urbanization in the history of the world. Its urban population, now 755 million people, has more than tripled since China initiated economic reforms in the 1980s.

• China's focus on export-oriented manufacturing based in urban areas has led to rural areas lagging behind cities in access to jobs and income, education, and medical care.

• China's SEZs and other economic zones have been spectacularly successful, becoming major growth poles that draw investment and migration.

• In 2013, China reached a major milestone that every rapidly developing country eventually reaches: its booming service sector began to create more jobs than did its industrial sector; however, this is also a harbinger of slower future growth.

• A British possession from the end of the first Opium War in 1842 until 1997, Hong Kong has long had a special role as a link to the global economy for China.

• Shanghai today is the largest city in China, and the region around the city is now responsible for as much as a quarter of China's GDP.

POPULATION AND GENDER

GEOGRAPHIC THEME 5

Population and Gender: Although East Asia remains a highly populous world region, families here are having far fewer children than in the past, resulting in slow population growth and populations that are aging. The legacy of China's now-abandoned one-child policy, combined with an enduring cultural preference in China for male children, has created a shortage of females.

From a global perspective, East Asia's rate of natural increase is low, 0.5 percent. This rate is very similar to that of North America (0.4 percent) but higher than in Europe (0.0 percent). In China, this low population growth is partially due to the legacy of government policies that harshly penalized families for having more than one child. But in China as well as elsewhere, urbanization and changing gender roles are also resulting in smaller families, regardless of official policy. Only in the two poorest countries—Mongolia (with 3 million people) and North Korea (with 25 million)—are women still averaging two or more children each, but even there family size is shrinking.

RESPONDING TO AN AGING POPULATION

Low birth rates mean that fewer young people are being added to the population, and improved living conditions mean that people are living longer across East Asia. The overall effect is that the average age of the populations is rising. Put another way, populations are aging. For Mongolia, South Korea, North Korea, and Taiwan, it will be several decades before the financial and social costs of supporting numerous elderly people will have to be addressed. China faces especially serious future problems with elder care because the one-child policy and urbanization have so drastically reduced family kin-groups. Japan, on the other hand, has already been dealing with the problem of having a large elderly population that requires support and a reduced number of young people to do the job.

Japan's Options

Japan's population has stopped growing altogether and the country is aging rapidly, raising concerns about economic productivity and humane ways to care for dependent people. The demographic transition (see Chapter 1) is in its late stages in this highly developed country, where 93 percent of the population lives in cities. Japan has a negative rate of natural increase (−0.2), the lowest in East Asia and the world, except for a few countries in Central Europe. If this trend continues, Japan's population is projected by the United Nations to plummet from the current 127 million to 108 million by 2050.

At the same time, the Japanese have among the world's longest life expectancy, at 83 years **(FIGURE 9.23)**. As a result, Japan also has the world's oldest population, 26 percent of which is over the age of 65. By 2050, the average age in Japan is projected to be 53, which can be compared with the projection for the United States, at 41 years. Retired people in Japan may then account for 40 percent of the population, and Japan's labor pool could be reduced by more than a third, but it would still need to produce enough to take care of more than twice as many retirees as it does now. Clearly, these demographic changes will have a momentous effect on Japan's economy, and the search for solutions is underway. One possibility is increasing immigration to bring in younger workers who will fill jobs and contribute to the tax rolls, as the United States and Europe have done.

In Japan, recruiting immigrant workers from other countries is a very unpopular solution to the aging crisis. Many Japanese people object to the presence of foreigners, and the few small minority populations with cultural connections to China or Korea have for years faced discrimination. Nonetheless, foreign workers are dribbling into Japan in a multitude of legal and illegal ways. Many are so-called guest workers from South Asia, brought in to fill the most dangerous and lowest-paying jobs with the understanding that they will eventually leave. Others are the descendants of Japanese people who once migrated to South

FIGURE 9.23 Japan's aging population. Elderly people in an adult day-care center on Gogo Island, Japan. Japan has the world's longest life expectancy (83 years) and also the world's oldest population: 26 percent of its people are over the age of 65.

America (Brazil and Peru). Regardless, Japan's foreign population remains tiny, making up only 1.5 percent (2 million people) of the total population. A recent UN report estimates that Japan would have to admit more than 640,000 immigrants per year just to maintain its present workforce and avoid a 6.7 percent annual drop in its GDP.

In a novel approach to Japan's demographic changes, government and industry have invested enormous sums of money in robotics over the past decade. Robots are already widespread in Japanese industries such as auto manufacturing, and their industrial use is growing. Now they are also being developed to care for the elderly, to guide patients through hospitals, to look after children, staff hotels, and even to make sushi. By 2025, the government plans to replace up to 15 percent of Japan's workforce with robots.

China's Options

The proportion of China's population over 65 is now only 10 percent, but this will change rapidly as conditions improve and life expectancies increase by 5 to 10 years to become more like those of China's affluent East Asian neighbors. China's current population pyramid shows that there are many people in their fifties who will be retiring in the not-too-distant future **(FIGURE 9.24A)**. However, a crisis in elder care is already upon China for two other reasons: the high rate of rural-to-urban migration and the shrinkage of family support systems because of the one-child policy (discussed below).

The lingering effects of the one-child policy have transformed Chinese families and Chinese society as a whole. For example, an only child has no siblings, so within two generations, the kinship categories of brother, sister, cousin, aunt, and uncle have disappeared from most families, meaning that any individual has very few, if any, related age peers with whom to share family responsibilities. Most children are doted on by several adults and children are not taught by siblings to share. Conscious efforts must be made to instill self-sufficiency in only children.

When hundreds of millions of young Chinese people were lured into cities to work, most thought rural areas would

benefit from remittances, and this has happened. However, few anticipated that the one-child family would mean that for every migrant, two aging parents would be left to fend for themselves, often in rural, underdeveloped areas. China's parliament passed a law in 2013 that requires family members to visit and support their elderly relatives. While such a law may be unenforceable, it shows how worried the government is about the social consequences of the prospect of an aging society, including the increasing spatial mismatch between the elderly who need care and their younger relatives who often have moved elsewhere in search of employment.

THE LEGACIES OF CHINA'S ONE-CHILD POLICY

In response to fears about overpopulation and environmental stress, China mandated a policy from 1979 to 2013 that each family could only have one child. The one-child policy was sometimes enforced brutally. At various times and in places, the government has waged a campaign of forced sterilizations and forced abortions for mothers who already have one child. The policy was also implemented with rewards for complying and with large fines for not complying. As a result, China's rate of natural increase (0.5 percent) is approximately the same as that of the United States, and less than half the world average (1.2 percent).

Recently, the policy has been relaxed. In 2013, when China also instituted broader economic reforms, couples were allowed to have two children, so long as one of the people in the couple is an only child, which is the case for most couples. Then in 2015, the Chinese government announced that the one-child policy would be replaced with a two-child policy. The step-by-step relaxation of the policy is due to growing concern about the forecasted shrinking of China's population, which should begin sometime between 2025 and 2050 (see Figure 9.24B), creating many of the same economic and social problems that Japan is now facing. But as China is an increasingly urban society where having more children is a net household cost rather than a benefit, many couples may choose to have only one child, which has become the new social norm. No baby boom is expected in the future.

Gender Imbalance and the Cultural Preference for Sons

The one-child policy, combined with an ancient cultural preference for male children, resulted in a severe gender imbalance. For many couples, the prospect of their only child being a daughter, without the possibility of having a son in the future, was devastating. The preference for sons relates to deeply patriarchal Confucian values that have prevailed throughout this region for millennia (see page 513). In fact, gender imbalance has emerged in the Koreas and Taiwan even without the one-child policy, which suggests that the influence of Confucian values is strong.

China's more severe gender imbalance is illustrated by its population pyramid (see Figure 9.24A). The average global sex ratio at birth is 107 boys to 100 girls, which evens out to 101 boys to 100 girls by the age of 5. In China, however, there are 116 boys for every 100 girls born. In fact, for nearly every category until age 70, males outnumber females. The census data show that there are already 50 million more men than women.

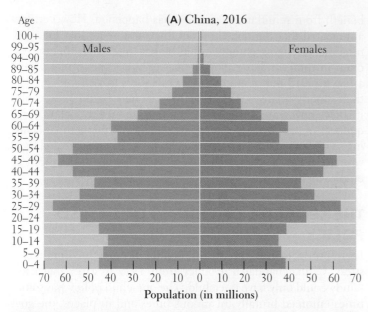

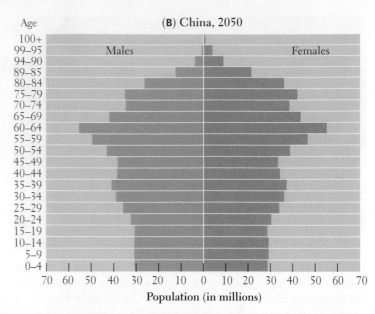

FIGURE 9.24 Population pyramids for China, 2016 and 2050 (projected). [Source consulted: U.S. Census Bureau, Population Division, International Data Base, 2016, at https://www.census.gov/population/international/data/idb/informationGateway.php]

What happened to the missing girls? There are several possible answers. Given the preference for male children, the births of these girls may simply have gone unreported by families hoping to conceal their daughters as they tried to conceive a male child. There are many anecdotes of girls being raised secretly or even disguised as boys. Also, adoption records indicate that girls are given up for international adoption much more often than boys are. Or the girls may have died in early infancy, either through neglect or infanticide. Finally, some parents conduct medical tests that can identify the sex of a fetus. There is evidence that in China, as elsewhere around the world, many parents choose to abort female fetuses.

There is some evidence that attitudes may be changing. For example, in Japan, South Korea, and Mongolia, the percentage of women receiving secondary education equals or exceeds that of men. In Japan, there is a slight gender imbalance *in favor of females*. In China, social policy makers have tried for decades to eliminate the old Confucian bias toward men by empowering women economically and socially. In some ways, they are succeeding, as Chinese women now participate in the workforce to a large degree. Nevertheless, the preference for sons persists.

A Shortage of Brides

A major side effect of the preference for sons is that there is now a growing shortage of women of marriageable age throughout East Asia. In 2012, China alone had an estimated deficit of 11 million women aged 20 to 35. Females are also effectively "missing" from the marriage rolls because many educated young women are too busy with career success to meet eligible young men.

Research suggests that at least 10 percent of young Chinese men will fail to find a mate; and poor, rural, uneducated men will have the most difficulty. The shortage of women will lower the birth rate yet further, which will contribute to the expected shrinkage of the population over the next century. Without spouses, children, or even siblings, there will be no one to care for single men when they age. Furthermore, China's growing millions of single young men are emerging as a potential threat to civil order, as they may be more prone to drug abuse, violent crime, HIV infection, and sex crimes. Cases of kidnapping and forced prostitution of young girls and women are already increasing.

POPULATION DISTRIBUTION

In East Asia, people are not evenly distributed across the land **(FIGURE 9.25)**. China, with 1.37 billion people, has almost one-fifth of the world's population. However, 90 percent of these people are clustered on the approximately one-sixth of the total land area that is suitable for agriculture, and roughly half of these live in urban areas. The population is concentrated especially densely in eastern China, including the North China Plain stretching from Beijing down to Shanghai; the coastal zone from Shanghai to Hong Kong that includes the delta of the Zhu Jiang (Pearl River) in the southeast; the Sichuan Basin; and the middle Chang Jiang (Yangtze) basin.

The west and south of the Korean Peninsula are also densely settled, as are northern and western Taiwan. In Japan, settlement is concentrated in a band that stretches from the cities of Tokyo and Yokohama on the island of Honshu facing the Pacific, south through the coastal zones of the Inland Sea to the islands of Shikoku and Kyushu. This urbanized region is one of the most extensive and heavily populated metropolitan zones in the world, accommodating 86 percent of Japan's total population. The rest of Japan is mountainous and more lightly settled. Mongolia is only lightly settled, with one modest urban area.

Human Well-Being

"The objective of development is to create an enabling environment for people to enjoy long, healthy and creative lives."

—MAHBUB UL HAQ, FOUNDER OF THE UN HUMAN DEVELOPMENT REPORT

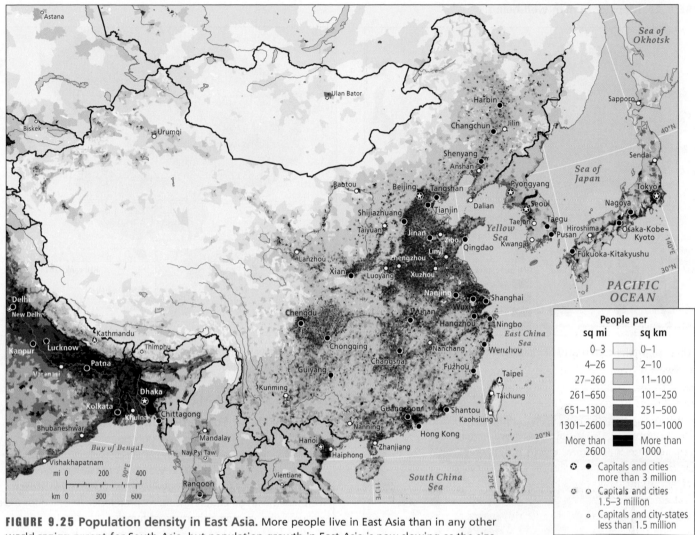

FIGURE 9.25 Population density in East Asia. More people live in East Asia than in any other world region except for South Asia, but population growth in East Asia is now slowing as the size of families is decreasing. Most of the region now faces the challenge of caring for large elderly populations. China is grappling also with an unexpected consequence of its one-child policy: a shortage of women.

There are different ways to measure how well specific countries enable their citizens to enjoy healthy and rewarding lives, but it is generally agreed that income per capita should not be the sole measure of well-being. In addition to gross national income (GNI) per capita (which is a variation of GDP), we include here maps on human development and on gender equality **(FIGURE 9.26)**. On these maps, Japan and South Korea (plus Taiwan in Map A) stand out as quite different from China and Mongolia (plus North Korea in Map A). Taiwan is not covered in Maps B and C because in UN data, Taiwan is included with China. North Korea is only included in Map A because it does not submit data to the United Nations.

Japan, South Korea, and Taiwan all have high GNI per capita, adjusted for purchasing power parity (PPP) (see Figure 9.26A). China and Mongolia have GNI per capita (PPP) that is slightly below the world average. What the income figures mask is the extent to which there is disparity of wealth in a country. Ironically, wealth disparity in Communist-governed China is larger than elsewhere in the region. The introduction of a market economy

during the last few decades has increased inequalities between individuals and between regions within China. In contrast, both South Korea and Japan have a relatively equitable distribution of economic resources. The GNI per capita for North Korea is an estimate, as data are hard to come by, but the country is undoubtedly very poor, due to severe government mismanagement.

Figure 9.26B depicts countries' rank on the Human Development Index (HDI), which is a calculation (based on adjusted real income, life expectancy, and educational attainment) of how adequately a country provides for the well-being of its citizens. Again, there are differences between the very high global ranking for Japan (20) and South Korea (17), and the high global rankings for China and Mongolia (both countries ranked 90 among countries in the world). China and Mongolia have made significant progress over the last 20 years and their global HDI rank is better than their medium GNI per capita (PPP). Both countries now provide equitable access to basic services for all of their people, and it is quite possible that their ranking—especially China's—will continue to improve in the years to come.

FIGURE 9.26 Maps of human well-being.

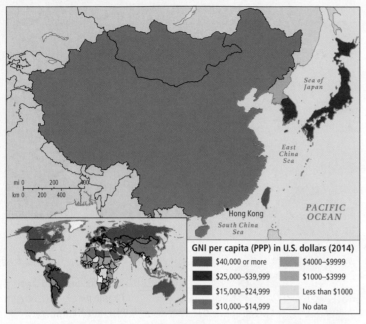

(A) Gross national income (GNI) per capita, adjusted for purchasing power parity (PPP).

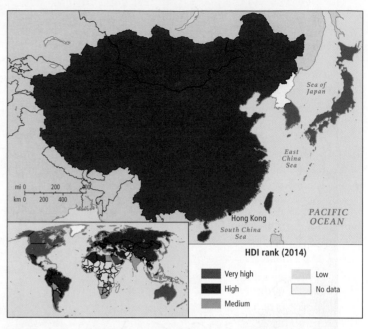

(B) Human Development Index (HDI).

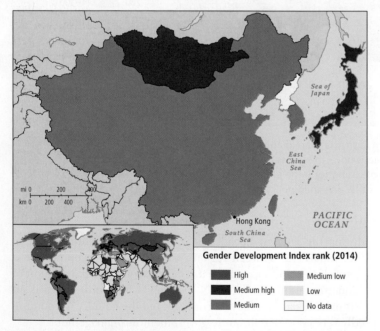

(C) Gender Development Index (GDI).

Figure 9.26C shows how well countries are ensuring gender equality. The Gender Development Index is based on the same variables as HDI—longevity, education, and income—but adjusted for gender differences. A high rank indicates that the genders are tending toward equality. Here, the differences among the countries don't align directly with the level of economic development. All countries rank medium or medium-high. Japan does relatively well and is in the medium-high GDI category. Affluent South Korea, however, only ranks in the medium category, together with China. Interestingly, Mongolia has a medium-high ranking, which may be a result of the emphasis on gender equity under communism and traditional cultural attitudes that promote some measures of gender equality.

THINGS TO REMEMBER

GEOGRAPHIC THEME 5 • **Population and Gender** Although East Asia remains a highly populous world region, families here are having far fewer children than in the past, resulting in slow population growth and populations that are aging. The legacy of China's now-abandoned one-child policy, combined with an enduring cultural preference in China for male children, has created a shortage of females.

• The Japanese have the world's longest life expectancy, at 83 years, and the world's oldest population—26 percent of Japanese people are over the age of 65. By 2050, this age group is projected to account for 40 percent of the population.

• China faces a crisis in elder care for two reasons: the high rate of rural-to-urban migration and the shrinkage of family support systems because of the legacy of the one-child policy.

• Research suggests that at least 10 percent of young Chinese men will fail to find a mate; and poor, rural, uneducated men will have the most difficulty.

• The general level of well-being varies in the region, from high (Japan, South Korea, Taiwan), to medium (China, Mongolia), to low (North Korea).

SOCIOCULTURAL ISSUES

Most countries in East Asia have one dominant ethnic group, but all countries have considerable cultural diversity. In China, for example, 93 percent of Chinese citizens call themselves *people of the Han*. The name harks back about 2000 years to the Han empire but it gained currency only in the early twentieth century, when nationalist leaders were trying to create a mass Chinese identity. Figure 9.13 shows how the Han empire extended to territories that are now populated by ethically Chinese people. The term *Han* simply connotes people who share a general way of life and pride in Chinese culture. The main language spoken by the Han is Mandarin, although it is only one of many Chinese dialects.

China's non-Han minorities number about 120 million people in more than 55 different ethnic or culture groups scattered across the country. Most live outside the Han heartland of eastern China **(FIGURE 9.27)**. Some of these areas have been designated *autonomous regions*, where minorities theoretically manage their own affairs. In practice, however, the Communist Party in Beijing controls these regions, especially those considered to have potentially rebellious populations and those that have resources of economic value. We profile a few of China's ethnic groups here.

Western China's Muslims Muslims of various ethnic origins have long been prominent minorities in China. All are originally of Central Asian origin and most are Turkic people who historically have been nomadic herders. Others specialized in trading. They tend to be concentrated in China's northwest and often think of themselves as quite separate from mainstream China.

Uygurs (pronounced WEE-gurs) and Kazakhs, who are Turkic-speaking Muslims, live in the autonomous region of Xinjiang in the far northwest (see the map in Figure 9.1). Though relatively few Uygurs or Kazakhs remain nomadic herders today, contact between them and similar peoples in Central Asia has been revived since China's market reforms began and the Soviet Union dissolved **(FIGURE 9.28)**.

The Beijing government claims this area's oil, other mineral resources, and irrigable agricultural land for national development. Accordingly, it has sent troops and many millions

FIGURE 9.28 The Silk Road market in Kashi. A Uygur man tests a horse before selling it at the Sunday market in Kashi, a city in China's Xinjiang Uygur Autonomous Region. Kashi has been an important trading center along the Silk Road for at least 2000 years.

of Han settlers to Xinjiang through what is known as the *Go West* policy. The Han settlers fill most managerial jobs in the bureaucracy, the military, and in the mineral extraction and power generation industries. An important secondary role of the Han is to dilute the power of Uygurs and Kazakhs within their own lands. In Xinjiang, there are now almost as many Han as Uygurs (9 versus 10 million), plus small numbers of other minorities.

The Beijing government has rushed to develop Xinjiang and its capital Urumqi by establishing economic development zones in the area, but the Uygurs have been left out of most policy-making roles and indeed have been excluded from participating in the economic boom. One young Uygur man in Urumqi writes of his discontent: "I am a strong man, and well-educated. But [Han] Chinese firms won't give me a job. Yet go down to the railroad station and you can see all the [Han] Chinese who've just arrived. They'll get jobs. It's a policy to swamp us."

Until recently, the Uygur people of Xinjiang expressed their resistance to Han dominance merely by reinvigorating their Islamic culture. Islamic prayers were increasingly heard publicly, more Muslim women were wearing Islamic dress, Uygur was spoken rather than Chinese, and Islamic architectural traditions were being revived. Then more active resistance groups, formed by Uygur separatists, began carrying out attacks on Chinese targets. In 2009, violence erupted in the streets of Urumqi between Uygurs and the Han, and about 200 people were killed. The Beijing government responded by harshly punishing the rioters and broadcasting the accusation that all Uygur separatists, even those committed to nonviolence, are Islamic fundamentalists bent on terrorism. This conflict is ongoing. As recently as 2015, Uygurs have attacked Chinese police and the government has cracked down on Uygur activists. The ancient city of Kashi, crisscrossed with walled courtyards and narrow alleys, is being demolished or modernized, depending on one's perspective. Many Uygurs view this as a form of cultural genocide.

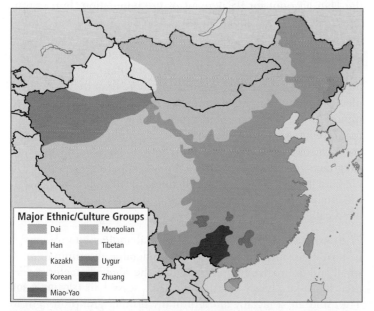

Major Ethnic/Culture Groups

- Dai
- Han
- Kazakh
- Korean
- Miao-Yao
- Mongolian
- Tibetan
- Uygur
- Zhuang

FIGURE 9.27 Major ethnic groups of China. This map shows the areas traditionally occupied by the Han and by ethnic minorities. It does not show the recent resettling of Han in Xinjiang and Tibet, nor does it show the Hui, people from many ethnic groups whose ancestors converted to Islam and who are found in disparate locations along the old Silk Road and in coastal southeastern China. [Source consulted: Chiao-min Hsieh and Jean Kan Hsieh, *China: A Provincial Atlas* (New York: Macmillan, 1995), p. 12. The website http://www.index-china.com/minority /minorityenglish.htm includes a comprehensive survey of minorities in China.]

Distinct from the Uygurs and Kazakhs are the Hui, who altogether number about 11 million. The original Hui people were descended from ancient Turkic Muslim traders who traveled the Silk Road from Europe across Central Asia to Kashgar (now Kashi) and on to Xian in the east (see the map of the Silk Road in Figure 5.9; see also Figure 9.28). Subgroups of Hui live in the Ningxia Huizu Autonomous Region and throughout northern, western, and southwestern China. There they continue as traders, farmers, and artisans. The tradition of commercial activity and adoption of the Chinese language among the Hui has facilitated their success in China. Many are active in the new free market economies of southeastern China as businesspeople, technicians, and financial managers, using their money not just to buy luxury goods but also to revive religious instruction and to fund their mosques, which are now more obvious in the landscape.

The Tibetans The Tibetans, in contrast to the prosperous and assimilated Hui, are an impoverished ethnic minority group of nearly 5 million people scattered thinly over a huge, high, mountainous region in western China. The history of Tibet's political status vis-à-vis China is long and complex, characterized by both cordial relations and conflict. During China's imperial era (prior to the twentieth century), Tibet maintained its own government but was controlled by China, as either a subordinate state or a province. In the early 1900s, Tibet declared itself separate and free from China and conducted its affairs as an independent country. It was able to maintain this status until 1949–1950, when the Chinese Communist army "liberated" Tibet, promising a "one country, two systems" structure, which suggested a great deal of regional self-governance. A Tibetan uprising in 1959 led China to abolish the Tibetan government and violently reorder Tibetan society. Since the 1950s, the Chinese government has referred to Tibet as the Xizang Autonomous Region. The Chinese government suppressed Tibetan Buddhism—a religion that Tibetans rally behind as a symbol of independence—by destroying thousands of temples and monasteries and massacring many thousands of monks and nuns. In 1959, the spiritual and political leader of Tibet, the Dalai Lama, was forced into exile in India along with thousands of his followers.

Since the 1990s, the Beijing government's strategy has been to overwhelm the Tibetans with secular social and economic modernization and with Han Chinese settlers rather than outright military force (though China maintains a military presence in Tibet). To attract trade and quell foreign criticism of its treatment of Tibetans, China now spends hundreds of millions of dollars on housing and on roads, railroads, and a tourism infrastructure that capitalizes on European and American interest in Tibetan culture. China presents its actions in Tibet as part of its overall strategy to integrate the entire country economically and socially. Schools are being built and jobs opened up to young Tibetans. A new railway link effectively connecting Tibet to the rest of

Ainu an indigenous minority group in Japan characterized by their light skin, heavy beards, and thick, wavy hair, who are thought to have migrated thousands of years ago from the northern Asian steppes

China was completed in 2006 **(FIGURE 9.29)**. The Han in Tibet see the railway as a public service that will promote Tibetan development, but Tibetan activists see it as a conveyor belt for more Han dominance over the Tibetan economy and culture.

Within Tibetan culture, women have held a somewhat higher position than in other cultures of East Asia. Buddhism did introduce many patriarchal attitudes to Tibet, but these did not curtail other more equitable traditions. Among nomadic herders, women could have more than one husband, just as men were free to have more than one wife. At marriage, a husband often joins the wife's family. By comparison, Han Chinese culture has typically regarded the women of Tibet and other western minorities as barbaric precisely because their roles were not circumscribed. They rode horses, they worked alongside the men in herding and agriculture, and they were generally more assertive in daily life than their Chinese counterparts.

Indigenous Diversity in Southern China In Yunnan Province in southern China, more than 20 groups of ancient native peoples live in remote areas of the deeply folded mountains that stretch into Southeast Asia. These groups speak many different languages, and many have cultural and language connections to the indigenous people of Tibet, Burma, Thailand, or Cambodia. Gender relations are different here than among the Han. A crucial difference may be that among several groups, most notably the Dai, a husband moves in with his wife's family at marriage and provides her family with labor or income. A husband inherits from his wife's family rather than from his birth family.

Taiwan's Many Minorities In Taiwan and the adjacent islands, the Han account for 95 percent of the population, but Taiwan is also home to 60 indigenous minorities. Some have cultural characteristics—languages, crafts, and agricultural and hunting customs—that indicate a strong connection to ancient cultures in Southeast Asia and the Pacific. The mountain dwellers among these groups have resisted assimilation more than the plains peoples. Both groups may live on reservations set aside for indigenous minorities if they choose, but most are now being absorbed into mainstream, urbanized, Han-influenced Taiwanese life. The Han are themselves not homogenous in Taiwan. They are divided into subgroups based on their ancestral home in China, and the people (and their descendants) who fled to Taiwan around the time of the establishment of the People's Republic of China in 1949 are often called "mainlanders."

The Ainu in Japan There are several indigenous minorities in Japan and most have suffered considerable discrimination. A small and distinctive minority group is the **Ainu,** characterized by their light skin, heavy beards, and thick, wavy hair **(FIGURE 9.30)**. The Ainu are a racially and culturally distinct group thought to have migrated many thousands of years ago from the northern Asian steppes. They once occupied Hokkaido and northern Honshu and lived by hunting, fishing, and some cultivation, but they are now being displaced by forestry and other development activities. Few full-blooded Ainu remain because, despite prejudice, they have been steadily assimilated into the mainstream Japanese population. The Ainu officially only number about 25,000, but

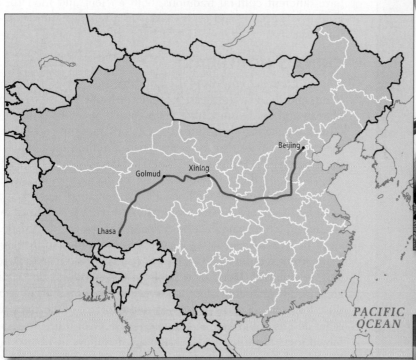

(A) Han Chinese passengers aboard the train combat altitude sickness by sleeping and using supplemental oxygen supplied to every passenger.

FIGURE 9.29 The Beijing-Tibet Railway. The western section of this rail-road opened in 2006. It is the world's highest railway, with nearly 600 miles (1000 kilometers) of line that are at an altitude above 13,000 feet (4000 meters). The trip from Beijing to Lhasa now takes 41 hours, and it costs much less to transport goods and people via the railway than it does using the existing highway. However, many Tibetans worry that the railway will become a conduit for more Han migration to Tibet. (Source consulted: "Beijing to Lhasa Train Travel Guide," http://www.tibettravel.org/tibet-train/beijing-to-lhasa-train.html)

(B) A view of the train as it passes through the mountains of the Plateau of Tibet.

there may be many more who don't identify themselves as Ainu in the Japanese census due to perceived discrimination. Somewhat belatedly, the Japanese parliament officially recognized the Ainu as an indigenous minority in 2008.

East Asia's Most Influential Cultural Export: The Overseas Chinese

China has had an impact on the rest of the world not only through its global trade, but also through the migration of its people to nearly all corners of the world. The first recorded emigration by the Chinese took place more than 2200 years ago. Following that, China's contacts then spread eastward to Korea and Japan, west-ward into Central and Southwest Asia via the Silk Road, and by the fifteenth century, to Southeast Asia, coastal India, the Arabian Peninsula, and even Africa.

Trade was probably the first impetus for Chinese emigra-tion. The early merchants, artisans, sailors, and laborers came mainly from China's southeastern coastal provinces. Taking their families with them, some settled permanently on the peninsulas and islands of what are now Indonesia, Thailand, Malaysia, and the Philippines. Today they form a prosperous urban commercial class known as the **Overseas Chinese.** The quintessential Overseas Chinese state in Southeast Asia is Singapore, where 77 percent of the population is ethnically Chinese (see Chapter 10).

In the nineteenth century, economic hardship in China and a growing international demand for labor spawned the migration of as many as 10 million Chinese people to countries all over the world. By the middle of the twentieth century, many others fleeing the repres-sion of China's Communist Revolution joined those from earlier migration. As a result, "Chinatowns" are found in most major world cities. The term Overseas Chinese has been extended to apply to Chinese emigrants and their descendants in all such locations.

Overseas Chinese Chinese emigrants and their descendants, especially those in Southeast Asia

THINGS TO REMEMBER

• There are several distinct minority groups in East Asia: the Uygurs, Kazakhs, and Tibetans in western China; the Hui, scat-tered throughout central China; and smaller indigenous groups in southern China, Taiwan, and Japan.

FIGURE 9.30 The Ainu of Japan. An Ainu man creates an item to be used in ancestor worship in Tokyo. In 2008, Japan's government formally recognized the Ainu as an indigenous people of Japan, more than 500 years after the Ainu began to be marginalized by mainstream Japanese culture.

- There is resentment and sometimes open resistance to Han Chinese domination in the various minority homelands.
- Throughout East Asia, minorities have experienced discrimination.
- Millions of Overseas Chinese live in cities and towns around the world, especially in Southeast Asia. Most have settled permanently in their foreign locations and many maintain relationships with China.

SUBREGIONS OF EAST ASIA

loess windblown material that forms deep soils in China, North America, and central Europe

Because China is so large and diverse, it is easiest to consider it by looking at its four major subregions—those of northeastern, central, southern,

and far north and west China (**FIGURE 9.31**). The other countries have different cultural traditions, both ancient and modern, as Figure 9.11 shows, and each makes up a separate subregion of East Asia.

CHINA'S NORTHEAST

China's northeast consists of the Loess Plateau, the North China Plain, and the Far Northeast subregion (**FIGURE 9.32**). The Loess Plateau and the North China Plain are the ancient heartland of China. By the eighth century, the city of Chang'an (now where the modern city of Xian is located) had 2 million inhabitants in the urban region and may have been the largest city in the world. After 900 C.E., the center of Chinese civilization shifted to the North China Plain, but the Loess Plateau remained a crucial part of China. Xian served as the eastern terminus of the Silk Road that connected China with Central Asia and Europe (see Figure 5.9).

The Loess Plateau (see Figure 9.32) and the North China Plain are linked physically as well as culturally. Both are covered by fine yellowish **loess,** or windblown soil particles. Loess soils are typically quite fertile but erode easily. For thousands of years, dust storms have picked up loess from the surface of the Gobi and other deserts to the west and carried it east. Over time, the windblown loess has filled up deep mountain valleys in Shanxi and Shaanxi provinces, creating an undulating plateau. The Huang He, in a second massive earth-moving process, transports millions of tons of loess sediment from the Loess Plateau to the coastal plain each year. Over the millennia, this river has created the North China Plain by depositing its heavy load of loess sediment in what was once a much larger Bo Hai Sea.

The Loess Plateau

An unusually diverse mixture of peoples from all over Central Asia and East Asia have found the Loess Plateau a fertile, though challenging, place to farm and herd. Among the many culture groups that share this densely inhabited and productive plateau are the Hui and the Han of China and, particularly in the western reaches, Mongols, Tibetans, Uygurs, and Kazakhs. Many of the plateau inhabitants farm cotton and millet—a tall-grass grain grown in many parts of Asia and Africa—in irrigated valley bottoms, or raise sheep on the drier grassy uplands. China's largest coal reserves are also found here.

Thousands of years of human occupation have stripped the plateau of its ancient forests, leaving only grasses to anchor the loess. Because the loess is so thick—hundreds of feet deep in many areas—and the winds persistent, some people have carved residences and animal sheds below the surface of the plateau (**FIGURE 9.33**). Torrential rains cut deep gullies into the loess, and such gullies now cover much of the landscape, causing people to abandon the agricultural terraces. For decades, the government has maintained a revegetation campaign to stabilize slopes (see the vignette on page 503).

The North China Plain

The North China Plain is the largest and most populous expanse of flat, arable land in China. Since the sixth century, it has been home to most of the imperial families, and from this region the Han Chinese have dominated the country.

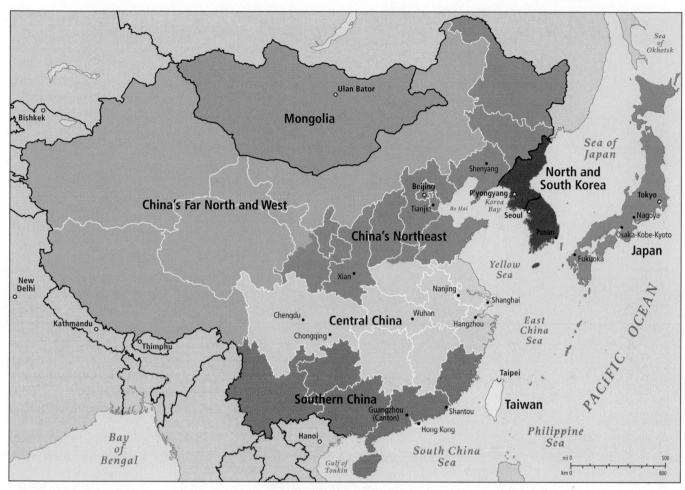

FIGURE 9.31 Subregions of East Asia. This map shows the four subregions of China as well as the other four subregions of East Asia: Taiwan, Japan, South and North Korea, and Mongolia.

Today the North China Plain is one of the most densely populated spaces in the country: 210 million people live in a space about the size of France. The majority are farmers who work the generally small fields that cover the plain. Most of them produce wheat, which grows well in the relatively dry climate. The forests that once blanketed the plain were cut down thousands of years ago, and now the only trees that survive are those surrounding temples and those planted as windbreaks.

The Huang He, the river that created the North China Plain, remains the plain's most important physical feature, serving as a major transportation artery and a source of irrigation water, but also producing disastrous floods. After the river descends from the Loess Plateau, its speed slows dramatically on the flat surface of the plain, and its load of sediment begins to settle out. This sediment raises the level of the riverbed year after year until, in some places, it lies higher than the plain. Occasionally, during the spring surge, the river breaks out of the levees that have been built to control it and rushes across the surrounding densely occupied plain. The spring surge has forced the Huang He to cut many new channels over time **(FIGURE 9.34)**, and it also endows the plain with a new layer of fertile soil as much as a yard (a meter) thick each year. Because it both enriches and destroys, the river is referred to as both the Mother of China and China's Sorrow. The Huang He

is also connected to the watershed of the Chang Jiang to the south by a canal. The Da Yunhe (Grand Canal), considered the oldest and longest canal in the world, was completed about 610 C.E. and originally was built to carry troops and supplies north. Later, in the eighteenth century, it was used to carry taxes (in the form of grains) to the capital, Beijing.

China's Far Northeast

The northeastern corner of China inland of the Korean peninsula was for centuries considered a peripheral region, partly because of its location and partly because of its harsh climate. The winters are long and bitterly cold, the summers short and hot. The growing season is less than 120 days. Nonetheless, China's Far Northeast subregion found an important niche in the country's economy in the post-revolution 1950s, when its rich mineral resources—oil, coal, gas, gold, copper, lead, and zinc—made possible major industrialization, and its fertile state farms began to produce wheat, corn, soybeans, sunflowers (for oil), and beets. The Far Northeast was known for its steel, coal, and petroleum production (it is still the country's leading oil-producing province), and for its "iron man" workers who labored heroically for the revolution in return for meager wages and barracks-like housing. The Heilong Jiang

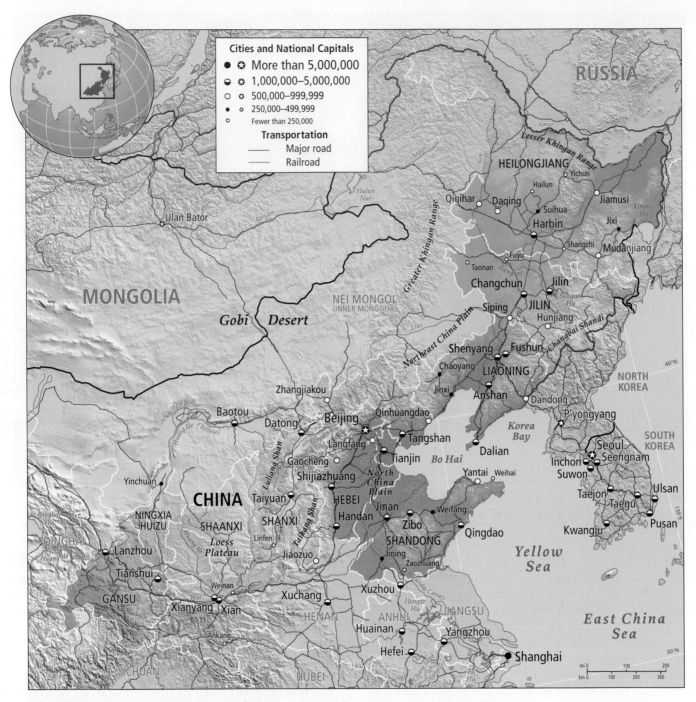

FIGURE 9.32 China's Northeast subregion.

(Amur River), which runs more than 1000 miles (1600 kilometers) along the border between Russia and China, became a conduit for trade with Russia, other parts of China, and Japan. The Far Northeast was slated to make the entire country industrially self-sufficient.

Beginning in the 1980s, however, when Communist Party policy introduced a market economy, the geographic focus of development shifted to Guangdong Province in Southern China, to the international banking center of Hong Kong (then a British colony), to the Chang Jiang (Yangtze) delta and Shanghai, and even to Xinjiang in far western China. By the 1990s, China's Far Northeast was considered a rust belt. Its famous steel production has been surpassed by facilities in provinces along the central coast. Its outdated industrial facilities are being dismantled to make way for new factories that can compete in the global economy. Its millions of laid-off workers are being evicted from slum dwellings that will be replaced with upscale apartments for more select, highly skilled technology workers who are being enticed from elsewhere in the country. Even so, the industrial heartland of the northeast, the Liaoning province, was the only province in China that did not increase its population from 2010 to 2014.

FIGURE 9.33 "Cave dwellings" of the Loess Plateau, China. In parts of Shanxi and Shaanxi provinces, the loess is so thick and firm that energy-efficient, cave-like houses can be dug out from it. First, a courtyard about 10 meters deep is dug out. Then rooms are excavated off the main courtyard.

Much of the revitalization is concentrated in cities on the coast, such as Dalian, and is financed by nearby Korean and Japanese firms that train young people to do telephone sales, information technology (IT), or manufacturing work. The Japanese occupied China's Far Northeast between 1932 and 1945. Their regime is remembered as brutal, but they were the first to introduce industrial development to the region. Now, because the Japanese are bringing investment cash and a chance for economic renewal, they are welcomed.

Beijing and Tianjin

There are 12 million residents in the urban core of Beijing, China's capital city, and 21 million in its metropolitan region. Beijing lies at the northern edge of the North China Plain. It is the administrative headquarters of the People's Republic of China and has several of the nation's most prestigious universities. Because Beijing is not located right at the coast, the nearby port city of Tianjin, an industrial and transportation center, is important for the regional economy. The selection of Beijing as host of the 2008 Summer Olympics established the city—and China—as a global leader in the modern era.

Beijing was once a grand imperial city full of architectural masterpieces—notably the Forbidden City, the former imperial headquarters, which is made up of approximately 1000 buildings. The Forbidden City, as well as the rest of central Beijing, is laid out as a square. The square is a traditional shape in Chinese urban design, as it represents Earth in ancient Chinese philosophy, and Chinese cities are planned as a microcosm of Earth. The rectangular city walls of historic Beijing are now gone and instead replaced by ring roads that follow the same pattern. In fact, much of Beijing's historic character was lost under the Communists, who tore down many monuments. Today, in the context of China's acceptance of capitalism, much of Beijing is again being reshaped, this time as a center for international commerce, with new neighborhoods of high-rise apartments, office towers, and hotels replacing Communist-era apartment blocks and older neighborhoods of nineteenth-century, tile-roofed, extended-family homes.

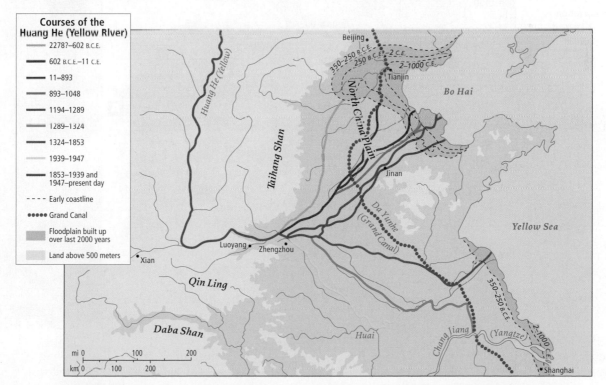

FIGURE 9.34 The Huang He's changing course. The lower course of the Huang He has changed its direction of flow many times over the last several thousand years. In 2000 B.C.E., it flowed north and entered the Bo Hai, south of Beijing. Then it repeatedly shifted to the east, then southeast, like the hand of a clock, until it joined another river, the Huai, and finally cut south all the way to the Chang Jiang delta at Shanghai. Now it once again flows into the Bo Hai.

THINGS TO REMEMBER

• The Loess Plateau has rich, thick soil, but agricultural practices have resulted in soil erosion, gullies within the landscape, and reduced agricultural productivity.

• After flowing through the Loess Plateau, the Huang He remains China's most important physical feature, serving as a major transportation artery and a source of irrigation water—as well as a source of disastrous floods and a conveyor or depositor of sediment.

• China's northeast is home to Beijing, the nation's capital, and the nearby industrial and port city of Tianjin.

CENTRAL CHINA

Central China consists of the upper, middle, and lower portions of the Chang Jiang (Yangtze River) basin **(FIGURE 9.35)**. Like many of Eurasia's rivers, the Chang Jiang starts in the Plateau of Tibet. It descends to skirt the south of Sichuan Province, then flows through the Witch Mountains (Wu Shan) and the Three Gorges

Dam, leaving the upper basin and flowing into the middle basin in China's densely occupied central plain. It then winds through the coastal plain (the lower basin) and enters the Pacific Ocean in a huge delta, the site of the famous trading city of Shanghai.

Sichuan Province

Sichuan, with 111 million people, has some of China's richest resources: fertile soil, a hospitable climate, inventive cultivators, and sufficient natural raw materials to support diversified industries. For years, Sichuan embodied the ideal of complementary agricultural and industrial sectors, but it could not keep pace with the coastal industries that attracted its young people. Now, old industrial cities like Chongqing are changing and are successfully recruiting foreign investors.

The heart of the province is the Sichuan Basin, also called the Red Basin because of its underlying red sandstone. It is a region of hills and plains crossed by many rivers that drain south toward the Chang Jiang. The basin is on a south-facing slope and is surrounded by mountains that are highest in the north and west. The mountains form a barrier against the arctic blasts of winter and trap the moist, warm air that moves up from the

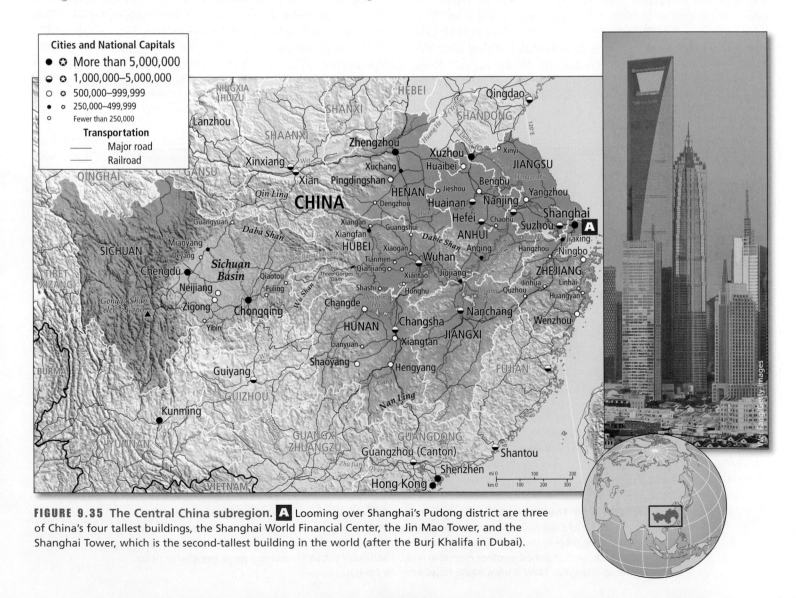

FIGURE 9.35 The Central China subregion. **A** Looming over Shanghai's Pudong district are three of China's four tallest buildings, the Shanghai World Financial Center, the Jin Mao Tower, and the Shanghai Tower, which is the second-tallest building in the world (after the Burj Khalifa in Dubai).

southeast. For these reasons, the Sichuan climate is generally mild and humid, and the basin is so often cloaked in fog or low cloud cover that it is said, "A Sichuan dog will bark at the sun."

Over thousands of years, Sichuan's relatively affluent farmers have cleared the native forests and manicured the landscape, so that only a few patches of old-growth forest are left in the uplands and mountains. The rivers have been channeled into an intricate system of irrigation streams for wet rice cultivation (see page 506) and, more recently, for growing the fine vegetables and fruits that affluent urban dwellers across China want.

The population density of Sichuan is more than 800 people per square mile (300 per square kilometer). Most of the people are farmers, growing rice, wheat, and corn, and as a sideline, raising animals such as silkworms, pigs, and poultry. In the new market economy, it is often these sidelines that earn them the most cash income. In the mountain pastures to the west of the basin, Tibetan herders raise cattle, yaks, sheep, and horses. (Sichuan has the largest population of Tibetans outside Tibet—close to 1 million.)

The two main cities of the Sichuan Basin are Chengdu, the provincial capital (with a population of 14 million in the metropolitan area), and Chongqing (with a population of 29 million in its larger metropolitan area). Chengdu is a transportation hub and is home to light industries, especially food processing and textile and precision instrument manufacturing. Some of its higher-quality products are sold in the global market. Chongqing, with its thousands of iron and steel manufacturers and machine-building industries, is an old industrial city that suffered from decline. Now, Chongqing is targeted for renewal as Sichuan attempts to attract workers back to the interior from the coastal provinces. As a result of the construction of the Three Gorges Dam (see pages 508–510), Chongqing, which lies at the western end of the dam reservoir, is also in line to become a hub for shipping and other new economic activities, especially as it benefits from an abundance of electricity from the Three Gorges Dam.

Sichuan's productivity and dense population has resulted in the degradation of its environment. The basin tends to be subject to temperature **inversion**—when the ground cools down more than the warm, wet air above does. This creates conditions where the cool air becomes "trapped" near the land surface; and along with it, industrial and vehicle emissions build up, leading to intolerable pollution levels. The inversion phenomenon also contributes to poor air quality in other places, such as Beijing.

The Central and Coastal Plains

After the Chang Jiang emerges from the Three Gorges Dam, it traverses an ancient, undulating lake bed that is interrupted in many places by low hills. This former lake bed is the middle basin of the Chang Jiang. It and the river's lower basin are rich agricultural regions dotted with industrial cities.

The middle basin and the lower basin have been filled with **alluvium** (river-borne sediments) carried down from the Plateau of Tibet and the Sichuan Basin. Other rivers entering the middle basin from the north and south also bring in loads of silt, which are added to the main river channel. The Chang Jiang carries a huge amount of sediment—as much as 186 million cubic yards

(142 million cubic meters) per year—past the large industrial city of Wuhan. This is particularly pronounced during the rainy season in the summer and early fall (the area is affected by a moderate monsoon—see page 501). This sediment, which under natural conditions was deposited on the basin floors during annual floods, enriched agricultural production. But because the floods often destroyed people, property, animals, and crops, the Chang Jiang, like the Huang He, now flows between levees. Still, it occasionally breaches the levees and floods both rural and urban areas, as it did most recently in 2010. As the river approaches the East China Sea, it deposits the last of its sediment load in a giant delta, at the outer limits of which lies the trading city of Shanghai.

The climate of the middle and lower basins, though not as pleasant as Sichuan's, is milder than that of the North China Plain. The Qin Ling range and other, lower hills that extend eastward across the northern limits of the basin block some of the cold northern winter winds. The mountains also trap warm, wet southern breezes, so the basins retain significant moisture during most of the year. The growing season is 9 to 10 months long. The natural forest cover has long since been removed to provide agricultural land for the rural population. Summer crops are rice, cotton, corn, and soybeans; winter crops include barley, wheat, rapeseed (canola), sesame seeds, and broad beans. There are more than 485 million people in the middle and lower basins of the Chang Jiang. In the past, many of them were farmers, but every year more people migrate to the many industrialized cities of the region. Shanghai overshadows all of these cities in importance.

For many centuries, Shanghai's location facing the East China Sea, with the whole of Central China at its back, positioned the city well to participate in whatever international trade was allowed. When the British forced trade on China in the 1800s, Shanghai became their base of operations. Before the Communist Revolution, Shanghai was among the most cosmopolitan cities on Earth, home to well-educated citizens, wealthy traders, and a goodly number of underworld figures as well. After the revolution, the Communists designated the city as the nexus of capitalist corruption, and it fell into disgrace.

When China began to open up to world trade in the 1980s, Shanghai entrepreneurs were well situated to take advantage of government incentives, such as reduced taxes, assistance in preparing building sites, and permission to take profits out of the country. These incentives helped motivate developers to build factories, workers' apartment blocks, international banks, luxury apartment houses, shopping malls, and elegant, extraordinarily tall skyscrapers (see Figure 9.35A). Today, some of the Overseas Chinese (see page 539; see also Chapter 10)—particularly those who fled the repression of the Communist Party to places such as Taiwan, Singapore, and Malaysia—have returned to invest in everything from television stations and discotheques to factories, shopping centers, and entertainment parks. Europeans and North and South Americans are attracted to Shanghai, too, hoping to find joint ventures with Chinese partners so that they can tap into the huge pool of consumers emerging in China.

inversion the unusual condition when warm air overlies cool air, impeding normal circulation

alluvium river-borne sediment

THINGS TO REMEMBER

• Sichuan is a densely populated basin in central China; rich in resources, it is a region of wet rice cultivation, fine fruits and vegetables, and silkworm production.

• The Chang Jiang brings sediment downstream, which makes its river valley agriculturally productive but also prone to flooding.

• The Chang Jiang empties into the ocean near Shanghai, China's largest city and its economic engine.

SOUTHERN CHINA

Southern China has two distinct sections **(FIGURE 9.36)**. The first is made up of the mountainous and mostly rural provinces of Yunnan and Guizhou to the southwest. The second includes the Guangxi Zhuangzu Autonomous Region and Guangdong and Fujian provinces on the southeastern coast, where the booming cities of China's evolving economy are located.

The Yunnan–Guizhou Plateau

The provinces of Yunnan and Guizhou share a plateau noted for its natural beauty and mild climate and for being the home of numerous indigenous groups that are culturally distinct from the Han Chinese. Some of them also practice Buddhism, which is a prominent religion in nearby Southeast Asia (see Figure 9.36A). The plateau is a rough land of deeply folded mountains that trend north-south, through which the Nu (Salween) and Mekong rivers flow. The heavily forested valleys are deeper than they are wide. Although people can call to one another across the valleys, it may take more than a day's difficult travel to reach the other side. The north-south orientation naturally connects the region more to Southeast Asia than to the Chinese coast. In some places, rope and bamboo bridges have been slung across the chasms. The landforms here are unstable, and earthquakes cause heavy damage to the stairstep terraces of rice paddies.

The valuable natural resources of the Yunnan–Guizhou Plateau are primarily biotic. Yunnan Province, with its fertile soil, is an important source of produce and meat for the bustling cities on the southeastern coast **(FIGURE 9.37)**. Yunnan is called "the national botanical garden" because many of China's plant species and one-third of its more than 400 bird species are native to the area's exotic landscapes. Guilin is noted for its fantastic formations of eroded limestone, known as *karst* **(FIGURE 9.38)**. Until the 1970s, the tropical forests of far southern Yunnan (close to Burma, Thailand, and Laos) harbored elephants, bears, porcupines,

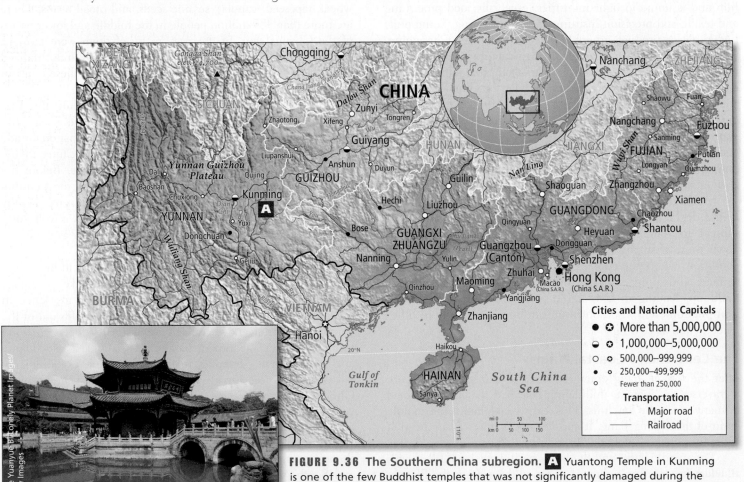

FIGURE 9.36 The Southern China subregion. **A** Yuantong Temple in Kunming is one of the few Buddhist temples that was not significantly damaged during the Cultural Revolution. Today it is a major destination for pilgrims and tourists. In surveys, between 11 and 16 percent of Chinese profess to be Buddhists, which makes Buddhism the largest organized religion in China (most Chinese, however, do not have a religious affiliation).

FIGURE 9.37 A restaurant in Yunnan. The vegetables that will make up much of the meal are displayed prominently at the entrance of a restaurant in Dali, Yunnan. A mild climate and fertile soils make Yunnan an important source of vegetables for much of China.

FIGURE 9.38 The fantastic karst landscapes of Yunnan's Stone Forest. The region is composed of karst (limestone) formations that have been eroded by water and wind to form jagged peaks jutting out from the surrounding land. This phenomenon is found in its most extreme form near Kunming in the Stone Forest (called the *Shilin* in Chinese), which has become a world-renowned tourist attraction.

gibbons, and boa constrictors. However, the flora, fauna, and geologic treasures of these provinces have been severely threatened by human disturbance of the environment. The region was also required to supply wood for China's industrialization. The historic range of the most famous forest dweller, the giant panda, once extended over much of southern China, but the animal is now limited to 40 panda preserves, mostly in the Sichuan province.

The mountains of Yunnan also lie at the heart of a booming heroin trade that flows from major producers in remote areas of Burma and Laos across the border into China. Kunming, the capital of Yunnan Province, is a key location in this heroin trade. From there, heroin is transported to China's northern and eastern coastal cities, where there is a market for the drug, and even onward to the global market.

The Southeastern Coast

The southeastern coastal zone of China has been a window to the outside world for centuries. During the Tang Dynasty (618–907 C.E.), its ports began launching ships that journeyed as far as the Persian Gulf and the eastern coasts of Africa. Around the same time, Arab traders began to visit this part of China. By the fifteenth century, some of the first Europeans in the region described a string of flourishing trading towns all along the coast. The overwhelming majority of Overseas Chinese have their roots along China's southeastern coast. The people of this area, as well as many Overseas Chinese, speak Cantonese, which is distinct from the Mandarin Chinese spoken in most other parts of China. In the 1980s, the central government decided to take advantage of this tradition of outside contact by designating several of the old coastal fishing towns as SEZs, with special rights to conduct business with the outside world and to attract foreign investors. The quick result was a chain of cities attracting millions of migrants to work in light industries that now supply the world with a variety of consumer products.

Much of this development is along the principal river of Southern China, the Zhu Jiang (Pearl River), which is joined by several tributaries to form one large delta south of the city of Guangzhou (Canton) in Guangdong Province. The lowlands along the rivers and delta have a subtropical climate and a perpetual growing season. The rich delta sediment and the interior hinterlands are used to cultivate sugarcane, tea, fruit, vegetables, herbs, timber, and mulberry trees for sericulture (the raising of silkworms), all of which are processed and packaged in the various SEZs and exported to global markets.

The city that epitomizes SEZ development more than any other is Shenzhen. When it was declared China's first SEZ in 1979, Shenzhen was little more than a fishing village situated next to Hong Kong. Surrounded by an 85-mile barbed-wire fence to control the movement of workers, modern Shenzhen emerged as a manufacturing hub that has grown to a metropolis of more than 10 million people. When the electronics companies hire temporary laborers to expand production for the Christmas gift season, millions more flow into the city. Shenzhen is still an urban landscape of rapid change. During the recession, growth ground to a halt, but like most boom-and-bust cities, it is again experiencing renewed growth. However, reduced competitiveness in low-wage manufacturing due to increasing real estate and labor costs has initiated a shift towards high-tech manufacturing and development. Shenzhen hopes to be a Chinese Silicon Valley.

Hong Kong

Densely populated Hong Kong is the economic heart of the southeastern coast. For 99 years, Hong Kong was controlled by Britain and served as a hub of the Asian colonial economy. In 1997, Britain's lease on Hong Kong ran out and the city reverted back to Chinese control. However, Hong Kong's economic importance has not diminished and the city, in itself, has the world's eleventh-largest trading economy. It used to have a large manufacturing base but much of that has migrated to mainland China; Hong Kong is instead a major financial center. The Hong Kong area also has the largest container port facilities in the world. Technically the ports of Shanghai and Singapore are larger, but the port in Hong Kong proper, plus two newer and more spacious ports in nearby Shenzhen and Guangzhou, together carry far more ocean-going goods than does any other urban area in the world. Basically, global investment is funneled into southeastern China via Hong Kong banks, and manufactured goods flow out of the region via its world-class ports. Given the very obvious success of many cities along China's southeastern coast, from Macao to Shanghai, Hong Kong will probably continue in its role as a financial hub in the development of this very rapidly growing region. As a de facto city-state, Hong Kong's position as a business center resembles that of Singapore in Southeast Asia.

Macao

Macao (also spelled Macau), built on a series of islands across the Zhu Jiang estuary from Hong Kong, is the oldest permanent European settlement in East Asia. Portuguese traders arrived in about 1516, and by 1557 had established a raucous colonial trading and gambling center. Macao remained a Portuguese colony until the Chinese government regained control of it in 1999, just like it did with Hong Kong in 1997. Today, this small territory has a population of about 600,000—primarily workers who earn a living in the tourism industry. Macao remains China's only gambling center; some of its casinos are managed by Las Vegas firms. Macao's GDP per capita is similar to Hong Kong's and many times that of the mainland.

THINGS TO REMEMBER

• The southwestern inland part of Southern China is home to several indigenous populations; the heavily forested mountainous terrain is also known for animals and plant diversity, as well as for its unique landforms.

• The southeastern coastal zone, which includes Hong Kong and Macao, is a major manufacturing, export, financial, and gambling center, and a window to the world.

FAR NORTH AND WEST CHINA

The Far North and West subregion of China **(FIGURE 9.39)** once occupied a central role in the global economic system. Traders carried Chinese and Central Asian products such as silks, rugs, spices and herbs, and ceramics over the Silk Road to Europe, where they exchanged them for gold and silver. Today, this route is cumbersome for long-term trade, but the subregion is recovering some of its ancient economic vitality as it connects the

yurts (or **gers**) round, heavy, felt tents stretched over collapsible willow lattice frames used by nomadic herders in northwestern China, Mongolia, and Central Asia

qanats underground tunnels, built by ancient cultures and still used today, that carry groundwater for irrigation in dry regions

consumers of eastern China with the resource economies of central Asia.

Despite its trading past, this large interior zone has historically been considered backward by the rulers of eastern China because of its dry, cold climate; its vast grasslands; its history of nomadic herding; and—in the lands of the Uygurs and Kazakhs—its persistent adherence to Islam. Settlements are widely dispersed, and most agriculture requires irrigation. Three of the subregion's four political divisions have been designated autonomous regions (not provinces) because of the high percentage of ethnic minority populations there (Nei Mongol, also called Inner Mongolia, is perhaps the best known of these autonomous regions), but their people do not enjoy real autonomy. The central government in Beijing retains control over political and economic policies because it sees this subregion as crucial to China's future: it has energy and other resources for industrial development, it is close to the emerging oil-rich economies of Central Asia and Russia, and it affords a place to resettle some of China's surplus population.

Xinjiang

The Xinjiang Uygur Autonomous Region in northwestern China is historically and physically part of Central Asia. Yet it is the largest of China's political divisions, accounting for one-sixth of China's territory.

The Uygur minority and its conflict with the central government in Beijing and resettled Han Chinese are discussed earlier. Xinjiang, which has 20 million inhabitants, consists of Central Asian ethnic groups, most of whom are Muslims, and recent Han immigrants (see Figure 9.39A). The peoples native to the subregion include Turkic-speaking Uygurs; there are also Mongols, Persian-speaking Tajiks, Kazakhs **(FIGURE 9.40)**, Kirghiz, Manchu-speaking Xibe, and Hui. They once made their living as nomadic herders and animal traders, moving with their herds and living in **yurts** (or **gers,** in Mongolia). These cozy houses—round, heavy, felt tents stretched over collapsible willow frames—can be folded and carried on horseback, in horse-drawn carts, or today, in trucks (see Figure 9.43A). In the last few decades, many nomadic people have urbanized or taken jobs in the emerging oil industry and now live in apartments provided for laborers.

Xinjiang consists of two dry basins: the Tarim Basin, occupied by the Taklimakan Desert, and the smaller Junggar Basin to the northeast. Both are virtually surrounded by 13,000-foot-high (4000-meter-high) mountains topped with snow and glaciers. In this distant corner of China, far from the world's oceans, rainfall is exceedingly sparse. Snow and glacial meltwater from the high mountain peaks are important sources of moisture. Much of the meltwater makes its way to underground rivers, where it is protected from the high rates of evaporation on the surface. Long ago, people built tunnels called **qanats** deep below the surface to carry groundwater dozens of miles to areas where it was needed. Qanats have made productive some of the hottest and driest places on Earth. The Turfan Depression, situated between the Junggar and the Tarim basins, is a case in point. In this basin that descends 500 feet (150 meters) below sea level, temperatures often reach 104°F (40°C), and evaporation rates are extremely high. But because the qanats bring irrigation water, this area produces some of China's best foods: melons, grapes, apples, and pears. The produce is sold to urban populations in eastern China.

In Xinjiang today, the local and the global, the very traditional and the very modern, confront each other daily. Herdspeople still living in yurts and gers dwell under high-voltage electric wires that supply new oil rigs. Tajik women weave traditional rugs that are sold to merchants who fly from faraway developed countries to the ancient trading city of Kashi (Kashgar) for the Sunday market (see Figure 9.28). With the breakup of the Soviet Union, citizens of the new republics of Central Asia have been eager to revive their trading heritage, and they have oil and gas to sell. China is welcoming them and attracting outside investors from Europe and the Americas by establishing free trade zones in cities such as Kashi and Urumqi. As discussed above, the Uygur and other ethnic leaders of Xinjiang are wary of Beijing, fearing that the

central government's singular intent is to exploit Xinjiang's oil and gas resources and to appropriate land for the resettling of eastern China's excess population.

The Plateau of Tibet

Situated in far western China, the Plateau of Tibet is the traditional home of the Tibetan people. Administratively, it includes the Xizang Autonomous Region and Qinghai Province. Tibet (Xizang) and Qinghai lie an average of 13,000 feet (4000 meters) and 10,000 feet (3000 meters) above sea level, respectively. They are surrounded by mountains that soar thousands of feet higher. They have cold, dry climates (late June can feel like March does on the American Great Plains) because of their high elevation and because the Himalayas to the south block warm, wet air from moving in from the Indian Ocean. Across the plateau, but especially along the northern foothills of the Himalayas, snowmelt and

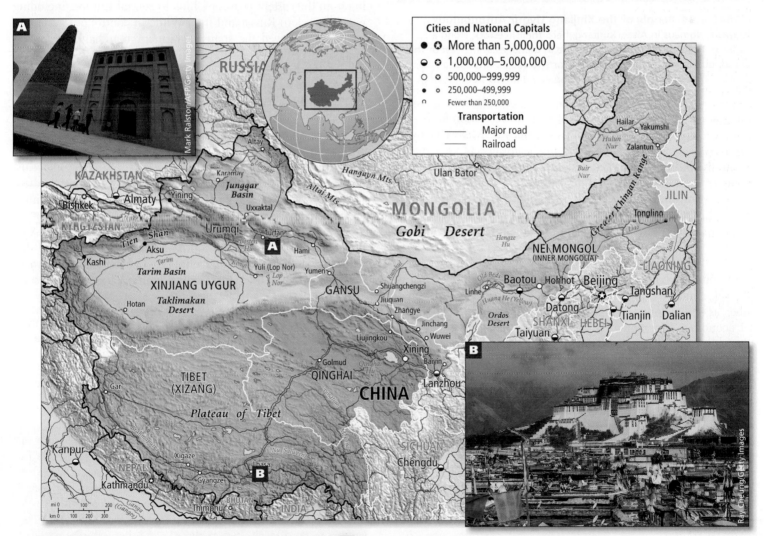

FIGURE 9.39 China's Far West and North subregion. A The Emin Minaret and Mosque in Turfan, Xinjian, China. Built in 1778 to honor a Uygur general who had suppressed a local rebellion against the emperor of the Chinese Qing dynasty, the structure is still a place of worship. B The Potala Palace rises over Lhasa, Tibet. Built in the seventeenth century as both a monastery and seat of government for much of Tibet, it was home to the last ten Dalai Lamas, a lineage of Buddhist teachers that goes back to 1391. Its construction marked their rise to power as heads of government. Since China's invasion of Tibet in 1959, and the subsequent exile of the fourteenth Dalai Lama to India, the Potala Palace has been made into a museum.

STR/AFP/Getty Images

FIGURE 9.40 People of the Xinjiang Uygur Autonomous Region. Nomads in Altay, Xinjiang, herd their livestock along a caravan. Nomadism is increasingly rare in China, as in the rest of the world, due in part to government efforts to secure international and provincial borders, which prevent people and goods from moving freely.

rainfall are sufficient to support a short growing season for barley and vegetables such as peas and broad beans. Meltwater from Himalayan snow and glaciers forms the headwaters of some major rivers: the Indus, Ganga (Ganges), and Brahmaputra begin in the western Himalayas, and the Nu (Salween), Irrawaddy, Mekong, Chang Jiang, and Huang He all begin along the eastern reaches of the Plateau of Tibet. Many of the larger settlements in Tibet are located along the valley of the Brahmaputra (called *Yarlung Zangbo* in Chinese) in the southern part of the province.

Traditionally, the economies of Tibet and Qinghai have been based on the raising of grazing animals. The main draft animal is the yak, which also provides meat, milk, butter, cheese, hides, and fiber, as well as dung and butterfat for fuel and light. Other animals of economic importance are sheep, horses, donkeys, cattle, and dogs. Animal husbandry on the sparse grasses of the plateau has required a mobile way of life so that the animals can be taken to the best available grasses at different times of the year. For several decades, though, the Chinese government has pressured Tibetan herds people to settle in permanent locations so that their wealth can be taxed, their children schooled, their sick cared for, and their dissidents curtailed. Still, throughout the plateau, many native (non-Han) peoples continue to live mobile yet solitary lifestyles, as they have for centuries, occasionally adopting some aspects of modern life and adapting to its restrictions. The political history of the Tibetan minority in China and suppression of Tibetan culture (**FIGURE 9.41**; see also Figure 9.39B) are discussed on page 538.

THINGS TO REMEMBER

• China's Far North and West subregion has historically been considered backward by the rulers of eastern China because of its dry, cold climate; its vast grasslands; its long history of nomadic herding; and its persistent adherence to Islam and Tibetan Buddhism.

• Three of the subregion's four political divisions have been designated autonomous regions because of the high percentage of ethnic minority populations there.

• The central government in Beijing retains control over political and economic policies because the subregion has energy and other resources for industrial development; it is close to the emerging oil-rich economies of Central Asia and Russia; and it affords a place to resettle some of China's surplus population.

MONGOLIA

The present country of Mongolia is the north-central part of what was once a much larger cultural area in eastern Central Asia known by the same name (**FIGURE 9.42**). Between 1206 and 1370, the Mongols, under Genghis Khan and his grandson Kublai Khan, created the largest-ever land-based empire, stretching from the eastern coast of China to central Europe, including parts of modern Russia and Iran. While in control of China, the Mongols improved the status of farmers, merchants, scientists, and engineers; fostered international trade; and broke the control of traditional elites by abolishing their automatic access to privilege. They refined the manufacture of textiles, jewelry, and blue and white porcelain, and trade in these wares flourished along the Silk Road. Although the Mongols brought many important reforms to China, they also practiced authoritarian rule and discrimination against ethnic Chinese. Deposed by the Chinese in 1366, the Mongols retreated to the north; over the next 300 years, their control of territory in Central Asia and Europe dwindled. Eventually, the southern part of Mongolia (somewhat confusingly called Inner Mongolia) became part of China.

The Mongolian Plateau lies in the heart of Central Asia, directly north of China and south of Siberia. It is high and dry, with an extreme continental climate. The temperature ranges from very cold in the winter to moderate in the summer, such

FIGURE 9.41 Tibet's religious traditions. A giant painting of the Buddha is unveiled by the monks of Drepung Monastery, Tibet's largest monastery, at the Yogurt Festival held there each August for the past 300 years. Located near Lhasa, Tibet, the monastery is one of three celebrated "university" monasteries for Tibetan Buddhists, each of which has a counterpart monastery in exile in India that houses monks who left Tibet when the Chinese invaded in 1959.

Hong Wu/Getty Images

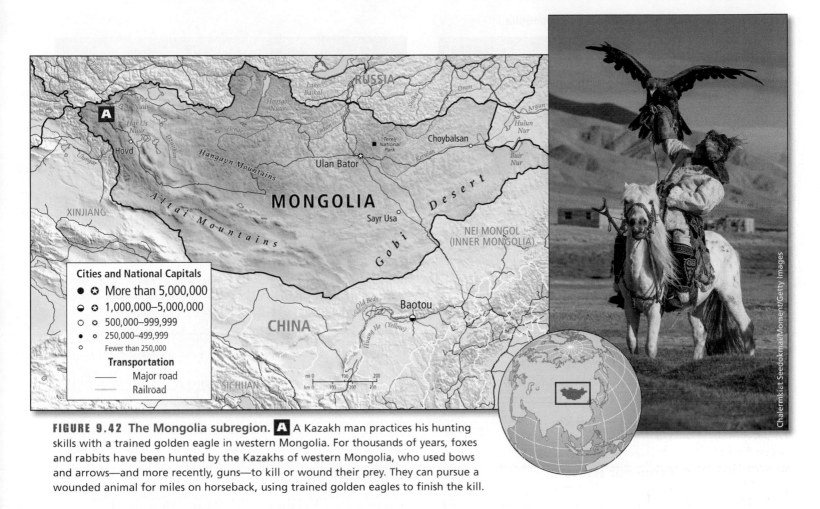

FIGURE 9.42 The Mongolia subregion. **A** A Kazakh man practices his hunting skills with a trained golden eagle in western Mongolia. For thousands of years, foxes and rabbits have been hunted by the Kazakhs of western Mongolia, who used bows and arrows—and more recently, guns—to kill or wound their prey. They can pursue a wounded animal for miles on horseback, using trained golden eagles to finish the kill.

as in the capital of Ulan Bator, where a summer day is often no warmer than 70°F (21°C). The physical geography of Mongolia can be broken down into four major zones. In the far southeast and extending across the border into China is the Gobi Desert—actually a very dry grassland that grades into true desert in particularly dry years or where it is overgrazed. It is likely to be especially prone to increased aridity with global climate change. To the west and northwest of the Gobi is a huge, rolling, somewhat moister grassland. The remaining two zones are Mongolia's two primary mountain ranges: the forested Hangayn Nuruu, in north-central Mongolia, and the grass- and shrub-covered Altai Mountains, which sweep around to the west and south and into the Gobi Desert (see the Figure 9.42 map).

With just 3 million people, Mongolia occupies a territory so large that its average population density is one of the lowest on Earth, at 4 people per square mile (1.5 per square kilometer). For thousands of years, the economy has been based on the nomadic herding of sheep, goats, camels, horses, and yaks. Nomadic pastoralism, one of the most complex agricultural traditions in the world, dates at least as far back as the domestication of plants (8000 to 10,000 years). Nomadic pastoralists must understand the intricate biological requirements of the animals they breed, and they must also know the ecology and seasonal cycles of the landscapes they traverse with their herds.

Today, only 32 percent of Mongolians are still engaged in rural activities, often related to this traditional lifestyle (**FIGURE 9.43A**;

see also Figure 9.42A). The other 68 percent now live in cities, primarily the capital, Ulan Bator (see Figure 9.43B). Urban workers are employed in a wide range of services and in industries related to the processing of minerals from Mongolia's mines and animal products such as hides, fur, and wool.

The Communist Era in Mongolia

Influenced by the Russian Revolution of 1917, and after considerable internal turmoil, Mongolia first declared independence from China in 1921; three years later it became a communist republic under the Soviet sphere of influence. From 1924 to 1989, Mongolia sought guidance from Soviet advisers and technicians. They helped set up a system of central planning that reorganized the nomadic pastoral economy according to socialist policies, but did so without drastically disrupting traditional lifeways. Rural households of extended families became economic collectives. Some households continued to herd and breed animals, others engaged in sedentary livestock production, and still others farmed crops. During this time, Mongolia also began to industrialize and urbanize. By 1985, nomadic herding and forestry were declining, accounting for less than 18 percent of the Mongolian GDP and less than 30 percent of the labor force, whereas employment in mining, industry, and services accounted for about half of the GDP and a third of the labor force.

During this period, the Mongolian economy was supported by the Soviets, who subsidized a wide array of social services.

FIGURE 9.43 Tradition and modernity in Mongolia.

(A) A family in western Mongolia moves their ger, a felt-covered, wood lattice–framed structure traditionally used by the nomadic peoples of Mongolia and Central Asia as their main dwelling. In the past, gers were moved by horses; trucks are now widely used for the seasonal moves that seminomadic herders frequently make to provide adequate pasture for their sheep.

(B) The outskirts of Ulan Bator, Mongolia, where the gers of recent migrants are overshadowed by large apartment blocks. Sixty-eight percent of Mongolians now live in cities.

By 1989, there was nearly universal education through middle school, and adult literacy had reached 93 percent. Health care was available to all, and life expectancy had risen dramatically. The previously high infant mortality rates dropped, and then fertility decreased.

Post-Communist Economic Issues

The collapse of the Soviet Union in 1991, coupled with several natural disasters and lower world prices for the minerals that Mongolia produces, led to immediate budget problems and declines in economic growth, human well-being, and social order. The Soviet advisers left without training Mongolian replacements to run the economy. Layoffs and factory closings caused a severe decline in living standards. Some people who were working in urban jobs were forced to return to the countryside. By 2003, about one-third of the population lived in poverty.

Some signs of a better future, however, were emerging. Mongolia's new membership in the WTO promised to bring increased international attention and development assistance, and private foreign investors took an interest in the country, especially in its minerals. China, with its booming economy just to the south, has been particularly interested in Mongolia's minerals, especially copper, coal, and gold. Compared to nearby post-Soviet states and China, Mongolia has also had a positive political development. It is ranked as a flawed democracy, but its elections and media are predominantly free and fair.

This period of increasing affluence was stopped dead in 2008 by the global recession, which was much more severe in Mongolia than in any of the other East Asian countries, primarily because Mongolia depended on the export of raw materials—fleece from sheep and goats, and mined minerals—for which market prices dropped precipitously. Prices have since turned upward again and because large-scale multinational mining companies are joining the fray, the environmental damage is escalating, while at the same time bringing renewed, post-recession economic growth. But Mongolia, as a landlocked country, is heavily dependent on raw material exports to China and energy imports from Russia, which means that recent economic shocks may be repeated in the future. Mongolia appears to be a boom-and-bust country that is at risk of developing a *resource curse*, which is when a country focuses its development strategy on resource extraction at the expense of other economic sectors. Such countries may then become overly dependent on commodity prices, suffer from economic volatility, and government corruption can be endemic as people in power benefit from the sudden influx of cash.

ON THE BRIGHT SIDE

In the last decade, Mongolia invested heavily in IT. Today, the national telecom company is upgrading to a 4G network. Most people, including nomads, have at least one cell phone, far out in the steppe. In just a 3-year period (from 2012 to 2015) the number of Internet users in Mongolia more than doubled. *[Source: Laura Pappano, "The Boy Genius of Ulan Bator," New York Times Magazine, September 15, 2013, p. 52.]* ■

Gender Roles in Mongolia

Traditionally, Mongolian women enjoyed a status approximately equal to that of men, but outside forces, such as Tibetan Buddhism in the sixteenth century and Chinese rule in the seventeenth and eighteenth centuries, eroded their status. The Communist era restored, or even enhanced, egalitarian gender attitudes. The daily work of women herders was valued by the government, and like their husbands, women were eligible for old-age pensions. A woman could leave an unsuccessful marriage because the government provided support for her and her children. A few women became teachers, judges, and party officials, and soon women were represented in institutions of higher education.

In today's modern nomadic pastoral economy, as in the past, women usually choose which stock to breed, tend the mother animals and their young, preserve meat and milk, and produce many essentials from animal hides and hair. Women provide the materials for the construction of gers and also dye and weave beautiful furnishings for the interiors. Both boys and girls are taught to ride horses at an early age, and both help with the herding. As they mature, women become more responsible than men for the routines of daily life in the home. Men engage primarily in pasturing the herds and arranging the marketing of the animals and animal products. Men also perform other tasks—such as the occasional planting of barley and wheat—that are carried out beyond the home compound.

When the Soviets left and the economy collapsed, women lost some of the status they had enjoyed, but they are now regaining legal protections and access to education, jobs, health care, housing, and credit. Demographically, Mongolian women have a life expectancy that is 8.5 years longer than men, they participate in the workforce to an extent similar to that of women in the United States, and they have surpassed Mongolian men in educational attainment. As in so many other countries that are moving from communism to a market economy, the informal economy—in which women are particularly active—has been an essential component of the Mongolian economic transition.

THINGS TO REMEMBER

- Sustainable management of Mongolia's sparse but valuable natural resources, such as minerals and grassland used for nomadic pastoralism, has proven to be difficult.

- Mongolia continues to move toward a better-functioning market economy, despite the 2007–2009 global recession and a recent economic slowdown in China.

- Mongolia has a better-than-average record on gender equality both for the region and globally

KOREA, NORTH AND SOUTH

Some of the most enduring international tensions in East Asia have been focused on the Korean Peninsula. As we have seen (page 518), after centuries of unity under one government, the peninsula was divided in 1953. The two resulting countries are dramatically different. Communist, poor, and inward-looking North Korea does not participate in the global economy, while more cosmopolitan, populous, newly democratic, and affluent South Korea has pursued a model of state-aided capitalist development.

Physically, Korea juts out from the Asian continent like a downturned thumb **(FIGURE 9.44)**. Two rivers, the Yalu (Amnok in Korean) and the Tumen, separate the peninsula from the Chinese mainland and a small area of Russian Siberia. Low-lying mountains cover much of North Korea and stretch along the eastern side of the peninsula into South Korea, covering nearly 70 percent of the peninsula. There is little level land for settlement in this mountainous zone. The rugged terrain disrupts ground communications from valley to valley. Along the western side of Korea, floodplains slope toward the Yellow Sea, and most people live on these western slopes and plains. Although the peninsula is surrounded by water on three sides, its climate is essentially continental because it lies so close to the huge Asian landmass. The same cyclical monsoons that "inhale" and "exhale" over the Asian continent (see page 501) bring hot, wet summers and cold, dry winters.

The Korean Peninsula was a unified country as early as 668 c.e. Scholars believe that present-day Koreans are descended primarily from people who migrated from the Altai Mountains in western Mongolia, because the Korean language appears to be most closely related to languages from that region. Other groups—Chinese, Manchurians, Japanese, and Mongols—invaded the peninsula, sometimes as settlers, other times as conquerors. The Buddhist and Chinese Confucian values brought by some of these groups have influenced Korea's educational, political, and legal systems. Korea is noted for its early advances in mathematics, medicine, and printing.

Contrasting Political Systems

Although an armistice ended the Korean War in 1953, North Korea and South Korea each assumed a strong nationalist stance, which has resulted in more than 50 years of hostile competition between the two. Both governments adopted the Korean concept of *juche*, which means "self-reliance" or "the right to govern yourself in your own way." In North Korea, juche was interpreted as unquestioned loyalty to the Great Leader, Kim Il Sung, and later to his son, Kim Jong Il, and since December 2011, his grandson Kim Jong Un, who is now president **(FIGURE 9.45A)**. There is little pretence of political freedom, and the government remains in essence a military dictatorship. Extraordinarily large amounts of money fund national defense, which is a very powerful institution in the country. North Korea limits its trade and other involvement with its continental neighbors, China and Russia. Its restrictive internal and external economic policies make it one of the poorest nations in the world, with a GDP per capita that is estimated to be as low as many African countries (see Figure 9.45B). Despite its apparent desire for isolation, North Korea has occasionally threatened South Korea and nearby Japan.

North Korea provoked global concern by producing enriched plutonium, forcing UN monitors to leave the country, and then withdrawing from the Treaty on the Non-Proliferation of Nuclear Weapons. Intelligence assessments confirm that North Korea tested its nuclear device in 2006. Other nuclear tests have followed, often as ploys to negotiate aid from the UN and the West. North Korea says it must prepare for an attack by the United States or Japan. It is generally agreed that the presence of nuclear arms in North Korea is cause for concern worldwide. The new leader, Kim Jong Un, has pursued a militaristic path similar to those of his predecessors.

The nuclear issue has diverted attention from massive mismanagement of the country's resources, especially food. Such mismanagement has resulted in very serious human rights abuses in North Korea. There are vast prisons for those who have deviated from the Communist Party line, engaged in entrepreneurialism (trading), or even just wandered the countryside looking for food. Eyewitness escapees have told the outside world of beatings, torture, and executions.

In South Korea, by contrast, juche has come to mean vigorous individualism, coupled with pride in and loyalty to one's own people and nation. Social criticism and even aggressive protests by labor unions and other social movements are allowed, with the understanding that one's ultimate loyalty is still to South Korea. Shortly after World War II, South Korea allied itself politically with the United States, Japan, and Europe, and sought

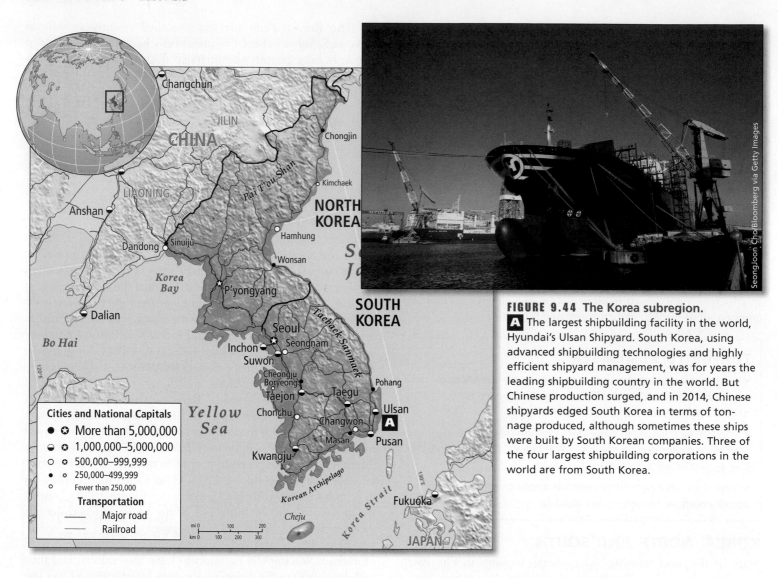

FIGURE 9.44 The Korea subregion.
A The largest shipbuilding facility in the world, Hyundai's Ulsan Shipyard. South Korea, using advanced shipbuilding technologies and highly efficient shipyard management, was for years the leading shipbuilding country in the world. But Chinese production surged, and in 2014, Chinese shipyards edged South Korea in terms of tonnage produced, although sometimes these ships were built by South Korean companies. Three of the four largest shipbuilding corporations in the world are from South Korea.

economic growth through foreign aid and capitalist development. But despite its economic success, South Korea was governed by a series of military dictatorships until the late 1980s, when democratic elections were held. South Korea even elected its first female president in 2013.

ON THE BRIGHT SIDE: Trending Toward Multiculturalism in South Korea

South Korea's economic success is also transforming the previously homogenous society, albeit slowly. Immigrant male workers and young immigrant women who marry Korean men—both groups typically originating in poorer Asian countries—are gradually altering the ethnic composition of South Korea. For example, every tenth marriage now involves a foreign spouse. This recent trend toward multiculturalism means that the country is inevitably moving toward a more inclusive notion of what it means to be Korean. ■

chaebol huge corporate conglomerates in South Korea that receive government support and protection

Contrasting Economies

The two Koreas differ dramatically in their economic resources. North Korea (with a population of 25 million) has better physical resources for industrial development, including forests and deposits of coal and iron ore. The many rivers descending from its mountains have considerable potential to generate hydroelectric power. South Korea (with a population of 51 million) has better resources for agriculture because its flatter terrain and slightly warmer climate make it possible to grow two crops of rice and millet per year. In recent decades, South Korea has surpassed North Korea in agricultural production many times over, and its electronic, automotive, chemical, and shipbuilding industries successfully compete with those in the Americas, Europe, and Japan (see Figure 9.44A). North Korea's industry is barely developed.

A major reason for South Korea's economic success has been the formation of huge corporations known as **chaebol.** These conglomerates—companies that are involved in a wide variety of economic sectors—include such internationally known companies as Samsung and Hyundai. South Korea's government has assisted the chaebol by making credit easily available to them and by helping them purchase foreign patents so that Koreans can focus on product quality and marketing rather than on inventing.

FIGURE 9.45 Life in North Korea.

(A) Hailed as the largest show of its kind in the world, the Arirang Festival celebrates the nation and its leaders. The 2-month-long celebration is held in P'yongyang in the largest stadium in the world. More than 100,000 people perform, and the festival features elaborately choreographed dances, gymnastics, and music. Among the highlights of the festival are images created in the bleachers by tens of thousands of schoolchildren who hold differently colored cards that they switch in rapid succession. Due to lack of funds, the North Korean government recently cancelled the latest installment of the festival.

(B) A man drives an ox cart through a street in P'yongyang that has a picture of former president Kim Il Sung hung in the background. The vast majority of North Koreans who labor in factories and on farms are plagued by outdated and often broken-down equipment. Virtually nobody owns a car, which is why the oversized street in the picture is devoid of traffic.

Nevertheless, the chaebol have come under increasing criticism in recent years because their close connections with the government have led to corruption and mismanagement.

Despite these problems, the South Korean economic system has worked well enough so that South Koreans, who were extremely poor at the end of the Korean War, can now afford the products they formerly only exported. For example, the GDP per capita is now close to the European Union average. The use of advanced telecommunication technologies and modern urban landscapes, especially in the capital Seoul, testify to South Korea's success in the global economy.

North Koreans have benefited from some communist social policies, particularly access to basic health care and education. In general, however, as the economy has deteriorated, these services have also declined. North Korea's industries are inefficient and its workers poorly motivated. Furthermore, through the demise of the Soviet Union, the country lost markets and access to raw materials, fuel, and technical training. Poor harvests on collective farms and recurrent cycles of floods and droughts have brought recurrent food shortages and even the extensive famines that happened from 1995 through 2001. Although the government has been unwilling to release information, as many as 2 million people may have died of hunger over that 6-year period. Today the situation is not quite as dire, but a 2013 UN report concluded that 84 percent of North Koreans are food insecure. The infant mortality rate in 2014 was 25 per 1000 in North Korea, compared to just 2 per 1000 in South Korea.

THINGS TO REMEMBER

• South Korea's model of state-aided capitalist development has made it a significant player in the global economy, both as a producer of a variety of goods and the headquarters of many important corporations.

• Communist and inward-looking North Korea is poverty-ridden, but its development of nuclear arms is a threat to South Korea and other countries in the region.

JAPAN

Japan consists of a chain of four main islands and hundreds of smaller ones **(FIGURE 9.46)**. Prone to severe earthquakes, tsunamis, and typhoons, and so mountainous that only 18 percent of its land can be cultivated, the Japanese archipelago might seem an unlikely place to find 127 million people living in affluent comfort. Yet its citizens have learned to cope with these limitations and have made the most of crowded conditions in cities and countryside alike.

A Wide Geographic Spectrum

Modern Japanese populations are descended from migrants from the Asian mainland, the Korean Peninsula, and the Pacific Islands. Ideas and material culture imported from these places include Buddhism, Confucian bureaucratic organization, architecture, the Chinese system of writing, the arts, and agricultural technology.

FIGURE 9.46 The Japan subregion.

A During rush hour in the Tokyo railway, "passenger arrangement staff" squeeze as many commuters onto a train as possible. They also make sure passengers can get off crowded trains safely, and signal train conductors when it is safe for the trains to depart.

B Dancers at a *yosakoi* festival in Sapporo, Hokkaido. Yosakoi is a modern and highly energetic form of a traditional dance that is performed by large costumed groups at festivals and sporting events.

C A 60-foot-high statue of a *Gundam*, or giant robot, in Odaiba, Tokyo. This statue represents a character from a popular anime (Japanese animation) TV series.

Cities and National Capitals

● ✪ More than 5,000,000
◕ ✪ 1,000,000–5,000,000
○ ✪ 500,000–999,999
• ✪ 250,000–499,999
○ Fewer than 250,000

Transportation

—— Major road
—— Railroad

Japan stretches over a range of latitudes roughly comparable to that between Canada's province of Nova Scotia and the Georgia coast (and even as far south as the Florida Keys, if one counts Japan's tiny, southernmost islands). Although half a world away, it has a similar range of climates to that of the North American Eastern Seaboard. The climate is cool on the far northern island of Hokkaido; temperate on the islands of Honshu, Kyushu, and Shikoku; and tropical in the southern Ryukyu Islands. Despite the moderating effect of the surrounding oceans, the great seasonal climate shifts of nearby continental East Asia give Japan a more extreme seasonal variation in temperature than it would have if it lay farther out to sea.

Honshu, the largest and most densely populated of Japan's islands, has the most mountains and forests. Because it has gained access to forest products from Southeast Asia and other parts of the world, Japan has been able to keep much of its own forest land in reserves. Today, the interior mountains of Honshu (and Hokkaido) remain largely forested, although many of the slopes are planted with a monoculture of *sugi* (Japanese cedar) that is harvested commercially as building material. The few flat lowlands and coastal areas of Honshu were once intensively cultivated and supported a dense rural population. While some farmland remains, industrial cities now fill many of these lowlands. Japan's largest flatland is the Kanto Plain on the east-central coast of Honshu, which is where the Tokyo–Yokohama urban agglomeration is located. Thirty percent of Japan's total population lives there. The flatlands that ring the Inland Sea are also heavily urbanized, especially the Kobe–Osaka–Kyoto metropolitan area, which is the second-largest concentration of people in Japan (see the Figure 9.46 map).

Japan's other three main islands are Hokkaido, the least populated island, referred to as Japan's northern frontier, and Kyushu and Shikoku, two smaller southern islands. The northern slopes of Kyushu and Shikoku, facing the Inland Sea, have narrow strips of land that are put to agricultural, industrial, and urban uses. The southern slopes are more rural and agricultural. Off Kyushu's southern tip, the very small, mountainous Ryukyu Islands stretch out in a 650-mile (1000-kilometer) chain, reaching almost to Taiwan. Although many areas are densely settled and are important for tourism and specialty agriculture, such as the growing of pineapples, the Ryukyu Islands are considered a poor, rural backwater of the four big islands.

The Challenge of Living Well on Limited Resources

Japan has among the highest ratios of people to farmland in the world: more than 7000 people depend on each square mile (more than 2700 per square kilometer) of cultivated (arable) land. Thus, Japanese agriculture relies heavily on high-yield varieties of rice and other crops, as well as on irrigation, fertilization, and mechanization. Because flat land is scarce, Japanese farmers have had to make efficient use of hill slopes by planting tea bushes and orchards and by terracing for rice paddies.

Although subsistence ways of life play a large part in Japanese national identity, today only 3 percent of the population fish or farm for a living. Japanese farms are highly productive in terms of output per acre; however, their food is some of the most expensive in the world. Foreign rice costs about half as much as domestically

grown rice does, but many Japanese people associate locally grown rice with quality and prefer it. Another farm product, Kobe beef, has a worldwide reputation for quality, but it sells for more than U.S.$150 per pound.

The seas surrounding Japan, where warm and cold ocean currents mix, are an especially important source of food for the country, especially considering the limited amount of farmland. There are some 4000 coastal fishing villages and tens of thousands of small craft that work these waters and bring home great quantities of tuna, halibut, mackerel, salmon, and other fish. In addition, a large aquaculture industry produces oysters, seaweed, and other foods in shallow bays, and freshwater fish in artificial ponds.

Despite all these efforts at food production, Japan is one of the world's largest consumers of imported agricultural products, especially seafood, grains, meat, and processed foods, and it has the lowest levels of **food self-sufficiency** (the ability to produce its own food supply) among developed countries. As shown in **FIGURE 9.47**, the average family spends more than 25 percent of its expenditures on food (about $616 per month in 2015). Americans, by contrast, spend about 13 percent.

Japan's Culture of Contrasts

Perhaps more than any other country in the world, Japan is renowned for its ability to confound foreigners. Here, in an attempt to penetrate some of Japan's cultural complexity, we look at a few of its unique cultural phenomena.

Japan has combined its long-standing traditions with customs from around the world, resulting in unique cultural contrasts. We see the mix in many ways: in the kinds of clothes people wear, in the foods they eat, in the music they listen to, in the sports they enjoy, and in the cultural impact they have abroad. Thus, among other contrasts, Japan is a land of kimonos and blue jeans, sushi and hamburgers, the *koto* (a traditional stringed instrument) and rap music, sumo wrestling and baseball. Japanese couples who marry in traditional wedding kimonos often

food self-sufficiency the ability of an area, such as a country, to produce food domestically to meet the needs of its people

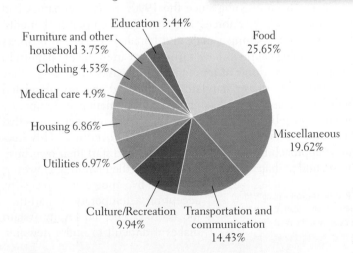

FIGURE 9.47 Average monthly expenditures for a Japanese worker's family, 2015. [Source consulted: Statistics Japan, http://www.stat.go.jp/english/data/kakei/156.htm]

change their clothing after the ceremony to a Western-style tuxedo and a white bridal dress for the reception. Japanese culture is a blend of modern influences and respect for tradition and authority (see Figure 9.46).

There is often a distinction in Japanese life between what is displayed on the outside, *omote*, and what is private on the inside, *ura*. The former often protects the latter, as in the contrasts between *omote-ji*, the outer layer of cloth in a kimono, and *ura-ji*, the layer closest to the skin. This cultural attitude is also expressed geographically. *Omote-Nippon*, the urban-industrial eastern (Pacific Ocean) side of Japan, trades with the world, while *Ura-Nippon* is the more traditional and secluded western (Sea of Japan) side of the country.

Working in Japan

Japan's distinctive culture is in part reflected in the traditional working lives of its people, as well as in their attitude toward the outsiders who perform many of the country's most menial jobs.

The Japanese are well known for their hard work, high-quality output, and devotion to their jobs. Since before World War II, the norm in Japan has been that an employer provides lifetime employment, regular paid vacations, a pension upon retirement, and often subsidized housing and perhaps even a graveyard for employees. Job-hopping is considered reprehensible. Personnel management instills loyalty to the employer. The prototypical Japanese corporate employee is a "salaryman"—a male, usually with a family he spends a lot of time away from, who lives to work. In sprawling Tokyo where the commute home is long and the employee is expected to work late, he can stay overnight at a capsule hotel, which is little more than a bed-in-a-tube made available at a low cost. Back at work, individualism is regarded as selfish, and making innovative suggestions is discouraged. Employees who have good ideas may not present them for several years, or may give credit to someone else. The Japanese language does not even have a word for "entrepreneur"; only recently has the English term become a buzzword in Japanese.

This norm is now receding in the face of larger economic and social changes. Since the slowdown of the Japanese economy, which has been ongoing since the 1990s, many companies had to reduce their workforces and cut back employee benefits. Moreover, some younger male and female graduates in business and technological fields have turned away from the old pattern in favor of the freedoms of temporary employment. Although they may settle for a lower standard of living and give up job-related benefits, women enjoy less gender discrimination in temporary or contract jobs, and temporary workers do not feel compelled to acquiesce to their superiors or to work overtime. This free-lance approach to work allows them time to start their own businesses and perhaps spend more time with their families. However, whether by choice or necessity, accepting a temporary or part-time job is risky. The old loyalty system makes it difficult to find a new job,

Asia–Pacific region a huge trading area that includes all of Asia and the countries around the Pacific Rim

and firms worried about industrial espionage are often unwilling to hire someone who has worked for a competing company.

Foreign Workers Foreign workers from poor countries come to Japan as guest workers to take some of the hard, dirty, dangerous, and low-paying jobs that today's Japanese workers avoid. They come from places such as China, Indonesia, the Philippines, Pakistan, Bangladesh, Sri Lanka, Iran, Peru, Bolivia, Brazil, and Africa. Some Latin American immigrants are descendants of Japanese people who went abroad in search of work some generations ago, when Japan was a poorer country. Foreign men often work in construction, in factories, and as dishwashers in restaurants; women work as hotel maids, as cleaners, and in other low-status jobs. Many foreign women are also employed as hostesses in bars and as sex workers. The recruitment of women for sex work has become a controversial issue because unscrupulous "agents" have been known to force women into sex-industry jobs after promising them other kinds of work. As the recession increased unemployment, some foreign workers, who make up only 1.5 percent of the Japanese population, have been paid by the Japanese government to go back home. This may be a short-sighted policy, as the Japanese workforce needs to expand in order to remain productive in the future.

THINGS TO REMEMBER

• Japan stretches over a range of latitudes from the cool north to the tropical south; much of the land is mountainous.

• Japan has among the highest ratios of people to farmland in the world: more than 7000 people depend on each square mile (more than 2700 per square kilometer) of cultivated land. Its food self-sufficiency level is very low by international standards.

• Japan has a culture of contrasts and complexities: long-standing indigenous customs and traditions mixed with aspects of global culture—in the home, in urban spaces, and in the workplace.

TAIWAN

Taiwan is located a little over 100 miles (160 kilometers) off China's southeastern coast **(FIGURE 9.48)**. With an area of 14,000 square miles (36,000 square kilometers) and more than 23 million people, Taiwan is a crowded place. A mountainous spine runs from the northeastern corner to the southern tip of the island, with a rather steep escarpment facing east and a long, gentler slope facing west. Most of the population lives on the western side, especially at lower elevations along the coastal plain. As farming has declined, populations have become concentrated in a few urban centers. The greatest concentration is in the far north, in the area surrounding Taipei, the capital.

Despite its small size and small population, Taiwan is one of the most prosperous countries of the **Asia–Pacific region,** a huge

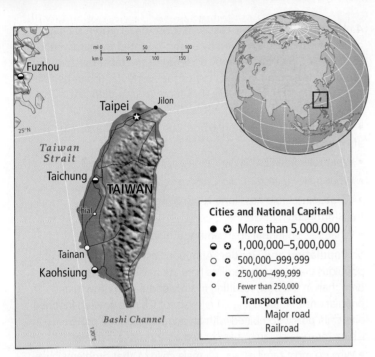

FIGURE 9.48 The Taiwan subregion.

trading area that includes all of Asia and the countries around the Pacific Rim (which consists of all the countries that border the Pacific Ocean). It ranks twentieth globally in the size of its exports, and its GDP per capita in 2015 was U.S.$47,500. After the Communist Revolution in China, Taiwan took advantage of its ardent anticommunist stance, its burgeoning refugee population, and its geographic location close to the mainland to draw aid and investment from Europe, America, and eventually, Japan. The island's economy quickly changed from overwhelmingly rural and agricultural to mostly urban and industrial. Green revolution technologies (see Chapter 1) allowed a surplus labor force in the countryside to be transferred to urban jobs, and today only 5 percent of Taiwanese people still work in agriculture. Many industries made products for Taiwan's impressive export markets. Then, slowly, the economic emphasis changed again, from labor-intensive industries to high-tech and service industries requiring education and technical skills.

By the mid-1990s, the domestic economy was expanding rapidly, and local consumers were absorbing a major proportion of Taiwan's own production: home appliances, electronics, automobiles, motorcycles, and synthetic textiles. The Taiwanese performed a neat pirouette to make possible this high rate of local consumption. As Taiwan lost its labor-intensive industries to cheaper labor markets throughout Asia—including mainland China—Taiwanese entrepreneurs built factories and made other investments in the very places that were giving them such stiff competition: Thailand, Indonesia, the Philippines, Malaysia, Vietnam, and the new SEZs in southern China. Taiwan was thus able to remain competitive as a rapidly growing economy with strong export markets; it also profited from dealing with the less advanced but emerging economies in the region, including especially China.

What Taiwan's future role should be in East Asia is a hotly debated question. The United States has had a strong relationship with Taiwan since the 1950s, when Taiwan was a symbol of anticommunism. More recently, depending on which political party is in charge, sometimes the Taiwanese government favors relaxed relations with China and at other times the tone is more nationalistic (see Figure 9.21C). China, on the other hand, regards Taiwan as an integral part of the mainland, and Taiwan could easily be militarily subdued, as China periodically demonstrates by flexing its military muscles in the Straits of Taiwan. At the same time, Taiwan is geographically and financially situated to enhance the development of mainland Chinese markets and the integration of the Asia–Pacific economy. China is currently accepting this reality.

THINGS TO REMEMBER

• Taiwan, a small crowded island off China, lives with political conditions that date back to the Communist Revolution in China—Taiwan claims sovereignty, while China considers Taiwan one of its provinces.

• Taiwan is relatively affluent and it plays a strong role in the Asia–Pacific economy, especially as a source of investment in China and other lesser-developed economies.

GEOGRAPHIC THEMES: East Asia Review and Self-Test

1. Environment: East Asia's most serious environmental problems result from its high population density combined with its rapid urbanization and environmentally unsustainable economic development. Climate change is intensifying the droughts and floods that have long plagued this region.

• Why has East Asia suffered for so many years from enormously destructive droughts and devastating floods? What has been done to protect the population from such hazards?

• How might the melting of China's highest glaciers affect the country's rivers? Which areas might be more affected than others?

• In which ways is China more vulnerable to global warming than Japan, the Koreas, or Taiwan?

2. Globalization and Development: East Asia pioneered a spectacularly successful economic development strategy that has transformed economies across the globe. Governments in the region intervened strategically to encourage the production of manufactured goods destined for sale abroad, primarily to the large economies of North America and Europe.

• East Asian countries are able to buy food on the global market but not produce the food they need at home. What problems might this create both for East Asia and the rest of the world?

• What is East Asia's, and especially Japan's, role in the global harvesting of wild-caught fish in the oceans?

3. Power and Politics: As the countries of East Asia develop economically, people have been pushing for more political freedom. Japan, South Korea, and Taiwan all have high levels of political freedom, and demands for political change in China are increasing, especially in urban areas.

• How is the Internet emerging as a tool to promote the expansion of political freedoms in China?

• How has China's government responded to protests and other efforts to promote political change by Chinese citizens?

• Why do so many countries claim control of uninhabited islands in or near the waters of East Asia?

4. Urbanization: Across East Asia, cities have grown rapidly over the last century, fueled by export-oriented manufacturing industries. China has recently undergone the most massive and rapid urbanization in the history of the world. Its urban population, now 755 million people, has more than tripled since China initiated economic reforms in the 1980s.

• How has China's focus on export-oriented manufacturing affected rural areas?

• What role did SEZs play in China's economic reforms?

• What major milestone did China reach in 2013?

• What special role has Hong Kong played in China's transformation?

• What sets Shanghai apart among China's major cities?

5. Population and Gender: Although East Asia remains a highly populous world region, families here are having far fewer children than in the past, resulting in slow population growth and populations that are aging. The legacy of China's now-abandoned one-child policy, combined with an enduring cultural preference in China for male children, has created a shortage of females.

• Why do most families want a male child? What problems have been created by the resulting shortage of women?

• What problems are arising in Japan that are related to its rapidly aging population? What might be the solutions to such problems?

Critical Thinking Questions

1. Why is the interior west of continental East Asia (western China and Mongolia) so dry and subject to extremes in temperature? What are the effects of this climate on local population size and forms of subsistence?

2. Historically, China has been afflicted with recurring famines. While famine is no longer a concern, China's access to food has changed with economic development. How?

3. How might a less authoritarian political system have resulted in changes in the overall conception and design of China's Three Gorges Dam?

4. Is there a contradiction between China's communist political system and its adoption of free markets? How do you see democracy and human rights evolving in China in the future?

5. In what ways might globalization have taken a different course if the voyages of Admiral Zheng He had inspired China to conquer a vast colonial empire, like that of Great Britain?

6. Contrast Japan's pre–World War II policies in East Asia with its current role in the region. Describe the principal similarities or differences.

7. In what ways has the one-child policy helped maintain stability in China? Even though the one child-policy is no longer in use, how might it have lingering effects on China in the future in terms of gender balance and economic growth?

8. How has East Asia's spatial pattern of urbanization been shaped by integration into the global economy?

9. What information technologies are now available to the Chinese and what role can such technologies have in the democratization of China?

Chapter Key Terms

Ainu 538
alluvium 545
Asia–Pacific region 558
chaebol 554
Confucianism 513
conurbation 527
Cultural Revolution 516
demilitarized zone 518
export-led growth 519
floating population 498

food security 505
food self-sufficiency 557
Great Leap Forward 516
growth poles 529
hukou system 498
inversion 545
just-in-time system 519
kaizen system 519
loess 540
Overseas Chinese 539

qanats 548
regional self-sufficiency 521
regional specialization 521
special economic zones (SEZs) 529
state-aided market economy 519
tsunami 500
typhoon 501
wet rice cultivation 506
yurts (or *gers*) 548

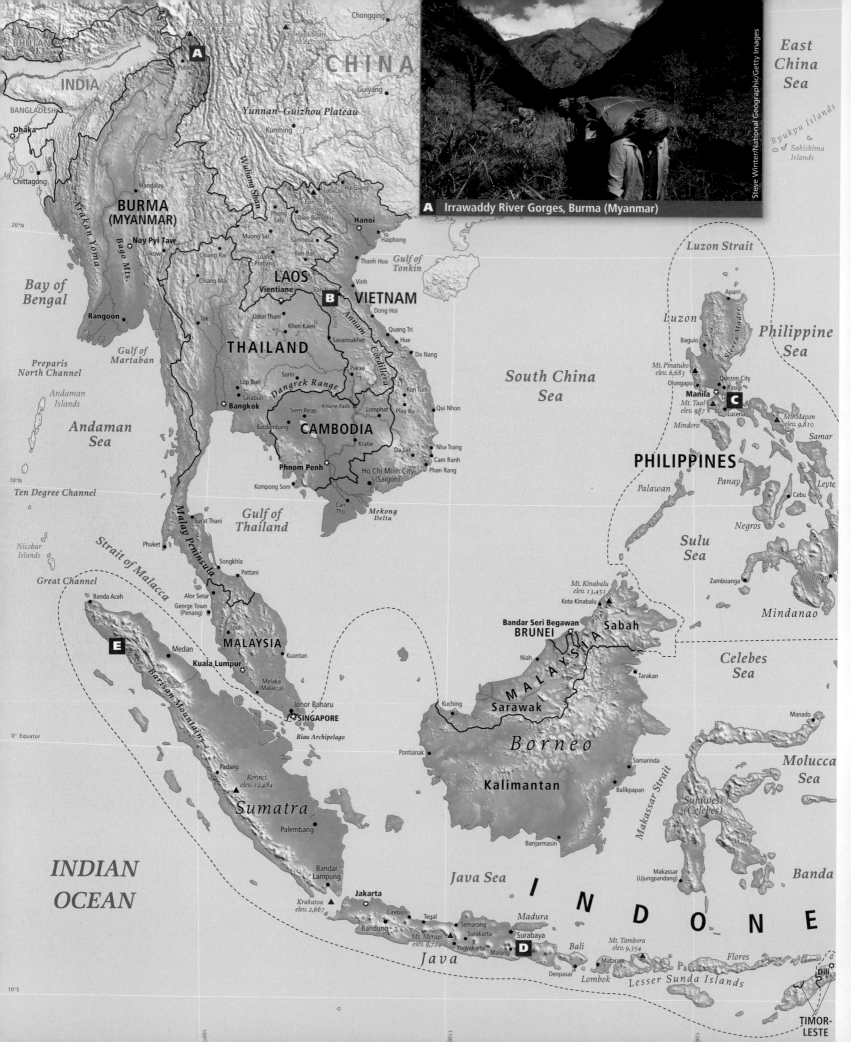

BHUTAN
INDIA
BANGLADESH
Dhaka
Chittagong

A

CHINA
Chongqing
Guiyang
Yunnan–Guizhou Plateau
Kunming

Hkakabo Razi elev. 19,295
Gonaga Shan elev. 24,700

Putao
Mandalay
BURMA (MYANMAR)
Arakan Yoma
Nay Pyi Taw
Bago Mts.
20°N
Loikow
Chiang Rai
Chiang Mai

Laa Cai
Ha Giang
Fan Si Pan elev. 10,312
Phong Saly
Dien Bien Phu
Muong Sai
Luang Prabang
Hanoi
Haiphong
Samneua
Ban Ban
LAOS
Vientiane
Ban Mana
Thanh Hoa
Vinh
VIETNAM

B

Gulf of Tonkin

Rangoon
Gulf of Martaban
Tak
Udon Thani
Khon Kaen
Savannakhet
Quang Tri
Hue
Da Nang
Annam Cordillera

Preparis North Channel
Bay of Bengal

THAILAND
Dangrek Range
Surin
Lop Buri
Saraburi
Bangkok
Siem Reap
Batdambang
Khone Falls
Lomphat
Kon Tum
Play Ku
Qui Nhon

CAMBODIA
Kratie
Da Lat
Nha Trang
Cam Ranh
Phan Rang

Phnom Penh
Kompong Som
Ho Chi Minh City (Saigon)

Andaman Islands
Andaman Sea

10°N
Ten Degree Channel

Surat Thani
Phuket
Malay Peninsula
Gulf of Thailand
Can Tho
Mekong Delta

Nicobar Islands
Great Channel

Songkhla
Pattani
Alor Setar
George Town (Penang)
Ipoh
MALAYSIA
Kuantan
Kuala Lumpur
Melaka (Malacca)
Johor Baharu
SINGAPORE
Riau Archipelago

Banda Aceh
E
Medan
Strait of Malacca
Barisan Mountains

0° Equator

Padang
Kerinci elev. 12,484
Sumatra
Palembang

INDIAN OCEAN

Bandar Lampung
Krakatoa elev. 2,667
Jakarta
Bandung
Cirebon
Tegal
Mt. Merapi elev. 9,724
Semarang
Surakarta
Yogyakarta
Malang
D
Java

10°S

East China Sea
Ryukyu Islands
Sakishima Islands

Steve Winter/National Geographic/Getty Images

A Irrawaddy River Gorges, Burma (Myanmar)

Luzon Strait
Luzon
Aparri
Baguio
Sierra Madre
Philippine Sea

South China Sea

Mt. Pinatubo elev. 6,683
Olongapo
Quezon City
Pasig
Manila
Mt. Taal elev. 987
Lucena
C
Mt. Mayon elev. 9,810
Samar

PHILIPPINES
Mindoro
Palawan
Panay
Cebu
Leyte
Negros
Sulu Sea

Mt. Kinabalu elev. 13,451
Kota Kinabalu
Sabah
Bandar Seri Begawan
BRUNEI
Niah
Tarakan
Celebes Sea

M A L A Y S I A
Kuching
Sarawak
Zamboanga
Davao
Mindanao

Borneo
Samarinda
Makassar Strait

Kalimantan
Balikpapan
Manado
Molucca Sea

Pontianak
Banjarmasin
Sulawesi (Celebes)
Makassar (Ujungpandang)
Java Sea
Madura
Surabaya
Bali
Mt. Tambora elev. 9,354
Banda Sea

Denpasar
Mataram
Lombok
Flores
Lesser Sunda Islands

I N D O N E

Dili
TIMOR-LESTE

10
Southeast Asia

Okinawa Islands
(Japan)

B Mekong River Plains, Laos

Wilbur E. Garrett/National Geographic/Getty Images

C Philippine Archipelago

Per-Andre Hoffmann / LOOK-foto/Getty Images

MICRONESIA

PALAU
Ngerulmud

Land Elevations

meters		feet
4877		16,000
3353		11,000
2134		7000
914		3000
305		1000
152		500
0		0

mi 0   100   200   300
km 0 100 200 300 400 500

1:17,000,000
Azimuthal Equidistant Projection

Halmahera

Sorong

M o l u c c a s

Buru   Ceram

West Papua
New Guinea

Puntjak Jaya
elev. 16,499 ▲

Maoke Mountains

Jayapura

PAPUA
NEW
GUINEA

Sea

S I A

Aru
Islands

S

Arafura
Sea

Merauke

Timor

Timor Sea

AUSTRALIA

D Volcanoes, Java, Indonesia

Oystein Lurd Andersen/E+/Getty Images

E Post-Earthquake/Tsunami, Sumatra, Indonesia

Kazuhiro Nogi/AFP/Getty Images

FIGURE 10.1 Regional map of Southeast Asia.

GEOGRAPHIC THEMES: Southeast Asia

After you read this chapter, you will be able to discuss the following geographic themes as they relate to the five thematic concepts:

1. Environment: Many of Southeast Asia's most critical environmental issues relate in some way to climate change. Deforestation, a major global source of greenhouse gas emissions, has been rapid; and the region is vulnerable to the climate change effects of flooding and droughts. Human manipulation of the land and resources is changing ecological systems and threatening food security. Rising ocean temperatures are straining aquatic ecosystems and storm surges threaten human settlements.

2. Globalization and Development: Globalization has brought both spectacular successes and occasional declines to the economies of Southeast Asia. Key to the economic development of this region are strategies that were pioneered earlier in East Asia: the formation of state-aided market economies, combined with export-led economic development. Regional economic cooperation is expanding slowly.

3. Power and Politics: There has been a general expansion of political freedoms throughout Southeast Asia in recent decades, but authoritarianism, corruption, and violence have at times reversed these gains.

4. Urbanization: Although Southeast Asia as a whole is only 47 percent urban, its cities are growing rapidly as agricultural employment declines and urban industries expand. The largest Southeast Asian cities, which are receiving most of the new rural-to-urban migrants, rarely have sufficient housing, water, sanitation, or jobs for all their people.

5. Population and Gender: Population dynamics vary considerably in this region because of differences in economic development, government policies, prescribed gender roles, and religious and cultural practices. Economic change has brought better job opportunities and increased status for women, who then often choose to have fewer children. Some countries also have gender imbalances because of a cultural preference for male children.

The Southeast Asia Region

The region of Southeast Asia (shown in **FIGURE 10.1**) has attracted global attention for several decades because so many of its countries emerged poverty-stricken from World War II, only to embark on a rapid journey to relative prosperity. There have been ups and downs: difficulties in finding smooth paths to stable democratic processes and institutions; urbanization at too rapid a pace; and a tendency to rely on development strategies that carry grave environmental and social side effects. Some of this region's transitions are now emulated by other developing regions: rapid industrialization, expansion of the middle class, education for the masses, the empowerment of women, public health care, improved food security, and slower population growth.

The five thematic concepts in this book are explored as they arise in the discussion of regional issues, with interactions between two or more themes featured. Vignettes, like the one that follows about the rights of a group of people indigenous to the state of Sarawak on the island of Borneo, illustrate one or more of the themes as they are experienced in individual lives.

GLOBAL PATTERNS LOCAL LIVES In December 2005, a group of indigenous people in the Malaysian state of Sarawak, on the island of Borneo (see the Figure 10.1 map), attended a public meeting wearing orangutan masks. They carried signs informing onlookers that, although the government protects natural areas for orangutan (large reddish-brown apes native to Borneo), it ignores the basic right of indigenous people to live on their own ancestral lands and practice their forest skills (FIGURE 10.2A).

FIGURE 10.2 Indigenous people in Sarawak, Borneo.

(A) An indigenous man in Sarawak hunts in the forests of Malaysian Borneo, which are being rapidly converted to palm oil plantations.

(B) Palm oil plantations in Sarawak cover vast areas that were once highly diverse tropical forests where many different groups of indigenous people lived, hunted, and cultivated tropical food gardens.

Over the years, before and since 2005, Sarawak forest dwellers have tried many tactics to save their lands from the deforestation that logging companies do in preparation for expanding oil palm plantations (see Figure 10.2B). Palm oil is a major food and cosmetic oil export; it is sold primarily to Europe, North America, Japan, and China, where many are unwary consumers. In the mid-twentieth century, the Sarawak state government began licensing logging companies to cut down tropical forests occupied by indigenous peoples (FIGURE 10.3) and export the timber.

By the 1980s, 90 percent of Sarawak's lowland forests had been degraded and 30 percent had been clear-cut. Much of the cleared area was replaced by oil palm plantations. As a consequence, indigenous people found that even on uncleared land, their hunts declined. Vegetative diversity was lost, streams and rivers became polluted with eroded sediment and by the fertilizers and pesticides applied to palm trees. Virtually none of the profits from palm oil were returned to the communities affected by the conversion of forests to plantations.

In the 1990s, a group of citizens in Berkeley, California, who were concerned about news reports of the deforestation in Sarawak, organized the Borneo Project. They offered to become a "sister city" to one indigenous group in Sarawak, the Uma Bawang, giving help wherever it was needed. With the help of the Borneo Project, the Uma Bawang began a community-based mapping project in 1995. Using rudimentary compass-and-tape techniques, they began mapping both the extent and the

FIGURE 10.3 Issues affecting indigenous groups worldwide. The Sarawak forest dwellers' mapping practices are spreading. There are about 5000 distinct indigenous groups in the world. On this map, each color represents one or more of these groups that are associated with a particular geographic location by their language, culture, and way of life. Many of these peoples have participated in community mapping projects, similar to those of the Sarawak forest dwellers, in order to identify and protect their rights to their traditional lands. The symbols reflect some of the global issues that affect a particular group. [Sources consulted: "Struggling Cultures," *National Geographic Atlas of the World*, 8th ed. (Washington, DC: National Geographic Society, 2005), p. 15; "Globalization: Effects on Indigenous Peoples" map, in Jerry Mander and Victoria Tauli-Corpuz, eds., *Paradigm Wars: Indigenous Peoples' Resistance to Economic Globalization* (San Francisco, CA: Sierra Club Books, 2006)]

biological content of their forest home. Since then, indigenous people from across Sarawak have learned how to use global positioning systems (GPS), geographic information systems (GIS), and satellite imagery to make more sophisticated maps. The power of these maps was demonstrated in 2001, when they were used to help win a precedent-setting court case that protected indigenous lands from an encroaching oil palm plantation.

This favorable court ruling was challenged in 2005, when the government appealed on behalf of the palm oil industry, but in 2009 the Malaysian federal court ruled in favor of the Uma Bawang. The decision stated that native customary land rights had been protected since 1939, when British colonial officials had directed the district lands and survey departments to map the boundaries of native lands. Now indigenous people have the right to sue the government for past illegal leases to logging companies, and as of 2014, some 8000 such cases were in litigation. The outcome of this litigation is as yet unclear, and in the Indonesian part of Borneo, deforestation for oil palm plantations is proceeding rapidly, as shown in the 2011 video "In Search of Forest, Part 1," at https://www.youtube .com/watch?v=WXVO0pxn-Kk. *[Sources: The Borneo Wire; http://www.aljazeera.com/indepth/features/2014/04/palm-oil-fuels-indonesia-deforestation-indigenous-displa-201443145636809366.html. For detailed source information, see Text Sources and Credits.]* ∎

It is not unusual for compliant or corrupt governments to help private developers take indigenous lands. The account in the vignette of the tactics used by an indigenous group to secure their rights to control and manage their ancestral lands mirrors the accounts of similar struggles in Middle and South America (see Chapter 3) and highlights a number of themes in this book: the environment, globalization and development, and power and politics. In particular, it shows how issues that may seem to have only local significance can gain global attention. This connection of the local to the global presents both challenges and opportunities for the world's indigenous people (see Figure 10.3). On one hand, the global market provides the demand for timber and palm oil; on the other, it is support from distant groups like the Borneo Project that help communities such as the Uma Bawang have their land claims validated. In fact, the use of mapping to help indigenous groups secure their legal rights to ancestral lands has now become a global phenomenon. Hundreds of indigenous groups throughout the world are collaborating with mapping specialists, many of them geographers, often resulting in their land rights being formally recognized by governments for the first time. In 2007, the United Nations Department of Economic and Social Affairs passed the Declaration on the Rights of Indigenous People and issued a report, *State of the World's Indigenous Peoples*, that documents several indigenous community mapping efforts. The 2015 report is available online at http://www.un.org /esa/socdev/unpfii/documents/2015/sowip2volume-ac.pdf.

What Makes Southeast Asia a Region?

Southeast Asia is physically a manifestation of tectonic forces that are described in the next section, called "Physical Geography." Aside from physical commonalities, the peninsula and island countries of Southeast Asia today share a deep cultural past,

related to but separate from the cultures of southwestern China. With the exception of Thailand, they all also share more recent experiences with European colonialism. During and after World War II, most countries went through severe political turmoil followed by the rapid modernization and industrialization that continues into the present.

Terms in This Chapter

The mainland Southeast Asian countries are Burma (Myanmar), Thailand, Laos, Cambodia, Vietnam, Malaysia, and Singapore; the island countries are Indonesia, Brunei, Timor-Leste (East Timor), and the Philippines. Malaysia occupies part of the peninsula and parts of the island of Borneo **(FIGURE 10.4)**.

Some governments in Southeast Asia choose to dispense with place-names that originated in their colonial past. Governments can also make name changes as a political strategy. Such is the case with Burma (see Figure 10.4), where a military government seized control in a coup d'état in 1990, shortly after the election of Aung San Suu Kyi. The military placed her under house arrest and changed the country's name to Myanmar. As discussed later in the chapter, the government of Burma is once again in transition, and in press reports, the names of both Burma and Myanmar are used. Although it seems that slowly the name of Myanmar is being accepted internally, Aung San Suu Kyi, who, after years of house arrest by the military government won a landslide election in November 2015, has stated that she prefers the traditional name of Burma. Therefore, in this text, we use the country's traditional name of Burma, with Myanmar in parentheses, a naming policy followed by the U.S. government.

Another potential point of confusion is Borneo, a large island that is shared by three countries. The part of the island known as Kalimantan is part of Indonesia; Sarawak and Sabah on the north coast are part of Malaysia; and Brunei is a very small, independent, oil-rich country, also on the north coast.

FIGURE 10.4 Political map of Southeast Asia. Please learn the relative locations of these countries and refer back to this map as you read.

THINGS TO REMEMBER

• Indigenous land claims are no longer just local issues. Indigenous groups have shifted their efforts to secure rights to ancestral lands to the global scale, thus highlighting a trend in which issues once considered local or national are now becoming global concerns.

PHYSICAL GEOGRAPHY

The physical patterns of Southeast Asia have a continuity that is not immediately obvious on a map. A map of the region shows a unified mainland region that is part of the Eurasian continent and a vast and complex series of islands arranged in chains and groups. These landforms are actually related in tectonic origin. Climate is another source of continuity; most of the region is tropical or subtropical.

LANDFORMS

Southeast Asia is a region of peninsulas and islands (see Figure 10.1). Although the region stretches over an area larger than the continental United States, most of that space is ocean; the area of all the region's landmass amounts to less than half that of the contiguous United States. Burma (Myanmar), Thailand, Laos, Cambodia, and Vietnam occupy the large Indochina peninsula that extends south of China. This peninsula itself sprouts the long, thin Malay Peninsula that is shared by outlying parts of Burma (Myanmar) and Thailand, a main part of Malaysia, and the city-state of Singapore, which is built on a series of islands at the southern tip. The **archipelago** (a series of large and small islands) that fans out to the south and east of the mainland is grouped into the countries of Indonesia, Malaysia, Brunei, Timor-Leste (East Timor), and the Philippines (see Figure 10.1C). Indonesia alone has some 17,000 islands, and the Philippines, 7000.

The irregular shapes and landforms of the Southeast Asian mainland and archipelago are the result of the same tectonic forces that were unleashed when India split off from the African Plate and gradually collided with Eurasia (see Figure 1.8). As a result of this collision, which is still underway, the mountainous folds of the Plateau of Tibet reach heights of almost 20,000 feet (6100 meters). These folded landforms, which bend out of the high plateau, turn south into Southeast Asia. There, they descend rapidly and then fan out to become the Indochina peninsula. Deep gorges widen out into valleys that stretch toward the sea, each containing mountains of 2000 to 3000 feet (600 to 900 meters) and a river or two flowing from the mountains of the Yunnan–Guizhou Plateau of China to the Andaman Sea, the Gulf of Thailand, and the South China Sea (see Figure 10.1A and the figure map). The major rivers of the Indochina peninsula are the Irrawaddy and the Salween in Burma (Myanmar); the Chao Phraya in Thailand; the Mekong, which flows through Laos, Cambodia, and Vietnam; and the Black and Red rivers of northern Vietnam. Several of these rivers have major delta formations—especially the Irrawaddy, Chao Phraya, and the Mekong—that are intensively cultivated and settled (see the discussion on pages 573–576; see also Figure 10.1B).

The curve formed by Sumatra, Java, the Lesser Sunda Islands (from Bali to Timor), and New Guinea conforms approximately to the shape of the Eurasian Plate's leading southern edge (see Figure 1.8). Where the Indian-Australian Plate plunges beneath the Eurasian Plate along this curve, hundreds of earthquakes occur and volcanoes abound, especially on the islands of Sumatra and Java (see Figure 10.1D, E). There are also volcanoes and earthquakes in the Philippines where the Philippine Plate pushes against the eastern edge of the Eurasian Plate and is in turn pushed by the Pacific Plate. The volcanoes of Southeast Asia are considered part of the Pacific Ring of Fire (see Figure 1.9; see also Figure 10.1C, D).

Volcanic eruptions and the mudflows and landslides that follow eruptions complicate and endanger the lives of many Southeast Asians. Earthquakes are especially problematic because of the tsunamis they can set off (see Figure 10.1E). The tsunami of December 2004, triggered by a giant earthquake (9.1 in magnitude) just north of Sumatra, swept east and west across the Indian Ocean, taking the lives of 230,000 people (170,000 in Aceh Province, Sumatra) and injuring many more. It was one of the deadliest natural disasters in recorded history. There has been a series of strong earthquakes since, with several along coastal Sumatra in August and September of 2009, April and May of 2010, April of 2012, and February of 2016, but they did not generate notable tsunamis. Over the long run, the volcanic material creates new land and provides minerals that enrich the soil for farming.

The now-submerged shelf of the Eurasian continent, known as Sundaland **(FIGURE 10.5)** extends under the Southeast Asian peninsulas and islands. It was above sea level during the recurring ice ages of the Pleistocene epoch, when much of the world's water was frozen in glaciers. The exposed shelf allowed ancient people (including *Homo erectus*) and Asian land animals (such as elephants, tigers, rhinoceroses, and orangutans; **FIGURE 10.6**) to travel south to what became the islands of Southeast Asia when sea levels rose.

CLIMATE AND VEGETATION

The largely tropical climate of Southeast Asia is distinguished by continuous warm temperatures in the lowlands—consistently above 65°F (18°C)—and heavy rain **(FIGURE 10.7)**. The rainfall is the result of two major processes: the monsoons (seasonally shifting winds) and the movement of the intertropical convergence zone (ITCZ), an area centered roughly at the equator, where surface winds converge from the northern and southern hemispheres and rise upward, resulting in rainfall. The wet summer (monsoon) season extends from May to October, when the warming of the Eurasian landmass sucks in moist air from the surrounding seas and pulls the ITCZ northward (see the Figure 10.7 map). Between November and April, there is a long dry season on the mainland, when the seasonal cooling of Eurasia causes dry, cool air from the interior of the continent to flow out toward the sea, pushing the ITCZ southward (see Figure 11.5). On the many islands, however, the winter can also be wet because the air that flows from the continent warms and picks up moisture as it passes south and east over the seas. The air releases its moisture as rain after ascending high enough over elevated landforms to cool. With rains coming from both the monsoon and the ITCZ, the island parts of Southeast Asia are among the wettest areas of the world.

archipelago a group, often a chain, of islands

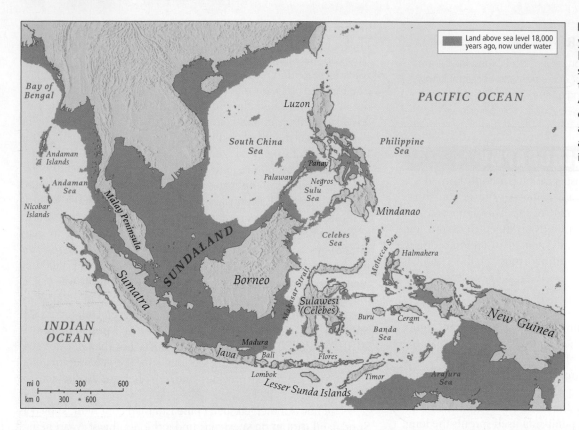

FIGURE 10.5 Sundaland 18,000 years ago, at the height of the last ice age. The now-submerged shelf of the Eurasian continent that extends under Southeast Asia's peninsulas and islands was exposed during the last ice age and remained above sea level until about 16,000 years ago, when that ice age was ending.

FIGURE 10.6 LOCAL LIVES: People and Animals in Southeast Asia

A A cat at a temple in Chiang Mai, Thailand. Considered the guardians of statues in temples throughout Thailand, cats are written about in ancient poetry and celebrated by Thailand's royal family. Stray cats are given refuge at certain temples.

B Clownfish swim amid sea anemones that capture prey with paralyzing toxins. Immune to these toxins, the clownfish eat tiny animals that could harm the anemones; in return, they are provided with a safe home. Clownfish, whose natural habitat is this region's saltwater, have also been successfully bred for aquarium tanks.

C A pig is decorated and paraded around a village in Vietnam during Tet, the Vietnamese New Year. Pork is especially prized in Vietnam and is an essential part of holiday meals. This pig will be served as an offering to village gods.

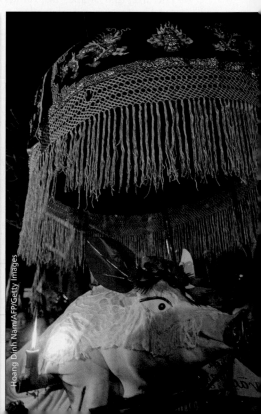

FIGURE 10.7 PHOTO ESSAY: Climates of Southeast Asia

The presence of so much wet, tropical air, combined with the movements of the intertropical convergence zone (ITCZ) across the region twice a year, makes Southeast Asia one of the wettest regions in the world.

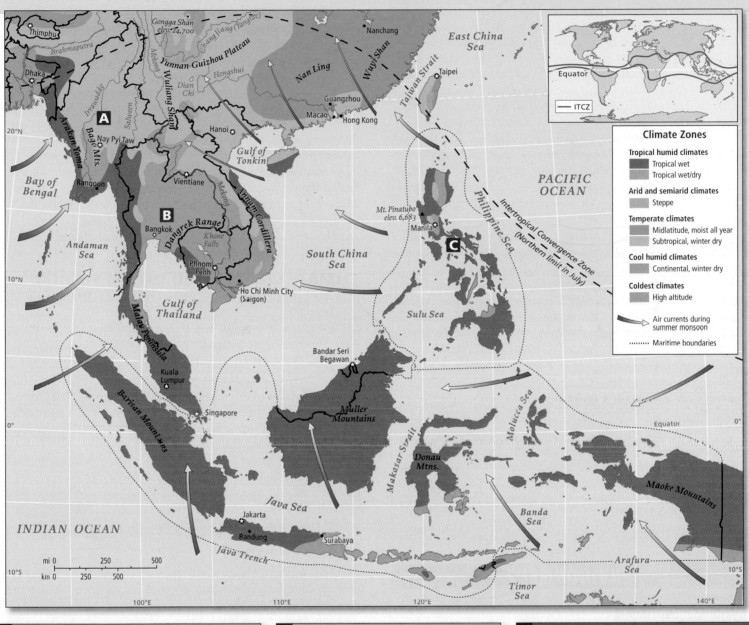

Climate Zones

Tropical humid climates
- Tropical wet
- Tropical wet/dry

Arid and semiarid climates
- Steppe

Temperate climates
- Midlatitude, moist all year
- Subtropical, winter dry

Cool humid climates
- Continental, winter dry

Coldest climates
- High altitude

→ Air currents during summer monsoon

∙∙∙∙ Maritime boundaries

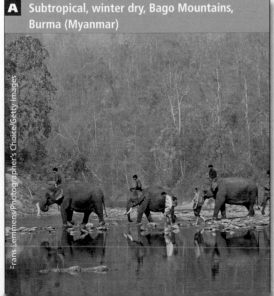

A Subtropical, winter dry, Bago Mountains, Burma (Myanmar)

Frans Lemmens/Photographer's Choice/Getty Images

B Tropical, wet/dry, Khao Yai, Thailand

Feargus Cooney/Lonely Planet Images/Getty Images

C Tropical, wet, Alaminos, Philippines

Danita Delimont/Gallo Images/Getty Images

Irregularly every 2 to 7 years, the normal patterns of rainfall are interrupted, especially in the islands, by the El Niño phenomenon (see Figure 11.6). In an El Niño event, the usual patterns of air and water circulation in the Pacific are reversed. Ocean temperatures are cooler than usual in the western Pacific near Southeast Asia. Instead of warm, wet air rising and condensing as rainfall, cool, dry air sits on the ocean surface. The result is severe drought, often with catastrophic effects for farmers and for tinder-dry forests that regularly catch fire. The cool El Niño air can trap smoke and other pollutants at Earth's surface, creating unusually toxic smog and such low visibility that planes sometimes crash.

The soils in Southeast Asia are typical of the tropics. Although not particularly fertile, they will support dense and prolific vegetation when left undisturbed for long periods. The warm temperatures and damp conditions promote the rapid decay of **detritus** (dead organic material) and the quick release of useful minerals. These minerals are taken up directly by the roots of the living forest rather than enriching the soil. Because rainfall is usually abundant during the summer wet season (when drought is only episodic), this region has some of the world's most impressive forests in both tropical and subtropical zones (see Figure 10.7). On the mainland, human interference, coupled with the long winter dry season, has reduced the forest cover. The relationships between humans and the environment that affect forests elsewhere in the region are discussed in the opening vignette and in the next section.

THINGS TO REMEMBER

- The irregular shapes and landforms of the Southeast Asian mainland and archipelago are the result of the same tectonic forces that were unleashed when India split off from the African Plate and crashed into Eurasia.

- Continuous warm temperatures in the lowlands and heavy rain distinguish the tropical climate of Southeast Asia, except during those irregular years of the El Niño phenomenon.

ENVIRONMENT

GEOGRAPHIC THEME 1

Environment: Many of Southeast Asia's most critical environmental issues relate in some way to climate change. Deforestation, a major global source of greenhouse gas emissions, has been rapid; and the region is vulnerable to the climate change effects of flooding and droughts. Human manipulation of the land and resources is changing ecological systems and threatening food security. Rising ocean temperatures are straining aquatic ecosystems and storm surges threaten human settlements.

DEFORESTATION

Southeast Asia has the second-highest rate of deforestation of any world region, after sub-Saharan Africa. With only 5 percent of the world's forests, Southeast Asia has accounted for nearly 25 percent of deforestation around the world in the

detritus dead organic material (such as plants and insects) that collects on the ground and in the soil

last 10 years. The forces driving deforestation in Southeast Asia are global. Photos A and B and the map in **FIGURE 10.8** show the supply chain that takes wood from various tropical regions to Europe and Asia, where it is turned into furniture and other products. Graphs C and D in Figure 10.8 show who is importing and exporting tropical logs. It is extremely difficult to know the extent of illegal deforestation, but in 2006 the World Wildlife Fund estimated that more than 40 percent of the timber production in Indonesia was illegal.

The regional environmental impacts of deforestation and the agriculture that often follows deforestation are explored in **FIGURE 10.9**. One effect is that indigenous people of the forest lose their living space and resources when the land is given over to export agriculture. Another, particular to Southeast Asia, is the loss of habitat for orangutans, the Sumatran tiger, the Sumatran rhinoceros, and tens of thousands of other lesser known species that are displaced when rain forests are converted to agricultural uses. Finally, deforestation processes (fires and vegetation decay) cause the release of greenhouse gases (GHGs), contributing to climate change.

CLIMATE CHANGE AND DEFORESTATION

Deforestation is a major contributor to global climate change because enormous amounts of carbon dioxide (CO_2) are released when forests are removed and the underbrush is burned. Fewer trees also mean that less CO_2 is absorbed from the atmosphere. Every day, 13 to 19 square miles (34 to 50 square kilometers) of Southeast Asia's rain forests are destroyed, much of it done illegally. A great deal of this deforestation takes place in Malaysia and Indonesia, where the logging of tropical hardwoods for sale on the global market is a major activity (see Figure 10.8). For decades, these activities have made Indonesia among the world's biggest contributors to global climate change. Since 1950, the tropical forest cover on the Indonesian part of the island of Borneo alone has decreased by about 60 percent, threatening megafauna and valuable food plants, like the Kalimantan mango. Oil palm plantations now constitute the predominant use of land there.

Oil Palm Plantations

Palm oil, which is used around the world in food, products, as a cooking oil, in soaps and cosmetics, and as a machine oil, is at the center of debates over global climate change in Southeast Asia. The oil palm originated in West Africa and was brought to Southeast Asia early in the colonial era. Today, Indonesia and Malaysia are the world's largest palm oil producers, and the expansion of their oil palm plantations has resulted in extensive deforestation. Much of the deforestation is in lowland peat swamps which, in a natural state, are capable of storing large amounts of carbon in their soils. To create space for oil palm plantations, the swamps are drained and most of their vegetation is burned (see Figure 10.9A). Often the fires spread underground to the peat beneath the forests, smoldering for years after the surface fires have been extinguished. These subsurface fires release enormous amounts of carbon into the atmosphere—the daily carbon emissions of fires in 2015 were so enormous that they surpassed the emission output of the entire United States. In recent years, smoke from burning forests and peat has periodically covered much of the region, dramatically reducing air quality and even making it necessary for people in rural areas to wear masks.

FIGURE 10.8 Trade in tropical timber. Most of the consumer demand for tropical timber products is in North America, Europe, and China. Increasingly large quantities of tropical timber sold to China are made into furniture, plywood, and flooring, which are then sold to consumers in the developed world. The trade shown in the graphics is mostly legal, but much of the wood that fuels China's wood-processing industries is actually harvested illegally in Southeast Asia. The data mapped are the latest trade flow data available (2015). [Sources consulted: Frances Maplesden, "Situation and Trends in World Timber Trade and the Tropical Timber Trade in Asia-Pacific." Presentation to Asia-Pacific Forestry Week, Clark Freeport Zone, Pampanga, Philippines, 22–26 February 2016; and Sam Lawson and Larry MacFaul, "Illegal Logging and Related Trade: Indicators of the Global Response" (London: Chatham House, 2010), at http://awsassets.panda.org/downloads /chatham_house_illegallogging_2010.pdf]

(A) A logging road in Sabah, Borneo.

(B) Tropical hardwoods are made into toy train tracks in Dongguan, China.

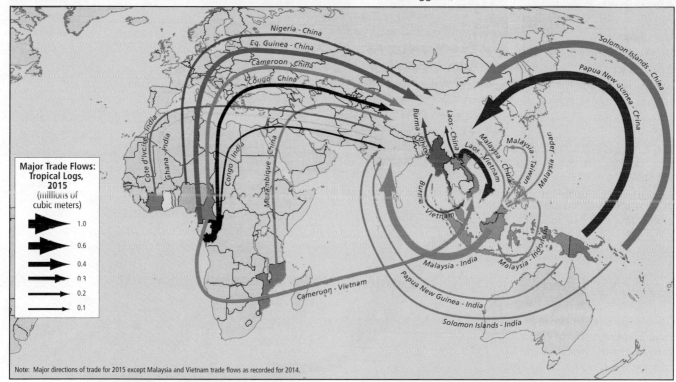

Note: Major directions of trade for 2015 except Malaysia and Vietnam trade flows as recorded for 2014.

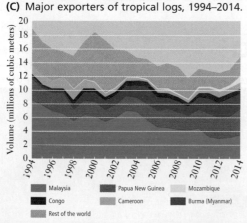

(C) Major exporters of tropical logs, 1994–2014.

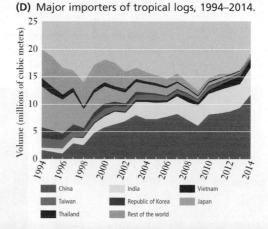

(D) Major importers of tropical logs, 1994–2014.

Thinking Geographically

Judging from the data in Figure 10.8, what is China's role in the global chain of tropical wood product consumption?

FIGURE 10.9 PHOTO ESSAY: Human Impacts on the Biosphere in Southeast Asia

Deforestation and the conversion of land to agricultural uses in Southeast Asia are having a profound impact both on regional ecosystems and on the entire biosphere. [Sources consulted: *United Nations Environment Programme, 2002, 2003, 2004, 2005, 2006* (New York: United Nations Development Program), at http://maps.grida.no/go/collection/globio-geo-3]

A A worker sprays a palm oil plantation in Puntianai, on the island of Sumatra in Indonesia, with pesticides. The expansion of oil palm plantations is now driving much of the deforestation in much of Malaysia and Indonesia.

Human Impact, 2002

Land cover
- Forests
- Grasslands
- Deserts
- Tundra
- Ice

Human impact on land
- High impact
- Medium–high impact
- Low–medium impact

Overfishing
- Threatened fisheries

Acid rain
- 5.5–4.9 pH

- National boundaries
- Maritime boundaries

mi 0   250   500
km 0   250   500

B A wild female orangutan and her baby eat flowers in Tanjung Puting National Park in Kalimantan, Indonesia. The deforestation and fragmentation of habitats caused by roads have rendered many plants, such as the Kalimantan mango, extinct; loss of such foods contributes to the endangerment of orangutan species.

C A satellite image of fires used to clear forests for agriculture in Indonesia and Malaysia. The CO_2 produced contributes to climate change and degrades local air quality.

D Rice terraces in the Philippines. Agricultural expansion is a major force behind deforestation in the region, and wet rice agriculture, because it releases methane, is a significant contributor of GHGs.

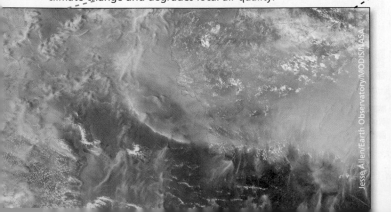

Thinking Geographically

After you have read about human impacts on the biosphere in Southeast Asia, you will be able to answer the following questions about photos in this chapter.

A How does this landscape of recently planted oil palms differ from that of the forest that was removed for this plantation?

B What are these orangutans eating? Do you think these animals would relish a meal of oil palm leaves? Why is this likely or unlikely?

C What are the geographic locations barely visible through the smoke?

Palm oil has been promoted as a potential solution to global climate change because it can be converted into fuel for automobiles. The claim is that because the palm trees absorb CO_2 from the atmosphere just as the original forest did, palm oil could be considered a "carbon neutral" fuel. This is a fallacy because palm trees do not absorb carbon at a rate equivalent to natural rain forests—and if forests are burned to clear land for plantations, it can take decades to offset the initial carbon emissions produced by deforestation. If a peat swamp is cleared for an oil palm plantation, more carbon is released through the burning of subsurface peat than centuries of CO_2 absorption by growing oil palms can counteract.

Even as deforestation proceeds across the region, many people are beginning to appreciate the costs of deforestation. Forests and peat swamps are now known to perform billions of dollars per year in services: Well-managed forests provide renewable plant and animal food, medicine and construction materials, and domestic fuel resources. The peat swamps act like giant sponges, soaking up water and thus mitigating floods. They also purify fresh water, preserve biodiversity, and absorb enormous amounts of CO_2.

CLIMATE CHANGE AND FOOD PRODUCTION

Food production contributes to global climate change in two ways. First, it is a major contributor to deforestation, and second, some types of cultivation actually produce significant GHGs.

Shifting Cultivation

Also known as *slash-and-burn* or *swidden* cultivation, shifting cultivation has been practiced sustainably for thousands of years in the hills and uplands of mainland Southeast Asia and in many parts of the islands **(FIGURE 10.10 MAP)**. To maintain soil fertility in these warm, wet environments where nutrients are quickly lost to decay, traditional farmers move their fields every 3 years or so, letting old plots lie fallow for 15 years or more. The regrowth of forest on once-cleared fields not only regenerates the soil and inhibits runoff, it also absorbs significant amounts of CO_2 from the atmosphere. However, if fallow periods are shortened or are disrupted by logging, soil fertility can collapse, making future cultivation impossible. In some cases, fertility can be temporarily restored through the use of chemical or organic fertilizers, but these can be too expensive for most farmers and chemical fertilizers may be ineffective over time or be too unhealthy for food cultivation. Tropical soils left bare of forest for too long eventually turn into hard, infertile, sunbaked clay called *laterite*.

Where population densities are relatively low, subsistence farmers can practice *sustainable shifting cultivation* indefinitely, but to allow for long fallow periods and still support human populations, this system requires larger areas than other types of agriculture. If population density increases, farmers are usually forced to shorten fallow periods, thus inhibiting forest regrowth and soil regeneration. Even though individual plots are small, the plots eventually become too close to one other because shifting cultivation requires clearing forest at each move. Where rural population densities are high, shifting cultivation now accounts for a significant portion of the region's deforestation. Moreover, because forest debris is usually burned, shifting cultivation can result in wildfires. This is especially true during an El Niño period, when rainfall is low. El Niño–induced wildfires have increased in recent years, particularly in Indonesia, further contributing to deforestation and carbon emissions there.

Wet Rice Cultivation

Another major indirect contributor to global climate change is Southeast Asia's most productive form of agriculture. *Wet rice cultivation* (sometimes called *paddy rice*) entails planting rice seedlings by hand in flooded and often terraced fields on land that was once forested (see Figure 10.9D). The fields are first cultivated with hand-guided plows pulled by water buffalo or tractors. Wet rice cultivation has transformed landscapes throughout Southeast Asia. It is practiced throughout this generally well-watered region, especially on rich volcanic soils and in places where rivers and streams bring a yearly supply of silt.

The flooding of rice fields also results in the production of methane, a powerful GHG that is responsible for about 20 percent of GHG emissions. It is estimated that up to one-third of the world's methane is released from flooded rice fields, where organic matter in soil undergoes fermentation as oxygen supplies are cut off. Wet rice has been cultivated for thousands of years, but the increase in human population in the last 25 years has driven a 17 percent expansion of the area devoted to wet rice. The map in Figure 10.10 shows patterns of field and forest crops across the region.

Commercial Agriculture

During the recent decades of relative prosperity, small farms once operated by families have been combined into large commercial farms owned by local or multinational corporations. These farms produce cash crops for export, such as rubber, palm oil, bananas, pineapples, tea, and rice (see Figure 10.10A and the figure map). Commercial farming on large tracts of deforested land reduces the need for labor by using mechanization, irrigation, and chemicals for fertilizer and pest control. The objectives of commercial farming are to generate big yields and quick profits, not to develop long-term sustainable agriculture. Many commercial farmers have achieved dramatic boosts in harvests (especially of rice) by using high-yield crop varieties, the result of green revolution research that has been applied in many parts of the world (see Chapter 8).

As we have noted in relation to other world regions, large-scale commercial (so-called *green revolution*) farming ultimately has significant negative environmental effects, including the loss of wildlife habitat and hence the loss of biodiversity (see Figure 10.9B); increases in soil erosion, flooding, and chemical pollution; and the depletion of groundwater resources. In

FIGURE 10.10 Agricultural patterns in Southeast Asia. Little virgin forest remains in this region. Tropical forestlands are still used for subsistence (food) and small-scale cash crops like coconut and spices. A wide variety of commercial food crops, such as rice, oil palm, rubber, bananas, citrus, and coffee, are grown for local consumption and export. [Source consulted: *Hammond Citation World Atlas* (Maplewood, NJ: Hammond, 1996), pp. 74, 83, 84]

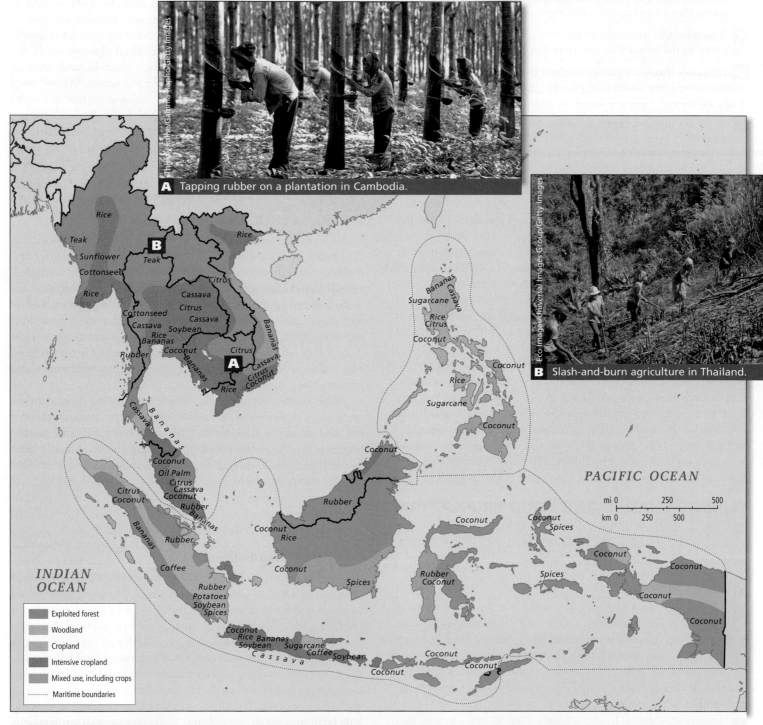

A Tapping rubber on a plantation in Cambodia.

B Slash-and-burn agriculture in Thailand.

Legend:
- Exploited forest
- Woodland
- Cropland
- Intensive cropland
- Mixed use, including crops
- ········ Maritime boundaries

Thinking Geographically

After you have read this entire section on agricultural patterns in Southeast Asia, you will able to answer the following questions.

- What happens to CO_2 in the atmosphere when there are fewer trees due to legal or illegal logging?

- On plantations of all types, native plants used for subsistence are often displaced by marketable crop plants. If this happens for long, the native plants may become extinct. What might be a way in which using slash-and-burn agriculture helps farmers avoid this particular impact on native plants?

- How might the different goals of the various agricultural systems affect climate change patterns?

addition, subsistence farmers unable to afford the new technologies find they cannot compete with large commercial farms and are forced to migrate to cities to look for work.

CLIMATE CHANGE IMPACTS RELATED TO WATER

Many of Southeast Asia's potential vulnerabilities to global climate change are related to water resources. Four areas of vulnerability to warming temperatures may affect the region's economy and food supply: glacial melting, increased evaporation, coral reef bleaching, and storm surge flooding (FIGURE 10.11). Also, human intervention in river systems, especially efforts to both control the water and to use it for electricity generation, increases vulnerability by interfering with ecological balance.

Melting Glaciers and Increased Evaporation

Like much of Asia, mainland Southeast Asia's largest rivers (the Irrawaddy, Salween, Mekong, and Red Rivers) are fed during the dry season (November through March) by glaciers high in the Himalayas. These glaciers are now melting at a rate that could result in their eventual disappearance. As glacial melting accelerates, the immediate risk is catastrophic flooding. While some areas have long been adapted to dramatic seasonal floods, which have taken place in this region for most of human history, many areas are unprepared for floods (see Figure 10.11D).

A longer-term concern is that dry-season flows in the rivers where glacial melt water has been the primary source of stream flow will be reduced to a trickle. As much as 15 percent of this region's rice harvest depends on the dry-season flows of the major rivers. The loss of these harvests would reduce many farmers' incomes and create food shortages for them and the cities they supply. Coming on top of a global rise in food prices in recent years, this would place further strain on the well-being of poor people throughout the region.

The higher temperatures associated with current trends in global climate change mean evaporation rates will rise, resulting in drier conditions in fields, lower lake levels, and lower fish catches because of changing habitats for aquatic animals. Evaporation that results in reduced river and groundwater levels can also cause saltwater intrusions into estuaries and freshwater aquifers.

Coral Reef Bleaching

Global climate change is expected to increase sea temperatures in Southeast Asia, threatening the coral reefs that sustain much of the region's fishing and tourism industries. A coral reef is an intricate structure that is composed of the calcium-rich skeletons of millions of tiny living creatures called coral polyps (see Figure 10.6B). The polyps are subject to **coral bleaching** (see Figure 10.11C), or color loss, which results when photosynthetic algae that live in the corals are expelled by a variety of human-instigated or naturally occurring changes. Both rising water temperature related to climate change and oceanic acidification resulting from the absorption of excess CO_2 in the atmosphere can cause bleaching. Ocean acidification is thought to have increased by about 30 percent over the last 340 years, a rate of change that may exceed the ability of ocean creatures to adapt. Bleaching also occurs in response to pollution from urban sources, such as sewage discharges, storm water runoff, and industrial water pollution.

Fish catches can be affected by bleaching as well. Many of the fish caught in Southeast Asia's seas depend on healthy coral reefs for their reproduction and survival. Under normal conditions, coral that is assaulted with just one of the bleaching situations recovers within weeks or months. However, severe or repeated bleaching can cause corals to die. Unprecedented global coral bleaching events affecting roughly half of the world's coral reefs took place in 1998, 2002, 2004, and 2006, and regional events occur somewhere every year. The thousands of rural communities throughout coastal Southeast Asia that rely on fish for food are thus also threatened by coral bleaching. So far, however, the greatest observable impacts on people have been in the tourism industry. In the Philippines, the coral bleaching of 1998 brought a dramatic decline in tourists who normally came to dive the country's usually spectacular reefs, resulting in a loss of about U.S.$30 million to the economy.

Storm Surges and Flooding

Although the relationship is not yet entirely understood, violent tropical storms seem to be increasing in both number and intensity as the climate warms. When Typhoon Haiyan hit in 2013, it became the strongest typhoon to make its way onto land in recorded history. Haiyan struck the Philippines, killing at least 5600 people and leaving millions homeless. Climatologists studying data on tropical storms predict that the entire Southeast Asian region will have a somewhat higher number of typhoons over the next few years and that the duration of peak winds along coastal zones will increase. This is significant because many poor, urban migrants have crowded into precarious dwellings in low-lying coastal cities such as Manila, Bangkok, Rangoon, and Jakarta, where coping with high winds, flooding, and the aftermath of storms will likely be a frequent experience (see Figure 10.11A).

RESPONSES TO CLIMATE CHANGE IN THE MEKONG RIVER SYSTEM

Many of the drivers of climate change lie beyond the policy control of Southeast Asians. Glacial melting is taking place beyond their borders. Temperature increases in oceans and lakes and increased evaporation rates must be addressed at the global scale. But Southeast Asians can drastically reduce CO_2 emissions by limiting the removal of forests for lumber exports and commercial agriculture.

Another way that the people of this region can moderate climate change is to reduce their own fossil fuel consumption by turning to alternative sources of energy. The demand for energy has doubled and doubled again just during the last decade as modernization increases. Both the Philippines and Indonesia have significant potential for generating electricity from geothermal energy (heat stored in Earth's crust). This energy is particularly accessible near active volcanoes, which both countries have in abundance. The Philippines already generates 27 percent of its electricity from geothermal energy (2015) and has 29 more projects under development. In Indonesia, by some estimates, geothermal energy could eventually provide a majority of energy needs; 62 projects are under development there. Solar energy is another attractive source, given that the entire region lies near the equator, the part of the planet

coral bleaching color loss that results when photosynthetic algae that live in corals are expelled

FIGURE 10.11 PHOTO ESSAY: Vulnerability to Climate Change in Southeast Asia

Much of Southeast Asia's vulnerability to climate change relates to water. Here we explore vulnerabilities related to tropical storms, flooding, and coral reefs.

A A father and his children on Mindanao, Philippines, sift through the wreckage of Typhoon Bopha in search of coconuts. Typhoons (known in the Atlantic as hurricanes) are projected to increase in frequency as a result of the warmer temperatures created by climate change.

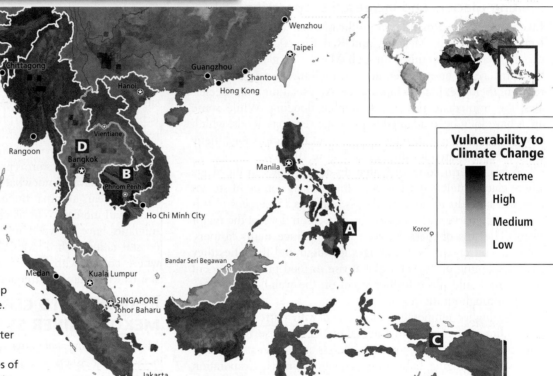

Vulnerability to Climate Change

Extreme
High
Medium
Low

B A husband and wife harvest shrimp and fish on Cambodia's Tonlé Sap Lake. Sixty percent of Cambodia's protein intake comes from this inland freshwater lake. Higher temperatures related to climate change may be increasing rates of evaporation from the lake, causing water levels to drop and fish catches to decline.

C A partially bleached coral in Cenderawasih Bay, West Papua, Indonesia. Climate change is raising ocean temperatures, resulting in more coral bleaching events that severely degrade reefs, which are home to many aquatic life forms.

D Residents of Bangkok, Thailand, paddle along a flooded street during catastrophic flooding, which is becoming more common in the city. Climate change could increase flooding along rivers as patterns of rainfall and glacial melting change.

Thinking Geographically

After you have read about vulnerability to climate change in Southeast Asia, you will able to answer the following questions about photos in this chapter.

A Typhoons are particularly significant for people living on which landforms?

C The text describes several causes of coral bleaching. Explain the link between coral bleaching and human activity in the case of rising ocean temperatures.

D How might flood-resistant cities help people adapt to climate change?

that receives the most solar energy. Wind energy may be cost-effective in Laos and Vietnam, where many population centers are in high-wind areas. But the energy source that is most popular and expanding the fastest is hydroelectric power.

Hydroelectric Dams and Their Consequences

The Mekong is one of Earth's longest and most complex rivers. Beginning high in the Himalayas as a tiny glacier-fed trickle, the Mekong (called the *Lancang* in China) flows through deep gorges, gaining speed and volume and picking up sediment for a thousand miles or more, through Yunnan, China. It skirts the borders of far-eastern India and Burma (Myanmar) before entering Laos, the halfway point on its journey to the sea. In this first half of its path, the Mekong, with its hundreds of tributaries, drains a huge area of China, bits of India, and northern Burma (Myanmar). The river system supports a remarkable amount of biodiversity, rivaled only by that of the Congo River and the Amazon: 20,000 species of plants and 2500 species of animals live here, including freshwater dolphins, giant catfish, and spiders the size of grapefruits. The ethnic diversity is also astonishing, with hundreds of cultures intricately attuned to the variable physical environment of the river basin. In its entirety, the Mekong sustains around 60 million people, mostly rural and poor.

As the Mekong flows southeast through Laos, its silt-laden waters descend from rugged mountain territory into rolling hills and then floodplains, where the silt is deposited as floodwaters slow down and spread out. Periodic floods pick up and redeposit the nutrient-rich silt over and over, providing farmers with natural fertilizer and fisheries with food. By the time the Mekong reaches Cambodia, it is a wide, slow, meandering river in a vast floodplain. In Vietnam it divides into **distributaries** that fan out to form the Mekong Delta.

Modernity is arriving late but bringing quick change. Most everyone now has access to cell phones, which need electricity to work. Electricity is essential to nearly every other aspect of development likely to come soon, hence the emphasis on hydro-electric power, which is especially appealing in China, where carbon emissions cause frequent traffic-stopping pollution. At least 86,000 hydroelectric dams have been built across China since the 1960s. By the time the Mekong leaves China at the Laotian border, it has already passed through 20 sites where dams are operational, under construction, or planned, with more on the river's many tributaries. Laos plans 9 Mekong dams; Cambodia, 2 more. Laos, whose main resources are its fertile soil and the water of the Mekong, plans to profit from the sales of the power it generates from dams. Thailand is already a major customer for

Laotian hydropower. The fact that soil fertility and river water quality will be inevitably changed by the dams has not yet gained much recognition by Laotian planners.

Dams have numerous consequences for the environment. Although dams can be used to control damage from flooding, floods serve a purpose. The natural seasonal ebb and flow of river water distributes moisture and nutrients downstream. It is now apparent that seasonal variations in flow are essential for the health of many plant and animal communities, and by extension, human communities. Dams restrict the movement and hence the reproduction of freshwater fish, and create vast reservoirs that cover villages and valuable cropland and produce hospitable habitat for disease-bearing mosquitoes.

As scientists have gained a better understanding of ecological nuances, they have warned that even expert planners tend to underestimate the damage done by the construction of dams and their operation over long periods of time. Planners, meanwhile, often overestimate the benefits that dams can produce. As the demand for energy increases, it is tempting to ignore the negative externalities of creating that energy with hydroelectric dams.

THINGS TO REMEMBER

GEOGRAPHIC THEME 1 • **Environment** Many of Southeast Asia's most critical environmental issues relate in some way to climate change. Deforestation, a major global source of greenhouse gas emissions, has been rapid; and the region is vulnerable to the climate change effects of flooding and droughts. Human manipulation of the land and resources is changing ecological systems and threatening food security. Rising ocean temperatures are straining aquatic ecosystems and storm surges threaten human settlements.

• The expansion of palm oil cultivation has resulted in extensive deforestation that has released enormous amounts of carbon into the atmosphere.

• Much of this region's rice harvest depends on the dry-season flows of the major rivers of mainland Southeast Asia for irrigation. But these rivers are threatened by glacial melting.

• Climate change poses special challenges for people in low-lying cities and in typhoon zones.

• Southeast Asians have a number of strategies available to ameliorate the impacts of climate change.

HUMAN PATTERNS OVER TIME

First settled in prehistory by migrants from the Eurasian continent, Southeast Asia was later influenced by Chinese, Indian, and Arab traders. Later still, it was colonized by Europe (from the 1500s to the early 1900s). The Philippines were acquired by the United States at the end of the Spanish-American War and were occupied by the United States from 1898 to 1946. During World War II, much of Southeast Asia was occupied by Japan. By the late twentieth century, occupation and domination by outsiders had ended and the region was profiting from selling manufactured goods to its former colonizers. Like Middle and South America, Africa,

distributaries branches of a river in its delta that fan out and carry water away from the main channel

and South Asia, Southeast Asia has had more success than other areas in achieving widespread, if modest, prosperity from its new links to the global economy. It has done so largely by following the example of some of East Asia's most successful countries, such as Japan, Korea, Taiwan, and more recently, China.

THE PEOPLING OF SOUTHEAST ASIA

The modern indigenous populations of Southeast Asia arose from two migrations widely separated in time. In the first migration, about 40,000 to 60,000 years ago, **Australo-Melanesians,** a group of hunters and gatherers from what are now northern India and Burma (Myanmar), moved to the exposed landmass of Sundaland and into present-day Australia. Their descendants still live in Indonesia's easternmost islands and in small, usually remote pockets on other islands and the Malay Peninsula, and in Australia, New Guinea, and other parts of Oceania (see the Figure 11.1 map).

In the second migration (from 6000 to 10,000 years ago, after the last ice age), people from southern China began moving into Southeast Asia. Their migration gained momentum about 5000 years ago, when a culture of skilled farmers and seafarers from southern China, the **Austronesians,** migrated first to what is now Taiwan, then to the Philippines, and then into the islands of Southeast Asia and the Malay Peninsula (see Figure 10.13A). Some of these adventuresome sea travelers eventually moved westward to southern India and to Madagascar (off the east coast of Africa; see Chapter 7), and eastward to the far reaches of the Pacific islands (see Figure 11.13).

DIVERSE CULTURAL INFLUENCES

Over the last several thousand years, Southeast Asia has been—and continues to be—shaped by a steady circulation of cultural influences, both internal and external. Merchants, religious teachers, and sometimes even invading armies from the Indian subcontinent, Southwest Asia, China, Japan, and finally Europe came to the region by overland trade routes and the surrounding seas. These newcomers, whether peaceful or aggressive, introduced religions (Buddhism, Hinduism, Islam, Daoism, Christianity), trade goods (such as cotton textiles), and food plants (such as mangoes and tamarinds) deep into the Indonesian and Philippine archipelagos and throughout the mainland. The monsoon winds, which blow from the west in the spring and summer, facilitated access for merchant ships from South Asia and the Persian Gulf. The ships sailed home on winds blowing from the east in the autumn and winter. These easterly winds carried ships with spices, bananas, sugarcane, silks, and other East and Southeast Asian items, as well as people, to the wider world.

Australo-Melanesians a group of hunters and gatherers who moved approximately 40,000 to 60,000 year ago from what are now northern India and Burma (Myanmar) to the exposed landmass of Sundaland and present-day Australia

Austronesians a group of skilled farmers and seafarers from southern China who migrated south to various parts of Southeast Asia between 6000 and 10,000 years ago

Religious Legacies

Spatial patterns of religion in Southeast Asia reveal an island–mainland division that reflects the history of influences from India, China, Southwest Asia, and Europe. Both Hinduism and Buddhism arrived thousands of years ago by monks and traders who traveled by sea and along overland trade routes

that connected India and China through Burma (see Figure 10.13B). Early Southeast Asian kingdoms and empires switched back and forth between Hinduism and Buddhism as their official religion. Spectacular ruins of these Hindu–Buddhist empires are scattered across the region; the most famous is the city of Angkor in what is now Cambodia. At its zenith in the 1100s, Angkor was among the largest cities in the world, and its ruins are now a World Heritage Site (see Figure 10.16A). Today, Buddhism dominates mainland Southeast Asia, while Hinduism remains dominant only on the Indonesian islands of Bali and Lombok.

In Vietnam, people practice a mix of Buddhist, Confucian, and Taoist customs that reflect the thousand years (ending in 938 C.E.) when Vietnam was part of various Chinese empires. China's traders and laborers also brought cultural influences to scattered coastal zones throughout Southeast Asia. Islam is now dominant in the islands of Southeast Asia. Islam came mainly through South Asia after India was conquered by Mughals in the fifteenth century. Muslim spiritual leaders and traders converted many formerly Hindu–Buddhist kingdoms in Indonesia, Malaysia, and parts of the southern Philippines. Islam remains dominant in Malaysia and in Indonesia, which has the largest Muslim population on Earth, and in the southern Philippines. Protestant Christianity was introduced by the Dutch but did not catch on. Roman Catholicism is the predominant religion in Timor-Leste and in most of the Philippines, which were colonized by Portugal and Spain, respectively.

EUROPEAN COLONIZATION

Between the sixteenth and the twentieth centuries, several European countries established colonies or quasi-colonies in Southeast Asia **(FIGURE 10.12)**. Drawn by the region's fabled spice trade, the Portuguese established the first permanent European settlement in Southeast Asia at the port of Malacca, Malaysia, in 1511. Although better ships and weapons gave the Portuguese an advantage, their anti-Islamic and pro-Catholic policies provoked strong resistance in Southeast Asia. Only on the small island of Timor-Leste did the Portuguese establish Catholicism as the dominant religion.

Arriving first in 1521, the Spanish had established trade links across the Pacific between the Philippines and their colonies in the Americas by 1540 **(FIGURE 10.13C)**. Like the Portuguese, they practiced a style of colonial domination grounded in Catholicism, but they met less resistance because of their greater tolerance of non-Christians. The Spanish ruled the Philippines for more than 350 years; as a result, except for the southernmost islands, which retain strong indigenous and long-standing Muslim influences, the Philippines is the most deeply Westernized and certainly the most Catholic part of Southeast Asia.

The Dutch were the most economically successful of the European colonial powers in Southeast Asia. From the sixteenth to the nineteenth centuries, they extended their control of trade over most of what is today called Indonesia, known at the time as the Dutch East Indies. The Dutch became interested in growing cash crops for export. Between 1830 and 1870, they forced indigenous farmers to leave their own fields and work part time without pay on Dutch coffee, sugar, and indigo plantations. The resulting disruption of local food production systems caused severe famines and provoked resistance that often took the form of Islamic religious movements. Such resistance movements hastened the

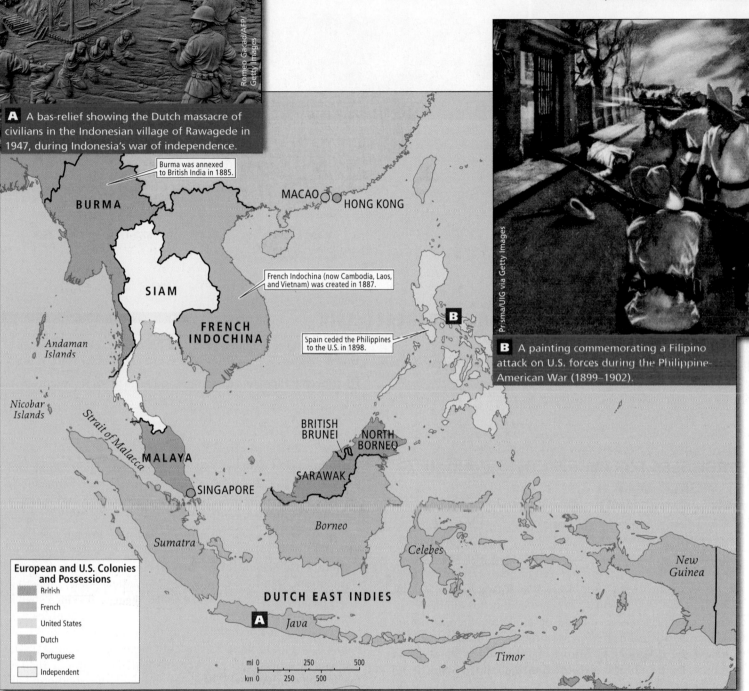

A A bas-relief showing the Dutch massacre of civilians in the Indonesian village of Rawagede in 1947, during Indonesia's war of independence.

Burma was annexed to British India in 1885.

French Indochina (now Cambodia, Laos, and Vietnam) was created in 1887.

Spain ceded the Philippines to the U.S. in 1898.

B A painting commemorating a Filipino attack on U.S. forces during the Philippine-American War (1899–1902).

MACAO   HONG KONG

BURMA

SIAM

FRENCH INDOCHINA

Andaman Islands

Nicobar Islands

Strait of Malacca

MALAYA

SINGAPORE

BRITISH BRUNEI   NORTH BORNEO

SARAWAK

Borneo

Sumatra

Celebes

New Guinea

European and U.S. Colonies and Possessions

- British
- French
- United States
- Dutch
- Portuguese
- Independent

DUTCH EAST INDIES

A *Java*

Timor

ml 0   250   500
km 0   250   500

FIGURE 10.12 European and U.S. colonies in Southeast Asia, 1914. Of the present-day countries in Southeast Asia, only Thailand (formerly called Siam) was never colonized. [Source consulted: *Hammond Times Concise Atlas of World History* (Maplewood, NJ: Hammond, 1994), p. 101]

spread of Islam throughout Indonesia, where the Dutch had made little effort to spread their Protestant version of Christianity.

Beginning in the late eighteenth century, the British established colonies at key ports on the Malay Peninsula. They held these ports both for their trade value and to protect the Strait of Malacca, the shortest passage for sea trade between Britain's empire in India and China. In the nineteenth century, Britain extended its rule over the rest of modern Malaysia in order to benefit from Malaysia's tin mines and plantations. Britain also added Burma to its empire, which provided access to forest resources and overland trade routes into southwest China.

The French first entered Southeast Asia as Catholic missionaries in the early seventeenth century. They worked mostly in the eastern mainland area in the modern states of Vietnam, Cambodia, and Laos. In the late nineteenth century, spurred by rivalry with Britain and other European powers for access to the markets of nearby China, the French formally colonized the area, which became known as French Indochina.

In all of Southeast Asia, the only country not to be colonized by Europe was Thailand (then known as Siam). Like Japan, it protected its sovereignty through both diplomacy and a vigorous drive toward European-style modernization.

A Ifugao tribal people, one of several Austronesian ethnic groups in the Philippines.

B The ruins of Bagan, capital of many Buddhist kingdoms since at least 874 C.E.

C A monument to the landing of Ferdinand Magellan in the Philippines, 1521.

| 60,000 B.C.E. | 40,000 B.C.E. | 20,000 B.C.E. | 8000 B.C.E. | 500 B.C.E. | 0 C.E. | 500 C.E. | 1000 C.E. | 1500 C.E. |

60,000–40,000 B.C.E. Austro-Melanesian migrations

8000 B.C.E. Austronesian migrations

c. 200 B.C.E.–Present Hindu–Buddhist kingdoms

874–1369 C.E.

1521

FIGURE 10.13 VISUAL HISTORY OF SOUTHEAST ASIA

Thinking Geographically

After you have read about the human history of Southeast Asia, you will be able to answer the following questions.

A From where did those who populated this region in prehistory come?

B How did Hinduism, Buddhism, and Islam arrive in Southeast Asia?

STRUGGLES FOR INDEPENDENCE AIDED BY WORLD WAR II

Agitation against colonial rule began in the late nineteenth century when Filipinos fought first against Spain in 1896. They then resisted control by the United States, which began in 1898 after the Spanish-American War (see the map in Figure 10.12). However, the Philippines and the rest of Southeast Asia did not win independence until after World War II (see Figure 10.12A, 10.13D, and 9.14). By then, Europe's ability to administer its colonies had been weakened by the devastation of the war, during which Japan had conquered most of the parts of Southeast Asia that had been controlled by Europe and the United States. Japan held these lands until it was defeated by the United States at the end of the war. By the mid-1950s, the colonial powers had granted self-government to most of the region, and all of Southeast Asia was independent by 1975.

The Vietnam War

The bitterest battle for independence took place in French Indochina (the territories of present-day Vietnam, Laos, and Cambodia), acquired by France in the nineteenth century. Although all three became nominally independent in 1949, France retained political and economic power over them. Various nationalist leaders, most notably Vietnam's Ho Chi Minh, headed resistance movements against continued French domination. After failed diplomatic efforts in Europe and the United States, the resistance leaders accepted military assistance from Communist China and the Soviet Union, despite ancient antipathies toward China for its previous millennia of domination. In this way, the Cold War was brought to mainland Southeast Asia.

In 1954, Ho Chi Minh defeated the French at Dien Bien Phu in northern Vietnam. The United States stepped in because it had become worried about the spread of international communism should the anticolonial resisters, now supported by communists, succeed. The **domino theory**—the idea that if one country "falls" to communism, other nearby countries will follow—was a major influence in this decision, because both North Korea and China had recently become communist. The Vietnamese resistance, which controlled the northern half of the country, attempted to wrest control of the southern half of Vietnam from the U.S.-supported and quite corrupt South Vietnamese government. The pace of the war accelerated in the mid-1960s. After many years of brutal conflict, public opinion in the United States forced a U.S. hasty withdrawal from the conflict in 1973. The civil war continued in Vietnam, finally ending in 1975, when the North defeated the South and established a new national government.

More than 4.5 million people died during the Vietnam War, including more than 58,000 U.S. soldiers. Another 4.5 million on both sides were wounded, and bombs, napalm, and chemical defoliants ruined much of the Vietnamese environment. Land mines continue to be a hazard to this day, and the effects of the highly toxic defoliant known as Agent Orange are still causing debilitating birth defects among many rural Vietnamese and Laotian people (see Figure 10.13E). The withdrawal from

domino theory a foreign policy theory that used the idea of the domino effect to suggest that if one country "fell" to communism, others in the neighboring region would also fall

D Japanese forces land in the Philippines during World War II.

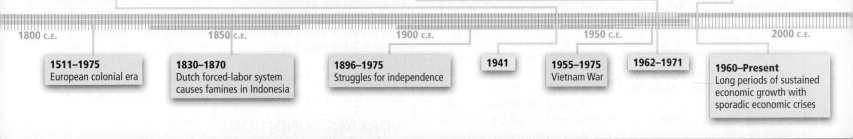

E The defoliant Agent Orange was sprayed over Vietnam and Laos between 1962 and 1971 to destroy crops and expose North Vietnamese troops.

F Bangkok, Thailand, is emerging as a global center for manufacturing and trade.

1800 C.E.   1850 C.E.   1900 C.E.   1950 C.E.   2000 C.E.

1511–1975
European colonial era

1830–1870
Dutch forced-labor system causes famines in Indonesia

1896–1975
Struggles for independence

1941

1955–1975
Vietnam War

1962–1971

1960–Present
Long periods of sustained economic growth with sporadic economic crises

C Which European countries established colonies in Southeast Asia; and what marked the end of the colonial era?

D How did this region experience World War II and its aftermath?

E How did rural Vietnamese and Laotian people get exposed to Agent Orange?

Vietnam in 1973 ranks as one of the most profound military defeats in U.S. history. After the war, the United States crippled Vietnam's recovery by imposing severe economic sanctions that lasted until 1993. Since then, however, the United States and Vietnam have developed a significant trade relationship.

The "Killing Fields" in Cambodia

In Cambodia, where the Vietnam War had spilled over the border, a particularly violent revolutionary faction called the Khmer Rouge seized control of the government in 1975. Inspired by the vision of a rural communist society, they attempted to destroy virtually all traces of European influence. They targeted Western-educated urbanites in particular, forcing them into labor camps where more than 2 million Cambodians—one-quarter of the population—starved or were executed in what became known as the "Killing Fields."

In 1979, Vietnam deposed the Khmer Rouge and, until 1989, ruled Cambodia through a puppet government. A 2-year civil war then ensued. Despite major UN efforts throughout the 1990s to establish a multiparty democracy in Cambodia, the country remained plagued by political tensions between rival factions and by government corruption.

In March 2009, Kang Kek Iew, the first of the Khmer Rouge leaders to be tried for war crimes and genocide, was forced to listen to and watch lengthy accounts of the torture of men, women, and children that he supervised. He was convicted in 2010 and sentenced to 35 years in prison. Most operatives in the killing fields will never be prosecuted.

THINGS TO REMEMBER

• The modern indigenous populations of Southeast Asia arose from two migrations, one about 40,000 to 60,000 years ago (the Australo-Melanesians), and one 6000 to 10,000 years ago (the Austronesians).

• Today, Buddhism dominates mainland Southeast Asia, while Hinduism remains dominant only on the Indonesian islands of Bali and Lombok.

• Islam is now dominant in some of the countries of Southeast Asia, especially Indonesia and Malaysia and in the southern Philippines.

• Between the sixteenth and the twentieth centuries, several European countries, the United States, and Japan established a series of colonies or quasi-colonies that covered almost all of Southeast Asia.

• Colonial rule in Southeast Asia began in 1511 with the Portuguese and came to a violent end in Vietnam and Cambodia, where war took the lives of more than 4.5 million people, including more than 58,000 U.S. soldiers.

GLOBALIZATION AND DEVELOPMENT

GEOGRAPHIC THEME 2

Globalization and Development: Globalization has brought both spectacular successes and occasional declines to the economies of Southeast Asia. Key to the economic development of this region are strategies that were pioneered earlier in East Asia: the formation of state-aided market economies, combined with export-led economic development. Regional economic cooperation is expanding slowly.

During and immediately after the era of European colonialism, trade among the countries within the region was inhibited by the fact that they all exported similar goods—primarily food and raw materials—and imposed tariffs against one another. They imported consumer products, industrial materials, machinery, and fossil fuels, mostly from the developed world.

BORROWED IDEAS FOR ECONOMIC GROWTH

Beginning in the 1960s, some national governments in Southeast Asia were able to create strong and sustained economic expansion by emulating two strategies for economic growth pioneered earlier by Japan, Taiwan, and South Korea (see Chapter 9). One was the formation of state-aided market economies. That is, national governments in Indonesia, Malaysia, Thailand, and to some extent the Philippines intervened strategically in financial institutions to make sure that certain economic sectors developed; in addition, investment by foreigners was limited so that the governments could have more control over the direction of the economy. The other strategy was export-led economic development, which focused investment on industries that manufactured products for export, primarily to developed countries. These strategies amounted to a limited and selective embrace of globalization in that global markets for the region's products were sought but foreign sources of capital were not.

These approaches were a dramatic departure from those used in other developing areas. In Middle and South America and parts of Africa, postcolonial governments relied on import substitution industries that produced low-quality manufactured goods mainly for local use. By contrast, export-led growth allowed Southeast Asia's industries to produce high-quality goods and to earn much more money in the vastly larger markets of the developed world. Standards of living increased markedly, especially in Malaysia, Singapore, Indonesia, and Thailand. Other important results of Southeast Asia's success were a decrease in wealth disparities and improvements in vital statistics—lower infant and maternal mortality, longer life expectancy, and lower population growth. By the 1990s, Southeast Asian countries had some of the highest economic growth rates in the world. The growth was based on an economic strategy that emphasized the export of manufactured goods—initially clothing and then more sophisticated technical products.

EXPORT PROCESSING ZONES

In the 1970s, some governments in the region began adopting an additional strategy for encouraging economic development. This time, they sought foreign sources of capital, but the places in which those sources could invest were limited to specially designated free trade areas. Such export processing zones (EPZs; see the discussion in Chapter 3) are places in which foreign companies can set up their facilities using inexpensive, locally available labor to produce items only for export. (Maquiladoras in Mexico are an example of companies set up by foreign firms.) Taxes are eliminated or greatly reduced as long as the products are re-exported (not sold within the country). Since the 1970s, EPZs have expanded economic development in Malaysia, Indonesia, Vietnam, and

feminization of labor the rising numbers of women in both the formal and the informal labor force

the Philippines, and are now used widely in China and Middle and South America, and to a lesser extent in sub-Saharan Africa.

THE FEMINIZATION OF LABOR

Between 80 and 90 percent of the workers in the EPZs are women, not only in Southeast Asia but in other world regions as well (FIGURE 10.14). The **feminization of labor** has been a distinct characteristic of globalization over the past three decades (see Chapter 9). Employers prefer to hire young, single women because they are perceived as the least expensive, least troublesome employees. Statistics do show that, generally, women across the globe will work for lower wages than men, will not complain about poor and unsafe working conditions, will accept being restricted to certain jobs on the basis of sex, and are not as likely as men to agitate for promotions; though, of course, there are many exceptions. The reasons for this are complex and are discussed throughout this book.

WORKING CONDITIONS

In general, the benefits of Southeast Asia's "economic miracle" have been unequally apportioned. In the region's new factories and other enterprises, it is not unusual for assembly-line employees to work 10 to 12 hours per day, 7 days per week for less than the legal minimum

FIGURE 10.14 The feminization of labor. Women assemble circuit boards in a factory in Vietnam.

Steve Raymer/Getty Images

Thinking Geographically
Why do many employers prefer workers to be young, single women?

wage and without benefits. Labor unions that typically would address working conditions and wage grievances are frequently repressed by governments, and international consumer pressure to improve working conditions at U.S.-based companies has been only partially effective. For example, the U.S.-based shoe company Nike has tried to sidestep the entire issue of customer complaints for years by outsourcing its manufacturing to non-U.S. contractors in the region. U.S. fair labor NGOs pursued one Nike subcontractor in Jakarta, Indonesia, where for 19 years workers were forced to work an extra hour every day without pay. The NGOs filed a lawsuit against the subcontractor, which they won, resulting in a $1 million settlement that will give the 4500 workers $222 each. However, given that there are more than 160,000 workers for Nike subcontractors in Indonesia alone, this settlement is but a drop in the bucket.

A much more powerful force that is improving working conditions and driving up pay is the growth of the service sector throughout the region. Many service sector jobs require at least a high school education, and competition for hiring these educated workers means that wages and working conditions are already better than in manufacturing and are likely to improve faster. The service sector now dominates the economy as a percentage of GDP and as percentage of employment in Singapore, Malaysia, Indonesia, the Philippines, and Thailand. In all other countries except Timor-Leste and Brunei, the service sector accounts for at least 40 percent of GDP, though agriculture still employs more than a majority of workers in Burma (Myanmar), Laos, and Timor-Leste. In other countries in the region, the service sector is approaching parity with the industrial sector as a percent of GDP and of employment. Only in Brunei is the industrial sector still dominant; there it is entirely based on oil production. In Timor-Leste, oil is the mainstay of GDP but employs only 10 percent of the population. Sixty-four percent of the people in Timor-Leste work in agriculture, which produces only 6 percent of the GDP.

ECONOMIC CRISIS AND RECOVERY: THE PERILS OF GLOBALIZATION

Periodically, economic crises have swept through this region, with varying effects on society. For example, in the late 1990s as part of an effort to open up national economies to the free market, and with a general push from the International Monetary Fund (IMF), Southeast Asian governments relaxed controls on the financial sector, which had traditionally been highly regulated. Soon, Southeast Asian banks were flooded with money from investors in the rich countries of the world who hoped to profit from the region's growing economies. Flush with cash and newfound freedoms, the banks often made reckless decisions. For example, bankers made risky loans to real estate developers, often for high-rise office building construction. As a result, many Southeast Asian cities soon had far too much office space, with millions of investment dollars tied up in projects that contributed little to economic development.

One of the forces that led bankers to make such bad decisions was a kind of corruption known as **crony capitalism.** In most Southeast Asian countries, as elsewhere, corruption is related to the close personal and family relationships between high-level politicians, bankers, and wealthy business owners. In Indonesia, for

example, the most lucrative government contracts and business opportunities were reserved for the children of former president Suharto, who ruled the country from 1967 to 1997. His children became some of the wealthiest people in Southeast Asia. This kind of corruption expanded considerably with the new foreign investment money, much of which was diverted to bribery or unnecessary projects that brought prestige to political leaders.

The cumulative effect of crony capitalism and the lifting of controls on banks was that many ventures failed to produce any profits at all. In response, foreign investors panicked, withdrawing their money on a massive scale. In 1996, before the crisis, there was a net inflow of U.S.$94 billion to Southeast Asia's leading economies. In 1997, inflows had ceased and there was a net outflow of U.S.$12 billion. This financial debacle forced millions of people into poverty and changed the political order in some countries. There were also some geographic aspects to the crisis: the banks involved were located primarily in Singapore and the major cities of Thailand, Malaysia, and Indonesia. These were the countries to feel the immediate effects of the crisis. While urban areas were hit hardest, eventually the effects filtered into the hinterland, and workers lost jobs even in remote areas.

The IMF Bailout and Its Aftermath

The IMF made a major effort to keep the region from sliding deeper into recession by instituting reforms designed to make banks more responsible in their lending practices. The IMF also required structural adjustment policies (SAPs; see Chapter 3), which required countries to cut government spending (especially on social services for poor people, such as health care and education) and abandon policies intended to protect domestic industries.

After several years of economic chaos and much debate over whether the IMF bailout and accompanying SAPs helped or hurt a majority of Southeast Asians, economies began to recover. In the largest of the region's economies, growth resumed by 1999; by 2006, the crisis had been more or less overcome. But the crisis of 1997–2006 remains an ominous reminder of the risks of globalization, and was the impetus to reform the SAP policies once so popular with the IMF (see discussion in Chapter 7).

The Impact of China's Growth

During the Southeast Asian financial crisis of 1997–2006, Singapore, Malaysia, and Thailand lost out to China in attracting new industries and foreign investors. By the early 2000s, China attracted more than twice as much foreign direct investment (FDI) as Southeast Asia. However, China's growth also became an opportunity for Southeast Asia. Singapore, Malaysia, Indonesia, Thailand, and the Philippines "piggy-backed" on China's growth by winning large contracts to upgrade China's infrastructure in areas such as wastewater treatment, gas distribution, and shopping mall development.

Southeast Asia has also used its comparative advantages over China to attract investment to itself. In poorer countries, such as Vietnam, wages have remained lower than in China. This has attracted investments in low-skill manufacturing that had been going to China. Some

crony capitalism a type of corruption in which politicians, bankers, and entrepreneurs, sometimes members of the same family, have close personal as well as business relationships

of the region's wealthier countries—Singapore, Malaysia, and Thailand—have been positioning themselves as locations that offer more highly skilled labor and better high-tech infrastructure than China does.

The Global Recession Beginning in 2007

The region was hit again by the effects of the global recession that began in 2007. Growth slowed markedly because high oil and food prices restricted disposable cash worldwide, and consumers in wealthy countries (especially in the United States and the EU countries, which are Southeast Asia's biggest trade partners) faced crippling debt that seriously curtailed their spending. By late 2009, East and Southeast Asia appeared to be recovering. In part, the recovery was based on the increased demand for goods and services within the domestic economies of China and Southeast Asia, where consumers now had some disposable cash. Southeast Asia's ability to respond to this demand was facilitated by policies that opened up intraregional trade and access to China.

But the positive developments of 2009 did not last. By 2012, financial troubles in Europe and slow recoveries in the United States and Japan all lowered demand for Southeast Asian products. Singapore, which has a highly sophisticated economy that is based on its banking services and the transfer of goods through its giant port, is particularly susceptible to slowdowns in rich countries. Manufacturing countries like Indonesia, Malaysia, and Thailand also fare best when there is demand for their products in international markets.

Cambodia, Laos, and Burma still rely on raw material exports to places like China. By 2015, China's economy had begun to slow, and in order to energize its economy and meet the consumer needs of its own people, China began discouraging imports in favor of Chinese-produced goods. Thus China's demand for raw materials and manufactured goods from Southeast Asia was reduced.

REGIONAL TRADE AND ASEAN

During the 1980s and 1990s, Southeast Asian countries traded more with China and the rich countries of the world than they did with each other. This issue of insufficient trade between Southeast Asian countries was the reason behind the creation of the **Association of Southeast Asian Nations (ASEAN),** an organization of Southeast Asian nations that has grown in strength over the years. The ASEAN member countries are Brunei, Burma (Myanmar), Cambodia, Indonesia, Laos, Malaysia, Philippines, Singapore, Thailand, and Vietnam; candidate members are Papua New Guinea and Timor-Leste. Close associates of ASEAN are China, Japan, South Korea, Australia, India, New Zealand, Russia, and the United States.

Association of Southeast Asian Nations (ASEAN) an organization of Southeast Asian governments that was established to further economic growth and political cooperation between member countries and with other areas of the world

social cohesion the willingness of members of a society to cooperate with each other in order for all to survive and prosper

Regional Integration

ASEAN started in 1967 as an anti-Communist, anti-China association, but it now focuses on agreements that strengthen regional cooperation, including agreements with China. One example is the Southeast Asian Nuclear Weapons–Free Zone Treaty signed in December 1995. Another is the ASEAN Economic Community, or AEC, a trade bloc patterned after the North American Free Trade Agreement and the European Union, which was formalized on December 31, 2015. Now, ASEAN focuses on increasing trade between members of the association. ASEAN has also ratified free trade agreements with Australia, New Zealand, China, India, Japan, and South Korea, and is a major participant in the Trans-Pacific Partnership (TPP) negotiations (not yet ratified by the United States and other countries, as of 2016).

The total intraregional trade between ASEAN countries in 2014 was more than trade with any other outside country or region; ASEAN countries represented 22.5 percent of total imports and 25.5 percent of total exports **(FIGURE 10.15)**. By 2015, ASEAN was addressing a variety of other issues that were aimed at encouraging such things as the free flow of services, skilled labor, investment, and capital between countries. ASEAN's overall stated goals are to develop a sustainable regional economic community that protects consumers and intellectual property rights and levels the playing field for competition.

Regional Inequalities

For all the emphasis on regional integration, the fact is that the 11 countries in the region of Southeast Asia vary widely in levels of economic development, as alluded to in the discussion above of the uneven performance in the growth of the service sector. Perhaps most telling are figures on gross national income (GNI) per capita, adjusted for purchasing power parity (PPP). Singapore, the smallest country, has a per capita GNI (PPP) of U.S.$80,270, one of the highest such figures in the world. The poorest country in the region is Cambodia, with a per capita GNI (PPP) of just U.S.$3080, one of the lowest such figures. Statistics on other aspects of human well-being are similarly divergent, as is discussed on pages 597–598 and shown in Figure 10.24. Part of the divergence in development and well-being is related to the variations in size and geographic features of countries in the region, but it is interesting to note that despite efforts at regional trade integration, there has been little region-wide effort to decrease the gross disparities in levels of development among ASEAN member states. There is no central bureaucracy similar to that in Brussels in the European Union, no agency like the European Commission that works toward **social cohesion** for the entire region; and there are no standards for financial transparency.

Piracy on the High Seas

An example of the weakness in regional solidarity that ASEAN has not managed to address is the reemergence of piracy in the waters of Southeast Asia. The narrow Malacca strait between Sumatra (Indonesia) and Malaysia has been the site of numerous hijackings (15 in 2014). Crews of small armed vessels take advantage of the crowded waters to sneak onto modest commercial cargo boats and tankers. More than 120,000 such small craft ply these waters every year. The pirates overpower the crew, hold the crewmembers captive, and quickly move the captured ship to one of a number of secluded small ports, from which they transfer or sell the cargo with the help of accomplices. Only a few deaths have resulted, but the losses are significant and governments as well as ASEAN officials have been unable to join forces to stop the phenomenon.

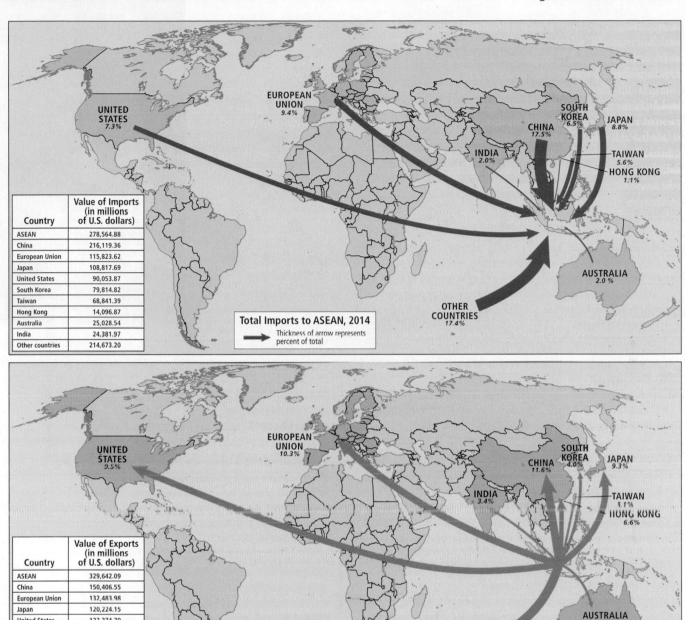

Country	Value of Imports (in millions of U.S. dollars)
ASEAN	278,564.88
China	216,119.36
European Union	115,823.62
Japan	108,817.69
United States	90,053.87
South Korea	79,814.82
Taiwan	68,841.39
Hong Kong	14,096.87
Australia	25,028.54
India	24,381.97
Other countries	214,673.20

Total Imports to ASEAN, 2014
Thickness of arrow represents percent of total

Country	Value of Exports (in millions of U.S. dollars)
ASEAN	329,642.09
China	150,406.55
European Union	132,483.98
Japan	120,224.15
United States	122,374.70
South Korea	51,624.21
India	43,325.81
Taiwan	39,472.10
Hong Kong	85,275.45
Australia	45,344.61
Other countries	172,226.10

Total Exports from ASEAN, 2014
Thickness of arrow represents percent of total

FIGURE 10.15 ASEAN imports and exports, 2014. The ASEAN countries are very active in world trade as importers and exporters. The goal of increasing intraregional trade is being reached. About one-quarter of the total regional trade is now between ASEAN countries, with imports at 22.5 percent and exports at 25.5 percent. [Source consulted: *ASEAN External Trade Statistics,* Table 20, "Top Ten ASEAN Trade Partner Countries/Regions, 2014," ASEAN, December 21, 2015, at http://www.asean.org/storage/2015/12/table20_asof121Dec15_R.pdf]

Tourism

International tourism is an important and rapidly growing economic activity in most Southeast Asian countries **(FIGURE 10.16)**. Between 1991 and 2014, the number of intra- and extra-ASEAN tourists went from about 20 million to 105 million. As in other trade matters, Southeast Asians are themselves increasingly engaging with neighboring countries **(FIGURE 10.17)**. This is a positive trend because familiarity between neighbors lays the groundwork for various forms of regional cooperation, including business ventures, cultural exchanges, law enforcement, and infrastructure improvements.

In response to its popularity with global and regional tourists, ASEAN members have been working to improve the region's transportation infrastructure. One such project is the

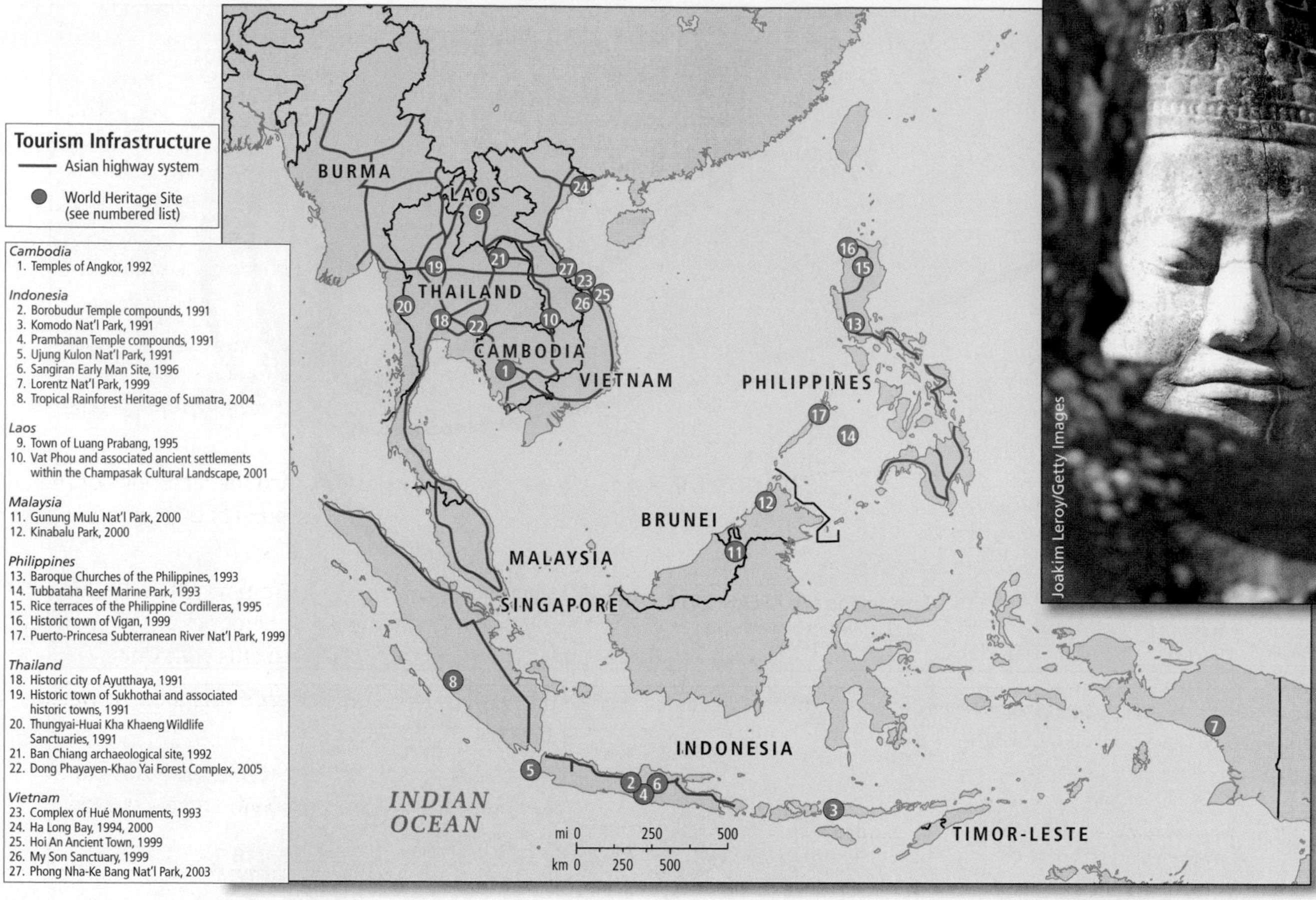
Tourism Infrastructure
Asian highway system
World Heritage Site (see numbered list)
Cambodia
1. Temples of Angkor, 1992
Indonesia
2. Borobudur Temple compounds, 1991
3. Komodo Nat'l Park, 1991
4. Prambanan Temple compounds, 1991
5. Ujung Kulon Nat'l Park, 1991
6. Sangiran Early Man Site, 1996
7. Lorentz Nat'l Park, 1999
8. Tropical Rainforest Heritage of Sumatra, 2004
Laos
9. Town of Luang Prabang, 1995
10. Vat Phou and associated ancient settlements within the Champasak Cultural Landscape, 2001
Malaysia
11. Gunung Mulu Nat'l Park, 2000
12. Kinabalu Park, 2000
Philippines
13. Baroque Churches of the Philippines, 1993
14. Tubbataha Reef Marine Park, 1993
15. Rice terraces of the Philippine Cordilleras, 1995
16. Historic town of Vigan, 1999
17. Puerto-Princesa Subterranean River Nat'l Park, 1999
Thailand
18. Historic city of Ayutthaya, 1991
19. Historic town of Sukhothai and associated historic towns, 1991
20. Thungyai-Huai Kha Khaeng Wildlife Sanctuaries, 1991
21. Ban Chiang archaeological site, 1992
22. Dong Phayayen-Khao Yai Forest Complex, 2005
Vietnam
23. Complex of Hué Monuments, 1993
24. Ha Long Bay, 1994, 2000
25. Hoi An Ancient Town, 1999
26. My Son Sanctuary, 1999
27. Phong Nha-Ke Bang Nat'l Park, 2003
BURMA
LAOS
THAILAND
CAMBODIA
VIETNAM
PHILIPPINES
BRUNEI
MALAYSIA
SINGAPORE
INDONESIA
TIMOR-LESTE
INDIAN OCEAN
mi 0 250 500
km 0 250 500

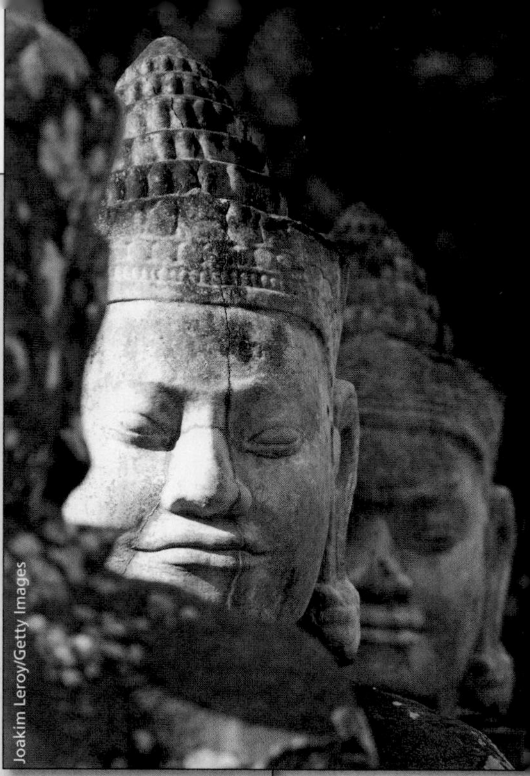
Joakim Leroy/Getty Images

FIGURE 10.16 Development of the Asian Highway: Transportation infrastructure for tourism. Many of the Southeast Asian World Heritage Sites are located along the partially completed Asian Highway, which will eventually connect Europe to Indonesia. The UNESCAP report, "Development of the Asian Highway," describes the project and includes a map of the entire system. The photo shows Buddha heads sculpted from stone at the Angkor Wat temple complex in Cambodia (number 1 on the map), the largest religious monument in the world. [Sources consulted: "World Heritage List," UN World Heritage Convention, at http://whc.unesco.org/en/list/; "Tourism Attractions Along the Asian Highway," UN Economic and Social Commission for Asia and the Pacific, 2004, at http://www.unescap.org/our-work/transport/asian-highway/about; and *Asia Times Online*, at http://www.atimes.com/atimes/Asian_Economy/images/highways.html]

Asian Highway, a web of standardized roads that loops through the mainland and connects it with Malaysia, Singapore, and Indonesia (the latter by ferry; see the Figure 10.16 map). Eventually, the Asian Highway will facilitate ground travel through 32 Eurasian countries, from Russia to Indonesia and from Turkey to Japan.

The surge of tourism in Southeast Asia is welcomed, but it also raises concerns about becoming too dependent on an industry that leaves economies vulnerable to events that precipitously stop the flow of visitors (natural disasters, political upheavals), or that leave local people vulnerable to the sometimes destructive demands of tourists (see the discussion of sex tourism on pages 603–604). Examples of disasters are the tsunami of December 2004 that killed several thousand people in Thailand and Indonesia, many of whom were international tourists; the giant typhoon, Haiyan, in the Philippines in 2013; and various human-made disasters, such as the terrorist bombings in Bali (2002, 2005) and in southern Thailand (2006, 2012). In addition, tourism can divert talent from contributing to national development, as the following vignette illustrates.

VIGNETTE Tan Phuc waits patiently for the mechanic to mount the new rear tire on his Chinese-made 125cc motorcycle. He is a motorcycle tour guide in Vietnam who likes to take his clients to places along Highway 14 in the Central Highlands, close to the historic Ho Chi Minh Trail. Tan's current client, a Dutch tourist, has spent 3 days filling his digital camera with images of vast coffee plantations covered in snowy blossoms, of silkworms frying in oil while their cocoons are unwound and spun into silk thread, and of indigenous minority children who met the gaze of his camera with casual curiosity. Tan finds his clients in bus stations and backpacker hostels and negotiates a daily rate for tours—usually between U.S.$50 and U.S.$75, not insignificant in a country that has an average per capita annual income of U.S.$2790.

Tan Phuc's life has always been affected by global forces. As a child in the Mekong Delta, he sold produce to U.S. soldiers at a nearby base. During the war, he lost his brother, a soldier for South Vietnam; after the war, his father spent 10 years in a Communist reeducation camp. Even with the hardships, Tan obtained an education. But when the annual inflation rate of

400 percent shrank his salary as a high school math teacher, Tan could not adequately meet his family's needs. Like other skilled Vietnamese people, Tan took advantage of Vietnam's transition to a market economy, which began in the 1980s, and established a business that caters to tourists. This means that he, along with many other educated Vietnamese people, no longer works in occupations crucial to Vietnam's future, like education. *[Source: The field work of Karl Russell Kirby. For detailed source information, see Text Sources and Credits.]* ■

FIGURE 10.17 Origin of international visitors to ASEAN countries, 2013. Southeast Asia remains the fastest-growing tourism market in the world. Arrivals have surged in Burma (Myanmar) (up 51 percent after recent political changes) and leveled off in Thailand, where riots and coups periodically scare off visitors. The biggest increase has come from intraregional travel—people in Southeast Asia visiting their neighbors. [Sources consulted: "ASEAN Yearbook, 2015": (A) http://www.asean.org/storage/images/2015/July /ASEAN-Yearbook/July%202015%20-%20ASEAN%20Statistical%20 Yearbook%202014.pdf; (B) http://www.asean.org/storage /images/2015/July/ASEAN-Yearbook/July%202015%20-%20 ASEAN%20Statistical%20Yearbook%202014.pdf]

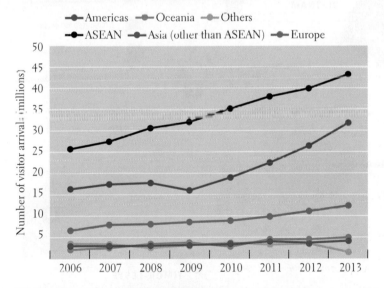

(A) Trends in visitor arrivals to ASEAN countries, 2006–2013

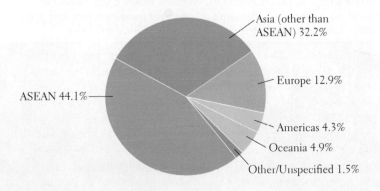

(B) Origin of international visitors to ASEAN countries, 2013

THINGS TO REMEMBER

GEOGRAPHIC THEME 2 • **Globalization and Development**
Globalization has brought both spectacular successes and occasional declines to the economies of Southeast Asia. Key to the economic development of this region are strategies that were pioneered earlier in East Asia: the formation of state-aided market economies, combined with export-led economic development. Regional economic cooperation is expanding slowly.

• The region was hit by a devastating recession in the late 1990s that lasted as late as 2006 in some places, and by a less severe slowdown related to the global recession that began in 2007.

• One of the forces that contributed to the recessions of the last two decades was a kind of corruption known as crony capitalism.

• ASEAN started in 1967 as an anti-Communist, anti-China association, but it now focuses on agreements that strengthen regional cooperation, including agreements with China.

• ASEAN has not yet addressed the wide disparities in wealth and well-being across the region.

• Regional integration, including the tourism infrastructure of the region, is expected to foster growth and encourage cooperation with China.

POWER AND POLITICS

GEOGRAPHIC THEME 3

Power and Politics: There has been a general expansion of political freedoms throughout Southeast Asia in recent decades, but authoritarianism, corruption, and violence have at times reversed these gains.

While there is a general shift away from authoritarianism and toward more political freedoms, the **FIGURE 10.18** map shows the great variation in political freedoms in the region. Some people argue that authoritarianism has deep cultural roots in Southeast Asia and should be given more respect. Others see a certain amount of authoritarianism as necessary to control corruption and political violence. Still others argue that respect for political freedoms is more likely to expose corruption and transform militant movements into peaceful political parties.

SOUTHEAST ASIA'S AUTHORITARIAN TENDENCIES

Some Southeast Asian leaders, such as Singapore's former prime minister, the late Lee Kuan Yew, have said that Asian values are not compatible with Western ideas of democracy and political freedom. Lee asserted that Asian values are grounded in the Confucian view that individuals should be submissive to authority; therefore Asian countries should avoid the highly contentious public debate of open electoral politics. Nevertheless, despite the

FIGURE 10.18 PHOTO ESSAY: Power and Politics in Southeast Asia

There are significant barriers to public participation in politics in this region, including political violence and authoritarian political cultures. However, people have been pushing for more access to the political process, and some countries have recently enacted political reforms.

A Rohingya refugees at a camp in Burma (Myanmar). The newly elected government in Burma (Myanmar) (which began operating in 2016) may or may not follow the former military government's policy of not allowing Rohingyas to become citizens. Burma (Myanmar) has more than 100 distinct ethnic minority groups, many of whom are denied basic political freedoms. For years the central government was dominated by the Burman ethnic group that forms more than two-thirds of the country's population. Several ethnically based violent insurgent groups are currently battling Burma's (Myanmar's) armed forces in the north of the country, along the Chinese border.

Paula Bronstein/Getty Images

Tjandjono Eranius/AFP/Getty Images

B The banned flag of the militant Free Papua Movement is carried at a demonstration in West Papua, Indonesia, shortly before police open fire. While political freedoms are growing in Indonesia, political violence and the repression of political movements have a long history and still occur.

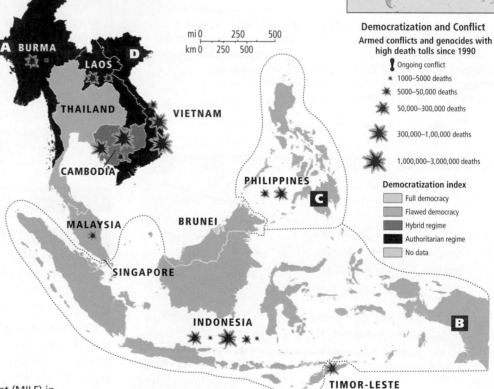

mi 0   250   500
km 0 250 500

Democratization and Conflict
Armed conflicts and genocides with high death tolls since 1990

- Ongoing conflict
- * 1000–5000 deaths
- 5000–50,000 deaths
- 50,000–300,000 deaths
- 300,000–1,00,000 deaths
- 1,000,000–3,000,000 deaths

Democratization index

- Full democracy
- Flawed democracy
- Hybrid regime
- Authoritarian regime
- No data

C Members of the Moro Islamic Liberation Front (MILF) in Mindanao, Philippines, celebrate an accord between their organization and the government. In return for the MILF's cessation of hostilities, the Muslim-dominated areas of Mindanao were granted more autonomy.

D Police and local militia members stop journalists from taking pictures outside a court in Ho Chi Minh City, Vietnam, where a pro-democracy activist was being tried. Vietnam's government remains communist and authoritarian and allows the population few political freedoms.

Karlos Manlupig/AFP/Getty Images

Ian Timberlake/AFP/Getty Images

Thinking Geographically

After you have read about the power and politics in Southeast Asia, you will able to answer the following questions.

A What details in this picture are clues that these people are refugees?

B Why would a government ban a flag?

C What does this picture suggest about the MILF?

D How might banning photos outside a courtroom inhibit public participation in politics?

high respect accorded Lee (the world's longest-serving prime minister), when confronted with governments that abuse their power, people throughout Southeast Asia have not submitted but rather have rebelled (see Figure 10.18A, B). Even in Singapore, the Western-educated son of Lee Kuan Yew, Lee Hsien Loong, who is now prime minister, has expressed more interest in the growth of political freedoms than his father did.

While authoritarianism seems to be giving way to more political freedom and regular elections, these processes are often complicated by corruption and violent state repression of political movements. For example, after being plagued with a history of colonialism under the Spanish, then military occupation by the United States from 1898 to 1946 (the U.S. maintained a large military base until 1991), and a long line of dictators throughout the twentieth century, the Philippines has now elected governments that have resolved some significant problems in the north. However, in a long-standing confrontation with Muslim militants in the southern islands, the government has been reluctant to compromise and the militants have continued their insurgency, periodically resorting to violence (see Figure 10.18C).

Thailand, a constitutional monarchy, was regarded for some years as the most stable democracy in the region, despite its record of numerous coups d'état. It was always a fragile democracy because of deep divisions between those who favored authoritarianism (including the monarchy, many among the economic elites, and the military) and those who were part of or supported the large **populist political movement** advocated by the political party Pheu Thai. In 2006, the military took over Thailand's government after a corrupt but charismatic prime minister, Thaksin Shinawatra, was implicated in several scandals. In 2011, Shinawatra's sister, Yingluck, was elected prime minister, with overwhelming support from her brother's supporters. But in 2014, in the twelfth coup d'état since 1932, she was removed from power. She was tried on corruption charges when the military once again seized control of the government. The military suspended the constitution, declared martial law, reduced the power of political parties, and imposed censorship of the Internet and the media. Urban prosperous elites, dominant in Bangkok and to the west and south of the country, favor the authoritarian military government, while a large and vocal rural majority north and east of Bangkok coalesces around what is best characterized as an uncompromising version of populist absolutism. Thailand remains in political transition: a new constitution has not yet been approved (2016) and legislative elections are not expected until the middle of 2017.

Other countries are also struggling to find a balance between meeting demands for more political freedoms and relying on authoritarianism to create stability. Cambodia's democracy is precarious because of widespread and entrenched corruption, and violence there is common. The wealthier and usually stable countries of Malaysia and Singapore continue to use authoritarian versions of democracy that impose severe limitations on freedom of the press and freedom of speech.

Across the region, when more democratically oriented governments fail to keep the peace, authoritarianism, often backed by a strong military, is seen as a justified counterforce to "too much" democracy. In virtually every country, the military has been called on to restore civil calm after periods of civil disorder. Military rank is highly regarded and many top elected officials have had military careers.

Very recently, authoritarianism has been challenged by political reform in Burma (Myanmar) where numerous regional ethnic minorities and pro-democracy movements have been repressed with an iron fist for decades. A gradual expansion of political freedoms has been taking place in Burma (Myanmar), as evidenced by the recent freeing of political prisoners such as Aung San Suu Kyi (who has led antimilitary protests and was under house arrest for 15 of the years between 1989 and 2010), a reduction in restrictions on the press, and more or less free elections in 2015. The military, calling these changes "disciplined democracy," still holds an unelected 25 percent of parliamentary seats, has the constitutional right to appoint key ministry positions (defense, home affairs, and border affairs), and must approve all changes to the constitution. But political change appears to be coming in Burma (Myanmar), and the country's economy is already anticipating transformation.

Little political reform is taking place in Laos and Vietnam (see Figure 10.18D), where authoritarian socialist regimes have a firm grip on power; or in Brunei, which is governed by an authoritarian sultanate. For these and other reasons, authoritarianism is likely to remain a powerful force in the politics of this region for some time.

MILITARISM'S ROOTS IN WORRIES ABOUT CHINA

While the countries of the region are unlikely to go to war against each other, they are keeping an eye on China, who often offers friendship through **soft power initiatives**—bits of aid that appear to have no ulterior geopolitical motives. Given that China has in recent years been encroaching on parts of the South China Sea previously considered to be Southeast Asian waters, it is understandable that Southeast Asian nations consider military preparedness to be prudent. Newly discovered oil resources in the South China Sea, as well as fishing resources and vital shipping routes, have resulted in competing territorial claims by China, Taiwan, Vietnam, Malaysia, Brunei, Indonesia, and the Philippines. One contention is over the Spratly Islands and associated maritime shoals, reefs, and cays, at least ten of which are occupied by the Philippines but others of which are physically claimed by China, Taiwan, and Vietnam.

populist political movement a coalition of usually rural or working class activists who claim that virtuous citizens are being exploited by a small group of elites who are presumed to be in illegitimate control of the government

soft power initiatives diplomatic overtures or offers of aid that appear to be only friendly with no ulterior geopolitical motives

Most countries have recently invested heavily in military equipment, primarily from Europe and the United States. In the midst of this buildup, countries with no claims to the South China Sea are arming themselves. Tiny, wealthy, trade-centered Singapore recently spent one quarter of its GDP on weapons. It has the largest military budget in the region and now produces armored troop carriers for use at home and for export.

CAN THE EXPANSION OF POLITICAL FREEDOMS BRING PEACE TO INDONESIA?

Indonesia is the largest country in Southeast Asia and the most fragmented—physically, culturally, and politically. It is comprised of more than 17,000 islands (3000 of which are inhabited), stretching over 3000 miles (8000 kilometers) of ocean. It is also the most culturally diverse, with hundreds of ethnic groups and multiple religions. Although Indonesia has the largest Muslim population in the world, there are also many Christians, Buddhists, Hindus, and followers of various local religions and spiritual traditions. With all these potentially divisive forces, many wonder whether this multi-island country of 250 million people might be headed for political disintegration, or whether it might instead prove to be a model of social integration.

Until the end of World War II, Indonesia was a loose assembly of distinct island cultures that Dutch colonists managed to hold together as the Dutch East Indies. When Indonesia became an independent country in 1945, its first president, Sukarno, hoped to forge a new nation out of these many parts, founded on a strong Communist ideology. To that end, he articulated a national philosophy known as *Pancasila,* which was aimed at holding the disparate nation together, primarily through nationalism and concepts of religious tolerance.

In 1965, during the height of the Cold War, Suharto, a staunchly anti-Communist general in the Indonesian army who was supported by the United States because he was against Communism, ousted Sukarno in a coup and ruled the country for another 33 years. It is now clear that Suharto's regime was responsible for a purge of ordinary citizens suspected of being communists, during which as many as a million people were killed.

Despite his abrupt removal from office by Suharto, Sukarno's unifying idea of **Pancasila** endures as a central theme of life in Indonesia (and less formally throughout the region). Pancasila embraces five precepts: *belief* in God, and the observance of *conformity, corporatism* (often defined as "organic social solidarity with the state"), *consensus,* and *harmony.* These last four precepts could be interpreted as discouraging dissent or even loyal opposition, and they seem to require a perpetual stance of national boosterism. Pancasila has been criticized by some Indonesians as being too secular, while others see it as not going far enough to protect the freedom of people to believe in multiple deities, as Indonesia's many Hindus do, or freedom to not believe in any deity at all. Some praise Pancasila's other precepts for counteracting the extreme ethnic diversity and geographic dispersion of the country, while others say that the precepts have had a chilling effect on participatory democracy and on criticism of the government, the president, and the army.

The first orderly democratic change of government in Indonesia did not take place until national elections in 2004. Since then, there have been several peaceful elections, and the government is stable enough to allow citizens to publicly protest some policies. But corruption remains a threat that encourages feelings of nostalgia for the Suharto model of authoritarian government.

Separatist movements have sprouted in four distinct areas in recent years, demonstrating Indonesia's fragility. The only rebellion to succeed was in Timor-Leste, which became an independent country in 2002. However, its case is unique in that this area was under Portuguese control until 1975, when it was forcibly integrated into Indonesia. Two other separatist movements (in the Molucca Islands and in West Papua; see Figure 10.18B) developed largely in response to Indonesia's forced **resettlement schemes,** which were begun in 1965. Also known as *transmigration schemes*, these programs have relocated approximately 20 million people from crowded islands such as Java to less densely settled islands. The policies were originally initiated under the Dutch in 1905 to relieve crowding and provide agricultural labor for plantations in thinly populated areas. After independence, Indonesia used resettlement schemes both to remove troublesome people and to bring outlying areas under closer control of the central government in Jakarta. Resettlement schemes continue today, though at a much smaller scale of roughly 60,000 people per year.

The far-western province of Aceh, in the north of Sumatra, has long been troubled by political violence and accusations of terrorism. However, the recent expansions of political freedoms in Aceh may have helped the province chart a course to peace. Conflicts there originally developed because most of the wealth yielded by Aceh's resources, primarily revenues from oil extraction, was going to the central government in Jakarta. The Acehnese people protested what they saw as the expropriation of oil without compensation, and Jakarta sent the military to silence them. Many who spoke out against the military presence or who supported the Free Aceh movement were accused of terrorism and arrested, jailed, forced into hiding, or even killed. The conflict seemed unresolvable. Then in 2004, the earthquake and tsunami in Aceh, which killed more than 170,000 Acehnese, suddenly brought many outside disaster relief workers to Sumatra because the Indonesian government was unable to give sufficient aid to the victims. Global press coverage of the relief effort mentioned the recent political violence; this created a powerful incentive for separatists and the government to cooperate in order to receive outside aid. A resulting peace accord signed in 2005 brought many former combatants into the political process as democratically elected local leaders. Virtually all the separatists laid down their arms; but underdevelopment and lack of opportunities for youth have instigated a revival of the Free Aceh movement.

Terrorism, Politics, and Economic Issues

Like authoritarianism, terrorism has long been a counterforce to political freedoms in this region. Terrorist violence short-circuits

Pancasila a national philosophy that embraces five precepts: *belief* in God and the observance of *conformity, corporatism, consensus,* and *harmony*

resettlement schemes government plans to move large numbers of people from one part of a country to another to relieve urban congestion, disperse political dissidents, or accomplish other social purposes; also called *transmigration*

the public debate that is at the heart of democratic processes and appears to reinforce the need for repressive authoritarian measures. However, it is important to recognize that terrorist movements often thrive in the context of both economic deprivation and political repression, two factors that are often interrelated. As was the case in Aceh and is the case in the southern Philippines (see Figure 10.18C), terrorism seems to draw support from people who feel shut out of opportunities for economic advancement. The peaceful solution to terrorism could be as simple as a willingness to listen to both the political and economic desires of those who might otherwise be attracted to terrorism.

THINGS TO REMEMBER

GEOGRAPHIC THEME 3 • **Power and Politics** There has been a general expansion of political freedoms throughout Southeast Asia in recent decades, but authoritarianism, corruption, and violence have at times reversed these gains.

• Some Southeast Asian leaders, such as Singapore's former prime minister, Lee Kuan Yew, have said that Asian values are not compatible with Western ideas of democracy and political freedom.

• All Southeast Asian countries have diverse multicultural populations that at times come into conflict with each other. Authoritarianism has been a tempting strategy for dealing with multiculturalism.

• Indonesia, a geographically dispersed and culturally complex country, has experimented with communism, authoritarianism, forced social cohesion through Pancasila, and democracy.

• Like authoritarianism, terrorism has been a counterforce to political freedoms in this region.

URBANIZATION

GEOGRAPHIC THEME 4

Urbanization: Although Southeast Asia as a whole is only 47 percent urban, its cities are growing rapidly as agricultural employment declines and urban industries expand. The largest Southeast Asian cities, which are receiving most of the new rural-to-urban migrants, rarely have sufficient housing, water, sanitation, or jobs for all their people.

Farmers who were once able to produce enough food for their families are now forced to migrate to cities, where they must purchase their food, and where the only affordable housing is in slums that lack essential services.

Southeast Asia as a whole is only 47 percent urban, but the rural–urban balance is shifting steadily in response to the economic transformation of the region into an industrial hot spot. The forces driving farmers into the cities are called the *push factors* in rural-to-urban migration. They include deforestation of subsistence lands for commercial crops like palm oil, the rising cost of farming, which is related to the use of new crops, technologies, and competition with agribusiness, and to the loss of farm labor opportunities. *Pull factors*, in contrast, are those that attract people to the city, such as abundant

manufacturing jobs, education opportunities, and the rumored excitement of urban living. In Southeast Asia, as in all other regions, these factors have come together to create steadily increasing urbanization. Malaysia is already 74 percent urban; the Philippines, 44 percent; Brunei, 77 percent; Burma (Myanmar), 34 percent; Cambodia, 21 percent; Vietnam, 33 percent; and Singapore, 100 percent.

Employment in agriculture has been declining throughout Southeast Asia since the introduction of new production methods that increased the use of labor-saving equipment and reduced the need for human labor. Chemical pesticide and fertilizer use has also grown. While such additives can increase harvests and the supply of food to cities, they also can drive the cost of production higher than what most farmers can afford. Many family farmers have sold their land to more prosperous local farmers or to agribusiness corporations and have moved to the cities. These people, skilled at traditional farming but with little formal education, often end up in the most menial of urban jobs **(FIGURE 10.19C)** and live in circumstances that do not allow them to grow their own food (see Figure 10.19A).

Labor-intensive manufacturing industries (garment and shoe making, for instance) are expanding in the cities and towns of the poorer countries, such as Cambodia, Vietnam, and parts of Indonesia and Timor-Leste. In the urban and suburban areas of the wealthier countries—Singapore, Malaysia, Thailand, and parts of Indonesia and the northern Philippines—technologically sophisticated manufacturing industries are also growing. These include automobile assembly, chemical and petroleum refining, and computer and other electronic equipment assembly. Riding on this growth in manufacturing are innumerable construction projects that often provide employment to recent migrants (see Figure 10.19B, D, and the vignette on the global economy in Chapter 1).

Cities like Jakarta, Manila, and Bangkok, among the most rapidly growing metropolitan areas in the world, are *primate cities*—cities that, with their suburbs, are vastly larger than all others in a country. Bangkok is more than 20 times larger than Thailand's next-largest metropolitan area, Udon Thani; and Manila is more than 10 times larger than Davao, the second-largest city in the Philippines. Thanks to their strong industrial base and political power and the massive immigration they attract, primate cities can dominate whole countries; and in the case of Singapore, the city constitutes the entire country.

Rarely can such cities provide sufficient housing, water, sanitation, or decent jobs for all the new rural-to-urban migrants. Many millions of urban residents in this region live in squalor, often on floating raft-villages on rivers and estuaries. Of all the cities in Southeast Asia, only Singapore provides well for nearly all of its citizens (see Figure 10.19B). Even there, however, a significant undocumented noncitizen population lives in poverty on islands surrounding the city. The experience of rural-to-urban migrants who go to Bangkok or Jakarta is more typical; migrants there often live in slums on the banks of polluted, trash-ridden bodies of water (see Figure 10.19A).

EMIGRATION RELATED TO URBANIZATION AND GLOBALIZATION

The same push and pull factors that convince migrants to go to the region's cities also tempt millions to continue out of Southeast Asia. These migrants are a major force of globalization, as they supply much of the world's growing demand for low- and middle-wage

FIGURE 10.19 PHOTO ESSAY: Urbanization in Southeast Asia

Southeast Asia is rapidly urbanizing, as manufacturing and service sector industries pull in people from rural areas and as changes in agriculture push farmers to the cities. Many cities are struggling to cope with rapid growth. [Source consulted: Population Reference Bureau, *2015 World Population Data Sheet*, at http://www.prb.org/pdf11/2015population-data-sheet_eng.pdf]

A A slum area in Jakarta, Indonesia, where 62 percent of the population lives in slums. Jakarta will grow by almost 50 percent by 2020.

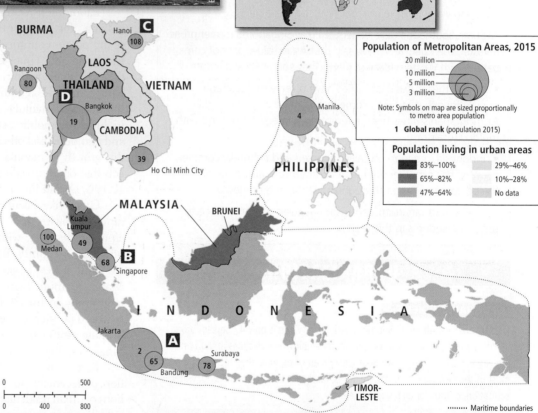

Population of Metropolitan Areas, 2015

20 million
10 million
5 million
3 million

Note: Symbols on map are sized proportionally to metro area population

1 Global rank (population 2015)

Population living in urban areas

- 83%–100%
- 65%–82%
- 47%–64%
- 29%–46%
- 10%–28%
- No data

········· Maritime boundaries

mi 0   500
km 0   400   800

B Publicly built housing in Singapore, where 85 percent of the population lives in apartment buildings designed, built, and managed by the government.

C Migrant workers from rural areas pull carts to a construction site in Hanoi, Vietnam. Many migrants start out in low-wage, manual labor jobs like this one.

D A woman works at a construction site in Bangkok, Thailand. Gender roles are changing as more women move to the cities.

Thinking Geographically

After you have read about urbanization in Southeast Asia, you will be able to answer the following questions.

A Why might these people be living on the water in Jakarta?

B If government-built apartment buildings in Singapore are intended for the middle class, what facilities house most of the undocumented, noncitizen population, and where are those facilities likely to be?

C What would be the consequences of these carts being replaced with one dump truck? Are the carts perhaps more efficient for some tasks?

D Discuss why the situation depicted in this photo might be viewed as a triumph for women's rights in Bangkok.

The Merchant Marines

Skilled male seamen from Southeast Asia make up a significant portion of the international merchant marines, where conditions are considerably better than they are for most migrant workers. Nonetheless, in the merchant marines, it is customary for workers to be paid according to their homeland's pay scales, which makes employing seamen from low-wage Southeast Asian countries attractive to ship owners. The seamen work aboard international freighters or on luxury cruise liners as deckhands, cooks, engine mechanics; a few become officers. Generally, seamen work for 6 months at a time—with only a few hours a day for breaks—saving nearly every penny. At the end of a tour of duty, they return home to their families for another 6 months, where they often contribute financially to the well-being of an extended group of kin and friends and send their children to receive an advanced education.

The Maid Trade

Women constitute well over 50 percent of the more than 8 million emigrants from Southeast Asia. Many skilled nurses and technicians from the Philippines work in European, North American, and Southwest Asian cities. About 3 million participate in the global "maid trade" (shown in **FIGURE 10.20A** and the figure map). Most are educated women from the Philippines and Indonesia who work under 2- to 4-year contracts in wealthy homes throughout Asia. An estimated 1 to 3 million Indonesian maids now work outside of the country, mostly in the Persian Gulf. In 2015 there were 80,000 housemaids living in the United Arab Emirates.

foreign exchange foreign currency that countries need to purchase imports

workers who are willing to travel or live temporarily in foreign countries. For example, 40 percent of the foreign workers in Taiwan are Indonesian. These migrants are also a globalizing force within their home countries because their remittances (monies sent home) boost family incomes and supply governments with badly needed **foreign exchange** (foreign currency) that countries use to purchase imports. Filipinos working abroad are their country's largest source of foreign exchange, sending home more than US $6 billion annually and increasing household annual income by an average of 40 percent. And when they return home, they bring new ways of thinking, some useful, some potentially disruptive.

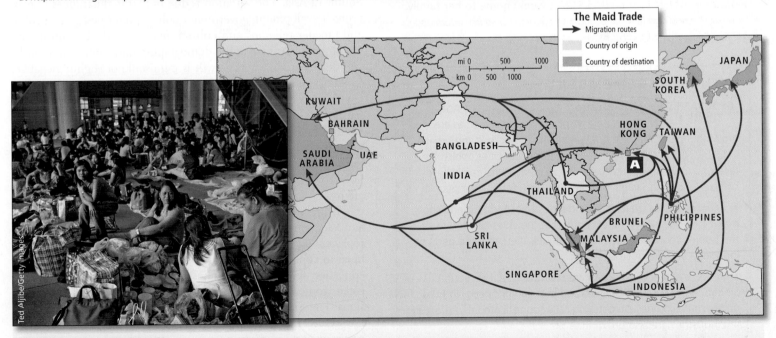

FIGURE 10.20 Globalization: The maid trade. By 2009, about 3 million women, primarily from the Philippines, Indonesia, and Sri Lanka, were working as domestic servants across Asia. The governments of Indonesia and the Philippines are now discouraging this migration, saying that it robs these countries of valuable human capital. **A** Filipina housemaids and nannies in Hong Kong relax on blankets on their day off. The Philippine government mandates that they must have Sundays off, so many public areas in central Hong Kong are occupied on Sundays by small groups of women talking, trading goods, playing games, and packing up things to send home. The map shows the origins and destinations of the maids. [Sources consulted: Joni Seager, *The Penguin Atlas of Women in the World* (New York: Penguin Books, 2003), p. 73, with updated information from the Migration Policy Institute, at http://www.migrationpolicy.org/regions/asia-and-pacific?qt-recent_program_activities=4#qt-recent_program_activities]

Thinking Geographically

How does this photo portray the inventiveness of women in the maid trade?

By 2000, governments in Southeast Asia were recognizing that migration constituted a loss for their countries as well as an advantage. They began stipulating minimum wages that their citizens had to be paid while working as housemaids and required a minimum number of days off. Resistance has grown to the entire trend of sending thousands of young women to work abroad as servants and nannies. When Joko Widodo was elected president of Indonesia (2014), he directed the labor department to begin retrieving all Indonesians working as housemaids in foreign countries, saying that the custom of sending domestic workers abroad undermined the country's "self-esteem and dignity." His position was grounded in the suspicion that the wages paid were too low and the working conditions, inhumane, perhaps including sexual exploitation. Indonesia has been joined in this opposition to housemaid emigration by the Philippines. Leaders in both countries are noting that the emigration of young people reveals their own poor development policies. Emigration flows result in a loss of human capital and are a reflection of the lack of satisfactory living and employment opportunities in their home countries, as the following vignette illustrates.

VIGNETTE Every Sunday is amah (nanny) day in Hong Kong. Gloria Cebu and her fellow Filipina maids and nannies stake out temporary geographic territory on the sidewalks and public spaces of the central business district (see Figure 10.20A). Informally arranging themselves according to the different dialects of Tagalog (the official language of the Philippines) they speak, they create room-like enclosures of cardboard boxes and straw mats, where they share food, play cards, give massages, and do each other's hair and nails. Gloria says it is the happiest time of her week, because for the other 6 days she works alone, caring for the children of two bankers.

Gloria, who is a trained law clerk, has a husband and two children back home in Manila. Because the economy of the Philippines has stagnated, she can earn more in Hong Kong as a nanny than in Manila in the legal profession. Every Sunday she sends most of her income (U.S.$125 a week) home to her family. *[Sources: Kirsty Vincin and the Economist. For detailed source information, see Text Sources and Credits.]* ∎

THINGS TO REMEMBER

GEOGRAPHIC THEME 4 • **Urbanization** Although Southeast Asia as a whole is only 47 percent urban, its cities are growing rapidly as agricultural employment declines and urban industries expand. The largest Southeast Asian cities, which are receiving most of the new rural-to-urban migrants, rarely have sufficient housing, water, sanitation, or jobs for all their people.

• Labor-intensive manufacturing industries (garment and shoe making, for instance) are expanding in the cities and towns of the poorer countries, such as Cambodia, Vietnam, and parts of Indonesia and Timor-Leste.

• In the urban and suburban areas of the wealthier countries—Singapore, Malaysia, Thailand, and parts of Indonesia and the northern Philippines—technologically sophisticated manufacturing industries are also growing.

• Cities like Jakarta, Manila, and Bangkok, among the most rapidly growing metropolitan areas in the world, are *primate cities*—cities that, with their suburbs, are vastly larger than all others in a country and are often the focus of rural-to-urban migration. These cities are unable to adequately house and employ the flood of migrants.

• Thousands of people emigrate from the region each year to seek temporary or long-term employment in the Middle East, Europe, North America, Asia, or on the world's oceans.

POPULATION AND GENDER

GEOGRAPHIC THEME 5

Population and Gender: Population dynamics vary considerably in this region because of differences in economic development, government policies, prescribed gender roles, and religious and cultural practices. Economic change has brought better job opportunities and increased status for women, who then often choose to have fewer children. Some countries also have gender imbalances because of a cultural preference for male children.

Southeast Asia is home to nearly 630 million people (almost double the U.S. population), who occupy a land area that is about one-half the size of the United States. At current rates, Southeast Asia's population, which is large and growing, is projected to reach more than 800 million by 2050, by which time much of this population will live in cities (**FIGURE 10.21**). However, population projections could be inaccurate, both because rates of natural increase are continuing to slow markedly and because many Southeast Asians are migrating to find employment outside the region.

POPULATION DYNAMICS

Population dynamics vary considerably among the countries of Southeast Asia. The variation is due in part to differences in economic development, government policy, prescribed gender roles, and broader religious and cultural practices. Most countries are nearing the last stage of the demographic transition, where births and deaths are low and growth is minuscule or slightly negative. In the last several decades, overall fertility rates in Southeast Asia have dropped rapidly (**FIGURE 10.22**). Whereas women formerly had 5 to 7 children, they now have 1 to 3, with Laos (3.9) and Timor-Leste (5.7) being the major exceptions. However, because in most countries populations are still quite young (regionally 27 percent of the population is aged 15 years or younger), population growth is likely for several decades because so many people are just coming into their reproductive years.

Brunei, Burma (Myanmar), Malaysia, Singapore, Thailand, and Vietnam have reduced their fertility rates so sharply—below replacement levels of 2.1 (see Figure 10.22)—that they will soon need to cope with aging and shrinking populations. Regionally, education levels are the highest in Singapore, where most educated women work outside the home at skilled jobs and professions. The Singapore government is now so concerned about the low fertility rate that it offers young couples various incentives for marrying and procreating.

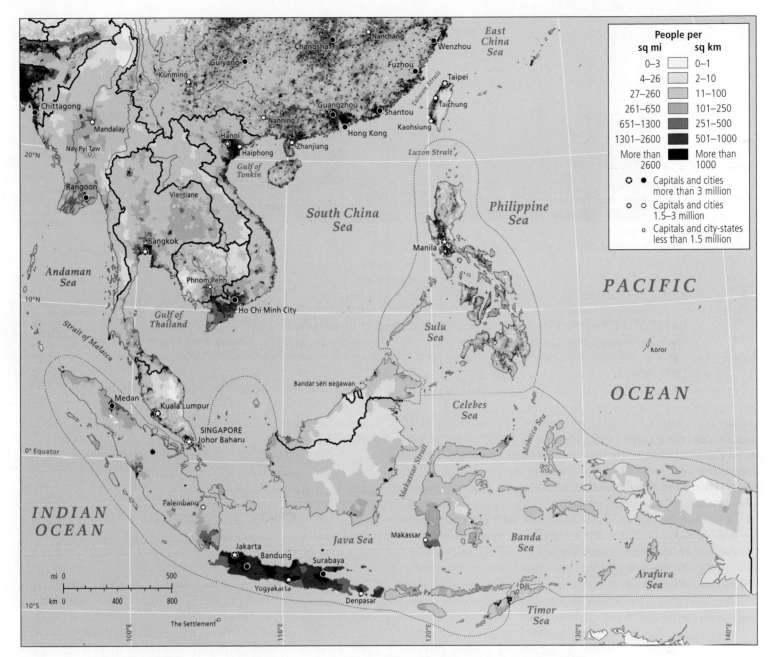

FIGURE 10.21 Population density in Southeast Asia. Population growth in Southeast Asia is slowing, largely related to economic development, urbanization, changing gender roles, and government policies. Fertility rates have declined sharply in all countries since the 1960s but are still high in the poorest areas.

An important source of population growth for Singapore is the steady stream of immigrants that its vibrant economy attracts. The country draws highly skilled workers from the wider region, as well as from the United States, Australia, and Europe. Singapore also attracts immigrants from elsewhere in Southeast Asia, primarily in the building trades and low-wage services. All countries in this region are losing population to emigration, with the exceptions of Singapore, Malaysia, Thailand, Vietnam, and Brunei.

Thailand's low fertility rate of 1.4 children per adult woman was achieved in part through a government-sponsored condom campaign, and in part by rapid economic development and urbanization, which made many couples feel that smaller families would be best. Also, as women gained more opportunities to work

and study outside the home, they decided to have fewer children. High literacy rates for both men and women, along with Buddhist attitudes that accept the use of contraception, have also been credited for the decline in Thailand's fertility rate.

The poorest and most rural countries in the region show the usual correlation between poverty, high fertility, and infant mortality. Timor-Leste, which is predominantly Roman Catholic due to the influence of its colonizer, Portugal, consists of the east half of the island of Timor, plus adjacent islands. Timor-Leste suffered a violent, impoverishing civil disturbance after declaring independence from Portugal in 1975. Indonesia immediately invaded and violently occupied it until 1999, when it became independent from Indonesia. Today Timor Leste is characterized by poverty,

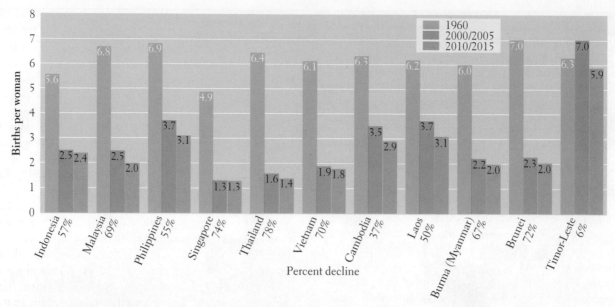

FIGURE 10.22 Total fertility rates, 1960, 2000–2005, and 2010–2015. Total fertility rates declined significantly for all Southeast Asian countries from 1960 to 2015. Only in Timor-Leste did they increase, and there they did so for a few years (between 2004 and 2011) and then began to drop. [Sources consulted: United Nations, *2015 Human Development Report*, Table 18.]

low use of birth control, a high fertility rate (5.7 children per adult woman), and a high infant mortality rate (45 per 1000 live births). Its oil resources bring some hope of reasonable prosperity but the volatility of oil prices make relying on oil risky.

On the mainland, Cambodia, Laos, Burma (Myanmar), and Vietnam are notably more poverty-stricken than are Thailand, Malaysia, and Singapore. Even Indonesia and the Philippines are less impoverished than these four (see Figure 10.24A). In Cambodia and Laos, fertility rates average 2.9 and 3.1, respectively. Infant mortality rates are 28 per 1000 live births for Cambodia and 68 per 1000 for Laos. Interestingly, Burma (Myanmar)—only recently emerging from a total military dictatorship into a partially democratic regime with a strong military component—has a literacy rate of 92 percent, a low fertility rate of 2.3, but a very high infant mortality rate of 62. Apparently the military government has done well looking after basic education but less well with health care. If democracy is allowed to flourish and women to participate to their fullest, we could expect rapid improvement in overall human well-being in Burma (Myanmar).

On the other hand, in Vietnam, where people are only slightly more prosperous and urbanized, the fertility rate has already dropped to 1.8 children per adult woman and the infant mortality rate is just 16 per 1000 births, far lower than those of Burma (Myanmar), Cambodia, and Laos. These low rates for Vietnam are explained by the fact that as a socialist state, Vietnam provides basic education and health care—including birth control—to all, regardless of income. In Vietnam, literacy rates are well over 90 percent for men and women, whereas they are only 76 percent for women in Cambodia and 68 percent in Laos. In addition, Vietnam's rapidly developing economy is pulling in foreign investment, which provides more employment for women, and

careers are replacing child rearing as the central focus of many women's lives.

In the 1990s, out of concern that a rapidly growing population would jeopardize Vietnam's upswing in economic development, the government of Vietnam reintroduced a two-child policy that it had previously had on and off since the 1960s. This policy has brought on gender imbalance, as some couples are choosing abortion if the fetus is female, and the birth rate dropped precipitously, below the replacement level. By 2015 all restrictions on fertility were removed.

The Philippines, which has a higher per capita income than Laos, Cambodia, Timor-Leste, or Vietnam, is an anomaly in population patterns, primarily because it is predominantly Roman Catholic (93 percent; see the Figure 10.26 map), a religion that officially does not allow birth control. Fertility there is among the highest in the region (3.2 per adult female); infant mortality is in the medium range (22 per 1000 live births); and maternal deaths from childbirth are high (120 per 100,000 births) despite female literacy rates also being high (93 percent). The following vignette offers some insights.

VIGNETTE In the Roman Catholic Philippines, Gina Judilla, who works outside the home, has had six children with her unemployed husband. They wanted only two, but because of the strong role of the Catholic Church and the political pressure it exerts, birth control was not available to them. Abortion is legal only to save the life of the mother, so with every succeeding pregnancy she tried folk methods of inducing an abortion. None worked. Now she can afford to send only two of her six children to school.

A move by family planners to provide national reproductive health services and sex education is under way. A recent survey showed that 48 percent of all pregnancies in the Philippines in 2010 were unintended, which often led to illegal "backstreet" abortions. In 2015, only 38 percent of Philippine women had access to modern birth control methods. *[Source: Likhaan Center for Women's Health, Inc., and the New York Times. For detailed source information, see Text Sources and Credits.]* ■

POPULATION PYRAMIDS

The youth and gender features of Southeast Asian populations are best appreciated by looking at the population pyramids for Indonesia. **FIGURE 10.23** shows the 2016 population and the projected population for 2050. The wide bottom and then narrowing of the 2016 pyramid indicates that most people are age 34 and under, but in the last 10 years births have begun to decline. The projections to 2050 show that eventually, with declining birth rates, those 34 and under will be outnumbered by those 35 and older. Indonesia will accumulate ever-larger numbers in the upper age groups, and the pyramid will eventually be more box-shaped, as those for Europe are now. The gender disparities (more males than females) that have developed over the last 20 years can be seen by carefully examining the length of the bars on the male and female sides of the pyramid for those aged 30–34 and under. The differences are slight but significant because there is a rather consistent deficit of females, which indicates that they were selected out before birth.

SOUTHEAST ASIA'S ENCOUNTER WITH HIV/AIDS

As in sub-Saharan Africa (see Chapter 7), HIV/AIDS is a significant public health issue in Southeast Asia, although infection and death rates are now declining across the region. Cambodia, Thailand, and Burma currently have the highest infection rates.

HIV rates are expected to increase rapidly in rural areas and in secondary cities where conservative religious leaders and faith-based international agencies restrict sex education and AIDS-prevention programs, such as the promotion of condom use. Sex education is widely viewed as promoting promiscuity. Aggressive prevention programs are even more essential because of popular customs: sexual experimentation (at least among men), the reluctance of women to insist that their husbands and boyfriends use condoms, intravenous drug use (primarily by men), and the high mobility of young adults. Also contributing to the spread of HIV among young people (male and female) are sex tourism and sex-related human trafficking (discussed further on page 603).

Thailand is one of the few countries that has been able to drastically reduce the incidence of HIV/AIDS. It did so through a well-funded program that increased the use of condoms, decreased STDs dramatically through health education, and reduced visits to sex workers by half. In Thailand in 2001, AIDS was the leading cause of death, overtaking stroke, heart disease, and cancer, but it is now just the third leading cause of death. Estimates are that in 2015 about 500,000 Thais are infected, down from 1 million 12 years ago. Men between the ages of 20 and 40 have the highest rates of infection; most vulnerable are men who have sex with men, and transgender sex workers.

GEOGRAPHIC PATTERNS OF HUMAN WELL-BEING

If human well-being is defined as the ability to enjoy long, healthy, and creative lives, as suggested by Mahbub ul Haq, the founder of the annual United Nations Human Development Report (UNHDR), then personal income statistics are not the best way to measure success. Nonetheless, income is relevant to

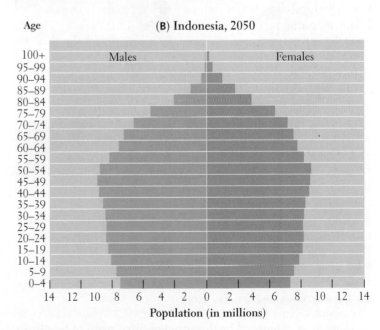

FIGURE 10.23 Population pyramids for Indonesia, 2016 and 2050 (projected). In 2016, Indonesia's population was 255.7 million. It is projected to be 366.5 million in 2050. [Source consulted: *International Data Base*, U.S. Census Bureau, 2016, at http://www.census.gov/population/international/data/idb/informationGateway.php]

FIGURE 10.24 Maps of human well-being.

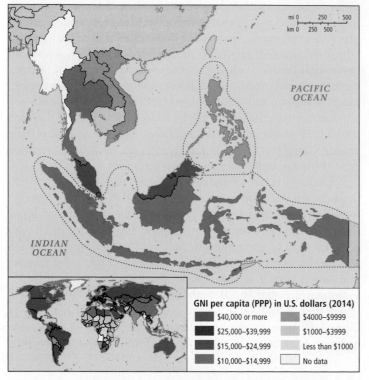

(A) Gross national income (GNI) per capita, adjusted for purchasing power parity (PPP).

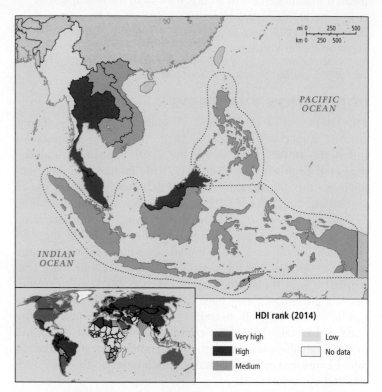

(B) Human Development Index (HDI).

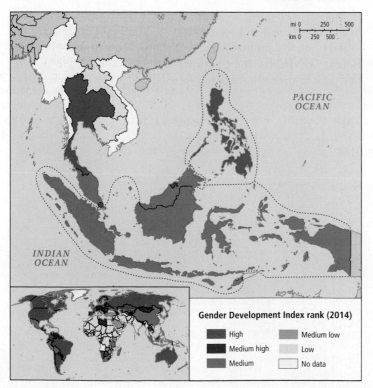

(C) Gender Development Index (GDI).

well-being, as evidenced by this series of three maps of Southeast Asia. **FIGURE 10.24A** shows gross national income (GNI) per capita (PPP) by country; Figure 10.24B shows each country's rank on the Human Development Index (HDI); and Figure 10.22C

depicts the gender development index (GDI) for the region, showing to what extent the genders are treated equally in society.

As shown in Figure 10.24A, two small countries, Singapore and Brunei, are the only places where annual per capita GNI (PPP) is in the highest category. For both countries, GNI (PPP) is well over U.S.$70,000, which is higher than in the United States and comparable to the richest countries in Europe (see the inset map of the world). Elsewhere in the region, per capita GNI (PPP) is considerably lower, with Malaysia the next highest, at U.S.$24,770. Thailand is near the top of the medium range (at $14,870); and below it are, in order, Indonesia, the Philippines, Vietnam, Laos, Timor-Leste, and Cambodia, with per capita GNI (PPP) rates varying from U.S.$10,190 to U.S.$3080. There are no data for Burma (Myanmar).

Figure 10.24B shows each country's rank on the Human Development Index (HDI), which is a calculation (based on adjusted real income, life expectancy, and educational attainment) of how adequately a country provides for the well-being of its citizens. Singapore and Brunei again do well at providing for the overall well-being of their citizens, as they rank in the very high category. Malaysia and Thailand rank high; Indonesia, the Philippines, Vietnam, Timor-Leste, Laos, and Cambodia rank medium; and Burma (Myanmar) ranks low.

Figure 10.24C shows how well countries ensure gender equality in three categories: reproductive health, political and education empowerment, and access to the labor market. The countries with rankings closest to 1 have the highest degree of gender equality. As might be expected, Singapore ranks high, as does Brunei. None rank medium high. Indonesia and Malaysia rank medium, and Cambodia, Laos, and Timor-Leste rank low. There are no data for Vietnam.

THINGS TO REMEMBER

GEOGRAPHIC THEME 5

• **Population and Gender** Population dynamics vary considerably in this region because of differences in economic development, government policies, prescribed gender roles, and religious and cultural practices. Economic change has brought better job opportunities and increased status for women, who then often choose to have fewer children. Some countries also have gender imbalances because of a cultural preference for male children.

• While population growth rates have slowed across the region, they are slowing the most in the developed economies of Singapore, Brunei, and Thailand.

• The variations in population growth are linked to relative prosperity, gender roles, and availability of birth control.

• An important source of population growth for Singapore is the steady stream of highly skilled immigrants that its vibrant economy attracts.

• HIV/AIDS is a threat, to varying degrees, throughout the region. Preventative measures have proven quite successful in Thailand.

SOCIOCULTURAL ISSUES

Because of the region's long and complex history, the people of Southeast Asia have a great diversity of cultures and religious traditions. Now globalization and urbanization are adding yet more diversity as new cultural influences come in from abroad and as urban women gain more independence and pursue careers outside the home.

CULTURAL AND RELIGIOUS PLURALISM

Southeast Asia is a place of **cultural pluralism** in that it is inhabited by groups of people from many different backgrounds. Over the past 40,000 years, migrants have come to the region from India, the Tibetan Plateau, the Himalayas, China, Southwest Asia, Europe, Japan, Korea, and the Pacific. Many of these groups have remained distinct, partly because they lived in isolated pockets separated by rugged topography or seas. However, the religious practices and traditions of many groups show diverse cultural influences (FIGURE 10.25).

As discussed in the "Religious Legacies" section (page 578), the major religious traditions of Southeast

> **cultural pluralism** the cultural identity characteristic of a region where groups of people from many different backgrounds have lived together for a long time but have remained distinct

FIGURE 10.25 LOCAL LIVES: Festivals in Southeast Asia

A Songkran, or New Year's Day, is celebrated in Thailand. Water is thrown on friends and passersby as a sign of respect and renewal. This developed from the tradition of gathering water used to wash statues of the Buddha and gently pouring that water on the shoulders of elderly relatives. However, since the festival happens during April, a very hot time of the year, a tradition developed of throwing water on anyone.

B A calligrapher at the Temple of Literature in Hanoi, Vietnam, elegantly writes down intentions for Tet (the New Year, celebrated in January or February). Exchanges of such calligraphy are hung in people's homes; the expressed intentions surround a particular idea or a quality such as happiness, wealth, virtue, knowledge, talent, or long life.

C An elaborately costumed dancer in the Ati-Atihan Festival in Kalibo, Aklan, in the Philippines. Originally celebrating the migration to the Philippines of a group of indigenous people from the island of Borneo, Ati-Atihan includes non-Christian traditions but is also celebrated by Christian Filipinos as a day honoring the infant Jesus.

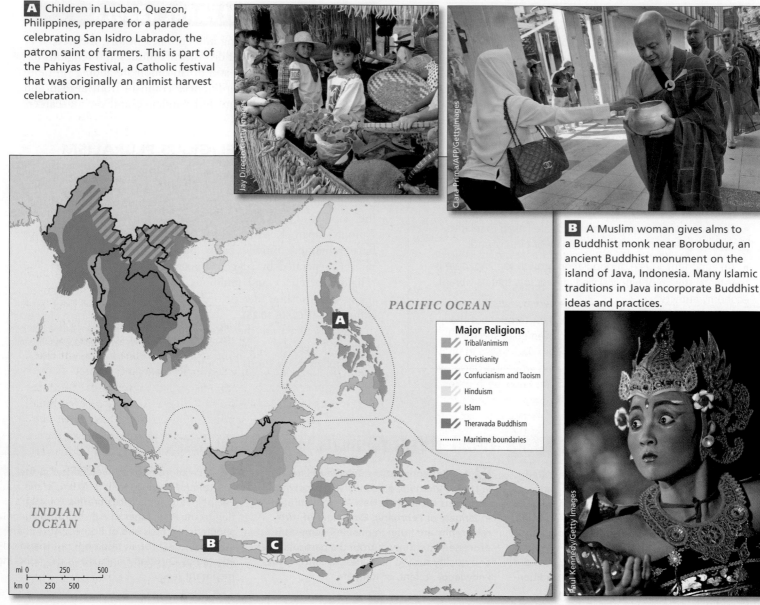

A Children in Lucban, Quezon, Philippines, prepare for a parade celebrating San Isidro Labrador, the patron saint of farmers. This is part of the Pahiyas Festival, a Catholic festival that was originally an animist harvest celebration.

PACIFIC OCEAN

Major Religions
- Tribal/animism
- Christianity
- Confucianism and Taoism
- Hinduism
- Islam
- Theravada Buddhism
- ········ Maritime boundaries

INDIAN OCEAN

mi 0   250   500
km 0   250   500

B A Muslim woman gives alms to a Buddhist monk near Borobudur, an ancient Buddhist monument on the island of Java, Indonesia. Many Islamic traditions in Java incorporate Buddhist ideas and practices.

FIGURE 10.26 Religions of Southeast Asia. Southeast Asia is religiously unusually diverse: five of the world's six major religions are practiced there. Animism, the oldest belief system, is found both in island and mainland locations and has many subtle influences. [Source consulted: *Oxford Atlas of the World* (New York: Oxford University Press, 1996), p. 27]

C A dancer in Bali, Indonesia, where traditional Hindu performances incorporate Buddhist and animist symbols and stories.

Asia include Hinduism, Buddhism, Confucianism, Taoism, Islam, Christianity, and animism **(FIGURE 10.26)**. In animism, a belief system common among many indigenous peoples, natural features such as rocks, trees, rivers, crop plants, and the rains all carry spiritual meaning. These natural phenomena are the focus of festivals and rituals to give thanks for bounty and to mark the passing of the seasons, and these ideas have permeated all the imported religious traditions of the region.

The patterns of religious practice are complex; the patterns of distribution reveal an island–mainland division. All but animism originated outside the region and were brought primarily by traders, priests (Brahmin and Christian), and colonists. Buddhism is dominant on the mainland, especially in Burma, Thailand, and Cambodia. In Vietnam, people practice a mix of Buddhism,

Confucianism, and Taoism that originated in China. Islam is dominant in Indonesia (the world's largest Muslim country), on the southern Malay Peninsula, and in Malaysia, and it is increasingly popular in the southern Philippines. Roman Catholicism is the predominant religion in Timor-Leste, the Molucca Islands, and in the middle and northern Philippines, where it was introduced by Portuguese and Spanish colonists. Hinduism first arrived with Indian traders thousands of years ago and was once much more widespread; now it is found only in small patches, chiefly on the islands of Bali and Lombok, east of Java. Recent Indian immigrants who came as laborers in the twentieth century (during the latter part of the European colonial period) have reintroduced Hinduism to Burma, Malaysia, and Singapore, but only as minority communities.

All of Southeast Asia's religions have changed as a result of exposure to one another. Many Muslims and Christians believe in spirits and practice rituals that have their roots in animism. Hindus and Christians in Indonesia, surrounded as they are by Muslims, have absorbed ideas from Islam, such as the seclusion of women. Muslims have absorbed ideas and customs from indigenous belief systems, especially ideas about kinship and marriage, as illustrated in the vignette below.

VIGNETTE Although arranged marriages have historically been the norm across Southeast Asia, in most urban and rural areas, marriages are now love matches. Such is the case for Harum and Adinda, who live in Tegal, on the island of Java in Indonesia. They met in high school and some 10 years later, after saving a considerable sum of money, decided to formally ask both sets of parents if they could marry. Harum, an accountant, would normally be expected to pay the wedding costs, which could run to many thousands of dollars; however, Adinda was able to contribute from her salary as a teacher. Like nearly all Javanese people, both are Muslim. Because Islam does not have elaborate marriage ceremonies, it is customary for colorful rituals from Christianity, Buddhism, and indigenous animism to enhance these elaborate and festive occasions.

In preparation, both bride and groom participate in unique Javanese rituals that remain from the days when marriages were arranged and the bride and groom did not know each other (FIGURE 10.27). A *pemaes*, a woman who prepares a bride for her wedding and whose role is to inject mystery and romance into the marriage relationship, bathes and perfumes the bride. She also puts on the bride's makeup and dresses her, all the while making offerings to the spirits of the bride's ancestors and counseling her about how to behave as a wife and how to avoid being dominated by her husband. The groom also takes part in ceremonies meant to prepare him for marriage. Both are counseled that their relationship is bound to change over the course of the decades as they mature and as their family grows older.

FIGURE 10.27 A Javanese wedding. In the *dahar klimah* phase of a traditional Javanese wedding, the bridegroom makes three small balls of the food and feeds them to the bride; she then does the same for him. The ritual reminds them that they should joyfully share whatever they have.

Despite the elaborate preparations for marriage, divorce in Indonesia (and also in Malaysia) is fairly common among Muslims, who often go through one or two marriages early in life before they settle into a stable relationship. Although the prevalence of divorce is lamented by society, it is not considered outrageous or disgraceful. Apparently, ancient indigenous customs predating Islam allowed for mating flexibility early in life, and this attitude is still tacitly accepted. *[Source: Jennifer W. Nourse and Walter Williams. For detailed source information, see Text Sources and Credits.]* ∎

Globalization Brings Cultural Diversity but Also Homogeneity

Southeast Asia's cities are extraordinarily culturally diverse, a fact that is illustrated by their food customs (FIGURE 10.28). But in some ways, this diversity decreases as the many groups continue to be exposed to each other and to global culture. For example, Malaysian teens of many different ethnicities (Chinese, Tamil, Malay, Bangladeshi) spend much of their spare time following the same European soccer teams, playing the same video games, visiting the same shopping centers, eating the same fast food, and talking to each other in English. While rural areas retain more traditional influences—of the world's 6000 or so actively spoken languages, 1000 can be found in rural Southeast Asia—in cities, one main language usually dominates trade and politics.

One group that has brought cultural diversity to Southeast Asia is the Overseas (or ethnic) Chinese (see Chapter 9). Small groups of traders from southern and coastal China have been active in Southeast Asia for thousands of years; over the centuries, there has been a constant trickle of immigrants from China. The forebearers of most of today's Overseas Chinese, however, began to arrive in large numbers during the nineteenth century, when the European colonizers needed labor for their plantations and mines. Later, those who fled China's Communist Revolution after 1949 joined them and all sought permanent homes in Southeast Asian trading centers. Today, more than 26 million Overseas Chinese live and work across Southeast Asia, primarily as shopkeepers and as small business owners. A few are wealthy financiers, and a significant number are still engaged in agricultural labor.

Chinese success at small-scale commercial and business activity throughout the region has reinforced the perception that the Chinese are diligent, clever, and extremely frugal, often working very long hours. Although few are actually wealthy, with their region-wide family and friendship connections and access to start-up money, some have been well positioned to take advantage of the new growth sectors in the globalizing economies of the region. Sometimes externally funded, new Chinese-owned enterprises have put out of business older, more traditional establishments that depend on local customers and employees of modest incomes (both ethnic Malay and ethnic Chinese).

In recent years, when low- and middle-income Southeast Asians were hurt by the recurring financial crises, some tended to blame their problems on the Overseas Chinese. Waves of violence resulted. Chinese people were assaulted, their temples desecrated, and their homes and businesses destroyed. Conflicts involving the Overseas Chinese have taken place in Vietnam, Malaysia, and in many parts of Indonesia (Sumatra, Java, Kalimantan, and

FIGURE 10.28 LOCAL LIVES: Food Diversity in Southeast Asia

A *Satay*, or marinated and grilled meat or fish often served with a peanut sauce, is prepared at a market in Chiang Mai, Thailand. Satay first became popular in this region in the nineteenth century on the island of Java, Indonesia, after the arrival of Muslim merchants and other immigrants from South and Southwest Asia. Some food scholars argue that satay is related to kebab, a food found throughout South and Southwest Asia.

B Throughout Southeast Asia, cold desserts feature shaved ice, coconut milk, evaporated or condensed milk, sweeteners, beans, noodles, corn, jelly, and other toppings. They are called *ais kacang* in Malaysia, *halo-halo* in the Philippines, *cendol* in Indonesia, and *ching bo leung* in Vietnam. Ice machines on European ships in the early twentieth century turned these desserts from beverages into ice cream–like sundaes.

C A *banh mi* sandwich, a street food in Vietnam and Laos, is a perfect example of the union of European (in this case, French) and Southeast Asian cuisine. European ingredients include French bread (a baguette), mayonnaise, and pâté. Vietnamese and Laotian ingredients include coriander, pickled vegetables, hot peppers, cucumber, sliced pork or headcheese, and fish sauce. This sandwich now has a U.S. audience and fans.

John Elk/Getty Images

Seet Ying Lai Photography/Getty Images

Rebecca Skinner/Getty Images

Sulawesi) as well. Some Overseas Chinese have attempted to diffuse tensions through public education about Chinese culture. Others have shown their civic awareness by financing economic and social aid projects to help their poorer neighbors, usually of local ethnic origins (Malay, Thai, Indonesian).

ON THE BRIGHT SIDE: LGBT Rights in Thailand and Vietnam

Lesbian, gay, bisexual, and transgender people are discriminated against throughout the world, and Southeast Asia is no exception. Both Thailand and Vietnam, though, are considering giving more legal recognition to same-sex couples in the arenas of property and child custody. Thailand is among the world's most open societies with regard to transgender people; transgender women are referred to as *kathoey*, and transgender men as *tom*, both of which are nonderogatory terms. Of the two, the more prominent are the kathoey, whose gender identity has become somewhat celebrated in recent years, though kathoeys often face discrimination in daily life. Both kathoeys and toms may soon benefit from legal protection as a "third gender." ∎

GENDER ROLES IN SOUTHEAST ASIA

Gender roles are being transformed across Southeast Asia by urbanization and the changes it brings to family organization and employment. Here we look at some surprising traditional patterns of gender roles in extended families. Moving to the city shifts people away from extended families and toward the nuclear family. The gains that women have made in political empowerment, educational achievement, and paid employment have only partially erased gender disparities.

Family Organization, Traditional and Modern

Throughout the region, it has been common for a newly married couple to reside with, or close to, the wife's parents. Along with this custom is a range of behavioral rules that empower the woman in a marriage, despite some basic patriarchal attitudes. For example, a family is headed by the oldest living male, usually the wife's father. When he dies, he passes on his wealth and power to the husband of his oldest daughter, not to his own son. (A son goes to live with his wife's parents and inherits from them.) Hence, a husband may live for many years as a subordinate in his father-in-law's home. Instead of the wife being the outsider, subject to the demands of her mother-in-law—as is the case, for example, in South Asia—it is the husband who must show deference. The inevitable tension between the wife's father and the son-in-law is resolved by the custom of *ritual avoidance*—in daily life they simply arrange to not encounter each other much. The wife manages communication between the two men by passing messages and even money back and forth. Consequently, she has access to a wealth of information crucial to the family and has the opportunity to influence each of the two men. Another traditional custom that empowers women is that women are often the family financial managers. Husbands turn over their pay to their wives, who then apportion the money to various household and family needs.

Urbanization and the shift to the nuclear family mean that young couples now frequently live apart from the extended family, an arrangement that takes the pressure to defer to his father-in-law off the husband. Because this nuclear family unit is often dependent entirely on itself for support, wives usually work for wages outside the home. Although married women lose the power they would have if they lived among their close kin, they are empowered by the opportunity to have a career and an income. The main drawback of this compact family structure, as many young families have discovered in Europe and the United States, is that there is no pool of relatives available to help working parents with child care and housework. Further, no one is left to help elderly parents maintain the rural family home.

Political and Economic Empowerment of Women

Women have made some impressive gains in politics in Southeast Asia. Economically, they still earn less money than men and work less outside the home, but this will likely change if their level of education in relation to that of men continues to increase (see "On the Bright Side," below).

Southeast Asia has had several prominent female leaders over the years, most of whom have risen to power in times of crisis as the leaders of movements opposing corrupt or undemocratic regimes. In the Philippines, Corazon Aquino, a member of a large and powerful family, became president in 1986 after leading the opposition to Ferdinand Marcos, whose 21-year presidency was infamous for its corruption and authoritarianism. She is credited with helping reinvigorate political freedoms in the Philippines. Gloria Macapagal-Arroyo, from another powerful family, became president in 2001 after opposing a similarly corrupt president, though she was then accused of corruption herself. Still, her administration kept the economy sound throughout the global recession starting in 2007. In 2010, Corazon's son, Benigno Aquino III, became president.

In Indonesia, Megawati Sukarnoputri became president in 2001 after decades of leading the opposition to Suharto's notoriously corrupt 31-year reign. In Burma (Myanmar), for more than two decades, a woman, Aung San Suu Kyi, often while confined in house arrest, has led opposition to the military dictatorship and largely succeeded in arranging for elections. In Thailand, Yingluck Shinawatra was elected prime minister in 2011 as part of a popular political movement started by her brother. She was deposed in a military coup.

All of these women leaders were wives, daughters, or siblings of powerful male political leaders, which raises some questions of nepotism. However, family favoritism cannot account for several countries where the percentage of female national legislators is well above the world average of 18 percent: Timor-Leste (39 percent), Laos (25 percent), Vietnam (24 percent), Singapore (25 percent), the Philippines (27 percent), and Cambodia (20 percent).

Despite their increasing successes in politics and their acknowledged role in managing family money, women still lag well behind men in terms of economic well-being (improved access to the labor market will help; see the GDI ranking in Figure 10.24C). Throughout the region, men have a higher rate of employment outside the home than women and are paid more for doing the same work. But changes may be on the way. In Brunei, Malaysia, the Philippines, and Thailand, significantly more women than men are completing training beyond secondary school. If training qualifications were the sole consideration for employment, women would appear to have an advantage over men. While women are still disadvantaged in relation to men despite their more advanced education, change is beginning. This advantage may be significant if service sector economies—which generally require more education—become dominant in more countries. The service economy already dominates in Singapore, the Philippines, Malaysia, Thailand, and Indonesia.

ON THE BRIGHT SIDE: Women's Improving Status in Southeast Asia

Building on strong traditional roles at the family level, women now occupy prominent political roles in many countries, and their rising educational levels suggest that even in the most conservative countries, changes in gender roles will be significant. One result of the rising political status of women is that there is more focus on controlling human trafficking and the abuse of women and girls in the sex industry. ■

GLOBALIZATION AND GENDER: THE SEX INDUSTRY

Southeast Asia has become one of several global centers for the sex industry, supported in large part by international visitors willing to pay for sex. **Sex tourism** in Southeast Asia grew out of the sexual entertainment industry that served foreign military troops stationed in Asia during World War II, the Korean War, and the Vietnam War. Now, primarily civilian men arrive from around the globe to live out their fantasies during a few weeks of vacation. The industry is found throughout the region but is most prominent in Thailand. In 2015, 29 million tourists visited Thailand alone, up from 250,000 in 1965. The largest number came from China, the fastest-growing segment of tourism arrivals. Some observers estimate that as many as 50 percent were looking for some kind of sexual experience. Even though the industry is officially illegal, some Thai government officials have publicly praised sex tourism for its role in helping the country weather economic crises, because it supports more than 2.2 million jobs. Some corrupt officials also favor sex tourism because it provides them with a source of untaxed income from bribes.

One result of the popularity of sex tourism is a high demand for sex workers, which has attracted organized crime. Estimates of the numbers of sex workers vary from 30,000 to more than a million in Thailand alone. Gangs often coerce girls and women into remaining in sex work once they have been tricked or forced into the trade. Demographers estimate that 20,000 to 30,000 Burmese girls taken from Burma (Myanmar) against their will—some as young as 12—are working in Thai brothels. Their wages are too low to enable them to buy their own freedom. In the course of their work, they must service more than 10 clients per day, and they are routinely exposed to physical abuse and sexually transmitted diseases, especially HIV.

sex tourism the sexual entertainment industry that serves primarily men who travel for the purpose of living out their fantasies during a few weeks of vacation

VIGNETTE Twenty-five-year-old Watsanah K. (not her real name) awakens at 11:00 every morning, attends afternoon classes in English and secretarial skills, and then goes to work at 4:00 P.M. in a bar in Patpong, Bangkok's red light district. There she will meet men from Europe, North America, Japan, Taiwan, China, Australia, Saudi Arabia, and elsewhere, who will pay to have sex with her. She leaves work at about 2:00 A.M., studies for a while, and then goes to sleep.

Watsanah was born in northern Thailand to an ethnic minority group made up of impoverished subsistence farmers. She married at 15 and had two children shortly thereafter. Several years later, her husband developed an opium addiction. She divorced him and left for Bangkok with her children. She found work at a factory that produced seatbelts for a nearby automobile plant. In 2008, Watsanah lost her job as the result of the global economic crisis. To feed her children, she became a sex worker.

Although the pay, between U.S.$400 and U.S.$800 a month, is much better than the U.S.$100 a month she earned in the factory, the work is dangerous and demeaning. Sex work, though widely practiced and generally accepted in Thailand, is illegal, and the women who do it are looked down on. As a result, Watsanah must live in constant fear of going to jail and losing her children. Moreover, she cannot always make her clients use condoms, which puts her at high risk of contracting AIDS and other sexually transmitted diseases. "I don't want my children to grow up and learn that their mother is a prostitute," says Watsanah. "That's why I am studying. Maybe by the time they are old enough to know, I will have a respectable job." [Source: Field notes of Alex Pulsipher and Debbi Hempel; Kaiser Family Foundation; BBC News. For detailed source information, see Text Sources and Credits.] ∎

THINGS TO REMEMBER

• The major religious traditions of Southeast Asia include Hinduism, Buddhism, Confucianism, Taoism, Islam, Christianity, and animism. All originated outside the region, with the exception of the animist belief systems.

• Lesbian, gay, bisexual, and transgender people are discriminated against throughout Southeast Asia, but both Thailand and Vietnam are considering giving more legal recognition to same-sex couples.

• Women have made some impressive gains in politics in Southeast Asia, but they still earn less money than men and work less outside the home.

• Gender roles are being transformed across Southeast Asia by urbanization and the changes it brings to family organization and employment.

• Some countries have become centers for the global sex industry, which puts many women at risk of violence and disease.

SUBREGIONS OF SOUTHEAST ASIA

The subregions of Southeast Asia share many similarities: tropical environments; ethnically diverse populations, with Overseas Chinese minorities that are especially active in commerce (see page 573); vestiges of European colonialism (except in Thailand); difficult political trade-offs between democracy and dictatorship; and the importance of religion. Some obvious differences among the subregions are in the pace of modernization and in the varying responses to cultural and ethnic diversity—from celebrating diversity to tolerating it to violently rejecting it.

MAINLAND SOUTHEAST ASIA: BURMA (MYANMAR) AND THAILAND

Burma (Myanmar) and Thailand occupy the major portion of the Southeast Asian mainland and share the long, slender peninsula that reaches south to Malaysia and Singapore (**FIGURE 10.29 MAP**). Although Burma and Thailand are adjacent and share similar physical environments, Burma is poor, depends on agriculture, and is just beginning to show signs of emerging from a repressive military government, whereas Thailand has rapidly industrialized and has had a more open, less repressive society, despite occasional coups d'état and military rule (the last in 2014). Both countries trade in the global economy, but in different ways: Burma supplies raw materials (including illegal drug components), while Thailand, though the world's largest exporter of rice (see Figure 10.29C), earns most of its income by providing low- to medium-wage labor, by hosting many multinational manufacturing firms, and through its thriving tourism industry that has a worldwide clientele.

The landforms of Burma (Myanmar) and northeastern Thailand consist of a series of ridges and gorges that bend out of the Plateau of Tibet and descend to the southeast, spreading across the Indochina peninsula. The Irrawaddy and Salween rivers originate from glaciers on the Plateau of Tibet and flow south through the narrow gorges. The Irrawaddy forms a huge delta at the southern tip of Burma (see Figure 10.29B). The Chao Phraya flows from Thailand's northern mountains through the large plain in central Thailand and enters the sea south of Bangkok. Most farming is done in the Burmese interior lowlands around Burma and in Thailand's central plain.

Ancient migrants from southern China, Tibet, and eastern India settled in the mountainous northern reaches of Burma and Thailand. The rugged topography has in the past protected these indigenous peoples from outside influences. The largest of these groups are the Shan, Karen, Mon, Chin, and Kachin, many of whom still follow traditional ways of life and practice animism. By contrast, the southern Burmese valleys and lowlands and Thailand's central plain are Buddhist, modernized, and have much urban development (Figure 10.29A). Burma is named for the Burmans, who constitute about 70 percent of the population and live primarily in the lowlands. Thailand is named for the Thais, a diverse group of indigenous people who originated in southern China. Both countries are predominantly Buddhist.

Burma (Myanmar)

Burma (Myanmar) is rich in natural resources. Between 70 and 80 percent of the world's remaining teakwood still grows in its interior uplands; a single tree can be worth U.S.$200,000. Other resources include oil, natural gas (of particular interest to India, China, and the United States), tin, antimony, zinc,

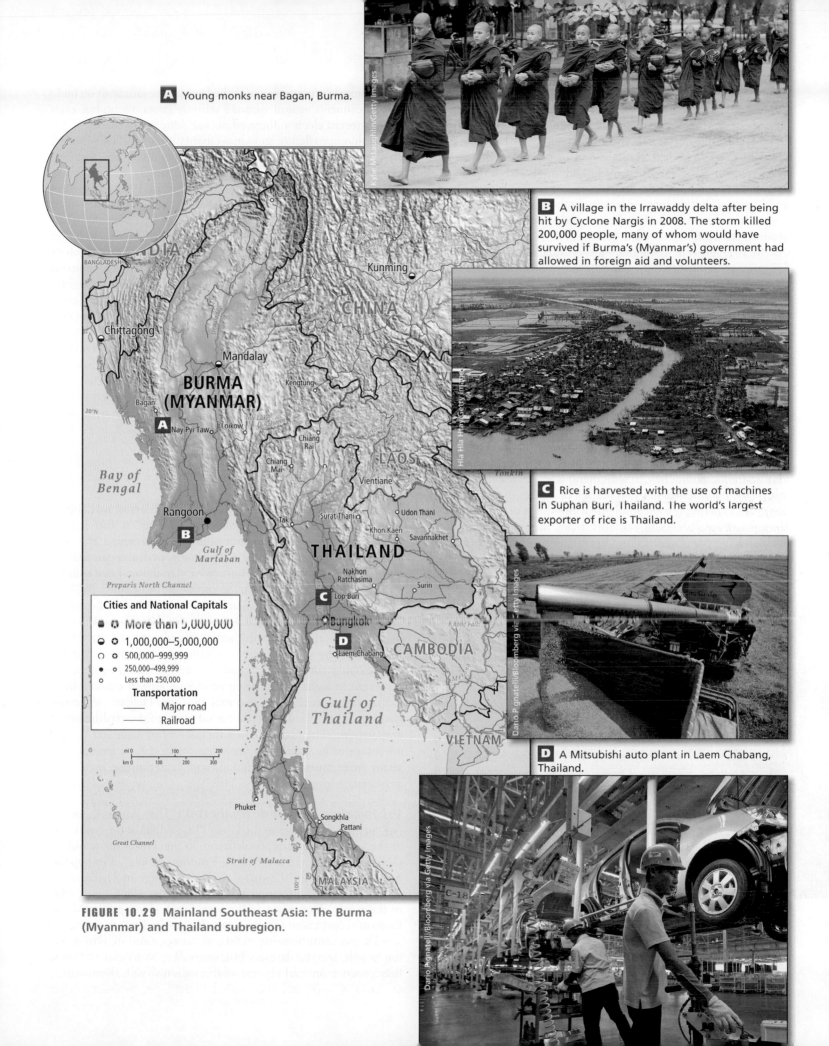

A Young monks near Bagan, Burma.

B A village in the Irrawaddy delta after being hit by Cyclone Nargis in 2008. The storm killed 200,000 people, many of whom would have survived if Burma's (Myanmar's) government had allowed in foreign aid and volunteers.

C Rice is harvested with the use of machines In Suphan Buri, Thailand. The world's largest exporter of rice is Thailand.

D A Mitsubishi auto plant in Laem Chabang, Thailand.

FIGURE 10.29 Mainland Southeast Asia: The Burma (Myanmar) and Thailand subregion.

copper, tungsten, limestone, marble, and gemstones. However, in part because of corruption and repression by a military junta, Burma (Myanmar) ranks as one of the region's poorest countries. Although data on per capita GNI are lacking, GNI (PPP) was estimated in 2011 to be U.S.$1535. About 70 percent of the people still live in rural villages; their livelihoods are based primarily on wet rice cultivation and the growing of corn, oilseeds, sugarcane, and legumes, and on logging teak and other tropical hardwoods. People in rural areas travel to nearby markets to sell or trade their produce for necessities **(FIGURE 10.30)**.

In 2011, Burma (Myanmar) was estimated to supply opium for more than 60 percent of the heroin market in the United States; it alternated with Afghanistan as the world's largest producer (the situation is now in flux, with production and transport under some level of control). Opium poppy cultivation, along with methamphetamine production, is the major source of income for a number of northern indigenous ethnic groups. The Wa, for example, who number about 700,000, depend on illegal drug production for 90 percent of their income and are said to protect their territory in northern Burma with surface-to-air missiles. The military government of Burma (Myanmar) has long encouraged, manipulated, and profited from the drug traffic, especially in the lightly settled uplands and mountains near the Thai border. When indigenous inhabitants protested, they were silenced by assassinations, plundering, resettlement, and the abduction and sale of their young women into the sex industries in neighboring countries. This repression resulted in hundreds of thousands of Burmese people becoming refugees; many now live in camps in Thailand and Bangladesh. The 730,000 mostly Muslim Rohingya migrants from Bangladesh taking refuge in Burma (Myanmar) have suffered violence from the Burmese. Since 1997, the

international community has applied economic sanctions on trade and investments against Burma's military government, but even after its recent election these efforts had little or no effect.

Unlike pro-democracy movements that resulted in real change elsewhere in the region, for more than two decades, people in Burma (Myanmar) protested futilely the rule of a corrupt and authoritarian military regime. In 1990 when, in a landslide, the people elected Aung San Suu Kyi to lead a civilian reformist government, the regime refused to step aside. Suu Kyi, who won the Nobel Peace Prize in 1991, spent 15 years under house arrest until 2010. Widespread pro-democracy protests were repeatedly and brutally repressed after 1990. Elections in 2010, still overwhelmingly controlled by the military, brought a modestly reformist former general, Thein Sein, to the presidency. Limited free elections in 2012 and 2015 gave the pro-democracy party another landslide victory, with Aung San Suu Kyi's party winning a majority in parliament; but the military retains an allotment of 25 percent of parliamentary seats. In 2012, Suu Kyi belatedly claimed her Nobel Prize in Norway and in 2016, her party's success in the election meant that she was poised to pick the president, a post she was constitutionally barred from because she had married a British citizen.

Thailand

During the boom years of the mid-1990s, Thailand was known as one of the Asian *tiger economies*—that is, it was rapidly approaching widespread modernization and prosperity. Thailand takes pride in having avoided colonization by European nations and in having transformed itself from a traditional agricultural society into a modern nation. According to the United Nations, between 1975 and 1998, Thailand had the world's fastest-growing economy, averaging an annual growth in GNI per capita of 4.9 percent (U.S. annual growth then averaged 1.9 percent). At the same time, the country kept unemployment relatively low for a developing country—at 6 percent—and inflation in check at 5 percent or lower. The soaring economic growth resulted from the rapid industrialization that took place in Thailand when the government provided attractive conditions for large multinational corporations. They had situated themselves in Thailand to take advantage of its literate yet low-wage workforce, its lenient laws about environmental degradation, and its permissive regulation of manufacturing and trade.

Urbanization and Industrialization The Bangkok metropolitan area, which lies in the Chao Phraya delta at the north end of the Gulf of Thailand, contributes 50 percent of the country's GNI and contains 16 percent (10.5 million people) of its population (see the Figure 10.19 map). Cities elsewhere in Thailand are also growing and attracting investment, especially Chiang Mai in the north, Khon Kaen in the east, Surat Thani on the Malay Peninsula, and Laem Chabang (see Figure 10.29D).

Despite its urbanizing trends, about one-third of Thai working people are still farmers but, according to official statistics, they produce only 10 percent of the nation's GNI. (Remember,

FIGURE 10.30 A market at Inle Lake, Burma (Myanmar). Several indigenous groups live in villages around the lake, growing vegetables, flowers, and rice. Men produce pottery and items made of silver and brass, and women weavers have made the area Burma's second-largest producer of silk products.

though, that much of what farmers, especially women, produce is not counted in the statistics.) Urbanization is proceeding rapidly; 46 percent of the population, or about 30 million people, live in cities, many in crowded, polluted, and often impoverished conditions. It is common for thousands of immigrants from the countryside to live in slums along urban riverbanks, as is shown for Indonesia in Figure 10.19A. Industrialization has had unanticipated detrimental side effects in Thailand. Many of the millions of rural people drawn to the cities seeking jobs, status, and an improved quality of life arrive only to find a difficult existence and meager earnings.

Political Instability Thailand, which is actually a monarchy where the much-revered king usually plays a low-profile role, had a reputation as a rapidly developing democracy that allowed for protest and provided mechanisms for constitutional adjustments and smooth transitions after regular elections. Nevertheless, in 2006, Thailand's democracy was weakened when a military coup d'état, professing loyalty to the king, overthrew an apparently corrupt but popular, and populist, prime minister, Thaksin Shinawatra. Red-shirted crowds of Thaksin supporters faced off against yellow-shirted supporters of the monarchy and military. Cycles of bloody street battles and arrests ensued well into 2012. Mr. Thaksin, still highly controversial because of corruption charges and a tendency to manipulate his red-shirt supporters, was followed in office by his younger sister, Yingluck Shinawatra, who was then overthrown by another coup (see the discussion on page 603). These blows to Thai democracy are likely to affect economic development because the civil unrest discourages tourism and investment.

While Thailand's infrastructure is not as developed as those of other middle-income countries, hopes are high: the emerging Asian Highway network (see the Figure 10.16 map) should facilitate trade with India and China; Thailand now has three international airports, making it easier for tourists to visit, and more than one-third of the population, or 24 million people, have cell phones, which elsewhere have helped facilitate entrepreneurship and general development; but in Thailand, the military government controls the Internet.

THINGS TO REMEMBER

• Burma (Myanmar) and Thailand differ significantly in political structure, economy, and sociocultural variables.

• Although rich in natural resources, Burma (Myanmar) is one of the region's poorest countries.

• Thailand had been regarded as a successful democracy on the fast track to development, but a recent series of coups d'état and civil unrest have slowed economic growth and cast a cloud over its future.

MAINLAND SOUTHEAST ASIA: LAOS, CAMBODIA, AND VIETNAM

On the map in **FIGURE 10.31**, the countries of Laos, Cambodia, and Vietnam on the Southeast Asian mainland might appear to be ideally suited for peaceful cooperation, sharing as they do a common heritage of Buddhism (see Figure 10.31C) and the Mekong River and its delta (see discussion above, page 577). Nonetheless, these three countries went through a disruptive half-century of war that pitted them against each other. Until the end of World War II, all three were colonies, known collectively as *French Indochina*, and all three fought long struggles to transform themselves into independent nations. And then came the Vietnam War and its aftermath (see the discussion on page 580). As of 2016, they remain essentially Communist states with measured amounts of free market capitalism. Visitors and investors began arriving in the 1990s, and by 2000, once-somber city streets were abuzz with people and enterprises, especially in Vietnam (see Figure 10.31D).

A long, curved spine of mountains runs through Laos and into Vietnam. In the southern part of the subregion, these mountains are flanked on the west by the broad floodplain of the Mekong that is occupied by Thailand and Cambodia (see the Figure 10.31 map), and on the east by the 1000-mile-long (1600-kilometer-long) coastline and fertile river deltas of Vietnam (see Figure 10.31A). Vietnam, with 91.7 million people, is by far the most populous of these three countries. Cambodia has 15.4 million people; Laos, almost entirely mountainous, has only 6.9 million. The rugged mountainous territory of northern Vietnam and Laos is the least densely occupied area in mainland Southeast Asia. Most of the subregion's population lives along the coastal zones of Vietnam and in the Mekong delta, where people accommodate the seasonal floods by building their homes on stilts, as in the Ganga-Brahmaputra delta in India. Farmers take advantage of the wet tropical climate and flat terrain to cultivate rice (see Figure 10.31B). The Red River delta in northern Vietnam is also an important rice-growing area. Throughout the subregion, about 65 percent of the people support themselves as subsistence farmers; in Cambodia, only 21 percent of the people live in cities. There, people farm small plots of land as they have for many years. Development experts concerned about sustainable livelihoods for Cambodians say that for the time being this is the wisest way to provide for the population, but to provide truly adequate nutrition and surplus food for cities, managed egalitarian agricultural reform is needed.

The main trading partners for Laos and Cambodia are Thailand and China (as sources of imports) and the United States and Thailand (as destinations for exports of timber, vegetables, and inexpensive manufactured goods). Thai investors wield considerable power in all of these countries because they have interests in the production of rice, vegetables, coffee, sugarcane, and cotton for export, in addition to such resources as timber, gypsum, tin, gold, gemstones, and hydroelectric power. As discussed on pages 575–577, a major environmental issue is the planned damming of the highly biodiverse Mekong River to create hydropower that would be sold primarily to Thailand, which is investing in the dams. Laos, which needs the money that would be generated by the dams, has been silent about the protests of Cambodia and Vietnam—both downstream of the dams—that the dams' environmental impact would be overwhelming.

B Farmers plant rice in Cambodia.

C A Buddhist statue in Vientiane, Laos.

D Traffic in Ho Chi Minh City.

A A woman rows a boat near Chau Doc, in the Mekong Delta region of Vietnam.

FIGURE 10.31 Mainland Southeast Asia: The Vietnam, Laos, and Cambodia subregion.

Cities and National Capitals
- ● More than 5,000,000
- ◐ 1,000,000–5,000,000
- ○ 500,000–999,999
- ● 250,000–499,999
- ○ Less than 250,000

Transportation
—— Major road
—— Railroad

Vietnam is the most developed of the eastern Southeast Asian mainland countries (see Figure 10.31D). After the United States withdrew its military forces in 1973, the communist government, assisted by the Soviet Union, began to invest aggressively in health care, basic nutrition, and basic education. By 2012, 87 percent of children reached the fifth grade, and adult literacy was above 94 percent. Forty-eight percent of the Vietnamese people are still cultivators (down from 56 percent just 7 years ago), and they have extensive knowledge of such practical matters as useful plants (including medicinal herbs) and animals, home building and maintenance, and fishing. The Vietnamese have also developed a national flair for excellent cuisine; when times are good, the dinner table is filled with artfully prepared fish, vegetables, herbs, rice, and fruits.

Vietnam has mineral resources (phosphates, coal, manganese, offshore oil) and at one time had a lush forest cover, much of which was destroyed by defoliants in the latter years of the Vietnam War. Despite its resources, Vietnam's economy languished, partly because of economic sanctions imposed by the United States and partly because of the inefficiencies of communism. In the mid-1980s, the Hanoi leadership began to introduce elements of a market economy in a program of economic and bureaucratic restructuring called *doi moi* (very similar to perestroika in the Soviet Union and the structural adjustment programs imposed on many countries by international lending agencies). With the lifting of the U.S. embargo in 1994, firms from all over the world began to open branches in Vietnam. By the late 1990s, about 90 percent of the country's industrial labor force worked in the private sector, producing such exportable products as textiles and clothing, cement, fertilizers, and processed food. Vietnam has entered preferential trade agreements with the European Union and with the United States via the Trans-Pacific Partnership.

The factors that brought growth during the 1990s and 2000s also caused considerable dislocation in some parts of society. One focus of doi moi has been the privatization of land and other publicly held assets. The farmers who own newly privatized land have more security and so pay more attention to conservation and efficient production. But land reform left out the poorest people, who have had to resort to the informal economy to make a living. And in the late 1990s, farmers felt the squeeze when some of their lands were confiscated for use in foreign-funded enterprises such as tourist hotels, golf courses, and oil refineries.

> **THINGS TO REMEMBER**
>
> • Laos, Cambodia, and Vietnam share the Mekong River and its delta on the Southeast Asian mainland; contentions about how to use this river sustainably are rising.
>
> • Laos and Cambodia are the poorest countries in the subregion and are still struggling after years of warfare associated with the Vietnam War.
>
> • Of the three countries in this subregion, Vietnam has the highest human well-being ranking, the largest population, and the largest economy.

ISLAND AND PENINSULAR SOUTHEAST ASIA: MALAYSIA, BRUNEI, AND SINGAPORE

Malaysia and neighboring Singapore and Brunei are the most economically successful countries in Southeast Asia. Malaysia was created in 1963 when the previously independent Federation of Malaya (the southern portion of the Malay Peninsula) was combined with Singapore (at the peninsula's southernmost tip) and the territories of Sarawak and Sabah (on the northern coast of the large island of Borneo to the east) **(FIGURE 10.32 MAP)**. All had been British colonies since the nineteenth century. Singapore became independent from Malaysia in 1965. The tiny and wealthy sultanate of Brunei, also on the northern coast of Borneo, refused to join Malaysia and remained a British colony until its independence in 1984. Singapore and Brunei

are two of the wealthiest countries in the world, both with annual per capita incomes of over $70,000 (PPP). Virtually all citizens have a high standard of living, with Brunei's wealth coming from its oil and natural gas and Singapore's from a diversified economy grounded in global trade. Malaysia also has a diversified economy, though it is much less wealthy than these two neighbors, with an annual GNI per capita (PPP) of about $24,000 (see Figure 10.24A).

Malaysia

Malaysia is home to 30.8 million ethnically diverse people. Slightly over 50 percent of Malaysia's people are Malays, and nearly all Malays are Muslims. Ethnic Chinese, most of whom are Buddhist, make up nearly 24 percent of the population. Just over 7 percent of people in Malaysia are Tamil- and English-speaking Indian Hindus, and 11 percent are indigenous peoples of Austronesian ancestry, who live primarily in Sarawak and Sabah. Until the 1970s, conflicts among these many groups divided the country socially and economically. Malaysians have worked hard to improve relationships among their diverse cultural groups, and they have had noteworthy success.

Today, most Malaysians (86 percent) live on the Malay Peninsula, which has 40 percent of the country's land area. Throughout most of its history, indigenous Malays and a small percentage of Tamil-speaking Indians inhabited peninsular Malaysia. The Indians were traders who plied the Strait of Malacca, the ancient route from India to the South China Sea. Even before the eastward spread of Islam in the thirteenth century, Arab traders began using the straits in the ninth and tenth centuries, stopping at small fishing villages along the way to replenish their ships. By 1400, virtually all of the ethnic Malay inhabitants of peninsular Malaysia had converted to Islam.

During the colonial era, the British brought in Chinese Buddhists and Indians to work as laborers on peninsular Malaysian plantations. These Overseas Chinese eventually became merchants and financiers, while Indians achieved success in the professions and in small businesses. The far more numerous Malays remained poor village farmers and plantation laborers. As economic disparities widened, antagonism between the groups increased. After independence, animosities exploded into widespread rioting by the poor in 1969. Political rights were suspended, and it took 2 years for the situation to calm down. The violence so shocked and frightened Malaysians that they agreed to address some of the fundamental social and economic causes.

After a decade of discussions among all groups, Malaysia launched a long-term affirmative action program, called *Bumiputra*, in the early 1980s, designed to help the Malays and indigenous peoples advance economically. The policy required Chinese business owners to have Malay partners. It set quotas that increased Malay access to schools and universities and to government jobs. As the program evolved in the

doi moi a program of economic and bureaucratic restructuring in Vietnam, similar to perestroika in the Soviet Union and structural adjustment programs imposed on many countries by international lending agencies

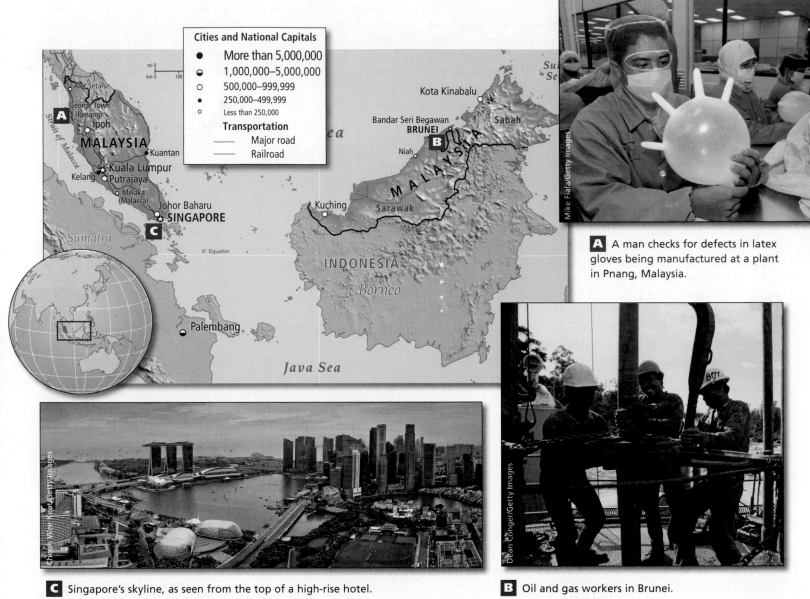

Cities and National Capitals

● More than 5,000,000
◒ 1,000,000–5,000,000
○ 500,000–999,999
• 250,000–499,999
○ Less than 250,000

Transportation
——— Major road
——— Railroad

A A man checks for defects in latex gloves being manufactured at a plant in Pnang, Malaysia.

C Singapore's skyline, as seen from the top of a high-rise hotel.

B Oil and gas workers in Brunei.

FIGURE 10.32 Island and peninsular Southeast Asia: The Malaysia, Singapore, and Brunei subregion.

1980s, the goal became to bring Malaysia to the status of a fully developed nation by the year 2020. The program has succeeded in narrowing some social inequalities fairly rapidly. In 2011, the United Nations ranked Malaysia sixty-second, in the high- (but not highest-) development category. As incomes in Malaysia have increased, wealth disparity (the gap between rich and poor), though still considerable, has narrowed rather than widened. This is an unusual pattern.

The economy of Malaysia has traditionally relied on raw materials export products such as rubber, palm oil, tin, oil and gas, and iron ore; and on some manufacturing (see Figure 10.32A, B, C). In contrast, much of the growth of the 1990s came from Japanese and U.S. investment in manufacturing, especially of electronics, and from offshore oil production

in the South China Sea. Timber exports are also important (especially in Sarawak and Sabah). The development of oil palm and rubber tree plantations on cleared forest land (see the vignette at the start of the chapter) has brought global criticism because of its multiple negative impacts: deforestation, burning of scrap vegetation, GHG emissions, river pollution, the demise of forest species (such as the orangutan, which in Malay means *man of the forest*), and the displacement of indigenous people. Because many countries in this region are collectively responsible for deforestation, Malaysia called for reinforcement of the ASEAN agreements on Transboundary Haze Pollution.

The Malaysian capital, Kuala Lumpur, boasts beautifully designed skyscrapers like those in Singapore, including what were in 2000 the two tallest buildings in the world—Petronas Towers I and II (buildings elsewhere have since surpassed the towers in height). Nonetheless, at least one-fifth of the city's residents are squatters, some on land in the shadows of the towers, some on raft houses on rivers or bays. Malaysia is hoping to benefit from a new, multibillion-dollar high-tech manufacturing corridor just south of Kuala Lumpur. Called the MSC Malaysia, it includes two new cities: the new federal administrative center, called Putrajaya, and another city named Cyberjaya. The MSC was expected to help the nation spring fully prepared into a prosperous twenty-first century, but growth has been undercut by various global recessions.

Like many on the Asian development fast track, Malaysia has volatile economic patterns. Currency values and stock market prices may fluctuate widely. Lavish building projects may be halted for years, as they were in Kuala Lumpur from 1998 through 2001 and again during the recession that began in 2007. Overall, however, it does seem that Malaysia is likely to have an increase in prosperity and to continue to foster cultural diversity while discouraging extremist factions, religious or ethnic.

Brunei

Brunei is a country, a little smaller than Delaware, on the north coast of Borneo, where it is situated between the Malaysian states of Sarawak and Sabah. It is a sultanate that has been ruled by the same family for more than 600 years. The sultan is both the head of government and the head of state, and the country became independent from Britain in 1984. Brunei has a small economy, with over half of its GNI from crude oil and natural gas production that make up 90 percent of the country's exports (see Figure 10.32B). Significant income from overseas investments, primarily in the petrochemical industries, helps to give Brunei the world's eighth-highest per capita GNI income ($72,190 in 2015). Medical care is free, as is education through the university level. Ethnically, Malays comprise two-thirds of the population, Chinese people about 11 percent, and a variety of others make up the rest. Sixty-seven percent of the people are Muslim, 13 percent are Buddhist, 10 percent Christian, and 10 percent practice other beliefs. More than 95 percent of the population is literate and the life expectancy is over 75 years.

Singapore

Singapore occupies one large and many small hot, humid, flat islands just off the southern tip of the Malay Peninsula. Its 5.5 million inhabitants, all of them urban, live at a density of more than 20,446 people per square mile (7890 per square kilometer), yet as one of the wealthiest countries in the world, Singapore's urban landscapes are elegant (see Figure 10.32C; see also Figure 10.19B). Singapore's wealth is based on pharmaceutical, biomedical, and electronic manufacturing; financial services; oil refining and petrochemical manufacturing; and oceanic transshipment services. Unemployment is low and poverty unusual, except among temporary and often undocumented immigrants—primarily from Indonesia—who live on little islands surrounding Singapore. Singaporeans seem to share the notion that financial prosperity is a worthy first priority, but although the free market economy reigns, the government has a strong regulating role and provides many services.

Ethnically, Singapore is overwhelmingly Chinese (74.2 percent). Malays account for 13 percent of the population, Indians for 9.2 percent, and mixed or "other" peoples for 3.3 percent. Singapore's people are diverse in religious beliefs: 33 percent are Buddhist or Taoist, 18 percent are Christian, 14.7 percent are Muslim, and 5 percent are Hindu; the rest include Sikhs, Jews, and Zoroastrians. As in Malaysia and Indonesia, the Singapore government officially subscribes to a national ethic not unlike *Pancasila* in Indonesia. The nation commands ultimate allegiance; loyalty is to be given next to the community and then to the family, which is recognized as the basic unit of society. Individual rights are respected but not championed. The emphasis is on shared values, racial and religious harmony, and community consensus rather than majority rule, which is thought to lead to conflict.

Singapore has a meticulously planned cityscape with safe, clean streets and little congestion because of an elaborate light rail system. Eighty percent of the people live in government-built housing estates (see Figure 10.32C), and workers are required to contribute up to 25 percent of their wages to a government-run pension fund. However, home care of the elderly is the responsibility of the family and is enforced by law. Each child, on average, receives 10 years of education and can continue further if his or her grades and exam scores are high enough. Literacy is close to 95 percent. The government strictly controls virtually all aspects of society. Permits are required for almost any activity that could have a public effect: having a car radio, owning a copier, working as a journalist, having a satellite dish, being a sex worker, or performing as a street artisan or entertainer. Law and order are strictly enforced. Some years ago, a visiting U.S. teenager was sentenced to a caning for spray-painting graffiti. Drug users are severely punished, and drug dealers are sentenced to life imprisonment or death. But virtually all citizens of Singapore seem to have accepted this control and strictness in return for a safe city and the fourth-highest GNI per capita in the world.

THINGS TO REMEMBER

• Singapore and Brunei are very small, relatively stable, and very wealthy authoritarian countries where the accepted wisdom is that democracy would be too chaotic.

• The largest and most populous country in this subregion, Malaysia, which has an increasingly diversified economy, has managed to control ethnic and political strife through the positive efforts on the part of the citizenry to painstakingly accommodate all interests. Democratic traditions are somewhat stronger in Malaysia, but authoritarian policies and official corruption are common.

• Malaysia is making some efforts to cope with its environmental issues caused mostly by unregulated development.

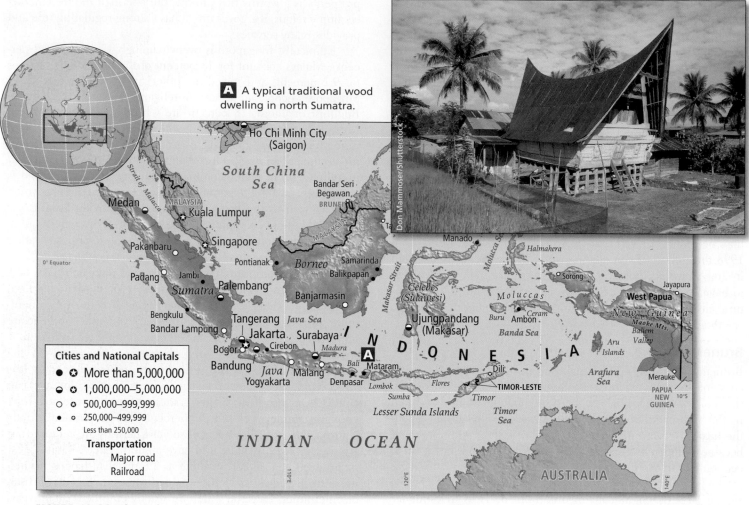

A A typical traditional wood dwelling in north Sumatra.

FIGURE 10.33 The Indonesia and Timor-Leste subregion.

INDONESIA AND TIMOR-LESTE

Indonesia, with the world's largest Muslim population, is a recent amalgamation of island groups, inhabited by people who, before the colonial era, never thought of themselves as a national unit **(FIGURE 10.33 MAP)**. Rivalry between island peoples remains strong, and many resent the dominance of the Javanese in government and business. This resentment and other burgeoning issues of identity and allegiance are serious threats to Indonesia's continued unity (see the discussion of Indonesia that begins with "Can the Expansion of Political Freedoms Bring Peace to Indonesia?" on pages 590–591). Emblematic of the fragile state of this subregion is the fact that Timor-Leste, the eastern half of the small island of Timor, is now independent. In 1999, after more than 20 years of armed conflict with Indonesia, in which the infrastructure was destroyed and as many as 250,000 people died (one-quarter of the population), the people of eastern Timor voted in a UN plebiscite to split from Indonesia and become known as Timor-Leste (East Timor). For 2 more years, the Indonesian military tried to enforce cohesion, in part because of

Timor-Leste's oil and gas deposits, but in 2002, after 1500 more deaths, Timor-Leste became an independent country.

The archipelago of Indonesia consists of the large islands of Sumatra, Java, and Sulawesi; the Lesser Sunda Islands east of Java (including Bali); Kalimantan, which shares the island of Borneo with Malaysia and Brunei; West Papua on the western half of New Guinea; and the Moluccas. In all, it contains some 17,000 islands, but some are small, uninhabited bits of coral reef. The term *Indonesia* was coined in 1850 by James Logan, a Singapore-residing Englishman, from two Greek words: *indos* (Indian) and *nesoi* (islands). Indonesians themselves now refer to the archipelago as *Tanah Air Kita*, meaning "Our Land and Water." This name conveys a sense of the archipelago environment, but falsely indicates a unified feeling of togetherness and environmental concern.

The island of Java, the most economically productive in Indonesia, is home to 57 percent of the country's population (145 million people in 2015) but only 7 percent of its available land. However, thousands of years of ash from Java's 17 volcanoes have made the soil rich and productive, capable of supporting large numbers of people. Another 21 percent of the population lives on the adjacent volcanic island of Sumatra. Indonesia frequently resettles large numbers of people on Sumatra to relieve population pressure elsewhere or to isolate dissidents.

The resettlement of Javanese people to other islands in Indonesia has been a contentious issue because indigenous people resent being inundated with Javanese culture and concepts of economic development. They see resettlement as a Javanese effort to gain access to, and profit from, indigenously held natural resources, such as oil, precious metals, and forestlands. Indigenous people, some of whom continue to live via subsistence agriculture as they have for millennia **(FIGURE 10.34)**, have been repeatedly devastated by the rapid changes brought to everyday life by the newcomers. The government's argument that resettlement provides jobs is true, strictly speaking. But often the habitat is degraded beyond repair. The central government tends to ignore these hardships, instead invoking the principles of *Pancasila*, arguing that resettlement strengthens Indonesian identity by spreading modernization and eliminating ways of life that differ from the government's vision of the norm.

FIGURE 10.34 Men from the Dani ethnic group take part in a mock battle in Baliem Valley, West Papua. Logging and mining interests, mostly owned by corporations from outside West Papua, as well as immigration from Java and other densely populated islands, have driven many Papuans off their land. Meanwhile, a large population of indigenous people still lives in the remote Baliem Valley of West Papua, where they practice subsistence agriculture.

THINGS TO REMEMBER

• Indonesia is densely settled and has the world's fourth-largest population. It also has the world's largest Muslim population.

• Timor-Leste, rich in oil and gas deposits yet with many impoverished residents, gained its independence from Indonesia in 2002.

• Indonesia is a country made up of numerous islands that are home to many ethnic groups. Now bound into one unit, it has had to construct a national identity.

• A contentious strategy for resolving the country's population and unemployment problems has been the resettlement of urban people to deforested lands, displacing indigenous peoples.

THE PHILIPPINES

The Philippines, lying at the northeastern reach of the Southeast Asian archipelago, comprises more than 7000 islands spread over about 500,000 square miles (1.3 million square kilometers) of ocean **(FIGURE 10.35 MAP)**. The two largest islands are Luzon in the north and Mindanao in the south. Together, they make up about two-thirds of the country's total land area, which is about the size of Arizona. The Philippine Islands, part of the Pacific Ring of Fire (see Figure 1.26), are volcanic. The violent eruption of Mount Pinatubo in June 1991 devastated 154 square miles (400 square kilometers) and blanketed most of Southeast Asia with ash (see Figure 10.35A). Volcanologists had predicted the eruption, and precautions were taken, so although about 3 million people were threatened, the death toll was less than 0.028 percent—still, 847 people died. Over time, volcanic eruptions have given the Philippines fertile soil and rich deposits of minerals (gold, copper, iron, chromate, and several other elements).

Population, Urbanization, and Disaffection When the Philippines became a virtual U.S. colony in 1898, after the Spanish-American War, it had 7 million people. Just over 110 years later, the population is nearly 15 times larger, at about 103 million. Today, close to 44 percent of Filipinos live in cities, about the same proportion for Southeast Asia as a whole (47 percent). The metropolitan area of Manila, the country's capital, with a population of more than 22 million, is one of the largest and most densely settled

A The eruption of Mount Pinatubo in 1991, which killed 847 people, affected about 3 million others, and impacted climate on a global scale for at least 3 years.

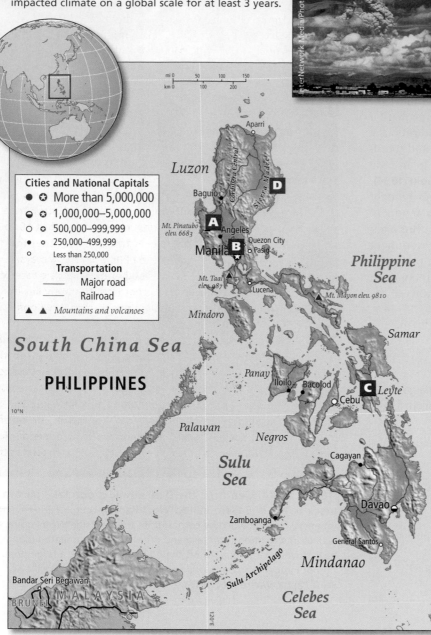

Cities and National Capitals

● ✪ More than 5,000,000
◒ ✪ 1,000,000–5,000,000
○ ✪ 500,000–999,999
• ○ 250,000–499,999
○ Less than 250,000

Transportation

—— Major road
—— Railroad
▲ ▲ *Mountains and volcanoes*

FIGURE 10.35 The Philippines subregion.

B Fire burns in a slum on the fringes of Manila's central business district.

C A 2006 landslide in Leyte, Philippines, that destroyed an entire village. Previous logging and mining are blamed for loosening the soil, leading to an avalanche after heavy rains.

D An indigenous woman fishes for squid near the Philippines Northern Sierra Madre National Park.

urban agglomerations in the world (15,400 people per square mile or 40,000 per square kilometer; see Figure 10.35B). Some urban residents were displaced to the city by the eruption of Mount Pinatubo, others by rapid deforestation (see Figure 10.35C), dam projects, and mechanized commercial agriculture.

Because of the forced nature of rural-to-urban migration, many people are unemployed urban squatters living in shelters they have built out of scraps. To the Philippine central government, the masses of urban poor represent a particular political and civil threat. Even though the excesses of the dictator Ferdinand Marcos ended in 1986 and the presidencies of Corazon Aquino, Gloria Macapagal-Aroyo, and Benigno Aquino III have filled the decades since with efforts to improve life, the people's faith in their government has not yet been restored. Marcos maintained a flamboyant and brutal regime beginning in 1965. Since his ouster in 1986, there has been insufficient economic progress, and violent social unrest has become very much a part of life in the Philippines.

Cultural and Economic Disparities The cultural complexity of Philippine society is illustrated by its ethnic profile. The majority of the population (96 percent) belongs to one or more of 60 distinct indigenous ethnic groups. The Chinese make up another 1.5 percent of the population, and the remaining 3 percent includes Europeans, Americans, other Asians, and people native to other Southeast Asian islands.

Most Filipinos are Roman Catholic (83 percent), but the southern and central portions of the island of Mindanao and the Sulu Archipelago have been predominantly Muslim for a long time. Over the last few decades, the central government has been resettling many thousands of Catholics in Mindanao, apparently to dilute the Muslim population. Rival family clans that have armed militias who do their bidding dominate local politics in Mindanao. In retaliation against those clans that are allied with Catholic elected national officials (President Gloria Arroyo, for one), Muslim fighters, some possibly linked to international terrorist movements, began operating in the southern Philippines, carrying out bombings that have killed scores of people (see Figure 10.18C and the figure map).

Wealth and poverty are also often associated with certain ethnic groups, as is the case elsewhere in Southeast Asia. The vast majority of the wealthy are descendants of Spanish and Spanish-Filipino plantation landowners or of Chinese financiers and businesspeople. It is estimated that 30 percent of the top 500 corporations in the islands are controlled today by ethnic Chinese Filipinos, who make up less than 1 percent of the population. Wealthy families control 78 percent of all corporate wealth in the Philippines. Meanwhile, the poor are overwhelmingly indigenous people.

In contrast to Malaysia, where ethnic disparities in wealth became an openly discussed national social issue and then were addressed with overt policies, the Philippine government under Marcos did not embark on an economic policy to help disadvantaged ethnic groups, or improve the country's infrastructure, or find strategies for national reconciliation. President Marcos's reaction to protests that erupted in 1972 was to declare martial law. He hung onto power for another 14 years, during which there was little progress in infrastructure development, education advancement, job creation, population control, or wealth redistribution. Successors to Ferdinand Marcos have been hampered in their efforts to settle social unrest and attract investment by the triple economic disasters of the Mount Pinatubo eruption, the closing of the U.S. military bases in 1991, and a drastic cut in U.S. foreign aid once the Americans left. By 2015, there was talk (unlikely to be fulfilled) that the United States might negotiate the reopening of the Subic Bay base as part of its new interest in playing a stronger role in the greater Asian region, a development that would have major economic impacts on the Philippines.

The case of the Philippines under Marcos and since shows how poor leadership and corrupt government can hold back a country. Despite the rampant sale abroad of its once-rich timber resources, the country's economy has been so lethargic that actual unemployment rates are thought to be as high as 30 percent, with 40 percent of the population living below the poverty line. While the older rural poor must scavenge for survival (Figure 10.35D), many of the best-educated young Filipinos have chosen to temporarily migrate to more vibrant economic zones, where they often work at jobs that are far below their qualifications (see the vignette about the maids from the Philippines on page 593).

ON THE BRIGHT SIDE: Filipinos Envision a Better Future

Ordinary Filipinos are keenly aware that a better social and political system is possible. In cooperation with the United Nations, the government of New Zealand, and several NGOs, a group of concerned Philippine citizens have produced 11 Human Development Reports over the last two decades. The report for 2009 is a model of frank talk about civic problems. Noting that human well-being has dropped dramatically over the last 10 to 15 years, the report states: "There is wide agreement that the weakness of political institutions in the Philippines is a major… hindrance to progress…. When all is said and done…the fate of [human development] policies and programs ultimately lies in the hands of the government" (Philippines Human Development Report 2008–2009, UNDP, p. 2).

The 2015 report advocates that "work" be defined broadly not only as making a living, but as a way to participate in the holistic project of human development— with social cohesion as a family and community goal. The report goes on to tutor the public on the multiple ways to measure human development and on just where the Philippines fits into the global picture. The careful marshaling of evidence by these citizens and their brave dedication to rational long-term change bodes well for the future of the Philippines ■

THINGS TO REMEMBER

• Uncontrolled population growth during the twentieth-century occupation by the United States led to unsustainable use of resources, heavy rural-to-urban migration, and deteriorating living standards and political order.

• Development has been lopsided, with the largest firms overwhelmingly in the hands of a tiny elite that pays little attention to human well-being.

• Ethnicity and clans still influence local politics in the Philippines, often with deadly consequences. Such rivalries are often conflated with unrest among minority Muslims in the southern provinces of the Philippines.

• In a series of frank Human Development Reports, the citizens themselves have identified enduring political problems as caused by a failure of leadership, especially in the realms of political institutions and education.

GEOGRAPHIC THEMES: Southeast Asia Review and Self-Test

1. Environment: Many of Southeast Asia's most critical environmental issues relate in some way to climate change. Deforestation, a major global source of greenhouse gas emissions, has been rapid; and the region is vulnerable to the climate change effects of flooding and droughts. Human manipulation of the land and resources is changing ecological systems and threatening food security. Rising ocean temperatures are straining aquatic ecosystems and storm surges threaten human settlements.

• How does deforestation lead to more GHGs?

• How is food production for global markets related to deforestation?

• How is deforestation related to increased risks of floods and droughts?

2. Globalization and Development: Globalization has brought both spectacular successes and occasional declines to the economies of Southeast Asia. Key to the economic development of this region are strategies that were pioneered earlier in East Asia: the formation of state-aided market economies, combined with export-led economic development. Regional economic cooperation is expanding slowly.

• How is globalization linked to the economic success of Southeast Asia?

• How is globalization linked to rapid urbanization and the growth of slums in this region?

• Why have global recessions caused such dramatic economic declines for the many workers in Southeast Asia?

3. Power and Politics: There has been a general expansion of political freedoms throughout Southeast Asia in recent decades, but authoritarianism, corruption, and violence have at times reversed these gains.

• How does the history of colonialism in this region influence the current dynamics of political power?

• What arguments have some Southeast Asian leaders made about Asian values not being compatible with Western ideas of democracy and political freedom?

• Why are some countries concerned about China's interest in the South China Sea?

• What are some of the explanations for the violence against ethnic or religious minorities that tends to occur during hard economic times?

• How is Indonesia's government working to alleviate ethnic tensions?

4. Urbanization: Although Southeast Asia as a whole is only 47 percent urban, its cities are growing rapidly as agricultural employment declines and urban industries expand. The largest Southeast Asian cities, which are receiving most of the new rural-to-urban migrants, rarely have sufficient housing, water, sanitation, or jobs for all their people.

• How have development strategies, such as mechanized agriculture and agribusiness, changed the degree of food self-sufficiency in this region?

• How are development and urbanization linked to the decision of many to migrate abroad for employment?

5. Population and Gender: Population dynamics vary considerably in this region because of differences in economic development, government policies, prescribed gender roles, and religious and cultural practices. Economic change has brought better job opportunities and increased status for women, who then often choose to have fewer children. Some countries also have gender imbalances because of a cultural preference for male children.

• Where have population growth rates slowed the most in this region?

• What is a major source of population growth for Singapore other than natural increase?

• What are variations in population growth linked to in this region?

• Where have efforts at HIV/AIDS prevention been more successful?

• Describe the circumstances that have led to better job opportunities for women.

• How are the improving roles and status of women changing attitudes toward female sex work?

Critical Thinking Questions

1. What do you think are the most serious threats to Southeast Asia posed by global climate change? What are the most promising responses to global climate change emerging from this region?

2. Which countries in the region have been most successful in controlling population growth? How do population issues differ among Southeast Asian countries? How do the population issues of the region as a whole compare with those of Europe or Africa?

3. Did those responsible for resolving the economic crisis of the 1990s opt for reregulation or deregulation of financial institutions? Why did they make this choice? Consider the role of crony capitalism in your response.

4. In what ways did state aid to market economies and export-led growth amount to a strategic and limited embrace of globalization?

5. ASEAN's overall stated goals are to develop a sustainable regional economic community. To what extent has the organization achieved these goals? Where has it fallen short?

6. Describe the spatial distribution of major religious traditions of this region. How have these religious traditions influenced each other over time?

7. What factors have produced major flows of refugees in Southeast Asia?

8. Relationships and gender roles in some traditional Southeast Asian families are not necessarily what an outsider might expect. Describe the characteristics of these relationships that interested you most. Why did they catch your attention? Discuss some of the ways in which gender roles vary from those typically found in South Asia or East Asia.

9. How has the role of the Overseas Chinese evolved over time? Why have some people in this region resented them?

10. Compare and contrast political freedoms in Singapore and Thailand or Singapore and Burma (Myanmar).

Chapter Key Terms

archipelago 567
Association of Southeast Asian Nations (ASEAN) 584
Australo-Melanesians 578
Austronesians 578
coral bleaching 575
crony capitalism 583

cultural pluralism 599
detritus 570
distributaries 577
doi moi 609
domino theory 580
feminization of labor 582
foreign exchange 593

Pancasila 590
populist political movement 589
resettlement schemes 590
sex tourism 603
social cohesion 584
soft power initiatives 589

C Australia's Eastern Highlands, Great Dividing Range

A Australian Desert

NORTH
PACIFIC
OCEAN

Land Elevations

meters	feet
4877	16,000
3353	11,000
2134	7000
914	3000
305	1000
152	500
0	0

Ocean Depths

meters	feet
0	0
300	984
3500	11,483
5000	16,404

mi 0   200   400   600   800
km 0   200   400   600   800 1000 1200

1:42,000,000
Mercator Projection

Kauai
Oahu
Molokai
Honolulu   Muui
Hawaii
(U.S.)   Hawaii

Line
Islands
Kiritimati

Polynesia

Marquesas
Islands

B Great Barrier Reef

Gonzalo Azumendi/Getty Images

0° Equa

E

Cook Islands
(New Zealand)

Papeete   Tahiti
F French Polynesia
(France)

Avatua
Rarotonga

SOUTH PACIFIC
OCEAN

Pitcai
Islan
(U.K.

160°W   150°W   140°W   130°W

D New Zealand, Milford Sound, South Island

Jean-Pierre Pieuchot/Getty Images

11

Oceania: Australia, New Zealand, and the Pacific

E Tūpai Atoll, French Polynesia

Danita Delimont/Gallo Images/Getty Images

F The High Island of Mo'orea, French Polynesia

Fotosearch/Getty Images

FIGURE 11.1 Regional map of Oceania

619

GEOGRAPHIC THEMES: Oceania

After you read this chapter, you will be able to discuss the following geographic themes as they relate to the five thematic concepts:

1. Environment: Oceania faces a host of environmental problems and public awareness of environmental issues is keen. Global climate change, primarily warming, has brought rising sea levels and increasingly variable rainfall. Other major threats to the region's unique ecology have come from the introduction of nonnative species and the expansion of herding, agriculture, fishing, fossil fuel extraction, waste disposal, and human settlements.

2. Globalization and Development: Globalization, coupled with Oceania's stronger focus on neighboring Asia (rather than long-time connections with Europe and North America), has transformed patterns of trade and economic development across Oceania. These changes are driven largely by Asia's growing affluence, its enormous demand for resources, and its similarly massive production of manufactured goods.

3. Power and Politics: Stark divisions have emerged in Oceania over definitions of democracy—the system of government that dominates in New Zealand, Australia, and Hawaii—versus the Pacific Way, a political and cultural philosophy based on the communitarian values of traditional cultures of the Pacific islands. The global refugee crisis tests this region's ability to maintain its reputation as a humanitarian refuge and zone of opportunity that maintains social cohesion.

4. Urbanization: Oceania is only lightly populated but it is highly urbanized. The shift from extractive economies to service economies is a major reason for the urbanization of the wealthiest parts of Oceania (Australia, New Zealand, Hawaii, Guam), where 80 to 100 percent of the population lives in cities. These trends are weakest in Papua New Guinea and many smaller Pacific islands.

5. Population and Gender: In this largest but least populated world region, there are two main patterns relevant to population and gender. Australia, New Zealand, and Hawaii have older and more slowly growing populations, and relatively more opportunities for women. The Pacific islands and Papua New Guinea have much more rural, younger, and rapidly growing populations, where women play a central role in family and community but enjoy fewer opportunities as economies modernize.

The Oceania Region

The region of Oceania, shown in **FIGURE 11.1**, is a vast and often idealized region that faces a number of difficult realities. The five thematic concepts in this book are explored as they arise in the discussion of these issues. Vignettes, like the one that follows about the Kiribati archipelago and the challenges its people face due to climate change, help illustrate the themes as they are experienced in individual lives.

GLOBAL PATTERNS, LOCAL LIVES Aurora (a pseudonym) is a nursing student in Brisbane, Australia, who is undergoing a wrenching personal transition due to climate change. Aurora is from Kiribati, a Pacific nation of 33 tiny islands barely 6 and a

(A) Inhabitants of a now-relocated Kiribati village gather where their village once stood. Kiribati's coastline is eroding as sea level rises.

(B) A swamp taro pit that was rendered infertile by intruding salt water. Swamp taro is a major food source for inhabitants of low-lying islands. It is cultivated in a carefully managed pit that makes use of a layer of fresh (not salty) groundwater that exists along the coastline. As sea level rises, salt water intrudes into this layer of fresh groundwater, killing the plants or stunting their growth.

FIGURE 11.2 Kiribati.

half feet above sea level. Rising seas are swamping the Kiribati archipelago, which straddles the equator and the International Date Line (IDL at 180° longitude).[1] Aurora pines for her island's blue lagoons. While she studies for a new future and prepares to become her household's primary income earner, she does so without the daily close company of her raucous extended family.

Several years ago, Aurora and her fellow citizens (100,000 people live in Kiribati) began to notice alarming environmental changes. Their drinking water was getting brackish (salty), several shoreline villages had sunk below sea level, and the climate had become so dry that skilled gardeners could no longer raise their customary cabbages, tomatoes, cucumbers, and swamp taro (FIGURE 11.2A, B). Drought conditions were being worsened by rising salty tides that saturated formerly fertile soil. Scientists

[1] The IDL is adjusted to the east so that all the islands of Kiribati are in the same time zone.

predicted that, with rising seas and more frequent droughts and storms, much of Kiribati will be uninhabitable in a few decades.

Aurora's nursing education, funded by a government program called AusAID, is one of several strategies developed by Kiribati President Anote Tong to encourage his people to gradually "migrate with dignity." Tong sees climate change as a huge challenge for his country, and he hopes that other South Pacific islands, as well as Australia and New Zealand, will allow Kiribati people to resettle there permanently. Through the AusAID program, Aurora and other young people receive an education that will help them find employment outside Kiribati and support their large extended families after their family members have joined them.

But it is a tough transition. Another nursing student says he feels as if he is losing his identity as his nation sinks into the sea. Some students' parents refuse to leave Kiribati. They acknowledge the changes in sea level and rainfall, but as Christians, they believe that "God is not so silly to allow people to perish just like that."

In 2014, the government of Kiribati under President Tong purchased the 5500-acre (2225-hectare) Natoavatu Estate on the multi-island nation of Fiji, 1300 miles (2092 kilometers) to the south (FIGURE 11.3). His plans are unclear: Does he plan to produce food to export to Kiribati, or to import Fijian soil to replace the soil lost to sea level rise in Kiribati, or to relocate Kiribati's people (100,000) to Fiji in the event of catastrophic sea rise? President Tong's purchase has triggered controversy in Fiji, as explored by geographer James Ellsmoor and engineer Zachary Rosen in a January 2016 essay on the Devpolicy blog. *[Sources consulted: Devpolicy Blog, http://devpolicy.org/kitibatis-land-purchase-in-fiji-does-it-make-sense-20160111/. For detailed source information, see Text Sources and Credits.]* ∎

To give the reader a sense of the size of the Pacific, the insert map A shows the distance from San Francisco to Melbourne, a flight that takes 16 hours, and crosses the International Date Line (IDL), an imaginary line at about 180 degrees east (or west) of the Greenwich Meridian. This line marks the change from one day to the next. If you leave San Francisco on a Tuesday, when you cross the IDL it will become Wednesday. Deviations in the dateline are made to avoid dividing countries, such as Kiribati.

Perhaps nowhere else on Earth are the effects of climate change as measurably real as they are in the low-lying islands of the Pacific, where changes in rainfall patterns and sea level are strikingly evident. These islands have contributed little to

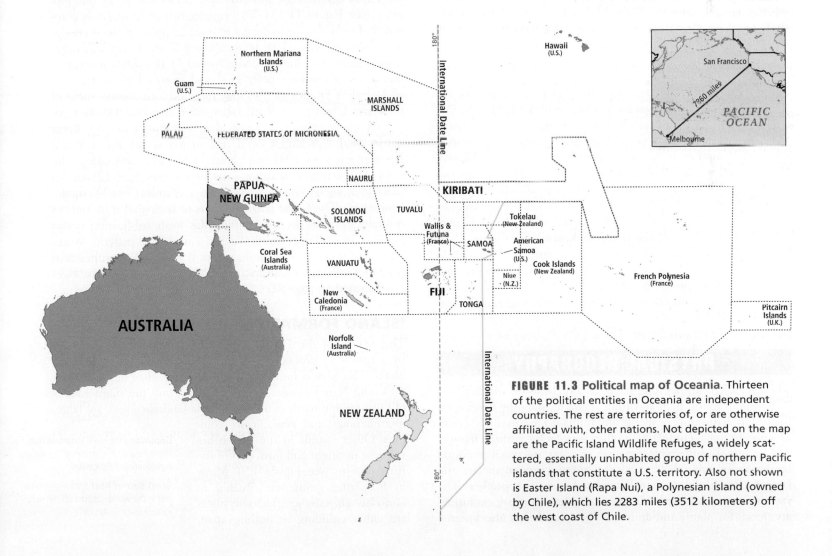

FIGURE 11.3 Political map of Oceania. Thirteen of the political entities in Oceania are independent countries. The rest are territories of, or are otherwise affiliated with, other nations. Not depicted on the map are the Pacific Island Wildlife Refuges, a widely scattered, essentially uninhabited group of northern Pacific islands that constitute a U.S. territory. Also not shown is Easter Island (Rapa Nui), a Polynesian island (owned by Chile), which lies 2283 miles (3512 kilometers) off the west coast of Chile.

the greenhouse gases that have triggered rapid climate change. Because many people in the islands have fairly simple lifestyles, most Pacific islands have relatively low emissions of greenhouse gases on a per capita basis. Their emissions are also low as a whole because the islands' total population is small. But seemingly remote places in the Pacific are anything but remote from general global influences. Beyond climate change, people in this region are dealing with changes brought on by tourism; environmental degradation caused by modern mining techniques; freshwater scarcity; ocean pollution; and rapidly changing cultural and trading patterns.

What Makes Oceania a Region?

Oceania is made up of Australia, New Zealand, Papua New Guinea, Hawaii, and the many small islands scattered across the Pacific Ocean (see Figure 11.3). It is a unique world region in that it is composed primarily of ocean and covers the largest area of Earth's surface of any region, yet it is home to only 40 million people, the vast majority of whom live in Australia (23.9 million), Papua New Guinea (7.7 million), and New Zealand (4.6 million). In this book, we also include the U.S. state of Hawaii (1.4 million) in Oceania. Altogether, the thousands of other Pacific islands are home to only 3 million people. The Pacific Ocean is both a uniting feature and a barrier: the vast ocean serves as a link that unites Oceania as a region, profoundly influencing life even on dry land, but across the region, the ocean also acts as a biological and cultural barrier.

Terms in This Chapter

Some maps in this chapter show different place-names for the same locations. This reflects the political evolution of the region, where some islands previously were grouped under one name that remains in everyday use but are now grouped into country units with another name. For example, the Caroline Islands, located north of New Guinea, still go by that name on maps and charts, but they have been divided into two countries: Palau (a small group of islands at the western end of the Caroline Islands) and the Federated States of Micronesia, which extends over 2000 miles west to east, from Yap to Kosrae (see Figure 11.3). In addition, there are three names commonly used to refer to large cultural groupings of islands: Micronesia, Melanesia, and Polynesia. These groupings are not political units; rather, they are based on ancient ethnic and cultural links and can be seen in Figure 11.13.

PHYSICAL GEOGRAPHY

The Pacific Ocean has long served as a highway on which plants and animals found their way from island to island by floating on the water, swimming, or flying, and humans have used the ocean as a way to make contact with other peoples for visiting, trading, and raiding. But the wide expanses of water are also a profound barrier to the natural diffusion of plant and animal species and keep Pacific Islanders spatially isolated from one another. The vast ocean has imposed solitude and fostered unique evolutionary trends for plants and animals. The solitude has also fostered

self-sufficiency and subsistence economies among its human occupants well into the modern era. Unfortunately, as we shall see, the ocean also serves as a conveyer of pollution, especially discarded plastics (see page 633).

CONTINENT FORMATION

The largest landmass in Oceania is the ancient continent of Australia at the southwestern edge of the region (see the Figure 11.1 map). The Australian continent is partially composed of some of the oldest rock on Earth and has been relatively stable for more than 200 million years, with very little volcanic activity and only an occasional mild earthquake. Australia was once a part of the great landmass called **Gondwana** that formed the southern part of the ancient supercontinent Pangaea (see Figure 1.8). What became present-day Australia broke free from Gondwana and drifted until it eventually collided with the part of the Eurasian Plate on which Southeast Asia sits. That impact created the mountainous island of New Guinea to the north of Australia.

Australia is shaped roughly like a dinner plate with an irregularly broken rim with two bites taken out of it: one in the north (the Gulf of Carpentaria) and one in the south (the Great Australian Bight). The center of the plate is the great lowland Australian desert, with only two hilly zones and rocky outcroppings (see Figure 11.1A). The eastern rim of Australia is composed of uplands; the highest and most complex of these are the long, curving Eastern Highlands (labeled "Great Dividing Range" in the Figure 11.1 map; see also Figure 11.1C). Over millennia, the forces of erosion—both wind and water—have worn most of Australia's landforms into low, rounded formations. Some of these, like Uluru (Ayers Rock), are quite spectacular **(FIGURE 11.4)**.

Off the northeastern coast of the continent lies the **Great Barrier Reef,** the largest coral reef in the world and a World Heritage Site since 1981 (see Figure 11.1B). It stretches along the coast of Queensland in an irregular arc for more than 1250 miles (2000 kilometers), covering 135,000 square miles (350,000 square kilometers). The Great Barrier Reef is so large that it influences Australia's climate by interrupting the westward-flowing ocean currents in the mid–South Pacific circulation pattern. Warm water is shunted to the south, where it warms the southeastern coast of Australia. Threats to the health of the Great Barrier Reef are discussed on page 627.

ISLAND FORMATION

The islands of the Pacific were (and are still being) created by a variety of processes related to the movement of tectonic plates. The islands found in the western reaches of Oceania—including New Guinea, New Caledonia, and the main islands of Fiji—are remnants of the Gondwana landmass; they are large, mountainous, and geologically complex. Other islands in the region are volcanic in origin and form part of the Ring of Fire (see Figure 1.9). Many in this latter group are situated in boundary zones where tectonic plates are either colliding or pulling apart.

Gondwana the great landmass that formed the southern part of the ancient supercontinent Pangaea

Great Barrier Reef the longest coral reef in the world, located off the northeastern coast of Australia

FIGURE 11.4 Uluru (Ayers Rock). This land formation, near the center of Australia, is a smooth remnant of ancient mountains. The site is held sacred by central Aboriginal Australians. It is also one of Australia's most popular tourist destinations.

For example, the Mariana Islands east of the Philippines are volcanoes that were formed when the Pacific Plate plunged beneath the Philippine Plate. The two islands of New Zealand were created when the eastern edge of the Indian-Australian Plate was thrust upward by its convergence with the Pacific Plate.

The Hawaiian Islands were produced through another form of volcanic activity associated with **hot spots,** places where particularly hot magma moving upward from Earth's core breaches the crust in tall plumes. Over the past 80 million years, the Pacific Plate has moved across these hot spots, creating a string of volcanic formations 3600 miles (5800 kilometers) long. The youngest volcanoes, only a few of which are active, are on or near the islands known as Hawaii.

Volcanic islands exist in three forms: volcanic high islands, low coral atolls, and **makatea** (sometimes called seamounts), which are coral platforms raised or uplifted by volcanism. High islands are usually volcanoes that rise above the sea into mountainous rocky formations which, because of their varying height and rugged landscapes, contain a rich variety of environments. New Zealand, the Hawaiian Islands, Mo'orea, and Easter Island are examples of high islands (see Figure 11.1D, F). An **atoll** is a low-lying island or chain of islets, formed of coral reefs that have built up on the rim of a submerged volcanic crater (see Figure 11.1E). These reefs are arranged around a central lagoon that was once the volcano's crater. Because of their low elevation, atoll islands tend to have only a small range of environments and very limited supplies of fresh water.

CLIMATE

Although the Pacific Ocean stretches almost from pole to pole, most of the land of Oceania is situated within the Pacific's tropical and subtropical latitudes. The tepid water temperatures of the central Pacific bring mild climates year-round to nearly all the inhabited parts of the region **(FIGURE 11.5)**. The southernmost reaches of Australia and New Zealand have the widest seasonal variations in temperature.

Moisture and Rainfall

With the exception of the vast arid interior of Australia, much of Oceania is warm and humid nearly all the time. New Zealand and the high islands of the Pacific receive copious rainfall; before human settlement, they supported dense forest vegetation. Now, after 1000 years of human impact, much of that forest is gone (see Figure 11.5B; see also Figure 11.9D).

Travelers approaching New Zealand, either by air or by sea, often notice a distinctive long white cloud that stretches above the north island. Seven hundred years ago, early **Maori** settlers (members of the Polynesian group) also noticed this phenomenon and they named that place *Aotearoa,* "land of the long white cloud," a name that is now applied to all of New Zealand. The cloud is the result of particularly high winds, complex landforms, and moist conditions.

The legendary **Roaring Forties** (named for the 40th parallel south) are powerful air and ocean currents that speed around the Southern Hemisphere (usually between the latitudes of 40 and 50 degrees) virtually unimpeded by landmasses. These *westerly winds* (winds that blow west to east), which are responsible for Aotearoa's distinctive moist cloud, deposit a drenching 130 inches (330 centimeters) of rain per year in the New Zealand highlands and more than 30 inches (76 centimeters) per year on the coastal lowlands (see Figure 11.5B). At the southern tip of New Zealand's North Island, the wind averages more than 40 miles per hour (64 kilometers per hour) nearly 120 days per year. Farmers in the area stake their cabbages to the ground so the plants will not blow away.

By contrast, two-thirds of Australia is overwhelmingly dry (see Figure 11.5A). The dominant winds affecting Australia are the north and

---

hot spots individual sites of upwelling material (magma) that originate deep in Earth's mantle and surface in a tall plume; hot spots tend to remain fixed relative to migrating tectonic plates

makatea coral platforms uplifted by volcanism, sometimes called seamounts

atoll a low-lying island or chain of islets, formed of coral reefs that have built up on the rim of a submerged volcano

Maori Polynesian people indigenous to New Zealand

Roaring Forties powerful air and ocean currents at about 40° S latitude that speed around the far Southern Hemisphere virtually unimpeded by landmasses

FIGURE 11.5 PHOTO ESSAY: Climates of Oceania

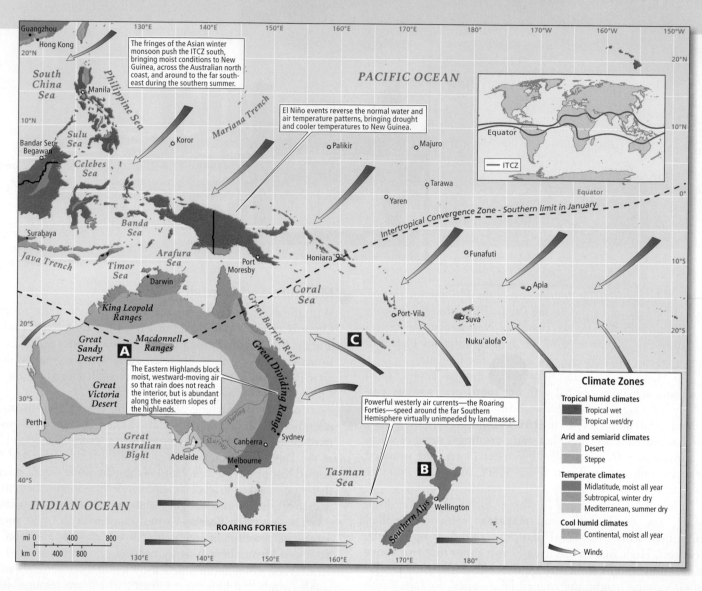

The fringes of the Asian winter monsoon push the ITCZ south, bringing moist conditions to New Guinea, across the Australian north coast, and around to the far southeast during the southern summer.

El Niño events reverse the normal water and air temperature patterns, bringing drought and cooler temperatures to New Guinea.

PACIFIC OCEAN

Equator

ITCZ

Equator

Intertropical Convergence Zone - Southern limit in January

The Eastern Highlands block moist, westward-moving air so that rain does not reach the interior, but is abundant along the eastern slopes of the highlands.

Powerful westerly air currents—the Roaring Forties—speed around the far Southern Hemisphere virtually unimpeded by landmasses.

INDIAN OCEAN

ROARING FORTIES

Climate Zones

Tropical humid climates
- Tropical wet
- Tropical wet/dry

Arid and semiarid climates
- Desert
- Steppe

Temperate climates
- Midlatitude, moist all year
- Subtropical, winter dry
- Mediterranean, summer dry

Cool humid climates
- Continental, moist all year

→ Winds

A Desert, Alice Springs, Australia

B Midlatitude, moist all year, New Zealand

C Tropical wet/dry, New Caledonia

south *easterlies* (winds that blow east to west) that converge east of the continent. The Great Dividing Range blocks the movement of moist, westward-moving air so that rain does not reach the interior (an orographic pattern; see Figure 1.12). As a result, a large portion of Australia receives less than 20 inches (50 centimeters) of rain per year, and humans have found rather limited uses for this interior territory. But the eastern (windward) slopes of the highlands receive more abundant moisture. This relatively moist eastern rim of Australia was favored as a habitat by both indigenous people and the Europeans who displaced them after 1800. During the southern summer, the fringes of the monsoon that passes over Southeast Asia and Eurasia bring moisture across Australia's northern coast. There, annual rainfall varies from 20 to 80 inches (50 to 200 centimeters).

Overall, Australia is so arid that it has only one major river system, which is in the temperate southeast where most Australians live. There, the Darling and Murray rivers drain one-seventh of the continent, flowing west and south into the Indian Ocean near Adelaide. One measure of Australia's overall dryness is that the entire average *annual* flow of the Murray-Darling river system is equal to just one day's average flow of the Amazon in Brazil.

In the island Pacific, mountainous high islands also exhibit *orographic rainfall* patterns, with a wet windward side and a dry leeward side (see Figure 11.5C). Rainfall amounts on the low-lying islands vary considerably across the region. Some of the islands lie directly in the path of trade winds, which usually deliver between 60 and 120 inches (152 to 305 centimeters) of rain per year. These islands support a remarkable variety of plants and animals. Other low-lying islands, particularly those near the equator, receive considerably less rainfall and are dominated by grasslands that support little animal life.

El Niño

Recall from Chapter 3 the *El Niño* phenomenon, a pattern of shifts in the circulation of air and water in the Pacific that occurs irregularly every 2 to 7 years. Although these cyclical shifts, or oscillations, are not yet well understood, scientists have worked out a model of how the oscillations may occur **(FIGURE 11.6)**.

The El Niño event of 1997–1998 illustrates the effects of this phenomenon. By December 1997, the island of New Guinea (north of Australia; see the Figure 11.1 map) had received very little rainfall for almost a year. Crops failed, springs and streams dried up, and fires broke out in tinder-dry forests. The cloudless sky allowed heat to radiate up and away from elevations above 7200 feet (2200 meters), so temperatures at high elevations dipped below freezing at night for stretches of a week or more. Tropical plants died, and people unaccustomed to chilly weather became ill. Meanwhile, at the other end of the system, along the Pacific coasts of North, Central, and South America, the warmer-than-usual weather brought unusually strong storms, high ocean surges, and damaging wind and rainfall (see Figure 11.6C).

In the 1980s, an opposite pattern in which normal weather conditions become intensified, was identified and named *La Niña*. It is now understood that La Niña patterns can bring unusually severe precipitation events (including tornadoes and blizzards) to places from the Indian Ocean to North America. La Niña is thought to have played a role in the major floods of 2010–2011 in Queensland, Australia, which were especially damaging because they followed a lengthy El Niño–connected drought.

FIGURE 11.6 A model of the El Niño phenomenon.

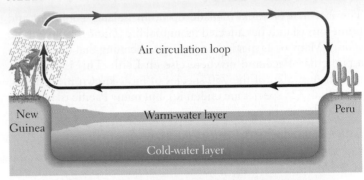

(A) Normal equatorial conditions. Water in the equatorial western Pacific (the New Guinea/Australia side) is warmer than water in the eastern Pacific (the Peru side). Due to prevailing wind patterns, the warm water piles up in the west. Warm air rises above the warm-water bulge in the western Pacific and forms rain clouds. The rising air cools, drops its moisture as rain, and, once in the higher atmosphere, moves in an easterly direction. In the east, the now dry, cool air descends, bringing little rainfall to Peru.

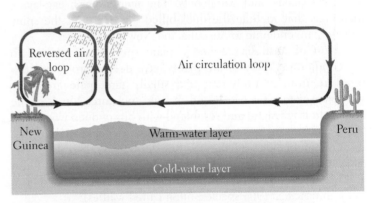

(B) Developing El Niño conditions. As an El Niño event develops, the ocean surface's warm-water bulge begins to move east. The air rising above it splits into two formations, one circling east to west in the upper atmosphere and one west to east.

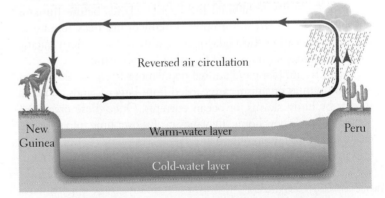

(C) Fully developed El Niño. Slowly, as the bulge of warm water at the surface of the ocean moves east, it forces the whole system into the fully developed El Niño, with air at the surface and in the upper atmosphere flowing in reverse of normal (A). Instead of warm, wet air rising over the mountains of New Guinea and condensing as rainfall, cool, dry, cloudless air descends to sit at Earth's surface in the west. Meanwhile, in the east, the normally dry, clear coast of Peru has clouds and rainfall. [*Sources consulted:* Environmental Dynamics Research, Inc., 1998; Ivan Cheung, George Washington University, Geography 137, Lecture 16, October 29, 2001.]

FAUNA AND FLORA

The fact that Oceania is made up of an isolated continent and numerous islands has affected its animal life (*fauna*) and plant life (*flora*). Many of its species are **endemic,** meaning that they exist in a particular place and nowhere else on Earth. This is especially true in Australia (of the 750 species of birds known in Australia, more than 325 species are endemic), but many Pacific islands also have endemic species.

Animal and Plant Life in Australia

The uniqueness of Australia's animal and plant life is the result of the continent's long physical isolation, large size, relatively homogeneous landforms, and arid climate. Since Australia broke away from Gondwana more than 65 million years ago, its animal and plant species have evolved in isolation. One spectacular result of this isolation is the presence of more than 144 living species of endemic marsupial animals. **Marsupials** are mammals whose babies at birth are still at a very immature stage; the marsupial then nurtures them in a pouch equipped with nipples. The best-known marsupials are kangaroos; other marsupials include wombats, koalas, and bandicoots. The **monotremes,** egg-laying mammals that include the duck-billed platypus and the spiny anteater, are endemic to Australia and New Guinea.

Most of Australia's endemic plant species are adapted to dry conditions. Many of the plants have deep taproots to draw moisture from groundwater, and small, hard, pale green, or shiny leaves to reflect heat and to hold moisture. Much of the continent is grassland and scrubland with bits of open woodland; there are only a few true forests, found in pockets along the Eastern Highlands and the southwestern tip and in Tasmania **(FIGURE 11.7)**. Two plant genera account for nearly all the forest and woodland plants: *Eucalyptus* (450 species, often called gum trees) and *Acacia* (900 species, often called wattles).

Plant and Animal Life in New Zealand and the Pacific Islands

Naturalists and evolutionary biologists have had a great interest in the species that inhabited the Pacific islands before humans arrived. Charles Darwin formulated many of his ideas about evolution after visiting the Galápagos Islands of the eastern Pacific (see the Figure 3.1 map) and the islands of Oceania.

Islands gain plant and animal populations from the sea and air around them as organisms are carried from larger islands and continents by birds, storms, or ocean currents. Once these organisms "colonize" their new home, they may evolve over time into new species that are unique to one island. High, wet islands generally contain more varied species because their more complex environments provide niches for a wider range of wayfarers and thus more opportunities for evolutionary change.

Once they arrive, human inhabitants modify the flora and fauna of islands. In prehistoric times, Asian explorers in oceangoing canoes brought plants such as bananas and breadfruit and animals such as pigs, chickens, and dogs to Oceania.

endemic belonging or restricted to a particular place

marsupials mammals whose babies at birth are still at a very immature stage; the marsupial then nurtures them in a pouch equipped with nipples

monotremes egg-laying mammals, such as the duck-billed platypus and the spiny anteater

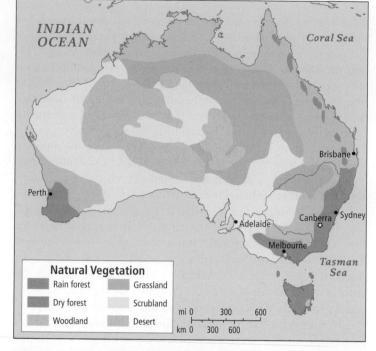

FIGURE 11.7 **Map of Australia's natural vegetation.** Much of Australia is grassland and scrubland. A few forests can be found in the Eastern Highlands, the far southwest, and Tasmania. [*Source consulted:* Tom L. McKnight, *Oceania* (Englewood Cliffs, NJ: Prentice Hall, 1995), p. 28.]

European settlers later brought grains, vegetables, fruits, invasive grasses, cattle, sheep, goats, rabbits, housecats, and rats. Today, human activities from tourism to military exercises to urbanization continue to change the flora and fauna of Oceania.

Generally, the diversity of land animals and plants is richest in the western Pacific, near the larger landmasses. It thins out to the east, where the islands are smaller and farther apart. The natural rain forest flora is rich and abundant in New Zealand and New Guinea, and also on the high islands of the Pacific. However, the natural fauna is much more limited on these islands. While New Guinea has fauna comparable to Australia, to which it was once connected via Sundaland (see Figure 10.5), New Zealand and the Pacific islands have no indigenous land mammals, almost no indigenous reptiles, and only a few indigenous species of frogs, as they were never connected to Australia and New Guinea by a land bridge that land animals could cross. Two indigenous birds in New Zealand, the kiwi and the huge moa (a bird that grew up to 12 feet [3.7 meters] tall), were a major source of food for the Maori people. The moa was hunted to extinction before the Europeans arrived. Today, New Zealand may well be the country with the most nonnative species of mammals, fish, and fowl, nearly all brought there by European settlers (see further discussion of this on page 629).

THINGS TO REMEMBER

- This largest region of the world is primarily water and has a very small population of just 40 million people.

- The largest land area in Oceania is the continent of Australia.

- The thousands of islands of the Pacific were created either as a result of plate tectonics or volcanic activity.

- The climate of most of the region is warm and humid almost all the time.

- Due to their isolation from the life forms found on other major landmasses, the fauna and flora of the region are unique in many distinctive ways that have informed our knowledge about evolution. Many species, such as marsupials and monotremes, are endemic to the region.

ENVIRONMENT

GEOGRAPHIC THEME 1

Environment: Oceania faces a host of environmental problems and public awareness of environmental issues is keen. Global climate change, primarily warming, has brought about rising sea levels and increasingly variable rainfall. Other major threats to the region's unique ecology have come from the introduction of nonnative species and the expansion of herding, agriculture, fishing, fossil fuel extraction, waste disposal, and human settlements.

GLOBAL CLIMATE CHANGE

Oceania, with the exception of Australia, is a minor contributor of greenhouse gases. Australia has some of the world's highest greenhouse gas emissions on a per capita basis (see Figure 1.14). Much of Australia's emissions, like those of the United States, result from the use of automobile-based transportation systems to connect a widely dispersed network of cities and towns. Additionally, Australia's heavy dependence on coal to generate electricity has led to high emissions, much as coal dependence has in the United States. Because Australia has a relatively small population (23.9 million people in 2015), it accounts for only slightly more than 1 percent of global emissions. However, despite the negligible contributions of greenhouse gases made by the islands of the Pacific, they, as well as Australia, are quite vulnerable to the effects of global climate change **(FIGURE 11.8)**.

Sea Level Rise and Storm Surges

The best scientific research indicates that global warming is raising sea levels mainly through thermal expansion as rising temperatures cause the water in the ocean to expand in size, but also by melting glaciers and polar ice caps. Obviously, this issue is of great concern to residents of islands that already barely rise above the waves (see Figure 11.8A). If sea levels rise 4 inches (10 centimeters) per decade, as predicted by the International Panel on Climate Change, many of the lowest-lying Pacific atolls, such as Tuvalu, will disappear under water within 50 years. Other islands, some with already very crowded coastal zones, will see these zones shrink and become more vulnerable to storm surges and cyclones. In late February 2016, the worst cyclone to hit the southern Pacific struck the Fiji island chain, killing 42 and leaving at least 60,000 without shelter, water, and electricity. Wind speeds exceeded 200 miles (325 kilometers) per hour and waves were 40 feet (13 meters) high.

Wildfires and Other Water-Related Vulnerabilities

Much of Oceania is particularly vulnerable to the droughts and floods that could result from global climate change. Parts of Australia and some low, dry Pacific islands are already undergoing prolonged droughts and freshwater shortages requiring changes in daily life and livelihoods. The fear is that the severe droughts are not the usual periodic dry spells but may represent permanent alterations in rainfall patterns that could also worsen wildfires. Such fires emerged as a major issue in Australia in February 2009, when 173 people died in a rural firestorm near Melbourne (see Figure 11.8B). In 2015 and 2016, repeated wildfires struck Western Australia, South Australia, Queensland, New South Wales, and Victoria—the last three lie in the most humid, and hence most fire-resistant, part of the country.

As fresh water becomes scarcer across the region, *virtual water* (see Chapter 1) becomes an issue. All export-related activities that permanently consume or degrade fresh water in their extraction or production processes (such as mineral mining; oil and gas extraction; meat, wool, and wheat production; and tourism-related construction and maintenance) are essentially extracting virtual water from places that are already under water stress. If the true costs of this freshwater depletion were counted and added to the price of the products, these exports might no longer be competitive on the world market—at least not until all global producers understood virtual water accountability to be in their best interests and raised their prices accordingly.

Another concern is that the warming of the oceans could not only result in stronger tropical storms, but also threaten coral reefs and the fisheries that depend on them by causing *coral bleaching* (see Figure 11.8C). When ocean waters warm just a few degrees, the living coral organisms expel the algae that live inside and give the coral its color. Coral bleaching is a phenomenon that has affected all reefs in this region in recent years, with scientists announcing, in 2016, Australia's "biggest ever environmental disaster": The Great Barrier Reef had lost 50 percent of its coral cover and was vulnerable to losing another 25 percent. Because so many fish depend on reefs, coral bleaching also threatens many fishing communities and ultimately the global food supply. This is especially true on some of the Pacific islands that have few other local food resources. Photos of Great Barrier coral bleaching can be seen at http://www.npr.org/sections/thetwo-way/2016/05/14/477963623/new-photos-show-the-rapid-pace-of-great-barrier-reef-bleaching.

Responses to Potential Climate Change Crises

Oceania is pursuing a number of alternative energy strategies to reduce greenhouse gas emissions. Although Australia remains dependent on fossil fuel sales (primarily the sale of coal, and also crude oil and natural gas) to Asia, it is pursuing renewable energy strategies for domestic use. These include increasing power production from geothermal, solar, and biomass sources and, because of water shortages, decreasing the emphasis on hydropower. New Zealand has set a goal of obtaining 95 percent of its energy from renewable sources by 2025. Much of this energy will come from wind power, which has a great deal of potential in this region, especially in areas near the Roaring Forties. On low Pacific islands, solar energy is now the most widely used alternative to costly and polluting imported fuel.

FIGURE 11.8 PHOTO ESSAY: Vulnerability to Climate Change in Oceania

Oceania is vulnerable to a wide variety of hazards related to climate change, including sea level rise, stronger tropical storm intensity, and the availability of water becoming less certain. Fortunately, several countries in this region are already implementing practices that are increasing resilience to climate hazards.

A A low-lying atoll in Tuvalu. With its highest point only 14.7 feet (4.5 meters) above sea level, Tuvalu is quite vulnerable to sea level rise. Combined with increased flooding during tropical storms, higher sea levels could make many low islands in this region uninhabitable. Tuvalu's government is already negotiating the future resettlement of parts of its population to nearby nations such as New Zealand.

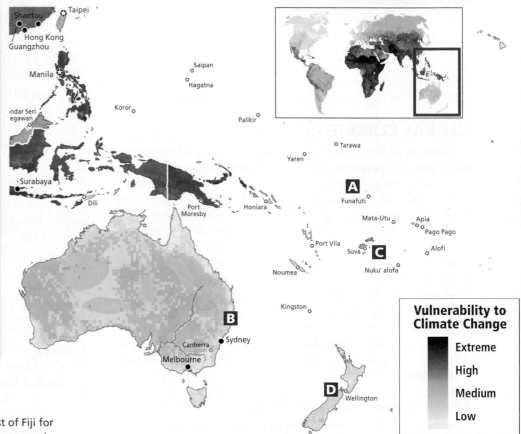

Vulnerability to Climate Change

- Extreme
- High
- Medium
- Low

B A wildfire outside of Brisbane, Australia. Higher temperatures are bringing stronger and more extensive wildfires to this region.

C A marine biologist monitors the reefs off the coast of Fiji for signs of coral bleaching. Caused by warmer temperatures, coral bleaching threatens many fish species that live on coral reefs. In turn, human communities on Pacific islands are threatened by the loss of fisheries.

D A vineyard fitted with a drip irrigation system, visible at the base of the vines, in Marlborough, New Zealand. These systems provide resilience in the face of drought and use substantially less water than other irrigation methods.

Thinking Geographically

After you have read about the vulnerability to climate change in Oceania, you will be able to answer the following questions.

A If sea levels rise 4 inches (10 centimeters) per decade, as predicted by the International Panel on Climate Change, name some places other than low-lying islands that will be affected.

B Check the location of Brisbane and then suggest a reason for why wildfires there are especially alarming.

C What kinds of actions could humans take to increase the resilience of coral reefs to climate change?

D Why would irrigation be needed in New Zealand, which has a wet climate?

ON THE BRIGHT SIDE: Water Conservation

Oceania is a world leader in implementing alternative water technologies. Some of these are simple but effective age-old methods, such as harvesting rainwater from roofs and the ground for household use. Most buildings in rural Australia, New Zealand, and many Pacific islands get at least part of their water this way, relieving surface and groundwater resources. Australia and New Zealand are now stretching water resources further by using very efficient drip irrigation technologies extensively in agriculture (see Figure 11.8D) and new low-cost water-filtration techniques. ■

INVASIVE SPECIES AND FOOD PRODUCTION

European systems for producing food and fiber introduced in Oceania have had a profound effect on its environments. Many of the unique endemic plants and animals of Oceania were displaced by **invasive species,** organisms that spread aggressively into regions outside their native range, adversely affecting economies or environments. Many exotic, or alien, plants and animals were brought to Oceania by Europeans to support their food production systems. Ironically, many of these same species are now major threats to food production **(FIGURE 11.9)**.

Australia

When Europeans first settled the continent, they brought many new animals and plants with them, sometimes intentionally, sometimes unintentionally. European rabbits are among the most destructive of these introduced species. Early British settlers who enjoyed eating rabbits brought them to Australia. Many were released for hunting, but with no natural predators, the rabbits multiplied quickly, consuming so much of the native vegetation that many indigenous animal species starved. Moreover, rabbits became a major source of agricultural crop loss and reduced the capacity of grasslands to support herds of introduced sheep and cattle. Attempts to control the rabbit population by introducing European foxes and cats backfired as these animals became major invasive species themselves **(FIGURE 11.10C)**. Foxes and cats have driven several native Australian predator species to extinction without having much effect on the rabbit population. Intentionally introduced diseases have proven more effective at controlling the rabbit population, though rabbits have repeatedly developed resistance to them.

Herding has also had a huge impact on Australian ecosystems. Because the climate is arid and soils in many areas are relatively infertile, the dominant land use in Australia is the grazing of introduced domesticated animals—primarily sheep, but also cattle. More than 15 percent of the land has been allocated for grazing and Australia leads the world in exports of sheep and cattle products.

Dingoes, the indigenous wild dogs of Australia (see Figure 11.10B), prey on sheep and young cattle. To separate the wild dogs from the herds, the Dingo Fence—the world's longest fence—was built. It extends 3488 miles (5614 kilometers) and is unfortunately a major ecological barrier to other wild species. Meanwhile, kangaroos (the natural prey of dingoes) have learned to live on the sheep side of the fence, where their population has boomed beyond sustainable levels.

New Zealand

New Zealand's environment has been transformed by introduced species and food production systems even more extensively than Australia's environment. No humans lived in New Zealand until about 700 years ago, when the Polynesian Maori people settled there. When they arrived, dense midlatitude rain forest covered 85 percent of the land. The Maori were cultivators who brought in yams and taro as well as other nonnative plants, rats, and birds. By the time of significant European settlement (after 1825), forest clearing and overhunting by the Maori had already degraded many environments and driven several bird species to extinction.

European settlement in New Zealand dramatically intensified environmental degradation. Attempts to re-create European farming and herding systems in New Zealand resulted in environments that today are actually hostile to many native species, a growing number of which are becoming extinct. Only 23 percent of the country remains forested, with ranches, farms, roads, and urban areas claiming more than 90 percent of the lowland area **(FIGURE 11.11)**.

The ordinary house cat (*Felis catus*), brought from Europe to control rabbits, mice, and rats, is an example of an interloper whose impact has been astonishing. In New Zealand, it is estimated that feral cats kill up to 100 million birds each year. Many of the victims are endemic birds such as *tuis* and *kukupa*, with little inborn wariness for predators. For some, the answer has been to eliminate all feral cats and sterilize all housecats—a project that gained steam in New Zealand in 2013. But this solution has not been popular with New Zealand's pet lovers or advocates; eliminating or reducing the number of cats may also give rise to a burgeoning rodent population. Despite the attempts to reduce the cat population, cats remain popular: in 2016, 45 percent of New Zealanders still owned one.

Most of New Zealand's cleared land is used for export-oriented farming and ranching. Grazing has become so widespread that today in New Zealand, there are 15 times as many sheep as people, and 3 times as many cattle. Both farming and ranching have severely degraded the environment. Soils exposed by the clearing of forests proved infertile, forcing farmers and ranchers to augment them with agricultural

invasive species organisms that spread into regions outside their native range, adversely affecting economies or environments

FIGURE 11.9 PHOTO ESSAY: Human Impacts on the Biosphere in Oceania

Scott Camazine/Getty Images

Despite its vast size and relatively small population, Oceania has been severely affected by human activity. Much of the damage has been done by people from distant countries, as well as countries in Oceania that export resources outside the region.

A In 1946, on Bikini Atoll in the Marshall Islands (then a U.S. territory), the United States conducted one of the first underwater tests of a nuclear weapon and its effects on naval vessels. Since then, the United States and France have conducted over 300 nuclear tests in Oceania, some without sufficient attention to nuclear contamination.

Daniele Delimont/Gallo Images/Getty Images

B Once hunted to near extinction to protect sheep herds introduced to the area, the Tasmanian devil is now threatened by low genetic diversity, which leaves it vulnerable to disease.

C An endangered great white shark tangled in a fish net off the coast of New Zealand. Fleets from around the world come to Oceania to take advantage of its fisheries, many of which are now overexploited. Many Pacific islands still sell fishing rights to foreign fleets because they need the income.

Map

Northern Mariana Islands (U.S.)

Hawaii (U.S.)

Guam (U.S.)

A

MARSHALL ISLANDS

PALAU

FEDERATED STATES OF MICRONESIA

NAURU

KIRIBATI

American Samoa (U.S.)

PAPUA NEW GUINEA

SOLOMON ISLANDS

TUVALU

Tokelau (New Zealand)

Wallis & Futuna (France)

SAMOA

Coral Sea Islands (Australia)

VANUATU

New Caledonia (France)

FIJI

TONGA

Niue (N.Z.)

AUSTRALIA

Norfolk Island (Australia)

NEW ZEALAND

D

B

C

Human Impact, 2002

Land cover
- Forests
- Grasslands
- Deserts
- Tundra
- Ice

Overfishing
- /// Threatened fisheries

Human impact on land
- High impact
- Medium–high impact
- Low–medium impact

Acid rain
- --- 5.5–4.9 pH

mi 0   400   800
km 0   400   800

D A herd of sheep in New Zealand, where ranches, farms, roads, and urban areas cover 90 percent of the lowlands. Forests once covered 85 percent of New Zealand, but after two centuries of export-oriented agriculture and forestry, only 23 percent of the country remains forested. In recent decades, there have been increased efforts to conserve the remaining forests.

Kim Westerskov/Getty Images

Raimund Linke/Getty Images

Thinking Geographically

After you read about human impacts on the biosphere in Oceania, you will be able to answer the following questions.

A Why did the nations that conducted nuclear tests in Oceania choose that part of the world?

B Why were so many plants and animals brought to Oceania from Europe and what are some negative results?

C Why might islands object to foreign fishing fleets in Oceania?

D What is the chief present impetus for deforestation in New Zealand?

chemicals. The chemicals, along with feces from sheep and cattle, have seriously polluted many waterways, causing the extinction of some aquatic species.

ON THE BRIGHT SIDE: Environmental Awareness in New Zealand

New Zealand is a global leader in environmental awareness. It formed the world's first environmentally focused national political party in 1973—the Green Party of Aotearoa—now the third largest, and spearheaded the world's first nuclear-free zone in 1984. New Zealand's government promotes a "clean and green" image internationally, but most New Zealanders acknowledge the severity of existing environmental problems and the need for further action. ■

GLOBALIZATION AND THE ENVIRONMENT IN THE PACIFIC ISLANDS

As the Pacific islands have become more connected to the global economy over the years, many unique species of plants and animals were driven to extinction as the islands were deforested and mined or converted to commercial agriculture. This has been the case in Hawaii, which is home to more threatened or endangered species than any other U.S. state, despite having less than 1 percent of the U.S. landmass. The extensive conversion of tropical Hawaiian forests to export crops, such as sugar cane and pineapples, has caused the extinction of numerous plant, bird, and land species.

Over the last century, across the span of the Pacific, flows of resources and pollutants have increased dramatically. Mining, nuclear pollution, commercial agriculture and fishing, and tourism are all examples of how globalization has transformed environments in the Pacific islands.

Mining in Papua New Guinea and Nauru

Mining has rendered the islanders of Oceania losers in three ways: foreign-owned mining companies that took advantage of poorly enforced or nonexistent environmental laws are responsible for major environmental damage. In the Ok Tedi Mine on Papua New Guinea, 80 million tons of mine waste devastated river systems **(FIGURE 11.12)**. The environmental degradation forced tens of thousands of indigenous subsistence cultivators into new mining market towns, where their horticulture skills were of little use

Figure 11.10 LOCAL LIVES: People and Pets in Oceania

A A rainbow lorikeet, a small parrot that is native in much of Oceania. Rainbow lorikeets are popular pets because of their plumage and curious disposition. However, in parts of Australia and New Zealand, they are considered pests because they consume fruit in orchards, are noisy, and leave behind large droppings.

B An Australian dingo, a type of wild dog that lives mainly in the Australian outback. Dingoes were likely brought to Australia by Aboriginal Australians, who used them as guard dogs and possibly as a food source. Shepherds consider dingoes to be pests, and in the 1880s, they built a 3488-mile (5614-kilometer) fence to keep dingoes out of southeastern Australia.

C A cat in New Zealand. Introduced throughout Oceania by Europeans who brought them as pets and to control rodent populations, cats quickly became feral. This had disastrous impacts on native species of marsupials and birds. Cats have been completely removed from several of New Zealand's smaller islands, including those designated as native bird sanctuaries.

Vanessa Mylett/Getty Images

Robin Smith/Getty Images

Copyright Kristen Gamble/NHPA/Getty Images

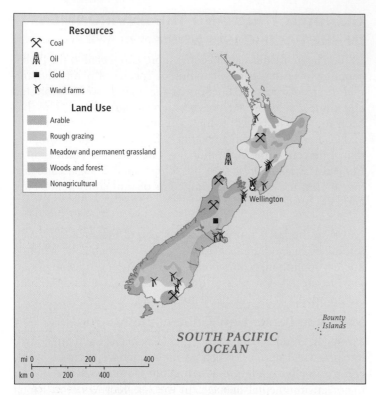

FIGURE 11.11 Land uses and natural resources of New Zealand. As a result of European settlement and the clearing of land for farming, only 23 percent of New Zealand remains forested. [*Source consulted:* Richard Nile and Christian Clerk, *Cultural Atlas of Australia, New Zealand, and the South Pacific* (New York: Facts on File, 1996), p. 194.]

and where they needed cash to buy food and pay rent. Also, most of the profits of mining go to foreign-owned mining companies, while the best-paying mining jobs go to outsiders.

In 2007, thirty thousand Papua New Guinea people sued the Australian parent mining company, which was then BHP Billiton, for U.S.$4 billion. Two villagers, Rex Dagi and Alex Maun, traveled to Europe and the United States to explain their cause and meet with international environmental groups. They and their supporters convinced U.S. and German partners in the Ok Tedi Mine to divest their shares.

However, the story of mining disasters in Oceania is extensive. The most extreme case of environmental damage caused by mining took place on the once densely forested Melanesian island of Nauru, which is one-third the size of Manhattan and located northeast of the Solomon Islands (see the Figure 11.1 map). Nauru's wealth was once legendary, based on proceeds from the strip-mining of high-grade phosphates derived from eons of bird droppings (guano) that are used to manufacture fertilizer. The phosphate mining companies were owned first by Germany, then Japan, and finally Australia. For a time in the early 1970s, Nauru had the highest per capita income in the world (although not distributed equitably). Today, the phosphate reserves are nearly depleted, the proceeds have been ill spent, and the environment destroyed. Junked mining equipment sits on miles of bleached white sand where forest once stood. Instead, Nauru now serves as a detention camp for more than 500 asylum seekers from Iraq, Iran, Afghanistan, Somalia, Cambodia, and Myanmar (Rohinga). Nauru has become part of Australia's

FIGURE 11.12 The Ok Tedi Mine.

(A) The Ok Tedi open-pit copper and gold mine. A large hole now exists where Mount Fubilan once stood.

(B) The Ok Tedi River as it flows downstream of the mine. Each year since its opening in 1984, the mine has discharged millions of tons of contaminated mine tailings and eroded sediments, resulting in a once-deep river becoming clogged with poisonous runoff that kills many trees along its banks. Sediment from the mine river has killed innumerable fish in the river, contaminated 500 square miles (1300 square kilometers) of farmland, and adversely affected 50,000 people in 120 villages. Litigation against the mine owners is ongoing.

Thinking Geographically

The products of the Ok Tedi Mine (gold and copper) are not used by the indigenous subsistence cultivators of Papua New Guinea, yet they were forced off their land and into new mining market towns as a result of sediment and chemical pollution. What are some possible ways to alert consumers of the gold and copper to the negative impacts of mining on indigenous peoples?

controversial offshoring policy for those seeking asylum, called the *Pacific Solution*, discussed on page 644.

Nuclear Pollution

The geopolitical aspects of globalization have hit Oceania especially hard. Nuclear weapons testing by France and the United States from the 1940s to the 1960s (during the Cold War), as well as the dumping of nuclear waste by various nuclear powers, have long been major environmental issues for the Pacific islands (see Figure 11.9A). In New Zealand in July 1985, the French secret service blew up the ship *Rainbow Warrior* owned by the antinuclear environmental group Greenpeace. In response, the 1986 Treaty of Rarotonga established the South Pacific Nuclear Free Zone, which was an expansion of a similar zone set up in New Zealand in 1984. Most independent countries in Oceania signed this treaty, which bans nuclear weapons testing and nuclear waste dumping on their lands. Because of political pressure from France and the United States, however, French Polynesia and U.S. territories such as the Marshall Islands have not signed the treaty.

Tourism

Even tourism, which until recently was considered a "clean" industry, can create environmental problems. Foreign-owned tourism enterprises not only take the profits from tourism home and leave the stress and many other social costs of tourism to be absorbed by local communities, they often accelerate the loss of wetlands and worsen beach erosion by clearing coastal vegetation for hotel construction, golf courses, and waterfront-related entertainment.

Tourism has also strained island water resources because of showering, laundering, and other services that consume fresh water. Furthermore, inadequate methods of disposing of sewage and trash from resorts have polluted many once-pristine areas. Ecotourism (see Chapter 3) aimed at reducing these impacts is now a common element of development throughout the Pacific, but environmental impacts from ecotourism are still generally high.

The Great Pacific Garbage Island

When oceanographer Charles Moore found himself surrounded by a massive floating island of degrading plastic garbage in the north Pacific in 1997, he thought it was an anomaly. But his investigations revealed that the disposable, throwaway aspect of modern living was responsible. The "island" included plastic beverage bottles and caps along with Lego blocks, trash bags, toothbrushes, footballs, and kayaks—indeed, virtually every consumer product made. By 2008, several such garbage masses were floating in the Pacific and Atlantic oceans. Modern plastics exposed to sunlight degrade into tiny bits that are then ingested by sea birds and animals or into microscopic fragments ingested by filter feeding organisms. The plastics also release carcinogenic chemicals as they break down. Millions of seabirds, sea mammals, and fish die each year after exposure to the residue of this trash. Humans are affected because, as marine scientists say, whatever goes into the ocean is likely to go into the food chain and eventually end up on someone's dinner plate.

The science of marine debris is young and the research needed must be done by a wide range of experts. In 2014, scientists in Europe and North America discovered a significant accounting error: from the vast amount of plastic manufactured, the five huge circulating masses of trash in the world ocean should have contained 10 to 100 times more plastic than they did. Where is the missing plastic, and what is the impact of all this plastic on marine animals and those who eat them? Efforts to clean up the floating islands of garbage are currently under study but have not yet been implemented.

The UN Convention on the Law of the Sea

Based on the idea that all the problems of the world's oceans are interrelated and need to be addressed as a whole, the United Nations Convention on the Law of the Sea (UNCLOS) established rules governing all uses of the world's oceans and seas; it has been ratified by 157 countries (although not the United States). The treaty allows islands to claim rights to ocean resources 200 miles (320 kilometers) out from their shores. Island countries can now make money by licensing privately owned fleets from Japan, South Korea, Russia, the United States, and elsewhere to fish within these offshore limits. As of yet, however, there is no overarching enforcement agency, and protecting the fisheries from overfishing by these rich and powerful licensees has turned out to be an enforcement nightmare for tiny island governments with few resources. Similarly, it has proven difficult to monitor and control the exploitation of seafloor mineral deposits by foreign mining companies.

THINGS TO REMEMBER

GEOGRAPHIC THEME 1 • **Environment:** Oceania faces a host of environmental problems and public awareness of environmental issues is keen. Global climate change, primarily warming, has brought about rising sea levels and increasingly variable rainfall. Other major threats to the region's unique ecology have come from the introduction of nonnative species and the expansion of herding, agriculture, fishing, fossil fuel extraction, waste disposal, and human settlements.

• In an effort to combat global warming, renewable energy alternatives to fossil fuels are being pursued across the region. New Zealand has set a goal of obtaining 95 percent of its energy from renewable sources by 2025.

• Throughout Oceania, the introduction of food production systems from elsewhere has resulted in the spread of ecologically and economically damaging invasive plant and animal species.

• Globalization and the patterns of consumption (from mining to vacationing) by people who live far from the Pacific are seriously affecting human and animal life in this region.

HUMAN PATTERNS OVER TIME

Oceania's past has been shaped by its ancient settlement from the Asian mainland and by the more recent arrival in Australia and New Zealand of Europeans. Oceania's present is increasingly being influenced by economic and geographic considerations, particularly its physical proximity to Asia.

THE PEOPLING OF OCEANIA

The longest-surviving inhabitants of Oceania are **Aboriginal Australians,** whose ancestors (the Australoids) migrated from Southeast Asia

Aboriginal Australians the longest-surviving inhabitants of Oceania, whose ancestors, the Australoids, migrated from Southeast Asia 50,000 to 70,000 years ago over the Sundaland landmass that was exposed during the ice ages

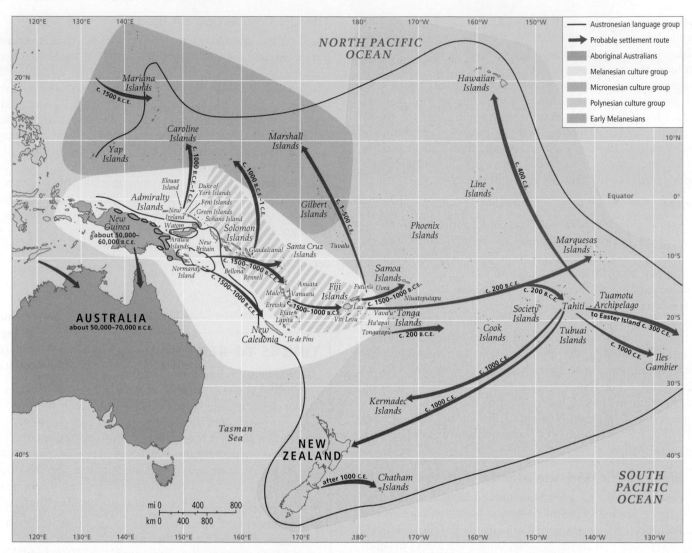

FIGURE 11.13 **Primary indigenous culture groups of Oceania**. By 50,000 to 70,000 years ago, humans had come to New Guinea and Australia. About 25,000 years ago, people began moving across the ocean to nearby Pacific islands. Movement into the more distant islands commenced with the arrival of Austronesians, who went on to inhabit the farthest reaches of Oceania. The largest culture group is Polynesia, which stretches from Hawaii to Easter Island (Rapa Nui, which is so far east that it is not on this map), to New Zealand. [*Source consulted:* Richard Nile and Christian Clerk, *Cultural Atlas of Australia, New Zealand, and the South Pacific* (New York: Facts on File, 1996), pp. 58–59.]

50,000 to 70,000 years ago (**FIGURE 11.13**; see also Figure 11.14A), at a time when sea level was somewhat lower. It is possible that some memory of this ancient journey may be preserved in Aboriginal oral traditions, which recall mountains and other geographic features that are now submerged under water. At about the same time that the Aboriginal Australians were set-

Melanesians a group of Australoids named for their relatively dark skin tones, a result of high levels of the protective pigment melanin; they settled throughout New Guinea and other nearby islands

Melanesia New Guinea and the islands south of the equator and west of Tonga (the Solomon Islands, New Caledonia, Fiji, and Vanuatu)

tling Australia and Tasmania, related groups were settling nearby areas. The distribution of these groups, the sequence of settlement, and the navigation skills necessary to explore and occupy this huge oceanic environment are complex. Figure 11.13 and the vignette about Mau Piailug, the modern-day Micronesian navigator, are attempts to clarify this story.

Melanesians, so named for their relatively dark skin tones, a result of high levels of the protective pigment *melanin*, migrated throughout New Guinea and other nearby islands, giving this area its name, **Melanesia.** Archaeological evidence indicates that they first arrived more than 50,000 to 60,000 years ago from Sundaland (see Figure 10.5), a now-submerged shelf exposed during the Pleistocene epoch. They lived in isolated pockets, which resulted in the evolution of hundreds of distinct yet related languages. Like the Aboriginal Australians, the early Melanesians survived mostly by hunting, gathering, and fishing, although some groups—especially those inhabiting the New Guinea highlands—eventually practiced agriculture.

VIGNETTE *With courage, you can travel anywhere in the world and never be lost. Because I have faith in the words of my ancestors, I'm a navigator.*

—Mau Piailug (1932–2010)

In 1976, Mau Piailug made history by sailing a reconstruction of a traditional double-hulled Pacific island voyaging canoe, the *Hōkūle'a,* across the 2400 miles (3860 kilometers) of deep ocean between Hawaii and Tahiti. He did so without a compass, charts, or other modern instruments, using only methods passed down through his family. To find his way, he relied mainly on observations of the stars, the Sun, and the Moon. When clouds covered the sky, he used the patterns of ocean waves and swells as well as the presence of seabirds to tell him of distant islands over the horizon.

Piailug reached Tahiti 33 days after leaving Hawaii and made the return trip in 22 days. His voyage resolved a major scholarly debate over how people settled the many remote islands of the Pacific without navigational instruments thousands of years before the arrival of Europeans. Some thought that navigation without instruments was impossible and argued that would-be settlers simply drifted about on their canoes at the mercy of the winds, most of them starving to death on the seas, with a few happening upon new islands by chance. It was hard to refute this argument because local navigational methods had died out almost everywhere. However, in isolated Micronesia, where Piailug lived, indigenous navigational traditions still survive.

After the successful 1976 voyage, Piailug trained several students in traditional navigational techniques. His efforts have become a symbol of cultural rebirth and a source of pride throughout the Pacific. In 2007, the protégés of Mau Piailug sailed from Hawaii through the Marshall Islands to Yokohama, Japan, to celebrate peace and the human need to stay connected with nature. In 2016, the *Hōkūle'a* arrived in the Potomac River in Washington, D.C., as part of a global traverse using Polynesian *wayfinding* navigation techniques. [Source: Facts on File, 2007; NPR, 2016.] ∎

Long after Melanesia was settled, **Micronesia** and **Polynesia** were settled between 5000 to 6000 years ago and other islands as recently as 1000 years ago, by linguistically related *Austronesians.* The Austronesians were a group of skilled farmers and seafarers originally from southern China who migrated through Southeast Asia and into the Pacific, sometimes mixing with the Melanesian peoples they encountered (see Chapter 10). Micronesia consists of the small islands that lie east of the Philippines and north of the equator (see the Figure 11.13 map). Polynesia is made up of numerous islands situated inside the large, irregular triangle formed by New Zealand, Hawaii, and Easter Island. (Easter Island, also called Rapa Nui, is a tiny speck of land in the far eastern Pacific, at 109° W 27° S, not shown in the figures in this chapter.) The voyages of Mau Piailug (see the vignette above) and recent experiments run by Polynesians have provided evidence that ancient sailors could navigate over vast distances using seasonal winds, astronomic calculations, bird and aquatic life, and wave patterns to reach the most far-flung islands of the Pacific. The Polynesians were fishers, hunter-gatherers, and cultivators who developed complex cultures and maintained trading relationships among their widely spaced islands.

In the millennia that have passed since first settlement, humans have continued to circulate throughout Oceania. Some apparently set out because their own space was overcrowded and full of conflict or because food reserves were declining. It is also likely that Pacific peoples were enticed to new locales by the same lures that later attracted some of the more romantic explorers from Europe and elsewhere: sparkling beaches, magnificent blue skies, aromatic breezes, and lovely landscapes.

THE SAGA OF RAPA NUI (EASTER ISLAND)

One of the most fascinating mysteries about the exploration and settlement of the Pacific by indigenous people is that regarding Easter Island (Rapa Nui), a 63-square-mile (163-square-kilometer) volcanic formation 2283 miles (3512 kilometers) off the west coast of Chile. Archeological evidence suggests that the island was settled by 100 or fewer Polynesian people about 1200 C.E. They arrived in double-hulled canoes. Over 900 monolithic human figures carved from local rock attest to a vibrant society that could support massive artworks, yet by the time Europeans arrived in 1722, decline was well under way, sparking a debate. Was environmental collapse the explanation for the decline, or political revolution, or diseases of some sort? Collaborative research by universities in California, Virginia, Spain, New Zealand, and Denmark have contributed to the developing understanding that Rapa Nui people experienced decline because of a range of environmental problems that prohibited continual production of sufficient food; this decline preceded the arrival of Europeans.

POSSIBILITIES OF EARLY CONTACT BETWEEN PACIFIC PEOPLES AND THE AMERICAS

Somehow, around 1300 C.E. (19 to 23 generations ago), the Rapa Nui were able to overcome their difficulties to a degree sufficient enough to continue exploration further to the east. Genomic data show that the Rapa Nui traveled to coastal South America long before Europeans appeared in the Americas or the Pacific, and there, they mixed with Native Americans. This evidence coincides with other studies that recently have found genomic proof of ancient Polynesian ancestry among indigenous Brazilians. And now, the long puzzling fact that sweet potatoes, genetically derived from sweet potatoes domesticated in South America, were being cultivated in the Pacific long before European contact, may have been explained. They were the result of very early contact between Polynesian people and Native Americans.

ARRIVAL OF THE EUROPEANS IN THE PACIFIC

The earliest recorded contact between Pacific peoples and Europeans took place in 1521, when the Portuguese navigator Ferdinand Magellan (exploring for Spain) landed on the island of Guam in Micronesia. The encounter ended badly. The islanders, intrigued by European vessels, tried to take a small skiff. For this crime, Magellan had his men kill the offenders and burn their village to the ground. A few months later, Magellan was himself killed by islanders in what later became the Philippines, which he had claimed for Spain. Nevertheless, by the 1560s, the Spanish had set up a lucrative Pacific trade route between Manila in the Philippines

Micronesia the small islands that lie east of the Philippines and north of the equator

Polynesia the numerous islands situated inside an irregular triangle formed by New Zealand, Hawaii, and Easter Island

A Aboriginal Australians with hunting tools.

B *Hokule'a 2*, a functioning replica of a traditional Hawaiian voyaging canoe.

C An etching of early British contact with Pacific Islanders in 1783.

| 75,000 B.C.E. | 50,000 B.C.E. | 25,000 B.C.E. | 1500 C.E. | 1550 C.E. | 1600 C.E. | 1650 C.E. |

70,000–50,000 B.C.E.
Aboriginal Australians' ancestors migrate from Southeast Asia

4000–3000 B.C.E.
Settlement of Micronesia and Polynesia

1500s C.E.
Spanish explorations

FIGURE 11.14 VISUAL HISTORY OF OCEANIA

Thinking Geographically

After you have read about the human history of Oceania, you will be able to answer the following questions.

A From where did the ancestors of Aboriginal Australians migrate to Oceania?

B Why did Europeans find it hard to believe that Polynesians could navigate boats like the one in this painting over the wide span of the Pacific and successfully find their way back and forth?

and Acapulco on the west coast of Mexico. Explorers from other European states followed, first taking an interest mainly in the region's valuable spices. The British and French explored Oceania extensively in the eighteenth century **(FIGURE 11.14C)**.

The Pacific was not formally divided among the colonial powers until the nineteenth century. By that time, the United States, Germany, and Japan had joined France and Britain in taking control of various island groups. As in other regions, the European colonizers of Oceania emphasized extractive agriculture and mining. Because native people were often displaced from their lands or exposed to exotic diseases to which they had no immunity, their populations declined sharply.

THE COLONIZATION OF AUSTRALIA AND NEW ZEALAND

Although all of Oceania has been under European or U.S. rule at some point, the most Westernized parts of the region are Australia and New Zealand. The colonization of these two countries by the British has resulted in Australia and New Zealand having many parallels with North America. In fact, the American Revolution was a major impetus for "settling" Australia because once the North American colonies became independent, the British needed another location where they could send their convicts and other outcasts. In early-nineteenth-century Britain, a relatively minor theft—for example, of a piglet—might be punished with 7 years of hard labor in Australia (see Figure 11.14D).

A steady flow of English and Irish convicts arrived in Australia until 1868. Most of the convicts chose to stay in the colony after their sentences were served and are given credit for Australia's rustic self-image and egalitarian spirit. They were joined by a much larger group of voluntary immigrants from the British Isles

who were attracted by the availability of inexpensive farmland. Waves of these immigrants continued to arrive until World War II. New Zealand was settled in the mid-1800s, somewhat later than Australia. Although its population also derives primarily from British immigrants, New Zealand was never a penal colony.

Another similarity among Australia, New Zealand, and North America was the treatment of indigenous peoples by European settlers. In both Australia and New Zealand, native peoples were killed outright, annihilated by infectious diseases, or shifted to the margins of society. The few who lived on territory the Europeans deemed undesirable were able to maintain their traditional ways of life. However, the vast majority of the survivors lived and worked in grinding poverty, either in urban slums or on cattle and sheep ranches. Today, native peoples still suffer from discrimination and maladies such as alcoholism and malnutrition. Even so, some progress is being made toward improving their lives (see pages 651–652). In 2008, the then newly elected prime minister of Australia, Kevin Rudd, officially apologized to Aboriginal people for the treatment they have received since the land was first colonized.

Closely related to attitudes toward indigenous people were attitudes toward immigrants of any color other than white. By 1901, a whites-only policy (called the "White Australia policy") governed Australian immigration, with favored migrants coming from the British Isles and (after World War II) from southern Europe. This discrimination persisted until the mid-1970s, when the White Australia policy on immigration was ended. In New Zealand, where similar racist attitudes prevailed, there was never an official whites-only policy, and by the 1970s, students and immigrants were arriving from Asia and the Pacific islands. But controversy over immigration in Australia and New Zealand continues. Recently,

D An etching of British convicts being transported to Australia in the mid-1800s.

E Silver mining in Australia, 1900.

F Chinese Australians participate in a parade celebrating the Chinese New Year.

1700 C.E.　　　　　　　1800 C.E.　　　1850 C.E.　　　1900 C.E.　　　1950 C.E.　　　2000 C.E.

1700s
Early French and British explorations

1788–1868
European population of Australia goes from 0 to 1.7 million

1788–1945
Era of European orientation

1945–1970s
Era of North American orientation

1970–Present
Era of increasing orientation to Asia

C What were the chief interests of early Europeans when they came to Oceania in the sixteenth, seventeenth, and eighteenth centuries?

D Summarize the sequence of immigrants to Australia from early years of European settlement to the present.

E Until World War II (approximately) on what did the economy in most parts of Oceania depend?

F What is the present role of Asians in the makeup of Australia's population?

debate has centered on the arrival of refugees by boat from various parts of Asia, including Iraq, Iran, Syria, Myanmar, and Cambodia (see Figure 11.17A and the discussion on page 644).

OCEANIA'S SHIFTING GLOBAL RELATIONSHIPS

During the twentieth century, Oceania's relationship with the rest of the world went through three phases: from a predominantly European focus, to identification with the United States and Canada, and finally to the currently emerging linkage with Asia.

Until roughly World War II, the colonial system gave the region a European orientation. In most places, the economy depended largely on the export of raw materials to Europe (see Figure 11.14E). Thus, even when a colony gained independence from Britain, as did Australia in 1901 and New Zealand in 1907, people remained strongly tied to their mother countries. Even today, the queen of England remains the titular head of state in both countries. During World War II, however, the European powers provided only token resistance to Japan's invasion of much of the Pacific and its bombing of northern Australia. This impotence on the part of Europe spurred a change in the region's political and economic orientation.

After World War II, the United States, which already had a strong foothold in the Philippines, became the dominant power in the Pacific and U.S. investment grew more important to the economies of Oceania. Australia and New Zealand joined the United States in a Cold War military alliance, and both fought alongside the United States in Korea and Vietnam, suffering considerable casualties and experiencing significant antiwar activity at home. U.S. cultural influences became strong, too, as North American products, technologies, movies, and pop music penetrated much of Oceania.

By the 1970s, another shift took place as many of the island groups were granted self-rule by their European colonizers, with Oceania steadily drawn into the growing economies of Asia. Since the 1960s, Australia's thriving mineral export sector has become increasingly geared toward supplying raw materials to Asian manufacturing industries (primarily Japan in the 1960s and China since the 1990s). Similarly, New Zealand's wool and dairy exports have gone mostly to Asian markets since the 1970s. Despite occasional backlashes against "Asianization," Australia, New Zealand, and the rest of Oceania are being transformed by Asian influences. Many Pacific islands have significant Chinese, Japanese, Filipino, and Indian minorities, and the small Asian minorities of Australia and New Zealand are growing in population (see Figure 11.14F). On some Pacific islands, such as Hawaii, Asians now constitute the largest portion (42 percent in Hawaii, for example) of the population.

THINGS TO REMEMBER

• Oceania can be divided into four distinct indigenous cultural regions: Australia and Tasmania, originally settled by Aboriginal Australians; Melanesia, settled by Melanesians; and Micronesia and Polynesia (includes New Zealand and Hawaii), settled by a variety of Austronesian peoples.

• Through colonization, Europeans were active in Oceania from the early sixteenth century (in New Zealand, after 1642) until the end of World War II. During the 50 years after the war, the United States, Australia, and New Zealand were the principal powers in the Pacific.

• Since the 1970s, the influence of Asian countries has grown throughout Oceania.

GLOBALIZATION AND DEVELOPMENT

GEOGRAPHIC THEME 2

Globalization and Development: Globalization, coupled with Oceania's stronger focus on neighboring Asia (rather than long-time connections with Europe and North America), has transformed patterns of trade and economic development across Oceania. These changes are driven largely by Asia's growing affluence, its enormous demand for resources, and its similarly massive production of manufactured goods.

GLOBALIZATION, DEVELOPMENT, AND OCEANIA'S NEW ASIAN ORIENTATION

One could say that globalization in Oceania began when the first European explorers arrived in the region, beginning the trend of influence by outsiders (primarily Europeans) on settlement, culture, and economics. More recently, the United States has exerted a powerful influence on trade and politics in the region. For the past several decades, however, globalization has reoriented this region toward Asia, which buys more than 70 percent of Australia's exports (mainly coal, iron ore, and other minerals). In 2011, China and India each purchased not only the output of mines, but also major shares of Australia's particularly high-quality coal deposits. Both countries use coal to generate energy and are trying to secure future access to more high-quality coal that produces relatively lower harmful emissions. Asia also buys nearly 40 percent of New Zealand's exports (primarily meat, wood, wool, wine, and dairy products), as well as many other products and services from islands across Oceania **(FIGURE 11.15)**.

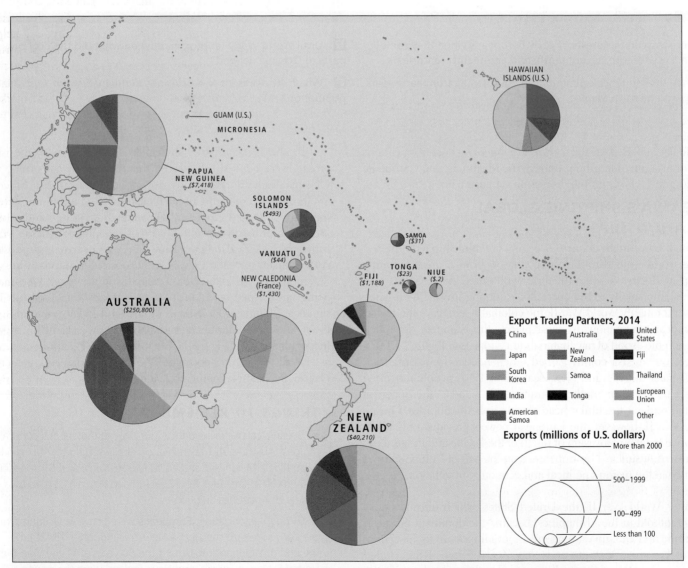

FIGURE 11.15 Exports from Oceania. The colors of each pie chart indicate a country's export trading partners. The "other" sections can include trade with Canada, Mexico, the Caribbean, non-EU Europe, sub-Saharan Africa, and other locales, some of them new trading partners. (Figures for Hawaii do not include exports to other parts of the United States.) [*Sources consulted: The World Factbook*, Central Intelligence Agency, https://www.cia.gov/library/publications /the-world-factbook/geos/xx.html; "East Asian and Pacific Affairs: Countries and Other Areas," U.S. Department of State, http://www.state.gov/p/eap/ci/index.htm.]

In addition, Asia is a major source of the region's imports. Because there is relatively little manufacturing in Oceania, most manufactured goods are imported from China, Japan, South Korea, Singapore, and Thailand—Oceania's leading trading partners. Both Australia and New Zealand have free trade agreements either completed or in continuous negotiation with Asia's two largest economies, China and Japan.

The Pacific islands are further along in their reorientation toward Asia than are Australia and New Zealand. Not only are coconut, forest, and fish products from the Pacific islands sold to Asian markets, but Asian companies own more and more of these industries on the various islands. Fleets from Asia regularly fish the offshore waters of Pacific island nations. Asians also dominate the Pacific island tourist trade, both as tourists and as investors in the tourism infrastructure. And growing numbers of Asians are taking up residence in the Pacific islands, exerting widespread economic and social influence.

THE STRESSES OF ASIA'S ECONOMIC DEVELOPMENT "MIRACLE" ON AUSTRALIA AND NEW ZEALAND

For Australia and New Zealand, Asia's global economic rise has generated both more trade and more competition with Asian economies in foreign markets. Throughout Oceania, local industries formerly enjoyed protected or preferential trade with Europe. They have lost that advantage because European Union (EU) regulations stemming from the EU's membership in the World Trade Organization (WTO) prohibit such arrangements. No longer protected in their trade with Europe, local industries now face stiff competition from larger companies in Asia that benefit from much cheaper labor.

Australia and New Zealand are somewhat unusual in having achieved broad prosperity not from manufacturing, but rather from the exporting of raw materials over a long period of time and to a wide range of customers (see Figures 11.14E and 11.15). Preferential trade with Europe allowed higher profits for many export industries. Meanwhile, strong labor movements in Australia and New Zealand meant that profits from these extractive industries were more equitably distributed throughout society than the profits in typical raw materials–based economies in Middle and South America and sub-Saharan Africa. Australian coal miners' unions successfully agitated not just for good wages, but also for the world's first 35-hour workweek. Other labor unions won a minimum wage, pensions, and aid to families with children long before such programs were enacted in many other industrialized countries. For decades, these arrangements were very successful. Both Australia and New Zealand had living standards comparable to those in North America but with a more egalitarian distribution of income. Since the 1970s, however, competition from Asian companies has seen many workers in Australia and New Zealand lose their jobs and their hard-won benefits scaled back or eliminated.

More recently, competition from Asian companies has also led to lower corporate profits in Oceania. As these profits have fallen, so have government tax revenues, which has resulted in cuts to previously high rates of social spending on welfare, health care, and education. The loss of social support, especially for those who have lost jobs, has contributed to rising poverty in recent years. Australia now has the second-highest poverty rate in the industrialized world. (The United States has the highest.)

Maintaining Raw Materials Exports as Service Economies Develop

Although their monetary contribution to national economies remains high, industries that export raw materials are a decreasing economic sector in the economies of Australia and New Zealand in that they now employ fewer people because of mechanization. This shift to replacing human labor with machinery has been essential for these industries to stay globally competitive with other countries where living standards are lower and workers are paid much less.

Today, the economies of both Australia and New Zealand are dominated by diverse and growing service sectors, which have links to the region's export sectors. Extracting minerals and managing herds and cropland have become technologically sophisticated enterprises that depend on many supporting services and an educated workforce rather than physical labor. Australia is now a world leader in providing technical and other services to mining and engineering companies, sheep farmers, and winemakers, and New Zealand's well-educated workforce and well-developed marketing infrastructure have helped it break into luxury markets for dairy products, meats, and fruits. Perhaps the most visible success has been New Zealand's global marketing of the indigenous kiwifruit (*Actinidia deliciosa*, a gooseberry native to southern China), now found in most U.S. food markets. (*Kiwi*, the slang term for anyone from New Zealand, originated with the flightless Kiwi bird, not the fruit.)

ECONOMIC CHANGE IN THE PACIFIC ISLANDS

In general, the Pacific islands are also shifting away from extractive industries, such as mining, farming, and fishing, and toward service industries, such as tourism and government. But, on many islands, self-sufficiency and resources from abroad cushion the stress of economic change, because many island households still construct their own homes and rely on fishing and subsistence cultivation for much of their food supply. On the islands of Fiji, for example, part-time subsistence agriculture engages more than 60 percent of the population, although it accounts for just under 17 percent of the gross national income. Remittances sent home from the thousands of Pacific Islanders working abroad are essential to many Pacific island economies and constitute more than half of all income on some islands. However, remittances are rarely reported as part of official statistics.

In the relatively poor nations and those with less-skilled populations (the Solomon Islands, Tuvalu, and parts of Papua New Guinea, for example), conditions typify what has been termed a **MIRAB economy**—one based on migration, remittances, aid, and bureaucracy. Foreign aid from former or present colonial powers supports government bureaucracies that supply employment for the educated and semiskilled. Although a MIRAB economy has little potential for growth, Islanders who can be self-sufficient in terms of food and shelter while saving extra cash for travel and occasional purchases of manufactured goods are sometimes said to have achieved **subsistence affluence.**

> **MIRAB economy** an economy based on migration, remittances, and an aid-supported bureaucracy
>
> **subsistence affluence** a lifestyle whereby people are self-sufficient with regard to most necessities and have some opportunities to save cash for travel and occasional purchases of manufactured goods

Where there is poverty in the Pacific, it is often related to geographic isolation, which means a lack of access to information and economic opportunity. Although computers and the new global communication networks (Internet and cell-phone service) are not yet widely available in the Pacific islands, they have the potential to significantly alleviate this isolation.

ON THE BRIGHT SIDE: Subsistence Affluence Practices Could Go Global

Those concerned with global sustainability have noted that many of the qualities of *subsistence affluence* practiced by Pacific Islanders have the potential to be expanded upon and adapted to other places. For example, local self-sufficiency based on home gardening and resource conservation strategies has been diffused to North America by Pacific Islander emigrants. The low-key optimism and accepting attitude toward life that are characteristic of Oceania have also been recognized as admirable and teachable qualities useful in many global locales. ■

The Advantages and Stresses of Tourism

Tourism is a growing part of the economy throughout Oceania. Tourists come largely from Japan, Korea, Taiwan, Southeast Asia, the Americas, and Europe (FIGURE 11.16). In 2013 (the latest year for which complete figures are available), 17.7 million tourists arrived in Oceania. Of those tourists, 23 percent came from Asia—16 percent from Japan alone; just 11 percent came from Europe, down from 17 percent in recent years; 33 percent came from North America; 19 percent were from within Oceania; and 12 percent traveled from other locations.

FIGURE 11.16 Tourism in Oceania. Tourism plays a major role in the economies of Oceania. Australia is the most popular destination, attracting about one-third of the total volume of tourists. The origins of the tourists reflect changing trade patterns in the region, with more and more journeying from Asia. Here, you see hotels on Honolulu's Waikiki Beach, which provide lodging for a majority of Hawaii's 6 million or more yearly visitors. Nearby is the giant Ala Moana shopping center geared to meet the tourists' shopping needs.

Large numbers of visitors expecting to be entertained and graciously accommodated can place a special stress on local inhabitants. In some Pacific island groups, the number of tourists far exceeds the island population. Guam, for example, receives tourists annually in numbers equivalent to five times its population. Palau and the Northern Mariana Islands annually receive more than four times their populations. And although they bring money to the islands' economies, these visitors create problems for island ecologies, place extra burdens on scarce resources (land, water, fuel, food, and waste management systems), and expect a standard of living that may be far out of reach for local people. Perhaps nowhere else in the region are the issues raised by tourism as clear as they are in Hawaii.

CASE STUDY: Conflict over Tourism in Hawaii

Since the 1950s, travel and tourism have been the largest industries in Hawaii, accounting for around 20 percent of the gross state product in 2015. Tourism is related in one way or another to nearly 75 percent of all jobs in the state. (By comparison, travel and tourism account for 9 percent of GDP worldwide.) In 2014, tourism employed one out of every six Hawaiians and accounted for 20 percent of state tax revenues.

Dramatic fluctuations in tourist visits, often driven by forces far removed from Oceania, can wreak havoc on local economies. Decreases in tourism affect not just tourist facilities but supporting industries as well. For example, the construction industry thrives by building condominiums, hotels, resorts, and retirement facilities. The Asian recession of the late 1990s, the terrorist attacks of September 11, 2001, and the global recession of 2008–2009 all affected Hawaii's economy by creating dramatic slumps in tourist visits. By 2015, however, Hawaii's tourism industry rebounded: more than 8.6 million visitors arrived that year, up 22 percent over 2009.

To ordinary citizens, mass tourism can sometimes seem like an invading force. For example, in 2015, Hawaii hosted eight tourists for every one Hawaiian. But what was the real net benefit of all these visitors, who had to be housed, fed, entertained (and cleaned up after)? This is a question Hawaiians continue to ask themselves. Furthermore, an important segment of the Honolulu tourist infrastructure—hotels, golf courses, specialty shopping centers, import shops, and nightclubs—is geared exclusively to foreign visitors and many such facilities are owned by foreign investors. Hawaiian citizens and even some vacationers can feel out of place in facilities focused on high-end foreign consumers.

Just the demand for golf courses imposed by mass tourism has resulted in what Native (indigenous) Hawaiians view as the desecration of sacred sites. Land that in precolonial times was communally owned, cultivated, and used for sacred rituals was first confiscated by the colonial government and more recently sold to Asian golf course developers. Now the only people with access to the sacred sites are fee-paying tourist golfers.

Nevertheless, the golf industry cannot be ignored. As of 2015, there were more than 90 golf courses in Hawaii, and the golf industry alone contributed $1.6 billion to the state's economy—more than twice the amount from agriculture. Golf's total economic impact is $2.5 billion, which represents about 12.5 percent of the state's income from the tourism sector, a fact that confuses the picture for those who oppose golf on environmental grounds.

The pressure to make land available for tourism of all kinds has decreased access to land by local citizens. For example, retired mainland Americans who relocate to Hawaii in search of a sunny spot—in what is often called *residential tourism*—have caused property values to rise steeply and thereby increased the costs of housing for local people. [*Sources: Hawaii Tourism Authority and a field report from Conrad M. Goodwin and Lydia Pulsipher. For detailed source information, see Text Sources and Credits.*] ∎

ON THE BRIGHT SIDE: Sustainable Tourism

Some Pacific islands have attempted to deal with the pressures of tourism by adopting the principle of *sustainable tourism,* which aims to decrease tourism's imprint and minimize disparities between hosts and visitors. Samoa, with financial aid from New Zealand, has developed sustainable tourism components (beaches, wetlands, forested island environments, and cultural attractions) and its islanders provide *knowledge-based tourism experiences* for the conscious traveler. These experiences are information-rich explanations of Samoa's political, social, and environmental issues. The initiative began in 2009, but by 2015, Samoa's tourism economy, while viable, was still not on a par with those of other Pacific islands. A locally published opinion piece, by Olivia Peterkin, noted that perhaps Samoa's sustainable tourism innovations could not result in big earnings, but they might indeed bring about long-term sustainability. [*Source consulted: "Samoa's Dwindling Tourism Industry," Think Global: Oceania, November 3, 2015. https://thinkglobaloceania.wordpress.com/2015/11/03/samoas-dwindling-tourism-industry/.*] ∎

THE FUTURE: DIVERSE GLOBAL ORIENTATIONS?

Despite the powerful forces pushing Oceania toward Asia, important factors still favor strong ties with Europe and North America. In spite of the trade links and China's recent efforts to expand diplomatic and cultural relations with Australia, both Australia and New Zealand remain staunch military allies of the United States. Over the years, both have participated in U.S.-led wars in Korea, Vietnam, Afghanistan, and Iraq. In 2012, in an apparent effort to check the growing influence of the Chinese military in the South China Sea and the Indian Ocean, the Australian government gave the U.S. Marine Corps access to a large tract of land near Darwin (located in Australia's Northern Territory). The United States and Australia also opened discussions regarding the use of the Cocos Islands (Australian possessions in the Indian Ocean) for reconnaissance purposes.

In some of the Pacific islands, strong links to Europe and North America are also upheld by Europe and the United States' continuing administrative control. In Micronesia, the United States governs Guam and the Northern Mariana Islands; in Polynesia, American Samoa is a U.S. territory. Just as the Hawaiian Islands are a U.S. state, the 120 islands of French Polynesia—including Tahiti and the rest of the Society Islands, the Marquesas Islands, and the Tuamotu Archipelago—are Overseas Lands of France. Any desire people in these possessions have for independence has not been sufficient to override the financial benefits of aid, subsidies, and investment money provided by France and the United States.

In 1989, the **Asia Pacific Economic Cooperation (APEC),** a coalition of 21 Pacific countries, was formed. Today, these states account for approximately 40 percent of the world's population, just over 50 percent of global production, and more than 40 percent of global trade. APEC members include Oceania's most populous countries (Australia, Papua New Guinea, and New Zealand), as well as Brunei, Canada, Chile, China, Hong Kong, Indonesia, Japan, South Korea, Malaysia, Mexico, Peru, the Philippines, Russia, Singapore, Taipei, Thailand, the United States, and Vietnam. APEC was organized to enhance economic prosperity and strengthen the Asia–Pacific community, and while much is made of its potential, its inability to compel its membership to act in any significant way has led some to dismiss the group as a pointless "talk shop."

An important effort to get APEC to evolve into a more potent force, perhaps to place it on a par with the European Union (after which it is partially patterned), is the **Trans-Pacific Partnership (TPP)** agreement brought forward in its final negotiated (but as of 2016, unapproved) form in October 2015. The TPP is seen by some as a way to smooth relations between Asia, Oceania, and the United States in matters of trade, food security, climate change, product regulations, and energy-efficient transportation. Others (surveyed among the public in the United States, Japan, and Mexico) view the TPP as mostly a give away to international business that would result in the movement of jobs to low-wage Asian countries and the limiting of competition, thus encouraging higher prices for consumers. The provision that would allow multinational corporations to challenge regulations before special tribunals, in particular, is intensely opposed by critics in the United States.

Asia Pacific Economic Cooperation (APEC) a group of 21 Pacific Rim countries organized in 1989 to increase trade and cooperation

Trans-Pacific Partnership (TPP) the largest regional trade accord in history that, if approved, would set new terms for trade between the United States and eleven other Pacific Rim countries

THINGS TO REMEMBER

GEOGRAPHIC THEME 2 • **Globalization and Development:** Globalization, coupled with Oceania's stronger focus on neighboring Asia (rather than long-time connections with Europe and North America), has transformed patterns of trade and economic development across Oceania. These changes are driven largely by Asia's growing affluence, its enormous demand for resources, and its similarly massive production of manufactured goods.

• Service industries are becoming the dominant source of income for most of Oceania's economies, although extractive industries remain important.

• Tourism is a significant and growing part of the economies in Oceania, but it can produce environmental and social stresses for the host countries.

• Oceania is at the center of a unique attempt by the Asia Pacific Economic Cooperation (APEC) alliance to forge international cooperation on trade, food security, climate change, and energy-efficient transportation, which may culminate in adoption of the Trans-Pacific Partnership (TPP) agreement.

POWER AND POLITICS

GEOGRAPHIC THEME 3

Power and Politics: Stark divisions have emerged in Oceania over definitions of democracy—the system of government that dominates in New Zealand, Australia, and Hawaii—versus the Pacific Way, a political and cultural philosophy based on the communitarian values of traditional cultures of the Pacific islands. The global refugee crisis tests this region's ability to maintain its reputation as a humanitarian refuge and zone of opportunity that values social cohesion.

In Australia and New Zealand, government is based on European-style parliamentary systems grounded on universal voting rights for adults, free speech, debate, and majority rule. Democratic principles influence debate and conflict resolution from the community level to the national level, and the internal and external political issues faced by Australia and New Zealand bear a strong similarity. Both have majority populations with a European heritage and large minority populations of indigenous people—Aboriginal people in Australia and Maori (Polynesian) in New Zealand. Political issues that drive internal debate in both countries include how to manage migration between the two countries and immigration from outside; how to maintain social cohesion in the face of increasing diversity; how to adjust their resource export–based economies to a greater focus on technical innovation and services; and how to reformulate international relationships as the region pivots from strong ethnic and economic connections to Europe to closer associations with Asia.

By contrast, the **Pacific Way** is based on traditional notions of power and problem solving and refers to a way of settling issues familiar to many Pacific Islanders. It was developed in small communities where decisions are reached in face-to-face meetings, and consensus and mutual understanding are favored rather than open confrontation and majority rule. High value is placed on respect for traditional leadership (especially the usual patriarchal leadership of families and villages) rather than on political freedoms such as free speech and individuality. As such, the Pacific Way can embody very different, but not invalid, definitions of fairness and corruption than do parliamentary systems.

The Pacific Way in Fiji

As a political and cultural philosophy, the Pacific Way was articulated as a formal concept in Fiji around the time of Fiji's independence from the United Kingdom in 1970. It subsequently gained popularity in many other Pacific islands, most of which gained independence in the 1970s and 1980s.

The Pacific Way carries a flavor of resistance to Europeanization and has often been invoked to uphold the notion of a regional identity shared by Pacific islands that grows out of their unique history and social experience. It was particularly influential among educators given the task of writing new textbooks to replace those used by the former colonial masters. The new texts focused students' attention away from Britain, France, and the United States and toward their

Pacific Way the regional identity and way of handling conflicts peacefully that grows out of Pacific Islanders' particular social experiences

Thinking Geographically

After you have read about power and politics in Oceania, you will be able to answer the following questions.

A When did Australia's whites-only immigration policy formally end?

B In Fiji, three coups d'état carried out by indigenous Fijians have removed legally elected governments headed by which ethnicity?

C Describe how the three coups in Fiji against elected governments could be described as a conflict between the Pacific Way and democracy.

D What was the colonial rationale, in place until 1993, that held the Aboriginal Australians had no prior claim to land in Australia?

own cultures. Appeals to the Pacific Way have also been used to uphold attempts by Pacific island governments to control their own economic development and solve their own political and social problems.

ON THE BRIGHT SIDE: The Value of the Pacific Way to Grassroots Sustainability

Regardless of its global political status, the Pacific Way is likely to endure, especially as a concept that upholds Pacific regional identity and traditional culture. Furthermore, some organizations now use the Pacific Way as the basis for an integrated approach to economic development and the resolution of environmental issues. For example, the Secretariat of the Pacific Regional Environmental Programme (SPREP) builds on traditional Pacific island economic activities—such as fishing and local traditions that require knowledge and awareness of the environment—to promote grassroots economic development and environmental sustainability. The strategic focuses of SPREP are climate change, biodiversity, and environmental policy design. Implementation of Pacific Way procedures by SPREP has the reputation of facilitating the involvement of the public perhaps more effectively than so-called democratic approaches. ■

In managing political conflict, the Pacific Way has proven to have a down side, at least from the perspective of those favoring Western-style democracy, because the Pacific Way has occasionally been invoked as a philosophical basis for overriding democratic elections if the results challenge the power of indigenous Pacific Islanders. In 1987, 2000, and 2006, indigenous Fijians used the Pacific Way to justify coups d'état against legally elected governments **(FIGURE 11.17B)**. All three of the overthrown governments were dominated by Indian Fijians, the descendants of people from India whom the British brought to Fiji more than a century ago to work on sugar plantations.

Fiji's population is now about evenly divided between indigenous Fijians and Indian Fijians. Indigenous Fijians are generally less prosperous and tend to live in rural areas where community affairs are still governed by traditional chiefs. Indian Fijians, in contrast, hold significant economic and political power, especially

FIGURE 11.17 PHOTO ESSAY: Power and Politics in Oceania

Politics in this region vary significantly between the more Europeanized areas of Australia, New Zealand, and Hawaii, and the more indigenous and traditional political cultures of New Guinea and many Pacific islands. Recently, the political culture of Australia and New Zealand coupled with their proximity to Asia has made these countries the focus of people fleeing poverty and persecution in Earth's trouble spots.

A A funeral at an immigration detention facility on Christmas Island, Australia, for refugees from Iraq and Iran who drowned while trying to reach Australia by boat. Immigration has been a political hot-button issue in Australia for many years. Particularly controversial is Australia's so-called *Pacific Solution*—a policy of paying Pacific islands, such as Nauru and Papua New Guinea, to take in refugee asylum-seekers, who arrive by smuggler's boats, and keep them indefinitely in detention facilities. Here, they may wait for years in prison-like conditions for their applications for asylum to be judged. Some Australians believe that because relatively few refugees come to their country, those who do should be treated better and accepted more quickly. Others see refugees as a security risk or an economic drain and insist that the long detentions will discourage other refugees from coming to Australia.

Mark Kolbe/Getty Images

William West/Getty Images

B A Fijian soldier during the military coup of 2006, which overturned the fair election of officials who were Indian-Fijian. Military takeovers also took place in Fiji in 1987 and 2000.

Democratization and Conflict Armed conflicts and genocides with high death tolls since 1990

✳ 13,000 deaths

Democratization index
- Full democracy
- Flawed democracy
- Hybrid regime
- Authoritarian regime
- No data

mi 0 · 250 · 500
km 0 · 250 · 500

HAWAIIAN ISLANDS (U.S.)

GUAM (U.S.)
NORTHERN MARIANA ISLANDS (U.S.)
MARSHALL ISLANDS
PALAU
FEDERATED STATES OF MICRONESIA
PAPUA NEW GUINEA
NAURU
KIRIBATI
SOLOMON ISLANDS
TUVALU
TOKELAU (N.Z.)
SAMOA
AMER. SAMOA (U.S.)
COOK ISLANDS (N.Z.)
FRENCH POLYNESIA
CHRISTMAS ISLAND (Australia)
VANUATU
WALLIS AND FUTUNA (Fr.)
TONGA
NIUE (N.Z.)
NEW CALEDONIA (France)
FIJI
AUSTRALIA
NEW ZEALAND
PITCAIRN (U.K.)

C Members of a violent criminal gang guard the entrance to their headquarters in Port Moresby, the capital of Papua New Guinea. Decades of rampant corruption have led to a breakdown of law and order in the capital and many other areas of the country.

D Corrie Bodney, an elder of the Ballaruk Aboriginal tribe, stages a sit-in at Perth International Airport, which is located on land the Ballaruk have occupied for thousands of years. In 2013, the government of Western Australia offered several Aboriginal tribes more than U.S.$1 billion to settle a larger claim, which included the entire city of Perth.

Torsten Blackwood/Getty Images

The West Australian/AFP/Getty Images

in the urban centers and in areas of tourism and sugar cultivation. In response to the coups, many Indian Fijians left Fiji, resulting in a loss of badly needed skilled workers that has slowed economic development.

Political responses around Oceania to the Fiji coups have been divided. Australia, New Zealand, and the United States (via APEC and the state government of Hawaii) have demanded that the election results stand and the Indian Fijians be returned to office. But much of Oceania has referenced the Pacific Way in arguments supporting the coup leaders. As in Fiji, those who govern many of the Pacific islands are leaders of indigenous descent who have not always had the strongest respect for political freedoms, especially when their hold on power is threatened. Their decisions have at times upheld traditional Pacific values such as stability, respect for authority, and certain kinds of environmental awareness, while at other times these decisions have contributed to corruption and civil disorder (see Figure 11.17C).

On the global stage, Fiji has been suspended from the Commonwealth of Nations (a union of former British colonies) for subverting majority-rule democracy. As a result, it is ineligible for Commonwealth aid and is not allowed to participate in Commonwealth sports events. And because sports play a central role in Pacific identity (see "Sports as a Unifying Force" on page 655), this latter sanction carries significant weight.

Migrants and Refugees from Abroad Present a Challenge for Oceania

Legal migrants with skills are welcomed every year to Australia and New Zealand in rather large numbers. In 2015, Australia's net immigration was about 184,000 and New Zealand's net immigration was about 50,000 (about 15 percent of these migrants are simply moving between Australia and New Zealand). Both countries have a high proportion of foreign-born residents. In Australia, about 30 percent of its 24 million people are foreign-born, and in New Zealand, approximately 20 percent of its 4.6 million were born abroad (in the United States, 13.7 percent were foreign-born in 2015). Both countries are just now revising their immigration rules, and both actively encourage immigration by professional workers from any country. Of those from outside Oceania, most are emigrating from India and China, with smaller percentages from the United Kingdom and Southeast Asia. Australia firmly contends that it has the right to decide who legally immigrates to its soil.

Unfortunately, ethnic and religious hostilities in places far from Oceania have unexpectedly exposed fissures in the region's reputation for generous and humane migration policies. Ruthless human traffickers, sensing that Australia and New Zealand can be counted on to accept people fleeing Earth's many conflict zones—Afghanistan, Pakistan, Syria, Iraq, Iran, Myanmar, Cambodia—offer to take people in rickety boats to the coasts of these two countries for outrageously high fees.

Both countries have humane policies for the orderly acceptance of refugees (defined as different from

ordinary immigrants). In 2015, New Zealand was prepared to resettle 750 refugees, Australia (a much larger country) 13,750, and both were prepared to help such refugees adjust to life in a new place (especially those Pacific Islanders fleeing the effects of climate change). Nonetheless, the huge and unexpected influx of impoverished and desperate people from distant regions far exceeded what the public in the two countries was willing to accept. Both countries were already dealing with right-wing political objections to cultural diversity, when they were faced, as was the EU, with thousands of more refugees than anticipated from Syria and elsewhere.

As Australia and New Zealand struggled to find a way to address the overflow and simultaneously maintain their own social cohesion, they devised the controversial **Pacific Solution** to handle undocumented asylum-seekers—a solution that, when it was first proposed in 2001, was regarded by the international community as racist and inappropriate. Similarly viewed in the present, the Pacific Solution mandates that undocumented asylum-seekers—those who have exceeded both countries' limited quotas—be held in detention centers in Nauru and on an outlying small island off Papua New Guinea **(FIGURE 11.18)**. Basic food and shelter are supplied and basic education is provided for the children, but for the adults, who range from physicians and teachers and technologists to unschooled farmhands, there is little to do and no future to look forward to. In April 2016, the Supreme Court of Papua New Guinea declared the detention center (then holding more than 1300 people) illegal and subject to closure.

Pacific Solution Australia and New Zealand's controversial approach to the excess number of undocumented immigrants

---

THINGS TO REMEMBER

GEOGRAPHIC THEME 3 • **Power and Politics:** Stark divisions have emerged in Oceania over definitions of democracy—the system of government that dominates in New Zealand, Australia, and Hawaii—versus the Pacific Way, a political and cultural philosophy based on the communitarian values of traditional cultures of the Pacific islands. The global refugee crisis tests this region's ability to maintain its reputation as a humanitarian refuge and zone of opportunity that maintains social cohesion.

• The Pacific Way, a political and cultural philosophy based on consensus, was first articulated in Fiji at the time of Fiji's independence from the United Kingdom in 1970, but gained popularity in many islands.

• In politics, the Pacific Way has occasionally been invoked as a philosophical basis for overriding democratic elections that challenge the power of indigenous Pacific Islanders.

• Governance in Oceania is, for the most part, based on democratic principles with regular elections; however, occasionally traditional power holders have negated or threatened political freedoms.

• The Pacific Solution is a highly controversial policy devised by Australia to address the overflow of impoverished asylum-seeking refugees.

FIGURE 11.18 Camp for refugees.

URBANIZATION

GEOGRAPHIC THEME 4

Urbanization: Oceania is only lightly populated but it is highly urbanized. The shift from extractive economies to service economies is a major reason for the urbanization of the wealthiest parts of Oceania (Australia, New Zealand, Hawaii, Guam), where 80 to 100 percent of the population lives in cities. These trends are weakest in Papua New Guinea and many smaller Pacific islands.

The global trend of migration from the countryside to cities is quite visible in Oceania, where 70 percent of the overall population now lives in urban areas. Australia and New Zealand have among the highest percentages of city dwellers outside Europe. More than 89 percent of Australians live in a string of cities along the country's relatively well-watered and fertile eastern and southeastern coasts. Similarly, 86 percent of New Zealanders live in urban areas. The vast majority of people in these two countries live in modern comfort, work in a range of occupations typical of highly industrialized societies, and have access to tax-supported health-care and leisure facilities (**FIGURE 11.19A**). Vibrant, urban-based service economies employ about three-quarters of the population in both countries. Declining employment in mining and agriculture, where mechanization has dramatically reduced the number of workers needed, has also contributed to urbanization.

Throughout the Pacific, urban centers have transformed natural landscapes. In some small densely populated countries, such as Guam, Palau, and the Marshall Islands, they have become the dominant landscape. Although cities are places of opportunity, they can also be sites of cultural change, conflict, and environmental hazards (see Figure 11.19C).

Most Pacific island towns and all the capital cities are located in ecologically fragile coastal settings. Many of these waterfront towns were established during the colonial era as ports or docking facilities and were situated in places suitable for only limited numbers of people. Consequently, little land is available for development and access to housing is inadequate. Squatter settlements have been a visible feature of the region's urban areas for decades. The coastal locations for towns and cities in Oceania make these settlements particularly vulnerable to rising sea levels and violent storms associated with climate change.

Multiculturalism has been enhanced by urbanization across the region. Many urban residents are letting go of the rural ways of their childhoods, as well as their ethnic identity and cultural commitments. As time goes on, more urban people are marrying across ethnic divisions, having only one or two children, and creating new patterns of social alliances and networks. Such cultural blending can ultimately result in enhanced social cohesion, but the result can also be new social tensions, which change the very nature of social life in the island Pacific. Urban unemployment and unrest are on the rise, and low rates of economic growth restrict the revenue available to governments to manage urban development.

THINGS TO REMEMBER

GEOGRAPHIC THEME 4 • **Urbanization:** Oceania is only lightly populated but it is highly urbanized. The shift from extractive economies to service economies is a major reason for the urbanization of the wealthiest parts of Oceania (Australia, New Zealand, Hawaii, Guam), where 80 to 100 percent of the population lives in cities. These trends are weakest in Papua New Guinea and many smaller Pacific islands.

• The global trend of migration from the countryside to cities is quite visible in Oceania, where 70 percent of the overall population lives in urban areas.

• Most Pacific island towns and all the capital cities are located in ecologically fragile coastal settings.

• Many urban residents are letting go of the rural ways of their childhoods, as well as their ethnic identity and cultural commitments.

POPULATION AND GENDER

GEOGRAPHIC THEME 5

Population and Gender: In this largest but least populated world region, there are two main patterns relevant to population and gender. Australia, New Zealand, and Hawaii have older and more slowly growing populations, and offer relatively more opportunities for women. The Pacific islands and Papua New Guinea have much more rural, younger, and rapidly growing populations, where women play a central role in family and community but enjoy fewer opportunities than men as economies modernize.

A number of factors are influencing population growth in Oceania; among them are important issues related to gender. While there is a trend toward equality across gender lines throughout Oceania, persistent gender inequality exists as well. A striking disparity is emerging between Australia, New Zealand,

FIGURE 11.19 PHOTO ESSAY: Urbanization in Oceania

There are two patterns of urbanization in Oceania. Australia, New Zealand, and Hawaii are very urbanized places and have high standards of living; Papua New Guinea and many Pacific islands have high rural densities and their cities tend to have lower standards of living and be characterized by coastal shantytowns. [*Source consulted: 2011 World Population Data Sheet*, Population Reference Bureau, http://www.prb.org/pdf11/2011population-data-sheet_eng.pdf.]

A Tourists climb the Harbor Bridge in Sydney, Australia, the largest city in Oceania and one that is consistently ranked among the most livable cities in the world, along with Melbourne and Perth, Australia, and Auckland, New Zealand.

B A free concert on Christmas Day in a public park in Auckland, New Zealand.

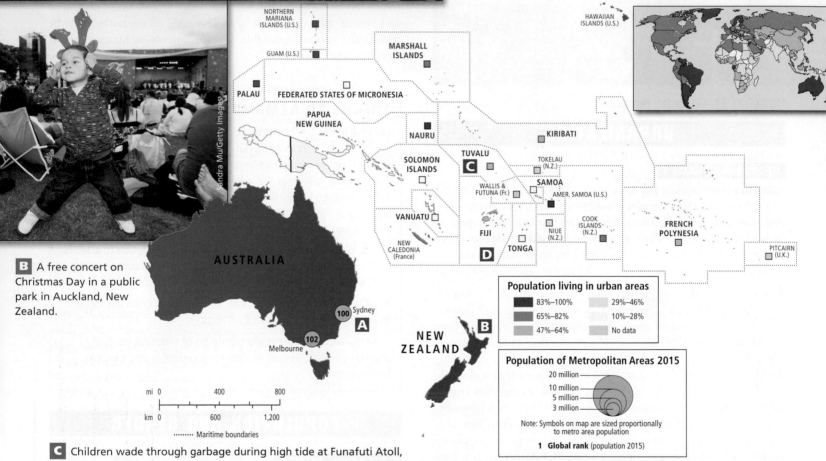

Population living in urban areas
- 83%–100%
- 65%–82%
- 47%–64%
- 29%–46%
- 10%–28%
- No data

Population of Metropolitan Areas 2015
- 20 million
- 10 million
- 5 million
- 3 million

Note: Symbols on map are sized proportionally to metro area population

1 Global rank (population 2015)

C Children wade through garbage during high tide at Funafuti Atoll, capital of Tuvalu. A very densely populated country, with 4847 people per square mile (1871 per square kilometer), Tuvalu is one of the poorest nations in Oceania, with a GDP (PPP) per capita of about U.S.$1600.

D A 4-year-old girl in a squatter settlement outside Suva, Fiji. Incomes are relatively low here, as is access to health care and education.

Thinking Geographically

After you have read about urbanization in Oceania, you will be able to answer the following questions.

A Describe the stereotypes about Australia that leave some surprised to learn that nearly 90 percent of its population lives in cities.

C and **D** What best explains why Pacific island cities have lower standards of living than Australia, New Zealand, and Hawaii?

and Hawaii on the one hand (where women are gaining political and economic power), and Papua New Guinea and the Pacific islands on the other hand (where change is much slower). In Australia, New Zealand, and Hawaii, where opportunities for women have improved, fertility rates are low at 1.9, well below replacement rates. In the islands, where opportunities for women are fewer, fertility rates range between 2.3 and 4.7.

Another and closely related factor affecting population growth is the age of the populations. On some of the poorer Pacific islands, close to 40 percent of the population is under the age of 15. So even if people decide to limit fertility, populations are likely to grow because a large proportion is just reaching reproductive age. This is not the case in Australia, New Zealand, and a few more wealthy islands, where just 20 percent is under age 15.

In Australia, New Zealand, and Hawaii, women's access to jobs and policy-making positions in government has improved, particularly over the last few decades. New Zealand and the Australian province of South Australia were among the first places in the world to grant European women full voting rights (in 1893 and 1895, respectively). New Zealand has elected two female prime ministers, and in 2010, Australia elected a woman prime minister. Moreover, according to the 2015 Inter-Parliamentary Union report on women in lower houses of parliament, in both countries, the proportion of women in national legislatures (31.4 percent in New Zealand, 26.7 percent in Australia) is well above the global average of 22 percent. In Papua New Guinea and the Pacific islands, women generally have far less political and economic power. No woman has yet been elected to a top-level national office, and women are a tiny minority in national legislatures when they are present at all. The one exception is Fiji, where 14 percent of those serving in Parliament are women, perhaps indicating that change is underway after years of political disruption in Fiji (see pages 642–644).

In both Australia and New Zealand, young women are pursuing higher education and professional careers and postponing marriage and childbearing until their thirties. (This is also a trend in Hawaii, Guam, and the islands with a French affiliation.) Nonetheless, both societies continue to reinforce the role of housewife for women in a variety of ways. For example, the expectation is that women, not men, will interrupt their careers to stay home to care for young or elderly family members. Women in Australia receive, on average, only about 70 percent of the pay that men receive for equivalent work. This represents, however, a smaller gender pay gap than in many other developed countries.

Throughout Papua New Guinea and the Pacific islands, gender roles and relationships vary greatly over the course of a lifetime. Because of the emphasis on community, male and female Pacific Islanders contribute to family assets through the formal and informal economies. Traditionally, men are the boat builders, navigators, fishers, and house builders. And men are usually the preparers of food, though women often supply some of the ingredients through their gathering and cultivating efforts (FIGURE 11.20). Most traders in marketplaces are women, and, aside from fish, the items they sell are also made and transported by women. Many young women today fulfill traditional roles as mates and mothers and practice a wide range of domestic crafts, such as weaving and basketry. In middle age, however, these same women may return to school and take up careers. With the aid of government scholarships, some Pacific Island women pursue higher education or job training that takes them far from the villages where they raised their children. Thus, the expectation that Aurora (in this chapter's opening vignette) will study far from home and then support her family and elders is in line with evolving gender roles in Pacific ways of life. Aurora can expect that as her credentials and experience accumulate, she will enjoy a position of considerable power in her community, an honor typically accorded to only elderly women on Pacific islands.

POPULATION NUMBERS AND DISTRIBUTION

Although Oceania occupies a huge portion of the planet, its total population is only 40 million people, close to that of the state of California (38 million). The people of Oceania live on a total land area slightly larger than the contiguous United States but spread out in bits and pieces across an ocean larger than the Eurasian landmass (FIGURE 11.21). The Pacific islands have nearly 4.75 million people, including Hawaii's 1.43 million; Australia has 23.9 million; Papua New Guinea, 7.7 million; and New Zealand, 4.6 million.

Population densities remain low in Australia, at 7.8 people per square mile (3 per square kilometer) for the country as a whole and about 130 per square mile (50 per square kilometer) on Australia's arable land. New Zealand's arable land density is quite a bit higher, at 2064.4 people per square mile (794 per square kilometer). In the Pacific islands, densities vary widely. Some are sparsely settled or uninhabited, while others—including some of the smallest, poorest, and lowest in elevation (which are thus some of the most exposed to rising sea levels; see Figure 11.19C)—are extremely densely populated. For example, French Polynesia, in the eastern Pacific, and Palau, in Micronesia (see Figure 11.1), have 26,689 and 4625 people, respectively, per square mile (10,265 and 1779, respectively, per square kilometer).

THINGS TO REMEMBER

GEOGRAPHIC THEME 5 • **Population and Gender:** In this largest but least populated world region, there are two main patterns relevant to population and gender. Australia, New Zealand, and Hawaii have older and more slowly growing populations, and offer relatively more opportunities for women. The Pacific islands and Papua New Guinea have much more rural, younger, and rapidly growing populations, where women play a central role in family and community but enjoy fewer opportunities than men as economies modernize.

FIGURE 11.20 LOCAL LIVES: Foodways in Oceania

A A winemaker in Australia's Hunter Valley assesses the shiraz grapes at her winery. Australia is the fourth-largest exporter of wine in the world, after Italy, France, and Spain.

B Taro is harvested in Hawaii. Grown in flooded fields, taro was originally brought to Oceania from Southeast Asia. It is now a major part of diets throughout the Pacific islands. While the leaves are also eaten, the cooked tuber is particularly prized as a source of calories when processed into *poi,* a paste of variable thickness that is eaten with the fingers.

C Food is removed from a Maori earth oven, or *hangi,* in New Zealand. First a pit is dug, and then a fire is made to heat stones placed in the pit. Baskets of food are placed over the hot stones (which are covered with cloth and then earth) for several hours until the food is cooked.

- New Zealand and the Australian province of South Australia were among the first places in the world to grant European women full voting rights (in 1893 and 1895, respectively).

- In Australia, New Zealand, and Hawaii, young women are pursuing higher education and professional careers and postponing marriage and childbearing until their thirties.

- In some Pacific islands, women can inherit or personally accrue considerable power in their own communities over the course of a lifetime, but compared to men, they enjoy fewer education and job opportunities.

- Although Oceania occupies a huge portion of the planet, its total population is only 40 million people, close to that of the state of California (38 million). Nonetheless, some islands are unusually densely occupied.

GEOGRAPHIC PATTERNS OF HUMAN WELL-BEING

Human well-being varies dramatically across Oceania as measured by the customary indicators used in this book: gross national income per capita (GNI per capita, adjusted for PPP), rank on the UN Human Development Index (HDI), and the Gender Development Index (GDI). Australia in 2015 ranked 2nd on the HDI. New Zealand ranked 9th in 2015. **FIGURE 11.22** offers maps delineating these three measures of human well-being across the region.

Figure 11.22A shows that, except for Australia, New Zealand, and Hawaii, Oceania has low levels of GNI per capita. Over the last decade, Australia typically has ranked among the top 20 countries in GNI per capita PPP (in 2015, it ranked 17th). New Zealand is usually among the top 25; in 2015, it ranked 23rd. Hawaii, part of the United States, typically has a

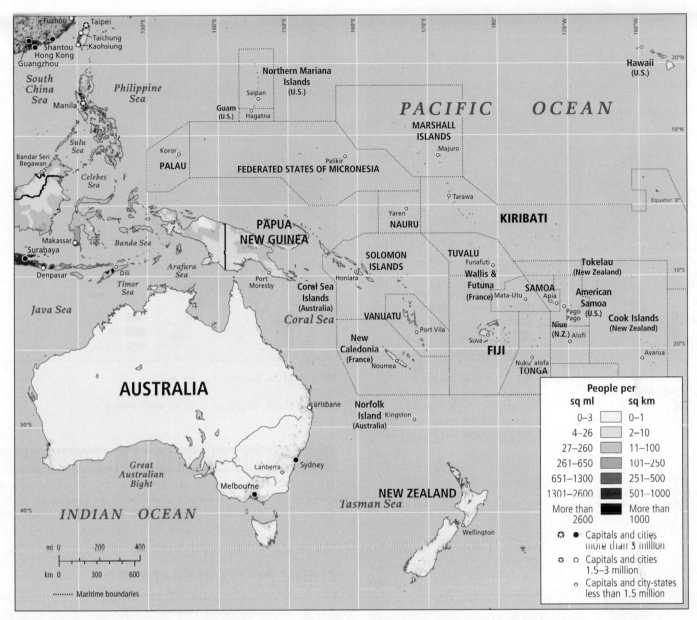

FIGURE 11.21 Population density in Oceania. Population density on this map is calculated for total area, much of which (especially in Australia, but also on many islands) is not usable for agriculture (arable). Population growth is slowing throughout Oceania as more people move to the cities, where health care is better and women are more likely to pursue careers—factors that make large families less likely. Australia, New Zealand, and Hawaii are furthest along in the urbanization process, with smaller families and increasingly older populations. Papua New Guinea and some small islands (French Polynesia and Palau, for example) are relatively densely occupied and still have high growth rates. The size of the region and the smallness of most islands make showing density difficult at printable scales.

GNI per capita PPP that is about $2000 higher than the average for the United States; in 2015, Hawaiian GNI per capita was $54,516.

Figure 11.22B shows each country's rank on the HDI, which is a calculation (based on adjusted real income, life expectancy, and educational attainment) of how adequately a country provides for the well-being of its citizens. Again with the exceptions of Australia (2), New Zealand (9), and Hawaii (because it is a state with an unusually well-developed social welfare system, Hawaii would rank higher than the overall U.S. level of 8 on the HDI),

the rest of Oceania ranks low on HDI or the data are not available. However, it should be remembered that, while the Pacific islands have low levels of income and human development as measured in official statistics, *subsistence affluence* (discussed on page 639) and the strong communitarian values of the *Pacific Way* (see the discussion on page 642) can result in higher-than-expected actual well-being.

Figure 11.22C shows the GDI, which is calculated on three sets of data: reproductive health, political and education empowerment, and access to the labor market. The countries

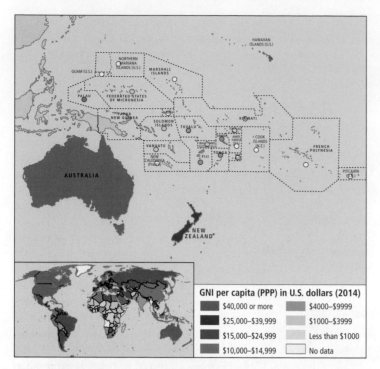

(A) Gross national income (GNI) per capita, adjusted for purchasing power parity (PPP)

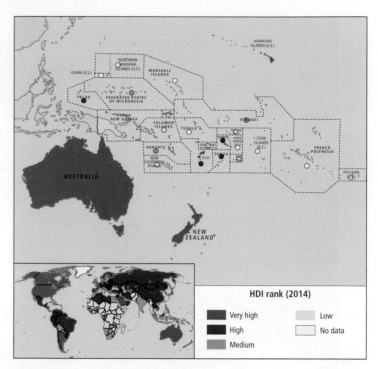

(B) Human Development Index (HDI)

FIGURE 11.22 Maps of human well-being.

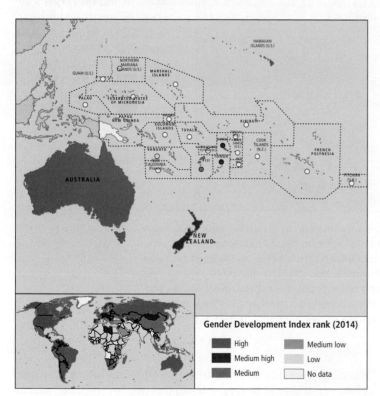

(C) Gender Development Index (GDI)

with rankings closest to 1 have the highest degree of gender equality, meaning that the genders are tending toward equality. Australia (1), New Zealand (2), Samoa (2), Tonga (2), Fiji (3), and Vanuatu (4) are the only countries for which data were available in 2015. Hawaii would share the rank of 1 with the United States.

SOCIOCULTURAL ISSUES

The cultural sea change in Oceania, away from Europe and toward Asia and the Pacific, has been accompanied by new respect for indigenous peoples, and increasing economic interdependence with Asia has diminished historic discrimination against Asians in this region.

ETHNIC ROOTS REEXAMINED

Until very recently, most people of European descent in Australia and New Zealand thought of themselves as Europeans in exile. Many considered their lives incomplete until they had made a pilgrimage to the British Isles or the European continent. In her book *An Australian Girl in London* (1902), Louise Mack wrote: "[We] Australians [are] packed away there at the other end of the world, shut off from all that is great in art and music, but born with a passionate craving to see, and hear and come close to these [European] great things and their home[land]s."

These longings for Europe were accompanied by racist attitudes toward both indigenous peoples and Asians. Most histories of Australia written in the early twentieth century failed to even mention the Aboriginal people, and later writings described them as amoral. At midcentury, there were numerous projects to take Aboriginal children from their parents and acculturate them to European ways in boarding schools known for abuse and brutality. From the 1920s to the 1960s, whites-only immigration policies barred Asians, Africans, and Pacific Islanders from migrating to Australia and discouraged them from entering New Zealand. As we have seen, trading patterns in that era further reinforced connections to Europe.

Weakening of the European Connection

When migration from the British Isles slowed after World War II, both Australia and New Zealand began to lure immigrants from southern and eastern Europe, many of whom had been displaced by the war. Hundreds of thousands came from Greece, Italy, and what was then Yugoslavia. The arrival of these non-English-speaking people initiated a shift toward a more multicultural society. Eventually, the whites-only immigration policy was abandoned and people began to arrive from many places. There was an influx of Vietnamese refugees in the early 1970s during the frantic exodus that followed the U.S. withdrawal from Vietnam. More recently, skilled workers from India and elsewhere in Asia have been helping to meet the growing demand for information technology (IT) specialists throughout the service sector.

As of 2015, 30 percent of the Australian population was foreign-born. The fastest-growing group was from India. Nevertheless, while new immigration policies are increasing the numbers of immigrants from China, Vietnam, and India, people of Asian birth or ancestry remain a small percentage of the total population in both Australia and New Zealand. In 2010, the latest year for which statistics are available, at least 26 percent of Australia's foreign-born residents were from Europe and 18 percent from Asia (**FIGURE 11.23**). New Zealand has similar proportions among its foreign-born residents, and most immigrants continue to come from Europe. Although (because of low birth rates and high immigration rates) Europeans are slowly decreasing as a percentage of the population in both Australia and New Zealand, they are projected to still constitute two-thirds or more of both countries' populations by 2021.

The Social Repositioning of Indigenous Peoples in Australia and New Zealand

Perhaps the most interesting population change in Australia and New Zealand is one of identity. For the first time in 200 years, the number of people in both countries who claim indigenous origins is increasing. Between 1991 and 1996, the number of Australians claiming Aboriginal origins rose by 33 percent. By 2009, the Aboriginal population was estimated at 528,600. In New Zealand, the number claiming Maori background rose by 20 percent (to 652,900) between 1991 and 2009.

These increases are mostly due to changing identities, not to a population boom. More positive attitudes toward indigenous peoples have encouraged the open acknowledgment of Aboriginal or Maori ancestry. Also, marriages between European and indigenous peoples are now more common. As a result, the number of people with a recognized mixed heritage is increasing.

As society has acknowledged that discrimination has been the main reason for the low social standing and impoverished state of indigenous peoples, respect for Aboriginal and Maori culture has also increased. The Aboriginal Australians base their way of life on the idea that the spiritual and physical worlds are intricately related (**FIGURE 11.24**). The dead are present everywhere in spirit, and they guide the living in how to relate to the physical environment. Much Aboriginal spirituality refers to the *Dreamtime*, the time of creation when the human spiritual connections to rocks, rivers, deserts, plants, and animals were made clear. However, very few Aboriginal people continue to practice their own cultural traditions or live close to ancient homelands. Instead, many live in impoverished urban conditions where the guidelines for living a respectful Aboriginal life are breaking down. In New Zealand, where the Maori constitute about 15 percent of the country's population and Auckland has the largest Polynesian population (including Native Maori) of any city in the world, there are now many efforts to bring Maori culture more into the mainstream of national life.

Aboriginal Land Claims

In 1988, during a bicentennial celebration of the founding of white Australia, a contingent of some 15,000 Aboriginal people protested that they had little reason to celebrate. During the same 200 years, they were assumed to have no prior claim to any land in Australia, they had lost basic civil rights, and they had effectively

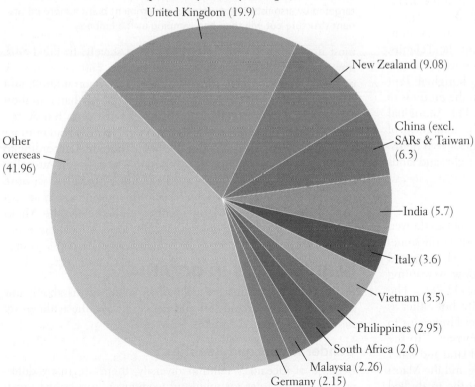

Australia's Cultural Diversity in 2010: Top 10 Countries of Birth (percent by country of origin)

United Kingdom (19.9)
New Zealand (9.08)
China (excl. SARs & Taiwan) (6.3)
India (5.7)
Italy (3.6)
Vietnam (3.5)
Philippines (2.95)
South Africa (2.6)
Malaysia (2.26)
Germany (2.15)
Other overseas (41.96)

FIGURE 11.23 Australia's cultural diversity in 2010. More than 30 percent (7.17 million) of Australia's people were born in other places, making Australia one of the world's most ethnically diverse nations. [*Source consulted:* Australian Bureau of Statistics, "Main Countries of Birth," *Year Book Australia,* 2008, Table 7.39, http://www.ausstats.abs.gov.au/ausstats/subscriber.nsf/0/8D6ED0E197 FE38A6CA2573E7000EC2AD/$File/13010_2008.pdf.]

FIGURE 11.24 Aboriginal rock art. A Mimi spirit painted on a rock at Kakadu National Park, Australia. To the Aboriginal Australians, Mimi spirits are teachers who pass between this world and another dimension via crevices in rocks. They are responsible for many teachings on hunting, food preparation, use of fire, dance, and sexuality.

FIGURE 11.25 A performer at the Aboriginal Tent Embassy in Canberra, Australia. Intermittently since 1972, and continuously since 1992, Aboriginal activists have camped out on the grounds of Australia's House of Parliament in Canberra. Considered by many to be the most effective political action ever taken by Aboriginal Australians, the first tent embassy was a response to the Australian government's denial of land ownership and other land rights to Aboriginal Australians in territories they had continuously occupied for thousands of years. As Aboriginal land rights have gained recognition, the tent embassy has championed other causes, including opposition to mining that threatens Aboriginal communities and cultural sites, as well as the plight of the Aboriginal urban poor, such as the community of Redfern in Sydney. The tent embassy remains controversial, and it has been targeted by arsonists. The Australian government plans a more permanent structure but will then ban camping at the embassy.

been erased from the Australian national consciousness. Into the 1960s, it was even illegal for Aboriginal Australians to drink alcohol.

British documents indicate that during colonial settlement, all Australian lands were deemed to be available for British use. The Aboriginal Australians were thought to be too primitive to have concepts of land ownership because their nomadic cultures had "no fixed abodes, fields or flocks, nor any internal hierarchical differentiation." After increasing pressure from Aboriginal activists, the Australian High Court declared this position void in 1993. After that, Aboriginal groups began to win some land claims, mostly for land in the arid interior previously controlled by the Australian government. **FIGURE 11.25** shows the Aboriginal Tent Embassy in 2012, versions of which have stood on the grounds of Parliament in Canberra for more than 40 years. The Aboriginal Tent Embassy was instrumental in raising public awareness of injustices and remains a national symbol of Aboriginal civil rights. Court cases and other efforts to restore Aboriginal rights and lands continue (see Figure 11.17D).

Maori Land Claims

In New Zealand, relations between the majority European-derived population and the indigenous Maori have proceeded only somewhat more amicably than in Australia. In 1840, the Maori signed the Waitangi Treaty with the British, assuming they were granting only rights of land usage, not ownership (see Figure 11.28B). The Maori did not regard land as a tradable commodity, but rather as an asset of the people, used by families and larger kin groups to fulfill their needs. The geographer Eric Pawson writes: "To the Maori the land was sacred . . . [and] the features of land and water bodies were woven through with spiritual meaning and the Maori creation myth." The British chose to assume that the treaty had given them *exclusive* rights to settle the land with British migrants and to extract wealth through farming, mining, and forestry.

By 1950, the Maori had lost all but 6.6 percent of their former lands to European settlers and the government. Maori numbers had shrunk from a probable 120,000 in the early 1800s to 42,000 in 1900, and the Maori came to occupy the lowest and

most impoverished rung of New Zealand society. In the 1990s, however, the Maori began to reclaim their culture, and they established a tribunal that forcefully advances Maori interests and land claims through the courts. Since then, nearly half a million acres of land and several major fisheries have been transferred back to Maori control. As of 2016, Maori in New Zealand number about 712,000 or 15 percent of the population, but, according to Australian censuses, about 100,000 more Maori live in Australia. Altogether, those who identify as Maori outnumber those who are officially ethnically Maori, which seems to indicate a shift toward popular acceptance of Maori identity. Nonetheless, the Maori still have notably higher unemployment, lower education levels, and poorer health than the New Zealand population as a whole.

GENDER ROLES IN OCEANIA

Perceptions of Oceania are colored by many myths about how men and women are and should be. As always, the realities are more complex than the myths.

Gender Myths and Realities

Because of Oceania's cultural diversity, there are many different acceptable roles for men and women. As mentioned previously, in the Pacific islands, men traditionally were cultivators, deepwater fishers, boat-builders, and masters of seafaring. In Polynesia, men also were responsible for many aspects of food preparation, including cooking. In the modern world, men fill many positions, but idealized male images continue to be associated with vigorous activities.

In Australia and New Zealand, the hypermasculine, white, working-class settler has long had prominence in the national mythologies. In New Zealand, he was a farmer and herdsman. In Australia, he was more often a many-skilled laborer—a stockman, sheep shearer, cane cutter, or digger (miner)—who possessed a laconic, laid-back sense of humor. Labeled a "swagman" for the pack he carried, he went from station (large farm) to station or mine to mine, working hard but sporadically, gambling and drinking, and then working again until he had enough money or experience to make it in the city **(FIGURE 11.26)**.

In cities, the swagman often felt ill at ease and chafed to return to the wilds. Now immortalized in songs ("Waltzing Matilda," for example), novels, and films, these men are portrayed as a rough and nomadic tribe whose social life is dominated by male camaraderie and frequent brawls. No small part of this characterization of males derived from the fact that many of Australia's first immigrants were convicts from the British Isles.

Today, as part of larger efforts to recognize the diversity of Australian society, new ways of life for men are emerging and are breaking down the national image of the tough male loner. Nonetheless, the old model persists and remains prominent in the public images of Australian businessmen, politicians, journalists, and movie stars.

Perhaps the most enduring myth Europeans created regarding Oceania was their characterization of the women of the Pacific islands as gentle, simple, compliant love objects. (Tourist brochures still promote this perception.) There is ample evidence to suggest that Pacific Islanders did have more sexual partners in a lifetime than Europeans did. However, the reports of unrestrained sexuality related by European sailors were no doubt influenced by the exaggerated fantasies one might expect from all-male crews living at sea for months at a time. The notes of Captain James Cook are typical: "No women I ever met were less reserved. Indeed, it appeared to me, that they visited us with no other view, than to make a surrender of their persons." Over the years, such notions about Pacific island women have been encouraged by the paintings and prints of Paul Gauguin **(FIGURE 11.27)**, the writings of novelist Herman Melville (*Typee*), and the studies of anthropologist Margaret Mead (*Coming of Age in Samoa*), as well as by movies and musicals such as *Mutiny on the Bounty* and *South Pacific*.

Historically, in the Pacific islands, women's roles varied considerably from those in Europe, but not in the ways early European explorers imagined. Women often exercised a good bit of power in family and clan, and their power increased with motherhood and advancing age. In Polynesia, a woman could achieve the rank of ruling chief in her own right, not just as the consort of a male chief. Women were primarily craftspeople, but they also contributed to subsistence by gathering fruits and nuts

FIGURE 11.26 Gender and national mythology. Australian wild horse hunter George Girdler epitomizes the hypermasculine, white, working-class settler that is central to the national mythologies of Australia and New Zealand and often serves as a role model for young men.

FIGURE 11.27 *Arearea* **("Amusement") by Paul Gauguin.** In this 1892 painting, Tahitian women are rendered in a European Romantic pastoral style that emphasizes their gentle, compliant demeanor.

and by fishing. And in some places—Micronesia, for example—lineage was established through women, not men (a custom that makes sense when a woman is likely to have more than one sexual partner).

Today, there are some trends toward equality across gender lines throughout Oceania, but the persistence of inequality is tenacious. A striking disparity is emerging: in Australia, New Zealand, Hawaii, and a very few other islands (see Figure 11.22C), women are gaining political and economic empowerment; in Papua New Guinea and most of the Pacific islands, change is much slower.

FORGING UNITY IN OCEANIA

Although wide ocean spaces and the great diversity of languages in the region sometimes make communication difficult, travel, sports, and festivals **(FIGURE 11.28)** are three forces that help bring the people of Oceania closer together.

Languages in Oceania

The linguist David Crystal tells us that each language through its vocabulary and structure offers a unique vision of the world. The Pacific islands—most notably Melanesia—have a rich variety of languages, each presenting a slightly different perspective. Some islands in a single chain can have several different languages. A case in point is Vanuatu, a chain of 80 mostly high volcanic islands to the east of northern Australia (see Figure 11.1). At least 108 languages are spoken by a population of just 180,000—an average of 1 language for every 1600 people. It is easy to see that such remote languages spoken by so few are endangered in a globalizing world.

While language can be both an important part of a community's cultural identity, it can also be a hindrance to cross-cultural understanding. In Melanesia and elsewhere in the Pacific, the need for communication with the wider world is served by a number of **pidgin**

> **pidgin** a language used for trading; one made up of words borrowed from the several languages of people involved in trading relationships

Figure 11.28 LOCAL LIVES: Festivals in Oceania

A A young Aboriginal dancer at the Garma Festival, which is held to encourage the practice of the traditional dance, singing, visual art, and ceremonies of the Yolngu people. The festival is held every year in Arnhem Land, which overlooks the Gulf of Carpentaria in Australia's Northern Territory.

B Waitangi Day in New Zealand, a national holiday that commemorates the signing of a treaty between the indigenous Maori of New Zealand and the British. The long boats shown here are Maori canoes, known as *waka*, and are part of a reenactment of the treaty's signing.

C The aerial theater and comedy troupe Dislocate performs at the Sydney Festival, a 3-week-long international arts festival that is held every January in Sydney, Australia.

languages that are similar enough to be mutually understood. Pidgins are made up of words borrowed from several languages by people involved in trading relationships. Over time, pidgins can grow into fairly complete languages, capable of fine nuances of expression. When a particular pidgin is in such common use that mothers talk to their children in it, then it can literally be called a "mother tongue." In Papua New Guinea, a version of pidgin English is the official language. Increasingly, English is the lingua franca for everyone in Oceania and, as such, it threatens the 1300 indigenous languages of the region.

Interisland Travel

One way in which unity is manifested in Oceania is interisland travel. Today, people travel in small planes from the outlying islands to hubs such as Fiji, where jumbo jets can be boarded for Auckland, Melbourne, and Honolulu. Cook Islanders call these little planes "the canoes of the modern age," and people travel for many reasons. Dancers from across the region attend the annual folk festival in Brisbane; businesspeople from Kiribati, Micronesia, can fly to Fiji to take a short course at the University of the South Pacific; a Cook Islands teacher can take graduate training in Hawaii; and sports fans can visit multiple locations over time.

Sports as a Unifying Force

Sports and games are a major feature of daily life throughout Oceania. The region has shared sports traditions with and borrowed them from cultures around the world. Surfing evolved in Hawaii and, like outrigger sailing and canoeing, derives from ancient navigational customs that matched human wits against the power of the ocean. On hundreds of Pacific islands and in Australia and New Zealand, rugby, volleyball, soccer, and cricket are important community-building activities. Baseball is a favorite in the parts of Micronesia that were U.S. trust territories. Women compete in the popular sport of netball (similar to basketball but without a backboard).

Pan-Oceania sports competitions are the single most common and resilient link among the countries of the region. Attendance at regional sports events is so desirable that low-income islanders will hold yard sales and raffles to amass the cash necessary to make the trip. The centrality of such competitions in daily life encourages regional identity and provides opportunities for ordinary citizens to travel extensively around the region and to other parts of the world.

The *haka* (FIGURE 11.29) is an example of how, in the post-colonial modern era, indigenous culture in Oceania is being revived, celebrated, and appropriated in new places by those who wish to project a multicultural image. The haka is a highly emotional and physical traditional dance performed by the Maori to motivate fellow participants and to intimidate opponents before a confrontation or major event, even an event such as a wedding. Dances like this have historically been a part of many cultures in the islands of Oceania, but the haka has now become an integral part of rugby, the region's most popular sport (FIGURE 11.30). Before almost every international match for the past century, the All Blacks (the New Zealand men's rugby

FIGURE 11.29 The haka, a Maori tradition. A haka performed by the New Zealand men's rugby team, the All Blacks, before a match against Australia.

team) have performed the haka: chanting, screaming, jumping, stomping their feet, poking out their tongues, widening their eyes to show the whites, and beating their thighs, arms, and chests.

Outside Oceania, those who perform the haka include the rugby teams at Jefferson High in Portland, Oregon, and Middlebury College in Vermont, and the football teams at Brigham Young University and the University of Hawaii. All these teams have players who are of Polynesian heritage. Most practitioners speak of the haka as filling them with the necessary exuberance, aggression, and spirituality to play a vigorous and successful game. To see Maori-created videos of the haka, go to http://www.youtube.com and type in "Wedding Haka - Subtitled & translated."

THINGS TO REMEMBER

- Oceania's long-standing cultural and economic links to Europe are being challenged by reinvigorated native traditions and identities and by economic globalization, which is strengthening the region's links to Asia.

- The number of Asian immigrants in Australia and New Zealand has been increasing over the past two decades, while the number of European immigrants has been on the decline. Asians, however, still make up only a small minority of the populations of both Australia and New Zealand.

- Indigenous people throughout the region are asserting their rights and finding empathy among their fellow citizens.

- Idealized gender roles have characterized perceptions of the region: the hypermasculine swagman or jack-of-all-trades workman for Australia and New Zealand, and the beautiful and compliant seductress for the Pacific islands. Both are limiting and untrue.

- Sports and festivals are unifying forces for the region, inspiring fundraisers that allow ordinary citizens to travel to games, thus reinforcing regional and ethnic identity and social cohesion.

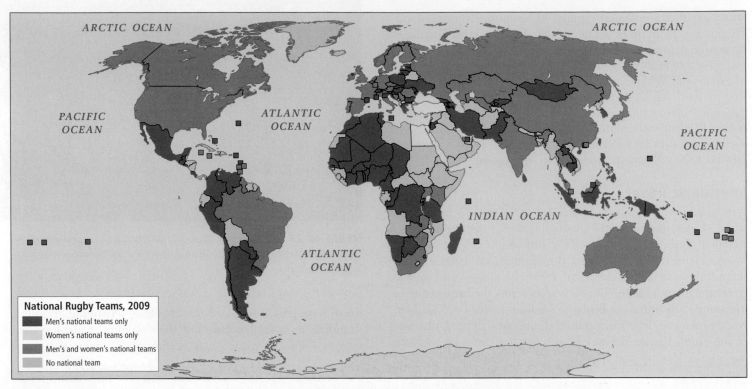

FIGURE 11.30 Rugby around the world. In more than 136 countries, women, men, boys, and girls play rugby. All of these countries have men's national rugby teams, and 58 women's national teams. The men's World Cup rugby competition began in 1987, the women's in 1991. In April 2010, New Zealand boasted the top team for both men and women, but the ranking of the men's teams can change weekly. [*Source consulted:* http://www.irb.com/aboutirb/organisation/index.html.]

GEOGRAPHIC THEMES: Oceania Review and Self-Test

1. Environment: Oceania faces a host of environmental problems and public awareness of environmental issues is keen. Global climate change, primarily warming, has brought about rising sea levels and increasingly variable rainfall. Other major threats to the region's unique ecology have come from the introduction of nonnative species and the expansion of herding, agriculture, fishing, fossil fuel extraction, waste disposal, and human settlements.

• Explain why some parts of Oceania are more vulnerable to the impacts of climate change than others? Which parts of the region contribute the most and least to greenhouse gas emissions?

• Why are coral reefs important ecologically, and how are they threatened by climate change?

• How have European farm animals and crops introduced in Australia and New Zealand affected the economies of those countries? How would you assess the purpose and success of the Dingo Fence? How do foreign patterns of consumption affect this region, and what are the threats to Oceania's food security?

2. Globalization and Development: Globalization, coupled with Oceania's stronger focus on neighboring Asia (rather than long-time connections with Europe and North America), has transformed patterns of trade and economic development across Oceania. These changes are driven largely by Asia's growing affluence, its enormous demand for resources, and its similarly massive production of manufactured goods.

• What does Asia buy from the Pacific islands, and what are Asia's trade connections to Australia and New Zealand?

• How does the loss of preferential trading ties with Europe affect the region's workers and tax revenues?

• What sector is the basis of most economies in this region?

• Explain why many Pacific Islanders can be said to have subsistence affluence despite having rather low monetary incomes.

3. Power and Politics: Stark divisions have emerged in Oceania over definitions of democracy—the system of government that dominates in New Zealand, Australia, and Hawaii—versus the Pacific Way, a political and cultural philosophy based on the traditional cultures of the Pacific islands. The global refugee crisis tests this region's ability to maintain its reputation as a humanitarian refuge and zone of opportunity that maintains social cohesion.

• How do ideas about the proper exercise of political power in the Pacific islands differ from those in Australia and New Zealand and some Europeanized islands?

• Why and when did the political philosophy known as the Pacific Way develop?

• In what situations has the Pacific Way been invoked? To what extent could the Pacific Way be construed as just a different version of democracy?

• How does the *Pacific Solution* to the asylum-seeker crisis in Australia fit philosophically with either Western democracy or the Pacific Way?

4. Urbanization: Oceania is only lightly populated but it is highly urbanized. The shift from extractive economies to service economies is a major reason for the urbanization of the wealthiest parts of Oceania (Australia, New Zealand, Hawaii, Guam), where 80 to 100 percent of the population lives in cities. These trends are weakest in Papua New Guinea and many smaller Pacific islands.

• Why, despite its low average population densities, is the region so highly urbanized?

• Which parts of Oceania are both densely populated and threatened by climate change?

• Explain how it can be said that parts of Oceania are both highly urbanized and lightly populated.

• What kind of cultural impact are immigration and emigration having in Oceania?

5. Population and Gender: In this largest but least populated world region, there are two main patterns relevant to population and gender. Australia, New Zealand, and Hawaii have older and more slowly growing populations, and offer relatively more opportunities for women. The Pacific islands and Papua New Guinea have much more rural, younger, and rapidly growing populations, where women play a central role in family and community but enjoy fewer opportunities than men as economies modernize.

• Contrast the population growth rates in Australia and New Zealand with those in Oceania.

• In which parts of this region is the aging of the population of most concern and why?

• Which parts of this region were among the first in the world to grant European women the right to vote?

• Describe the possible changing gender roles of Pacific island women over the course of their lifetimes.

• What are some physical, social, and economic reasons for Oceania's relatively small overall population?

Critical Thinking Questions

1. What has been the impact of European colonialism on the demography and settlement patterns of Aboriginal Australians and Maori peoples?

2. What do you find most remarkable about the navigation skills of Pacific Islanders? How might recognition of these skills contribute to general multicultural understanding?

3. Describe and explain the changing orientation of Oceania to Europe, the Americas, and Asia over the years since the 1500s.

4. As Australia and New Zealand have moved away from their intense cultural and economic involvement with Europe, new policies and attitudes have evolved to facilitate their deeper involvement with Asia. If you were a college student in Australia or New Zealand, how might you experience these changes? Think about fellow students, career choices, language learning, and travel choices.

5. Discuss the emerging cultural identity of the Pacific islands, taking note of the extent to which Australia and New Zealand share or do not share in this identity. What factors are helping to forge a sense of unity across Polynesia and beyond? (First, review the spatial extent of Polynesia.)

6. Discuss the many ways in which Asia has historic, and now increasingly economic, ties to Oceania. In your discussion, include patterns of population distribution, mineral exports and imports, technological interactions, and tourism.

7. To what extent can the countries of Oceania exercise control over the future as the climate changes?

8. Australia and New Zealand differ from each other physically. Compare and contrast the two countries in relation to water, vegetation, and prehistoric and modern animal populations.

9. Indigenous peoples worldwide are taking action to safeguard their cultures, rights, and access to land and resources. Discuss how the indigenous peoples of Australia, New Zealand, and the Pacific islands are serving as leaders in this movement and what measures they are taking to reconstitute a sense of cultural heritage.

10. How is tourism both boosting economies and straining environments and societies throughout the Pacific islands? Describe the solutions that are being proposed to reduce the negative impacts of tourism.

11. Compare how women do and do not have political and economic power in Australia, New Zealand, Papua New Guinea, and the Pacific islands.

12. Compared with other regions, Australia and New Zealand are somewhat unusual in having become broadly prosperous on the basis of raw materials exports. How would you explain this achievement?

Chapter Key Terms

Aboriginal Australians 633
Asia Pacific Economic Cooperation (APEC) 641
atoll 623
endemic 626
Gondwana 622
Great Barrier Reef 622
hot spots 623

invasive species 629
makatea 623
Maori 623
marsupials 626
Melanesia 634
Melanesians 634
Micronesia 635
MIRAB economy 639

monotremes 626
Pacific Solution 644
Pacific Way 642
pidgin 654
Polynesia 635
Roaring Forties 623
subsistence affluence 639
Trans-Pacific Partnership (TPP) 641

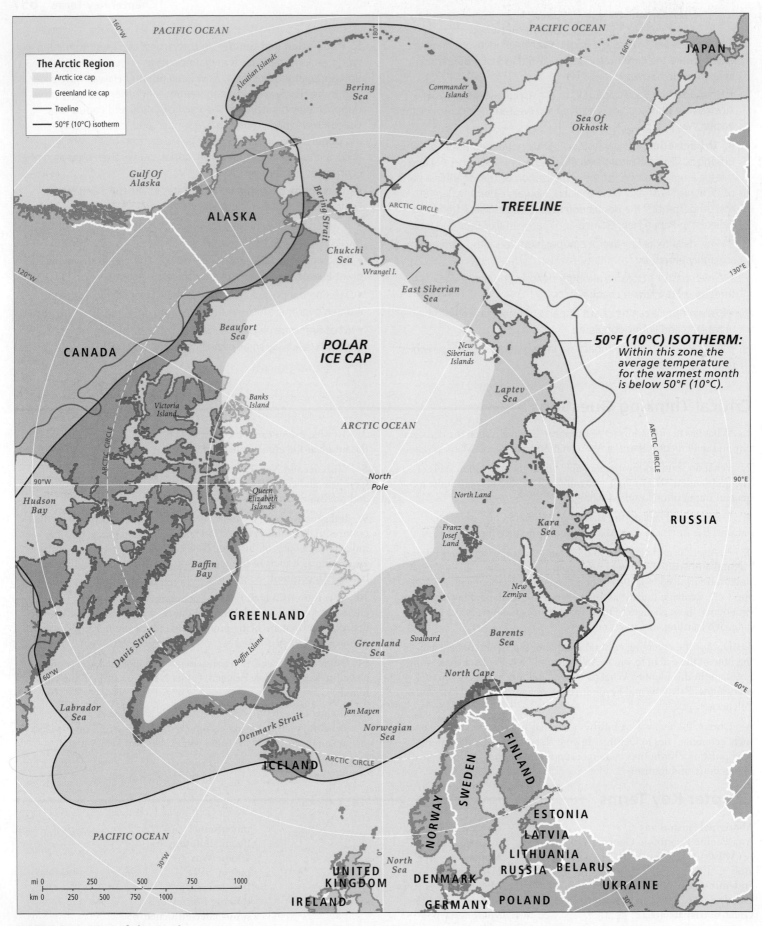

The Arctic Region

Arctic ice cap
Greenland ice cap
Treeline
50°F (10°C) isotherm

PACIFIC OCEAN

PACIFIC OCEAN

JAPAN

Aleutian Islands

Bering Sea

Commander Islands

Sea Of Okhostk

160°W

180°

160°E

Gulf Of Alaska

ALASKA

ARCTIC CIRCLE

TREELINE

Bering Strait

Chukchi Sea

Wrangel I.

East Siberian Sea

130°E

120°W

Beaufort Sea

POLAR ICE CAP

New Siberian Islands

50°F (10°C) ISOTHERM:
Within this zone the average temperature for the warmest month is below 50°F (10°C).

CANADA

Banks Island

Laptev Sea

Victoria Island

ARCTIC CIRCLE

ARCTIC OCEAN

North Land

ARCTIC CIRCLE

90°W

90°E

Hudson Bay

Queen Elizabeth Islands

North Pole

Kara Sea

RUSSIA

Franz Josef Land

Baffin Bay

New Zemlya

Baffin Island

GREENLAND

Svalbard

Barents Sea

Davis Strait

Greenland Sea

Labrador Sea

Jan Mayen

North Cape

60°W

60°E

Denmark Strait

Norwegian Sea

FINLAND

ICELAND

ARCTIC CIRCLE

SWEDEN

NORWAY

PACIFIC OCEAN

ESTONIA

LATVIA

LITHUANIA

30°W

North Sea

RUSSIA

BELARUS

0°

UNITED KINGDOM

DENMARK

POLAND

UKRAINE

30°E

IRELAND

GERMANY

mi 0   250   500   750   1000

km 0   250   500   750   1000

FIGURE E1.1 Map of the Arctic.

Polar Regions

GEOGRAPHIC THEMES

After you read this epilogue, you will be able to discuss the following geographic themes as they relate to the five thematic concepts:

1. Environment: Climate change is transforming environments more profoundly at the poles than at any other place on Earth. Temperatures far above normal are melting polar glaciers and sea ice rapidly, resulting in habitat loss for many species.

2. Globalization and Development: In the Arctic, military spending, mining, fishing, and hunting have made up the economic base for a long time; however, a new range of industries could appear in the Arctic because the Arctic Ocean will likely be ice-free in the summer by as early as 2030. The international fishing industry has already begun to exert pressure on Antarctica, and the human population of Antarctica may increase.

3. Power and Politics: The polar regions are a growing source of conflict. Multiple nations with competing territorial claims are interested in controlling the rapidly changing and increasingly valuable lands, seas, and resources of this region.

4. Urbanization: Urban centers are few and far between in the Arctic and completely absent in the Antarctic. Rural areas are generally depopulating in the Arctic, as people move to the cities or leave the region. Meanwhile, the largest Arctic cities are also in decline due to changes in Russia's economy; however, several smaller Scandinavian cities are growing because of renewed interest in the area.

5. Population and Gender: Polar populations are stable on the whole, with population growth in Scandinavia and North America balancing population decline in Russia. Human development is generally lower in the Arctic regions of countries, relative to national averages. There is also a complex gender imbalance at the poles.

The Polar Regions

Despite their remote locations and few human inhabitants, the polar regions (including the Arctic and Antarctica) are undergoing changes of all types at an accelerating rate **(FIGURE E1.1)**. Climate change is transforming areas once covered in ice into open ocean and land, bringing interest both in trade across potentially more navigable seas and in resource extraction from newly exposed land. Politics are also shifting as nations compete for control of these rapidly changing regions. The indigenous populations, long accustomed to adapting to harsh environments, face not only shifting climates but a greater presence of national governments and corporations.

GLOBAL PATTERNS, LOCAL LIVES Mikkel scans the hills of Finnmark County, Norway, for signs of his herd of 600 reindeer **(FIGURE E1.2)**. Some of the young have not been getting enough to eat, and with this February's temperatures dropping down to 5°F (−15°C) in the mornings, he may have to bring some home with him to nurse them back to health. The cold brings more work but is actually a blessing after years of unusually warm winters, which have reduced the availability of lichen and moss that reindeer graze on. Spying a group of reindeer about to enter into some woods, Mikkel jumps on his snowmobile, hoping to head them off before they encounter the lynx and wolverines, hiding among the trees, who have killed 20 of his herd just this year.

Lately, Mikkel worries less about climate change and predators than about all the new development in Finnmark County. The government recently approved a new open-pit copper mine, the owners of which plan to dump mine waste into a nearby fjord where some of Mikkel's friends fish for salmon. Meanwhile, new windmill farms are occupying areas where reindeer used to give birth to calves and graze. There are also more tourists eager to experience the Arctic (which has brought new roads and resorts that take away herding land), retirees who gather wild berries that Mikkel's family depends on, and hunters who kill wild birds that Mikkel must now work harder to hunt.

FIGURE E1.2 Sami reindeer herding in northern Norway. With warmer temperatures reducing their herd's access to food, and with more land taken for new industrial development and tourism, many Sami reindeer herders are concerned that their way of life is threatened.

Mikkel takes a dim view of the tourist industry's appropriation of the culture of his people, the indigenous Sami, who spread across northern Norway, Sweden, Finland, and Russia. While Mikkel and his kin struggle through the cold, barely getting by in some years, tours advertise a happy, carefree image of Sami life, with trips on sleds drawn by reindeer driven by Sami guides, some of whom aren't actually Sami. Christmas brings further annoyance, as the town of Rovaniemi, Finland (located just south of the Arctic Circle), advertises itself as the "Official Hometown of Santa Claus," with pageantry in which Santa's helpers are dressed in imitations of Sami dress (worn by non-Sami) and imitations of Sami crafts are sold in gift shops, alongside playing cards depicting Sami as drunken fools. [Sources: *Intelligent Life Magazine;* IPinCH; the *Guardian.* For detailed source information, see Text Sources and Credits.] ■

What Makes the Arctic and Antarctic Regions?

The Arctic is most often defined as the areas north of the Arctic Circle, where the Sun never sets for part of the summer and never rises for part of the winter. Some consider the Arctic to be the area where the average temperature for the warmest month is below 50°F (10°C), as indicated by the red line on the map (see E1.1). Humans have occupied the Arctic for more than 40,000 years, mostly as nomadic hunters and gatherers operating on land, sea, or both. The Arctic lands are all part of larger countries with populations and governments centered in the south. Only a few large Arctic cities were built as military outposts, as service providers for extractive industries, and as centers for government administration. The Antarctic, the Southern Hemisphere's counterpart to the Arctic, is usually defined by the continent of Antarctica. There is no indigenous human population, and its current inhabitants are a few thousand scientists who live at scattered research outposts for a few months or years at a time.

PHYSICAL GEOGRAPHY

The intense cold of the polar regions result in ice caps that cover much of the Arctic Ocean and Antarctica. Arctic landforms are highly varied, centered on an ocean surrounded by coastal lowlands and a few mountain ranges. The tallest of the mountains are in Greenland, where they reach as high as 12,119 feet (3694 meters). There are 12 major river systems that drain into the Arctic Ocean, some of which drain vast areas south of the Arctic Circle. The largest Arctic rivers are found in Russia, where the Yenisei River has a larger discharge than the Mississippi River. Parts of the coastal Arctic are known for a landform called the **fjord,** a long, narrow inlet formed by glacial erosion **(FIGURE E1.3).** Fjords are most common in northern Scandinavia, Greenland, Alaska, and northern Canada; there are also a few in Antarctica.

Antarctica has a similarly wide range of landforms. Underneath a sheet of ice that averages about 1 mile (1.6 kilometers) thick, Antarctica is a complex archipelago of islands. There are a few ice-free areas near the coast that

fjord a long, narrow inlet formed by glacial erosion

FIGURE E1.3 A fjord in Norway during the winter.

together make up about 2 percent of Antarctica. The Antarctic ice cap holds roughly 90 percent of the world's ice and 70 percent of its fresh water. With so much water locked in the form of ice, there are few rivers in Antarctica, and the longest runs for just under 20 miles (32 kilometers). Antarctica's tallest mountain, Mount Vinson, reaches 16,050 feet (4892 meters).

Polar climates are characterized by extremely low solar radiation and relatively little precipitation. Because the waters of the Arctic Ocean never get below 28°F (−2°C) and rarely above 46°F (8°C), they give the Arctic a more moderate climate, with warmer winters relative to some places to the south. In fact, the lowest temperatures in the Northern Hemisphere are in the area between Verkhoyansk and Oymyakon in northern Russia (well south of the Arctic Circle), where the average daily low temperatures are below −53°F (−47°C) from December through February. In Antarctica, the continent's landmass and ice cap cool down more severely in the winter, resulting in the lowest temperatures ever recorded on Earth (−128°F, −89°C).

CLIMATE AND VEGETATION

Bitter cold, snow and ice, lack of sunlight, and lack of moisture severely limit vegetation in the polar regions. Arctic vegetation varies between mixed taiga and tundra (see Chapter 5) to pure tundra and Arctic desert. Soil that is permanently frozen a few feet below the surface (permafrost) limits the growth of trees and other plants with deep roots. The vegetation is mostly lichen and mosses that stay close to the ground to conserve warmth and moisture and can survive more than a year of being covered by snow. Antarctica has vegetation only on the 2 percent of its area that is free from snow and ice and where only a few grasses, lichens, and mosses survive. Despite holding so much of Earth's water, Antarctica is a desert; most of its interior areas receive only 2 inches (5 centimeters) of precipitation each year, less than the Sahara.

ENVIRONMENT

Environment: Climate change is transforming environments more profoundly at the poles than at any other place on Earth. Temperatures far above normal are melting polar glaciers and sea ice rapidly, resulting in habitat loss for many species.

The increase in the intensity of climate change at the poles is called **polar amplification,** which scientists think is related mainly to reductions in sea ice and snow cover. Perhaps the most noticeable effect of climate change is the increase in Arctic temperatures and the resultant melting of Arctic sea ice, which has decreased by roughly 8 percent in the past 30 years. Recent research suggests that the Arctic Ocean may be ice-free during summers by as early as 2030. These changes are already challenging some Arctic species that hunt or give birth on the ice, such as polar bears, walruses, certain kinds of seals, and some seabirds. Algae are also being affected; some native species are giving way to invaders from the south, which has some oceanographers concerned because algae are the foundation of Arctic food chains.

With regard to temperatures and melting, the situation is more complex in Antarctica than in the Arctic. Recent studies suggest that while overall temperatures are increasing on land and in the seas, ice and snow are melting in some places and accumulating in others **(FIGURE E1.4)**. Temperatures are increasing in the seas surrounding Antarctica, in west Antarctica, and on the Antarctic peninsula, where coastal ice over land is melting and sea ice is now present for shorter times each year. However, in eastern Antarctica, temperatures on land are actually declining and sea ice is increasing. Recent research by NASA suggests that temperatures are rising in the continental interior of Antarctica. This is increasing the potential for the air to hold moisture,

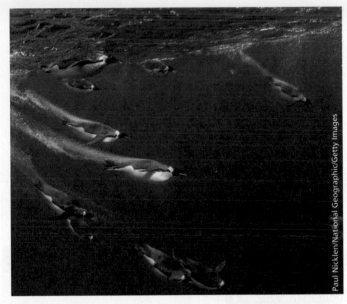

FIGURE E1.5 Emperor penguins dive beneath sea ice in Antarctica. Emperor penguins live only in Antarctica and are proving extremely sensitive to climate change. In warmer years, declines in sea ice—the penguins' ideal hunting habitat—have resulted in widespread starvation among the penguins. In colder-than-normal years, few penguin chicks hatch.

resulting in higher snowfall that may be offsetting the loss of ice along the coast, contributing to an overall gain of ice on the continent.

As a result of these changes, Antarctic species that breed and hunt on sea ice, such as several types of penguin, will decline in some areas and possibly increase in others **(FIGURE E1.5)**. In the seas surrounding Antarctica, rising temperatures have more uniform impacts, such as reductions in **krill,** the tiny crustaceans on which whales, seals, penguins, and fish feed. This is already resulting in declining seal populations in some areas.

Climate change at the poles may also result in increased greenhouse gas (GHG) emissions because methane, a powerful GHG, is released when permafrost thaws and polar oceans warm. Some studies estimate that the amount of methane trapped in permafrost under the polar oceans and frozen in the oceans (as methane hydrates) could be large enough to trigger much more rapid warming of Earth's climate if released. Dramatic increases in methane release from the Arctic have been documented in recent years, though some studies suggest that certain bacteria in permafrost soils on land can consume some of this methane.

polar amplification the increase in the intensity of climate change at the poles

krill tiny crustaceans that whales, seals, penguins, and fish feed on

FIGURE E1.4 Scientists measure the sea level at Antarctica's Ross Sea. While most sea level rises predicted for the near future are caused by thermal expansion of the oceans, sea levels could rise 200 feet (60 meters) if Antarctica's massive glaciers were all to melt.

THINGS TO REMEMBER

GEOGRAPHIC THEME 1 • **Environment** Climate change is transforming environments more profoundly at the poles than at any other place on Earth. Temperatures far above normal are melting polar glaciers and sea ice rapidly, resulting in habitat loss for many species.

- The Arctic is most often defined as the areas north of the Arctic Circle, where the Sun never sets for part of the summer and never rises for part of the winter. Some consider the Arctic to be the area where the average temperature for the warmest month is below 50°F (10°C).

- The Antarctic, the Southern Hemisphere's counterpart to the Arctic, is usually defined by the continent of Antarctica.

- Bitter cold, snow and ice, lack of sunlight, and lack of moisture severely limit vegetation in the polar regions.

- The increase in the intensity of climate change at the poles is called polar amplification, which scientists think is related mainly to reductions in sea ice and snow cover.

GLOBALIZATION AND DEVELOPMENT

GEOGRAPHIC THEME 2

Globalization and Development: In the Arctic, military spending, mining, fishing, and hunting have made up the economic base for a long time; however, a new range of industries could appear in the Arctic because the Arctic Ocean will likely be ice-free in the summer by as early as 2030. The international fishing industry has already begun to exert pressure on Antarctica, and the human population of Antarctica may increase.

Warming at the poles may bring more potential for economic development from large-scale operations, particularly new mining or energy (mainly oil and natural gas) operations that will be better able to navigate polar oceans that will not be clogged by sea ice and build roads to areas that will no longer have impassable deep snow or glaciers. Ice-free shipping routes through the Arctic Ocean have long been sought after because they offer shorter travel distances between the Atlantic Ocean and Pacific Ocean, requiring as much as 40 percent less fuel than current sea routes involving the Suez or Panama canals. However, the persistence of summer sea ice until at least 2030, along with the infrastructure needed to support Arctic Ocean shipping routes, means that the routes are not likely to be in service until 2040 **(FIGURE E1.6)**.

FIGURE E1.6 **A nuclear-powered ice breaker cuts a path through the Kara Sea for a convoy of ships.**

Sovfoto/UIG via Getty Images

Paul Sutherland/National Geographic/Getty Images

FIGURE E1.7 **One of the thousands of fishing vessels active in the Southern Ocean.** Species like the Patagonian toothfish (marketed as Chilean sea bass) are severely overfished. Tiny shrimp known as krill, which are used in fish farms and health food supplements, are also being overfished.

Climate change is also making development and even survival more difficult for smaller-scale operations. Many people who hunt, fish, or herd in the Arctic are having a much more difficult time as species move to new areas and as new techniques to exploit them are required by warmer temperatures. For example, some hunting grounds once accessible year-round by snowmobiles now require expensive paved roads for part of the year, especially during the spring thaw that turns unpaved roads into impassable mud pits (see Chapter 5 for a discussion of the rasputitsa).

Antarctica is also experiencing more pressure for resource extraction. The international fishing industry is rapidly increasing its catches in the Southern Ocean, though this is mostly due to a growing global demand for fish, not due to changes related to the rise in temperatures **(FIGURE E1.7)**. Meanwhile, larger human populations in Antarctica could develop with the renegotiation of the Antarctic Treaty of 1959, which currently bans all resource extraction. The treaty won't be formally reviewed until 2048, but more than 30 countries are already building up research presences on the continent, hoping to secure claims to mineral and energy resources once they become available. These resources include oil, natural gas, and a wide variety of metallic ore deposits located along the Antarctic coastline.

THINGS TO REMEMBER

GEOGRAPHIC THEME 2 • **Globalization and Development** In the Arctic, military spending, mining, fishing, and hunting have made up the economic base for a long time; however, a new range of industries could appear in the Arctic because the Arctic Ocean will likely be ice-free in the summer by as early as 2030. The international fishing industry has already begun to exert pressure on Antarctica, and the human population of Antarctica may increase.

• Larger human populations could develop in Antarctica with the renegotiation of the Antarctic Treaty of 1959, which currently bans all resource extraction but will be formally reviewed in 2048.

• Climate change is making development and even survival more difficult for smaller-scale operations.

POWER AND POLITICS

GEOGRAPHIC THEME 3

Power and Politics: The polar regions are a growing source of conflict. Multiple nations with competing territorial claims are interested in controlling the rapidly changing and increasingly valuable lands, seas, and resources of this region.

Eight states have territorial claims in the Arctic, and all are members of an intergovernmental forum known as the *Arctic Council*. Established in 1996 to promote cooperation, coordination, and interaction, the Arctic Council specifically avoids geopolitical and security issues, which may limit its utility in the future. One source of contention is the provisions of the United Nations Convention on the Law of the Sea (UNCLOS, see Chapter 11), which allows for countries to claim exclusive economic development rights for 200 miles (322 kilometers) out from their coastlines. Currently there are disputes between Canada, Denmark, Russia, and the United States over several islands and the surrounding seas to which UNCLOS would grant owner's rights.

The Arctic has a long history of strategic and war-related activity, and all Arctic nations are expanding their military presence here. The largest Arctic cities all have a significant military component to their economy, and even the most remote parts of the Arctic had some level of military activity during the Cold War between the United States and the Soviet Union. This included testing of nuclear weapons, constant patrols by nuclear-armed jets and submarines, and extensive military exercises.

Antarctica has a whole different set of nations with territorial claims, several of which overlap **(FIGURE E1.8)**. While formal military operations are not allowed, according to the Antarctic Treaty of 1959, more than 30 national militaries help maintain their home countries' research stations **(FIGURE E1.9)**.

THINGS TO REMEMBER

GEOGRAPHIC THEME 3 • **Power and Politics** The polar regions are a growing source of conflict. Multiple nations with competing territorial claims are interested in controlling the rapidly changing and increasingly valuable lands, seas, and resources of this region.

• All Arctic nations are expanding their polar military presence. The Arctic has a long history of war-related activity.

URBANIZATION

GEOGRAPHIC THEME 4

Urbanization: Urban centers are few and far between in the Arctic and completely absent in the Antarctic. Rural areas are generally depopulating in the Arctic, as people move to the cities or leave the region. Meanwhile, the largest Arctic cities are also in decline due to changes in Russia's economy; however, several smaller Scandinavian cities are growing because of renewed interest in the area.

Throughout the Arctic, many indigenous people are moving to cities as their livelihoods, which have been based on hunting, fishing, or herding, have become more difficult because of climate change.

The largest Arctic cities are in Russia, and were founded at the direction of the government in Moscow in the twentieth century to fulfill both strategic and resource-extraction needs. The biggest is Murmansk (population 300,000), which was created by the Russian Empire in 1915 as an ice-free port that would enable Russia's allies in World War I to supply it with military equipment and ammunition **(FIGURE E1.10)**. Still a major military outpost, Murmansk's population has plummeted since the fall of the Soviet Union, as the government has been less able to support the city's industries. People are also leaving Norilsk (population 175,000), a city that was founded in the late 1920s as a state-run mining **gulag camp** that depended on prisoners supplied by the government for much of its labor. Vorkuta (population 70,000) was started in the 1930s as a coal-mining gulag camp and is also shrinking.

The largest Arctic city outside of Russia is Tromsø, Norway (population 72,000), which was a fortress town at least 1000 years ago, populated by Vikings and Sami. In contrast to the Arctic cities in Russia, Tromsø and other Arctic cities in Scandinavia are growing moderately due to increased government and private investment in Arctic resource extraction. The Finnish city of Rovaniemi (population 61,000), which lies just 6 miles (10 kilometers) south of the Arctic Circle, is an administrative and commercial center for Finland's Arctic. Both Tromsø and Rovaniemi have major universities and government agencies. Outside of Russia and Scandinavia, the next largest Arctic town is Barrow, Alaska (population 4000), which serves the surrounding oil and gas industry on the north slope of Alaska.

gulag camp a settlement set up by the central government of the USSR, in which prisoners were sent to state-run enterprises to provide labor

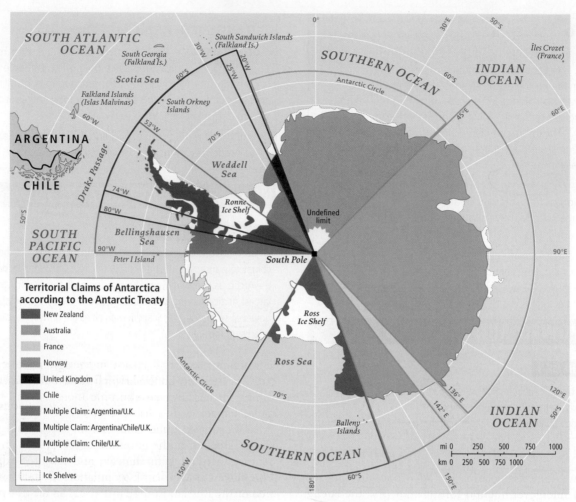

FIGURE E1.8 Territorial claims on Antarctica.

THINGS TO REMEMBER

GEOGRAPHIC THEME 4 • **Urbanization** Urban centers are few and far between in the Arctic and completely absent in the Antarctic. Rural areas are generally depopulating in the Arctic, as people move to the cities or leave the region. Meanwhile, the largest Arctic cities are also in decline due to changes in Russia's economy; however, several smaller Scandinavian cities are growing because of renewed interest in the area.

POPULATION AND GENDER

GEOGRAPHIC THEME 5

Population and Gender: Polar populations are stable on the whole, with population growth in Scandinavia and North America balancing population decline in Russia. Human development is generally lower in the Arctic regions of countries, relative to national averages. There is also a complex gender imbalance at the poles.

The Arctic is home to roughly 2.5 million people, most of whom live in Russia. Here populations are declining due to emigration of urbanites who came to the Arctic in the twentieth century

(mainly from western Russia). Indigenous populations, though, are growing throughout the Arctic because of their relatively high childbirth rate. For example, women in Nunavut, Canada, give birth to 3 children, on average.

The gender imbalance in the polar regions is complex. Arctic North America, Greenland, and Scandinavia generally have more men than women, while the Russian Arctic has more women than men. Antarctica has many more men than women, due to male dominance in the scientific disciplines practiced at the research stations there.

In almost every case, Arctic places have an above-average ratio of men to women with respect to national averages. For example, the gender ratio is 94:100 (94 males for every 100 females) in Murmansk, Russia, while in Russia as a whole the gender ratio is 86:100. The tendency for Arctic areas to have above-average male populations is usually explained by its economic base in resource-extraction industries, which tend to employ more men, as well as by more traditional patriarchal power structures, which extend male dominance into higher-paying government and service sector jobs. These factors tend to encourage women seeking more opportunities for education and career advancement to migrate (Russia's lower ratio of men to women, relative to other parts of the Arctic, is related to higher mortality of men due to cardiovascular disease, work-related hazards, and higher rates of alcoholism and suicide).

FIGURE E1.9 **The Amundsen-Scott South Pole Station.** Just outside the dome-shaped Amundsen-Scott station—a U.S. research station—fly the flags of the first countries to sign the Antarctic Treaty of 1959, which bans all military and resource extraction–related activity, making the continent a scientific and nature reserve.

Among indigenous populations that consume large amounts of Arctic fish, environmental pollution may also be introducing a gender imbalance, with twice as many girls as boys being born in some villages. Studies suggest that chemicals used all over the world in pesticides, electrical transformers, and flame retardants act as *endocrine disruptors* that, when they reach high enough concentrations, mimic human hormones that switch the gender of a fetus from male to female during pregnancy. These chemicals become concentrated in Arctic fish through several linked processes. First, they make their way to the Arctic through wind and ocean currents. Then, a process of **bioaccumulation** takes place as organisms absorb the toxins faster than their bodies can excrete them. On an ecosystem scale, a process called **biomagnification** occurs as the concentration of toxins increases at successive levels of the food chain. For example, PCBs, a toxic class of chemicals used in electrical equipment, exist in fairly low concentrations in the bodies of Arctic plankton, but the smaller fish that eat the plankton have much higher concentrations, the larger fish

FIGURE E1.10 **Reindeer racing at a winter festival in Murmansk, Russia.**

that eat the smaller fish have even higher concentrations, and the humans who eat the larger fish have the highest concentrations of PCBs. The gender imbalance that endocrine disruptors are introducing to some indigenous Arctic communities is renewing efforts to make governments regulate the use of chemicals more carefully.

> **bioaccumulation** a process in which organisms absorb toxins faster than their bodies can excrete them
>
> **biomagnification** a process in which the concentration of toxins increases at successive levels of the food chain

---

THINGS TO REMEMBER

GEOGRAPHIC THEME 5 • **Population and Gender** Polar populations are stable on the whole, with population growth in Scandinavia and North America balancing population decline in Russia. Human development is generally lower in the Arctic regions of countries, relative to national averages. There is also a complex gender imbalance at the poles.

• The Arctic is home to roughly 2.5 million people, most of whom live in Russia, where populations are declining due to the emigration of urbanites who came to the Arctic in the twentieth century, mainly from western Russia.

SOCIOCULTURAL ISSUES

The vignette at the beginning of this epilogue hints at some of the issues facing indigenous people in this vast, sparsely populated region. Climate change is just one of many threats to indigenous cultures, which for more than 500 years have endured many unwelcome influences of national governments, corporations, tourists, and other outsiders. Visitors often come to the Arctic with grand designs that overlook or are openly hostile to the people who have lived there for tens of thousands of years. For example, Norway had a policy of *Norwegianization* from 1850 to the 1980s that was aimed at destroying Sami culture. The Sami languages were made illegal, Sami lands were confiscated, and knowledge of the Norwegian language was required to buy or lease land. Norwegianization was gradually abandoned in the 1980s, and a Sami parliament was established in 1989 with authority to distribute funds designated for the preservation of Sami culture. Even so, discrimination against Sami persists and many Norwegians strongly oppose efforts to elevate the Sami languages to an equal status with Norwegian even in areas where Sami live in large numbers.

Russia's indigenous populations have suffered brutal repression, displacement, and forced abandonment of their nomadic lifestyle at the hands of successive Russian governments over the past 500 years. During Imperial Russia's colonization of Siberia (1580 to the late 1700s), several czars waged genocidal wars against Arctic peoples, often wiping out entire cultures. The groups that were left were severely reduced in number and had much of their land taken from them by Russian colonists from the west. Only the most remote and warlike, such as the Chukchi of the far eastern Arctic, were able to maintain their independence. During the Soviet era, many nomadic hunters and gatherers were displaced by mining and other extractive industries and forced onto collective farms and reindeer ranches. Their children were sent to boarding schools in Moscow and St. Petersburg, where many of them died of illness. In the post-Soviet era, many collective farms have closed as the oil and gas industry has expanded further into reindeer herding lands.

Arctic indigenous peoples in North America have a similar history of facing systematic abuse and repression by successive governments. Imperial Russia claimed possession of Alaska from the 1780s until 1867, during which time many native peoples were forced into slavery by Russian fur traders. After the United States purchased Alaska from Russia in 1867, whaling in the North American Arctic increased, reducing access to an important source of food for many Arctic indigenous peoples, who faced periodic starvation. Influxes of foreign whalers, miners, fur traders, and missionaries also introduced diseases such as measles and influenza, which were deadly to Inuit populations in Alaska

and the Canadian Arctic and could kill more than 30 percent of a community in a single epidemic.

Missionaries considered most indigenous cultures to be savage and in need of "civilizing." With the assistance of the government, hundreds of thousands of children, both in the Arctic and the rest of North America, were forcibly removed from their families and sent to boarding schools, where they were encouraged to abandon all aspects of their native culture. These schools, established in the late nineteenth and early twentieth centuries, were renowned for physical abuse as well as for having such low-quality health care that deadly epidemics killed thousands of children.

The long history of oppression and dehumanization of indigenous people suggests that they will remain on a semi-impoverished periphery in the era of heightened interest in Arctic resources. In addition, the many changes that climate change is already bringing are stressing Arctic indigenous cultures to the breaking point. As Aili Keskitalo, the first female president of Norway's Sami Parliament put it, "We are used to having to adapt. But we cannot adapt ourselves to death."

ON THE BRIGHT SIDE Conservation efforts at the poles are being strengthened. In the Arctic, the United States, Canada, Greenland, Norway, and Russia created a conservation plan for polar bears in 2013 that addresses threats from shipping, oil and gas exploration, and military conflict. In 2015, Russia took steps to protect wild salmon fishing areas in its far east. In 2013, the U.S. government protected large areas of bird habitats in northern Alaska from oil and gas development. In Nunavut, Canada, a new marine reserve for whales was created in 2011. Additions to the Antarctic Treaty that have been in effect since 1998 are designed to protect the seas around Antarctica from overfishing and pollution. ■

THINGS TO REMEMBER

• The long history of oppression and dehumanization of indigenous people suggests that they will remain on a semi-impoverished periphery in the era of heightened interest in Arctic resources.

Critical Thinking Questions

1. How could climate change at the poles influence your life, and how could your life influence climate change at the poles?

2. What new industries are increasing their presence in the Arctic and why?

3. Why are conflicts over the Arctic growing?

4. What are the differences between large cities in Russia and Scandinavia?

5. What factors contribute to the complex polar gender imbalance?

6. What common experiences have Arctic indigenous people endured over the past several centuries?

Epilogue 1 Key Terms

bioaccumulation 665
biomagnification 665

fjord 660
gulag camp 663

krill 661
polar amplification 661

Space

GEOGRAPHIC THEMES

After you read this epilogue, you will be able to discuss the following geographic themes as they relate to the five thematic concepts:

1. Environment: The proliferation of human-made satellites orbiting Earth provides a greater understanding of this planet and the solar system, but also brings hazards that are complicating further space use and exploration. Over the long term, there are also significant threats to Earth's ecosystems from space, such as impacts from large asteroids or comets, which have caused mass extinctions of species throughout Earth's history.

2. Globalization and Development: The range of commercial activity in space is expanding from satellite-based enterprises to include space launching and asteroid mining. These activities will continue to propel space exploration for the foreseeable future.

3. Power and Politics: While all spacefaring nations have agreed that no parts of outer space can be claimed as national territory, there is disagreement over the rights of individuals or corporations to own property in space. National militaries have played a central role in space exploration, and agreements to limit the use of weapons of mass destruction (including nuclear weapons) have not limited the militarization of space.

4. Urbanization: The challenges of life in the harsh environment of space mean that urban settlements not on Earth are distant prospects. However, the use of space-based remote sensing technologies to better understand urban processes on Earth and to help cities adapt to change is a well-developed and growing field within geography and related disciplines.

5. Population and Gender: While it is not yet clear that a long-term, sustained human population is possible in space, observation of Earth from space has been providing insights into population issues for decades. There is also a significant gender gap in space exploration and research.

Outer Space

Given the increasing environmental and social stresses on Earth, it's no surprise that space exploration has captured the imagination of generations of humans. Research suggests that there are many valuable resources in space and possibly billions of planets in our galaxy that can support life. However, vast regions of space are hostile to human life, and a wide range of obstacles complicate adaptation to this new frontier. Nevertheless the parts of outer space closest to Earth are already highly utilized, with thousands of satellites having been launched into orbit. This region is now central to our understanding of processes on Earth, the management of global communications, and the conduct of war **(FIGURE E2.1)**.

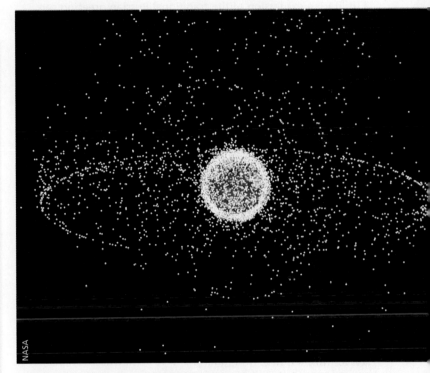

NASA

FIGURE E2.1 (A) Objects orbiting Earth. Ninety-five percent of the objects orbiting Earth that are currently being tracked by NASA are space junk, or nonfunctional satellites. In this computer-generated image, the white dots represent the objects. (The dots are not to scale and are much smaller relative to Earth in real life.) The crowded region closest to Earth, is called low Earth orbit (LEO), and lies between 99 miles (160 kilometers) and 1243 miles (2000 kilometers) above Earth.

STR/AFP/Getty Images

(B) A rocket booster that fell to Earth without burning up and landed in Vietnam. Most pieces of space junk are much smaller.

GLOBAL PATTERNS, LOCAL LIVES On June 8, 1985, two days after being hurtled into space in their *Soyuz* spacecraft, Russian cosmonauts Vladimir Dzhanibekov and Viktor Savinikh found themselves floating into the dark and frozen *Salyut 7* space station (FIGURE E2.2). Searching around with their flashlights, they found the crackers and salt tablets left by the previous crew, a utilitarian expression of the traditional Russian greeting ceremony in which important or beloved guests are welcomed with a large round loaf of bread, symbolizing wealth, and a dish of salt, symbolizing protection. The salt was particularly relevant in this situation. Of the 181 people who had ventured into space by 1985, five had died, making a cosmonaut's job more dangerous than service in most military combat units.

Dzhanibekov and Savinikh faced a particularly precarious mission this day. *Salyut 7* had lost all power and communications systems 4 months ago and was drifting out of its orbit. Their job was to find out what had gone wrong, fix it, restore the station to operation, and prepare it for the next crew. While inspecting the frozen station, the cosmonauts had to work short shifts due to the absence of ventilation, which created a risk of carbon dioxide poisoning from their own exhaled breath. Eventually, the cause of the power loss was found—a faulty battery charger for *Salyut 7*'s solar power system—which was easily mended.

Despite its mechanical problems, *Salyut 7* was the most advanced and comfortable of all the Soviet space stations, with a dining table, an electric oven, a refrigerator, utensils and dishes, and the best food of any Soviet mission, consisting mostly of freeze-dried dishes that could be rehydrated with hot or cold water. Nevertheless, living quarters were somewhat cramped, with 3178 cubic feet (90 cubic meters) of living space (roughly half the size of a small mobile home), especially compared to the current International Space Station (ISS), which is the size of a six-bedroom house. With no laundry, cosmonauts wore the same underwear for several days in a row, after which it was jettisoned out the waste hatch and descended toward Earth, burning up on reentry. Bathing was restricted to monthly sponge baths, during which elaborate precautions were taken to contain water droplets in zero gravity, as the droplets could easily damage the station's electronics.

As with all Soviet stations, *Salyut 7* had elaborate cameras for photographing phenomena on the surface of Earth and relatively little equipment for observing astronomic phenomena. Its many observations of Earth were celebrated in the Soviet media for their contributions to oil and gas exploration, agriculture, Earth science, and mapping, and were estimated to have had a substantial economic impact. Direct observations of Earth by the cosmonauts themselves were also considered highly valuable, and the crew kept an elaborate log of everything they saw from the station.

Salyut 7 achieved many "firsts" in space. Shortly after its launch in 1982, it became home to the first flowering plants in space, the first Indian in space, and the first woman to go on a

geospace the region of outer space closest to Earth, including the upper atmosphere and extending out 40,000 miles (65,000 km), to the farthest reaches of Earth's magnetic field

interplanetary space the region of space between the Sun and planets of our solar system

interstellar space the region of space between our solar system and others within the Milky Way galaxy

intergalactic space the region of space between the Milky Way galaxy and other galaxies

FIGURE E2.2 Cosmonauts aboard the *Salyut 7* spacecraft in 1982.

spacewalk (who also performed the first welding in space). *Salyut 7* was the first Soviet space station to have a system for digitally sending images back to Earth from orbit, instead of relying on cosmonauts to bring back exposed film, which made its observational capabilities much more useful.

After its last manned mission in 1986, *Salyut 7* was placed into a high "storage" orbit, from which it would be retrieved for later use. However, the economic and political disruptions surrounding the fall of the USSR resulted in several cancelled missions. Meanwhile, higher-than-expected discharges of solar particles from the Sun created *solar wind drag* that started *Salyut 7* on a descent toward Earth, where it burned up in the atmosphere, scattering debris over the town of Capitán Bermúdez, Argentina, in 1991.

The story of *Salyut 7* captures the difficulties of life in space and the persistence of humans in attempting to overcome these difficulties. It highlights the importance of space as a platform for observing Earth, as well as the ways that events on Earth can wreak havoc with even the best-laid plans for space exploration. [Sources: The Space Review; astronautix.com; and arstechnica.com. For detailed source information, see Text Sources and Credits. ∎

What Makes Outer Space a Region?

Outer space is the area between Earth's atmosphere and all other objects in the universe. Strictly speaking, asteroids, planets, and stars are not part of outer space, but rather are contained within it. However, these and other celestial objects are treated as part of outer space in this epilogue.

Outer space is usually broken down into several regions. **Geospace** is the region of outer space closest to Earth. Geospace includes the upper atmosphere and extends out 40,000 miles (65,000 km), to the farthest reaches of Earth's magnetic field. **Interplanetary space** exists between the Sun and planets of our solar system; **interstellar space** exists between our solar system and others within our galaxy (the Milky Way); and **intergalactic space** exists between our galaxy and other galaxies. So far, humans have left geospace only on the missions to the Moon.

PHYSICAL GEOGRAPHY

A less hospitable environment than outer space is hard to imagine. Intense cold of −454°F (−270°C) is combined with extremely low pressure, approaching a perfect vacuum, that would suck all the air from a human being's lungs, causing them to pass out within 15 seconds and most likely die within 60 seconds. And yet direct sunlight can create temperatures of 250°F (121°C) on the surface of objects such as the Moon or a space vehicle facing the Sun. Outside a space station orbiting Earth, radiation from the Sun is so intense that bare skin would burn immediately in space after microseconds, likely resulting in skin cancer. Tiny bits of rock, metal, and other solid material, called *meteoroids*, are another hazard in outer space. Often created by collisions between larger objects, tiny meteoroids can travel at thousands of miles per hour in space, potentially ripping through human skin and leaving space vehicles covered with tiny impact craters. There is also very little gravity in space, resulting in a bewildering array of complications to daily life, even when one is shielded from other space hazards.

SPACE WEATHER

The Sun and other stars are constantly ejecting charged particles that form a gaslike plasma that can conduct electricity and respond to magnetic fields. The Sun's emission of these particles is known as **solar wind,** which is thrown out at millions of miles per hour, reaching far beyond the orbits of the most distant planets in the solar system. Earth is largely shielded from the solar wind by its magnetic field, also called the *magnetosphere* **(FIGURE E2.3)**. Some particles get through the magnetosphere and light up the sky near poles, creating the *northern* and *southern lights* and occasionally disrupting electric power-delivery systems. Solar wind is lethal to terrestrial life, and all celestial objects without strong magnetic fields are bombarded by this form of intense radiation. All spacecraft are designed to provide shielding, but the most intense solar winds still penetrate, increasing the risk of radiation sickness and cancer among space travelers.

FIGURE E2.3 Artist's rendition of Earth's magnetic field. Satellite is shown in orbit and is not to scale.

ENVIRONMENT

GEOGRAPHIC THEME 1

Environment: The proliferation of human-made satellites orbiting Earth provides a greater understanding of this planet and the solar system, but also brings hazards that are complicating further space use and exploration. Over the long term, there are also significant threats to Earth's ecosystems from space, such as impacts from large asteroids or comets, which have caused mass extinctions of species throughout Earth's history.

A **satellite** is any object in space that orbits another object of greater mass. Earth's moon is a natural satellite, and over 8000 artificial satellites have been launched into orbit in order to provide important information to people on Earth. Satellites and the information they can provide about Earth itself, and to a lesser degree about outer space, have been the primary focus of all space programs. About half of all satellites launched since 1957, when the USSR launched the first satellite into orbit, have been for military purposes; the other half have been for civilian purposes. Until the late 1990s, most satellites were launched for military and civilian government purposes, but civilian commercial satellite launches have increased dramatically in the last few years and may outnumber all other types of launches in coming decades **(FIGURE E2.4)**. The Soviet Union (and later, Russia) has launched just under half of all satellites, and the United States has launched around a quarter.

TYPES OF SATELLITES

Roughly a quarter of all satellites launched provide *remote sensing* services to national militaries. These "spy satellites" gather a wide array of visual, photographic, and electronic signal data, and the vast majority belong to the United States and Russia. The extent of secret surveillance that these satellites are capable of is not publicly available information, and controversies have raged over the last few years as information from these satellites has been leaked to the Internet by people and organizations that object to what they see as a breach of their right to privacy (see Chapter 2).

In recent years, there has been a dramatic growth of civilian remote sensing satellites, many of which geographers can access to help analyze natural and human-induced processes on Earth's surface. Some of the ways the data are used are to estimate deforestation, to monitor oil spills and forest fires, and to track many processes related to climate change, such as glacial melting, variations in surface temperature, and sea level rise. Numerous other processes can be monitored by remote sensing and are discussed below.

After remote sensing satellites, communications satellites are the next-largest group, comprising about 16 percent of all satellites launched. Most are used to extend the range of telephone, television, radio, Internet, and other communications signals that must traverse long distances. Because most broadcasted signals travel in a straight line, they are limited in range by the curvature of Earth as

solar wind tiny charged particles emitted by the Sun that travel at millions of miles per hour and form a gaslike plasma that can conduct electricity and respond to magnetic fields

satellite any object in space that orbits another object of greater mass

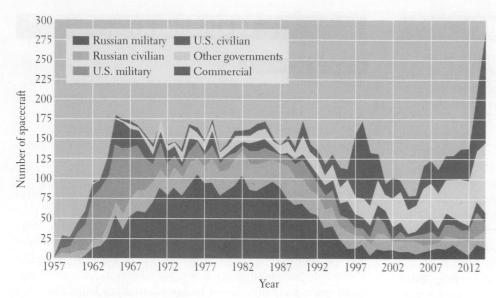

FIGURE E2.4 Space launches from 1957 to 2014. Notice the dominance of the Russian and U.S. military space launches until the late 1980s, and the recent growth of other governmental and commercial space launches since then. (Source: http://claudelafleur.qc.ca/images/YR-5714.jpg)

well as by hills and mountain ranges. Communications satellites receive signals from the surface of the planet and bounce them back, thus allowing them to avoid these obstacles **(FIGURE E2.5)**.

Other satellites include weather satellites, which assist forecasting by monitoring cloud formations, temperature, and precipitation; navigation satellites, such as the Global Positioning System, that provide accurate latitude and longitude as well as elevation data to receivers on Earth; and astronomic and solar observatories such as the Hubble Space Telescope.

SPACE JUNK

Thousands of satellites no longer function and are considered **space junk.** Most of these pose relatively little risk of becoming dangerous space junk, as they are located in **geostationary orbit (GEO),** meaning that they are located at least 22,236 miles (35,786 kilometers) above a fixed point on Earth and they also do not move relative to each other. However, as can be seen in Figure E2.1, many satellites are also located in **low Earth orbit (LEO),** the region which lies between 99 miles (160 kilometers) and 1243 miles (2000 kilometers) above Earth, in which satellites do move relative to each other. Often they travel at speeds of around 17,500 miles per hour, roughly ten times faster than a speeding bullet. Hundreds of millions of pieces of space junk have accumulated in LEO, most of them tiny pieces broken off of nonfunctional satellites or parts of rockets used to put spacecraft

space junk nonfunctioning satellites

geostationary orbit (GEO) an orbit located at least 22,236 miles (35,786 kilometers) above a fixed point on Earth, in which satellites do not move relative to each other

low Earth orbit (LEO) an orbit that lies between 99 miles (160 kilometers) and 1243 miles (2000 kilometers) above Earth, in which satellites move relative to each other

Kessler scenario a situation where so much space junk accumulates around Earth that travel through low Earth orbit becomes impossible

in orbit. Some of the debris is created when satellites are hit by micrometeors or other pieces of space junk. Collisions between satellites have been rare, but when they occur they create thousands of pieces of debris. An increasing number of satellites are hit by other pieces of debris each day. Some have been destroyed by Earth-based militaries of the United States, Russia, and China, as a demonstration of their capabilities.

In the lower levels of Earth orbit, at an altitude of 250 miles (400 kilometers), gravity pulls most debris

out of orbit and into the atmosphere, where it burns up due to friction with air molecules. Fortunately, this is also the zone where most crewed spaceflights currently take place. Even so, solar panels on space stations and windows on space vehicles are often damaged by space junk, necessitating shielding. Expeditions to the Moon, Mars, or asteroids require passing through the much more dangerous higher zones of LEO.

Donald Kessler of the U.S. National Aeronautics and Space Administration (NASA) cautions that if something isn't done about space junk we may reach a point where so much space junk surrounds Earth that travel through LEO will become impossible, a situation now known as the **Kessler scenario.** While there is currently no international agreement limiting further accumulation of space junk, all spacefaring nations require satellite designers, builders, and launchers to take precautions. Most now require that satellites be able to relocate themselves after their useful life is complete, either to an out-of-the-way "graveyard orbit" or back toward Earth, where they will burn up on reentry. There are also efforts underway to remove defunct satellites and rocket parts or other pieces of debris that weigh more than several thousand pounds before they collide with other pieces of space junk, creating even more debris. Proposals to accomplish this include netting, harpooning, or otherwise capturing and then towing defunct satellites into the lower levels of Earth orbit, where they will incinerate due to friction with Earth's atmosphere.

ASTEROID RISKS

A long-term threat to Earth from space is the risk of an asteroid or comet impact, such as the asteroid that was between 200 and 620 feet wide that flattened 2000 square kilometers of forest in Tunguska, Siberia, in 1908; or the 6-mile-wide asteroid that landed just off the coast of the Yucatán Peninsula 66 million years ago, possibly causing the extinction of the dinosaurs and numerous other life-forms. While a global-scale "civilization killing" asteroid is predicted to hit Earth once every 100 million years, four Tunguska-sized asteroids are predicted to pass near Earth over the next 50 years. Strategies for dealing with asteroid impact mostly involve early warning, though asteroid-related technology is developing rapidly.

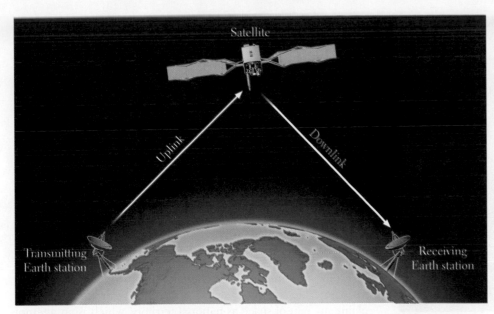

FIGURE E2.5 Satellites and signals. Because of their high altitude, satellites can extend the range of communications signals on Earth that would otherwise be blocked by topography and the curvature of Earth. (Source: http://www.radio-electronics.com/info/satellite/communications_satellite/communications-satellite-technology.php)

THINGS TO REMEMBER

GEOGRAPHIC THEME 1 • **Environment** The proliferation of human-made satellites orbiting Earth provides a greater understanding of this planet and the solar system, but also brings hazards that are complicating further space use and exploration. Over the long term, there are also significant threats to Earth's ecosystems from space, such as impacts from large asteroids or comets, which have caused mass extinctions of species throughout Earth's history.

• Outer space is the area between Earth's atmosphere and all other objects in the universe.

• The Sun and other stars are constantly ejecting charged particles that form a gaslike plasma that can conduct electricity, respond to magnetic fields, and occasionally disrupt electric power-delivery systems on Earth.

• In recent years, there has been a dramatic growth of civilian remote sensing satellites, many of which geographers can access to help analyze natural and human-induced processes on Earth's surface.

• We may reach a point where so much space junk surrounds Earth that travel through LEO will become impossible, a situation now known as the Kessler scenario.

GLOBALIZATION AND DEVELOPMENT

GEOGRAPHIC THEME 2

Globalization and Development: The range of commercial activity in space is expanding from satellite-based enterprises to include space launching and asteroid mining. These activities will continue to propel space exploration for the foreseeable future.

Space-based commercial activity began in 1962 with the launch of the first telecommunications satellites. Since then, there have been over 8000 satellites launched into space, 3600 of which are still in orbit. Of the roughly 1300 satellites that are still operational, more than half are communications satellites, launched primarily by the United States and Russia (or the former Soviet Union). A major increase in the number of communications satellites in orbit is planned for the near future, with several companies planning on putting up hundreds or even thousands more communications satellites to provide global wireless Internet access. Commercial activity is now expanding into the building and launching of rockets and other spacecraft, as private companies now provide launch services at significantly lower costs than those from government-sponsored space programs. SpaceX, a U.S. company founded in 2002 by former PayPal entrepreneur and Tesla Motors CEO Elon Musk (of South Africa), plans to drop prices by an order of magnitude, from its current price for launching a 29,000-pound payload into LEO of U.S.$56 million (already the lowest price in the business) to U.S.$5 million, in part by using reusable rockets.

ASTEROID MINING

There are more than 100 million asteroids in the solar system, some of which contain precious metals such as gold and platinum in quantities valued in the trillions of dollars. In 2010, Japan's space program initiated the first major feasibility test of asteroid mining by landing a probe on an asteroid, taking samples, and returning to Earth with them. A growing number of companies are currently planning to mine asteroids, primarily with robotic spacecraft operated from Earth's surface. The enormous profits that could be reaped by asteroid mining will likely drive space exploration beyond geospace for the foreseeable future **(FIGURE E2.6)**.

Asteroid mining could transform Earth in ways that are as yet barely understood. Successful asteroid mining would likely collapse the prices of many metals and jeopardize millions of mining jobs across the planet. This is especially true for metals such as gold or platinum that are rare on Earth but abundant enough in space that a single asteroid could easily contain more of these metals than has ever been mined in human history. But prices for more economically important metals, such as iron or nickel, could also decline due to the presence of asteroids relatively close to Earth that contain vast amounts of these elements. Proponents of asteroid mining argue that a larger supply of these

FIGURE E2.6 Mining outpost. An artist's rendition of a mining outpost on an asteroid near Earth that is about 2000 feet (610 meters) wide and 1000 feet (305 meters) long.

Walter Myers/Stocktrek Images/Getty Images

metals will increase their use by, for example, making platinum more available for use in fuel cells and gold more available for use in electronics, potentially creating millions more jobs and raising living standards in the process. Even without the trillion-dollar profits that are predicted, asteroid mining will probably take place in order to supply space exploration missions, given that some asteroids hold large amounts of water, which can be split into hydrogen for fuel and oxygen for respiration.

While tests and plans for asteroid mining are now well advanced, no one is yet mining asteroids. Deep Space Industries of Mountain View, California, plans to start mining by 2023, and other companies are on track to follow closely behind. In 2016, Luxembourg began actively recruiting asteroid-mining companies to locate within its borders.

Outer Space Treaty of 1967 treaty that the United States and all spacefaring nations have signed and which forbids nations from claiming territory in space

THINGS TO REMEMBER

GEOGRAPHIC THEME 2 • **Globalization and Development** The range of commercial activity in space is expanding from satellite-based enterprises to include space launching and asteroid mining. These activities will continue to propel space exploration for the foreseeable future.

• Commercial activity is now expanding into the building and launching of rockets and other spacecraft, with private companies now providing launch services at significantly lower costs than those from government-sponsored space programs.

• Successful asteroid mining would likely collapse the prices of many metals and jeopardize millions of mining jobs across the planet. Proponents of asteroid mining argue that a larger supply of these metals will increase their use, creating millions more jobs and raising living standards in the process.

POWER AND POLITICS

GEOGRAPHIC THEME 3

Power and Politics: While all spacefaring nations have agreed that no parts of outer space can be claimed as national territory, there is disagreement over the rights of individuals or corporations to own property in space. National militaries have played a central role in space exploration, and agreements to limit the use of weapons of mass destruction (including nuclear weapons) in space have not limited the militarization of space.

SPACE AND THE LAW

Following intense lobbying by several asteroid-mining companies, the Space Act of 2015 was passed by the U.S. Congress and signed into law by President Obama, giving companies the rights to exploit any minerals or other nonliving resources they find in space. The law also states that the U.S. government will not claim any part of space as national territory, which is an attempt to avoid a clear break with the **Outer Space Treaty of 1967.** This treaty, which the United States and all spacefaring nations have signed and which forbids nations from claiming territory in space, declares that the exploration of all celestial objects "shall be carried out for the benefit and the interests of all countries and shall be the province of all mankind." The U.S. government seemed to uphold this principle when it cited the treaty during its successful defense in court in 2001 against a parking ticket issued it by beef jerky entrepreneur and space enthusiast Gregory Nemitz, who had registered a claim to the asteroid Eros, which NASA had just landed a probe on.

The U.S. government has so far made no move to withdraw from the Outer Space Treaty, possibly due to other provisions of the treaty which ban the use or deployment of weapons of mass destruction in space. While the treaty doesn't ban weapons from space (and the United States, Russia, and China have already deployed weapons there), many argue that it reduces the potential for armed conflict in space by preventing nations and corporations from making any broad territorial claims.

THE MILITARIZATION OF SPACE

National militaries have played a pivotal role in space exploration, and the first object to enter space was a German V2 rocket during World War II. After the war, former Nazi military scientists played key roles in both the U.S. and Soviet space programs, which were contained within military agencies or evolved out of them. Much space exploration has been shaped by competing military interests, such as the Cold War between the United States and the Soviet Union, and the more recent U.S. military rivalry with China. Much of the hardware that makes space exploration possible was developed by national militaries, including the intercontinental ballistic missiles (originally designed to carry nuclear warheads) that have propelled most launches into space, as well as the many space-based systems for remote sensing, communication, and navigation. Every major military power on Earth now considers these space-based technologies to be an essential part of warfare, and all the major spacefaring nations now have militarized "space forces" that constantly train for war in space to protect their satellites and, if deemed necessary, disable those of other countries. Of all Earth's military "space powers," the United States has by far the most currently functioning satellites in orbit (between 150 and 200). Russia follows with about 75, and then China, with 35.

THINGS TO REMEMBER

GEOGRAPHIC THEME 3 • **Power and Politics** While all spacefaring nations have agreed that no parts of outer space can be claimed as national territory, there is disagreement over the rights of individuals or corporations to own property in space. National militaries have played a central role in space exploration, and agreements to limit the use of weapons of mass destruction (including nuclear weapons) in space have not limited the militarization of space.

• Following intense lobbying by several asteroid mining companies, the Space Act of 2015 was signed into law by President Obama, giving U.S. companies the rights to exploit any minerals or other nonliving resources they find in space.

• Every major military power on Earth now considers space-based technologies to be an essential part of warfare, and all the major spacefaring nations now have militarized "space forces" that constantly train for war in space to protect their satellites and, if necessary, disable those of other countries.

URBANIZATION

GEOGRAPHIC THEME 4

Urbanization: The challenges of life in the harsh environment of space mean that urban settlements not on Earth are distant prospects. However, the use of space-based remote sensing technologies to better understand urban processes on Earth and to help cities adapt to change is a well-developed and growing field within geography and related disciplines.

The physical challenges posed to humans by prolonged habitation in space include radiation and space debris, the effects of microgravity, and the need to minimize the use of materials from Earth. Shielding against radiation and space debris has long been included in spacecraft and space stations, but long-term settlements would need more extensive protection. Human settlements on asteroids, moons, and some planets will likely take advantage of underground architecture, which provides shielding of all kinds as well as insulation from the huge temperature swings found in interplanetary space. But these protections will do little to mitigate the effects of long-term exposure to microgravity, which has brought about a wide range of physical changes in humans who spend even short periods in space. Despite extensive exercise regimes, humans regularly have declines in bone density, loss of muscle mass, and permanently impaired vision after visits to space. In addition to these biological obstacles to urbanization in space, the high costs of launching material into space will require extensive capabilities in materials recycling and space-based resource acquisition. Unless there are dramatic advances in technological capability, these complex challenges will likely delay the development of long-term space-based settlements for several decades.

REMOTE SENSING OF URBAN AREAS

While actual urban settlements in space may be decades or more away, space-based remotely sensed observations of Earth have been contributing to our understanding of urbanization since the 1970s. Remotely sensed data is often made into spatially precise maps that are particularly useful for understanding urban change over time. An increasingly important use for remotely sensed data is in disaster management, where the large amount of detailed spatial information that can be gathered rapidly by satellites is very valuable. For example, during the aftermath of Hurricane Katrina, residents and emergency response personnel in New Orleans, Louisiana, needed maps of flooding in the city. Remote sensing scientists at the U.S. Geological Survey's Center for Earth Resources Observations and Science (EROS) were able to provide detailed and comprehensive maps showing the exact location and depth of the flooding within days. This information proved invaluable to emergency responders as they struggled to assist residents trapped in their homes and on rooftops. Over the course of the next weeks and months, scientists at EROS were able to provide detailed maps and digital images of the disaster **(FIGURE E2.7)**, which became essential tools for city planners and disaster management personnel as they tried to understand why

DigitalGlobe/Getty Images

FIGURE E2.7 Satellite image of New Orleans after Hurricane Katrina. A satellite image of the 800-foot-wide (240-meter) break in East New Orleans' Industrial Canal in Louisiana and surrounding flooding on August 31, 2005, two days after Hurricane Katrina hit the city. The image was collected from Quickbird, a commercial satellite capable of capturing objects larger than 5.2 feet (1.6 meters) from an orbit 280 miles (450 kilometers) above the surface of Earth.

the city's flood-protection systems had failed. Looking forward, geographers and urban planners will be using remotely sensed satellite data on a daily basis to help cities across the world adapt to climate change, growing urban populations, and decaying infrastructure.

THINGS TO REMEMBER

GEOGRAPHIC THEME 4
• **Urbanization** The challenges of life in the harsh environment of space mean that urban settlements not on Earth are distant prospects. However, the use of space-based remote sensing technologies to better understand urban processes on Earth and to help cities adapt to change is a well-developed and growing field within geography and related disciplines.

• An increasingly important use for remotely sensed data is in disaster management, where the large amount of detailed spatial information that can be gathered rapidly by satellites can prove very valuable.

POPULATION AND GENDER

GEOGRAPHIC THEME 5

Population and Gender: While it is not yet clear that a long-term, sustained human population is possible in space, observation of Earth from space has been providing insights into population issues for decades. There is also a significant gender gap in space exploration and research.

At present, human abilities to produce healthy populations in space appear uncertain. Studies with animals show that while reproduction is possible in space, exposure to radiation and zero gravity significantly reduces sperm and egg counts and may result in infertile offspring. Studies with animals born in space show that they have many birth defects, including poor equilibrium due to delayed development of cells that sense gravity. NASA offers astronauts the option of having samples of their sperm and eggs frozen for later use before they go into space, though many have conceived and given birth to healthy children after they have returned.

REMOTE SENSING AND POPULATION ISSUES

Remote sensing satellites have been providing important data for estimating population size and location on Earth, as well as for understanding related issues such as food security and disease. Satellite images can be analyzed to detect changes in land use that help researchers estimate population growth, especially around urban areas. Satellites with infrared sensors can monitor the health of crops as they are growing, allowing analysts to improve forecasts of harvests; and provide advance warning of food scarcity and famine in rural areas. Satellites can also collect data on factors that relate to economic activity and income levels, such as the use of electric lights at night, that can help provide estimates of a population's overall risk of famine. Conditions that contribute to disease can also be detected by using satellites. For example, remotely sensed images of algal

blooms in the waters of coastal Bangladesh have helped predict the locations of outbreaks of cholera, and satellite data on vegetation cover and precipitation patterns have been used to guide mosquito-eradication efforts aimed at containing outbreaks of malaria in Thailand.

GENDER IN SPACE RESEARCH

There is a significant gender gap in space-related research and exploration. In 1963, two years after the Soviet Union sent the first man (Yuri Gagarin) into space, they sent the first woman (Valentina Tereshkova) into space. She was the twelfth human in space, and it was another 19 years and 76 crewed space missions before the Soviet Union sent the second woman into space. The United States had a 20-year ban on women in the space program, despite early studies that showed that women had particular advantages for space travel, such as smaller size, a lower metabolic rate, and excellent performance under stress. The perception by NASA engineers that women were unsuited to the stresses of spaceflight delayed women's participation in the U.S. space program until the late 1970s, and the first U.S. woman wasn't launched into space until 1983. Since the 1980s, over 60 women have been into space, more than 50 of them from the United States. Women still represent less than 14 percent of human spaceflights since 1983, but that share has increased to 31 percent of spaceflights since 2013. The gender gap in space exploration continues, in part because of the requirement of most space programs that astronauts have at least a bachelor's degree in a STEM (science, technology, engineering, or mathematics) field. Women are historically underrepresented in these fields due to a variety of cultural influences, including teachers, friends, and families who discourage girls and young women from taking advanced math and science courses.

THINGS TO REMEMBER

GEOGRAPHIC THEME 5
• **Population and Gender** While it is not yet clear that a long-term, sustained human population is possible in space, observation of Earth from space has been providing insights into population issues for decades. There is also a significant gender gap in space exploration and research.

• Remote sensing satellites have been providing important data for estimating population size and location on Earth, as well as for understanding related issues such as food security and disease.

• The perception by NASA engineers that women were unsuited to the stresses of spaceflight delayed women's participation in the U.S. space program until the late 1970s, and the first U.S. woman wasn't launched into space until 1983.

SOCIOCULTURAL ISSUES

All of the risks and uncertainties of space travel and exploration are dwarfed by the potentially profound implications for humanity of finding even just a single new planet hospitable to humans, of a single form of intelligent life. What will human occupation of other worlds mean for life on Earth? If the past history of European colonialism is any example, the effects will be complex and not completely understood, even hundreds of years after the first space colonies are established.

The most habitable extraterrestrial environment within current reach is Mars, where data collected over 50 years of exploration (NASA launched the first successful probe in 1964) indicate that the planet has water in ice caps, subsurface lakes, and in its atmosphere; and has enough sunlight that solar power can be generated. While Mars's climate is generally colder and more extreme than Earth's (Mars has an average high at the equator of 70°F (20°C) and an average low of −100°F (−73°C)), it is still within the range of human adaptability. After Mars, there are a few other candidates for human habitability within the solar system, such as Jupiter's moon Europa, which may contain a marine environment beneath its icy surface. Beyond the solar system, the possibilities for finding habitable planets improve, as suggested by data from NASA's Kepler space observatory, a Sun-orbiting space telescope launched in 2009 and designed to find Earth-sized planets in the Milky Way galaxy that have similar temperature ranges as those on Earth, given their location in their solar systems. By 2016, scientists using Kepler had confirmed the existence of more than 20 such planets and could extrapolate, based on previous data, that there may be as many as 40 billion habitable Earth-sized planets in the Milky Way galaxy.

WHAT WILL SPACE EXPLORATION DO TO LIFE ON EARTH?

The extension of human civilization into outer space will bring complex changes to Earth, some of which could parallel the transformations brought about by European colonialism in the Americas. Already the focus on asteroid mining mirrors the objectives of early Spanish colonialists, who came to the Americas seeking gold and other precious metals, often leaving chaos and disorder in their wake. Could the exploration of space bring new biological resources, similar to the exchange of food crops between Europe and the Americas that eventually transformed the entire world? Hopefully space-age humanity will fare better than indigenous American cultures, who were nearly wiped out by diseases brought by European explorers. Perhaps spacefarers will learn from our current predicament with invasive species, which are the greatest threat to ecosystems after habitat loss caused by human activity, and strictly control introductions of new life-forms to Earth.

The potential for huge profits to be reaped from space exploration offers many interesting scenarios as well. With a single asteroid capable of delivering almost unimaginable riches, disparities in wealth could increase as mines on Earth close while asteroid-mining operations expand, bringing in enormous profits. On the other hand, resources from space could also provide the basis for a new age of technological advancement and higher living standards, much like the resources of the Americas fueled Europe's Industrial Revolution of the 1800s.

Perhaps the biggest implications of human space exploration involve potential encounters with intelligent life, a subject that has permeated popular culture for more than a century. Humans have been searching for signals of extraterrestrial life using radio antennae since the 1890s, and from the 1970s to the 1990s NASA funded the Search for Extraterrestrial Intelligence (SETI), a program that adapted these techniques to scientific standards. Despite a few intriguing signals, these efforts have produced relatively little. NASA has also sent out spacecraft containing messages for potential extraterrestrial civilizations, such as the *Voyager I* and *II* spacecraft, launched in 1977 to gather data about Jupiter and Saturn, and which are now traveling in interstellar space.

People have reported unidentified flying objects (UFOs) across cultures for all of recorded human history. The subject of UFOs is hugely popular today, though most sightings can be explained without reference to extraterrestrial visitors. After several waves of sightings in the 1950s and 1960s, the U.S. Air Force commissioned a study of UFO reports from the University of Colorado. The report both dismissed the subject as unscientific and left many of the cases contained within the study unexplained. Since then, notions of a wide-ranging cover-up of an extraterrestrial presence on Earth have become ubiquitous, and multiple public opinion surveys in the United States over the past several decades have reported that as much as 80 percent of the general public believes that the U.S. government is hiding information on this subject. Meanwhile, reported sightings of UFOs have risen steadily for more than two decades. Explanations for this range from the greater ease of reporting sightings via the Internet, to the expansion of the use of drone aircraft, to more widely available cell phones with video cameras, to a general shift away from ridicule of the UFO phenomenon and toward social acceptance of it as a genuine mystery.

ON THE BRIGHT SIDE

At a cost of more than U.S.$150 billion, the International Space Station (ISS) is the largest peacetime international collaboration in history. Its two largest participants in terms of funding and personnel, the United States and Russia, were once adversaries in the Cold War and are still occasional rivals. Since its first components were lifted into orbit in 1998, the ISS has served as a platform for experiments on microgravity, radiation, and other space-related subjects. As long as a football field from end to end, the ISS is the largest human-made object in orbit and is visible with the naked eye. Crews often do not share a common language, are isolated from family and peers, and adapt to a noisy and sometimes hazardous indoor environment plagued by mold and fungi. The ISS's record of unbroken collaboration in these difficult circumstances has earned it multiple peace-related awards, including a nomination for the Nobel Peace Prize, and encouraged future international cooperation on other space projects, such as a Moon base. ∎

THINGS TO REMEMBER

- The question of what human habitation on other planets will do to life on Earth is broad and complex, but some parallels can be drawn from the history of European colonialism in the Americas.

Critical Thinking Questions

1. If the Kessler scenario occurred, how would it impact your life?

2. How has your life been influenced by the use of data from satellites (remote sensing)?

3. If asteroid mining became a major industry, what places described earlier in this book would be the most affected?

4. What factors make war in space more likely or unlikely?

5. Given what you have learned about gender from earlier chapters of this book, do you think that the gender gap in space exploration and research will persist, or will it shrink until men and women are equally represented in these fields?

6. What problems and opportunities for Earth could human habitation of other worlds bring?

Epilogue 2 Key Terms

geospace 668
geostationary orbit (GEO) 670
intergalactic space 668
interplanetary space 668

interstellar space 668
Kessler scenario 670
low Earth orbit (LEO) 670
Outer Space Treaty of 1967 672

satellite 669
solar wind 669
space junk 670

GLOSSARY

Note: These definitions are only partial and are to be understood in the context in which they appear in the text.

Aboriginal Australians (p. 633) the longest-surviving inhabitants of Oceania, whose ancestors, the Australoids, migrated from Southeast Asia 50,000 to 70,000 years ago over the Sundaland landmass that was exposed during the ice ages

acculturation (p. 169) adaptation of a minority culture to the host culture enough to function effectively and be self-supporting; culture borrowing

acid rain (p. 70) precipitation that has formed through the interaction of rainwater or moisture in the air with sulfur dioxide and nitrogen oxides emitted during the burning of fossil fuels, making it acidic

active aging (p. 232) a three-pronged EU policy aimed at changing society's attitudes toward the abilities and needs of people as they age

agribusiness (p. 84) the business of farming conducted by large-scale operations that purchase, produce, finance, package, and distribute agricultural products

agriculture (p. 30) the practice of producing food and other products through animal husbandry, or the raising of animals, and the cultivation of plants

agroecology (p. 460) the practice of traditional, nonchemical methods of crop fertilization and the use of natural predators to control pests

agroforestry (p. 383) growing economically useful crops of trees on farms, in conjunction with the usual plants and crops, to reduce dependence on trees from non-farmed forests and to provide income to the farmer

Ainu (p. 538) an indigenous cultural minority group in Japan characterized by their light skin, heavy beards, and thick, wavy hair who are thought to have migrated thousands of years ago from the northern Asian steppes

alluvium (p. 545) river-borne sediment

Altiplano (p. 184) an area of high plains in the central Andes of South America

animism (p. 419) a belief system in which spirits, including those of the deceased, are thought to exist everywhere and to offer protection to those who pay their respects

apartheid (p. 393) a system of laws mandating racial segregation; South Africa was an apartheid state from 1948 until 1994

aquifers (p. 66) ancient natural underground reservoirs of water

archipelago (p. 567) a group, often a chain, of islands

Asia-Pacific region (p. 558) a huge trading area that includes all of Asia and the countries around the Pacific Rim

Asia Pacific Economic Cooperation (APEC) (p. 641) a group of 21 Pacific Rim countries organized in 1989 to increase trade and cooperation

assimilation (pp. 169, 235) the loss of old ways of life and the adoption of the lifestyle of another culture

Association of Southeast Asian Nations (ASEAN) (p. 584) an organization of Southeast Asian governments that was established to further economic growth and political cooperation between member countries and with other areas of the world

asylum seekers (p. 235) refugees who left their homes because of violent persecution and seek new legal status in an adopted country

atoll (p. 623) a low-lying island or chain of islets, formed of coral reefs that have built up on the circular or oval rims of a submerged volcano

Australo-Melanesians (p. 578) a group of hunters and gatherers who moved approximately 40,000 to 60,000 years ago from what are now northern Indian and Burma (Myanmar) to the exposed landmass of Sundaland and present-day Australia

Austronesians (p. 578) a group of skilled farmers and seafarers from southern China who migrated south to various parts of Southeast Asia between 6000 and 10,000 years ago

authoritarianism (p. 40) a political system that subordinates individual freedom to the power of the state or of elite regional and local leaders

Aztecs (p. 145) indigenous people of high-central Mexico noted for their advanced civilization before the Spanish conquest

Berber (p. 359) an indigenous group of people in North Africa that are most numerous in Morocco and Algeria

bioaccumulation (p. 665) a process in which organisms absorb toxins faster than their bodies can excrete them

biodiversity (p. 135) the variety of life forms to be found in a given area

biomagnification (p. 665) a process in which the concentration of toxins increases at successive levels of the food chain

biosphere (p. 19) the entirety of Earth's integrated physical spheres, with humans and other impacts included as part of nature

birth rate (p. 49) the number of births per 1000 people in a given population, per unit of time (usually per year)

Bolsheviks (p. 275) a faction of Communists who came to power during the Russian Revolution

brain drain (p. 161) the migration of educated and ambitious young adults to cities or foreign countries, depriving the communities from which the young people come of talented youth in whom they have invested years of nurturing and education

brownfields (p. 99) old industrial sites whose degraded conditions pose obstacles to redevelopment

Buddhism (p. 454) a religion of Asia that originated in India in the sixth century B.C.E. as a reinterpretation of Hinduism; it emphasizes modest living and peaceful self-reflection leading to enlightenment

buffer state (p. 304) a country that is situated between two rival and more powerful political entities and serves as a neutral territory, limiting direct contact and potential conflict between the two powers

capitalism (pp. 41, 214) an economic system based on the private ownership of the means of production and distribution of goods, driven by the profit motive and characterized by a competitive marketplace; an economic system characterized by privately owned businesses and industrial firms that adjust prices and output to match the demands of the market

capitalists (p. 275) usually a wealthy minority that owns the majority of factories, farms, businesses, and other means of production

carbon sequestration (p. 381) the removal and storage of carbon taken from the atmosphere

carrying capacity (p. 32) the maximum number of people that a given territory can support sustainably with food, water, and other essential resources

cartel (p. 334) a group of producers strong enough to control production and set prices for products

cartographer (p. 4) geographer who specializes in depicting geographic information on maps

cash economy (p. 52) an economic system that tends to be urban but may be rural, in which skilled workers, well-trained specialists, and even farm laborers are paid in money

caste system (p. 453) a complex, ancient Hindu system for dividing society into hereditary hierarchical classes

Caucasia (p. 263) the mountainous region between the Black Sea and the Caspian Sea

central planning (p. 214) a communist economic model in which a central bureaucracy dictates prices and output with the stated aim of allocating goods equitably across society according to need

centrally planned, or socialist, economy (p. 276) an economic system in which the state owns all land and means of production, while government officials direct all economic activity, including the locating of factories, residences, and transportation infrastructure

chaebol (p. 554) huge corporate conglomerates in South Korea that receive government support and protection

Christianity (p. 327) a monotheistic religion based on the belief in the teachings of Jesus of Nazareth, a Jew, who described God's relationship to humans as primarily one of love and support, as exemplified by the Ten Commandments

cisgender (p. 54) someone who has a gender identity that aligns with the gender that they were assigned at birth

civil disobedience (p. 456) protesting of laws or policies by peaceful direct action

civil society (p. 41) the social groups and traditions that function independently of the state and its institutions to foster a sense of unity and informed common purpose among the general population

clear-cutting (p. 74) a method of logging that involves cutting down all trees on a given plot of land regardless of age, health, or species

climate (p. 15) the long-term balance of temperature and precipitation that characteristically prevails in a particular region

climate change (p. 19) a slow shifting of climate patterns due to the general cooling or warming of the atmosphere

Cohesion Policy (p. 215) EU guidelines for programs aimed at removing social and economic disparities across Europe

Cold War (pp. 214, 276) a period of conflict, tension, and competition between the United States and the Soviet Union that lasted from 1945 to 1991; the contest that pitted the United States and western Europe, who were espousing free market capitalism and democracy, against the USSR and its allies, who were promoting a centrally planned economy and a socialist state

commercial agriculture (p. 384) farming in which crops are grown deliberately for cash rather than solely as food for the farm family

commodities (p. 396) raw materials that are traded, usually to other countries, for processing or manufacturing into more valuable goods

commodity dependence (p. 396) economic dependence on exports of agricultural and mineral raw materials

Common Agricultural Program (CAP) (p. 220) an EU program meant to guarantee secure and safe food supplies at affordable prices, that are produced sustainably by farmers who earn a fair income

communal conflict (p. 463) a euphemism for religiously based violence in South Asia

communism (pp. 41, 214, 275) an ideology, based largely on the writings of the German revolutionary Karl Marx, that calls on workers to unite to overthrow capitalism and establish an egalitarian society in which workers share what they produce; as practiced, communism was actually a socialized system of public services and a centralized government and economy in which citizens participated only indirectly through Communist Party representatives

Communist Party (p. 276) the political organization that ruled the USSR from 1917 to 1991; other communist countries, such as China, Mongolia, North Korea, and Cuba, also have communist parties

Confucianism (p. 513) a Chinese philosophy that teaches the importance of stability and social order based on traditional institutions such as the patriarchal family, community, and the state

continental climate (p. 199) a midlatitude climate pattern in which summers are fairly hot and moist, and winters become longer and colder the deeper into the interior of the continent one goes

contested space (p. 160) any area that several groups claim or want to use in different and often conflicting ways, such as the Amazon or Palestine

conurbation (pp. 98, 527) an area formed when several cities expand so that their edges meet and coalesce

coral bleaching (p. 575) color loss that results when photosynthetic algae that live in the corals are expelled

coup d'état (p. 157) a military- or civilian-led forceful takeover of a government

Creoles (p. 148) people mostly of European descent born in the Americas

crony capitalism (p. 583) a type of corruption in which politicians, bankers, and entrepreneurs, sometimes members of the same family, have close personal as well as business relationships

cultural homogenization (p. 198) the tendency toward uniformity of ideas, values, technologies, and institutions among associated culture groups; the loss of ethnic distinctiveness

cultural pluralism (p. 599) the cultural identity characteristic of a region where groups of people from many different backgrounds have lived together for a long time but have remained distinct

Cultural Revolution (p. 516) a political movement launched in 1966 to force the entire population of China to support the continuing revolution

culture (p. 55) all the ideas, materials, and institutions that people have invented to use to live on Earth that are not directly part of our biological inheritance

czar (p. 274) the ruler of the Russian empire

decolonization (p. 214) the process of dissolving or ending colonial relationships, institutions, and mind-sets

death rate (p. 49) the ratio of total deaths to total population in a specified community, usually expressed in numbers per 1000 or in percentages

delta (p. 14) the triangular-shaped plain of sediment that forms where a river meets the sea

demilitarized zone (p. 518) a border area between rival states where military activities are prohibited; usually refers to the border between North and South Korea

democratization (p. 40) the transition toward political systems guided by competitive elections

demographic transition (p. 52) the change from high birth and death rates to low birth and death rates that usually accompanies a cluster of other changes, such as change from a subsistence to a cash economy, increased education rates, and urbanization

desertification (p. 322) a set of ecological changes that converts arid lands into deserts

detritus (p. 570) dead organic material (such as plants and insects) that collects on the ground and in the soil

development (p. 33) an inexact concept typically referring to the shift from primary activities to secondary, tertiary, and quaternary activities

diaspora (p. 326) the dispersion of Jews around the globe after they were expelled from the Eastern Mediterranean by the Roman

Empire, beginning in 73 C.E.; the term can now refer to other dispersed culture groups

dictator (p. 157) a ruler who claims absolute authority, governing with little respect for the law or the rights of citizens

diffusion (p. 329) the process by which ideas, things, and people move across space and time

digital divide (p. 80) the discrepancy in access to information technology between small, rural, and poor areas and large, wealthy cities that contain major government research laboratories and universities

distributaries (p. 577) branches of a river in its delta that fan out and carry water away from the main channel

divide and rule (p. 403) the deliberate intensification of divisions and conflicts by potential rulers; in the case of sub-Saharan Africa, by European colonial powers

doi moi (p. 609) a program of economic and bureaucratic restructuring in Vietnam, very similar to perestroika in the Soviet Union and the structural adjustment programs imposed on many countries by international lending agencies

domestication (p. 30) the process whereby humans used selective breeding to develop plants and animals for use as food, construction materials, and sources of energy, or to improve life in other ways

domino theory (p. 580) a foreign policy theory that used the idea of the domino effect to suggest that if one country "fell" to communism, others in the neighboring region were also likely to fall

double day (p. 237) the longer workday of women with jobs outside the home who also work as caretakers, housekeepers, and/or cooks for their families

dowry (p. 478) a price paid by the family of the bride to the groom (the opposite of bride price); formerly, a custom practiced only by the rich

dual economy (p. 396) an economy in which the population is divided by economic disparities into two groups, one prosperous and the other near or below the poverty level

ecological footprint (p. 19) the amount of biologically productive land and sea area needed to sustain a person at the current average standard of living for a given population

economic core (p. 77) the dominant economic region within a larger region

economic diversification (p. 336) the expansion of an economy to include a wider array of activities

economic migrants (p. 235) those who seek employment opportunities better than what they have at home

economies of scale (p. 218) reductions in the unit cost of production that occur when goods or services are efficiently mass produced, resulting in increased profits per unit

ecotourism (p. 141) nature-oriented vacations, often taken in endangered and remote areas usually by travelers from affluent nations

El Niño (p. 136) periodic climate-altering changes, especially in the circulation of the Pacific Ocean, now understood to operate on a global scale

endemic (p. 626) belonging or restricted to a particular place

erosion (p. 14) the process by which fragmented rock and soil are moved over a distance, primarily by wind and water

ethnic cleansing (p. 42) the deliberate removal of an ethnic group from a particular area by forced migration

ethnic group (p. 55) a group of people who share a common ancestry and sense of common history, a set of beliefs, a way of life, a technology, and usually a common geographic location of origin

ethnicity (p. 106) the quality of belonging to a particular culture group

euro (€) (p. 218) the official (but not required) currency of the European Union as of January 1999

Euro zone (p. 37) those countries in the European Union that use the Euro currency

European Union (EU) (p. 196) a supranational organization that unites most of the countries of West, South, North, and Central Europe

evangelical Protestantism (p. 171) a Christian movement that focuses on personal salvation and empowerment of the individual through miraculous healing and transformation; some practitioners preach to the poor the gospel of success—that a life dedicated to Christ will result in prosperity for the believer

exclave (p. 253) a portion of a country that is separated from the main part

export processing zones (EPZs) (p. 152) specially created legal spaces or industrial parks within a country where, to attract foreign owned factories, duties and taxes are not charged

export-led growth (p. 519) an economic development strategy that relies heavily on the production of manufactured goods destined for sale abroad

extended family (p. 170) a family that consists of related individuals beyond the nuclear family of parents and children

extraction (p. 33) mining, forestry, and agriculture

failed state (p. 359) a country whose government has lost control and is no longer able to fulfill its basic responsibilities as a defender of a sovereign state

fair trade (p. 36) trade that values equity throughout the international trade system, not just market forces; now proposed as an alternative to free trade

fallow period (p. 383) the time (ideally, a decade or more) during which an agricultural plot is allowed to rest and regrow a natural vegetation cover

favelas (p. 163) Brazilian urban slums and shantytowns built by the poor; called colonias, barrios, or barriadas in other countries

female genital mutilation (FGM) (p. 418) removing the labia and clitoris and sometimes stitching the vulva nearly shut

female seclusion (p. 354) the requirement that women stay out of public view

feminization of labor (p. 582) the rising numbers of women in both the formal and the informal labor force

Fertile Crescent (p. 325) an arc of lush, fertile land formed by the uplands of the Tigris and Euphrates river systems and the Zagros Mountains, where nomadic peoples began the earliest known agricultural communities

fjord (p. 660) a long, narrow inlet formed by glacial erosion

floating population (p. 498) the Chinese term for people who live and work in a place other than their household registration location; many are people who have left economically depressed rural areas for the cities

floodplain (p. 14) the flat land around a river where sediment is deposited during flooding

food security (pp. 30, 505) the ability of a society to consistently supply a sufficient amount of basic food to the entire population; when people consistently have access to sufficient amounts of food to maintain a healthy life

food self-sufficiency (p. 557) the ability of an area, such as a country, to produce food domestically to meet the needs of its people

foreign exchange (p. 593) foreign currency that countries need to purchase imports

formal economy (p. 37) all aspects of the economy that take place in official channels

forward capital (p. 191) a capital city built in the hinterland to draw migrants and investment for economic development and sometimes for political/strategic reasons

fossil fuel (p. 315) a source of energy formed from the remains of dead plants and animals; fossil fuel includes oil, coal, and natural gas

fracking (p. 74) a drilling technology that uses high pressure injection of water to fracture rock formations, releasing natural gas and oil that they contain

free trade (p. 35) the unrestricted international exchange of goods, services, and capital

Gazprom (p. 278) in Russia, the state-owned energy company; it is the largest gas company in the world

gender (p. 53) the ways a particular social group defines the differences between the sexes

gender roles (p. 53) the socially assigned roles for males and females although exceptions and variations exist

genetic modification (GMO) (p. 32) in agriculture, the practice of splicing together the genes from widely divergent species to achieve particular desirable characteristics

genocide (pp. 42, 407) the deliberate destruction of an ethnic, racial, religious, or political group

gentrification (p. 100) the renovation of old urban districts by affluent investors, a process that often displaces poorer residents

Geographic Information Science (GISc) (p. 7) the body of science that supports multiple spatial analysis technologies and keeps them at the cutting edge

geopolitics (p. 41) the strategies that countries use to ensure that their own interests are served in relations with other countries

geospace (p. 668) the region of outer space closest to Earth, including the upper atmosphere and extending out to the furthest reaches of Earth's magnetic field to 40,000 miles (65,000 km)

geostationary orbit (p. 670) an orbit located at least 22,236 miles (35,786 kilometers) above a fixed point on Earth, in which satellites do not move relative to each other

glasnost (p. 276) literally, "openness"; the policies instituted in the late 1980s under Mikhail Gorbachev that encouraged greater transparency and openness in Soviet government and society

global economy (p. 33) the worldwide system in which goods, services, and labor are exchanged

globalization (p. 32) the growth of worldwide linkages and the changes these linkages are bringing about in daily life, especially in economies, but also in cultural and biological spheres

global scale (p. 11) the level of geography that encompasses the entire world as a single unified area

global warming (p. 20) the warming of Earth's climate as atmospheric levels of greenhouse gases increase

Gondwana (p. 622) the great landmass that formed the southern part of the ancient supercontinent Pangaea

grassroots economic development (p. 402) economic development projects designed to provide sustainable livelihoods in rural and urban areas; these often use simple technology that requires minimal or no investment in imported materials

Great Barrier Reef (p. 622) the longest coral reef in the world, located off the northeastern coast of Australia

Great Leap Forward (p. 516) a failed economic reform program under Mao Zedong intended to quickly raise China's industrial level

Green (p. 201) an adjective indicating a person or group that is environmentally conscious

greenhouse gases (GHG) (p. 20) gases, such as carbon dioxide and methane, released into the atmosphere by human activities, which become harmful when released in excessive amounts

green revolution (p. 31) increases in food production brought about through the use of new seeds, fertilizers, mechanized equipment, irrigation, pesticides, and herbicides

gross domestic product (GDP) per capita (p. 37) the total market value of all goods and services produced within a particular country's borders and within a given year, divided by the number of people in the country

gross national income (GNI) per capita (p. 37) the sum of a country's gross domestic product plus all net income received from overseas, divided by the mid-year population

groundwater (p. 385) water naturally stored in aquifers as long as 5000 years ago during wetter climate conditions

Group of Eight (G8) (p. 280) an organization of eight countries with large economies: France, the United States, Britain, Germany, Japan, Italy, Canada, and Russia

growth poles (p. 529) zones of development whose success draws more investment and migration to a region

guest workers (p. 235) legal workers from outside a country who help fulfill the need for temporary workers but who are expected to return home when they are no longer needed

gulag camp (p. 663) a settlement set up by the central government of the USSR, in which prisoners were sent to state-run enterprises to provide labor

Gulf states (p. 315) countries that border the Persian Gulf; Saudi Arabia, Kuwait, Bahrain, Oman, Qatar, and the United Arab Emirates

hacienda (p. 150) a large agricultural estate in Middle or South America, more common in the past; usually not specialized by crop and not focused on market production

hajj (p. 328) the pilgrimage to the city of Makkah (Mecca) that all Muslims are encouraged to undertake at least once in a lifetime

Harappa culture (p. 451) see *Indus Valley civilization*

Hinduism (p. 453) a major world religion practiced by approximately 900 million people, 800 million of whom live in India; a complex belief system, with roots both in ancient literary texts (known as the Great Tradition) and in highly localized folk traditions (known as the Little Tradition)

housing segregation (p. 100) a common phenomenon in cities, in which ethnic or racial or economic classes live in separate neighborhoods and the segregation is maintained by discriminatory customs, not law

Holocaust (p. 214) during World War II, a massive execution by the Nazis of 6 million Jews and 5 million gentiles (non-Jews), including ethnic Poles and other Slavs, Roma (Gypsies), disabled and mentally ill people, gays, lesbians, transgendered people, and political dissidents

Horn of Africa (p. 379) the triangular peninsula that juts out from northeastern Africa below the Red Sea and wraps around the Arabian Peninsula

hot spots (p. 623) individual sites of upwelling material (magma) that originate deep in Earth's mantle and surface in a tall plume; hot spots tend to remain fixed relative to migrating tectonic plates

horticulture (p. 383) hands-on farming that pays close attention to the specific needs of individual plants

hukou system (p. 498) the system in China where each person's permanent residence is registered and any person who wants to migrate must obtain permission from authorities to do so; the system has recently been liberalized

human geography (p. 3) the study of patterns and processes that have shaped human understanding, use, and alteration of Earth's surface

humanism (p. 209) a philosophy and value system that emphasizes the dignity and worth of the individual, regardless of wealth or social status

human well-being (p. 38) various measures of the extent to which people are able to obtain a healthy and socially rewarding standard of living in an environment that is safe and sustainable

immigration (p. 49) in-migration to a place or country (see also *migration*)

import substitution industrialization (ISI) (p. 150) policies that encourage local production of machinery and other items that previously had been imported at great expense from abroad

Incas (p. 145) indigenous people who ruled the largest pre-Columbian state in the Americas, with a domain stretching from southern Colombia to northern Chile and Argentina

income disparity (p. 149) the gap in income between rich and poor

indigenous (p. 134) native to a particular place or region

Indus Valley civilization (p. 451) the first substantial settled agricultural communities, which appeared about 4500 years ago along the Indus River in modern-day Pakistan and northwest India

industrial production (p. 33) processing, manufacturing, and construction

Industrial Revolution (p. 30) a series of innovations and ideas that occurred broadly between 1750 and 1850, which changed the way goods were manufactured

informal economy (p. 37) all aspects of the economy that take place outside official channels

intergalactic space (p. 668) the region of space between the Milky Way Galaxy and other galaxies

interplanetary space (p. 668) the region of space between the Sun and planets of our solar system

interstellar space (p. 668) the region of space between our solar system and others within the Milky Way Galaxy

infrastructure (p. 76) road, rail, and communication networks and other facilities necessary for economic activity

intertropical convergence zone (ITCZ) (p. 379) a band of atmospheric currents that circle the globe roughly around the equator; warm winds from both north and south converge at the ITCZ, pushing air upward and causing copious rainfall

intifada (p. 343) a prolonged Palestinian uprising against Israel

invasive species (p. 629) organisms that spread into regions outside of their native range, adversely affecting economies or environments

inversion (p. 545) the unusual condition when warm air overlies cool air, impeding normal circulation

Iron Curtain (p. 214) a long, fortified border zone that separated western Europe from (then) eastern Europe during the Cold War

Islam (p. 314) a monotheistic religion that emerged in the seventh century C.E. when, according to tradition, God revealed the tenets of the religion to the Prophet Muhammad

Islamic jihadism (p. 224) a personal or group struggle to promote Islamic revivalism, often with the threat of or actual use of force

Islamism (p. 314) a grassroots religious revival in Islam that seeks political power to curb what are seen as dangerous secular influences; also seeks to replace secular governments and civil laws with governments and laws guided by Islamic principles

isthmus (p. 134) a narrow strip of land that joins two larger land areas

Jainism (p. 454) a religion of Asia that originated as a reformist movement within Hinduism more than 2000 years ago; Jains are found mainly in western India and large urban centers throughout the region and are known for their educational achievements, nonviolence, and strict vegetarianism

jati (p. 453) in Hindu India, the subcaste into which a person is born, which traditionally defined the individual's experience for a lifetime

jihadists (p. 340) militant Islamists engaging in jihad, especially in opposition to Western influence

Judaism (p. 326) a monotheistic religion characterized by the belief in one god (Yahweh), a strong ethical code summarized in the Ten Commandments, and an enduring ethnic identity

just-in-time system (p. 519) the system pioneered in Japanese manufacturing that clusters companies that are part of the same production system close together so that they can deliver parts to each other precisely when they are needed

kaizen system (p. 519) a system of continuous manufacturing improvement, pioneered in Japan, in which production lines are constantly adjusted, improved, and surveyed for errors to save time and money and ensure that fewer defective parts are produced

Kessler scenario (p. 670) a situation where so much space junk accumulates around Earth that travel through low Earth orbit becomes impossible

krill (p. 661) tiny crustaceans that whales, seals, penguins, and fish feed on

Kyoto Protocol (p. 26) an amendment to a United Nations treaty on global warming, the Protocol is an international agreement, adopted in 1997 and in force in 2005, that sets binding targets for industrialized countries for the reduction of emissions of greenhouse gases

ladino (p. 178) a local term for mestizo, used in Central America

landforms (p. 12) physical features of Earth's surface, such as mountain ranges, river valleys, basins, and cliffs

Latino (p. 65) a term used to refer to all Spanish-speaking people from Middle and South America, although their ancestors may have been European, African, Asian, or Native American

latitude (p. 4) the distance in degrees north or south of the equator; lines of latitude run parallel to the equator, and are also called parallels

legend (p. 4) a small box somewhere on a map that provides basic information about how to read the map, such as the meaning of the symbols and colors used

less developed countries (LDCs) (p. 33) countries that have labor-intensive and low-wage, often agricultural, economies

liberation theology (p. 171) a movement within the Roman Catholic Church that uses the teachings of Jesus to encourage the poor to organize to change their own lives and to encourage the rich to promote social and economic equity

lingua franca (p. 421) language of trade

living wages (p. 36) minimum wages high enough to support a healthy life

local scale (p. 11) the level of geography that describes the space where an individual lives or works; a city, town, or rural area

loess (p. 540) windblown material that forms deep soils in China, North America, and central Europe

longitude (p. 4) the distance in degrees east and west of Greenwich, England; lines of longitude, also called meridians, run from pole to pole (the line of longitude at Greenwich is 0° and is known as the prime meridian)

low Earth orbit (LEO) (p. 670) an orbit that lies between 99 miles (160 kilometers) and 1243 miles (2000 kilometers) above Earth, in which satellites move relative to each other

machismo (p. 170) a set of values that defines manliness in Middle and South America

makatea (p. 623) coral platforms uplifted by volcanism, sometimes called seamounts

Maori (p. 623) Polynesian people indigenous to New Zealand

map projections (p. 7) the various ways of showing the spherical Earth on a flat surface

maquiladoras (p. 152) foreign-owned, tax-exempt factories, often located in Mexican towns just across the border from U.S. towns, that hire workers at low wages to assemble manufactured goods which are then exported for sale

marianismo (p. 170) a set of values based on the life of the Virgin Mary, the mother of Jesus, that defines the proper social roles for women in Middle and South America

marketization (p. 152) the development of a free market economy in support of free trade

marsupials (p. 626) mammals whose babies at birth are still at a very immature stage; the marsupial then nurtures them in a pouch equipped with nipples

Mediterranean climate (p. 199) a climate pattern of warm, dry summers and mild, rainy winters

Melanesia (p. 634) New Guinea and the islands south of the equator and west of Tonga (the Solomon Islands, New Caledonia, Fiji, and Vanuatu)

Melanesians (p. 634) a group of Australoids named for their relatively dark skin tones, a result of high levels of the protective pigment melanin; they settled throughout New Guinea and other nearby islands

mercantilism (pp. 149, 209) the policy by which Europeans sought to increase the power and wealth of their mother countries by managing all aspects of production, transport, and commerce in their colonies, as well as trade between colonies and the mother countries; a strategy for increasing a country's power and wealth by acquiring colonies and managing all aspects of their production, transport, and trade for the colonizer's benefit

mestizos (p. 148) people of mixed European, African, and indigenous descent

metropolitan areas (p. 96) cities of 50,000 or more and their surrounding suburbs and towns

microcredit (p. 461) a program based on peer support that makes very small loans available to very low-income entrepreneurs

Micronesia (p. 635) the small islands that lie east of the Philippines and north of the equator

Middle America (p. 135) in this book, a region that includes Mexico, Central America, and the islands of the Caribbean

midlatitude temperate climate (p. 199) as in south-central North America, China, and Europe, a climate that is moist all year with relatively mild winters and long, mild-to-hot summers

migration (p. 49) the movement of people from one place or country to another, often for safety or economic reasons

MIRAB economy (p. 639) an economy based on migration, remittances, and aid-supported bureaucracy

mixed agriculture (p. 383) farming that involves raising a variety of crops and animals on a single farm, often to take advantage of several environmental riches

Mongols (p. 273) a loose confederation of nomadic pastoral people centered in East and Central Asia, who by the thirteenth century had established by conquest an empire that stretched from Europe to the Pacific

mono-crop agriculture (p. 384) farming in which a single crop species is grown in a large field

monotheism (p. 326) the belief system based on the idea that there is only one god

monotremes (p. 626) egg-laying mammals, such as the duck-billed platypus and the spiny anteater

monsoon (p. 18) a wind pattern in which in summer months, warm, wet air coming from the ocean brings copious rainfall, and in winter, cool, dry air moves from the continental interior toward the ocean

more developed countries (MDCs) (p. 33) countries that have economies often generated by tertiary and quaternary sectors and that generally provide adequate education, health care, and other social services to help their people contribute to economic development

Mughals (p. 452) a dynasty of Central Asian origin that ruled India from the sixteenth century to the nineteenth century

multicultural society (p. 55) a society in which many culture groups live in close association

multinational corporation (p. 34) a business organization that operates extraction, production, and/or distribution facilities in multiple countries

Muslims (p. 328) followers of Islam

nationalism (p. 213) devotion to the interests or culture of a particular country, nation, or cultural group; the idea that a group of people living in a specific territory and sharing cultural traits should be united in a single country to which they are loyal and obedient

nationalize (p. 152) to seize private property and place under government ownership, with some compensation

nomadic pastoralists (p. 271) people whose way of life and economy are centered on tending grazing animals that are moved seasonally to gain access to the best pasture

nongovernmental organizations (NGOs) (p. 44) associations outside the formal institutions of government in which individuals, often from widely differing backgrounds and locations, share views and activism on political, social, economic, or environmental issues

nonpoint sources of pollution (p. 267) diffuse sources of environmental contamination, such as untreated automobile exhaust, raw sewage, and agricultural chemicals, that drain from fields into water supplies

North American Free Trade Agreement (NAFTA) (pp. 81, 153) a free trade agreement made in 1994 that added Mexico to the 1989 economic arrangement between the United States and Canada

North Atlantic Drift (p. 199) the easternmost end of the Gulf Stream, a broad, warm-water current that brings large amounts of warm water to the coasts of Europe

North Atlantic Treaty Organization (NATO) (p. 223) a military alliance between European (including Turkey) and North American countries, developed during the Cold War to counter the influence of the Soviet Union; after the breakup of the Soviet Union in 1991, NATO expanded to include much of eastern Europe. NATO now focuses on providing the international security and cooperation needed to expand the European Union

nuclear family (p. 109) a family consisting of a married father and mother and their children

occupied Palestinian Territories (oPT) (p. 315) Palestinian lands occupied by Israel since 1967

offshore outsourcing (p. 460) the contracting of certain business functions or production functions to providers in areas where labor and other costs are lower

oligarchs (p. 278) in Russia, those who acquired great wealth during the privatization of Russia's resources and who use that wealth to exercise power

OPEC (Organization of the Petroleum Exporting Countries) (p. 334) a cartel of oil-producing countries—currently, Algeria, Angola, Iran, Iraq, Kuwait, Libya, Nigeria, Qatar, Saudi Arabia, the United Arab Emirates, Indonesia, Ecuador, and Venezuela—that was established to regulate the production and price of oil

organically grown (p. 85) products produced without chemical fertilizers and pesticides

orographic rainfall (p. 18) rainfall produced when a moving moist air mass encounters a mountain range, rises, cools, and releases condensed moisture that falls as rain

Ottoman Empire (p. 330) influential Islamic empire centered in today's Turkey that lasted from the 1200s to the early 1900s

Outer Space Treaty of 1967 (p. 672) treaty that the United States and all spacefaring nations have signed and which forbids nations from claiming territory in space

Overseas Chinese (p. 539) Chinese emigrants and their descendants, especially those in Southeast Asia

Pacific Rim (p. 102) a term that refers to all the countries that border the Pacific Ocean

Pacific Solution (p. 644) Australia's and New Zealand's controversial approach to the excess number of undocumented immigrants

Pacific Way (p. 642) the regional identity and way of handling conflicts peacefully that grows out of Pacific Islanders' particular social experiences

Pancasila (p. 590) a national philosophy that embraces five precepts: belief in God and the observance of conformity, corporatism, consensus, and harmony

Partition (p. 457) the breakup following Indian independence that resulted in the establishment of Hindu India and Muslim Pakistan

pastoralism (p. 387) a way of life based on herding; practiced primarily on savannas, on desert margins, or in the mixture of grass and shrubs called open bush

patriarchal (p. 353) relating to a social organization in which the father is supreme in the clan or family

perestroika (p. 276) literally, "restructuring"; the restructuring of the Soviet economic system that was done in the late 1980s in an attempt to revitalize the economy

permafrost (p. 265) permanently frozen soil that lies just beneath the surface

physical geography (p. 3) the study of Earth's physical processes: how they work and interact, how they affect humans, and how they are affected by humans

pidgin (p. 654) a language used for trading; one made up of words borrowed from the several languages of people involved in trading relationships

plantation (p. 150) a large factory farm that grows and partially processes a single cash crop

plate tectonics (p. 12) the scientific theory that Earth's surface is composed of large plates that float on top of an underlying layer of molten rock; the movement and interaction of the plates create many of the large features of Earth's surface, particularly mountains

polar amplification (p. 661) the increase in the intensity of climate change at the poles

political ecology (p. 38) the study of the interactions among development, politics, human well-being, and the environment

political freedoms (p. 40) the rights and capacities that support individual and collective liberty and public participation in political decision making

polygyny (p. 418) the practice of having multiple wives

Polynesia (p. 635) the numerous islands situated inside an irregular triangle formed by New Zealand, Hawaii, and Easter Island

population pyramid (p. 49) a graph that depicts the age and sex structures of a political unit, usually a country

populist political movement (p. 589) a coalition of usually rural or working class activists who claim that virtuous citizens are being exploited by a small group of elites who are presumed to be in illegitimate control of the government

post-colonial (p. 226) the conditions and attitudes that persist in a society after colonization is over

precipitation (p. 15) dew, rain, sleet, hail, and snow

primary sector (p. 33) economic activity based on extracting resources (mining, forestry, and agriculture)

primate city (p. 161) a city, plus its suburbs, that is vastly larger than all others in a country and in which economic and political activity is centered

privatization (pp. 152, 278) the selling of formerly government-owned industries and firms to private companies or individuals; the sale of industries that were formerly owned and operated by the government to private companies or individuals

pull factors (p. 104) positive factors that draw in migrants

purchasing power parity (PPP) (p. 37) the amount that the local currency equivalent of U.S.$1 will purchase in a given country

purdah (p. 477) the practice of concealing women from the eyes of nonfamily men

push factors (p. 104) negative factors that cause people to leave home, family, and friends

push/pull phenomenon of urbanization (p. 45) conditions, such as political instability or economic changes, that encourage (push) people to leave rural areas, and urban factors, such as job opportunities, that encourage (pull) people to move to urban areas

qanats (p. 548) underground tunnels, built by ancient cultures and still used today, that carry groundwater for irrigation in dry regions

quaternary sector (p. 33) economic activities based on intellectual pursuits in education, research, and information technology

Qur'an (or **Koran**) (p. 318) the holy book of Islam, believed by Muslims to contain the words Allah revealed to Muhammad through the archangel Gabriel

race (p. 58) a social or political construct that is based on apparent characteristics such as skin color, hair texture, and face and body shape, but that is of no biological significance

rate of natural increase (RNI) (p. 49) the rate of population growth measured as the excess of births over deaths per 1000 individuals per year without regard for the effects of migration

redlining (p. 100) The practice used by banks of drawing red lines on urban maps to show where they will not approve loans for home buyers regardless of qualifications

refugees (p. 235) those who have fled from their homes because of war or violent social unrest or discrimination

region (p. 9) a unit of Earth's surface that contains distinct patterns of physical features and/or distinct patterns of human development

regional conflict (p. 464) a conflict created by the resistance of a regional ethnic or religious minority to the authority of a national or state government; currently, these are the most intense armed conflicts in South Asia

regional self-sufficiency (p. 521) an economic policy in Communist China that encouraged each region to develop independently in the hope of evening out the wide disparities in the national distribution of production and income

regional specialization (p. 521) specialization (rather than self-sufficiency) that takes advantage of regional variations in climate, natural resources, and location

religious nationalism (p. 463) the association of a particular religion with a particular territory or political unit to the exclusion of other religions

resettlement schemes (p. 590) government plans to move large numbers of people from one part of a country to another to relieve urban congestion, disperse political dissidents, or accomplish other social purposes; also called *transmigration*

Ring of Fire (p. 13) the tectonic plate junctures around the edges of the Pacific Ocean; characterized by volcanoes and earthquakes

Roaring Forties (p. 623) powerful air and ocean currents at about 40° S latitude that speed around the far Southern Hemisphere virtually unimpeded by landmasses

Roma (p. 214) the now-preferred term in Europe for Gypsies; some prefer to be called Gypsies as a point of ethnic identity

Russian Federation (p. 264) Russia and its political subunits, which include 21 internal republics

Russification (p. 285) the czarist and Soviet policy of encouraging ethnic Russians to settle in non-Russian areas as a way to assert political control

Sahel (p. 377) a band of arid grassland, where steppe and savanna grasses grow, that runs east-west along the southern edge of the Sahara

Salafism (p. 340) a religious movement that started in Egypt and that has influenced the ideology of radical Islamism

salinization (p. 321) a process that occurs when large quantities of water are used to irrigate areas where evaporation rates are high, leaving behind salts and other minerals

satellite (p. 669) is any object in space that orbits another object of greater mass

scale (of a map) (p. 4) the proportion that relates the dimensions of the map to the dimensions of the area it represents; also, variable-sized units of geographical analysis from the local scale to the regional scale to the global scale

Schengen Agreement (p. 218) an agreement first signed in 1985 and implemented in 1995 that allows for free movement across common borders in the European Union

seawater desalination (p. 322) the removal of salt from seawater—usually accomplished through the use of expensive and energy-intensive technologies—to make the water suitable for drinking or irrigating

secondary sector (p. 33) economic activity based on processing, manufacturing, and construction

secular states (p. 340) countries that have no state religion and in which religion has no direct influence on affairs of state or civil law

Security Council (p. 44) a division of the UN with five permanent members (the United States, Britain, China, France, and Russia) and ten rotating members

self-reliant development (p. 402) small-scale development in rural areas that is focused on developing local skills, creating local jobs, producing products or services for local consumption, and maintaining local control so that participants retain a sense of ownership over the process

services (p. 33) sales, entertainment, and financial services

sex (p. 53) the biological category of male, female, intersex; does not indicate how people may behave or identify themselves

sex tourism (p. 603) the sexual entertainment industry that services primarily men who travel for the purpose of living out their fantasies during a few weeks of vacation

sex work (p. 470) the provision of sexual acts for a fee

shari'a (p. 329) Islamic religious law that guides daily life according to the interpretations of the Qur'an

sheikhs (p. 363) traditionally, patriarchal leaders who controlled the use of lands shared among several ancestral tribal groups; today, the title is still used to denote someone in a leadership position

Shi'ite (or **Shi'a**) (p. 329) the smaller of two major groups of Muslims; Shi'ites are found primarily in Iran and southern Iraq

shifting cultivation (pp. 144, 383) a productive system of agriculture in which small plots are cleared in forestlands, the dried brush is burned to release nutrients, and the clearings are planted with multiple species; each plot is used for only 2 or 3 years and then abandoned for many years of regrowth

Siberia (p. 264) the territories of Russia that are located east of the Ural Mountains

Sikhism (p. 454) a religion of South Asia that combines beliefs of Islam and Hinduism

silt (p. 136) fine soil particles

Slavs (p. 273) a group of people who originated in what are now Poland, Ukraine, and Belarus

slum (p. 45) a densely populated area characterized by crowding, run-down housing, and inadequate access to food, clean water, education, and social services

smog (p. 70) a combination of industrial emissions, car exhaust, and water vapor that frequently hovers as a yellow-brown haze over many cities, causing a variety of health problems

social cohesion (pp. 226, 584) an expression of solidarity within the EU-28, involving balanced and sustainable development, reduced structural disparities between regions and countries, and the promotion of equal opportunities for all people; the willingness of members of a society to cooperate with each other in order for all to survive and prosper

social exclusion (p. 226) the condition of being systematically left out of important opportunities for participation in society

social protection (p. 226) in Europe, tax-supported systems that provide citizens with benefits such as health care, long-term care, pensions, and relief from poverty and social exclusion

social safety net (p. 93) the services provided by the government—such as welfare, unemployment benefits, and health care—that prevent people from falling into extreme poverty

socialism (p. 214) primarily an economic system in which all have access to basic goods and services and all large-scale industries are collectively owned, with any profits benefitting society as a whole

soft power initiatives (p. 589) diplomatic overtures or offers of aid that appear to be only friendly with no ulterior geopolitical motives

solar wind (p. 669) tiny charged particles emitted by the Sun that travel at millions of miles per hour and form a gas-like plasma that can conduct electricity and respond to magnetic fields

South America (p. 135) the continent south of Central America

Soviet Union (p. 262) see *Union of Soviet Socialist Republics*

space junk (p. 670) non-functioning satellites

spatial distribution (p. 3) the arrangement of a phenomenon across Earth's surface

spatial interaction (p. 3) the flow of goods, people, services, or information across space and among places

special economic zones (SEZs) (p. 529) free trade zones within China, which are commonly called export processing zones (EPZs) elsewhere

state (p. 40) an organized political community in a defined territory and under one government

state-aided market economy (p. 519) an economic system based on market principles but with strong government guidance; in contrast to the limited government, free market economic system of the United States and, to a lesser degree, Europe

steppes (p. 264) semiarid, grass-covered plains

structural adjustment (p. 36) belt-tightening measures imposed to force a government to make its economy more open to market forces; often involves closing or privatizing government enterprises and reducing education, health-care programs, and other social services that benefit the poor

structural adjustment programs (SAPs) (p. 151) policies that require economic reorganization toward less government involvement in industry, agriculture, and social services; sometimes imposed by the World Bank and the International Monetary Fund as conditions for receiving loans

subcontinent (p. 440) a term often used to refer to the entire Indian peninsula, including Nepal, Bhutan, India, Pakistan, and Bangladesh

subduction zone (p. 135) a zone where one tectonic plate slides under another

subsidies (p. 220) monetary assistance granted by a government to an individual or group in support of an activity, such as farming, that is viewed as being in the public interest

subsistence affluence (p. 639) a lifestyle whereby people are self-sufficient with regard to most necessities and have some opportunities to save cash for travel and occasional purchases of manufactured goods

subsistence agriculture (p. 383) farming that provides food for only the farmer's family and is usually done on small farms

subsistence economy (p. 52) an economy in which families produce most of their own food, clothing, and shelter

suburbs (p. 96) populated areas along the peripheries of cities

summer monsoon (p. 441) rains that begin every June when the warm, moist ITCZ air first reaches the mountainous Western Ghats

Sunni (p. 329) the larger of two major groups of Muslims

sustainable agriculture (p. 32) farming that meets human needs without poisoning the environment or using up water and soil resources

sustainable development (p. 38) the improvement of current standards of living in ways that will not jeopardize those of future generations

taiga (p. 265) subarctic coniferous forests

Taliban (p. 466) an archconservative Islamist movement that gained control of the government of Afghanistan for a while in the mid-1990s

temperature-altitude zones (p. 136) regions of the same latitude that vary in climate according to altitude

tertiary sector (p. 33) economic activity based on sales, entertainment, and financial services

theocratic states (p. 340) countries that require all government leaders to subscribe to a state religion and all citizens to follow rules decreed by that religion

total fertility rate (TFR) (p. 49) the average number of children that women in a country are likely to have at the present rate of natural increase

trade deficit (p. 81) the extent to which the money earned by exports is exceeded by the money spent on imports

trade winds (p. 136) winds that blow from the northeast and the southeast toward the equator

trafficking (p. 296) the recruiting, transporting, and harboring of people through coercion for the purpose of exploiting them

Trans-Pacific Partnership (TPP) (pp. 82, 641) a trade agreement signed by 12 Pacific Rim countries, including the United States, Japan, Mexico, Canada, Australia, Malaysia, Vietnam, Singapore, Chile, and Peru

transgender (p. 54) someone who has a gender identity that differs from the gender they were assigned at birth

transparency (p. 403) in politics, the state of being open to observation and participation by the public

tsunami (p. 500) a large sea wave caused by an earthquake

tundra (pp. 121, 265) a region of winters so long and cold that the ground is permanently frozen several feet below the surface; a treeless area, between the ice cap and the tree line of arctic regions, where the subsoil is permanently frozen

typhoon (p. 501) a tropical storm in the western Pacific Ocean; called a cyclone or hurricane elsewhere

UNASUR (p. 154) a union of South American nations that was organized in May of 2008; it emulates the European Union, linking the economies of Argentina, Bolivia, Brazil, Chile, Colombia, Ecuador, Guyana, Paraguay, Peru, Suriname, Uruguay, and Venezuela

underemployment (p. 281) the condition in which people are working too few hours to make a decent living or are highly trained but working at menial jobs

Union of Soviet Socialist Republics (USSR) (p. 262) the multinational union formed from the Russian empire in 1922 and dissolved in 1991; commonly known as the *Soviet Union*

United Nations (UN) (p. 44) an assembly of 193 member states that sponsors programs and agencies that focus on economic development, security, general health and well-being, democratization, peacekeeping assistance in hot spots around the world, humanitarian aid, and scientific research

United Nations Human Development Index (HDI) (p. 38) an index that calculates a country's level of well-being, based on a formula of factors that considers income adjusted to PPP, data on life expectancy at birth, and data on educational attainment

United Nations Gender Development Index (GDI) (p. 38) a composite measure of disparity in human development (longevity, education, and income) by gender. The closer the rank is to 1, the smaller the gap between women and men; that is, the more that the genders are tending toward equality

urban growth poles (p. 164) locations within cities that are attractive to investment, innovative immigrants, and trade, and thus attract economic development like a magnet

urban sprawl (p. 72) the encroachment of suburbs on agricultural land

urbanization (p. 45) the process whereby cities, towns, and suburbs grow as populations shift from rural to urban livelihoods

value chains (p. 402) links between various aspects of a production line to maximize efficiency and profits

varna (p. 453) the four hierarchically ordered divisions of society in Hindu India underlying the caste system: Brahmins (priests), Kshatriyas (warriors/kings), Vaishyas (merchants/landowners), and Sudras (laborers/artisans)

veil (p. 355) the custom of covering the body with a loose dress and/or of covering the head—and in some places the face—with a scarf

virtual water (p. 27) the water used to produce a product, such as an apple or a pair of shoes

Wahhabism (p. 340) a religious movement that started in Saudi Arabia and that has influenced the ideology of radical Islamism

water footprint (p. 27) all the water a person consumes, including both virtual water and the water they consume directly

water scarcity (p. 446) a lack of sufficient water resources to meet water needs

weather (p. 15) the short-term and spatially limited expression of climate that can change in a matter of minutes

weathering (p. 14) the physical or chemical decomposition of rocks by sunlight, rain, snow, ice, and the effects of life-forms

welfare state (p. 213) a government that accepts responsibility for the well-being of its people, guaranteeing basic necessities such as education, affordable food, employment, housing, and health care for all citizens

West Bank barrier (p. 345) an Israeli-built concrete wall or fence that now surrounds much of the West Bank and encompasses many of the Jewish settlements there

wet rice cultivation (p. 506) a prolific type of rice production that requires the plant roots to be submerged in water for part of the growing season

winter monsoon (p. 442) a weather pattern that begins by November, when the cooling Eurasian landmass sends the cooler, drier, heavier air over South Asia

World Trade Organization (WTO) (p. 36) a global institution made up of member countries whose stated mission is to lower trade barriers and to establish ground rules for international trade

yurts (or **gers**) (p. 548) round, heavy, felt tents stretched over collapsible willow lattice frames used by nomadic herders in northwestern China, Mongolia, and Central Asia

Zionism (p. 342) a movement that started in nineteenth-century Europe to create a Jewish homeland in Palestine

Facts have been gathered from the following sources for text on the pages indicated.

CHAPTER 1

35 Adapted from Lydia Pulsipher's field notes, 1992–2000 (Olivia and Tanya), and Alex Pulsipher's field notes, 1999–2008 (Setiya).
56 Adapted from Lydia Pulsipher's field notes in Honolulu, 1995.

CHAPTER 2

65 Adapted from Jesse McKinley, "Drought Adds to Hardships in California," *New York Times*, February 21, 2009; David Leonhardt, "Job Losses Show Breadth of Recession," *New York Times*, March 3, 2009; Catherine Rampell, "As Layoffs Surge, Women May Pass Men in Job Force," *New York Times*, February 5, 2009; Alissa Figueroa, *PBS Newshour*, "In Calif. Town, Prison May Fix One Problem, but Create Another," May 5, 2011, at http://www.pbs.org/newshour/bb/social_issues-jan-june11-mendota_05-05/.
87 Adapted from NPR Staff, *All Things Considered*, "What Recession? It's Boom Time for Nebraska Farms," February 25, 2011, at http://tinyurl.com/6jyljoa.

CHAPTER 3

132 Oxfam America, 2005; Juan Forero, "Rain Forest Residents, Texaco Face Off In Ecuador," *Morning Edition*, National Public Radio, April 30, 2009, at http://www.npr.org/templates/story/story.php?storyId=103233560; Gonzalo Solano, "Damages in Chevron Ecuador Suit Jump Billions," *San Francisco Chronicle*, September 18, 2010, at http://www.sfgate.com/default/article/Damages-in-Chevron-Ecuador-suit-jump-billions-3173946.php; Patrick Radden Keefe, "Reversal of Fortune," *New Yorker*, January 9, 2012, pp. 38–49; Reuters, "Ecuador: Villagers Try Again to Hold Chevron to $18bn Ruling," June 29, 2012, at http://www.independent.co.uk/news/world/americas/ecuador-villagers-try-again-to-hold-chevron-to-18bn-ruling-7897095.html.
152 NPR reports by John Idste, August 14, 2001, and August 16 and 25, 2003, and by Gary Hadden, August 27, 2003; David Bacon, "Anti-China Campaign Hides Maquiladora Wage Cuts," ZNet, February 3, 2003, at http://zcomm.org/znetarticle/anti-china-campaign-hides-maquiladora-wage-cuts-by-david-bacon; Maquila Portal, April 2006, at http://www.maquilaportal.com/cgi-bin/public/index.pl; "How Rising Wages Are Changing the Game in China,"

Bloomberg Businessweek, March 26, 2006, at http://www.businessweek.com/stories/2006-03-26/how-rising-wages-are-changing-the-game-in-china; Pete Engardio, "So Much for the Cheap 'China Price,'" *Bloomberg Businessweek*, June 4, 2009, at http://www.businessweek.com/magazine/content/09_24/b4135054963557.htm; "Avalanche of Chinese Investment in Mexico," Maquila Portal, December 1, 2011, at http://www.maquilaportal.com/index.php?blog/show/Avalanche-of-Chinese-investment-in-Mexico.html; Gustavo de Lima Palhares and Higor Uzzun Sales, "An Economic Analysis of the Role of Foreign Investments in Brazil," April 18, 2012, Council on Hemispheric Affairs, at http://www.coha.org/an-economic-analysis-of-the-role-of-foreign-investments-in-brazil.
154 Adapted from Andrew Wheat, "Toxic Bananas," *Multinational Monitor* 17: 6–7, September 1996 (updated 2007); David Magney, "Costa Rican Bananas," David Magney Environmental Consulting, December 16, 2005 (updated July 9, 2007), at http://www.magney.org/photofiles/CostaRica-Bananas1.htm.
164 Lydia Pulsipher's field notes on Brazil, updated with the help of John Mueller, Fortaleza, Brazil, 2006.

CHAPTER 4

203 Adapted from Rob Gifford, "Gardeners Brighten London Under Cover of Dark," *Morning Edition*, National Public Radio, May 15, 2006, at http://www.npr.org/templates/story/story.php?storyId=5404229.
222 Lydia Pulsipher's conversations with Vera Kuzmic and Dusan Kramberger, 1993–2012.
224 Source consulted: Democratization Index, adapted from "The Democracy Index 2014: Democracy Under Stress," Economist Intelligence Unit, at http://www.eiu.com/Handlers/Whitepaper Handler.ashx?fi=Democracy-index-2014.pdf&mode=wp&campaignid=Democracy0115]
257 Conversations with geographer Margareta Lelea, a Romanian specialist and post-doctoral researcher, Departments of Entomology and Human and Community Development, University of California, Davis; Doreen Carvajal and Stephen Castle, "A U.S. Hog Giant Transforms Eastern Europe," *New York Times*, May 5, 2009, at http://www.nytimes.com/2009/05/06/business/global/06smithfield.html?

pagewanted=all&module=Search&mabReward=relbias%3Ar.

CHAPTER 5

262 Ellen Barry and Andrew E. Kramer, "In Biting Cold, Protesters Pack the Center of Moscow," *New York Times*, February 4, 2012, at http://www.nytimes.com/2012/02/05/world/europe/tens-of-thousands-protest-putin-in-moscow-russia.html?pagewanted=all; Simon Saradzhyan and Nabi Abdullaev, "Putin Election Victory Doesn't Pave an Easy Path Through His Third Presidential Term," *Christian Science Monitor*, March 5, 2012, at http://www.csmonitor.com/Commentary/Opinion/2012/0305/Putin-election-victory-doesn-t-pave-an-easy-path-through-his-third-presidential-term; Ellen Barry and Michael Schwirtz, "After Election, Putin Faces Challenges to Legitimacy," *New York Times*, March 5, 2012, at http://www.nytimes.com/2012/03/06/world/europe/observers-detail-flaws-in-russian-election.html?pagewanted=all; Michael Schwirtz, "Fear of Return to '90s Hardship Fuels Support for Putin," *New York Times*, March 3, 2012, at http://www.nytimes.com/2012/03/04/world/europe/in-russia-vote-fear-of-hardship-fuels-putin-support.html?pagewanted=all; Amanda Walker, "Last Anti-Putin Rally Before Russian Election," *Sky News*, February 26, 2012, at http://news.sky.com/story/929127/last-anti-putin-rally-before-russian-election.
270 "Shrinking Aral Sea," NASA Earth Observatory, August 25, 2000, at http://earthobservatory.nasa.gov/Features/WorldOfChange/aral_sea.php; "Central Asia: Uzbekistan," *World Factbook*, Central Intelligence Agency, at https://www.cia.gov/library/publications/the-world-factbook/geos/uz.html; "Cotton Fact Sheet: Uzbekistan," ICAC, at http://www.icac.org/econ_stats/country_fact_sheets/fact_sheet_uzbekistan_2011.pdf.
285 Source consulted: "Ukraine Crisis in Maps," *New York Times*, http://www.nytimes.com/interactive/2014/02/27/world/europe/ukraine-divisions-crimea.html?_r=0,
288 Sources consulted: 2015 World Population Data Sheet, Population Reference Bureau, at http://www.prb.org/pdf15/2015-world-population-data-sheet_eng.pdf; Demographia World Urban Areas, 11th Annual Edition: 2015, at http://www.demographia.com/

db-worldua.pdf; Timothy Heleniak, "Russia's Demographic Decline Continues," Population Reference Bureau, June 2002, at http://www.prb.org/Publications/Articles/2002/RussiasDemographicDeclineContinues.aspx; "Half of All Premature Deaths of Russian Adults Down to Alcohol," University of Oxford, June 26, 2009, at http://www.ox.ac.uk/media/news_stories/2009/090626.html.
290 John Lancaster, "Tomorrowland," *National Geographic*, February 2012, at http://ngm.nationalgeographic.com/2012/02/astana/lancaster-text.
297 "Overcoming Barriers: Human Mobility and Development," Human Development Report 2009, United Nations Development Program, at http://hdr.undp.org/en/content/human-development-report-2009.
298 Source consulted: Aftonbladet 2015 (http://www.aftonbladet.se/sportbladet/kronikorer/erikniva/article21361496.ab)

CHAPTER 6

316 Mona Eltahawy, "Why Do They Hate Us?" *Foreign Policy*, April 23, 2012, at http://www.foreignpolicy.com/articles/2012/04/23/why_do_they_hate_us; Allison Good, "Debating the War on Women," *Foreign Policy*, April 24, 2012, at http://www.foreignpolicy.com/articles/2012/04/24/debating_the_war_on_women; "Talk to Al Jazeera: Why Arab Women Still 'Have No Voice,'" at http://www.aljazeera.com/programmes/talktojazeera/2012/04/201242111373249723.html.
321 A. M. MacDonald, H. C. Bonsor, B. E. O. Dochartaigh, and R. G. Taylor, "Quantitative Maps of Groundwater Resources in Africa," *Environmental Research Letters*, April 19, 2012, at http://iopscience.iop.org/1748-9326/7/2/024009/pdf/1748-9326_7_2_024009.pdf.

CHAPTER 7

376 Adapted from "Juliana Rotich— This Is What I See! Crowdsourced Crisis Mapping in Real Time," 99 *Faces*, March 17, 2011, at http://99faces.tv/julianarotich/; Ushahidi Downloads, at http://download.ushahidi.com/; "Africa: The Next Chapter," *TED Radio Hour*, National Public Radio, at http://www.npr.org/2012/06/29/155904209/africa-the-next-chapter.
381 Adapted from "Silas Kpanan-'Ayoung Siakor: A Voice for the Forest

and Its People," Goldman Environmental Prize, at http://www.goldmanprize.org/node/442; Scott Simon, "Reflections of a Liberian Environmental Activist," *Weekend Edition Saturday*, National Public Radio, April 29, 2006, at http://www.npr.org/templates/story/story.php?storyId=5370987; Lydia Pulsipher's personal communication with Siakor, September 2006.
385 *Scientific American*, January, 2016, p. 48.
392 http://www.bbc.com/news/world-africa-33259003
396 Kaoru Kimura, Duncan Wambogo Omole, and Mark Williams, "ICT in Sub-Saharan Africa: Success Stories," Engaging the Private Sector to Upgrade Infrastructure, World Bank, at http://siteresources.worldbank.org/AFRICAEXT/Resources/258643-1271798012256/ICT-19.pdf.
396 http://www.bbc.com/news/world-africa-20138322.
399 "Africa: Five Years Later, Airtel's Africa Operations Still Struggling to Hit Targets," *Wall Street Journal*, All Africa website, September 2015, http://allafrica.com/stories/201509230480.html.
400 Emeka Umejei, "China's Engagement with Nigeria: Opportunity or Opportunist?" African East Asian Affairs, December 2015, issues 3 and 4, p. 56ff.
401 Deputy Directors-General, "'Africa Should Trade More with Africa to Secure Future Growth'—DDG Rugwabiza," World Trade Organization, April 12, 2012, at http://www.wto.org/english/news_e/news12_e/ddg_12apr12_e.htm.
407 "Topics: Zimbabwe," *New York Times*, at http://www.nytimes.com/topic/destination/zimbabwe.
411 "Growing Greener Cities in Africa," Food and Agriculture Organization of the United Nations (Rome: UNFAO, 2012), at http://www.fao.org/docrep/016/i3002e/i3002e.pdf.
415 "Issues," Eastern and Southern Africa: HIV and AIDS, UNICEF, at http://www.unicef.org/esaro/5482_HIV_AIDS.html.
418 James Fenske, "African Polygamy: Past and Present," February 15, 2012, at https://editorialexpress.com/cgi-bin/conference/download.cgi?db_name=CSAE2012&paper_id=115.
426 Kevin Sullivan, "Bridging the Digital Divide," *Washington Post National Weekly Edition*, July 17–23, 2006, pp. 10–11; *Africa The Good News*, February 11, 2010, at http://www.africagoodnews.com/ict/africa-448-million-mobile-

network-subscribers.html; Tolu Ogunlesi and Stephanie Busari, "Seven Ways Mobile Phones Have Changed Lives in Africa," CNN, September 14, 2012, at http://www.cnn.com/2012/09/13/world/africa/mobile-phones-change-africa/; Christopher Schuetze, "Sustainable Innovation: The Ethanol Stove," *International Herald Tribune*, October 22, 2012, at http://rendezvous.blogs.nytimes.com/2012/10/22/sustainable-innovation-the-ethanol-stove/.
429 Christina Russo, "Shining a Bright Light on Africa's Elephant Slaughter," *Yale Environment 360*, September 20, 2012, at http://e360.yale.edu/feature/shining_a_bright_light_on_africas_elephant_slaughter/2575/#.UFx7s0lMklk.twitter.
432 "Despite Grim Headlines, Africa Is Booming," *Talk of the Nation*, National Public Radio, July 11, 2012, at http://www.npr.org/2012/07/11/156621234/despite-grim-headlines-africa-is-booming.
432 Famine Early Warning System Network, "Malawi Food Security Update," USAID, February 17, 2010, at http://www.reliefweb.int/rw/rwb.nsf/db900sid/SMAR-82S447?OpenDocument; Bob Carr, "Australia Boosts Food Security Commitment to Africa," Australian Minister for Foreign Affairs, October 9, 2012, at http://reliefweb.int/report/kenya/australia-boosts-food-security-commitment-africa.

CHAPTER 8
438 Adapted from "Modi Goes on Fast over Narmada Dam," *India eNews*, April 16, 2006; Rahul Kumar, "Medha Patkar Ends Fast After Court Order on Rehabilitation," One World South Asia.
449 Lydia and Alex Pulsipher's field notes, Nilgiri Hills, June 2000; Government of Tamil Nadu, *Tamil Nadu Human Development Report* (Delhi: Social Sciences Press, 2003), at http://hdr.undp.org/sites/default/files/tn_nhdr_2003.pdf.
463 Beth Roy, *Some Trouble with Cows—Making Sense of Social Conflict* (Berkeley: University of California Press, 1994), pp. 18–19.
463 Jim Yardley, "Protests Awaken a Goliath in India," India's Way, *New York Times*, October 29, 2011, at http://www.nytimes.com/2011/10/30/world/asia/indias-middle-class-appears-to-shed-political-apathy.html.
468 "India Sees Explosion of Mobile Technology," Tell Me More, National Public Radio, July 20, 2010, at http://www.npr.

org/2010/07/20/128645261/india-see-explosion-of-mobile-technology.
479 Internews Afghanistan, March 2008, at http://www.internews.org/bulletin/afghanistan/Afghan_200803.html; Maria Fantappie and Brittany Tanasa, "Fearless Women Use Radio to Make Ripples of Change in Herat, Afghanistan," September 7, 2011, at http://cima.ned.org/fearless-women-use-radio-make-ripples-change-herat-afghanistan.
485 Adapted from Eric Valli, "Himalayan Caravans," *National Geographic*, December 1993, L 5–35; see also the movie *Himalaya*.

CHAPTER 9
498 Adapted from Peter S. Goodman, "In China's Cities, a Turn from Factories," *Washington Post*, September 25, 2004, at http://www.washingtonpost.com/wp-dyn/articles/A48818-2004Sep24.html; Louisa Lim, "The End of Agriculture in China," *Reporter's Notebook*, National Public Radio, May 19, 2006, at http://www.npr.org/templates/story/story.php?storyId=5411325; background information from Kathy Chen, "Boom-Town Bound," *Wall Street Journal*, October 29, 1996, p. A6; "Life Lessons," *Wall Street Journal*, July 9, 1997.
498 "China's 'Floating Population' Exceeds 221 mln," China.org.cn, March 1, 2011, at http://www.china.org.cn/china/2011-03/01/content_22025827.htm.
503 Adapted from Maria Siow, "Desertification: One of the Challenges Faced by China," *Asia Pacific News*, October 29, 2009, at http://www.channelnewsasia.com/stories/eastasia/view/1014326/1/.html.
503 Associated Press, "China Flooding Has Killed 701 in 2010; Worst Toll Since 1998," *USA Today*, July 21, 2010, at http://usatoday30.usatoday.com/news/world/2010-07-21-china-flooding_N.htm; "China Faces Worst Flooding in 12 Years on Yangtze," Reuters, July 14, 2010, at http://www.reuters.com/article/2010/07/15/us-asia-weather-china-idUSTRE66E0ES20100715.
507 http://www.st.nmfs.noaa.gov/st1/fus/fus10/08_perita2010.pdf; "FAO State of World Fisheries, Aquaculture Report—Fish Consumption," *The Fish Site*, September 10, 2012, at http://www.thefishsite.com/articles/1447/fao-state-of-world-fisheries-aquaculture-report-fish-consumption; Mike Stones, "Fish Consumption Hits All Time High: FAO Report," *Food Navigator*, February 2, 2011, at http://www.food-navigator.com/Financial-Industry/

Fish-consumption-hits-all-time-high-FAO-report.
520 Mark Manyin and Mary Beth Nikitin, "Foreign Assistance to North Korea," Congressional Research Service 7-5700, June 11, 2013, at http://www.fas.org/sgp/crs/row/R40095.pdf; Michael Wines, "Beijing's Air Is Cleaner, but Far from Clean," *New York Times*, October 16, 2009, at http://www.nytimes.com/2009/10/17/world/asia/17beijing.html?partner=rss&emc=rss&_r=0.
525 "World Report 2012: China," Human Rights Watch, at http://www.hrw.org/world-report-2012/world-report-2012-china; "Chinese Social Network Renren Sees Losses Increase," *News Business*, BBC, May 14, 2012, at http://www.bbc.co.uk/news/business-18068128.
529 "America's Transport Infrastructure: Life in the Slow Lane," *Economist*, April 28, 2011, at http://www.economist.com/node/18620944; Yukon Huang, "Reinterpreting China's Success Through the New Economic Geography," Carnegie Endowment for International Peace, November 10, 2010, at http://carnegieendowment.org/2010/11/10/reinterpreting-china-s-success-through-new-economic-geography/1lyq; Matthew Stadlen, "China Opens World's Longest High-Speed Rail Route," *News China*, BBC, December 25, 2012, at http://www.bbc.co.uk/news/world-asia-china-20842836.
533–534 "Gender Statistics," World Bank, at http://data.worldbank.org/data-catalog/gender-statistics; "Field Listing: Sex Ratio," *World Factbook*, Central Intelligence Agency, at https://www.cia.gov/library/publications/the-world-factbook/fields/2018.html; Rob Brooks, "China's Biggest Problem? Too Many Men," *CNN Opinion*, March 4, 2013, at http://www.cnn.com/2012/11/14/opinion/china-challenges-one-child-brooks/index.html; http://www.census.gov/population/international/data/idb/region.php.
537 Preeti Bhattacharji,"Uighurs and China's Xinjiang Region," *Backgrounder*, Council on Foreign Relations, May 29, 2012, at http://www.cfr.org/china/uighurs-chinas-xinjiang-region/p16870.
538–539, 540 "World Directory of Minorities and Indigenous Peoples—Japan: Ainu," *RefWorld*, UNHCR, June 2008, at http://www.refworld.org/docid/49749cfe23.html.
539 "East & Southeast Asia: Singapore," *World Factbook*, Central Intelligence Agency, at

https://www.cia.gov/library/publications/the-world-factbook/geos/sn.html.

541 "China's Five Largest Oil Producing Provinces and Cities," *China Oil Trader*, October 4, 2012, at http://www.chinaoiltrader.com/?p=202.

545 "Chengdu Population Fourth Largest in Chinese Cities," *China Daily*, July 7, 2011, at http://www.chinadaily.com.cn/china/2011-07/07/content_12855869.htm.

545 "Yangtze River Floods," *Encyclopædia Britannica*, at http://www.britannica.com/EBchecked/topic/1503369/Yangtze-River-floods.

546 "Saving the Giant Panda," *Los Angeles Times*, July 30, 2010, at http://articles.latimes.com/2010/jul/30/opinion/la-ed-panda-20100730.

547 "Top 50 World Container Ports," *About the Industry*, World Shipping Council, at http://www.worldshipping.org/about-the-industry/global-trade/top-50-world-container-ports; *2012 International Trade Statistics Yearbook*, International Merchandise Trade Statistics, United Nations, July 2, 2013, at http://comtrade.un.org/pb/.

551 Five reports on Mongolia by Louisa Lim, *Morning Edition*, National Public Radio, September 7–11, 2009; *Mongolia Human Development Report 2007* (Ulan Bator, Mongolia: United Nations Development Programme, 2007), at http://www.mn.undp.org/content/mongolia/en/home/library/National-Human-Development-Reports/mongolia-human-development-report-2007.html.

557 A.B., "Worldwide Cost of Living: The Expenses of Japan," *Economist*, July 7, 2011, at http://www.economist.com/blogs/gulliver/2011/07/worldwide-cost-living.

558 Coco Masters, "Japan to Immigrants: Thanks, but You Can Go Home Now," *Time*, April 20, 2009, at http://www.time.com/time/world/article/0,8599,1892469,00.html.

CHAPTER 10

564–566 Adapted from the *Borneo Wire*, Jessica Lawrence, ed., Spring 2006, at http://borneoproject.org/article.php?id=623; "Community Stops Illegal Logging and Bulldozing Towards Protected Rainforests," Jessica Lawrence, ed., *Borneo Wire*, Fall 2006, at http://borneoproject.org/article.php?list=type&type=39; Mark Bujang, "A Community Initiative: Mapping Dayak's Customary Lands in Sarawak," presented at the Regional Community Mapping Network Workshop, November 8–10, 2004, Diliman, Quezon City, Philippines; Julia Zappei, "Malaysia Highest Court Affirms Tribes' Land Rights," Associated Press, May 10, 2009, at http://borneoproject.org/article.php?id=762; "US Announces Its Support of the UN Declaration of the Rights of Indigenous Peoples (UNDRIP), December 16, 2010," International Forum on Globalization, at http://ifg.org/programs/indig/USUNDRIP.html; video of deforestation in Indonesian Borneo, at http://news.mongabay.com/2012/0326-muaratae-video.html.

584 Coco Masters, "Japan to Immigrants: Thanks, but You Can Go Home Now," *Time*, April 20, 2009, at http://www.time.com/time/world/article/0,8599,1892469,00.html.

584 "ASEAN Economic Community Scorecard: Charting Progress Toward Regional Economic Integration," Jakarta: ASEAN Secretariat, March 2012, at http://www.aseansec.org/documents/scorecard_final.pdf.

586–587 The fieldwork of Karl Russell Kirby, Department of Geography, University of Tennessee, 2010.

594 Adapted from notes by Kirsty Vincin, an Australian teacher in Hong Kong, November 17, 2009, and "The Filipina Sisterhood: An Anthropology of Happiness," *Economist*, December 20, 2001, at http://www.economist.com/node/883909.

596–597 Likhaan Center for Women's Health, 2010, at http://www.likhaan.org/; Carlos H. Conde, "Bill to Increase Access to Contraception Is Dividing Filipinos," *New York Times*, October 25, 2009, at http://www.nytimes.com/2009/10/26/world/asia/26iht-phils.html.

601 Adapted from personal communications with anthropologist Jennifer W. Nourse, University of Richmond, a specialist in Southeast Asia, 2010; Walter Williams, *Javanese Lives* (Piscataway, NJ: Rutgers University Press, 1991), pp. 128–134.

604 Adapted from the field notes of Alex Pulsipher and Debbi Hempel, 2000; coverage of the HIV-AIDS conference in Thailand, July 11–16, 2004, by the Kaiser Family Foundation; "Life as a Thai Sex Worker," *BBC News*, February 22, 2007, at http://news.bbc.co.uk/2/hi/asia-pacific/6360603.stm.

CHAPTER 11

620–621 Adapted from audio segments and articles written by Brian Reed for National Public Radio, "Preparing for Sea Level Rise, Islanders Leave Home," at http://www.npr.org/2011/02/17/133681251/preparing-for-sea-level-rise-islanders-leave-home; "Could People from Kiribati Be 'Climate Change Refugees?'" February 17, 2011, at http://www.npr.org/blogs/the-two-way/2011/02/17/133848076/could-people-from-kiribati-be-climate-change-refugees; "Kiribati Mulls Fiji Land Purchase in Battle Against the Sea," *BBC News Asia*, March 8, 2012, at http://www.bbc.co.uk/news/world-asia-17295862.

640–641 Richard Nile and Christian Clerk, *Cultural Atlas of Australia, New Zealand, and the South Pacific* (New York: Facts on File, 1996), pp. 63–65; *Voyage Weblog*, Polynesian Voyaging Society, 2007, at http://pvs.kcc.hawaii.edu/2007voyage/index.html.

655 Adapted from articles by Phil Wilkins, "Tonga Can Only Match the Kiwis in the Haka," Brisbane, Australia, October 25, 2003; Mark Falcous, "The Decolonizing National Imaginary: Promotional Media Constructions During the 2005 Lions Tour of Aotearoa," *New Zealand Journal of Sport and Social Issues*, November 2007, 31(4): 374–393.

EPILOGUES

660 http://www.intelligentlife-magazine.com/content/features/among-reindeer, http://www.sfu.ca/ipinch/outputs/blog/appropriation-month-when-did-sami-sign-santa-s-helpers, http://www.theguardian.com/global-development/2016/feb/21/sami-people-reindeer-herders-arctic-culture.

668 http://www.astronautix.com/flights/salt7eo1.htm, http://arstechnica.com/science/2014/09/the-little-known-soviet-mission-to-rescue-a-dead-space-station/3/, http://www.thespacereview.com/article/1137/1.

Text Sources and Credits   TC-3

Index

Page numbers in *italic* indicate figures; page numbers in **bold** indicate key terms.

WITHDRAWN

PHYSICAL GEOGRAPHY MAP

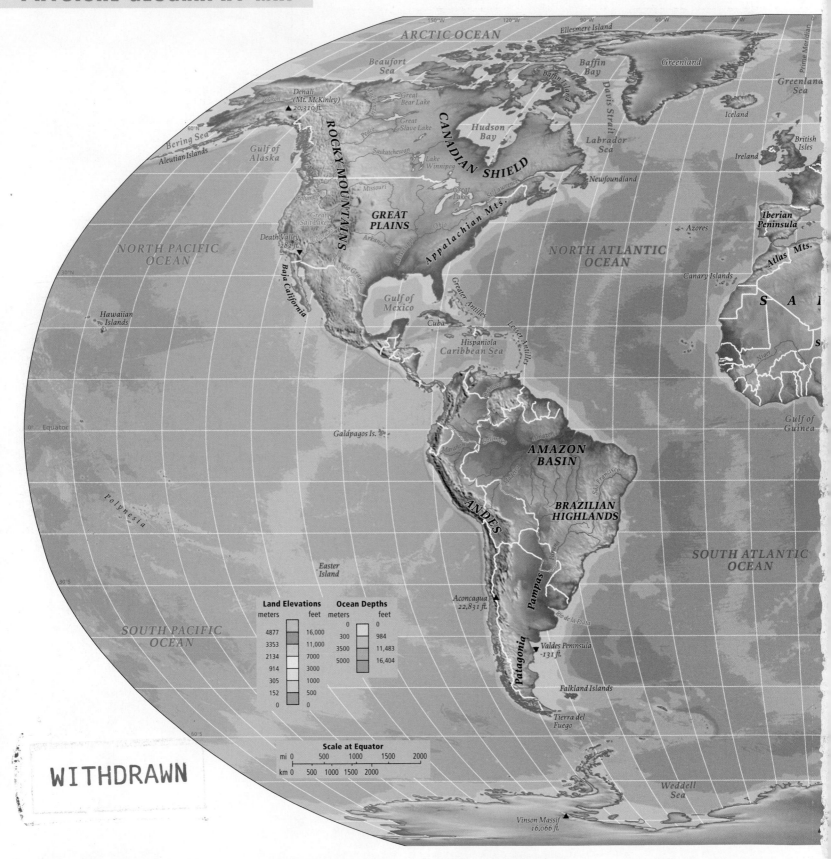

ARCTIC OCEAN

Beaufort
Sea

Ellesmere Island

Baffin
Bay

Greenland

Greenland
Sea

Denali
(Mt. McKinley)
▲ 20,310 ft.

Yukon

Great
Bear Lake

Mackenzie

Baffin Island

Davis Strait

Iceland

Bering Sea

Gulf of
Alaska

Aleutian Islands

ROCKY MOUNTAINS

Peace

Great
Slave Lake

CANADIAN SHIELD

Hudson
Bay

Labrador
Sea

Ireland

British
Isles

Prime Meridian

Saskatchewan

Lake
Winnipeg

Great
Lakes

St. Lawrence

Newfoundland

Columbia

Missouri

GREAT
PLAINS

Ohio

Appalachian Mts.

NORTH ATLANTIC
OCEAN

Iberian
Peninsula

NORTH PACIFIC
OCEAN

Great
Salt Lake

Arkansas

Mississippi

Azores

Death Valley
-282 ft. ▼

Colorado

Rio Grande

Atlas Mts.

Canary Islands

Baja California

Gulf of
Mexico

Greater Antilles

S   A

Hawaiian
Islands

Cuba

Hispaniola

Lesser Antilles

Caribbean Sea

Niger

Orinoco

Galápagos Is.

Gulf of
Guinea

Equator

Marañón

Solimões

Amazon

AMAZON
BASIN

São Francisco

Polynesia

ANDES

Madeira

BRAZILIAN
HIGHLANDS

Easter
Island

SOUTH ATLANTIC
OCEAN

SOUTH PACIFIC
OCEAN

Aconcagua
22,831 ft. ▼

Pampas

Rio de la Plata

Land Elevations
meters	feet
4877	16,000
3353	11,000
2134	7000
914	3000
305	1000
152	500
0	0

Ocean Depths
meters	feet
0	0
300	984
3500	11,483
5000	16,404

Valdés Peninsula
-131 ft. ▼

Patagonia

Falkland Islands

Tierra del
Fuego

Scale at Equator
mi 0   500   1000   1500   2000
km 0 500 1000 1500 2000

Weddell
Sea

Vinson Massif ▲
16,066 ft.

WITHDRAWN